Cutnell & Johnson Physics

Twelfth Edition

DAVID YOUNG

SHANE STADLER
Louisiana State University

VICE PRESIDENT AND GENERAL MANAGER	Aurora Martinez
EXECUTIVE EDITOR	John LaVacca
SENIOR EDITOR	Jennifer Yee
ASSOCIATE EDITOR	Georgia Larsen
ASSISTANT EDITOR	Samantha Hart
SENIOR MANAGING EDITOR	Mary Donovan
MARKETING MANAGER	Sean Willey
SENIOR MANAGER, COURSE DEVELOPMENT AND PRODUCTION	Svetlana Barskaya
SENIOR COURSE PRODUCTION OPERATIONS SPECIALIST	Patricia Gutierrez
COURSE CONTENT DEVELOPER	Corrina Santos
DESIGNER	Wendi Lai
COVER IMAGES CREDIT	Digital Composite Image Of Iris © Erwin Matuschat/ EyeEm/Getty Images, DNA structure © KTSDESIGN/ SCIENCE PHOTO LIBRARY/Getty Images

This book was set in 9.5/12 STIX Two Text by Lumina Datamatics®, Inc.

Founded in 1807, John Wiley & Sons, Inc. has been a valued source of knowledge and understanding for more than 200 years, helping people around the world meet their needs and fulfill their aspirations. Our company is built on a foundation of principles that include responsibility to the communities we serve and where we live and work. In 2008, we launched a Corporate Citizenship Initiative, a global effort to address the environmental, social, economic, and ethical challenges we face in our business. Among the issues we are addressing are carbon impact, paper specifications and procurement, ethical conduct within our business and among our vendors, and community and charitable support. For more information, please visit our Web site: www.wiley.com/go/citizenship.

Evaluation copies are provided to qualified academics and professionals for review purposes only, for use in their courses during the next academic year. These copies are licensed and may not be sold or transferred to a third party. Upon completion of the review period, please return the evaluation copy to Wiley. Return instructions and a free-of-charge return shipping label are available at: www.wiley.com/go/returnlabel. If you have chosen to adopt this textbook for use in your course, please accept this book as your complimentary desk copy. Outside of the United States, please contact your local representative.

Main Text binder Edition: 9781119773610
(Volume 1): 9781119803690
(Volume 2): 9781119803775
ePDF: 9781119788591
epub: 9781119773535

The inside back cover will contain printing identification and country of origin if omitted from this page. In addition, if the ISBN on the cover differs from the ISBN on this page, the one on the cover is correct.

Printed in the United States of America

SKY10029038_081121

Courtesy David Young

DAVID YOUNG received his Ph.D. in experimental condensed matter physics from Florida State University in 1998. He then held a post-doc position in the Department of Chemistry and the Princeton Materials Institute at Princeton University before joining the faculty in the Department of Physics and Astronomy at Louisiana State University in 2000. His research focuses on the synthesis and characterization of high-quality single crystals of novel electronic and magnetic materials. The goal of his research group is to understand the physics of electrons in materials under extreme conditions, i.e., at temperatures close to absolute zero, in high magnetic fields, and under high pressure. He is the coauthor of over 250 research publications that have appeared in peer-reviewed journals, such as *Physical Review B, Physical Review Letters*, and *Nature*. Professor Young has taught introductory physics with the Cutnell & Johnson text since he was a senior undergraduate almost 30 years ago. He routinely lectures to large sections, often in excess of 300 students. To engage such a large number of students, he uses *WileyPLUS*, electronic response systems, tutorial-style recitation sessions, and in-class demonstrations. Professor Young has received multiple awards for outstanding teaching of undergraduates. David enjoys spending his free time with his family, playing basketball, and working on his house.

I would like to thank my family for their continuous love and support.   —David Young

Courtesy Shane Stadler

SHANE STADLER Shane Stadler earned a Ph.D. in experimental condensed matter physics from Tulane University in 1998. Afterwards, he accepted a National Research Council Postdoctoral Fellowship with the Naval Research Laboratory in Washington, DC, where he conducted research on artificially structured magnetic materials. Three years later, he joined the faculty in the Department of Physics at Southern Illinois University (the home institution of John Cutnell and Ken Johnson, the original authors of this textbook), before joining the Department of Physics and Astronomy at Louisiana State University in 2008. His research group studies novel magnetic materials for applications in the areas of spintronics and magnetic cooling.

Over the past twenty years, Professor Stadler has taught the full spectrum of physics courses, from physics for students outside the sciences, to graduate-level physics courses, such as classical electrodynamics. He teaches classes that range from fewer than ten students to those with enrollments of over 300. His educational interests are focused on developing teaching tools and methods that apply to both small and large classes, and which are applicable to emerging teaching strategies, such as "flipping the classroom."

In his spare time, Shane writes science fiction/thriller novels.

I would like to thank my parents, George and Elissa, for their constant support and encouragement.   —Shane Stadler

Dear Students and Instructors:

Welcome to college physics! To the students: We know there is a negative stigma associated with physics, and you yourself may harbor some trepidation as you begin this course. But fear not! We are here to help. Whether you are worried about your math proficiency, understanding the concepts, or developing your problem-solving skills, the resources available to you are designed to address all of these areas and more. When we were students and had to take introductory physics, we had a printed textbook, a pencil, and some paper. That was it! Can you learn physics this way? You bet! We did! But research has shown that learning styles vary greatly among students. Maybe some of you have a more visual preference, or auditory preference, or some other preferred learning modality. In any case, the resources available to you in this course will satisfy all of these preferences and improve your chance of success. Take a moment to explore below what the textbook and online course have to offer. We suspect that, as you continue to improve throughout the course, some of that initial trepidation will be replaced with confidence and excitement.

To start, 12e will continue to offer a new learning medium unique to this book in the form of a comprehensive set of lecture videos—one for every section (259 in all). These animated lectures (created and narrated by the authors) are 2–10 minutes in length, and explain the basic concepts and learning objectives of each section. They are assignable within WileyPLUS and can be paired with follow-up questions that are gradable. In addition to supplementing traditional lecturing, the videos can be used in a variety of ways, including, flipping the classroom, lectures for online courses, and reviewing for exams.

Next, we created a new end-of-chapter section of problems called *Physics in Biology, Medicine, and Sports*. The text has always offered real-world examples of physics applications in the life sciences, not only to generate interest for the many students in one of the premedical fields, but also to underscore the fact that the human body, like any biological system, is governed by the laws of physics. Overall, in this edition, we have more than doubled the number of bio-inspired examples and problems to more than 300 total for the text, which reflects our commitment to showing students how relevant physics is in their lives. Although they are of general instructional value, many of the end-of-chapter problems are also similar to what premed students will encounter in the Chemical and Physical Foundations of Biological Systems Passages section of the MCAT.

Building on the success of the *Team Problems* introduced in the previous edition, we have doubled their number in each chapter. These are context-rich end-of-chapter problems of variable difficulty designed for group cooperation, but they may also be tackled by the individual student. We have also added a "worksheet" activity to each new Team Problem, which helps to guide the team of students along the way through the solution. These problems offer a great way to promote peer-to-peer instruction in the classroom, as well as serve as the focal point of a group activity during a breakout session as part of an online course.

One of the great strengths of this text is the synergistic relationship it develops between problem solving and conceptual understanding. For instance, available in WileyPLUS are animated *Chalkboard Videos*, which consist of short (2–3 min) videos narrated by the authors that demonstrate step-by-step practical solutions to typical homework problems. We have more than doubled the number of *Chalkboard Videos* in the new edition, so that every section of the book has at least one, which provides comprehensive coverage of the entire course material in this format.

One of the most important techniques developed in the text for solving problems involving multiple forces is the *free-body diagram (FBD)*. Many problems in the force-intensive chapters, such as Chapters 4 and 18, take advantage of the new FBD drawing tools now available online in WileyPLUS, where students can construct the FBDs for a select number of problems and be graded on them. Also available are numerous *Guided Online (GO) Tutorials* that implement a step-by-step pedagogical approach, which provides students a low-stakes environment for refining their problem-solving skills. Finally, new to this edition will be *Adaptive Assignments*. Seamlessly integrated into WileyPLUS, *Adaptive Assignments,* which are powered by Knewton's world-class recommendation engine, offer the most targeted, efficient way for instructors to tailor practice and preparation. This new assignment type pinpoints learner knowledge gaps and offers personalized, just-in-time support, targeted instruction, and detailed answer explanations.

Over the last 20 years nothing has had a more significant impact on the way students learn than the World Wide Web. Students essentially have 24/7 access to countless sources of digital multimedia. They complete homework assignments online with their PC's, tablets, and smart phones. Online homework systems are no longer "in the future," but are now the norm. Physics is no exception. Unfortunately, having all of this information readily available comes at a price. Students have fundamentally changed the way in which they approach their homework assignments. Instead of struggling through the entire solution to a problem from scratch, where much of the learning process takes place, they default to online resources, where they can pay for access to written solutions to the end-of-chapter problems. Alternatively, many students find solutions by simply searching the questions on Google, for example. As a result, a student's online-homework grade has become a rather poor measure of their knowledge of the course material. What's even worse is the false sense of security the students feel as a result of their inflated homework grades. They feel confident and prepared for exams, because they have 95%–100% on their homework. Unfortunately, a poor performance on the first exam is often the initial indicator of their level of misunderstanding. The content outlined below, and the functionality of WileyPLUS and the *Adaptive Assignments,* will provide students with all the resources they need to be successful in the course:

- The *Lecture Videos* created by the authors for each section include questions with intelligent feedback when a student enters the wrong answer.
- The multistep *GO Tutorial* problems created in WileyPLUS are designed to provide targeted, intelligent feedback.
- The *Free-body Diagram* vector drawing tools provide students an easy way to enter answers requiring vector drawing, and also provide enhanced feedback.
- *Chalkboard Video Solutions* take the students step-by-step through the solution and the thought process of the authors. Problem-solving strategies are discussed, and common misconceptions and potential pitfalls are addressed. The students can then apply these techniques to solve similar, but different, problems.

All of these features are designed to encourage students to remain within the WileyPLUS environment, as opposed to pursuing the "pay-for solutions" websites that short circuit the learning process.

To the students—We strongly recommend that you take this honest approach to the course. Take full advantage of the many features and learning resources that accompany the text and the online content. Be engaged with the material and push yourself to work through the exercises. Physics may not be the easiest subject to understand initially, but with the Wiley resources at your disposal, and your hard work, you CAN be successful.

We are immensely grateful to all of you who have provided feedback as we've worked on this new edition, and to our students who have taught us how to teach. Thank you for your guidance, and keep the feedback coming. Best wishes for success in this course and wherever your major may take you!

Brief Contents

Contents

12 Temperature and Heat 356

13 The Transfer of Heat 394

14 The Ideal Gas Law and Kinetic Theory 416

15 Thermodynamics 441

16 Waves and Sound 478

17 The Principle of Linear Superposition and Interference Phenomena   513

Appendixes   A-1

Note: Chapter sections marked with an asterisk (*) can be omitted with little impact to the overall development of the material.

Our Vision and WileyPLUS with Adaptive Assignments

Our Vision

Our goal is to provide students with the skills they need to succeed in this course, and instructors with the tools they need to develop those skills.

Skills Development

One of the great strengths of this text is the synergistic relationship between conceptual understanding, problem solving, and establishing relevance. We identify here some of the core features of the text that support these synergies.

Conceptual Understanding Students often regard physics as a collection of equations that can be used blindly to solve problems. However, a good problem-solving technique does not begin with equations. It starts with a firm grasp of physics concepts and how they fit together to provide a coherent description of natural phenomena. Helping students develop a conceptual understanding of physics principles is a primary goal of this text. The features in the text that work toward this goal are:

- *Lecture Videos (one for each section of the text)*
- *Conceptual Examples*
- *Concepts & Calculations* problems (now with video solutions)
- *Focus on Concepts* homework material
- *Check Your Understanding* questions
- *Concept Simulations* (an online feature)

Problem Solving The ability to reason in an organized and mathematically correct manner is essential to solving problems, and helping students to improve their reasoning skills is also one of our primary goals. To this end, we have included the following features:

- *Math Skills boxes* for just-in-time delivery of math support
- *Explicit reasoning steps* in all examples
- *Reasoning Strategies* for solving certain classes of problems
- *Analyzing Multiple-Concept Problems*
- *Video Support and Tutorials* (in *WileyPLUS*)
    Chalkboard Video Solutions
    Physics Demonstration Videos
    Video Help
    Concept Simulations
- *Problem Solving Insights*

Relevance Since it is always easier to learn something new if it can be related to day-to-day living, we want to show students that physics principles come into play over and over again in their lives. To emphasize this goal, we have included a wide range of applications of physics principles. Many of these applications are biomedical in nature (for example, wireless capsule endoscopy). Others deal with modern technology (for example, 3-D movies). Still others focus on things that we take for granted in our lives (for example, household plumbing). To call attention to the applications we have used the label The Physics of.

WileyPLUS with Adaptive Assignments

WileyPLUS is an innovative, research-based online environment for effective teaching and learning. The hallmark of *WileyPLUS* with Adaptive Assignments for this text is that the media- and text-based resources are all created by the authors of the project, providing a seamless presentation of content.

WileyPLUS builds students' confidence because it takes the guesswork out of studying by providing students with a clear roadmap: **what to do, how to do it, if they did it right**.

With *WileyPLUS*, our efficacy research shows that students improve their outcomes by as much as one letter grade. *WileyPLUS* helps students take more initiative, so you'll have greater impact on their achievement in the classroom and beyond.

With WileyPLUS, instructors receive:

- **Breadth and Depth of Assessment:** *WileyPLUS* contains a wealth of online questions and problems for creating online homework and assessment including:
    - ALL end-of-chapter questions, plus favorites from past editions not found in the printed text, coded algorithmically, each with at least one form of instructor-controlled question assistance (GO tutorials, hints, link to text, video help)
    - Simulation, animation, and video-based questions
    - Free body and vector drawing questions
    - Test bank questions

- **Gradebook:** *WileyPLUS* provides instant access to reports on trends in class performance, student use of course materials, and progress toward learning objectives, thereby helping instructors' decisions and driving classroom discussion.

With WileyPLUS, students receive:

- The complete digital textbook, including the Lecture Videos, saving students up to 60% off the cost of a printed text
- Question assistance, including links to relevant sections in the online digital textbook
- Immediate feedback and proof of progress, 24/7
- Integrated, multimedia resources—including animations, simulations, video demonstrations, and much more—that provide multiple study paths and encourage more active learning
- GO Tutorials
- Chalkboard Videos
- Free Body Diagram/Vector Drawing Questions

New to WileyPLUS for the Twelfth Edition

Team Problems In each chapter, the end-of-chapter problems contain four *Team Problems* that are designed for group problem-solving exercises. These are context-rich problems of medium difficulty designed for group cooperation, but may also be tackled by the individual student. Accompanying some of these team problems are "worksheets" that serve to guide the students through the solution in a tutorial fashion. Many of these problems read like parts of an adventure story, where the student (or their team) is the main character. The motivation for each problem is clear and personal—the pronoun "you" is used throughout, and the problem statements often start with "You and your team need to ...". Pictures and diagrams are not given with these problems except in rare cases. Students must visualize the problems and discuss strategies with their team members to solve them. The problems require two or more steps/multiple concepts (hence the "medium" difficulty level) and may require basic principles

learned earlier. Sometimes, there is no specific target variable given, but rather questions like *Will it work?* or *Is it safe?* Suggested solutions are given in the Instructor Solutions Manual.

"The Physics of . . ." Examples The text now contains over 300 real-world application examples that reflect our commitment to showing students how relevant physics is in their lives. Each application is identified in the text with the label The Physics of. A subset of these examples focuses on biomedical applications, and we have more than doubled their number in the new edition. In fact, we have created a new end-of-chapter problem section called the *Physics in Biology, Medicine, and Sports*. Students majoring in biomedical and life sciences will find new examples in every chapter covering topics such as cooling the human brain, abdominal aortic aneurysms, the mechanical properties of bone, and many more! The application of physics principles to biomedical problems in these examples is similar to what premed students will encounter in the *Chemical and Physical Foundations of Biological Systems Passages* section of the MCAT. All biomedical examples and end-of-chapter problems will be marked with the **BIO** icon.

EXAMPLE 8 | BIO The Physics of Hearing Loss—Standing Waves in the Ear

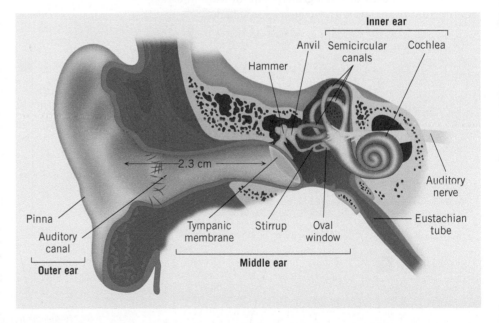

Interactive Graphics The online reading experience within WileyPLUS has been enhanced with the addition of "Interactive Graphics." Several static figures in each chapter have been transformed to include interactive elements. These graphics drive students to be more engaged with the extensive art program and allow them to more easily absorb complex and/or long multi-part figures.

Use the checkboxes to select the correct vectors as instructed above. Move the vectors to the correct starting points and orient them in the correct direction as instructed.

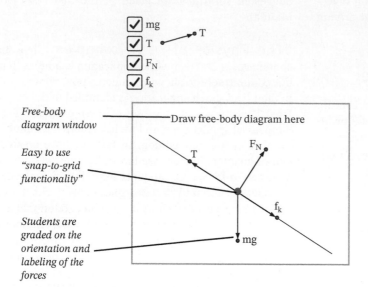

Key to Force Labels

mg = Gravitational force
T = Tension force
F_N = Normal force
f_k = Kinetic friction force

Free-body diagram window

Draw free-body diagram here

Easy to use "snap-to-grid functionality"

Students are graded on the orientation and labeling of the forces

Also Available in WileyPLUS

Free-Body Diagram (FBD) Tools
For many problems involving multiple forces, an interactive free-body diagram tool in *WileyPLUS* is used to construct the diagram. It is essential for students to practice drawing FBDs, as that is the critical first step in solving many equilibrium and non-equilibrium problems with Newton's second law.

GO Tutorial Problems
Some of the homework problems found in the collection at the end of each chapter are marked with a special GO icon. All of these problems (550 of them) are available for assignment online via *WileyPLUS*. Each of these problems in *WileyPLUS* includes a guided tutorial option that instructors can make available for student access with or without penalty.

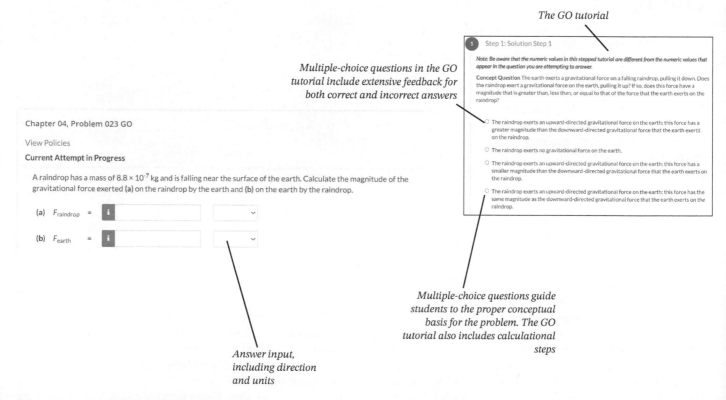

The GO tutorial

Multiple-choice questions in the GO tutorial include extensive feedback for both correct and incorrect answers

Chapter 04, Problem 023 GO

View Policies

Current Attempt in Progress

A raindrop has a mass of 8.8×10^{-7} kg and is falling near the surface of the earth. Calculate the magnitude of the gravitational force exerted (a) on the raindrop by the earth and (b) on the earth by the raindrop.

(a) $F_{raindrop}$ =

(b) F_{earth} =

Answer input, including direction and units

Multiple-choice questions guide students to the proper conceptual basis for the problem. The GO tutorial also includes calculational steps

Adaptive Assignments

Seamlessly integrated into WileyPLUS and based on learning science, Adaptive Assignments are the most targeted, efficient way for instructors to tailor practice and preparation. This new assignment type pinpoints learner knowledge gaps and offers personalized, just-in-time support, targeted instruction and detailed answer explanations.

Acknowledgments

The effective revision of a textbook in this age goes well beyond the efforts of the authors and requires the expertise of a great number of people. Their abilities must range from brainstorming revision concepts to editing written and digital content to creating marketing strategies. They have to learn how the needs of our readers have changed and to anticipate how they might transform in the future. As the authors, we are extremely grateful to the people who have provided direction, talent, and effort as a part of the team that made this revision possible. Well done and thank you!

We would especially like to acknowledge John LaVacca III, our Senior Editor. His knowledge of the textbook landscape has offered crucial guidance in developing this revision in order to meet the needs of students and instructors alike. He has provided us ample encouragement and flexibility to pursue the creative aspects of this work within the constraints of a well-planned strategy.

It was a great pleasure working with Jennifer Yee, Senior Editor, on this edition, and we would like to thank her for keeping this edition on track and coordinating our efforts with the other facets of this production. Keeping us informed of the timelines for all aspects of the project made the process smooth and devoid of stress (at least on the part of authors!). We wish her well at her new position in Wiley's Professional Learning Group! Congratulations Jen!

We owe our gratitude to Corrina Santos, Course Content Developer, and Samantha Hart, Editorial Assistant for their careful quality control and editing of the new content, and Anju Joshi and Wendi Lai for layout and design, as well as keeping the production process running smoothly. With the lack of direct interaction due to travel restrictions during the past year, we were impressed with the cohesive team effort, despite the lack of in-person interactions, and we are grateful for and recognize the individual contributions from Georgia Larson, Assistant Editor, Svetlana Barskaya, Senior Manager, Course Development and Production, Mary Donovan, Senior Managing Editor, and Patricia Gutierrez, Senior Course Production Specialist. We would also like to welcome our new Marketing Manager, Sean Willey, who has hit the ground running on getting the word out about the new features of this edition.

We think all would agree that the sales representatives of John Wiley & Sons, Inc. form the backbone of this entire effort. Without you, and your expertise and understanding of the book's features, its online content, its integration into WileyPLUS, and the needs of students and instructors in higher education, it would never have had the success that is has over the past 30 years! And you should know that it was one of your own, Dayna Leaman, our good friend, who got us into this business in the first place! Thank you, Dayna!

Finally, we would like to thank our physics colleagues and their students who have kindly shared their ideas about effective pedagogy, and improved the text by identifying errors. We are grateful for all of their suggestions as they have helped us to produce a more clear and accurate work. To the reviewers of this and previous editions, we especially owe a large debt of gratitude. Specifically, we thank:

Lai Cao, *Baton Rouge Magnet High School*
Candee Chambers-Colbeck, *Maryville University*
Diana Driscoll, *Case Western Reserve University*
Costas Efthimiou, *University of Central Florida*
Robert Egler, *North Carolina State University*
Sambandamurthy Ganapathy, *The State University of New York at Buffalo*
Joseph Ganem, *Loyola University, Maryland*
Jasper Halekas, *University of Iowa*
Lilit Haroyan, *East Los Angeles College*
Richard Holland, *Rend Lake College and SIUC*
Klaus Honscheid, *Ohio State University*
Shyang Huang, *Missouri State University*
Craig Kleitzing, *University of Iowa*
Wayne Mathe, *University of South Florida, Sarasota-Manatee Campus*
Wayne Mathé, *University of South Florida Tampa*
Samuel Mensah, *University of Memphis*
Mark Morgan-Tracy, *University of New Mexico*
Kriton Papavasiliou, *Virginia Polytechnic Institute and State University*
Kriton Papavasiliou, *Virginia Tech*
Payton Parker, *Midlothian Heritage High School*
Christian Prewitt, *Midlothian Heritage High School*
Joshua Ravenscraft, *Vernon Hills High School and College of Lake County*
Bill Schmidt, *Meredith College*
Brian Schuft, *North Carolina A&T State University*
Andreas Shalchi, *University of Manitoba*
Deepshikha Shukla, *Rockford University*
Jennifer Snyder, *San Diego Mesa College*
Paul Sokol, *Indiana University Bloomington*
Richard Taylor, *University of Oregon*
Beth Thacker, *Texas Tech University*
Anne Topper, *Queen's University*
David Ulrich, *Portland Community College*
Luc T. Wille, *Florida Atlantic University*

About the cover: One of the great successes of physics as a science is its ability to accurately predict the behavior of the natural world across all length scales, from the atomically small to the astronomically large. The cover image represents this achievement with an artist's rendition of a human eye drifting in the cosmos. At the center of the eye's pupil is the famous double helix of DNA. The motion of the planets and stars is largely determined by the gravitational force. The formation of an image by our eyes is governed by the laws of optics, and the amino acid base pairs in DNA are held together by electrostatic forces. In this course, you will study all of these topics and learn how to apply them to solve a variety of problems.

In spite of our best efforts to produce an error-free book, errors no doubt remain. They are solely our responsibility, and we would appreciate hearing of any that you find. We hope that this text makes learning and teaching physics easier and more enjoyable, and we look forward to hearing about your experiences with it. Please feel free to write us care of Physics Editor, Global Education, John Wiley & Sons, Inc., 111 River Street, Hoboken, NJ 07030, or contact the authors at **dyoun14@gmail.com** or **sstadler23@gmail.com**.

The animation techniques and special effects used in the film *Avengers: Infinity War* rely on computers and mathematical concepts such as trigonometry and vectors. Such mathematical concepts will be useful throughout this book in our discussion of physics.

Lifestyle pictures/Alamy Stock Photo

Introduction and Mathematical Concepts

1.1 The Nature of Physics

Physics is the most basic of the sciences, and it is at the very root of subjects like chemistry, engineering, astronomy, and even biology. The discipline of physics has developed over many centuries, and it continues to evolve. It is a mature science, and its laws encompass a wide scope of phenomena that range from the formation of galaxies to the interactions of particles in the nuclei of atoms. Perhaps the most visible evidence of physics in everyday life is the eruption of new applications that have improved our quality of life, such as new medical devices, and advances in computers and high-tech communications.

The exciting feature of physics is its capacity for predicting how nature will behave in one situation on the basis of experimental data obtained in another situation. Such predictions place physics at the heart of modern technology and, therefore, can have a tremendous

LEARNING OBJECTIVES

After reading this module, you should be able to...

1.1 Describe the fundamental nature of physics.

1.2 Describe different systems of units.

1.3 Solve unit conversion problems.

1.4 Solve trigonometry problems.

1.5 Distinguish between vectors and scalars.

1.6 Solve vector addition and subtraction problems by graphical methods.

1.7 Calculate vector components.

1.8 Solve vector addition and subtraction problems using components.

impact on our lives. Rocketry and the development of space travel have their roots firmly planted in the physical laws of Galileo Galilei (1564–1642) and Isaac Newton (1642–1727). The transportation industry relies heavily on physics in the development of engines and the design of aerodynamic vehicles. Entire electronics and computer industries owe their existence to the invention of the transistor, which grew directly out of the laws of physics that describe the electrical behaviors of solids. The telecommunications industry depends extensively on electromagnetic waves, whose existence was predicted by James Clerk Maxwell (1831–1879) in his theory of electricity and magnetism. The medical profession uses X-ray, ultrasonic, and magnetic resonance methods for obtaining images of the interior of the human body, and physics lies at the core of all of these. Perhaps the most widespread impact in modern technology is that due to the laser. Fields ranging from space exploration to medicine benefit from this incredible device, which is a direct application of the principles of atomic physics.

Because physics is so fundamental, it is a required course for students in a wide range of major areas. We welcome you to the study of this fascinating topic. You will learn how to see the world through the "eyes" of physics and to reason as a physicist does. In the process, you will learn how to apply physics principles to a wide range of problems. We hope that you will come to recognize that physics has important things to say about your environment.

1.2 Units

Physics experiments involve the measurement of a variety of quantities, and a great deal of effort goes into making these measurements as accurate and reproducible as possible. The first step toward ensuring accuracy and reproducibility is defining the units in which the measurements are made.

In this text, we emphasize the system of units known as **SI units,** which stands for the French phrase "Le **S**ystème **I**nternational d'Unités." By international agreement, this system employs the **meter** (m) as the unit of length, the **kilogram** (kg) as the unit of mass, and the **second** (s) as the unit of time. Two other systems of units are also in use, however. The CGS system utilizes the centimeter (cm), the gram (g), and the second for length, mass, and time, respectively, and the BE or British Engineering system (the gravitational version) uses the foot (ft), the slug (sl), and the second. **Table 1.1** summarizes the units used for length, mass, and time in the three systems.

Originally, the meter was defined in terms of the distance measured along the earth's surface between the north pole and the equator. Eventually, a more accurate measurement standard was needed, and by international agreement the meter became the distance between two marks on a bar of platinum–iridium alloy (see **Figure 1.1**) kept at a temperature of 0 °C. Today, to meet further demands for increased accuracy, the meter is defined as the distance that light travels in a vacuum in a time of 1/299 792 458 second. This definition arises because the speed of light is a universal constant that is defined to be 299 792 458 m/s.

Courtesy NIST Archives

FIGURE 1.1 The standard platinum–iridium meter bar.

TABLE 1.1 **Units of Measurement**

	System		
	SI	CGS	BE
Length	Meter (m)	Centimeter (cm)	Foot (ft)
Mass	Kilogram (kg)	Gram (g)	Slug (sl)
Time	Second (s)	Second (s)	Second (s)

The definition of a kilogram as a unit of mass has also undergone changes over the years. As Chapter 4 discusses, the mass of an object indicates the tendency of the object to continue in motion with a constant velocity. Originally, the kilogram was expressed in terms of a specific amount of water. Since 1879, and until very recently, one kilogram was defined to be the mass of a standard cylinder of platinum–iridium alloy, like the one in **Figure 1.2**. Currently, one kilogram is defined in terms of two other **base** units (see below)—the meter and the second, which are based on fixed fundamental constants of nature. The kilogram is defined by setting Planck's constant h exactly equal to $6.62607015 \times 10^{-34}$ J · s. We will see in Section 29.2 that Planck's constant is important in the field of quantum mechanics. This new definition of the kilogram took effect on May 20, 2019.

As with the units for length and mass, the present definition of the second as a unit of time is different from the original definition. Originally, the second was defined according to the average time for the earth to rotate once about its axis, one day being set equal to 86 400 seconds. The earth's rotational motion was chosen because it is naturally repetitive, occurring over and over again. Today, we still use a naturally occurring repetitive phenomenon to define the second, but of a very different kind. We use the electromagnetic waves emitted by cesium-133 atoms in an atomic clock like that in **Figure 1.3**. One second is defined as the time needed for 9 192 631 770 wave cycles to occur.*

The units for length, mass, and time, along with a few other units that will arise later, are regarded as **base** SI units. The word "base" refers to the fact that these units are used along with various laws to define additional units for other important physical quantities, such as force and energy. The units for such other physical quantities are referred to as **derived** units, since they are combinations of the base units. Derived units will be introduced from time to time, as they arise naturally along with the related physical laws.

The value of a quantity in terms of base or derived units is sometimes a very large or very small number. In such cases, it is convenient to introduce larger or smaller units that are related to the normal units by multiples of ten. **Table 1.2** summarizes the prefixes that are used to denote multiples of ten. For example, 1000 or 10^3 meters are referred to as 1 kilometer (km), and 0.001 or 10^{-3} meter is called 1 millimeter (mm). Similarly, 1000 grams and 0.001 gram are referred to as 1 kilogram (kg) and 1 milligram (mg), respectively. Appendix A contains a discussion of scientific notation and powers of ten, such as 10^3 and 10^{-3}.

Science Source

FIGURE 1.2 The standard platinum–iridium kilogram is kept at the International Bureau of Weights and Measures in Sèvres, France. This copy of it was assigned to the United States in 1889 and is housed at the National Institute of Standards and Technology.

Geoffrey Wheeler

FIGURE 1.3 This atomic clock, the NIST-F1, keeps time with an uncertainty of about one second in sixty million years.

TABLE 1.2	Standard Prefixes Used to Denote Multiples of Ten	
Prefix	**Symbol**	**Factor[a]**
tera	T	10^{12}
giga	G	10^{9}
mega	M	10^{6}
kilo	k	10^{3}
hecto	h	10^{2}
deka	da	10^{1}
deci	d	10^{-1}
centi	c	10^{-2}
milli	m	10^{-3}
micro	μ	10^{-6}
nano	n	10^{-9}
pico	p	10^{-12}
femto	f	10^{-15}

1.3 The Role of Units in Problem Solving

The Conversion of Units

Since any quantity, such as length, can be measured in several different units, it is important to know how to convert from one unit to another. For instance, the foot can be used to express the distance between the two marks on the standard platinum–iridium meter bar. There are 3.281 feet in one meter, and this number can be used to convert from meters to feet, as the following example demonstrates.

*See Chapter 16 for a discussion of waves in general and Chapter 24 for a discussion of electromagnetic waves in particular.

[a]Appendix A contains a discussion of powers of ten and scientific notation.

EXAMPLE 1 | The World's Highest Waterfall

The highest waterfall in the world is Angel Falls in Venezuela, with a total drop of 979.0 m (see **Figure 1.4**). Express this drop in feet.

Reasoning   When converting between units, we write down the units explicitly in the calculations and treat them like any algebraic quantity. In particular, we will take advantage of the following algebraic fact: Multiplying or dividing an equation by a factor of 1 does not alter an equation.

Solution   Since 3.281 feet = 1 meter, it follows that (3.281 feet)/(1 meter) = 1. Using this factor of 1 to multiply the equation "Length = 979.0 meters," we find that

$$\text{Length} = (979.0 \text{ m})(1) = (979.0 \text{ meters})\left(\frac{3.281 \text{ feet}}{1 \text{ meter}}\right) = \boxed{3212 \text{ feet}}$$

The colored lines emphasize that the units of meters behave like any algebraic quantity and cancel when the multiplication is performed, leaving only the desired unit of feet to describe the answer. In this regard, note that 3.281 feet = 1 meter also implies that (1 meter)/(3.281 feet) = 1. However, we chose not to multiply by a factor of 1 in this form, because the units of meters would not have canceled.

A calculator gives the answer as 3212.099 feet. Standard procedures for significant figures, however, indicate that the answer should be rounded off to four significant figures, since the value of 979.0 meters is accurate to only four significant figures. In this regard, the "1 meter" in the denominator does not limit the significant figures of the answer, because this number is precisely one meter by definition of the conversion factor. Appendix B contains a review of significant figures.

Andoni Canela/Age Fotostock

FIGURE 1.4   Angel Falls in Venezuela is the highest waterfall in the world.

Problem-Solving Insight   In any conversion, if the units do not combine algebraically to give the desired result, the conversion has not been carried out properly.

With this in mind, the next example stresses the importance of writing down the units and illustrates a typical situation in which several conversions are required.

EXAMPLE 2 | BIO The World's Fastest Land Animal

The cheetah can accelerate for a short time to a top speed of 75 miles/hour, making it the fastest animal on land (see **Figure 1.5**). Express 75 miles/hour in meters/second.

Reasoning   As in Example 1, it is important to write down the units explicitly in the calculations and treat them like any algebraic quantity. Here, we take advantage of two well-known relationships—namely, 5280 feet = 1 mile and 3600 seconds = 1 hour. As a result, (5280 feet)/(1 mile) = 1 and (3600 seconds)/(1 hour) = 1. In our solution we will use the fact that multiplying and dividing by these factors of unity does not alter an equation.

Solution   Multiplying and dividing by factors of unity, we find the cheetah's speed in feet per second as shown here:

$$\text{Speed} = \left(75 \frac{\text{miles}}{\text{hour}}\right)(1)(1) =$$

$$\left(75 \frac{\text{miles}}{\text{hour}}\right)\left(\frac{5280 \text{ feet}}{1 \text{ mile}}\right)\left(\frac{1 \text{ hour}}{3600 \text{ seconds}}\right) = 110 \frac{\text{feet}}{\text{second}}$$

To convert feet into meters, we use the fact that (1 meter)/(3.281 feet) = 1:

$$\text{Speed} = \left(110 \frac{\text{feet}}{\text{second}}\right)(1) =$$

$$\left(110 \frac{\text{feet}}{\text{second}}\right)\left(\frac{1 \text{ meter}}{3.281 \text{ feet}}\right) = \boxed{34 \frac{\text{meters}}{\text{second}}}$$

DrZoltan/Pixabay

FIGURE 1.5   The cheetah is the world's fastest land animal.

In addition to their role in guiding the use of conversion factors, units serve a useful purpose in solving problems. They can provide an internal check to eliminate errors, if they are carried along during each step of a calculation and treated like any algebraic factor.

> **Problem-Solving Insight**   In particular, remember that only quantities with the same units can be added or subtracted.

Thus, at one point in a calculation, if you find yourself adding 12 miles to 32 kilometers, stop and reconsider. Either miles must be converted into kilometers or kilometers must be converted into miles before the addition can be carried out. A collection of useful conversion factors is given on the page facing the inside of the front cover.

The reasoning strategy that we have followed in Examples 1 and 2 for converting between units is outlined as follows:

> **REASONING STRATEGY**   **Converting Between Units**
>
> 1. In all calculations, write down the units explicitly.
> 2. Treat all units as algebraic quantities. In particular, when identical units are divided, they are eliminated algebraically.
> 3. Use the conversion factors located on the page facing the inside of the front cover of the textbook. Be guided by the fact that multiplying or dividing an equation by a factor of 1 does not alter the equation. For instance, the conversion factor of 3.281 feet = 1 meter might be applied in the form (3.281 feet)/(1 meter) = 1. This factor of 1 would be used to multiply an equation such as "Length = 5.00 meters" in order to convert meters to feet.
> 4. Check to see that your calculations are correct by verifying that the units combine algebraically to give the desired unit for the answer. Only quantities with the same units can be added or subtracted.

Sometimes an equation is expressed in a way that requires specific units to be used for the variables in the equation. In such cases it is important to understand why only certain units can be used in the equation, as the following example illustrates.

EXAMPLE 3 | **BIO** The Physics of the Body Mass Index

The body mass index (BMI) takes into account your mass in kilograms (kg) and your height in meters (m) and is defined as follows:

$$\text{BMI} = \frac{\text{Mass in kg}}{(\text{Height in m})^2}$$

However, the BMI is often computed using the weight* of a person in pounds (lb) and his or her height in inches (in.). Thus, the expression for the BMI incorporates these quantities, rather than the mass in kilograms and the height in meters. Starting with the definition above, determine the expression for the BMI that uses pounds and inches.

Reasoning   We will begin with the BMI definition and work separately with the numerator and the denominator. We will determine the mass in kilograms that appears in the numerator from the weight in pounds by using the fact that 1 kg corresponds to 2.205 lb. Then, we will determine the height in meters that appears in the denominator from the height in inches with the aid of the facts that 1 m = 3.281 ft and 1 ft = 12 in. These conversion factors are located on the page facing the inside of the front cover of the text.

Solution   Since 1 kg corresponds to 2.205 lb, the mass in kilograms can be determined from the weight in pounds in the following way:

$$\text{Mass in kg} = (\text{Weight in lb})\left(\frac{1\ \text{kg}}{2.205\ \text{lb}}\right)$$

Since 1 ft = 12 in. and 1 m = 3.281 ft, we have

$$\text{Height in m} = (\text{Height in in.})\left(\frac{1\ \text{ft}}{12\ \text{in.}}\right)\left(\frac{1\ \text{m}}{3.281\ \text{ft}}\right)$$

Substituting these results into the numerator and denominator of the BMI definition gives

$$\text{BMI} = \frac{\text{Mass in kg}}{(\text{Height in m})^2} = \frac{(\text{Weight in lb})\left(\dfrac{1\ \text{kg}}{2.205\ \text{lb}}\right)}{(\text{Height in in.})^2\left(\dfrac{1\ \text{ft}}{12\ \text{in.}}\right)^2\left(\dfrac{1\ \text{m}}{3.281\ \text{ft}}\right)^2}$$

$$= \left(\frac{1\ \text{kg}}{2.205\ \text{lb}}\right)\left(\frac{12\ \text{in.}}{1\ \text{ft}}\right)^2\left(\frac{3.281\ \text{ft}}{1\ \text{m}}\right)^2\frac{(\text{Weight in lb})}{(\text{Height in in.})^2}$$

$$\boxed{\text{BMI} = \left(703.0\ \frac{\text{kg}\cdot\text{in.}^2}{\text{lb}\cdot\text{m}^2}\right)\frac{(\text{Weight in lb})}{(\text{Height in in.})^2}}$$

For example, if your weight and height are 180 lb and 71 in., your body mass index is 25 kg/m². The BMI can be used to assess approximately whether your weight is normal for your height (see Table 1.3).

TABLE 1.3	**The Body Mass Index**
BMI (kg/m²)	**Evaluation**
Below 18.5	Underweight
18.5–24.9	Normal
25.0–29.9	Overweight
30.0–39.9	Obese
40 and above	Morbidly obese

---

*Weight and mass are different concepts, and the relationship between them will be discussed in Section 4.7.

Dimensional Analysis

We have seen that many quantities are denoted by specifying both a number and a unit. For example, the distance to the nearest telephone may be 8 meters, or the speed of a car might be 25 meters/second. Each quantity, according to its physical nature, requires a certain *type* of unit. Distance must be measured in a length unit such as meters, feet, or miles, and a time unit will not do. Likewise, the speed of an object must be specified as a length unit divided by a time unit. In physics, the term **dimension** is used to refer to the physical nature of a quantity and the type of unit used to specify it. Distance has the dimension of length, which is symbolized as [L], while speed has the dimensions of length [L] divided by time [T], or [L/T]. Many physical quantities can be expressed in terms of a combination of fundamental dimensions such as length [L], time [T], and mass [M]. Later on, we will encounter certain other quantities, such as temperature, which are also fundamental. A fundamental quantity like temperature cannot be expressed as a combination of the dimensions of length, time, mass, or any other fundamental dimension.

Dimensional analysis is used to check mathematical relations for the consistency of their dimensions. As an illustration, consider a car that starts from rest and accelerates to a speed v in a time t. Suppose we wish to calculate the distance x traveled by the car but are not sure whether the correct relation is $x = \frac{1}{2}vt^2$ or $x = \frac{1}{2}vt$. We can decide by checking the quantities on both sides of the equals sign to see whether they have the same dimensions. If the dimensions are not the same, the relation is incorrect. For $x = \frac{1}{2}vt^2$, we use the dimensions for distance [L], time [T], and speed [L/T] in the following way:

$$x = \tfrac{1}{2}vt^2$$

Dimensions $\qquad\qquad [\text{L}] \overset{?}{=} \left[\dfrac{\text{L}}{T}\right][\text{T}]^2 = [\text{L}][\text{T}]$

Dimensions cancel just like algebraic quantities, and pure numerical factors like $\frac{1}{2}$ have no dimensions, so they can be ignored. The dimension on the left of the equals sign does not match those on the right, so the relation $x = \frac{1}{2}vt^2$ cannot be correct. On the other hand, applying dimensional analysis to $x = \frac{1}{2}vt$, we find that

$$x = \tfrac{1}{2}vt$$

Dimensions $\qquad\qquad [\text{L}] \overset{?}{=} \left[\dfrac{\text{L}}{T}\right][T] = [\text{L}]$

> **Problem-Solving Insight**   You can check for errors that may have arisen during algebraic manipulations by performing a dimensional analysis on the final expression.

The dimension on the left of the equals sign matches that on the right, so this relation is dimensionally correct. If we know that one of our two choices is the right one, then $x = \frac{1}{2}vt$ is it. In the absence of such knowledge, however, dimensional analysis cannot identify the correct relation. It can only identify which choices *may be* correct, since it does not account for numerical factors like $\frac{1}{2}$ or for the manner in which an equation was derived from physics principles.

Check Your Understanding

(The answers are given at the end of the book.)

1. **(a)** Is it possible for two quantities to have the same dimensions but different units?

   (b) Is it possible for two quantities to have the same units but different dimensions?

2. You can always add two numbers that have the same units (such as 6 meters + 3 meters). Can you always add two numbers that have the same dimensions, such as two numbers that have the dimensions of length [L]?

3. The following table lists four variables, along with their units:

Variable	Units
x	Meters (m)
v	Meters per second (m/s)
t	Seconds (s)
a	Meters per second squared (m/s^2)

These variables appear in the following equations, along with a few numbers that have no units. In which of the equations are the units on the left side of the equals sign consistent with the units on the right side?

(a) $x = vt$

(b) $x = vt + \frac{1}{2}at^2$

(c) $v = at$

(d) $v = at + \frac{1}{2}at^3$

(e) $v^3 = 2ax^2$

(f) $t = \sqrt{\frac{2x}{a}}$

4. In the equation $y = c^n at^2$ you wish to determine the integer value (1, 2, etc.) of the exponent n. The dimensions of y, a, and t are known. It is also known that c has no dimensions. Can dimensional analysis be used to determine n?

1.4 Trigonometry

Scientists use mathematics to help them describe how the physical universe works, and trigonometry is an important branch of mathematics. Three trigonometric functions are utilized throughout this text. They are the sine, the cosine, and the tangent of the angle θ (Greek theta), abbreviated as sin θ, cos θ, and tan θ, respectively. These functions are defined below in terms of the symbols given along with the right triangle in **Interactive Figure 1.6**.

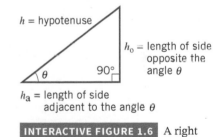

h = hypotenuse
h_o = length of side opposite the angle θ
h_a = length of side adjacent to the angle θ

INTERACTIVE FIGURE 1.6 A right triangle.

DEFINITION OF SIN θ, COS θ, AND TAN θ

$$\sin \theta = \frac{h_o}{h} \qquad (1.1)$$

$$\cos \theta = \frac{h_a}{h} \qquad (1.2)$$

$$\tan \theta = \frac{h_o}{h_a} \qquad (1.3)$$

h = length of the **hypotenuse** of a right triangle
h_o = length of the side **opposite** the angle θ
h_a = length of the side **adjacent** to the angle θ

The sine, cosine, and tangent of an angle are numbers without units, because each is the ratio of the lengths of two sides of a right triangle. Example 4 illustrates a typical application of Equation 1.3.

EXAMPLE 4 | Using Trigonometric Functions

On a sunny day, a tall building casts a shadow that is 67.2 m long. The angle between the sun's rays and the ground is $\theta = 50.0°$, as **Figure 1.7** shows. Determine the height of the building.

Reasoning We want to find the height of the building. Therefore, we begin with the colored right triangle in **Figure 1.7** and identify the height as the length h_o of the side opposite the angle θ. The length of the shadow is the length h_a of the side that is adjacent to the angle θ. The ratio of the length of the opposite side to the length of the adjacent side is the tangent of the angle θ, which can be used to find the height of the building.

Solution We use the tangent function in the following way, with $\theta = 50.0°$ and $h_a = 67.2$ m:

$$\tan \theta = \frac{h_o}{h_a} \tag{1.3}$$

$$h_o = h_a \tan \theta = (67.2 \text{ m})(\tan 50.0°) = (67.2 \text{ m})(1.19) = \boxed{80.0 \text{ m}}$$

The value of tan 50.0° is found by using a calculator.

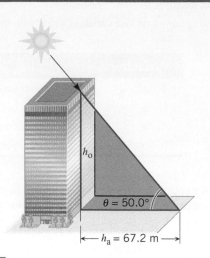

FIGURE 1.7 From a value for the angle θ and the length h_a of the shadow, the height h_o of the building can be found using trigonometry.

The sine, cosine, or tangent may be used in calculations such as that in Example 4, depending on which side of the triangle has a known value and which side is asked for.

> **Problem-Solving Insight** However, the choice of which side of the triangle to label h_o (opposite) and which to label h_a (adjacent) can be made only after the angle θ is identified.

Often the values for two sides of the right triangle in **Interactive Figure 1.6** are available, and the value of the angle θ is unknown. The concept of **inverse trigonometric functions** plays an important role in such situations. Equations 1.4–1.6 give the inverse sine, inverse cosine, and inverse tangent in terms of the symbols used in the drawing. For instance, Equation 1.4 is read as "θ equals the angle whose sine is h_o/h."

$$\theta = \sin^{-1}\left(\frac{h_o}{h}\right) \tag{1.4}$$

$$\theta = \cos^{-1}\left(\frac{h_a}{h}\right) \tag{1.5}$$

$$\theta = \tan^{-1}\left(\frac{h_o}{h_a}\right) \tag{1.6}$$

The use of -1 as an exponent in Equations 1.4–1.6 *does not mean* "take the reciprocal." For instance, $\tan^{-1}(h_o/h_a)$ does not equal $1/\tan(h_o/h_a)$. Another way to express the inverse trigonometric functions is to use arc sin, arc cos, and arc tan instead of $\sin^{-1}$, $\cos^{-1}$, and $\tan^{-1}$. Example 5 illustrates the use of an inverse trigonometric function.

EXAMPLE 5 | Using Inverse Trigonometric Functions

A lakefront drops off gradually at an angle θ, as **Figure 1.8** indicates. For safety reasons, it is necessary to know how deep the lake is at various distances from the shore. To provide some information about the depth, a lifeguard rows straight out from the shore a distance of 14.0 m and drops a weighted fishing line. By measuring the length of the line, the lifeguard determines the depth to be 2.25 m. **(a)** What is the value of θ? **(b)** What would be the depth d of the lake at a distance of 22.0 m from the shore?

Reasoning Near the shore, the lengths of the opposite and adjacent sides of the right triangle in **Figure 1.8** are $h_o = 2.25$ m and $h_a = 14.0$ m, relative to the angle θ. Having made this identification, we can use the inverse tangent to find the angle in part (a). For part (b) the opposite and adjacent sides farther from the shore become $h_o = d$ and $h_a = 22.0$ m. With the value for θ obtained in part (a), the tangent function can be used to find the unknown depth. Considering the way in which the lake bottom drops off in **Figure 1.8**, we expect the unknown depth to be greater than 2.25 m.

Solution **(a)** Using the inverse tangent given in Equation 1.6, we find that

$$\theta = \tan^{-1}\left(\frac{h_o}{h_a}\right) = \tan^{-1}\left(\frac{2.25 \text{ m}}{14.0 \text{ m}}\right) = \boxed{9.13°}$$

(b) With $\theta = 9.13°$, the tangent function given in Equation 1.3 can be used to find the unknown depth farther from the shore, where $h_o = d$ and $h_a = 22.0$ m. Since $\tan \theta = h_o/h_a$, it follows that

$$h_o = h_a \tan \theta$$

$$d = (22.0 \text{ m})(\tan 9.13°) = \boxed{3.54 \text{ m}}$$

which is greater than 2.25 m, as expected.

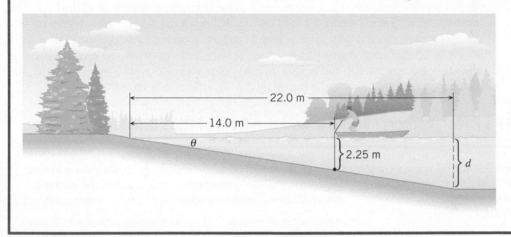

FIGURE 1.8 If the distance from the shore and the depth of the water at any one point are known, the angle θ can be found with the aid of trigonometry. Knowing the value of θ is useful, because then the depth d at another point can be determined.

The right triangle in **Interactive Figure 1.6** provides the basis for defining the various trigonometric functions according to Equations 1.1–1.3. These functions always involve an angle and two sides of the triangle. There is also a relationship among the lengths of the three sides of a right triangle. This relationship is known as the **Pythagorean theorem** and is used often in this text.

PYTHAGOREAN THEOREM

The square of the length of the hypotenuse of a right triangle is equal to the sum of the squares of the lengths of the other two sides:

$$h^2 = h_o^2 + h_a^2 \qquad\qquad (1.7)$$

1.5 | Scalars and Vectors

The volume of water in a swimming pool might be 50 cubic meters, or the winning time of a race could be 11.3 seconds. In cases like these, only the size of the numbers matters. In other words, *how much* volume or time is there? The 50 specifies the amount of water in units of cubic meters, while the 11.3 specifies the amount of time in seconds. Volume and time are examples of scalar quantities. A **scalar quantity** is one that can be described with a single number (including any units) giving its size or magnitude. Some other common scalars are temperature (e.g., 20 °C) and mass (e.g., 85 kg).

While many quantities in physics are scalars, there are also many that are not, and for these quantities the magnitude tells only part of the story. Consider **Figure 1.9**, which depicts a car that has moved 2 km along a straight line from start to finish. When describing the motion, it is incomplete to say that "the car moved a distance of 2 km." This statement would indicate only that the car ends up somewhere on a circle whose center is at the starting point and whose radius is 2 km. A complete description must include the direction along with the distance, as in the statement "the car moved a distance of 2 km in a direction 30° north of east." A quantity that deals inherently with *both magnitude and direction* is called a **vector quantity**. Because direction is an important characteristic of vectors, arrows are used to represent them; ***the direction of the arrow gives the direction of the vector***. The colored arrow in **Figure 1.9**, for example, is called the *displacement vector*, because it shows how the car is displaced from its starting point. Chapter 2 discusses this particular vector.

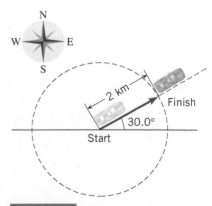

FIGURE 1.9 A vector quantity has a magnitude and a direction. The colored arrow in this drawing represents a displacement vector.

The length of the arrow in **Figure 1.9** represents the magnitude of the displacement vector. If the car had moved 4 km instead of 2 km from the starting point, the arrow would have been drawn twice as long. ***By convention, the length of a vector arrow is proportional to the magnitude of the vector***.

In physics there are many important kinds of vectors, and the practice of using the length of an arrow to represent the magnitude of a vector applies to each of them. All forces, for instance, are vectors. In common usage a force is a push or a pull, and the direction in which a force acts is just as important as the strength or magnitude of the force. The magnitude of a force is measured in SI units called newtons (N). An arrow representing a force of 20 newtons is drawn twice as long as one representing a force of 10 newtons.

The fundamental distinction between scalars and vectors is the characteristic of direction. Vectors have it, and scalars do not. Conceptual Example 6 helps to clarify this distinction and explains what is meant by the "direction" of a vector.

CONCEPTUAL EXAMPLE 6 | Vectors, Scalars, and the Role of Plus and Minus Signs

There are places where the temperature is +20 °C at one time of the year and −20 °C at another time. Do the plus and minus signs that signify positive and negative temperatures imply that temperature is a vector quantity? **(a)** Yes **(b)** No

Reasoning   A hallmark of a vector is that there is both a magnitude and a physical direction associated with it, such as 20 meters due east or 20 meters due west.

Answer (a) is incorrect.   The plus and minus signs associated with +20 °C and −20 °C do not convey a physical direction, such as due east or due west. Therefore, temperature cannot be a vector quantity.

Answer (b) is correct.   On a thermometer, the algebraic signs simply mean that the temperature is a number less than or greater than zero on the temperature scale being used and have nothing to do with east, west, or any other physical direction. Temperature, then, is not a vector. It is a scalar, and scalars can sometimes be negative.

Often, for the sake of convenience, quantities such as volume, time, displacement, velocity, and force are represented in physics by symbols. In this text, we write vectors in boldface symbols (**this is boldface**) with arrows above them* and write scalars in italic symbols (*this is italic*). Thus, a displacement vector is written as "$\vec{\mathbf{A}}$ = 750 m, due east," where the $\vec{\mathbf{A}}$ is a boldface symbol. By itself, however, separated from the direction, the magnitude of this vector is a scalar quantity. Therefore, the magnitude is written as "A = 750 m," where the A is an italic symbol without an arrow.

Check Your Understanding

(*The answer is given at the end of the book.*)

5. Which of the following statements, if any, involves a vector? **(a)** I walked 2 miles along the beach. **(b)** I walked 2 miles due north along the beach. **(c)** I jumped off a cliff and hit the water traveling at 17 miles per hour. **(d)** I jumped off a cliff and hit the water traveling straight down at a speed of 17 miles per hour. **(e)** My bank account shows a negative balance of −25 dollars.

1.6 Vector Addition and Subtraction

Addition

Often it is necessary to add one vector to another, and the process of addition must take into account both the magnitude and the direction of the vectors. The simplest situation occurs when the vectors point along the same direction—that is, when they are colinear,

---

*Vectors are also sometimes written in other texts as boldface symbols without arrows above them.

as in **Figure 1.10**. Here, a car first moves along a straight line, with a displacement vector $\vec{A}$ of 275 m, due east. Then the car moves again in the same direction, with a displacement vector $\vec{B}$ of 125 m, due east. These two vectors add to give the total displacement vector $\vec{R}$, which would apply if the car had moved from start to finish in one step. The symbol $\vec{R}$ is used because the total vector is often called the **resultant vector**. With the tail of the second arrow located at the head of the first arrow, the two lengths simply add to give the length of the total displacement. This kind of vector addition is identical to the familiar addition of two scalar numbers ($2 + 3 = 5$) *and can be carried out here only because the vectors point along the same direction.* In such cases we add the individual magnitudes to get the magnitude of the total, knowing in advance what the direction must be. Formally, the addition is written as follows:

$$\vec{R} = \vec{A} + \vec{B}$$

$$\vec{R} = 275 \text{ m, due east} + 125 \text{ m, due east} = 400 \text{ m, due east}$$

Perpendicular vectors are frequently encountered, and **Figure 1.11** indicates how they can be added. This figure applies to a car that first travels with a displacement vector $\vec{A}$ of 275 m, due east, and then with a displacement vector $\vec{B}$ of 125 m, due north. The two vectors add to give a resultant displacement vector $\vec{R}$. Once again, the vectors to be added are arranged in a tail-to-head fashion, and the resultant vector points from the tail of the first to the head of the last vector added. The resultant displacement is given by the vector equation

$$\vec{R} = \vec{A} + \vec{B}$$

The addition in this equation cannot be carried out by writing $R = 275 \text{ m} + 125 \text{ m}$, because the vectors have different directions. Instead, we take advantage of the fact that the triangle in **Figure 1.11** is a right triangle and use the Pythagorean theorem (Equation 1.7). According to this theorem, the magnitude of $\vec{R}$ is

$$R = \sqrt{(275 \text{ m})^2 + (125 \text{ m})^2} = 302 \text{ m}$$

The angle θ in **Figure 1.11** gives the direction of the resultant vector. Since the lengths of all three sides of the right triangle are now known, $\sin \theta$, $\cos \theta$, or $\tan \theta$ can be used to determine θ. Noting that $\tan \theta = B/A$ and using the inverse trigonometric function, we find that:

$$\theta = \tan^{-1}\left(\frac{B}{A}\right) = \tan^{-1}\left(\frac{125 \text{ m}}{275 \text{ m}}\right) = 24.4°$$

Thus, the resultant displacement of the car has a magnitude of 302 m and points north of east at an angle of 24.4°. This displacement would bring the car from the start to the finish in **Figure 1.11** in a single straight-line step.

When two vectors to be added are not perpendicular, the tail-to-head arrangement does not lead to a right triangle, and the Pythagorean theorem cannot be used. **Figure 1.12a** illustrates such a case for a car that moves with a displacement $\vec{A}$ of 275 m, due east, and then with a displacement $\vec{B}$ of 125 m, in a direction 55.0° north of west. As usual, the resultant displacement vector $\vec{R}$ is directed from the tail of the first to the head of the last vector added. The vector addition is still given according to

$$\vec{R} = \vec{A} + \vec{B}$$

However, the magnitude of $\vec{R}$ is not $R = A + B$, because the vectors $\vec{A}$ and $\vec{B}$ do not have the same direction, and neither is it $R = \sqrt{A^2 + B^2}$, because the vectors are not perpendicular, so the Pythagorean theorem does not apply. Some other means must be used to find the magnitude and direction of the resultant vector.

One approach uses a graphical technique. In this method, a diagram is constructed in which the arrows are drawn tail to head. The lengths of the vector arrows are drawn to scale, and the angles are drawn accurately (with a protractor, perhaps). Then the length of the arrow representing the resultant vector is measured with a ruler. This length is converted to the magnitude of the resultant vector by using the scale factor with which

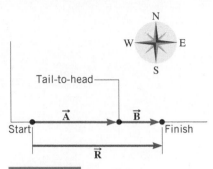

FIGURE 1.10 Two colinear displacement vectors $\vec{A}$ and $\vec{B}$ add to give the resultant displacement vector $\vec{R}$.

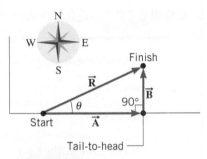

FIGURE 1.11 The addition of two perpendicular displacement vectors $\vec{A}$ and $\vec{B}$ gives the resultant vector $\vec{R}$.

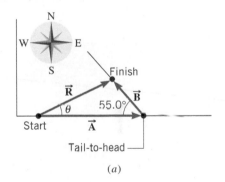

(a)

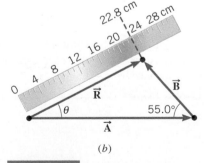

(b)

FIGURE 1.12 (a) The two displacement vectors $\vec{A}$ and $\vec{B}$ are neither colinear nor perpendicular, but even so they add to give the resultant vector $\vec{R}$. (b) In one method for adding them together, a graphical technique is used.

the drawing is constructed. In **Figure 1.12b**, for example, a scale of one centimeter of arrow length for each 10.0 m of displacement is used, and it can be seen that the length of the arrow representing $\vec{R}$ is 22.8 cm. Since each centimeter corresponds to 10.0 m of displacement, the magnitude of $\vec{R}$ is 228 m. The angle θ, which gives the direction of $\vec{R}$, can be measured with a protractor to be $\theta = 26.7°$ north of east.

Subtraction

The subtraction of one vector from another is carried out in a way that depends on the following fact. *When a vector is multiplied by −1, the magnitude of the vector remains the same, but the direction of the vector is reversed.* Conceptual Example 7 illustrates the meaning of this statement.

CONCEPTUAL EXAMPLE 7 | Multiplying a Vector by –1

Consider two vectors described as follows:

1. A woman climbs 1.2 m up a ladder, so that her displacement vector $\vec{D}$ is 1.2 m, upward along the ladder, as in **Figure 1.13a**.

2. A man is pushing with 450 N of force on his stalled car, trying to move it eastward. The force vector $\vec{F}$ that he applies to the car is 450 N, due east, as in **Figure 1.14a**.

What are the physical meanings of the vectors $-\vec{D}$ and $-\vec{F}$?
(a) $-\vec{D}$ points upward along the ladder and has a magnitude of −1.2 m; $-\vec{F}$ points due east and has a magnitude of −450 N.
(b) $-\vec{D}$ points downward along the ladder and has a magnitude of −1.2 m; $-\vec{F}$ points due west and has a magnitude of −450 N.

(c) $-\vec{D}$ points downward along the ladder and has a magnitude of 1.2 m; $-\vec{F}$ points due west and has a magnitude of 450 N.

Reasoning A displacement vector of $-\vec{D}$ is $(-1)\vec{D}$. The presence of the (−1) factor reverses the direction of the vector, but does not change its magnitude. Similarly, a force vector of $-\vec{F}$ has the same magnitude as the vector $\vec{F}$ but has the opposite direction.

Answer (a) and (b) are incorrect. While scalars can sometimes be negative, magnitudes of vectors are never negative.

Answer (c) is correct. The vectors $-\vec{D}$ and $-\vec{F}$ have the same magnitudes as $\vec{D}$ and $\vec{F}$, but point in the opposite direction, as indicated in **Figures 1.13b** and **1.14b**.

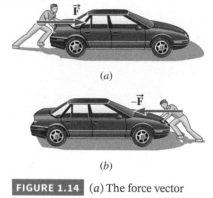

FIGURE 1.14 (a) The force vector for a man pushing on a car with 450 N of force in a direction due east is $\vec{F}$. (b) The force vector for a man pushing on a car with 450 N of force in a direction due west is $-\vec{F}$.

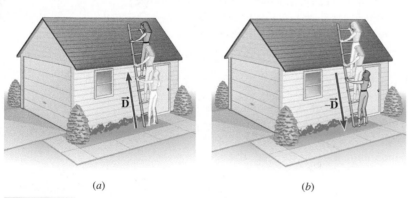

FIGURE 1.13 (a) The displacement vector for a woman climbing 1.2 m up a ladder is $\vec{D}$. (b) The displacement vector for a woman climbing 1.2 m down a ladder is $-\vec{D}$.

In practice, vector subtraction is carried out exactly like vector addition, except that one of the vectors added is multiplied by a scalar factor of −1. To see why, look at the two vectors $\vec{A}$ and $\vec{B}$ in **Figure 1.15a**. These vectors add together to give a third vector $\vec{C}$, according to $\vec{C} = \vec{A} + \vec{B}$. Therefore, we can calculate vector $\vec{A}$ as $\vec{A} = \vec{C} - \vec{B}$, which is an example of vector subtraction. However, we can also write this result as $\vec{A} = \vec{C} + (-\vec{B})$ and treat it as vector addition. **Figure 1.15b** shows how to calculate vector $\vec{A}$ by adding the vectors $\vec{C}$ and $-\vec{B}$. Notice that vectors $\vec{C}$ and $-\vec{B}$ are arranged tail to head and that any suitable method of vector addition can be employed to determine $\vec{A}$.

Check Your Understanding

(The answers are given at the end of the book.)

6. Two vectors $\vec{A}$ and $\vec{B}$ are added together to give a resultant vector $\vec{R}$: $\vec{R} = \vec{A} + \vec{B}$. The magnitudes of $\vec{A}$ and $\vec{B}$ are 3 m and 8 m, respectively, but the vectors can have any orientation. What are **(a)** the maximum possible value and **(b)** the minimum possible value for the magnitude of $\vec{R}$?

7. Can two nonzero perpendicular vectors be added together so their sum is zero?

8. Can three or more vectors with unequal magnitudes be added together so their sum is zero?

9. In preparation for this question, review Conceptual Example 7. Vectors $\vec{A}$ and $\vec{B}$ satisfy the vector equation $\vec{A} + \vec{B} = 0$. **(a)** How does the magnitude of $\vec{B}$ compare with the magnitude of $\vec{A}$? **(b)** How does the direction of $\vec{B}$ compare with the direction of $\vec{A}$?

10. Vectors $\vec{A}$, $\vec{B}$, and $\vec{C}$ satisfy the vector equation $\vec{A} + \vec{B} = \vec{C}$, and their magnitudes are related by the scalar equation $A^2 + B^2 = C^2$. How is vector $\vec{A}$ oriented with respect to vector $\vec{B}$?

11. Vectors $\vec{A}$, $\vec{B}$, and $\vec{C}$ satisfy the vector equation $\vec{A} + \vec{B} = \vec{C}$, and their magnitudes are related by the scalar equation $A + B = C$. How is vector $\vec{A}$ oriented with respect to vector $\vec{B}$?

1.7 | The Components of a Vector

Vector Components

Suppose a car moves along a straight line from start to finish, as in **Figure 1.16**, the corresponding displacement vector being $\vec{r}$. The magnitude and direction of the vector $\vec{r}$ give the distance and direction traveled along the straight line. However, the car could also arrive at the finish point by first moving due east, turning through 90°, and then moving due north. This alternative path is shown in the drawing and is associated with the two displacement vectors $\vec{x}$ and $\vec{y}$. The vectors $\vec{x}$ and $\vec{y}$ are called the *x* vector component and the *y* vector component of $\vec{r}$.

Vector components are very important in physics and have two basic features that are apparent in **Figure 1.16**. One is that the components add together to equal the original vector:

$$\vec{r} = \vec{x} + \vec{y}$$

The components $\vec{x}$ and $\vec{y}$, when added vectorially, convey exactly the same meaning as does the original vector $\vec{r}$: they indicate how the finish point is displaced relative to the starting point. The other feature of vector components that is apparent in **Figure 1.16** is that $\vec{x}$ and $\vec{y}$ are not just any two vectors that add together to give the original vector $\vec{r}$: they are perpendicular vectors. This perpendicularity is a valuable characteristic, as we will soon see.

Any type of vector may be expressed in terms of its components, in a way similar to that illustrated for the displacement vector in **Figure 1.16**. **Interactive Figure 1.17** shows an arbitrary vector $\vec{A}$ and its vector components $\vec{A}_x$ and $\vec{A}_y$. The components are drawn parallel to convenient *x* and *y* axes and are perpendicular. They add vectorially to equal the original vector $\vec{A}$:

$$\vec{A} = \vec{A}_x + \vec{A}_y$$

There are times when a drawing such as **Interactive Figure 1.17** is not the most convenient way to represent vector components, and **Figure 1.18** presents an alternative method. The disadvantage of this alternative is that the tail-to-head arrangement of $\vec{A}_x$ and $\vec{A}_y$ is missing, an arrangement that is a nice reminder that $\vec{A}_x$ and $\vec{A}_y$ add together to equal $\vec{A}$.

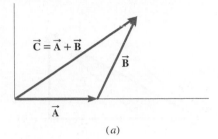

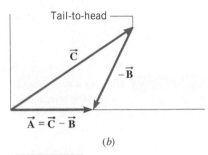

FIGURE 1.15 (*a*) Vector addition according to $\vec{C} = \vec{A} + \vec{B}$. (*b*) Vector subtraction according to $\vec{A} = \vec{C} - \vec{B} = \vec{C} + (-\vec{B})$.

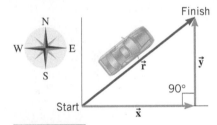

FIGURE 1.16 The displacement vector $\vec{r}$ and its vector components $\vec{x}$ and $\vec{y}$.

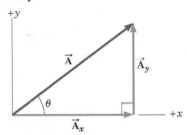

INTERACTIVE FIGURE 1.17 An arbitrary vector $\vec{A}$ and its vector components $\vec{A}_x$ and $\vec{A}_y$.

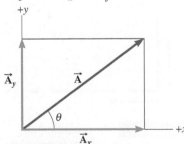

FIGURE 1.18 This alternative way of drawing the vector $\vec{A}$ and its vector components is completely equivalent to that shown in **Interactive Figure 1.17**.

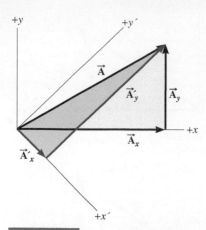

FIGURE 1.19 The vector components of the vector depend on the orientation of the axes used as a reference.

The definition that follows summarizes the meaning of vector components:

> **DEFINITION OF VECTOR COMPONENTS**
>
> In two dimensions, the vector components of a vector $\vec{A}$ are two perpendicular vectors $\vec{A}_x$ and $\vec{A}_y$ that are parallel to the x and y axes, respectively, and add together vectorially according to $\vec{A} = \vec{A}_x + \vec{A}_y$.

> **Problem-Solving Insight** In general, the components of any vector can be used in place of the vector itself in any calculation where it is convenient to do so.

The values calculated for vector components depend on the orientation of the vector relative to the axes used as a reference. **Figure 1.19** illustrates this fact for a vector $\vec{A}$ by showing two sets of axes, one set being rotated clockwise relative to the other. With respect to the black axes, vector $\vec{A}$ has perpendicular vector components $\vec{A}_x$ and $\vec{A}_y$; with respect to the colored rotated axes, vector $\vec{A}$ has different vector components $\vec{A}'_x$ and $\vec{A}'_y$. The choice of which set of components to use is purely a matter of convenience.

Scalar Components

It is often easier to work with the **scalar components**, A_x and A_y (note the italic symbols), rather than the vector components $\vec{A}_x$ and $\vec{A}_y$. Scalar components are positive or negative numbers (with units) that are defined as follows: The scalar component A_x has a magnitude equal to that of $\vec{A}_x$ and is given a positive sign if $\vec{A}_x$ points along the $+x$ axis and a negative sign if $\vec{A}_x$ points along the $-x$ axis. The scalar component A_y is defined in a similar manner. The following table shows an example of vector and scalar components:

Vector Components	Scalar Components	Unit Vectors
$\vec{A}_x$ = 8 meters, directed along the $+x$ axis	A_x = +8 meters	$\vec{A}_x$ = (+8 meters)$\hat{x}$
$\vec{A}_y$ = 10 meters, directed along the $-y$ axis	A_y = −10 meters	$\vec{A}_y$ = (−10 meters)$\hat{y}$

In this text, when we use the term "component," we will be referring to a scalar component, unless otherwise indicated.

Another method of expressing vector components is to use unit vectors. A **unit vector** is a vector that has a magnitude of 1, but no dimensions. We will use a caret (^) to distinguish it from other vectors. Thus,

$\hat{x}$ is a dimensionless unit vector of length l that points in the positive x direction, and

$\hat{y}$ is a dimensionless unit vector of length l that points in the positive y direction.

These unit vectors are illustrated in **Figure 1.20**. With the aid of unit vectors, the vector components of an arbitrary vector $\vec{A}$ can be written as $\vec{A}_x = A_x\hat{x}$ and $\vec{A}_y = A_y\hat{y}$, where A_x and A_y are its scalar components (see the drawing and the third column of the table above). The vector $\vec{A}$ is then written as $\vec{A} = A_x\hat{x} + A_y\hat{y}$.

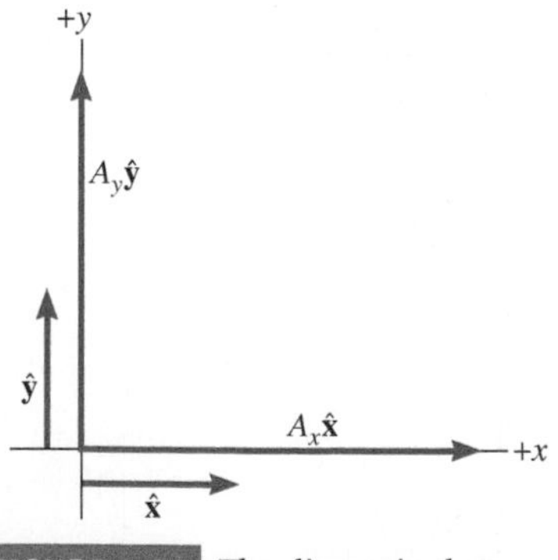

Resolving a Vector into Its Components

FIGURE 1.20 The dimensionless unit vectors $\hat{x}$ and $\hat{y}$ have magnitudes equal to 1, and they point in the $+x$ and $+y$ directions, respectively. Expressed in terms of unit vectors, the vector components of the vector $\vec{A}$ are $A_x\hat{x}$ and $A_y\hat{y}$.

If the magnitude and direction of a vector are known, it is possible to find the components of the vector. The process of finding the components is called "resolving the vector into its components." As Example 8 illustrates, this process can be carried out with the aid of trigonometry, because the two perpendicular vector components and the original vector form a right triangle.

EXAMPLE 8 | Finding the Components of a Vector

A displacement vector $\vec{r}$ has a magnitude of $r = 175$ m and points at an angle of 50.0° relative to the x axis in **Figure 1.21**. Find the x and y components of this vector.

Reasoning We will base our solution on the fact that the triangle formed in **Figure 1.21** by the vector $\vec{r}$ and its components $\vec{x}$ and $\vec{y}$ is a right triangle. This fact enables us to use the trigonometric sine and cosine functions, as defined in Equations 1.1 and 1.2.

> **Problem-Solving Insight** You can check to see whether the components of a vector are correct by substituting them into the Pythagorean theorem in order to calculate the magnitude of the original vector.

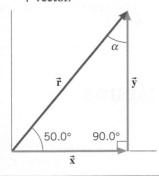

FIGURE 1.21 The x and y components of the displacement vector $\vec{r}$ can be found using trigonometry.

Solution The y component can be obtained using the 50.0° angle and Equation 1.1, $\sin \theta = y/r$:

$$y = r \sin \theta = (175 \text{ m})(\sin 50.0°) = \boxed{134 \text{ m}}$$

In a similar fashion, the x component can be obtained using the 50.0° angle and Equation 1.2, $\cos \theta = x/r$:

$$x = r \cos \theta = (175 \text{ m})(\cos 50.0°) = \boxed{112 \text{ m}}$$

Math Skills Either acute angle of a right triangle can be used to determine the components of a vector. The choice of angle is a matter of convenience. For instance, instead of the 50.0° angle, it is also possible to use the angle α in **Figure 1.21**. Since $\alpha + 50.0° = 90.0°$, it follows that $\alpha = 40.0°$. The solution using α yields the same answers as the solution using the 50.0° angle:

$$\cos \alpha = \frac{y}{r}$$

$$y = r \cos \alpha = (175 \text{ m})(\cos 40.0°) = \boxed{134 \text{ m}}$$

$$\sin \alpha = \frac{x}{r}$$

$$x = r \sin \alpha = (175 \text{ m})(\sin 40.0°) = \boxed{112 \text{ m}}$$

Since the vector components and the original vector form a right triangle, the Pythagorean theorem can be applied to check the validity of calculations such as those in Example 8. Thus, with the components obtained in Example 8, the theorem can be used to verify that the magnitude of the original vector is indeed 175 m, as given initially:

$$r = \sqrt{(112 \text{ m})^2 + (134 \text{ m})^2} = 175 \text{ m}$$

It is possible for one of the components of a vector to be zero. This does not mean that the vector itself is zero, however.

> **Problem-Solving Insight** For a vector to be zero, every vector component must individually be zero.

Thus, in two dimensions, saying that $\vec{A} = 0$ is equivalent to saying that $\vec{A}_x = 0$ and $\vec{A}_y = 0$. Or, stated in terms of scalar components, if $\vec{A} = 0$, then $A_x = 0$ and $A_y = 0$.

> **Problem-Solving Insight** Two vectors are equal if, and only if, they have the same magnitude and direction.

Thus, if one displacement vector points east and another points north, they are *not* equal, even if each has the same magnitude of 480 m. In terms of vector components, two vectors $\vec{A}$ and $\vec{B}$ are equal if, and only if, each vector component of one is equal to the corresponding vector component of the other. In two dimensions, if $\vec{A} = \vec{B}$, then $\vec{A}_x = \vec{B}_x$ and $\vec{A}_y = \vec{A}_y$. Alternatively, using scalar components, we write that $A_x = B_x$ and $A_y = B_y$.

Check Your Understanding

(*The answers are given at the end of the book.*)

12. Which of the following displacement vectors (if any) are equal?

Variable	Magnitude	Direction
$\vec{A}$	100 m	30° north of east
$\vec{B}$	100 m	30° south of west
$\vec{C}$	50 m	30° south of west
$\vec{D}$	100 m	60° east of north

13. Two vectors, $\vec{A}$ and $\vec{B}$, are shown in **CYU Figure 1.1**. **(a)** What are the signs (+ or −) of the scalar components, A_x and A_y, of the vector $\vec{A}$? **(b)** What are the signs of the scalar components, B_x and B_y, of the vector $\vec{B}$? **(c)** What are the signs of the scalar components, R_x and R_y, of the vector $\vec{R}$, where $\vec{R} = \vec{A} + \vec{B}$?

14. Are two vectors with the same magnitude necessarily equal?

15. The magnitude of a vector has doubled, its direction remaining the same. Can you conclude that the magnitude of each component of the vector has doubled?

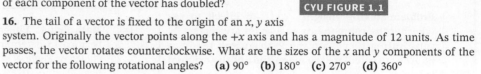

CYU FIGURE 1.1

16. The tail of a vector is fixed to the origin of an x, y axis system. Originally the vector points along the $+x$ axis and has a magnitude of 12 units. As time passes, the vector rotates counterclockwise. What are the sizes of the x and y components of the vector for the following rotational angles? **(a)** 90° **(b)** 180° **(c)** 270° **(d)** 360°

17. A vector has a component of zero along the x axis of a certain axis system. Does this vector necessarily have a component of zero along the x axis of another (rotated) axis system?

1.8 Addition of Vectors by Means of Components

The components of a vector provide the most convenient and accurate way of adding (or subtracting) any number of vectors. For example, suppose that vector $\vec{A}$ is added to vector $\vec{B}$. The resultant vector is $\vec{C}$, where $\vec{C} = \vec{A} + \vec{B}$. **Interactive Figure 1.22a** illustrates this vector addition, along with the x and y vector components of $\vec{A}$ and $\vec{B}$. In part b of the drawing, the vectors $\vec{A}$ and $\vec{B}$ have been removed, because we can use the vector components of these vectors in place of them. The vector component $\vec{B}_x$ has been shifted downward and arranged tail to head with vector component $\vec{A}_x$. Similarly, the vector component $\vec{A}_y$ has been shifted to the right and arranged tail to head with the vector component $\vec{B}_y$. The x components are colinear and add together to give the x component of the resultant vector $\vec{C}$. In like fashion, the y components are colinear and add together to give the y component of $\vec{C}$. In terms of scalar components, we can write

$$C_x = A_x + B_x \quad \text{and} \quad C_y = A_y + B_y$$

The vector components $\vec{C}_x$ and $\vec{C}_y$ of the resultant vector form the sides of the right triangle shown in **Interactive Figure 1.22c**. Thus, we can find the magnitude of $\vec{C}$ by using the Pythagorean theorem:

$$C = \sqrt{C_x^2 + C_y^2}$$

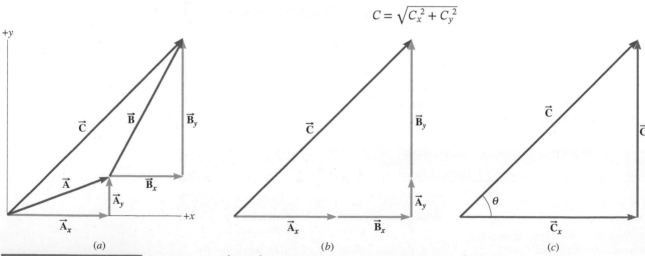

INTERACTIVE FIGURE 1.22 (a) The vectors $\vec{A}$ and $\vec{B}$ add together to give the resultant vector $\vec{C}$. The x and y components of $\vec{A}$ and $\vec{B}$ are also shown. (b) The drawing illustrates that $\vec{C}_x = \vec{A}_x + \vec{B}_x$ and $\vec{C}_y = \vec{A}_y + \vec{B}_y$. ($c$) Vector $\vec{C}$ and its components form a right triangle.

The angle θ that $\vec{C}$ makes with the x axis is given by $\theta = \tan^{-1}(C_y/C_x)$. Example 9 illustrates how to add several vectors using the component method.

Analyzing Multiple-Concept Problems

EXAMPLE 9 | The Component Method of Vector Addition

A jogger runs 145 m in a direction 20.0° east of north (displacement vector $\vec{A}$) and then 105 m in a direction 35.0° south of east (displacement vector $\vec{B}$). Using components, determine the magnitude and direction of the resultant vector $\vec{C}$ for these two displacements.

Reasoning **Figure 1.23** shows the vectors $\vec{A}$ and $\vec{B}$, assuming that the y axis corresponds to the direction due north. The vectors are arranged in a tail-to-head fashion, with the resultant vector $\vec{C}$ drawn from the tail of $\vec{A}$ to the head of $\vec{B}$. The components of the vectors are also shown in the figure. Since $\vec{C}$ and its components form a right triangle (red in the drawing), we will use the Pythagorean theorem and trigonometry to express the magnitude and directional angle θ for $\vec{C}$ in terms of its components. The components of $\vec{C}$ will then be obtained from the components of $\vec{A}$ and $\vec{B}$ and the data given for these two vectors.

Knowns and Unknowns The data for this problem are listed in the table that follows:

Description	Symbol	Value	Comment
Magnitude of vector $\vec{A}$		145 m	
Direction of vector $\vec{A}$		20.0° east of north	See **Figure 1.23**.
Magnitude of vector $\vec{B}$		105 m	
Direction of vector $\vec{B}$		35.0° south of east	See **Figure 1.23**.
Unknown Variable			
Magnitude of resultant vector	C	?	
Direction of resultant vector	θ	?	

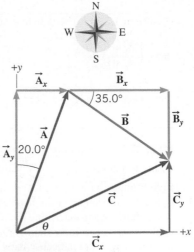

FIGURE 1.23 The vectors $\vec{A}$ and $\vec{B}$ add together to give the resultant vector $\vec{C}$. The vector components of $\vec{A}$ and $\vec{B}$ are also shown. The resultant vector $\vec{C}$ can be obtained once its components have been found.

Modeling the Problem

STEP 1 Magnitude and Direction of $\vec{C}$ In **Figure 1.23** the vector $\vec{C}$ and its components $\vec{C}_x$ and $\vec{C}_y$ form a right triangle, as the red arrows show. Applying the Pythagorean theorem to this right triangle shows that the magnitude of $\vec{C}$ is given by Equation 1a at the right. From the red triangle it also follows that the directional angle θ for the vector $\vec{C}$ is given by Equation 1b at the right.

$$C = \sqrt{C_x^2 + C_y^2} \qquad \text{(1a)}$$

$$\theta = \tan^{-1}\left(\frac{C_y}{C_x}\right) \qquad \text{(1b)}$$

STEP 2 Components of $\vec{C}$ Since vector $\vec{C}$ is the resultant of vectors $\vec{A}$ and $\vec{B}$, we have $\vec{C} = \vec{A} + \vec{B}$ and can write the scalar components of $\vec{C}$ as the sum of the scalar components of $\vec{A}$ and $\vec{B}$:

$$\boxed{C_x = A_x + B_x} \quad \text{and} \quad \boxed{C_y = A_y + B_y}$$

These expressions can be substituted into Equations 1a and 1b for the magnitude and direction of $\vec{C}$, as shown at the right.

$$C = \sqrt{C_x^2 + C_y^2} \qquad \text{(1a)}$$

$$\boxed{C_x = A_x + B_x} \quad \boxed{C_y = A_y + B_y} \qquad \text{(2)}$$

$$\theta = \tan^{-1}\left(\frac{C_y}{C_x}\right) \qquad \text{(1b)}$$

$$\boxed{C_x = A_x + B_x} \quad \boxed{C_y = A_y + B_y} \qquad \text{(2)}$$

Solution Algebraically combining the results of each step, we find that

$$\overset{\text{STEP 1}}{} \qquad \overset{\text{STEP 2}}{}$$

$$C = \sqrt{C_x^2 + C_y^2} = \sqrt{(A_x + B_x)^2 + (A_y + B_y)^2}$$

$$\theta = \tan^{-1}\left(\frac{C_y}{C_x}\right) = \tan^{-1}\left(\frac{A_y + B_y}{A_x + B_x}\right)$$

$$\underset{\text{STEP 1}}{} \qquad \underset{\text{STEP 2}}{}$$

To use these results we need values for the individual components of $\vec{A}$ and $\vec{B}$. Referring to **Figure 1.23**, we find these values to be

$A_x = (145 \text{ m}) \sin 20.0° = 49.6 \text{ m}$ and
$A_y = (145 \text{ m}) \cos 20.0° = 136 \text{ m}$

$B_x = (105 \text{ m}) \cos 35.0° = 86.0 \text{ m}$ and
$B_y = -(105 \text{ m}) \sin 35.0° = -60.2 \text{ m}$

Note that the component B_y is negative, because $\vec{B}_y$ points downward, in the negative y direction in the drawing. Substituting these values into the results for C and θ gives

$$C = \sqrt{(A_x + B_x)^2 + (A_y + B_y)^2}$$

$$= \sqrt{(49.6 \text{ m} + 86.0 \text{ m})^2 + (136 \text{ m} - 60.2 \text{ m})^2} = \boxed{155 \text{ m}}$$

$$\theta = \tan^{-1}\left(\frac{A_y + B_y}{A_x + B_x}\right)$$

$$= \tan^{-1}\left(\frac{136 \text{ m} - 60.2 \text{ m}}{49.6 \text{ m} + 86.0 \text{ m}}\right) = \boxed{29°}$$

Math Skills   According to the definitions given in Equations 1.1 and 1.2, the sine and cosine functions are $\sin \phi = \dfrac{h_o}{h}$ and $\cos \phi = \dfrac{h_a}{h}$, where h_o is the length of the side of a right triangle that is opposite the angle ϕ, h_a is the length of the side adjacent to the angle ϕ, and h is the length of the hypotenuse (see **Figure 1.24a**). Applications of the sine and cosine functions to determine the scalar components of a vector occur frequently. In such applications we begin by identifying the angle ϕ. **Figure 1.24b** shows the relevant portion of **Figure 1.23** and indicates that $\phi = 20.0°$ for the vector $\vec{A}$. In this case we have $h_o = A_x$, $h_a = A_y$, and $h = A = 145$ m; it follows that

$$\sin 20.0° = \frac{h_o}{h} = \frac{A_x}{A} \text{ or } A_x = A \sin 20.0° = (145 \text{ m}) \sin 20.0° = 49.6 \text{ m}$$

$$\cos 20.0° = \frac{h_a}{h} = \frac{A_y}{A} \text{ or } A_y = A \cos 20.0° = (145 \text{ m}) \cos 20.0° = 136 \text{ m}$$

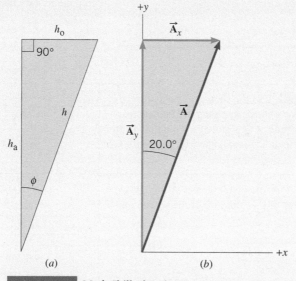

FIGURE 1.24   Math Skills drawing.

Related Homework: Problems 42, 44, 47, 50

Check Your Understanding

(*The answer is given at the end of the book.*)

18. Two vectors, $\vec{A}$ and $\vec{B}$, have vector components that are shown (to the same scale) in **CYU Figure 1.2**. The resultant vector is labeled $\vec{R}$. Which drawing shows the correct vector sum of $\vec{A} + \vec{B}$? **(a)** 1, **(b)** 2, **(c)** 3, **(d)** 4

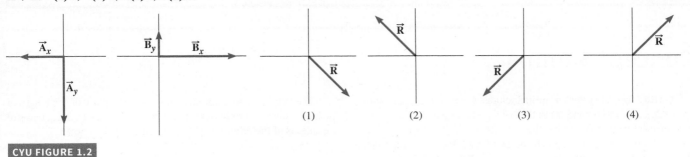

(1)   (2)   (3)   (4)

CYU FIGURE 1.2

In later chapters we will often use the component method for vector addition. For future reference, the main features of the reasoning strategy used in this technique are summarized below.

> REASONING STRATEGY   **The Component Method of Vector Addition**
>
> 1. **For each vector to be added, determine the x and y components relative to a conveniently chosen x, y coordinate system. Be sure to take into account the directions of the components by using plus and minus signs to denote whether the components point along the positive or negative axes.**
> 2. **Find the algebraic sum of the x components, which is the x component of the resultant vector. Similarly, find the algebraic sum of the y components, which is the y component of the resultant vector.**
> 3. **Use the x and y components of the resultant vector and the Pythagorean theorem to determine the magnitude of the resultant vector.**
> 4. **Use the inverse sine, inverse cosine, or inverse tangent function to find the angle that specifies the direction of the resultant vector.**

EXAMPLE 10 | BIO Multi-joint Movements

Figure 1.25 shows an example of a multi-joint movement involving the shoulder and elbow joints. The view from above shows a person holding a ball in a position that involves both shoulder flexion and elbow extension. Vector $\vec{A}$ represents the position of the elbow joint relative to the shoulder joint, and vector $\vec{B}$ represents the position of the ball relative to the elbow joint. Use the component method of vector addition and the angles given in the figure to find the magnitude and direction (θ) of vector $\vec{C}$, which represents the position of the ball relative to the shoulder joint. The magnitude of vector $\vec{A}$ is 35.6 cm, and the magnitude of vector $\vec{B}$ is 31.2 cm. The angle θ is measured relative to a vertical anatomical plane known as the frontal or coronal plane.

Reasoning Similar to Example 9, the vectors $\vec{A}$ and $\vec{B}$ in **Figure 1.25** are drawn tail-to-head. Thus, the resultant vector $\vec{C} = \vec{A} + \vec{B}$. The components of vector $\vec{C}$ will be obtained from the components of vectors $\vec{A}$ and $\vec{B}$. Once C_x and C_y are known, we can calculate the magnitude of vector $\vec{C}$ using the Pythagorean theorem. The directional angle θ will be determined from the components of vector $\vec{C}$.

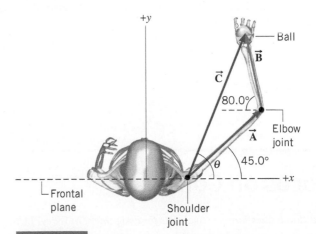

FIGURE 1.25 Top view of a multi-joint movement involving the shoulder and elbow joints. The vectors $\vec{A}$ and $\vec{B}$ add together to give the resultant vector $\vec{C}$, which represents the position vector of the ball with respect to the shoulder joint. The resultant vector $\vec{C}$ can be obtained, once its components are determined.

Solution   Applying the component method of vector addition, we know that $C_x = A_x + B_x$, and $C_y = A_y + B_y$. From **Figure 1.25**, we see that $A_x = (35.6 \text{ cm}) \cos 45° = 25.2 \text{ cm}$, $B_x = -(31.2) \cos 80° = -5.42 \text{ cm}$, $A_y = (35.6 \text{ cm}) \sin 45° = 25.2 \text{ cm}$, and $B_y = (31.2 \text{ cm}) \sin 80° = 30.7 \text{ cm}$. Therefore, $C_x = 25.2 \text{ cm} + (-5.42 \text{ cm}) = 19.8 \text{ cm}$, and $C_y = 25.2 \text{ cm} + 30.7 \text{ cm} = 55.9 \text{ cm}$. The magnitude of vector $\vec{C}$ is now found by the Pythagorean theorem:

$C = \sqrt{C_x^2 + C_y^2} = \sqrt{(19.8 \text{ cm})^2 + (55.9 \text{ cm})^2} = \boxed{59.3 \text{ cm}}$ The directional angle θ is found by using the tangent function:

$$\theta = \tan^{-1}\left(\frac{C_y}{C_x}\right) = \tan^{-1}\left(\frac{55.9 \text{ cm}}{19.8 \text{ cm}}\right) = \boxed{70.5°}.$$

Concept Summary

1.2 Units   The SI system of units includes the meter (m), the kilogram (kg), and the second (s) as the base units for length, mass, and time, respectively. One meter is the distance that light travels in a vacuum in a time of 1/299 792 458 second. One kilogram is the mass of a standard cylinder of platinum–iridium alloy kept at the International Bureau of Weights and Measures. One second is the time for a certain type of electromagnetic wave emitted by cesium-133 atoms to undergo 9 192 631 770 wave cycles.

1.3 The Role of Units in Problem Solving   To convert a number from one unit to another, multiply the number by the ratio of the two units. For instance, to convert 979 meters to feet, multiply 979 meters by the factor (3.281 foot/1 meter).

The dimension of a quantity represents its physical nature and the type of unit used to specify it. Three such dimensions are length [L], mass [M], time [T]. Dimensional analysis is a method for checking mathematical relations for the consistency of their dimensions.

1.4 Trigonometry   The sine, cosine, and tangent functions of an angle θ are defined in terms of a right triangle that contains θ, as in Equations 1.1–1.3, where h_o and h_a are, respectively, the lengths of the sides opposite and adjacent to the angle θ, and h is the length of the hypotenuse.

$$\sin \theta = \frac{h_o}{h} \qquad (1.1)$$

$$\cos \theta = \frac{h_a}{h} \qquad (1.2)$$

$$\tan \theta = \frac{h_o}{h_a} \qquad (1.3)$$

The inverse sine, inverse cosine, and inverse tangent functions are given in Equations 1.4–1.6.

$$\theta = \sin^{-1}\left(\frac{h_o}{h}\right) \qquad (1.4)$$

$$\theta = \cos^{-1}\left(\frac{h_a}{h}\right) \qquad (1.5)$$

$$\theta = \tan^{-1}\left(\frac{h_o}{h_a}\right) \qquad (1.6)$$

The Pythagorean theorem states that the square of the length of the hypotenuse of a right triangle is equal to the sum of the squares of the lengths of the other two sides, according to Equation 1.7.

$$h^2 = h_o^2 + h_a^2 \qquad (1.7)$$

1.5 Scalars and Vectors   A scalar quantity is described by its size, which is also called its magnitude. A vector quantity has both a magnitude and a direction. Vectors are often represented by arrows, the length of the arrow being proportional to the magnitude of the vector and the direction of the arrow indicating the direction of the vector.

1.6 Vector Addition and Subtraction   One procedure for adding vectors utilizes a graphical technique, in which the vectors to be added are arranged in a tail-to-head fashion. The resultant vector is drawn from the tail of the first vector to the head of the last vector. The subtraction of a vector is treated as the addition of a vector that has been multiplied by a scalar factor of −1. Multiplying a vector by −1 reverses the direction of the vector.

1.7 The Components of a Vector   In two dimensions, the vector components of a vector $\vec{A}$ are two perpendicular vectors $\vec{A}_x$ and $\vec{A}_y$ that are parallel to the x and y axes, respectively, and that add together vectorially so that $\vec{A} = \vec{A}_x + \vec{A}_y$. The scalar component A_x has a magnitude that is equal to that of $\vec{A}_x$ and is given a positive sign if $\vec{A}_x$ points along the $+x$ axis and a negative sign if $\vec{A}_x$ points along the $-x$ axis. The scalar component A_y is defined in a similar manner.

Two vectors are equal if, and only if, they have the same magnitude and direction. Alternatively, two vectors are equal in two dimensions if the x vector components of each are equal and the y vector components of each are equal. A vector is zero if, and only if, each of its vector components is zero.

1.8 Addition of Vectors by Means of Components   If two vectors $\vec{A}$ and $\vec{B}$ are added to give a resultant $\vec{C}$ such that $\vec{C} = \vec{A} + \vec{B}$, then $C_x = A_x + B_x$ and $C_y = A_y + B_y$, where C_x, A_x, and B_x are the scalar components of the vectors along the x direction, and C_y, A_y, and B_y are the scalar components of the vectors along the y direction.

Focus on Concepts

Online

Additional questions are available for assignment in WileyPLUS.

Section 1.6   Vector Addition and Subtraction

1. During a relay race, runner A runs a certain distance due north and then hands off the baton to runner B, who runs for the same distance in a direction south of east. The two displacement vectors $\vec{A}$ and $\vec{B}$ can be added together to give a resultant vector $\vec{R}$. Which drawing correctly shows the resultant vector? (a) 1 (b) 2 (c) 3 (d) 4

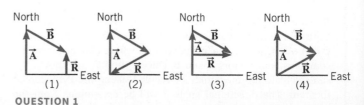

QUESTION 1

2. How is the magnitude R of the resultant vector $\vec{R}$ in the drawing related to the magnitudes A and B of the vectors $\vec{A}$ and $\vec{B}$?

(a) The magnitude of the resultant vector $\vec{R}$ is equal to the sum of the magnitudes of $\vec{A}$ and $\vec{B}$, or $R = A + B$. (b) The magnitude of the resultant vector $\vec{R}$ is greater than the sum of the magnitudes of $\vec{A}$ and $\vec{B}$, or $R > A$ $+ B$. (c) The magnitude of the resultant vector $\vec{R}$ is less than the sum of the magnitudes of $\vec{A}$ and $\vec{B}$, or $R < A + B$.

QUESTION 2

3. The first drawing shows three displacement vectors, $\vec{A}$, $\vec{B}$, and $\vec{C}$, which are added in a tail-to-head fashion. The resultant vector is labeled $\vec{R}$. Which of the following drawings shows the correct resultant vector for $\vec{A} + \vec{B} - \vec{C}$? (a) 1 (b) 2 (c) 3

(1)   (2)   (3)

QUESTION 3

4. The first drawing shows the sum of three displacement vectors, $\vec{A}$, $\vec{B}$, and $\vec{C}$. The resultant vector is labeled $\vec{R}$. Which of the following drawings shows the correct resultant vector for $\vec{A} - \vec{B} - \vec{C}$? (a) 1 (b) 2 (c) 3

(1)   (2)   (3)

QUESTION 4

Section 1.7   The Components of a Vector

5. A person is jogging along a straight line, and her displacement is denoted by the vector $\vec{A}$ in the drawings. Which drawing represents the correct vector components, $\vec{A}_x$ and $\vec{A}_y$, for the vector $\vec{A}$? (a) 1 (b) 2 (c) 3 (d) 4

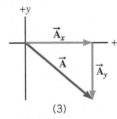 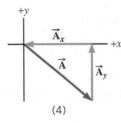

(1)   (2)

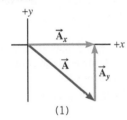 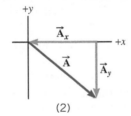

(3)   (4)

QUESTION 5

6. A person drives a car for a distance of 450.0 m. The displacement $\vec{A}$ of the car is illustrated in the drawing. What are the *scalar* components of this displacement vector?

(a) $A_x = 0$ m and $A_y = +450.0$ m

(b) $A_x = 0$ m and $A_y = -450.0$ m

(c) $A_x = +450.0$ m and $A_y = +450.0$ m

(d) $A_x = -450.0$ m and $A_y = 0$ m

(e) $A_x = -450.0$ m and $A_y = +450.0$ m

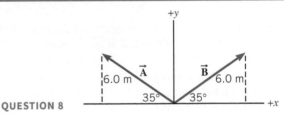

QUESTION 6

7. Drawing a shows a displacement vector $\vec{A}$ (450.0 m along the $-y$ axis). In this x, y coordinate system the scalar components are $A_x = 0$ m and $A_y = -450.0$ m. Suppose that the coordinate system is rotated counterclockwise by 35.0°, but the magnitude (450.0 m) and direction of vector $\vec{A}$ remain unchanged, as in drawing

b. What are the scalar components, $A_{x'}$ and $A_{y'}$, of the vector $\vec{A}$ in the rotated x', y' coordinate system?

(a) $A_{x'} = -369$ m and $A_{y'} = -258$ m

(b) $A_{x'} = +369$ m and $A_{y'} = -258$ m

(c) $A_{x'} = +258$ m and $A_{y'} = +369$ m

(d) $A_{x'} = +258$ m and $A_{y'} = -369$ m

(e) $A_{x'} = -258$ m and $A_{y'} = -369$ m

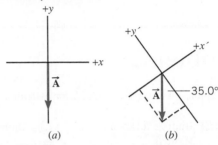

QUESTION 7   (a)   (b)

8. Suppose the vectors $\vec{A}$ and $\vec{B}$ in the drawing have magnitudes of 6.0 m and are directed as shown. What are A_x and B_x, the scalar components of $\vec{A}$ and $\vec{B}$ along the x axis?

	A_x	B_x
(a)	$+(6.0 \text{ m}) \cos 35° = +4.9$ m	$-(6.0 \text{ m}) \cos 35° = -4.9$ m
(b)	$+(6.0 \text{ m}) \sin 35° = +3.4$ m	$-(6.0 \text{ m}) \cos 35° = -4.9$ m
(c)	$-(6.0 \text{ m}) \cos 35° = -4.9$ m	$+(6.0 \text{ m}) \sin 35° = +3.4$ m
(d)	$-(6.0 \text{ m}) \cos 35° = -4.9$ m	$+(6.0 \text{ m}) \cos 35° = +4.9$ m
(e)	$-(6.0 \text{ m}) \sin 35° = -3.4$ m	$+(6.0 \text{ m}) \sin 35° = +3.4$ m

QUESTION 8

Section 1.8   Addition of Vectors by Means of Components

9. Drawing a shows two vectors $\vec{A}$ and $\vec{B}$, and drawing b shows their components. The scalar components of these vectors are as follows:

$$A_x = -4.9 \text{ m} \qquad A_y = +3.4 \text{ m}$$
$$B_x = +4.9 \text{ m} \qquad B_y = +3.4 \text{ m}$$

When the vectors $\vec{A}$ and $\vec{B}$ are added, the resultant vector is $\vec{R}$, so that $\vec{R} = \vec{A} + \vec{B}$. What are the values of R_x and R_y, the x and y components of $\vec{R}$?

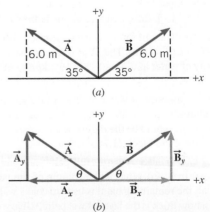

QUESTION 9   (b)

10. The displacement vectors $\vec{A}$ and $\vec{B}$, when added together, give the resultant vector $\vec{R}$, so that $\vec{R} = \vec{A} + \vec{B}$. Use the data in the drawing to find the magnitude R of the resultant vector and the angle θ that it makes with the $+x$ axis.

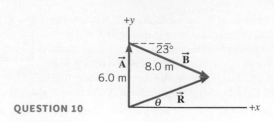

QUESTION 10

Problems

Online

Additional questions are available for assignment in WileyPLUS.

SSM Student Solutions Manual	**BIO** Biomedical application
MMH Problem-solving help	**E** Easy
GO Guided Online Tutorial	**M** Medium
V-HINT Video Hints	**H** Hard
CHALK Chalkboard Videos	**WS** Worksheet
	T Team Problem

Section 1.2 Units

Section 1.3 The Role of Units in Problem Solving

1. **E** **GO** A student sees a newspaper ad for an apartment that has 1330 square feet (ft^2) of floor space. How many square meters of area are there?

2. **E** Bicyclists in the Tour de France reach speeds of 34.0 miles per hour (mi/h) on flat sections of the road. What is this speed in **(a)** kilometers per hour (km/h) and **(b)** meters per second (m/s)?

3. **E** **SSM** Vesna Vulovic survived the longest fall on record without a parachute when her plane exploded and she fell 6 miles, 551 yards. What is this distance in meters?

4. **E** Suppose a man's scalp hair grows at a rate of 0.35 mm per day. What is this growth rate in feet per century?

5. **E** Given the quantities $a = 9.7$ m, $b = 4.2$ s, $c = 69$ m/s, what is the value of the quantity $d = a^3/(cb^2)$?

6. **E** Consider the equation $v = \frac{1}{3}zxt^2$. The dimensions of the variables v, x, and t are [L]/[T], [L], and [T], respectively. The numerical factor 3 is dimensionless. What must be the dimensions of the variable z, such that both sides of the equation have the same dimensions? Show how you determined your answer.

7. **E** **SSM** A bottle of wine known as a magnum contains a volume of 1.5 liters. A bottle known as a jeroboam contains 0.792 U.S. gallons. How many magnums are there in one jeroboam?

8. **E** The CGS unit for measuring the viscosity of a liquid is the poise (P): 1 P = 1 g/(s · cm). The SI unit for viscosity is the kg/(s · m). The viscosity of water at 0 °C is 1.78×10^{-3} kg/(s · m). Express this viscosity in poise.

9. **E** The amount of water on the Earth's surface is 326 million trillion gallons, or 3.26×10^{20} gallons. If all of this water was placed in a sphere, what would be the radius of this sphere in m? Compare this to the radius of the Moon (1.74×10^6 m).

10. **M** **GO** **CHALK** A partly full paint can has 0.67 U.S. gallons of paint left in it. **(a)** What is the volume of the paint in cubic meters? **(b)** If all the remaining paint is used to coat a wall evenly (wall area = 13 m^2), how thick is the layer of wet paint? Give your answer in meters.

11. **M** **SSM** **CHALK** A spring is hanging down from the ceiling, and an object of mass m is attached to the free end. The object is pulled down, thereby stretching the spring, and then released. The object oscillates up and down, and the time T required for one complete up-and-down oscillation is given by the equation $T = 2\pi\sqrt{m/k}$, where k is known as the spring constant. What must be the dimension of k for this equation to be dimensionally correct?

Section 1.4 Trigonometry

12. **E** You are driving into St. Louis, Missouri, and in the distance you see the famous Gateway to the West arch. This monument rises to a height of 192 m. You estimate your line of sight with the top of the arch to be 2.0° above the horizontal. Approximately how far (in kilometers) are you from the base of the arch?

13. **E** **SSM** A highway is to be built between two towns, one of which lies 35.0 km south and 72.0 km west of the other. What is the shortest length of highway that can be built between the two towns, and at what angle would this highway be directed with respect to due west?

14. **E** **GO** A hill that has a 12.0% grade is one that rises 12.0 m vertically for every 100.0 m of distance in the horizontal direction. At what angle is such a hill inclined above the horizontal?

15. **E** **GO** The corners of a square lie on a circle of diameter $D = 0.35$ m. Each side of the square has a length L. Find L.

16. **E** **GO** The drawing shows a person looking at a building on top of which an antenna is mounted. The horizontal distance between the person's eyes and the building is 85.0 m. In part a the person is looking at the base of the antenna, and his line of sight makes an angle of 35.0° with the horizontal. In part b the person is looking at the top of the antenna, and his line of sight makes an angle of 38.0° with the horizontal. How tall is the antenna?

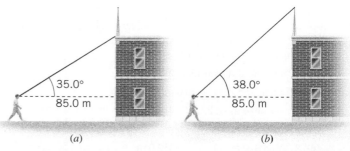

(a)   (b)

PROBLEM 16

17. **E** **CHALK** The two hot-air balloons in the drawing are 48.2 and 61.0 m above the ground. A person in the left balloon observes that the right balloon is 13.3° above the horizontal. What is the horizontal distance x between the two balloons?

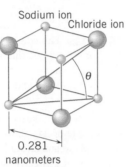

PROBLEM 17

18. **M** **MMH** The drawing shows sodium and chloride ions positioned at the corners of a cube that is part of the crystal structure of sodium chloride (common table salt). The edges of the cube are each 0.281 nm (1 nm = 1 nanometer = 10^{-9} m) in length. What is the value of the angle θ in the drawing?

19. **M** **GO** A person is standing at the edge of the water and looking out at the ocean (see the drawing). The height of the person's eyes above the water is $h = 1.6$ m, and the radius of the earth is $R = 6.38 \times 10^6$ m. **(a)** How far is it to the horizon? In other words, what is the distance d from the person's eyes to the horizon? *(Note: At the horizon the angle between the line of sight and the radius of the earth is 90°.)* **(b)** Express this distance in miles.

Sodium ion
Chloride ion

0.281 nanometers

PROBLEM 18

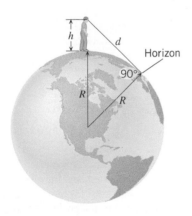

PROBLEM 19

20. **M** **SSM** Three deer, A, B, and C, are grazing in a field. Deer B is located 62 m from deer A at an angle of 51° north of west. Deer C is located 77° north of east relative to deer A. The distance between deer B and C is 95 m. What is the distance between deer A and C? *(Hint: Consider the law of cosines given in Appendix E.)*

21. **H** An aerialist on a high platform holds on to a trapeze attached to a support by an 8.0-m cord. (See the drawing.) Just before he jumps off the platform, the cord makes an angle of 41° with the vertical. He jumps, swings down, then back up, releasing the trapeze at the instant it is 0.75 m below its initial height. Calculate the angle θ that the trapeze cord makes with the vertical at this instant.

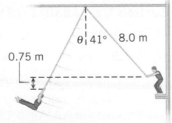

PROBLEM 21

Section 1.6   Vector Addition and Subtraction

22. **E** **SSM** **(a)** Two workers are trying to move a heavy crate. One pushes on the crate with a force $\vec{A}$, which has a magnitude of 445 newtons and is directed due west. The other pushes with a force $\vec{B}$, which has a magnitude of 325 newtons and is directed due north. What are the magnitude and direction of the resultant force $\vec{A} + \vec{B}$ applied to the crate? **(b)** Suppose that the second worker applies a force $-\vec{B}$ instead of $\vec{B}$. What then are the magnitude and direction of the resultant force $\vec{A} - \vec{B}$ applied to the crate? In both cases express the direction relative to due west.

23. **E** A force vector $\vec{F}_1$ points due east and has a magnitude of 200 newtons. A second force $\vec{F}_2$ is added to $\vec{F}_1$. The resultant of the two vectors has a magnitude of 400 newtons and points along the east/west line. Find the magnitude and direction of $\vec{F}_2$. Note that there are two answers.

24. **E** **SSM** Consider the following four force vectors:

$$\vec{F}_1 = 50.0 \text{ newtons, due east}$$
$$\vec{F}_2 = 10.0 \text{ newtons, due east}$$
$$\vec{F}_3 = 40.0 \text{ newtons, due west}$$
$$\vec{F}_4 = 30.0 \text{ newtons, due west}$$

Which two vectors add together to give a resultant with the smallest magnitude, and which two vectors add to give a resultant with the largest magnitude? In each case specify the magnitude and direction of the resultant.

25. **E** **GO** Vector $\vec{A}$ has a magnitude of 63 units and points due west, while vector $\vec{B}$ has the same magnitude and points due south. Find the magnitude and direction of **(a)** $\vec{A} + \vec{B}$ and **(b)** $\vec{A} - \vec{B}$. Specify the directions relative to due west.

26. **E** Two bicyclists, starting at the same place, are riding toward the same campground by two different routes. One cyclist rides 1080 m due east and then turns due north and travels another 1430 m before reaching the campground. The second cyclist starts out by heading due north for 1950 m and then turns and heads directly toward the campground. **(a)** At the turning point, how far is the second cyclist from the campground? **(b)** In what direction (measured relative to due east) must the second cyclist head during the last part of the trip?

27. **E** **GO** The drawing shows a triple jump on a checkerboard, starting at the center of square A and ending on the center of square B. Each side of a square measures 4.0 cm. What is the magnitude of the displacement of the colored checker during the triple jump?

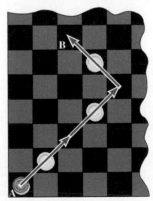

PROBLEM 27

28. **E** Given the vectors $\vec{P}$ and $\vec{Q}$ shown on the grid, sketch and calculate the magnitudes of the vectors **(a)** $\vec{M} = \vec{P} + \vec{Q}$ and **(b)** $\vec{K} = 2\vec{P} - \vec{Q}$. Use the tail-to-head method and express the magnitudes in centimeters with the aid of the grid scale shown in the drawing.

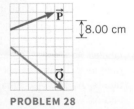

PROBLEM 28

29. ⬛M ⬛MMH ⬛CHALK Vector $\vec{A}$ has a magnitude of 12.3 units and points due west. Vector $\vec{B}$ points due north. **(a)** What is the magnitude of $\vec{B}$ if $\vec{A} + \vec{B}$ has a magnitude of 15.0 units? **(b)** What is the direction of $\vec{A} + \vec{B}$ relative to due west? **(c)** What is the magnitude of $\vec{B}$ if $\vec{A} - \vec{B}$ has a magnitude of 15.0 units? **(d)** What is the direction of $\vec{A} - \vec{B}$ relative to due west?

30. ⬛M ⬛SSM A car is being pulled out of the mud by two forces that are applied by the two ropes shown in the drawing. The dashed line in the drawing bisects the 30.0° angle. The magnitude of the force applied by each rope is 2900 newtons. Arrange the force vectors tail to head and use the graphical technique to answer the following questions. **(a)** How much force would a single rope need to apply to accomplish the same effect as the two forces added together? **(b)** How would the single rope be directed relative to the dashed line?

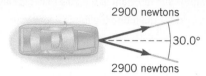

2900 newtons

30.0°

2900 newtons

PROBLEM 30

31. ⬛M ⬛GO A jogger travels a route that has two parts. The first is a displacement $\vec{A}$ of 2.50 km due south, and the second involves a displacement $\vec{B}$ that points due east. **(a)** The resultant displacement $\vec{A} + \vec{B}$ has a magnitude of 3.75 km. What is the magnitude of $\vec{B}$, and what is the direction of $\vec{A} + \vec{B}$ relative to due south? **(b)** Suppose that $\vec{A} - \vec{B}$ had a magnitude of 3.75 km. What then would be the magnitude of $\vec{B}$, and what is the direction of $\vec{A} - \vec{B}$ relative to due south?

32. ⬛M At a picnic, there is a contest in which hoses are used to shoot water at a beach ball from three directions. As a result, three forces act on the all, $\vec{F}_1$, $\vec{F}_2$, and $\vec{F}_3$ (see the drawing). The magnitudes of $\vec{F}_1$ and $\vec{F}_2$ are $F_1 = 50.0$ newtons and $F_2 = 90.0$ newtons. Using a scale drawing and the graphical technique, determine **(a)** the magnitude of $\vec{F}_3$ and **(b)** the angle θ such that the resultant force acting on the ball is zero.

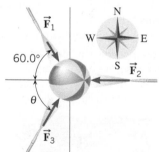

PROBLEM 32

Section 1.7   The Components of a Vector

33. ⬛E ⬛GO A force vector has a magnitude of 575 newtons and points at an angle of 36.0° below the positive x axis. What are **(a)** the x scalar component and **(b)** the y scalar component of the vector?

34. ⬛E ⬛SSM Vector $\vec{A}$ points along the $+y$ axis and has a magnitude of 100.0 units. Vector $\vec{B}$ points at an angle of 60.0° above the $+x$ axis and has a magnitude of 200.0 units. Vector $\vec{C}$ points along the $+x$ axis and has a magnitude of 150.0 units. Which vector has **(a)** the largest x component and **(b)** the largest y component?

35. ⬛E Soccer player #1 is 8.6 m from the goal (see the drawing). If she kicks the ball directly into the net, the ball has a displacement labeled $\vec{A}$. If, on the other hand, she first kicks it to player #2, who then kicks it into the net, the ball undergoes two successive displacements, $\vec{A}_y$ and $\vec{A}_x$. What are the magnitudes and directions of $\vec{A}_x$ and $\vec{A}_y$?

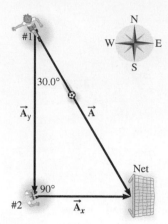

PROBLEM 35

36. ⬛E ⬛GO The components of vector $\vec{A}$ are A_x and A_y (both positive), and the angle that it makes with respect to the positive x axis is θ. Find the angle θ if the components of the displacement vector $\vec{A}$ are **(a)** $A_x = 12$ m and $A_y = 12$ m, **(b)** $A_x = 17$ m and $A_y = 12$ m, and **(c)** $A_x = 12$ m and $A_y = 17$ m.

37. ⬛E During takeoff, an airplane climbs with a speed of 180 m/s at an angle of 34° above the horizontal. The speed and direction of the airplane constitute a vector quantity known as the velocity. The sun is shining directly overhead. How fast is the shadow of the plane moving along the ground? (That is, what is the magnitude of the horizontal component of the plane's velocity?)

38. ⬛E ⬛SSM The x vector component of a displacement vector $\vec{r}$ has a magnitude of 125 m and points along the negative x axis. The y vector component has a magnitude of 184 m and points along the negative y axis. Find the magnitude and direction of $\vec{r}$. Specify the direction with respect to the negative x axis.

39. ⬛E Your friend has slipped and fallen. To help her up, you pull with a force $\vec{F}$, as the drawing shows. The vertical component of this force is 130 newtons, and the horizontal component is 150 newtons. Find **(a)** the magnitude of $\vec{F}$ and **(b)** the angle θ.

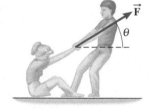

PROBLEM 39

40. ⬛M ⬛GO Two racing boats set out from the same dock and speed away at the same constant speed of 101 km/h for half an hour (0.500 h), the blue boat headed 25.0° south of west, and the green boat headed 37.0° south of west. During this half hour **(a)** how much farther west does the blue boat travel, compared to the green boat, and **(b)** how much farther south does the green boat travel, compared to the blue boat? Express your answers in km.

41. ⬛M ⬛SSM ⬛MMH ⬛CHALK The magnitude of the force vector $\vec{F}$ is 82.3 newtons. The x component of this vector is directed along the $+x$ axis and has a magnitude of 74.6 newtons. The y component points along the $+y$ axis. **(a)** Find the direction of $\vec{F}$ relative to the $+x$ axis. **(b)** Find the component of $\vec{F}$ along the $+y$ axis.

Section 1.8   Addition of Vectors by Means of Components

42. ⬛E ⬛SSM Consult Multiple-Concept Example 9 in preparation for this problem. A golfer, putting on a green, requires three strokes to "hole the ball." During the first putt, the ball rolls 5.0 m due east. For the second putt, the ball travels 2.1 m at an angle of 20.0° north of east. The third putt is 0.50 m due north. What displacement (magnitude and direction relative to due east) would have been needed to "hole the ball" on the very first putt?

43. **E** **CHALK** The three displacement vectors in the drawing have magnitudes of $A = 5.00$ m, $B = 5.00$ m, and $C = 4.00$ m. Find the resultant (magnitude and directional angle) of the three vectors by means of the component method. Express the directional angle as an angle above the positive or negative x axis.

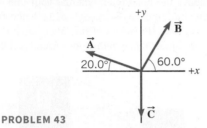

PROBLEM 43

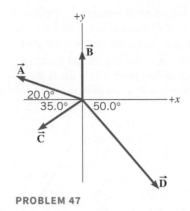

PROBLEM 47

44. **E** **MMH** Multiple-Concept Example 9 reviews the concepts that play a role in this problem. Two forces are applied to a tree stump to pull it out of the ground. Force $\vec{F}_A$ has a magnitude of 2240 newtons and points 34.0° south of east, while force $\vec{F}_B$ has a magnitude of 3160 newtons and points due south. Using the component method, find the magnitude and direction of the resultant force $\vec{F}_A + \vec{F}_B$ that is applied to the stump. Specify the direction with respect to due east.

45. **E** **GO** A baby elephant is stuck in a mud hole. To help pull it out, game keepers use a rope to apply a force $\vec{F}_A$, as part a of the drawing shows. By itself, however, force $\vec{F}_A$ is insufficient. Therefore, two additional forces $\vec{F}_B$ and $\vec{F}_C$ are applied, as in part b of the drawing. Each of these additional forces has the same magnitude F. The magnitude of the resultant force acting on the elephant in part b of the drawing is k times larger than that in part a. Find the ratio F/F_A when $k = 2.00$.

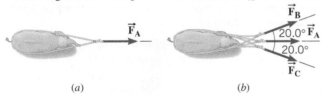

(a)   (b)

PROBLEM 45

46. **E** Displacement vector $\vec{A}$ points due east and has a magnitude of 2.00 km. Displacement vector $\vec{B}$ points due north and has a magnitude of 3.75 km. Displacement vector $\vec{C}$ points due west and has a magnitude of 2.50 km. Displacement vector $\vec{D}$ points due south and has a magnitude of 3.00 km. Find the magnitude and direction (relative to due west) of the resultant vector $\vec{A} + \vec{B} + \vec{C} + \vec{D}$.

47. **E** Multiple-Concept Example 9 provides background pertinent to this problem. The magnitudes of the four displacement vectors shown in the drawing are $A = 16.0$ m, $B = 11.0$ m, $C = 12.0$ m, and $D = 26.0$ m. Determine the magnitude and directional angle for the resultant that occurs when these vectors are added together.

48. **M** **GO** Two geological field teams are working in a remote area. A global positioning system (GPS) tracker at their base camp shows the location of the first team as 38 km away, 19° north of west, and the second team as 29 km away, 35° east of north. When the first team uses its GPS to check the position of the second team, what does the GPS give for the second team's (a) distance from them and (b) direction, measured from due east?

49. **M** **SSM** A sailboat race course consists of four legs, defined by the displacement vectors $\vec{A}$, $\vec{B}$, $\vec{C}$, and $\vec{D}$, as the drawing indicates. The magnitudes of the first three vectors are $A = 3.20$ km, $B = 5.10$ km, and $C = 4.80$ km. The finish line of the course coincides with the starting line. Using the data in the drawing, find the distance of the fourth leg and the angle θ.

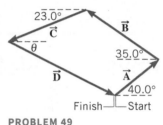

PROBLEM 49

50. **M** Multiple-Concept Example 9 deals with the concepts that are important in this problem. A grasshopper makes four jumps. The displacement vectors are (1) 27.0 cm, due west; (2) 23.0 cm, 35.0° south of west; (3) 28.0 cm, 55.0° south of east; and (4) 35.0 cm, 63.0° north of east. Find the magnitude and direction of the resultant displacement. Express the direction with respect to due west.

51. **M** The route followed by a hiker consists of three displacement vectors $\vec{A}$, $\vec{B}$, and $\vec{C}$. Vector $\vec{A}$ is along a measured trail and is 1550 m in a direction 25.0° north of east. Vector $\vec{B}$ is not along a measured trail, but the hiker uses a compass and knows that the direction is 41.0° east of south. Similarly, the direction of vector $\vec{C}$ is 35.0° north of west. The hiker ends up back where she started. Therefore, it follows that the resultant displacement is zero, or $\vec{A} + \vec{B} + \vec{C} = 0$. Find the magnitudes of (a) vector $\vec{B}$ and (b) vector $\vec{C}$.

Additional Problems   Online

52. **E** A monkey is chained to a stake in the ground. The stake is 3.00 m from a vertical pole, and the chain is 3.40 m long. How high can the monkey climb up the pole?

53. **M** A car drives from its initial position for 150 km in a direction that is 20° north of east and stops. The car then drives due north for 125 km and stops again. After the car comes to rest the second time, what is the displacement (a) magnitude and (b) direction of the car from its initial position? State your direction relative to due East.

54. **E** An ocean liner leaves New York City and travels 18.0° north of east for 155 km. How far east and how far north has it gone? In other words, what are the magnitudes of the components of the ship's displacement vector in the directions (a) due east and (b) due north?

55. **E** **GO** A pilot flies her route in two straight-line segments. The displacement vector $\vec{A}$ for the first segment has a magnitude of 244 km and a direction 30.0° north of east. The displacement vector $\vec{B}$ for the second segment has a magnitude of 175 km and a direction

due west. The resultant displacement vector is $\vec{R} = \vec{A} + \vec{B}$ and makes an angle θ with the direction due east. Using the component method, find the magnitude of $\vec{R}$ and the directional angle θ.

56. **M** **SSM** Vector $\vec{A}$ has a magnitude of 6.00 units and points due east. Vector $\vec{B}$ points due north. **(a)** What is the magnitude of $\vec{B}$, if the vector $\vec{A} + \vec{B}$ points 60.0° north of east? **(b)** Find the magnitude of $\vec{A} + \vec{B}$.

57. **M** **GO** Three forces act on an object, as indicated in the drawing. Force $\vec{F}_1$ has a magnitude of 21.0 newtons (21.0 N) and is directed 30.0° to the left of the $+y$ axis. Force $\vec{F}_2$ has a magnitude of 15.0 N and points along the $+x$ axis. What must be the magnitude and direction (specified by the angle θ in the drawing) of the third force $\vec{F}_3$ such that the vector sum of the three forces is 0 N?

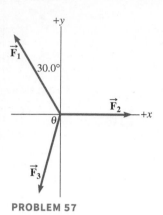

PROBLEM 57

58. **M** **CHALK** You live in the building on the left in the drawing, and a friend lives in the other building. The two of you are having a discussion about the heights of the buildings, and your friend claims that the height of his building is more than 1.50 times the height of yours. To resolve the issue you climb to the roof of your building and estimate that your line of sight to the top edge of the other building makes an angle of 21° above the horizontal, whereas your line of sight to the base of the other building makes an angle of 52° below the horizontal. Determine the ratio of the height of the taller building to the height of the shorter building. State whether your friend is right or wrong.

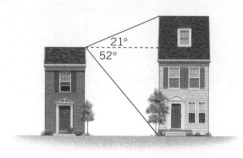

PROBLEM 58

Physics in Biology, Medicine, and Sports

59. **E** **BIO** **1.3** Azelastine hydrochloride is an antihistamine nasal spray. A standard size container holds one fluid ounce (oz) of the liquid. You are searching for this medication in a European drugstore and are asked how many milliliters (mL) there are in one fluid ounce. Using the following conversion factors, determine the number of milliliters in a volume of one fluid ounce: 1 gallon (gal) = 128 oz, 3.785×10^{-3} cubic meters (m³) = 1 gal, and 1 mL = 10^{-6} m³.

60. **E** **BIO** **1.3** The volume of liquid flowing per second is called the volume flow rate Q and has the dimensions of $[L]^3/[T]$. The flow rate of a liquid through a hypodermic needle during an injection can be estimated with the following equation:

$$Q = \frac{\pi R^n (P_2 - P_1)}{8\eta L}$$

The length and radius of the needle are L and R, respectively, both of which have the dimension $[L]$. The pressures at opposite ends of the needle are P_2 and P_1, both of which have the dimensions of $[M]/\{[L][T]^2\}$. The symbol η represents the viscosity of the liquid and has the dimensions of $[M]/\{[L][T]\}$. The symbol π stands for pi and, like the number 8 and the exponent n, has no dimensions. Using dimensional analysis, determine the value of n in the expression for Q.

61. **E** **BIO** **1.3** During their lifetime, a person will consume approximately 1.2×10^4 gallons of water. An Olympic-size swimming pool, which has a volume of 2.5 million liters (L), can provide a lifetime of water for how many people? Recall that 1 gallon (gal) = 3.785 $\times 10^{-3}$ m³, and 1 mL = 10^{-6} m³.

62. **E** **BIO** **1.3** Deoxyribonucleic acid, or DNA, is a molecule composed of two chains that wrap around each other to form a double helix, as the figure shows. DNA carries the genetic information for the growth and development of all known organisms. If you were to uncoil all of the DNA chains in a human body and lay them end-to-end, they would stretch for a distance of 1.6×10^{13} m. What is this distance in miles? For comparison, the average distance from Earth to Pluto is about 3.7 billion miles.

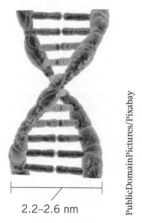

PROBLEM 62 2.2–2.6 nm

PublicDomainPictures/Pixabay

63. **M** **BIO** **1.3** An owl's keen eyesight is not the only ability that assists them in locating prey. They also have exceptional hearing. Take the barn owl, for example. Their entire head is designed to focus in on low frequency sounds. They have a dish-shaped face that collects and amplifies sound in much the same way a dish antenna collects electromagnetic signals from orbiting satellites (see images below). Their ears are hidden behind feathers on each side of their head behind their eyes. The position of their ear openings is not symmetrical, as their left ear is slightly higher than their right one. This asymmetry allows them to determine if a source of sound is above or below their line of sight. Furthermore, they use a technique known as *triangulation* to pinpoint the location of prey in front of them. Consider the distance between the owl's ears to correspond to the base of a triangle, with the long sides extending from each ear to a mouse located some distance away (see figure). If the mouse is directly in front of

the owl, then the time it takes the sound to travel from the mouse to its left and right ear will be the same. However, if the mouse is not directly in front, such as in the figure, then the sound will reach the owl's left and right ears at different times. The owl's auditory system is extremely sensitive to this time difference. In fact, it can detect a time difference between its left and right ear of 30 millionths of a second! The owl turns its head until the sound difference is eliminated, which indicates its prey is directly in front.

Pexels/Pixabay

Wikilmages/Pixabay

The figure shows an example of the mouse being located to the left of the owl's forward direction (dashed line). In this case, the sound created by the mouse would reach the owl's left ear a fraction of a second sooner than it reaches the right ear. Based on the information shown in the figure (not to scale) calculate the distance between the owl's ears (x).

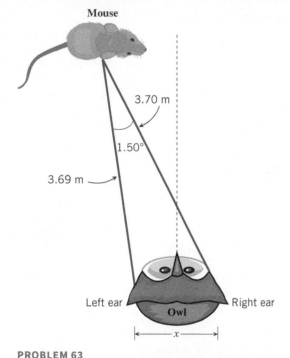

PROBLEM 63

64. **E** **BIO** **1.3**   The total surface area of the United States is 9.6×10^6 km^2. Seventy-four percent of this land can support spiders. Given that there are approximately 5.0×10^4 spiders per acre, how many spiders are there in the United States? Recall that 1 acre = 0.004 km^2.

65. **M** **BIO** **1.4**   Scoliosis is a common disease of the spine, where the spine is not straight, but forms an "S" or "C" shape while viewing a patient from the front. One of the most widely used indicators of scoliosis is a measurement of the Cobb angle (a), as it is a quantitative measure of the spinal curvature. The Cobb angle is calculated directly on X-ray radiographs of the spine (see photo). To determine the Cobb angle, the health professional must decide which two vertebrae are at each end of the deformity. They correspond to the two vertebrae that are rotated the most from a vertical position. In the following sample X-ray, these are shown with red arrows. Lines are then drawn (solid blue) along the endplates of these vertebrae until they intersect as shown, forming the Cobb angle a. In this example, an isosceles triangle is formed, where the two equivalent sides have a length h, and the third side has a length of $0.65h$. If the angle θ in the figure has a value of 38°, what is the value of the Cobb angle, a?

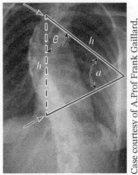

Case courtesy of A.Prof Frank Gaillard, Radiopaedia.org, rID: 49374

PROBLEM 65

Concepts and Calculations Problems

Online

This chapter has presented an introduction to the mathematics of trigonometry and vectors, which will be used throughout the text. In this section we apply some of the important features of this mathematics, and review some concepts that can help in anticipating some of the characteristics of the numerical answers.

66. [M] [CHALK] The figure shows two displacement vectors $\vec{A}$ and $\vec{B}$. Vector $\vec{A}$ points at an angle of 22° above the x axis and has an unknown magnitude. Vector $\vec{B}$ has an x component $B_x = 35.0$ m and has an unknown y component B_y. These two vectors are equal. *Concepts:* (i) What does the condition that vector $\vec{A}$ equals $\vec{B}$ imply about the magnitudes and directions of the vectors? (ii) What does the condition that vector $\vec{A}$ equals $\vec{B}$ imply about the x and y components of the vectors? *Calculations:* Find the magnitude of $\vec{A}$ and the value of B_y.

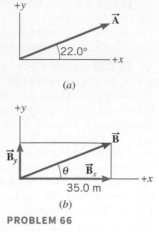

(a)

(b)

PROBLEM 66

67. [M] [CHALK] [SSM] The figure shows three displacement vectors $\vec{A}$, $\vec{B}$, and $\vec{C}$. These vectors are arranged in tail-to-head fashion, and they

add together to give a resultant displacement $\vec{R}$, which lies along the x axis. Note that the vector $\vec{B}$ is parallel to the x axis. *Concepts:* (i) How is the magnitude of $\vec{A}$ related to its scalar components A_x and A_y? (ii) Do any of the vectors in the figure have a zero value for either their x or y components? (If so, which ones?) (iii) What does the fact that $\vec{A}$, $\vec{B}$, and $\vec{C}$ add together to give $\vec{R}$ tell you about the components of these vectors? *Calculations:* What is the magnitude of the vector $\vec{A}$ and its directional angle θ?

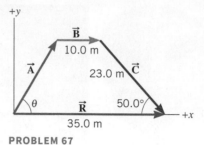

PROBLEM 67

Team Problems

Online

68. [E] [BIO] [WS] [T] **Recovering a Downed Eagle.** You are on a road that runs east and west, and another member of your team is on the same road, 1.20 km east of you. Someone reported seeing an eagle strike a wire and eventually go down somewhere in a dense forest north of the road. Your objective is to rescue the downed eagle. The bird is fitted with a transmitter, and you have a detector that indicates that it is located at a distance of 6.00×10^2 m from you. The team member located 1.20 km east of you has a similar detector that indicates that the eagle is 9.00×10^2 m away from them. Note that neither detector gives directional information. **(a)** Assuming a straight-line path, determine the direction you must go in order to find the eagle. **(b)** Assuming you want to minimize your walking distance in the thick forest, how should you proceed? Find the minimum distance traveled in the forest.

69. [E] [BIO] [WS] [T] **Focusing a Gamma Knife.** A gamma knife is a medical device used to deliver high intensity radiation to a tumor inside the human body while minimizing damage to the surrounding tissue. The device directs multiple radiation beams through the same point (e.g., at the position of a tumor) so that the intensity in the volume where the beams intersect is extremely high. Suppose the position of a tumor relative to the origin of an xy coordinate system is represented by a vector of magnitude 32.5 cm that makes an angle of $\theta_0 = +76.4°$ with respect to the $+x$ axis. If two radiation sources are located on the x axis at positions $x_1 = +10.7$ cm and $x_2 = -14.3$ cm, at what angles relative to the $+x$ axis must the radiation beams from these two sources be directed so that they intersect at the tumor?

70. [M] [T] **The Waterfall.** You and your team are exploring a river in South America when you come to the bottom of a tall waterfall. You estimate the cliff over which the water flows to be about 100 feet tall. You have to choose between climbing the cliff or backtracking and taking another route, but climbing the cliff would cut two hours off of your trip. There is only one experienced climber in the group: she would climb the cliff alone and drop a rope over the edge to lift supplies and allow the others to climb without packs. The climber estimates it will take her 45 minutes to get to the top. However, you

are concerned that the rope might be too short to reach the bottom of the cliff (it is exactly 30.0 m long). If it is too short, she'll have to climb back down (another 45 minutes) and you will be too far behind schedule to get to your destination before dark. As you contemplate how to determine whether the rope is long enough, you notice that the late afternoon shadow of the cliff grows as the sun descends over its edge. You suddenly remember your trigonometry. You measure the length of the shadow from the base of the cliff to the shadow's edge (144 ft), and the angle subtended between the base and top of the cliff measured from the shadow's edge. The angle is 38.1°. Do you send the climber, or start backtracking to take another route?

71. [M] [T] **The Weather Monitor.** Your South American expedition splits into two groups: one that stays at home base, and yours that goes off to set up a sensor that will monitor precipitation, temperature, and sunlight through the upcoming winter. The sensor must link up to a central communications system at base camp that simultaneously uploads the data from numerous sensors to a satellite. In order to set up and calibrate the sensor, you will have to communicate with base camp to give them specific location information. Unfortunately, the group's communication and navigation equipment has dwindled to walkie-talkies and a compass due to a river-raft mishap, which means your group must not exceed the range of the walkie-talkies (3.0 miles). However, you do have a laser rangefinder to help you measure distances as you navigate with the compass. After a few hours of hiking, you find the perfect plateau on which to mount the sensor. You have carefully mapped your path from base camp around lakes and other obstacles: 550.0 m West (W), 275 m S, 750.0 m W, 900.0 m NE, 800.0 m W, and 400.0 m 30.0° W of S. The final leg is due south, 2.20 km up a constant slope and ending at a plateau that is 320.0 m above the level of base camp. **(a)** How far are you from base camp? Will you be able to communicate with home base using the walkie-talkies? **(b)** What is the geographical direction from base camp to the sensor (expressed in the form $\theta°$ south of west, etc.)? **(c)** What is the angle of inclination from base camp to the detector?

The pilots in the United States Navy's Blue Angels can perform high-speed maneuvers in perfect unison. They do so by controlling the displacement, velocity, and acceleration of their jet aircraft. These three concepts and the relationships among them are the focus of this chapter.

erniedecker/iStockphoto

Kinematics in One Dimension

2.1 | Displacement

There are two aspects to any motion. In a purely descriptive sense, there is the movement itself. Is it rapid or slow, for instance? Then, there is the issue of what causes the motion or what changes it, which requires that forces be considered. **Kinematics** deals with the concepts that are needed to describe motion, without any reference to forces. The present chapter discusses these concepts as they apply to motion in one dimension, and the next chapter treats two-dimensional motion. **Dynamics** deals with the effect that forces have on motion, a topic that is considered in Chapter 4. Together, kinematics and dynamics form the branch of physics known as **mechanics**. We turn now to the first of the kinematics concepts to be discussed, which is displacement.

To describe the motion of an object, we must be able to specify the location of the object at all times, and **Figure 2.1** shows how to do this for one-dimensional motion. In this drawing, the initial position of a car is indicated by the vector labeled $\vec{x}_0$. The length of $\vec{x}_0$ is the distance of the car from an arbitrarily chosen origin. At a later time the car has moved

LEARNING OBJECTIVES

After reading this module, you should be able to...

2.1 Define one-dimensional displacement.

2.2 Discriminate between speed and velocity.

2.3 Define one-dimensional acceleration.

2.4 Use one-dimensional kinematic equations to predict future or past values of variables.

2.5 Solve one-dimensional kinematic problems.

2.6 Solve one-dimensional free-fall problems.

2.7 Predict kinematic quantities using graphical analysis.

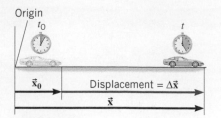

FIGURE 2.1 The displacement $\Delta\vec{x}$ is a vector that points from the initial position $\vec{x}_0$ to the final position $\vec{x}$.

to a new position, which is indicated by the vector $\vec{x}$. The **displacement** of the car $\Delta\vec{x}$ (read as "delta x" or "the change in x") is a vector drawn from the initial position to the final position. Displacement is a vector quantity in the sense discussed in Section 1.5, for it conveys both a magnitude (the distance between the initial and final positions) and a direction. The displacement can be related to $\vec{x}_0$ and $\vec{x}$ by noting from the drawing that

$$\vec{x}_0 + \Delta\vec{x} = \vec{x} \qquad \text{or} \qquad \Delta\vec{x} = \vec{x} - \vec{x}_0$$

Thus, the displacement $\Delta\vec{x}$ is the difference between $\vec{x}$ and $\vec{x}_0$, and the Greek letter delta (Δ) is used to signify this difference. It is important to note that the change in any variable is always the final value minus the initial value.

> **DEFINITION OF DISPLACEMENT**
>
> The displacement is a vector that points from an object's initial position to its final position and has a magnitude that equals the shortest distance between the two positions.
>
> **SI Unit of Displacement:** meter (m)

The SI unit for displacement is the meter (m), but there are other units as well, such as the centimeter and the inch. When converting between centimeters (cm) and inches (in.), remember that 2.54 cm = 1 in.

Often, we will deal with motion along a straight line. In such a case, a displacement in one direction along the line is assigned a positive value, and a displacement in the opposite direction is assigned a negative value. For instance, assume that a car is moving along an east/west direction and that a positive (+) sign is used to denote a direction due east. Then, $\Delta\vec{x} = +500$ m represents a displacement that points to the east and has a magnitude of 500 meters. Conversely, $\Delta\vec{x} = -500$ m is a displacement that has the same magnitude but points in the opposite direction, due west.

The magnitude of the displacement vector is the shortest distance between the initial and final positions of the object. However, this does not mean that displacement and distance are the same physical quantities. In **Figure 2.1**, for example, the car could reach its final position after going forward and backing up several times. In that case, the total distance traveled by the car would be greater than the magnitude of the displacement vector.

Check Your Understanding

(The answer is given at the end of the book.)

1. A honeybee leaves the hive and travels a total distance of 2 km before returning to the hive. What is the magnitude of the displacement vector of the bee?

2.2 | Speed and Velocity

Average Speed

One of the most obvious features of an object in motion is how fast it is moving. If a car travels 200 meters in 10 seconds, we say its average speed is 20 meters per second, the *average speed* being the distance traveled divided by the time required to cover the distance:

$$\text{Average speed} = \frac{\text{Distance}}{\text{Elapsed time}} \tag{2.1}$$

Equation 2.1 indicates that the unit for average speed is the unit for distance divided by the unit for time, or meters per second (m/s) in SI units. Example 1 illustrates how the idea of average speed is used.

EXAMPLE 1 | Distance Run by a Jogger

How far does a jogger run in 1.5 hours (5400 s) if his average speed is 2.22 m/s?

Reasoning The average speed of the jogger is the average distance per second that he travels. Thus, the distance covered by the jogger is equal to the average distance per second (his average speed) multiplied by the number of seconds (the elapsed time) that he runs.

Solution To find the distance run, we rewrite Equation 2.1 as

$$\text{Distance} = (\text{Average speed})(\text{Elapsed time}) = (2.22 \text{ m/s})(5400 \text{ s})$$
$$= \boxed{12\,000 \text{ m}}$$

Speed is a useful idea, because it indicates how fast an object is moving. However, speed does not reveal anything about the direction of the motion. To describe both how fast an object moves and the direction of its motion, we need the vector concept of velocity.

Average Velocity

To define the velocity of an object, we will use two concepts that we have already encountered, displacement and time. The building of new concepts from more basic ones is a common theme in physics. In fact the great strength of physics as a science is that it builds a coherent understanding of nature through the development of interrelated concepts.

Suppose that the initial position of the car in **Figure 2.1** is $\vec{x}_0$ when the time is t_0. A little later that car arrives at the final position $\vec{x}$ at the time t. The difference between these times is the time required for the car to travel between the two positions. We denote this difference by the shorthand notation Δt (read as "delta t"), where Δt represents the final time t minus the initial time t_0:

$$\underbrace{\Delta t = t - t_0}_{\text{Elapsed time}}$$

Note that Δt is defined in a manner analogous to $\Delta\vec{x}$, which is the final position minus the initial position ($\Delta\vec{x} = \vec{x} - \vec{x}_0$). Dividing the displacement $\Delta\vec{x}$ of the car by the elapsed time Δt gives the **average velocity** of the car. It is customary to denote the average value of a quantity by placing a horizontal bar above the symbol representing the quantity. The average velocity, then, is written as $\overline{\vec{v}}$, as specified in Equation 2.2:

DEFINITION OF AVERAGE VELOCITY

$$\text{Average velocity} = \frac{\text{Displacement}}{\text{Elapsed time}}$$

$$\overline{\vec{v}} = \frac{\vec{x} - \vec{x}_0}{t - t_0} = \frac{\Delta\vec{x}}{\Delta t} \tag{2.2}$$

SI Unit of Average Velocity: meter per second (m/s)

Equation 2.2 indicates that the unit for average velocity is the unit for length divided by the unit for time, or meters per second (m/s) in SI units. Velocity can also be expressed in other units, such as kilometers per hour (km/h) or miles per hour (mi/h).

Average velocity is a vector that points in the same direction as the displacement in Equation 2.2. **Figure 2.2** illustrates that the velocity of an object confined to move along a line can point either in one direction or in the opposite direction. As with displacement, we will use plus and minus signs to indicate the two possible directions. If the displacement points in the positive direction, the average velocity is positive. Conversely, if the displacement points in the negative direction, the average velocity is negative. Example 2 illustrates these features of average velocity.

Michael Dietrich/Age Fotostock

FIGURE 2.2 The boats in this photograph are traveling in opposite directions; in other words, the velocity of one boat points in a direction that is opposite to the velocity of the other boat.

EXAMPLE 2 | The World's Fastest Jet-Engine Car

Andy Green in the car *ThrustSSC* set a world record of 341.1 m/s (763 mi/h) in 1997. The car was powered by two jet engines, and it was the first one officially to exceed the speed of sound. To establish such a record, the driver makes two runs through the course, one in each direction, to nullify wind effects. **Figure 2.3a** shows that the car first travels from left to right and covers a distance of 1609 m (1 mile) in a time of 4.740 s. **Figure 2.3b** shows that in the

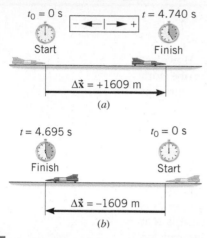

FIGURE 2.3   The arrows in the box at the top of the drawing indicate the positive and negative directions for the displacements of the car, as explained in Example 2.

reverse direction, the car covers the same distance in 4.695 s. From these data, determine the average velocity for each run.

Reasoning   Average velocity is defined as the displacement divided by the elapsed time. In using this definition we recognize that the displacement is not the same as the distance traveled. Displacement takes the direction of the motion into account, and distance does not. During both runs, the car covers the same distance of 1609 m. However, for the first run the displacement is $\Delta\vec{x} = +1609$ m, while for the second it is $\Delta\vec{x} = -1609$ m. The plus and minus signs are essential, because the first run is to the right, which is the positive direction, and the second run is in the opposite or negative direction.

Solution   According to Equation 2.2, the average velocities are

Run 1 $\bar{\vec{v}} = \dfrac{\Delta\vec{x}}{\Delta t} = \dfrac{+1609 \text{ m}}{4.740 \text{ s}} = \boxed{+339.5 \text{ m/s}}$

Run 2 $\bar{\vec{v}} = \dfrac{\Delta\vec{x}}{\Delta t} = \dfrac{-1609 \text{ m}}{4.695 \text{ s}} = \boxed{-342.7 \text{ m/s}}$

In these answers the algebraic signs convey the directions of the velocity vectors. In particular, for run 2 the minus sign indicates that the average velocity, like the displacement, points to the left in **Figure 2.3b**. The magnitudes of the velocities are 339.5 and 342.7 m/s. The average of these numbers is 341.1 m/s, and this is recorded in the record book.

Instantaneous Velocity

Suppose the magnitude of your average velocity for a long trip was 20 m/s. This value, being an average, does not convey any information about how fast you were moving or the direction of the motion at any instant during the trip. Both can change from one instant to another. Surely there were times when your car traveled faster than 20 m/s and times when it traveled more slowly. The **instantaneous velocity** $\vec{v}$ of the car indicates how fast the car moves and the direction of the motion at each instant of time. The magnitude of the instantaneous velocity is called the **instantaneous speed**, and it is the number (with units) indicated by the speedometer.

The instantaneous velocity at any point during a trip can be obtained by measuring the time interval Δt for the car to travel a *very small* displacement $\Delta\vec{x}$. We can then compute the average velocity over this interval. If the time Δt is small enough, the instantaneous velocity does not change much during the measurement. Then, the instantaneous velocity $\vec{v}$ at the point of interest is approximately equal to ($\approx$) the average velocity $\bar{\vec{v}}$ computed over the interval, or $\vec{v} \approx \bar{\vec{v}} = \Delta\vec{x}/\Delta t$ (for sufficiently small Δt). In fact, in the limit that Δt becomes infinitesimally small, the instantaneous velocity and the average velocity become equal, so that

$$\vec{v} = \lim_{\Delta t \to 0} \frac{\Delta\vec{x}}{\Delta t} \tag{2.3}$$

The notation $\lim_{\Delta t \to 0} \dfrac{\Delta\vec{x}}{\Delta t}$ means that the ratio $\Delta\vec{x}/\Delta t$ is defined by a limiting process in which smaller and smaller values of Δt are used, so small that they approach zero. As smaller values of Δt are used, $\Delta\vec{x}$ also becomes smaller. However, the ratio $\Delta\vec{x}/\Delta t$ does *not* become zero but, rather, approaches the value of the instantaneous velocity. For brevity, we will use the word *velocity* to mean "instantaneous velocity" and *speed* to mean "instantaneous speed."

Check Your Understanding

(The answers are given at the end of the book.)

2. Is the average speed of a vehicle a vector or a scalar quantity?

3. Two buses depart from Chicago, one going to New York and one to San Francisco. Each bus travels at a speed of 30 m/s. Do they have equal velocities?

4. One of the following statements is incorrect. **(a)** The car traveled around the circular track at a constant velocity. **(b)** The car traveled around the circular track at a constant speed. Which statement is incorrect?

5. A straight track is 1600 m in length. A runner begins at the starting line, runs due east for the full length of the track, turns around and runs halfway back. The time for this run is five minutes. What is the runner's average velocity, and what is his average speed?

6. The average velocity for a trip has a positive value. Is it possible for the instantaneous velocity at a point during the trip to have a negative value?

2.3 | Acceleration

In a wide range of motions, the velocity changes from moment to moment, such as in the case of the sprinter in **Photo 2.1**. To describe the manner in which it changes, the concept of acceleration is needed. The velocity of a moving object may change in a number of ways. For example, it may increase, as it does when the driver of a car steps on the gas pedal to pass the car ahead. Or it may decrease, as it does when the driver applies the brakes to stop at a red light. In either case, the change in velocity may occur over a short or a long time interval. To describe how the velocity of an object changes during a given time interval, we now introduce the new idea of acceleration. This idea depends on two concepts that we have previously encountered, velocity and time. Specifically, the notion of acceleration emerges when the *change* in the velocity is combined with the time during which the change occurs.

The meaning of **average acceleration** can be illustrated by considering a plane during takeoff. **Figure 2.4** focuses attention on how the plane's velocity changes along the runway. During an elapsed time interval $\Delta t = t - t_0$, the velocity changes from an initial value of $\vec{v}_0$ to a final velocity of $\vec{v}$. The change $\Delta\vec{v}$ in the plane's velocity is its final velocity minus its initial velocity, so that $\Delta\vec{v} = \vec{v} - \vec{v}_0$. The average acceleration $\overline{\vec{a}}$ is defined in the following manner, to provide a measure of how much the velocity changes per unit of elapsed time.

DEFINITION OF AVERAGE ACCELERATION

$$\text{Average acceleration} = \frac{\text{Change in velocity}}{\text{Elapsed time}}$$

$$\overline{\vec{a}} = \frac{\vec{v} - \vec{v}_0}{t - t_0} = \frac{\Delta\vec{v}}{\Delta t} \qquad (2.4)$$

SI Unit of Average Acceleration: meter per second squared (m/s²)

The average acceleration $\overline{\vec{a}}$ is a vector that points in the same direction as $\Delta\vec{v}$, the change in the velocity. Following the usual custom, plus and minus signs indicate the two possible directions for the acceleration vector when the motion is along a straight line.

Kaliva/Shutterstock.com

PHOTO 2.1 As this sprinter explodes out of the starting block, his velocity is changing, which means that he is accelerating.

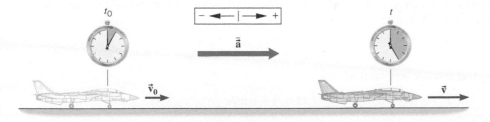

FIGURE 2.4 During takeoff, the plane accelerates from an initial velocity $\vec{v}_0$ to a final velocity $\vec{v}$ during the time interval $\Delta t = t - t_0$.

We are often interested in an object's acceleration at a particular instant of time. The **instantaneous acceleration** $\vec{a}$ can be defined by analogy with the procedure used in Section 2.2 for instantaneous velocity:

$$\vec{a} = \lim_{\Delta t \to 0} \frac{\Delta \vec{v}}{\Delta t} \tag{2.5}$$

Equation 2.5 indicates that the instantaneous acceleration is a limiting case of the average acceleration. When the time interval Δt for measuring the acceleration becomes extremely small (approaching zero in the limit), the average acceleration and the instantaneous acceleration become equal. Moreover, in many situations the acceleration is constant, so the acceleration has the same value at any instant of time. In the future, we will use the word *acceleration* to mean "instantaneous acceleration." Example 3 deals with the acceleration of a plane during takeoff.

EXAMPLE 3 | Acceleration and Increasing Velocity

Suppose the plane in **Figure 2.4** starts from rest ($\vec{v}_0 = 0$ m/s) when $t_0 = 0$ s. The plane accelerates down the runway and at $t = 29$ s attains a velocity of $\vec{v} = +260$ km/h, where the plus sign indicates that the velocity points to the right. Determine the average acceleration of the plane.

Reasoning The average acceleration of the plane is defined as the change in its velocity divided by the elapsed time. The change in the plane's velocity is its final velocity $\vec{v}$ minus its initial velocity $\vec{v}_0$, or $\vec{v} - \vec{v}_0$. The elapsed time is the final time t minus the initial time t_0, or $t - t_0$.

Problem-Solving Insight The change in any variable is the final value minus the initial value: for example, the change in velocity is $\Delta \vec{v} = \vec{v} - \vec{v}_0$, and the change in time is $\Delta t = t - t_0$.

Solution The average acceleration is expressed by Equation 2.4 as

$$\bar{\vec{a}} = \frac{\vec{v} - \vec{v}_0}{t - t_0} = \frac{260 \text{ km/h} - 0 \text{ km/h}}{29 \text{ s} - 0 \text{ s}} = \boxed{+9.0 \frac{\text{km/h}}{\text{s}}}$$

The average acceleration calculated in Example 3 is read as "nine kilometers per hour per second." Assuming the acceleration of the plane is constant, a value of $+9.0 \frac{\text{km/h}}{\text{s}}$ means the velocity changes by +9.0 km/h during each second of the motion. During the first second, the velocity increases from 0 to 9.0 km/h; during the next second, the velocity increases by another 9.0 km/h to 18 km/h, and so on. **Figure 2.5** illustrates how the velocity changes during the first two seconds. By the end of the 29th second, the velocity is 260 km/h.

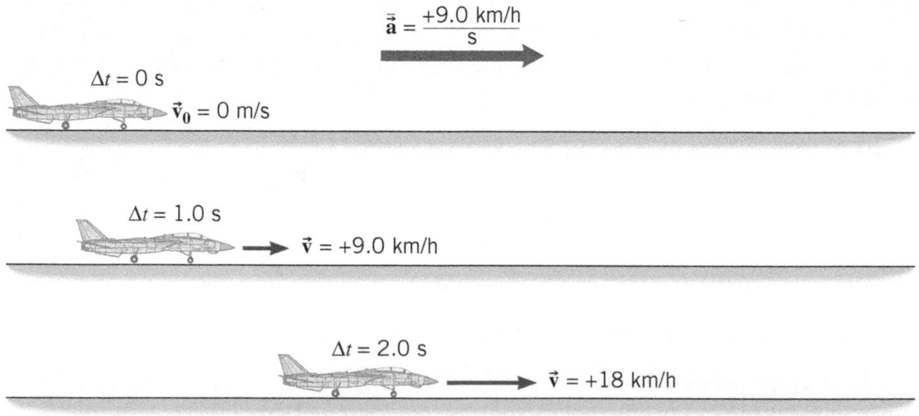

$$\bar{\vec{a}} = \frac{+9.0 \text{ km/h}}{\text{s}}$$

$\Delta t = 0$ s

$\vec{v}_0 = 0$ m/s

$\Delta t = 1.0$ s

$\vec{v} = +9.0$ km/h

$\Delta t = 2.0$ s

$\vec{v} = +18$ km/h

FIGURE 2.5 An acceleration of $+9.0 \frac{\text{km/h}}{\text{s}}$ means that the velocity of the plane changes by +9.0 km/h during each second of the motion. The "+" direction for $\vec{a}$ and $\vec{v}$ is to the right.

It is customary to express the units for acceleration solely in terms of SI units. One way to obtain SI units for the acceleration in Example 3 is to convert the velocity units from km/h to m/s:

$$\left(260 \frac{\text{km}}{\text{h}}\right)\left(\frac{1000 \text{ m}}{1 \text{ km}}\right)\left(\frac{1 \text{ h}}{3600 \text{ s}}\right) = 72 \frac{\text{m}}{\text{s}}$$

The average acceleration then becomes

$$\bar{\vec{a}} = \frac{72 \text{ m/s} - 0 \text{ m/s}}{29 \text{ s} - 0 \text{ s}} = +2.5 \text{ m/s}^2$$

where we have used $2.5 \frac{m/s}{s} = 2.5 \frac{m}{s \cdot s} = 2.5 \frac{m}{s^2}$. An acceleration of $2.5 \frac{m}{s^2}$ is read as "2.5 meters per second per second" (or "2.5 meters per second squared") and means that the velocity changes by 2.5 m/s during each second of the motion.

Example 4 deals with a case where the motion becomes slower as time passes.

EXAMPLE 4 | Acceleration and Decreasing Velocity

A drag racer crosses the finish line, and the driver deploys a parachute and applies the brakes to slow down, as **Figure 2.6** illustrates. The driver begins slowing down when $t_0 = 9.0$ s and the car's velocity is $\vec{v}_0 = +28$ m/s. When $t = 12.0$ s, the velocity has been reduced to $\vec{v} = +13$ m/s. What is the average acceleration of the dragster?

Reasoning The average acceleration of an object is always specified as its change in velocity, $\vec{v} - \vec{v}_0$, divided by the elapsed time, $t - t_0$. This is true whether the final velocity is less than the initial velocity or greater than the initial velocity.

Solution The average acceleration is, according to Equation 2.4,

$$\bar{\vec{a}} = \frac{\vec{v} - \vec{v}_0}{t - t_0} = \frac{13 \text{ m/s} - 28 \text{ m/s}}{12.0 \text{ s} - 9.0 \text{ s}} = \boxed{-5.0 \text{ m/s}^2}$$

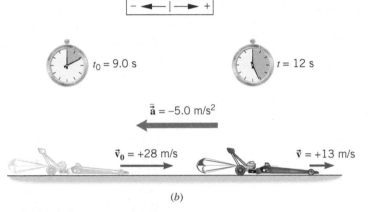

$t_0 = 9.0$ s

$t = 12$ s

$\bar{\vec{a}} = -5.0$ m/s^2

$\vec{v}_0 = +28$ m/s

$\vec{v} = +13$ m/s

(a)

(b)

Icon Sportswire/Getty Images

FIGURE 2.6 (a) To slow down, a drag racer deploys a parachute and applies the brakes. (b) The velocity of the car is decreasing, giving rise to an average acceleration $\bar{\vec{a}}$ that points opposite to the velocity.

Figure 2.7 shows how the velocity of the dragster in Example 4 changes during the braking, assuming that the acceleration is constant throughout the motion. The acceleration calculated in Example 4 is negative, indicating that the acceleration points to the left in the drawing. As a result, the acceleration and the velocity point in *opposite* directions.

| **Problem-Solving Insight** Whenever the acceleration and velocity vectors have opposite directions, the object slows down and is said to be "decelerating."

In contrast, the acceleration and velocity vectors in **Figure 2.5** point in the *same* direction, and the object speeds up.

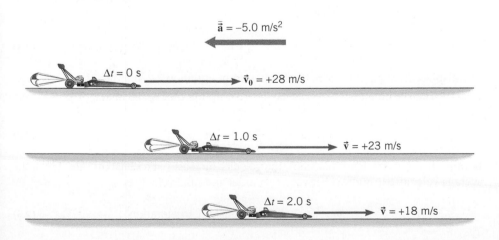

$\bar{\vec{a}} = -5.0$ m/s^2

$\Delta t = 0$ s

$\vec{v}_0 = +28$ m/s

$\Delta t = 1.0$ s

$\vec{v} = +23$ m/s

$\Delta t = 2.0$ s

$\vec{v} = +18$ m/s

FIGURE 2.7 Here, an acceleration of -5.0 m/s^2 means the velocity decreases by 5.0 m/s during each second of elapsed time.

Check Your Understanding

(The answers are given at the end of the book.)

7. At one instant of time, a car and a truck are traveling side by side in adjacent lanes of a highway. The car has a greater velocity than the truck has. Does the car necessarily have the greater acceleration?

8. Two cars are moving in the same direction (the positive direction) on a straight road. The acceleration of each car also points in the positive direction. Car 1 has a greater acceleration than car 2 has. Which one of the following statements is true? **(a)** The velocity of car 1 is always greater than the velocity of car 2. **(b)** The velocity of car 2 is always greater than the velocity of car 1. **(c)** In the same time interval, the velocity of car 1 changes by a greater amount than the velocity of car 2 does. **(d)** In the same time interval, the velocity of car 2 changes by a greater amount than the velocity of car 1 does.

9. An object moving with a constant acceleration slows down if the acceleration points in the direction opposite to the direction of the velocity. But can an object ever come to a permanent halt if its acceleration truly remains constant?

10. A runner runs half the remaining distance to the finish line every ten seconds. She runs in a straight line and does not ever reverse her direction. Does her acceleration have a constant magnitude?

2.4 | Equations of Kinematics for Constant Acceleration

It is now possible to describe the motion of an object traveling with a constant acceleration along a straight line. To do so, we will use a set of equations known as the equations of kinematics for constant acceleration. These equations entail no new concepts, because they will be obtained by combining the familiar ideas of displacement, velocity, and acceleration. However, they will provide a very convenient way to determine certain aspects of the motion, such as the final position and velocity of a moving object.

In discussing the equations of kinematics, it will be convenient to assume that the object is located at the origin $\vec{x}_0 = 0$ m when $t_0 = 0$ s. With this assumption, the displacement $\Delta\vec{x} = \vec{x} - \vec{x}_0$ becomes $\Delta\vec{x} = \vec{x}$. Furthermore, in the equations that follow, as is customary, we dispense with the use of boldface symbols overdrawn with small arrows for the displacement, velocity, and acceleration vectors. We will, however, continue to convey the directions of these vectors with plus or minus signs.

Consider an object that has an initial velocity of v_0 at time $t_0 = 0$ s and moves for a time t with a constant acceleration a. For a complete description of the motion, it is also necessary to know the final velocity and displacement at time t. The final velocity v can be obtained directly from Equation 2.4:

$$\bar{a} = a = \frac{v - v_0}{t} \quad \text{or} \quad v = v_0 + at \quad \text{(constant acceleration)} \qquad (2.4)$$

The displacement x at time t can be obtained from Equation 2.2, if a value for the average velocity $\bar{v}$ can be obtained. Considering the assumption that $x_0 = 0$ m at $t_0 = 0$ s, we have

$$\bar{v} = \frac{x - x_0}{t - t_0} = \frac{x}{t} \quad \text{or} \quad x = \bar{v}t \qquad (2.2)$$

Because the acceleration is constant, the velocity increases at a constant rate. Thus, the average velocity $\bar{v}$ is midway between the initial and final velocities:

$$\bar{v} = \tfrac{1}{2}(v_0 + v) \quad \text{(constant acceleration)} \qquad (2.6)$$

Equation 2.6, like Equation 2.4, applies only if the acceleration is constant and cannot be used when the acceleration is changing. The displacement at time t can now be determined as

$$x = \bar{v}t = \tfrac{1}{2}(v_0 + v)t \qquad \text{(constant acceleration)} \qquad \text{(2.7)}$$

Notice in Equations 2.4 ($v = v_0 + at$) and 2.7 [$x = \tfrac{1}{2}(v_0 + v)t$] that there are five kinematic variables:

1. x = displacement
2. $a = \bar{a}$ = acceleration (constant)
3. v = final velocity at time t
4. v_0 = initial velocity at time $t_0 = 0$ s
5. t = time elapsed since $t_0 = 0$ s

Each of the two equations contains four of these variables, so if three of them are known, the fourth variable can always be found. Example 5 illustrates how Equations 2.4 and 2.7 are used to describe the motion of an object.

Analyzing Multiple-Concept Problems

EXAMPLE 5 | The Displacement of a Speedboat

The speedboat in **Figure 2.8** has a constant acceleration of $+2.0 \text{ m/s}^2$. If the initial velocity of the boat is $+6.0$ m/s, find the boat's displacement after 8.0 seconds.

Reasoning As the speedboat accelerates, its velocity is changing. The displacement of the speedboat during a given time interval is equal to the product of its average velocity during that interval and the time. The average velocity, on the other hand, is just one-half the sum of the boat's initial and final velocities. To obtain the final velocity, we will use the definition of constant acceleration, as given in Equation 2.4.

Knowns and Unknowns The numerical values for the three known variables are listed in the table:

Description	Symbol	Value	Comment
Acceleration	a	$+2.0 \text{ m/s}^2$	Positive, because the boat is accelerating to the right, which is the positive direction. See Figure 2.8b.
Initial velocity	v_0	$+6.0$ m/s	Positive, because the boat is moving to the right, which is the positive direction. See Figure 2.8b.
Time interval	t	8.0 s	
Unknown Variable			
Displacement of boat	x	?	

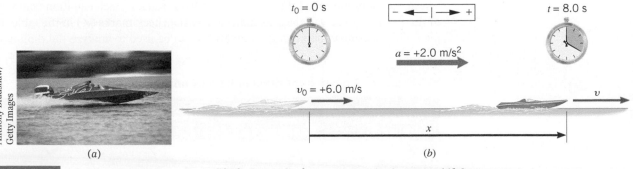

FIGURE 2.8 (a) An accelerating speedboat. (b) The boat's displacement x can be determined if the boat's acceleration, initial velocity, and time of travel are known.

Anthony Bradshaw/ Getty Images

Modeling the Problem

STEP 1 Displacement Since the acceleration is constant, the displacement x of the boat is given by Equation 2.7, in which v_0 and v are the initial and final velocities, respectively. In this equation, two of the variables, v_0 and t, are known (see the Knowns and Unknowns table), but the final velocity v is not. However, the final velocity can be determined by employing the definition of acceleration, as Step 2 shows.

$$x = \tfrac{1}{2}(v_0 + v)t \qquad (2.7)$$

$$\boxed{?}$$

STEP 2 Acceleration According to Equation 2.4, which is just the definition of constant acceleration, the final velocity v of the boat is

$$\boxed{v = v_0 + at} \qquad (2.4)$$

All the variables on the right side of the equals sign are known, and we can substitute this relation for v into Equation 2.7, as shown at the right.

$$x = \tfrac{1}{2}(v_0 + v)t \qquad (2.7)$$

$$\boxed{v = v_0 + at} \qquad (2.4)$$

Solution Algebraically combining the results of Steps 1 and 2, we find that

$$\overset{\text{STEP 1}}{\downarrow} \qquad \overset{\text{STEP 2}}{\downarrow}$$

$$x = \tfrac{1}{2}(v_0 + v)t = \tfrac{1}{2}[v_0 + (v_0 + at)]t = v_0 t + \tfrac{1}{2}at^2$$

The displacement of the boat after 8.0 s is

$$x = v_0 t + \tfrac{1}{2}at^2 = (+6.0 \text{ m/s})(8.0 \text{ s}) + \tfrac{1}{2}(+2.0 \text{ m/s}^2)(8.0 \text{ s})^2 = \boxed{+112 \text{ m}}$$

Related Homework: Problems 36, 50

In Example 5, we combined two equations [$x = \tfrac{1}{2}(v_0 + v)t$ and $v = v_0 + at$] into a single equation by algebraically eliminating the final velocity v of the speedboat (which was not known). The result was the following expression for the displacement x of the speedboat:

$$x = v_0 t + \tfrac{1}{2}at^2 \qquad \text{(constant acceleration)} \qquad (2.8)$$

The first term ($v_0 t$) on the right side of this equation represents the displacement that would result if the acceleration were zero and the velocity remained constant at its initial value of v_0. The second term ($\tfrac{1}{2}at^2$) gives the additional displacement that arises because the velocity changes (a is not zero) to values that are different from its initial value.

Equation 2.8 allows us to calculate the displacement of an object undergoing constant acceleration, given its acceleration (a), the initial velocity (v_0), and the time (t). However, even if the time interval is not given, we can still calculate the displacement, provided we also know the final velocity (v) of the object. We can solve Equation 2.4 for the time t, and then make this substitution into Equation 2.8. The result is $x = (v^2 - v_0^2)/2a$. Solving for v^2 gives

$$v^2 = v_0^2 + 2ax \qquad \text{(constant acceleration)} \qquad (2.9)$$

Equation 2.9 is often used when the time involved in the motion is unknown.

Table 2.1 presents a summary of the equations that we have been considering. These equations are called the **equations of kinematics**. Each equation contains four of the five kinematic variables, as indicated by the check marks (✓) in the table. Example 6 illustrates how the kinematic equations can be used to analyze the motion of the

TABLE 2.1 **Equations of Kinematics for Constant Acceleration**

Equation Number	Equation	Variables				
		x	a	v	v_0	t
(2.4)	$v = v_0 + at$	—	✓	✓	✓	✓
(2.7)	$x = \tfrac{1}{2}(v + v_0)t$	✓	—	✓	✓	✓
(2.8)	$x = v_0 t + \tfrac{1}{2}at^2$	✓	✓	—	✓	✓
(2.9)	$v^2 = v_0^2 + 2ax$	✓	✓	✓	✓	—

fastest animal in the world, the peregrine falcon. The next section shows how to apply the equations of kinematics.

EXAMPLE 6 | BIO The Peregrine Falcon

The peregrine falcon (see **Figure 2.9**) is the fastest animal on the planet. Although it flies at speeds less than 60.0 mph during normal flight, its wing shape and aerodynamic features allow it to attain typical speeds of up to 225 mph during dives. The highest velocity recorded for a stooping (diving) peregrine falcon is 242 mph! Other birds cannot even breathe at much lesser speeds. Suppose a peregrine falcon reaches 230.0 mph while performing a vertical dive to intercept a starling that is in straight and level flight. The starling sees the falcon when it is 22.0 m above, and tries to make a move to avoid it. **(a)** How much time (in seconds) does the starling have to avoid a direct hit? **(b)** If the falcon is capable of decelerating with an average magnitude of 15.0g, where $g = 9.80$ m/s² is the acceleration due to gravity, what is its stopping distance in meters?

Reasoning **(a)** The speed of the falcon and its distance from the starling are given. We can therefore apply a kinematic equation that relates the time, speed, and distance, that is, Equation 2.7. **(b)** We will define the $+y$ direction to be the same as that of the initial velocity, which is downward. Therefore the sign of the acceleration will be

Jerry Ting/Flickr

FIGURE 2.9 The peregrine falcon is the fastest animal on the planet.

negative since the falcon is decelerating. Assuming a constant deceleration of 15.0g, given the initial speed (230.0 mph) and the final speed (0 mph), we can use Equation 2.9 to find the stopping distance.

Solution **(a)** Since we want to know the time in seconds, we want to express the speed in m/s: 230.0 mph = 102.8 m/s. Knowing the falcon's initial speed in the y direction and its distance to the starling, we use Equation 2.7 (with y in place of x):

$$y = \tfrac{1}{2}(v_0 + v)t$$

Since the speed of the falcon is constant, $v_0 = v$, and the equation becomes

$$y = \tfrac{1}{2}(v_0 + v_0)t = v_0 t$$

Solving for t, and substituting the known values, we obtain

$$t = \frac{y}{v_0} = \frac{22.0 \text{ m}}{102.8 \text{ m/s}} = \boxed{0.214 \text{ s}}$$

(b) We can use Equation 2.9 to find the stopping distance, since the acceleration is given:

$$v^2 = v_0{}^2 + 2ay$$

Solving for y, and setting the final velocity $v = 0$ m/s and $a = -15.0g$, we get

$$y = -\frac{v_0{}^2}{a} = -\frac{v_0{}^2}{(-15.0g)} = \frac{(102.8 \text{ m/s})^2}{(15.0)(9.80 \text{ m/s}^2)} = \boxed{71.9 \text{ m}}$$

The acceleration a is negative since it points in a direction opposite to that of the velocity vector. By this calculation, if the falcon cannot maneuver to intercept the starling as it tries to escape, the falcon will overshoot its prey's position by nearly 50 meters. It turns out, however, that peregrine falcons can accelerate *laterally* with magnitudes that exceed 15.0g during their dives, and have the innate ability to tweak their intercept angles to outmaneuver their prey.

Check Your Understanding

(The answers are given at the end of the book.)

11. The muzzle velocity of a gun is the velocity of the bullet when it leaves the barrel. The muzzle velocity of one rifle with a short barrel is greater than the muzzle velocity of another rifle that has a longer barrel. In which rifle is the acceleration of the bullet larger?

12. A motorcycle starts from rest and has a constant acceleration. In a time interval t, it undergoes a displacement x and attains a final velocity v. Then t is increased so that the displacement is $3x$. In this same increased time interval, what final velocity does the motorcycle attain?

2.5 | Applications of the Equations of Kinematics

The equations of kinematics can be applied to any moving object, as long as the acceleration of the object is constant. However, remember that each equation contains four variables. Therefore, numerical values for three of the four must be available if an equation is to be used to calculate the value of the remaining variable. To avoid errors when using these equations, it helps to follow a few sensible guidelines and to be alert for a few situations that can arise during your calculations.

> **Problem-Solving Insight**   Decide at the start which directions are to be called positive (+) and negative (−) relative to a conveniently chosen coordinate origin.

This decision is arbitrary, but important because displacement, velocity, and acceleration are vectors, and their directions must always be taken into account. In the examples that follow, the positive and negative directions will be shown in the drawings that accompany the problems. It does not matter which direction is chosen to be positive. However, once the choice is made, it should not be changed during the course of the calculation.

> **Problem-Solving Insight**   As you reason through a problem before attempting to solve it, be sure to interpret the terms "decelerating" or "deceleration" correctly, should they occur in the problem statement.

These terms are the source of frequent confusion, and Conceptual Example 7 offers help in understanding them.

CONCEPTUAL EXAMPLE 7 | Deceleration Versus Negative Acceleration

A car is traveling along a straight road and is decelerating. Which one of the following statements correctly describes the car's acceleration? **(a)** It must be positive. **(b)** It must be negative. **(c)** It could be positive or negative.

Reasoning The term "decelerating" means that the acceleration vector points opposite to the velocity vector and indicates that the car is slowing down. One possibility is that the velocity vector of the car points to the right, in the positive direction, as **Figure 2.10a** shows. The term "decelerating" implies, then, that the acceleration vector of the car points to the left, which is the negative direction. Another possibility is that the car is traveling to the left, as in **Figure 2.10b**. Now, since the velocity vector points to the left, the acceleration vector would point opposite, or to the right, which is the positive direction.

Answers (a) and (b) are incorrect.   The term "decelerating" means only that the acceleration vector points opposite to the velocity vector. It is not specified whether the velocity vector of the car points in the positive or negative direction. Therefore, it is not possible to know whether the acceleration is positive or negative.

Answer (c) is correct.   As shown in **Figure 2.10**, the acceleration vector of the car could point in the positive or the negative

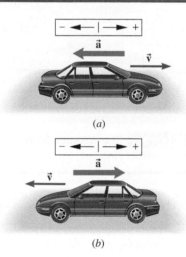

(a)

(b)

FIGURE 2.10   When a car decelerates along a straight road, the acceleration vector points opposite to the velocity vector, as Conceptual Example 7 discusses.

direction, so that the acceleration could be either positive or negative, depending on the direction in which the car is moving.

Related Homework: Problem 13

> **Problem-Solving Insight**   Sometimes there are two possible answers to a kinematics problem, each answer corresponding to a different situation.

Example 8 discusses one such case.

EXAMPLE 8 | The Physics of Spacecraft Retrorockets

The spacecraft shown in **Figure 2.11a** is traveling with a velocity of +3250 m/s. Suddenly the retrorockets are fired, and the spacecraft begins to slow down with an acceleration whose magnitude is 10.0 m/s². What is the velocity of the spacecraft when the displacement of the craft is +215 km, relative to the point where the retrorockets began firing?

Spacecraft Data				
x	a	v	v_0	t
+215 000 m	−10.0 m/s²	?	+3250 m/s	

Reasoning Since the spacecraft is slowing down, the acceleration must be opposite to the velocity. The velocity points to the right in the drawing, so the acceleration points to the left, in the negative direction; thus, $a = -10.0$ m/s². The three known variables are listed as follows:

The final velocity v of the spacecraft can be calculated using Equation 2.9, since it contains the four pertinent variables.

Solution From Equation 2.9 $(v^2 = v_0^2 + 2ax)$, we find that

$$v = \pm \sqrt{v_0^2 + 2ax}$$

$$= \pm\sqrt{(3250 \text{ m/s})^2 + 2(-10.0 \text{ m/s}^2)(215\,000 \text{ m})}$$

$$= \boxed{+2500 \text{ m/s}} \text{ and } \boxed{-2500 \text{ m/s}}$$

Both of these answers correspond to the *same* displacement ($x = +215$ km), but each arises in a different part of the motion. The answer $v = +2500$ m/s corresponds to the situation in **Figure 2.11a**, where the spacecraft has slowed to a speed of 2500 m/s, but is still traveling to the right. The answer $v = -2500$ m/s arises because the retrorockets eventually bring the spacecraft to a momentary halt and cause it to reverse its direction. Then it moves to the left, and its speed increases due to the continually firing rockets. After a time, the velocity of the craft becomes $v = -2500$ m/s, giving rise to the situation in **Figure 2.11b**. In both parts of the drawing the spacecraft has the same displacement, but a greater travel time is required in part b compared to part a.

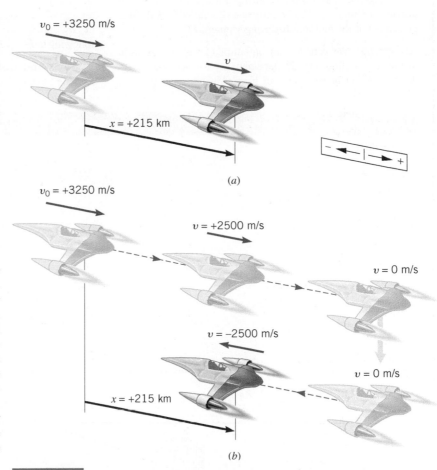

(a)

(b)

FIGURE 2.11 (a) Because of an acceleration of −10.0 m/s², the spacecraft changes its velocity from v_0 to v. (b) Continued firing of the retrorockets changes the direction of the craft's motion.

Problem-Solving Insight The motion of two objects may be interrelated, so that they share a common variable. The fact that the motions are interrelated is an important piece of information. In such cases, data for only two variables need be specified for each object.

Problem-Solving Insight Often the motion of an object is divided into segments, each with a different acceleration. When solving such problems, it is important to realize that the final velocity for one segment is the initial velocity for the next segment, as Example 9 illustrates.

Analyzing Multiple-Concept Problems

EXAMPLE 9 | A Motorcycle Ride

A motorcycle ride consists of two segments, as shown in **Interactive Figure 2.12a**. During segment 1, the motorcycle starts from rest, has an acceleration of $+2.6$ m/s^2, and has a displacement of $+120$ m. Immediately after segment 1, the motorcycle enters segment 2 and begins slowing down with an acceleration of -1.5 m/s^2 until its velocity is $+12$ m/s. What is the displacement of the motorcycle during segment 2?

Reasoning   We can use an equation of kinematics from Table 2.1 to find the displacement x_2 for segment 2. To do this, it will be necessary to have values for three of the

variables that appear in the equation. Values for the acceleration ($a_2 = -1.5$ m/s^2) and final velocity ($v_2 = +12$ m/s) are given. A value for a third variable, the initial velocity v_{02}, can be obtained by noting that it is also the final velocity of segment 1. The final velocity of segment 1 can be found by using an appropriate equation of kinematics, since three variables (x_1, a_1, and v_{01}) are known for this part of the motion, as the following table reveals.

Knowns and Unknowns   The data for this problem are listed in the table:

Description	Symbol	Value	Comment
Explicit Data			
Displacement for segment 1	x_1	$+120$ m	
Acceleration for segment 1	a_1	$+2.6$ m/s^2	Positive, because the motorcycle moves in the $+x$ direction and speeds up.
Acceleration for segment 2	a_2	-1.5 m/s^2	Negative, because the motorcycle moves in the $+x$ direction and slows down.
Final velocity for segment 2	v_2	$+12$ m/s	
Implicit Data			
Initial velocity for segment 1	v_{01}	0 m/s	The motorcycle starts from rest.
Unknown Variable			
Displacement for segment 2	x_2	?	

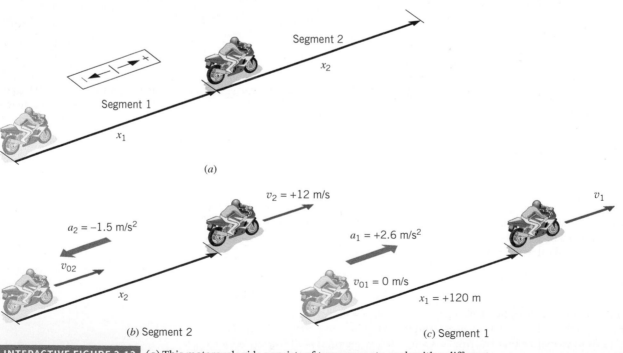

(a)

(b) Segment 2

(c) Segment 1

INTERACTIVE FIGURE 2.12   (a) This motorcycle ride consists of two segments, each with a different acceleration. (b) The variables for segment 2. (c) The variables for segment 1.

Modeling the Problem

STEP 1 Displacement During Segment 2   Interactive Figure 2.12*b* shows the situation during segment 2. Two of the variables—the final velocity v_2 and the acceleration a_2—are known, and for convenience we choose Equation 2.9 to find the displacement x_2 of the motorcycle:

$$v_2^2 = v_{02}^2 + 2a_2x_2 \qquad (2.9)$$

Solving this relation for x_2 yields Equation 1 at the right. Although the initial velocity v_{02} of segment 2 is not known, we will be able to determine its value from a knowledge of the motion during segment 1, as outlined in Steps 2 and 3.

$$x_2 = \frac{v_2^2 - v_{02}^2}{2a_2} \qquad (1)$$

$$\boxed{?}$$

STEP 2 Initial Velocity of Segment 2   Since the motorcycle enters segment 2 immediately after leaving segment 1, the initial velocity v_{02} of segment 2 is equal to the final velocity v_1 of segment 1, or $v_{02} = v_1$. Squaring both sides of this equation gives

$$\boxed{v_{02}^2 = v_1^2}$$

This result can be substituted into Equation 1 as shown at the right. In the next step v_1 will be determined.

$$x_2 = \frac{v_2^2 - v_{02}^2}{2a_2} \qquad (1)$$

$$\boxed{v_{02}^2 = v_1^2} \qquad (2)$$

$$\boxed{?}$$

STEP 3 Final Velocity of Segment 1   Interactive Figure 2.12*c* shows the motorcycle during segment 1. Since we know the initial velocity v_{01}, the acceleration a_1, and the displacement x_1, we can employ Equation 2.9 to find the final velocity v_1 at the end of segment 1:

$$\boxed{v_1^2 = v_{01}^2 + 2a_1x_1}$$

This relation for v_1^2 can be substituted into Equation 2, as shown at the right.

$$x_2 = \frac{v_2^2 - v_{02}^2}{2a_2} \qquad (1)$$

$$\boxed{v_{02}^2 = v_1^2} \qquad (2)$$

$$\boxed{v_1^2 = v_{01}^2 + 2a_1x_1} \qquad (3)$$

Solution   Algebraically combining the results of each step, we find that

$$\overset{\text{STEP 1}}{\quad} \quad \overset{\text{STEP 2}}{\quad} \quad \overset{\text{STEP 3}}{\quad}$$

$$x_2 = \frac{v_2^2 - v_{02}^2}{2a_2} = \frac{v_2^2 - v_1^2}{2a_2} = \frac{v_2^2 - (v_{01}^2 + 2a_1x_1)}{2a_2}$$

The displacement x_2 of the motorcycle during segment 2 is

$$x_2 = \frac{v_2^2 - (v_{01}^2 + 2a_1x_1)}{2a_2}$$

$$= \frac{(+12 \text{ m/s})^2 - [(0 \text{ m/s})^2 + 2(+2.6 \text{ m/s}^2)(+120 \text{ m})]}{2(-1.5 \text{ m/s}^2)} = \boxed{+160 \text{ m}}$$

Math Skills   Interactive Figure 2.12*a* shows which direction has been chosen as the positive direction, and the choice leads to a value of $x_2 = +160$ m for the displacement of the motorcycle during segment 2. This answer means that displacement is 160 m in the positive direction, which is upward and to the right in the drawing. The choice for the positive direction is completely arbitrary, however. The meaning of the answer to the problem will be the same no matter what choice is made. Suppose that the choice for the positive direction were opposite to that shown in **Interactive Figure 2.12*a***. Then, all of the data listed in the Knowns and Unknowns table would appear with algebraic signs opposite to those specified, and the calculation of the displacement x_2 would appear as follows:

$$x_2 = \frac{(-12 \text{ m/s})^2 - [(0 \text{ m/s})^2 + 2(-2.6 \text{ m/s}^2)(-120 \text{ m})]}{2(+1.5 \text{ m/s}^2)} = \boxed{-160 \text{ m}}$$

The value for x_2 has the same magnitude of 160 m as before, but it is now negative. The negative sign means that the displacement of the motorcycle during segment 2 is 160 m ***in the negative direction***. But the negative direction is now upward and to the right in **Interactive Figure 2.12*a***, so the ***meaning*** of the result is exactly the same as it was before.

Now that we have seen how the equations of kinematics are applied to various situations, it's a good idea to summarize the reasoning strategy that has been used. This strategy, which is outlined below, will also be used when we consider freely falling bodies in Section 2.6 and two-dimensional motion in Chapter 3.

REASONING STRATEGY **Applying the Equations of Kinematics**

1. **Make a drawing to represent the situation being studied. A drawing helps us to see what's happening.**

2. **Decide which directions are to be called positive (+) and negative (−) relative to a conveniently chosen coordinate origin. Do not change your decision during the course of a calculation.**

3. **In an organized way, write down the values (with appropriate plus and minus signs) that are given for any of the five kinematic variables (x, a, v, v_0, and t). Be on the alert for implicit data, such as the phrase "starts from rest," which means that the value of the initial velocity is $v_0 = 0$ m/s. The data summary tables used in the examples in the text are a good way to keep track of this information. In addition, identify the variables that you are being asked to determine.**

4. **Before attempting to solve a problem, verify that the given information contains values for at least three of the five kinematic variables. Once the three known variables are identified along with the desired unknown variable, the appropriate relation from Table 2.1 can be selected. Remember that the motions of two objects may be interrelated, so they may share a common variable. The fact that the motions are interrelated is an important piece of information. In such cases, data for only two variables need be specified for each object.**

5. **When the motion of an object is divided into segments, as in Example 9, remember that the final velocity of one segment is the initial velocity for the next segment.**

6. **Keep in mind that there may be two possible answers to a kinematics problem as, for instance, in Example 8. Try to visualize the different physical situations to which the answers correspond.**

2.6 Freely Falling Bodies

Everyone has observed the effect of gravity as it causes objects to fall downward. In the absence of air resistance, it is found that all bodies at the same location above the earth fall vertically with the same acceleration. Furthermore, if the distance of the fall is small compared to the radius of the earth, the acceleration remains essentially constant throughout the descent. This idealized motion, in which air resistance is neglected and the acceleration is nearly constant, is known as **free-fall**. Since the acceleration is constant in free-fall, the equations of kinematics can be used.

The acceleration of a freely falling body is called the **acceleration due to gravity**, and its magnitude (without any algebraic sign) is denoted by the symbol g. The acceleration due to gravity is directed downward, toward the center of the earth. Near the earth's surface, g is approximately

$$g = 9.80 \text{ m/s}^2 \quad \text{or} \quad 32.2 \text{ ft/s}^2$$

Unless circumstances warrant otherwise, we will use either of these values for g in subsequent calculations. In reality, however, g decreases with increasing altitude and varies slightly with latitude.

Figure 2.13a shows the well-known phenomenon of a rock falling faster than a sheet of paper. The effect of air resistance is responsible for the slower fall of the paper, for when air is removed from the tube, as in **Figure 2.13b**, the rock and the paper have exactly the same acceleration due to gravity. In the absence of air, the rock and the paper both exhibit free-fall motion. Free-fall is closely approximated for objects falling near the surface of the moon, where there is no air to retard the motion. A nice demonstration of free-fall was performed on the moon by astronaut David Scott, who dropped a hammer and a feather simultaneously from the same height. Both experienced the same

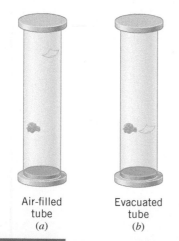

Air-filled   Evacuated
tube   tube
(a)   (b)

FIGURE 2.13 (a) In the presence of air resistance, the acceleration of the rock is greater than that of the paper. (b) In the absence of air resistance, both the rock and the paper have the same acceleration.

acceleration due to lunar gravity and consequently hit the ground at the same time. The acceleration due to gravity near the surface of the moon is approximately one-sixth as large as that on the earth.

When the equations of kinematics are applied to free-fall motion, it is natural to use the symbol y for the displacement, since the motion occurs in the vertical or y direction. Thus, when using the equations in Table 2.1 for free-fall motion, we will simply replace x with y. There is no significance to this change. The equations have the same algebraic form for either the horizontal or vertical direction, provided that the acceleration remains constant during the motion. We now turn our attention to several examples that illustrate how the equations of kinematics are applied to freely falling bodies.

EXAMPLE 10 | A Falling Stone

A stone is dropped from rest from the top of a tall building, as Animated Figure 2.14 indicates. After 3.00 s of free-fall, what is the displacement y of the stone?

Reasoning The upward direction is chosen as the positive direction. The three known variables are shown in the table below. The initial velocity v_0 of the stone is zero, because the stone is dropped from rest. The acceleration due to gravity is negative, since it points downward in the negative direction.

Stone Data				
y	a	v	v_0	t
?	-9.80 m/s^2		0 m/s	3.00 s

Equation 2.8 contains the appropriate variables and offers a direct solution to the problem. Since the stone moves downward, and upward is the positive direction, we expect the displacement y to have a negative value.

> **Problem-Solving Insight** It is only when values are available for at least three of the five kinematic variables (y, a, v, v_0, and t) that the equations in Table 2.1 can be used to determine the fourth and fifth variables.

Solution Using Equation 2.8, we find that

$$y = v_0 t + \tfrac{1}{2}at^2 = (0 \text{ m/s})(3.00 \text{ s}) + \tfrac{1}{2}(-9.80 \text{ m/s}^2)(3.00 \text{ s})^2$$

$$= \boxed{-44.1 \text{ m}}$$

The answer for y is negative, as expected.

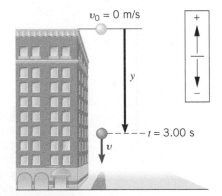

ANIMATED FIGURE 2.14 The stone, starting with zero velocity at the top of the building, is accelerated downward by gravity.

EXAMPLE 11 | The Velocity of a Falling Stone

After 3.00 s of free-fall, what is the velocity v of the stone in Animated Figure 2.14?

Reasoning Because of the acceleration due to gravity, the magnitude of the stone's downward velocity increases by 9.80 m/s during each second of free-fall. The data for the stone are the same as in Example 10, and Equation 2.4 offers a direct solution for the

final velocity. Since the stone is moving downward in the negative direction, the value determined for v should be negative.

Solution Using Equation 2.4, we obtain

$$v = v_0 + at = 0 \text{ m/s} + (-9.80 \text{ m/s}^2)(3.00 \text{ s}) = \boxed{-29.4 \text{ m/s}}$$

The velocity is negative, as expected.

The acceleration due to gravity is always a downward-pointing vector. It describes how the speed increases for an object that is falling freely downward. This same acceleration also describes how the speed decreases for an object moving upward under the influence of gravity alone, in which case the object eventually comes to a momentary halt and then falls back to earth. Examples 12 and 13 show how the equations of kinematics are applied to an object that is moving upward under the influence of gravity.

EXAMPLE 12 | How High Does It Go?

A football game customarily begins with a coin toss to determine who kicks off. The referee tosses the coin up with an initial speed of 5.00 m/s. In the absence of air resistance, how high does the coin go above its point of release?

Reasoning The coin is given an upward initial velocity, as in **Figure 2.15**. But the acceleration due to gravity points downward. Since the velocity and acceleration point in opposite directions, the coin slows down as it moves upward. Eventually, the velocity of the coin becomes $v = 0$ m/s at the highest point. Assuming that the upward direction is positive, the data can be summarized as shown below:

Coin Data				
y	a	v	v_0	t
?	-9.80 m/s^2	0 m/s	$+5.00$ m/s	

With these data, we can use Equation 2.9 ($v^2 = v_0^2 + 2ay$) to find the maximum height y.

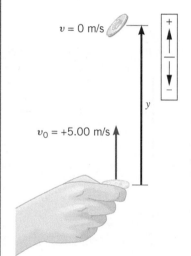

$v = 0$ m/s

$v_0 = +5.00$ m/s

y

FIGURE 2.15 At the start of a football game, a referee tosses a coin upward with an initial velocity of $v_0 = +5.00$ m/s. The velocity of the coin is momentarily zero when the coin reaches its maximum height.

Problem-Solving Insight Implicit data are important. In Example 12, for instance, the phrase "how high does the coin go" refers to the maximum height, which occurs when the final velocity v in the vertical direction is $v = 0$ m/s.

Solution Rearranging Equation 2.9, we find that the maximum height of the coin above its release point is

$$y = \frac{v^2 - v_0^2}{2a} = \frac{(0 \text{ m/s})^2 - (5.00 \text{ m/s})^2}{2(-9.80 \text{ m/s}^2)} = \boxed{1.28 \text{ m}}$$

Math Skills The rearrangement of algebraic equations is a problem-solving step that occurs often. The guiding rule in such a procedure is that whatever you do to one side of an equation, you must also do to the other side. Here in Example 12, for instance, we need to rearrange $v^2 = v_0^2 + 2ay$ (Equation 2.9) in order to determine y. The part of the equation that contains y is $2ay$, and we begin by isolating that part on one side of the equals sign. To do this we subtract v_0^2 from each side of the equation:

$$v^2 - v_0^2 = \cancel{v_0^2} + 2ay - \cancel{v_0^2} = 2ay$$

Thus, we see that $2ay = v^2 - v_0^2$. Then, we eliminate the term $2a$ from the left side of the equals sign by dividing both sides of the equation by $2a$.

$$\frac{2\cancel{a}y}{2\cancel{a}} = \frac{v^2 - v_0^2}{2a} \quad \text{or} \quad y = \frac{v^2 - v_0^2}{2a}$$

The term $2a$ occurs both in the numerator and the denominator on the left side of this result and can be eliminated algebraically, leaving the desired expression for y.

EXAMPLE 13 | How Long Is It in the Air?

In **Figure 2.15**, what is the total time the coin is in the air before returning to its release point?

Reasoning During the time the coin travels upward, gravity causes its speed to decrease to zero. On the way down, however, gravity causes the coin to regain the lost speed. Thus, the time for the coin to go up is equal to the time for it to come down. In other words, the total travel time is twice the time for the upward motion. The data for the coin during the upward trip are the same as in Example 12. With these data, we can use Equation 2.4 ($v = v_0 + at$) to find the upward travel time.

Solution Rearranging Equation 2.4, we find that

$$t = \frac{v - v_0}{a} = \frac{0 \text{ m/s} - 5.00 \text{ m/s}}{-9.80 \text{ m/s}^2} = 0.510 \text{ s}$$

The total up-and-down time is twice this value, or $\boxed{1.02 \text{ s}}$.

It is possible to determine the total time by another method. When the coin is tossed upward and returns to its release point, the displacement for the *entire trip* is $y = 0$ m. With this value for the displacement, Equation 2.8 ($y = v_0 t + \frac{1}{2}at^2$) can be used to find the time for the entire trip directly.

Examples 12 and 13 illustrate that the expression "freely falling" does not necessarily mean an object is falling down. A freely falling object is any object moving either upward or downward under the influence of gravity alone. In either case, the object always experiences the same *downward acceleration* due to gravity, a fact that is the focus of the next example.

CONCEPTUAL EXAMPLE 14 | Acceleration Versus Velocity

There are three parts to the motion of the coin in **Figure 2.15**, in which air resistance is being ignored. On the way up, the coin has an upward-pointing velocity vector with a decreasing magnitude. At the top of its path, the velocity vector of the coin is momentarily zero. On the way down, the coin has a downward-pointing velocity vector with an increasing magnitude. Compare the acceleration vector of the coin with the velocity vector. **(a)** Do the direction and magnitude of the acceleration vector behave in the same fashion as the direction and magnitude of the velocity vector or **(b)** does the acceleration vector have a constant direction and a constant magnitude throughout the motion?

Reasoning   Since air resistance is being ignored, the coin is in free-fall motion. This means that the acceleration vector of the coin is the acceleration due to gravity. Acceleration is the rate at which velocity *changes* and is not the same concept as velocity itself.

Answer (a) is incorrect.   During the upward and downward parts of the motion, and also at the top of the path, the acceleration due to gravity has a constant downward direction and a constant magnitude of 9.80 m/s². In other words, the acceleration vector of the coin does not behave as the velocity vector does. In particular, the acceleration vector is not zero at the top of the motional path just because the velocity vector is zero there. Acceleration is the rate at which the velocity is changing, and the velocity is changing at the top even though at one instant it is zero.

Answer (b) is correct.   The acceleration due to gravity has a constant downward direction and a constant magnitude of 9.80 m/s² at all times during the motion.

The motion of an object that is thrown upward and eventually returns to earth has a symmetry that is useful to keep in mind from the point of view of problem solving. The calculations just completed indicate that a time symmetry exists in free-fall motion, in the sense that the time required for the object to reach maximum height equals the time for it to return to its starting point.

A type of symmetry involving the speed also exists. **Figure 2.16** shows the coin considered in Examples 12 and 13. At any displacement *y* above the point of release, the coin's speed during the upward trip equals the speed at the same point during the downward trip. For instance, when $y = +1.04$ m, Equation 2.9 gives two possible values for the final velocity v, assuming that the initial velocity is $v_0 = +5.00$ m/s:

$$v^2 = v_0^2 + 2ay = (5.00 \text{ m/s})^2 + 2(-9.80 \text{ m/s}^2)(1.04 \text{ m}) = 4.62 \text{ m}^2/\text{s}^2$$

$$v = \pm 2.15 \text{ m/s}$$

The value $v = +2.15$ m/s is the velocity of the coin on the upward trip, and $v = -2.15$ m/s is the velocity on the downward trip. The speed in both cases is identical and equals 2.15 m/s. Likewise, the speed just as the coin returns to its point of release is 5.00 m/s, which equals the initial speed. This symmetry involving the speed arises because the coin loses 9.80 m/s in speed each second on the way up and gains back the same amount each second on the way down. In Conceptual Example 15, we use just this kind of symmetry to guide our reasoning as we analyze the motion of a pellet shot from a gun.

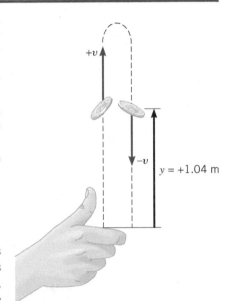

FIGURE 2.16   For a given displacement along the motional path, the upward speed of the coin is equal to its downward speed, but the two velocities point in opposite directions.

CONCEPTUAL EXAMPLE 15 | Taking Advantage of Symmetry

Interactive Figure 2.17a shows a pellet that has been fired straight upward from a gun at the edge of a cliff. The initial speed of the pellet is 30 m/s. It goes up and then falls back down, eventually hitting the ground beneath the cliff. In **Interactive Figure 2.17b** the pellet has been fired straight downward at the same initial speed. In the absence of air resistance, would the

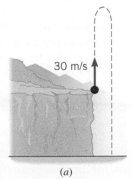

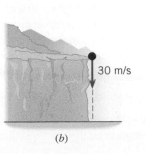

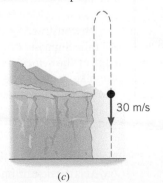

INTERACTIVE FIGURE 2.17
(*a*) From the edge of a cliff, a pellet is fired straight upward from a gun. The pellet's initial speed is 30 m/s. (*b*) The pellet is fired straight downward with an initial speed of 30 m/s. (*c*) In Conceptual Example 15 this drawing plays the central role in reasoning that is based on symmetry.

(*a*)   (*b*)   (*c*)

pellet in **Interactive Figure 2.17b** strike the ground with **(a)** a smaller speed than, **(b)** the same speed as, or **(c)** a greater speed than the pellet in **Interactive Figure 2.17a**?

Reasoning In the absence of air resistance, the motion is that of free-fall, and the symmetry inherent in free-fall motion offers an immediate answer.

Answers (a) and (c) are incorrect. These answers are incorrect, because they are inconsistent with the symmetry that is discussed next in connection with the correct answer.

Answer (b) is correct. Interactive Figure 2.17c shows the pellet after it has been fired upward and has fallen back down to its starting point. Symmetry indicates that the speed in **Interactive Figure 2.17c** is the same as in **Interactive Figure 2.17a**—namely, 30 m/s, as is also the case when the pellet has been actually fired downward. Consequently, whether the pellet is fired as in **Interactive Figure 2.17a** or **b**, it begins to move downward from the cliff edge at a speed of 30 m/s. In either case, there is the same acceleration due to gravity and the same displacement from the cliff edge to the ground below. Given these conditions, when the pellet reaches the ground, it has the same speed in both **Interactive Figures 2.17a** and **b**.

Related Homework: Problems 42, 49

Check Your Understanding

(The answers are given at the end of the book.)

13. An experimental vehicle slows down and comes to a halt with an acceleration whose magnitude is 9.80 m/s^2. After reversing direction in a negligible amount of time, the vehicle speeds up with an acceleration of 9.80 m/s^2. Except for being horizontal, is this motion **(a)** the same as or **(b)** different from the motion of a ball that is thrown straight upward, comes to a halt, and falls back to earth? Ignore air resistance.

14. A ball is thrown straight upward with a velocity $\vec{v}_0$ and in a time t reaches the top of its flight path, which is a displacement $\vec{y}$ above the launch point. With a launch velocity of $2\vec{v}_0$, what would be the time required to reach the top of its flight path and what would be the displacement of the top point above the launch point? **(a)** $4t$ and $2\vec{y}$ **(b)** $2t$ and $4\vec{y}$ **(c)** $2t$ and $2\vec{y}$ **(d)** $4t$ and $4\vec{y}$ **(e)** t and $2\vec{y}$.

15. Two objects are thrown vertically upward, first one, and then, a bit later, the other. Is it **(a)** possible or **(b)** impossible that both objects reach the same maximum height at the same instant of time?

16. A ball is dropped from rest from the top of a building and strikes the ground with a speed v_f. From ground level, a second ball is thrown straight upward at the same instant that the first ball is dropped. The initial speed of the second ball is $v_0 = v_f$, the same speed with which the first ball eventually strikes the ground. Ignoring air resistance, decide whether the balls cross paths **(a)** at half the height of the building, **(b)** above the halfway point, or **(c)** below the halfway point.

2.7 Graphical Analysis of Velocity and Acceleration

Graphical techniques are helpful in understanding the concepts of velocity and acceleration. Suppose a bicyclist is riding with a constant velocity of $v = +4$ m/s. The position x of the bicycle can be plotted along the vertical axis of a graph, while the time t is plotted along the horizontal axis. Since the position of the bike increases by 4 m every second, the graph of x versus t is a straight line. Furthermore, if the bike is assumed to be at $x = 0$ m when $t = 0$ s, the straight line passes through the origin, as **Figure 2.18** shows. Each point on this line gives the position of the bike at a particular time. For instance, at $t = 1$ s the position is 4 m, while at $t = 3$ s the position is 12 m.

In constructing the graph in **Figure 2.18**, we used the fact that the velocity was +4 m/s. Suppose, however, that we were given this graph, but did not have prior knowledge of the velocity. The velocity could be determined by considering what happens to the bike between the times of 1 and 3 s, for instance. The change in time is $\Delta t = 2$ s. During this time interval, the position of the bike changes from +4 to +12 m, and the change in position is $\Delta x = +8$ m. The ratio $\Delta x/\Delta t$ is called the **slope** of the straight line.

$$\text{Slope} = \frac{\Delta x}{\Delta t} = \frac{+8\text{ m}}{2\text{ s}} = +4\text{ m/s}$$

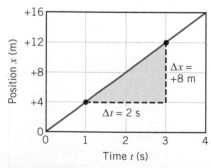

FIGURE 2.18 Position–time graph for an object moving with a constant velocity of $v = \Delta x/\Delta t = +4$ m/s.

Notice that the slope is equal to the velocity of the bike. This result is no accident, because $\Delta x/\Delta t$ is the definition of average velocity (see Equation 2.2). ***Thus, for an object moving*** *with a constant velocity, the slope of the straight line in a position–time graph gives the velocity.* Since the position–time graph is a straight line, any time interval Δt can be chosen to calculate the velocity. Choosing a different Δt will yield a different Δx, but the velocity $\Delta x/\Delta t$ will not change. In the real world, objects rarely move with a constant velocity at all times, as the next example illustrates.

EXAMPLE 16 | A Bicycle Trip

A bicyclist maintains a constant velocity on the outgoing leg of a trip, zero velocity while stopped, and another constant velocity on the way back. **Interactive Figure 2.19** shows the corresponding position–time graph. Using the time and position intervals indicated in the drawing, obtain the velocities for each segment of the trip.

Reasoning   The average velocity $\overline{v}$ is equal to the displacement Δx divided by the elapsed time Δt, $\overline{v} = \Delta x/\Delta t$. The displacement is the final position minus the initial position, which is a positive number for segment 1 and a negative number for segment 3. Note for segment 2 that $\Delta x = 0$ m, since the bicycle is at rest. The drawing shows values for Δt and Δx for each of the three segments.

Solution   The average velocities for the three segments are

Segment 1 $\overline{v} = \dfrac{\Delta x}{\Delta t} = \dfrac{800 \text{ m} - 400 \text{ m}}{400 \text{ s} - 200 \text{ s}} = \dfrac{+400 \text{ m}}{200 \text{ s}} = \boxed{+2 \text{ m/s}}$

Segment 2 $\overline{v} = \dfrac{\Delta x}{\Delta t} = \dfrac{1200 \text{ m} - 1200 \text{ m}}{1000 \text{ s} - 600 \text{ s}} = \dfrac{0 \text{ m}}{400 \text{ s}} = \boxed{0 \text{ m/s}}$

Segment 3 $\overline{v} = \dfrac{\Delta x}{\Delta t} = \dfrac{400 \text{ m} - 800 \text{ m}}{1800 \text{ s} - 1400 \text{ s}} = \dfrac{-400 \text{ m}}{400 \text{ s}} = \boxed{-1 \text{ m/s}}$

In the second segment of the journey the velocity is zero, reflecting the fact that the bike is stationary. Since the position of the bike does not change, segment 2 is a horizontal line that has a zero slope. In the third part of the motion the velocity is negative, because the position of the bike decreases from $x = +800$ m to $x = +400$ m during the 400-s interval shown in the graph. As a result, segment 3 has a negative slope, and the velocity is negative.

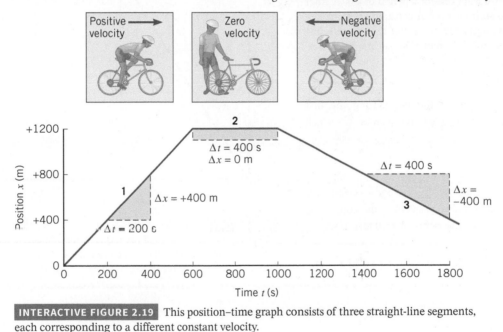

INTERACTIVE FIGURE 2.19   This position–time graph consists of three straight-line segments, each corresponding to a different constant velocity.

If the object is accelerating, its velocity is changing. When the velocity is changing, the position–time graph is not a straight line, but is a curve, perhaps like that in **Figure 2.20**. This curve was drawn using Equation 2.8 ($x = v_0 t + \frac{1}{2}at^2$), assuming an acceleration of $a = 0.26$ m/s^2 and an initial velocity of $v_0 = 0$ m/s. The velocity at any instant of time can be determined by measuring the slope of the curve at that instant. The slope at any point along the curve is defined to be the slope of the tangent line drawn to the curve at that point. For instance, in **Figure 2.20** a tangent line is drawn at $t = 20.0$ s. To determine the slope of the tangent line, a triangle is constructed using an arbitrarily chosen time interval of $\Delta t = 5.0$ s. The change in x associated with this time interval can be read from the tangent line as $\Delta x = +26$ m. Therefore,

$$\text{Slope of tangent line} = \frac{\Delta x}{\Delta t} = \frac{+26 \text{ m}}{5.0 \text{ s}} = +5.2 \text{ m/s}$$

The slope of the tangent line is the instantaneous velocity, which in this case is $v = +5.2$ m/s. This graphical result can be verified by using Equation 2.4 with $v_0 = 0$ m/s: $v = at = (+0.26 \text{ m/s}^2)(20.0 \text{ s}) = +5.2$ m/s.

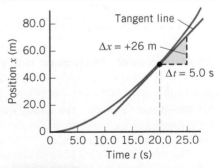

FIGURE 2.20   When the velocity is changing, the position–time graph is a curved line. The slope $\Delta x/\Delta t$ of the tangent line drawn to the curve at a given time is the instantaneous velocity at that time.

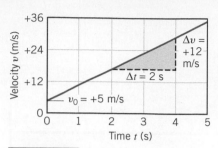

FIGURE 2.21 A velocity–time graph that applies to an object with an acceleration of $\Delta v/\Delta t = +6$ m/s^2. The initial velocity is $v_0 = +5$ m/s when $t = 0$ s.

Insight into the meaning of acceleration can also be gained with the aid of a graphical representation. Consider an object moving with a constant acceleration of $a = +6$ m/s^2. If the object has an initial velocity of $v_0 = +5$ m/s, its velocity at any time is represented by Equation 2.4 as

$$v = v_0 + at = 5 \text{ m/s} + (6 \text{ m/s}^2)t$$

This relation is plotted as the velocity–time graph in **Figure 2.21**. The graph of v versus t is a straight line that intercepts the vertical axis at $v_0 = 5$ m/s. The slope of this straight line can be calculated from the data shown in the drawing:

$$\text{Slope} = \frac{\Delta v}{\Delta t} = \frac{+12 \text{ m/s}}{2 \text{ s}} = +6 \text{ m/s}^2$$

The ratio $\Delta v/\Delta t$ is, by definition, equal to the average acceleration (Equation 2.4), so **the slope of the straight line in a velocity–time graph is the average acceleration.**

EXAMPLE 17 | BIO The Average Acceleration of a Sprinter

Figure 2.22 shows the velocity-time graph for a sprinter competing in a 100-m race. At the beginning of the race, usually initiated by the report of a starter's pistol, the sprinter accelerates from rest for a period of 4.00 s. After accelerating, the sprinter runs at constant speed, before slowing (decelerating) near the end of the race. Use the velocity and time intervals shown in the graph to determine the sprinter's average acceleration during the first 4.00 s of the race.

Reasoning The slope of the line during the first 4-second interval of the race will be given by $\frac{\Delta v}{\Delta t}$, which is equivalent to the sprinter's average acceleration, $\bar{a} = \frac{\Delta v}{\Delta t}$.

Solution Using the fact that the average acceleration is given by $\bar{a} = \frac{\Delta v}{\Delta t}$, we just need to calculate the slope of the velocity-time graph over the period from 0 to 4.00 s.

$$\bar{a} = \frac{\Delta v}{\Delta t} = \frac{+12.5 \text{ m/s} - 0}{4.00 \text{ s} - 0} = \boxed{+3.13 \text{ m/s}^2}$$

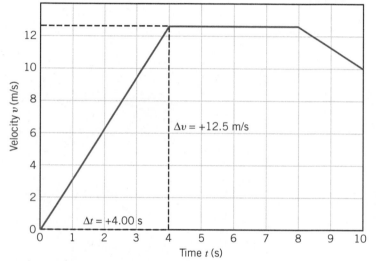

FIGURE 2.22 The velocity-time graph for a track and field athlete running a 100-m sprint. At the start of the race, the sprinter accelerates from rest for 4.00 s and then maintains a constant speed of 12.5 m/s for approximately four more seconds, before decelerating near the end of the race.

Concept Summary

2.1 Displacement Displacement is a vector that points from an object's initial position to its final position. The magnitude of the displacement is the shortest distance between the two positions.

2.2 Speed and Velocity The average speed of an object is the distance traveled by the object divided by the time required to cover the distance, as shown in Equation 2.1.

The average velocity $\bar{\vec{v}}$ of an object is the object's displacement $\Delta \vec{x}$ divided by the elapsed time Δt, as shown in Equation 2.2. Average velocity is a vector that has the same direction as the displacement. When the elapsed time becomes infinitesimally small, the average velocity becomes equal to the instantaneous velocity $\vec{v}$, the velocity at an instant of time, as indicated in Equation 2.3.

$$\text{Average speed} = \frac{\text{Distance}}{\text{Elapsed time}} \qquad (2.1)$$

$$\bar{\vec{v}} = \frac{\Delta \vec{x}}{\Delta t} \qquad (2.2)$$

$$\vec{v} = \lim_{\Delta t \to 0} \frac{\Delta \vec{x}}{\Delta t} \qquad (2.3)$$

2.3 Acceleration The average acceleration $\bar{\vec{a}}$ is a vector. It equals the change $\Delta \vec{v}$ in the velocity divided by the elapsed time Δt, the change in the velocity being the final minus the initial velocity; see Equation 2.4. When Δt becomes infinitesimally small, the average acceleration becomes equal to the instantaneous acceleration $\vec{a}$, as indicated in Equation 2.5. Acceleration is the rate at which the velocity is changing.

$$\vec{\bar{a}} = \frac{\Delta \vec{v}}{\Delta t} \tag{2.4}$$

$$\vec{a} = \lim_{\Delta t \to 0} \frac{\Delta \vec{v}}{\Delta t} \tag{2.5}$$

2.4 Equations of Kinematics for Constant Acceleration/ 2.5 Applications of the Equations of Kinematics The equations of kinematics apply when an object moves with a constant acceleration along a straight line. These equations relate the displacement $x - x_0$, the acceleration a, the final velocity v, the initial velocity v_0, and the elapsed time $t - t_0$. Assuming that $x_0 = 0$ m at $t_0 = 0$ s, the equations of kinematics are as shown in Equations 2.4 and 2.7–2.9.

$$v = v_0 + at \tag{2.4}$$

$$x = \tfrac{1}{2}(v_0 + v)t \tag{2.7}$$

$$x = v_0 t + \tfrac{1}{2}at^2 \tag{2.8}$$

$$v^2 = v_0^2 + 2ax \tag{2.9}$$

2.6 Freely Falling Bodies In free-fall motion, an object experiences negligible air resistance and a constant acceleration due to gravity. All objects at the same location above the earth have the same acceleration due to gravity. The acceleration due to gravity is directed toward the center of the earth and has a magnitude of approximately 9.80 m/s^2 near the earth's surface.

2.7 Graphical Analysis of Velocity and Acceleration The slope of a plot of position versus time for a moving object gives the object's velocity. The slope of a plot of velocity versus time gives the object's acceleration.

Focus on Concepts

Online

Additional questions are available for assignment in WileyPLUS.

Section 2.1 Displacement

1. What is the difference between distance and displacement? **(a)** Distance is a vector, while displacement is not a vector. **(b)** Displacement is a vector, while distance is not a vector. **(c)** There is no difference between the two concepts; they may be used interchangeably.

Section 2.2 Speed and Velocity

2. A jogger runs along a straight and level road for a distance of 8.0 km and then runs back to her starting point. The time for this round-trip is 2.0 h. Which one of the following statements is true? **(a)** Her average speed is 8.0 km/h, but there is not enough information to determine her average velocity. **(b)** Her average speed is 8.0 km/h, and her average velocity is 8.0 km/h. **(c)** Her average speed is 8.0 km/h, and her average velocity is 0 km/h.

Section 2.3 Acceleration

3. The velocity of a train is 80.0 km/h, due west. One and a half hours later its velocity is 65.0 km/h, due west. What is the train's average acceleration? **(a)** 10.0 km/h^2, due west **(b)** 43.3 km/h^2, due west **(c)** 10.0 km/h^2, due east **(d)** 43.3 km/h^2, due east **(e)** 53.3 km/h^2, due east.

Section 2.4 Equations of Kinematics for Constant Acceleration

4. In which one of the following situations can the equations of kinematics *not* be used? **(a)** When the velocity changes from moment to moment, **(b)** when the velocity remains constant, **(c)** when the acceleration changes from moment to moment, **(d)** when the acceleration remains constant.

5. In a race two horses, Silver Bullet and Shotgun, start from rest and each maintains a constant acceleration. In the same elapsed time Silver Bullet runs 1.20 times farther than Shotgun. According to the equations of kinematics, which one of the following is true concerning the accelerations of the horses? **(a)** $a_{\text{Silver Bullet}} = 1.44\, a_{\text{Shotgun}}$ **(b)** $a_{\text{Silver Bullet}} = a_{\text{Shotgun}}$ **(c)** $a_{\text{Silver Bullet}} = 2.40\, a_{\text{Shotgun}}$ **(d)** $a_{\text{Silver Bullet}} = 1.20\, a_{\text{Shotgun}}$ **(e)** $a_{\text{Silver Bullet}} = 0.72\, a_{\text{Shotgun}}$

Section 2.6 Freely Falling Bodies

6. A rocket is sitting on the launch pad. The engines ignite, and the rocket begins to rise straight upward, picking up speed as it goes. At about 1000 m above the ground the engines shut down, but the rocket continues straight upward, losing speed as it goes. It reaches the top of its flight path and then falls back to earth. Ignoring air resistance, decide which one of the following statements is true. **(a)** All of the rocket's motion, from the moment the engines ignite until just before the rocket lands, is free-fall. **(b)** Only part of the rocket's motion, from just after the engines shut down until just before it lands, is free-fall. **(c)** Only the rocket's motion while the engines are firing is free-fall. **(d)** Only the rocket's motion from the top of its flight path until just before landing is free-fall. **(e)** Only part of the rocket's motion, from just after the engines shut down until it reaches the top of its flight path, is free-fall.

7. The top of a cliff is located a distance H above the ground. At a distance $H/2$ there is a branch that juts out from the side of the cliff, and on this branch a bird's nest is located. Two children throw stones at the nest with the same initial speed, one stone straight downward from the top of the cliff and the other stone straight upward from the ground. In the absence of air resistance, which stone hits the nest in the least amount of time? **(a)** There is insufficient information for an answer. **(b)** Both stones hit the nest in the same amount of time. **(c)** The stone thrown from the ground. **(d)** The stone thrown from the top of the cliff.

Section 2.7 Graphical Analysis of Velocity and Acceleration

8. The graph accompanying this problem shows a three-part motion. For each of the three parts, A, B, and C, identify the direction of the motion. A positive velocity denotes motion to the right. **(a)** A right, B left, C right **(b)** A right, B right, C left **(c)** A right, B left, C left **(d)** A left, B right, C left **(e)** A left, B right, C right

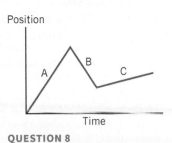

QUESTION 8

Problems

Online

Additional questions are available for assignment in WileyPLUS.

SSM Student Solutions
Manual

MMH Problem-solving help

GO Guided Online Tutorial

V-HINT Video Hints

CHALK Chalkboard Videos

BIO Biomedical application

E Easy

M Medium

H Hard

WS Worksheet

T Team Problem

Section 2.1 Displacement,

Section 2.2 Speed and Velocity

1. **E** **SSM** The space shuttle travels at a speed of about 7.6×10^3 m/s. The blink of an astronaut's eye lasts about 110 ms. How many football fields (length = 91.4 m) does the shuttle cover in the blink of an eye?

2. **E** **GO** For each of the three pairs of positions listed in the following table, determine the magnitude and direction (positive or negative) of the displacement.

	Initial position x_0	Final position x
(a)	+2.0 m	+6.0 m
(b)	+6.0 m	+2.0 m
(c)	−3.0 m	+7.0 m

3. **E** **SSM** Due to continental drift, the North American and European continents are drifting apart at an average speed of about 3 cm per year. At this speed, how long (in years) will it take for them to drift apart by another 1500 m (a little less than a mile)?

4. **E** **GO** The data in the following table describe the initial and final positions of a moving car. The elapsed time for each of the three pairs of positions listed in the table is 0.50 s. Review the concept of average velocity in Section 2.2 and then determine the average velocity (magnitude and direction) for each of the three pairs. Note that the algebraic sign of your answers will convey the direction.

	Initial position x_0	Final position x
(a)	+2.0 m	+6.0 m
(b)	+6.0 m	+2.0 m
(c)	−3.0 m	+7.0 m

5. **E** **CHALK** One afternoon, a couple walks three-fourths of the way around a circular lake, the radius of which is 1.50 km. They start at the west side of the lake and head due south to begin with. **(a)** What is the distance they travel? **(b)** What are the magnitude and direction (relative to due east) of the couple's displacement?

6. **E** The three-toed sloth is the slowest-moving land mammal. On the ground, the sloth moves at an average speed of 0.037 m/s, considerably slower than the giant tortoise, which walks at 0.076 m/s. After 12 minutes of walking, how much further would the tortoise have gone relative to the sloth?

7. **E** **GO** An 18-year-old runner can complete a 10.0-km course with an average speed of 4.39 m/s. A 50-year-old runner can cover the same distance with an average speed of 4.27 m/s. How much later (in seconds) should the younger runner start in order to finish the course *at the same time* as the older runner?

8. **E** A tourist being chased by an angry bear is running in a straight line toward his car at a speed of 4.0 m/s. The car is a distance d away. The bear is 26 m behind the tourist and running at 6.0 m/s. The tourist reaches the car safely. What is the maximum possible value for d?

9. **M** **V-HINT** In reaching her destination, a backpacker walks with an average velocity of 1.34 m/s, due west. This average velocity results because she hikes for 6.44 km with an average velocity of 2.68 m/s, due west, turns around, and hikes with an average velocity of 0.447 m/s, due east. How far east did she walk?

10. **M** **SSM** **CHALK** A bicyclist makes a trip that consists of three parts, each in the same direction (due north) along a straight road. During the first part, she rides for 22 minutes at an average speed of 7.2 m/s. During the second part, she rides for 36 minutes at an average speed of 5.1 m/s. Finally, during the third part, she rides for 8.0 minutes at an average speed of 13 m/s. **(a)** How far has the bicyclist traveled during the entire trip? **(b)** What is her average velocity for the trip?

11. **M** **V-HINT** A car makes a trip due north for three-fourths of the time and due south one-fourth of the time. The average northward velocity has a magnitude of 27 m/s, and the average southward velocity has a magnitude of 17 m/s. What is the average velocity (magnitude and direction) for the entire trip?

12. **H** You are on a train that is traveling at 3.0 m/s along a level straight track. Very near and parallel to the track is a wall that slopes upward at a 12° angle with the horizontal. As you face the window (0.90 m high, 2.0 m wide) in your compartment, the train is moving to the left, as the drawing indicates. The top edge of the wall first appears at window corner A and eventually disappears at window corner B. How much time passes between appearance and disappearance of the upper edge of the wall?

PROBLEM 12

Section 2.3 Acceleration

13. **E** Review Conceptual Example 7 as background for this problem. A car is traveling to the left, which is the negative direction. The direction of travel remains the same throughout this problem. The car's initial speed is 27.0 m/s, and during a 5.0-s interval, it changes to a final speed of **(a)** 29.0 m/s and **(b)** 23.0 m/s. In each case, find the acceleration (magnitude and algebraic sign) and state whether or not the car is decelerating.

14. **E** **GO** **(a)** Suppose that a NASCAR race car is moving to the right with a constant velocity of +82 m/s. What is the average acceleration of the car? **(b)** Twelve seconds later, the car is halfway around the track and traveling in the opposite direction with the same speed. What is the average acceleration of the car?

15. **E** Over a time interval of 2.16 years, the velocity of a planet orbiting a distant star reverses direction, changing from +20.9 km/s to −18.5 km/s. Find **(a)** the total change in the planet's velocity (in m/s) and **(b)** its average acceleration (in m/s²) during this interval. Include the correct algebraic sign with your answers to convey the directions of the velocity and the acceleration.

16. **E** **SSM** A motorcycle has a constant acceleration of 2.5 m/s². Both the velocity and acceleration of the motorcycle point in the same

direction. How much time is required for the motorcycle to change its speed from (a) 21 to 31 m/s, and (b) 51 to 61 m/s?

17. 🇪 A sprinter explodes out of the starting block with an acceleration of $+2.3$ m/s^2, which she sustains for 1.2 s. Then, her acceleration drops to zero for the rest of the race. What is her velocity (a) at $t = 1.2$ s and (b) at the end of the race?

18. 🇪 **GO** The initial velocity and acceleration of four moving objects at a given instant in time are given in the following table. Determine the final *speed* of each of the objects, assuming that the time elapsed since $t = 0$ s is 2.0 s.

	Initial velocity v_0	Acceleration a
(a)	$+12$ m/s	$+3.0$ m/s^2
(b)	$+12$ m/s	-3.0 m/s^2
(c)	-12 m/s	$+3.0$ m/s^2
(d)	-12 m/s	-3.0 m/s^2

19. 🇪 **GO** An Australian emu is running due north in a straight line at a speed of 13.0 m/s and slows down to a speed of 10.6 m/s in 4.0 s. (a) What is the direction of the bird's acceleration? (b) Assuming that the acceleration remains the same, what is the bird's velocity after an additional 2.0 s has elapsed?

20. 🇪 **SSM** **CHALK** For a standard production car, the highest road-tested acceleration ever reported occurred in 1993, when a Ford RS200 Evolution went from zero to 26.8 m/s (60 mi/h) in 3.275 s. Find the magnitude of the car's acceleration.

21. 🇲 **MMH** A car is traveling along a straight road at a velocity of $+36.0$ m/s when its engine cuts out. For the next twelve seconds the car slows down, and its average acceleration is $\bar{a}_1$. For the next six seconds the car slows down further, and its average acceleration is $\bar{a}_2$. The velocity of the car at the end of the eighteen-second period is $+28.0$ m/s. The ratio of the average acceleration values is $\bar{a}_1/\bar{a}_2 = 1.50$. Find the velocity of the car at the end of the initial twelve-second interval.

22. 🇭 Two motorcycles are traveling due east with different velocities. However, four seconds later, they have the same velocity. During this four-second interval, cycle A has an average acceleration of 2.0 m/s^2 due east, while cycle B has an average acceleration of 4.0 m/s^2 due east. By how much did the speeds *differ* at the beginning of the four-second interval, and which motorcycle was moving faster?

Section 2.4   Equations of Kinematics for Constant Acceleration, Section 2.5   Applications of the Equations of Kinematics

23. 🇪 In getting ready to slam-dunk the ball, a basketball player starts from rest and sprints to a speed of 6.0 m/s in 1.5 s. Assuming that the player accelerates uniformly, determine the distance he runs.

24. 🇪 **SSM** A jogger accelerates from rest to 3.0 m/s in 2.0 s. A car accelerates from 38.0 to 41.0 m/s also in 2.0 s. (a) Find the acceleration (magnitude only) of the jogger. (b) Determine the acceleration (magnitude only) of the car. (c) Does the car travel farther than the jogger during the 2.0 s? If so, how much farther?

25. 🇪 **GO** A VW Beetle goes from 0 to 60.0 mi/h with an acceleration of $+2.35$ m/s^2. (a) How much time does it take for the Beetle to reach this speed? (b) A top-fuel dragster can go from 0 to 60.0 mi/h in 0.600 s. Find the acceleration (in m/s^2) of the dragster.

26. 🇪 (a) What is the magnitude of the average acceleration of a skier who, starting from rest, reaches a speed of 8.0 m/s when going down a slope for 5.0 s? (b) How far does the skier travel in this time?

27. 🇪 **SSM** A jetliner, traveling northward, is landing with a speed of 69 m/s. Once the jet touches down, it has 750 m of runway in which to

reduce its speed to 6.1 m/s. Compute the average acceleration (magnitude and direction) of the plane during landing.

28. 🇪 **V-HINT** **MMH** The Kentucky Derby is held at the Churchill Downs track in Louisville, Kentucky. The track is one and one-quarter miles in length. One of the most famous horses to win this event was Secretariat. In 1973 he set a Derby record that would be hard to beat. His average acceleration during the last four quarter-miles of the race was $+0.0105$ m/s^2. His velocity at the start of the final mile ($x = +1609$ m) was about $+16.58$ m/s. The acceleration, although small, was very important to his victory. To assess its effect, determine the difference between the time he would have taken to run the final mile at a constant velocity of $+16.58$ m/s and the time he actually took. Although the track is oval in shape, assume it is straight for the purpose of this problem.

29. 🇪 **SSM** **MMH** A cart is driven by a large propeller or fan, which can accelerate or decelerate the cart. The cart starts out at the position $x = 0$ m, with an initial velocity of $+5.0$ m/s and a constant acceleration due to the fan. The direction to the right is positive. The cart reaches a maximum position of $x = +12.5$ m, where it begins to travel in the negative direction. Find the acceleration of the cart.

30. 🇪 **MMH** Two rockets are flying in the same direction and are side by side at the instant their retrorockets fire. Rocket A has an initial velocity of $+5800$ m/s, while rocket B has an initial velocity of $+8600$ m/s. After a time t both rockets are again side by side, the displacement of each being zero. The acceleration of rocket A is -15 m/s^2. What is the acceleration of rocket B?

31. 🇲 **CHALK** **MMH** A car is traveling at 20.0 m/s, and the driver sees a traffic light turn red. After 0.530 s (the reaction time), the driver applies the brakes, and the car decelerates at 7.00 m/s^2. What is the stopping distance of the car, as measured from the point where the driver first sees the red light?

32. 🇲 **GO** **MMH** A race driver has made a pit stop to refuel. After refueling, he starts from rest and leaves the pit area with an acceleration whose magnitude is 6.0 m/s^2; after 4.0 s he enters the main speedway. At the same instant, another car on the speedway and traveling at a constant velocity of 70.0 m/s overtakes and passes the entering car. The entering car maintains its acceleration. How much time is required for the entering car to catch the other car?

33. 🇲 **V-HINT** In a historical movie, two knights on horseback start from rest 88.0 m apart and ride directly toward each other to do battle. Sir George's acceleration has a magnitude of 0.300 m/s^2, while Sir Alfred's has a magnitude of 0.200 m/s^2. Relative to Sir George's starting point, where do the knights collide?

34. 🇲 **GO** Two soccer players start from rest, 48 m apart. They run directly toward each other, both players accelerating. The first player's acceleration has a magnitude of 0.50 m/s^2. The second player's acceleration has a magnitude of 0.30 m/s^2. (a) How much time passes before the players collide? (b) At the instant they collide, how far has the first player run?

35. 🇲 **MMH** **CHALK** A car is traveling at a constant speed of 33 m/s on a highway. At the instant this car passes an entrance ramp, a second car enters the highway from the ramp. The second car starts from rest and has a constant acceleration. What acceleration must it maintain, so that the two cars meet for the first time at the next exit, which is 2.5 km away?

36. 🇲 **GO** Refer to Multiple-Concept Example 5 to review a method by which this problem can be solved. You are driving your car, and the traffic light ahead turns red. You apply the brakes for 3.00 s, and the velocity of the car decreases to $+4.50$ m/s. The car's deceleration has a magnitude of 2.70 m/s^2 during this time. What is the car's displacement?

37. 🇭 A Boeing 747 "Jumbo Jet" has a length of 59.7 m. The runway on which the plane lands intersects another runway. The width of the intersection is 25.0 m. The plane decelerates through the intersection at a rate of 5.70 m/s^2 and clears it with a final speed of 45.0 m/s. How much time is needed for the plane to clear the intersection?

38. **H** **SSM** A locomotive is accelerating at 1.6 m/s². It passes through a 20.0-m-wide crossing in a time of 2.4 s. After the locomotive leaves the crossing, how much time is required until its speed reaches 32 m/s?

Section 2.6 Freely Falling Bodies

39. **E** **SSM** The greatest height reported for a jump into an airbag is 99.4 m by stuntman Dan Koko. In 1948 he jumped from rest from the top of the Vegas World Hotel and Casino. He struck the airbag at a speed of 39 m/s (88 mi/h). To assess the effects of air resistance, determine how fast he would have been traveling on impact had air resistance been absent.

40. **E** A dynamite blast at a quarry launches a chunk of rock straight upward, and 2.0 s later it is rising at a speed of 15 m/s. Assuming air resistance has no effect on the rock, calculate its speed **(a)** at launch and **(b)** 5.0 s after launch.

41. **E** **GO** **MMH** A ball is thrown vertically upward, which is the positive direction. A little later it returns to its point of release. The ball is in the air for a total time of 8.0 s. What is its initial velocity? Neglect air resistance.

42. **E** Review Conceptual Example 15 before attempting this problem. Two identical pellet guns are fired simultaneously from the edge of a cliff. These guns impart an initial speed of 30.0 m/s to each pellet. Gun A is fired straight upward, with the pellet going up and then falling back down, eventually hitting the ground beneath the cliff. Gun B is fired straight downward. In the absence of air resistance, how long after pellet B hits the ground does pellet A hit the ground?

43. **E** **MMH** An astronaut on a distant planet wants to determine its acceleration due to gravity. The astronaut throws a rock straight up with a velocity of +15 m/s and measures a time of 20.0 s before the rock returns to his hand. What is the acceleration (magnitude and direction) due to gravity on this planet?

44. **E** **SSM** A hot-air balloon is rising upward with a constant speed of 2.50 m/s. When the balloon is 3.00 m above the ground, the balloonist accidentally drops a compass over the side of the balloon. How much time elapses before the compass hits the ground?

45. **E** **GO** A ball is thrown straight upward and rises to a maximum height of 16 m above its launch point. At what height above its launch point has the speed of the ball decreased to one-half of its initial value?

46. **E** A diver springs upward with an initial speed of 1.8 m/s from a 3.0-m board. **(a)** Find the velocity with which he strikes the water. *[Hint: When the diver reaches the water, his displacement is y = −3.0 m (measured from the board), assuming that the downward direction is chosen as the negative direction.]* **(b)** What is the highest point he reaches above the water?

47. **E** **GO** A ball is thrown straight upward. At 4.00 m above its launch point, the ball's speed is one-half its launch speed. What maximum height above its launch point does the ball attain?

48. **E** **SSM** From her bedroom window a girl drops a water-filled balloon to the ground, 6.0 m below. If the balloon is released from rest, how long is it in the air?

49. **E** Before working this problem, review Conceptual Example 15. A pellet gun is fired straight downward from the edge of a cliff that is 15 m above the ground. The pellet strikes the ground with a speed of 27 m/s. How far above the cliff edge would the pellet have gone had the gun been fired straight upward?

50. **E** **V-HINT** Consult Multiple-Concept Example 5 in preparation for this problem. The velocity of a diver just before hitting the water is −10.1 m/s, where the minus sign indicates that her motion is directly downward. What is her displacement during the last 1.20 s of the dive?

51. **M** **V-HINT** A golf ball is dropped from rest from a height of 9.50 m. It hits the pavement, then bounces back up, rising just 5.70 m before falling back down again. A boy then catches the ball on the way down when it is 1.20 m above the pavement. Ignoring air resistance, calculate the total amount of time that the ball is in the air, from drop to catch.

52. **M** **MMH** A woman on a bridge 75.0 m high sees a raft floating at a constant speed on the river below. Trying to hit the raft, she drops a stone from rest when the raft has 7.00 m more to travel before passing under the bridge. The stone hits the water 4.00 m in front of the raft. Find the speed of the raft.

53. **M** **GO** **CHALK** Two stones are thrown simultaneously, one straight upward from the base of a cliff and the other straight downward from the top of the cliff. The height of the cliff is 6.00 m. The stones are thrown with the same speed of 9.00 m/s. Find the location (above the base of the cliff) of the point where the stones cross paths.

54. **M** **GO** Two arrows are shot vertically upward. The second arrow is shot after the first one, but while the first is still on its way up. The initial speeds are such that both arrows reach their maximum heights at the same instant, although these heights are different. Suppose that the initial speed of the first arrow is 25.0 m/s and that the second arrow is fired 1.20 s after the first. Determine the initial speed of the second arrow.

55. **M** **CHALK** **SSM** A cement block accidentally falls from rest from the ledge of a 53.0-m-high building. When the block is 14.0 m above the ground, a man, 2.00 m tall, looks up and notices that the block is directly above him. How much time, at most, does the man have to get out of the way?

56. **M** **V-HINT** A model rocket blasts off from the ground, rising straight upward with a constant acceleration that has a magnitude of 86.0 m/s² for 1.70 seconds, at which point its fuel abruptly runs out. Air resistance has no effect on its flight. What maximum altitude (above the ground) will the rocket reach?

57. **H** While standing on a bridge 15.0 m above the ground, you drop a stone from rest. When the stone has fallen 3.20 m, you throw a second stone straight down. What initial velocity must you give the second stone if they are both to reach the ground at the same instant? Take the downward direction to be the negative direction.

Section 2.7 Graphical Analysis of Velocity and Acceleration

58. **E** **SSM** A person who walks for exercise produces the position–time graph given with this problem. **(a)** Without doing any calculations, decide which segments of the graph (A, B, C, or D) indicate positive, negative, and zero average velocities. **(b)** Calculate the average velocity for each segment to verify your answers to part (a).

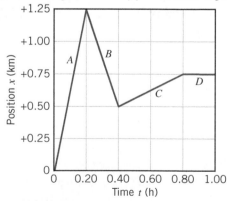

PROBLEM 58

59. **E** Starting at x = −16 m at time t = 0 s, an object takes 18 s to travel 48 m in the +x direction at a constant velocity. Make a position–time graph of the object's motion and calculate its velocity.

60. **E** **MMH** A snowmobile moves according to the velocity–time graph shown in the drawing. What is the snowmobile's average acceleration during each of the segments A, B, and C?

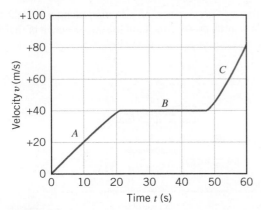

PROBLEM 60

61. **E** A bus makes a trip according to the position–time graph shown in the drawing. What is the average velocity (magnitude and direction) of the bus during each of the segments A, B, and C? Express your answers in km/h.

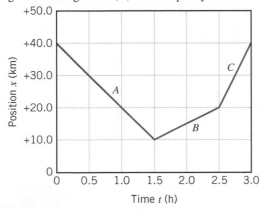

PROBLEM 61

62. **M GO CHALK** A bus makes a trip according to the position–time graph shown in the illustration. What is the average acceleration (in km/h^2) of the bus for the entire 3.5-h period shown in the graph?

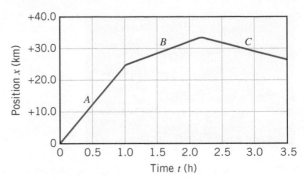

PROBLEM 62

63. **M** A runner is at the position $x = 0$ m when time $t = 0$ s. One hundred meters away is the finish line. Every ten seconds, this runner runs half the remaining distance to the finish line. During each ten-second segment, the runner has a constant velocity. For the first forty seconds of the motion, construct (a) the position–time graph and (b) the velocity–time graph.

Additional Problems

Online

64. **F GO** The data in the following table represent the initial and final velocities for a boat traveling along the x axis. The elapsed time for each of the four pairs of velocities in the table is 2.0 s. Review the concept of average acceleration in Section 2.3 and then determine the average acceleration (magnitude and direction) for each of the four pairs. Note that the algebraic sign of your answers will convey the direction.

	Initial velocity v_0	Final velocity v
(a)	+2.0 m/s	+5.0 m/s
(b)	+5.0 m/s	+2.0 m/s
(c)	−6.0 m/s	−3.0 m/s
(d)	+4.0 m/s	−4.0 m/s

65. **E** In 1954 the English runner Roger Bannister broke the four-minute barrier for the mile with a time of 3:59.4 s (3 min and 59.4 s). In 1999 the Moroccan runner Hicham el-Guerrouj set a record of 3:43.13 s for the mile. If these two runners had run in the same race, each running the entire race at the average speed that earned him a place in the record books, el-Guerrouj would have won. By how many meters?

66. **E MMH** At the beginning of a basketball game, a referee tosses the ball straight up with a speed of 4.6 m/s. A player cannot touch the ball until after it reaches its maximum height and begins to fall down. What is the minimum time that a player must wait before touching the ball?

67. **E** Electrons move through a certain electric circuit at an average speed of 1.1×10^{-2} m/s. How long (in minutes) does it take an electron to traverse 1.5 m of wire in the filament of a light bulb?

68. **E V-HINT MMH** A cheetah is hunting. Its prey runs for 3.0 s at a constant velocity of +9.0 m/s. Starting from rest, what constant acceleration must the cheetah maintain in order to run the same distance as its prey runs in the same time?

69. **M GO** The leader of a bicycle race is traveling with a constant velocity of +11.10 m/s and is 10.0 m ahead of the second-place cyclist. The second-place cyclist has a velocity of +9.50 m/s and an acceleration of +1.20 m/s^2. How much time elapses before he catches the leader?

70. **M V-HINT** A golfer rides in a golf cart at an average speed of 3.10 m/s for 28.0 s. She then gets out of the cart and starts walking at an average speed of 1.30 m/s. For how long (in seconds) must she walk if her average speed for the entire trip, riding and walking, is 1.80 m/s?

71. **M GO** Two cars cover the same distance in a straight line. Car A covers the distance at a constant velocity. Car B starts from rest and maintains a constant acceleration. Both cars cover a distance of 460 m in 210 s. Assume that they are moving in the +x direction. Determine (a) the constant velocity of car A, (b) the final velocity of car B, and (c) the acceleration of car B.

72. **M V-HINT MMH** A police car is traveling at a velocity of 18.0 m/s due north, when a car zooms by at a constant velocity of 42.0 m/s due north. After a reaction time of 0.800 s the policeman begins to pursue the speeder with an acceleration of 5.00 m/s^2. Including the reaction time, how long does it take for the police car to catch up with the speeder?

73. **M** **GO** **SSM** A jet is taking off from the deck of an aircraft carrier, as shown in the image. Starting from rest, the jet is catapulted with a constant acceleration of $+31$ m/s^2 along a straight line and reaches a velocity of $+62$ m/s. Find the displacement of the jet.

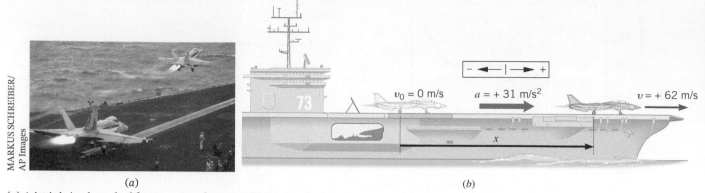

(*a*) A jet is being launched from an aircraft carrier. (*b*) During the launch, a catapult accelerates the jet down the flight deck.

PROBLEM 73

Physics in Biology, Medicine, and Sports

74. **E** **BIO** 2.6 The drawing shows a device that you can make with a piece of cardboard, which can be used to measure a person's reaction time. Hold the card at the top and suddenly drop it. Ask a friend to try to catch the card between his or her thumb and index finger. Initially, your friend's fingers must be level with the asterisks at the bottom. By noting where your friend catches the card, you can determine his or her reaction time in milliseconds (ms). Calculate the distances d_1, d_2, and d_3.

PROBLEM 74

75. **E** **BIO** 2.2 You step onto a hot beach with your bare feet. A nerve impulse, generated in your foot, travels through your nervous system at an average speed of 110 m/s. How much time does it take for the impulse, which travels a distance of 1.8 m, to reach your brain?

76. **E** **BIO** 2.4 The left ventricle of the heart accelerates blood from rest to a velocity of $+26$ cm/s. (**a**) If the displacement of the blood during the acceleration is $+2.0$ cm, determine its acceleration (in cm/s^2). (**b**) How much time does the blood take to reach its final velocity?

77. **M** **BIO** 2.5 During the late 1940s, Colonel John Paul Stapp was a pioneer in studying the effects of acceleration and deceleration on the human body. He made multiple runs strapped to a rocket sled that quickly accelerated him to high speeds along a straight track (see the images below). His research led to improvements in restraining harnesses and seatbelts for pilots and automobile occupants. During his final run, he reached a maximum speed of 632 mph. When the sled's braking system brought it to rest, Colonel Stapp experienced a deceleration of magnitude $46.2g$, or 46.2 times the acceleration due to gravity at the Earth's surface. Although he survived, he did sustain injuries, such as a fractured wrist, broken ribs, and bleeding in his eyes. Calculate how long it took to bring the rocket sled to rest. Assume the deceleration was constant during the braking period.

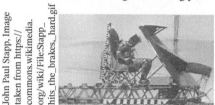

PROBLEM 77

78. **E** **BIO** 2.3 What is the acceleration of a thoroughbred racehorse? Starting from rest out of the gate, thoroughbreds can reach a maximum speed of 19 m/s (~43 mph.) over a distance of 35 m.

PROBLEM 78

79. **E** **BIO** 2.3 What is the acceleration of the fastest human sprinters? Olympic sprinter Usain Bolt reaches his maximum acceleration during the first 20.0 m of the sprint, at which time he is moving with a speed of 6.94 m/s. In reality, Bolt continues to accelerate for the first 60.0 m into the race, reaching a top speed near 28 mph.

PROBLEM 79

80. **M** **BIO** **2.5** Almost 80% of all traffic accidents involve distracted driving. Looking at a cell phone text message for even a couple of seconds is very dangerous. A typical human reaction time is 190 milliseconds. Assume a car is traveling at 71 mph. If the maximum deceleration of the car is a constant 8.0 m/s^2, what will be the stopping distance of the car? Take into account the reaction time.

81. **M** **BIO** **2.6** In 2012, Austrian skydiver Felix Baumgartner set a record for the highest exit altitude of 127 852 feet. He jumped from a helium balloon located in the earth's stratosphere (see the photo). At that altitude the atmosphere is too thin to breathe, so Baumgartner wore a special pressurized suit. During his free-fall toward the earth, he reached a maximum speed of 843.6 mph., which is greater than the speed of sound. This set a record for greatest speed achieved during free-fall and also set off a small sonic boom heard by his friends and family on the ground. Assume the effect of air resistance can be neglected and the acceleration due to gravity remains constant during his fall with a value of 9.807 m/s^2, and calculate the distance he fell before reaching 843.6 mph.

https://www.redbull.com/int-en/projects/red-bull-stratos

PROBLEM 81

Concepts and Calculations Problems

Online

In this chapter we have studied the displacement, velocity, and acceleration vectors. The concept questions are designed to review some of the important concepts that can help in anticipating some of the characteristics of the numerical answers.

82. **M** **CHALK** A skydiver is falling straight down, along the negative y direction. **(a)** During the initial part of the fall, her speed increases from 16 to 28 m/s in 1.5 s, as in part a of the figure. **(b)** Later, her parachute opens, and her speed decreases from 48 to 26 m/s in 11 s, as in part b of the drawing. *Concepts:* **(i)** Is her average acceleration positive or negative when her speed is increasing in part a of the figure? **(ii)** Is her average acceleration positive or negative when her speed is decreasing in part b of the figure? *Calculations:* In both instances (parts a and b of the figure) determine the magnitude and direction of her average acceleration.

83. **M** **CHALK** **SSM** A top-fuel dragster starts from rest and has a constant acceleration of 40.0 m/s^2. *Concepts:* **(i)** At time t the dragster has a certain velocity. Keeping in mind that the dragster starts from rest, when the time doubles to $2t$, does the velocity also double? **(ii)** When the time doubles to $2t$, does the displacement of the dragster also double? *Calculations:* What are the **(a)** final velocities and **(b)** displacements of the dragster at the end of 2.0 s and at the end of twice this time, or 4.0 s?

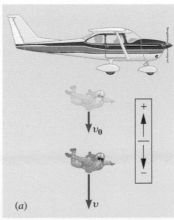

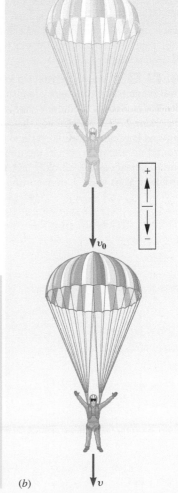

PROBLEM 82

(a)

(b)

Team Problems

84. **E** **T** **WS** **BIO** **A Fall on the Moon.** You and your team are stationed at an earth-based mission control facility and have been tasked with assisting astronauts via radio communication as they exit a spacecraft that has recently landed on the moon. The astronauts intended to explore the lunar surface but, unfortunately, they hit a snag even before they opened the hatch. Their craft has landed right on the edge of a 15-ft cliff, so that there is no way off the lander except to drop that distance from the bottom rung of a ladder on the side of the ship. You look up the structural parameters of the space suits, the gear they are wearing, and the physical limitations of the crew, and find that, on earth, the maximum drop they can safely endure is 4.0 ft. Is it safe for them to make the 15-ft drop on the moon? Explain your reasoning using quantitative information. Note that the acceleration due to gravity on the moon is one-sixth as large as that on earth.

85. **E** **T** **WS** **BIO** **Helmets to Prevent Concussions** In contact sports such as American football, hockey, and rugby, players routinely experience collisions in which considerable contact is made with the head, and therefore the brain can be subjected to significant forces. Repetitive collisions in which the brain experiences accelerations of 10 times the acceleration due to gravity (i.e., 10g) or greater can result in accumulative permanent brain damage. With current helmet design, the effective stopping distance for a football player undergoing a head-to-head collision with an initial speed of 8.00 mph. is 2.25 inches. **(a)** What is the average magnitude of the acceleration during the collision (in terms of g)? Is it a safe value? **(b)** The effective stopping distance can be increased by adding better cushioning features. You and your team are assigned to determine the increase in the effective stopping distance required for a new helmet design that will result in an effective average acceleration of less than 10.0g (with an initial speed of 8.00 mph.), thereby reducing the accumulative effects of head contact.

86. **M** **T** **A Deadly Virus.** On your flight out of South America your cargo plane emergency-lands on a desert island somewhere in the South Pacific. Because it was a beach landing, the pilot skids the plane on its belly rather than lowering the landing gear. The landing is rough: an initial vertical drop followed by an abrupt horizontal deceleration. When the dust settles, the pilot and your scientist colleagues are all okay, but you know there might be a big problem. Your team is transporting samples of a deadly virus from a recent breakout in a small village. The samples were packed in a cryogenically cooled container that has a special seal that may leak if it exceeds an acceleration greater than 10.0g. The container has a mechanical accelerometer gauge that measures the maximum vertical deceleration in case the container is dropped. It reads a value of $a = 7.30g$. However, you are still worried: there was another component of the acceleration. You get out of the plane and measure the length of the skid the plane made in the sand as it landed: 154 ft. The pilot claims the plane's horizontal speed on impact was 160.0 mph. Assuming the worst case scenario (i.e., that the vertical and horizontal accelerations occurred simultaneously), and that the horizontal deceleration was uniform, is it likely that the seal on the virus container is compromised? Explain.

87. **M** **T** **The Lost Drone.** You and your team are exploring the edge of an Antarctic mountain range and you send a drone ahead to help navigate. After takeoff you lose sight of the drone and, a few seconds later, the controls malfunction and the drone stops sending visual images and navigational information except for speed and directional data. Changing speeds erratically, the drone heads west until it makes a drastic turn at the 5-minute mark to 35.0° east of south. After nearly ten minutes, the speed drops to zero and the drone stops sending data. It has crashed. Using the speed/directional data, the team draws up the graph shown in the drawing. How far is the drone from you, and in what direction must you go to retrieve it? Express your result as a geographical direction (i.e., in the form 30° north of east, etc.).

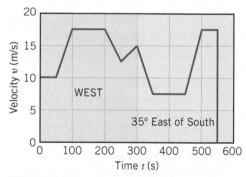

PROBLEM 87

The beautiful water fountain at the World War II Memorial in Washington, D.C., is illuminated at night. The arching streams of water follow parabolic paths whose sizes depend on the launch velocity of the water and the acceleration due to gravity, assuming that the effects of air resistance are negligible.

threespeedjones/iStockphoto

Kinematics in Two Dimensions

LEARNING OBJECTIVES

After reading this module, you should be able to...

3.1 Define two-dimensional kinematic variables.

3.2 Use two-dimensional kinematic equations to predict future or past values of variables.

3.3 Analyze projectile motion to predict future or past values of variables.

3.4 Apply relative velocity equations.

3.1 Displacement, Velocity, and Acceleration

In Chapter 2 the concepts of displacement, velocity, and acceleration are used to describe an object moving in one dimension. There are also situations in which the motion is along a curved path that lies in a plane. Such two-dimensional motion can be described using the same concepts. In Grand Prix racing, for example, the course follows a curved road, and **Figure 3.1** shows race car at two different positions along it. These positions are identified by the vectors $\vec{r}$ and $\vec{r}_0$, which are drawn from an arbitrary coordinate origin. The **displacement** $\Delta\vec{r}$ of the car is the vector drawn from the initial position $\vec{r}_0$ at time t_0 to the final position $\vec{r}$ at time t. The magnitude of $\Delta\vec{r}$ is the shortest distance between the two positions. In the drawing, the vectors $\vec{r}_0$ and $\Delta\vec{r}$ are drawn tail to head, so it is evident that $\vec{r}$ is the vector sum of $\vec{r}_0$ and $\Delta\vec{r}$. (See Sections 1.5 and 1.6 for a review of vectors and vector addition.) This means that $\vec{r} = \vec{r}_0 + \Delta\vec{r}$, or

$$\text{Displacement} = \Delta\vec{r} = \vec{r} - \vec{r}_0$$

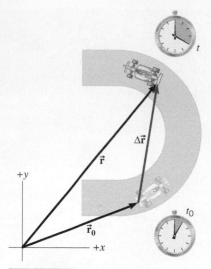

FIGURE 3.1 The displacement $\Delta\vec{r}$ of the car is a vector that points from the initial position of the car at time t_0 to the final position at time t. The magnitude of $\Delta\vec{r}$ is the shortest distance between the two positions.

The displacement here is defined as it is in Chapter 2. Now, however, the displacement vector may lie anywhere in a plane, rather than just along a straight line.

The **average velocity** $\overline{\vec{v}}$ of the car in **Figure 3.1** between the two positions is defined in a manner similar to that in Equation 2.2, as the displacement $\Delta\vec{r} = \vec{r} - \vec{r}_0$ divided by the elapsed time $\Delta t = t - t_0$:

$$\overline{\vec{v}} = \frac{\vec{r} - \vec{r}_0}{t - t_0} = \frac{\Delta\vec{r}}{\Delta t} \qquad (3.1)$$

Since both sides of Equation 3.1 must agree in direction, the average velocity vector has the same direction as the displacement $\Delta\vec{r}$. The velocity of the car at an instant of time is its **instantaneous velocity** $\vec{v}$. The average velocity becomes equal to the instantaneous velocity $\vec{v}$ in the limit that Δt becomes infinitesimally small ($\Delta t \rightarrow 0$ s):

$$\vec{v} = \lim_{\Delta t \rightarrow 0} \frac{\Delta\vec{r}}{\Delta t}$$

Figure 3.2 illustrates that the instantaneous velocity $\vec{v}$ is tangent to the path of the car. The drawing also shows the vector components $\vec{v}_x$ and $\vec{v}_y$ of the velocity, which are parallel to the x and y axes, respectively.

The **average acceleration** $\overline{\vec{a}}$ is defined just as it is for one-dimensional motion—namely, as the change in velocity, $\Delta\vec{v} = \vec{v} - \vec{v}_0$, divided by the elapsed time Δt:

$$\overline{\vec{a}} = \frac{\vec{v} - \vec{v}_0}{t - t_0} = \frac{\Delta\vec{v}}{\Delta t} \qquad (3.2)$$

The average acceleration has the same direction as the change in velocity $\Delta\vec{v}$. In the limit that the elapsed time becomes infinitesimally small, the average acceleration becomes equal to the **instantaneous acceleration** $\vec{a}$:

$$\vec{a} = \lim_{\Delta t \rightarrow 0} \frac{\vec{v}}{\Delta t}$$

The acceleration has a vector component $\vec{a}_x$ along the x direction and a vector component $\vec{a}_y$ along the y direction.

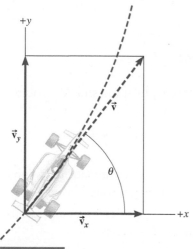

FIGURE 3.2 The instantaneous velocity $\vec{v}$ and its two vector components $\vec{v}_x$ and $\vec{v}_y$.

Check Your Understanding

(The answer is given at the end of the book.)

1. Suppose you are driving due east, traveling a distance of 1500 m in 2 minutes. You then turn due north and travel the same distance in the same time. What can be said about the average speeds and the average velocities for the two segments of the trip? **(a)** The average speeds are the same, and the average velocities are the same. **(b)** The average speeds are the same, but the average velocities are different. **(c)** The average speeds are different, but the average velocities are the same.

3.2 | Equations of Kinematics in Two Dimensions

To understand how displacement, velocity, and acceleration are applied to two-dimensional motion, consider a spacecraft equipped with two engines that are mounted perpendicular to each other. These engines produce the only forces that the craft experiences, and the spacecraft is assumed to be at the coordinate origin when $t_0 = 0$ s, so that $\vec{r}_0 = 0$ m. At a later time t, the spacecraft's displacement is $\Delta\vec{r} = \vec{r} - \vec{r}_0 = \vec{r}$. Relative to the x and y axes, the displacement $\vec{r}$ has vector components of $\vec{x}$ and $\vec{y}$, respectively.

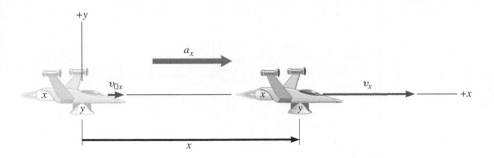

FIGURE 3.3 The spacecraft is moving with a constant acceleration a_x parallel to the x axis. There is no motion in the y direction, and the y engine is turned off.

In **Figure 3.3** only the engine oriented along the x direction is firing, and the vehicle accelerates along this direction. It is assumed that the velocity in the y direction is zero, and it remains zero, since the y engine is turned off. The motion of the spacecraft along the x direction is described by the five kinematic variables x, a_x, v_x, v_{0x}, and t. Here the symbol "x" reminds us that we are dealing with the x components of the displacement, velocity, and acceleration vectors. (See Sections 1.7 and 1.8 for a review of vector components.) The variables x, a_x, v_x, and v_{0x} are scalar components (or "components," for short). As Section 1.7 discusses, these components are positive or negative numbers (with units), depending on whether the associated vector components point in the $+x$ or the $-x$ direction. If the spacecraft has a constant acceleration along the x direction, the motion is exactly like that described in Chapter 2, and the equations of kinematics can be used. For convenience, these equations are written in the first four rows of **Table 3.1**, where each equation contains four of the five kinematic variables.

TABLE 3.1 Equations of Kinematics for Constant Acceleration in Two-Dimensional Motion

Equation Number	Equation	x	a_x	v_x	v_{0x}	t
(3.3a)	$v_x = v_{0x} + a_x t$	—	✓	✓	✓	✓
(3.4a)	$x = \frac{1}{2}(v_{0x} + v_x)t$	✓	—	✓	✓	✓
(3.5a)	$x = v_{0x}t + \frac{1}{2}a_x t^2$	✓	✓	—	✓	✓
(3.6a)	$v_x^2 = v_{0x}^2 + 2a_x x$	✓	✓	✓	✓	—

Equation Number	Equation	y	a_y	v_y	v_{0y}	t
(3.3b)	$v_y = v_{0y} + a_y t$	—	✓	✓	✓	✓
(3.4b)	$y = \frac{1}{2}(v_{0y} + v_y)t$	✓	—	✓	✓	✓
(3.5b)	$y = v_{0y}t + \frac{1}{2}a_y t^2$	✓	✓	—	✓	✓
(3.6b)	$v_y^2 = v_{0y}^2 + 2a_y y$	✓	✓	✓	✓	—

Figure 3.4 is analogous to **Figure 3.3**, except that now only the y engine is firing, and the spacecraft accelerates along the y direction. Such a motion can be described in terms of the kinematic variables y, a_y, v_y, v_{0y}, and t. And if the acceleration along the y direction is constant, these variables are related by the equations of kinematics, as written in the last four rows of **Table 3.1**. Like their counterparts in the x direction, the scalar components, y, a_y, v_y, and v_{0y}, may be positive (+) or negative (−) numbers (with units).

If both engines of the spacecraft are firing *at the same time,* the resulting motion takes place in part along the x axis and in part along the y axis, as **Interactive Figure 3.5** illustrates. The thrust of each engine gives the vehicle a corresponding acceleration component. The x engine accelerates the craft in the x direction and causes a change in the x component of the velocity. Likewise, the y engine causes a change in the y component of the velocity.

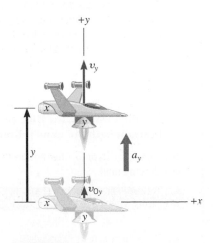

FIGURE 3.4 The spacecraft is moving with a constant acceleration a_y parallel to the y axis. There is no motion in the x direction, and the x engine is turned off.

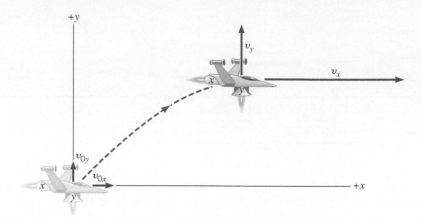

The two-dimensional motion of the spacecraft can be viewed as the combination of the separate x and y motions.

> **Problem-Solving Insight** It is important to realize that the x part of the motion occurs exactly as it would if the y part did not occur at all. Similarly, the y part of the motion occurs exactly as it would if the x part of the motion did not exist.

In other words, the x and y motions are independent of each other.

The independence of the x and y motions lies at the heart of two-dimensional kinematics. It allows us to treat two-dimensional motion as two distinct one-dimensional motions, one for the x direction and one for the y direction. Everything that we have learned in Chapter 2 about kinematics in one dimension will now be applied separately to each of the two directions. In so doing, we will be able to describe the x and y variables separately and then bring these descriptions together to understand the two-dimensional picture. Examples 1 and 2 take this approach in dealing with a moving spacecraft.

EXAMPLE 1 | The Displacement of a Spacecraft

In **Interactive Figure 3.5**, the directions to the right and upward are the positive directions. In the x direction, the spacecraft has an initial velocity component of $v_{0x} = +22$ m/s and an acceleration component of $a_x = +24$ m/s^2. In the y direction, the analogous quantities are $v_{0y} = +14$ m/s and $a_y = +12$ m/s^2. At a time of $t = 7.0$ s, find the x and y components of the spacecraft's displacement.

Reasoning The motion in the x direction and the motion in the y direction can be treated separately, each as a one-dimensional motion subject to the equations of kinematics for constant acceleration (see **Table 3.1**). By following this procedure we will be able to determine x and y, which specify the spacecraft's location after an elapsed time of 7.0 s.

> **Problem-Solving Insight** When the motion is two-dimensional, the time variable t has the same value for both the x and y directions.

Solution The data for the motion in the x direction are listed in the following table:

x-Direction Data				
x	a_x	v_x	v_{0x}	t
?	+24 m/s^2		+22 m/s	7.0 s

The x component of the craft's displacement can be found by using Equation 3.5a:

$$x = v_{0x}t + \tfrac{1}{2}a_xt^2 = (22 \text{ m/s})(7.0 \text{ s}) + \tfrac{1}{2}(24 \text{ m/s}^2)(7.0 \text{ s})^2 = \boxed{+740 \text{ m}}$$

The data for the motion in the y direction are listed in the following table:

y-Direction Data				
y	a_y	v_y	v_{0y}	t
?	+12 m/s^2		+14 m/s	7.0 s

The y component of the craft's displacement can be found by using Equation 3.5b:

$$y = v_{0y}t + \tfrac{1}{2}a_yt^2 = (14 \text{ m/s})(7.0 \text{ s}) + \tfrac{1}{2}(12 \text{ m/s}^2)(7.0 \text{ s})^2 = \boxed{+390 \text{ m}}$$

After 7.0 s, the spacecraft is 740 m to the right and 390 m above the origin.

Analyzing Multiple-Concept Problems

EXAMPLE 2 | The Velocity of a Spacecraft

This example also deals with the spacecraft in **Interactive Figure 3.5**. As in Example 1, the x components of the craft's initial velocity and acceleration are $v_{0x} = +22$ m/s and $a_x = +24$ m/s^2, respectively. The corresponding y components are $v_{0y} = +14$ m/s and $a_y = +12$ m/s^2. At a time of $t = 7.0$ s, find the spacecraft's final velocity (magnitude and direction).

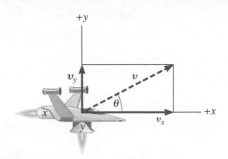

Reasoning **Figure 3.6** shows the final velocity vector, which has components v_x and v_y and a magnitude v. The final velocity is directed at an angle θ above the $+x$ axis. The vector and its components form a right triangle, the hypotenuse being the magnitude of the velocity and the components being the other two sides. Thus, we can use the Pythagorean theorem to determine the magnitude v from values for the components v_x and v_y. We can also use trigonometry to determine the directional angle θ.

FIGURE 3.6 The velocity vector has components v_x and v_y and a magnitude v. The magnitude gives the speed of the spacecraft, and the angle θ gives the direction of travel relative to the positive x direction.

Knowns and Unknowns The data for this problem are listed in the table that follows:

Description	Symbol	Value	Comment
x component of acceleration	a_x	$+24$ m/s^2	
x component of initial velocity	v_{0x}	$+22$ m/s	
y component of acceleration	a_y	$+12$ m/s^2	
y component of initial velocity	v_{0y}	$+14$ m/s	
Time	t	7.0 s	Same time for x and y directions
Unknown Variables			
Magnitude of final velocity	v	?	
Direction of final velocity	θ	?	

Modeling the Problem

STEP 1 Final Velocity In **Figure 3.6** the final velocity vector and its components v_x and v_y form a right triangle. Applying the Pythagorean theorem to this right triangle shows that the magnitude v of the final velocity is given in terms of the components by Equation 1a at the right. From the right triangle in **Figure 3.6** it also follows that the directional angle θ is given by Equation 1b at the right.

$$v = \sqrt{v_x^{\,2} + v_y^{\,2}} \qquad \text{(1a)}$$

$$\theta = \tan^{-1}\!\left(\frac{v_y}{v_x}\right) \qquad \text{(1b)}$$

STEP 2 The Components of the Final Velocity Values are given for the kinematic variables a_x, v_{0x}, and t in the x direction and for the corresponding variables in the y direction (see the table of knowns and unknowns). For each direction, then, these values allow us to calculate the final velocity components v_x and v_y by using Equations 3.3a and 3.3b from the equations of kinematics.

$$\boxed{v_x = v_{0x} + a_x t} \qquad \text{(3.3a)}$$

$$\boxed{v_y = v_{0y} + a_y t} \qquad \text{(3.3b)}$$

These expressions can be substituted into Equations 1a and 1b for the magnitude and direction of the final velocity, as shown at the right.

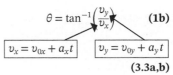

$$v = \sqrt{v_x^{\,2} + v_y^{\,2}} \qquad \text{(1a)}$$

$$\boxed{v_x = v_{0x} + a_x t} \qquad \boxed{v_y = v_{0y} + a_y t}$$
$$\text{(3.3a,b)}$$

$$\theta = \tan^{-1}\!\left(\frac{v_y}{v_x}\right) \qquad \text{(1b)}$$

$$\boxed{v_x = v_{0x} + a_x t} \qquad \boxed{v_y = v_{0y} + a_y t}$$
$$\text{(3.3a,b)}$$

Solution Algebraically combining the results of each step, we find that

STEP 1   STEP 2

$$v = \sqrt{v_x{}^2 + v_y{}^2} = \sqrt{(v_{0x} + a_x t)^2 + (v_{0y} + a_y t)^2}$$

$$\theta = \tan^{-1}\left(\frac{v_y}{v_x}\right) = \tan^{-1}\left(\frac{v_{0y} + a_y t}{v_{0x} + a_x t}\right)$$

STEP 1   STEP 2

With the data given for the kinematic variables in the x and y directions, we find that the magnitude and direction of the final velocity of the spacecraft are

$$v = \sqrt{(v_{0x} + a_x t)^2 + (v_{0y} + a_y t)^2}$$

$$= \sqrt{[(22 \text{ m/s}) + (24 \text{ m/s}^2)(7.0 \text{ s})]^2 + [(14 \text{ m/s}) + (12 \text{ m/s}^2)(7.0 \text{ s})]^2} = \boxed{210 \text{ m/s}}$$

$$\theta = \tan^{-1}\left(\frac{v_{0y} + a_y t}{v_{0x} + a_x t}\right) = \tan^{-1}\left[\frac{(14 \text{ m/s}) + (12 \text{ m/s}^2)(7.0 \text{ s})}{(22 \text{ m/s}) + (24 \text{ m/s}^2)(7.0 \text{ s})}\right] = \boxed{27°}$$

After 7.0 s, the spacecraft, at the position determined in Example 1, has a velocity of 210 m/s in a direction of 27° above the positive x axis.

Related Homework: Problem 59

The following Reasoning Strategy gives an overview of how the equations of kinematics are applied to describe motion in two dimensions, as in Examples 1 and 2.

REASONING STRATEGY   **Applying the Equations of Kinematics in Two Dimensions**

1. **Make a drawing to represent the situation being studied.**

2. **Decide which directions are to be called positive (+) and negative (−) relative to a conveniently chosen coordinate origin. Do not change your decision during the course of a calculation.**

3. **Remember that the time variable t has the same value for the part of the motion along the x axis and the part along the y axis.**

4. **In an organized way, write down the values (with appropriate + and − signs) that are given for any of the five kinematic variables associated with the x direction and the y direction. Be on the alert for implied data, such as the phrase "starts from rest," which means that the values of the initial velocity components are zero: $v_{0x} = 0$ m/s and $v_{0y} = 0$ m/s. The data summary tables that are used in the examples are a good way of keeping track of this information. In addition, identify the variables that you are being asked to determine.**

5. **Before attempting to solve a problem, verify that the given information contains values for at least three of the kinematic variables. Do this for the x and the y direction of the motion. Once the three known variables are identified along with the desired unknown variable, the appropriate relations from Table 3.1 can be selected.**

6. **When the motion is divided into segments, remember that the final velocity for one segment is the initial velocity for the next segment.**

7. **Keep in mind that a kinematics problem may have two possible answers. Try to visualize the different physical situations to which the answers correspond.**

Check Your Understanding

(The answer is given at the end of the book.)

2. A power boat, starting from rest, maintains a constant acceleration. After a certain time t, its displacement and velocity are $\vec{r}$ and $\vec{v}$. At time $2t$, what would be its displacement and velocity, assuming the acceleration remains the same? **(a)** $2\vec{r}$ and $2\vec{v}$ **(b)** $2\vec{r}$ and $4\vec{v}$ **(c)** $4\vec{r}$ and $2\vec{v}$ **(d)** $4\vec{r}$ and $4\vec{v}$

3.3 Projectile Motion

The biggest thrill in baseball is a home run. The motion of the ball on its curving path into the stands is a common type of two-dimensional motion called "projectile motion." A good description of such motion can often be obtained with the assumption that air resistance is absent.

Using the equations in **Table 3.1**, we consider the horizontal and vertical parts of the motion separately. In the horizontal or x direction, the moving object (the projectile) does not slow down in the absence of air resistance. Thus, the x component of the velocity remains constant at its initial value or $v_x = v_{0x}$, and the x component of the acceleration is $a_x = 0 \text{ m/s}^2$. In the vertical or y direction, however, the projectile experiences the effect of gravity. As a result, the y component of the velocity v_y is not constant and changes. The y component of the acceleration a_y is the downward acceleration due to gravity. If the path or trajectory of the projectile is near the earth's surface, a_y has a magnitude of 9.80 m/s^2. In this text, then, the phrase "projectile motion" means that $a_x = 0 \text{ m/s}^2$ and a_y equals the acceleration due to gravity. Example 3 and other examples in this section illustrate how the equations of kinematics are applied to projectile motion.

EXAMPLE 3 | A Falling Care Package

Figure 3.7 shows an airplane moving horizontally with a constant velocity of $+115 \text{ m/s}$ at an altitude of 1050 m. The directions to the right and upward have been chosen as the positive directions. The plane releases a "care package" that falls to the ground along a curved trajectory. Ignoring air resistance, determine the time required for the package to hit the ground.

Reasoning The time required for the package to hit the ground is the time it takes for the package to fall through a vertical distance of 1050 m. In falling, it moves to the right, as well as downward, but these two parts of the motion occur independently. Therefore, we can focus solely on the vertical part. We note that the package is moving initially in the horizontal or x direction, not in the y direction, so that $v_{0y} = 0 \text{ m/s}$. Furthermore, when the package hits the ground, the y component of its displacement is $y = -1050 \text{ m}$, as the drawing shows. The acceleration is that

due to gravity, so $a_y = -9.80 \text{ m/s}^2$. These data are summarized as follows:

y-Direction Data				
y	a_y	v_y	v_{0y}	t
-1050 m	-9.80 m/s^2		0 m/s	?

With these data, Equation 3.5b ($y = v_{0y}t + \frac{1}{2}a_y t^2$) can be used to find the fall time.

> **Problem-Solving Insight** The variables y, a_y, v_y, and v_{0y} are scalar components. Therefore, an algebraic sign (+ or −) must be included with each one to denote direction.

Solution Since $v_{0y} = 0 \text{ m/s}$, it follows from Equation 3.5b that $y = \frac{1}{2}a_y t^2$ and

$$t = \sqrt{\frac{2y}{a_y}} = \sqrt{\frac{2(-1050 \text{ m})}{-9.80 \text{ m/s}^2}} = \boxed{14.6 \text{ s}}$$

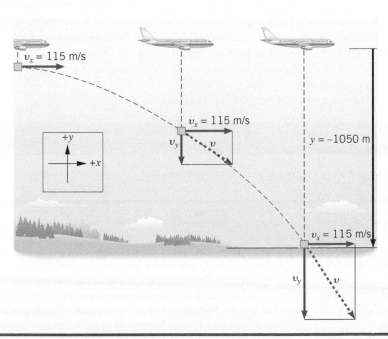

$v_x = 115 \text{ m/s}$

$+y$

$+x$

$v_x = 115 \text{ m/s}$

v_y v

$y = -1050 \text{ m}$

$v_x = 115 \text{ m/s}$

v_y v

FIGURE 3.7 The package falling from the plane is an example of projectile motion, as Examples 3 and 4 discuss.

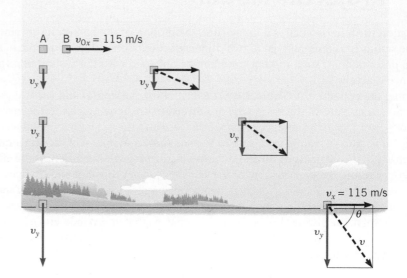

FIGURE 3.8 Package A and package B are released simultaneously at the same height and strike the ground at the same time because their y variables (y, a_y, and v_{0y}) are the same.

The freely falling package in Example 3 picks up vertical speed on the way downward. The horizontal component of the velocity, however, retains its initial value of $v_{0x} = +115$ m/s throughout the entire descent. Since the plane also travels at a constant horizontal velocity of $+115$ m/s, it remains directly above the falling package. The pilot always sees the package directly beneath the plane, as the dashed vertical lines in **Figure 3.7** show. This result is a direct consequence of the fact that the package has no acceleration in the horizontal direction. In reality, air resistance would slow down the package, and it would not remain directly beneath the plane during the descent.

Figure 3.8 illustrates what happens to two packages that are released simultaneously from the same height, in order to emphasize that the vertical and horizontal parts of the motion in Example 3 occur independently. Package A is dropped from a stationary balloon and falls straight downward toward the ground, since it has no horizontal velocity component ($v_{0x} = 0$ m/s). Package B, on the other hand, is given an initial velocity component of $v_{0x} = +115$ m/s in the horizontal direction, as in Example 3, and follows the path shown in the figure. Both packages hit the ground at the same time. Not only do the packages in **Figure 3.8** reach the ground at the same time, but the y components of their velocities are also equal at all points on the way down. However, package B does hit the ground with a greater speed than does package A. Remember, speed is the magnitude of the velocity vector, and the velocity of B has an x component, whereas the velocity of A does not. The magnitude and direction of the velocity vector for package B at the instant just before the package hits the ground is computed in Example 4.

Analyzing Multiple-Concept Problems

EXAMPLE 4 | The Velocity of the Care Package

Figure 3.7 shows a care package falling from a plane, and **Figure 3.8** shows this package as package B. As in Example 3, the directions to the right and upward are chosen as the positive directions, and the plane is moving horizontally with a constant velocity of $+115$ m/s at an altitude of 1050 m. Ignoring air resistance, find the magnitude v and the directional angle θ of the final velocity vector that the package has just before it strikes the ground.

Reasoning Figures 3.7 and 3.8 show the final velocity vector, which has components v_x and v_y and a magnitude v. The vector is directed at an angle θ below the horizontal or x direction. We note the right triangle formed by the vector and its components. The hypotenuse of the triangle is the magnitude of the velocity, and the components are the other two sides. As in Example 2, we can use the Pythagorean theorem to express the magnitude or speed v in terms of the components v_x and v_y, and we can use trigonometry to determine the directional angle θ.

Knowns and Unknowns The data for this problem are listed in the table that follows:

Description	Symbol	Value	Comment
Explicit Data, x Direction			
x component of initial velocity	v_{0x}	+115 m/s	Package has plane's horizontal velocity at instant of release
Implicit Data, x Direction			
x component of acceleration	a_x	0 m/s^2	No horizontal acceleration, since air resistance is ignored
Explicit Data, y Direction			
y component of displacement	y	−1050 m	Negative, since upward is positive and package falls downward
Implicit Data, y Direction			
y component of initial velocity	v_{0y}	0 m/s	Package traveling horizontally in x direction at instant of release, not in y direction
y component of acceleration	a_y	−9.80 m/s^2	Acceleration vector for gravity points downward in the negative direction
Unknown Variables			
Magnitude of final velocity	v	?	
Direction of final velocity	θ	?	

Modeling the Problem

STEP 1 Final Velocity Using the Pythagorean theorem to express the speed v in terms of the components v_x and v_y (see **Figure 3.7** or **3.8**), we obtain Equation 1a at the right. Furthermore, in a right triangle, the cosine of an angle is the side adjacent to the angle divided by the hypotenuse. With this in mind, we see in **Figure 3.7** or **Figure 3.8** that the directional angle θ is given by Equation 1b at the right.

$$v = \sqrt{v_x^2 + v_y^2} \qquad (1a)$$

$$\theta = \cos^{-1}\left(\frac{v_x}{v}\right) = \cos^{-1}\left(\frac{v_x}{\sqrt{v_x^2 + v_y^2}}\right)$$
$$(1b)$$

STEP 2 The Components of the Final Velocity Reference to the table of knowns and unknowns shows that, in the x direction, values are available for the kinematic variables v_{0x} and a_x. Since the acceleration a_x is zero, the final velocity component v_x remains unchanged from its initial value of v_{0x}, so we have

$$\boxed{v_x = v_{0x}}$$

In the y direction, values are available for y, v_{0y}, and a_y, so that we can determine the final velocity component v_y using Equation 3.6b from the equations of kinematics:

$$\boxed{v_y^2 = v_{0y}^2 + 2a_y y} \qquad (3.6b)$$

These results for v_x and v_y can be substituted into Equations 1a and 1b, as shown at the right.

$$v = \sqrt{v_x^2 + v_y^2} \qquad (1a)$$

$$\boxed{v_x = v_{0x}} \quad \boxed{v_y^2 = v_{0y}^2 + 2a_y y}$$
$$(3.6b)$$

$$\theta = \cos^{-1}\left(\frac{v_x}{\sqrt{v_x^2 + v_y^2}}\right) \qquad (1b)$$

$$\boxed{v_x = v_{0x}} \quad \boxed{v_y^2 = v_{0y}^2 + 2a_y y} \quad (3.6b)$$

Solution Algebraically combining the results of each step, we find that

$$\overset{\text{STEP 1}}{v} = \overset{}{\sqrt{v_x^2 + v_y^2}} \overset{\text{STEP 2}}{=} \sqrt{v_{0x}^2 + v_{0y}^2 + 2a_y y}$$

$$\theta = \overset{\text{STEP 1}}{\cos^{-1}\left(\frac{v_x}{\sqrt{v_x^2 + v_y^2}}\right)} = \overset{\text{STEP 2}}{\cos^{-1}\left(\frac{v_{0x}}{\sqrt{v_{0x}^2 + v_{0y}^2 + 2a_y y}}\right)}$$

With the data given for the kinematic variables in the x and y directions, we find that the magnitude and direction of the final velocity of the package are

$$v = \sqrt{v_{0x}^2 + v_{0y}^2 + 2a_y y} = \sqrt{(115 \text{ m/s})^2 + (0 \text{ m/s})^2 + 2(-9.80 \text{ m/s}^2)(-1050 \text{ m})} = \boxed{184 \text{ m/s}}$$

$$\theta = \cos^{-1}\left(\frac{v_{0x}}{\sqrt{v_{0x}^2 + v_{0y}^2 + 2a_y y}}\right) = \cos^{-1}\left[\frac{115 \text{ m/s}}{\sqrt{(115 \text{ m/s})^2 + (0 \text{ m/s})^2 + 2(-9.80 \text{ m/s}^2)(-1050 \text{ m})}}\right] = \boxed{51.3°}$$

Problem-Solving Insight
The speed of a projectile at any location along its path is the magnitude v of its velocity at that location: $v = \sqrt{v_x^2 + v_y^2}$. Both the horizontal and vertical velocity components contribute to the speed.

Related Homework: Problems 32, 37, 64

An important feature of projectile motion is that there is no acceleration in the horizontal, or *x*, direction. Conceptual Example 5 discusses an interesting implication of this feature.

CONCEPTUAL EXAMPLE 5 | Shot a Bullet Into the Air . . .

Suppose you are driving in a convertible with the top down. The car is moving to the right at a constant velocity. As **Figure 3.9** illustrates, you point a rifle straight upward and fire it. In the absence of air resistance, would the bullet land **(a)** behind you, **(b)** ahead of you, or **(c)** in the barrel of the rifle?

FIGURE 3.9 The car is moving with a constant velocity to the right, and the rifle is pointed straight up. In the absence of air resistance, a bullet fired from the rifle has no acceleration in the horizontal direction. Example 5 discusses what happens to the bullet.

Reasoning Because there is no air resistance to slow it down, the bullet experiences no horizontal acceleration. Thus, the bullet's horizontal velocity component does not change, and it stays the same as that of the rifle and the car.

Answers (a) and (b) are incorrect. If air resistance were present, it would slow down the bullet and cause it to land behind you, toward the rear of the car. However, air resistance is absent. If the bullet were to land ahead of you, its horizontal velocity component would have to be greater than that of the rifle and the car. This cannot be, since the bullet's horizontal velocity component never changes.

Answer (c) is correct. Since the bullet's horizontal velocity component does not change, it retains its initial value, and remains matched to that of the rifle and the car. As a result, the bullet remains directly above the rifle at all times and would fall directly back into the barrel of the rifle. This situation is analogous to that in **Figure 3.7**, where the care package, as it falls, remains directly below the plane.

Related Homework: Problem 38

Often projectiles, such as footballs and baseballs, are sent into the air at an angle with respect to the ground. From a knowledge of the projectile's initial velocity, a wealth of information can be obtained about the motion. For instance, Example 6 demonstrates how to calculate the maximum height reached by the projectile.

EXAMPLE 6 | The Height of a Kickoff

A placekicker kicks a football at an angle of $\theta = 40.0°$ above the horizontal axis, as **Animated Figure 3.10** shows. The initial speed of the ball is $v_0 = 22$ m/s. Ignore air resistance, and find the maximum height H that the ball attains.

Reasoning The maximum height is a characteristic of the vertical part of the motion, which can be treated separately from the horizontal part. In preparation for making use of this fact, we calculate the vertical component of the initial velocity:

$$v_{0y} = v_0 \sin \theta = +(22 \text{ m/s}) \sin 40.0° = +14 \text{ m/s}$$

The vertical component of the velocity, v_y, decreases as the ball moves upward. Eventually, $v_y = 0$ m/s at the maximum height H. The data below can be used in Equation 3.6b ($v_y^2 = v_{0y}^2 + 2a_y y$) to find the maximum height:

y-Direction Data				
y	a_y	v_y	v_{0y}	t
$H = ?$	-9.80 m/s^2	0 m/s	$+14$ m/s	

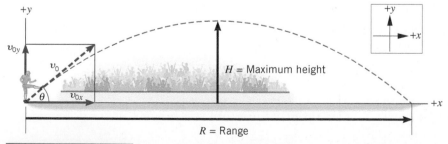

ANIMATED FIGURE 3.10 A football is kicked with an initial speed of v_0 at an angle of θ above the ground. The ball attains a maximum height H and a range R.

Problem-Solving Insight When a projectile reaches maximum height, the vertical component of its velocity is momentarily zero ($v_y = 0$ m/s). However, the horizontal component of its velocity is not zero.

The height H depends only on the y variables; the same height would have been reached had the ball been thrown *straight up* with an initial velocity of $v_{0y} = +14$ m/s.

Solution From Equation 3.6b, we find that

$$y = H = \frac{v_y^2 - v_{0y}^2}{2a_y} = \frac{(0 \text{ m/s})^2 - (14 \text{ m/s})^2}{2(-9.80 \text{ m/s}^2)} = \boxed{+10 \text{ m}}$$

It is also possible to find the total time or "hang time" during which the football in **Animated Figure 3.10** is in the air. Example 7 shows how to determine this time.

EXAMPLE 7 | The Physics of the "Hang Time" of a Football

For the motion illustrated in **Animated Figure 3.10**, ignore air resistance and use the data from Example 6 to determine the time of flight between kickoff and landing.

Reasoning Given the initial velocity, it is the acceleration due to gravity that determines how long the ball stays in the air. Thus, to find the time of flight we deal with the vertical part of the motion. Since the ball starts at and returns to ground level, the displacement in the y direction is zero. The initial velocity component in the y direction is the same as that in Example 6; that is, $v_{0y} = +14$ m/s. Therefore, we have

y-Direction Data				
y	a_y	v_y	v_{0y}	t
0 m	-9.80 m/s^2		$+14$ m/s	?

The time of flight can be determined from Equation 3.5b ($y = v_{0y}t + \frac{1}{2}a_y t^2$).

Solution Using Equation 3.5b and the fact that $y = 0$ m, we find that

$$y = 0 = v_{0y}t + \tfrac{1}{2}a_y t^2 = \left(v_{0y} + \tfrac{1}{2}a_y t\right)t$$

There are two solutions to this equation. One is given by $v_{0y} + \frac{1}{2}a_y t = 0$, with the result that

$$t = \frac{-2v_{0y}}{a_y} = \frac{-2(14 \text{ m/s})}{-9.80 \text{ m/s}^2} = 2.9 \text{ s}$$

The other is given by $t = 0$ s. The solution we seek is $\boxed{t = 2.9 \text{ s}}$, because $t = 0$ s corresponds to the initial kickoff.

Another important feature of projectile motion is called the "range." The range, as **Animated Figure 3.10** shows, is the horizontal distance traveled between launching and landing, assuming the projectile returns to the *same vertical level* at which it was fired. Example 8 shows how to obtain the range.

EXAMPLE 8 | The Range of a Kickoff

For the motion shown in **Animated Figure 3.10** and discussed in Examples 6 and 7, ignore air resistance and calculate the range R of the projectile.

Reasoning The range is a characteristic of the horizontal part of the motion. Thus, our starting point is to determine the horizontal component of the initial velocity:

$$v_{0x} = v_0 \cos \theta = +(22 \text{ m/s}) \cos 40.0° = +17 \text{ m/s}$$

Recall from Example 7 that the time of flight is $t = 2.9$ s. Since there is no acceleration in the x direction, v_x remains constant, and the range is simply the product of $v_x = v_{0x}$ and the time.

Solution The range is

$$x = R = v_{0x}t = +(17 \text{ m/s})(2.9 \text{ s}) = \boxed{+49 \text{ m}}$$

The range in the previous example depends on the angle θ at which the projectile is fired above the horizontal. When air resistance is absent, the maximum range results when $\theta = 45°$.

Math Skills   To show that a projectile launched from and returning to ground level has its maximum range when $\theta = 45°$, we begin with the expression for the range R from Example 8:

$$R = v_{0x}t = (v_0 \cos \theta)t$$

Example 7 shows that the time of flight of the projectile is

$$t = \frac{-2v_{0y}}{a_y} = \frac{-2v_{0y}}{-9.80 \text{ m/s}^2} = \frac{2v_{0y}}{9.80 \text{ m/s}^2}$$

According to Example 6, the velocity component v_{0y} in this result is $v_{0y} = v_0 \sin \theta$, so that

$$t = \frac{2v_0 \sin \theta}{9.80 \text{ m/s}^2}$$

Substituting this expression for t into the range expression gives

$$R = (v_0 \cos \theta)t = (v_0 \cos \theta)\frac{2v_0 \sin \theta}{9.80 \text{ m/s}^2} = \frac{v_0^2(2 \sin \theta \cos \theta)}{9.80 \text{ m/s}^2}$$

Equation 6 (Other Trigonometric Identities) in Appendix E.2 shows that $2 \sin \theta \cos \theta = \sin 2\theta$, so the range expression becomes

$$R = \frac{v_0^2(2 \sin \theta \cos \theta)}{9.80 \text{ m/s}^2} = \frac{v_0^2 \sin 2\theta}{9.80 \text{ m/s}^2}$$

In this result R has its maximum value when $\sin 2\theta$ has its maximum value of 1. This occurs when $2\theta = 90°$, or when $\theta = 45°$.

In projectile motion, the magnitude of the acceleration due to gravity affects the trajectory in a significant way. For example, a baseball or a golf ball would travel much farther and higher on the moon than on the earth, when launched with the same initial velocity. The reason is that the moon's gravity is only about one-sixth as strong as the earth's.

Section 2.6 points out that certain types of symmetry with respect to time and speed are present for freely falling bodies. These symmetries are also found in projectile motion, since projectiles are falling freely in the vertical direction. In particular, the time required for a projectile to reach its maximum height H is equal to the time spent returning to the ground. In addition, **Figure 3.11** shows that the speed v of the object at any height above the ground on the upward part of the trajectory is equal to the speed v at the same height on the downward part. Although the two speeds are the same, the velocities are different, because they point in different directions. Conceptual Example 9 shows how to use this type of symmetry in your reasoning.

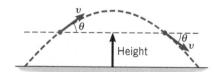

FIGURE 3.11   The speed v of a projectile at a given height above the ground is the same on the upward and downward parts of the trajectory. The velocities are different, however, since they point in different directions.

CONCEPTUAL EXAMPLE 9 | Two Ways to Throw a Stone

From the top of a cliff overlooking a lake, a person throws two stones. The stones have identical initial speeds v_0, but stone 1 is thrown downward at an angle θ below the horizontal, while stone 2 is thrown upward at the same angle above the horizontal, as

Figure 3.12 shows. Neglect air resistance and decide which stone, if either, strikes the water with the greater velocity: **(a)** both stones strike the water with the same velocity, **(b)** stone 1 strikes with the greater velocity, **(c)** stone 2 strikes with the greater velocity.

Reasoning Note point P in the drawing, where stone 2 returns to its initial height; here the speed of stone 2 is v_0 (the same as its initial speed), but its velocity is directed at an angle θ below the horizontal. This is exactly the type of projectile symmetry illustrated in **Figure 3.11**, and this symmetry will lead us to the correct answer.

Answers (b) and (c) are incorrect. You might guess that stone 1, being hurled downward, would strike the water with the greater velocity. Or, you might think that stone 2, having reached a greater height than stone 1, would hit the water with the greater velocity. To understand why neither of these answers is correct, see the response for answer (a) below.

Answer (a) is correct. Let's follow the path of stone 2 as it rises to its maximum height and falls back to earth. When it reaches point P in the drawing, stone 2 has a velocity that is identical to the velocity with which stone 1 is thrown downward from the top of the cliff (see the drawing). From this point on, the velocity of stone 2 changes in exactly the same way as that for stone 1, so both stones strike the water with the same velocity.

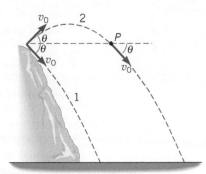

FIGURE 3.12 Two stones are thrown off the cliff with identical initial speeds v_0, but at equal angles θ that are below and above the horizontal. Conceptual Example 9 compares the velocities with which the stones hit the water below.

Related Homework: Problems 23, 67

In all the examples in this section, the projectiles follow a curved trajectory. In general, if the only acceleration is that due to gravity, the shape of the path can be shown to be a *parabola*.

Check Your Understanding

(*The answers are given at the end of the book.*)

3. A projectile is fired into the air, and it follows the parabolic path shown in **CYU Figure 3.1**, landing on the right. There is no air resistance. At any instant, the projectile has a velocity $\vec{v}$ and an acceleration $\vec{a}$. Which one or more of the drawings could *not* represent the directions for $\vec{v}$ and $\vec{a}$ at any point on the trajectory?

   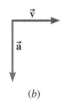

(a)   (b)   (c)   (d)

CYU FIGURE 3.1

4. An object is thrown upward at an angle θ above the ground, eventually returning to earth. **(a)** Is there any place along the trajectory where the velocity and acceleration are perpendicular? If so, where? **(b)** Is there any place where the velocity and acceleration are parallel? If so, where?

5. Is the acceleration of a projectile equal to zero when the projectile reaches the top of its trajectory?

6. In baseball, the pitcher's mound is raised to compensate for the fact that the ball falls downward as it travels from the pitcher toward the batter. If baseball were played on the moon, would the pitcher's mound have to be **(a)** higher than, **(b)** lower than, or **(c)** the same height as it is on earth?

7. A tennis ball is hit upward into the air and moves along an arc. Neglecting air resistance, where along the arc is the speed of the ball **(a)** a minimum and **(b)** a maximum?

8. A wrench is accidentally dropped from the top of the mast on a sailboat. Air resistance is negligible. Will the wrench hit at the same place on the deck whether the sailboat is at rest or moving with a constant velocity?

9. A rifle, at a height H above the ground, fires a bullet parallel to the ground. At the same instant and at the same height, a second bullet is dropped from rest. In the absence of air resistance, which bullet, if either, strikes the ground first?

10. A stone is thrown horizontally from the top of a cliff and eventually hits the ground below. A second stone is dropped from rest from the same cliff, falls through the same height, and also hits the ground below. Ignore air resistance. Is each of the following quantities different or the same in the two cases? **(a)** Displacement **(b)** Speed just before impact with the ground **(c)** Time of flight

11. A leopard springs upward at a 45° angle and then falls back to the ground. Air resistance is negligible. Does the leopard, at any point on its trajectory, ever have a speed that is one-half its initial value?

12. Two balls are launched upward from the same spot at different angles with respect to the ground. Both balls rise to the same maximum height. Ball A, however, follows a trajectory that has a greater range than that of ball B. Ignoring air resistance, decide which ball, if either, has the greater launch speed.

3.4 | Relative Velocity

To someone hitchhiking along a highway, two cars speeding by in adjacent lanes seem like a blur. But if the cars have the same velocity, each driver sees the other remaining in place, one lane away. The hitchhiker observes a velocity of perhaps 30 m/s, but each driver observes the other's velocity to be zero. Clearly, the velocity of an object is relative to the observer who is making the measurement.

Figure 3.13 illustrates the concept of relative velocity by showing a passenger walking toward the front of a moving train. The people sitting on the train see the passenger walking with a velocity of +2.0 m/s, where the plus sign denotes a direction to the right. Suppose the train is moving with a velocity of +9.0 m/s relative to an observer standing on the ground. Then the ground-based observer would see the passenger moving with a velocity of +11 m/s, due in part to the walking motion and in part to the train's motion. As an aid in describing relative velocity, let us define the following symbols:

$$\vec{v}_{\boxed{PT}} = \text{velocity of the } \boxed{\text{Passenger}} \text{ relative to the } \boxed{\text{Train}} = +2.0 \text{ m/s}$$
$$\vec{v}_{\boxed{TG}} = \text{velocity of the } \boxed{\text{Train}} \text{ relative to the } \boxed{\text{Ground}} = +9.0 \text{ m/s}$$
$$\vec{v}_{\boxed{PG}} = \text{velocity of the } \boxed{\text{Passenger}} \text{ relative to the } \boxed{\text{Ground}} = +11 \text{ m/s}$$

In terms of these symbols, the situation in **Figure 3.13** can be summarized as follows:

$$\vec{v}_{PG} = \vec{v}_{PT} + \vec{v}_{TG} \tag{3.7}$$

or

$$\vec{v}_{PG} = (2.0 \text{ m/s}) + (9.0 \text{ m/s}) = +11 \text{ m/s}$$

According to Equation 3.7*, $\vec{v}_{PG}$ is the vector sum of $\vec{v}_{PT}$ and $\vec{v}_{TG}$, and this sum is shown in the drawing. Had the passenger been walking toward the rear of the train, rather than

FIGURE 3.13 The velocity of the passenger relative to the ground-based observer is $\vec{v}_{PG}$. It is the vector sum of the velocity $\vec{v}_{PT}$ of the passenger relative to the train and the velocity $\vec{v}_{TG}$ of the train relative to the ground: $\vec{v}_{PG} = \vec{v}_{PT} + \vec{v}_{TG}$.

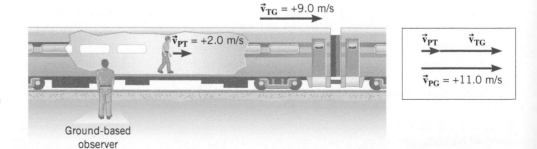

*This equation assumes that the train and the ground move in a straight line relative to one another.

toward the front, the velocity relative to the ground-based observer would have been $\vec{v}_{PG} = (-2.0 \text{ m/s}) + (9.0 \text{ m/s}) = +7.0 \text{ m/s}$.

Each velocity symbol in Equation 3.7 contains a two-letter subscript. The first letter in the subscript refers to the body that is moving, while the second letter indicates the object relative to which the velocity is measured. For example, $\vec{v}_{TG}$ and $\vec{v}_{PG}$ are the velocities of the **T**rain and **P**assenger measured relative to the **G**round. Similarly, $\vec{v}_{PT}$ is the velocity of the **P**assenger measured by an observer sitting on the **T**rain.

The ordering of the subscript symbols in Equation 3.7 follows a definite pattern. The first subscript (P) on the left side of the equation is also the first subscript on the right side of the equation. Likewise, the last subscript (G) on the left side is also the last subscript on the right side. The third subscript (T) appears only on the right side of the equation as the two "inner" subscripts. The colored boxes below emphasize the pattern of the symbols in the subscripts:

$$\vec{v}_{\boxed{PG}} = \vec{v}_{\boxed{P}T} + \vec{v}_{T\boxed{G}}$$

PHOTO 3.1 In landing on a moving aircraft carrier, the pilot of the helicopter must match the helicopter's horizontal velocity to the carrier's velocity, so that the relative velocity of the helicopter and the carrier is zero.

In other situations such as in the case of a helicopter landing on an aircraft carrier (**Photo 3.1**), the subscripts will not necessarily be P, G, and T, but will be compatible with the names of the objects involved in the motion.

Equation 3.7 has been presented in connection with one-dimensional motion, but the result is also valid for two-dimensional motion. **Figure 3.14** depicts a common situation that deals with relative velocity in two dimensions. Part *a* of the drawing shows a boat being carried downstream by a river; the engine of the boat is turned off. In part *b*, the engine is turned on, and now the boat moves across the river in a diagonal fashion because of the combined motion produced by the current and the engine. The list below gives the velocities for this type of motion and the objects relative to which they are measured:

$$\vec{v}_{\boxed{BW}} = \text{velocity of the } \boxed{\text{Boat}} \text{ relative to the } \boxed{\text{Water}}$$
$$\vec{v}_{\boxed{WS}} = \text{velocity of the } \boxed{\text{Water}} \text{ relative to the } \boxed{\text{Shore}}$$
$$\vec{v}_{\boxed{BS}} = \text{velocity of the } \boxed{\text{Boat}} \text{ relative to the } \boxed{\text{Shore}}$$

The velocity $\vec{v}_{BW}$ of the boat relative to the water is the velocity measured by an observer who, for instance, is floating on an inner tube and drifting downstream with the current. When the engine is turned off, the boat also drifts downstream with the current, and $\vec{v}_{BW}$ is zero. When the engine is turned on, however, the boat can move relative to the water, and $\vec{v}_{BW}$ is no longer zero. The velocity $\vec{v}_{WS}$ of the water relative to the shore is the velocity of the current measured by an observer on the shore. The velocity $\vec{v}_{BS}$ of the boat relative to the shore is due to the combined motion of the boat relative to the water and the motion of the water relative to the shore. In symbols,

$$\vec{v}_{\boxed{BS}} = \vec{v}_{\boxed{B}W} + \vec{v}_{W\boxed{S}}$$

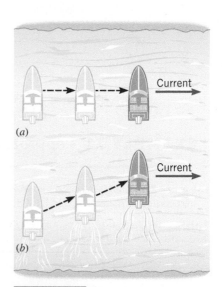

FIGURE 3.14 (*a*) A boat with its engine turned off is carried along by the current. (*b*) With the engine turned on, the boat moves across the river in a diagonal fashion.

The ordering of the subscripts in this equation is identical to that in Equation 3.7, although the letters have been changed to reflect a different physical situation. Example 10 illustrates the concept of relative velocity in two dimensions.

EXAMPLE 10 | Crossing a River

The engine of a boat drives it across a river that is 1800 m wide. The velocity $\vec{v}_{BW}$ of the boat relative to the water is 4.0 m/s, directed perpendicular to the current, as in **Interactive Figure 3.15**. The velocity $\vec{v}_{WS}$ of the water relative to the shore is 2.0 m/s. **(a)** What is the velocity $\vec{v}_{BS}$ of the boat relative to the shore? **(b)** How long does it take for the boat to cross the river?

Reasoning **(a)** The velocity of the boat relative to the shore is $\vec{v}_{BS}$. It is the vector sum of the velocity $\vec{v}_{BW}$ of the boat relative to the water and the velocity $\vec{v}_{WS}$ of the water relative to the shore: $\vec{v}_{BS} = \vec{v}_{BW} + \vec{v}_{WS}$. Since $\vec{v}_{BW}$ and $\vec{v}_{WS}$ are both known, we can use this relation among the velocities, with the aid of trigonometry, to find the magnitude and directional angle of $\vec{v}_{BS}$. **(b)** The component of $\vec{\mathbf{v}}_{BS}$ that is parallel to the width of the river (see **Interactive Figure 3.15**) determines how fast the boat crosses the river; this parallel component is $v_{BS} \sin \theta = v_{BW} = 4.0$ m/s. The time for the boat to cross the river is equal to the width of the river divided by the magnitude of this velocity component.

Solution   **(a)** Since the vectors $\vec{\mathbf{v}}_{BW}$ and $\vec{\mathbf{v}}_{WS}$ are perpendicular (see **Interactive Figure 3.15**), the magnitude of $\vec{\mathbf{v}}_{BS}$ can be determined by using the Pythagorean theorem:

$$v_{BS} = \sqrt{(v_{BW})^2 + (v_{WS})^2} = \sqrt{(4.0 \text{ m/s})^2 + (2.0 \text{ m/s})^2} = \boxed{4.5 \text{ m/s}}$$

Thus, the boat moves at a speed of 4.5 m/s with respect to an observer on shore. The direction of the boat relative to the shore is given by the angle θ in the drawing:

$$\tan \theta = \frac{v_{BW}}{v_{WS}} \quad \text{or} \quad \theta = \tan^{-1}\left(\frac{v_{BW}}{v_{WS}}\right) = \tan^{-1}\left(\frac{4.0 \text{ m/s}}{2.0 \text{ m/s}}\right) = \boxed{63°}$$

(b) The time t for the boat to cross the river is

$$t = \frac{\text{Width}}{v_{BS} \sin \theta} = \frac{1800 \text{ m}}{4.0 \text{ m/s}} = \boxed{450 \text{ s}}$$

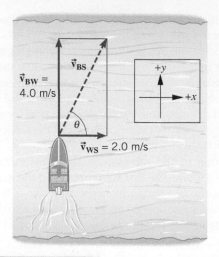

INTERACTIVE FIGURE 3.15   The velocity of the boat relative to the shore is $\vec{\mathbf{v}}_{BS}$. It is the vector sum of the velocity $\vec{\mathbf{v}}_{BW}$ of the boat relative to the water and the velocity $\vec{\mathbf{v}}_{WS}$ of the water relative to the shore: $\vec{\mathbf{v}}_{BS} = \vec{\mathbf{v}}_{BW} + \vec{\mathbf{v}}_{WS}$.

Sometimes, situations arise when two vehicles are in relative motion, and it is useful to know the relative velocity of one with respect to the other. Example 11 considers this type of relative motion.

EXAMPLE 11 | Approaching an Intersection

Figure 3.16*a* shows two cars approaching an intersection along perpendicular roads. The cars have the following velocities:

$\vec{\mathbf{v}}_{\boxed{AG}}$ = velocity of $\boxed{\text{car A}}$ relative to the $\boxed{\text{Ground}}$ = 25.0 m/s, eastward

$\vec{\mathbf{v}}_{\boxed{BG}}$ = velocity of $\boxed{\text{car B}}$ relative to the $\boxed{\text{Ground}}$ = 15.8 m/s, northward

Find the magnitude and direction of $\vec{\mathbf{v}}_{AB}$, where

$\vec{\mathbf{v}}_{\boxed{AB}}$ = velocity of $\boxed{\text{car A}}$ as measured by a passenger in $\boxed{\text{car B}}$

Reasoning   To find $\vec{\mathbf{v}}_{AB}$, we use an equation whose subscripts follow the order outlined earlier. Thus,

$$\vec{\mathbf{v}}_{\boxed{AB}} = \vec{\mathbf{v}}_{\boxed{A}G} + \vec{\mathbf{v}}_{G\boxed{B}}$$

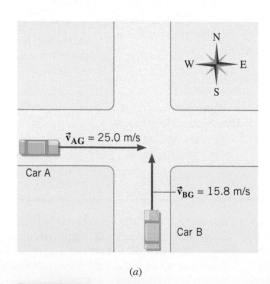

(*a*)

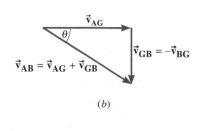

(*b*)

FIGURE 3.16   Two cars are approaching an intersection along perpendicular roads.

In this equation, the term $\vec{v}_{GB}$ is the velocity of the ground relative to a passenger in car B, rather than $\vec{v}_{BG}$, which is given as 15.8 m/s, northward. In other words, the subscripts are reversed. However, $\vec{v}_{GB}$ is related to $\vec{v}_{BG}$ according to

$$\vec{v}_{GB} = -\vec{v}_{BG}$$

This relationship reflects the fact that a passenger in car B, moving northward relative to the ground, looks out the car window and sees the ground moving southward, in the opposite direction. Therefore, the equation $\vec{v}_{AB} = \vec{v}_{AG} + \vec{v}_{GB}$ may be used to find $\vec{v}_{AB}$, provided we recognize $\vec{v}_{GB}$ as a vector that points opposite to the given velocity $\vec{v}_{BG}$. With this in mind, **Figure 3.16b** illustrates how $\vec{v}_{AG}$ and $\vec{v}_{GB}$ are added vectorially to give $\vec{v}_{AB}$.

Problem-Solving Insight In general, the velocity of object R relative to object S is always the negative of the velocity of object S relative to R: $\vec{v}_{RS} = -\vec{v}_{SR}$.

Solution From the vector triangle in **Figure 3.16b**, the magnitude and direction of $\vec{v}_{AB}$ can be calculated as

$$v_{AB} = \sqrt{(v_{AG})^2 + (v_{GB})^2} = \sqrt{(25.0 \text{ m/s})^2 + (-15.8 \text{ m/s})^2}$$
$$= \boxed{29.6 \text{ m/s}}$$

and

$$\cos\theta = \frac{v_{AG}}{v_{AB}} \quad \text{or} \quad \theta = \cos^{-1}\left(\frac{v_{AG}}{v_{AB}}\right) = \cos^{-1}\left(\frac{25.0 \text{ m/s}}{29.6 \text{ m/s}}\right) = \boxed{32.4°}$$

THE PHYSICS OF . . . raindrops falling on car windows.   While driving a car, have you ever noticed that the rear window sometimes remains dry, even though rain is falling? This phenomenon is a consequence of relative velocity, as **Figure 3.17** helps to explain. Part *a* shows a car traveling horizontally with a velocity of $\vec{v}_{CG}$ and a raindrop falling vertically with a velocity of $\vec{v}_{RG}$. Both velocities are measured relative to the ground. To determine whether the raindrop hits the window, however, we need to consider the velocity of the raindrop relative to the car, not to the ground. This velocity is $\vec{v}_{RC}$, and we know that

$$\vec{v}_{RC} = \vec{v}_{RG} + \vec{v}_{GC} = \vec{v}_{RG} - \vec{v}_{CG}$$

Here, we have used the fact that $\vec{v}_{GC} = -\vec{v}_{CG}$. Part *b* of the drawing shows the tail-to-head arrangement corresponding to this vector subtraction and indicates that the direction of $\vec{v}_{RC}$ is given by the angle θ_R. In comparison, the rear window is inclined at an angle θ_W with respect to the vertical (see the blowup in part *a*). When θ_R is greater than θ_W, the raindrop will miss the window. However, θ_R is determined by the speed v_{RG} of the raindrop and the speed v_{CG} of the car, according to $\theta_R = \tan^{-1}(v_{CG}/v_{RG})$. At higher car speeds, the angle θ_R becomes too large for the drop to hit the window. At a high enough speed, then, the car simply drives out from under each falling drop!

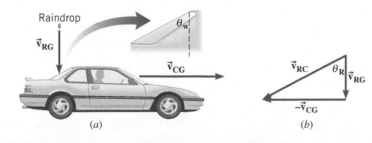

(a)   (b)

FIGURE 3.17 (*a*) With respect to the ground, a car is traveling at a velocity of $\vec{v}_{CG}$ and a raindrop is falling at a velocity of $\vec{v}_{RG}$. The rear window of the car is inclined at an angle θ_W with respect to the vertical. (*b*) This tail-to-head arrangement of vectors corresponds to the equation $\vec{v}_{RC} = \vec{v}_{RG} - \vec{v}_{CG}$.

Check Your Understanding

(The answers are given at the end of the book.)

13. Three cars, A, B, and C, are moving along a straight section of a highway. The velocity of A relative to B is $\vec{v}_{AB}$, the velocity of A relative to C is $\vec{v}_{AC}$, and the velocity of C relative to B is $\vec{v}_{CB}$. Fill in the missing velocities in the table.

	$\vec{v}_{AB}$	$\vec{v}_{AC}$	$\vec{v}_{CB}$
(a)	?	+40 m/s	+30 m/s
(b)	?	+50 m/s	−20 m/s
(c)	+60 m/s	+20 m/s	?
(d)	−50 m/s	?	+10 m/s

14. On a riverboat cruise, a plastic bottle is accidentally dropped overboard. A passenger on the boat estimates that the boat pulls ahead of the bottle by 5 meters each second. Is it possible to conclude that the magnitude of the velocity of the boat with respect to the shore is 5 m/s?

15. A plane takes off at St. Louis, flies straight to Denver, and then returns the same way. The plane flies at the same speed with respect to the ground during the entire flight, and there are no head winds or tail winds. Since the earth revolves around its axis once a day, you might expect that the times for the outbound trip and the return trip differ, depending on whether the plane flies against the earth's rotation or with it. Is this true, or are the two times the same?

16. A child is playing on the floor of a recreational vehicle (RV) as it moves along the highway at a constant velocity. He has a toy cannon, which shoots a marble at a fixed angle and speed with respect to the floor. The cannon can be aimed toward the front or the rear of the RV. Is the range toward the front the same as, less than, or greater than the range toward the rear? Answer this question **(a)** from the child's point of view and **(b)** from the point of view of an observer standing still on the ground.

17. Three swimmers can swim equally fast relative to the water. They have a race to see who can swim across a river in the least time. Swimmer A swims perpendicular to the current and lands on the far shore downstream, because the current has swept him in that direction. Swimmer B swims upstream at an angle to the current, choosing the angle so that he lands on the far shore directly opposite the starting point. Swimmer C swims downstream at an angle to the current in an attempt to take advantage of the current. Who crosses the river in the least time?

EXAMPLE 12 | BIO The Physics of Blood Splatter

The shape of the splatter made by a falling drop of blood on a surface can provide information on the velocity of the drop at impact. By measuring the physical dimensions of the pattern, forensic scientists can determine the drop's angle of impact with the surface. If a drop of blood falls straight downward and strikes a horizontal surface, the pattern it leaves is fairly circular (**Figure 3.18a**). However, if the drop strikes the surface at some angle less than 90°, then the pattern will be elongated with a tail (**Figure 3.18a**). Consider the following situation where a chef at a restaurant accidentally cuts her finger while slicing vegetables. She quickly gets in her car to drive to the closest emergency room for stitches, but she keeps her hand outside the window, so as to not drip blood inside the car. Based on the elongation of the drops of blood on the roadway, the impact angle was determined to be 15°. Assume the drops of blood behave like a projectile released from a height of 1.1 m and determine the speed of the car.

Reasoning This problem is similar to the physics of falling raindrops. We can analyze the blood splatter by using relative velocity. Here, we assume the chef's car is traveling horizontally with a velocity relative to ground of $\vec{v}_{CG}$, and the blood drop is falling vertically and impacts the ground with a velocity relative to the car of $\vec{v}_{BC}$ (**Figure 3.18b**). The velocity of the blood drop relative to the ground, which determines its elongation on impact, will be given by:

$$\vec{v}_{BG} = \vec{v}_{BC} + \vec{v}_{CG}$$

Notice how the subscripts follow the order outlined earlier. This equation can be solved for the velocity of the car relative to ground ($\vec{v}_{CG}$). The velocity of the blood drop relative to the car ($\vec{v}_{BC}$) at impact can be determined using the kinematic equations for freely falling bodies.

Solution From the vector triangle in **Figure 3.18c**, $\tan 15° = \frac{v_{BC}}{v_{CG}}$, therefore, $v_{CG} = \frac{v_{BC}}{\tan 15°}$. To calculate this we have to find $\vec{v}_{BC}$. This represents the speed of the blood drop relative to the car at impact. The speed of the blood drop at impact will be equal to the speed of an object that falls from rest through a distance of

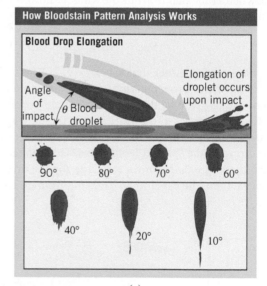

How Bloodstain Pattern Analysis Works

Blood Drop Elongation

Angle of impact
θ Blood droplet

Elongation of droplet occurs upon impact

90°   80°   70°   60°

40°   20°   10°

(a)

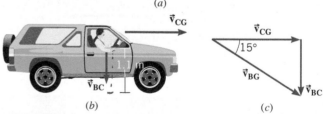

(b)   (c)

FIGURE 3.18 (a) The shape of a drop of blood splatter on a horizontal surface depends on the angle of impact (θ) of the drop. (b) The velocity of the car relative to the ground is $\vec{v}_{CG}$, and the velocity of the blood drop relative to the car is $\vec{v}_{BC}$. (c) This figure illustrates how $\vec{v}_{BC}$ and $\vec{v}_{CG}$ add vectorially to give $\vec{v}_{BG}$.

1.1 m. Using Equation 2.9, we have $v_{BC}^2 = v_{BC_0}^2 + 2ay$ or $v_{BC} = \sqrt{v_{BC_0}^2 + 2ay}$. Therefore,

$$v_{BC} = \sqrt{(0 \text{ m/s})^2 + 2(-9.8 \text{ m/s}^2)(-1.1 \text{ m})} = 4.6 \text{ m/s}$$

Finally, $v_{CG} = \frac{v_{BC}}{\tan 15°} = \frac{4.6 \text{ m/s}}{\tan 15°} = \boxed{17 \text{ m/s}}$

Concept Summary

3.1 Displacement, Velocity, and Acceleration The position of an object is located with a vector $\vec{r}$ drawn from the coordinate origin to the object. The displacement $\Delta\vec{r}$ of the object is defined as $\Delta\vec{r} = \vec{r} - \vec{r}_0$, where $\vec{r}$ and $\vec{r}_0$ specify its final and initial positions, respectively.

The average velocity $\overline{\vec{v}}$ of an object moving between two positions is defined as its displacement $\Delta\vec{r} = \vec{r} - \vec{r}_0$ divided by the elapsed time $\Delta t = t - t_0$, as in Equation 3.1.

$$\overline{\vec{v}} = \frac{\vec{r} - \vec{r}_0}{t - t_0} = \frac{\Delta\vec{r}}{\Delta t} \tag{3.1}$$

The instantaneous velocity $\vec{v}$ is the velocity at an instant of time. The average velocity becomes equal to the instantaneous velocity in the limit that Δt becomes infinitesimally small ($\Delta t \rightarrow 0$ s), as shown in Equation 1.

$$\vec{v} = \lim_{\Delta t \rightarrow 0} \frac{\Delta\vec{r}}{\Delta t} \tag{1}$$

The average acceleration $\overline{\vec{a}}$ of an object is the change in its velocity, $\Delta\vec{v} = \vec{v} - \vec{v}_0$ divided by the elapsed time $\Delta t = t - t_0$, as in Equation 3.2.

$$\overline{\vec{a}} = \frac{\vec{v} - \vec{v}_0}{t - t_0} = \frac{\Delta\vec{v}}{\Delta t} \tag{3.2}$$

The instantaneous acceleration $\vec{a}$ is the acceleration at an instant of time. The average acceleration becomes equal to the instantaneous acceleration in the limit that the elapsed time Δt becomes infinitesimally small, as shown in Equation 2.

$$\vec{a} = \lim_{\Delta t \rightarrow 0} \frac{\Delta\vec{v}}{\Delta t} \tag{2}$$

3.2 Equations of Kinematics in Two Dimensions Motion in two dimensions can be described in terms of the time t and the x and y components of four vectors: the displacement, the acceleration, and the initial and final velocities. The x part of the motion occurs exactly as it would if the y part did not occur at all. Similarly, the y part of the motion occurs exactly as it would if the x part of the motion did not exist. The motion can be analyzed by treating the x and y components of the four vectors separately and realizing that the time t is the same for each component.

When the acceleration is constant, the x components of the displacement, the acceleration, and the initial and final velocities are related by the equations of kinematics, and so are the y components:

x Component		y Component	
$v_x = v_{0x} + a_x t$	(3.3a)	$v_y = v_{0y} + a_y t$	(3.3b)
$x = \frac{1}{2}(v_{0x} + v_x)t$	(3.4a)	$y = \frac{1}{2}(v_{0y} + v_y)t$	(3.4b)
$x = v_{0x}t + \frac{1}{2}a_x t^2$	(3.5a)	$y = v_{0y}t + \frac{1}{2}a_y t^2$	(3.5b)
$v_x^2 = v_{0x}^2 + 2a_x x$	(3.6a)	$v_y^2 = v_{0y}^2 + 2a_y y$	(3.6b)

The directions of the components of the displacement, the acceleration, and the initial and final velocities are conveyed by assigning a plus $(+)$ or minus $(-)$ sign to each one.

3.3 Projectile Motion Projectile motion is an idealized kind of motion that occurs when a moving object (the projectile) experiences only the acceleration due to gravity, which acts vertically downward. If the trajectory of the projectile is near the earth's surface, the vertical component a_y of the acceleration has a magnitude of 9.80 m/s^2. The acceleration has no horizontal component ($a_x = 0 \text{ m/s}^2$), the effects of air resistance being negligible.

There are several symmetries in projectile motion: (1) The time to reach maximum height from any vertical level is equal to the time spent returning from the maximum height to that level. (2) The speed of a projectile depends only on its height above its launch point, and not on whether it is moving upward or downward.

3.4 Relative Velocity The velocity of object A relative to object B is $\vec{v}_{AB}$, and the velocity of object B relative to object C is $\vec{v}_{BC}$. The velocity of A relative to C is shown in Equation 3 (note the ordering of the subscripts). While the velocity of object A relative to object B is $\vec{v}_{AB}$, the velocity of B relative to A is $\vec{v}_{BA} = -\vec{v}_{AB}$.

$$\vec{v}_{AC} = \vec{v}_{AB} + \vec{v}_{BC} \tag{3}$$

Focus on Concepts

Additional questions are available for assignment in WileyPLUS.

Section 3.3 Projectile Motion

1. The drawing shows projectile motion at three points along the trajectory. The speeds at the points are v_1, v_2, and v_3. Assume there is no air resistance and rank the speeds, largest to smallest. (Note that the symbol $>$ means "greater than.") (a) $v_1 > v_3 > v_2$ (b) $v_1 > v_2 > v_3$ (c) $v_2 > v_3 > v_1$ (d) $v_2 > v_1 > v_3$ (e) $v_3 > v_2 > v_1$

QUESTION 1

2. Two balls are thrown from the top of a building, as in the drawing. Ball 1 is thrown straight down, and ball 2 is thrown with the same speed, but upward at an angle θ with respect to the horizontal. Consider the motion of the balls after they are released. Which one of the following statements is true? (a) The acceleration of ball 1 becomes larger and larger as it falls, because the ball is going faster and faster. (b) The acceleration of ball 2 decreases as it rises, becomes zero at the top of the trajectory, and then increases as the ball begins to fall toward the ground. (c) Both balls have the same acceleration at all times. (d) Ball 2 has an acceleration in both the horizontal and vertical directions, but ball 1 has an acceleration only in the vertical direction.

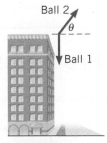

QUESTION 2

3. Each drawing shows three points along the path of a projectile, one on its way up, one at the top, and one on its way down. The launch point is on the left in each drawing. Which drawing correctly represents the acceleration $\vec{a}$ of the projectile at these three points? (a) 1 (b) 2 (c) 3 (d) 4 (e) 5

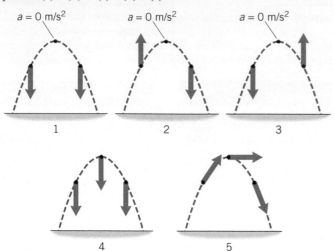

1   2   3

4   5

QUESTION 3

4. Ball 1 is thrown into the air and it follows the trajectory for projectile motion shown in the drawing. At the instant that ball 1 is at the top of its trajectory, ball 2 is dropped from rest from the same height. Which ball reaches the ground first? (a) Ball 1 reaches the ground first, since it is moving at the top of the trajectory, while ball 2 is dropped from rest. (b) Ball 2 reaches the ground first, because it has the shorter distance to travel. (c) Both balls reach the ground at the same time. (d) There is not enough information to tell which ball reaches the ground first.

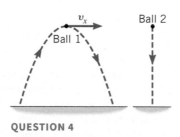

QUESTION 4

5. Two objects are fired into the air, and the drawing shows the projectile motions. Projectile 1 reaches the greater height, but projectile 2 has the greater range. Which one is in the air for the greater amount of time? (a) Projectile 1, because it travels higher than projectile 2. (b) Projectile 2, because it has the greater range. (c) Both projectiles spend the same amount of time in the air. (d) Projectile 2, because it has the smaller initial speed and, therefore, travels more slowly than projectile 1.

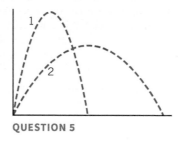

QUESTION 5

Section 3.4   Relative Velocity

6. A slower-moving car is traveling behind a faster-moving bus. The velocities of the two vehicles are as follows:

$\vec{v}_{CG}$ = velocity of the **C**ar relative to the **G**round = +12 m/s

$\vec{v}_{BG}$ = velocity of the **B**us relative to the **G**round = +16 m/s

A passenger on the bus gets up and walks toward the front of the bus with a velocity of $\vec{v}_{PB}$, where $\vec{v}_{PB}$ = velocity of the **P**assenger relative to the **B**us = +2 m/s. What is $\vec{v}_{PC}$, the velocity of the **P**assenger relative to the **C**ar?

(a) +2 m/s + 16 m/s + 12 m/s = +30 m/s

(b) −2 m/s + 16 m/s + 12 m/s = +26 m/s

(c) +2 m/s + 16 m/s − 12 m/s = +6 m/s

(d) −2 m/s + 16 m/s − 12 m/s = +2 m/s

7. Your car is traveling behind a jeep. Both are moving at the same speed, so the velocity of the jeep relative to you is zero. A spare tire is strapped to the back of the jeep. Suddenly the strap breaks, and the tire falls off the jeep. Will your car hit the spare tire before the tire hits the road? Assume that air resistance is absent. (a) Yes. As long as the car doesn't slow down, it will hit the tire. (b) No. The car will not hit the tire before the tire hits the ground, no matter how close you are to the jeep. (c) If the tire falls from a great enough height, the car will hit the tire. (d) If the car is far enough behind the jeep, the car will not hit the tire.

8. The drawing shows two cars traveling in different directions with different speeds. Their velocities are:

$\vec{v}_{AG}$ = velocity of car **A** relative to the **G**round = 27.0 m/s, due east

$\vec{v}_{BG}$ = velocity of car **B** relative to the **G**round = 21.0 m/s, due north

The passenger of car B looks out the window and sees car A. What is the velocity (magnitude and direction) of car A as observed by the passenger of car B? In other words, what is the velocity $\vec{v}_{AB}$ of car A relative to car B? Give the directional angle of $\vec{v}_{AB}$ with respect to due east.

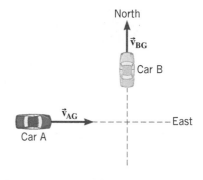

QUESTION 8

Problems   Online

Additional questions are available for assignment in WileyPLUS.

Section 3.1   Displacement, Velocity, and Acceleration

1. **E** **SSM** Two trees have perfectly straight trunks and are both growing perpendicular to the flat horizontal ground beneath them. The sides of the trunks that face each other are separated by 1.3 m. A frisky squirrel makes three jumps in rapid succession. First, he leaps from the foot of one tree to a spot that is 1.0 m above the ground on the other tree. Then, he jumps back to the first tree, landing on it

at a spot that is 1.7 m above the ground. Finally, he leaps back to the other tree, now landing at a spot that is 2.5 m above the ground. What is the magnitude of the squirrel's displacement?

2. **E** A meteoroid is traveling east through the atmosphere at 18.3 km/s while descending at a rate of 11.5 km/s. What is its speed, in km/s?

3. **E** **GO** In a football game a kicker attempts a field goal. The ball remains in contact with the kicker's foot for 0.050 s, during which time it experiences an acceleration of 340 m/s^2. The ball is launched at an angle of 51° above the ground. Determine the horizontal and vertical components of the launch velocity.

4. **E** A baseball player hits a triple and ends up on third base. A baseball "diamond" is a square, each side of length 27.4 m, with home plate and the three bases on the four corners. What is the magnitude of the player's displacement?

5. **E** **SSM** In diving to a depth of 750 m, an elephant seal also moves 460 m due east of his starting point. What is the magnitude of the seal's displacement?

6. **E** A mountain-climbing expedition establishes two intermediate camps, labeled A and B in the drawing, above the base camp. What is the magnitude Δr of the displacement between camp A and camp B?

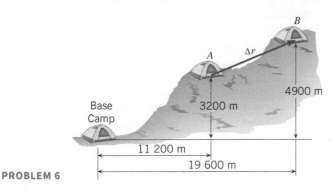

PROBLEM 6

7. **E** **SSM** A radar antenna is tracking a satellite orbiting the earth. At a certain time, the radar screen shows the satellite to be 162 km away. The radar antenna is pointing upward at an angle of 62.3° from the ground. Find the x and y components (in km) of the position vector of the satellite, relative to the antenna.

8. **E** **GO** In a mall, a shopper rides up an escalator between floors. At the top of the escalator, the shopper turns right and walks 9.00 m to a store. The magnitude of the shopper's displacement from the bottom of the escalator to the store is 16.0 m. The vertical distance between the floors is 6.00 m. At what angle is the escalator inclined above the horizontal?

9. **E** **SSM** A skateboarder, starting from rest, rolls down a 12.0-m ramp. When she arrives at the bottom of the ramp her speed is 7.70 m/s. (a) Determine the magnitude of her acceleration, assumed to be constant. (b) If the ramp is inclined at 25.0° with respect to the ground, what is the component of her acceleration that is parallel to the ground?

10. **M** **V-HINT** **MMH** A bird watcher meanders through the woods, walking 0.50 km due east, 0.75 km due south, and 2.15 km in a direction 35.0° north of west. The time required for this trip is 2.50 h. Determine the magnitude and direction (relative to due west) of the bird watcher's (a) displacement and (b) average velocity. Use kilometers and hours for distance and time, respectively.

11. **M** **MMH** **CHALK** The earth moves around the sun in a nearly circular orbit of radius 1.50×10^{11} m. During the three summer months (an elapsed time of 7.89×10^6 s), the earth moves one-fourth of the distance around the sun. (a) What is the average

speed of the earth? (b) What is the magnitude of the average velocity of the earth during this period?

Section 3.2   Equations of Kinematics in Two Dimensions/Section 3.3   Projectile Motion

12. **E** A spacecraft is traveling with a velocity of $v_{0x} = 5480$ m/s along the +x direction. Two engines are turned on for a time of 842 s. One engine gives the spacecraft an acceleration in the +x direction of $a_x = 1.20$ m/s^2, while the other gives it an acceleration in the +y direction of $a_y = 8.40$ m/s^2. At the end of the firing, find (a) v_x and (b) v_y.

13. **E** **SSM** A volleyball is spiked so that it has an initial velocity of 15 m/s directed downward at an angle of 55° below the horizontal. What is the horizontal component of the ball's velocity when the opposing player fields the ball?

14. **E** **CHALK** As a tennis ball is struck, it departs from the racket horizontally with a speed of 28.0 m/s. The ball hits the court at a horizontal distance of 19.6 m from the racket. How far above the court is the tennis ball when it leaves the racket?

15. **E** A skateboarder shoots off a ramp with a velocity of 6.6 m/s, directed at an angle of 58° above the horizontal. The end of the ramp is 1.2 m above the ground. Let the x axis be parallel to the ground, the +y direction be vertically upward, and take as the origin the point on the ground directly below the top of the ramp. (a) How high above the ground is the highest point that the skateboarder reaches? (b) When the skateboarder reaches the highest point, how far is this point horizontally from the end of the ramp?

16. **E** **GO** A puck is moving on an air hockey table. Relative to an x, y coordinate system at time $t = 0$ s, the x components of the puck's initial velocity and acceleration are $v_{0x} = +1.0$ m/s and $a_x = +2.0$ m/s^2. The y components of the puck's initial velocity and acceleration are $v_{0y} = +2.0$ m/s and $a_y = -2.0$ m/s^2. Find the magnitude and direction of the puck's velocity at a time of $t = 0.50$ s. Specify the direction relative to the +x axis.

17. **E** **SSM** A spider crawling across a table leaps onto a magazine blocking its path. The initial velocity of the spider is 0.870 m/s at an angle of 35.0° above the table, and it lands on the magazine 0.0770 s after leaving the table. Ignore air resistance. How thick is the magazine? Express your answer in millimeters.

18. **E** A horizontal rifle is fired at a bull's-eye. The muzzle speed of the bullet is 670 m/s. The gun is pointed directly at the center of the bull's-eye, but the bullet strikes the target 0.025 m below the center. What is the horizontal distance between the end of the rifle and the bull's-eye?

19. **E** **MMH** A golfer imparts a speed of 30.3 m/s to a ball, and it travels the maximum possible distance before landing on the green. The tee and the green are at the same elevation. (a) How much time does the ball spend in the air? (b) What is the longest hole in one that the golfer can make, if the ball does not roll when it hits the green?

20. **E** **GO** A golfer hits a shot to a green that is elevated 3.0 m above the point where the ball is struck. The ball leaves the club at a speed of 14.0 m/s at an angle of 40.0° above the horizontal. It rises to its maximum height and then falls down to the green. Ignoring air resistance, find the speed of the ball just before it lands.

21. **E** **SSM** In the aerials competition in skiing, the competitors speed down a ramp that slopes sharply upward at the end. The sharp upward slope launches them into the air, where they perform acrobatic maneuvers. The end of a launch ramp is directed 63° above the horizontal. With this launch angle, a skier attains a height of 13 m above the end of the ramp. What is the skier's launch speed?

22. 🅴 🆖 A space vehicle is coasting at a constant velocity of 21.0 m/s in the $+y$ direction relative to a space station. The pilot of the vehicle fires a RCS (reaction control system) thruster, which causes it to accelerate at 0.320 m/s² in the $+x$ direction. After 45.0 s, the pilot shuts off the RCS thruster. After the RCS thruster is turned off, find (a) the magnitude and (b) the direction of the vehicle's velocity relative to the space station. Express the direction as an angle measured from the $+y$ direction.

23. 🅴 SSM As preparation for this problem, review Conceptual Example 9. The drawing shows two planes each about to drop an empty fuel tank. At the moment of release each plane has the same speed of 135 m/s, and each tank is at the same height of 2.00 km above the ground. Although the speeds are the same, the velocities are different at the instant of release, because one plane is flying at an angle of 15.0° above the horizontal and the other is flying at an angle of 15.0° below the horizontal. Find the magnitude and direction of the velocity with which the fuel tank hits the ground if it is from (a) plane A and (b) plane B. In each part, give the directional angles with respect to the horizontal.

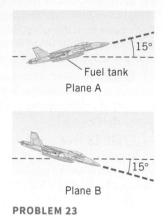
Fuel tank
Plane A

Plane B

PROBLEM 23

24. 🅴 🆖 A criminal is escaping across a rooftop and runs off the roof horizontally at a speed of 5.3 m/s, hoping to land on the roof of an adjacent building. Air resistance is negligible. The horizontal distance between the two buildings is D, and the roof of the adjacent building is 2.0 m below the jumping-off point. Find the maximum value for D.

25. 🅴 On a spacecraft, two engines are turned on for 684 s at a moment when the velocity of the craft has x and y components of $v_{0x} = 4370$ m/s and $v_{0y} = 6280$ m/s. While the engines are firing, the craft undergoes a displacement that has components of $x = 4.11 \times 10^6$ m and $y = 6.07 \times 10^6$ m. Find the x and y components of the craft's acceleration.

26. 🅴 🆖 In the absence of air resistance, a projectile is launched from and returns to ground level. It follows a trajectory similar to that shown in **Animated Figure 3.10** and has a range of 23 m. Suppose the launch speed is doubled, and the projectile is fired at the same angle above the ground. What is the new range?

27. 🅴 SSM A fire hose ejects a stream of water at an angle of 35.0° above the horizontal. The water leaves the nozzle with a speed of 25.0 m/s. Assuming that the water behaves like a projectile, how far from a building should the fire hose be located to hit the highest possible fire?

28. 🅴 A major-league pitcher can throw a baseball in excess of 41.0 m/s. If a ball is thrown horizontally at this speed, how much will it drop by the time it reaches a catcher who is 17.0 m away from the point of release?

29. 🅴 A quarterback claims that he can throw the football a horizontal distance of 183 m (200 yd). Furthermore, he claims that he can do this by launching the ball at the relatively low angle of 30.0° above the horizontal. To evaluate this claim, determine the speed with which this quarterback must throw the ball. Assume that the ball is launched and caught at the same vertical level and that air resistance can be ignored. For comparison, a baseball pitcher who can accurately throw a fastball at 45 m/s (100 mph) would be considered exceptional.

30. 🅴 SSM An eagle is flying horizontally at 6.0 m/s with a fish in its claws. It accidentally drops the fish. (a) How much time passes before the fish's speed doubles? (b) How much additional time would be required for the fish's speed to double again?

31. 🅴 🆖 The highest barrier that a projectile can clear is 13.5 m, when the projectile is launched at an angle of 15.0° above the horizontal. What is the projectile's launch speed?

32. 🅴 Consult Multiple-Concept Example 4 for background before beginning this problem. Suppose the water at the top of Niagara Falls has a horizontal speed of 2.7 m/s just before it cascades over the edge of the falls. At what vertical distance below the edge does the velocity vector of the water point downward at a 75° angle below the horizontal?

33. 🅴 MMH On a distant planet, golf is just as popular as it is on earth. A golfer tees off and drives the ball 3.5 times as far as he would have on earth, given the same initial velocities on both planets. The ball is launched at a speed of 45 m/s at an angle of 29° above the horizontal. When the ball lands, it is at the same level as the tee. On the distant planet, what are (a) the maximum height and (b) the range of the ball?

34. 🅴 CHALK A rocket is fired at a speed of 75.0 m/s from ground level, at an angle of 60.0° above the horizontal. The rocket is fired toward an 11.0-m-high wall, which is located 27.0 m away. The rocket attains its launch speed in a negligibly short period of time, after which its engines shut down and the rocket coasts. By how much does the rocket clear the top of the wall?

35. 🅴 A rifle is used to shoot twice at a target, using identical cartridges. The first time, the rifle is aimed parallel to the ground and directly at the center of the bull's-eye. The bullet strikes the target at a distance of H_A below the center, however. The second time, the rifle is similarly aimed, but from twice the distance from the target. This time the bullet strikes the target at a distance of H_B below the center. Find the ratio H_B/H_A.

36. 🅼 SSM An airplane with a speed of 97.5 m/s is climbing upward at an angle of 50.0° with respect to the horizontal. When the plane's altitude is 732 m, the pilot releases a package. (a) Calculate the distance along the ground, measured from a point directly beneath the point of release, to where the package hits the earth. (b) Relative to the ground, determine the angle of the velocity vector of the package just before impact.

37. 🅼 V-HINT Multiple-Concept Example 4 deals with a situation similar to that presented here. A marble is thrown horizontally with a speed of 15 m/s from the top of a building. When it strikes the ground, the marble has a velocity that makes an angle of 65° with the horizontal. From what height above the ground was the marble thrown?

38. 🅼 V-HINT MMH Review Conceptual Example 5 before beginning this problem. You are traveling in a convertible with the top down. The car is moving at a constant velocity of 25 m/s, due east along flat ground. You throw a tomato straight upward at a speed of 11 m/s. How far has the car moved when you get a chance to catch the tomato?

39. 🅼 V-HINT A diver springs upward from a diving board. At the instant she contacts the water, her speed is 8.90 m/s, and her body is extended at an angle of 75.0° with respect to the horizontal surface of the water. At this instant her vertical displacement is −3.00 m, where downward is the negative direction. Determine her initial velocity, both magnitude and direction.

40. 🅼 🆖 MMH A soccer player kicks the ball toward a goal that is 16.8 m in front of him. The ball leaves his foot at a speed of 16.0 m/s and an angle of 28.0° above the ground. Find the speed of the ball when the goalie catches it in front of the net.

41. **M** **V-HINT** In the javelin throw at a track-and-field event, the javelin is launched at a speed of 29 m/s at an angle of 36° above the horizontal. As the javelin travels upward, its velocity points above the horizontal at an angle that decreases as time passes. How much time is required for the angle to be reduced from 36° at launch to 18°?

42. **M** **SSM** An airplane is flying with a velocity of 240 m/s at an angle of 30.0° with the horizontal, as the drawing shows. When the altitude of the plane is 2.4 km, a flare is released from the plane. The flare hits the target on the ground. What is the angle θ?

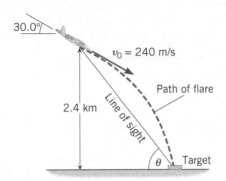

PROBLEM 42

43. **M** **V-HINT** A child operating a radio-controlled model car on a dock accidentally steers it off the edge. The car's displacement 1.1 s after leaving the dock has a magnitude of 7.0 m. What is the car's speed at the instant it drives off the edge of the dock?

44. **M** **MMH** After leaving the end of a ski ramp, a ski jumper lands downhill at a point that is displaced 51.0 m horizontally from the end of the ramp. His velocity, just before landing, is 23.0 m/s and points in a direction 43.0° below the horizontal. Neglecting air resistance and any lift he experiences while airborne, find his initial velocity (magnitude and direction) when he left the end of the ramp. Express the direction as an angle relative to the horizontal.

45. **M** **GO** Stones are thrown horizontally with the same velocity from the tops of two different buildings. One stone lands twice as far from the base of the building from which it was thrown as does the other stone. Find the ratio of the height of the taller building to the height of the shorter building.

46. **H** **SSM** The drawing shows an exaggerated view of a rifle that has been "sighted in" for a 91.4-meter target. If the muzzle speed of the bullet is $v_0 = 427$ m/s, what are the two possible angles θ_1 and θ_2 between the rifle barrel and the horizontal such that the bullet will hit the target? One of these angles is so large that it is never used in target shooting. (*Hint: The following trigonometric identity may be useful:* $2 \sin \theta \cos \theta = \sin 2\theta$.)

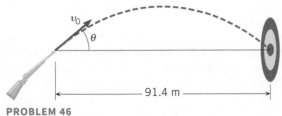

PROBLEM 46

47. **H** **MMH** A projectile is launched from ground level at an angle of 12.0° above the horizontal. It returns to ground level. To what value should the launch angle be adjusted, without changing the launch speed, so that the range doubles?

48. **H** **SSM** From the top of a tall building, a gun is fired. The bullet leaves the gun at a speed of 340 m/s, parallel to the ground. As the drawing shows, the bullet puts a hole in a window of another building and hits the wall that faces the window. Using the data in the drawing,

determine the distances D and H, which locate the point where the gun was fired. Assume that the bullet does not slow down as it passes through the window.

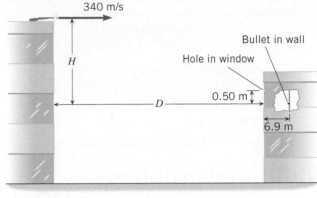

PROBLEM 48

Section 3.4 Relative Velocity

49. **E** In a marathon race Chad is out in front, running due north at a speed of 4.00 m/s. John is 95 m behind him, running due north at a speed of 4.50 m/s. How long does it take for John to pass Chad?

50. **E** **SSM** A swimmer, capable of swimming at a speed of 1.4 m/s in still water (i.e., the swimmer can swim with a speed of 1.4 m/s relative to the water), starts to swim directly across a 2.8-km-wide river. However, the current is 0.91 m/s, and it carries the swimmer downstream. **(a)** How long does it take the swimmer to cross the river? **(b)** How far downstream will the swimmer be upon reaching the other side of the river?

51. **E** **GO** Two friends, Barbara and Neil, are out rollerblading. With respect to the ground, Barbara is skating due south at a speed of 4.0 m/s. Neil is in front of her. With respect to the ground, Neil is skating due west at a speed of 3.2 m/s. Find Neil's velocity (magnitude and direction relative to due west), as seen by Barbara.

52. **E** At some airports there are speed ramps to help passengers get from one place to another. A speed ramp is a moving conveyor belt on which you can either stand or walk. Suppose a speed ramp has a length of 105 m and is moving at a speed of 2.0 m/s relative to the ground. In addition, suppose you can cover this distance in 75 s when walking on the ground. If you walk at the same rate with respect to the speed ramp that you walk on the ground, how long does it take for you to travel the 105 m using the speed ramp?

53. **E** You are in a hot-air balloon that, relative to the ground, has a velocity of 6.0 m/s in a direction due east. You see a hawk moving directly away from the balloon in a direction due north. The speed of the hawk relative to you is 2.0 m/s. What are the magnitude and direction of the hawk's velocity relative to the ground? Express the directional angle relative to due east.

54. **E** **GO** On a pleasure cruise a boat is traveling relative to the water at a speed of 5.0 m/s due south. Relative to the boat, a passenger walks toward the back of the boat at a speed of 1.5 m/s. **(a)** What are the magnitude and direction of the passenger's velocity relative to the water? **(b)** How long does it take for the passenger to walk a distance of 27 m on the boat? **(c)** How long does it take for the passenger to cover a distance of 27 m on the water?

55. **E** **GO** The captain of a plane wishes to proceed due west. The cruising speed of the plane is 245 m/s relative to the air. A weather report indicates that a 38.0-m/s wind is blowing from the south to the north. In what direction, measured with respect to due west, should the pilot head the plane?

56. **M** **V-HINT** A person looking out the window of a stationary train notices that raindrops are falling vertically down at a speed of 5.0 m/s relative to the ground. When the train moves at a constant velocity, the raindrops make an angle of 25° when they move past the window, as the drawing shows. How fast is the train moving?

PROBLEM 56

57. **M** **SSM** **CHALK** Mario, a hockey player, is skating due south at a speed of 7.0 m/s relative to the ice. A teammate passes the puck to him. The puck has a speed of 11.0 m/s and is moving in a direction of 22° west of south, relative to the ice. What are the magnitude and direction (relative to due south) of the puck's velocity, as observed by Mario?

58. **H** A jetliner can fly 6.00 hours on a full load of fuel. Without any wind it flies at a speed of 2.40×10^2 m/s. The plane is to make a round-trip by heading due west for a certain distance, turning around, and then heading due east for the return trip. During the entire flight, however, the plane encounters a 57.8-m/s wind from the jet stream, which blows from west to east. What is the maximum distance that the plane can travel due west and just be able to return home?

Additional Problems

Online

59. **E** Useful background for this problem can be found in Multiple-Concept Example 2. On a spacecraft two engines fire for a time of 565 s. One gives the craft an acceleration in the x direction of $a_x = 5.10$ m/s², while the other produces an acceleration in the y direction of $a_y = 7.30$ m/s². At the end of the firing period, the craft has velocity components of $v_x = 3775$ m/s and $v_y = 4816$ m/s. Find the magnitude and direction of the initial velocity. Express the direction as an angle with respect to the +x axis.

60. **E** A dolphin leaps out of the water at an angle of 35° above the horizontal. The horizontal component of the dolphin's velocity is 7.7 m/s. Find the magnitude of the vertical component of the velocity.

61. **E** A hot-air balloon is rising straight up with a speed of 3.0 m/s. A ballast bag is released from rest relative to the balloon at 9.5 m above the ground. How much time elapses before the ballast bag hits the ground?

62. **E** **CHALK** A golf ball rolls off a horizontal cliff with an initial speed of 11.4 m/s. The ball falls a vertical distance of 15.5 m into a lake below. **(a)** How much time does the ball spend in the air? **(b)** What is the speed v of the ball just before it strikes the water?

63. **E** **GO** When chasing a hare along a flat stretch of ground, a greyhound leaps into the air at a speed of 10.0 m/s, at an angle of 31.0° above the horizontal. **(a)** What is the range of his leap and **(b)** for how much time is he in the air?

64. **E** **SSM** Multiple-Concept Example 4 provides useful background for this problem. A diver runs horizontally with a speed of 1.20 m/s off a platform that is 10.0 m above the water. What is his speed just before striking the water?

65. **E** **GO** A ball is thrown upward at a speed v_0 at an angle of 52° above the horizontal. It reaches a maximum height of 7.5 m. How high would this ball go if it were thrown straight upward at speed v_0?

66. **M** **MMH** A golfer, standing on a fairway, hits a shot to a green that is elevated 5.50 m above the point where she is standing. If the ball leaves her club with a velocity of 46.0 m/s at an angle of 35.0° above the ground, find the time that the ball is in the air before it hits the green.

67. **M** **SSM** As preparation for this problem, review Conceptual Example 9. The two stones described there have identical initial speeds of $v_0 = 13.0$ m/s and are thrown at an angle $\theta = 30.0°$, one below the horizontal and one above the horizontal. What is the distance *between* the points where the stones strike the ground?

68. **M** **GO** Relative to the ground, a car has a velocity of 16.0 m/s, directed due north. Relative to this car, a truck has a velocity of 24.0 m/s, directed 52.0° north of east. What is the magnitude of the truck's velocity relative to the ground?

69. **M** **V-HINT** The lob in tennis is an effective tactic when your opponent is near the net. It consists of lofting the ball over his head, forcing him to move quickly away from the net (see the drawing). Suppose that you lob the ball with an initial speed of 15.0 m/s, at an angle of 50.0° above the horizontal. At this instant your opponent is 10.0 m away from the ball. He begins moving away from you 0.30 s later, hoping to reach the ball and hit it back at the moment that it is 2.10 m above its launch point. With what minimum average speed must he move? (Ignore the fact that he can stretch, so that his racket can reach the ball before he does.)

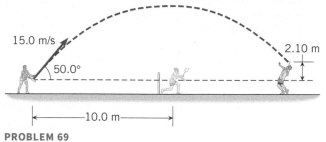

PROBLEM 69

70. **M** **GO** A baseball player hits a home run, and the ball lands in the left-field seats, 7.5 m above the point at which it was hit. It lands with a velocity of 36 m/s at an angle of 28° below the horizontal (see the drawing). The positive directions are upward and to the right in the drawing. Ignoring air resistance, find the magnitude and direction of the initial velocity with which the ball leaves the bat.

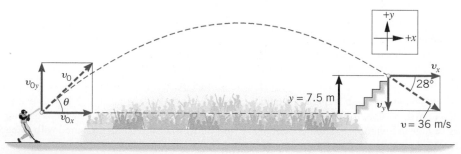

PROBLEM 70

Physics in Biology, Medicine, and Sports

71. **M** **BIO** **3.3** The Pacific white-sided dolphin (see photo) can jump straight out of the water to incredible heights near 9.10 m (~30 feet). If a dolphin jumps out of the water at a launch angle of 45.0°, **(a)** what will be its maximum height, and **(b)** how far will the dolphin jump horizontally? Assume the dolphin leaves the water at the same speed as when it jumps straight upward.

Terry Hardie

PROBLEM 71

72. **M** **BIO** **3.3** A spear-throwing lever, called an atlatl, is a device used in creating greater velocity of a thrown spear or dart (see the figure) compared to when thrown with just the arm. Used primarily in hunting, as well as a weapon, the atlatl can produce spear speeds near 90 mph. Historical evidence suggests that the atlatl was used by *Homo sapiens* dating back to 30 000 to 42 000 years ago. Suppose an atlatl is used to launch a spear at 40.0 m/s with a launch angle of 10.0° above the horizontal. What is the speed of the spear 1.00 s after it is thrown? Assume air resistance is negligible.

PROBLEM 72

73. **M** **BIO** **3.4** One of the world's greatest spectacles of animal migration is the annual circular trek of the wildebeest through the Serengeti National Park in Tanzania, Africa. Each year the availability of grazing land drives nearly 1.7 million wildebeest into river crossings, such as that shown in the photo. The wildebeest begin by walking across the river in the shallow waters close to the bank but end up swimming across in the middle of the fast-moving river, where the water is deeper. Eventually they reach the shallow area near the other bank, where they can walk out on the other side. This produces the characteristic zigzag pattern seen in the photo. The drawing shows a view of the river crossing from above and the path traced out by the wildebeest. Assume that the wildebeest walk at 1.00 m/s and can swim at 2.00 m/s with respect to still water. Use the distances in the drawing to calculate **(a)** how long it takes them to cross the river and **(b)** the speed of the river current.

Julian W/Shutterstock.com

PROBLEM 73

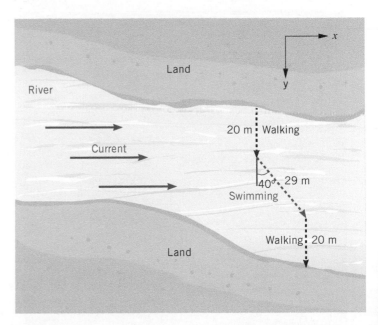

PROBLEM 73

74. Ⓜ ⒷⒾⓄ **3.3**   On average out of every 1000 newborn babies, one to two will suffer from a condition known as *pyloric stenosis*. This is characterized by a narrowing of the opening between the stomach and small intestine. One of the symptoms of this condition can be extremely forceful vomiting after feeding, known as projectile vomiting (see the drawing). The condition can usually be treated by opening the gastric outlet through laparoscopic surgery, and once corrected, it rarely

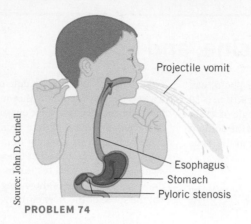

Source: John D. Cutnell

PROBLEM 74

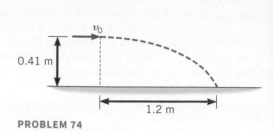

PROBLEM 74

causes any long-term side effects. Imagine an infant is sitting upright, so that his mouth is 41 cm above the floor. He suddenly vomits horizontally, with it landing 1.2 m on the floor in front of him, as shown in the drawing. As the name suggests, treat the vomit as a projectile and calculate the velocity at which it leaves his mouth.

75. Ⓜ ⒷⒾⓄ **3.3**   Archerfish are one of the few fish that prey on land insects and other small animals. They do this by shooting a jet of water from their specialized mouths at an insect resting on a leaf or plant stem above the water (see the photo). This often stuns the insect, causing it to fall into the water, where it can be eaten. In the drawing, an archerfish is shooting water toward an unsuspecting horsefly at rest on a blade of marsh grass above the surface of a pond. Use the information in the drawing to calculate how high (h) the horsefly is above the water. The initial velocity of the water (v_0) is 3.2 m/s, and assume that the projected water reaches its maximum height above the pond's surface when it strikes the horsefly.

Warren Photographic

PROBLEM 75

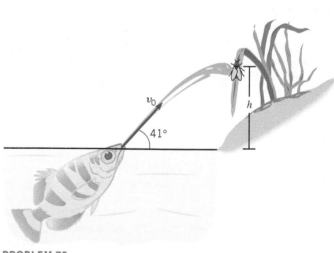

PROBLEM 75

Concepts and Calculations Problems   Online

A primary focus of this chapter has been projectile motion. These additional problems serve as a review of the basic features of this type of motion.

76. Ⓜ ⒸⒽⒶⓁⓀ ⓈⓈⓂ   Three balls are launched simultaneously, as in the figure. The red ball is launched at an initial velocity $\vec{v}_0$ directed at an angle θ above the horizontal. The blue ball is launched vertically with a speed of 10.0 m/s, and the yellow ball is launched horizontally with a speed of 4.6 m/s. The blue and red balls reach identical maximum heights and return to the ground at the same time, and the red ball returns to the ground to meet the yellow ball at the same place along the horizontal. *Concepts:* (i) Since the blue and red balls reach the same maximum height, what can be said about the y components of their velocities? (ii) Since the red and yellow balls travel equal distances along the x direction in the same amount of time, what can be said about the x components of their initial velocities? *Calculations:* Find the launch speed v_0 and launch angle θ for the red ball.

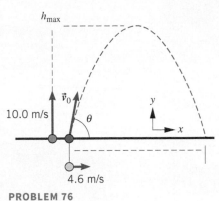

PROBLEM 76

77. **M** **CHALK** A projectile is launched from and returns to ground level, as the figure shows. There is no air resistance. The horizontal range of the projectile is measured to be $R = 175$ m, and the horizontal component of the launch velocity is $v_{0x} = +25$ m/s. *Concepts:* (i) What is the final value of the horizontal component v_x of the projectile's velocity? (ii) Can the time be determined for the horizontal part of the motion? (iii) Is the time for the horizontal part of the motion the same as that for the vertical part? (iv) For the vertical part of the motion, what is the displacement of the projectile? *Calculations:* Find the vertical component v_{0y} of the projectile.

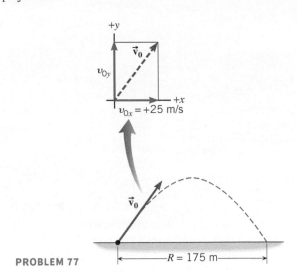

PROBLEM 77

Team Problems

Online

78. **E** **T** **WS** **A Starship Getaway.** You and your crew are on a spaceship that is being pursued by an enemy craft. Your ship is currently hiding behind an enormous space station, looking to make a run for the gate of a wormhole that will take you to safety. As the enemy ship passes to the other side of the space station, you will have exactly 5.00 s to burn your x and y thrusters, but must then turn them off so that you are not detected by the enemy craft as it comes around the other side of the space station. The gate is located at coordinates $x = 455$ m, $y = 633$ m, and your ship is located at the origin. The other requirement is that you enter the gate with a speed of exactly 220.0 m/s. Assuming that you keep your x and y thrusters on for the full 5.00 s (they turn on and off simultaneously), and that the acceleration is constant during this time, determine **(a)** the angle that your acceleration vector must make with the x axis, **(b)** the magnitude of the acceleration, and **(c)** the x and y components of the acceleration so that you know how much acceleration that each thruster should provide. **(d)** How far are you from the gate when the thrusters are turned off?

79. **E** **T** **WS** **A Water Slide.** You and your team are designing a water slide that launches a rider at a maximum speed of 30.0 mph. **(a)** To be safe, the maximum vertical component of the velocity of impact must be less than or equal to 20.0 mph. What launch angle results in a maximum vertical impact velocity of 20.0 mph? **(b)** Next, you have to design the pool in which the rider lands. For safety, the rider should land in the center of the pool when launched at maximum speed. Assuming one edge of the pool is at the launch position of the slide, how long should you make the pool (in meters)?

(c) What is the maximum height the rider can reach with this design (in meters)? Assume that the launch height is at water level.

80. **M** **T** **A Water Balloon Battle.** You are launching water balloons at a rival team using a large slingshot. The other team is set up on the opposite side of a flat-topped building that is 30.0 ft tall and 50.0 ft wide. Your reconnaissance team has reported that the opposition is set up 10.0 m from the wall of the building. Your balloon launcher is calibrated for launch speeds that can reach as high as 100 mph at angles between 0 and 80.0° from the horizontal. Since a direct shot is not possible (the opposing team is on the opposite side of the building), you plan to splash the other team by making a balloon explode on the ground near them. If your launcher is located 55.0 m from the building (opposite side as the opposing team), what should your launch velocity be (magnitude and direction) to land a balloon 5.0 meters beyond the opposing team with maximum impact (i.e., maximum vertical speed)?

81. **M** **T** **A Motorcycle Jump.** You are planning to make a jump with your motorcycle by driving over a ramp that will launch you at an angle of 30.0° with respect to the horizontal. The front edge of the ramp on which you are supposed to land, however, is 25.0 ft lower than the edge of the launch ramp (i.e., your launch height). **(a)** Assuming a launch speed of 65.0 mph, at what horizontal distance from your launch point should the landing ramp be placed? **(b)** In order to land smoothly, the angle of the landing ramp should match the direction of your velocity vector when you touch down. What should be the angle of the landing ramp?

CHAPTER 4

In order to successfully land the planetary rover, *Curiosity*, on the surface of Mars, NASA scientists and engineers had to take into account many forces, such as the thrust from the "sky crane" rocket engines, the tension in its towing cables, air resistance, and the weight of the rover due to the gravity on Mars, just to name a few. This chapter will discuss how forces influence the motion of objects.

NASA/JPL-Caltec

Forces and Newton's Laws of Motion

4.1 The Concepts of Force and Mass

In common usage, a **force** is a push or a pull, as the examples in **Figure 4.1** illustrate. In football, an offensive lineman pushes against his opponent. The tow bar attached to a speeding boat pulls a water skier. Forces such as those that push against the football player or pull the skier are called **contact forces**, because they arise from the physical contact between two objects. There are circumstances, however, in which two objects exert forces on one another even though they are not touching. Such forces are referred to as **noncontact forces** or **action-at-a-distance forces**. One example of such a noncontact force occurs when

86

a diver is pulled toward the earth because of the force of gravity. The earth exerts this force even when it is not in direct contact with the diver. In **Figure 4.1**, arrows are used to represent the forces. It is appropriate to use arrows, because a force is a vector quantity and has both a magnitude and a direction. The direction of the arrow gives the direction of the force, and the length is proportional to its strength or magnitude.

The word **mass** is just as familiar as the word *force*. A massive supertanker, for instance, is one that contains an enormous amount of mass. As we will see in the next section, it is difficult to set such a massive object into motion and difficult to bring it to a halt once it is moving. In comparison, a penny does not contain much mass. The emphasis here is on the amount of mass, and the idea of direction is of no concern. Therefore, mass is a scalar quantity.

During the seventeenth century, Isaac Newton, building on the work of Galileo, developed three important laws that deal with force and mass. Collectively they are called "Newton's laws of motion" and provide the basis for understanding the effect that forces have on an object. Because of the importance of these laws, a separate section will be devoted to each one.

4.2 Newton's First Law of Motion

The First Law

To gain some insight into Newton's first law, think about the game of ice hockey (**Figure 4.2**). If a player does not hit a stationary puck, it will remain at rest on the ice. After the puck is struck, however, it coasts on its own across the ice, slowing down only slightly because of friction. Since ice is very slippery, there is only a relatively small amount of friction to slow down the puck. In fact, if it were possible to remove all friction and wind resistance, and if the rink were infinitely large, the puck would coast forever in a straight line at a constant speed. Left on its own, the puck would lose none of the velocity imparted to it at the time it was struck. This is the essence of Newton's first law of motion:

> **NEWTON'S FIRST LAW OF MOTION**
>
> **An object continues in a state of rest or in a state of motion at a constant velocity (constant speed in a constant direction), unless compelled to change that state by a net force.**

In the first law the phrase "net force" is crucial. Often, several forces act simultaneously on a body, and ***the net force is the vector sum of all of them***. Individual forces matter only to the extent that they contribute to the total. For instance, if friction and other opposing forces were absent, a car could travel forever at 30 m/s in a straight line, without using any gas after it had come up to speed. In reality gas is needed, but only so that the engine can produce the necessary force to cancel opposing forces such as friction. This cancellation ensures that there is no net force to change the state of motion of the car.

(a)

(b)

(c)

FIGURE 4.1 The arrow labeled $\vec{F}$ represents the force that acts on (a) the football player on the left, (b) the water skier, and (c) the cliff diver.

FIGURE 4.2 The game of ice hockey can give some insight into Newton's laws of motion.

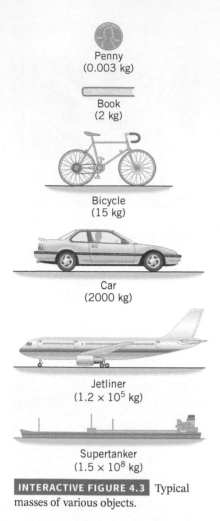

Penny
(0.003 kg)

Book
(2 kg)

Bicycle
(15 kg)

Car
(2000 kg)

Jetliner
(1.2×10^5 kg)

Supertanker
(1.5×10^8 kg)

INTERACTIVE FIGURE 4.3 Typical masses of various objects.

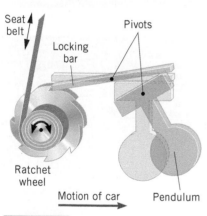

FIGURE 4.4 Inertia plays a central role in one seat-belt mechanism. The gray part of the drawing applies when the car is at rest or moving at a constant velocity. The colored parts show what happens when the car suddenly slows down, as in an accident.

When an object moves at a constant speed in a constant direction, its velocity is constant. Newton's first law indicates that a state of rest (zero velocity) and a state of constant velocity are completely equivalent, in the sense that neither one requires the application of a net force to sustain it. *The purpose served when a net force acts on an object is not to sustain the object's velocity, but, rather, to change it*.

Inertia and Mass

A greater net force is required to change the velocity of some objects than of others. For instance, a net force that is just enough to cause a bicycle to pick up speed will cause only an imperceptible change in the motion of a freight train. In comparison to the bicycle, the train has a much greater tendency to remain at rest. Accordingly, we say that the train has more **inertia** than the bicycle. Quantitatively, the inertia of an object is measured by its **mass**. The following definition of inertia and mass indicates why Newton's first law is sometimes called the law of inertia:

> **DEFINITION OF INERTIA AND MASS**
>
> **Inertia is the natural tendency of an object to remain at rest or in motion at a constant velocity. The mass of an object is a quantitative measure of inertia.**
>
> **SI Unit of Inertia and Mass: kilogram (kg)**

The SI unit for mass is the kilogram (kg), whereas the units in the CGS system and the BE system are the gram (g) and the slug (sl), respectively. Conversion factors between these units are given on the page facing the inside of the front cover. **Interactive Figure 4.3** gives the typical masses of various objects, ranging from a penny to a supertanker. The larger the mass, the greater is the inertia. Often the words "mass" and "weight" are used interchangeably, but this is incorrect. Mass and weight are different concepts, and Section 4.7 will discuss the distinction between them.

THE PHYSICS OF . . . seat belts. **Figure 4.4** shows a useful application of inertia. Automobile seat belts unwind freely when pulled gently, so they can be buckled. But in an accident, they hold you safely in place. One seat-belt mechanism consists of a ratchet wheel, a locking bar, and a pendulum. The belt is wound around a spool mounted on the ratchet wheel. While the car is at rest or moving at a constant velocity, the pendulum hangs straight down, and the locking bar rests horizontally, as the gray part of the drawing shows. Consequently, nothing prevents the ratchet wheel from turning, and the seat belt can be pulled out easily. When the car suddenly slows down in an accident, however, the relatively massive lower part of the pendulum keeps moving forward because of its inertia. The pendulum swings on its pivot into the position shown in color and causes the locking bar to block the rotation of the ratchet wheel, thus preventing the seat belt from unwinding.

An Inertial Reference Frame

Newton's first law (and also the second law) can appear to be invalid to certain observers. Suppose, for instance, that you are a passenger riding in a friend's car. While the car moves at a constant speed along a straight line, you do not feel the seat pushing against your back to any unusual extent. This experience is consistent with the first law, which indicates that in the absence of a net force you should move with a constant velocity. Suddenly the driver floors the gas pedal. Immediately you feel the seat pressing against your back as the car accelerates. Therefore, you sense that a force is being applied to you. The first law leads you to believe that your motion should change, and, relative to the ground outside, your motion does change. But *relative to the car,* you can see that your motion does *not* change, because you remain stationary with respect to the car. Clearly, Newton's first law does not hold for observers who use the accelerating car as a frame of reference. As a result, such a reference frame is said to be *noninertial*. All accelerating reference frames are noninertial. In contrast, observers for whom the law of inertia is valid are said to be using **inertial reference frames** for their observations, as defined below:

| **DEFINITION OF AN INERTIAL REFERENCE FRAME**
| **An inertial reference frame is one in which Newton's law of inertia is valid.**

The acceleration of an inertial reference frame is zero, so it moves with a constant velocity. All of Newton's laws of motion are valid in inertial reference frames, and when we apply these laws, we will be assuming such a reference frame. In particular, the earth itself is a good approximation of an inertial reference frame.

Check Your Understanding

(*The answer is given at the end of the book.*)

1. Which of the following statements can be explained by Newton's first law? (A): When your car suddenly comes to a halt, you lunge forward. (B): When your car rapidly accelerates, you are pressed backward against the seat. **(a)** Neither A nor B **(b)** Both A and B **(c)** A but not B **(d)** B but not A

4.3 Newton's Second Law of Motion

Newton's first law indicates that if no net force acts on an object, then the velocity of the object remains unchanged. The second law deals with what happens when a net force does act. Consider a hockey puck once again. When a player strikes a stationary puck, he causes the velocity of the puck to change. In other words, he makes the puck accelerate. The cause of the acceleration is the force that the hockey stick applies. As long as this force acts, the velocity increases, and the puck accelerates. Now, suppose another player strikes the puck and applies twice as much force as the first player does. The greater force produces a greater acceleration. In fact, if the friction between the puck and the ice is negligible, and if there is no wind resistance, the acceleration of the puck is directly proportional to the force. Twice the force produces twice the acceleration. Moreover, the acceleration is a vector quantity, just as the force is, and points in the same direction as the force.

Often, several forces act on an object simultaneously. Friction and wind resistance, for instance, do have some effect on a hockey puck. In such cases, it is the net force, or the vector sum of all the forces acting, that is important. Mathematically, the net force is written as $\Sigma\vec{F}$, where the Greek capital letter Σ (sigma) denotes the vector sum. Newton's second law states that the acceleration is proportional to the net force acting on the object.

In Newton's second law, the net force is only one of two factors that determine the acceleration. The other is the inertia or mass of the object. After all, the same net force that imparts an appreciable acceleration to a hockey puck (small mass) will impart very little acceleration to a semitrailer truck (large mass). Newton's second law states that for a given net force, the magnitude of the acceleration is inversely proportional to the mass. Twice the mass means one-half the acceleration, if the same net force acts on both objects. Thus, the second law shows how the acceleration depends on both the net force and the mass, as given in Equation 4.1.

| **NEWTON'S SECOND LAW OF MOTION**
| When a net external force $\Sigma\vec{F}$ acts on an object of mass m, the acceleration $\vec{a}$
| that results is directly proportional to the net force and has a magnitude that is
| inversely proportional to the mass. The direction of the acceleration is the same
| as the direction of the net force.
|
| $$\vec{a} = \frac{\Sigma\vec{F}}{m} \qquad \text{or} \qquad \Sigma\vec{F} = m\vec{a} \qquad\qquad (4.1)$$
|
| **SI Unit of Force:** $\text{kg} \cdot \text{m/s}^2 = \text{newton (N)}$

TABLE 4.1   **Units for Mass, Acceleration, and Force**

System	Mass	Acceleration	Force
SI	kilogram (kg)	meter/second^2 (m/s^2)	newton (N)
CGS	gram (g)	centimeter/second^2 (cm/s^2)	dyne (dyn)
BE	slug (sl)	foot/second^2 (ft/s^2)	pound (lb)

Note that the net force in Equation 4.1 includes only the forces that the environment exerts on the object of interest. Such forces are called **external forces**. In contrast, **internal forces** are forces that one part of an object exerts on another part of the object and are not included in Equation 4.1.

According to Equation 4.1, the SI unit for force is the unit for mass (kg) times the unit for acceleration (m/s^2), or

$$\text{SI unit for force} = (\text{kg})\left(\frac{\text{m}}{\text{s}^2}\right) = \frac{\text{kg} \cdot \text{m}}{\text{s}^2}$$

The combination of kg · m/s^2 is called a *newton* (N) and is a derived SI unit, not a base unit; 1 newton = 1 N = 1 kg · m/s^2.

In the CGS system, the procedure for establishing the unit of force is the same as with SI units, except that mass is expressed in grams (g) and acceleration in cm/s^2. The resulting unit for force is the *dyne*; 1 dyne = 1 g · cm/s^2.

In the BE system, the unit for force is defined to be the pound (lb),* and the unit for acceleration is ft/s^2. With this procedure, Newton's second law can then be used to obtain the BE unit for mass:

$$\text{BE unit for force} = \text{lb} = (\text{unit for mass})\left(\frac{\text{ft}}{\text{s}^2}\right)$$

$$\text{Unit for mass} = \frac{\text{lb} \cdot \text{s}^2}{\text{ft}}$$

The combination of lb · s^2/ft is the unit for mass in the BE system and is called the *slug* (sl); 1 slug = 1 sl = 1 lb · s^2/ft.

Table 4.1 summarizes the various units for mass, acceleration, and force. Conversion factors between force units from different systems are provided on the page facing the inside of the front cover.

When using the second law to calculate the acceleration, it is necessary to determine the net force that acts on the object. In this determination a **free-body diagram** helps enormously. A free-body diagram is a diagram that represents the object and the forces that act on it. Only the forces that *act on the object* appear in a free-body diagram. Forces that the object exerts on its environment are not included. Example 1 illustrates the use of a free-body diagram.

EXAMPLE 1 | Pushing a Stalled Car

Two people are pushing a stalled car, as **Animated Figure 4.5a** indicates. The mass of the car is 1850 kg. One person applies a force of 275 N to the car, while the other applies a force of 395 N. Both forces act in the same direction. A third force of 560 N also acts on the car, but in a direction opposite to that in which the people are pushing. This force arises because of friction and the extent to which the pavement opposes the motion of the tires. Find the acceleration of the car.

Reasoning   According to Newton's second law, the acceleration is the net force divided by the mass of the car. To determine the net force, we use the free-body diagram in **Animated Figure 4.5b**. In this diagram, the car is represented as a dot, and its motion is along the $+x$ axis. The diagram makes it clear that the forces all act along one direction. Therefore, they can be added as colinear vectors to obtain the net force.

---

*We refer here to the gravitational version of the BE system, in which a force of one pound is defined to be the pull of the earth on a certain standard body at a location where the acceleration due to gravity is 32.174 ft/s^2.

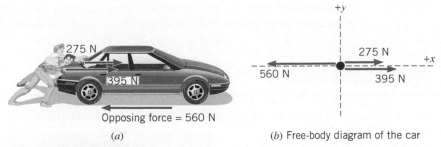

(a)

(b) Free-body diagram of the car

ANIMATED FIGURE 4.5 (a) Two people push a stalled car. A force created by friction and the pavement opposes their efforts. (b) A free-body diagram that shows the horizontal forces acting on the car.

> **Problem-Solving Insight** The direction of the acceleration is always the same as the direction of the net force.

The acceleration can now be obtained:

$$a = \frac{\Sigma F}{m} = \frac{+110\text{ N}}{1850\text{ kg}} = \boxed{+0.059\text{ m/s}^2} \qquad \textbf{(4.1)}$$

Solution From Equation 4.1, the acceleration is $a = (\Sigma F)/m$. The net force is

$$\Sigma F = +275\text{ N} + 395\text{ N} - 560\text{ N} = +110\text{ N}$$

The plus sign indicates that the acceleration points along the $+x$ axis, in the same direction as the net force.

Check Your Understanding

(The answers are given at the end of the book.)

2. The net external force acting on an object is zero. Which one of the following statements is true? **(a)** The object can only be stationary. **(b)** The object can only be traveling with a constant velocity. **(c)** The object can be either stationary or traveling with a constant velocity. **(d)** The object can only be traveling with a velocity that is changing.

3. In Case A an object is moving straight downward with a constant speed of 9.80 m/s, while in Case B an object is moving straight downward with a constant acceleration of magnitude 9.80 m/s². Which one of the following is true? **(a)** A nonzero net external force acts on the object in both cases. **(b)** A nonzero net external force acts on the object in neither case. **(c)** A nonzero net external force acts on the object in Case A only. **(d)** A nonzero net external force acts on the object in Case B only.

4.4 The Vector Nature of Newton's Second Law of Motion

When a football player throws a pass, the direction of the force he applies to the ball is important. Both the force and the resulting acceleration of the ball are vector quantities, as are all forces and accelerations. The directions of these vectors can be taken into account in two dimensions by using x and y components. The net force $\Sigma\vec{F}$ in Newton's second law has components ΣF_x and ΣF_y, while the acceleration $\vec{a}$ has components a_x and a_y. Consequently, Newton's second law, as expressed in Equation 4.1, can be written in an equivalent form as two equations, one for the x components and one for the y components:

$$\Sigma F_x = ma_x \qquad \textbf{(4.2a)}$$

$$\Sigma F_y = ma_y \qquad \textbf{(4.2b)}$$

This procedure is similar to that employed in Chapter 3 for the equations of two-dimensional kinematics (see **Table 3.1**). The components in Equations 4.2a and 4.2b are scalar components and will be either positive or negative numbers, depending on whether they point along the positive or negative x or y axis. The remainder of this section deals with examples that show how these equations are used.

EXAMPLE 2 | Applying Newton's Second Law Using Components

A man is stranded on a raft (mass of man and raft = 1300 kg), as shown in **Figure 4.6a**. By paddling, he causes an average force $\vec{P}$ of 17 N to be applied to the raft in a direction due east (the +x direction). The wind also exerts a force $\vec{A}$ on the raft. This force has a magnitude of 15 N and points 67° north of east. Ignoring any resistance from the water, find the x and y components of the raft's acceleration.

Reasoning Since the mass of the man and the raft is known, Newton's second law can be used to determine the acceleration components from the given forces. According to the form of the second law in Equations 4.2a and 4.2b, the acceleration component in a given direction is the component of the net force in that direction divided by the mass. As an aid in determining the components ΣF_x and ΣF_y of the net force, we use the free-body diagram in **Figure 4.6b**. In this diagram, the directions due east and due north are the +x and +y directions, respectively.

Solution **Figure 4.6b** shows the force components:

Force	x Component	y Component
$\vec{P}$	+17 N	0 N
$\vec{A}$	$+(15\text{ N})\cos 67° = +6\text{ N}$	$+(15\text{ N})\sin 67° = +14\text{ N}$
	$\Sigma F_x = +17\text{ N} + 6\text{ N} = +23\text{ N}$	$\Sigma F_y = +14\text{ N}$

The plus signs indicate that ΣF_x points in the direction of the +x axis and ΣF_y points in the direction of the +y axis. The x and y components of the acceleration point in the directions of ΣF_x and ΣF_y, respectively, and can now be calculated:

$$a_x = \frac{\Sigma F_x}{m} = \frac{+23\text{ N}}{1300\text{ kg}} = \boxed{+0.018\text{ m/s}^2} \quad (4.2a)$$

$$a_y = \frac{\Sigma F_y}{m} = \frac{+14\text{ N}}{1300\text{ kg}} = \boxed{+0.011\text{ m/s}^2} \quad (4.2b)$$

These acceleration components are shown in **Figure 4.6c**.

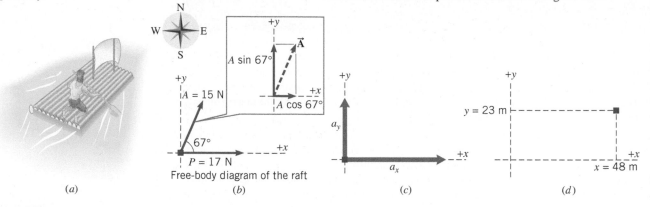

(a)　　(b)　　(c)　　(d)

FIGURE 4.6 (a) A man is paddling a raft, as in Examples 2 and 3. (b) The free-body diagram shows the forces $\vec{P}$ and $\vec{A}$ that act on the raft. Forces acting on the raft in a direction perpendicular to the surface of the water play no role in the examples and are omitted for clarity. (c) The raft's acceleration components a_x and a_y. (d) In 65 s, the components of the raft's displacement are x = 48 m and y = 23 m.

EXAMPLE 3 | The Displacement of a Raft

At the moment that the forces $\vec{P}$ and $\vec{A}$ begin acting on the raft in Example 2, the velocity of the raft is 0.15 m/s, in a direction due east (the +x direction). Assuming that the forces are maintained for 65 s, find the x and y components of the raft's displacement during this time interval.

Reasoning Once the net force acting on an object and the object's mass have been used in Newton's second law to determine the acceleration, it becomes possible to use the equations of kinematics to describe the resulting motion. We know from Example 2 that the acceleration components are a_x = +0.018 m/s² and a_y = +0.011 m/s², and it is given here that the initial velocity

components are v_{0x} = +0.15 m/s and v_{0y} = 0 m/s. Thus, Equation 3.5a ($x = v_{0x}t + \frac{1}{2}a_x t^2$) and Equation 3.5b ($y = v_{0y}t + \frac{1}{2}a_y t^2$) can be used with t = 65 s to determine the x and y components of the raft's displacement.

Solution According to Equations 3.5a and 3.5b, the x and y components of the displacement are

$$x = v_{0x}t + \tfrac{1}{2}a_x t^2 = (0.15\text{ m/s})(65\text{ s}) + \tfrac{1}{2}(0.018\text{ m/s}^2)(65\text{ s})^2 = \boxed{48\text{ m}}$$

$$y = v_{0y}t + \tfrac{1}{2}a_y t^2 = (0\text{ m/s})(65\text{ s}) + \tfrac{1}{2}(0.011\text{ m/s}^2)(65\text{ s})^2 = \boxed{23\text{ m}}$$

Figure 4.6d shows the final location of the raft.

Check Your Understanding

(The answers are given at the end of the book.)

4. Newton's second law indicates that when a net force acts on an object, it must accelerate. Does this mean that when two or more forces are applied to an object simultaneously, it must accelerate?

5. All of the following, except one, cause the acceleration of an object to double. Which one is the exception? **(a)** All forces acting on the object double. **(b)** The net force acting on the object doubles. **(c)** Both the net force acting on the object and the mass of the object double. **(d)** The net force acting on the object remains the same, while the mass of the object is reduced by a factor of two.

4.5 | Newton's Third Law of Motion

Imagine you are in a football game. You line up facing your opponent, the ball is snapped, and the two of you crash together. No doubt, you feel a force. But think about your opponent. He too feels something, for while he is applying a force to you, you are applying a force to him. In other words, there isn't just one force on the line of scrimmage; there is a pair of forces. Newton was the first to realize that all forces occur in pairs and there is no such thing as an isolated force, existing all by itself. His third law of motion deals with this fundamental characteristic of forces.

> **NEWTON'S THIRD LAW OF MOTION**
>
> **Whenever one object exerts a force on a second object, the second object exerts an oppositely directed force of equal magnitude on the first object.**

The third law is often called the "action–reaction" law, because it is sometimes quoted as follows: "For every action (force) there is an equal, but opposite, reaction."

Figure 4.7 illustrates how the third law applies to an astronaut who is drifting just outside a spacecraft and who pushes on the spacecraft with a force $\vec{\mathbf{P}}$. According to the third law, the spacecraft pushes back on the astronaut with a force $-\vec{\mathbf{P}}$ that is equal in magnitude but opposite in direction. In Example 4, we examine the accelerations produced by each of these forces.

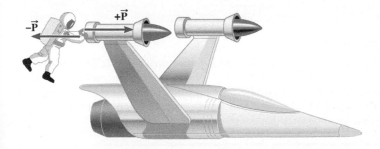

FIGURE 4.7 The astronaut pushes on the spacecraft with a force $+\vec{\mathbf{P}}$. According to Newton's third law, the spacecraft simultaneously pushes back on the astronaut with a force $-\vec{\mathbf{P}}$.

EXAMPLE 4 | The Accelerations Produced by Action and Reaction Forces

Suppose that the mass of the spacecraft in **Figure 4.7** is $m_S = 11\ 000$ kg and that the mass of the astronaut is $m_A = 92$ kg. In addition, assume that the astronaut pushes with a force of $\vec{\mathbf{P}} = +36$ N on the spacecraft. Find the accelerations of the spacecraft and the astronaut.

Reasoning Consistent with Newton's third law, when the astronaut applies the force $\vec{\mathbf{P}} = +36$ N to the spacecraft, the spacecraft applies a reaction force $-\vec{\mathbf{P}} = -36$ N to the astronaut. As a result, the spacecraft and the astronaut accelerate in opposite directions. Although the action and reaction forces have the same magnitude, they do not create accelerations of the same magnitude, because the spacecraft and the astronaut have different masses. According to Newton's second law, the astronaut, having a much smaller mass, will experience a much larger acceleration. In applying the second law, we note that the net force acting on the spacecraft is $\Sigma \vec{F} = \vec{P}$, while the net force acting on the astronaut is $\Sigma \vec{F} = -\vec{P}$.

> **Problem-Solving Insight** Even though the magnitudes of the action and reaction forces are always equal, these forces do not necessarily produce accelerations that have equal magnitudes, since each force acts on a different object that may have a different mass.

Solution Using the second law, we find that the acceleration of the spacecraft is

$$\vec{a}_S = \frac{\vec{P}}{m_S} = \frac{+36\ \text{N}}{11\ 000\ \text{kg}} = \boxed{+0.0033\ \text{m/s}^2}$$

The acceleration of the astronaut is

$$\vec{a}_A = \frac{-\vec{P}}{m_A} = \frac{-36\ \text{N}}{92\ \text{kg}} = \boxed{-0.39\ \text{m/s}^2}$$

THE PHYSICS OF . . . automatic trailer brakes. There is a clever application of Newton's third law in some rental trailers. As **Figure 4.8** illustrates, the tow bar connecting the trailer to the rear bumper of a car contains a mechanism that can automatically actuate brakes on the trailer wheels. This mechanism works without the need for electrical connections between the car and the trailer. When the driver applies the car brakes, the car slows down. Because of inertia, however, the trailer continues to roll forward and begins pushing against the bumper. In reaction, the bumper pushes back on the tow bar. The reaction force is used by the mechanism in the tow bar to "push a brake pedal" for the trailer.

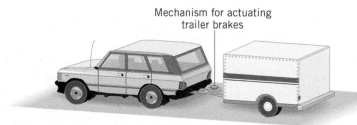

Mechanism for actuating
trailer brakes

FIGURE 4.8 Some rental trailers include an automatic brake-actuating mechanism.

Check Your Understanding

(The answer is given at the end of the book.)

6. A father and his seven-year-old daughter are facing each other on ice skates. With their hands, they push off against one another. Which one or more of the following statements is (are) true? **(a)** Each experiences an acceleration that has a different magnitude. **(b)** Each experiences an acceleration of the same magnitude. **(c)** Each experiences a pushing force that has a different magnitude. **(d)** Each experiences a pushing force that has the same magnitude.

4.6 | Types of Forces: An Overview

Newton's three laws of motion make it clear that forces play a central role in determining the motion of an object. In the next four sections some common forces will be discussed: the gravitational force (Section 4.7), the normal force (Section 4.8), frictional forces (Section 4.9), and the tension force (Section 4.10). In later chapters, we will encounter still others, such as electric and magnetic forces. It is important to realize that Newton's second law is always valid, regardless of which of these forces may act on an object. One does not have a different law for every type of common force. Thus, we need only to determine what forces are acting on an object, add them together to form the net force, and then use Newton's second law to determine the object's acceleration.

In nature there are two general types of forces, fundamental and nonfundamental. Fundamental forces are the ones that are truly unique, in the sense that all other forces can be explained in terms of them. Only three fundamental forces have been discovered:

1. Gravitational force
2. Strong nuclear force
3. Electroweak force

The gravitational force is discussed in the next section. The strong nuclear force plays a primary role in the stability of the nucleus of the atom (see Section 31.2). The electroweak force is a single force that manifests itself in two ways (see Section 32.6). One manifestation is the electromagnetic force that electrically charged particles exert on one another (see Sections 18.5, 21.2, and 21.8). The other manifestation is the so-called weak nuclear force that plays a role in the radioactive disintegration of certain nuclei (see Section 31.5).

Except for the gravitational force, all of the forces discussed in this chapter are non-fundamental, because they are related to the electromagnetic force. They arise from the interactions between the electrically charged particles that comprise atoms and molecules. Our understanding of which forces are fundamental, however, is continually evolving. For instance, in the 1860s and 1870s James Clerk Maxwell showed that the electric force and the magnetic force could be explained as manifestations of a single electromagnetic force. Then, in the 1970s, Sheldon Glashow (1932–), Abdus Salam (1926–1996), and Steven Weinberg (1933–) presented the theory that explains how the electromagnetic force and the weak nuclear force are related to the electroweak force. They received a Nobel Prize in 1979 for their achievement. Today, efforts continue that have the goal of further reducing the number of fundamental forces.

4.7 | **The Gravitational Force**

Newton's Law of Universal Gravitation

Objects fall downward because of gravity, and Chapters 2 and 3 discuss how to describe the effects of gravity by using a value of $g = 9.80$ m/s^2 for the downward acceleration it causes. However, nothing has been said about why g is 9.80 m/s^2. The reason is fascinating, as we will see later in this section.

The acceleration due to gravity is like any other acceleration, and Newton's second law indicates that it must be caused by a net force. In addition to his famous three laws of motion, Newton also provided a coherent understanding of the **gravitational force**. His "law of universal gravitation" is stated as follows:

> **NEWTON'S LAW OF UNIVERSAL GRAVITATION**
>
> **Every particle in the universe exerts an attractive force on every other particle. A particle is a piece of matter, small enough in size to be regarded as a mathematical point. For two particles that have masses m_1 and m_2 and are separated by a distance r, the force that each exerts on the other is directed along the line joining the particles (see Figure 4.9) and has a magnitude given by**
>
> $$F = G \frac{m_1 m_2}{r^2} \qquad (4.3)$$
>
> **The symbol G denotes the universal gravitational constant, whose value is found experimentally to be**
>
> $$G = 6.674 \times 10^{-11} \, \text{N} \cdot \text{m}^2/\text{kg}^2$$

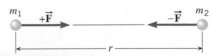

FIGURE 4.9 The two particles, whose masses are m_1 and m_2, are attracted by gravitational forces $+\vec{F}$ and $-\vec{F}$.

The constant G that appears in Equation 4.3 is called the **universal gravitational constant**, because it has the same value for all pairs of particles anywhere in the universe, no matter what their separation. The value for G was first measured in an experiment by the English scientist Henry Cavendish (1731–1810), more than a century after Newton proposed his law of universal gravitation.

To see the main features of Newton's law of universal gravitation, look at the two particles in **Figure 4.9**. They have masses m_1 and m_2 and are separated by a distance r. In the picture, it is assumed that a force pointing to the right is positive. The gravitational forces point along the line joining the particles and are

$+\vec{\mathbf{F}}$, the gravitational force exerted on particle 1 by particle 2

$-\vec{\mathbf{F}}$, the gravitational force exerted on particle 2 by particle 1

These two forces have equal magnitudes and opposite directions. They act on different bodies, causing them to be mutually attracted. In fact, these forces are an action–reaction pair, as required by Newton's third law. Example 5 shows that the magnitude of the gravitational force is extremely small for ordinary values of the masses and the distance between them.

EXAMPLE 5 | Gravitational Attraction

What is the magnitude of the gravitational force that acts on each particle in **Figure 4.9**, assuming $m_1 = 12$ kg (approximately the mass of a bicycle), $m_2 = 25$ kg, and $r = 1.2$ m?

Reasoning and Solution The magnitude of the gravitational force can be found using Equation 4.3:

$$F = G\frac{m_1 m_2}{r^2} = (6.67 \times 10^{-11}\,\text{N}\cdot\text{m}^2/\text{kg}^2)\frac{(12\,\text{kg})(25\,\text{kg})}{(1.2\,\text{m})^2}$$
$$= \boxed{1.4 \times 10^{-8}\,\text{N}}$$

For comparison, you exert a force of about 1 N when pushing a doorbell, so that the gravitational force is exceedingly small in circumstances such as those here. This result is due to the fact that G itself is very small. However, if one of the bodies has a large mass, like that of the earth (5.98×10^{24} kg), the gravitational force can be large.

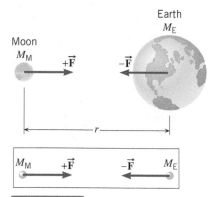

FIGURE 4.10 The gravitational force that each uniform sphere of matter exerts on the other is the same as if each sphere were a particle with its mass concentrated at its center. The earth (mass M_E) and the moon (mass M_M) approximate such uniform spheres.

As expressed by Equation 4.3, Newton's law of gravitation applies only to particles. However, most familiar objects are too large to be considered particles. Nevertheless, the law of universal gravitation can be applied to such objects with the aid of calculus. Newton was able to prove that an object of finite size can be considered to be a particle for purposes of using the gravitation law, provided the mass of the object is distributed with spherical symmetry about its center. Thus, Equation 4.3 can be applied when each object is a sphere whose mass is spread uniformly over its entire volume. **Figure 4.10** shows this kind of application, assuming that the earth and the moon are such uniform spheres of matter. In this case, r is the distance *between the centers of the spheres* and not the distance between the outer surfaces. The gravitational forces that the spheres exert on each other are the same as if the entire mass of each were concentrated at its center. Even if the objects are not uniform spheres, Equation 4.3 can be used to a good degree of approximation if the sizes of the objects are small relative to the distance of separation r.

Weight

The weight of an object exists because of the gravitational pull of the earth, according to the following definition:

DEFINITION OF WEIGHT

The weight of an object on or above the earth is the gravitational force that the earth exerts on the object. The weight always acts downward, toward the center of the earth. On or above another astronomical body, the weight is the gravitational force exerted on the object by that body.

SI Unit of Weight: newton (N)

Using W for the magnitude of the weight,* m for the mass of the object, and M_E for the mass of the earth, it follows from Equation 4.3 that

$$W = G\frac{M_E m}{r^2} \qquad (4.4)$$

Equation 4.4 and **Figure 4.11** both emphasize that an object has weight whether or not it is resting on the earth's surface, because the gravitational force is acting even when the distance r is not equal to the radius R_E of the earth. However, the gravitational force becomes weaker as r increases, since r is in the denominator of Equation 4.4. **Figure 4.12**, for example, shows how the weight of the Hubble Space Telescope becomes smaller as the distance r from the center of the earth increases. In Example 6 the telescope's weight is determined when it is on earth and in orbit.

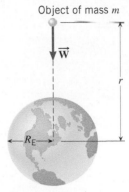

Object of mass m

Mass of earth = M_E

FIGURE 4.11 On or above the earth, the weight $\overrightarrow{W}$ of an object is the gravitational force exerted on the object by the earth.

EXAMPLE 6 | The Hubble Space Telescope

The mass of the Hubble Space Telescope is 11 600 kg. Determine the weight of the telescope **(a)** when it was resting on the earth and **(b)** as it is in its orbit 598 km above the earth's surface.

Reasoning The weight of the Hubble Space Telescope is the gravitational force exerted on it by the earth. According to Equation 4.4, the weight varies inversely as the square of the radial distance r. Thus, we expect the telescope's weight on the earth's surface (r smaller) to be greater than its weight in orbit (r larger).

> **Problem-Solving Insight** When applying Newton's gravitation law to uniform spheres of matter, remember that the distance r is between the centers of the spheres, not between the surfaces.

Solution **(a)** On the earth's surface, the weight is given by Equation 4.4 with $r = 6.38 \times 10^6$ m (the earth's radius):

$$W = G\frac{M_E m}{r^2} = \frac{(6.67 \times 10^{-11} \text{ N} \cdot \text{m}^2/\text{kg}^2)(5.98 \times 10^{24} \text{ kg})(11\,600 \text{ kg})}{(6.38 \times 10^6 \text{ m})^2}$$

$$\boxed{W = 1.14 \times 10^5 \text{N}}$$

(b) When the telescope is 598 km above the surface, its distance from the center of the earth is

$$r = 6.38 \times 10^6 \text{ m} + 598 \times 10^3 \text{ m} = 6.98 \times 10^6 \text{m}$$

The weight now can be calculated as in part (a), except that the new value of r must be used: $\boxed{W = 0.950 \times 10^5 \text{N}}$. As expected, the weight is less in orbit.

The space age has forced us to broaden our ideas about weight. For instance, an astronaut weighs only about one-sixth as much on the moon as on the earth. To obtain his weight on the moon from Equation 4.4, it is only necessary to replace M_E by M_M (the mass of the moon) and let $r = R_M$ (the radius of the moon).

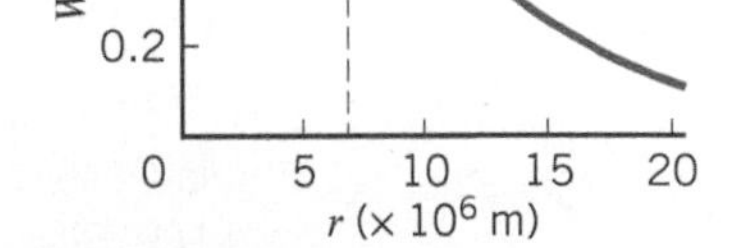

Relation Between Mass and Weight

Although massive objects weigh a lot on the earth, mass and weight are not the same quantity. As Section 4.2 discusses, mass is a quantitative measure of inertia. As such, mass is an intrinsic property of matter and does not change as an object is moved from one location to another. Weight, in contrast, is the gravitational force acting on the object and can vary, depending on how far the object is above the earth's surface or whether it is located near another body such as the moon.

The relation between weight W and mass m can be written in two ways:

$$W = \boxed{G\frac{M_E}{r^2}} m \qquad (4.4)$$

$$W = m\boxed{g} \qquad (4.5)$$

$R_E = 6.38 \times 10^6$ m

FIGURE 4.12 The weight of the Hubble Space Telescope decreases as the telescope gets farther from the earth. The distance from the center of the earth to the telescope is r.

*Often, the word "weight" and the phrase "magnitude of the weight" are used interchangeably, even though weight is a vector. Generally, the context makes it clear when the direction of the weight vector must be taken into account.

Equation 4.4 is Newton's law of universal gravitation, and Equation 4.5 is Newton's second law (net force equals mass times acceleration) incorporating the acceleration g due to gravity. These expressions make the distinction between mass and weight stand out. The weight of an object whose mass is m depends on the values for the universal gravitational constant G, the mass M_E of the earth, and the distance r. These three parameters together determine the acceleration g due to gravity. The specific value of $g = 9.80 \text{ m/s}^2$ applies only when r equals the radius R_E of the earth. For larger values of r, as would be the case on top of a mountain, the effective value of g is less than 9.80 m/s^2. The fact that g decreases as the distance r increases means that the weight likewise decreases. The mass of the object, however, does not depend on these effects and does not change. Conceptual Example 7 further explores the difference between mass and weight.

CONCEPTUAL EXAMPLE 7 | Mass Versus Weight

A vehicle designed for exploring the moon's surface (see **Photo 4.1**) is being tested on earth, where it weighs roughly six times more than it will on the moon. The acceleration of the vehicle along the ground is measured. To achieve the same acceleration on the moon, will the required net force be **(a)** the same as, **(b)** greater than, or **(c)** less than that on earth?

Reasoning   Do not be misled by the fact that the vehicle weighs more on earth. The greater weight occurs only because the mass and radius of the earth are different from the mass and radius of the moon. In any event, in Newton's second law the net force is proportional to the vehicle's mass, not its weight.

Answers (b) and (c) are incorrect.   According to Newton's second law, for a given acceleration, the net force depends only on the mass. If the required net force were greater or smaller on the moon than it is on the earth, the implication would be that the vehicle's mass is different on the moon than it is on earth, which is contrary to fact.

Answer (a) is correct.   The net force $\Sigma \vec{F}$ required to accelerate the vehicle is specified by Newton's second law as $\Sigma \vec{F} = m\vec{a}$, where m is the vehicle's mass and $\vec{a}$ is the acceleration. For a given acceleration, the net force depends only on the mass, which is the same on the moon as it is on the earth. Therefore, the required net force is the same on the moon as it is on the earth.

Related Homework: Problems 24, 84

NASA/Johnson Space Center

PHOTO 4.1 On the moon the Lunar Roving Vehicle that astronaut Eugene Cernan is driving and the Lunar Excursion Module (behind the Roving Vehicle) have the same masses that they have on the earth. However, their weights are different on the moon than on the earth, as Conceptual Example 7 discusses.

Check Your Understanding

(The answers are given at the end of the book.)

7. When a body is moved from sea level to the top of a mountain, what changes: **(a)** the body's mass, **(b)** its weight, or **(c)** both its mass and its weight?

8. Object A weighs twice as much as object B at the same spot on the earth. Would the same be true at a given spot on Mars?

9. Three particles have identical masses. Each particle experiences only the gravitational forces due to the other two particles. How should the particles be arranged so each one experiences a net gravitational force that has the same magnitude? **(a)** On the corners of an equilateral triangle **(b)** On three of the four corners of a square **(c)** On the corners of a right triangle

10. Two objects with masses m_1 and m_2 are separated by a distance $2d$ (see CYU Figure 4.1). Mass m_2 is greater than mass m_1. A third object has a mass m_3. All three objects are located on the same straight line. The net gravitational force acting on the third object is zero. Which one of the drawings correctly represents the locations of the objects?

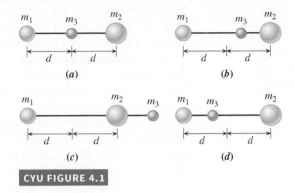

CYU FIGURE 4.1

4.8 # The Normal Force

The Definition and Interpretation of the Normal Force

In many situations, an object is in contact with a surface, such as a tabletop. Because of the contact, there is a force acting on the object. The present section discusses only one component of this force, the component that acts perpendicular to the surface. The next section discusses the component that acts parallel to the surface. The perpendicular component is called the **normal force**.

> **DEFINITION OF THE NORMAL FORCE**
>
> The normal force $\vec{F}_N$ is one component of the force that a surface exerts on an object with which it is in contact—namely, the component that is perpendicular to the surface.

Figure 4.13 shows a block resting on a horizontal table and identifies the two forces that act on the block, the weight $\vec{W}$ and the normal force $\vec{F}_N$. To understand how an inanimate object, such as a tabletop, can exert a normal force, think about what happens when you sit on a mattress. Your weight causes the springs in the mattress to compress. As a result, the compressed springs exert an upward force (the normal force) on you. In a similar manner, the weight of the block causes invisible "atomic springs" in the surface of the table to compress, thus producing a normal force on the block.

Newton's third law plays an important role in connection with the normal force. In Figure 4.13, for instance, the block exerts a force on the table by pressing down on it. Consistent with the third law, the table exerts an oppositely directed force of equal magnitude on the block. This reaction force is the normal force. The magnitude of the normal force indicates how hard the two objects press against each other.

If an object is resting on a horizontal surface and there are no vertically acting forces except the object's weight and the normal force, the magnitudes of these two forces are equal; that is, $F_N = W$. This is the situation in Figure 4.13. The weight must be balanced by the normal force for the object to remain at rest on the table. If the magnitudes of these forces were not equal, there would be a net force acting on the block, and the block would accelerate either upward or downward, in accord with Newton's second law.

If other forces in addition to $\vec{W}$ and $\vec{F}_N$ act in the vertical direction, the magnitudes of the normal force and the weight are no longer equal. In Figure 4.14a, for instance, a box whose weight is 15 N is being pushed downward against a table. The pushing force has a magnitude of 11 N. Thus, the total downward force exerted on the box is 26 N, and this must be balanced by the upward-acting normal force if the box is to remain at rest. In this situation, then, the normal force is 26 N, which is considerably larger than the weight of the box.

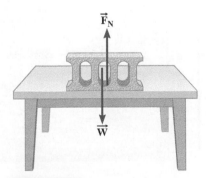

FIGURE 4.13 Two forces act on the block, its weight $\vec{W}$ and the normal force $\vec{F}_N$ exerted by the surface of the table.

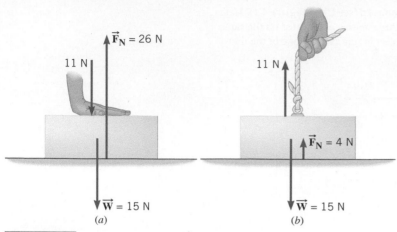

$\vec{\mathbf{F}}_N$ = 26 N

11 N

$\vec{\mathbf{W}}$ = 15 N

(a)

11 N

$\vec{\mathbf{F}}_N$ = 4 N

$\vec{\mathbf{W}}$ = 15 N

(b)

FIGURE 4.14 (a) The normal force $\vec{\mathbf{F}}_N$ is greater than the weight of the box, because the box is being pressed downward with an 11-N force. (b) The normal force is smaller than the weight, because the rope supplies an upward force of 11 N that partially supports the box.

Figure 4.14b illustrates a different situation. Here, the box is being pulled upward by a rope that applies a force of 11 N. The net force acting on the box due to its weight and the rope is only 4 N, downward. To balance this force, the normal force needs to be only 4 N. It is not hard to imagine what would happen if the force applied by the rope were increased to 15 N—exactly equal to the weight of the box. In this situation, the normal force would become zero. In fact, the table could be removed, since the block would be supported entirely by the rope. The situations in **Figure 4.14** are consistent with the idea that the magnitude of the normal force indicates how hard two objects press against each other. Clearly, the box and the table press against each other harder in part a of the picture than in part b.

Like the box and the table in **Figure 4.14**, various parts of the human body press against one another and exert normal forces. Example 8 illustrates the remarkable ability of the human skeleton to withstand a wide range of normal forces.

EXAMPLE 8 | BIO The Physics of the Human Skeleton

In a circus balancing act, a woman performs a headstand on top of a standing performer's head, as **Figure 4.15a** illustrates. The woman weighs 490 N, and the standing performer's head and neck weigh 50 N. It is primarily the seventh cervical vertebra in the spine that supports all the weight above the shoulders. What is the normal force that this vertebra exerts on the neck and head of the standing performer **(a)** before the act and **(b)** during the act?

Reasoning To begin, we draw a free-body diagram for the neck and head of the standing performer. Before the act, there are only two forces, the weight of the standing performer's head and neck, and the normal force. During the act, an additional force is present due to the woman's weight. In both cases, the upward and downward

forces must balance for the head and neck to remain at rest. This condition of balance will lead us to values for the normal force.

Solution **(a)** **Figure 4.15b** shows the free-body diagram for the standing performer's head and neck before the act. The only forces acting are the normal force $\vec{\mathbf{F}}_N$ and the 50-N weight. These two forces must balance for the standing performer's head and neck to remain at rest. Therefore, the seventh cervical vertebra exerts a normal force of $\boxed{F_N = 50 \text{ N}}$.

(b) **Figure 4.15c** shows the free-body diagram that applies during the act. Now, the total downward force exerted on the standing performer's head and neck is 50 N + 490 N = 540 N, which must be balanced by the upward normal force, so that $\boxed{F_N = 540 \text{ N}}$.

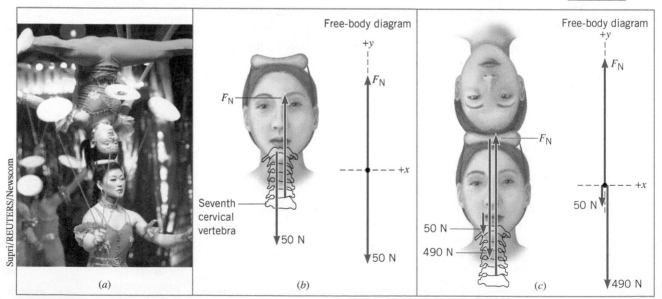

(a)

Free-body diagram

F_N

Seventh cervical vertebra

+y

F_N

+x

50 N

50 N

(b)

Free-body diagram

F_N

50 N

490 N

+y

F_N

+x

50 N

490 N

(c)

FIGURE 4.15 (a) A young woman keeps her balance during a performance by China's Sichuan Acrobatic group. A free-body diagram is shown for the standing performer's body above the shoulders (b) before the act and (c) during the act. For convenience, the scales used for the vectors in parts b and c are different.

In summary, the normal force does not necessarily have the same magnitude as the weight of the object. The value of the normal force depends on what other forces are present. It also depends on whether the objects in contact are accelerating. In one situation that involves accelerating objects, the magnitude of the normal force can be regarded as a kind of "apparent weight," as we will now see.

Apparent Weight

Usually, the weight of an object can be determined with the aid of a scale. However, even if a scale is working properly, there are situations in which it does not give the correct weight. In such situations, the reading on the scale gives only the "apparent" weight, rather than the gravitational force or "true" weight. The apparent weight is the force that the object exerts on the scale with which it is in contact.

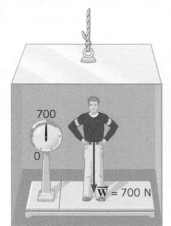 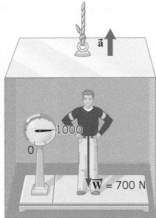 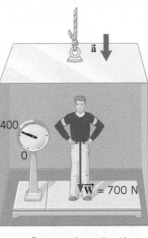 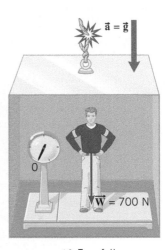

(a) No acceleration ($\vec{\mathbf{v}}$ = constant)　　(b) Upward acceleration　　(c) Downward acceleration　　(d) Free-fall

INTERACTIVE FIGURE 4.16　(a) When the elevator is not accelerating, the scale registers the true weight (W = 700 N) of the person. (b) When the elevator accelerates upward, the apparent weight (1000 N) exceeds the true weight. (c) When the elevator accelerates downward, the apparent weight (400 N) is less than the true weight. (d) The apparent weight is zero if the elevator falls freely—that is, if it falls with the acceleration due to gravity.

To see the discrepancies that can arise between true weight and apparent weight, consider the scale in the elevator in **Interactive Figure 4.16**. The reasons for the discrepancies will be explained shortly. A person whose true weight is 700 N steps on the scale. If the elevator is at rest or moving with a constant velocity (either upward or downward), the scale registers the true weight, as **Interactive Figure 4.16a** illustrates.

If the elevator is accelerating, the apparent weight and the true weight are not equal. When the elevator accelerates upward, the apparent weight is greater than the true weight, as **Interactive Figure 4.16b** shows. Conversely, if the elevator accelerates downward, as in part c, the apparent weight is less than the true weight. In fact, if the elevator falls freely, so its acceleration is equal to the acceleration due to gravity, the apparent weight becomes zero, as part d indicates. In a situation such as this, where the apparent weight is zero, the person is said to be "weightless." The apparent weight, then, does not equal the true weight if the scale and the person on it are accelerating.

The discrepancies between true weight and apparent weight can be understood with the aid of Newton's second law. **Figure 4.17** shows a free-body diagram of the person in the elevator. The two forces that act on him are the true weight $\vec{\mathbf{W}} = m\vec{\mathbf{g}}$ and the normal force $\vec{\mathbf{F}}_N$ exerted by the platform of the scale. Applying Newton's second law in the vertical direction gives

$$\Sigma F_y = +F_N - mg = ma$$

where a is the acceleration of the elevator and person. In this result, the symbol g stands for the magnitude of the acceleration due to gravity and can never be a negative quantity. However, the acceleration a may be either positive or negative, depending on whether

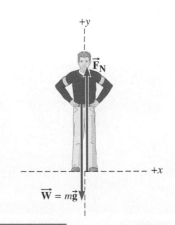

FIGURE 4.17　A free-body diagram showing the forces acting on the person riding in the elevator of **Interactive Figure 4.16**. $\vec{\mathbf{W}}$ is the true weight, and $\vec{\mathbf{F}}_N$ is the normal force exerted on the person by the platform of the scale.

the elevator is accelerating upward (+) or downward (−). Solving for the normal force F_N shows that

$$\underbrace{F_N}_{\substack{\text{Apparent}\\\text{weight}}} = \underbrace{mg}_{\substack{\text{True}\\\text{weight}}} + ma \tag{4.6}$$

In Equation 4.6, F_N is the magnitude of the normal force exerted on the person by the scale. But in accord with Newton's third law, F_N is also the magnitude of the downward force that the person exerts on the scale—namely, the apparent weight.

Equation 4.6 contains all the features shown in **Interactive Figure 4.16**. If the elevator is not accelerating, $a = 0$ m/s^2, and the apparent weight equals the true weight. If the elevator accelerates upward, a is positive, and the equation shows that the apparent weight is greater than the true weight. If the elevator accelerates downward, a is negative, and the apparent weight is less than the true weight. If the elevator falls freely, $a = -g$, and the apparent weight is zero. The apparent weight is zero because when both the person and the scale fall freely, they cannot push against one another. In this text, when the weight is given, it is assumed to be the true weight, unless stated otherwise.

Check Your Understanding

(The answers are given at the end of the book.)

11. A stack of books whose true weight is 165 N is placed on a scale in an elevator. The scale reads 165 N. From this information alone, can you tell whether the elevator is moving with a constant velocity of 2 m/s upward, is moving with a constant velocity of 2 m/s downward, or is at rest?

12. A 10-kg suitcase is placed on a scale that is in an elevator. In which direction is the elevator accelerating when the scale reads 75 N and when it reads 120 N? **(a)** Downward when it reads 75 N and upward when it reads 120 N **(b)** Upward when it reads 75 N and downward when it reads 120 N **(c)** Downward in both cases **(d)** Upward in both cases

13. You are standing on a scale in an elevator that is moving upward with a constant velocity. The scale reads 600 N. The following table shows five options for what the scale reads when the elevator slows down as it comes to a stop, when it is stopped, and when it picks up speed on its way back down. Which one of the five options correctly describes the scale's readings? Note that the symbol < means "less than" and > means "greater than."

Option	Elevator slows down as it comes to a halt	Elevator is stopped	Elevator picks up speed on its way back down
(a)	> 600 N	> 600 N	> 600 N
(b)	< 600 N	600 N	< 600 N
(c)	> 600 N	600 N	< 600 N
(d)	< 600 N	< 600 N	< 600 N
(e)	< 600 N	600 N	> 600 N

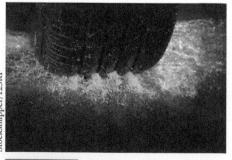

FIGURE 4.18 This photo, shot parallel to the road surface, shows a tire rolling under wet conditions. The channels in the tire collect and divert water away from the regions where the tire contacts the surface, thus providing better traction.

Stocksnapper/123RF

4.9 Static and Kinetic Frictional Forces

When an object is in contact with a surface, there is a force acting on the object. The previous section discusses the component of this force that is perpendicular to the surface, which is called the normal force. When the object moves or attempts to move along the surface, there is also a component of the force that is parallel to the surface. This parallel force component is called the **frictional force**, or simply **friction**.

In many situations considerable engineering effort is expended trying to reduce friction. For example, oil is used to reduce the friction that causes wear and tear in the pistons and cylinder walls of an automobile engine. Sometimes, however, friction is absolutely necessary. Without friction, car tires could not provide the traction needed to move the car. In fact, the raised tread on a tire is designed to maintain friction. On a wet road, the spaces in the tread pattern (see **Figure 4.18**) provide channels for the water to collect and be diverted away. Thus, these channels largely prevent the water from coming between the tire surface and the road surface, where it would reduce friction and allow the tire to skid.

Surfaces that appear to be highly polished can actually look quite rough when examined under a microscope. Such an examination reveals that two surfaces in contact touch only at relatively few spots, as **Figure 4.19** illustrates. The microscopic area of contact for these spots is substantially less than the apparent macroscopic area of contact between the surfaces—perhaps thousands of times less. At these contact points the molecules of the different bodies are close enough together to exert strong attractive intermolecular forces on one another, leading to what are known as "cold welds." Frictional forces are associated with these welded spots, but the exact details of how frictional forces arise are not well understood. However, some empirical relations have been developed that make it possible to account for the effects of friction.

Interactive Figure 4.20 helps to explain the main features of the type of friction known as **static friction**. The block in this drawing is initially at rest on a table, and as long as there is no attempt to move the block, there is no static frictional force. Then, a horizontal force $\vec{F}$ is applied to the block by means of a rope. If $\vec{F}$ is small, as in part *a*, experience tells us that the block still does not move. Why? It does not move because the static frictional force $\vec{f}_s$ exactly cancels the effect of the applied force. The direction of $\vec{f}_s$ is opposite to that of $\vec{F}$ and the magnitude of $\vec{f}_s$ equals the magnitude of the applied force, $f_s = F$. Increasing the applied force in **Interactive Figure 4.20** by a small amount still does not cause the block to move. There is no movement because the static frictional force also increases by an amount that cancels out the increase in the applied force (see part *b* of the drawing). If the applied force continues to increase, however, there comes a point when the block finally "breaks away" and begins to slide. The force just before breakaway represents the *maximum static frictional force* $\vec{f}_s^{\text{MAX}}$ that the table can exert on the block (see part *c* of the drawing). Any applied force that is greater than $\vec{f}_s^{\text{MAX}}$ cannot be balanced by static friction, and the resulting net force accelerates the block to the right.

Experimental evidence shows that, to a good degree of approximation, the maximum static frictional force between a pair of dry, unlubricated surfaces has two main characteristics. It is independent of the apparent macroscopic area of contact between the objects, provided that the surfaces are hard or nondeformable. For instance, in **Figure 4.21** the maximum static frictional force that the surface of the table can exert on a block is the same, whether the block is resting on its largest or its smallest side. The other main characteristic of $\vec{f}_s^{\text{MAX}}$ is that its magnitude is proportional to the magnitude of the normal force $\vec{F}_N$. As Section 4.8 points out, the magnitude of the normal force indicates how hard two surfaces are being pressed together. The harder they are pressed, the larger is f_s^{MAX}, presumably because the number of "cold-welded," microscopic contact points is increased. Equation 4.7 expresses the proportionality between f_s^{MAX} and F_N with the aid of a proportionality constant μ_s, which is called the **coefficient of static friction**.

> **STATIC FRICTIONAL FORCE**
>
> The magnitude f_s of the static frictional force can have any value from zero up to a maximum value of f_s^{MAX}, depending on the applied force. In other words, $f_s \leq f_s^{\text{MAX}}$, where the symbol $\leq$ is read as "less than or equal to." The equality holds only when f_s attains its maximum value, which is
>
> $$f_s^{\text{MAX}} = \mu_s F_N \qquad (4.7)$$
>
> In Equation 4.7, μ_s is the coefficient of static friction, and F_N is the magnitude of the normal force.

It should be emphasized that Equation 4.7 relates only the *magnitudes* of $\vec{f}_s^{\text{MAX}}$ and $\vec{F}_N$, *not the vectors themselves*. This equation does not imply that the directions of the vectors are the same. In fact, $\vec{f}_s^{\text{MAX}}$ is parallel to the surface, while $\vec{F}_N$ is perpendicular to it.

The coefficient of static friction, being the ratio of the magnitudes of two forces $(\mu_s = f_s^{\text{MAX}}/F_N)$, has no units. Also, it depends on the type of material from which each surface is made (steel on wood, rubber on concrete, etc.), the condition of the

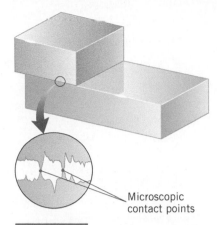

FIGURE 4.19 Even when two highly polished surfaces are in contact, they touch only at relatively few points.

No movement
(a)

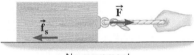

No movement
(b)

When movement just begins
(c)

INTERACTIVE FIGURE 4.20 (*a*) and (*b*) Applying a small force $\vec{F}$ to the block produces no movement, because the static frictional force $\vec{f}_s$ exactly balances the applied force. (*c*) The block just begins to move when the applied force is slightly greater than the maximum static frictional force $\vec{f}_s^{\text{MAX}}$.

FIGURE 4.21 The maximum static frictional force $\vec{f}_s^{\text{MAX}}$ would be the same, no matter which side of the block is in contact with the table.

TABLE 4.2 Approximate Values of the Coefficients of Friction for Various Surfaces[a]

Materials	Coefficient of Static Friction, μ_s	Coefficient of Kinetic Friction, μ_k
Glass on glass (dry)	0.94	0.4
Ice on ice (clean, 0 °C)	0.1	0.02
Rubber on dry concrete	1.0	0.8
Rubber on wet concrete	0.7	0.5
Steel on ice	0.1	0.05
Steel on steel (dry hard steel)	0.78	0.42
Teflon on Teflon	0.04	0.04
Wood on wood	0.35	0.3

[a]The last column gives the coefficients of kinetic friction, a concept that will be discussed shortly.

surfaces (polished, rough, etc.), and other variables such as temperature. Table 4.2 gives some typical values of μ_s for various surfaces. Example 9 illustrates the use of Equation 4.7 for determining the maximum static frictional force.

Analyzing Multiple-Concept Problems

EXAMPLE 9 | The Force Needed to Start a Skier Moving

A skier is standing motionless on a horizontal patch of snow. She is holding onto a horizontal tow rope, which is about to pull her forward (see **Figure 4.22a**). The skier's mass is 59 kg, and the coefficient of static friction between the skis and snow is 0.14. What is the magnitude of the maximum force that the tow rope can apply to the skier without causing her to move?

Reasoning When the rope applies a relatively small force, the skier does not accelerate. The reason is that the static frictional force opposes the applied force and the two forces have the same magnitude. We can apply Newton's second law in the horizontal direction to this situation. In order for the rope to pull the skier forward, it must exert a force large enough to overcome the *maximum* static frictional force acting on the skis. The magnitude of the maximum static frictional force depends on the coefficient of static friction (which is known) and on the magnitude of the normal force. We can determine the magnitude of the normal force by using Newton's second law, along with the fact that the skier does not accelerate in the vertical direction.

Knowns and Unknowns The data for this problem are as follows:

Description	Symbol	Value
Mass of skier	m	59 kg
Coefficient of static friction	μ_s	0.14
Unknown Variable		
Magnitude of maximum horizontal force that tow rope can apply	F	?

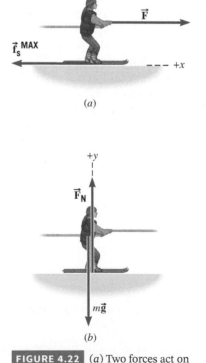

(a)

(b)

FIGURE 4.22 (a) Two forces act on the skier in the horizontal direction just before she begins to move. (b) Two vertical forces act on the skier.

Modeling the Problem

STEP 1 Newton's Second Law (Horizontal Direction)   **Figure 4.22a** shows the two horizontal forces that act on the skier just before she begins to move: the force $\vec{F}$ applied by the tow rope and the maximum static frictional force $\vec{f}_s^{MAX}$. Since the skier is standing motionless, she is not accelerating in the horizontal or x direction, so $a_x = 0$ m/s^2. Applying Newton's second law (Equation 4.2a) to this situation, we have

$$\Sigma F_x = ma_x = 0$$

Since the net force ΣF_x in the x direction is $\Sigma F_x = +F - f_s^{MAX}$, Newton's second law can be written as $+F - f_s^{MAX} = 0$. Thus,

$$F = f_s^{MAX}$$

We do not know f_s^{MAX}, but its value will be determined in Steps 2 and 3.

$$F = f_s^{MAX} \qquad (1)$$
$$\uparrow$$
$$\boxed{?}$$

STEP 2 The Maximum Static Frictional Force   The magnitude f_s^{MAX} of the maximum static frictional force is related to the coefficient of static friction μ_s and the magnitude F_N of the normal force by Equation 4.7:

$$\boxed{f_s^{MAX} = \mu_s F_N} \qquad (4.7)$$

We now substitute this result into Equation 1, as indicated in the right column. The coefficient of static friction is known, but F_N is not. An expression for F_N will be obtained in the next step.

$$F = f_s^{MAX} \qquad (1)$$
$$\uparrow$$
$$\boxed{f_s^{MAX} = \mu_s F_N} \qquad (4.7)$$
$$\uparrow$$
$$\boxed{?}$$

STEP 3 Newton's Second Law (Vertical Direction)   We can find the magnitude F_N of the normal force by noting that the skier does not accelerate in the vertical or y direction, so $a_y = 0$ m/s^2. **Figure 4.22b** shows the two vertical forces that act on the skier: the normal force $\vec{F}_N$ and her weight $m\vec{g}$. Applying Newton's second law (Equation 4.2b) in the vertical direction gives

$$\Sigma F_y = ma_y = 0$$

The net force in the y direction is $\Sigma F_y = +F_N - mg$, so Newton's second law becomes $+F_N - mg = 0$. Thus,

$$\boxed{F_N = mg}$$

We now substitute this result into Equation 4.7, as shown at the right.

$$F = f_s^{MAX} \qquad (1)$$
$$\uparrow$$
$$\boxed{f_s^{MAX} = \mu_s F_N} \qquad (4.7)$$
$$\uparrow$$
$$\boxed{F_N = mg}$$

Solution   Algebraically combining the results of the three steps, we have

$$\overset{\textbf{STEP 1}}{\downarrow} \qquad \overset{\textbf{STEP 2}}{\downarrow} \qquad \overset{\textbf{STEP 3}}{\downarrow}$$
$$F \;=\; f_s^{MAX} \;=\; \mu_s F_N \;=\; \mu_s mg$$

The magnitude F of the maximum force is

$$F = \mu_s mg = (0.14)(59 \text{ kg})(9.80 \text{ m/s}^2) = \boxed{81 \text{ N}}$$

If the force exerted by the tow rope exceeds this value, the skier will begin to accelerate forward.

Related Homework: Problem 39

BIO   **THE PHYSICS OF . . . rock climbing.**   Static friction is often essential, as it is to the rock climber in **Figure 4.23**, for instance. She presses outward against the walls of the rock formation with her hands and feet to create sufficiently large normal forces, so that the static frictional forces help support her weight.

Once two surfaces begin sliding over one another, the static frictional force is no longer of any concern. Instead, a type of friction known as **kinetic* friction** comes

---

*The word "kinetic" is derived from the Greek word *kinetikos*, meaning "pertaining to motion."

FIGURE 4.23 In maneuvering her way up Devil's Tower at Devil's Tower National Monument in Wyoming, this rock climber uses the static frictional forces between her hands and feet and the vertical rock walls to help support her weight.

into play. The kinetic frictional force opposes the relative sliding motion. If you have ever pushed an object across a floor, you may have noticed that it takes less force to keep the object sliding than it takes to get it going in the first place. In other words, the kinetic frictional force is usually less than the static frictional force.

Experimental evidence indicates that the kinetic frictional force $\vec{f}_k$ has three main characteristics, to a good degree of approximation. It is independent of the apparent area of contact between the surfaces (see **Figure 4.21**). It is independent of the speed of the sliding motion, if the speed is small. And last, the magnitude of the kinetic frictional force is proportional to the magnitude of the normal force. Equation 4.8 expresses this proportionality with the aid of a proportionality constant μ_k, which is called the **coefficient of kinetic friction**.

KINETIC FRICTIONAL FORCE

The magnitude f_k of the kinetic frictional force is given by

$$f_k = \mu_k F_N \tag{4.8}$$

In Equation 4.8, μ_k is the coefficient of kinetic friction, and F_N is the magnitude of the normal force.

Equation 4.8, like Equation 4.7, is a relationship between only the magnitudes of the frictional and normal forces. The directions of these forces are perpendicular to each other. Moreover, like the coefficient of static friction, the coefficient of kinetic friction is a number without units and depends on the type and condition of the two surfaces that are in contact. As indicated in **Table 4.2**, values for μ_k are typically less than those for μ_s, reflecting the fact that kinetic friction is generally less than static friction. The next example illustrates the effect of kinetic friction.

Analyzing Multiple-Concept Problems

EXAMPLE 10 | Sled Riding

A sled and its rider are moving at a speed of 4.0 m/s along a horizontal stretch of snow, as **Figure 4.24a** illustrates. The snow exerts a kinetic frictional force on the runners of the sled, so the sled slows down and eventually comes to a stop. The coefficient of kinetic friction is 0.050. What is the displacement x of the sled?

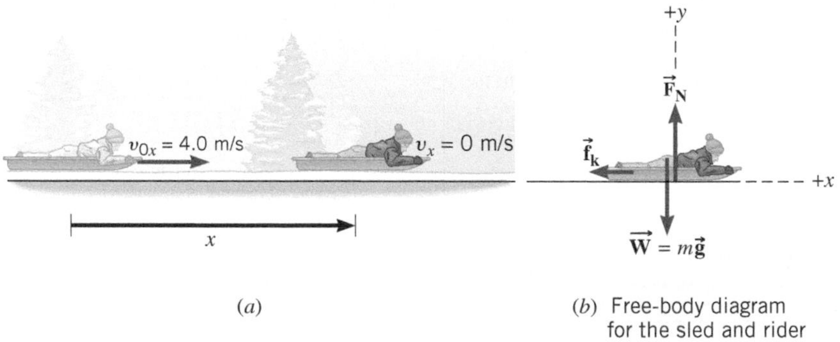

FIGURE 4.24 (a) The moving sled decelerates because of the kinetic frictional force. (b) Three forces act on the moving sled: the weight $\vec{W}$ of the sled and its rider, the normal force $\vec{F}_N$, and the kinetic frictional force $\vec{f}_k$. The free-body diagram for the sled and rider shows these forces.

Reasoning As the sled slows down, its velocity is decreasing. As our discussions in Chapters 2 and 3 indicate, the changing velocity is described by an acceleration (which in this case is a deceleration since the sled is slowing down). Assuming that the acceleration is constant, we can use one of the equations of kinematics from Chapter 3 to relate the displacement to the initial and final velocities and to the acceleration. The acceleration of the sled is not given directly. However, we can determine it by using Newton's second law of motion, which relates the acceleration to the net force (which is the kinetic frictional force in this case) acting on the sled and to the mass of the sled (plus rider).

Knowns and Unknowns The data for this problem are listed in the table:

Description	Symbol	Value	Comment
Explicit Data			
Initial velocity	v_{0x}	+4.0 m/s	Positive, because the velocity points in the +x direction. See the drawing.
Coefficient of kinetic friction	μ_k	0.050	
Implicit Data			
Final velocity	v_x	0 m/s	The sled comes to a stop.
Unknown Variable			
Displacement	x	?	

Modeling the Problem

STEP 1 Displacement To obtain the displacement x of the sled we will use Equation 3.6a from the equations of kinematics:

$$v_x^2 = v_{0x}^2 + 2a_x x$$

Solving for the displacement x gives the result shown at the right. This equation is useful because two of the variables, v_{0x} and v_x, are known and the acceleration a_x can be found by applying Newton's second law to the accelerating sled (see Step 2).

$$x = \frac{v_x^2 - v_{0x}^2}{2a_x} \qquad (1)$$

STEP 2 Newton's Second Law Newton's second law, as given in Equation 4.2a, states that the acceleration a_x is equal to the net force ΣF_x divided by the mass m:

$$a_x = \frac{\Sigma F_x}{m}$$

The free-body diagram in **Figure 4.24b** shows that the only force acting on the sled in the horizontal or x direction is the kinetic frictional force $\vec{f}_k$. We can write this force as $-f_k$, where f_k is the magnitude of the force and the minus sign indicates that it points in the $-x$ direction. Since the net force is $\Sigma F_x = -f_k$, Equation 4.2a becomes

$$a_x = \frac{-f_k}{m}$$

This result can now be substituted into Equation 1, as shown at the right.

$$x = \frac{v_x^2 - v_{0x}^2}{2a_x} \qquad (1)$$

$$a_x = \frac{-f_k}{m} \qquad (2)$$

STEP 3 Kinetic Frictional Force We do not know the magnitude f_k of the kinetic frictional force, but we do know the coefficient of kinetic friction μ_k. According to Equation 4.8, the two are related by

$$f_k = \mu_k F_N \qquad (4.8)$$

where F_N is the magnitude of the normal force. This relation can be substituted into Equation 2, as shown at the right. An expression for F_N will be obtained in the next step.

$$x = \frac{v_x^2 - v_{0x}^2}{2a_x} \qquad (1)$$

$$a_x = \frac{-f_k}{m} \qquad (2)$$

$$f_k = \mu_k F_N \qquad (4.8)$$

STEP 4 Normal Force The magnitude F_N of the normal force can be found by noting that the sled does not accelerate in the vertical or y direction ($a_y = 0$ m/s^2). Thus, Newton's second law, as given in Equation 4.2b, becomes

$$\Sigma F_y = ma_y = 0$$

There are two forces acting on the sled in the y direction: the normal force $\vec{F}_N$ and its weight $\vec{W}$ (see **Figure 4.24b**). Therefore, the net force in the y direction is

$$\Sigma F_y = +F_N - W$$

where $W = mg$ (Equation 4.5). Thus, Newton's second law becomes

$$+F_N - mg = 0 \quad \text{or} \quad F_N = mg$$

This result for F_N can be substituted into Equation 4.8, as shown at the right.

$$x = \frac{v_x^2 - v_{0x}^2}{2a_x} \qquad (1)$$

$$a_x = \frac{-f_k}{m} \qquad (2)$$

$$f_k = \mu_k F_N \qquad (4.8)$$

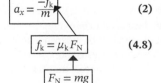

$$F_N = mg$$

Solution Algebraically combining the results of each step, we find that

STEP 1   STEP 2   STEP 3   STEP 4

$$x = \frac{v_x^2 - v_{0x}^2}{2a_x} = \frac{v_x^2 - v_{0x}^2}{2\left(\frac{-f_k}{m}\right)} = \frac{v_x^2 - v_{0x}^2}{2\left(\frac{-\mu_k F_N}{m}\right)} = \frac{v_x^2 - v_{0x}^2}{2\left(\frac{-\mu_k mg}{m}\right)} = \frac{v_x^2 - v_{0x}^2}{2(-\mu_k g)}$$

Note that the mass m of the sled and rider is algebraically eliminated from the final result. Thus, the displacement of the sled is

$$x = \frac{v_x^2 - v_{0x}^2}{2(-\mu_k g)} = \frac{(0 \text{ m/s})^2 - (+4.0 \text{ m/s})^2}{2[-(0.050)(9.80 \text{ m/s}^2)]} = \boxed{+16 \text{ m}}$$

Related Homework: Problems 42, 87

Static friction opposes the impending relative motion between two objects, while kinetic friction opposes the relative sliding motion that actually does occur. In either case, *relative motion* is opposed. However, this opposition to relative motion does not mean that friction prevents or works against the motion of *all* objects. For instance, consider what happens when you walk.

BIO **THE PHYSICS OF . . . walking.** Your foot exerts a force on the earth, and the earth exerts a reaction force on your foot. This reaction force is a static frictional force, and it opposes the impending backward motion of your foot, propelling you forward in the process. Kinetic friction can also cause an object to move, all the while opposing relative motion, as it does in Example 10. In this example the kinetic frictional force acts on the sled and opposes the relative motion of the sled and the earth. Newton's third law indicates, however, that since the earth exerts the kinetic frictional force on the sled, the sled must exert a reaction force on the earth. In response, the earth accelerates, but because of the earth's huge mass, the motion is too slight to be noticed.

Example 11 discusses the low-friction conditions on the inside of an important structural component of the human body.

EXAMPLE 11 | **BIO** The Amazing Synovial Joint

A synovial joint is a joint between bones that rub against each other, such as in elbows, hips, and knees. In a healthy joint, the inner contact surfaces are covered with cartilage and immersed in a lubricant called synovial fluid, which provides an extremely low coefficient of kinetic friction estimated to be $\mu_k \approx 2.00 \times 10^{-3}$, a value that has been difficult to match in the development of artificial joints. To get some intuition as to how "slippery" this is, consider a hockey puck coated with a material such that the coefficient of kinetic friction between it and the surface on which it slides is equal to 2.00×10^{-3}. Neglecting air resistance and assuming the surface on which the puck slides is horizontal and level, **(a)** how far will the puck travel and **(b)** how long will it take before it comes to rest if its initial speed is 15.0 mi/h (6.71 m/s)?

REASONING Knowing the coefficient of kinetic friction, we can determine the force of kinetic friction and then the acceleration (which will be negative). Since the mass of the puck was not given, we should expect it to be eliminated from the problem, and our solution should therefore be independent of the puck's mass. Once we know the (constant) acceleration, we can apply the kinematic equations to get both the distance it travels and the time it takes to come to rest.

SOLUTION **(a)** Here the force of kinetic friction acts opposite to the direction of motion and has a magnitude given by Equation 4.8:

$$f_k = \mu_k F_N$$

Since, in this situation, the weight and the normal force are equal and opposite in magnitude, $F_N = mg$, and the frictional force becomes

$$f_k = \mu_k mg$$

From Newton's second law (Equation 4.2a), the net force is equal to the mass times the acceleration, and we have

$$\Sigma F_x = \mu_k mg = ma_x$$

And therefore, since the masses cancel, the magnitude of the acceleration is

$$a_x = \mu_k g = (2.00 \times 10^{-3})(9.80 \text{ m/s}^2) = 0.0196 \text{ m/s}^2$$

This acceleration will be opposite to the direction of motion, which we will define to be in the $+x$ direction, and thus, the acceleration will be negative, that is, $a_x = -0.0196 \text{ m/s}^2$.

Now that the acceleration is known, as well as the initial velocity and final velocity (0 m/s), we can apply the kinematic equations to find the distance traveled. In this case, we use Equation 3.6a:

$$v_x^2 = v_{0x}^2 + 2a_x x$$

Setting the final velocity equal to zero and solving for x gives

$$x = -\frac{v_{0x}^2}{2a_x} = -\frac{(6.71 \text{ m/s})^2}{2(-0.0196 \text{ m/s}^2)} = \boxed{1150 \text{ m}}$$

(b) The time it takes for the puck to come to rest can be found from Equation 3.3a:

$$v_x = v_{0x} + a_x t$$

Setting the final velocity equal to zero and solving for the time gives

$$t = \frac{-v_{0x}}{a_x} = \frac{-6.71 \text{ m/s}}{-0.0196 \text{ m/s}^2} = \boxed{342 \text{ s}}$$

So the puck will travel a distance of over a kilometer and will remain in motion for 5 minutes and 42 seconds!

Although low-friction materials have been developed that perform better than the synovial joint, materials that go into the human body must satisfy strict criteria in regard to toxicity and compatibility with the human immune system. In addition, the durability of artificial joints must be exceptional, as they are subjected to large forces and repetitive motion.

Check Your Understanding

(The answers are given at the end of the book.)

14. Suppose that the coefficients of static and kinetic friction have values such that $\mu_s = 1.4\mu_k$ for a crate in contact with a cement floor. Which one of the following statements is true? **(a)** The magnitude of the static frictional force is always 1.4 times the magnitude of the kinetic frictional force. **(b)** The magnitude of the kinetic frictional force is always 1.4 times the magnitude of the static frictional force. **(c)** The magnitude of the maximum static frictional force is 1.4 times the magnitude of the kinetic frictional force.

15. A person has a choice of either pushing or pulling a sled at a constant velocity, as **CYU Figure 4.2** illustrates. Friction is present. If the angle θ is the same in both cases, does it require less force to push or to pull the sled?

CYU FIGURE 4.2

16. A box has a weight of 150 N and is being pulled across a horizontal floor by a force that has a magnitude of 110 N. The pulling force can point horizontally, or it can point above the horizontal at an angle θ. When the pulling force points horizontally, the kinetic frictional force acting on the box is twice as large as when the pulling force points at the angle θ. Find θ.

17. A box rests on the floor of an elevator. Because of static friction, a force is required to start the box sliding across the floor when the elevator is **(a)** stationary, **(b)** accelerating upward, and **(c)** accelerating downward. Rank the forces required in these three situations in ascending order—that is, smallest first.

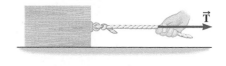

(a)

(b)

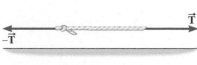

(c)

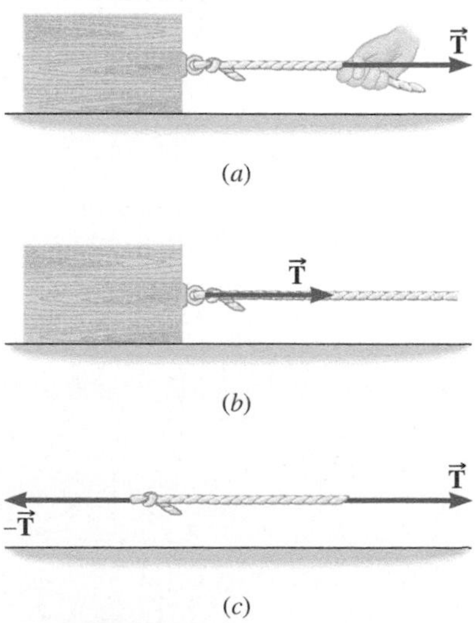

FIGURE 4.25 (a) A force $\vec{T}$ is being applied to the right end of a rope. (b) The force is transmitted to the box. (c) Forces are applied to both ends of the rope. These forces have equal magnitudes and opposite directions.

4.10 The Tension Force

Forces are often applied by means of cables or ropes that are used to pull an object. For instance, **Figure 4.25a** shows a force $\vec{T}$ being applied to the right end of a rope attached to a box. Each particle in the rope in turn applies a force to its neighbor. As a result, the force is applied to the box, as part *b* of the drawing shows.

In situations such as that in **Figure 4.25**, we say that the force $\vec{T}$ is applied to the box because of the tension in the rope, meaning that the tension and the force applied to the box have the same magnitude. However, the word "tension" is commonly used to mean the tendency of the rope to be pulled apart. To see the relationship between these two uses of the word "tension," consider the left end of the rope, which applies the force $\vec{T}$ to the box. In accordance with Newton's third law, the box applies a reaction force to the rope. The reaction force has the same magnitude as $\vec{T}$ but is oppositely directed. In other words, a force $-\vec{T}$ acts on the left end of the rope. Thus, forces of equal magnitude act on opposite ends of the rope, as in **Figure 4.25c**, and tend to pull it apart.

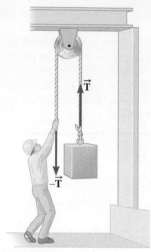

FIGURE 4.26 The force $\vec{T}$ applied at one end of a massless rope is transmitted undiminished to the other end, even when the rope bends around a pulley, provided the pulley is also massless and friction is absent.

In the previous discussion, we have used the concept of a "massless" rope ($m = 0$ kg) without saying so. In reality, a massless rope does not exist, but it is useful as an idealization when applying Newton's second law. According to the second law, a net force is required to accelerate an object that has mass. In contrast, no net force is needed to accelerate a massless rope, since $\Sigma\vec{F} = m\vec{a}$ and $m = 0$ kg. Thus, when a force $\vec{T}$ is applied to one end of a massless rope, none of the force is needed to accelerate the rope. As a result, the force $\vec{T}$ is also applied undiminished to the object attached at the other end, as we assumed in **Figure 4.25**.* If the rope had mass, however, some of the force $\vec{T}$ would have to be used to accelerate the rope. The force applied to the box would then be less than $\vec{T}$, and the tension would be different at different locations along the rope. In this text we will assume that a rope connecting one object to another is massless, unless stated otherwise. The ability of a massless rope to transmit tension undiminished from one end to the other is not affected when the rope passes around objects such as the pulley in **Figure 4.26** (provided the pulley itself is massless and frictionless).

Check Your Understanding

(*The answer is given at the end of the book.*)

18. A rope is used in a tug-of-war between two teams of five people each. Both teams are equally strong, so neither team wins. An identical rope is tied to a tree, and the same ten people pull just as hard on the loose end as they did in the contest. In both cases, the people pull steadily with no jerking. Which rope sustains the greater tension, **(a)** the rope tied to the tree or **(b)** the rope in the tug-of-war, or **(c)** do the ropes sustain the same tension?

4.11 Equilibrium Applications of Newton's Laws of Motion

Have you ever been so upset that it took days to recover your "equilibrium"? In this context, the word "equilibrium" refers to a balanced state of mind, one that is not changing wildly. In physics, the word "equilibrium" also refers to a lack of change, but in the sense that the velocity of an object isn't changing. If its velocity doesn't change, an object is not accelerating. Our definition of equilibrium, then, is as follows:

> **DEFINITION OF EQUILIBRIUM†**
>
> **An object is in equilibrium when it has zero acceleration.**

Newton's laws of motion apply whether or not an object is in equilibrium. For an object in equilibrium the acceleration is zero ($\vec{a} = 0$ m/s²) in Newton's second law, and the present section presents several examples of this type. In the nonequilibrium case the acceleration of the object is not zero ($\vec{a} \neq 0$ m/s²) in the second law, and Section 4.12 deals with these kinds of situations.

Since the acceleration is zero for an object in equilibrium, all of the acceleration components are also zero. In two dimensions, this means that $a_x = 0$ m/s² and $a_y = 0$ m/s². Substituting these values into the second law ($\Sigma F_x = ma_x$ and $\Sigma F_y = ma_y$) shows that the x component and the y component of the net force must each be zero. In other words,

---

*If a rope is not accelerating, $\vec{a}$ is zero in the second law, and $\Sigma\vec{F} = m\vec{a} = 0$, regardless of the mass of the rope. Then, the rope can be ignored, no matter what mass it has.

†In this discussion of equilibrium we ignore rotational motion, which is discussed in Chapters 8 and 9. In Section 9.2 a more complete treatment of the equilibrium of a rigid object is presented and takes into account the concept of torque and the fact that objects can rotate.

the forces acting on an object in equilibrium must balance. Thus, in two dimensions, the equilibrium condition is expressed by two equations:

$$\Sigma F_x = 0 \qquad\qquad (4.9a)$$
$$\Sigma F_y = 0 \qquad\qquad (4.9b)$$

In using Equations 4.9a and 4.9b to solve equilibrium problems, we will use the following five-step reasoning strategy:

REASONING STRATEGY **Analyzing Equilibrium Situations**

1. Select the object (often called the "system") to which Equations 4.9a and 4.9b are to be applied. It may be that two or more objects are connected by means of a rope or a cable. If so, it may be necessary to treat each object separately according to the following steps.

2. Draw a free-body diagram for each object chosen above. Be sure to include only forces that act on the object. *Do not include forces that the object exerts on its environment.*

3. Choose a set of x, y axes for each object and resolve all forces in the free-body diagram into components that point along these axes. Select the axes so that as many forces as possible point along one or the other of the two axes. Such a choice minimizes the calculations needed to determine the force components.

4. Apply Equations 4.9a and 4.9b by setting the sum of the x components and the sum of the y components of the forces each equal to zero.

5. Solve the two equations obtained in Step 4 for the desired unknown quantities, remembering that two equations can yield answers for only two unknowns at most.

Example 12 illustrates how these steps are followed. It deals with a traction device in which three forces act together to bring about the equilibrium.

EXAMPLE 12 | BIO The Physics of Traction for the Foot

Figure 4.27a shows a traction device used with a foot injury. The weight of the 2.2-kg object creates a tension in the rope that passes around the pulleys. Therefore, tension forces $\vec{T}_1$ and $\vec{T}_2$ are applied to the pulley on the foot. (It may seem surprising that the rope applies a force to either side of the foot pulley. A similar effect occurs when you place a finger inside a rubber band and push downward. You can feel each side of the rubber band pulling upward on the finger.) The foot pulley is kept in equilibrium because the foot also applies a force $\vec{F}$ to it. This force arises in reaction (Newton's third law) to the pulling effect of the forces $\vec{T}_1$ and $\vec{T}_2$. Ignoring the weight of the foot, find the magnitude of $\vec{F}$.

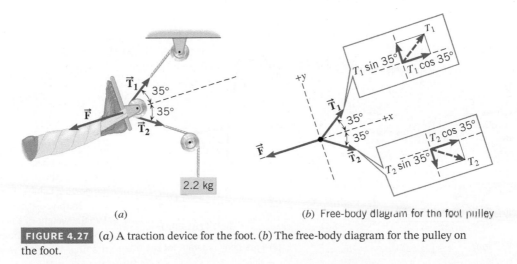

(a)

(b) Free-body diagram for the foot pulley

FIGURE 4.27 (a) A traction device for the foot. (b) The free-body diagram for the pulley on the foot.

Reasoning The forces $\vec{T}_1$, $\vec{T}_2$, and $\vec{F}$ keep the pulley on the foot at rest. The pulley, therefore, has no acceleration and is in equilibrium. As a result, the sum of the x components and the sum of the y components of the three forces must each be zero. **Figure 4.27b** shows the free-body diagram of the pulley on the foot. The x axis is chosen to be along the direction of force $\vec{F}$, and the components of the forces $\vec{T}_1$ and $\vec{T}_2$ are indicated in the drawing. (See Section 1.7 for a review of vector components.)

> **Problem-Solving Insight** Choose the orientation of the x, y axes for convenience. In Example 12, the axes have been rotated so the force $\vec{F}$ points along the x axis. Since $\vec{F}$ does not have a component along the y axis, the analysis is simplified.

Solution Since the sum of the y components of the forces is zero, it follows that

$$\Sigma F_y = +T_1 \sin 35° - T_2 \sin 35° = 0 \qquad \text{(4.9b)}$$

or $T_1 = T_2$. In other words, the magnitudes of the tension forces are equal. In addition, the sum of the x components of the forces is zero, so we have that

$$\Sigma F_x = +T_1 \cos 35° + T_2 \cos 35° - F = 0 \qquad \text{(4.9a)}$$

Solving for F and letting $T_1 = T_2 = T$, we find that $F = 2T \cos 35°$. However, the tension T in the rope is determined by the weight of the 2.2-kg object: $T = mg$, where m is its mass and g is the acceleration due to gravity. Therefore, the magnitude of $\vec{F}$ is

$$F = 2T \cos 35° = 2mg \cos 35° = 2(2.2 \text{ kg})(9.80 \text{ m/s}^2) \cos 35° = \boxed{35 \text{ N}}$$

Example 13 presents another situation in which three forces are responsible for the equilibrium of an object. However, in this example all the forces have different magnitudes.

EXAMPLE 13 | Replacing an Engine

An automobile engine has a weight $\vec{W}$, whose magnitude is $W = 3150$ N. This engine is being positioned above an engine compartment, as **Figure 4.28a** illustrates. To position the engine, a worker is using a rope. Find the tension $\vec{T}_1$ in the supporting cable and the tension $\vec{T}_2$ in the positioning rope.

Reasoning Under the influence of the forces $\vec{W}$, $\vec{T}_1$, and $\vec{T}_2$ the ring in **Figure 4.28a** is at rest and, therefore, in equilibrium. Consequently, the sum of the x components and the sum of the y components of these forces must each be zero; $\Sigma F_x = 0$ and $\Sigma F_y = 0$. By using these relations, we can find T_1 and T_2. **Figure 4.28b** shows the free-body diagram of the ring and the force components for a suitable x, y axis system.

> **Problem-Solving Insight** When an object is in equilibrium, as here in Example 13, the net force is zero, $\Sigma \vec{F} = 0$. This does not mean that each individual force is zero. It means that the vector sum of all the forces is zero.

Solution The free-body diagram shows the components for each of the three forces, and the components are listed in the following table:

Force	x Component	y Component
$\vec{T}_1$	$-T_1 \sin 10.0°$	$+T_1 \cos 10.0°$
$\vec{T}_2$	$+T_2 \sin 80.0°$	$-T_2 \cos 80.0°$
$\vec{W}$	0	$-W$

The plus signs in the table denote components that point along the positive axes, and the minus signs denote components that point along the negative axes. Setting the sum of the x components and the sum of the y components equal to zero leads to the following equations:

$$\Sigma F_x = -T_1 \sin 10.0° + T_2 \sin 80.0° = 0 \qquad \text{(4.9a)}$$

$$\Sigma F_y = +T_1 \cos 10.0° - T_2 \cos 80.0° - W = 0 \qquad \text{(4.9b)}$$

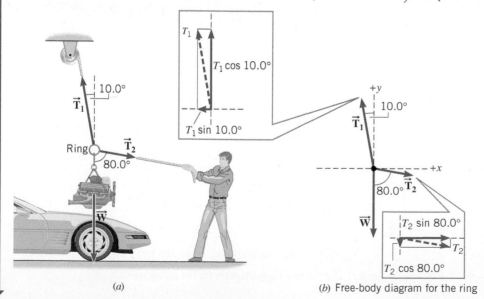

(a)

(b) Free-body diagram for the ring

FIGURE 4.28 (a) The ring is in equilibrium because of the three forces $\vec{T}_1$ (the tension force in the supporting cable), $\vec{T}_2$ (the tension force in the positioning rope), and $\vec{W}$ (the weight of the engine). (b) The free-body diagram for the ring.

Solving the first of these equations for T_1 shows that

$$T_1 = \left(\frac{\sin 80.0°}{\sin 10.0°}\right)T_2$$

Substituting this expression for T_1 into the second equation gives

$$\left(\frac{\sin 80.0°}{\sin 10.0°}\right)T_2 \cos 10.0° - T_2 \cos 80.0° - W = 0$$

$$T_2 = \frac{W}{\left(\dfrac{\sin 80.0°}{\sin 10.0°}\right)\cos 10.0° - \cos 80.0°}$$

Setting $W = 3150$ N in this result yields $\boxed{T_2 = 582 \text{ N}}$.

Since $T_1 = \left(\dfrac{\sin 80.0°}{\sin 10.0°}\right)T_2$ and $T_2 = 582$ N, it follows that

$$\boxed{T_1 = 3.30 \times 10^3 \text{ N}}.$$

An object can be moving and still be in equilibrium, provided there is no acceleration. Example 14 illustrates such a case, and the solution is again obtained using the five-step reasoning strategy summarized at the beginning of the section.

EXAMPLE 14 | Equilibrium at Constant Velocity

A jet plane is flying with a constant speed along a straight line, at an angle of 30.0° above the horizontal, as **Figure 4.29a** indicates. The plane has a weight $\vec{W}$ whose magnitude is $W = 86\,500$ N, and its engines provide a forward thrust $\vec{T}$ of magnitude $T = 103\,000$ N. In addition, the lift force $\vec{L}$ (directed perpendicular to the wings) and the force $\vec{R}$ of air resistance (directed opposite to the motion) act on the plane. Find $\vec{L}$ and $\vec{R}$.

Reasoning   **Figure 4.29b** shows the free-body diagram of the plane, including the forces $\vec{W}, \vec{L}, \vec{T}$, and $\vec{R}$. Since the plane is flying with a constant speed along a straight line, it is not accelerating, it is in equilibrium, and the sum of the x components and the sum of the y components of these forces must be zero. The weight $\vec{W}$ and the thrust $\vec{T}$ are known, so the lift force $\vec{L}$ and the force $\vec{R}$ of air resistance can be obtained from these equilibrium conditions. To calculate the components, we have chosen axes in the free-body diagram that are rotated by 30.0° from their usual horizontal–vertical positions. This has been done purely for convenience, since the weight $\vec{W}$ is then the only force that does not lie along either axis.

Problem-Solving Insight   A moving object is in equilibrium if it moves with a constant velocity; then its acceleration is zero. A zero acceleration is the fundamental characteristic of an object in equilibrium.

Solution   When determining the components of the weight, it is necessary to realize that the angle β in **Figure 4.29a** is 30.0°. Part c of the drawing focuses attention on the geometry that is responsible for this fact. There it can be seen that $\alpha + \beta = 90°$ and $\alpha + 30.0° = 90°$, with the result that $\beta = 30.0°$. The following table lists the components of the forces acting on the jet.

Force	x Component	y Component
$\vec{W}$	$-W \sin 30.0°$	$-W \cos 30.0°$
$\vec{L}$	0	$+L$
$\vec{T}$	$+T$	0
$\vec{R}$	$-R$	0

Setting the sum of the x component of the forces to zero gives

$$\Sigma F_x = -W \sin 30.0° + T - R = 0 \qquad \textbf{(4.9a)}$$

$$R = T - W \sin 30.0° = 103\,000 \text{ N} - (86\,500 \text{ N}) \sin 30.0° = \boxed{59\,800 \text{ N}}$$

Setting the sum of the y component of the forces to zero gives

$$\Sigma F_y = -W \cos 30.0° + L = 0 \qquad \textbf{(4.9b)}$$

$$L = W \cos 30.0° = (86\,500 \text{ N}) \cos 30.0° = \boxed{74\,900 \text{ N}}$$

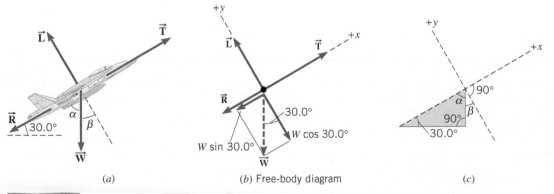

(a)   (b) Free-body diagram   (c)

FIGURE 4.29   (a) A plane moves with a constant velocity at an angle of 30.0° above the horizontal due to the action of four forces, the weight $\vec{W}$, the lift $\vec{L}$, the engine thrust $\vec{T}$, and the air resistance $\vec{R}$. (b) The free-body diagram for the plane. (c) This geometry occurs often in physics.

Check Your Understanding

(The answers are given at the end of the book.)

19. In which one of the following situations could an object possibly be in equilibrium? **(a)** Three forces act on the object; the forces all point along the same line but may have different directions. **(b)** Two perpendicular forces act on the object. **(c)** A single force acts on the object. **(d)** In none of the situations described in (a), (b), and (c) could the object possibly be in equilibrium.

20. A stone is thrown from the top of a cliff. Air resistance is negligible. As the stone falls, is it **(a)** in equilibrium or **(b)** not in equilibrium?

21. During the final stages of descent, a sky diver with an open parachute approaches the ground with a constant velocity. There is no wind to blow him from side to side. Which one of the following statements is true? **(a)** The sky diver is not in equilibrium. **(b)** The force of gravity is the only force acting on the sky diver, so that he is in equilibrium. **(c)** The sky diver is in equilibrium because no forces are acting on him. **(d)** The sky diver is in equilibrium because two forces act on him, the downward-acting force of gravity and the upward-acting force of the parachute.

22. A crate hangs from a ring at the middle of a rope, as **CYU Figure 4.3** illustrates. A person is pulling on the right end of the rope to keep the crate in equilibrium. Can the rope ever be made to be perfectly horizontal?

CYU FIGURE 4.3

$\boxed{4.12}$ Nonequilibrium Applications of Newton's Laws of Motion

When an object is accelerating, it is not in equilibrium. The forces acting on it are not balanced, so the net force is not zero in Newton's second law. However, with one exception, the reasoning strategy followed in solving nonequilibrium problems is identical to that used in equilibrium situations. The exception occurs in Step 4 of the five steps outlined at the beginning of the previous section. Since the object is now accelerating, the representation of Newton's second law in Equations 4.2a and 4.2b applies instead of Equations 4.9a and 4.9b:

$$\Sigma F_x = ma_x \quad \textbf{(4.2a)} \qquad \text{and} \qquad \Sigma F_y = ma_y \quad \textbf{(4.2b)}$$

Example 15 uses these equations in a situation where the forces are applied in directions similar to those in Example 12, except that now an acceleration is present.

EXAMPLE 15 | Towing a Supertanker

A supertanker of mass $m = 1.50 \times 10^8$ kg is being towed by two tugboats, as in **Figure 4.30a**. The tensions in the towing cables apply the forces $\vec{T}_1$ and $\vec{T}_2$ at equal angles of 30.0° with respect to the tanker's axis. In addition, the tanker's engines produce a forward drive force $\vec{D}$, whose magnitude is $D = 75.0 \times 10^3$ N. Moreover, the water applies an opposing force $\vec{R}$, whose magnitude is $R = 40.0 \times 10^3$ N. The tanker moves forward with an acceleration that points along the tanker's axis and has a magnitude of 2.00×10^{-3} m/s². Find the magnitudes of the tensions $\vec{T}_1$ and $\vec{T}_2$.

Reasoning The unknown forces $\vec{T}_1$ and $\vec{T}_2$ contribute to the net force that accelerates the tanker. To determine $\vec{T}_1$ and $\vec{T}_2$, therefore, we analyze the net force, which we will do using components. The various force components can be found by referring to the free-body diagram for the tanker in **Figure 4.30b**, where the ship's axis is chosen as the x axis. We will then use Newton's second law in its component form, $\Sigma F_x = ma_x$ and $\Sigma F_y = ma_y$, to obtain the magnitudes of $\vec{T}_1$ and $\vec{T}_2$.

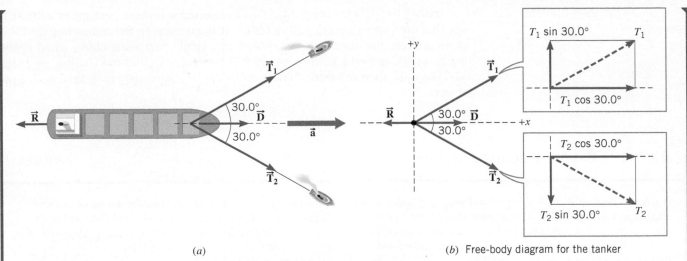

(a)

(b) Free-body diagram for the tanker

FIGURE 4.30 (a) Four forces act on a supertanker: $\vec{T}_1$ and $\vec{T}_2$ are the tension forces due to the towing cables, $\vec{D}$ is the forward drive force produced by the tanker's engines, and $\vec{R}$ is the force with which the water opposes the tanker's motion. (b) The free-body diagram for the tanker.

Solution   The individual force components are summarized as follows:

Force	x Component	y Component
$\vec{T}_1$	$+T_1 \cos 30.0°$	$+T_1 \sin 30.0°$
$\vec{T}_2$	$+T_2 \cos 30.0°$	$-T_2 \sin 30.0°$
$\vec{D}$	$+D$	0
$\vec{R}$	$-R$	0

Since the acceleration points along the x axis, there is no y component of the acceleration ($a_y = 0$ m/s^2). Consequently, the sum of the y components of the forces must be zero:

$$\Sigma F_y = +T_1 \sin 30.0° - T_2 \sin 30.0° = 0$$

This result shows that the magnitudes of the tensions in the cables are equal, $T_1 = T_2$. Since the ship accelerates along the x direction, the sum of the x components of the forces is not zero. The second law indicates that

$$\Sigma F_x = T_1 \cos 30.0° + T_2 \cos 30.0° + D - R = ma_x$$

Since $T_1 = T_2$, we can replace the two separate tension symbols by a single symbol T, the magnitude of the tension. Solving for T gives

$$T = \frac{ma_x + R - D}{2 \cos 30.0°}$$

$$= \frac{(1.50 \times 10^8 \text{ kg})(2.00 \times 10^{-3} \text{ m/s}^2) + 40.0 \times 10^3 \text{ N} - 75.0 \times 10^3 \text{ N}}{2 \cos 30.0°}$$

$$= \boxed{1.53 \times 10^5 \text{ N}}$$

Math Skills   The sine and cosine functions are defined in Equations 1.1 and 1.2 as $\sin \theta = \dfrac{h_o}{h}$ and $\cos \theta = \dfrac{h_a}{h}$, where h_o is the length of the side of a right triangle that is opposite the angle θ, h_a is the length of the side adjacent to the angle θ, and h is the length of the hypotenuse (see **Figure 4.31a**). When using the sine and cosine functions to determine the scalar components of a vector, we begin by identifying the angle θ. **Figure 4.31b** indicates that $\theta = 30.0°$ for the vector $\vec{T}_1$. The components of $\vec{T}_1$ are T_{1x} and T_{1y}. Comparing the shaded triangles in **Figure 4.31**, we can see that $h_o = T_{1y}$, $h_a = T_{1x}$, and $h = T_1$. Therefore, we have

$$\cos 30.0° = \frac{h_a}{h} = \frac{T_{1x}}{T_1} \quad \text{or} \quad T_{1x} = T_1 \cos 30.0°$$

$$\sin 30.0° = \frac{h_o}{h} = \frac{T_{1y}}{T_1} \quad \text{or} \quad T_{1y} = T_1 \sin 30.0°$$

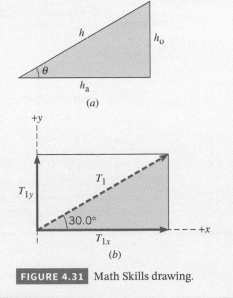

(a)

(b)

FIGURE 4.31   Math Skills drawing.

It often happens that two objects are connected somehow, perhaps by a drawbar like that used when a truck pulls a trailer. If the tension in the connecting device is of no interest, the objects can be treated as a single composite object when applying Newton's second law. However, if it is necessary to find the tension, as in the next example, then the second law must be applied separately to at least one of the objects.

EXAMPLE 16 | Hauling a Trailer

A truck is hauling a trailer along a level road, as **Figure 4.32a** illustrates. The mass of the truck is $m_1 = 8500$ kg and that of the trailer is $m_2 = 27\,000$ kg. The two move along the x axis with an acceleration of $a_x = 0.78$ m/s². Ignoring the retarding forces of friction and air resistance, determine **(a)** the tension $\vec{T}$ in the horizontal drawbar between the trailer and the truck and **(b)** the force $\vec{D}$ that propels the truck forward.

Reasoning   Since the truck and the trailer accelerate along the horizontal direction and friction is being ignored, only forces that have components in the horizontal direction are of interest. Therefore, **Figure 4.32** omits the weight and the normal force, which act vertically. To determine the tension force $\vec{T}$ in the drawbar, we draw the free-body diagram for the trailer and apply Newton's second law, $\Sigma F_x = ma_x$. Similarly, we can determine the propulsion force $\vec{D}$ by drawing the free-body diagram for the truck and applying Newton's second law.

> **Problem-Solving Insight**   A free-body diagram is very helpful when applying Newton's second law. Always start a problem by drawing the free-body diagram.

Solution   **(a)** The free-body diagram for the trailer is shown in **Figure 4.32b**. There is only one horizontal force acting on the trailer, the tension force $\vec{T}$ due to the drawbar. Therefore, it is straightforward to obtain the tension from $\Sigma F_x = m_2 a_x$, since the mass of the trailer and the acceleration are known:

$$\Sigma F_x = T = m_2 a_x = (27\,000 \text{ kg})(0.78 \text{ m/s}^2) = \boxed{21\,000 \text{ N}}$$

(b) Two horizontal forces act on the truck, as the free-body diagram in **Figure 4.32b** shows. One is the desired force $\vec{D}$. The other is the force $\vec{T}'$. According to Newton's third law, $\vec{T}'$ is the force with which the trailer pulls back on the truck, in reaction to the truck pulling forward. If the drawbar has negligible mass, the magnitude of $\vec{T}'$ is equal to the magnitude of $\vec{T}$—namely, 21 000 N. Since the magnitude of $\vec{T}'$, the mass of the truck, and the acceleration are known, $\Sigma F_x = m_1 a_x$ can be used to determine the drive force:

$$\Sigma F_x = +D - T' = m_1 a_x$$

$$D = m_1 a_x + T' = (8500 \text{ kg})(0.78 \text{ m/s}^2) + 21\,000 \text{ N} = \boxed{28\,000 \text{ N}}$$

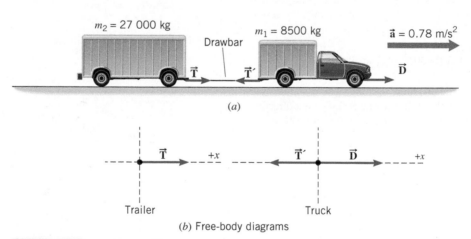

(a)

(b) Free-body diagrams

FIGURE 4.32   (a) The force $\vec{D}$ acts on the truck and propels it forward. The drawbar exerts the tension force $\vec{T}'$ on the truck and the tension force $\vec{T}$ on the trailer. (b) The free-body diagrams for the trailer and the truck, ignoring the vertical forces.

In Section 4.11 we examined situations where the net force acting on an object is zero, and in the present section we have considered two examples where the net force is not zero. Conceptual Example 17 illustrates a common situation where the net force is zero at certain times but is not zero at other times.

CONCEPTUAL EXAMPLE 17 | The Motion of a Water Skier

Figure 4.33 shows a water skier at four different moments:

a. The skier is floating motionless in the water.

b. The skier is being pulled out of the water and up onto the skis.

c. The skier is moving at a constant speed along a straight line.

d. The skier has let go of the tow rope and is slowing down.

For each moment, explain whether the net force acting on the skier is zero.

Reasoning and Solution According to Newton's second law, if an object has zero acceleration, the net force acting on it is zero. In such a case, the object is in equilibrium. In contrast, if the object has an acceleration, the net force acting on it is not zero. Such an object is not in equilibrium. We will consider the acceleration in each of the four phases of the motion to decide whether the net force is zero.

a. The skier is floating motionless in the water, so her velocity and acceleration are both zero. Therefore, the net force acting on her is zero, and she is in equilibrium.

b. As the skier is being pulled up and out of the water, her velocity is increasing. Thus, she is accelerating, and the net force acting on her is not zero. The skier is not in equilibrium. The direction of the net force is shown in **Figure 4.33b**.

c. The skier is now moving at a constant speed along a straight line (**Figure 4.33c**), so her velocity is constant. Since her velocity is constant, her acceleration is zero. Thus, the net force acting on her is zero, and she is again in equilibrium, even though she is moving.

d. After the skier lets go of the tow rope, her speed decreases, so she is decelerating. Thus, the net force acting on her is not zero, and she is not in equilibrium. The direction of the net force is shown in **Figure 4.33d**.

Related Homework: Problem 60

(a)

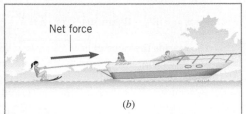
(b)

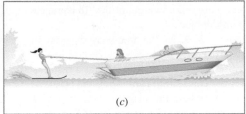
(c)

(d)

FIGURE 4.33 A water skier (a) floating in the water, (b) being pulled up by the boat, (c) moving at a constant velocity, and (d) slowing down.

The force of gravity is often present among the forces that affect the acceleration of an object. Examples 18 and 19 deal with typical situations.

EXAMPLE 18 | Accelerating Blocks

Block 1 (mass $m_1 = 8.00$ kg) is moving on a frictionless $30.0°$ incline. This block is connected to block 2 (mass $m_2 = 22.0$ kg) by a massless cord that passes over a massless and frictionless pulley (see **Figure 4.34a**). Find the acceleration of each block and the tension in the cord.

Reasoning Since both blocks accelerate, there must be a net force acting on each one. The key to this problem is to realize that Newton's second law can be used separately for each block to relate the net force and the acceleration. Note also that both blocks have accelerations of the same magnitude a, since they move as a

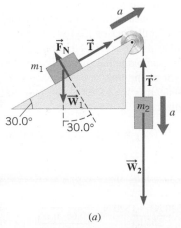
(a)

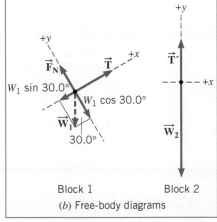
Block 1   Block 2
(b) Free-body diagrams

FIGURE 4.34 (a) Three forces act on block 1: its weight $\vec{W}_1$, the normal force $\vec{F}_N$, and the force $\vec{T}$ due to the tension in the cord. Two forces act on block 2: its weight $\vec{W}_2$ and the force $\vec{T}'$ due to the tension. The acceleration is labeled according to its magnitude a. (b) Free-body diagrams for the two blocks.

unit. We assume that block 1 accelerates up the incline and choose this direction to be the $+x$ axis. If block 1 in reality accelerates down the incline, then the value obtained for the acceleration will be a negative number.

> **Problem-Solving Insight** Mass and weight are different quantities. They cannot be interchanged when solving problems.

Solution Three forces act on block 1: (1) $\vec{\mathbf{W}}_1$ is its weight [$W_1 = m_1g = (8.00\ \text{kg})(9.80\ \text{m/s}^2) = 78.4\ \text{N}$], (2) $\vec{\mathbf{T}}$ is the force applied because of the tension in the cord, and (3) $\vec{\mathbf{F}}_N$ is the normal force the incline exerts. **Figure 4.34b** shows the free-body diagram for block 1. The weight is the only force that does not point along the x, y axes, and its x and y components are given in the diagram. Applying Newton's second law ($\Sigma F_x = m_1a_x$) to block 1 shows that

$$\Sigma F_x = -W_1 \sin 30.0° + T = m_1a$$

where we have set $a_x = a$. This equation cannot be solved as it stands, since both T and a are unknown quantities. To complete the solution, we next consider block 2.

Two forces act on block 2, as the free-body diagram in **Figure 4.34b** indicates: (1) $\vec{\mathbf{W}}_2$ is its weight [$W_2 = m_2g = (22.0\ \text{kg})(9.80\ \text{m/s}^2) = 216\ \text{N}$] and (2) $\vec{\mathbf{T}}'$ is exerted as a result of block 1 pulling back on the connecting cord. Since the cord and the frictionless pulley are massless, the magnitudes of $\vec{\mathbf{T}}'$ and $\vec{\mathbf{T}}$ are the same: $T' = T$. Applying Newton's second law ($\Sigma F_y = m_2a_y$) to block 2 reveals that

$$\Sigma F_y = T - W_2 = m_2(-a)$$

The acceleration a_y has been set equal to $-a$ since block 2 moves downward along the $-y$ axis in the free-body diagram, consistent with the assumption that block 1 moves up the incline. Now

there are two equations in two unknowns, and they may be solved simultaneously (see Appendix C) to give T and a:

$$\boxed{T = 86.3\ \text{N}} \quad \text{and} \quad \boxed{a = 5.89\ \text{m/s}^2}$$

Math Skills The two equations containing the unknown quantities T and a are

$$-W_1 \sin 30.0° + T = m_1a \quad (1)$$
and
$$T - W_2 = m_2(-a) \quad (2)$$

Neither equation by itself can yield numerical values for T and a. However, the two equations can be solved simultaneously in the following manner. To begin with, we rearrange Equation (1) to give $T = m_1a + W_1 \sin 30.0°$. Next, we substitute this result into Equation (2) and obtain a result containing only the unknown acceleration a:

$$\underbrace{m_1a + W_1 \sin 30.0°}_{T} - W_2 = m_2(-a)$$

Rearranging this equation so that the terms m_1a and m_2a stand alone on the left of the equals sign, we have

$$m_1a + m_2a = W_2 - W_1 \sin 30.0° \quad \text{or} \quad a = \frac{W_2 - W_1 \sin 30.0°}{m_1 + m_2}$$

This result yields the value of $\boxed{a = 5.89\ \text{m/s}^2}$. We can also substitute the expression for a into either one of the two starting equations and obtain a result containing only the unknown tension T. We choose Equation (1) and find that

$$-W_1 \sin 30.0° + T = m_1\underbrace{\left(\frac{W_2 - W_1 \sin 30.0°}{m_1 + m_2}\right)}_{a}$$

Solving for T gives

$$T = W_1 \sin 30.0° + m_1\left(\frac{W_2 - W_1 \sin 30.0°}{m_1 + m_2}\right)$$

This expression yields the value of $\boxed{T = 86.3\ \text{N}}$.

EXAMPLE 19 | Hoisting a Scaffold

A window washer on a scaffold is hoisting the scaffold up the side of a building by pulling downward on a rope, as in **Figure 4.35a**. The magnitude of the pulling force is 540 N, and the combined mass of the worker and the scaffold is 155 kg. Find the upward acceleration of the unit.

Reasoning The worker and the scaffold form a single unit, on which the rope exerts a force in three places. The left end of the rope exerts an upward force $\vec{\mathbf{T}}$ on the worker's hands. This force arises because he pulls downward with a 540-N force, and the rope exerts an oppositely directed force of equal magnitude on him, in accord with Newton's third law. Thus, the magnitude T of the upward force is $T = 540\ \text{N}$ and is the magnitude of the tension in the rope. If the masses of the rope and each pulley are negligible and if the pulleys are friction-free, the tension is transmitted undiminished along the rope. Then, a 540-N tension force $\vec{\mathbf{T}}$ acts

upward on the left side of the scaffold pulley (see part a of the drawing). A tension force is also applied to the point P, where the rope attaches to the roof. The roof pulls back on the rope in accord with the third law, and this pull leads to the 540-N tension force $\vec{\mathbf{T}}$ that acts on the right side of the scaffold pulley. In addition to the three upward forces, the weight of the unit must be taken into account [$W = mg = (155\ \text{kg})(9.80\ \text{m/s}^2) = 1520\ \text{N}$]. Part b of the drawing shows the free-body diagram.

Solution Newton's second law ($\Sigma F_y = ma_y$) can be applied to calculate the acceleration a_y:

$$\Sigma F_y = +T + T + T - W = ma_y$$

$$a_y = \frac{3T - W}{m} = \frac{3(540\ \text{N}) - 1520\ \text{N}}{155\ \text{kg}} = \boxed{0.65\ \text{m/s}^2}$$

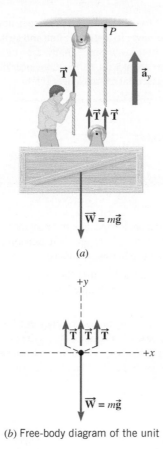

(a)

$$\vec{W} = m\vec{g}$$

+y

+x

$$\vec{W} = m\vec{g}$$

(b) Free-body diagram of the unit

FIGURE 4.35 (a) A window washer pulls down on the rope to hoist the scaffold up the side of a building. The force $\vec{T}$ results from the effort of the window washer and acts on him and the scaffold in three places, as discussed in Example 19. (b) The free-body diagram of the unit comprising the man and the scaffold.

EXAMPLE 20 | BIO The Importance of Seatbelts

Police departments around the country use a device, called *The Convincer*, to educate the public on the safety of seatbelts (see **Figure 4.36a**). The device consists of a sled that is free to move on rails that are inclined at 6.5° above the horizontal. A passenger is placed in the sled and secured with a seatbelt. The sled is released from rest and travels a distance of 2.3 m along the rails before slamming into a stop at the bottom, thereby simulating a car collision. Assume the rails are frictionless and calculate the speed of the sled at impact.

Reasoning The free-body diagram for the sled is shown in **Figure 4.36b**. The x-component of the sled's weight is the only force acting on the sled in the x-direction. Once released, the sled will accelerate down the rails. Thus, this is a nonequilibrium

problem, and the sum of the forces acting on the sled in the x-direction will be equal to its mass times its acceleration in that direction. Since the sled's acceleration down the rails will be constant, we can use our relationships from kinematics to calculate its speed when it strikes the stop.

Solution Applying Newton's second law to the sled in the x-direction, we have $\Sigma F_x = W \sin 6.5° = ma_x$. We can solve this expression for the acceleration of the sled in the x-direction: $a_x = \dfrac{W \sin 6.5°}{m} = \dfrac{mg \sin 6.5°}{m} = g \sin 6.5° = (9.8 \text{ m/s}^2)(\sin 6.5°) = 1.1 \text{ m/s}^2$. Now that we have the acceleration of the sled, we can use Equation 3.6a to calculate its speed when it strikes the stop: $v_x^2 = v_{0x}^2 + 2a_xx$. Solving for v_x, and using the fact that the sled begins

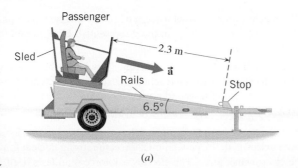

(a)

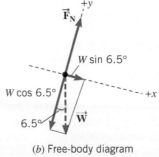

(b) Free-body diagram

FIGURE 4.36 (a) The passenger in the sled accelerates down the rails and collides with a stop in order to simulate a car collision. (b) Free-body diagram showing the forces acting on the sled. The only force providing its acceleration down the rails is the x-component of its weight.

from rest ($v_{0x} = 0$), we get the following result for its speed at impact: $v_x = \sqrt{2a_xx} = \sqrt{2(1.1 \text{ m/s}^2)(2.3 \text{ m})} = \boxed{2.2 \text{ m/s}}$. Consider the magnitude of this speed. This is approximately equal to 5 mi/h, which is a relatively low speed for a car. Even so, the passenger still receives quite a jolt when the sled strikes the stop. Without

the seatbelt to decelerate the passenger upon impact, she would continue to move in a straight line and strike the front of the sled (Newton's first law). Since many car collisions occur at speeds much larger than 5 mi/h, *The Convincer* demonstrates that seatbelt use is essential for passenger safety.

Check Your Understanding

(The answers are given at the end of the book.)

23. A circus performer hangs stationary from a rope. She then begins to climb upward by pulling herself up, hand over hand. When she starts climbing, is the tension in the rope **(a)** less than, **(b)** equal to, or **(c)** greater than it is when she hangs stationary?

24. A freight train is accelerating on a level track. Other things being equal, would the tension in the coupling between the engine and the first car change if some of the cargo in the last car were transferred to any one of the other cars?

25. Two boxes have masses m_1 and m_2, and m_2 is greater than m_1. The boxes are being pushed across a frictionless horizontal surface (see **CYU Figure 4.4**). As the drawing shows, there are two possible arrangements, and the pushing force is the same in each. In which arrangement, **(a)** or **(b)**, does the force that the left box applies to the right box have a greater magnitude, or **(c)** is the magnitude the same in both cases?

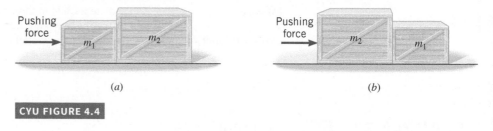

(a)   *(b)*

CYU FIGURE 4.4

Concept Summary

4.1 The Concepts of Force and Mass A force is a push or a pull and is a vector quantity. Contact forces arise from the physical contact between two objects. Noncontact forces are also called action-at-a-distance forces, because they arise without physical contact between two objects.

Mass is a property of matter that determines how difficult it is to accelerate or decelerate an object. Mass is a scalar quantity.

4.2 Newton's First Law of Motion Newton's first law of motion, sometimes called the law of inertia, states that an object continues in a state of rest or in a state of motion at a constant velocity unless compelled to change that state by a net force.

Inertia is the natural tendency of an object to remain at rest or in motion at a constant velocity. The mass of a body is a quantitative measure of inertia and is measured in an SI unit called the kilogram (kg). An inertial reference frame is one in which Newton's law of inertia is valid.

4.3 Newton's Second Law of Motion/4.4 The Vector Nature of Newton's Second Law of Motion Newton's second law of motion states that when a net force $\Sigma\vec{F}$ acts on an object of mass m, the acceleration $\vec{a}$ of the object can be obtained from Equation 4.1. This is a

vector equation and, for motion in two dimensions, is equivalent to Equations 4.2a and 4.2b. In these equations the x and y subscripts refer to the scalar components of the force and acceleration vectors. The SI unit of force is the newton (N).

$$\Sigma\vec{F} = m\vec{a} \qquad (4.1)$$
$$\Sigma F_x = ma_x \qquad (4.2a)$$
$$\Sigma F_y = ma_y \qquad (4.2b)$$

When determining the net force, a free-body diagram is helpful. A free-body diagram is a diagram that represents the object and the forces acting on it.

4.5 Newton's Third Law of Motion Newton's third law of motion, often called the action–reaction law, states that whenever one object exerts a force on a second object, the second object exerts an oppositely directed force of equal magnitude on the first object.

4.6 Types of Forces: An Overview Only three fundamental forces have been discovered: the gravitational force, the strong nuclear force, and the electroweak force. The electroweak force manifests itself as either the electromagnetic force or the weak nuclear force.

4.7 The Gravitational Force   Newton's law of universal gravitation states that every particle in the universe exerts an attractive force on every other particle. For two particles that are separated by a distance r and have masses m_1 and m_2, the law states that the magnitude of this attractive force is as given in Equation 4.3. The direction of this force lies along the line between the particles. The constant G has a value of $G = 6.674 \times 10^{-11}$ N · m²/kg² and is called the universal gravitational constant.

$$F = G\frac{m_1 m_2}{r^2} \tag{4.3}$$

The weight W of an object on or above the earth is the gravitational force that the earth exerts on the object and can be calculated from the mass m of the object and the magnitude g of the acceleration due to the earth's gravity, according to Equation 4.5.

$$W = mg \tag{4.5}$$

4.8 The Normal Force   The normal force $\vec{F}_N$ is one component of the force that a surface exerts on an object with which it is in contact—namely, the component that is perpendicular to the surface.

The apparent weight is the force that an object exerts on the platform of a scale and may be larger or smaller than the true weight mg if the object and the scale have an acceleration a (+ if upward, – if downward). The apparent weight is given by Equation 4.6.

$$\text{Apparent weight} = mg + ma \tag{4.6}$$

4.9 Static and Kinetic Frictional Forces   A surface exerts a force on an object with which it is in contact. The component of the force perpendicular to the surface is called the normal force. The component parallel to the surface is called friction.

The force of static friction between two surfaces opposes any impending relative motion of the surfaces. The magnitude of the static frictional force depends on the magnitude of the applied force and can assume any value up to the maximum specified in Equation 4.7,

where μ_s is the coefficient of static friction and F_N is the magnitude of the normal force.

$$f_s^{\text{MAX}} = \mu_s F_N \tag{4.7}$$

The force of kinetic friction between two surfaces sliding against one another opposes the relative motion of the surfaces. This force has a magnitude given by Equation 4.8, where μ_k is the coefficient of kinetic friction.

$$f_k = \mu_k F_N \tag{4.8}$$

4.10 The Tension Force   The word "tension" is commonly used to mean the tendency of a rope to be pulled apart due to forces that are applied at each end. Because of tension, a rope transmits a force from one end to the other. When a rope is accelerating, the force is transmitted undiminished only if the rope is massless.

4.11 Equilibrium Applications of Newton's Laws of Motion   An object is in equilibrium when the object has zero acceleration, or, in other words, when it moves at a constant velocity. The constant velocity may be zero, in which case the object is stationary. The sum of the forces that act on an object in equilibrium is zero. Under equilibrium conditions in two dimensions, the separate sums of the force components in the x direction and in the y direction must each be zero, as in Equations 4.9a and 4.9b.

$$\Sigma F_x = 0 \tag{4.9a}$$
$$\Sigma F_y = 0 \tag{4.9b}$$

4.12 Nonequilibrium Applications of Newton's Laws of Motion   If an object is not in equilibrium, then Newton's second law, as expressed in Equations 4.2a and 4.2b, must be used to account for the acceleration.

$$\Sigma F_x = ma_x \tag{4.2a}$$
$$\Sigma F_y = ma_y \tag{4.2b}$$

Focus on Concepts

Addtional questions are available for assignment in WileyPLUS

Section 4.2   Newton's First Law of Motion

1. An object is moving at a constant velocity. All but one of the following statements could be true. Which one cannot be true? (a) No forces act on the object. (b) A single force acts on the object. (c) Two forces act simultaneously on the object. (d) Three forces act simultaneously on the object.

2. A cup of coffee is sitting on a table in a recreational vehicle (RV). The cup slides toward the rear of the RV. According to Newton's first law, which one or more of the following statements could describe the motion of the RV? (A) The RV is at rest, and the driver suddenly accelerates. (B) The RV is moving forward, and the driver suddenly accelerates. (C) The RV is moving backward, and the driver suddenly hits the brakes. (a) A (b) B (c) C (d) A and B (e) A, B, and C

Section 4.4   The Vector Nature of Newton's Second Law of Motion

3. Two forces act on a moving object that has a mass of 27 kg. One force has a magnitude of 12 N and points due south, while the other force has a magnitude of 17 N and points due west. What is the acceleration of the object? (a) 0.63 m/s² directed 55° south of west (b) 0.44 m/s² directed 24° south of west (c) 0.77 m/s² directed 35° south of west (d) 0.77 m/s² directed 55° south of west (e) 1.1 m/s² directed 35° south of west

Section 4.5   Newton's Third Law of Motion

4. Which one of the following is true, according to Newton's laws of motion? Ignore friction. (a) A sports utility vehicle (SUV) hits a stationary motorcycle. Since it is stationary, the motorcycle sustains a greater force than the SUV does. (b) A semitrailer truck crashes all the way through a wall. Since the wall collapses, the wall sustains a greater force than the truck does. (c) Sam (18 years old) and his sister (9 years old) go ice skating. They push off against each other and fly apart. Sam flies off with the greater acceleration. (d) Two astronauts on a space walk are throwing a ball back and forth between each other. In this game of catch the distance between them remains constant. (e) None of the above is true, according to the third law.

5. Two ice skaters, Paul and Tom, are each holding on to opposite ends of the same rope. Each pulls the other toward him. The magnitude of

Paul's acceleration is 1.25 times greater than the magnitude of Tom's acceleration. What is the ratio of Paul's mass to Tom's mass? **(a)** 0.67 **(b)** 0.80 **(c)** 0.25 **(d)** 1.25 **(e)** 0.50

Section 4.7 The Gravitational Force

6. In another solar system a planet has twice the earth's mass and three times the earth's radius. Your weight on this planet is _____ times your earth-weight. Assume that the masses of the earth and of the other planet are uniformly distributed. **(a)** 0.667 **(b)** 2.000 **(c)** 0.111 **(d)** 0.444 **(e)** 0.222

7. What is the mass on Mercury of an object that weighs 784 N on the earth's surface? **(a)** 80.0 kg **(b)** 48.0 kg **(c)** 118 kg **(d)** 26.0 kg **(e)** There is not enough information to calculate the mass.

Section 4.8 The Normal Force

8. The apparent weight of a passenger in an elevator is greater than his true weight. Which one of the following is true? **(a)** The elevator is either moving upward with an increasing speed or moving upward with a decreasing speed. **(b)** The elevator is either moving upward with an increasing speed or moving downward with an increasing speed. **(c)** The elevator is either moving upward with a decreasing speed or moving downward with a decreasing speed. **(d)** The elevator is either moving upward with an increasing speed or moving downward with a decreasing speed. **(e)** The elevator is either moving upward with a decreasing speed or moving downward with an increasing speed.

9. The drawings show three examples of the force with which someone pushes against a vertical wall. In each case the magnitude of the pushing force is the same. Rank the normal forces that the wall applies to the pusher in ascending order (smallest first). **(a)** C, B, A **(b)** B, A, C **(c)** A, C, B **(d)** B, C, A **(e)** C, A, B

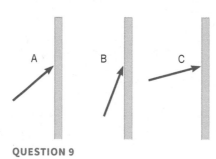

QUESTION 9

Section 4.9 Static and Kinetic Frictional Forces

10. The drawing shows three blocks, each with the same mass, stacked one upon the other. The bottom block rests on a frictionless horizontal surface and is being pulled by a force $\vec{F}$ that is parallel to this surface. The surfaces where the blocks touch each other have identical coefficients of static friction. Which one of the following correctly describes the magnitude of the *net* force of static friction f_s that acts on each block?

(a) $f_{s,A} = f_{s,B} = f_{s,C}$

(b) $f_{s,A} = f_{s,B} = \frac{1}{2}f_{s,C}$

(c) $f_{s,A} = 0$ and $f_{s,B} = \frac{1}{2}f_{s,C}$

(d) $f_{s,C} = 0$ and $f_{s,A} = \frac{1}{2}f_{s,B}$

(e) $f_{s,A} = f_{s,C} = \frac{1}{2}f_{s,B}$

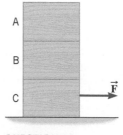

QUESTION 10

11. Three identical blocks are being pulled or pushed across a horizontal surface by a force $\vec{F}$, as shown in the drawings. The force $\vec{F}$ in each case has the same magnitude. Rank the kinetic frictional forces that act on the blocks in ascending order (smallest first).

(a) B, C, A (b) C, A, B (c) B, A, C (d) C, B, A (e) A, C, B

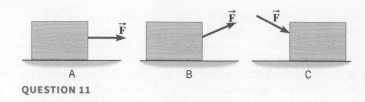

QUESTION 11

Section 4.10 The Tension Force

12. A heavy block is suspended from a ceiling using pulleys in three different ways, as shown in the drawings. Rank the tension in the rope that passes over the pulleys in ascending order (smallest first).

(a) B, A, C (b) C, B, A (c) A, B, C (d) C, A, B (e) B, C, A

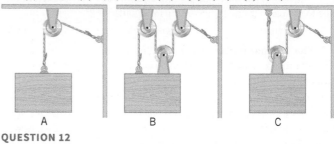

QUESTION 12

Section 4.11 Equilibrium Applications of Newton's Laws of Motion

13. A certain object is in equilibrium. Which one of the following statements is *not* true? **(a)** The object must be at rest. **(b)** The object has a constant velocity. **(c)** The object has no acceleration. **(d)** No net force acts on the object.

14. Two identical boxes are being pulled across a horizontal floor at a constant velocity by a horizontal pulling force of 176 N that is applied to one of the boxes, as the drawing shows. There is kinetic friction between each box and the floor. Find the tension in the rope between the boxes. **(a)** 176 N **(b)** 88.0 N **(c)** 132 N **(d)** 44.0 N **(e)** There is not enough information to calculate the tension.

QUESTION 14

Section 4.12 Nonequilibrium Applications of Newton's Laws of Motion

15. A man is standing on a platform that is connected to a pulley arrangement, as the drawing shows. By pulling upward on the rope with a force $\vec{P}$ the man can raise the platform and himself. The total mass of the man plus the platform is 94.0 kg. What pulling force should the man apply to create an upward acceleration of 1.20 m/s²?

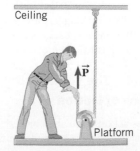

QUESTION 15

Problems

Note to Instructors: Most of the homework problems in this chapter are available for assignment via WileyPLUS. See the Preface for additional details. Additional problems are available for assignment in WileyPLUS

SSM Student Solutions Manual

MMH Problem-solving help

GO Guided Online Tutorial

V-HINT Video Hints

CHALK Chalkboard Videos

BIO Biomedical application

E Easy

M Medium

H Hard

WS Worksheet

T Team Problem

Section 4.3   Newton's Second Law of Motion

1. **E** An airplane has a mass of 3.1×10^4 kg and takes off under the influence of a constant net force of 3.7×10^4 N. What is the net force that acts on the plane's 78-kg pilot?

2. **E** **MMH** A boat has a mass of 6800 kg. Its engines generate a drive force of 4100 N due west, while the wind exerts a force of 800 N due east and the water exerts a resistive force of 1200 N due east. What are the magnitude and direction of the boat's acceleration?

3. **E** **GO** Two horizontal forces, $\vec{F}_1$ and $\vec{F}_2$, are acting on a box, but only $\vec{F}_1$ is shown in the drawing. $\vec{F}_2$ can point either to the right or to the left. The box moves only along the x axis. There is no friction between the box and the surface. Suppose that $\vec{F}_1 = +9.0$ N and the mass of the box is 3.0 kg. Find the magnitude and direction of $\vec{F}_2$ when the acceleration of the box is (a) $+5.0$ m/s^2, (b) -5.0 m/s^2, and (c) 0 m/s^2.

PROBLEM 3

4. **E** In the amusement park ride known as Magic Mountain Superman, powerful magnets accelerate a car and its riders from rest to 45 m/s (about 100 mi/h) in a time of 7.0 s. The combined mass of the car and riders is 5.5×10^3 kg. Find the average net force exerted on the car and riders by the magnets.

5. **E** **SSM** A person in a kayak starts paddling, and it accelerates from 0 to 0.60 m/s in a distance of 0.41 m. If the combined mass of the person and the kayak is 73 kg, what is the magnitude of the net force acting on the kayak?

6. **E** Scientists are experimenting with a kind of gun that may eventually be used to fire payloads directly into orbit. In one test, this gun accelerates a 5.0-kg projectile from rest to a speed of 4.0×10^3 m/s. The net force accelerating the projectile is 4.9×10^5 N. How much time is required for the projectile to come up to speed?

7. **E** **SSM** **MMH** A 1580-kg car is traveling with a speed of 15.0 m/s. What is the magnitude of the horizontal net force that is required to bring the car to a halt in a distance of 50.0 m?

8. **E** **GO** The space probe *Deep Space 1* was launched on October 24, 1998. Its mass was 474 kg. The goal of the mission was to test a new kind of engine called an ion propulsion drive. This engine generated only a weak thrust, but it could do so over long periods of time with the consumption of only small amounts of fuel. The mission was spectacularly successful. At a thrust of 56 mN how many days were required for the probe to attain a velocity of 805 m/s (1800 mi/h), assuming that the probe started from rest and that the mass remained nearly constant?

9. **M** **SSM** Two forces $\vec{F}_A$ and $\vec{F}_B$ are applied to an object whose mass is 8.0 kg. The larger force is $\vec{F}_A$. When both forces point due east, the object's acceleration has a magnitude of 0.50 m/s^2. However, when $\vec{F}_A$ points due east and $\vec{F}_B$ points due west, the acceleration is 0.40 m/s^2, due east. Find (a) the magnitude of $\vec{F}_A$ and (b) the magnitude of $\vec{F}_B$.

10. **M** **V-HINT** An electron is a subatomic particle ($m = 9.11 \times 10^{-31}$ kg) that is subject to electric forces. An electron moving in the $+x$ direction accelerates from an initial velocity of $+5.40 \times 10^5$ m/s to a final velocity of $+2.10 \times 10^6$ m/s while traveling a distance of 0.038 m. The electron's acceleration is due to two electric forces parallel to the x axis: $\vec{F}_1 = +7.50 \times 10^{-17}$ N, and $\vec{F}_2$, which points in the $-x$ direction. Find the magnitudes of (a) the net force acting on the electron and (b) the electric force $\vec{F}_2$.

Section 4.4   The Vector Nature of Newton's Second Law of Motion, Section 4.5   Newton's Third Law of Motion

11. **E** Only two forces act on an object (mass = 3.00 kg), as in the drawing. Find the magnitude and direction (relative to the x axis) of the acceleration of the object.

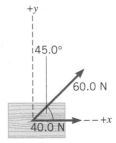

PROBLEM 11

12. **E** **SSM** **MMH** A rocket of mass 4.50×10^5 kg is in flight. Its thrust is directed at an angle of 55.0° above the horizontal and has a magnitude of 7.50×10^6 N. Find the magnitude and direction of the rocket's acceleration. Give the direction as an angle above the horizontal.

13. **E** When a parachute opens, the air exerts a large drag force on it. This upward force is initially greater than the weight of the sky diver and, thus, slows him down. Suppose the weight of the sky diver is 915 N and the drag force has a magnitude of 1027 N. The mass of the sky diver is 93.4 kg. What are the magnitude and direction of his acceleration?

14. **E** **GO** **CHALK** Two skaters, a man and a woman, are standing on ice. Neglect any friction between the skate blades and the ice. The mass of the man is 82 kg, and the mass of the woman is 48 kg. The woman pushes on the man with a force of 45 N due east. Determine the acceleration (magnitude and direction) of (a) the man and (b) the woman.

15. **M** **V-HINT** **CHALK** A space probe has two engines. Each generates the same amount of force when fired, and the directions of these forces can be independently adjusted. When the engines are fired simultaneously and each applies its force in the same direction, the probe, starting from rest, takes 28 s to travel a certain distance. How long does it take to travel the same distance, again starting from rest, if the engines are fired simultaneously and the forces that they apply to the probe are perpendicular?

16. **H** At a time when mining asteroids has become feasible, astronauts have connected a line between their 3500-kg space tug and a 6200-kg asteroid. Using their tug's engine, they pull on the asteroid with a force of 490 N. Initially the tug and the asteroid are at rest, 450 m apart. How much time does it take for the tug and the asteroid to meet?

17. **H** **SSM** A 325-kg boat is sailing 15.0° north of east at a speed of 2.00 m/s. Thirty seconds later, it is sailing 35.0° north of east at a speed of 4.00 m/s. During this time, three forces act on the boat: a 31.0-N force directed 15.0° north of east (due to an auxiliary engine), a

23.0-N force directed 15.0° south of west (resistance due to the water), and $\vec{F}_W$ (due to the wind). Find the magnitude and direction of the force $\vec{F}_W$. Express the direction as an angle with respect to due east.

Section 4.7   The Gravitational Force

18. **E** **GO** A 5.0-kg rock and a 3.0×10^{-4}-kg pebble are held near the surface of the earth. (a) Determine the magnitude of the gravitational force exerted on each by the earth. (b) Calculate the magnitude of the acceleration of each object when released.

19. **E** Mars has a mass of 6.46×10^{23} kg and a radius of 3.39×10^6 m. (a) What is the acceleration due to gravity on Mars? (b) How much would a 65-kg person weigh on this planet?

20. **E** On earth, two parts of a space probe weigh 11 000 N and 3400 N. These parts are separated by a center-to-center distance of 12 m and may be treated as uniform spherical objects. Find the magnitude of the gravitational force that each part exerts on the other out in space, far from any other objects.

21. **E** **GO** A raindrop has a mass of 5.2×10^{-7} kg and is falling near the surface of the earth. Calculate the magnitude of the gravitational force exerted (a) on the raindrop by the earth and (b) on the earth by the raindrop.

22. **E** The weight of an object is the same on two different planets. The mass of planet A is only sixty percent of that of planet B. Find the ratio r_A/r_B of the radii of the planets.

23. **E** **SSM** A bowling ball (mass = 7.2 kg, radius = 0.11 m) and a billiard ball (mass = 0.38 kg, radius = 0.028 m) may each be treated as uniform spheres. What is the magnitude of the maximum gravitational force that each can exert on the other?

24. **E** Review Conceptual Example 7 in preparation for this problem. In tests on earth a lunar surface exploration vehicle (mass = 5.90×10^3 kg) achieves a forward acceleration of 0.220 m/s². To achieve this same acceleration on the moon, the vehicle's engines must produce a drive force of 1.43×10^3 N. What is the magnitude of the frictional force that acts on the vehicle on the moon?

25. **E** **SSM** Synchronous communications satellites are placed in a circular orbit that is 3.59×10^7 m above the surface of the earth. What is the magnitude of the acceleration due to gravity at this distance?

26. **E** The drawing (not to scale) shows one alignment of the sun, earth, and moon. The gravitational force $\vec{F}_{SM}$ that the sun exerts on the moon is perpendicular to the force $\vec{F}_{EM}$ that the earth exerts on the moon. The masses are: mass of sun = 1.99×10^{30} kg, mass of earth = 5.98×10^{24} kg, mass of moon = 7.35×10^{22} kg. The distances shown in the drawing are $r_{SM} = 1.50 \times 10^{11}$ m and $r_{EM} = 3.85 \times 10^8$ m. Determine the magnitude of the net gravitational force on the moon.

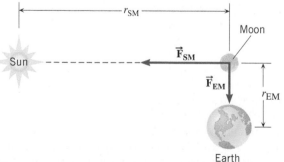

PROBLEM 26

27. **E** The drawing shows three particles far away from any other objects and located on a straight line. The masses of these particles are $m_A = 363$ kg, $m_B = 517$ kg, and $m_C = 154$ kg. Find the magnitude and direction of the net gravitational force acting on (a) particle A, (b) particle B, and (c) particle C.

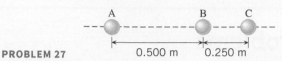

PROBLEM 27

28. **E** A space traveler weighs 540.0 N on earth. What will the traveler weigh on another planet whose radius is twice that of earth and whose mass is three times that of earth?

29. **M** **CHALK** A spacecraft is on a journey to the moon. At what point, as measured from the center of the earth, does the gravitational force exerted on the spacecraft by the earth balance that exerted by the moon? This point lies on a line between the centers of the earth and the moon. The distance between the earth and the moon is 3.85×10^8 m, and the mass of the earth is 81.4 times as great as that of the moon.

30. **M** **GO** A neutron star has a mass of 2.0×10^{30} kg (about the mass of our sun) and a radius of 5.0×10^3 m (about the height of a good-sized mountain). Suppose an object falls from rest near the surface of such a star. How fast would this object be moving after it had fallen a distance of 0.010 m? (Assume that the gravitational force is constant over the distance of the fall and that the star is not rotating.)

31. **M** **SSM** The sun is more massive than the moon, but the sun is farther from the earth. Which one exerts a greater gravitational force on a person standing on the earth? Give your answer by determining the ratio F_{sun}/F_{moon} of the magnitudes of the gravitational forces. Use the data on the inside of the front cover.

32. **H** Two particles are located on the x axis. Particle 1 has a mass m and is at the origin. Particle 2 has a mass $2m$ and is at $x = +L$. A third particle is placed between particles 1 and 2. Where on the x axis should the third particle be located so that the magnitude of the gravitational force on *both* particle 1 and particle 2 doubles? Express your answer in terms of L.

Section 4.8   The Normal Force, Section 4.9   Static and Kinetic Frictional Forces

33. **E** A 35-kg crate rests on a horizontal floor, and a 65-kg person is standing on the crate. Determine the magnitude of the normal force that (a) the floor exerts on the crate and (b) the crate exerts on the person.

34. **E** **SSM** A 60.0-kg crate rests on a level floor at a shipping dock. The coefficients of static and kinetic friction are 0.760 and 0.410, respectively. What horizontal pushing force is required to (a) just start the crate moving and (b) slide the crate across the dock at a constant speed?

35. **E** **CHALK** A rocket blasts off from rest and attains a speed of 45 m/s in 15 s. An astronaut has a mass of 57 kg. What is the astronaut's apparent weight during takeoff?

36. **E** **GO** A car is traveling up a hill that is inclined at an angle θ above the horizontal. Determine the ratio of the magnitude of the normal force to the weight of the car when (a) $\theta = 15°$ and (b) $\theta = 35°$.

37. **E** A woman stands on a scale in a moving elevator. Her mass is 60.0 kg, and the combined mass of the elevator and scale is an additional 815 kg. Starting from rest, the elevator accelerates upward. During the acceleration, the hoisting cable applies a force of 9410 N. What does the scale read during the acceleration?

38. **E** A Mercedes-Benz 300SL ($m = 1700$ kg) is parked on a road that rises 15° above the horizontal. What are the magnitudes of (a) the normal force and (b) the static frictional force that the ground exerts on the tires?

39. **E** **GO** Consult Multiple-Concept Example 9 to explore a model for solving this problem. A person pushes on a 57-kg refrigerator with a horizontal force of -267 N; the minus sign indicates that the force points in the $-x$ direction. The coefficient of static friction is 0.65. (a) If the refrigerator does not move, what are the magnitude and direction of the static frictional force that the floor exerts on the

refrigerator? (b) What is the magnitude of the largest pushing force that can be applied to the refrigerator before it just begins to move?

40. **E** **SSM** A 6.00-kg box is sliding across the horizontal floor of an elevator. The coefficient of kinetic friction between the box and the floor is 0.360. Determine the kinetic frictional force that acts on the box when the elevator is (a) stationary, (b) accelerating upward with an acceleration whose magnitude is 1.20 m/s^2, and (c) accelerating downward with an acceleration whose magnitude is 1.20 m/s^2.

41. **E** **GO** A cup of coffee is on a table in an airplane flying at a constant altitude and a constant velocity. The coefficient of static friction between the cup and the table is 0.30. Suddenly, the plane accelerates forward, its altitude remaining constant. What is the maximum acceleration that the plane can have without the cup sliding backward on the table?

42. **M** **V-HINT** Consult Multiple-Concept Example 10 in preparation for this problem. Traveling at a speed of 16.1 m/s, the driver of an automobile suddenly locks the wheels by slamming on the brakes. The coefficient of kinetic friction between the tires and the road is 0.720. What is the speed of the automobile after 1.30 s have elapsed? Ignore the effects of air resistance.

43. **M** **SSM** **CHALK** A person is trying to judge whether a picture (mass = 1.10 kg) is properly positioned by temporarily pressing it against a wall. The pressing force is perpendicular to the wall. The coefficient of static friction between the picture and the wall is 0.660. What is the minimum amount of pressing force that must be used?

44. **M** **V-HINT** Air rushing over the wings of high-performance race cars generates unwanted horizontal air resistance but also causes a vertical *downforce*, which helps the cars hug the track more securely. The coefficient of static friction between the track and the tires of a 690-kg race car is 0.87. What is the magnitude of the maximum acceleration at which the car can speed up without its tires slipping when a 4060-N downforce and an 1190-N horizontal-air-resistance force act on it?

45. **H** While moving in, a new homeowner is pushing a box across the floor at a constant velocity. The coefficient of kinetic friction between the box and the floor is 0.41. The pushing force is directed downward at an angle θ below the horizontal. When θ is greater than a certain value, it is not possible to move the box, no matter how large the pushing force is. Find that value of θ.

Section 4.10   The Tension Force,
Section 4.11   Equilibrium Applications
of Newton's Laws of Motion

46. **E** **MMH** The helicopter in the drawing is moving horizontally to the right at a constant velocity $\vec{V}$. The weight of the helicopter is $W =$ 53 800 N. The lift force $\vec{L}$ generated by the rotating blade makes an angle of 21.0° with respect to the vertical. (a) What is the magnitude of the lift force? (b) Determine the magnitude of the air resistance $\vec{R}$ that opposes the motion.

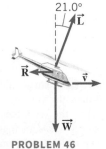

PROBLEM 46

47. **E** **SSM** Three forces act on a moving object. One force has a magnitude of 80.0 N and is directed due north. Another has a magnitude of 60.0 N and is directed due west. What must be the magnitude and direction of the third force, such that the object continues to move with a constant velocity?

48. **E** **SSM** A worker stands still on a roof sloped at an angle of 36° above the horizontal. He is prevented from slipping by a static frictional force of 390 N. Find the mass of the worker.

49. **E** A stuntman is being pulled along a rough road at a constant velocity by a cable attached to a moving truck. The cable is parallel to the ground. The mass of the stuntman is 109 kg, and the coefficient of kinetic friction between the road and him is 0.870. Find the tension in the cable.

50. **E** **GO** The drawing shows a circus clown who weighs 890 N. The coefficient of static friction between the clown's feet and the ground is 0.53. He pulls vertically downward on a rope that passes around three pulleys and is tied around his feet. What is the minimum pulling force that the clown must exert to yank his feet out from under himself?

PROBLEM 50

51. **M** **V-HINT** **MMH** **CHALK** During a storm, a tree limb breaks off and comes to rest across a barbed wire fence at a point that is not in the middle between two fence posts. The limb exerts a downward force of 151 N on the wire. The left section of the wire makes an angle of 14.0° relative to the horizontal and sustains a tension of 447 N. Find the magnitude and direction of the tension that the right section of the wire sustains.

52. **M** **GO** **MMH** **CHALK** A toboggan slides down a hill and has a constant velocity. The angle of the hill is 8.00° with respect to the horizontal. What is the coefficient of kinetic friction between the surface of the hill and the toboggan?

53. **H** **SSM** A bicyclist is coasting straight down a hill at a constant speed. The combined mass of the rider and bicycle is 80.0 kg, and the hill is inclined at 15.0° with respect to the horizontal. Air resistance opposes the motion of the cyclist. Later, the bicyclist climbs the same hill at the same constant speed. How much force (directed parallel to the hill) must be applied to the bicycle in order for the bicyclist to climb the hill?

54. **H** **MMH** A kite is hovering over the ground at the end of a straight 43-m line. The tension in the line has a magnitude of 16 N. Wind blowing on the kite exerts a force of 19 N, directed 56° above the horizontal. Note that the line attached to the kite is *not* oriented at an angle of 56° above the horizontal. Find the height of the kite, relative to the person holding the line.

Section 4.12   Nonequilibrium Applications
of Newton's Laws of Motion

55. **E** A 1450-kg submarine rises straight up toward the surface. Seawater exerts both an upward buoyant force of 16 140 N on the submarine and a downward resistive force of 1030 N. What is the submarine's acceleration?

56. **E** **SSM** A 15-g bullet is fired from a rifle. It takes 2.50×10^{-3} s for the bullet to travel the length of the barrel, and it exits the barrel with a speed of 715 m/s. Assuming that the acceleration of the bullet is constant, find the average net force exerted on the bullet.

57. **E** A fisherman is fishing from a bridge and is using a "45-N test line." In other words, the line will sustain a maximum force of 45 N without breaking. What is the weight of the heaviest fish that can be pulled up vertically when the line is reeled in (a) at a constant speed and (b) with an acceleration whose magnitude is 2.0 m/s^2?

58. **E** **SSM** Only two forces act on an object (mass = 4.00 kg), as in the drawing. Find the magnitude and direction (relative to the *x* axis) of the acceleration of the object.

59. **E** A helicopter flies over the arctic ice pack at a constant altitude, towing an airborne 129-kg laser sensor that measures the thickness of the ice (see the drawing). The helicopter and the sensor both move only in the horizontal direction and have a horizontal acceleration of magnitude 2.84 m/s^2. Ignoring air resistance, find the tension in the cable towing the sensor.

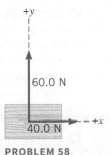

PROBLEM 58

PROBLEM 59

60. **E** Review Conceptual Example 17 as background for this problem. The water skier there has a mass of 73 kg. Find the magnitude of the net force acting on the skier when **(a)** she is accelerated from rest to a speed of 11 m/s in 8.0 s and **(b)** she lets go of the tow rope and glides to a halt in 21 s.

61. **E** A rescue helicopter is lifting a man (weight = 822 N) from a capsized boat by means of a cable and harness. **(a)** What is the tension in the cable when the man is given an initial upward acceleration of 1.10 m/s²? **(b)** What is the tension during the remainder of the rescue when he is pulled upward at a constant velocity?

62. **E** A car is towing a boat on a trailer. The driver starts from rest and accelerates to a velocity of +11 m/s in a time of 28 s. The combined mass of the boat and trailer is 410 kg. The frictional force acting on the trailer can be ignored. What is the tension in the hitch that connects the trailer to the car?

63. **E** **GO** A 292-kg motorcycle is accelerating up along a ramp that is inclined 30.0° above the horizontal. The propulsion force pushing the motorcycle up the ramp is 3150 N, and air resistance produces a force of 250 N that opposes the motion. Find the magnitude of the motorcycle's acceleration.

64. **E** **SSM** A student is skateboarding down a ramp that is 6.0 m long and inclined at 18° with respect to the horizontal. The initial speed of the skateboarder at the top of the ramp is 2.6 m/s. Neglect friction and find the speed at the bottom of the ramp.

65. **E** A man seeking to set a world record wants to tow a 109 000-kg airplane along a runway by pulling horizontally on a cable attached to the airplane. The mass of the man is 85 kg, and the coefficient of static friction between his shoes and the runway is 0.77. What is the greatest acceleration the man can give the airplane? Assume that the airplane is on wheels that turn without any frictional resistance.

66. **M** **V-HINT** A 205-kg log is pulled up a ramp by means of a rope that is parallel to the surface of the ramp. The ramp is inclined at 30.0° with respect to the horizontal. The coefficient of kinetic friction between the log and the ramp is 0.900, and the log has an acceleration of magnitude 0.800 m/s². Find the tension in the rope.

67. **M** **GO** To hoist himself into a tree, a 72.0-kg man ties one end of a nylon rope around his waist and throws the other end over a branch of the tree. He then pulls downward on the free end of the rope with a force of 358 N. Neglect any friction between the rope and the branch, and determine the man's upward acceleration.

68. **M** **CHALK** **SSM** Two objects (45.0 and 21.0 kg) are connected by a massless string that passes over a massless, frictionless pulley. The pulley hangs from the ceiling. Find **(a)** the acceleration of the objects and **(b)** the tension in the string.

69. **M** **GO** A train consists of 50 cars, each of which has a mass of 6.8×10^3 kg. The train has an acceleration of $+8.0 \times 10^{-2}$ m/s². Ignore friction and determine the tension in the coupling **(a)** between the 30th and 31st cars and **(b)** between the 49th and 50th cars.

70. **M** In Problem 65, an 85-kg man plans to tow a 109 000-kg airplane along a runway by pulling horizontally on a cable attached to it. Suppose that he instead attempts the feat by pulling the cable at an angle of 9.0° above the horizontal. The coefficient of static friction

between his shoes and the runway is 0.77. What is the greatest acceleration the man can give the airplane? Assume that the airplane is on wheels that turn without any frictional resistance.

71. **M** **MMH** The drawing shows a large cube (mass = 25 kg) being accelerated across a horizontal frictionless surface by a horizontal force $\vec{P}$. A small cube (mass = 4.0 kg) is in contact with the front surface of the large cube and will slide downward unless $\vec{P}$ is sufficiently large. The coefficient of static friction between the cubes is 0.71. What is the smallest magnitude that $\vec{P}$ can have in order to keep the small cube from sliding downward?

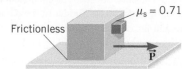

PROBLEM 71

72. **M** **V-HINT** The alarm at a fire station rings and an 86-kg fireman, starting from rest, slides down a pole to the floor below (a distance of 4.0 m). Just before landing, his speed is 1.4 m/s. What is the magnitude of the kinetic frictional force exerted on the fireman as he slides down the pole?

73. **M** **GO** Two blocks are sliding to the right across a horizontal surface, as the drawing shows. In Case A the mass of each block is 3.0 kg. In Case B the mass of block 1 (the block behind) is 6.0 kg, and the mass of block 2 is 3.0 kg. No frictional force acts on block 1 in either Case A or Case B. However, a kinetic frictional force of 5.8 N does act on block 2 in both cases and opposes the motion. For both Case A and Case B determine **(a)** the magnitude of the forces with which the blocks push against each other and **(b)** the magnitude of the acceleration of the blocks.

Case A   Case B

PROBLEM 73

74. **M** **SSM** A person whose weight is 5.20×10^2 N is being pulled up vertically by a rope from the bottom of a cave that is 35.1 m deep. The maximum tension that the rope can withstand without breaking is 569 N. What is the shortest time, starting from rest, in which the person can be brought out of the cave?

75. **M** **CHALK** A girl is sledding down a slope that is inclined at 30.0° with respect to the horizontal. The wind is aiding the motion by providing a steady force of 105 N that is parallel to the motion of the sled. The combined mass of the girl and the sled is 65.0 kg, and the coefficient of kinetic friction between the snow and the runners of the sled is 0.150. How much time is required for the sled to travel down a 175-m slope, starting from rest?

76. **H** A small sphere is hung by a string from the ceiling of a van. When the van is stationary, the sphere hangs vertically. However, when the van accelerates, the sphere swings backward so that the string makes an angle of θ with respect to the vertical. **(a)** Derive an expression for the magnitude a of the acceleration of the van in terms of the angle θ and the magnitude g of the acceleration due to gravity. **(b)** Find the acceleration of the van when $\theta = 10.0°$. **(c)** What is the angle θ when the van moves with a constant velocity?

77. **H** **SSM** A penguin slides at a constant velocity of 1.4 m/s down an icy incline. The incline slopes above the horizontal at an angle of 6.9°. At the bottom of the incline, the penguin slides onto a horizontal patch of ice. The coefficient of kinetic friction between the penguin and the ice is the same for the incline as for the horizontal patch. How much time is required for the penguin to slide to a halt after entering the horizontal patch of ice?

Additional Problems

Online

78. **E** **GO** A person with a black belt in karate has a fist that has a mass of 0.70 kg. Starting from rest, this fist attains a velocity of 8.0 m/s in 0.15 s. What is the magnitude of the average net force applied to the fist to achieve this level of performance?

79. **E** **SSM** A 95.0-kg person stands on a scale in an elevator. What is the apparent weight when the elevator is **(a)** accelerating upward with an acceleration of 1.80 m/s², **(b)** moving upward at a constant speed, and **(c)** accelerating downward with an acceleration of 1.30 m/s²?

80. **E** **MMH** Two forces, $\vec{F}_1$ and $\vec{F}_2$, act on the 7.00-kg block shown in the drawing. The magnitudes of the forces are $F_1 = 59.0$ N and $F_2 = 33.0$ N. What is the horizontal acceleration (magnitude and direction) of the block?

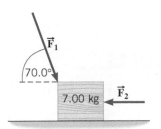

PROBLEM 80

81. **E** **SSM** A student presses a book between his hands, as the drawing indicates. The forces that he exerts on the front and back covers of the book are perpendicular to the book and are horizontal. The book weighs 31 N. The coefficient of static friction between his hands and the book is 0.40. To keep the book from falling, what is the magnitude of the minimum pressing force that each hand must exert?

PROBLEM 81

82. **E** In a European country a bathroom scale displays its reading in kilograms. When a man stands on this scale, it reads 92.6 kg. When he pulls down on a chin-up bar installed over the scale, the reading decreases to 75.1 kg. What is the magnitude of the force he exerts on the chin-up bar?

83. **E** **SSM** A 1380-kg car is moving due east with an initial speed of 27.0 m/s. After 8.00 s the car has slowed down to 17.0 m/s. Find the magnitude and direction of the net force that produces the deceleration.

84. **E** In preparation for this problem, review Conceptual Example 7. A space traveler whose mass is 115 kg leaves earth. What are his weight and mass **(a)** on earth and **(b)** in interplanetary space where there are no nearby planetary objects?

85. **E** **(a)** Calculate the magnitude of the gravitational force exerted on a 425-kg satellite that is a distance of two earth radii from the center of the earth. **(b)** What is the magnitude of the gravitational force exerted on the earth by the satellite? **(c)** Determine the magnitude of the satellite's acceleration. **(d)** What is the magnitude of the earth's acceleration?

86. **M** **SSM** **MMH** The drawing shows Robin Hood (mass = 77.0 kg) about to escape from a dangerous situation. With one hand, he is gripping the rope that holds up a chandelier (mass = 195 kg). When he cuts the rope where it is tied to the floor, the chandelier will fall, and he will be pulled up toward a balcony above. Ignore the friction between the rope and the beams over which it slides, and find **(a)** the acceleration with which Robin is pulled upward and **(b)** the tension in the rope while Robin escapes.

87. **M** **V-HINT** Consult Multiple-Concept Example 10 for insight into solving this type of problem. A box is sliding up an incline that makes an angle of 15.0° with respect to the horizontal. The coefficient of kinetic friction between the box and the surface of the incline is 0.180. The initial speed of the box at the bottom of the incline is 1.50 m/s. How far does the box travel along the incline before coming to rest?

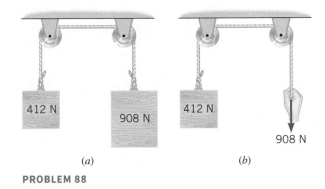

PROBLEM 86

88. **H** As part *a* of the drawing shows, two blocks are connected by a rope that passes over a set of pulleys. One block has a weight of 412 N, and the other has a weight of 908 N. The rope and the pulleys are massless and there is no friction. **(a)** What is the acceleration of the lighter block? **(b)** Suppose that the heavier block is removed, and a downward force of 908 N is provided by someone pulling on the rope, as part *b* of the drawing shows. Find the acceleration of the remaining block. **(c)** Explain why the answers in (a) and (b) are different.

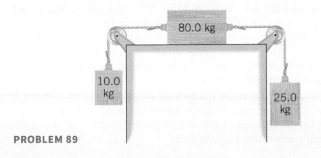

(a)   *(b)*

PROBLEM 88

89. **H** **SSM** The three objects in the drawing are connected by strings that pass over massless and friction-free pulleys. The objects move, and the coefficient of kinetic friction between the middle object and the surface of the table is 0.100. **(a)** What is the acceleration of the three objects? **(b)** Find the tension in each of the two strings.

80.0 kg

10.0 kg

25.0 kg

PROBLEM 89

90. M GO A flatbed truck is carrying a crate up a hill of angle of inclination $\theta = 10.0°$, as the figure illustrates. The coefficient of static friction between the truck bed and the crate is $\mu_s = 0.350$. Find the maximum acceleration that the truck can attain before the crate begins to slip backward relative to the truck.

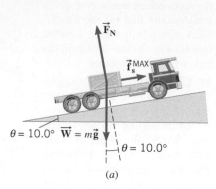

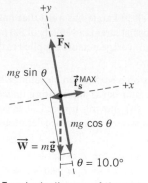

(b) Free-body diagram of the crate

PROBLEM 90

Physics in Biology, Medicine, and Sports

91. M BIO **4.3** To this point, we have neglected the effects of air resistance (or drag) on falling bodies, which acts in the opposite direction of the object's weight. However, drag can significantly affect the motion of a falling object. The drag force (F_d) due to air resistance can be represented by the following relationship: $F_d = CAv^n$, where C is the coefficient of air friction, A is the cross-sectional area of the falling object, and v is the object's speed relative to the air. The value of n will vary depending on the object's speed. It typically has a value near 1 at low speeds and 2 at high speeds. The important point is that, as the speed of the object increases, the drag force increases, and the acceleration decreases. The figure shows the free-body diagram for a skydiver, where drag has been included. Eventually, the drag force will be equal to the weight, and the sum of the forces acting on the skydiver will be zero. Thus, the acceleration will be zero, and the skydiver will fall at constant speed. This speed is referred to as the *terminal velocity*. What is the terminal velocity of a 95-kg skydiver? Assume the cross-sectional area of the skydiver is 0.20 m², the coefficient of air friction is 0.90 kg/m³, and $n = 2$.

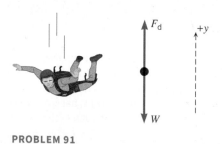

PROBLEM 91

92. E BIO **4.8** Walking up an inclined surface can have tremendous health benefits. Your leg muscles and hip flexors work harder to propel you up the incline than when you walk across level terrain. Likewise, your abdominal muscles must contract to keep your body upright and stable, since it is not perpendicular to the inclined surface, which helps to strengthen your core. Imagine a person in leather boots standing on an inclined oak plank, as the figure shows. If the coefficient of static friction between her boots and the plank is 0.60, what is the angle of the incline, assuming she is at the maximum angle before she begins to slide?

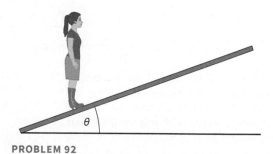

PROBLEM 92

93. M BIO **4.10** The longest bone in the human body is the femur. It is the only bone located in the upper leg. Fractures, or breaking of the femur bone, often involve high-speed accidents, such as from skiing or in an automobile collision. An alternative way to apply skeletal traction to the injured leg, as compared to Example 12, is shown in the figure. **(a)** What is the tension in the rope, and **(b)** what is the traction force (F_T) that acts on the leg?

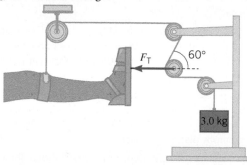

PROBLEM 93

94. M BIO **4.3** The sport of American football often involves athletes running at high speeds and smashing into each other. While they wear protective pads and helmets, their bodies still experience large decelerations and forces. In fact, some of the collisions between players, or between a player and the ground, are comparable to what a person might endure in an automobile accident. In 2017, a large study found that 110 out of 111 deceased NFL players had *chronic traumatic encephalopathy* (CTE), a degenerative brain disorder associated with repetitive head trauma (see photo). Other studies have linked CTE to higher rates of early onset dementia and memory loss. Consider an NFL player running at 3.92 m/s. During a collision and tackle, his head is brought to a stop in a distance of 0.125 m. (a) Calculate the magnitude of the deceleration of the player's head (assume the deceleration to be constant and treat this as a one-dimensional problem). How does it compare to the acceleration of gravity near the surface of Earth? (b) Assume the mass of the player's brain is 1.35 kg and calculate the force experienced by the brain.

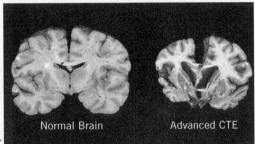

Boston University Center for the Study of Traumatic Encephalopathy, Image taken from https://commons.wikimedia.org/wiki/File:Chronic_Traumatic_Encephalopathy.png. Licensed under CC by 4.0

PROBLEM 94

95. **M** **BIO** **4.8** Doctor Doom's Fearfall ride at Universal Studios in Orlando, Florida launches thrill seekers straight upward with accelerations greater than what astronauts experienced during the launch of the Space Shuttle. Assume a 75.0-kg rider undergoes a constant acceleration and reaches a maximum speed of 40.0 mph during the first 5.10 m of the trip. What is the apparent weight of the rider during this time?

PROBLEM 95

96. **E** **BIO** **4.4** At an instant when a soccer ball is in contact with the foot of a player kicking it, the horizontal or x component of the ball's acceleration is 810 m/s^2 and the vertical or y component of its acceleration is 1100 m/s^2. The ball's mass is 0.43 kg. What is the magnitude of the net force acting on the soccer ball at this instant?

97. **E** **GO** **BIO** **4.4** A billiard ball strikes and rebounds from the cushion of a pool table perpendicularly. The mass of the ball is 0.38 kg. The ball approaches the cushion with a velocity of +2.1 m/s and rebounds with a velocity of −2.0 m/s. The ball remains in contact with the cushion for a time of 3.3 × 10^{-3} s. What is the average net force (magnitude and direction) exerted on the ball by the cushion?

98. **E** **MMH** **BIO** **4.9** An 81-kg baseball player slides into second base. The coefficient of kinetic friction between the player and the ground is 0.49. (a) What is the magnitude of the frictional force? (b) If the player comes to rest after 1.6 s, what was his initial velocity?

99. **E** **MMH** **BIO** **4.9** The speed of a bobsled is increasing because it has an acceleration of 2.4 m/s^2. At a given instant in time, the forces resisting the motion, including kinetic friction and air resistance, total 450 N. The combined mass of the bobsled and its riders is 270 kg. (a) What is the magnitude of the force propelling the bobsled forward? (b) What is the magnitude of the net force that acts on the bobsled?

100. **E** **BIO** **4.3** When a 58-g tennis ball is served, it accelerates from rest to a speed of 45 m/s. The impact with the racket gives the ball a constant acceleration over a distance of 44 cm. What is the magnitude of the net force acting on the ball?

101. **M** **GO** **BIO** **4.9** A skater with an initial speed of 7.60 m/s stops propelling himself and begins to coast across the ice, eventually coming to rest. Air resistance is negligible. (a) The coefficient of kinetic friction between the ice and the skate blades is 0.100. Find the deceleration caused by kinetic friction. (b) How far will the skater travel before coming to rest?

102. **M** **GO** **BIO** **4.11** A mountain climber, in the process of crossing between two cliffs by a rope, pauses to rest. She weighs 535 N. As the drawing shows, she is closer to the left cliff than to the right cliff, with the result that the tensions in the left and right sides of the rope are not the same. Find the tensions in the rope to the left and to the right of the mountain climber.

PROBLEM 102

103. **M** **BIO** **V-HINT** **4.11** The person in the drawing is standing on crutches. Assume that the force exerted on each crutch by the ground is directed along the crutch, as the force vectors in the drawing indicate. If the coefficient of static friction between a crutch and the ground is 0.90, determine the largest angle θ^{MAX} that the crutch can have just before it begins to slip on the floor.

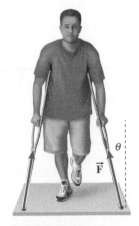

PROBLEM 103

Concepts and Calculations Problems

Online

Newton's three laws of motion provide the basis for understanding the effect of forces on the motion of an object. The second law is especially important, because it provides the quantitative relationship between net force and acceleration. The following problems serve as a review of the essential features of this relationship.

104. **M** **CHALK** **SSM** The figure shows two forces, $\vec{F}_1 = +3000$ N and $\vec{F}_2 = +5000$ N, acting on a spacecraft; the plus signs indicate that the forces are directed along the +x axis. A third force $\vec{F}_3$ also acts on the spacecraft but is not shown in the drawing. *Concepts:* (i) Suppose the spacecraft were stationary. What would be the

direction of $\vec{F}_3$? (ii) When the spacecraft is moving at a constant velocity of +850 m/s, what is the direction of $\vec{F}_3$? *Calculations:* Find the direction and magnitude of $\vec{F}_3$.

PROBLEM 104

105. **M** **CHALK** On earth a block has a weight of 88 N. This block is sliding on a horizontal surface on the moon, where the acceleration due to gravity is 1.60 m/s². As the figure shows, the block is being pulled by a horizontal rope in which the tension is $T = 24$ N. The coefficient of kinetic friction between the block and the surface is $\mu_k = 0.20$. *Concepts:* (i) Which of Newton's laws of motion provides the best way to determine the acceleration of the block? (ii) Does the net force in the *x* direction equal the tension *T*? *Calculations:* Determine the acceleration of the block.

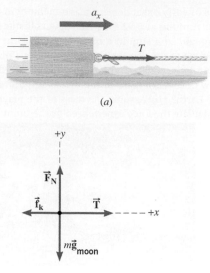

(a)

(b) Free-body diagram for the block

PROBLEM 105

Team Problems

Online

106. **E** **T** **WS** **A Water Balloon Launcher.** Two metal poles sticking vertically out of the ground are separated by 1.50 m. A giant slingshot is formed by attaching two pieces of elastic surgical tubing to the poles (one each) with their opposite ends attached to a launch basket that holds a water balloon. The two elastic tubes are the same length, but are different (one is harder to stretch than the other). The launching mechanism requires that the basket be pulled along a line that bisects, and is perpendicular to, the line connecting the two poles. You and your team are trying to determine how to aim this balloon launcher taking into consideration that the two sides of the slingshot provide forces of different magnitudes. When the launch basket is pulled 2.00 meters straight back (parallel to the ground) and connected to the triggering mechanism, you find that the magnitude of the pulling force exerted on the basket by the tubing on the left pole is 1.33 times larger than that exerted by the tubing connected to the right pole. How will the direction of the initial acceleration (the instant after the trigger is pulled) of the launch basket deviate from the straight-line path, that is, from the line that bisects the poles? Express your answer in terms of the angle of deviation to the left or right of the bisecting line.

107. **M** **T** **WS** **Extracting a Truck Engine.** You and your team removed a 175-kg engine from the chassis of a truck and hung it from a chain of length $L = 2.80$ m from the 4.00-m high ceiling of a garage. You now want to move the truck out of the garage and move in another vehicle into which you intend to place the engine. To make the exchange, you must pull the hanging engine to the side so that the chain makes an angle of $\theta = 30.0°$ with respect to the vertical. Located on the wall directly to the right of the engine (i.e., in the direction you want to pull it) is a pulley mounted 1.20 m below the ceiling. You measure a horizontal distance of 2.35 m between the wall and the point where the chain is connected to the engine while it is hanging vertically. You get the idea to thread a rope through the pulley on the wall, connect it to the engine at the same place where the chain is connected, and then pull the rope to move the engine to the side. However, you need a rope that can safely do the job. What is the minimum tension the rope should be rated to handle? Give your answer in both newtons and pounds.

108. **E** **T** **Towing a Detector.** You are in a helicopter towing a 129-kg laser detector that is mapping out the thickness of the Brunt Ice Shelf along the coast of Antarctica. The original cable used to suspend the detector was damaged and replaced by a lighter one with a maximum tension rating of 310.0 pounds, not much more than the weight of the detector. The replacement cable would work without question in the case that the detector and helicopter were not accelerating. However, some acceleration of the helicopter is inevitable. In order to monitor the tension force on the cable to make sure the maximum is not exceeded (and therefore to not lose the very expensive detector) you calculate the maximum angle the cable can make with the vertical without the cable exceeding the tension limit. (a) Assuming straight and level flight of the helicopter, what is that maximum angle? (b) What is the corresponding acceleration? (c) Your colleague wants to add a 10.0-kg infrared camera to the detector. What is the maximum allowable angle now?

109. **M** **T** **Stuck on a Ledge.** You and your friends pull your car off a mountain road onto an overlook to take pictures of the frozen lake below. You lose your footing on the packed snow, fall over the cliff, and land in a soft snow bank on a ledge, 10 feet below the edge. The cliff's wall is too smooth to climb upwards and too steep and dangerous to climb the remaining 300 feet to the bottom. While trying to think your way out of your predicament, you realize one reason you had slipped was that you had misjudged how the cliff sloped towards the edge (a decline of about 5–10 degrees). There is nothing near the edge to which your friends can tie a rope, and they fear the footing is too slippery for them to pull you to the top. However, they can get the car close enough to the edge to tie the rope to the bumper and hang it over the edge. You are worried, however, that the car might slip over the edge as well. On your phone (which has survived the fall) you look up the coefficient of static friction of snow tires on packed snow and find that it ranges from 0.22 to 0.26, and that the model of your friends' car weighs 3856 lb. Your own weight fluctuates between 120–130 lb. (a) What is the worst-case scenario in terms of the parameters given? (b) Assuming the worst-case scenario, should your friends bring the car close to the edge? (c) Next, should you risk tying the rope to the trailer hitch and trying to climb out? You assume the hitch is low enough to the ground so that the tension force is parallel to the slope near the edge of the cliff and that you climb up the rope at a constant velocity.

Crew members are refueling a V-22 Osprey helicopter. The green lights near the tips of its rotating blades experience a net force and acceleration that points toward the center of the circle. We will now see how and why this net force and acceleration arise.

Chief Petty Officer Joe Kane (U.S. Navy), Image taken from https://commons.wikimedia.org/wiki/File:20080406165033%21V-22_Osprey_refueling_edit1.jpg

Dynamics of Uniform Circular Motion

5.1 Uniform Circular Motion

There are many examples of motion on a circular path. Of the many possibilities, we single out those that satisfy the following definition:

> **DEFINITION OF UNIFORM CIRCULAR MOTION**
> **Uniform circular motion is the motion of an object traveling at a constant (uniform) speed on a circular path.**

As an example of uniform circular motion, **Figure 5.1** shows a model airplane on a guideline. The speed of the plane is the magnitude of the velocity vector $\vec{v}$, and since the speed is constant, the vectors in the drawing have the same magnitude at all points on the circle.

Sometimes it is more convenient to describe uniform circular motion by specifying the period of the motion, rather than the speed. The **period** T is the time required to travel once around the circle—that is, to make one complete revolution. There is a relationship between

LEARNING OBJECTIVES

After reading this module, you should be able to...

5.1 Define uniform circular motion.

5.2 Solve uniform circular motion kinematic problems.

5.3 Solve uniform circular motion dynamic problems.

5.4 Solve problems involving banked curves.

5.5 Analyze circular gravitational orbits.

5.6 Solve application problems involving gravity and uniform circular motion.

5.7 Analyze vertical circular motion.

FIGURE 5.1 The motion of a model airplane flying at a constant speed on a horizontal circular path is an example of uniform circular motion.

period and speed, since speed v is the distance traveled (here, the circumference of the circle = $2\pi r$) divided by the time T:

$$v = \frac{2\pi r}{T} \tag{5.1}$$

If the radius is known, as in Example 1, the speed can be calculated from the period, or vice versa.

EXAMPLE 1 | A Tire-Balancing Machine

The wheel of a car has a radius of $r = 0.29$ m and is being rotated at 830 revolutions per minute (rpm) on a tire-balancing machine. Determine the speed (in m/s) at which the outer edge of the wheel is moving.

Reasoning The speed v can be obtained directly from $v = 2\pi r/T$, but first the period T is needed. The period is the time for one revolution, and it must be expressed in seconds, because the problem asks for the speed in meters per second.

Solution Since the tire makes 830 revolutions in one minute, the number of minutes required for a single revolution is

$$\frac{1}{830 \text{ revolutions/min}} = 1.2 \times 10^{-3} \text{ min/revolution}$$

Therefore, the period is $T = 1.2 \times 10^{-3}$ min, which corresponds to 0.072 s. Equation 5.1 can now be used to find the speed:

$$v = \frac{2\pi r}{T} = \frac{2\pi(0.29 \text{ m})}{0.072 \text{ s}} = \boxed{25 \text{ m/s}}$$

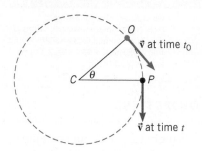

(a)

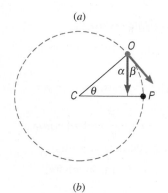

(b)

FIGURE 5.2 (a) For an object (•) in uniform circular motion, the velocity $\vec{v}$ has different directions at different places on the circle. (b) The velocity vector has been removed from point P, shifted parallel to itself, and redrawn with its tail at point O.

The definition of uniform circular motion emphasizes that the speed, or the magnitude of the velocity vector, is constant. It is equally significant that the direction of the vector is *not constant*. In **Figure 5.1**, for instance, the velocity vector changes direction as the plane moves around the circle. Any change in the velocity vector, even if it is only a change in direction, means that an acceleration is occurring. This particular acceleration is called "centripetal acceleration," because it points toward the center of the circle, as the next section explains.

5.2 Centripetal Acceleration

In this section we determine how the magnitude a_c of the centripetal acceleration depends on the speed v of the object and the radius r of the circular path. We will see that $a_c = v^2/r$.

In **Figure 5.2a** an object (symbolized by a dot •) is in uniform circular motion. At a time t_0 the velocity is tangent to the circle at point O, and at a later time t the velocity is tangent at point P. As the object moves from O to P, the radius traces out the angle θ, and the velocity vector changes direction. To emphasize the change, part b of the picture shows the velocity vector removed from point P, shifted parallel to itself, and redrawn with its tail at point O. The angle β between the two vectors indicates the change in direction. Since the radii CO and CP are perpendicular to the tangents at points O and P, respectively, it follows that $\alpha + \beta = 90°$ and $\alpha + \theta = 90°$. Therefore, angle β and angle θ are equal.

As always, acceleration is the change $\Delta\vec{v}$ in velocity divided by the elapsed time Δt, or $\vec{a} = \Delta\vec{v}/\Delta t$. **Figure 5.3a** shows the two velocity vectors oriented at the angle θ with respect to one another, together with the vector $\Delta\vec{v}$ that represents the change

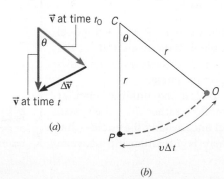

(a)

(b)

FIGURE 5.3 (a) The directions of the velocity vector at times t and t_0 differ by the angle θ. (b) When the object moves along the circle from O to P, the radius r traces out the same angle θ. Here, the sector COP has been rotated clockwise by 90° relative to its orientation in **Figure 5.2**.

in velocity. The change $\Delta\vec{v}$ is the increment that must be added to the velocity at time t_0, so that the resultant velocity has the new direction after an elapsed time $\Delta t = t - t_0$. **Figure 5.3b** shows the sector of the circle *COP*. In the limit that Δt is very small, the arc length *OP* is approximately a straight line whose length is the distance $v\,\Delta t$ traveled by the object. In this limit, *COP* is an isosceles triangle, as is the triangle in part *a* of the drawing. Note that both triangles have equal angles θ. This means that they are similar triangles, so that

$$\frac{\Delta v}{v} = \frac{v\,\Delta t}{r}$$

This equation can be solved for $\Delta v/\Delta t$, to show that the magnitude a_c of the centripetal acceleration is given by $a_c = v^2/r$.

Centripetal acceleration is a vector quantity and, therefore, has a direction as well as a magnitude. The direction is toward the center of the circle, and Conceptual Example 2 helps us to set the stage for explaining this important fact.

CONCEPTUAL EXAMPLE 2 | Which Way Will the Object Go?

In **Figure 5.4** an object, such as a model airplane on a guideline, is in uniform circular motion. The object is symbolized by a dot (•), and at point *O* it is released suddenly from its circular path. For instance, suppose that the guideline for a model plane is cut suddenly. Does the object move **(a)** along the straight tangent line between points *O* and *A* or **(b)** along the circular arc between points *O* and *P*?

Reasoning Newton's first law of motion (see Section 4.2) guides our reasoning. This law states that an object continues in a state of rest or in a state of motion at a constant velocity (i.e., at a constant speed along a straight line) unless compelled to change that state by a net force. When an object is suddenly released from its circular path, there is no longer a net force being applied to the object. In the case of the model airplane, the guideline cannot apply a force, since it is cut. Gravity certainly acts on the plane, but the wings provide a lift force that balances the weight of the plane.

Answer (b) is incorrect. An object such as a model airplane will remain on a circular path only if a net force keeps it there. Since there is no net force, it cannot travel on the circular arc.

Answer (a) is correct. In the absence of a net force, the plane or any object would continue to move at a constant speed along a straight line in the direction it had at the time of release, consistent with Newton's first law. This speed and direction are given in **Figure 5.4** by the velocity vector $\vec{v}$.

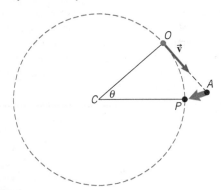

FIGURE 5.4 If an object (•) moving on a circular path were released from its path at point *O*, it would move along the straight tangent line *OA* in the absence of a net force.

As Example 2 discusses, the object in **Figure 5.4** would travel on a tangent line if it were released from its circular path suddenly at point *O*. It would move in a straight line to point *A* in the time it would have taken to travel on the circle to point *P*. It is as if, in the process of remaining on the circle, the object drops through the distance *AP*, and *AP* is directed toward the center of the circle in the limit that the angle θ is small. Thus, the object in uniform circular motion accelerates toward the center of the circle at every moment. Since the word "centripetal" means "moving toward a center," the acceleration is called **centripetal acceleration.**

CENTRIPETAL ACCELERATION

Magnitude: The centripetal acceleration of an object moving with a speed v on a circular path of radius r has a magnitude a_c given by

$$a_c = \frac{v^2}{r} \tag{5.2}$$

Direction: The centripetal acceleration vector always points toward the center of the circle and continually changes direction as the object moves.

The following example illustrates the effect of the radius r on the centripetal acceleration.

EXAMPLE 3 | The Physics of a Bobsled Track

The bobsled track at the 1994 Olympics in Lillehammer, Norway, contained turns with radii of 33 m and 24 m, as **Interactive Figure 5.5** illustrates. Find the centripetal acceleration at each turn for a speed of 34 m/s, a speed that was achieved in the two-man event. Express the answers as multiples of $g = 9.8$ m/s^2.

Reasoning   In each case, the magnitude of the centripetal acceleration can be obtained from $a_c = v^2/r$. Since the radius r is in the denominator on the right side of this expression, we expect the acceleration to be smaller when r is larger.

Solution   From $a_c = v^2/r$ it follows that

Radius = **33 m** $a_c = \dfrac{(34 \text{ m/s})^2}{33 \text{ m}} = 35$ m/s^2 = $\boxed{3.6\ g}$

Radius = **24 m** $a_c = \dfrac{(34 \text{ m/s})^2}{24 \text{ m}} = 48$ m/s^2 = $\boxed{4.9\ g}$

The centripetal acceleration is indeed smaller when the radius is larger. In fact, with r in the denominator on the right of $a_c = v^2/r$, the acceleration approaches zero when the radius becomes very large. Uniform circular motion along the arc of an infinitely large circle entails no acceleration, because it is just like motion at a constant speed along a straight line.

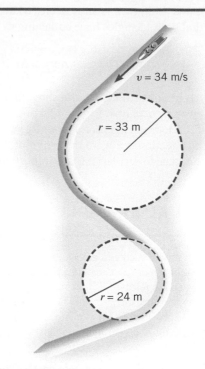

INTERACTIVE FIGURE 5.5   This bobsled travels at the same speed around two curves with different radii. For the turn with the larger radius, the sled has a smaller centripetal acceleration.

In Section 4.11 we learned that an object is in equilibrium when it has zero acceleration. Conceptual Example 4 discusses whether an object undergoing uniform circular motion can ever be at equilibrium.

CONCEPTUAL EXAMPLE 4 | Uniform Circular Motion and Equilibrium

A car moves at a constant speed along a straight line as it approaches a circular turn. In which of the following parts of the motion is the car in equilibrium? **(a)** As it moves along the straight line toward the circular turn, **(b)** as it is going around the turn, **(c)** as it moves away from the turn along a straight line.

Reasoning   An object is in equilibrium when it has no acceleration, according to the definition given in Section 4.11. If the object's velocity remains constant, both in magnitude and direction, its acceleration is zero.

Answer (b) is incorrect.   As the car goes around the turn, the direction of travel changes, so the car has a centripetal acceleration that is characteristic of uniform circular motion. Because of this acceleration, the car is not in equilibrium during the turn.

Answers (a) and (c) are correct.   As the car either approaches the turn or moves away from the turn it is traveling along a straight line, and both the speed and direction of the motion are constant. Thus, the velocity vector does not change, and there is no acceleration. Consequently, for these parts of the motion, the car is in equilibrium.

Related Homework: Problem 10

We have seen that going around tight turns (smaller r) and gentle turns (larger r) at the same speed entails different centripetal accelerations. And most drivers know that such turns "feel" different. This feeling is associated with the force that is present in uniform circular motion, and we turn to this topic in the next section.

Check Your Understanding

(The answers are given at the end of the book.)

1. The car in **CYU Figure 5.1** is moving clockwise around a circular section of road at a constant speed. What are the directions of its velocity and acceleration at **(a)** position 1 and **(b)** position 2? Specify your responses as north, east, south, or west.

2. The speedometer of your car shows that you are traveling at a constant speed of 35 m/s. Is it possible that your car is accelerating?

3. Consider two people, one on the earth's surface at the equator and the other at the north pole. Which has the larger centripetal acceleration?

4. Which of the following statements about centripetal acceleration is true? **(a)** An object moving at a constant velocity cannot have a centripetal acceleration. **(b)** An object moving at a constant speed may have a centripetal acceleration.

5. A car is traveling at a constant speed along the road *ABCDE* shown in **CYU Figure 5.2.** Sections *AB* and *DE* are straight. Rank the accelerations in each of the four sections according to magnitude, listing the smallest first.

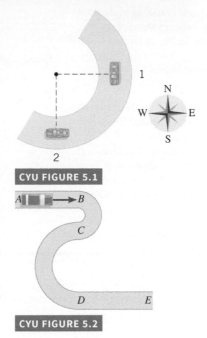

CYU FIGURE 5.1

CYU FIGURE 5.2

5.3 | Centripetal Force

Newton's second law indicates that whenever an object accelerates, there must be a net force to create the acceleration. Thus, in uniform circular motion there must be a net force to produce the centripetal acceleration. The second law gives this net force as the product of the object's mass m and its acceleration v^2/r. The net force causing the centripetal acceleration is called the **centripetal force** $\vec{F}_c$ and points in the same direction as the acceleration—that is, toward the center of the circle.

> **CENTRIPETAL FORCE**
>
> *Magnitude:* The centripetal force is the name given to the net force required to keep an object of mass m, moving at a speed v, on a circular path of radius r, and it has a magnitude of
>
> $$F_c = \frac{mv^2}{r} \qquad (5.3)$$
>
> *Direction:* The centripetal force always points toward the center of the circle and continually changes direction as the object moves.

The phrase "centripetal force" does not denote a new and separate force created by nature. The phrase merely labels the net force pointing toward the center of the circular path, and this net force is the vector sum of all the force components that point along the radial direction. Sometimes the centripetal force consists of a single force such as tension (see Example 5), friction (see Example 7), the normal force or a component thereof (see Examples 9 and 14), or the gravitational force (see Examples 10–12). However, there are circumstances when a number of different forces contribute simultaneously to the centripetal force (see Section 5.7).

In some cases, it is easy to identify the source of the centripetal force, as when a model airplane on a guideline flies in a horizontal circle. The only force pulling the plane inward is the tension in the line, so this force alone (or a component of it) is the centripetal force. Example 5 illustrates the fact that higher speeds require greater tensions.

EXAMPLE 5 | The Effect of Speed on Centripetal Force

The model airplane in **Figure 5.6** has a mass of 0.90 kg and moves at a constant speed on a circle that is parallel to the ground. The path of the airplane and its guideline lie in the same horizontal plane, because the weight of the plane is balanced by the lift generated by its wings. Find the tension in the guideline (length = 17 m) for speeds of 19 and 38 m/s.

Reasoning Since the plane flies on a circular path, it experiences a centripetal acceleration that is directed toward the center of the circle. According to Newton's second law of motion, this acceleration is produced by a net force that acts on the plane, and this net force is called the centripetal force. The centripetal force is also directed toward the center of the circle. Since the tension in the guideline is the only force pulling the plane inward, it must be the centripetal force.

Solution Equation 5.3 gives the tension directly: $F_c = T = mv^2/r$.

$$\textbf{Speed = 19 m/s} \quad T = \frac{(0.90 \text{ kg})(19 \text{ m/s})^2}{17 \text{ m}} = \boxed{19 \text{ N}}$$

$$\textbf{Speed = 38 m/s} \quad T = \frac{(0.90 \text{ kg})(38 \text{ m/s})^2}{17 \text{ m}} = \boxed{76 \text{ N}}$$

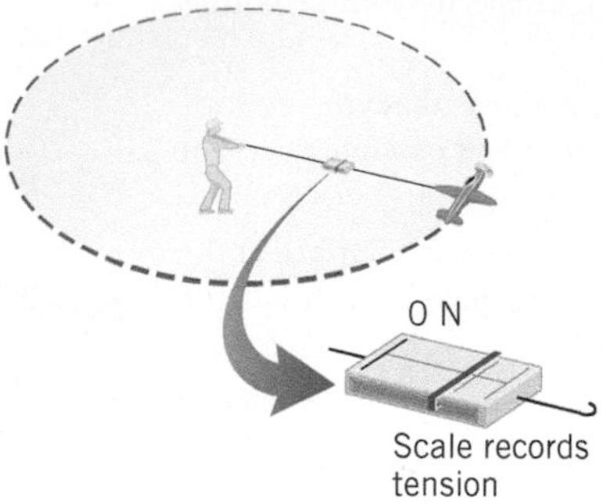

FIGURE 5.6 The scale records the tension in the guideline. See Example 5.

Scale records tension

0 N

Conceptual Example 6 deals with another case where it is easy to identify the source of the centripetal force.

CONCEPTUAL EXAMPLE 6 | The Physics of a Trapeze Act

In a circus, a man hangs upside down from a trapeze, legs bent over the bar and arms downward, holding his partner (see **Figure 5.7**). Is it harder for the man to hold his partner **(a)** when the partner hangs straight down and is stationary or **(b)** when the partner is swinging through the straight-down position?

Reasoning Whenever an object moves on a circular path, it experiences a centripetal acceleration that is directed toward the center of the path. A net force, known as the centripetal force, is required to produce this acceleration.

Answer (a) is incorrect. When the man and his partner are stationary, the man's arms must support only his partner's weight. When the two are swinging, however, the man's arms must provide the additional force required to produce the partner's centripetal acceleration. Thus, it is easier, not harder, for the man to hold his partner when the partner hangs straight down and is stationary.

Answer (b) is correct. When the two are swinging, the partner is moving on a circular arc and, therefore, has a centripetal acceleration. The man's arms must support the partner's weight and must

simultaneously exert an additional pull to provide the centripetal force that produces this acceleration. Because of the additional pull, it is harder for the man to hold his partner while swinging.

Related Homework: Problems 14, 16

FIGURE 5.7 See Example 6 for a discussion of the role of centripetal force in this trapeze act.

INTERCHROME/SuperStock

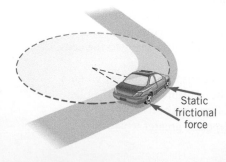

FIGURE 5.8 When the car moves without skidding around a curve, static friction between the road and the tires provides the centripetal force to keep the car on the road.

Static frictional force

When a car moves at a steady speed around an unbanked curve, the centripetal force keeping the car on the curve comes from the static friction between the road and the tires, as **Figure 5.8** indicates. It is static, rather than kinetic friction, because the tires are not slipping with respect to the radial direction. If the static frictional force is insufficient, given the speed and the radius of the turn, the car will skid off the road. Example 7 shows how an icy road can limit safe driving.

Analyzing Multiple-Concept Problems

EXAMPLE 7 | Centripetal Force and Safe Driving

At what maximum speed can a car safely negotiate a horizontal unbanked turn (radius = 51 m) in dry weather (coefficient of static friction = 0.95) and icy weather (coefficient of static friction = 0.10)?

Reasoning The speed v at which the car of mass m can negotiate a turn of radius r is related to the centripetal force F_c that is available, according to $F_c = mv^2/r$ (Equation 5.3). Static friction provides the centripetal force and can provide a maximum force f_s^{MAX} given by $f_s^{MAX} = \mu_s F_N$ (Equation 4.7), in which μ_s is the coefficient of static friction and F_N is the normal force. Thus, we will use the fact that $F_c = f_s^{MAX}$ to obtain the maximum speed from Equation 5.3. It will also be necessary to evaluate the normal force, which we will do by considering that it must balance the weight of the car. Experience indicates that the maximum speed should be greater for the dry road than for the icy road.

Knowns and Unknowns The data for this problem are as follows:

Description	Symbol	Value	Comment
Radius of turn	r	51 m	
Coefficient of static friction	μ_s	0.95	Dry conditions
Coefficient of static friction	μ_s	0.10	Icy conditions
Unknown Variable			
Speed of car	v	?	

Modeling the Problem

STEP 1 Speed and Centripetal Force According to Equation 5.3, we have

$$F_c = \frac{mv^2}{r}$$

Solving for the speed v gives Equation 1 at the right. To use this result, it is necessary to consider the centripetal force F_c, which we do in Step 2.

$$v = \sqrt{\frac{rF_c}{m}} \quad (1)$$

STEP 2 Static Friction The force of static friction is the centripetal force, and the greater its value, the greater is the speed at which the car can negotiate the turn. The maximum force f_s^{MAX} of static friction is given by $f_s^{MAX} = \mu_s F_N$ (Equation 4.7). Thus, the maximum available centripetal force is

$$\boxed{F_c = f_s^{MAX} = \mu_s F_N}$$

which we can substitute into Equation 1, as shown at the right. The next step considers the normal force F_N.

$$v = \sqrt{\frac{rF_c}{m}} \quad (1)$$
$$F_c = \mu_s F_N \quad (2)$$

STEP 3 The Normal Force The fact that the car does not accelerate in the vertical direction tells us that the normal force balances the car's weight mg, or

$$\boxed{F_N = mg}$$

This result for the normal force can now be substituted into Equation 2, as indicated at the right.

$$v = \sqrt{\frac{rF_c}{m}} \quad (1)$$
$$F_c = \mu_s F_N \quad (2)$$
$$F_N = mg$$

Solution Algebraically combining the results of each step, we find that

STEP 1   STEP2   STEP 3

$$v = \sqrt{\frac{rF_c}{m}} = \sqrt{\frac{r\mu_s F_N}{m}} = \sqrt{\frac{r\mu_s mg}{m}}$$

The mass m of the car is eliminated algebraically from this result, and we find that the maximum speeds are

Dry road ($\mu_s = 0.95$) $v = \sqrt{r\mu_s g} = \sqrt{(51\text{ m})(0.95)(9.8\text{ m/s}^2)} = \boxed{22\text{ m/s}}$

Icy road ($\mu_s = 0.10$) $v = \sqrt{r\mu_s g} = \sqrt{(51\text{ m})(0.10)(9.8\text{ m/s}^2)} = \boxed{7.1\text{ m/s}}$

As expected, the dry road allows a greater maximum speed.

Problem-Solving Insight
When using an equation to obtain a numerical answer, algebraically solve for the unknown variable in terms of the known variables. Then substitute in the numbers for the known variables, as this example shows.

Related Homework: Problems 12, 17, 22

A passenger in **Figure 5.8** must also experience a centripetal force to remain on the circular path. However, if the upholstery is very slippery, there may not be enough static friction to keep him in place as the car makes a tight turn at high speed. Then, when viewed from inside the car, he appears to be thrown toward the outside of the curve. What really happens is that the passenger slides off on a tangent to the circle, until he encounters a source of centripetal force to keep him in place while the car turns. This occurs when the passenger bumps into the side of the car, which pushes on him with the necessary force. While rounding a curve in a car can be exciting, moving on a circular path at high speed can have significant physiological effects, as the next example illustrates.

THE PHYSICS OF . . . flying an airplane in a banked turn.   Sometimes the source of the centripetal force is not obvious. A pilot making a turn, for instance, banks or tilts the plane at an angle to create the centripetal force. As a plane flies, the air pushes upward on the wing surfaces with a net lifting force $\vec{L}$ that is perpendicular to the wing surfaces, as **Animated Figure 5.9a** shows. Part *b* of the drawing illustrates that when the plane is banked at an angle θ, a component $L \sin \theta$ of the lifting force is directed toward the center of the turn. It is this component that provides the centripetal force. Greater speeds and/or tighter turns require greater centripetal forces. In such situations, the pilot must bank the plane at a larger angle, so that a larger component of the lift points toward the center of the turn. The technique of banking into a turn also has an application in the construction of high-speed roadways, where the road itself is banked to achieve a similar effect, as the next section discusses.

ANIMATED FIGURE 5.9 (*a*) The air exerts an upward lifting force $\frac{1}{2}\vec{L}$ on each wing. (*b*) When a plane executes a circular turn, the plane banks at an angle θ. The lift component $L \sin \theta$ is directed toward the center of the circle and provides the centripetal force.

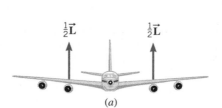
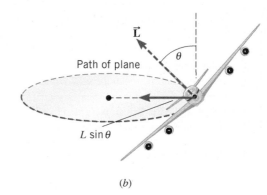

(*a*)   (*b*)

EXAMPLE 8 | BIO The Physics of *g*-LOC

The National Aeronautics and Space Administration (NASA) studies the physiological effects of large accelerations on astronauts. Some of these studies use a machine known as a centrifuge. This machine consists of a long arm, rotating about an axis through its center. At the two ends of the arm, and other distances from the center, are chambers in which the astronauts sit. The astronauts move on a circular path, much like a model airplane flying in a circle on a guideline (**Figure 5.10**). One setup inside the chamber is a seat that is free to pivot, so that when the centrifuge is rotating, the astronaut's head will point toward the center of the circular motion. The normal force from the astronaut's chair provides the centripetal force. The greater the astronaut's speed, the greater the normal force needs to be. The normal force is responsible for the astronaut's apparent weight. As the speed increases, the astronaut will feel heavier. This is referred to as "pulling g's", where *g* represents the acceleration due to gravity near earth's surface. If the centripetal acceleration is 1g, the astronaut will feel her normal weight. If it is 2g, she will feel twice her normal weight, and so on. One side effect of pulling a high number of g's is the potential loss of consciousness known as g-LOC. This occurs because

the fluids, primarily the blood, in an astronaut's body can move around. Just like a spinning washing machine drains the water from wet clothes, the astronaut's blood drains away from her head, which causes the blood to collect in her lower extremities, such as the legs and feet. The heart has to work against this motion, and at some point, is unable to maintain sufficient blood flow, and thus oxygen to the brain, which results in the astronaut blacking out or losing consciousness. The centrifuge helps astronauts and fighter jet pilots to train in techniques to prevent g-LOC. Other techniques to combat g-LOC include the use of special suits called anti-g suits (see Example 15 and Problem 52). If an astronaut in the centrifuge is rotating at 22.5 rpm, how heavy does she feel, compared to her real weight? The radius of the centrifuge is 8.84 m.

Reasoning   How heavy the astronaut feels will be determined by the magnitude of the normal force from the chair. The normal force also provides the centripetal force in this problem, so we can apply Equation 5.3. This requires us to know the speed of the astronaut, which we can determine from the period of the circular motion and Equation 5.1.

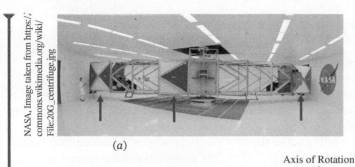

(a)

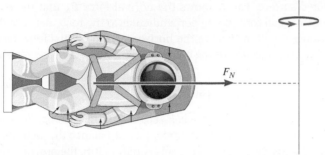

Axis of Rotation

F_N

(b)

FIGURE 5.10 (a) The 20-g centrifuge at the NASA Ames Research Center. The arrows in the picture show the locations of chambers, called cabs, where the astronaut can sit. (b) Orientation of the astronaut while the centrifuge is rotating. Notice that the normal force from the chair that acts on the astronaut points inward, toward the center of the circular motion.

Solution We begin by converting the rotation rate to the period:

$$22.5 \text{ rpm} = \frac{22.5 \text{ revolutions}}{\text{min}} \times \frac{1 \text{ min}}{60 \text{ s}} = 0.375 \text{ rev/s}$$

The period (T) is simply the reciprocal of this value: $T = \frac{1}{0.375 \text{ rev/s}} = 2.67$ s. **Figure 5.10b** shows the normal force acting on the astronaut is equal to the centripetal force: $F_N = F_C = \frac{mv^2}{r}$, by applying Equation 5.3. We can write the speed (v) in terms of the period by using Equation 5.1: $v = \frac{2\pi r}{T}$. Substituting this into the expression for F_N, we find

$$F_N = \frac{4m\pi^2 r^2}{r T^2} = \frac{4m\pi^2 r}{T^2} = m\left(\frac{4\pi^2 r}{T^2}\right).$$

Since F_N determines the astronaut's apparent weight, the term in parentheses represents "g effective." Comparing this to $g = 9.80$ m/s², we have

$$\text{"g effective"} = \frac{4\pi^2 r}{T^2} = \frac{4\pi^2 (8.84 \text{ m})}{(2.67 \text{ s})^2} = 48.9 \text{ m/s}^2 = \boxed{5g}$$

Thus, the astronaut will feel as if she weighs five times her true weight. We would say the astronaut is pulling 5 g's.

Related Homework: Problems 51, 52, and 53

Check Your Understanding

(The answers are given at the end of the book.)

6. A car is traveling in uniform circular motion on a section of road whose radius is r (see **CYU Figure 5.3**). The road is slippery, and the car is just on the verge of sliding. **(a)** If the car's speed were doubled, what would be the smallest radius at which the car does not slide? Express your answer in terms of r. **(b)** What would be your answer to part (a) if the car were replaced by one that weighed twice as much, the car's speed still being doubled?

r

CYU FIGURE 5.3

7. Other things being equal, would it be easier to drive at high speed around an unbanked horizontal curve on the moon than to drive around the same curve on the earth?

8. What is the chance of a light car safely rounding an unbanked curve on an icy road as compared to that of a heavy car: worse, the same, or better? Assume that both cars have the same speed and are equipped with identical tires.

9. A penny is placed on a rotating turntable. Where on the turntable does the penny require the largest centripetal force to remain in place: at the center of the turntable or at the edge of the turntable?

When a car travels without skidding around an unbanked curve, the static frictional force between the tires and the road provides the centripetal force. The reliance on friction can be eliminated completely for a given speed, however, if the curve is banked at an angle relative to the horizontal, much in the same way that a plane is banked while making a turn.

Interactive Figure 5.11a shows a car going around a friction-free banked curve. The radius of the curve is r, where r is measured parallel to the horizontal and not to the slanted surface. Part b shows the normal force $\vec{F}_N$ that the road applies to the car, the normal force being perpendicular to the road. Because the roadbed makes an angle θ with respect to the horizontal, the normal force has a component $F_N \sin \theta$ that points toward the center C of the circle and provides the centripetal force:

$$F_c = F_N \sin \theta = \frac{mv^2}{r}$$

The vertical component of the normal force is $F_N \cos \theta$ and, since the car does not accelerate in the vertical direction, this component must balance the weight mg of the car. Therefore, $F_N \cos \theta = mg$. Dividing this equation into the previous one shows that

$$\frac{F_N \sin \theta}{F_N \cos \theta} = \frac{mv^2/r}{mg}$$

$$\tan \theta = \frac{v^2}{rg} \qquad (5.4)$$

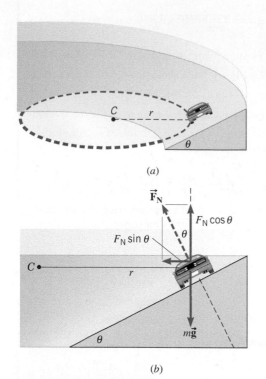

(a)

(b)

INTERACTIVE FIGURE 5.11 (a) A car travels on a circle of radius r on a frictionless banked road. The banking angle is θ, and the center of the circle is at C. (b) The forces acting on the car are its weight $m\vec{g}$ and the normal force $\vec{F}_N$. A component $F_N \sin \theta$ of the normal force provides the centripetal force.

Equation 5.4 indicates that, for a given speed v, the centripetal force needed for a turn of radius r can be obtained from the normal force by banking the turn at an angle θ, independent of the mass of the vehicle. Greater speeds and smaller radii require more steeply banked curves—that is, larger values of θ. At a speed that is too small for a given θ, a car would slide down a frictionless banked curve; at a speed that is too large, a car would slide off the top. The next example deals with a famous banked curve.

EXAMPLE 9 | The Physics of the Daytona International Speedway

The Daytona 500 is the major event of the NASCAR (National Association for Stock Car Auto Racing) season. It is held at the Daytona International Speedway in Daytona, Florida. The turns in this oval track have a maximum radius (at the top) of $r = 316$ m and are banked steeply, with $\theta = 31°$ (see **Interactive Figure 5.11**). Suppose these maximum-radius turns were frictionless. At what speed would the cars have to travel around them?

Reasoning In the absence of friction, the horizontal component of the normal force that the track exerts on the car must provide the centripetal force. Therefore, the speed of the car is given by Equation 5.4.

Solution From Equation 5.4, it follows that

$$v = \sqrt{rg \tan \theta} = \sqrt{(316 \text{ m})(9.80 \text{ m/s}^2) \tan 31°}$$

$$= \boxed{43 \text{ m/s (96 mph)}}$$

Drivers actually negotiate the turns at speeds up to 195 mph, however, which requires a greater centripetal force than that implied by Equation 5.4 for frictionless turns. Static friction provides the additional force.

Check Your Understanding

(The answer is given at the end of the book.)

10. Go to **Concept Simulation 5.2** at **www.wiley.com/college/cutnell** to review the concepts involved in this question. Two cars are identical, except for the type of tread design on their tires. The cars are driven at the same speed and enter the same unbanked horizontal turn. Car A cannot negotiate the turn, but car B can. Which tread design, the one on car A or the one on car B, yields a larger coefficient of static friction between the tires and the road?

5.5 | Satellites in Circular Orbits

Today there are many satellites in orbit about the earth. The ones in circular orbits are examples of uniform circular motion. Like a model airplane on a guideline, each satellite is kept on its circular path by a centripetal force. The gravitational pull of the earth provides the centripetal force and acts like an invisible guideline for the satellite.

There is only one speed that a satellite can have if the satellite is to remain in an orbit with a fixed radius. To see how this fundamental characteristic arises, consider the gravitational force acting on the satellite of mass m in **Figure 5.12**. Since the gravitational force is the only force acting on the satellite in the radial direction, it alone provides the centripetal force. Therefore, using Newton's law of gravitation (Equation 4.3), we have

$$F_c = G\frac{mM_E}{r^2} = \frac{mv^2}{r}$$

where G is the universal gravitational constant, M_E is the mass of the earth, and r is the distance from the center of the earth to the satellite. Solving for the speed v of the satellite gives

$$v = \sqrt{\frac{GM_E}{r}} \tag{5.5}$$

If the satellite is to remain in an orbit of radius r, the speed must have precisely this value. Note that the radius r of the orbit is in the denominator in Equation 5.5. This means that the closer the satellite is to the earth, the smaller is the value for r and the greater the orbital speed must be.

The mass m of the satellite does not appear in Equation 5.5, having been eliminated algebraically. ***Consequently, for a given orbit, a satellite with a large mass has exactly the same orbital speed as a satellite with a small mass.*** However, more effort is certainly required to lift the larger-mass satellite into orbit. The orbital speed of one famous artificial satellite is determined in the following example.

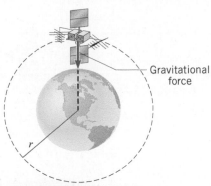

FIGURE 5.12 For a satellite in circular orbit around the earth, the gravitational force provides the centripetal force.

EXAMPLE 10 | The Physics of the Hubble Space Telescope

Determine the speed of the Hubble Space Telescope (see **Figure 5.13**) orbiting at a height of 598 km above the earth's surface.

Reasoning Before Equation 5.5 can be applied, the orbital radius r must be determined *relative to the center of the earth*. Since the radius of the earth is approximately 6.38×10^6 m, and the height of the telescope above the earth's surface is 0.598×10^6 m, the orbital radius is $r = 6.98 \times 10^6$ m.

> **Problem-Solving Insight** The orbital radius r that appears in the relation $v = \sqrt{GM_E/r}$ is the distance from the satellite to the center of the earth (not to the surface of the earth).

Solution The orbital speed is

$$v = \sqrt{\frac{GM_E}{r}} = \sqrt{\frac{(6.67 \times 10^{-11}\text{ N} \cdot \text{m}^2/\text{kg}^2)(5.98 \times 10^{24}\text{ kg})}{6.98 \times 10^6\text{ m}}}$$

$$v = \boxed{7.56 \times 10^3\text{ m/s (16 900 mi/h)}}$$

FIGURE 5.13 The Hubble Space Telescope orbiting the earth.

Frank Whitney/The Image Bank/Getty Images

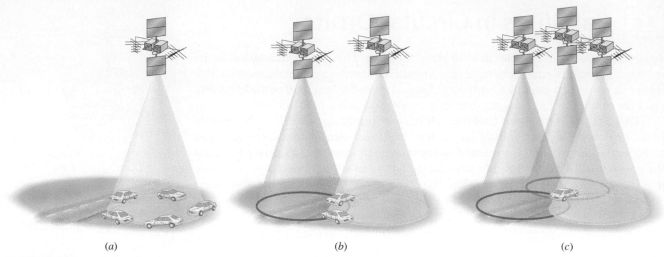

(a)   *(b)*   *(c)*

FIGURE 5.14 The Global Positioning System (GPS) of satellites can be used with a GPS receiver to locate an object, such as a car, on the earth. (*a*) One satellite identifies the car as being somewhere on a circle. (*b*) A second places it on another circle, which identifies two possibilities for the exact spot. (*c*) A third provides the means for deciding where the car is.

THE PHYSICS OF . . . the Global Positioning System. Many applications of satellite technology affect our lives. An increasingly important application is the network of 24 satellites called the Global Positioning System (GPS), which can be used to determine the position of an object to within 5 m or less. **Figure 5.14** illustrates how the system works. Each GPS satellite carries a highly accurate atomic clock, whose time is transmitted to the ground continually by means of radio waves. In the drawing, a car carries a computerized GPS receiver that can detect the waves and is synchronized to the satellite clock. The receiver can therefore determine the distance between the car and a satellite from a knowledge of the travel time of the waves and the speed at which they move. This speed, as we will see in Chapter 24, is the speed of light and is known with great precision. A measurement using a single satellite locates the car somewhere on a circle, as **Figure 5.14a** shows, while a measurement using a second satellite locates the car on another circle. The intersection of the circles reveals two possible positions for the car, as in **Figure 5.14b**. With the aid of a third satellite, a third circle can be established, which intersects the other two and identifies the car's exact position, as in **Figure 5.14c**. The use of ground-based radio beacons to provide additional reference points leads to a system called Differential GPS, which can locate objects even more accurately than the satellite-based system alone. Navigational systems for automobiles and portable systems that tell hikers and people with visual impairments where they are located are two of the many uses for the GPS technique. GPS applications are so numerous that they have developed into a multibillion dollar industry.

Equation 5.5 applies to human-made satellites or to natural satellites like the moon. It also applies to circular orbits about any astronomical object, provided M_E is replaced by the mass of the object on which the orbit is centered. Example 11, for instance, shows how scientists have applied this equation to conclude that a supermassive black hole is probably located at the center of the galaxy known as M87. This galaxy is located at a distance of about 50 million light-years away from the earth. (One light-year is the distance that light travels in a year, or 9.5×10^{15} m.)

EXAMPLE 11 | The Physics of Locating a Black Hole

The Hubble Telescope has detected the light being emitted from different regions of galaxy M87, which is shown in **Figure 5.15**. The black circle identifies the center of the galaxy. From the characteristics of this light, astronomers have determined that the orbiting speed is 7.5×10^5 m/s for matter located at a distance of 5.7×10^{17} m from the center. Find the mass M of the object located at the galactic center.

Reasoning and Solution Replacing M_E in Equation 5.5 with M gives $v = \sqrt{GM/r}$, which can be solved to show that

$$M = \frac{v^2 r}{G} = \frac{(7.5 \times 10^5 \text{ m/s})^2 (5.7 \times 10^{17} \text{ m})}{6.67 \times 10^{-11} \text{ N} \cdot \text{m}^2/\text{kg}^2}$$

$$= \boxed{4.8 \times 10^{39} \text{ kg}}$$

Since the mass of our sun is 2.0×10^{30} kg, matter equivalent to 2.4 billion suns is located at the center of galaxy M87. The volume of space in which this matter is located contains relatively few visible stars, so researchers believe that the data provide strong evidence for the existence of a supermassive black hole. The term "black hole" is used because the tremendous mass prevents even light from escaping. The light that forms the image in **Figure 5.15** comes not from the black hole itself, but from matter that surrounds it.*

FIGURE 5.15 This image of the ionized gas (yellow) at the heart of galaxy M87 was obtained by the Hubble Space Telescope. The circle identifies the center of the galaxy, at which a black hole is thought to exist.

The period T of a satellite is the time required for one orbital revolution. As in any uniform circular motion, the period is related to the speed of the motion by $v = 2\pi r/T$. Substituting v from Equation 5.5 shows that

$$\sqrt{\frac{GM_E}{r}} = \frac{2\pi r}{T}$$

Solving this expression for the period T gives

$$T = \frac{2\pi r^{3/2}}{\sqrt{GM_E}} \tag{5.6}$$

Although derived for earth orbits, Equation 5.6 can also be used for calculating the periods of those planets in nearly circular orbits about the sun, if M_E is replaced by the mass M_S of the sun and r is interpreted as the distance between the center of the planet and the center of the sun. The fact that the period is proportional to the three-halves power of the orbital radius is known as Kepler's third law, and it is one of the laws discovered by Johannes Kepler (1571–1630) during his studies of planetary motion. Kepler's third law also holds for elliptical orbits, which will be discussed in Chapter 9.

An important application of Equation 5.6 occurs in the field of communications, where "synchronous satellites" are put into a circular orbit that is in the plane of the equator, as **Figure 5.16** shows. The orbital period is chosen to be one day, which is also the time it takes for the earth to turn once about its axis. Therefore, these satellites move around their orbits in a way that is synchronized with the rotation of the earth. For earth-based observers, synchronous satellites have the useful characteristic of appearing in fixed positions in the sky and can serve as "stationary" relay stations for communication signals sent up from the earth's surface.

THE PHYSICS OF . . . digital satellite system TV. This is exactly what is done in the digital satellite systems that are a popular alternative to cable TV. As the blowup in **Figure 5.16** indicates, a small "dish" antenna on your house picks up the digital TV signals relayed back to earth by the satellite. After being decoded, these signals are delivered to your TV set. All synchronous satellites are in orbit at the same height above the earth's surface, as Example 12 shows.

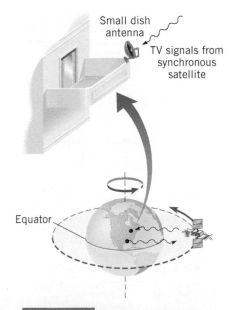

FIGURE 5.16 A synchronous satellite orbits the earth once per day on a circular path in the plane of the equator. Digital satellite system television uses such satellites as relay stations. TV signals are sent up from the earth's surface and then rebroadcast down to your own small dish antenna.

*More complete analysis shows that the mass of the black hole is equivalent to 6.6 billion suns (K. Gebhardt et al., *The Astrophysical Journal*, Vol. 729, 2011, p. 119).

EXAMPLE 12 | The Orbital Radius for Synchronous Satellites

What is the height H above the earth's surface at which all synchronous satellites (regardless of mass) must be placed in orbit?

Reasoning The period T of a synchronous satellite is one day, so we can use Equation 5.6 to find the distance r from the satellite to the center of the earth. To find the height H of the satellite above the earth's surface we will have to take into account the fact that the earth itself has a radius of 6.38×10^6 m.

Solution A period of one day* corresponds to $T = 8.64 \times 10^4$ s. In using this value it is convenient to rearrange the equation $T = 2\pi r^{3/2}/\sqrt{GM_E}$ as follows:

$$r^{3/2} = \frac{T\sqrt{GM_E}}{2\pi}$$

$$= \frac{(8.64 \times 10^4 \text{ s})\sqrt{(6.67 \times 10^{-11} \text{ N} \cdot \text{m}^2/\text{kg}^2)(5.98 \times 10^{24} \text{ kg})}}{2\pi}$$

By squaring and then taking the cube root, we find that $r = 4.23 \times 10^7$ m. Since the radius of the earth is approximately 6.38×10^6 m, the height of the satellite above the earth's surface is

$$H = 4.23 \times 10^7 \text{ m} - 0.64 \times 10^7 \text{ m} = \boxed{3.59 \times 10^7 \text{ m (22 300 mi)}}$$

Math Skills In solving $T = \frac{2\pi r^{3/2}}{\sqrt{GM_E}}$ (Equation 5.6) for r, we remember that whatever is done to one side of the equals sign must also be done to the other side. To isolate the term $r^{3/2}$ on one side we multiply both sides by $\sqrt{GM_E}$ and divide both sides by 2π:

$$T\frac{\sqrt{GM_E}}{2\pi} = \frac{2\pi r^{3/2}}{\sqrt{GM_E}}\frac{\sqrt{GM_E}}{2\pi} \quad \text{or} \quad r^{3/2} = \frac{T\sqrt{GM_E}}{2\pi}$$

Next, we square both sides of the result for $r^{3/2}$:

$$(r^{3/2})^2 = \left(\frac{T\sqrt{GM_E}}{2\pi}\right)^2 \quad \text{or} \quad r^3 = \frac{T^2 GM_E}{4\pi^2}$$

Finally, taking the cube root of the result for r^3 gives

$$\sqrt[3]{r^3} = \sqrt[3]{\frac{T^2 GM_E}{4\pi^2}} \quad \text{or} \quad (r^3)^{1/3} = \left(\frac{T^2 GM_E}{4\pi^2}\right)^{1/3} \quad \text{or}$$

$$r = \frac{T^{2/3} G^{1/3} M_E^{1/3}}{4^{1/3}\pi^{2/3}}$$

Using values for T, G, and M_E in the result for r shows that $r = 4.23 \times 10^7$ m.

Courtesy NASA

FIGURE 5.17 In a state of apparent weightlessness in orbit, astronaut and pilot Guy S. Gardner and mission specialist William M. Shepherd (in chair) float, with Gardner appearing to "balance" the chair on his nose.

Check Your Understanding

(The answer is given at the end of the book.)

11. Two satellites are placed in orbit, one about Mars and the other about Jupiter, such that the orbital speeds are the same. Mars has the smaller mass. Is the radius of the satellite in orbit about Mars less than, greater than, or equal to the radius of the satellite orbiting Jupiter?

5.6 | Apparent Weightlessness and Artificial Gravity

The idea of life on board an orbiting satellite conjures up visions of astronauts floating around in a state of "weightlessness," as in **Figure 5.17**. Actually, this state should be called "apparent weightlessness," because it is similar to the condition of zero apparent weight that occurs in an elevator during free-fall. Conceptual Example 13 explores this similarity.

*Successive appearances of the sun define the solar day of 24 h or 8.64×10^4 s. The sun moves against the background of the stars, however, and the time required for the earth to turn once on its axis relative to the fixed stars is 23 h 56 min, which is called the sidereal day. The sidereal day should be used in Example 12, but the neglect of this effect introduces an error of less than 0.4% in the answer.

CONCEPTUAL EXAMPLE 13 | The Physics of Apparent Weightlessness

Figure 5.18 shows a person on a scale in a freely falling elevator and in a satellite in a circular orbit. Assume that when the person is standing stationary on the earth, his weight is 800 N (180 lb). In each case, what apparent weight is recorded by the scale? **(a)** The scale in the elevator records 800 N while that in the satellite records 0 N. **(b)** The scale in the elevator records 0 N while that in the satellite records 800 N. **(c)** Both scales record 0 N.

Reasoning As Section 4.8 discusses, apparent weight is the force that an object exerts on the platform of a scale. This force depends on whether or not the object and the platform are accelerating together.

Answer (a) is incorrect. The scale and the person in the free-falling elevator are accelerating toward the earth at the same rate. Therefore, they cannot push against one another, and so the scale does not record an apparent weight of 800 N.

Answer (b) is incorrect. The scale and the person in the satellite are accelerating toward the center of the earth at the same rate (they have the same centripetal acceleration). Therefore, they cannot push against one another, and so the scale does not record an apparent weight of 800 N.

Answer (c) is correct. The scale and the person in the elevator fall together and, therefore, they cannot push against one another. Therefore, the scale records an apparent weight of 0 N. In the orbiting satellite in **Figure 5.18b**, both the person and the scale are in uniform circular motion. Objects in uniform circular motion continually accelerate or "fall" toward the center of the circle in order to remain on the circular path. Consequently, both the person and the scale "fall" with the same acceleration toward the center of the earth and cannot push against one another. Thus, the apparent weight in the satellite is zero, just as it is in the freely falling elevator.

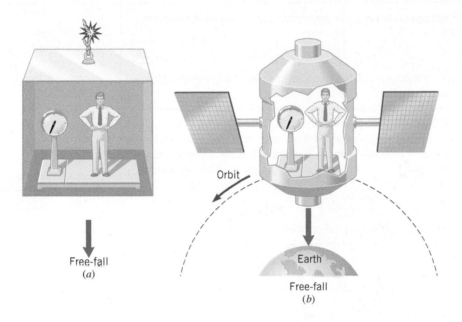

Orbit

Free-fall
(a)

Earth

Free-fall
(b)

FIGURE 5.18 (*a*) During free-fall, the elevator accelerates downward with the acceleration due to gravity, and the apparent weight of the person is zero. (*b*) The orbiting space station is also in free-fall toward the center of the earth.

THE PHYSICS OF . . . artificial gravity. The physiological effects of prolonged apparent weightlessness are only partially known. To minimize such effects, it is likely that artificial gravity will be provided in large space stations of the future. To help explain artificial gravity, **Figure 5.19** shows a space station rotating about an axis. Because of the rotational motion, any object located at a point P on the interior surface of the station experiences a centripetal force directed toward the axis. The surface of the station provides this force by pushing on the feet of an astronaut, for instance. The centripetal force can be adjusted to match the astronaut's earth-weight by properly selecting the rotational speed of the space station, as Example 14 illustrates.

FIGURE 5.19 The surface of the rotating space station pushes on an object with which it is in contact and thereby provides the centripetal force that keeps the object moving on a circular path.

Analyzing Multiple-Concept Problems

EXAMPLE 14 | Artificial Gravity

At what speed must the interior surface of the space station ($r = 1700$ m) move in **Figure 5.19**, so that the astronaut located at point P experiences a push on his feet that equals his weight on earth?

Reasoning The floor of the rotating space station exerts a normal force on the feet of the astronaut. This is the centripetal force ($F_c = mv^2/r$) that keeps the astronaut moving on a circular path. Since the magnitude of the normal force equals the astronaut's weight on earth, we can determine the speed v of the space station's floor.

Knowns and Unknowns The data for this problem are given in the following table:

Description	Symbol	Value
Radius of space station	r	1700 m
Unknown Variable		
Speed of space station's floor	v	?

Modeling the Problem

STEP 1 Speed and Centripetal Force The centripetal force acting on the astronaut is given by Equation 5.3 as

$$F_c = \frac{mv^2}{r}$$

Solving for the speed v gives Equation 1 at the right. Step 2 considers the centripetal force F_c that appears in this result.

$$v = \sqrt{\frac{rF_c}{m}} \qquad (1)$$

$$\boxed{?}$$

STEP 2 Magnitude of the Centripetal Force The normal force applied to the astronaut's feet by the floor is the centripetal force and has a magnitude equal to the astronaut's earth-weight. This earth-weight is given by Equation 4.5 as the astronaut's mass m times the magnitude g of the acceleration due to the earth's gravity. Thus, we have for the centripetal force that

$$\boxed{F_c = mg}$$

which can be substituted into Equation 1, as indicated at the right.

$$v = \sqrt{\frac{rF_c}{m}} \qquad (1)$$

$$\boxed{F_c = mg}$$

Solution Algebraically combining the results of each step, we find that

$$\overset{\text{STEP 1}}{\downarrow}\quad\overset{\text{STEP 2}}{\downarrow}$$
$$v = \sqrt{\frac{rF_c}{m}} = \sqrt{\frac{rm g}{m}}$$

The astronaut's mass m is eliminated algebraically from this result, so the speed of the space station floor is

$$v = \sqrt{rg} = \sqrt{(1700 \text{ m})(9.80 \text{ m/s}^2)} = \boxed{130 \text{ m/s}}$$

Check Your Understanding

(The answer is given at the end of the book.)

12. The acceleration due to gravity on the moon is one-sixth that on earth. **(a)** Is the true weight of a person on the moon less than, greater than, or equal to the true weight of the same person on earth? **(b)** Is the apparent weight of a person in orbit about the moon less than, greater than, or equal to the apparent weight of the same person in orbit about the earth?

5.7 | *Vertical Circular Motion

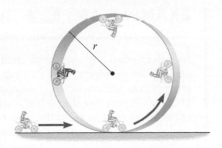

Motorcycle stunt drivers perform a feat in which they drive their cycles around a vertical circular track, as in **Interactive Figure 5.20a**. Usually, the speed varies in this stunt. When the speed of travel on a circular path changes from moment to moment, the motion is said to be nonuniform. Nonetheless, we can use the concepts that apply to uniform circular motion to gain considerable insight into the motion that occurs on a vertical circle.

There are four points on a vertical circle where the centripetal force can be identified easily, as **Interactive Figure 5.20b** indicates. As you look at **Interactive Figure 5.20b**, keep in mind that the centripetal force is not a new and separate force of nature. Instead, at each point the centripetal force is the net sum of all the force components oriented along the radial direction, and it points toward the center of the circle. The drawing shows only the weight of the cycle plus rider (magnitude $= mg$) and the normal force pushing on the cycle (magnitude $= F_N$). The propulsion and braking forces are omitted for simplicity, because they do not act in the radial direction. The magnitude of the centripetal force at each of the four points is given as follows in terms of mg and F_N:

(1) $\underbrace{F_{N1} - mg = \dfrac{mv_1^2}{r}}_{= F_{c1}}$ (3) $\underbrace{F_{N3} + mg = \dfrac{mv_3^2}{r}}_{= F_{c3}}$

(2) $\underbrace{F_{N2} = \dfrac{mv_2^2}{r}}_{= F_{c2}}$ (4) $\underbrace{F_{N4} = \dfrac{mv_4^2}{r}}_{= F_{c4}}$

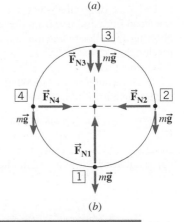

(b)

INTERACTIVE FIGURE 5.20 (a) A vertical loop-the-loop motorcycle stunt. (b) The normal force $\vec{F}_N$ and the weight $m\vec{g}$ of the cycle and the rider are shown here at four locations.

> **Problem-Solving Insight** Centripetal force is the name given to the net force that points toward the center of a circular path. As shown here, there may be more than one force that contributes to this net force.

THE PHYSICS OF . . . the loop-the-loop motorcycle stunt.   As the cycle goes around, the magnitude of the normal force changes. It changes because the speed changes and because the weight does not have the same effect at every point. At the bottom, the normal force and the weight oppose one another, giving a centripetal force of magnitude $F_{N1} - mg$. At the top, the normal force and the weight reinforce each other to provide a centripetal force whose magnitude is $F_{N3} + mg$. At points 2 and 4 on either side, only F_{N2} and F_{N4} provide the centripetal force. The weight is tangent to the circle at points 2 and 4 and has no component pointing toward the center. If the speed at each of the four places is known, along with the mass and the radius, the normal forces can be determined.

Riders who perform the loop-the-loop trick must have at least a minimum speed at the top of the circle to remain on the track. This speed can be determined by considering the centripetal force at point 3. The speed v_3 in the equation $F_{N3} + mg = mv_3^2/r$ is a minimum when F_{N3} is zero. Then, the speed is given by $v_3 = \sqrt{rg}$. At this speed, the track does not exert a normal force to keep the cycle on the circle, because the weight mg provides all the centripetal force. Under these conditions, the rider experiences an apparent weightlessness like that discussed in Section 5.6, because for an instant the rider and the cycle are falling freely toward the center of the circle.

Check Your Understanding

(The answers are given at the end of the book.)

13. Would a change in the earth's mass affect **(a)** the banking of airplanes as they turn, **(b)** the banking of roadbeds, **(c)** the speeds with which satellites are put into circular orbits, and **(d)** the performance of the loop-the-loop motorcycle stunt?

14. A stone is tied to a string and whirled around in a circle at a constant speed. Is the string more likely to break when the circle is horizontal or when it is vertical? Assume that the constant speed is the same in each case.

EXAMPLE 15 | BIO The Physics of *g*-Suits

Jet fighter pilots and astronauts often experience large accelerations from centripetal forces during flight. This is commonly referred to as "pulling *g*'s", as mentioned in Example 8. For example, pilots flying along the path of a vertical circle are susceptible to the loss of consciousness (blackout) due to blood flowing from the top of their head and pooling in their lower body. To help prevent this potentially dangerous situation, the pilot wears a device called a *g*-suit. The suit contains air-filled bladders that expand and squeeze on a pilot's legs and abdomen. This helps prevent the blood from draining from the upper part of the body, thereby allowing the pilot to remain conscious during higher levels of acceleration. Consider the following situation shown in **Figure 5.21**, where a fighter pilot is flying upside down along a vertical circle whose radius is 1.2 km. What must be the speed of the plane, so that the normal force acting on the pilot is equal to his weight?

Reasoning The forces acting on the pilot are shown in **Figure 5.21**. Since the jet is undergoing circular motion, we should be able to identify the centripetal force. We can then apply Equation 5.3 to calculate the speed of the plane.

Solution Since both forces on the pilot in **Figure 5.21** point in toward the center of the circular motion, the centripetal force will be equal to the sum of the normal force and the weight: $F_C = F_N + mg$. Equation 5.3 tells us that the centripetal force is equal to mv^2/r. Therefore, we find that $\frac{mv^2}{r} = F_N + mg$. The problem tells us we want the normal force on the pilot to be equal to his weight. Thus, we make that substitution and find the following: $\frac{mv^2}{r} = mg + mg = 2mg$. Notice that the mass of the pilot drops out, and we can solve for the speed of the plane:

$$v = \sqrt{2rg} = \sqrt{2(1200 \text{ m})(9.8 \text{ m/s}^2)} = \boxed{150 \text{ m/s}}$$

In addition to the importance of *g*-suits in this type of maneuver, this particular combination of forces can be very alarming for young pilots. Since the normal force on the pilot is equal to his weight, he feels as if he is sitting right side up, when in fact, he is upside down. This can lead to spatial disorientation.

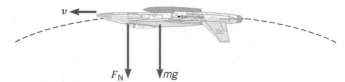

FIGURE 5.21 The normal force and weight of a pilot both point down toward the center of the circular motion while flying upside down at the top of a vertical circle. The speed of the plane will change the magnitude of the normal force (i.e., how much the seat pushes against the pilot).

Concept Summary

5.1 Uniform Circular Motion Uniform circular motion is the motion of an object traveling at a constant (uniform) speed on a circular path. The period T is the time required for the object to travel once around the circle. The speed v of the object is related to the period and the radius r of the circle by Equation 5.1.

$$v = \frac{2\pi r}{T} \qquad (5.1)$$

5.2 Centripetal Acceleration An object in uniform circular motion experiences an acceleration, known as centripetal acceleration. The magnitude a_c of the centripetal acceleration is given by Equation 5.2, where v is the speed of the object and r is the radius of the circle. The direction of the centripetal acceleration vector always points toward the center of the circle and continually changes as the object moves.

$$a_c = \frac{v^2}{r} \qquad (5.2)$$

5.3 Centripetal Force To produce a centripetal acceleration, a net force pointing toward the center of the circle is required. This net force is called the centripetal force, and its magnitude F_c is given by Equation 5.3, where m and v are the mass and speed of the object, and r is the radius of the circle. The direction of the centripetal force vector, like that of the centripetal acceleration vector, always points toward the center of the circle.

$$F_c = \frac{mv^2}{r} \qquad (5.3)$$

5.4 Banked Curves A vehicle can negotiate a circular turn without relying on static friction to provide the centripetal force, provided the turn is banked at an angle relative to the horizontal. The angle θ at which a friction-free curve must be banked is related to the speed v of the vehicle, the radius r of the curve, and the magnitude g of the acceleration due to gravity by Equation 5.4.

$$\tan \theta = \frac{v^2}{rg} \qquad (5.4)$$

5.5 Satellites in Circular Orbits When a satellite orbits the earth, the gravitational force provides the centripetal force that keeps the satellite moving in a circular orbit. The speed v and period T of a satellite depend on the mass M_E of the earth and the radius r of the orbit according to Equations 5.5 and 5.6, where G is the universal gravitational constant.

$$v = \sqrt{\frac{GM_E}{r}} \qquad (5.5)$$

$$T = \frac{2\pi r^{3/2}}{\sqrt{GM_E}} \qquad (5.6)$$

5.6 Apparent Weightlessness and Artificial Gravity The apparent weight of an object is the force that it exerts on a scale with which it is in contact. All objects, including people, on board an orbiting satellite are in free-fall, since they experience negligible air resistance and they have an acceleration that is equal to the acceleration due to gravity. When a person is in free-fall, his or her apparent weight is zero, because both the person and the scale fall freely and cannot push against one another.

5.7 Vertical Circular Motion Vertical circular motion occurs when an object, such as a motorcycle, moves on a vertical circular path. The speed of the object often varies from moment to moment, and so do the magnitudes of the centripetal acceleration and centripetal force.

Focus on Concepts

Online

Additional questions are available for assignment in WileyPLUS.

Section 5.2 Centripetal Acceleration

1. Two cars are traveling at the same constant speed v. As the drawing indicates, car A is moving along a straight section of the road, while car B is rounding a circular turn. Which statement is true about the accelerations of the cars? (a) The acceleration of both cars is zero, since they are traveling at a constant speed. (b) Car A is accelerating, but car B is not accelerating. (c) Car A is not accelerating, but car B is accelerating. (d) Both cars are accelerating.

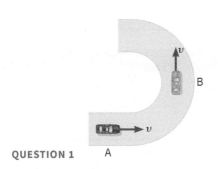

QUESTION 1

2. Two cars are driving at the same constant speed v around a racetrack. However, they are traveling through turns that have different radii, as shown in the drawing. Which statement is true about the magnitude of the centripetal acceleration of each car? (a) The magnitude of the centripetal acceleration of each car is the same, since the cars are moving at the same speed. (b) The magnitude of the centripetal acceleration of the car at A is greater than that of the car at B, since the radius of the circular track is smaller at A. (c) The magnitude of the centripetal acceleration of the car at A is greater than that of the car at B, since the radius of the circular track is greater at A. (d) The magnitude of the centripetal acceleration of the car at A is less than that of the car at B, since the radius of the circular track is smaller at A.

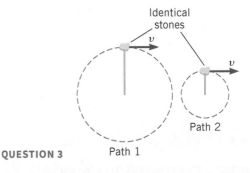

QUESTION 2

Section 5.3 Centripetal Force

3. The drawing shows two identical stones attached to cords that are being whirled on a tabletop at the same speed. The radius of the larger circle is twice that of the smaller circle. How is the tension T_1 in the longer cord related to the tension T_2 in the shorter cord? (a) $T_1 = T_2$ (b) $T_1 = 2T_2$ (c) $T_1 = 4T_2$ (d) $T_1 = \frac{1}{2}T_2$ (e) $T_1 = \frac{1}{4}T_2$

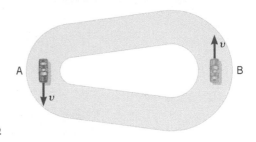

QUESTION 3

4. Three particles have the following masses (in multiples of m_0) and move on three different circles with the following speeds (in multiples of v_0) and radii (in multiples of r_0):

Particle	Mass	Speed	Radius
1	m_0	$2v_0$	r_0
2	m_0	$3v_0$	$3r_0$
3	$2m_0$	$2v_0$	$4r_0$

Rank the particles according to the magnitude of the centripetal force that acts on them, largest first. (**a**) 1, 2, 3 (**b**) 1, 3, 2 (**c**) 2, 1, 3 (**d**) 2, 3, 1 (**e**) 3, 2, 1

Section 5.4   Banked Curves

5. Two identical cars, one on the moon and one on the earth, have the same speed and are rounding banked turns that have the same radius r. There are two forces acting on each car, its weight mg and the normal force F_N exerted by the road. Recall that the weight of an object on the moon is about one-sixth of its weight on earth. How does the centripetal force on the moon compare with that on the earth? (**a**) The centripetal forces are the same. (**b**) The centripetal force on the moon is less than that on the earth. (**c**) The centripetal force on the moon is greater than that on the earth.

Section 5.5   Satellites in Circular Orbits

6. Two identical satellites are in orbit about the earth. As the drawing shows, one orbit has a radius r and the other $2r$. The centripetal force on the satellite in the larger orbit is _____ as that on the satellite in the smaller orbit. (**a**) the same (**b**) twice as great (**c**) four times as great (**d**) half as great (**e**) one-fourth as great

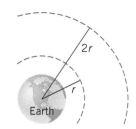

QUESTION 6

Section 5.7   Vertical Circular Motion

7. The drawing shows an extreme skier at the bottom of a ski jump. At this point the track is circular with a radius r. Two forces act on the skier, her weight mg and the normal force F_N. Which relation describes how the net force acting on her is related to her mass m and speed v and to the radius r? Assume that "up" is the positive direction.

(a) $F_N + mg = \dfrac{mv^2}{r}$ (b) $F_N - mg = \dfrac{mv^2}{r}$

(c) $F_N = \dfrac{mv^2}{r}$ (d) $-mg = \dfrac{mv^2}{r}$

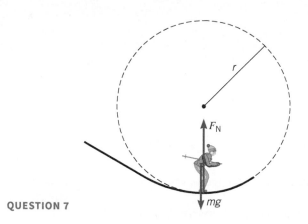

QUESTION 7

Problems

Online

Additional questions are available for assignment in WileyPLUS.

SSM	Student Solutions Manual	**BIO**	Biomedical application
MMH	Problem-solving help	**E**	Easy
GO	Guided Online Tutorial	**M**	Medium
V-HINT	Video Hints	**H**	Hard
CHALK	Chalkboard Videos	**WS**	Worksheet
		T	Team Problem

	Radius	Speed
Example 1	0.50 m	12 m/s
Example 2	Infinitely large	35 m/s
Example 3	1.8 m	2.3 m/s

Section 5.1   Uniform Circular Motion,

Section 5.2   Centripetal Acceleration

1. **E** **SSM** A car travels at a constant speed around a circular track whose radius is 2.6 km. The car goes once around the track in 360 s. What is the magnitude of the centripetal acceleration of the car?

2. **E** **GO** The following table lists data for the speed and radius of three examples of uniform circular motion. Find the magnitude of the centripetal acceleration for each example.

3. **E** **GO** Speedboat A negotiates a curve whose radius is 120 m. Speedboat B negotiates a curve whose radius is 240 m. Each boat experiences the same centripetal acceleration. What is the ratio v_A/v_B of the speeds of the boats?

4. **E** **SSM** How long does it take a plane, traveling at a constant speed of 110 m/s, to fly once around a circle whose radius is 2850 m?

5. **E** **MMH** The blade of a windshield wiper moves through an angle of $90.0°$ in 0.40 s. The tip of the blade moves on the arc of a circle that has a radius of 0.45 m. What is the magnitude of the centripetal acceleration of the tip of the blade?

6. **E** **V-HINT** There is a clever kitchen gadget for drying lettuce leaves after you wash them. It consists of a cylindrical container mounted so that it can be rotated about its axis by turning a hand crank. The outer wall of the cylinder is perforated with small holes. You put the wet leaves in the container and turn the crank to spin off the water. The radius of the container is 12 cm. When the cylinder is rotating at 2.0 revolutions per second, what is the magnitude of the centripetal acceleration at the outer wall?

7. **E** **SSM** Computer-controlled display screens provide drivers in the Indianapolis 500 with a variety of information about how their cars are performing. For instance, as a car is going through a turn, a speed of 221 mi/h (98.8 m/s) and centripetal acceleration of 3.00 g (three times the acceleration due to gravity) are displayed. Determine the radius of the turn (in meters).

8. **M** **V-HINT** **CHALK** A computer is reading data from a rotating CD-ROM (an antiquated disc-shaped data storage device). At a point that is 0.030 m from the center of the disc, the centripetal acceleration is 120 m/s². What is the centripetal acceleration at a point that is 0.050 m from the center of the disc?

9. **M** **CHALK** The earth rotates once per day about an axis passing through the north and south poles, an axis that is perpendicular to the plane of the equator. Assuming the earth is a sphere with a radius of 6.38×10^6 m, determine the speed and centripetal acceleration of a person situated **(a)** at the equator and **(b)** at a latitude of 30.0° north of the equator.

Section 5.3   Centripetal Force

10. **E** **SSM** Review Example 3, which deals with the bobsled in **Interactive Figure 5.5**. Also review Conceptual Example 4. The mass of the sled and its two riders in **Interactive Figure 5.5** is 350 kg. Find the magnitude of the centripetal force that acts on the sled during the turn with a radius of **(a)** 33 m and **(b)** 24 m.

11. **E** At an amusement park there is a ride in which cylindrically shaped chambers spin around a central axis. People sit in seats facing the axis, their backs against the outer wall. At one instant the outer wall moves at a speed of 3.2 m/s, and an 83-kg person feels a 560-N force pressing against his back. What is the radius of the chamber?

12. **E** **MMH** Multiple-Concept Example 7 reviews the concepts that play a role in this problem. Car A uses tires for which the coefficient of static friction is 1.1 on a particular unbanked curve. The maximum speed at which the car can negotiate this curve is 25 m/s. Car B uses tires for which the coefficient of static friction is 0.85 on the same curve. What is the maximum speed at which car B can negotiate the curve?

13. **E** A speed skater goes around a turn that has a radius of 31 m. The skater has a speed of 14 m/s and experiences a centripetal force of 460 N. What is the mass of the skater?

14. **E** For background pertinent to this problem, review Conceptual Example 6. In **Figure 5.7** the man hanging upside down is holding a partner who weighs 475 N. Assume that the partner moves on a circle that has a radius of 6.50 m. At a swinging speed of 4.00 m/s, what force must the man apply to his partner in the straight-down position?

15. **E** **MMH** A car is safely negotiating an unbanked circular turn at a speed of 21 m/s. The road is dry, and the maximum static frictional force acts on the tires. Suddenly a long wet patch in the road decreases the maximum static frictional force to one-third of its dry-road value. If the car is to continue safely around the curve, to what speed must the driver slow the car?

16. **E** See Conceptual Example 6 to review the concepts involved in this problem. A 9.5-kg monkey is hanging by one arm from a branch and is swinging on a vertical circle. As an approximation, assume a radial distance of 85 cm between the branch and the point where the monkey's mass is located. As the monkey swings through the lowest point on the circle, it has a speed of 2.8 m/s. Find **(a)** the magnitude of the centripetal force acting on the monkey and **(b)** the magnitude of the tension in the monkey's arm.

17. **E** **GO** Multiple-Concept Example 7 deals with the concepts that are important in this problem. A penny is placed at the outer edge of a disk (radius = 0.150 m) that rotates about an axis perpendicular to the plane of the disk at its center. The period of the rotation is 1.80 s. Find the minimum coefficient of friction necessary to allow the penny to rotate along with the disk.

18. **M** **CHALK** **SSM** The hammer throw is a track-and-field event in which a 7.3-kg ball (the "hammer") is whirled around in a circle several times and released. It then moves upward on the familiar curving path of projectile motion and eventually returns to earth some distance away. The world record for this distance is 86.75 m, achieved in 1986 by Yuriy Sedykh. Ignore air resistance and the fact that the ball is released above the ground rather than at ground level. Furthermore, assume that the ball is whirled on a circle that has a radius of 1.8 m and that its velocity at the instant of release is directed 41° above the horizontal. Find the magnitude of the centripetal force acting on the ball just prior to the moment of release.

19. **M** **V-HINT** An 830-kg race car can drive around an unbanked turn at a maximum speed of 58 m/s without slipping. The turn has a radius of curvature of 160 m. Air flowing over the car's wing exerts a downward-pointing force (called the *downforce*) of 11 000 N on the car. **(a)** What is the coefficient of static friction between the track and the car's tires? **(b)** What would be the maximum speed if no downforce acted on the car?

20. **M** **CHALK** **MMH** A "swing" ride at a carnival consists of chairs that are swung in a circle by 15.0-m cables attached to a vertical rotating pole, as the drawing shows. Suppose the total mass of a chair and its occupant is 179 kg. **(a)** Determine the tension in the cable attached to the chair. **(b)** Find the speed of the chair.

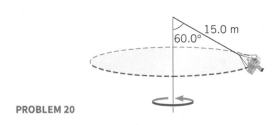

PROBLEM 20

Section 5.4   Banked Curves

21. **E** On a banked race track, the smallest circular path on which cars can move has a radius of 112 m, while the largest has a radius of 165 m, as the drawing illustrates. The height of the outer wall is 18 m. Find **(a)** the smallest and **(b)** the largest speed at which cars can move on this track without relying on friction.

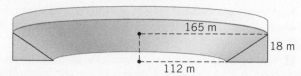

22. **E** **V-HINT** Before attempting this problem, review Examples 7 and 9. Two curves on a highway have the same radii. However, one is unbanked and the other is banked at an angle θ. A car can safely travel along the unbanked curve at a maximum speed v_0 under conditions when the coefficient of static friction between the tires and the road is $\mu_s = 0.81$. The banked curve is frictionless, and the car can negotiate it at the same maximum speed v_0. Find the angle θ of the banked curve.

23. **E** **GO** A woman is riding a Jet Ski at a speed of 26 m/s and notices a seawall straight ahead. The farthest she can lean the craft in order to make a turn is 22°. This situation is like that of a car on a curve that is banked at an angle of 22°. If she tries to make the turn without slowing down, what is the minimum distance from the seawall that she can begin making her turn and still avoid a crash?

24. **E** Two banked curves have the same radius. Curve A is banked at an angle of 13°, and curve B is banked at an angle of 19°. A car can travel around curve A without relying on friction at a speed of 18 m/s. At what speed can this car travel around curve B without relying on friction?

25. **M** **MMH** **CHALK** A racetrack has the shape of an inverted cone, as the drawing shows. On this surface the cars race in circles that are parallel to the ground. For a speed of 34.0 m/s, at what value of the distance d should a driver locate his car if he wishes to stay on a circular path without depending on friction?

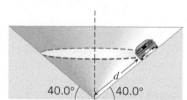

26. **M** **GO** A jet flying at 123 m/s banks to make a horizontal circular turn. The radius of the turn is 3810 m, and the mass of the jet is 2.00×10^5 kg. Calculate the magnitude of the necessary lifting force.

27. **H** The drawing shows a baggage carousel at an airport. Your suitcase has not slid all the way down the slope and is going around at a constant speed on a circle ($r = 11.0$ m) as the carousel turns. The coefficient of static friction between the suitcase and the carousel is 0.760, and the angle θ in the drawing is 36.0°. How much time is required for your suitcase to go around once?

Section 5.5 Satellites in Circular Orbits,

Section 5.6 Apparent Weightlessness and Artificial Gravity

28. **E** **GO** Two satellites are in circular orbits around the earth. The orbit for satellite A is at a height of 360 km above the earth's surface, while that for satellite B is at a height of 720 km. Find the orbital speed for each satellite.

29. **E** A rocket is used to place a synchronous satellite in orbit about the earth. What is the speed of the satellite in orbit?

30. **E** **SSM** A satellite is in a circular orbit around an unknown planet. The satellite has a speed of 1.70×10^4 m/s, and the radius of the orbit is 5.25×10^6 m. A second satellite also has a circular orbit around this same planet. The orbit of this second satellite has a radius of 8.60×10^6 m. What is the orbital speed of the second satellite?

31. **E** **GO** A satellite is in a circular orbit about the earth ($M_E = 5.98 \times 10^{24}$ kg). The period of the satellite is 1.20×10^4 s. What is the speed at which the satellite travels?

32. **E** A satellite circles the earth in an orbit whose radius is twice the earth's radius. The earth's mass is 5.98×10^{24} kg, and its radius is 6.38×10^6 m. What is the period of the satellite?

33. **M** **GO** **CHALK** A satellite has a mass of 5850 kg and is in a circular orbit 4.10×10^5 m above the surface of a planet. The period of the orbit is 2.00 hours. The radius of the planet is 4.15×10^6 m. What would be the true weight of the satellite if it were at rest on the planet's surface?

34. **M** **V-HINT** Two newly discovered planets follow circular orbits around a star in a distant part of the galaxy. The orbital speeds of the planets are determined to be 43.3 km/s and 58.6 km/s. The slower planet's orbital period is 7.60 years. **(a)** What is the mass of the star? **(b)** What is the orbital period of the faster planet, in years?

Section 5.7 Vertical Circular Motion

35. **E** **SSM** A motorcycle has a constant speed of 25.0 m/s as it passes over the top of a hill whose radius of curvature is 126 m. The mass of the motorcycle and driver is 342 kg. Find the magnitudes of **(a)** the centripetal force and **(b)** the normal force that acts on the cycle.

36. **E** **SSM** For the normal force in **Interactive Figure 5.20** to have the same magnitude at all points on the vertical track, the stunt driver must adjust the speed to be different at different points. Suppose, for example, that the track has a radius of 3.0 m and that the driver goes past point 1 at the bottom with a speed of 15 m/s. What speed must she have at point 3, so that the normal force at the top has the same magnitude as it did at the bottom?

37. **E** **GO** A special electronic sensor is embedded in the seat of a car that takes riders around a circular loop-the-loop ride at an amusement park. The sensor measures the magnitude of the normal force that the seat exerts on a rider. The loop-the-loop ride is in the vertical plane and its radius is 21 m. Sitting on the seat before the ride starts, a rider is level and stationary, and the electronic sensor reads 770 N. At the top of the loop, the rider is upside down and moving, and the sensor reads 350 N. What is the speed of the rider at the top of the loop?

38. **E** **GO** A 0.20-kg ball on a stick is whirled on a vertical circle at a constant speed. When the ball is at the three o'clock position, the stick tension is 16 N. Find the tensions in the stick when the ball is at the twelve o'clock and at the six o'clock positions.

39. **M** **GO** **MMH** A stone is tied to a string (length = 1.10 m) and whirled in a circle at the same constant speed in two different ways. First, the circle is horizontal and the string is nearly parallel to the ground. Next, the circle is vertical. In the vertical case the maximum tension in the string is 15.0% larger than the tension that exists when the circle is horizontal. Determine the speed of the stone.

40. **M** **GO** **CHALK** A motorcycle is traveling up one side of a hill and down the other side. The crest of the hill is a circular arc with a radius of 45.0 m. Determine the maximum speed that the cycle can have while moving over the crest without losing contact with the road.

Additional Problems

Online

41. **E** **SSM** In a skating stunt known as crack-the-whip, a number of skaters hold hands and form a straight line. They try to skate so that the line rotates about the skater at one end, who acts as the pivot. The skater farthest out has a mass of 80.0 kg and is 6.10 m from the pivot. He is skating at a speed of 6.80 m/s. Determine the magnitude of the centripetal force that acts on him.

42. **E** **GO** **MMH** The second hand and the minute hand on one type of clock are the same length. Find the ratio ($a_{c, \text{second}}/a_{c, \text{minute}}$) of the centripetal accelerations of the tips of the second hand and the minute hand.

43. **E** **MMH** A child is twirling a 0.0120-kg plastic ball on a string in a horizontal circle whose radius is 0.100 m. The ball travels once around the circle in 0.500 s. (a) Determine the centripetal force acting on the ball. (b) If the speed is doubled, does the centripetal force double? If not, by what factor does the centripetal force increase?

44. **E** **GO** **MMH** Two cars are traveling at the same speed of 27 m/s on a curve that has a radius of 120 m. Car A has a mass of 1100 kg, and car B has a mass of 1600 kg. Find the magnitude of the centripetal acceleration and the magnitude of the centripetal force for each car.

45. **E** **SSM** A roller coaster at an amusement park has a dip that bottoms out in a vertical circle of radius r. A passenger feels the seat of the car pushing upward on her with a force equal to twice her weight as she goes through the dip. If $r = 20.0$ m, how fast is the roller coaster traveling at the bottom of the dip?

46. **M** **V-HINT** The large blade of a helicopter is rotating in a horizontal circle. The length of the blade is 6.7 m, measured from its tip to the center of the circle. Find the ratio of the centripetal acceleration at the end of the blade to that which exists at a point located 3.0 m from the center of the circle.

47. **M** A block is hung by a string from the inside roof of a van. When the van goes straight ahead at a speed of 28 m/s, the block hangs vertically down. But when the van maintains this same speed around an unbanked curve (radius = 150 m), the block swings toward the outside of the curve. Then the string makes an angle θ with the vertical. Find θ.

48. **M** **GO** A rigid massless rod is rotated about one end in a horizontal circle. There is a particle of mass m_1 attached to the center of the rod and a particle of mass m_2 attached to the outer end of the rod. The inner section of the rod sustains a tension that is three times as great as the tension that the outer section sustains. Find the ratio m_1/m_2.

49. **M** **GO** A space laboratory is rotating to create artificial gravity, as the figure indicates. Its period of rotation is chosen so that the outer ring ($r_O = 2150$ m) simulates the acceleration due to gravity on earth (9.80 m/s^2). What should be the radius r_1 of the inner ring, so it simulates the acceleration due to gravity on the surface of Mars (3.72 m/s^2)?

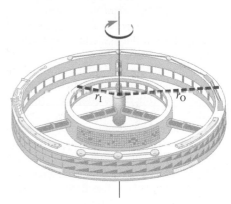

PROBLEM 49

Physics in Biology, Medicine, and Sports

50. **E** **BIO** **5.2** The aorta is a major artery, rising upward from the left ventricle of the heart and curving down to carry blood to the abdomen and lower half of the body. The curved artery can be approximated as a semicircular arch whose diameter is 5.0 cm. If blood flows through the aortic arch at a speed of 0.32 m/s, what is the magnitude (in m/s^2) of the blood's centripetal acceleration?

51. **M** **BIO** **5.2** A centrifuge is a device in which a small container of material is rotated at a high speed on a circular path. Such a device is used in medical laboratories, for instance, to cause the more dense red blood cells to settle through the less dense blood serum and collect at the bottom of the container. Suppose the centripetal acceleration of the sample is 6.25×10^3 times as large as the acceleration due to gravity. How many revolutions per minute is the sample making, if it is located at a radius of 5.00 cm from the axis of rotation?

52. **E** **BIO** **V-HINT** **5.7** Pilots of high-performance fighter planes can be subjected to large centripetal accelerations during high-speed turns (see Example 8). Because of these accelerations, the pilots are subjected to forces that can be much greater than their body weight, leading to an accumulation of blood in the abdomen and legs. As a result, the brain becomes starved for blood, and the pilot can lose consciousness ("black out"). The pilots wear "anti-g suits" to help keep the blood from draining out of the brain. To appreciate the forces that a fighter pilot must endure, consider the magnitude F_N of the normal force that the pilot's seat exerts on him at the bottom of a dive. The magnitude of the pilot's weight is W. The plane is traveling at 230 m/s on a vertical circle of radius 690 m. Determine the ratio F_N/W. For comparison, note that blackout can occur for values of F_N/W as small as 2 if the pilot is not wearing an anti-G suit.

53. **E** **BIO** **5.3** Consider again the "human centrifuge" in Example 8. If the chamber is located 15 m from the center of the circle. At what speed must the chamber move so that an astronaut is subjected to 7.5 times the acceleration due to gravity?

54. **M** **BIO** **5.6** Astronauts in the International Space Station (ISS) are in free-fall as they orbit the earth, so their apparent weight is zero. While floating around weightless may appear exciting and fun, there can be significant negative health effects on astronauts who are in space for long periods of time. Muscles atrophy (including the heart), bone density decreases, and flattening of the eyeballs occurs along with changes to the retina. On earth, gravity pulls downward on the blood and other fluids in the body. However, in orbit, the fluids redistribute themselves, which leads to more fluids in the upper half of the body and an increase in intercranial pressure. Perhaps one of the most surprising effects of long-term weightlessness is a change at the genetic level. In 2015 and 2016, former NASA astronaut Scott Kelly participated in a year-long "twins experiment" with his twin brother Mark. Scott spent a year onboard the ISS, while Mark remained on the ground. After returning to earth, medical examination of the brothers showed that certain genes turn on and off in space. Telomeres, which form the end of chromosomes and slow down their deterioration, temporarily got longer in Scott while he was in space. If the human race is to survive and we are to travel from earth to other planets, we will need to live and work in space. One possible solution to the negative physiological effects of space travel is to simulate gravity, as discussed in Example 13. In the 2015 movie *The Martian* the Hermes spaceship that took the crew to Mars and back had a rotating section to create artificial gravity of 1g (see photo, highlighted section). If the speed of the outer wall of the rotating section of the Hermes is 14 m/s, what is the diameter (d) of the section?

PROBLEM 54

55. **M** **BIO** **5.3** Adhesive capsulitis, also known as a frozen shoulder, is a condition that affects the motion of the shoulder joint. The articular shoulder capsule becomes inflamed, stiff, and restricts a person's mobility. Physical therapy provides one course of treatment for a frozen shoulder. One of the exercises often performed is the pendulum. Here, the patient bends forward and lets the injured arm hang downward and swing freely like a pendulum (see figure). The patient slowly swings their arm in a circular motion, for say 10 cycles, and then switches direction and completes 10 more. The patient may hold a light weight dumbbell while performing this motion, or just use an empty hand. Imagine the situation shown in the figure. A patient is moving a 7.0-lb (3.2-kg) dumbbell at constant speed and completes one circular motion in 1.2 s. What is the radius (r) of the circular motion?

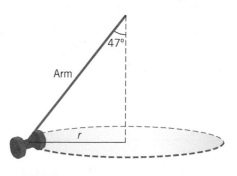

PROBLEM 55

56. **E** **BIO** **5.2** A dental drill operates at incredibly high speeds. It uses compressed air to do this, and it can accommodate multiple drilling heads. The head shown in the figure is called a dental burr, and it has a diameter of 0.30 mm. If the drill operates at 1.8×10^5 rpm, what is the centripetal acceleration of a point on the outside edge of the burr?

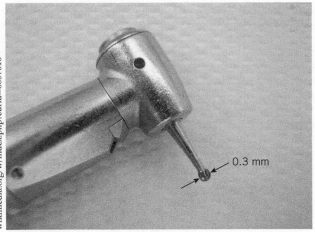

0.3 mm

PROBLEM 56

57. **M** **BIO** **5.3**   Does your apparent weight depend on where you are located on the earth's surface? Yes! Your apparent weight on a scale standing at the equator will be less than your apparent weight while standing on a scale at the North Pole (see figure). This difference occurs due to the earth's rotation. How much extra mass would a 65.0-kg person have to hold while weighing himself at the equator, so that his apparent weight there is equal to his apparent weight at the North Pole?

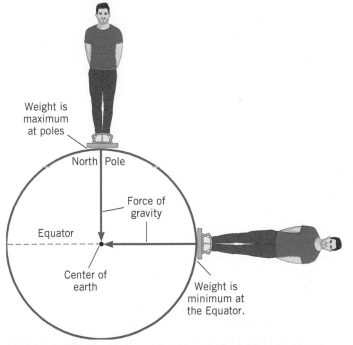

Weight is maximum at poles

North Pole

Force of gravity

Equator

Center of earth

Weight is minimum at the Equator.

Source: Md. Khaja Shareef, Measure the Mass of the Earth, Science Inspiration.
http://scienceinspiration.blogspot.com/2012/05/measure-mass-of-earth.html?

PROBLEM 57

58. **M** **BIO** **5.3**   A variation of the amusement park ride discussed in Problem 11 is the Gravitron, which consists of a rotating cylinder, where riders stand with their backs against the inner surface of the cylinder (see image). As the cylinder rotates faster, riders get stuck to the wall. The effect is great enough that on some rides, the floor even drops away. Let's model the Gravitron as a rotating cylinder with vertical walls. The radius of the cylinder is 5.5 m. At one point during the ride, the speed of the Gravitron is 24 rpm, at which point a rider slides down the wall at constant speed. What is the coefficient of kinetic friction between the rider and the wall?

PROBLEM 58

Concepts and Calculations Problems

Online

In uniform circular motion the concepts of acceleration and force play central roles. Problem 59 deals with acceleration, particularly in connection with the equations of kinematics. Problem 60 deals with force, and it stresses that centripetal force is the net force along the radial direction. This means that all forces along the radial direction must be taken into account when identifying the centripetal force.

59. **M** **CHALK** **SSM** At time $t = 0$ s, automobile A is traveling at a speed of 18 m/s along a straight road and is picking up speed with an acceleration that has a magnitude of 3.5 m/s² as in part *a* of the figure. At $t = 0$ s, automobile B is traveling at a speed of 18 m/s in uniform circular motion as it negotiates a turn (part *b* of the figure). It has a centripetal acceleration whose magnitude is also 3.5 m/s². *Concepts:* (i) Which automobile has a constant acceleration? (ii) For which automobile do the equations of kinematics apply? *Calculations:* Determine the speed of each automobile when $t = 2$ s.

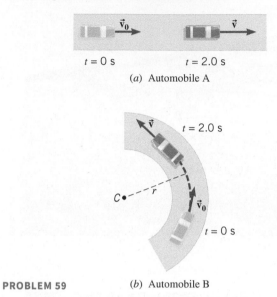

PROBLEM 59   (*b*) Automobile B

60. **M** **CHALK** Ball A is attached to one end of a rigid massless rod, while an identical ball B is attached to the center of the rod, as shown in the figure. Each ball has a mass of $m = 0.50$ kg, and the length of each half of the rod is $L = 0.40$ m. This arrangement is held by the empty end and is whirled around in a horizontal circle at a constant rate, so that each ball is in uniform circular motion. Ball A travels at a constant speed of $v_A = 5.0$ m/s. *Concepts:* (i) How many tension forces contribute to the centripetal force that acts on ball A? (ii) How many tension forces contribute to the centripetal force that acts on ball B? (iii) Is the speed of ball A the same as that of ball B? *Calculations:* Find the tension in each half of the rod.

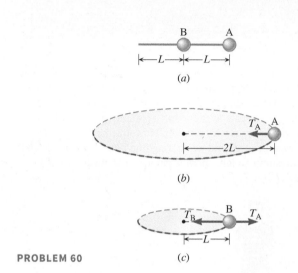

PROBLEM 60   (*c*)

Team Problems

Online

61. **E** **T** **WS** **A Ball Bearing Launch System.** A steel ball bearing of mass $m = 0.100$ kg has a tiny loop on its surface through which is connected a hook that protrudes from the end of a light cable. On the opposite end of the cable (length $L = 0.850$ m) is a trigger that extracts the hook from the loop, and thereby releases the ball. Built into the cable is an in-line digital scale that reports the tension in the cable when the trigger is pulled. You and your team take turns launching the ball bearing by twirling it in a horizontal circle above your head and pulling the release trigger. (**a**) If the largest value recorded by the scale is 28.5 lb (the units given by the digital in-line scale), what is the speed of the ball bearing at launch? Assume that the mass of the cable is negligible. (**b**) What is the maximum horizontal range with this launch speed? For simplicity, you can assume that the ball bearing is launched from ground level, and you can ignore air resistance.

62. **E** **T** **WS** **Not Breaking an Egg.** The House of Fabergé in Russia created the first Fabergé egg in the late 1800s. It is an intricate golden egg, laden with jewels, the first of which was crafted for Tsar Alexander III as a gift for his wife. You and your team are transporting an expensive Fabergé egg that is so fragile that it may be damaged if it experiences an apparent weight of magnitude greater than 2.50 times that of its actual weight on the surface of the earth. The egg is to be

shipped in a military transport aircraft that travels at 370.0 mi/h. Calculate the minimum allowable radius (in meters) of a horizontal turn of the plane. Assume that the plane banks so that the normal force acting on the egg from the internal surface of the plane is the source of the centripetal force required for the turn.

63. **M** **T** **Multilevel Artificial Gravity.** You and your team have the task of designing a space station that will mimic the force due to gravity on the earth, the moon, and Mars. The station's purpose is to physically prepare astronauts for the transition to the different gravitational forces they will experience on Mars and the moon, but also to acclimate them to earth's gravity when they return. The space station should be constructed of concentric cylinders, with a maximum allowed diameter of 300 m. (**a**) Draw a sketch of the space station and label the levels (i.e., the cylinders) according to the celestial body whose gravitational force each cylinder simulates. (**b**) Determine the required radii of each cylinder. Assume the outermost cylinder has the maximum allowable diameter. (**c**) What is the period of rotation of the station? ($g_{earth} = 9.80$ m/s², $g_{Mars} = 3.76$ m/s², and $g_{moon} = 1.62$ m/s²)

64. **M** **T** **A Dangerous Ride.** You and your exploration team are stuck on a steep slope in the Andes Mountains in Argentina. A deadly winter storm is approaching and you must get down the mountain before the storm hits. Your path leads you around an extremely slippery, horizontal curve with a diameter of 90.0 m and banked at an angle of 40.0° relative to the horizontal. You get the idea to unpack the toboggan that you have been using to haul supplies, load your team upon it, and ride it down the mountain to get enough speed to get around the banked curve. You must be extremely careful, however, not to slide down the bank: At the bottom of the curve is a steep cliff. (a) Neglecting friction and air resistance, what must be the speed of your toboggan in order to get around the curve without sliding up or down its bank? Express your answer in m/s and mph.

(b) You will need to climb up the mountain and ride the toboggan down in order to attain the speed you need to safely navigate the curve (from part (a)). The mountain slope leading into the curve is at an angle of 30.0° relative to the horizontal, and the coefficient of kinetic friction between the toboggan and the surface of the slope is ($\mu_{mountain} = 0.150$). How far up the mountain (distance along the slope, not elevation) from the curve should you start your ride? Note: the path down the mountain levels off at the bottom so that the toboggan enters the curve moving in the horizontal plane (i.e., in the same plane as the curve).

CHAPTER 6

On the popular Behemoth roller coaster at Canada's Wonderland amusement park, riders cresting the top of the hill will pick up speed on the way down, as they convert gravitational potential energy into kinetic energy. This conversion is governed by the principle of conservation of energy, a central topic in this chapter.

Oleksiy Maksymenko Photography/Alamy Images

LEARNING OBJECTIVES

After reading this module, you should be able to...

6.1 Determine the work done by a constant force.

6.2 Relate work and kinetic energy.

6.3 Use gravitational potential energy.

6.4 Identify conservative and nonconservative forces.

6.5 Use the law of conservation of mechanical energy.

6.6 Solve problems involving nonconservative forces.

6.7 Solve problems involving power.

6.8 Employ the law of conservation of energy.

6.9 Determine the work done by a variable force.

Work and Energy

6.1 Work Done by a Constant Force

Work is a familiar concept. For example, it takes work to push a stalled car. In fact, more work is done when the pushing force is greater or when the displacement of the car is greater. Force and displacement are, in fact, the two essential elements of work, as **Figure 6.1** illustrates. The drawing shows a constant pushing force $\vec{F}$ that points in the same direction as the resulting displacement $\vec{s}$.* In such a case, the work W is defined

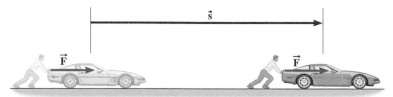

FIGURE 6.1 Work is done when a force $\vec{F}$ pushes a car through a displacement $\vec{s}$.

*When discussing work, it is customary to use the symbol $\vec{s}$ for the displacement, rather than $\vec{x}$ or $\vec{y}$.

as the magnitude F of the force times the magnitude s of the displacement: $W = Fs$. The work done to push a car is the same whether the car is moved north to south or east to west, provided that the amount of force used and the distance moved are the same. Since work does not convey directional information, it is a scalar quantity.

The equation $W = Fs$ indicates that the unit of work is the unit of force times the unit of distance, or the newton · meter in SI units. One newton · meter is referred to as a *joule* (J) (rhymes with "cool"), in honor of James Joule (1818–1889) and his research into the nature of work, energy, and heat. **Table 6.1** summarizes the units for work in several systems of measurement.

TABLE 6.1 Units of Measurement for Work

System	Force	×	Distance	=	Work
SI	newton (N)		meter (m)		joule (J)
CGS	dyne (dyn)		centimeter (cm)		erg
BE	pound (lb)		foot (ft)		foot · pound (ft · lb)

The definition of work as $W = Fs$ does have one surprising feature: If the distance s is zero, the work is zero, even if a force is applied. Pushing on an immovable object, such as a brick wall, may tire your muscles, but there is no work done of the type we are discussing. In physics, the idea of work is intimately tied up with the idea of motion. If the object does not move, the force acting on the object does no work.

Often, the force and displacement do not point in the same direction. For instance, **Figure 6.2a** shows a suitcase-on-wheels being pulled to the right by a force that is applied along the handle. The force is directed at an angle θ relative to the displacement. In such a case, only the component of the force along the displacement is used in defining work. As **Figure 6.2b** shows, this component is $F \cos \theta$, and it appears in the general definition below:

DEFINITION OF WORK DONE BY A CONSTANT* FORCE

The work done on an object by a constant force $\vec{F}$ is

$$W = (F \cos \theta)s \qquad \qquad (6.1)$$

where F is the magnitude of the force, s is the magnitude of the displacement, and θ is the angle between the force and the displacement.

SI Unit of Work: newton · meter = joule (J)

When the force points in the same direction as the displacement, then $\theta = 0°$, and Equation 6.1 reduces to $W = Fs$. Example 1 shows how Equation 6.1 is used to calculate work.

EXAMPLE 1 | Pulling a Suitcase-on-Wheels

Find the work done by a 45.0-N force in pulling the suitcase in **Figure 6.2a** at an angle $\theta = 50.0°$ for a distance $s = 75.0$ m.

Reasoning The pulling force causes the suitcase to move a distance of 75.0 m and does work. However, the force makes an angle of 50.0° with the displacement, and we must take this

angle into account by using the definition of work given by Equation 6.1.

Solution The work done by the 45.0-N force is

$$W = (F \cos \theta)s = [(45.0 \text{ N}) \cos 50.0°](75.0 \text{ m}) = \boxed{2170 \text{ J}}$$

The answer is expressed in newton · meters or joules (J).

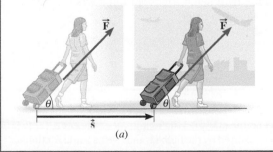

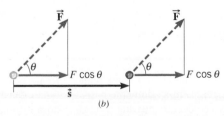

FIGURE 6.2 (*a*) Work can be done by a force $\vec{F}$ that points at an angle θ relative to the displacement $\vec{s}$. (*b*) The force component that points along the displacement is $F \cos \theta$.

---

*Section 6.9 considers the work done by a variable force.

The definition of work in Equation 6.1 takes into account only the component of the force in the direction of the displacement. The force component perpendicular to the displacement does no work. To do work, there must be a force *and* a displacement, and since there is no displacement in the perpendicular direction, there is no work done by the perpendicular component of the force. If the entire force is perpendicular to the displacement, the angle θ in Equation 6.1 is 90°, and the force does no work at all.

Work can be either positive or negative, depending on whether a component of the force points in the same direction as the displacement or in the opposite direction. Example 2 illustrates how positive and negative work arise.

EXAMPLE 2 | BIO The Physics of Weight Lifting

The weight lifter in **Interactive Figure 6.3a** is bench-pressing a barbell whose weight is 710 N. In part *b* of the figure, he raises the barbell a distance of 0.65 m above his chest, and in part *c* he lowers it the same distance. The weight is raised and lowered at a constant velocity. Determine the work done on the barbell by the weight lifter during **(a)** the lifting phase and **(b)** the lowering phase.

Reasoning To calculate the work, it is necessary to know the force exerted by the weight lifter. The barbell is raised and lowered at a constant velocity and, therefore, is in equilibrium. Consequently, the force $\vec{F}$ exerted by the weight lifter must balance the weight of the barbell, so $F = 710$ N. During the lifting phase, the force $\vec{F}$ and displacement $\vec{s}$ are in the same direction, as **Interactive Figure 6.3b** shows. The angle between them is $\theta = 0°$. When the barbell is lowered, however, the force and displacement are in opposite directions, as in **Interactive Figure 6.3c**. The angle

between the force and the displacement is now $\theta = 180°$. With these observations, we can find the work.

Solution **(a)** During the lifting phase, the work done by the force $\vec{F}$ is given by Equation 6.1 as

$$W = (F \cos \theta)s = [(710 \text{ N}) \cos 0°](0.65 \text{ m}) = \boxed{460 \text{ J}}$$

(b) The work done during the lowering phase is

$$W = (F \cos \theta)s$$

$$= [(710 \text{ N}) \cos 180°](0.65 \text{ m}) = \boxed{-460 \text{ J}}$$

since $\cos 180° = -1$. The work is negative, because the force is opposite to the displacement. Weight lifters call each complete up-and-down movement of the barbell a repetition, or "rep." The lifting of the weight is referred to as the positive part of the rep, and the lowering is known as the negative part.

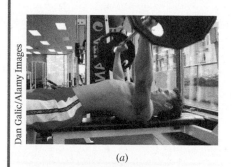 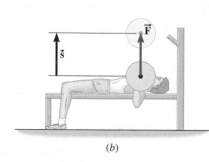 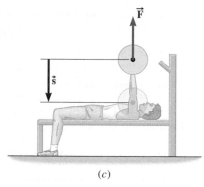

(a)   (b)   (c)

INTERACTIVE FIGURE 6.3 (*a*) In the bench press, work is done during both the lifting and lowering phases. (*b*) During the lifting phase, the force $\vec{F}$ does positive work. (*c*) During the lowering phase, the force does negative work.

Example 3 deals with the work done by a static frictional force when it acts on a crate that is resting on the bed of an accelerating truck.

EXAMPLE 3 | Accelerating a Crate

Figure 6.4a shows a 120-kg crate on the flatbed of a truck that is moving with an acceleration of $a = +1.5$ m/s² along the positive *x* axis. The crate does not slip with respect to the truck as the truck undergoes a displacement whose magnitude is $s = 65$ m. What is the total work done on the crate by all of the forces acting on it?

Reasoning The free-body diagram in **Figure 6.4b** shows the forces that act on the crate: (1) the weight $\vec{W}$ of the crate, (2) the normal force $\vec{F}_N$ exerted by the flatbed, and (3) the static frictional force $\vec{f}_s$, which is exerted by the flatbed in the forward direction and keeps the crate from slipping backward. The weight and the normal force are perpendicular to the displacement, so they do no work. Only the static frictional force does work, since it acts in the x direction. To determine the frictional force, we note that the crate does not slip and, therefore, must have the same acceleration of a = +1.5 m/s² as does the truck. The force creating this acceleration is the static frictional force, and, knowing the mass of the crate and its acceleration, we can use Newton's second law to obtain its magnitude. Then, knowing the frictional force and the displacement, we can determine the total work done on the crate.

Solution From Newton's second law, we find that the magnitude f_s of the static frictional force is

$$f_s = ma = (120 \text{ kg})(1.5 \text{ m/s}^2) = 180 \text{ N}$$

The total work is that done by the static frictional force and is

$$W = (f_s \cos\theta)s = (180 \text{ N})(\cos 0°)(65 \text{ m}) = \boxed{1.2 \times 10^4 \text{ J}} \qquad (6.1)$$

The work is positive, because the frictional force is in the same direction as the displacement ($\theta = 0°$).

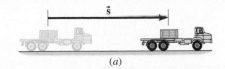
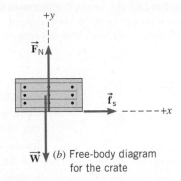

(b) Free-body diagram for the crate

FIGURE 6.4 (a) The truck and crate are accelerating to the right for a distance of s = 65 m. (b) The free-body diagram for the crate.

Check Your Understanding

(*The answers are given at the end of the book.*)

1. Two forces $\vec{F}_1$ and $\vec{F}_2$ are acting on the box shown in **CYU Figure 6.1**, causing the box to move across the floor. The two force vectors are drawn to scale. Which one of the following statements is correct? **(a)** $\vec{F}_2$ does more work than $\vec{F}_1$ does. **(b)** $\vec{F}_1$ does more work than $\vec{F}_2$ does. **(c)** Both forces do the same amount of work. **(d)** Neither force does any work.

2. A box is being moved with a velocity $\vec{v}$ by a force $\vec{P}$ (in the same direction as $\vec{v}$) along a level horizontal floor. The normal force is $\vec{F}_N$, the kinetic frictional force is $\vec{f}_k$, and the weight is $m\vec{g}$. Which one of the following statements is correct? **(a)** $\vec{P}$ does positive work, $\vec{F}_N$ and $\vec{f}_k$ do zero work, and $m\vec{g}$ does negative work. **(b)** $\vec{F}_N$ does positive work, $\vec{P}$ and $\vec{f}_k$ do zero work, and $m\vec{g}$ does negative work. **(c)** $\vec{f}_k$ does positive work, $\vec{F}_N$ and $m\vec{g}$ do zero work, and $\vec{P}$ does negative work. **(d)** $\vec{P}$ does positive work, $\vec{F}_N$ and $m\vec{g}$ do zero work, and $\vec{f}_k$ does negative work.

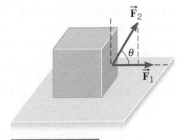

CYU FIGURE 6.1

3. A force does positive work on a particle that has a displacement pointing in the +x direction. This same force does negative work on a particle that has a displacement pointing in the +y direction. In which quadrant of the x, y coordinate system does the force lie? **(a)** First **(b)** Second **(c)** Third **(d)** Fourth

4. A suitcase is hanging straight down from your hand as you ride an escalator. Your hand exerts a force on the suitcase, and this force does work. This work is **(a)** positive when you ride up and negative when you ride down, **(b)** negative when you ride up and positive when you ride down, **(c)** positive when you ride up or down, **(d)** negative when you ride up or down.

6.2 The Work–Energy Theorem and Kinetic Energy

Most people expect that if you do work, you get something as a result. In physics, when a net force performs work on an object, there is always a result from the effort. The result is a change in the **kinetic energy** of the object. As we will now see, the relationship that relates work to the change in kinetic energy is known as the **work–energy theorem**. This theorem is obtained by bringing together three basic concepts that we've already learned

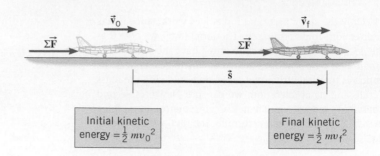

A constant net external force $\Sigma\vec{F}$ acts over a displacement $\vec{s}$ and does work on the plane. As a result of the work done, the plane's kinetic energy changes.

about. First, we'll apply Newton's second law of motion, $\Sigma F = ma$, which relates the net force ΣF to the acceleration a of an object. Then, we'll determine the work done by the net force when the object moves through a certain distance. Finally, we'll use Equation 2.9, one of the equations of kinematics, to relate the distance and acceleration to the initial and final speeds of the object. The result of this approach will be the work–energy theorem.

To gain some insight into the idea of kinetic energy and the work–energy theorem, look at **Interactive Figure 6.5**, where a constant net external force $\Sigma\vec{F}$ acts on an airplane of mass m. This net force is the vector sum of all the external forces acting on the plane, and, for simplicity, it is assumed to have the same direction as the displacement $\vec{s}$. According to Newton's second law, the net force produces an acceleration a, given by $a = \Sigma F/m$. Consequently, the speed of the plane changes from an initial value of v_0 to a final value of v_f.* Multiplying both sides of $\Sigma F = ma$ by the distance s gives

$$\underbrace{(\Sigma F)s}_{\substack{\text{Work done by}\\ \text{net ext. force}}} = mas$$

The left side of this equation is the work done by the net external force. The term as on the right side can be related to v_0 and v_f by using $v_f^2 = v_0^2 + 2as$ (Equation 2.9) from the equations of kinematics. Solving this equation gives

$$as = \tfrac{1}{2}\left(v_f^2 - v_0^2\right)$$

Substituting this result into $(\Sigma F)s = mas$ shows that

$$\underbrace{(\Sigma F)s}_{\substack{\text{Work done by}\\ \text{net ext. force}}} = \underbrace{\tfrac{1}{2}mv_f^2}_{\substack{\text{Final}\\ \text{KE}}} - \underbrace{\tfrac{1}{2}mv_0^2}_{\substack{\text{Initial}\\ \text{KE}}}$$

This expression is the work–energy theorem. Its left side is the work W done by the net external force, whereas its right side involves the difference between two terms, each of which has the form $\tfrac{1}{2}(\text{mass})(\text{speed})^2$. The quantity $\tfrac{1}{2}(\text{mass})(\text{speed})^2$ is called kinetic energy (KE) and plays a significant role in physics, as we will see in this chapter and later on in other chapters as well.

DEFINITION OF KINETIC ENERGY

The kinetic energy KE of an object with mass m and speed v is given by

$$\text{KE} = \tfrac{1}{2}mv^2 \tag{6.2}$$

SI Unit of Kinetic Energy: joule (J)

The SI unit of kinetic energy is the same as the unit for work, the joule. Kinetic energy, like work, is a scalar quantity. These are not surprising observations, because work and kinetic energy are closely related, as is clear from the following statement of the work–energy theorem.

THE WORK–ENERGY THEOREM

When a net external force does work W on an object, the kinetic energy of the object changes from its initial value of KE_0 to a final value of KE_f, the difference between the two values being equal to the work:

$$W = \text{KE}_f - \text{KE}_0 = \tfrac{1}{2}mv_f^2 - \tfrac{1}{2}mv_0^2 \tag{6.3}$$

*For extra emphasis, the final speed is now represented by the symbol v_f, rather than v.

The work–energy theorem may be derived for any direction of the force relative to the displacement, not just the situation in **Interactive Figure 6.5**. In fact, the force may even vary from point to point along a path that is curved rather than straight, and the theorem remains valid. According to the work–energy theorem, a moving object has kinetic energy, because work was done to accelerate the object from rest to a speed v_f.[†] Conversely, an object with kinetic energy can perform work, if it is allowed to push or pull on another object. Example 4 illustrates the work–energy theorem and considers a single force that does work to change the kinetic energy of a space probe.

Analyzing Multiple-Concept Problems

EXAMPLE 4 | The Physics of an Ion Propulsion Drive

The space probe *Deep Space 1* was launched October 24, 1998, and it used a type of engine called an ion propulsion drive. An ion propulsion drive generates only a weak force (or thrust), but can do so for long periods of time using only small amounts of fuel. Suppose the probe, which has a mass of 474 kg, is traveling at an initial speed of 275 m/s. No forces act on it except the 5.60×10^{-2}-N thrust of its engine. This external force $\vec{F}$ is directed parallel to the displacement $\vec{s}$, which has a magnitude of 2.42×10^9 m (see **Figure 6.6**). Determine the final speed of the probe, assuming that its mass remains nearly constant.

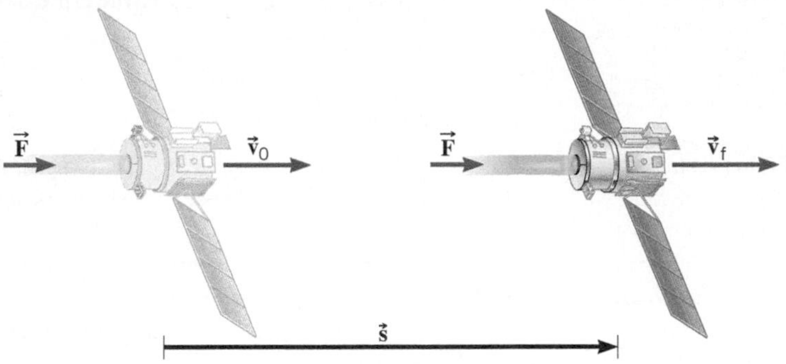

Reasoning   If we can determine the final kinetic energy of the space probe, we can determine its final speed, since kinetic energy is related to mass and speed according to Equation 6.2 and the mass of the probe is known. We will use the work–energy theorem ($W = KE_f - KE_0$), along with the definition of work, to find the final kinetic energy.

FIGURE 6.6   An ion propulsion drive generates a single force $\vec{F}$ that points in the same direction as the displacement $\vec{s}$. The force performs positive work, causing the space probe to gain kinetic energy.

Knowns and Unknowns   The following list summarizes the data for this problem:

Description	Symbol	Value	Comment
Explicit Data			
Mass	m	474 kg	
Initial speed	v_0	275 m/s	
Magnitude of force	F	5.60×10^{-2} N	
Magnitude of displacement	s	2.42×10^9 m	
Implicit Data			
Angle between force $\vec{F}$ and displacement $\vec{s}$	θ	0°	The force is parallel to the displacement.
Unknown Variable			
Final speed	v_f	?	

[†]Strictly speaking, the work–energy theorem, as given by Equation 6.3, applies only to a single particle, which occupies a mathematical point in space. A macroscopic object, however, is a collection or system of particles and is spread out over a region of space. Therefore, when a force is applied to a macroscopic object, the point of application of the force may be anywhere on the object. To take into account this and other factors, a discussion of work and energy is required that is beyond the scope of this text. The interested reader may refer to A. B. Arons, *The Physics Teacher*, October 1989, p. 506.

Modeling the Problem

STEP 1 Kinetic Energy   An object of mass m and speed v has a kinetic energy KE given by Equation 6.2 as $KE = \frac{1}{2}mv^2$. Using the subscript f to denote the final kinetic energy and the final speed of the probe, we have that

$$KE_f = \frac{1}{2}mv_f^2$$

Solving for v_f gives Equation 1 at the right. The mass m is known but the final kinetic energy KE_f is not, so we will turn to Step 2 to evaluate it.

$$v_f = \sqrt{\frac{2(KE_f)}{m}} \qquad (1)$$

$?$

STEP 2 The Work–Energy Theorem   The work–energy theorem relates the final kinetic energy KE_f of the probe to its initial kinetic energy KE_0 and the work W done by the force of the engine. According to Equation 6.3, this theorem is $W = KE_f - KE_0$. Solving for KE_f shows that

$$KE_f = KE_0 + W$$

The initial kinetic energy can be expressed as $KE_0 = \frac{1}{2}mv_0^2$, so the expression for the final kinetic energy becomes

$$\boxed{KE_f = \frac{1}{2}mv_0^2 + W}$$

This result can be substituted into Equation 1, as indicated at the right. Note from the data table that we know the mass m and the initial speed v_0. The work W is not known and will be evaluated in Step 3.

$$v_f = \sqrt{\frac{2(KE_f)}{m}} \qquad (1)$$

$$\boxed{KE_f = \frac{1}{2}mv_0^2 + W} \qquad (2)$$

$?$

STEP 3 Work   The work W is that done by the net external force acting on the space probe. Since there is only the one force $\vec{F}$ acting on the probe, it is the net force. The work done by this force is given by Equation 6.1 as

$$\boxed{W = (F\cos\theta)s}$$

where F is the magnitude of the force, θ is the angle between the force and the displacement, and s is the magnitude of the displacement. All the variables on the right side of this equation are known, so we can substitute it into Equation 2, as shown in the right column.

$$v_f = \sqrt{\frac{2(KE_f)}{m}} \qquad (1)$$

$$\boxed{KE_f = \frac{1}{2}mv_0^2 + W} \qquad (2)$$

$$\boxed{W = (F\cos\theta)s}$$

Solution   Algebraically combining the results of the three steps, we have

STEP 1   STEP 2   STEP 3

$$v_f = \sqrt{\frac{2(KE_f)}{m}} = \sqrt{\frac{2\left(\frac{1}{2}mv_0^2 + W\right)}{m}} = \sqrt{\frac{2\left[\frac{1}{2}mv_0^2 + (F\cos\theta)s\right]}{m}}$$

The final speed of the space probe is

$$v_f = \sqrt{\frac{2\left[\frac{1}{2}mv_0^2 + (F\cos\theta)s\right]}{m}}$$

$$= \sqrt{\frac{2\left[\frac{1}{2}(474\text{ kg})(275\text{ m/s})^2 + (5.60\times10^{-2}\text{N})(\cos 0°)(2.42\times10^9\text{m})\right]}{474\text{ kg}}} = \boxed{805\text{ m/s}}$$

Related Homework: Problem 22

In Example 4 only the force of the engine does work. If several external forces act on an object, they must be added together vectorially to give the net force. The work done by the net force can then be related to the change in the object's kinetic energy by using the work–energy theorem, as in the next example.

Analyzing Multiple-Concept Problems

EXAMPLE 5 | Downhill Skiing

A 58-kg skier is coasting down a 25° slope, as **Figure 6.7a** shows. Near the top of the slope, her speed is 3.6 m/s. She accelerates down the slope because of the gravitational force, even though a kinetic frictional force of magnitude 71 N opposes her motion. Ignoring air resistance, determine the speed at a point that is displaced 57 m downhill.

Reasoning The skier's speed at a point 57 m downhill (her final speed) depends on her final kinetic energy. According to the work–energy theorem, her final kinetic energy is related to her initial kinetic energy (which we can calculate directly) and the work done by the net external force that acts on her. The work can be evaluated directly from its definition.

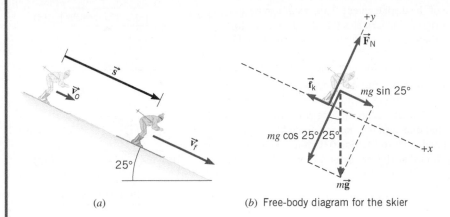

FIGURE 6.7 (a) A skier coasting downhill. (b) The free-body diagram for the skier.

(a)

(b) Free-body diagram for the skier

Knowns and Unknowns   The data for this problem are listed below:

Description	Symbol	Value	Comment
Explicit Data			
Mass	m	58 kg	
Initial speed	v_0	3.6 m/s	
Magnitude of kinetic frictional force	f_k	71 N	
Magnitude of skier's displacement	s	57 m	
Angle of slope above horizontal		25°	
Implicit Data			
Angle between net force acting on skier and her displacement	θ	0°	The skier is accelerating down the slope, so the direction of the net force is parallel to her displacement.
Unknown Variable			
Final speed	v_f	?	

Modeling the Problem

STEP 1 Kinetic Energy   The final speed v_f of the skier is related to her final kinetic energy KE_f and mass m by Equation 6.2:

$$KE_f = \tfrac{1}{2} m v_f^2$$

Solving for v_f gives Equation 1 at the right. Her mass is known, but her final kinetic energy is not, so we turn to Step 2 to evaluate it.

$$v_f = \sqrt{\frac{2(KE_f)}{m}} \qquad (1)$$

$\boxed{?}$

STEP 2 The Work–Energy Theorem   The final kinetic energy KE_f of the skier is related to her initial kinetic energy KE_0 and the work W done by the net external force acting on her. This is according to the work–energy theorem (Equation 6.3): $KE_f = KE_0 + W$.

The initial kinetic energy can be written as $KE_0 = \frac{1}{2}mv_0^2$, so the expression for the final kinetic energy becomes

$$KE_f = \frac{1}{2}mv_0^2 + W$$

This result can be substituted into Equation 1, as indicated at the right. The mass m and the initial speed v_0 are known. The work W is not known, and will be evaluated in Steps 3 and 4.

$$v_f = \sqrt{\frac{2(KE_f)}{m}} \qquad (1)$$

$$KE_f = \frac{1}{2}mv_0^2 + W \qquad (2)$$

$$\boxed{?}$$

STEP 3 Work The work W done by the net external force acting on the skier is given by Equation 6.1 s

$$W = (\Sigma F \cos\theta)s$$

where ΣF is the magnitude of the net force, θ is the angle between the net force and the displacement, and s is the magnitude of the displacement. This result can be substituted into Equation 2, as shown at the right. The variables θ and s are known (see the table of knowns and unknowns), and the net external force will be determined in Step 4.

$$v_f = \sqrt{\frac{2(KE_f)}{m}} \qquad (1)$$

$$KE_f = \frac{1}{2}mv_0^2 + W \qquad (2)$$

$$W = (\Sigma F \cos\theta)s \qquad (3)$$

$$\boxed{?}$$

STEP 4 The Net External Force **Figure 6.7b** is a free-body diagram for the skier and shows the three external forces acting on her: the gravitational force $m\vec{g}$, the kinetic frictional force $\vec{f}_k$ and the normal force $\vec{F}_N$. The net external force along the y axis is zero, because there is no acceleration in that direction (the normal force balances the component $mg\cos 25°$ of the weight perpendicular to the slope). Using the data from the table of knowns and unknowns, we find that the net external force along the x axis is

$$\Sigma F = mg\sin 25° - f_k$$

$$= (58 \text{ kg})(9.80 \text{ m/s}^2)\sin 25° - 71 \text{ N} = 170 \text{ N}$$

$$v_f = \sqrt{\frac{2(KE_f)}{m}} \qquad (1)$$

$$KE_f = \frac{1}{2}mv_0^2 + W \qquad (2)$$

$$W = (\Sigma F \cos\theta)s \qquad (3)$$

$$\Sigma F = mg\sin 25° - f_k = 170 \text{ N}$$

All the variables on the right side of this equation are known, so we substitute this expression into Equation 3 for the net external force (see the right column).

Math Skills In **Figure 6.7b** the angle between the gravitational force and the $-y$ axis equals the 25° angle of the slope. **Figure 6.8** uses β for this angle and helps to explain why $\beta = 25°$. It shows the vertical directional line along which the gravitational force acts, as well as the slope and the x, y axes. Since this directional line is perpendicular to the ground, the triangle ABC is a right triangle. It follows, then, that $\alpha + 25° = 90°$ or $\alpha = 65°$. Since the y axis is perpendicular to the slope, it is also true that $\alpha + \beta = 90°$. Solving this equation for the angle β and substituting $\alpha = 65°$ gives

$$\beta = 90° - \alpha = 90° - 65° = 25°$$

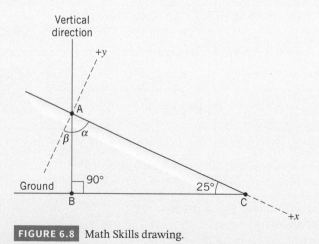

FIGURE 6.8 Math Skills drawing.

Solution   Algebraically combining the four steps, we arrive at the following expression for the final speed of the skier:

STEP 1   STEP 2   STEP 3 and 4

$$v_f = \sqrt{\frac{2(KE_f)}{m}} = \sqrt{\frac{2\left(\frac{1}{2}mv_0^2 + W\right)}{m}} = \sqrt{\frac{2\left[\frac{1}{2}mv_0^2 + (170\ N)(\cos\theta)s\right]}{m}}$$

Referring to the table of knowns and unknowns for the values of the symbols in this result, we find that the final speed of the skier is

$$v_f = \sqrt{\frac{2\left[\frac{1}{2}mv_0^2 + (170\ N)(\cos\theta)s\right]}{m}}$$

$$= \sqrt{\frac{2\left[\frac{1}{2}(58\ kg)(3.6\ m/s)^2 + (170\ N)(\cos 0°)(57\ m)\right]}{58\ kg}} = \boxed{19\ m/s}$$

Related Homework: Problem 22

Problem-Solving Insight   **Example 5 emphasizes that the *work–energy theorem* deals with the work done by the net external force. The work–energy theorem does not apply to the work done by an individual force,** unless that force happens to be the only one present, in which case it is the net force.

If the work done by the net force is *positive,* as in Example 5, the kinetic energy of the object *increases.* If the work done is *negative,* the kinetic energy *decreases.* If the work is zero, the kinetic energy remains the same. Conceptual Example 6 explores these ideas further in a situation where the component of the net force along the direction of the displacement changes during the motion.

CONCEPTUAL EXAMPLE 6 | Work and Kinetic Energy

Figure 6.9 illustrates a satellite moving about the earth in a circular orbit and in an elliptical orbit. The only external force that acts on the satellite is the gravitational force. In which orbit does the kinetic energy of the satellite change, **(a)** the circular orbit or **(b)** the elliptical orbit?

Reasoning   The gravitational force is the only force acting on the satellite, so it is the net force. With this fact in mind, we will apply the work–energy theorem, which states that the work done by the net force equals the change in the kinetic energy.

Answer (a) is incorrect.   For the circular orbit in **Figure 6.9a**, the gravitational force $\vec{F}$ does no work on the satellite since the force is perpendicular to the instantaneous displacement $\vec{s}$ at all times. Thus, the work done by the net force is zero, and according to the work–energy theorem (Equation 6.3), the kinetic energy of the satellite (and, hence, its speed) remains the same everywhere on the orbit.

Answer (b) is correct.   For the elliptical orbit in **Figure 6.9b**, the gravitational force $\vec{F}$ does do work. For example, as the satellite moves toward the earth in the top part of **Figure 6.9b**, there is a component of $\vec{F}$ that points in the same direction as the

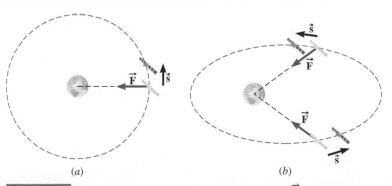

FIGURE 6.9   (a) In a circular orbit, the gravitational force $\vec{F}$ is always perpendicular to the displacement $\vec{s}$ of the satellite and does no work. (b) In an elliptical orbit, there can be a component of the force along the displacement, and, consequently, work can be done.

displacement. Consequently, $\vec{F}$ does positive work during this part of the orbit, and the kinetic energy of the satellite increases. When the satellite moves away from the earth, as in the lower part of **Figure 6.9b**, $\vec{F}$ has a component that points opposite to the displacement and, therefore, does negative work. As a result, the kinetic energy of the satellite decreases.

Related Homework: Problem 18

(The answers are given at the end of the book.)

5. A sailboat is moving at a constant velocity. Is work being done by a net external force acting on the boat?

6. A ball has a speed of 15 m/s. Only one external force acts on the ball. After this force acts, the speed of the ball is 7 m/s. Has the force done **(a)** positive, **(b)** zero, or **(c)** negative work on the ball?

7. A rocket is at rest on the launch pad. When the rocket is launched, its kinetic energy increases. Consider all of the forces acting on the rocket during the launch, and decide whether the following statement is true or false: The amount by which the kinetic energy of the rocket increases during the launch is equal to the work done by the force generated by the rocket's engine.

8. A net external force acts on a particle that is moving along a straight line. This net force is not zero. Which one of the following statements is correct? **(a)** The velocity, but not the kinetic energy, of the particle is changing. **(b)** The kinetic energy, but not the velocity, of the particle is changing. **(c)** Both the velocity and the kinetic energy of the particle are changing.

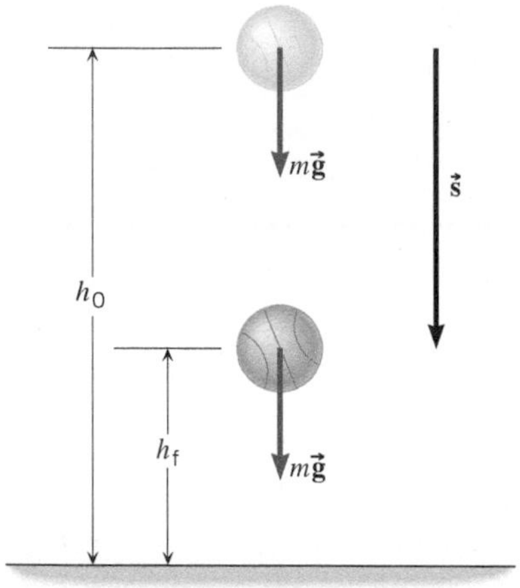

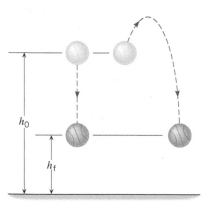

FIGURE 6.10 Gravity exerts a force $m\vec{\mathbf{g}}$ on the basketball. Work is done by the gravitational force as the basketball falls from a height of h_0 to a height of h_f.

6.3 Gravitational Potential Energy

Work Done by the Force of Gravity

The gravitational force is a well-known force that can do positive or negative work, and **Figure 6.10** helps to show how the work can be determined. This drawing depicts a basketball of mass m moving vertically downward, the force of gravity $m\vec{\mathbf{g}}$ being the only force acting on the ball. The initial height of the ball is h_0, and the final height is h_f, both distances measured from the earth's surface. The displacement $\vec{\mathbf{s}}$ is downward and has a magnitude of $s = h_0 - h_f$. To calculate the work W_{gravity} done on the ball by the force of gravity, we use $W = (F \cos \theta)s$ with $F = mg$ and $\theta = 0°$, since the force and displacement are in the same direction:

$$W_{\text{gravity}} = (mg \cos 0°)(h_0 - h_f) = mg(h_0 - h_f) \tag{6.4}$$

Equation 6.4 is valid for *any path* taken between the initial and final heights, and not just for the straight-down path shown in **Figure 6.10**. For example, the same expression can be derived for both paths shown in **Figure 6.11**. Thus, only the *difference in vertical distances* $(h_0 - h_f)$ need be considered when calculating the work done by gravity. Since the difference in the vertical distances is the same for each path in the drawing, the work done by gravity is the same in each case. We are assuming here that the difference in heights is small compared to the radius of the earth, so that the magnitude g of the acceleration due to gravity is the same at every height. Moreover, for positions close to the earth's surface, we can use the value of $g = 9.80 \text{ m/s}^2$.

Only the difference $h_0 - h_f$ appears in Equation 6.4, so h_0 and h_f themselves need not be measured from the earth. For instance, they could be measured relative to a level that is one meter above the ground, and $h_0 - h_f$ would still have the same value. Example 7 illustrates how the work done by gravity is used with the work–energy theorem.

FIGURE 6.11 An object can move along different paths in going from an initial height of h_0 to a final height of h_f. In each case, the work done by the gravitational force is the same [$W_{\text{gravity}} = mg(h_0 - h_f)$], since the change in vertical distance $(h_0 - h_f)$ is the same.

EXAMPLE 7 | A Gymnast on a Trampoline

A gymnast springs vertically upward from a trampoline as in **Figure 6.12a**. The gymnast leaves the trampoline at a height of 1.20 m and reaches a maximum height of 4.80 m before falling back down. All heights are measured with respect to the ground. Ignoring air resistance, determine the initial speed v_0 with which the gymnast leaves the trampoline.

Reasoning We can find the initial speed of the gymnast (mass = m) by using the work–energy theorem, provided the work done by the net external force can be determined. Since only the gravitational force acts on the gymnast in the air, it is the net force, and we can evaluate the work by using the relation $W_{\text{gravity}} = mg(h_0 - h_f)$.

Solution **Figure 6.12b** shows the gymnast moving upward. The initial and final heights are $h_0 = 1.20$ m and $h_f = 4.80$ m, respectively. The initial speed is v_0 and the final speed is $v_f = 0$ m/s, since the gymnast comes to a momentary halt at the highest point. Since $v_f = 0$ m/s, the final kinetic energy is $KE_f = 0$ J, and the work–energy theorem becomes $W = KE_f -$ $KE_0 = -\overline{KE_0}$. The work W is that due to gravity, so this theorem reduces to

$$W_{gravity} = mg(h_0 - h_f) = -\tfrac{1}{2}mv_0^2$$

Solving for v_0 gives

$$v_0 = \sqrt{-2g(h_0 - h_f)} = \sqrt{-2(9.80 \text{ m/s}^2)(1.20 \text{ m} - 4.80 \text{ m})}$$

$$= \boxed{8.40 \text{ m/s}}$$

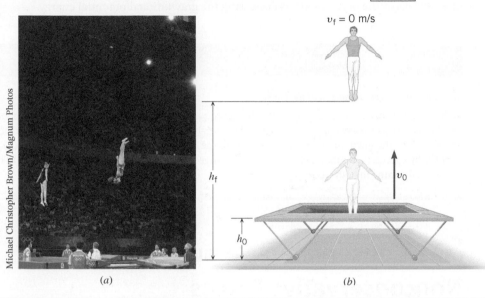

$v_f = 0$ m/s

h_f

v_0

h_0

FIGURE 6.12 (*a*) A gymnast bounces on a trampoline. (*b*) The gymnast moves upward with an initial speed v_0 and reaches maximum height with a final speed of zero.

(*a*)

(*b*)

Michael Christopher Brown/Magnum Photos

Gravitational Potential Energy

An object in motion has kinetic energy. There are also other types of energy. For example, an object may possess energy by virtue of its position relative to the earth and is said to have gravitational potential energy. A pile driver, for instance, is used to pound "piles," or structural support beams, into the ground. The device contains a massive hammer that is raised to a height h and dropped (see **Figure 6.13**), so the hammer has the potential to do the work of driving the pile into the ground. The greater the height h, the greater is the potential for doing work, and the greater is the gravitational potential energy.

Now, let's obtain an expression for the gravitational potential energy. Our starting point is Equation 6.4 for the work done by the gravitational force as an object moves from an initial height h_0 to a final height h_f:

$$W_{gravity} = \underbrace{mgh_0}_{\substack{\text{Gravitational} \\ \text{potential energy} \\ \text{PE}_0 \text{ (initial)}}} - \underbrace{mgh_f}_{\substack{\text{Gravitational} \\ \text{potential energy} \\ \text{PE}_f \text{ (final)}}} \qquad (6.4)$$

This shows that the work done by the gravitational force is equal to the difference between the initial and final values of the quantity mgh. The value of mgh is larger when the height is larger and smaller when the height is smaller. We identify the quantity mgh as the **gravitational potential energy**. The concept of potential energy is associated only with a type of force known as a "conservative" force, as we will discuss in Section 6.4.

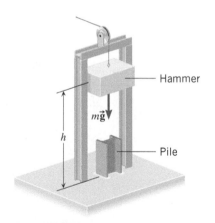

Hammer

$m\vec{g}$

h

Pile

FIGURE 6.13 In a pile driver, the gravitational potential energy of the hammer relative to the ground is $PE = mgh$.

DEFINITION OF GRAVITATIONAL POTENTIAL ENERGY

The gravitational potential energy PE is the energy that an object of mass m has by virtue of its position relative to the surface of the earth. That position is measured by the height h of the object relative to an arbitrary zero level:

$$PE = mgh \qquad (6.5)$$

SI Unit of Gravitational Potential Energy: joule (J)

Gravitational potential energy, like work and kinetic energy, is a scalar quantity and has the same SI unit as they do—the joule. It is the *difference* between two potential energies that is related by Equation 6.4 to the work done by the force of gravity. Therefore, the zero level for the heights can be taken anywhere, as long as both h_0 and h_f are measured relative to the same zero level. The gravitational potential energy depends on both the object and the earth (m and g, respectively), as well as the height h. Therefore, the gravitational potential energy belongs to the object and the earth as a system, although one often speaks of the object alone as possessing the gravitational potential energy.

Check Your Understanding

(The answer is given at the end of the book.)

9. In a simulation on earth, an astronaut in his space suit climbs up a vertical ladder. On the moon, the same astronaut makes the exact same climb. Which one of the following statements correctly describes how the gravitational potential energy of the astronaut changes during the climb? **(a)** It changes by a greater amount on the earth. **(b)** It changes by a greater amount on the moon. **(c)** The change is the same in both cases.

6.4 | Conservative Versus Nonconservative Forces

The gravitational force has an interesting property that when an object is moved from one place to another, the work done by the gravitational force does not depend on the choice of path. In **Figure 6.11**, for instance, an object moves from an initial height h_0 to a final height h_f along two different paths. As Section 6.3 discusses, the work done by gravity depends only on the initial and final heights, and not on the path between these heights. For this reason, the gravitational force is called a **conservative force**, according to version 1 of the following definition:

> **DEFINITION OF A CONSERVATIVE FORCE**
>
> **Version 1** A force is conservative when the work it does on a moving object is independent of the path between the object's initial and final positions.
>
> **Version 2** A force is conservative when it does no net work on an object moving around a closed path, starting and finishing at the same point.

Figure 6.14 helps us to illustrate version 2 of the definition of a conservative force. The picture shows a roller coaster car racing through dips and double dips, ultimately

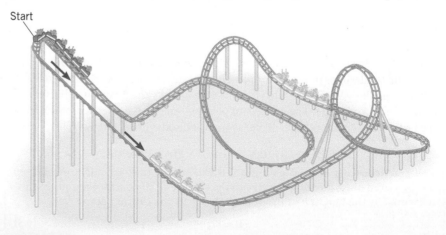
Start

FIGURE 6.14 A roller coaster track is an example of a closed path.

returning to its starting point. This kind of path, which begins and ends at the same place, is called a *closed* path. Gravity provides the only force that does work on the car, assuming that there is no friction or air resistance. Of course, the track exerts a normal force, but this force is always directed perpendicular to the motion and does no work. On the downward parts of the trip, the gravitational force does positive work, increasing the car's kinetic energy. Conversely, on the upward parts of the motion, the gravitational force does negative work, decreasing the car's kinetic energy. Over the entire trip, the gravitational force does as much positive work as negative work, so the net work is zero, and the car returns to its starting point with the same kinetic energy it had at the start. Therefore, consistent with version 2 of the definition of a conservative force, $W_{\text{gravity}} = 0$ J for a closed path.

The gravitational force is our first example of a conservative force. Later, we will encounter others, such as the elastic force of a spring and the electrical force of electrically charged particles. With each conservative force we will associate a potential energy, as we have already done in the gravitational case (see Equation 6.5). For other conservative forces, however, the algebraic form of the potential energy will differ from that in Equation 6.5.

Not all forces are conservative. A force is nonconservative if the work it does on an object moving between two points depends on the path of the motion between the points. The kinetic frictional force is one example of a nonconservative force. When an object slides over a surface and kinetic friction is present, the frictional force points opposite to the sliding motion and does negative work. Between any two points, greater amounts of work are done over longer paths between the points, so that the work depends on the choice of path. Thus, the kinetic frictional force is nonconservative. Air resistance is another nonconservative force. ***The concept of potential energy is not defined for a nonconservative force.***

For a closed path, the total work done by a nonconservative force is not zero as it is for a conservative force. In **Figure 6.14**, for instance, a frictional force would oppose the motion and slow down the car. Unlike gravity, friction would do negative work on the car throughout the entire trip, on *both* the up and down parts of the motion. Assuming that the car makes it back to the starting point, the car would have *less* kinetic energy than it had originally. **Table 6.2** gives some examples of conservative and nonconservative forces.

In normal situations, conservative forces (such as gravity) and nonconservative forces (such as friction and air resistance) act simultaneously on an object. Therefore, we write the work W done by the net external force as $W = W_{\text{c}} + W_{\text{nc}}$, where W_{c} is the work done by the conservative forces and W_{nc} is the work done by the nonconservative forces. According to the work–energy theorem, the work done by the net external force is equal to the change in the object's kinetic energy, or $W_{\text{c}} + W_{\text{nc}} = \frac{1}{2}mv_{\text{f}}^2 - \frac{1}{2}mv_0^2$. If the only conservative force acting is the gravitational force, then $W_{\text{c}} = W_{\text{gravity}} = mg(h_0 - h_{\text{f}})$, and the work–energy theorem becomes

$$mg(h_0 - h_{\text{f}}) + W_{\text{nc}} = \frac{1}{2}mv_{\text{f}}^2 - \frac{1}{2}mv_0^2$$

The work done by the gravitational force can be moved to the right side of this equation, with the result that

$$W_{\text{nc}} = \left(\tfrac{1}{2}mv_{\text{f}}^2 - \tfrac{1}{2}mv_0^2\right) + (mgh_{\text{f}} - mgh_0) \tag{6.6}$$

In terms of kinetic and potential energies, we find that

$$\underbrace{W_{\text{nc}}}_{\substack{\text{Net work done by} \\ \text{nonconservative} \\ \text{forces}}} = \underbrace{(\text{KE}_{\text{f}} - \text{KE}_0)}_{\substack{\text{Change in kinetic} \\ \text{energy}}} + \underbrace{(\text{PE}_{\text{f}} - \text{PE}_0)}_{\substack{\text{Change in} \\ \text{gravitational} \\ \text{potential energy}}} \tag{6.7a}$$

Equation 6.7a states that the net work W_{nc} done by all the external nonconservative forces equals the change in the object's kinetic energy plus the change in its gravitational potential energy. It is customary to use the delta symbol (Δ) to denote such changes;

TABLE 6.2	Some Conservative and Nonconservative Forces

Conservative Forces

Gravitational force (Ch. 4)

Elastic spring force (Ch. 10)

Electric force (Ch. 18, 19)

Nonconservative Forces

Static and kinetic frictional forces

Air resistance

Tension

Normal force

Propulsion force of a rocket

thus, $\Delta\text{KE} = (\text{KE}_f - \text{KE}_0)$ and $\Delta\text{PE} = (\text{PE}_f - \text{PE}_0)$. With the delta notation, the work–energy theorem takes the form

$$W_{nc} = \Delta\text{KE} + \Delta\text{PE} \qquad (6.7b) \leftarrow$$

In the next two sections, we will show why the form of the work–energy theorem expressed by Equations 6.7a and 6.7b is useful.

Math Skills   The change in both the kinetic and the potential energy is *always* the final value minus the initial value: $\Delta\text{KE} = \text{KE}_f - \text{KE}_0$ and $\Delta\text{PE} = \text{PE}_f - \text{PE}_0$. For example, consider a skier starting from rest at the top of a steep ski jumping ramp, gathering speed on the way down, and leaving the end of the ramp at a high speed. The initial kinetic energy is $\text{KE}_0 = 0$ J, since the skier is at rest at the top. The final kinetic energy, however, is $\text{KE}_f = \frac{1}{2}mv_f^2$, since the skier leaves the ramp at a high speed v_f. The kinetic energy increases from zero to a nonzero value, so the *change* in kinetic energy has a positive value of

$$\Delta\text{KE} = \text{KE}_f - \text{KE}_0 = \tfrac{1}{2}mv_f^2 - 0\text{ J} = \tfrac{1}{2}mv_f^2$$

If ΔKE were written incorrectly as the initial value minus the final value or $\Delta\text{KE} = \text{KE}_0 - \text{KE}_f$, the skier's kinetic energy would appear to change by the following amount:

Incorrect Equation

$$\Delta\text{KE} = \text{KE}_0 - \text{KE}_f = 0\text{ J} - \tfrac{1}{2}mv_f^2 - \tfrac{1}{2}mv_f^2$$

This change is negative. It would imply that the skier, although starting at the top with zero kinetic energy, nevertheless loses kinetic energy on the way down! This is nonsense and would lead to serious errors in problem solving.

6.5 The Conservation of Mechanical Energy

The concept of work and the work–energy theorem have led us to the conclusion that an object can possess two kinds of energy: kinetic energy, KE, and gravitational potential energy, PE. The sum of these two energies is called the **total mechanical energy** E, so that $E = \text{KE} + \text{PE}$. Later on we will update this definition to include other types of potential energy in addition to the gravitational form. The concept of total mechanical energy will be extremely useful in describing the motion of objects.

By rearranging the terms on the right side of Equation 6.7a, the work–energy theorem can be expressed in terms of the total mechanical energy:

$$W_{nc} = (\text{KE}_f - \text{KE}_0) + (\text{PE}_f - \text{PE}_0) \qquad (6.7a)$$
$$= \underbrace{(\text{KE}_f + \text{PE}_f)}_{E_f} - \underbrace{(\text{KE}_0 + \text{PE}_0)}_{E_0}$$

or

$$W_{nc} = E_f - E_0 \qquad (6.8)$$

Remember: Equation 6.8 is just another form of the work–energy theorem. It states that W_{nc}, the net work done by external nonconservative forces, changes the total mechanical energy from an initial value of E_0 to a final value of E_f.

The conciseness of the work–energy theorem in the form $W_{nc} = E_f - E_0$ allows an important basic principle of physics to stand out. This principle is known as the conservation of mechanical energy. Suppose that the net work W_{nc} done by external nonconservative forces is zero, so $W_{nc} = 0$ J. Then, Equation 6.8 reduces to

$$E_f = E_0 \qquad (6.9a)$$
$$\underbrace{\tfrac{1}{2}mv_f^2 + mgh_f}_{E_f} = \underbrace{\tfrac{1}{2}mv_0^2 + mgh_0}_{E_0} \qquad (6.9b)$$

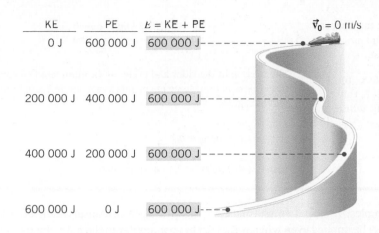

KE	PE	$E = KE + PE$
0 J	600 000 J	600 000 J
200 000 J	400 000 J	600 000 J
400 000 J	200 000 J	600 000 J
600 000 J	0 J	600 000 J

$\vec{v}_0 = 0$ m/s

INTERACTIVE FIGURE 6.15 If friction and wind resistance are ignored, a bobsled run illustrates how kinetic and potential energy can be interconverted, while the total mechanical energy remains constant. The total mechanical energy is 600 000 J, being all potential energy at the top and all kinetic energy at the bottom.

Equation 6.9a indicates that the final mechanical energy is equal to the initial mechanical energy. Consequently, the total mechanical energy *remains constant all along the path between the initial and final points,* never varying from the initial value of E_0. A quantity that remains constant throughout the motion is said to be "conserved." The fact that the total mechanical energy is conserved when $W_{nc} = 0$ J is called the **principle of conservation of mechanical energy**.

> **THE PRINCIPLE OF CONSERVATION OF MECHANICAL ENERGY**
>
> The total mechanical energy ($E = KE + PE$) of an object remains constant as the object moves, provided that the net work done by external nonconservative forces is zero, $W_{nc} = 0$ J.

The principle of conservation of mechanical energy offers keen insight into the way in which the physical universe operates. While the sum of the kinetic and potential energies at any point is conserved, the two forms may be interconverted or transformed into one another. Kinetic energy of motion is converted into potential energy of position, for instance, when a moving object coasts up a hill. Conversely, potential energy is converted into kinetic energy when an object is allowed to fall. **Interactive Figure 6.15** shows such transformations of energy for a bobsled run, assuming that nonconservative forces, such as friction and wind resistance, can be ignored. The normal force, being directed perpendicular to the path, does no work. Only the force of gravity does work, so the total mechanical energy E remains constant at all points along the run. The conservation principle is well known for the ease with which it can be applied, as in the following example.

EXAMPLE 8 | A Daredevil Motorcyclist

A motorcyclist is trying to leap across the canyon shown in **Figure 6.16** by driving horizontally off the cliff at a speed of 38.0 m/s. Ignoring air resistance, find the speed with which the cycle strikes the ground on the other side.

Reasoning Once the cycle leaves the cliff, no forces other than gravity act on the cycle, since air resistance is being ignored. Thus, the work done by external nonconservative forces is zero, $W_{nc} = 0$ J. Accordingly, the principle of conservation of mechanical energy

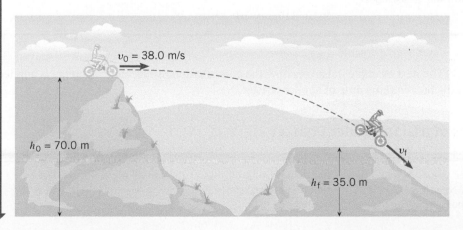

$v_0 = 38.0$ m/s

$h_0 = 70.0$ m

v_f

$h_f = 35.0$ m

FIGURE 6.16 A daredevil jumping a canyon.

holds, so the total mechanical energy is the same at the final and initial positions of the motorcycle. We will use this important observation to determine the final speed of the cyclist.

> **Problem-Solving Insight** Be on the alert for factors, such as the mass m here in Example 8, that sometimes can be eliminated algebraically when using the conservation of mechanical energy.

Solution   The principle of conservation of mechanical energy is written as

$$\underbrace{\tfrac{1}{2}mv_f^2 + mgh_f}_{E_f} = \underbrace{\tfrac{1}{2}mv_0^2 + mgh_0}_{E_0} \qquad (6.9b)$$

The mass m of the rider and cycle can be eliminated algebraically from this equation, since m appears as a factor in every term. Solving for v_f gives

$$v_f = \sqrt{v_0^2 + 2g(h_0 - h_f)}$$

$$v_f = \sqrt{(38.0\ \text{m/s})^2 + 2(9.80\ \text{m/s}^2)(70.0\ \text{m} - 35.0\ \text{m})} = \boxed{46.2\ \text{m/s}}$$

Examples 9 and 10 emphasize that the principle of conservation of mechanical energy can be applied even when forces act perpendicular to the path of a moving object.

CONCEPTUAL EXAMPLE 9 | The Favorite Swimming Hole

A rope is tied to a tree limb. It is used by a swimmer who, starting from rest, swings down toward the water below, as in **Figure 6.17**. Only two forces act on him during his descent, the nonconservative force $\vec{T}$, which is due to the tension in the rope, and his weight, which is due to the conservative gravitational force. There is no air resistance. His initial height h_0 and final height h_f are known. Can we use the principle of conservation of mechanical energy to find his speed v_f at the point where he lets go of the rope, even though a nonconservative external force is present? **(a)** Yes. **(b)** No.

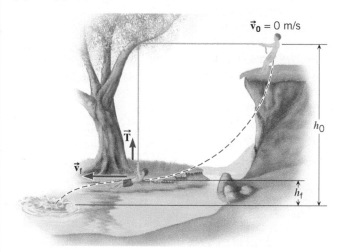

FIGURE 6.17   During the downward swing, the tension $\vec{T}$ in the rope acts perpendicular to the circular arc and, hence, does no work on the person.

Reasoning   The principle of conservation of mechanical energy can be used even in the presence of nonconservative forces, provided that the net work W_{nc} done by the nonconservative forces is zero. $W_{nc} = 0$ J.

Answer (b) is incorrect.   The mere presence of nonconservative forces does not prevent us from applying the principle of conservation of mechanical energy. The deciding factor is whether the net work done by the nonconservative forces is zero. In this case, the tension force $\vec{T}$ is always perpendicular to the circle (see **Figure 6.17**). Thus, the angle θ between $\vec{T}$ and the instantaneous displacement of the person is always 90°. According to Equation 6.1, the work is proportional to the cosine of this angle, or cos 90°, which is zero, with the result that the work is also zero. As a result, the tension here does not prevent us from using the conservation principle.

> **Problem-Solving Insight**   When nonconservative forces are perpendicular to the motion, we can still use the principle of conservation of mechanical energy, because such "perpendicular" forces do no work.

Answer (a) is correct.   As the person swings downward, he follows a circular path, as shown in **Figure 6.17**. Since the tension force is the only nonconservative force acting and is always perpendicular to this path, the corresponding net work is $W_{nc} = 0$ J, and we can apply the principle of conservation of mechanical energy to find the speed v_f.

Related Homework: Problem 38

The next example illustrates how the conservation of mechanical energy is applied to the breathtaking drop of a roller coaster.

EXAMPLE 10 | The Physics of a Giant Roller Coaster

The Kingda Ka is a giant among roller coasters (see **Figure 6.18**). Located in Jackson Township, New Jersey, the ride includes a vertical drop of 127 m. Suppose that the coaster has a speed of 6.00 m/s at the top of the drop. Neglect friction and air resistance and find the speed of the riders at the bottom.

Reasoning Since we are neglecting friction and air resistance, we may set the work done by these forces equal to zero. A normal force from the seat acts on each rider, but this force is perpendicular to the motion, so it does not do any work. Thus, the work done by external nonconservative forces is zero, $W_{nc} = 0$ J, and we may use the principle of conservation of mechanical energy to find the speed of the riders at the bottom.

Solution The principle of conservation of mechanical energy states that

$$\underbrace{\tfrac{1}{2}mv_f^2 + mgh_f}_{E_f} = \underbrace{\tfrac{1}{2}mv_0^2 + mgh_0}_{E_0} \qquad \text{(6.9b)}$$

The mass m of the rider appears as a factor in every term in this equation and can be eliminated algebraically. Solving for the final speed gives

$$v_f = \sqrt{v_0^2 + 2g(h_0 - h_f)}$$

$$v_f = \sqrt{(6.00 \text{ m/s})^2 + 2(9.80 \text{ m/s}^2)(127 \text{ m})}$$

$$= \boxed{50.3 \text{ m/s (112 mi/h)}}$$

where the vertical drop is $h_0 - h_f = 127$ m.

FIGURE 6.18 The Kingda Ka roller coaster in Six Flags Great Adventure, located in Jackson Township, New Jersey, is a giant. It includes a vertical drop of 127 m.

Six Flags/Splash News/Newscom

When applying the principle of conservation of mechanical energy in solving problems, we have been using the following reasoning strategy:

> **REASONING STRATEGY** **Applying the Principle of Conservation of Mechanical Energy**
>
> 1. Identify the external conservative and nonconservative forces that act on the object. For this principle to apply, the total work done by nonconservative forces must be zero, $W_{nc} = 0$ J. A nonconservative force that is perpendicular to the displacement of the object does no work, for example.
>
> 2. Choose the location where the gravitational potential energy is taken to be zero. This location is arbitrary but must not be changed during the course of solving a problem.
>
> 3. Set the final total mechanical energy of the object equal to the initial total mechanical energy, as in Equations 6.9a and 6.9b. The total mechanical energy is the sum of the kinetic and potential energies.

Check Your Understanding

(The answers are given at the end of the book.)

10. Suppose the total mechanical energy of an object is conserved. Which one or more of the following statements is/are true? **(a)** If the kinetic energy decreases, the gravitational potential energy increases. **(b)** If the gravitational potential energy decreases, the kinetic energy increases. **(c)** If the kinetic energy does not change, the gravitational potential energy also does not change.

11. In Example 10 the Kingda Ka roller coaster starts with a speed of 6.0 m/s at the top of the drop and attains a speed of 50.3 m/s when it reaches the bottom. If the roller coaster were then to start up an identical hill, what speed would it attain when it reached the top? Assume that friction and air resistance are absent. **(a)** Greater than 6.0 m/s **(b)** Exactly 6.0 m/s **(c)** Between 0 m/s and 6.0 m/s **(d)** 0 m/s

12. CYU Figure 6.2 shows an empty fuel tank about to be released by three different jet planes. At the moment of release, each plane has the same speed, and each tank is at the same height above the ground. However, the direc-

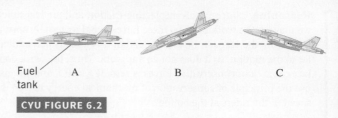

Fuel tank

A   B   C

CYU FIGURE 6.2

tions of the velocities of the planes are different. Which tank has the largest speed upon hitting the ground? Ignore friction and air resistance. (a) A (b) B (c) C (d) Each tank hits the ground with the same speed.

13. In which one or more of the following situations is the principle of conservation of mechanical energy obeyed? (a) An object moves uphill with an increasing speed. (b) An object moves uphill with a decreasing speed. (c) An object moves uphill with a constant speed. (d) An object moves downhill with an increasing speed. (e) An object moves downhill with a decreasing speed. (f) An object moves downhill with a constant speed.

6.6 | Nonconservative Forces and the Work–Energy Theorem

Most moving objects experience nonconservative forces, such as friction, air resistance, and propulsive forces, and the work W_{nc} done by the net external nonconservative force is not zero. In these situations, the difference between the final and initial total mechanical energies is equal to W_{nc} according to $W_{nc} = E_f - E_0$ (Equation 6.8). Consequently, the total mechanical energy is not conserved. The next two examples illustrate how Equation 6.8 is used when nonconservative forces are present and do work.

EXAMPLE 11 | The Kingda Ka Revisited

In Example 10, we ignored nonconservative forces, such as friction. In reality, however, such forces are present when the roller coaster descends. The actual speed of the riders at the bottom is 45.0 m/s, which is less than that determined in Example 10. Assuming again that the coaster has a speed of 6.00 m/s at the top, find the work done by nonconservative forces on a 55.0-kg rider during the descent from a height h_0 to a height h_f, where $h_0 - h_f = 127$ m.

Reasoning Since the speed at the top, the final speed, and the vertical drop are given, we can determine the initial and final total mechanical energies of the rider. The work–energy theorem, $W_{nc} = E_f - E_0$, can then be used to determine the work W_{nc} done by the nonconservative forces.

> **Problem-Solving Insight** As illustrated here and in Example 3, a nonconservative force such as friction can do negative or positive work. It does

negative work when it has a component opposite to the displacement and slows down the object. It does positive work when it has a component in the direction of the displacement and speeds up the object.

Solution The work–energy theorem is

$$W_{nc} = \underbrace{\left(\tfrac{1}{2}mv_f^2 + mgh_f\right)}_{E_f} - \underbrace{\left(\tfrac{1}{2}mv_0^2 + mgh_0\right)}_{E_0} \qquad \textbf{(6.8)}$$

Rearranging the terms on the right side of this equation gives

$$W_{nc} = \tfrac{1}{2}m\left(v_f^2 - v_0^2\right) - mg(h_0 - h_f)$$

$$W_{nc} = \tfrac{1}{2}(55.0 \text{ kg})[(45.0 \text{ m/s})^2 - (6.00 \text{ m/s})^2] -$$

$$(55.0 \text{ kg})(9.80 \text{ m/s}^2)(127 \text{ m}) = \boxed{-1.38 \times 10^4 \text{J}}$$

EXAMPLE 12 | Fireworks

A 0.20-kg rocket in a fireworks display is launched from rest and follows an erratic flight path to reach the point P, as **Figure 6.19** shows. Point P is 29 m above the starting point. In the process, 425 J of work is done on the rocket by

the nonconservative force generated by the burning propellant. Ignoring air resistance and the mass lost due to the burning propellant, find the speed v_f of the rocket at the point P.

Reasoning The only nonconservative force acting on the rocket is the force generated by the burning propellant, and the work done by this force is $W_{nc} = 425$ J. Because work is done by a non-conservative force, we use the work–energy theorem in the form $W_{nc} = E_f - E_0$ to find the final speed v_f of the rocket.

Solution From the work–energy theorem we have

$$W_{nc} = \left(\tfrac{1}{2}mv_f^2 + mgh_f\right) - \left(\tfrac{1}{2}mv_0^2 + mgh_0\right) \qquad (6.8)$$

Solving this expression for the final speed of the rocket and noting that the initial speed of the rocket at rest is $v_0 = 0$ m/s, we get

$$v_f = \sqrt{\frac{2[W_{nc} + \tfrac{1}{2}mv_0^2 - mg(h_f - h_0)]}{m}}$$

$$v_f = \sqrt{\frac{2[425\ \text{J} + \tfrac{1}{2}(0.20\ \text{kg})(0\ \text{m/s})^2 - (0.20\ \text{kg})(9.80\ \text{m/s}^2)(29\ \text{m})]}{0.20\ \text{kg}}}$$

$$= \boxed{61\ \text{m/s}}$$

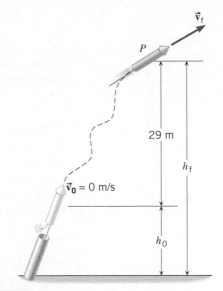

FIGURE 6.19 A fireworks rocket, moving along an erratic flight path, reaches a point P that is 29 m above the launch point.

Check Your Understanding

(*The answers are given at the end of the book.*)

14. A net external nonconservative force does positive work on a particle. Based solely on this information, you are justified in reaching only one of the following conclusions. Which one is it? **(a)** The kinetic and potential energies of the particle both decrease. **(b)** The kinetic and potential energies of the particle both increase. **(c)** Neither the kinetic nor the potential energy of the particle changes. **(d)** The total mechanical energy of the particle decreases. **(e)** The total mechanical energy of the particle increases.

15. In one case, a sports car, its engine running, is driven up a hill at a constant speed. In another case, a truck approaches a hill, and its driver turns off the engine at the bottom of the hill. The truck then coasts up the hill. Which vehicle is obeying the principle of conservation of mechanical energy? Ignore friction and air resistance. **(a)** Both the sports car and the truck **(b)** Only the sports car **(c)** Only the truck

6.7 | Power

In many situations, the time it takes to do work is just as important as the amount of work that is done. Consider two automobiles that are identical in all respects (e.g., both have the same mass), except that one has a "souped-up" engine. The car with the souped-up engine can go from 0 to 27 m/s (60 mph) in 4 seconds, while the other car requires 8 seconds to achieve the same speed. Each engine does work in accelerating its car, but one does it more quickly. Where cars are concerned, we associate the quicker performance with an engine that has a larger horsepower rating. A large horsepower rating means that the engine can do a large amount of work in a short time. In physics, the horsepower rating is just one way to measure an engine's ability to generate power. The idea of **power** incorporates both the concepts of work and time, for power is work done per unit time.

DEFINITION OF AVERAGE POWER

Average power $\bar{P}$ is the average rate at which work W is done, and it is obtained by dividing W by the time t required to perform the work:

$$\bar{P} = \frac{\text{Work}}{\text{Time}} = \frac{W}{t} \qquad (6.10a)$$

SI Unit of Power: joule/s = watt (W)

The definition of average power presented in Equation 6.10a involves work. However, the work–energy theorem relates the work done by a net external force to the change in the energy of the object (see, for example, Equations 6.3 and 6.8). Therefore, we can also define average power as the rate at which the energy is changing, or as the change in energy divided by the time during which the change occurs:

$$\overline{P} = \frac{\text{Change in energy}}{\text{Time}} \tag{6.10b}$$

Since work, energy, and time are scalars, power is also a scalar. The unit in which power is expressed is that of work divided by time, or a joule per second in SI units. One joule per second is called a watt (W), in honor of James Watt (1736–1819), developer of the steam engine. The unit of power in the BE system is the foot · pound per second (ft · lb/s), although the familiar horsepower (hp) unit is used frequently for specifying the power generated by electric motors and internal combustion engines:

$$1 \text{ horsepower} = 550 \text{ foot} \cdot \text{pounds/second} = 745.7 \text{ watts}$$

Table 6.3 summarizes the units for power in the various systems of measurement.

TABLE 6.3   Units of Measurement for Power

System	Work	÷	Time	=	Power
SI	joule (J)		second (s)		watt (W)
CGS	erg		second (s)		erg per second (erg/s)
BE	foot · pound (ft · lb)		second (s)		foot · pound per second (ft · lb/s)

BIO **THE PHYSICS OF . . . human metabolism.**   Equation 6.10b provides the basis for understanding the production of power in the human body. In this context the "Change in energy" on the right-hand side of the equation refers to the energy produced by metabolic processes, which, in turn, is derived from the food we eat. Table 6.4 gives typical metabolic rates of energy production needed to sustain various activities. Running at 15 km/h (9.3 mi/h), for example, requires metabolic power sufficient to operate eighteen 75-watt light bulbs, and the metabolic power used in sleeping would operate a single 75-watt bulb.

An alternative expression for power can be obtained from Equation 6.1, which indicates that the work W done when a constant net force of magnitude F points in the same direction as the displacement is $W = (F\cos 0°)s = Fs$. Dividing both sides of this equation by the time t it takes for the force to move the object through the distance s, we obtain

$$\frac{W}{t} = \frac{Fs}{t}$$

But W/t is the average power $\overline{P}$, and s/t is the average speed $\overline{v}$, so that

$$\overline{P} = F\overline{v} \tag{6.11}$$

Example 13 applies the concepts of work and power to the performance of the human heart.

TABLE 6.4   Human Metabolic Rates[a]

Activity	Rate (watts)
Running (15 km/h)	1340 W
Skiing	1050 W
Biking	530 W
Walking (5 km/h)	280 W
Sleeping	77 W

[a]For a young 70-kg male.

EXAMPLE 13 | BIO Power Output of the Human Heart

The total average work done during one contraction and relaxation (i.e., during one beat) of the heart of a person at rest is about 1.33 J. This is split between the work done by the right ventricle (0.22 J) and the left ventricle (1.11 J). **(a)** If the efficiency of the heart is 10.0% (i.e., its work output is 1/10th of its energy input), what is the total energy expended in one day by the heart of someone who has an average heart rate of 70.0 beats/minute? **(b)** What is the average output power of the heart during this time?

Reasoning   **(a)** Knowing the total work done during a single beat of the heart and the heart rate, the total work done by the heart in one day can be determined. Since we know the efficiency, we can

then determine the total energy expended. **(b)** The average power output is equal to the total work done divided by the total time.

Solution   **(a)** The total work done in one day is the total number of heartbeats (N) in one day multiplied by the work done during one heartbeat (W_B):

$$W_T = NW_B$$

The total number of heartbeats is given by the number of beats per minute multiplied by the number of minutes in one day:

$$N = \left(\frac{70 \text{ beats}}{\text{min}}\right)\left(\frac{60 \text{ min}}{\text{hr}}\right)\left(\frac{24 \text{ hr}}{\text{day}}\right) = 100\,804 \frac{\text{beats}}{\text{day}}$$

The total work done by the heart in one day is then given by

$$W_T = NW_B = (100\ 804\ \text{beats})\left(1.33\ \frac{\text{J}}{\text{beats}}\right) = 1.34 \times 10^5\ \text{J}$$

Since the energy input is 10 times the work output, the total energy (E_T) expended by the heart in one day is given by

$$E_T = (10)W_T = \boxed{1.34 \times 10^6\ \text{J}}$$

The energy expended by the heart ultimately comes from food, the energy of which is measured in units of kcal (or the Calorie). Since 1 kcal = 4186 J, the heart burns 320.0 kcal per day in this example.

(b) The average power is equal to the work done divided by the time (Equation 6.10a). In order to determine the output power in watts (W), we divide the total work done by the heart in one day by the number of seconds in one day (86 400 s) to obtain

$$P = \frac{W_T}{t} = \frac{1.34 \times 10^5\ \text{J}}{8.64 \times 10^4\ \text{s}} = \boxed{1.55\ \text{W}}$$

This is the output power. Considering that the energy efficiency of the heart is about 10%, the required input power delivered to the heart by the body is 15.5 W. The total metabolic rate for a resting person is about 100 W, that is, about the same value as that of the electrical power delivered to a bright light bulb. The heart therefore accounts for about 15% of the body's total power output.

Check Your Understanding

(The answers are given at the end of the book.)

16. Engine A has a greater power rating than engine B. Which one of the following statements correctly describes the abilities of these engines to do work? **(a)** Engines A and B can do the same amount of work, but engine A can do it more quickly. **(b)** Engines A and B can do the same amount of work in the same amount of time. **(c)** In the same amount of time, engine B can do more work than engine A.

17. Is it correct to conclude that one engine is doing twice the work that another is doing just because it is generating twice the power? **(a)** Yes **(b)** No

6.8 | Other Forms of Energy and the Conservation of Energy

Up to now, we have considered only two types of energy, kinetic energy and gravitational potential energy. There are many other types, however. Electrical energy is used to run electrical appliances. Energy in the form of heat is utilized in cooking food. Moreover, the work done by the kinetic frictional force often appears as heat, as you can experience by rubbing your hands back and forth. When gasoline is burned, some of the stored chemical energy is released and does the work of moving cars, airplanes, and boats. The chemical energy stored in food provides the energy needed for metabolic processes.

The research of many scientists, most notably Albert Einstein, led to the discovery that mass itself is one manifestation of energy. Einstein's famous equation, $E_0 = mc^2$, describes how mass m and energy E_0 are related, where c is the speed of light in a vacuum and has a value of 3.00×10^8 m/s. Because the speed of light is so large, this equation implies that very small masses are equivalent to large amounts of energy. The relationship between mass and energy will be discussed further in Chapter 28.

We have seen that kinetic energy can be converted into gravitational potential energy and vice versa. In general, energy of all types can be converted from one form to another. **BIO** THE PHYSICS OF . . . **transforming chemical energy in food into mechanical energy.** Part of the chemical energy stored in food, for example, is transformed into gravitational potential energy when a hiker climbs a mountain. Suppose a 65-kg hiker eats a 250-Calorie* snack, which contains 1.0×10^6 J of chemical energy. If this were 100% converted into potential energy $mg(h_f - h_0)$, the change in height would be

$$h_f - h_0 = \frac{1.0 \times 10^6\ \text{J}}{(65\ \text{kg})(9.8\ \text{m/s}^2)} = 1600\ \text{m}$$

*Energy content in food is typically given in units called Calories, which we will discuss in Section 12.7.

At a more realistic conversion efficiency of 50%, the change in height would be 800 m. Similarly, in a moving car the chemical energy of gasoline is converted into kinetic energy, as well as into electrical energy and heat.

Whenever energy is transformed from one form to another, it is found that no energy is gained or lost in the process; the total of all the energies before the process is equal to the total of the energies after the process. This observation leads to the following important principle:

THE PRINCIPLE OF CONSERVATION OF ENERGY

Energy can neither be created nor destroyed, but can only be converted from one form to another.

Learning how to convert energy from one form to another more efficiently is one of the main goals of modern science and technology.

6.9 | Work Done by a Variable Force

The work W done by a constant force (constant in both magnitude and direction) is given by Equation 6.1 as $W = (F \cos \theta)s$. Quite often, situations arise in which the force is not constant but changes with the displacement of the object. For instance, **Animated Figure 6.20a** shows an archer using a high-tech compound bow. This type of bow consists of a series of pulleys and strings that produce a force–displacement graph like that in **Animated Figure 6.20b**. One of the key features of the compound bow is that the force rises to a maximum as the string is drawn back, and then falls to 60% of this maximum value when the string is fully drawn. The reduced force at $s = 0.500$ m makes it much easier for the archer to hold the fully drawn bow while aiming the arrow.

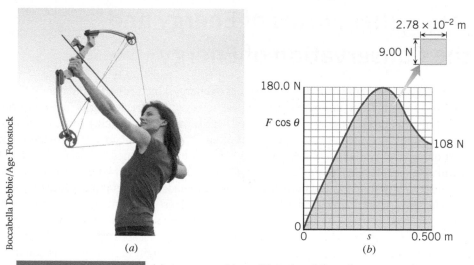

ANIMATED FIGURE 6.20 (a) A compound bow. (b) A plot of $F \cos \theta$ versus s as the bowstring is drawn back.

When the force varies with the displacement, as in **Animated Figure 6.20b**, we cannot use the relation $W = (F \cos \theta)s$ to find the work, because this equation is valid only when the force is constant. However, we can use a graphical method. In this method we divide the total displacement into very small segments, Δs_1, Δs_2, and so on (see **Figure 6.21a**). For each segment, the *average value* of the force component is indicated by a short horizontal line. For example, the short horizontal line for segment Δs_1 is labeled $(F \cos \theta)_1$ in **Figure 6.21a**. We can then use this average value as the constant-force component in Equation 6.1 and determine an approximate value for the work ΔW_1 done during the first segment: $\Delta W_1 = (F \cos \theta)_1 \Delta s_1$. But this work is just the area of the colored rectangle in the drawing. The word "area" here refers to the area of a rectangle that has a width of Δs_1 and a height of $(F \cos \theta)_1$; it does not mean an area in square meters, such as the area of a parcel of land. In a like manner, we can calculate an approximate

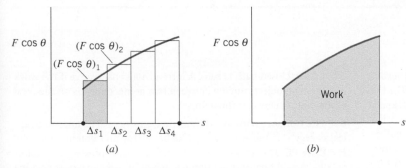

FIGURE 6.21 (*a*) The work done by the average-force component $(F \cos \theta)_1$ during the small displacement Δs_1 is $(F \cos \theta)_1 \Delta s_1$, which is the area of the colored rectangle. (*b*) The work done by a variable force is equal to the colored area under the graph of $F \cos \theta$ versus *s*.

value for the work for each segment. Then we add the results for the segments to get, approximately, the work *W* done by the variable force:

$$W \approx (F \cos \theta)_1 \, \Delta s_1 + (F \cos \theta)_2 \, \Delta s_2 + \cdots$$

The symbol $\approx$ means "approximately equal to." The right side of this equation is the sum of all the rectangular areas in **Figure 6.21a** and is an approximate value for the area shaded in color under the graph in **Figure 6.21b**. If the rectangles are made narrower and narrower by decreasing each Δs, the right side of this equation eventually becomes equal to the area under the graph.

> **Problem-Solving Insight**   The work done by a variable force in moving an object is equal to the area under the graph of $F \cos \theta$ versus *s*.

Example 14 illustrates how to use this graphical method to determine the approximate work done when a high-tech compound bow is drawn.

EXAMPLE 14 | The Physics of a Compound Bow

Find the work that the archer must do in drawing back the string of the compound bow in **Animated Figure 6.20a** from 0 to 0.500 m.

Reasoning   The work is equal to the colored area under the curved line in **Animated Figure 6.20b**. For convenience, this area is divided into a number of small squares, each having an area of $(9.00 \text{ N})(2.78 \times 10^{-2} \text{ m}) = 0.250 \text{ J}$. The area can be found by counting the number of squares under the curve and multiplying by the area per square.

Solution   We estimate that there are 242 colored squares in the drawing. Since each square represents 0.250 J of work, the total work done is

$$W = (242 \text{ squares})\left(0.250 \, \frac{\text{J}}{\text{square}}\right) = \boxed{60.5 \text{ J}}$$

When the arrow is fired, part of this work is imparted to it as kinetic energy.

EXAMPLE 15 | BIO The Physics of Seatbelts

Seatbelts are the most important safety feature for automobile drivers and passengers. Roughly 50% of passengers killed in auto accidents were not wearing seatbelts. In modern cars, seatbelts, along with sensors and airbags, function as an advanced safety restraining system. The graph below shows the force applied to the chest wall as a function of displacement during a frontend collision. Modern seatbelts use a pretensioner effect when the collision is high-speed, and therefore particularly dangerous. This reduces the maximum force experienced by the chest wall from the seatbelt during the impact. Using the data in the graph, calculate the work done on the chest wall from the variable force applied by the seatbelt. Assume the force from the seatbelt is applied parallel to the chest wall displacement.

Reasoning   The work is equal to the area under the solid curve. We see from the graph that this will consist of the combined area of two rectangles (Areas I and III), and one triangle (Area II).

Solution   The total area under the curve will be equal to Area I + Area II + Area III. From the data in **Figure 6.22**, we see that Area I = $(5000 \text{ N} - 0 \text{ N})(0.20 \text{ m} - 0.00 \text{ m}) = 1000 \text{ J}$. Area II = ½(0.20 m −

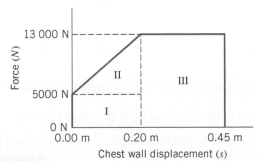

FIGURE 6.22 The variable force applied by a seatbelt to the chest wall as a function of the wall's displacement. The work done by the seatbelt will be equal to the total area under the solid curve.

$0.00 \text{ m})(13\,000 \text{ N} - 5000 \text{ N}) = 800 \text{ J}$. And finally, Area III = $(13\,000 \text{ N} - 0 \text{ N})(0.45 \text{ m} - 0.20 \text{ m}) = 3300 \text{ J}$. The work done is therefore equal to $1000 \text{ J} + 800 \text{ J} + 3300 \text{ J} = \boxed{5100 \text{ J}}$

Concept Summary

6.1 Work Done by a Constant Force The work W done by a constant force acting on an object is given by Equation 6.1, where F is the magnitude of the force, s is the magnitude of the displacement, and θ is the angle between the force and the displacement vectors. Work is a scalar quantity and can be positive or negative, depending on whether the force has a component that points, respectively, in the same direction as the displacement or in the opposite direction. The work is zero if the force is perpendicular ($\theta = 90°$) to the displacement.

$$W = (F \cos \theta)s \tag{6.1}$$

6.2 The Work–Energy Theorem and Kinetic Energy The kinetic energy KE of an object of mass m and speed v is given by Equation 6.2. The work–energy theorem states that the work W done by the net external force acting on an object equals the difference between the object's final kinetic energy KE_f and initial kinetic energy KE_0, according to Equation 6.3. The kinetic energy increases when the net force does positive work and decreases when the net force does negative work.

$$KE = \tfrac{1}{2}mv^2 \tag{6.2}$$

$$W = KE_f - KE_0 \tag{6.3}$$

6.3 Gravitational Potential Energy Work done by the force of gravity on an object of mass m is given by Equation 6.4, where h_0 and h_f are the initial and final heights of the object, respectively.

$$W_{\text{gravity}} = mg(h_0 - h_f) \tag{6.4}$$

$$PE = mgh \tag{6.5}$$

Gravitational potential energy PE is the energy that an object has by virtue of its position. For an object near the surface of the earth, the gravitational potential energy is given by Equation 6.5, where h is the height of the object relative to an arbitrary zero level.

6.4 Conservative Versus Nonconservative Forces A conservative force is one that does the same work in moving an object between two points, independent of the path taken between the points. Alternatively, a force is conservative if the work it does in moving an object

around any closed path is zero. A force is nonconservative if the work it does on an object moving between two points depends on the path of the motion between the points.

6.5 The Conservation of Mechanical Energy The total mechanical energy E is the sum of the kinetic energy and potential energy: $E = KE + PE$. The work–energy theorem can be expressed in an alternate form as shown in Equation 6.8, where W_{nc} is the net work done by the external nonconservative forces, and E_f and E_0 are the final and initial total mechanical energies, respectively.

$$W_{nc} = E_f - E_0 \tag{6.8}$$

The principle of conservation of mechanical energy states that the total mechanical energy E remains constant along the path of an object, provided that the net work done by external nonconservative forces is zero. Whereas E is constant, KE and PE may be transformed into one another.

6.6 Nonconservative Forces and the Work–Energy Theorem/

6.7 Power Average power $\overline{P}$ is the work done per unit time or the rate at which work is done, as shown in Equation 6.10a. It is also the rate at which energy changes, as shown in Equation 6.10b. When a force of magnitude F acts on an object moving with an average speed $\overline{v}$, the average power is given by Equation 6.11.

$$\overline{P} = \frac{\text{Work}}{\text{Time}} \tag{6.10a}$$

$$\overline{P} = \frac{\text{Change in energy}}{\text{Time}} \tag{6.10b}$$

$$\overline{P} = F\overline{v} \tag{6.11}$$

6.8 Other Forms of Energy and the Conservation of Energy The principle of conservation of energy states that energy can neither be created nor destroyed, but can only be transformed from one form to another.

6.9 Work Done by a Variable Force The work done by a variable force of magnitude F in moving an object through a displacement of magnitude s is equal to the area under the graph of $F \cos \theta$ versus s. The angle θ is the angle between the force and displacement vectors.

Focus on Concepts

Online

Additional questions are available for assignment in WileyPLUS.

Section 6.1 Work Done by a Constant Force

1. The same force $\vec{F}$ pushes in three different ways on a box moving with a velocity $\vec{v}$, as the drawings show. Rank the work done by the force $\vec{F}$ in ascending order (smallest first): **(a)** A, B, C **(b)** A, C, B **(c)** B, A, C **(d)** C, B, A **(e)** C, A, B

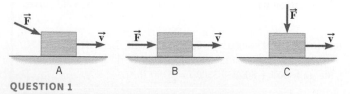

2. Consider the force $\vec{F}$ shown in the drawing. This force acts on an object that can move along the positive or negative x axis, or along the positive or negative y axis. The work done by this force is positive when the displacement of the object is along the _____ axis or along the _____ axis: **(a)** $-x$, $-y$ **(b)** $-x$, $+y$ **(c)** $+x$, $+y$ **(d)** $+x$, $-y$

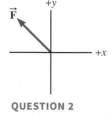

Section 6.2 The Work–Energy Theorem and Kinetic Energy

3. Two forces $\vec{F}_1$ and $\vec{F}_2$ act on a particle. As a result the speed of the particle increases. Which one of the following is NOT possible?

(a) The work done by $\vec{F}_1$ is positive, and the work done by $\vec{F}_2$ is zero.
(b) The work done by $\vec{F}_1$ is zero, and the work done by $\vec{F}_2$ is positive.
(c) The work done by each force is positive. (d) The work done by each force is negative. (e) The work done by $\vec{F}_1$ is positive, and the work done by $\vec{F}_2$ is negative.

4. Force $\vec{F}_1$ acts on a particle and does work W_1. Force $\vec{F}_2$ acts simultaneously on the particle and does work W_2. The speed of the particle does not change. Which one of the following must be true? (a) W_1 is zero and W_2 is positive (b) $W_1 = -W_2$ (c) W_1 is positive and W_2 is zero (d) W_1 is positive and W_2 is positive.

Section 6.4   Conservative Versus Nonconservative Forces

5. A person is riding on a Ferris wheel. When the wheel makes one complete turn, the net work done on the person by the gravitational force _____. (a) is positive (b) is negative (c) is zero (d) depends on how fast the wheel moves (e) depends on the diameter of the wheel

Section 6.5   The Conservation of Mechanical Energy

6. In which one of the following circumstances could mechanical energy not possibly be conserved, even if friction and air resistance are absent? (a) A car moves up a hill, its velocity continually decreasing along the way. (b) A car moves down a hill, its velocity continually increasing along the way. (c) A car moves along level ground at a constant velocity. (d) A car moves up a hill at a constant velocity.

7. A ball is fixed to the end of a string, which is attached to the ceiling at point P. As the drawing shows, the ball is projected downward at A with the launch speed v_0. Traveling on a circular path, the ball comes to a halt at point B. What enables

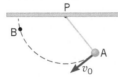
QUESTION 7

the ball to reach point B, which is above point A? Ignore friction and air resistance. (a) The work done by the tension in the string (b) The ball's initial gravitational potential energy (c) The ball's initial kinetic energy (d) The work done by the gravitational force

Section 6.6   Nonconservative Forces and the Work–Energy Theorem

8. In which one of the following circumstances does the principle of conservation of mechanical energy apply, even though a nonconservative force acts on the moving object? (a) The nonconservative force points in the same direction as the displacement of the object. (b) The nonconservative force is perpendicular to the displacement of the object. (c) The nonconservative force has a direction that is opposite to the displacement of the object. (d) The nonconservative force has a component that points in the same direction as the displacement of the object. (e) The nonconservative force has a component that points opposite to the displacement of the object.

9. A 92.0-kg skydiver with an open parachute falls straight downward through a vertical height of 325 m. The skydiver's velocity remains constant. What is the work done by the nonconservative force of air resistance, which is the only nonconservative force acting? (a) -2.93×10^5 J (b) 0 J (c) $+2.93 \times 10^5$ J (d) The answer is not obtainable, because insufficient information about the skydiver's speed is given.

Section 6.7   Power

10. The power needed to accelerate a projectile from rest to its launch speed v in a time t is 43.0 W. How much power is needed to accelerate the same projectile from rest to a launch speed of $2v$ in a time of $\frac{1}{2}t$?

Problems
Online

Additional questions are available for assignment in WileyPLUS.

SSM   Student Solutions Manual
MMH   Problem-solving help
GO   Guided Online Tutorial
V-HINT   Video Hints
CHALK   Chalkboard Videos
BIO   Biomedical application
E   Easy
M   Medium
H   Hard
WS   Worksheet
T   Team Problem

Section 6.1   Work Done by a Constant Force

1. E SSM During a tug-of-war, team A pulls on team B by applying a force of 1100 N to the rope between them. The rope remains parallel to the ground. How much work does team A do if they pull team B toward them a distance of 2.0 m?

2. E GO You are moving into an apartment and take the elevator to the 6th floor. Suppose your weight is 685 N and that of your belongings is 915 N. (a) Determine the work done by the elevator in lifting you and your belongings up to the 6th floor (15.2 m) at a constant velocity. (b) How much work does the elevator do on you alone (without belongings) on the downward trip, which is also made at a constant velocity?

3. E The brakes of a truck cause it to slow down by applying a retarding force of 3.0×10^3 N to the truck over a distance of 850 m. What is the work done by this force on the truck? Is the work positive or negative? Why?

4. E A 75.0-kg man is riding an escalator in a shopping mall. The escalator moves the man at a constant velocity from ground level to the floor above, a vertical height of 4.60 m. What is the work done on the man by (a) the gravitational force and (b) the escalator?

5. E SSM Suppose in Figure 6.2 that $+1.10 \times 10^3$ J of work is done by the force $\vec{F}$ (magnitude = 30.0 N) in moving the suitcase a distance of 50.0 m. At what angle θ is the force oriented with respect to the ground?

6. E A person pushes a 16.0-kg shopping cart at a constant velocity for a distance of 22.0 m. She pushes in a direction 29.0° below the horizontal. A 48.0-N frictional force opposes the motion of the cart. (a) What is the magnitude of the force that the shopper exerts? Determine the work done by (b) the pushing force, (c) the frictional force, and (d) the gravitational force.

7. E SSM MMH The drawing shows a plane diving toward the ground and then climbing back upward. During each of these motions, the lift force $\vec{L}$ acts perpendicular to the displacement $\vec{s}$, which has the

same magnitude, 1.7×10^3 m, in each case. The engines of the plane exert a thrust $\vec{T}$, which points in the direction of the displacement and has the same magnitude during the dive and the climb. The weight $\vec{W}$ of the plane has a magnitude of 5.9×10^4 N. In both motions, net work is performed due to the combined action of the forces $\vec{L}$, $\vec{T}$, and $\vec{W}$. **(a)** Is more net work done during the dive or the climb? Explain. **(b)** Find the difference between the net work done during the dive and the climb.

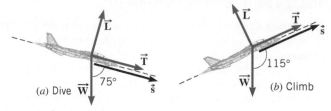

PROBLEM 7

8. **E** A person pulls a toboggan for a distance of 35.0 m along the snow with a rope directed 25.0° above the snow. The tension in the rope is 94.0 N. **(a)** How much work is done on the toboggan by the tension force? **(b)** How much work is done if the same tension is directed parallel to the snow?

9. **M** **V-HINT** As a sailboat sails 52 m due north, a breeze exerts a constant force $\vec{F}_1$ on the boat's sails. This force is directed at an angle west of due north. A force $\vec{F}_2$ of the same magnitude directed due north would do the same amount of work on the sailboat over a distance of just 47 m. What is the angle between the direction of the force $\vec{F}_1$ and due north?

10. **M** **GO** A 55-kg box is being pushed a distance of 7.0 m across the floor by a force $\vec{P}$ whose magnitude is 160 N. The force $\vec{P}$ is parallel to the displacement of the box. The coefficient of kinetic friction is 0.25. Determine the work done on the box by each of the *four* forces that act on the box. Be sure to include the proper plus or minus sign for the work done by each force.

11. **M** **V-HINT** A 1.00×10^2-kg crate is being pushed across a horizontal floor by a force $\vec{P}$ that makes an angle of 30.0° below the horizontal. The coefficient of kinetic friction is 0.200. What should be the magnitude of $\vec{P}$, so that the net work done by it and the kinetic frictional force is zero?

12. **H** A 1200-kg car is being driven up a 5.0° hill. The frictional force is directed opposite to the motion of the car and has a magnitude of $f = 524$ N. A force $\vec{F}$ is applied to the car by the road and propels the car forward. In addition to these two forces, two other forces act on the car: its weight $\vec{W}$ and the normal force $\vec{F}_N$ directed perpendicular to the road surface. The length of the road up the hill is 290 m. What should be the magnitude of $\vec{F}$, so that the net work done by all the forces acting on the car is +150 kJ?

Section 6.2   The Work–Energy Theorem and Kinetic Energy

13. **E** A fighter jet is launched from an aircraft carrier with the aid of its own engines and a steam-powered catapult. The thrust of its engines is 2.3×10^5 N. In being launched from rest it moves through a distance of 87 m and has a kinetic energy of 4.5×10^7 J at lift-off. What is the work done on the jet by the catapult?

14. **E** **MMH** A golf club strikes a 0.045-kg golf ball in order to launch it from the tee. For simplicity, assume that the average net force applied to the ball acts parallel to the ball's motion, has a magnitude of 6800 N, and is in contact with the ball for a distance of 0.010 m. With what speed does the ball leave the club?

15. **E** **SSM** It takes 185 kJ of work to accelerate a car from 23.0 m/s to 28.0 m/s. What is the car's mass?

16. **E** Starting from rest, a 1.9×10^{-4}-kg flea springs straight upward. While the flea is pushing off from the ground, the ground exerts an average upward force of 0.38 N on it. This force does $+2.4 \times 10^{-4}$ J of work on the flea. **(a)** What is the flea's speed when it leaves the ground? **(b)** How far upward does the flea move while it is pushing off? Ignore both air resistance and the flea's weight.

17. **E** **GO** A water-skier is being pulled by a tow rope attached to a boat. As the driver pushes the throttle forward, the skier accelerates. A 70.3-kg water-skier has an initial speed of 6.10 m/s. Later, the speed increases to 11.3 m/s. Determine the work done by the net external force acting on the skier.

18. **E** As background for this problem, review Conceptual Example 6. A 7420-kg satellite has an elliptical orbit, as in **Figure 6.9b**. The point on the orbit that is farthest from the earth is called the *apogee* and is at the far right side of the drawing. The point on the orbit that is closest to the earth is called the *perigee* and is at the left side of the drawing. Suppose that the speed of the satellite is 2820 m/s at the apogee and 8450 m/s at the perigee. Find the work done by the gravitational force when the satellite moves from **(a)** the apogee to the perigee and **(b)** the perigee to the apogee.

19. **E** **GO** A 16-kg sled is being pulled along the horizontal snow-covered ground by a horizontal force of 24 N. Starting from rest, the sled attains a speed of 2.0 m/s in 8.0 m. Find the coefficient of kinetic friction between the runners of the sled and the snow.

20. **E** **GO** An asteroid is moving along a straight line. A force acts along the displacement of the asteroid and slows it down. The asteroid has a mass of 4.5×10^4 kg, and the force causes its speed to change from 7100 to 5500 m/s. **(a)** What is the work done by the force? **(b)** If the asteroid slows down over a distance of 1.8×10^6 m, determine the magnitude of the force.

21. **M** The concepts in this problem are similar to those in Multiple-Concept Example 4, except that the force doing the work in this problem is the tension in the cable. A rescue helicopter lifts a 79-kg person straight up by means of a cable. The person has an upward acceleration of 0.70 m/s^2 and is lifted from rest through a distance of 11 m. **(a)** What is the tension in the cable? How much work is done by **(b)** the tension in the cable and **(c)** the person's weight? **(d)** Use the work–energy theorem and find the final speed of the person.

22. **M** **V-HINT** Consult Multiple-Concept Example 5 for insight into solving this problem. A skier slides horizontally along the snow for a distance of 21 m before coming to rest. The coefficient of kinetic friction between the skier and the snow is $\mu_k = 0.050$. Initially, how fast was the skier going?

23. **M** **GO** **CHALK** Under the influence of its drive force, a snowmobile is moving at a constant velocity along a horizontal patch of snow. When the drive force is shut off, the snowmobile coasts to a halt. The snowmobile and its rider have a mass of 136 kg. Under the influence of a drive force of 205 N, it is moving at a constant velocity whose magnitude is 5.50 m/s. The drive force is then shut off. Find **(a)** the distance in which the snowmobile coasts to a halt and **(b)** the time required to do so.

24. **H** The model airplane in Figure 5.6 is flying at a speed of 22 m/s on a horizontal circle of radius 16 m. The mass of the plane is 0.90 kg. The person holding the guideline pulls it in until the radius becomes 14 m. The plane speeds up, and the tension in the guideline becomes four times greater. What is the net work done on the plane?

Section 6.3   Gravitational Potential Energy,
Section 6.4   Conservative Versus Nonconservative Forces

25. **E** **CHALK** A 75.0-kg skier rides a 2830-m-long lift to the top of a mountain. The lift makes an angle of 14.6° with the horizontal. What is the change in the skier's gravitational potential energy?

26. **E** Juggles and Bangles are clowns. Juggles stands on one end of a teeter-totter at rest on the ground. Bangles jumps off a platform 2.5 m above the ground and lands on the other end of the teeter-totter, launching Juggles into the air. Juggles rises to a height of 3.3 m above the ground, at which point he has the same amount of gravitational potential energy as Bangles had before he jumped, assuming both potential energies are measured using the ground as the reference level. Bangles' mass is 86 kg. What is Juggles' mass?

27. **E** A 0.60-kg basketball is dropped out of a window that is 6.1 m above the ground. The ball is caught by a person whose hands are 1.5 m above the ground. (a) How much work is done on the ball by its weight? What is the gravitational potential energy of the basketball, relative to the ground, when it is (b) released and (c) caught? (d) How is the change $(PE_f - PE_0)$ in the ball's gravitational potential energy related to the work done by its weight?

28. **E** A pole-vaulter just clears the bar at 5.80 m and falls back to the ground. The change in the vaulter's potential energy during the fall is -3.70×10^3 J. What is his weight?

29. **E** **SSM** A bicyclist rides 5.0 km due east, while the resistive force from the air has a magnitude of 3.0 N and points due west. The rider then turns around and rides 5.0 km due west, back to her starting point. The resistive force from the air on the return trip has a magnitude of 3.0 N and points due east. (a) Find the work done by the resistive force during the round trip. (b) Based on your answer to part (a), is the resistive force a conservative force? Explain.

30. **E** **GO** **CHALK** "Rocket Man" has a propulsion unit strapped to his back. He starts from rest on the ground, fires the unit, and accelerates straight upward. At a height of 16 m, his speed is 5.0 m/s. His mass, including the propulsion unit, has the approximately constant value of 136 kg. Find the work done by the force generated by the propulsion unit.

31. **E** **SSM** A 55.0-kg skateboarder starts out with a speed of 1.80 m/s. He does +80.0 J of work on himself by pushing with his feet against the ground. In addition, friction does −265 J of work on him. In both cases, the forces doing the work are nonconservative. The final speed of the skateboarder is 6.00 m/s. (a) Calculate the change $(\Delta PE = PE_f − PE_0)$ in the gravitational potential energy. (b) How much has the vertical height of the skater changed, and is the skater above or below the starting point?

Section 6.5   The Conservation of Mechanical Energy

32. **E** A 35-kg girl is bouncing on a trampoline. During a certain interval after she leaves the surface of the trampoline, her kinetic energy decreases to 210 J from 440 J. How high does she rise during this interval? Neglect air resistance.

33. **E** **SSM** **MMH** A gymnast is swinging on a high bar. The distance between his waist and the bar is 1.1 m, as the drawing shows. At the top of the swing his speed is momentarily zero. Ignoring friction and treating the gymnast as if all of his mass is located at his waist, find his speed at the bottom of the swing.

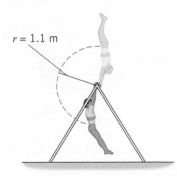

$r = 1.1$ m

PROBLEM 33

34. **E** **CHALK** **GO** The skateboarder in the drawing starts down the left side of the ramp with an initial speed of 5.4 m/s. Neglect nonconservative forces, such as friction and air resistance, and find the height h of the highest point reached by the skateboarder on the right side of the ramp.

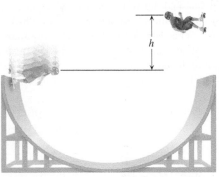

h

PROBLEM 34

35. **E** **MMH** A slingshot fires a pebble from the top of a building at a speed of 14.0 m/s. The building is 31.0 m tall. Ignoring air resistance, find the speed with which the pebble strikes the ground when the pebble is fired (a) horizontally, (b) vertically straight up, and (c) vertically straight down.

36. **E** **GO** The drawing shows two boxes resting on frictionless ramps. One box is relatively light and sits on a steep ramp. The other box is heavier and rests on a ramp that is less steep. The boxes are released from rest at A and allowed to slide down the ramps. The two boxes have masses of 11 and 44 kg. If A and B are 4.5 and 1.5 m, respectively, above the ground, determine the speed of (a) the lighter box and (b) the heavier box when each reaches B. (c) What is the ratio of the kinetic energy of the heavier box to that of the lighter box at B?

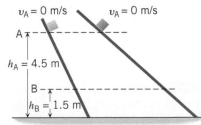

$v_A = 0$ m/s $v_A = 0$ m/s

A

$h_A = 4.5$ m

B

$h_B = 1.5$ m

PROBLEM 36

37. **E** A 47.0-g golf ball is driven from the tee with an initial speed of 52.0 m/s and rises to a height of 24.6 m. (a) Neglect air resistance and determine the kinetic energy of the ball at its highest point. (b) What is its speed when it is 8.0 m below its highest point?

38. **E** **MMH** Consult Conceptual Example 9 in preparation for this problem. The drawing shows a person who, starting from rest at the top of a cliff, swings down at the end of a rope, releases it, and falls into the water below. There are two paths by which the person can enter the water. Suppose he enters the water at a speed of 13.0 m/s via path 1. How fast is he moving on path 2 when he releases the rope at a height of 5.20 m above the water? Ignore the effects of air resistance.

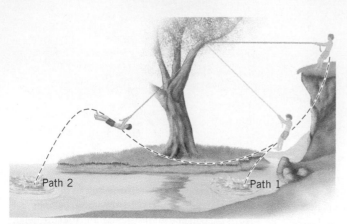

Path 2   Path 1

PROBLEM 38

39. **M** **GO** **SSM** The drawing shows a skateboarder moving at 5.4 m/s along a horizontal section of a track that is slanted upward by 48° above the horizontal at its end, which is 0.40 m above the ground. When she leaves the track, she follows the characteristic path of projectile motion. Ignoring friction and air resistance, find the maximum height H to which she rises above the end of the track.

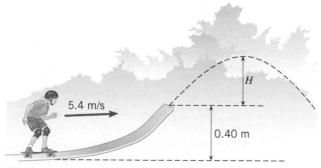

5.4 m/s

H

0.40 m

PROBLEM 39

40. **M** **CHALK** **GO** A small lead ball, attached to a 0.75-m rope, is being whirled in a circle that lies in the vertical plane. The ball is whirled at a constant rate of three revolutions per second and is released on the upward part of the circular motion when it is 1.5 m above the ground. The ball travels straight upward. In the absence of air resistance, to what maximum height above the ground does the ball rise?

41. **M** **CHALK** A pendulum consists of a small object hanging from the ceiling at the end of a string of negligible mass. The string has a length of 0.75 m. With the string hanging vertically, the object is given an initial velocity of 2.0 m/s parallel to the ground and swings upward on a circular arc. Eventually, the object comes to a momentary halt at a point where the string makes an angle θ with its initial vertical orientation and then swings back downward. Find the angle θ.

42. **M** **V-HINT** A semitrailer is coasting downhill along a mountain highway when its brakes fail. The driver pulls onto a runaway-truck ramp that is inclined at an angle of 14.0° above the horizontal. The semitrailer coasts to a stop after traveling 154 m along the ramp. What was the truck's initial speed? Neglect air resistance and friction.

43. **M** **GO** The drawing shows two frictionless inclines that begin at ground level ($h = 0$ m) and slope upward at the same angle θ. One track is longer than the other, however. Identical blocks are projected up each track with the same initial speed v_0. On the longer track the block slides upward until it reaches a maximum height H above the ground. On the shorter track the block slides upward, flies off the end of the track at a height H_1 above the ground, and then follows the familiar parabolic trajectory of projectile motion. At the highest point of this trajectory, the block is a height H_2 above the end of the track. The initial total mechanical energy of each block is the same and is

all kinetic energy. The initial speed of each block is $v_0 = 7.00$ m/s, and each incline slopes upward at an angle of $\theta = 50.0°$. The block on the shorter track leaves the track at a height of $H_1 = 1.25$ m above the ground. Find **(a)** the height H for the block on the longer track and **(b)** the total height $H_1 + H_2$ for the block on the shorter track.

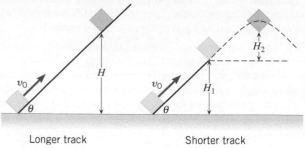

v_0

H

θ

Longer track

v_0

H_1

H_2

θ

Shorter track

PROBLEM 43

44. **H** A skier starts from rest at the top of a hill. The skier coasts down the hill and up a second hill, as the drawing illustrates. The crest of the second hill is circular, with a radius of $r = 36$ m. Neglect friction and air resistance. What must be the height h of the first hill so that the skier just loses contact with the snow at the crest of the second hill?

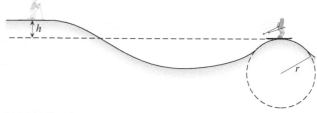

h

r

PROBLEM 44

45. **H** A person starts from rest at the top of a large frictionless spherical surface, and slides into the water below (see the drawing). At what angle θ does the person leave the surface? *(Hint: When the person leaves the surface, the normal force is zero.)*

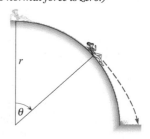

r

θ

PROBLEM 45

Section 6.6 Nonconservative Forces and the Work–Energy Theorem

46. **E** **SSM** A projectile of mass 0.750 kg is shot straight up with an initial speed of 18.0 m/s. **(a)** How high would it go if there were no air resistance? **(b)** If the projectile rises to a maximum height of only 11.8 m, determine the magnitude of the average force due to air resistance.

47. **E** A basketball player makes a jump shot. The 0.600-kg ball is released at a height of 2.00 m above the floor with a speed of 7.20 m/s. The ball goes through the net 3.10 m above the floor at a speed of 4.20 m/s. What is the work done on the ball by air resistance, a nonconservative force?

48. **E** Starting from rest, a 93-kg firefighter slides down a fire pole. The average frictional force exerted on him by the pole has a magnitude of 810 N, and his speed at the bottom of the pole is 3.4 m/s. How far did he slide down the pole?

49. **E** **GO** A student, starting from rest, slides down a water slide. On the way down, a kinetic frictional force (a nonconservative force) acts on her. The student has a mass of 83.0 kg, and the height of the water slide is 11.8 m. If the kinetic frictional force does -6.50×10^3 J of work, how fast is the student going at the bottom of the slide?

50. **E** **SSM** The (nonconservative) force propelling a 1.50×10^3-kg car up a mountain road does 4.70×10^6 J of work on the car. The car starts from rest at sea level and has a speed of 27.0 m/s at an altitude of 2.00×10^2 m above sea level. Obtain the work done on the car by the combined forces of friction and air resistance, both of which are nonconservative forces.

51. **M** **V-HINT** In attempting to pass the puck to a teammate, a hockey player gives it an initial speed of 1.7 m/s. However, this speed is inadequate to compensate for the kinetic friction between the puck and the ice. As a result, the puck travels only one-half the distance between the players before sliding to a halt. What minimum initial speed should the puck have been given so that it reached the teammate, assuming that the same force of kinetic friction acted on the puck everywhere between the two players?

52. **M** **MMH** A 67.0-kg person jumps from rest off a 3.00-m-high tower straight down into the water. Neglect air resistance. She comes to rest 1.10 m under the surface of the water. Determine the magnitude of the average force that the water exerts on the diver. This force is nonconservative.

53. **M** At a carnival, you can try to ring a bell by striking a target with a 9.00-kg hammer. In response, a 0.400-kg metal piece is sent upward toward the bell, which is 5.00 m above. Suppose that 25.0% of the hammer's kinetic energy is used to do the work of sending the metal piece upward. How fast must the hammer be moving when it strikes the target so that the bell just barely rings?

Section 6.7   Power

54. **E** A person is making homemade ice cream. She exerts a force of magnitude 22 N on the free end of the crank handle on the ice-cream maker, and this end moves on a circular path of radius 0.28 m. The force is always applied parallel to the motion of the handle. If the handle is turned once every 1.3 s, what is the average power being expended?

55. **E** **MMH** A car accelerates uniformly from rest to 20.0 m/s in 5.6 s along a level stretch of road. Ignoring friction, determine the average power required to accelerate the car if (a) the weight of the car is 9.0×10^3 N and (b) the weight of the car is 1.4×10^4 N.

56. **E** **GO** A helicopter, starting from rest, accelerates straight up from the roof of a hospital. The lifting force does work in raising the helicopter. An 810-kg helicopter rises from rest to a speed of 7.0 m/s in a time of 3.5 s. During this time it climbs to a height of 8.2 m. What is the average power generated by the lifting force?

57. **M** **V-HINT** The cheetah is one of the fastest-accelerating animals, because it can go from rest to 27 m/s (about 60 mi/h) in 4.0 s. If its

mass is 110 kg, determine the average power developed by the cheetah during the acceleration phase of its motion. Express your answer in (a) watts and (b) horsepower.

58. **M** **V-HINT** **CHALK** In 2.0 minutes, a ski lift raises four skiers at constant speed to a height of 140 m. The average mass of each skier is 65 kg. What is the average power provided by the tension in the cable pulling the lift?

Section 6.9   Work Done by a Variable Force

59. **E** The graph shows how the force component $F \cos \theta$ along the displacement varies with the magnitude s of the displacement. Find the work done by the force. *(Hint: Recall how the area of a triangle is related to the triangle's base and height.)*

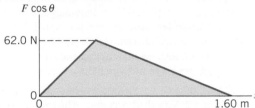

PROBLEM 59

60. **E** **CHALK** The force component along the displacement varies with the magnitude of the displacement, as shown in the graph. Find the work done by the force in the interval from (a) 0 to 1.0 m, (b) 1.0 to 2.0 m, and (c) 2.0 to 4.0 m. *(Note: In the last interval the force component is negative, so the work is negative.)*

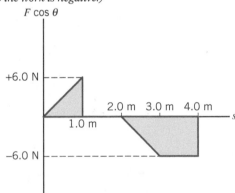

PROBLEM 60

61. **E** **GO** A net external force is applied to a 6.00-kg object that is initially at rest. The net force component along the displacement of the object varies with the magnitude of the displacement as shown in the drawing. What is the speed of the object at $s = 20.0$ m?

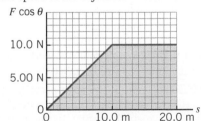

PROBLEM 61

Additional Problems

Online

62. **E** A cable lifts a 1200-kg elevator at a constant velocity for a distance of 35 m. What is the work done by (a) the tension in the cable and (b) the elevator's weight?

63. **E** **SSM** A 2.00-kg rock is released from rest at a height of 20.0 m. Ignore air resistance and determine the kinetic energy, gravitational potential energy, and total mechanical energy at each of the following heights: 20.0, 10.0, and 0 m.

64. **E** **GO** When an 81.0-kg adult uses a spiral staircase to climb to the second floor of his house, his gravitational potential energy increases by 2.00×10^3 J. By how much does the potential energy of an 18.0-kg child increase when the child climbs a normal staircase to the second floor?

65. **M** **GO** A husband and wife take turns pulling their child in a wagon along a horizontal sidewalk. Each exerts a constant force and pulls the wagon through the same displacement. They do the same amount of work, but the husband's pulling force is directed 58° above the horizontal, and the wife's pulling force is directed 38° above the horizontal. The husband pulls with a force whose magnitude is 67 N. What is the magnitude of the pulling force exerted by his wife?

66. **M** **V-HINT** Some gliders are launched from the ground by means of a winch, which rapidly reels in a towing cable attached to the glider. What average power must the winch supply in order to accelerate a 184-kg ultralight glider from rest to 26.0 m/s over a horizontal distance of 48.0 m? Assume that friction and air resistance are negligible, and that the tension in the winch cable is constant.

67. **M** **V-HINT** A 63-kg skier coasts up a snow-covered hill that makes an angle of 25° with the horizontal. The initial speed of the skier is 6.6 m/s. After coasting 1.9 m up the slope, the skier has a speed of 4.4 m/s. **(a)** Find the work done by the kinetic frictional force that acts on the skis. **(b)** What is the magnitude of the kinetic frictional force?

68. **M** **V-HINT** A water slide is constructed so that swimmers, starting from rest at the top of the slide, leave the end of the slide traveling horizontally. As the drawing shows, one person hits the water 5.00 m from the end of the slide in a time of 0.500 s after leaving the slide. Ignoring friction and air resistance, find the height H in the drawing.

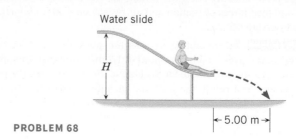

Water slide

H

←— 5.00 m —→

PROBLEM 68

69. **M** **GO** A car, starting from rest, accelerates in the $+x$ direction as in the figure. It has a mass of 1.10×10^3 kg and maintains an acceleration of $+4.60$ m/s^2 for 5.00 s. Assume that a single horizontal force (not shown) accelerates the vehicle. Determine the average power generated by this force.

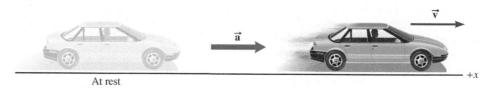

$\vec{a}$

$\vec{v}$

At rest

$+x$

PROBLEM 69

Physics in Biology, Medicine, and Sports

70. **E** **SSM** **BIO** **6.2** The hammer throw is a track-and-field event in which a 7.3 kg ball (the "hammer"), starting from rest, is whirled around in a circle several times and released. It then moves upward on the familiar curving path of projectile motion. In one throw, the hammer is given a speed of 29 m/s. For comparison, a .22 caliber bullet has a mass of 2.6 g and, starting from rest, exits the barrel of a gun at a speed of 410 m/s. Determine the work done to launch the motion of **(a)** the hammer and **(b)** the bullet.

71. **E** **GO** **BIO** **6.6** In the sport of skeleton a participant jumps onto a sled (known as a skeleton) and proceeds to slide down an icy track, belly down and head first. In the 2010 Winter Olympics, the track had sixteen turns and dropped 126 m in elevation from top to bottom. **(a)** In the absence of nonconservative forces, such as friction and air resistance, what would be the speed of a rider at the bottom of the track? Assume that the speed at the beginning of the run is relatively small and can be ignored. **(b)** In reality, the gold-medal winner (Canadian Jon Montgomery) reached the bottom in one heat with a speed of 40.5 m/s (about 91 mi/h). How much work was done on him and his sled (assuming a total mass of 118 kg) by nonconservative forces during this heat?

PROBLEM 71

Richard Heathcote / Staff / Getty Images

72. **E** **BIO** **SSM** **6.7** Bicyclists in the Tour de France do enormous amounts of work during a race. For example, the average power per kilogram generated by seven-time-winner Lance Armstrong ($m = 75.0$ kg) is 6.50 W per kilogram of his body mass. **(a)** How much work does he do during a 135-km race in which his average speed is 12.0 m/s? **(b)** Often, the work done is expressed in nutritional Calories rather than in joules. Express the work done in part (a) in terms of nutritional Calories, noting that 1 joule = 2.389×10^{-4} nutritional Calories.

73. **E** **BIO** **GO** **6.7** You are working out on a rowing machine. Each time you pull the rowing bar (which simulates the oars) toward you, it moves a distance of 1.2 m in a time of 1.5 s. The readout on the display indicates that the average power you are producing is 82 W. What is the magnitude of the force that you exert on the handle?

74. **M** **BIO** **6.3** A 65.0-kg mountain climber just ate a Johnny Rocket's bacon-cheddar-double-burger meal along with two peanut butter milkshakes, which have 5810 Calories (5810 kilocalories). Assume she can convert these calories into energy with 100% efficiency and use the fact that 1 kcal = 4186 J. Calculate how many times she could climb up Mt. Everest—the world's tallest mountain. The height of Mt. Everest is 8840 m.

75. **M** **BIO** **6.2** Which has more kinetic energy: the world's fastest man or a 9-mm bullet fired from a handgun? Usain Bolt holds the world record in the 100-m run. His top speed was recorded at 27.4 mph (12.3 m/s), and his mass is 94.0 kg. In comparison, a 9-mm bullet is fired at a speed of 791 mph (354 m/s), and it has a mass of 7.45×10^{-3} kg. Calculate **(a)** the kinetic energy of Usain Bolt, and **(b)** the kinetic energy of the 9-mm bullet.

76. **M** **BIO** **6.7** A typical human brain has a mass of 1.4 kg, which makes up about 2% of a person's mass. However, the brain requires 20% of the body's power consumption. In essence, the brain requires 10 times as much power as any other organ in the body. The average power output of the human brain is about 2.0×10^1 W. Assume that the brain's power output could be converted into mechanical work and calculate how high we could use this power to lift an apple ($m = 0.20$ kg) in one minute.

77. **M** **BIO** **6.1** The barbell curl (see the image) is an exercise that can strengthen and enlarge the biceps muscle in the upper arm. During a single contraction, the biceps muscle in each arm can exert a force of 850 N, as the muscle contracts through a distance of 7.2 cm. How much work can the biceps muscle perform in one contraction?

PROBLEM 77

78. **M** **BIO** **6.7** In the motion picture *Iron Man*, Robert Downey, Jr.'s character, Tony Stark, creates the original arc reactor to power his Iron Man suit (see the image). Tony calculated the energy output of the arc reactor to be 3.00 GJ/s (gigajoules per second), and his math is always right. In comparison, calculate how much energy is produced by a human heart over an 80.0-year lifetime. The average output power of the heart is 1.33 W.

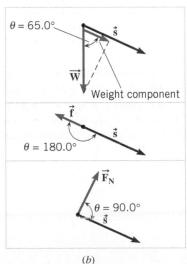

PROBLEM 78

Concepts and Calculations Problems

Online

Problem 79 reviews the important concept of work, and illustrates how forces can give rise to positive, negative, and zero work. Problem 80 examines the all-important conservation of mechanical energy.

79. **M** **CHALK** **SSM** The skateboarder in the figure is coasting down a ramp, and there are three forces acting on her: her weight $\vec{W}$ (magnitude = 675 N), a frictional force $\vec{f}$ (magnitude = 125 N) that opposes her motion, and a normal force $\vec{F}_N$ (magnitude = 612 N) *Concepts:* Part b of the figure shows each force, along with the displacement $\vec{s}$ of the skateboarder. By examining these diagrams and without doing any numerical calculations, determine whether the work done by each force is positive, negative, or zero. Provide a reason for each answer. *Calculations:* Determine the net work done by the three forces when she coasts for a distance of 9.2 m.

80. **M** **CHALK** The figure shows a 0.41-kg block sliding from A to B along a frictionless surface. When the block reaches B, it continues to slide along the horizontal surface BC where the kinetic frictional force acts. As a result, the block slows down, coming to rest at C. The kinetic energy of

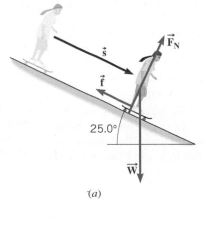

PROBLEM 79

the block at A is 37 J, and the heights of A and B are 12.0 and 7.0 m above the ground, respectively. *Concepts:* (i) Is the total mechanical energy of the block conserved as the block goes from A to B? Why or why not? (ii) When the block reaches point B, has its kinetic energy increased, decreased, or remained the same relative to what it was at A? Give a reason for your answer. (iii) Is the total mechanical energy of the block conserved as the block goes from B to C? Justify your answer. *Calculations:* (a) What is the value of the kinetic energy of the block when it reaches B? (b) How much work does the kinetic frictional force do during the BC segment of the trip?

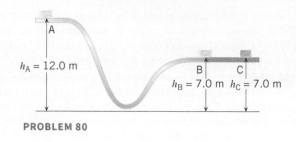

$h_A = 12.0$ m

$h_B = 7.0$ m $h_C = 7.0$ m

PROBLEM 80

Team Problems   Online

81. **E T WS** **A Runaway Sled.** You and your team are on the top of an icy slope that has a steep grade (i.e., angle of ascent) of 24.0°, and need to pull up a heavy sled loaded with supplies. You fasten a pulley into the ice at the top of the slope through which you thread a 150.0-ft rope (which is just long enough to reach the sled) and tie it to the sled. The rope has a strict tension rating of 450.0 lb, and a warning sticker on one end claims that it will break if this limit is exceeded by just 10.0 lb. Lined up like a tug-o-war team, you and six others in your group steadily pull on the rope with a nearly constant force, resulting in a constant rate of ascent. However, after exactly 45.0 s of pulling, the rope suddenly snaps, and the sled and its crucial supplies plummet down the hill and disappear into a snow bank. You measure the length of rope pulled through the pulley before it failed to be 85.0 ft. (a) Neglecting friction, how fast was the sled going when it reached the bottom of the slope? (b) For the next attempt to pull the sled up the slope, you get a stronger rope and want to use a motor-driven winch (it was tiring work by hand). What should be the minimum power rating of the motor (in horsepower) in order that the winch pulls the sled at the same rate as your team had been pulling it by hand? (*Note: 1 horsepower = 745.7 watts.*)

82. **E T WS** **A Tire Swing.** As background for this problem, review Conceptual Example 9. You and your team are designing a tire swing that hangs from a sturdy branch that is located 22.0 feet above the base of a tree. You tie one end of a rope to the branch and the other to a 35.0-lb tire, and adjust the length of the swing (i.e., the distance from the branch to the center of the tire) so that the center of the tire is 1.20 m above the ground at its lowest point. Near the tree is a grassy hill, the crest of which is 12.0 feet above the base of the tree, which makes for a perfect takeoff point. (a) If starting from rest at the top of the hill (i.e., the center of the swing is 12.0 feet above the base of the tree), how fast would a rider be moving when they reached the bottom? (b) If the maximum allowable weight of a rider is 160.0 lb, what should be the maximum tension rating for the rope?

83. **M T** **A Makeshift Elevator.** While exploring an elaborate tunnel system, you and your team get lost and find yourselves at the bottom of a 450-m vertical shaft. Suspended from a thick rope (near the floor) is a large rectangular bucket that looks like it had been used

to transport tools and debris up and down the tunnel. Mounted on the floor near one of the walls is a gasoline engine (3.5 hp) that turns a pulley and rope, and a sign that reads "Emergency Lift." It is clear that the engine is used to drive the bucket up the shaft. On the wall next to the engine is a sign indicating that a full tank of gas will last exactly 15 minutes when the engine is running at full power. You open the engine's gas tank and estimate that it is ¼ full, and there are no other sources of gasoline. (a) Assuming zero friction, if you send your team's lightest member (who weighs 125 lb), and the bucket weighs 150 lb when empty, how far up the shaft will the engine take her (and the bucket)? Will it get her out of the mine? (b) Assuming an effective collective friction (from the pulleys, etc.) of $\mu_{eff} = 0.10$ (so that $F_f = \mu_{eff} M g$, where M is the total mass of the bucket plus team member), will the engine (with a ¼-full tank of gas) lift her to the top of the shaft?

84. **M T** **A Sledding Contest.** You are in a sledding contest where you start at a height of 40.0 m above the bottom of a valley and slide down a hill that makes an angle of 25.0° with respect to the horizontal. When you reach the valley, you immediately slide up a second hill that makes an angle of 15.0° with respect to the horizontal. The winner of the contest will be the contestant who travels the greatest distance up the second hill. You must now choose between using your flat-bottomed plastic sled, or your "Blade Runner," which glides on two steel rails. The hill you will ride down is covered with loose snow. However, the hill you will climb on the other side is a popular sledding hill, and is packed hard and is slick. The two sleds perform very differently on the two surfaces, the plastic one performing better on loose snow, and the Blade Runner doing better on hard-packed snow or ice. The performances of each sled can be quantified in terms of their respective coefficients of kinetic friction on the two surfaces. For the plastic sled: $\mu = 0.17$ on loose snow, and $\mu = 0.15$ on packed snow or ice. For the Blade Runner, $\mu = 0.19$ on loose snow, and $\mu = 0.070$ on packed snow or ice. Assuming the two hills are shaped like inclined planes, and neglecting air resistance, (a) how far does each sled make it up the second hill before stopping? (b) Assuming the total mass of the sled plus rider is 55.0 kg in both cases, how much work is done by nonconservative forces (over the total trip) in each case?

Traffic accidents often involve the collision between two or more motor vehicles. Although increased seat belt use, air bags, and other technologies have greatly improved the safety of modern automobiles, almost 90 Americans lose their lives everyday in traffic accidents. Much of the kinetic energy in the collision goes into permanently damaging the vehicles. In physics such a collision is classified as being inelastic. Inelastic collisions are one of the two basic types that this chapter introduces.

Marcel Langthim/Pixabay

Impulse and Momentum

7.1 The Impulse–Momentum Theorem

There are many situations in which the force acting on an object is not constant, but varies with time. For instance, **Figure 7.1*a*** shows a baseball being hit, and part *b* of the figure illustrates approximately how the force applied to the ball by the bat changes during the time of contact. The magnitude of the force is zero at the instant t_0 just before the bat touches the ball. During contact, the force rises to a maximum and then returns to zero at the time t_f when the ball leaves the bat. The time interval $\Delta t = t_f - t_0$ during which the bat and ball are in contact is quite short, being only a few thousandths of a second, although the maximum force can be very large, often exceeding thousands of newtons. For comparison, the graph also shows the magnitude $\overline{F}$ of the average force exerted on the ball during the time of contact. **Figure 7.2** depicts other situations in which a time-varying force is applied to a ball.

LEARNING OBJECTIVES

After reading this module, you should be able to...

7.1 Apply the impulse–momentum theorem.

7.2 Apply the law of conservation of linear momentum.

7.3 Analyze one-dimensional collisions.

7.4 Analyze two-dimensional collisions.

7.5 Determine the location and the velocity of the center of mass.

191

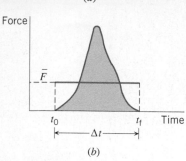

FIGURE 7.1 (a) The collision time between a bat and a ball is very short, often less than a millisecond, but the force can be quite large. (b) When the bat strikes the ball, the magnitude of the force exerted on the ball rises to a maximum and then returns to zero when the ball leaves the bat. The time interval during which the force acts is Δt, and the magnitude of the average force is $\overline{F}$.

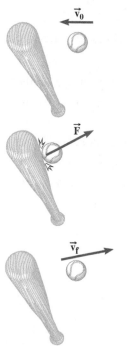

FIGURE 7.3 When a bat hits a ball, an average force $\vec{\overline{F}}$ is applied to the ball by the bat. As a result, the ball's velocity changes from an initial value of $\vec{v}_0$ (top drawing) to a final value of $\vec{v}_f$ (bottom drawing).

FIGURE 7.2 In each of these situations, the force applied to the ball varies with time. The time of contact is small, but the maximum force can be large.

To describe how a time-varying force affects the motion of an object, we will introduce two new ideas: the impulse of a force and the linear momentum of an object. These ideas will be used with Newton's second law of motion to produce an important result known as the impulse–momentum theorem. This theorem plays a central role in describing collisions, such as that between a ball and a bat. Later on, we will see also that the theorem leads in a natural way to one of the most fundamental laws in physics, the conservation of linear momentum.

If a baseball is to be hit well, both the magnitude of the force and the time of contact are important. When a large average force acts on the ball for a long enough time, the ball is hit solidly. To describe such situations, we bring together the average force and the time of contact, calling the product of the two the **impulse** of the force.

DEFINITION OF IMPULSE

The impulse $\vec{J}$ of a force is the product of the average force $\vec{\overline{F}}$ and the time interval Δt during which the force acts:

$$\vec{J} = \vec{\overline{F}}\,\Delta t \tag{7.1}$$

Impulse is a vector quantity and has the same direction as the average force.

SI Unit of Impulse: newton · second (N · s)

When a ball is hit, it responds to the value of the impulse. A large impulse produces a large response; that is, the ball departs from the bat with a large velocity. However, we know from experience that the more massive the ball, the less velocity it has after leaving the bat. Both mass and velocity play a role in how an object responds to a given impulse, and the effect of each of them is included in the concept of **linear momentum**, which is defined as follows:

DEFINITION OF LINEAR MOMENTUM

The linear momentum $\vec{p}$ of an object is the product of the object's mass m and velocity $\vec{v}$:

$$\vec{p} = m\vec{v} \tag{7.2}$$

Linear momentum is a vector quantity that points in the same direction as the velocity.

SI Unit of Linear Momentum: kilogram · meter/second (kg · m/s)

Newton's second law of motion can now be used to reveal a relationship between impulse and momentum. **Figure 7.3** shows a ball with an initial velocity $\vec{v}_0$ approaching a bat, being struck by the bat, and then departing with a final velocity $\vec{v}_f$. When the velocity of an object changes from $\vec{v}_0$ to $\vec{v}_f$ during a time interval Δt, the average acceleration $\vec{\overline{a}}$ is given by Equation 2.4 as

$$\vec{\overline{a}} = \frac{\vec{v}_f - \vec{v}_0}{\Delta t}$$

According to Newton's second law, $\Sigma\vec{\mathbf{F}} = m\vec{\mathbf{a}}$, the average acceleration is produced by the net average force $\Sigma\vec{\mathbf{F}}$. Here $\Sigma\vec{\mathbf{F}}$ represents the vector sum of all the average forces that act on the object. Thus,

$$\Sigma\vec{\mathbf{F}} = m\left(\frac{\vec{\mathbf{v}}_f - \vec{\mathbf{v}}_0}{\Delta t}\right) = \frac{m\vec{\mathbf{v}}_f - m\vec{\mathbf{v}}_0}{\Delta t} \qquad (7.3)$$

In this result, the numerator on the far right is the final momentum minus the initial momentum, which is the change in momentum. Thus, the net average force is given by the change in momentum per unit of time.* Multiplying both sides of Equation 7.3 by Δt yields Equation 7.4, which is known as the **impulse–momentum theorem**.

IMPULSE–MOMENTUM THEOREM

When a net average force $\Sigma\vec{\mathbf{F}}$ acts on an object during a time interval Δt the impulse of this force is equal to the change in momentum of the object:

$$\underbrace{(\Sigma\vec{\mathbf{F}})\Delta t}_{\text{Impulse}} = \underbrace{m\vec{\mathbf{v}}_f}_{\substack{\text{Final} \\ \text{momentum}}} - \underbrace{m\vec{\mathbf{v}}_0}_{\substack{\text{Initial} \\ \text{momentum}}} \qquad (7.4)$$

Impulse = Change in momentum

During a collision, it is often difficult to measure the net average force $\Sigma\vec{\mathbf{F}}$, so it is not easy to determine the impulse $(\Sigma\vec{\mathbf{F}})\,\Delta t$ directly. On the other hand, it is usually straightforward to measure the mass and velocity of an object, so that its momentum $m\vec{\mathbf{v}}_f$ just after the collision and $m\vec{\mathbf{v}}_0$ just before the collision can be found. Thus, the impulse–momentum theorem allows us to gain information about the impulse indirectly by measuring the change in momentum that the impulse causes (see **Photo 7.1**). Then, armed with a knowledge of the contact time Δt, we can evaluate the net average force. Examples 1 and 2 illustrate how the theorem is used in this way.

PHOTO 7.1 During the launch of SpaceX's *Falcon Heavy*, the engines fire and create a force that applies an impulse to the rocket. This impulse causes the momentum of the rocket to increase in the upward direction.

John Raoux/AP Images

EXAMPLE 1 | BIO The Physics of Sneezing—A Violent Respiratory Event

Contagious respiratory diseases, such as SARS (severe acute respiratory syndrome), H5N1 and H1N1 influenza, and more recently coronavirus disease 2019 (COVID-19), have motivated researchers to study the physics of human sneezing, as water droplets ejected into the air from a person's nose and mouth are a common mode of viral transmission. One of the factors that determines social distancing guidelines during a pandemic is how far the droplets travel in the air after sneezing. Studies have shown that this distance can vary greatly (from less than 1 m to over 8 m), and it depends on many variables, such as the force on the droplets during the sneeze, sneeze velocity, droplet size, wind conditions, and so on (see **Figure 7.4**). A typical sneeze can contain as many as 40 000 droplets of various sizes, with diameters ranging from <10 μm to >1000 μm. Calculate the magnitude of the average force applied to a single droplet during a sneeze that takes place over a time duration of 0.200 s. Assume the diameter of an average droplet is 15.0 μm (mass of 1.77×10^{-12} kg), and it is ejected with a speed of 100 mph (45.0 m/s).

Reasoning We can use the impulse—momentum theorem (Equation 7.4) to calculate the average net force on the droplet. During the sneeze, the forces acting on the droplet include its weight, the propelling force (due to a difference in air pressure, $\vec{\mathbf{F}}_P$), and possibly forces between droplets and the walls of the nasal cavity. During the sneeze, the propelling force is much greater than these other forces, so we neglect them here. Therefore, the net force $\Sigma\vec{\mathbf{F}} = \vec{\mathbf{F}}_P$ will be equal to the propelling force. The change in momentum of the droplet can be calculated from its mass and change in speed.

Solution From Equation 7.4, $\vec{\mathbf{F}}\Delta t = m\vec{\mathbf{v}}_f - m\vec{\mathbf{v}}_0$. Here, the initial velocity of the droplet is zero, and the final velocity is given in the problem. Since we want to calculate the magnitude of the average force, we drop the vector notation. Solving for $\overline{F}$ we find:

$$\overline{F}_P = \overline{F} = \frac{mv_f}{\Delta t} = \frac{(1.77 \times 10^{-12}\text{ kg})(45.0\text{ m/s})}{(0.200\text{ s})} = \boxed{3.98 \times 10^{-10}\text{ N}}$$

Even though the average force applied to the droplet is quite small, its acceleration will be large (>200 m/s^2) due to its very small mass.

---

*The equality between the net force and the change in momentum per unit time is the version of the second law of motion presented originally by Newton.

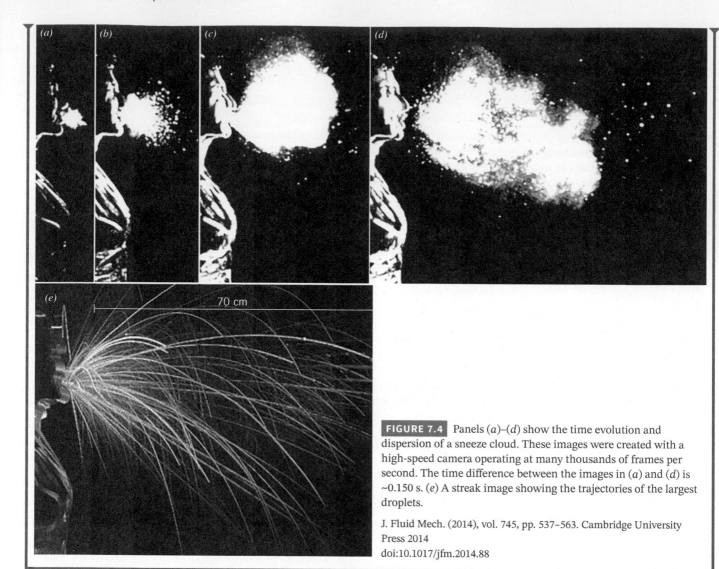

FIGURE 7.4 Panels (*a*)–(*d*) show the time evolution and dispersion of a sneeze cloud. These images were created with a high-speed camera operating at many thousands of frames per second. The time difference between the images in (*a*) and (*d*) is ~0.150 s. (*e*) A streak image showing the trajectories of the largest droplets.

J. Fluid Mech. (2014), vol. 745, pp. 537–563. Cambridge University Press 2014

doi:10.1017/jfm.2014.88

EXAMPLE 2 | A Well-Hit Ball

A baseball ($m = 0.14$ kg) has an initial velocity of $\vec{\mathbf{v}}_0 = -38$ m/s as it approaches a bat. We have chosen the direction of approach as the negative direction. The bat applies an average force $\overrightarrow{\mathbf{F}}$ that is much larger than the weight of the ball, and the ball departs from the bat with a final velocity of $\vec{\mathbf{v}}_f = +58$ m/s. **(a)** Determine the impulse applied to the ball by the bat. **(b)** Assuming that the time of contact is $\Delta t = 1.6 \times 10^{-3}$ s, find the average force exerted on the ball by the bat.

Reasoning Two forces act on the ball during impact, and together they constitute the net average force: the average force $\overrightarrow{\mathbf{F}}$ exerted by the bat, and the weight of the ball. As in Example 1, since $\overrightarrow{\mathbf{F}}$ is much greater than the weight of the ball, we neglect the weight. Thus, the net average force is equal to $\overrightarrow{\mathbf{F}}$, or $\Sigma\overrightarrow{\mathbf{F}} = \overrightarrow{\mathbf{F}}$. In hitting the ball, the bat imparts an impulse to it. We cannot use Equation 7.1 ($\overrightarrow{\mathbf{J}} = \overrightarrow{\mathbf{F}}\,\Delta t$) to determine the impulse $\overrightarrow{\mathbf{F}}$ directly, since $\overrightarrow{\mathbf{F}}$ is not known. We can find the impulse indirectly, however, by turning to the impulse–momentum theorem, which states that the impulse is equal to the ball's final momentum minus its initial momentum. With values for the impulse and the time of contact, Equation 7.1 can be used to determine the average force applied by the bat to the ball.

> **Problem-Solving Insight** Momentum is a vector and, as such, has a magnitude and a direction. For

motion in one dimension, be sure to indicate the direction by assigning a plus or minus sign to it, as in this example.

Solution (a) According to the impulse–momentum theorem, the impulse $\overrightarrow{\mathbf{J}}$ applied to the ball is

$$\overrightarrow{\mathbf{J}} = m\vec{\mathbf{v}}_f - m\vec{\mathbf{v}}_0$$

$$= \underbrace{(0.14\ \text{kg})(+58\ \text{m/s})}_{\text{Final momentum}} - \underbrace{(0.14\ \text{kg})(-38\ \text{m/s})}_{\text{Initial momentum}}$$

$$= \boxed{+13.4\ \text{kg}\cdot\text{m/s}}$$

(b) Now that the impulse is known, the contact time can be used in Equation 7.1 to find the average force $\overrightarrow{\mathbf{F}}$ exerted by the bat on the ball:

$$\overrightarrow{\mathbf{F}} = \frac{\overrightarrow{\mathbf{J}}}{\Delta t} = \frac{+13.4\ \text{kg}\cdot\text{m/s}}{1.6 \times 10^{-3}\,\text{s}}$$

$$= \boxed{+8400\ \text{N}}$$

The force is positive, indicating that it points opposite to the velocity of the approaching ball. A force of 8400 N corresponds to 1900 lb, such a large value being necessary to change the ball's momentum during the brief contact time.

EXAMPLE 3 | A Rainstorm

During a storm, rain comes straight down with a velocity of $\vec{v}_0 = -15$ m/s and hits the roof of a car perpendicularly (see **Figure 7.5**). The mass of rain per second that strikes the car roof is 0.060 kg/s. Assuming that the rain comes to rest upon striking the car ($\vec{v}_f = 0$ m/s), find the average force exerted by the rain on the roof.

Reasoning   This example differs from Example 2 in an important way. Example 2 gives information about the ball and asks for the force applied to the ball. In contrast, the present example gives information about the rain but asks for the force acting on the roof. However, the force exerted on the roof by the rain and the force exerted on the rain by the roof have equal magnitudes and opposite directions, according to Newton's law of action and reaction (see Section 4.5). Thus, we will find the force exerted on the rain and then apply the law of action and reaction to obtain the force on the roof. Two forces act on the rain while it impacts with the roof: the average force $\vec{F}$ exerted by the roof and the weight of the rain. These two forces constitute the net average force. By comparison, however, the force $\vec{F}$ is much greater than the weight, so we may neglect the weight. Thus, the net average force becomes equal to $\vec{F}$, or $\Sigma\vec{F} = \vec{F}$. The value of $\vec{F}$ can be obtained by applying the impulse–momentum theorem to the rain.

Solution   The average force $\vec{F}$ needed to reduce the rain's velocity from $\vec{v}_0 = -15$ m/s to $\vec{v}_f = 0$ m/s is given by Equation 7.4 as

$$\vec{F} = \frac{m\vec{v}_f - m\vec{v}_0}{\Delta t} = -\left(\frac{m}{\Delta t}\right)\vec{v}_0$$

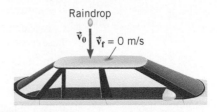

Raindrop

FIGURE 7.5   A raindrop falling on a car roof has an initial velocity of $\vec{v}_0$ just before striking the roof. The final velocity of the raindrop is $\vec{v}_f = 0$ m/s, because it comes to rest on the roof.

The term $m/\Delta t$ is the mass of rain per second that strikes the roof, so that $m/\Delta t = 0.060$ kg/s. Thus, the average force exerted on the rain by the roof is

$$\vec{F} = -(0.060 \text{ kg/s})(-15 \text{ m/s}) = +0.90 \text{ N}$$

This force is in the positive or upward direction, which is reasonable since the roof exerts an upward force on each falling drop in order to bring it to rest. According to the action–reaction law, the force exerted on the roof by the rain also has a magnitude of 0.90 N but points downward: Force on roof = $\boxed{-0.90 \text{ N}}$.

As you reason through problems such as those in Examples 1, 2, and 3, take advantage of the impulse–momentum theorem. It is a powerful statement that can lead to significant insights. The following Conceptual Example further illustrates its use.

CONCEPTUAL EXAMPLE 4 | Hailstones Versus Raindrops

In Example 3 rain is falling on the roof of a car and exerts a force on it. Instead of rain, suppose hail is falling. The hail comes straight down at a mass rate of $m/\Delta t = 0.060$ kg/s and an initial velocity of $\vec{v}_0 = -15$ m/s and strikes the roof perpendicularly, just as the rain does in Example 3. However, unlike rain, hail usually does not come to rest after striking a surface. Instead, the hailstones bounce off the roof of the car. If hail fell instead of rain, would the force on the roof be **(a)** smaller than, **(b)** equal to, or **(c)** greater than that calculated in Example 3?

Reasoning   The raindrops and the hailstones fall in the same manner. That is, both fall with the same initial velocity and mass rate, and both strike the roof perpendicularly. However, there is an important difference: the raindrops come to rest (see **Figure 7.5**) after striking the roof, whereas the hailstones bounce upward (see **Figure 7.6**). According to the impulse–momentum theorem (Equation 7.4), the impulse that acts on an object is equal to the change in the object's momentum. This change is $m\vec{v}_f - m\vec{v}_0 = m \, \Delta\vec{v}$ and is proportional to the change $\Delta\vec{v}$ in the velocity. For a raindrop, the

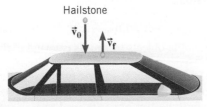

Hailstone

FIGURE 7.6   Hailstones have a downward velocity of $\vec{v}_0$ just before striking this car roof. They rebound with an upward velocity of $\vec{v}_f$.

change in velocity is from $\vec{v}_0$ (downward) to zero. For a hailstone, the change is from $\vec{v}_0$ (downward) to $\vec{v}_f$ (upward). Thus, a raindrop and a hailstone experience different changes in velocity, and, hence, different changes in momentum and different impulses.

Answers (a) and (b) are incorrect.   The change $\Delta\vec{v}$ in the velocity of a raindrop is smaller than that of a hailstone, since a raindrop does not rebound after striking the roof. According to the impulse-momentum theorem, a smaller impulse acts on a raindrop. But impulse is the product of the average force and the time interval Δt. Since the same amount of mass falls in the same time interval in either case, Δt is the same for both a raindrop and a hailstone. The smaller impulse acting on a raindrop, then, means that the car roof exerts a smaller force on it. By Newton's third law, this implies that a raindrop, not a hailstone, exerts a smaller force on the roof.

Answer (c) is correct.   A hailstone experiences a larger change in momentum than does a raindrop, since a hailstone rebounds after striking the roof. Therefore, according to the impulse–momentum theorem, a greater impulse acts on a hailstone. An impulse is the product of the average force and the time interval Δt. Since the same amount of mass falls in the same time interval in either case, Δt is the same for a hailstone as for a raindrop. The greater impulse acting on a hailstone means that the car roof exerts a greater force on a hailstone than on a raindrop. According to Newton's action–reaction law (see Section 4.5) then, the car roof experiences a greater force from the hail than from the rain.

Related Homework: Problems 3, 7

Check Your Understanding

(The answers are given at the end of the book.)

1. Two identical automobiles have the same speed, one traveling east and one traveling west. Do these cars have the same momentum?

2. In Times Square in New York City, people celebrate on New Year's Eve. Some just stand around, but many move about randomly. Consider a group consisting of all of these people. Approximately, what is the total linear momentum of this group at any given instant?

3. Two objects have the same momentum. Do the velocities of these objects necessarily have **(a)** the same directions and **(b)** the same magnitudes?

4. (a) Can a single object have a kinetic energy but no momentum? **(b)** Can a group of two or more objects have a total kinetic energy that is not zero but a total momentum that is zero?

5. Suppose you are standing on the edge of a dock and jump straight down. If you land on sand your stopping time is much shorter than if you land on water. Using the impulse–momentum theorem as a guide, determine which one of the following statements is correct. **(a)** In bringing you to a halt, the sand exerts a greater impulse on you than does the water. **(b)** In bringing you to a halt, the sand and the water exert the same impulse on you, but the sand exerts a greater average force. **(c)** In bringing you to a halt, the sand and the water exert the same impulse on you, but the sand exerts a smaller average force.

6. An airplane is flying horizontally with a constant momentum during a time interval Δt. **(a)** Is there a *net* impulse acting on the plane during this time? Use the impulse–momentum theorem to guide your thinking. **(b)** In the horizontal direction, both the thrust generated by the engines and air resistance act on the plane. Considering your answer to part (a), how is the impulse of the thrust related (in magnitude and direction) to the impulse of the force due to the air resistance?

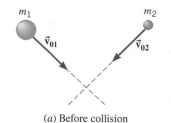

(a) Before collision

(b) During collision

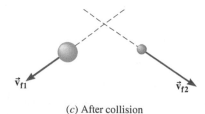

(c) After collision

INTERACTIVE FIGURE 7.7 *(a)* The velocities of the two objects before the collision are $\vec{\mathbf{v}}_{01}$ and $\vec{\mathbf{v}}_{02}$. *(b)* During the collision, each object exerts a force on the other. These forces are $\vec{\mathbf{F}}_{12}$ and $\vec{\mathbf{F}}_{21}$. *(c)* The velocities after the collision are $\vec{\mathbf{v}}_{f1}$ and $\vec{\mathbf{v}}_{f2}$.

7.2 | The Principle of Conservation of Linear Momentum

It is worthwhile to compare the impulse–momentum theorem to the work–energy theorem discussed in Chapter 6. The impulse–momentum theorem states that the impulse produced by a net force is equal to the change in the object's momentum, whereas the work–energy theorem states that the work done by a net force is equal to the change in the object's kinetic energy. The work–energy theorem leads directly to the principle of conservation of mechanical energy (see Section 6.5), and, as we will shortly see, the impulse–momentum theorem also leads to a conservation principle, known as the conservation of linear momentum.

We begin by applying the impulse–momentum theorem to a midair collision between two objects. The two objects (masses m_1 and m_2) are approaching each other with initial velocities $\vec{\mathbf{v}}_{01}$ and $\vec{\mathbf{v}}_{02}$, as **Interactive Figure 7.7a** shows. The collection of objects being studied is referred to as the "system." In this case, the system contains only the two objects. They interact during the collision in part *b* and then depart with the final velocities $\vec{\mathbf{v}}_{f1}$ and $\vec{\mathbf{v}}_{f2}$ shown in part *c*. Because of the collision, the initial and final velocities are not the same.

Two types of forces act on the system:

1. *Internal forces:* Forces that the objects within the system exert on each other.

2. *External forces:* Forces exerted on the objects by agents external to the system.

During the collision in **Interactive Figure 7.7b**, $\vec{\mathbf{F}}_{12}$ is the force exerted on object 1 by object 2, while $\vec{\mathbf{F}}_{21}$ is the force exerted on object 2 by object 1. These forces are action–reaction forces that are equal in magnitude but opposite in direction, so $\vec{\mathbf{F}}_{12} = -\vec{\mathbf{F}}_{21}$. They are internal forces, since they are forces that the two objects within the system exert on each other. The force of gravity also acts on the objects, their weights being $\vec{\mathbf{W}}_1$ and $\vec{\mathbf{W}}_2$. These weights, however, are external forces, because they are applied by the earth, which is outside the system. Friction and air resistance would also be considered external forces, although these forces are ignored here for the sake of simplicity. The impulse–momentum theorem, as applied to each object, gives the following results:

Object 1
$$(\underbrace{\vec{\mathbf{W}}_1}_{\substack{\text{External}\\\text{force}}} + \underbrace{\vec{\mathbf{F}}_{12}}_{\substack{\text{Internal}\\\text{force}}})\,\Delta t = m_1\vec{\mathbf{v}}_{f1} - m_1\vec{\mathbf{v}}_{01}$$

Object 2
$$(\underbrace{\vec{\mathbf{W}}_2}_{\substack{\text{External}\\\text{force}}} + \underbrace{\vec{\mathbf{F}}_{21}}_{\substack{\text{Internal}\\\text{force}}})\,\Delta t = m_2\vec{\mathbf{v}}_{f2} - m_2\vec{\mathbf{v}}_{02}$$

Adding these equations produces a single result for the system as a whole:

$$(\underbrace{\vec{\mathbf{W}}_1 + \vec{\mathbf{W}}_2}_{\substack{\text{External}\\\text{forces}}} + \underbrace{\vec{\mathbf{F}}_{12} + \vec{\mathbf{F}}_{21}}_{\substack{\text{Internal}\\\text{forces}}})\,\Delta t = \underbrace{(m_1\vec{\mathbf{v}}_{f1} + m_2\vec{\mathbf{v}}_{f2})}_{\substack{\text{Total final}\\\text{momentum }\vec{\mathbf{P}}_f}} - \underbrace{(m_1\vec{\mathbf{v}}_{01} + m_2\vec{\mathbf{v}}_{02})}_{\substack{\text{Total initial}\\\text{momentum }\vec{\mathbf{P}}_0}}$$

On the right side of this equation, the quantity $m_1\vec{\mathbf{v}}_{f1} + m_2\vec{\mathbf{v}}_{f2}$ is the vector sum of the final momenta for each object, or the total final momentum $\vec{\mathbf{P}}_f$ of the system. Likewise, the quantity $m_1\vec{\mathbf{v}}_{01} + m_2\vec{\mathbf{v}}_{02}$ is the total initial momentum $\vec{\mathbf{P}}_0$. Therefore, the result above becomes

$$\left(\begin{array}{c}\textbf{Sum of average}\\\textbf{external forces}\end{array} + \begin{array}{c}\textbf{Sum of average}\\\textbf{internal forces}\end{array}\right)\Delta t = \vec{\mathbf{P}}_f - \vec{\mathbf{P}}_0 \qquad (7.5)$$

The advantage of the internal/external force classification is that the internal forces always add together to give zero, as a consequence of Newton's law of action–reaction; $\vec{\mathbf{F}}_{12} = -\vec{\mathbf{F}}_{21}$ so that $\vec{\mathbf{F}}_{12} + \vec{\mathbf{F}}_{21} = 0$. Cancellation of the internal forces occurs no matter how many parts there are to the system and allows us to ignore the internal forces, as Equation 7.6 indicates:

$$\textbf{(Sum of average external forces) } \Delta t = \vec{\mathbf{P}}_f - \vec{\mathbf{P}}_0 \qquad (7.6)$$

We developed this result with gravity as the only external force. But, in general, the sum of the external forces on the left includes *all* external forces.

With the aid of Equation 7.6, it is possible to see how the conservation of linear momentum arises. Suppose that the sum of the external forces is zero. A system for which this is true is called an **isolated system**. Then Equation 7.6 indicates that

$$0 = \vec{\mathbf{P}}_f - \vec{\mathbf{P}}_0 \quad \text{or} \quad \vec{\mathbf{P}}_f = \vec{\mathbf{P}}_0 \qquad (7.7a)$$

Thus, the final total momentum of the isolated system after the objects in **Interactive Figure 7.7** collide is the same as the initial total momentum.* Explicitly writing out the final and initial momenta for the two-body collision, we obtain from Equation 7.7a that

$$\underbrace{m_1\vec{\mathbf{v}}_{f1} + m_2\vec{\mathbf{v}}_{f2}}_{\vec{\mathbf{P}}_f} = \underbrace{m_1\vec{\mathbf{v}}_{01} + m_2\vec{\mathbf{v}}_{02}}_{\vec{\mathbf{P}}_0} \qquad (7.7b)$$

This result is an example of a general principle known as the **principle of conservation of linear momentum**.

PRINCIPLE OF CONSERVATION OF LINEAR MOMENTUM

The total linear momentum of an isolated system remains constant (is conserved). An isolated system is one for which the vector sum of the average external forces acting on the system is zero.

---

*Technically, the initial and final momenta are equal when the impulse of the sum of the external forces is zero—that is, when the left-hand side of Equation 7.6 is zero. Sometimes, however, the initial and final momenta are very nearly equal even when the sum of the external forces is not zero. This occurs when the time Δt during which the forces act is so short that it is effectively zero. Then, the left-hand side of Equation 7.6 is approximately zero.

The conservation-of-momentum principle applies to a system containing any number of objects, regardless of the internal forces, provided the system is isolated. Whether the system is isolated depends on whether the vector sum of the external forces is zero. Judging whether a force is internal or external depends on which objects are included in the system, as Conceptual Example 5 illustrates.

CONCEPTUAL EXAMPLE 5 | Is the Total Momentum Conserved?

Imagine two balls colliding on a billiard table that is friction-free. Using the momentum-conservation principle as a guide, decide which statement is correct: **(a)** The total momentum of the system that contains only one of the two balls is the same before and after the collision. **(b)** The total momentum of the system that contains both of the two balls is the same before and after the collision.

Reasoning The total momentum of an isolated system is the same before and after the collision; in such a situation the total momentum is said to be conserved. An isolated system is one for which the vector sum of the average external forces acting on the system is zero. To decide whether statement (a) or (b) is correct, we need to examine the one-ball and two-ball systems and see if they are, in fact, isolated.

Answer (a) is incorrect. In **Figure 7.8a** only one ball is included in the system, as indicated by the rectangular dashed box. The forces acting on this system are all external and include the weight $\vec{W}_1$ of the ball and the normal force $\vec{F}_{N1}$ due to the table. Since the ball does not accelerate in the vertical direction, the normal force must balance the weight, so the vector sum of these two vertical forces is zero. However, there is a third external force to consider. Ball 2 is outside the system, so the force $\vec{F}_{12}$ that it applies to the system (ball 1) during the collision is an external force. As a result, the vector sum of the three external forces is not zero. Thus, the one-ball system is not an isolated system, and its momentum is not the same before and after the collision.

Answer (b) is correct. The rectangular dashed box in **Figure 7.8b** shows that both balls are included in the system. The collision forces are not shown, because they are internal forces and cannot cause the total momentum of the two-ball system to change. The external forces include the weights $\vec{W}_1$ and $\vec{W}_2$

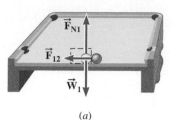

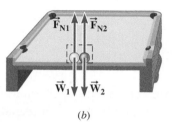

(b)

FIGURE 7.8 Two billiard balls collide on a pool table. (a) The rectangular dashed box emphasizes that only one of the balls (ball 1) is included in the system; $\vec{W}_1$ is its weight, $\vec{F}_{N1}$ is the normal force exerted on ball 1 by the pool table, and $\vec{F}_{12}$ is the force exerted on ball 1 by ball 2. (b) Now both balls are included in the system; $\vec{W}_2$ is the weight of ball 2, and $\vec{F}_{N2}$ is the normal force acting on it.

of the balls and the upward-pointing normal forces $\vec{F}_{N1}$ and $\vec{F}_{N2}$. Since the balls do not accelerate in the vertical direction, $\vec{F}_{N1}$ balances $\vec{W}_1$, and $\vec{F}_{N2}$ balances $\vec{W}_2$. Furthermore, the table is friction-free. Thus, there is no net external force to change the total momentum of the two-ball system, and, as a result, the total momentum is the same before and after the collision.

Next, we apply the principle of conservation of linear momentum to the problem of assembling a freight train.

EXAMPLE 6 | Assembling a Freight Train

A freight train is being assembled in a switching yard, and **Animated Figure 7.9** shows two boxcars. Car 1 has a mass of $m_1 = 65 \times 10^3$ kg and moves at a velocity of $v_{01} = +0.80$ m/s. Car 2, with a mass of $m_2 = 92 \times 10^3$ kg and a velocity of $v_{02} = +1.3$ m/s,

overtakes car 1 and couples to it. Neglecting friction, find the common velocity v_f of the cars after they become coupled.

Reasoning The two boxcars constitute the system. The sum of the external forces acting on the system is zero, because

(a) Before coupling

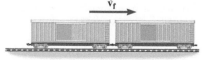

(b) After coupling

ANIMATED FIGURE 7.9 (a) The boxcar on the left eventually catches up with the other boxcar and (b) couples to it. The coupled cars move together with a common velocity after the collision.

the weight of each car is balanced by a corresponding normal force, and friction is being neglected. Thus, the system is isolated, and the principle of conservation of linear momentum applies. The coupling forces that each car exerts on the other are internal forces and do not affect the applicability of this principle.

> **Problem-Solving Insight** The conservation of linear momentum is applicable only when the net external force acting on the system is zero. Therefore, the first step in applying momentum conservation is to be sure that the net external force is zero.

Solution   Momentum conservation indicates that

$$\underbrace{(m_1 + m_2)v_f}_{\substack{\text{Total momentum} \\ \text{after collision}}} = \underbrace{m_1 v_{01} + m_2 v_{02}}_{\substack{\text{Total momentum} \\ \text{before collision}}}$$

This equation can be solved for v_f, the common velocity of the two cars after the collision:

$$v_f = \frac{m_1 v_{01} + m_2 v_{02}}{m_1 + m_2}$$

$$= \frac{(65 \times 10^3 \text{ kg})(0.80 \text{ m/s}) + (92 \times 10^3 \text{ kg})(1.3 \text{ m/s})}{(65 \times 10^3 \text{ kg} + 92 \times 10^3 \text{ kg})}$$

$$= \boxed{+1.1 \text{ m/s}}$$

In the previous example it can be seen that the velocity of car 1 increases, while the velocity of car 2 decreases as a result of the collision. The acceleration and deceleration arise at the moment the cars become coupled, because the cars exert internal forces on each other. The powerful feature of the momentum-conservation principle is that it allows us to determine the changes in velocity without knowing what the internal forces are. Example 7 further illustrates this feature.

> **Problem-Solving Insight**   It is important to realize that the total linear momentum may be conserved even when the kinetic energies of the individual parts of a system change.

EXAMPLE 7 | Ice Skaters

Starting from rest, two skaters push off against each other on smooth level ice, where friction is negligible. As **Interactive Figure 7.10a** shows, one is a woman ($m_1 = 54$ kg), and one is a man ($m_2 = 88$ kg). Part b of the drawing shows that the woman moves away with a velocity of $v_{f1} = +2.5$ m/s. Find the "recoil" velocity v_{f2} of the man.

Reasoning   For a system consisting of the two skaters on level ice, the sum of the external forces is zero. This is because the weight of each skater is balanced by a corresponding normal force and friction is negligible. The skaters, then, constitute an isolated system, and the principle of conservation of linear momentum applies. We expect the man to have a smaller recoil speed for the following reason: The internal forces that the man and woman exert on each other during pushoff have equal magnitudes but opposite directions, according to Newton's action–reaction law. The man, having the larger mass, experiences a smaller acceleration according to Newton's second law. Hence, he acquires a smaller recoil speed.

Solution   The total momentum of the skaters before they push on each other is zero, since they are at rest. Momentum conservation requires that the total momentum remains zero after the skaters have separated, as in **Interactive Figure 7.10b**:

$$\underbrace{m_1 v_{f1} + m_2 v_{f2}}_{\substack{\text{Total momentum} \\ \text{after pushing}}} = \underbrace{0}_{\substack{\text{Total momentum} \\ \text{before pushing}}}$$

Solving for the recoil velocity of the man gives

$$v_{f2} = \frac{-m_1 v_{f1}}{m_2} = \frac{-(54 \text{ kg})(+2.5 \text{ m/s})}{88 \text{ kg}} = \boxed{-1.5 \text{ m/s}}$$

The minus sign indicates that the man moves to the left in the drawing. After the skaters separate, the total momentum of the system remains zero, because momentum is a vector quantity, and the momenta of the man and the woman have equal magnitudes but opposite directions.

> **Math Skills**   When you use the momentum-conservation principle, you must choose which direction to call positive. The choice is arbitrary, but it is important. *The results of using the principle can only be interpreted with respect to your choice of the positive direction.* In Example 7, the direction to the right has been selected as the positive direction (see **Interactive Figure 7.10**), with the result that the velocity v_{f1} of the woman is given as $v_{f1} = +2.5$ m/s, and the answer for the velocity v_{f2} of the man is $v_{f2} = -1.5$ m/s. This answer means that the man moves off in the negative direction, which is to the left in the drawing. Suppose, however, that the positive direction in **Interactive Figure 7.10** had been chosen to be to the left. Then, the velocity v_{f1} of the woman would have been given as $v_{f1} = -2.5$ m/s, and the calculation for the velocity of the man would have revealed that the answer for v_{f2} is
>
> $$v_{f2} = \frac{-m_1 v_{f1}}{m_2} = \frac{-(54 \text{ kg})(-2.5 \text{ m/s})}{88 \text{ kg}} = +1.5 \text{ m/s}$$
>
> This answer means that the man moves off in the positive direction, which is now the direction to the left. Thus, the momentum-conservation principle leads to the same conclusion for the man. He moves off to the left, irrespective of the arbitrary choice of the positive direction.

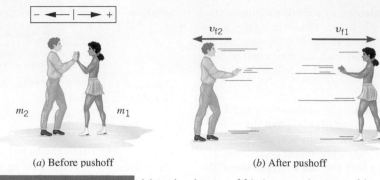

(a) Before pushoff   (b) After pushoff

INTERACTIVE FIGURE 7.10 (a) In the absence of friction, two skaters pushing on each other constitute an isolated system. (b) As the skaters move away, the total linear momentum of the system remains zero, which is what it was initially.

In Example 7, for instance, the initial kinetic energy is zero, since the skaters are stationary. But after they push off, the skaters are moving, so each has kinetic energy. The kinetic energy changes because work is done by the internal force that each skater exerts on the other. However, internal forces cannot change the total linear momentum of a system, since the total linear momentum of an isolated system is conserved in the presence of such forces.

When applying the principle of conservation of linear momentum, we have been following a definite reasoning strategy that is summarized as follows:

REASONING STRATEGY   Applying the Principle of Conservation of Linear Momentum

1. Decide which objects are included in the system.

2. Relative to the system that you have chosen, identify the internal forces and the external forces.

3. Verify that the system is isolated. In other words, verify that the sum of the external forces applied to the system is zero. Only if this sum is zero can the conservation principle be applied. If the sum of the average external forces is not zero, consider a different system for analysis.

4. Set the total final momentum of the isolated system equal to the total initial momentum. Remember that linear momentum is a vector. If necessary, apply the conservation principle separately to the various vector components.

Check Your Understanding

(*The answers are given at the end of the book.*)

7. An object slides along the surface of the earth and slows down because of kinetic friction. If the object alone is considered as the system, the kinetic frictional force must be identified as an external force that, according to Equation 7.4, decreases the momentum of the system. **(a)** If *both* the object and the earth are considered to be the system, is the force of kinetic friction still an external force? **(b)** Can the frictional force change the total linear momentum of the two-body system?

8. A satellite explodes in outer space, far from any other body, sending thousands of pieces in all directions. Is the linear momentum of the satellite before the explosion less than, equal to, or greater than the total linear momentum of all the pieces after the explosion?

9. On a distant asteroid, a large catapult is used to throw chunks of stone into space. Could such a device be used as a propulsion system to move the asteroid closer to the earth?

10. A canoe with two people aboard is coasting with an initial momentum of +110 kg · m/s. Then, one of the people (person 1) dives off the back of the canoe. During this time, the net average external force acting on the system (the canoe and the two people) is zero. The table lists four possibilities for the final momentum of person 1 and the final momentum of person 2 plus the canoe, immediately after person 1 dives off. Only one possibility could be correct. Which one is it?

	Final Momenta	
	Person 1	**Person 2 and Canoe**
(a)	$-60 \text{ kg} \cdot \text{m/s}$	$+170 \text{ kg} \cdot \text{m/s}$
(b)	$-30 \text{ kg} \cdot \text{m/s}$	$+110 \text{ kg} \cdot \text{m/s}$
(c)	$-40 \text{ kg} \cdot \text{m/s}$	$-70 \text{ kg} \cdot \text{m/s}$
(d)	$+80 \text{ kg} \cdot \text{m/s}$	$-30 \text{ kg} \cdot \text{m/s}$

11. You are a passenger on a jetliner that is flying at a constant velocity. You get up from your seat and walk toward the front of the plane. Because of this action, your forward momentum increases. Does the forward momentum of the plane itself decrease, remain the same, or increase?

12. An ice boat is coasting on a frozen lake. Friction between the ice and the boat is negligible, and so is air resistance. Nothing is propelling the boat. From a bridge someone jumps straight down into the boat, which continues to coast straight ahead. **(a)** Does the total horizontal momentum of the boat plus the jumper change? **(b)** Does the speed of the boat itself increase, decrease, or remain the same?

13. **Concept Simulation 7.1** at **www.wiley.com/college/cutnell** reviews the concepts that are pertinent in this question. In movies, Superman hovers in midair, grabs a villain by the neck, and throws him forward. Superman, however, remains stationary. This is not possible, because it violates which one or more of the following: **(a)** The law of conservation of energy **(b)** Newton's second law **(c)** Newton's third law **(d)** The principle of conservation of linear momentum

14. The energy released by the exploding gunpowder in a cannon propels the cannonball forward. Simultaneously, the cannon recoils. The mass of the cannonball is less than that of the cannon. Which has the greater kinetic energy, the launched cannonball or the recoiling cannon? Assume that momentum conservation applies.

7.3 Collisions in One Dimension

As discussed in the previous section, the total linear momentum is conserved when two objects collide, provided they constitute an isolated system. When the objects are atoms or subatomic particles, the total kinetic energy of the system is often conserved also. In other words, the total kinetic energy of the particles before the collision equals the total kinetic energy of the particles after the collision, so that kinetic energy gained by one particle is lost by another.

In contrast, when two macroscopic objects collide, such as two cars, the total kinetic energy after the collision is generally less than that before the collision. Kinetic energy is lost mainly in two ways: (1) It can be converted into heat because of friction, and (2) it is spent in creating permanent distortion or damage, as in an automobile collision. With very hard objects, such as a solid steel ball and a marble floor, the permanent distortion suffered upon collision is much smaller than with softer objects and, consequently, less kinetic energy is lost.

Collisions are often classified according to whether the total kinetic energy changes during the collision:

1. *Elastic collision:* One in which the total kinetic energy of the system after the collision is equal to the total kinetic energy before the collision.

2. *Inelastic collision:* One in which the total kinetic energy of the system is *not* the same before and after the collision; if the objects stick together after colliding, the collision is said to be *completely inelastic.*

> **Problem-Solving Insight** As long as the net external force is zero, the conservation of linear momentum applies to any type of collision. This is true whether the collision is elastic or inelastic.

The boxcars coupling together in **Animated Figure 7.9** provide an example of a completely inelastic collision. When a collision is completely inelastic, the greatest amount of kinetic energy is lost. Example 8 shows how one particular elastic collision is described using the conservation of linear momentum and the fact that no kinetic energy is lost.

Analyzing Multiple-Concept Problems

EXAMPLE 8 | A Head-On Collision

Figure 7.11 illustrates an elastic head-on collision between two balls. One ball has a mass of $m_1 = 0.250$ kg and an initial velocity of +5.00 m/s. The other has a mass of $m_2 = 0.800$ kg and is initially at rest. No external forces act on the balls. What are the velocities of the balls after the collision?

Reasoning Three facts will guide our solution. The first is that the collision is elastic, so that the kinetic energy of the two-ball system is the same before and after the balls collide. The second fact is that the collision occurs head-on. This means that the velocities before and after the balls collide all point along the same line. In other words, the collision occurs in one dimension. Last, no external forces act on the balls, with the result that the two-ball system is isolated and its total linear momentum is conserved. We expect that ball 1, having the smaller mass, will rebound to the left after striking ball 2, which is more massive. Ball 2 will be driven to the right in the process.

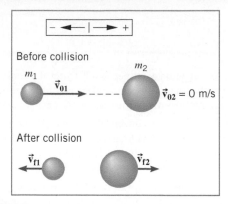

FIGURE 7.11 A 0.250-kg ball, traveling with an initial velocity of $v_{01} = +5.00$ m/s, undergoes an elastic collision with a 0.800-kg ball that is initially at rest.

Knowns and Unknowns The data for this problem are given in the following table:

Description	Symbol	Value	Comment
Explicit Data			
Mass of ball 1	m_1	0.250 kg	
Initial velocity of ball 1	v_{01}	+5.00 m/s	Before collision occurs.
Mass of ball 2	m_2	0.800 kg	
Implicit Data			
Initial velocity of ball 2	v_{02}	0 m/s	Before collision occurs, ball is initially at rest.
Unknown Variables			
Final velocity of ball 1	v_{f1}	?	After collision occurs.
Final velocity of ball 2	v_{f2}	?	After collision occurs.

Modeling the Problem

STEP 1 Elastic Collision Since the collision is elastic, the kinetic energy of the two-ball system is the same before and after the balls collide:

$$\underbrace{\tfrac{1}{2}m_1 v_{f1}^2 + \tfrac{1}{2}m_2 v_{f2}^2}_{\substack{\text{Total kinetic energy} \\ \text{after collision}}} = \underbrace{\tfrac{1}{2}m_1 v_{01}^2 + 0}_{\substack{\text{Total kinetic energy} \\ \text{before collision}}}$$

$$v_{f1}^2 = v_{01}^2 - \frac{m_2}{m_1} v_{f2}^2 \qquad (1)$$

Here we have utilized the fact that ball 2 is at rest before the collision. Thus, its initial velocity v_{02} is zero and so is its initial kinetic energy. Solving this equation for v_{f1}^2, the square of the velocity of ball 1 after the collision, gives Equation 1 at the right. To use this result, we need a value for v_{f2}, which we will obtain in Step 2.

STEP 2 Conservation of Linear Momentum Total linear momentum of the two-ball system is conserved since no external forces act on the system. This fact does not depend on whether the collision is elastic. Conservation of linear momentum indicates that

$$\underbrace{m_1 v_{f1} + m_2 v_{f2}}_{\substack{\text{Total momentum} \\ \text{after collision}}} = \underbrace{m_1 v_{01} + 0}_{\substack{\text{Total momentum} \\ \text{before collision}}}$$

We have again utilized the fact that ball 2 is at rest before the collision; since its initial velocity v_{02} is zero, so is its initial momentum. Solving the expression above for v_{f2} gives

$$v_{f2} = \frac{m_1}{m_2}(v_{01} - v_{f1})$$

which can be substituted into Equation 1, as shown at the right.

$$v_{f1}^2 = v_{01}^2 - \frac{m_2}{m_1}v_{f2}^2 \qquad \text{(1)}$$

$$v_{f2} = \frac{m_1}{m_2}(v_{01} - v_{f1}) \qquad \text{(2)}$$

Solution   Algebraically combining the results of each step, we find that

$$\overset{\textbf{STEP 1}}{\underset{\downarrow}{}} \qquad \overset{\textbf{STEP 2}}{\underset{\downarrow}{}}$$

$$v_{f1}^2 = v_{01}^2 - \frac{m_2}{m_1}v_{f2}^2 = v_{01}^2 - \frac{m_2}{m_1}\left[\frac{m_1}{m_2}(v_{01} - v_{f1})\right]^2$$

Solving for v_{f1} shows that

$$v_{f1} = \left(\frac{m_1 - m_2}{m_1 + m_2}\right)v_{01} = \left(\frac{0.250\text{ kg} - 0.800\text{ kg}}{0.250\text{ kg} + 0.800\text{ kg}}\right)(+5.00\text{ m/s}) = \boxed{-2.62\text{ m/s}} \qquad \textbf{(7.8a)}$$

Math Skills   The equation $v_{f1}^2 = v_{01}^2 - \frac{m_2}{m_1}\left[\frac{m_1}{m_2}(v_{01} - v_{f1})\right]^2$ is a quadratic equation, because it contains the square of the unknown variable v_{f1}. Such equations can always be solved using the quadratic formula (see Appendix C.4, Equation C-2), although it is often awkward and time consuming to do so. A more sensible approach in this case is to begin by rearranging the equation as follows:

$$v_{f1}^2 = v_{01}^2 - \frac{m_2}{m_1}\frac{m_1}{m_2}\frac{m_1}{m_2}(v_{01} - v_{f1})^2 \quad \text{or} \quad v_{f1}^2 = v_{01}^2 - \frac{m_1}{m_2}(v_{01} - v_{f1})^2$$

Further rearrangement and use of the fact that $v_{01}^2 - v_{f1}^2 = (v_{01} - v_{f1})(v_{01} + v_{f1})$ produces

$$v_{01}^2 - v_{f1}^2 = \frac{m_1}{m_2}(v_{01} - v_{f1})^2 \quad \text{or} \quad (\cancel{v_{01} - v_{f1}})(v_{01} + v_{f1}) = \frac{m_1}{m_2}(\cancel{v_{01} - v_{f1}})(v_{01} - v_{f1})$$

This result is equivalent to

$$v_{01} + v_{f1} = \frac{m_1}{m_2}v_{01} - \frac{m_1}{m_2}v_{f1} \quad \text{or} \quad \left(\frac{m_1}{m_2} + 1\right)v_{f1} = \left(\frac{m_1}{m_2} - 1\right)v_{01}$$

Thus, we find that

$$v_{f1} = \frac{\left(\dfrac{m_1}{m_2} - 1\right)}{\left(\dfrac{m_1}{m_2} + 1\right)}v_{01} \quad \text{or} \quad v_{f1} = \left(\frac{m_1 - m_2}{m_1 + m_2}\right)v_{01}$$

Substituting the expression for v_{f1} into Equation 2 gives

$$v_{f2} = \left(\frac{2m_1}{m_1 + m_2}\right)v_{01} = \left[\frac{2(0.250\text{ kg})}{0.250\text{ kg} + 0.800\text{ kg}}\right](+5.00\text{ m/s}) = \boxed{+2.38\text{ m/s}} \qquad \textbf{(7.8b)}$$

The negative value for v_{f1} indicates that ball 1 rebounds to the left after the collision in **Figure 7.11**, while the positive value for v_{f2} indicates that ball 2 moves to the right, as expected.

Related Homework: Problem 35

We can get a feel for an elastic collision by dropping a steel ball onto a hard surface, such as a marble floor. If the collision were elastic, the ball would rebound to its original height, as **Figure 7.12a** illustrates. In contrast, a partially deflated basketball exhibits little rebound from a relatively soft asphalt surface, as in part *b*, indicating that a fraction of the ball's kinetic energy is dissipated during the inelastic collision. The very deflated basketball in part *c* has no bounce at all, and a maximum amount of kinetic energy is lost during the completely inelastic collision.

The next example illustrates a completely inelastic collision in a device called a "ballistic pendulum." This device can be used to measure the speed of a bullet.

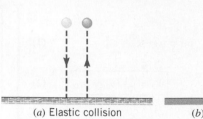

(a) Elastic collision

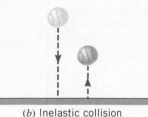

(b) Inelastic collision

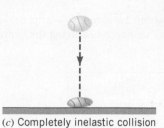

(c) Completely inelastic collision

FIGURE 7.12 (a) A hard steel ball would rebound to its original height after striking a hard marble surface if the collision were elastic. (b) A partially deflated basketball has little bounce on a soft asphalt surface. (c) A very deflated basketball has no bounce at all.

Analyzing Multiple-Concept Problems

EXAMPLE 9 | The Physics of Measuring the Speed of a Bullet

A ballistic pendulum can be used to measure the speed of a projectile, such as a bullet. The ballistic pendulum shown in **Figure 7.13a** consists of a stationary 2.50-kg block of wood suspended by a wire of negligible mass. A 0.0100-kg bullet is fired into the block, and the block (with the bullet in it) swings to a maximum height of 0.650 m above the initial position (see part *b* of the drawing). Find the speed with which the bullet is fired, assuming that air resistance is negligible.

Reasoning The physics of the ballistic pendulum can be divided into two parts. The first is the completely inelastic collision between the bullet and the block. The total linear momentum of the system (block plus bullet) is conserved during the collision, because the suspension wire supports the system's weight, which means that the sum of the external forces acting on the system is nearly zero. The second part of the physics is the resulting motion of the block and bullet as they swing upward after the collision. As the system swings upward, the principle of conservation of mechanical energy applies, since nonconservative forces do no work (see Section 6.5). The tension force in the wire does no work because it acts perpendicular to the motion. Since air resistance is negligible, we can ignore the work it does. The conservation principles for linear momentum and mechanical energy provide the basis for our solution.

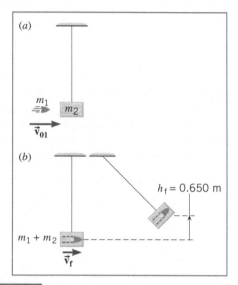

FIGURE 7.13 (a) A bullet approaches a ballistic pendulum. (b) The block and bullet swing upward after the collision.

Knowns and Unknowns The data for this problem are as follows:

Description	Symbol	Value	Comment
Explicit Data			
Mass of bullet	m_1	0.0100 kg	
Mass of block	m_2	2.50 kg	
Height to which block plus bullet swings	h_f	0.650 m	Maximum height of swing.
Implicit Data			
Initial velocity of block	v_{02}	0 m/s	Before collision, block is stationary.
Unknown Variable			
Speed with which bullet is fired	v_{01}	?	Before collision with block.

Modeling the Problem

STEP 1 Completely Inelastic Collision Just after the bullet collides with it, the block (with the bullet in it) has a speed v_f. Since linear momentum is conserved, the total momentum of the block–bullet system after the collision is the same as it is before the collision:

$$\underbrace{(m_1 + m_2)v_f}_{\substack{\text{Total momentum} \\ \text{after collision}}} = \underbrace{m_1 v_{01} + 0}_{\substack{\text{Total momentum} \\ \text{before collision}}}$$

$$v_{01} = \left(\frac{m_1 + m_2}{m_1}\right)v_f \qquad (1)$$

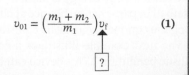

Note that the block is at rest before the collision ($v_{02} = 0$ m/s), so its initial momentum is zero. Solving this equation for v_{01} gives Equation 1 at the right. To find a value for the speed v_f in this equation, we turn to Step 2.

STEP 2 Conservation of Mechanical Energy The speed v_f immediately after the collision can be obtained from the maximum height h_f to which the system swings, by using the principle of conservation of mechanical energy:

$$\underbrace{(m_1 + m_2)gh_f}_{\substack{\text{Total mechanical energy} \\ \text{at top of swing, all} \\ \text{potential}}} = \underbrace{\tfrac{1}{2}(m_1 + m_2)v_f^2}_{\substack{\text{Total mechanical energy} \\ \text{at bottom of swing, all} \\ \text{kinetic}}}$$

This result can be solved for v_f to show that

$$\boxed{v_f = \sqrt{2gh_f}}$$

which can be substituted into Equation 1, as shown at the right. In applying the energy-conservation principle, it is tempting to say that the total potential energy at the top of the swing is equal to the total kinetic energy of the bullet before it strikes the block $[(m_1 + m_2)gh_f = \tfrac{1}{2}m_1v_{01}^2]$ and solve directly for v_{01}. This is incorrect, however, because the collision between the bullet and the block is inelastic, so that some of the bullet's initial kinetic energy is dissipated during the collision (due to friction and structural damage to the block and bullet).

$$v_{01} = \left(\frac{m_1 + m_2}{m_1}\right)v_f \qquad (1)$$

$$\boxed{v_f = \sqrt{2gh_f}}$$

Solution Algebraically combining the results of each step, we find that

$$\overset{\textbf{STEP 1}}{\underset{\downarrow}{}} \qquad \overset{\textbf{STEP 2}}{\underset{\downarrow}{}}$$

$$v_{01} = \left(\frac{m_1 + m_2}{m_1}\right)v_f = \left(\frac{m_1 + m_2}{m_1}\right)\sqrt{2gh_f}$$

$$v_{01} = \left(\frac{0.0100\text{ kg} + 2.50\text{ kg}}{0.0100\text{ kg}}\right)\sqrt{2(9.80\text{ m/s}^2)(0.650\text{ m})} = \boxed{+896\text{ m/s}}$$

Related Homework: Problem 27

Check Your Understanding

(The answers are given at the end of the book.)

15. Two balls collide in a one-dimensional elastic collision. The two balls constitute a system, and the net external force acting on them is zero. The table shows four possible sets of values for the initial and final momenta of the two balls, as well as their initial and final kinetic energies. Only one set of values could be correct. Which set is it?

		Initial (Before Collision)		Final (After Collision)	
		Momentum	Kinetic Energy	Momentum	Kinetic Energy
(a)	Ball 1:	+4 kg · m/s	12 J	−5 kg · m/s	10 J
	Ball 2:	−3 kg · m/s	5 J	−1 kg · m/s	7 J
(b)	Ball 1:	+7 kg · m/s	22 J	+5 kg · m/s	18 J
	Ball 2:	+2 kg · m/s	8 J	+4 kg · m/s	15 J
(c)	Ball 1:	−5 kg · m/s	12 J	−6 kg · m/s	15 J
	Ball 2:	−8 kg · m/s	31 J	−9 kg · m/s	25 J
(d)	Ball 1:	+9 kg · m/s	25 J	+6 kg · m/s	18 J
	Ball 2:	+4 kg · m/s	15 J	+7 kg · m/s	22 J

16. In an elastic collision, is the kinetic energy of *each* object the same before and after the collision?

17. **Concept Simulation 7.2** at **www.wiley.com/college/cutnell** illustrates the concepts that are involved in this question. Also review Multiple-Concept Example 8. Suppose two objects collide head on, as in Example 8, where initially object 1 (mass = m_1) is moving and object 2 (mass = m_2) is stationary. Now assume that they have the same mass, so $m_1 = m_2$. Which one of the following statements is true? **(a)** Both objects have the same velocity (magnitude and direction) after the collision. **(b)** Object 1 rebounds with one-half its initial speed, while object 2 moves to the right, as in **Figure 7.11**, with one-half the speed that object 1 had before the collision. **(c)** Object 1 stops completely, while object 2 acquires the same velocity (magnitude and direction) that object 1 had before the collision.

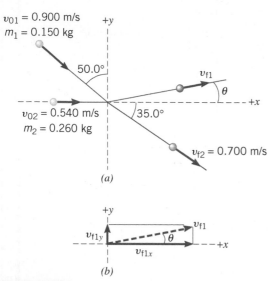

FIGURE 7.14 (*a*) Top view of two balls colliding on a horizontal frictionless table. (*b*) This part of the drawing shows the *x* and *y* components of the velocity of ball 1 after the collision.

7.4 | Collisions in Two Dimensions

The collisions discussed so far have been head-on, or one-dimensional, because the velocities of the objects all point along a single line before and after contact. Collisions often occur, however, in two or three dimensions. **Figure 7.14** shows a two-dimensional case in which two balls collide on a horizontal frictionless table.

For the system consisting of the two balls, the external forces include the weights of the balls and the corresponding normal forces produced by the table. Since each weight is balanced by a normal force, the sum of the external forces is zero, and the total momentum of the system is conserved, as Equation 7.7b indicates. Momentum is a vector quantity, however, and in two dimensions the *x* and *y* components of the total momentum are conserved separately. In other words, Equation 7.7b is equivalent to the following two equations:

x Component
$$\underbrace{m_1 v_{f1x} + m_2 v_{f2x}}_{P_{fx}} = \underbrace{m_1 v_{01x} + m_2 v_{02x}}_{P_{0x}} \qquad (7.9a)$$

y Component
$$\underbrace{m_1 v_{f1y} + m_2 v_{f2y}}_{P_{fy}} = \underbrace{m_1 v_{01y} + m_2 v_{02y}}_{P_{0y}} \qquad (7.9b)$$

These equations are written for a system that contains two objects. If a system contains more than two objects, a mass-times-velocity term must be included for each additional object on either side of Equations 7.9a and 7.9b.

7.5 | Center of Mass

In previous sections, we have encountered situations in which objects interact with one another, such as the two skaters pushing off in Example 7. In these situations, the mass of the system is located in several places, and the various objects move relative to each other before, after, and even during the interaction. It is possible, however, to speak of a kind of average location for the total mass by introducing a concept known as the **center of mass** (abbreviated as "cm"). With the aid of this concept, we will be able to gain additional insight into the principle of conservation of linear momentum.

The center of mass is a point that represents the average location for the total mass of a system. **Figure 7.15**, for example, shows two particles of mass m_1 and m_2 that are located on the *x* axis at the positions x_1 and x_2, respectively. The position x_{cm} of the center-of-mass point from the origin is defined to be

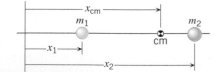

FIGURE 7.15 The center of mass cm of the two particles is located on a line between them and lies closer to the more massive particle.

Center of mass
$$x_{cm} = \frac{m_1 x_1 + m_2 x_2}{m_1 + m_2} \qquad (7.10)$$

Each term in the numerator of this equation is the product of a particle's mass and position, while the denominator is the total mass of the system. If the two masses are equal, we expect the average location of the total mass to be midway between the particles. With $m_1 = m_2 = m$, Equation 7.10 becomes $x_{cm} = (mx_1 + mx_2)/(m + m) = \frac{1}{2}(x_1 + x_2)$, which indeed corresponds to the point midway between the particles. Alternatively, suppose that $m_1 = 5.0$ kg and $x_1 = 2.0$ m, while $m_2 = 12$ kg and $x_2 = 6.0$ m. Then we expect the average location of the total mass to be located closer to particle 2, since it is more massive. Equation 7.10 is also consistent with this expectation, for it gives

$$x_{cm} = \frac{(5.0 \text{ kg})(2.0 \text{ m}) + (12 \text{ kg})(6.0 \text{ m})}{5.0 \text{ kg} + 12 \text{ kg}} = 4.8 \text{ m}$$

If a system contains more than two particles, the center-of-mass point can be determined by generalizing Equation 7.10. For three particles, for instance, the numerator would contain a third term m_3x_3, and the total mass in the denominator would be $m_1 + m_2 + m_3$. For a macroscopic object, which contains many, many particles, the center-of-mass point is located at the geometric center of the object, provided that the mass is distributed symmetrically about the center. Such would be the case for a billiard ball. For objects such as a golf club, the mass is not distributed symmetrically, and the center-of-mass point is not located at the geometric center of the club. The driver used to launch a golf ball from the tee, for instance, has more mass in the club head than in the handle, so the center-of-mass point is closer to the head than to the handle.

Equation 7.10 (and its generalization to more than two particles) deals with particles that lie along a straight line. A system of particles, however, may include particles that do not all lie along a single straight line. For particles lying in a plane, an equation like Equation 7.10 uses the x coordinates of each particle to give the x coordinate of the center of mass. A similar equation uses the y coordinates of each particle to give the y coordinate of the center of mass. If a system contains rigid objects, each of them may be treated as if it were a particle with all of the mass located at the object's own center of mass. In this way, a collection of rigid objects becomes a collection of particles, and the x and y coordinates of the center of mass of the collection can be determined as described earlier in this paragraph.

To see how the center-of-mass concept is related to momentum conservation, suppose that the two particles in a system are moving, as they would be during a collision. With the aid of Equation 7.10, we can determine the velocity v_{cm} of the center-of-mass point. During a time interval Δt, the particles experience displacements of Δx_1 and Δx_2, as **Interactive Figure 7.16** shows. They have different displacements during this time because they have different velocities. Equation 7.10 can be used to find the displacement Δx_{cm} of the center of mass by replacing x_{cm} by Δx_{cm}, x_1 by Δx_1, and x_2 by Δx_2:

$$\Delta x_{cm} = \frac{m_1 \Delta x_1 + m_2 \Delta x_2}{m_1 + m_2}$$

Now we divide both sides of this equation by the time interval Δt. In the limit as Δt becomes infinitesimally small, the ratio $\Delta x_{cm}/\Delta t$ becomes equal to the instantaneous velocity v_{cm} of the center of mass. (See Section 2.2 for a review of instantaneous velocity.) Likewise, the ratios $\Delta x_1/\Delta t$ and $\Delta x_2/\Delta t$ become equal to the instantaneous velocities v_1 and v_2, respectively. Thus, we have

Velocity of center of mass $\quad v_{cm} = \dfrac{m_1 v_1 + m_2 v_2}{m_1 + m_2}$ (7.11)

The numerator $(m_1 v_1 + m_2 v_2)$ on the right-hand side in Equation 7.11 is the momentum of particle 1 $(m_1 v_1)$ plus the momentum of particle 2 $(m_2 v_2)$, which is the total linear momentum of the system. In an isolated system, the total linear momentum does not change because of an interaction such as a collision. Therefore, Equation 7.11 indicates that the velocity v_{cm} of the center of mass does not change either. To emphasize this important point, consider the collision discussed in Example 8. With the data from that example, we can apply Equation 7.11 to determine the velocity of the center of mass before and after the collision:

Before collision

$$v_{cm} = \frac{(0.250 \text{ kg})(+5.00 \text{ m/s}) + (0.800 \text{ kg})(0 \text{ m/s})}{0.250 \text{ kg} + 0.800 \text{ kg}} = +1.19 \text{ m/s}$$

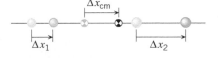

INTERACTIVE FIGURE 7.16 During a time interval Δt, the displacements of the particles are Δx_1 and Δx_2, while the displacement of the center of mass is Δx_{cm}.

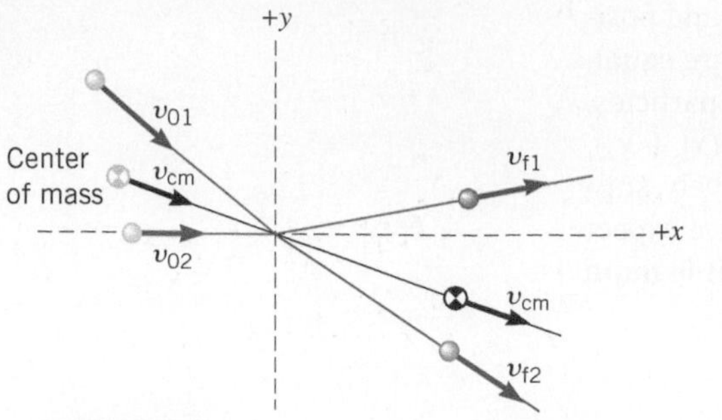

FIGURE 7.17 This drawing shows the two balls in Figure 7.14*a* and the path followed by the center-of-mass point as the balls approach the collision point and then move away from it. Because of momentum conservation, the velocity of the center of mass of the balls is the same before and after the collision (see the vectors labeled v_{cm}). As a result, the center of mass moves along the same straight-line path before and after the collision.

After collision

$$v_{cm} = \frac{(0.250 \text{ kg})(-2.62 \text{ m/s}) + (0.800 \text{ kg})(+2.38 \text{ m/s})}{0.250 \text{ kg} + 0.800 \text{ kg}} = +1.19 \text{ m/s}$$

Thus, the velocity of the center of mass is the same before and after the objects interact during a collision in which the total linear momentum is conserved.

In a two-dimensional collision, the velocity of the center of mass is also the same before and after the collision, provided that the total linear momentum is conserved. **Figure 7.17** illustrates this fact. The drawing reproduces **Figure 7.14*a***, except that the initial and final center-of-mass locations are indicated, along with the velocity vector for the center of mass at these two places (see the vectors labeled v_{cm}). As the drawing shows, this velocity vector is the same before and after the balls collide. Thus, the center of mass moves along a straight line as the two balls approach the collision point and continues along the same straight line following the collision.

Check Your Understanding

(The answers are given at the end of the book.)

18. Would you expect the center of mass of a baseball bat to be located halfway between the ends of the bat, nearer the lighter end, or nearer the heavier end?

19. A sunbather is lying on a floating raft that is stationary. She then gets up and walks to one end of the raft. Consider the sunbather and raft as an isolated system. **(a)** What is the velocity of the center of mass of this system while she is walking? **(b)** Does the raft itself move while she is walking? If so, what is the direction of the raft's velocity relative to that of the sunbather?

20. Water, dripping at a constant rate from a faucet, falls to the ground. At any instant there are many drops in the air between the faucet and the ground. Where does the center of mass of the drops lie relative to the halfway point between the faucet and the ground? **(a)** Above it **(b)** Below it **(c)** Exactly at the halfway point

EXAMPLE 10 | BIO The Physics of High Jumpers—The "Fosbury Flop"

Track athletes participating in the high jump often utilize a technique known as the Fosbury Flop, which is named for Dick Fosbury, who first brought attention to the technique in the 1968 Summer Olympics. Here, the athlete positions the body in the air so as to actually travel "backwards" over the bar, as shown in **Figure 7.18*a***. We can see why this is such an effective technique by calculating the position of the athlete's center of mass, as the body maintains this curved shape while passing over the bar. In **Figure 7.18*b*** we approximate the athlete's body as consisting of three different linear mass segments. Segment A represents the mass of the head, neck, trunk, and arms. Segment B represents the mass from the upper and lower legs, and Segment C represents the mass from the feet. The masses and positions of the centers of mass of each segment are indicated in **Figure 7.18*b***. Use this information to calculate the position of the overall center of mass of the three segments.

Reasoning This is a 2-dimensional center of mass problem, where we can find the x and y coordinates of the position of the overall center of mass by applying Equation 7.10 twice, once for the x-direction and once for the y-direction.

Solution We have three mass segments, so we will have three terms in the numerator and denominator of Equation 7.10. Let's begin with the x-direction:

$$x_{cm} = \frac{m_A x_A + m_B x_B + m_C x_C}{m_A + m_B + m_C}$$

$$= \frac{(39 \text{ kg})(99 \text{ cm}) + (20 \text{ kg})(37 \text{ cm}) + (1.9 \text{ kg})(0 \text{ cm})}{39 \text{ kg} + 20 \text{ kg} + 1.9 \text{ kg}} = \boxed{76 \text{ cm}}$$

And now for the y-direction:

$$y_{cm} = \frac{m_A y_A + m_B y_B + m_C y_C}{m_A + m_B + m_C}$$

$$= \frac{(39 \text{ kg})(36 \text{ cm}) + (20 \text{ kg})(37 \text{ cm}) + (1.9 \text{ kg})(0 \text{ cm})}{39 \text{ kg} + 20 \text{ kg} + 1.9 \text{ kg}} = \boxed{35 \text{ cm}}$$

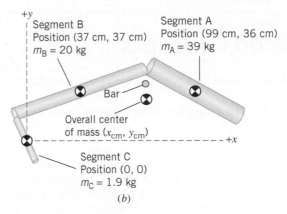

+y

Segment B
Position (37 cm, 37 cm)
m_B = 20 kg

Segment A
Position (99 cm, 36 cm)
m_A = 39 kg

Bar

Overall center
of mass (x_{cm}, y_{cm})

+x

Segment C
Position (0, 0)
m_C = 1.9 kg

(a)   (b)

FIGURE 7.18   (a) A high jumper clearing the bar by performing the Fosbury Flop. This technique allows her to keep her center of mass closer to the ground. (b) The approximate distribution of the high jumper's mass as she passes over the bar. The masses and locations of different parts of her body are represented by the three segments in the figure.

Therefore, the position of the overall center of mass is given by (x_{cm} = 76 cm, y_{cm} = 35 cm). What makes this technique so effective is that it keeps the high jumper's center of mass closer to the ground. In fact, her center of mass actually passes below the bar. Keeping her center of mass closer to the ground reduces the total energy she needs to clear the bar.

Concept Summary

7.1 The Impulse–Momentum Theorem   The impulse $\vec{J}$ of a force is the product of the average force $\vec{F}$ and the time interval Δt during which the force acts, according to Equation 7.1. Impulse is a vector that points in the same direction as the average force.

The linear momentum $\vec{p}$ of an object is the product of the object's mass m and velocity $\vec{v}$, according to Equation 7.2. Linear momentum is a vector that points in the same direction as the velocity. The total linear momentum of a system of objects is the vector sum of the momenta of the individual objects.

The impulse–momentum theorem states that when a net average force $\Sigma\vec{F}$ acts on an object, the impulse of this force is equal to the change in momentum of the object, as in Equation 7.4.

$$\vec{J} = \vec{F}\Delta t \tag{7.1}$$

$$\vec{p} = m\vec{v} \tag{7.2}$$

$$(\Sigma\vec{F})\Delta t = m\vec{v}_f - m\vec{v}_0 \tag{7.4}$$

7.2 The Principle of Conservation of Linear Momentum   External forces are those forces that agents external to the system exert on objects within the system. An isolated system is one for which the vector sum of the external forces acting on the system is zero.

The principle of conservation of linear momentum states that the total linear momentum of an isolated system remains constant. For a two-body system, the conservation of linear momentum can be written as in Equation 7.7b, where m_1 and m_2 are the masses, $\vec{v}_{f1}$ and $\vec{v}_{f2}$ are the final velocities, and $\vec{v}_{01}$ and $\vec{v}_{02}$ are the initial velocities of the objects.

$$\underbrace{m_1\vec{v}_{f1} + m_2\vec{v}_{f2}}_{\text{Final total linear momentum}} = \underbrace{m_1\vec{v}_{01} + m_2\vec{v}_{02}}_{\text{Initial total linear momentum}} \tag{7.7b}$$

7.3 Collisions in One Dimension   An elastic collision is one in which the total kinetic energy of the system after the collision is equal to the total kinetic energy of the system before the collision. An inelastic collision is one in which the total kinetic energy of the system is not the same before and after the collision. If the objects stick together after the collision, the collision is said to be completely inelastic.

7.4 Collisions in Two Dimensions   When the total linear momentum is conserved in a two-dimensional collision, the x and y components of the total linear momentum are conserved separately. For a collision between two objects, the conservation of total linear momentum can be written as in Equations 7.9a and 7.9b.

$$\underbrace{m_1v_{f1x} + m_2v_{f2x}}_{\substack{x \text{ component of final} \\ \text{total linear momentum}}} = \underbrace{m_1v_{01x} + m_2v_{02x}}_{\substack{x \text{ component of initial} \\ \text{total linear momentum}}} \tag{7.9a}$$

$$\underbrace{m_1v_{f1y} + m_2v_{f2y}}_{\substack{y \text{ component of final} \\ \text{total linear momentum}}} = \underbrace{m_1v_{01y} + m_2v_{02y}}_{\substack{y \text{ component of initial} \\ \text{total linear momentum}}} \tag{7.9b}$$

7.5 Center of Mass   The location of the center of mass of two particles lying on the x axis is given by Equation 7.10, where m_1 and m_2 are the masses of the particles and x_1 and x_2 are their positions relative to the coordinate origin. If the particles move with velocities v_1 and v_2, the velocity v_{cm} of the center of mass is given by Equation 7.11. If the total linear momentum of a system of particles remains constant during an interaction such as a collision, the velocity of the center of mass also remains constant.

$$x_{cm} = \frac{m_1x_1 + m_2x_2}{m_1 + m_2} \tag{7.10}$$

$$v_{cm} = \frac{m_1v_1 + m_2v_2}{m_1 + m_2} \tag{7.11}$$

Focus on Concepts

Online

Additional questions are available for assignment in WileyPLUS.

Section 7.1   The Impulse–Momentum Theorem

1. Two identical cars are traveling at the same speed. One is heading due east and the other due north, as the drawing shows. Which statement is true regarding the kinetic energies and momenta of the cars? **(a)** They both have the same kinetic energies and the same momenta. **(b)** They have the same kinetic energies, but different momenta. **(c)** They have different kinetic energies, but the same momenta. **(d)** They have different kinetic energies and different momenta.

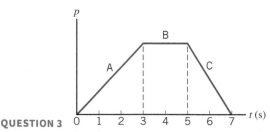

QUESTION 1

2. Six runners have the mass (in multiples of m_0), speed (in multiples of v_0), and direction of travel that are indicated in the table. Which two runners have identical momenta? **(a)** B and C **(b)** A and C **(c)** C and D **(d)** A and E **(e)** D and F

Runner	Mass	Speed	Direction of Travel
A	$\frac{1}{2}m_0$	v_0	Due north
B	m_0	v_0	Due east
C	m_0	$2v_0$	Due south
D	$2m_0$	v_0	Due west
E	m_0	$\frac{1}{2}v_0$	Due north
F	$2m_0$	$2v_0$	Due west

3. A particle is moving along the $+x$ axis, and the graph shows its momentum p as a function of time t. Use the impulse–momentum theorem to rank (largest to smallest) the three regions according to the *magnitude* of the impulse applied to the particle. **(a)** A, B, C **(b)** A, C, B **(c)** A and C (a tie), B **(d)** C, A, B **(e)** B, A, C

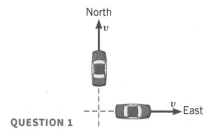

QUESTION 3

4. A particle moves along the $+x$ axis, and the graph shows its momentum p as a function of time t. In each of the four regions a force, which may or may not be nearly zero, is applied to the particle. In which region is the *magnitude* of the force largest and in which region is it smallest? **(a)** B largest, D smallest **(b)** C largest, B smallest

(c) A largest, D smallest **(d)** C largest, A smallest **(e)** A largest, C smallest

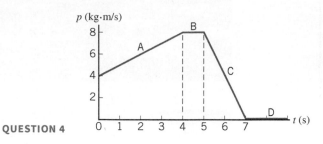

QUESTION 4

Section 7.2   The Principle of Conservation of Linear Momentum

5. As the text discusses, the conservation of linear momentum is applicable only when the system of objects is an isolated system. Which of the systems listed below are isolated systems?

  1. A ball is dropped from the top of a building. The system is the ball.

  2. A ball is dropped from the top of a building. The system is the ball and the earth.

  3. A billiard ball collides with a stationary billiard ball on a frictionless pool table. The system is the moving ball.

  4. A car slides to a halt in an emergency. The system is the car.

  5. A space probe is moving in deep space where gravitational and other forces are negligible. The system is the space probe.

(a) Only 2 and 5 are isolated systems. **(b)** Only 1 and 3 are isolated systems. **(c)** Only 3 and 5 are isolated systems. **(d)** Only 4 and 5 are isolated systems. **(e)** Only 5 is an isolated system.

Section 7.3   Collisions in One Dimension

6. Two objects are involved in a completely inelastic one-dimensional collision. The net external force acting on them is zero. The table lists four possible sets of the initial and final momenta and kinetic energies of the two objects. Which is the only set that could occur?

		Initial (Before Collision)		Final (After Collision)	
		Momentum	Kinetic Energy	Momentum	Kinetic Energy
a.	Object 1:	$+6 \text{ kg} \cdot \text{m/s}$	15 J	$+8 \text{ kg} \cdot \text{m/s}$	9 J
	Object 2:	$0 \text{ kg} \cdot \text{m/s}$	0 J		
b.	Object 1:	$+8 \text{ kg} \cdot \text{m/s}$	5 J	$+6 \text{ kg} \cdot \text{m/s}$	12 J
	Object 2:	$-2 \text{ kg} \cdot \text{m/s}$	7 J		
c.	Object 1:	$-3 \text{ kg} \cdot \text{m/s}$	1 J	$+1 \text{ kg} \cdot \text{m/s}$	4 J
	Object 2:	$+4 \text{ kg} \cdot \text{m/s}$	6 J		
d.	Object 1:	$0 \text{ kg} \cdot \text{m/s}$	3 J	$-8 \text{ kg} \cdot \text{m/s}$	11 J
	Object 2:	$-8 \text{ kg} \cdot \text{m/s}$	8 J		

Section 7.4   Collisions in Two Dimensions

7. Object 1 is moving along the x axis with an initial momentum of $+16 \text{ kg} \cdot \text{m/s}$, where the $+$ sign indicates that it is moving to the right. As the drawing shows, object 1 collides with a second object that is initially at rest. The collision is not head-on, so the objects move off in different directions after the collision. The net external force acting on the two-object system is zero. After the collision, object 1 has a momentum whose y component is $-5 \text{ kg} \cdot \text{m/s}$. What is the y component of the momentum of object 2 after the collision? (a) $0 \text{ kg} \cdot \text{m/s}$ (b) $+16 \text{ kg} \cdot \text{m/s}$ (c) $+5 \text{ kg} \cdot \text{m/s}$ (d) $-16 \text{ kg} \cdot \text{m/s}$ (e) The y component of the momentum of object 2 cannot be determined.

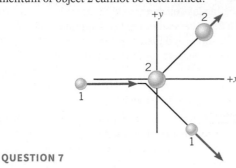

QUESTION 7

Section 7.5   Center of Mass

8. The drawing shows three particles that are moving with different velocities. Two of the particles have mass m, and the third has a mass $2m$. At the instant shown, the center of mass (cm) of the three particles is at the coordinate origin. What is the velocity v_{cm} (magnitude and direction) of the center of mass?

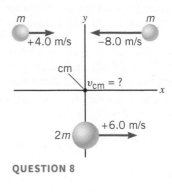

QUESTION 8

Problems

Online

Additional questions are available for assignment in WileyPLUS.

SSM Student Solutions Manual	**BIO** Biomedical application
MMH Problem-solving help	**E** Easy
GO Guided Online Tutorial	**M** Medium
V-HINT Video Hints	**H** Hard
CHALK Chalkboard Videos	**WS** Worksheet
	T Team Problem

Section 7.1   The Impulse–Momentum Theorem

1. **E** **SSM** A 46-kg skater is standing still in front of a wall. By pushing against the wall she propels herself backward with a velocity of -1.2 m/s. Her hands are in contact with the wall for 0.80 s. Ignore friction and wind resistance. Find the magnitude and direction of the average force she exerts on the wall (which has the same magnitude as, but opposite direction to, the force that the wall applies to her).

2. **E** A model rocket is constructed with a motor that can provide a total impulse of $29.0 \text{ N} \cdot \text{s}$. The mass of the rocket is 0.175 kg. What is the speed that this rocket achieves when launched from rest? Neglect the effects of gravity and air resistance.

3. **E** Before starting this problem, review Conceptual Example 4. Suppose that the hail described there bounces off the roof of the car with a velocity of $+15 \text{ m/s}$. Ignoring the weight of the hailstones, calculate the force exerted by the hail on the roof. Compare your answer to that obtained in Example 2 for the rain, and verify that your answer is consistent with the conclusion reached in Conceptual Example 4.

4. **E** **GO** In a performance test, each of two cars takes 9.0 s to accelerate from rest to 27 m/s. Car A has a mass of 1400 kg, and car B has a mass of 1900 kg. Find the net average force that acts on each car during the test.

5. **E** **SSM** A volleyball is spiked so that its incoming velocity of $+4.0 \text{ m/s}$ is changed to an outgoing velocity of -21 m/s. The mass of the volleyball is 0.35 kg. What impulse does the player apply to the ball?

6. **E** Two arrows are fired horizontally with the same speed of 30.0 m/s. Each arrow has a mass of 0.100 kg. One is fired due east and the other due south. Find the magnitude and direction of the total momentum of this two-arrow system. Specify the direction with respect to due east.

7. **E** Refer to Conceptual Example 4 as an aid in understanding this problem. A hockey goalie is standing on ice. Another player fires a puck ($m = 0.17 \text{ kg}$) at the goalie with a velocity of $+65 \text{ m/s}$. (a) If the goalie catches the puck with his glove in a time of $5.0 \times 10^{-3} \text{ s}$, what is the average force (magnitude and direction) exerted on the goalie by the puck? (b) Instead of catching the puck, the goalie slaps it with his stick and returns the puck straight back to the player with a velocity of -65 m/s. The puck and stick are in contact for a time of $5.0 \times 10^{-3} \text{ s}$. Now what is the average force exerted on the goalie by the puck? Verify that your answers to parts (a) and (b) are consistent with the conclusion of Conceptual Example 4.

8. **E** A space probe is traveling in outer space with a momentum that has a magnitude of $7.5 \times 10^{7} \text{ kg} \cdot \text{m/s}$. A retrorocket is fired to slow down the probe. It applies a force to the probe that has a magnitude of $2.0 \times 10^{6} \text{ N}$ and a direction opposite to the probe's motion. It fires for a period of 12 s. Determine the momentum of the probe after the retrorocket ceases to fire.

9. **M** **MMH** A stream of water strikes a stationary turbine blade horizontally, as the drawing illustrates. The incident water stream has a velocity of $+16.0 \text{ m/s}$, while the exiting water stream has a velocity of -16.0 m/s. The mass of water per second that strikes the blade is 30.0 kg/s. Find the magnitude of the average force exerted on the water by the blade.

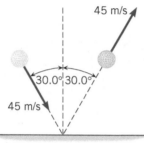

$v_0 = +16.0$ m/s

Stationary
turbine
blade

$v_f = -16.0$ m/s

PROBLEM 9

10. **M** **GO** A student ($m = 63$ kg) falls freely from rest and strikes the ground. During the collision with the ground, he comes to rest in a time of 0.040 s. The average force exerted on him by the ground is $+18\,000$ N, where the upward direction is taken to be the positive direction. From what height did the student fall? Assume that the only force acting on him during the collision is that due to the ground.

11. **M** **V-HINT** **CHALK** A golf ball strikes a hard, smooth floor at an angle of 30.0° and, as the drawing shows, rebounds at the same angle. The mass of the ball is 0.047 kg, and its speed is 45 m/s just before and after striking the floor. What is the magnitude of the impulse applied to the golf ball by the floor? (*Hint: Note that only the vertical component of the ball's momentum changes during impact with the floor, and ignore the weight of the ball.*)

45 m/s

30.0° 30.0°

45 m/s

PROBLEM 11

12. **M** **V-HINT** An 85-kg jogger is heading due east at a speed of 2.0 m/s. A 55-kg jogger is heading 32° north of east at a speed of 3.0 m/s. Find the magnitude and direction of the sum of the momenta of the two joggers.

13. **M** **GO** A basketball ($m = 0.60$ kg) is dropped from rest. Just before striking the floor, the ball has a momentum whose magnitude is 3.1 kg · m/s. At what height was the basketball dropped?

14. **H** **SSM** A dump truck is being filled with sand. The sand falls straight downward from rest from a height of 2.00 m above the truck bed, and the mass of sand that hits the truck per second is 55.0 kg/s. The truck is parked on the platform of a weight scale. By how much does the scale reading exceed the weight of the truck and sand?

Section 7.2 The Principle of Conservation of Linear Momentum

15. **E** **MMH** In a science fiction novel two enemies, Bonzo and Ender, are fighting in outer space. From stationary positions they push against each other. Bonzo flies off with a velocity of $+1.5$ m/s, while Ender recoils with a velocity of -2.5 m/s. (a) Without doing any calculations, decide which person has the greater mass. Give your reasoning. (b) Determine the ratio $m_{\text{Bonzo}}/m_{\text{Ender}}$ of the masses of these two enemies.

16. **E** A 2.3-kg cart is rolling across a frictionless, horizontal track toward a 1.5-kg cart that is held initially at rest. The carts are loaded with strong magnets that cause them to attract one another. Thus, the speed of each cart increases. At a certain instant before the carts collide, the first cart's velocity is $+4.5$ m/s, and the second cart's velocity is -1.9 m/s. (a) What is the total momentum of the system of the two carts at this instant? (b) What was the velocity of the first cart when the second cart was still at rest?

17. **E** **GO** As the drawing illustrates, two disks with masses m_1 and m_2 are moving horizontally to the right at a speed of v_0. They

are on an air-hockey table, which supports them with an essentially frictionless cushion of air. They move as a unit, with a compressed spring between them, which has a negligible mass. Then the spring is released and allowed to push the disks outward. Consider the situation where disk 1 comes to a momentary halt shortly after the spring is released. Assuming that $m_1 = 1.2$ kg, $m_2 = 2.4$ kg, and $v_0 = +5.0$ m/s, find the velocity of disk 2 at that moment.

m_1 m_2 v_0

PROBLEM 17

18. **E** **SSM** A lumberjack (mass = 98 kg) is standing at rest on one end of a floating log (mass = 230 kg) that is also at rest. The lumberjack runs to the other end of the log, attaining a velocity of $+3.6$ m/s relative to the shore, and then hops onto an identical floating log that is initially at rest. Neglect any friction and resistance between the logs and the water. (a) What is the velocity of the first log just before the lumberjack jumps off? (b) Determine the velocity of the second log if the lumberjack comes to rest on it.

19. **E** **GO** **MMH** An astronaut in his space suit and with a propulsion unit (empty of its gas propellant) strapped to his back has a mass of 146 kg. The astronaut begins a space walk at rest, with a completely filled propulsion unit. During the space walk, the unit ejects some gas with a velocity of $+32$ m/s. As a result, the astronaut recoils with a velocity of -0.39 m/s. After the gas is ejected, the mass of the astronaut (now wearing a partially empty propulsion unit) is 165 kg. What percentage of the gas was ejected from the completely filled propulsion unit?

20. **E** **SSM** A two-stage rocket moves in space at a constant velocity of 4900 m/s. The two stages are then separated by a small explosive charge placed between them. Immediately after the explosion the velocity of the 1200-kg upper stage is 5700 m/s in the same direction as before the explosion. What is the velocity (magnitude and direction) of the 2400-kg lower stage after the explosion?

21. **M** **V-HINT** A 40.0-kg boy, riding a 2.50-kg skateboard at a velocity of $+5.30$ m/s across a level sidewalk, jumps forward to leap over a wall. Just after leaving contact with the board, the boy's velocity relative to the sidewalk is 6.00 m/s, 9.50° above the horizontal. Ignore any friction between the skateboard and the sidewalk. What is the skateboard's velocity relative to the sidewalk at this instant? Be sure to include the correct algebraic sign with your answer.

22. **M** **GO** The lead female character in the movie *Diamonds Are Forever* is standing at the edge of an offshore oil rig. As she fires a gun, she is driven back over the edge and into the sea. Suppose the mass of a bullet is 0.010 kg and its velocity is $+720$ m/s. Her mass (including the gun) is 51 kg. (a) What recoil velocity does she acquire in response to a single shot from a stationary position, assuming that no external force keeps her in place? (b) Under the same assumption, what would be her recoil velocity if, instead, she shoots a blank cartridge that ejects a mass of 5.0×10^{-4} kg at a velocity of $+720$ m/s?

23. **M** **MMH** A 0.015-kg bullet is fired straight up at a falling wooden block that has a mass of 1.8 kg. The bullet has a speed of 810 m/s when it strikes the block. The block originally was dropped from rest from the top of a building and has been falling for a time t when the collision with the bullet occurs. As a result of the collision, the block (with the bullet in it) reverses direction, rises, and comes to a momentary halt at the top of the building. Find the time t.

24. **M** **SSM** By accident, a large plate is dropped and breaks into three pieces. The pieces fly apart parallel to the floor. As the plate falls, its momentum has only a vertical component and no component parallel to the floor. After the collision, the component of the total momentum parallel to the floor must remain zero, since the net external force acting on the plate has no component parallel to the floor. Using the data shown in the drawing, find the masses of pieces 1 and 2.

3.00 m/s

25.0°

m_1

1.79 m/s

m_2

45.0°

$m_3 = 1.30$ kg

3.07 m/s

PROBLEM 24

25. **H** Adolf and Ed are wearing harnesses and are hanging at rest from the ceiling by means of ropes attached to them. Face to face, they push off against one another. Adolf has a mass of 120 kg, and Ed has a mass of 78 kg. Following the push, Adolf swings upward to a height of 0.65 m above his starting point. To what height above his own starting point does Ed rise?

Section 7.3 Collisions in One Dimension,

Section 7.4 Collisions in Two Dimensions

26. **E** After sliding down a snow-covered hill on an inner tube, Ashley is coasting across a level snowfield at a constant velocity of +2.7 m/s. Miranda runs after her at a velocity of +4.5 m/s and hops on the inner tube. How fast do the two slide across the snow together on the inner tube? Ashley's mass is 71 kg and Miranda's is 58 kg. Ignore the mass of the inner tube and any friction between the inner tube and the snow.

27. **E** Consult Multiple-Concept Example 9 for background pertinent to this problem. A 2.50-g bullet, traveling at a speed of 425 m/s, strikes the wooden block of a ballistic pendulum, such as that in Figure 7.13. The block has a mass of 215 g. **(a)** Find the speed of the bullet–block combination immediately after the collision. **(b)** How high does the combination rise above its initial position?

28. **E GO** One object is at rest, and another is moving. The two collide in a one-dimensional, completely inelastic collision. In other words, they stick together after the collision and move off with a common velocity. Momentum is conserved. The speed of the object that is moving initially is 25 m/s. The masses of the two objects are 3.0 and 8.0 kg. Determine the final speed of the two-object system after the collision for the case when the large-mass object is the one moving initially and the case when the small-mass object is the one moving initially.

29. **E SSM** Batman (mass = 91 kg) jumps straight down from a bridge into a boat (mass = 510 kg) in which a criminal is fleeing. The velocity of the boat is initially +11 m/s. What is the velocity of the boat after Batman lands in it?

30. **E** A car (mass = 1100 kg) is traveling at 32 m/s when it collides head-on with a sport utility vehicle (mass = 2500 kg) traveling in the opposite direction. In the collision, the two vehicles come to a halt. At what speed was the sport utility vehicle traveling?

31. **E SSM** A 5.00-kg ball, moving to the right at a velocity of +2.00 m/s on a frictionless table, collides head-on with a stationary 7.50-kg ball. Find the final velocities of the balls if the collision is **(a)** elastic and **(b)** completely inelastic.

32. **E CHALK** The drawing shows a collision between two pucks on an air-hockey table. Puck A has a mass of 0.025 kg and is moving along the x axis with a velocity of +5.5 m/s. It makes a collision with puck B, which has a mass of 0.050 kg and is initially at rest. The collision is not head-on. After the collision, the two pucks fly apart with

the angles shown in the drawing. Find the final speeds of **(a)** puck A and **(b)** puck B.

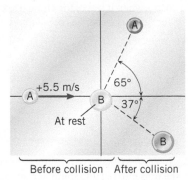

+5.5 m/s

A

B

At rest

65°

37°

A

B

PROBLEM 32 Before collision   After collision

33. **E SSM** A projectile (mass = 0.20 kg) is fired at and embeds itself in a stationary target (mass = 2.50 kg). With what percentage of the projectile's incident kinetic energy does the target (with the projectile in it) fly off after being struck?

34. **E GO** Object A is moving due east, while object B is moving due north. They collide and stick together in a completely inelastic collision. Momentum is conserved. Object A has a mass of $m_A = 17.0$ kg and an initial velocity of $\vec{v}_{0A} = 8.00$ m/s, due east. Object B, however, has a mass of $m_B = 29.0$ kg and an initial velocity of $\vec{v}_{0B} = 5.00$ m/s, due north. Find the magnitude and direction of the total momentum of the two-object system after the collision.

35. **E** Multiple-Concept Example 8 deals with some of the concepts that are used to solve this problem. A cue ball (mass = 0.165 kg) is at rest on a frictionless pool table. The ball is hit dead center by a pool stick, which applies an impulse of +1.50 N · s to the ball. The ball then slides along the table and makes an elastic head-on collision with a second ball of equal mass that is initially at rest. Find the velocity of the second ball just after it is struck.

36. **M CHALK GO** A ball is attached to one end of a wire, the other end being fastened to the ceiling. The wire is held horizontal, and the ball is released from rest (see the drawing). It swings downward and strikes a block initially at rest on a horizontal frictionless surface. Air resistance is negligible, and the collision is elastic. The masses of the ball and block are, respectively, 1.60 kg and 2.40 kg, and the length of the wire is 1.20 m. Find the velocity (magnitude and direction) of the ball **(a)** just before the collision, and **(b)** just after the collision.

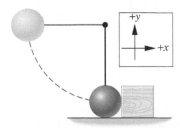

+y

+x

PROBLEM 36

37. **M V-HINT** A girl is skipping stones across a lake. One of the stones accidentally ricochets off a toy boat that is initially at rest in the water (see the drawing). The 0.072-kg stone strikes the boat at a velocity of 13 m/s, 15° below due east, and ricochets off at a velocity of 11 m/s, 12° above due east. After being struck by the stone, the boat's velocity is 2.1 m/s, due east. What is the mass of the boat? Assume the water offers no resistance to the boat's motion.

East

15°

13 m/s

11 m/s

East

12°

PROBLEM 37

38. [M] [GO] A mine car (mass = 440 kg) rolls at a speed of 0.50 m/s on a horizontal track, as the drawing shows. A 150-kg chunk of coal has a speed of 0.80 m/s when it leaves the chute. Determine the speed of the car–coal system after the coal has come to rest in the car.

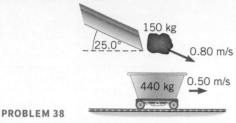

150 kg

25.0°

0.80 m/s

440 kg   0.50 m/s

PROBLEM 38

39. [M] [SSM] A 50.0-kg skater is traveling due east at a speed of 3.00 m/s. A 70.0-kg skater is moving due south at a speed of 7.00 m/s. They collide and hold on to each other after the collision, managing to move off at an angle θ south of east, with a speed of v_f. Find (a) the angle θ and (b) the speed v_f, assuming that friction can be ignored.

40. [M] [CHALK] [GO] A 4.00-g bullet is moving horizontally with a velocity of +355 m/s, where the +sign indicates that it is moving to the right (see part *a* of the drawing). The bullet is approaching two blocks resting on a horizontal frictionless surface. Air resistance is negligible. The bullet passes completely through the first block (an inelastic collision) and embeds itself in the second one, as indicated in part *b*. Note that both blocks are moving after the collision with the bullet. The mass of the first block is 1150 g, and its velocity is +0.550 m/s after the bullet passes through it. The mass of the second block is 1530 g. (a) What is the velocity of the second block after the bullet embeds itself? (b) Find the ratio of the total kinetic energy after the collisions to that before the collisions.

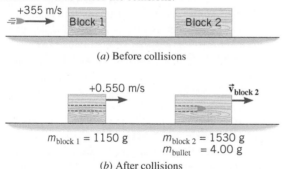

+355 m/s

Block 1   Block 2

(*a*) Before collisions

+0.550 m/s $\vec{v}_{block\ 2}$

$m_{block\ 1}$ = 1150 g $m_{block\ 2}$ = 1530 g
                           m_{bullet} = 4.00 g

(*b*) After collisions

PROBLEM 40

41. [M] [V-HINT] An electron collides elastically with a stationary hydrogen atom. The mass of the hydrogen atom is 1837 times that of the electron. Assume that all motion, before and after the collision, occurs along the same straight line. What is the ratio of the kinetic energy of the hydrogen atom after the collision to that of the electron before the collision?

42. [M] [GO] A 60.0-kg person, running horizontally with a velocity of +3.80 m/s, jumps onto a 12.0-kg sled that is initially at rest. (a) Ignoring the effects of friction during the collision, find the velocity of the

sled and person as they move away. (b) The sled and person coast 30.0 m on level snow before coming to rest. What is the coefficient of kinetic friction between the sled and the snow?

43. [H] [SSM] Starting with an initial speed of 5.00 m/s at a height of 0.300 m, a 1.50-kg ball swings downward and strikes a 4.60-kg ball that is at rest, as the drawing shows. (a) Using the principle of conservation of mechanical energy, find the speed of the 1.50-kg ball just before impact. (b) Assuming that the collision is elastic, find the velocities (magnitude and direction) of both balls just after the collision. (c) How high does each ball swing after the collision, ignoring air resistance?

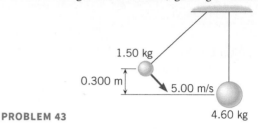

1.50 kg

0.300 m

5.00 m/s

4.60 kg

PROBLEM 43

Section 7.5   Center of Mass

44. [E] Two particles are moving along the *x* axis. Particle 1 has a mass m_1 and a velocity $v_1 = +4.6$ m/s. Particle 2 has a mass m_2 and a velocity $v_2 = -6.1$ m/s. The velocity of the center of mass of these two particles is zero. In other words, the center of mass of the particles remains stationary, even though each particle is moving. Find the ratio m_1/m_2 of the masses of the particles.

45. [E] [GO] John's mass is 86 kg, and Barbara's is 55 kg. He is standing on the *x* axis at $x_J = +9.0$ m, while she is standing on the *x* axis at $x_B = 12.0$ m. They switch positions. How far and in which direction does their center of mass move as a result of the switch?

46. [E] [GO] Two stars in a binary system orbit around their center of mass. The centers of the two stars are 7.17×10^{11} m apart. The larger of the two stars has a mass of 3.70×10^{30} kg, and its center is 2.08×10^{11} m from the system's center of mass. What is the mass of the smaller star?

47. [M] [GO] [CHALK] The drawing shows a sulfur dioxide molecule. It consists of two oxygen atoms and a sulfur atom. A sulfur atom is twice as massive as an oxygen atom. Using this information and the data provided in the drawing, find (a) the *x* coordinate and (b) the *y* coordinate of the center of mass of the sulfur dioxide molecule. Express your answers in nanometers (1 nm = 10^{-9} m).

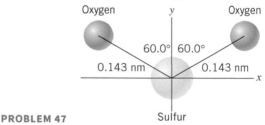

Oxygen   *y*   Oxygen

60.0° 60.0°

0.143 nm   0.143 nm

x

Sulfur

PROBLEM 47

Additional Problems

Online

48. [E] [GO] A golf ball bounces down a flight of steel stairs, striking several steps on the way down, but never hitting the edge of a step. The ball starts at the top step with a vertical velocity component of zero. If all the collisions with the stairs are elastic, and if the vertical height of the staircase is 3.00 m, determine the bounce height when the ball reaches the bottom of the stairs. Neglect air resistance.

49. [E] [MMH] A baseball (*m* = 149 g) approaches a bat horizontally at a speed of 40.2 m/s (90 mi/h) and is hit straight back at a speed of 45.6 m/s (102 mi/h). If the ball is in contact with the bat for a time of 1.10 ms, what is the average force exerted on the ball by the bat? Neglect the weight of the ball, since it is so much less than the force of the bat. Choose the direction of the incoming ball as the positive direction.

50. **E** **GO** A wagon is rolling forward on level ground. Friction is negligible. The person sitting in the wagon is holding a rock. The total mass of the wagon, rider, and rock is 95.0 kg. The mass of the rock is 0.300 kg. Initially the wagon is rolling forward at a speed of 0.500 m/s. Then the person throws the rock with a speed of 16.0 m/s. Both speeds are relative to the ground. Find the speed of the wagon after the rock is thrown directly forward in one case and directly backward in another.

51. **M** **V-HINT** Two ice skaters have masses m_1 and m_2 and are initially stationary. Their skates are identical. They push against one another, as in **Interactive Figure 7.10**, and move in opposite directions with different speeds. While they are pushing against each other, any kinetic frictional forces acting on their skates can be ignored. However, once the skaters separate, kinetic frictional forces eventually bring them to a halt. As they glide to a halt, the magnitudes of their accelerations are equal, and skater 1 glides twice as far as skater 2. What is the ratio m_1/m_2 of their masses?

52. **M** **GO** For the situation depicted in the figure, use momentum conservation to determine the magnitude and direction of the final velocity of ball 1 after the collision.

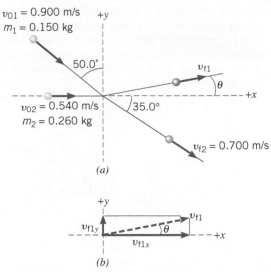

(a) Top view of two balls colliding on a horizontal surface. (b) This part of the drawing shows the x and y components of the velocity of ball 1 after the

PROBLEM 52 collision.

Physics in Biology, Medicine, and Sports

53. **E** **BIO** **7.1** When jumping straight down, you can be seriously injured if you land stiff-legged. One way to avoid injury is to bend your knees upon landing to reduce the force of the impact. A 75-kg man just before contact with the ground has a speed of 6.4 m/s. **(a)** In a stiff-legged landing he comes to a halt in 2.0 ms. Find the average net force that acts on him during this time. **(b)** When he bends his knees, he comes to a halt in 0.10 s. Find the average net force now. **(c)** During the landing, the force of the ground on the man points upward, while the force due to gravity points downward. The average net force acting on the man includes both of these forces. Taking into account the directions of the forces, find the force of the ground on the man in parts (a) and (b).

54. **E** **BIO** **MMH** **7.2** For tests using a *ballistocardiograph*, a patient lies on a horizontal platform that is supported on jets of air. Because of the air jets, the friction impeding the horizontal motion of the platform is negligible. Each time the heart beats, blood is pushed out of the heart in a direction that is nearly parallel to the platform. Since momentum must be conserved, the body and the platform recoil, and this recoil can be detected to provide information about the heart. For each beat, suppose that 0.050 kg of blood is pushed out of the heart with a velocity of +0.25 m/s and that the mass of the patient and platform is 85 kg. Assuming that the patient does not slip with respect to the platform, and that the patient and platform start from rest, determine the recoil velocity.

55. **M** **BIO** **V-HINT** **7.5** The drawing shows a human figure in a sitting position. For purposes of this problem, there are three parts to the figure, and the center of mass of each one is shown in the drawing. These parts are: (1) the torso, neck, and head (total mass = 41 kg) with a center of mass located on the y axis at a point 0.39 m above the origin, (2) the upper legs (mass = 17 kg) with a center of mass located on the x axis at a point 0.17 m to the right of the origin, and (3) the lower legs and feet (total mass = 9.9 kg) with a center of mass located 0.43 m to the right of and 0.26 m below the origin. Find the x and y coordinates of the center of mass of the human figure. Note that the mass of the arms and hands (approximately 12% of the whole-body mass) has been ignored to simplify the drawing.

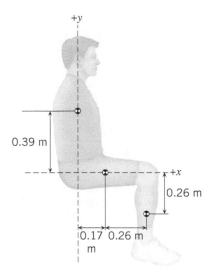

PROBLEM 55

56. **M** **BIO** **7.1** As of 2008, 42 people had survived a fall from a height above 10 000 feet without a parachute. Without air resistance, a person would hit the ground with a speed of 240 m/s (540 mph!). However, by placing your body in a skydiving position (see photo), you can maximize your cross-sectional area and take advantage of the drag force to limit your speed. In this position, your terminal speed will only be 54 m/s (120 mph). This is still a high rate of speed. Just imagine being in an automobile accident at 120 mph! Nevertheless, there is a higher probability of survival, depending upon what you land on. After all, it is not the fall that will kill you, but the impact. If you strike concrete at 54 m/s, we will say nice things about you at your funeral. You can take comfort in the fact that you won't even feel it. Contrary to popular belief, hitting water at this speed is not much better. Your body will have to displace and compress the water as you enter it. But water, and most liquids, are difficult to compress. The surface of the liquid does not "give" nearly enough to cushion your fall. It is no surprise that the survivors fell into deep snow, trees and bushes, and marsh mud. **(a)** Calculate the average force exerted on a 75-kg person who strikes concrete at 54 m/s and comes to rest in 0.0052 s. **(b)** In comparison, calculate the average force exerted on the same person who strikes deep snow at 54 m/s and comes to rest 5.7 m deep in the snow. Assume the person's deceleration is constant and the only force that acts on him when he is brought to rest is the snow.

PROBLEM 56

skeeze/Pixabay

57. **M** **BIO** **7.3** Cephalopods, such as octopi and squids, can use a form of jet propulsion to move quickly through the water. This technique of locomotion is useful in providing short bursts of speed to catch prey and to avoid predators, and it gives them the title of fastest marine invertebrates. Oxygenated water is taken in by the cephalopod and then expelled by muscular contraction through an opening called the hyponome. Assume a 4.2-kg squid draws in 0.32 kg of water and then quickly (in 0.12 s) expels this water at a speed of 11 m/s. **(a)** What is the speed of the squid just after the water is expelled, and **(b)** what is the average force exerted on the expelled water?

PROBLEM 57

joakant/Pixabay

58. **M** **BIO** **7.3** Frank "Cannonball" Richards was an American entertainer and vaudeville performer. He was famous for having incredibly strong abdominal muscles. In feats of strength, he would often let people punch him in the stomach, jump on him, and even hit him with sledgehammers. However, his greatest claim to fame was being shot in the stomach with a 47-kg (104 lb) cannonball (see the photo). While performing, Richards limited this demonstration to just twice a day, since it was quite painful. He would live a long life and pass away at age 81 in 1969. If the speed of the cannonball was 11 m/s right before it collided

with Richards and the speed of the cannonball was 0.75 m/s right after the collision, what was Richards's speed immediately after being struck by the cannonball? Assume Richards's mass was 86 kg.

Unknown stock footage film-maker; Image taken from https://en.wikipedia.org/wiki/File:Frank_%-22Cannonball%22_Richards_shot_in_stomach_with_cannonball.jpg

PROBLEM 58

59. **M** **BIO** **7.1** A professional golfer hits a golf ball off the tee a horizontal distance of 310 m (see the photo). The ball leaves the tee at an angle of 45° with respect to the horizontal. If the head of the driver is in contact with the 45-g ball for 1.0 ms, calculate the average force exerted on the ball by the driver.

Leonard Kamsler/Popperfoto/Getty Images

PROBLEM 59

60. **M** **BIO** **7.3** One common inaccuracy in action movies occurs when someone gets shot. Oftentimes, when struck by the bullet, the person goes flying backward. Let's consider an average person ($m = 75$ kg) standing at rest on frictionless ice who is shot at point-blank range with a high-energy handgun—a .357 Magnum. The mass of the bullet is 8.0 g and its muzzle velocity is approximately 1450 ft/s (440 m/s). **(a)** What is the velocity of the person after he is shot? Assume the bullet remains lodged in his body. You should find that your answer to part (a) is quite small—far too small to send the body flying backward, as is often portrayed in the movies. Does this mean getting shot is not that dangerous? No! The bullet transfers a tremendous amount of kinetic energy to a very small area of the body. This energy can do damaging work on the body's internal organs and structures. **(b)** Calculate the kinetic energy of the bullet just before it strikes the body. Now assume the person gets "shot" with a baseball ($m = 140$ g) that has the same kinetic energy as the bullet. **(c)** What would be the speed of the baseball in mph?

Concepts and Calculations Problems

Online

Momentum and energy are two of the most fundamental concepts in physics. As we have seen in this chapter, momentum is a vector and, like all vectors, has a magnitude and a direction. In contrast, energy is a scalar quantity, as Chapter 6 discusses, and does not have a direction associated with it. Problem 61 provides the opportunity to review how the vector nature of momentum and the scalar nature of kinetic energy influence calculations using these quantities. Problem 62 explores some further differences between momentum and kinetic energy.

61. **M** **CHALK** **SSM** Two joggers, Jim and Tom, are both running at a speed of 4.00 m/s (see the figure). Jim has a mass of 90.0 kg, and Tom has a mass of 55.0 kg. *Concepts:* (i) Does the total kinetic energy of the two-jogger system have a smaller value in part *a* or *b* of the figure, or is it the same in both cases? (ii) Does their total momentum have a smaller magnitude in part *a* or *b* of the figure, or is it the same for each? *Calculations:* Find the kinetic energy and momentum of the two-jogger system when **(a)** Jim and Tom are both running due north as in part *a* of the figure, and **(b)** Jim is running due north and Tom is running due south, as in part *b* of the figure.

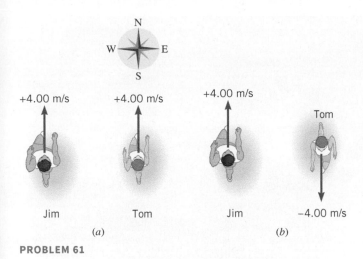

62. **M** **CHALK** The table gives mass and speed data for the two objects in the figure. *Concepts:* (i) Is it possible for two objects to have different speeds when their momenta have the same amplitude? Explain your answer. (ii) If two objects have the same momenta do they necessarily have the same kinetic energy? Explain. *Calculations:* Find the magnitude of the momentum and the kinetic energy for each object.

	Mass	**Speed**
Object A	2.0 kg	6.0 m/s
Object B	6.0 kg	2.0 m/s

A 6.0 m/s
2.0 kg

B 2.0 m/s
6.0 kg

PROBLEM 62

PROBLEM 61

Team Problems

Online

63. **E** **T** **WS** **A Balancing Act.** A heavy object that is part of a gigantic particle detector that will be used in a new particle accelerator is composed of four identically shaped disks and an aluminum plate. Each disk is composed of a single pure element: one each of lead (Pb), copper (Cu), tungsten (W), and titanium (Ti). The disks are 60.0 cm in diameter and 25.0 cm thick, and are bolted through their centers to the corners of a square piece of aluminum (Al) plate that is 12.0 cm thick. When bolted in place, the curved edges of the disks touch those of their neighbors mounted on adjacent corners. If a typical right-handed xy coordinate system is defined such that its origin is at the center of the square plate, the tungsten disk is located in the first quadrant (i.e., the $+x, +y$ quadrant), titanium in the second ($-x, +y$), copper in the third ($-x, -y$), and, finally, the lead disk is in the fourth quadrant ($+x, -y$). It is crucial, in order to place the object where it belongs on the detector, that the object be lifted with the aluminum plate parallel to the ground using a single chain. Your team is tasked with locating the position at which a hook should be placed in order to connect the chain that will be used to lift the awkward object with a crane. Determine the x and y coordinates of the hook position in centimeters. The densities of the materials in g/cm^3 are as follows: lead (11.3), copper (8.96), tungsten (19.3), titanium (4.51), and aluminum (2.70). (*Hint: you can treat each object as a point mass with its total mass concentrated at its geometric center.*)

64. **E** **T** **WS** **Which one is heavier?** Two smooth disks about the size of drink coasters are identical in size and overall appearance, except that one is composed of radioactive cobalt (Co) and the other of iron (Fe), which is not radioactive. Since the two disks are exactly the same size, their weights are slightly different because they are made from different materials. However, no scale is available to measure the weight and you are unable to discern which one is heavier by handling them (something you must avoid since one is radioactive). However, there is an air hockey table available (i.e., an air table on which the disks can move with nearly zero friction). You and your team have been given the task of determining which of the two disks is radioactive (a radiation detector is not available). Describe a one-dimensional collision experiment that does not require measuring the masses or speeds of the disks and that uses the air hockey table to reveal which disk is heavier. Explain in detail (with mathematical reasoning) how it will provide the information that you seek. The densities of the materials in g/cm^3 are: Fe (7.86) and Co (8.90).

65. **M** **T** **Docking a Spaceship.** You and your crew must dock your 2.50×10^4 kg spaceship at Spaceport Alpha, which is orbiting Mars. In the process, Alpha's control tower has requested that you ram another vessel, a freight ship of mass 1.65×10^4 kg, latch onto it, and use your combined momentum to bring it into dock. The freight ship is not moving with respect to the colossal Spaceport Alpha, which has a mass of 1.85×10^7 kg. Alpha's automated system that guides incoming spacecraft into dock requires that the incoming speed is less than 2.00 m/s. **(a)** Assuming a perfectly linear alignment of your ship's velocity vector with the freight ship (which is stationary with respect to Alpha) and Alpha's docking port, what must be your ship's speed (before colliding with the freight ship) so that the combination of the freight ship and your ship arrives at Alpha's docking port with a speed of 1.50 m/s? **(b)** How does the velocity of Spaceport Alpha change when the combination of your vessel and the freight ship successfully docks with it? **(c)** Suppose you made a mistake while maneuvering your vessel in an attempt to ram the freight ship and, rather than latching on to it and making a perfectly inelastic collision, you strike it and knock it in the direction of the spaceport with a perfectly elastic collision. What is the speed of the freight ship in that case (assuming your ship had the same initial velocity as that calculated in part (a))?

66. **M** **T** **Spaceship Malfunction.** You and your team get a distress call from another ship indicating that there has been an explosion and their propulsion and navigation systems are now off line. A large piece of their ship has been jettisoned into space, and their life support systems are failing. They need help as soon as possible. You try to track them, but are only able to pick up the piece of the craft that had been projected away from them in the explosion. You look up the make and model of their ship, a Galaxy Explorer 5000 (GE 5000), which has a total mass of 5.50×10^4 kg. The piece heading in your direction is the ship's storage module, and has a mass of 7.50×10^3 kg. The GE 5000 was at rest with respect to your ship at the time of the explosion. **(a)** Assuming you received the distress call 20.0 minutes before the jettisoned storage module passed by your ship at a speed of 750.0 m/s, what is the speed of the damaged craft? **(b)** How far away is the Galaxy Explorer 5000 when its storage module passes by? **(c)** In which direction, and how fast must you go, in order to catch up with the GE 5000 before its life support systems fail. (The crew had 80.0 minutes of life support left when they made the distress call, so you have 60.0 minutes to get to them.)

CHAPTER 8

The figure shows the front view of a turbine jet engine on a commercial aircraft. The rotating fan blades collect air into the engine before it is compressed, mixed with fuel, and ignited to produce thrust. The rotational motion of the blades can be described using the concepts of angular displacement, angular velocity, and angular acceleration within the framework of rotational kinematics.

Mikhail Starodubov/Shutterstock.com

LEARNING OBJECTIVES

After reading this module, you should be able to...

8.1 Define angular displacement.

8.2 Define angular velocity and angular acceleration.

8.3 Solve rotational kinematics problems.

8.4 Relate angular and tangential variables.

8.5 Distinguish between centripetal and tangential accelerations.

8.6 Analyze rolling motion.

8.7 Use the right-hand rule to determine the direction of angular vectors.

Rotational Kinematics

8.1 Rotational Motion and Angular Displacement

In the simplest kind of rotation, points on a rigid object move on circular paths. In **Figure 8.1**, for example, we see the circular paths for points A, B, and C on a spinning skater. The centers of all such circular paths define a line, called the **axis of rotation**.

The angle through which a rigid object rotates about a fixed axis is called the **angular displacement**. **Figure 8.2** shows how the angular displacement is measured for a thin rotating disc. Here, the axis of rotation passes through the center of the disc and is perpendicular to its surface. On the surface of the disc we draw a radial line, which is a line that intersects the axis of rotation perpendicularly. As the disc turns, we observe the angle through which this line moves relative to a convenient reference line that does not rotate. The radial line moves from its initial orientation at angle θ_0 to a final orientation at angle θ (Greek letter theta). In the process, the line sweeps out the angle $\theta - \theta_0$. As with other differences that we have encountered ($\Delta x = x - x_0$, $\Delta v = v - v_0$, $\Delta t = t - t_0$), it is customary to denote the difference between the final and initial angles by the notation $\Delta\theta$ (read as "delta theta"): $\Delta\theta = \theta - \theta_0$. The angle $\Delta\theta$ is the angular displacement. A rotating object may rotate either counterclockwise or clockwise, and standard convention calls a counterclockwise displacement positive and a clockwise displacement negative.

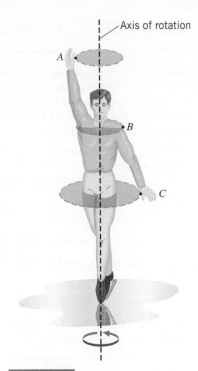

FIGURE 8.1 When a rigid object rotates, points on the object, such as A, B, or C, move on circular paths. The centers of the circles form a line that is the axis of rotation.

DEFINITION OF ANGULAR DISPLACEMENT

When a rigid body rotates about a fixed axis, the angular displacement is the angle $\Delta\theta$ swept out by a line passing through any point on the body and intersecting the axis of rotation perpendicularly. By convention, the angular displacement is positive if it is counterclockwise and negative if it is clockwise.

SI Unit of Angular Displacement: radian (rad)*

Angular displacement is often expressed in one of three units. The first is the familiar **degree**, and it is well known that there are 360 degrees in a circle. The second unit is the **revolution (rev)**, one revolution representing one complete turn of 360°. The most useful unit from a scientific viewpoint, however, is the SI unit called the **radian (rad)**. **Figure 8.3** shows how the radian is defined, again using a disc as an example. The picture focuses attention on a point P on the disc. This point starts out on the stationary reference line, so that $\theta_0 = 0$ rad, and the angular displacement is $\Delta\theta = \theta - \theta_0 = \theta$. As the disc rotates, the point traces out an arc of length s, which is measured along a circle of radius r. Equation 8.1 defines the angle θ in radians:

$$\theta(\text{in radians}) = \frac{\text{Arc length}}{\text{Radius}} = \frac{s}{r} \tag{8.1}$$

According to this definition, an angle in radians is the ratio of two lengths; for example, meters/meters. In calculations, therefore, the radian is treated as a number without units and has no effect on other units that it multiplies or divides.

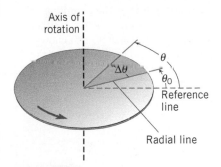

FIGURE 8.2 The angular displacement of a rotating disc is the angle $\Delta\theta$ swept out by a radial line as the disc turns about its axis of rotation.

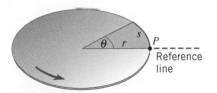

FIGURE 8.3 In radian measure, the angle θ is defined to be the arc length s divided by the radius r.

To convert between degrees and radians, it is only necessary to remember that the arc length of an entire circle of radius r is the circumference $2\pi r$. Therefore, according to Equation 8.1, *the number of radians that corresponds to 360°, or one revolution, is*

$$\theta = \frac{2\pi r}{r} = 2\pi \text{ rad}$$

Since 2π rad corresponds to 360°, the number of degrees in one radian is

$$1 \text{ rad} = \frac{360°}{2\pi} = 57.3°$$

It is useful to express an angle θ in radians, because then the arc length s subtended at any radius r can be calculated by multiplying θ by r. Example 1 illustrates this point and also shows how to convert between degrees and radians.

---

*The radian is neither a base SI unit nor a derived one. It is regarded as a supplementary SI unit.

EXAMPLE 1 | The Physics of Synchronous Communications Satellites

Synchronous or "stationary" communications satellites are put into an orbit whose radius is $r = 4.23 \times 10^7$ m. The orbit is in the plane of the equator, and two adjacent satellites have an angular separation of $\theta = 2.00°$, as **Figure 8.4** illustrates. Find the arc length s (see the drawing) that separates the satellites.

Reasoning Since the radius r and the angle θ are known, we may find the arc length s by using the relation θ (in radians) $= s/r$. But first, the angle must be converted to radians from degrees.

Solution To convert 2.00° into radians, we use the fact that 2π radians is equivalent to 360°:

$$2.00° = (2.00 \text{ degrees})\left(\frac{2\pi \text{ radians}}{360 \text{ degrees}}\right) = 0.0349 \text{ radians}$$

From Equation 8.1, it follows that the arc length between the satellites is

$$s = r\theta = (4.23 \times 10^7 \text{ m})(0.0349 \text{ rad}) = \boxed{1.48 \times 10^6 \text{ m (920 miles)}}$$

The radian, being a unitless quantity, is dropped from the final result, leaving the answer expressed in meters.

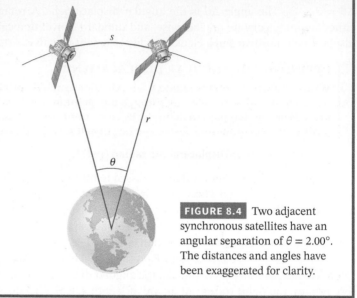

FIGURE 8.4 Two adjacent synchronous satellites have an angular separation of $\theta = 2.00°$. The distances and angles have been exaggerated for clarity.

Conceptual Example 2 takes advantage of the radian as a unit for measuring angles and explains the spectacular phenomenon of a total solar eclipse.

CONCEPTUAL EXAMPLE 2 | The Physics of a Total Solar Eclipse

The diameter of the sun is about 400 times greater than that of the moon. By coincidence, the sun is also about 400 times farther from the earth than is the moon. For an observer on earth, compare the angles subtended by the sun and the moon. **(a)** The angle subtended by the sun is much greater than that subtended by the moon. **(b)** The angle subtended by the sun is much smaller than that subtended by the moon. **(c)** The angles subtended by the sun and the moon are approximately equal.

Reasoning Equation 8.1 ($\theta = s/r$), which is the definition of an angle in radians, will guide us. The distance r between either the sun and the earth or the moon and the earth is great enough that the arc length s is very nearly equal to the diameter of the sun or the moon.

Answers (a) and (b) are incorrect. Interactive Figure 8.5a shows a person on earth viewing the sun and the moon.

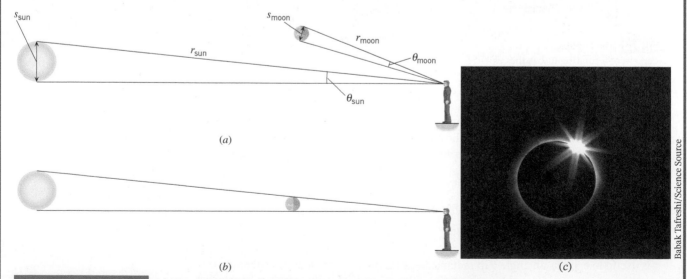

(a)

(b)

(c)

INTERACTIVE FIGURE 8.5 (a) The angles subtended by the moon and sun at the eyes of the observer are θ_{moon} and θ_{sun}. (The distances and angles are exaggerated for the sake of clarity.) (b) Since the moon and sun subtend approximately the same angle, the moon blocks nearly all the sun's light from reaching the observer's eyes. (c) The result is a total solar eclipse.

Babak Tafreshi/Science Source

For the diameters s of the sun and the moon, we know that $s_{sun} \approx 400 s_{moon}$ (where the symbol $\approx$ means "approximately equal to"). For the distances r between the sun and the earth or the moon and the earth, we know that $r_{sun} \approx 400 r_{moon}$. Because of these facts, the ratio s/r is approximately the same for the sun and for the moon. Therefore, the subtended angle ($\theta = s/r$) is approximately the same for the sun and the moon.

Answer (c) is correct. Applying Equation 8.1 to the case of the sun and the moon gives $\theta_{sun} = s_{sun}/r_{sun}$ and $\theta_{moon} = s_{moon}/r_{moon}$.

We know, however, that $s_{sun} \approx 400 s_{moon}$ and $r_{sun} \approx 400 r_{moon}$. Substitution into the expression for θ_{sun} gives

$$\theta_{sun} = \frac{s_{sun}}{r_{sun}} \approx \frac{400 \, s_{moon}}{400 \, r_{moon}} = \theta_{moon}$$

Interactive Figure 8.5b shows what happens when the moon comes between the sun and the earth. Since the angles subtended by the sun and the moon are nearly equal, the moon blocks most of the sun's light from reaching the observer's eyes, and a total solar eclipse like that in **Interactive Figure 8.5c** occurs.

Related Homework: Problem 7, 10

Check Your Understanding

(The answers are given at the end of the book.)

1. In **CYU Figure 8.1**, the flat triangular sheet *ABC* is lying in the plane of the paper. This sheet is going to rotate about first one axis and then another axis. Both of these axes lie in the plane of the paper and pass through point *A*. For each of the axes the points *B* and *C* move on separate circular paths that have the same radii. Identify these two axes.

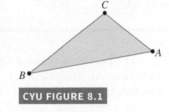

CYU FIGURE 8.1

2. Three objects are visible in the night sky. They have the following diameters (in multiples of d) and subtend the following angles (in multiples of θ_0) at the eye of the observer. Object A has a diameter of $4d$ and subtends an angle of $2\theta_0$. Object B has a diameter of $3d$ and subtends an angle of $\theta_0/2$. Object C has a diameter of $d/2$ and subtends an angle of $\theta_0/8$. Rank them in descending order (greatest first) according to their distance from the observer.

8.2 | Angular Velocity and Angular Acceleration

Angular Velocity

In Section 2.2 we introduced the idea of linear velocity to describe how fast an object moves and the direction of its motion. The average linear velocity $\vec{v}$ was defined as the linear displacement $\Delta \vec{x}$ of the object divided by the time Δt required for the displacement to occur, or $\vec{v} = \Delta \vec{x}/\Delta t$ (see Equation 2.2). We now introduce the analogous idea of angular velocity to describe the motion of a rigid object rotating about a fixed axis. The **average angular velocity** $\bar{\omega}$ (Greek letter omega) is defined as the angular displacement $\Delta \theta = \theta - \theta_0$ divided by the elapsed time Δt during which the displacement occurs.

> **DEFINITION OF AVERAGE ANGULAR VELOCITY**
>
> $$\text{Average angular velocity} = \frac{\text{Angular displacement}}{\text{Elapsed time}}$$
>
> $$\bar{\omega} = \frac{\theta - \theta_0}{t - t_0} = \frac{\Delta \theta}{\Delta t} \qquad (8.2)$$
>
> **SI Unit of Angular Velocity: radian per second (rad/s)**

The SI unit for angular velocity is the radian per second (rad/s), although other units such as revolutions per minute (rev/min or rpm) are also used. In agreement with the sign convention adopted for angular displacement, angular velocity is positive when the rotation is counterclockwise and negative when it is clockwise. Example 3 shows how the concept of average angular velocity is applied to a gymnast.

EXAMPLE 3 | Gymnast on a High Bar

A gymnast on a high bar swings through two revolutions in a time of 1.90 s, as **Figure 8.6** suggests. Find the average angular velocity (in rad/s) of the gymnast.

Reasoning   The average angular velocity of the gymnast in rad/s is the angular displacement in radians divided by the elapsed time. However, the angular displacement is given as two revolutions, so we begin by converting this value into radian measure.

Solution   The angular displacement (in radians) of the gymnast is

$$\Delta\theta = -2.00 \text{ revolutions} \left(\frac{2\pi \text{ radians}}{1 \text{ revolution}} \right) = -12.6 \text{ radians}$$

where the minus sign denotes that the gymnast rotates clockwise (see the drawing). The average angular velocity is

$$\overline{\omega} = \frac{\Delta\theta}{\Delta t} = \frac{-12.6 \text{ rad}}{1.90 \text{ s}} = \boxed{-6.63 \text{ rad/s}} \qquad (8.2)$$

FIGURE 8.6   Swinging on a high bar.

The **instantaneous angular velocity** ω is the angular velocity that exists at any given instant. To measure it, we follow the same procedure used in Chapter 2 for the instantaneous linear velocity. In this procedure, a small angular displacement $\Delta\theta$ occurs during a small time interval Δt. The time interval is so small that it approaches zero ($\Delta t \to 0$), and in this limit, the measured average angular velocity, $\overline{\omega} = \Delta\theta/\Delta t$, becomes the instantaneous angular velocity ω:

$$\omega = \lim_{\Delta t \to 0} \overline{\omega} = \lim_{\Delta t \to 0} \frac{\Delta\theta}{\Delta t} \qquad (8.3)$$

The magnitude of the instantaneous angular velocity, without reference to whether it is a positive or negative quantity, is called the **instantaneous angular speed**. If a rotating object has a constant angular velocity, the instantaneous value and the average value are the same.

Angular Acceleration

In linear motion, a changing velocity means that an acceleration is occurring. Such is also the case in rotational motion; a changing angular velocity means that an **angular acceleration** is occurring. There are many examples of angular acceleration. For instance, when the push buttons of an electric blender are changed from a lower setting to a higher setting, the angular velocity of the blades increases. We will define the average angular acceleration in a fashion analogous to that used for the average linear acceleration. Recall that the average linear acceleration $\vec{\overline{a}}$ is equal to the change $\Delta\vec{v}$ in the linear velocity of an object divided by the elapsed time Δt: $\vec{\overline{a}} = \Delta\vec{v}/\Delta t$ (Equation 2.4). When the angular velocity changes from an initial value of ω_0 at time t_0 to a final value of ω at time t, the average angular acceleration $\overline{\alpha}$ (Greek letter alpha) is defined similarly:

> **DEFINITION OF AVERAGE ANGULAR ACCELERATION**
>
> $$\text{Average angular acceleration} = \frac{\text{Change in angular velocity}}{\text{Elapsed time}}$$
>
> $$\overline{\alpha} = \frac{\omega - \omega_0}{t - t_0} = \frac{\Delta\omega}{\Delta t} \qquad (8.4)$$
>
> **SI Unit of Average Angular Acceleration: radian per second squared (rad/s²)**

The SI unit for average angular acceleration is the unit for angular velocity divided by the unit for time, or (rad/s)/s = rad/s². An angular acceleration of +5 rad/s², for example, means that the angular velocity of the rotating object increases by +5 radians per second during each second of acceleration.

The **instantaneous angular acceleration** α is the angular acceleration at a given instant. In discussing linear motion, we assumed a condition of constant acceleration, so that the average and instantaneous accelerations were identical ($\vec{\bar{a}} = \vec{a}$). Similarly, we assume that the angular acceleration is constant, so that the instantaneous angular acceleration α and the average angular acceleration $\bar{\alpha}$ are the same ($\bar{\alpha} = \alpha$). The next example illustrates the concept of angular acceleration.

EXAMPLE 4 | A Jet Revving Its Engines

A jet awaiting clearance for takeoff is momentarily stopped on the runway. As seen from the front of one engine, the fan blades are rotating with an angular velocity of -110 rad/s, where the negative sign indicates a clockwise rotation (see **Figure 8.7**). As the plane takes off, the angular velocity of the blades reaches -330 rad/s in a time of 14 s. Find the angular acceleration, assuming it to be constant.

Reasoning Since the angular acceleration is constant, it is equal to the average angular acceleration. The average acceleration is the change in the angular velocity, $\omega - \omega_0$, divided by the elapsed time, $t - t_0$.

Solution Applying the definition of average angular acceleration given in Equation 8.4, we find that

$$\bar{\alpha} = \frac{\omega - \omega_0}{t - t_0} = \frac{(-330 \text{ rad/s}) - (-110 \text{ rad/s})}{14 \text{ s}} = \boxed{-16 \text{ rad/s}^2}$$

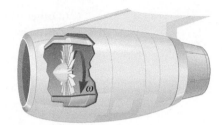

FIGURE 8.7 The fan blades of a jet engine have an angular velocity ω in a clockwise direction.

Thus, the magnitude of the angular velocity increases by 16 rad/s during each second that the blades are accelerating. The negative sign in the answer indicates that the direction of the angular acceleration is also in the clockwise direction.

Check Your Understanding

(The answers are given at the end of the book.)

3. A pair of scissors is being used to cut a piece of paper in half. Does each blade of the scissors have the same angular velocity (both magnitude and direction) at a given instant?

4. An electric clock is hanging on a wall. As you are watching the second hand rotate, the clock's battery stops functioning, and the second hand comes to a halt over a brief period of time. Which one of the following statements correctly describes the angular velocity ω and angular acceleration α of the second hand as it slows down? **(a)** ω and α are both negative. **(b)** ω is positive and α is negative. **(c)** ω is negative and α is positive. **(d)** ω and α are both positive.

8.3 The Equations of Rotational Kinematics

In Chapters 2 and 3 the concepts of displacement, velocity, and acceleration were introduced. We then combined these concepts and developed a set of equations called the equations of kinematics for constant acceleration (see Tables 2.1 and 3.1). These equations are a great aid in solving problems involving linear motion in one and two dimensions.

We now take a similar approach for rotational motion by bringing together the ideas of angular displacement, angular velocity, and angular acceleration to produce a set of equations called the equations of rotational kinematics for constant angular acceleration. These equations, like those developed in Chapters 2 and 3 for linear motion, will prove very useful in solving problems that involve rotational motion.

A complete description of rotational motion requires values for the angular displacement $\Delta\theta$, the angular acceleration α, the final angular velocity ω, the initial angular velocity ω_0, and the elapsed time Δt. In Example 4, for instance, only the angular displacement of the fan blades during the 14-s interval is missing. Such missing information can be calculated, however. For convenience in the calculations, we assume that the orientation of the rotating object is given by $\theta_0 = 0$ rad at time $t_0 = 0$ s. Then, the angular displacement becomes $\Delta\theta = \theta - \theta_0 = \theta$, and the time interval becomes $\Delta t = t - t_0 = t$.

In Example 4, the angular velocity of the fan blades changes at a constant rate from an initial value of $\omega_0 = -110$ rad/s to a final value of $\omega = -330$ rad/s. Therefore, the average angular velocity is midway between the initial and final values:

$$\overline{\omega} = \tfrac{1}{2}[(-110 \text{ rad/s}) + (-330 \text{ rad/s})] = -220 \text{ rad/s}$$

In other words, when the angular acceleration is constant, the average angular velocity is given by

$$\overline{\omega} = \tfrac{1}{2}(\omega_0 + \omega) \tag{8.5}$$

With a value for the average angular velocity, Equation 8.2 can be used to obtain the angular displacement of the fan blades:

$$\theta = \overline{\omega}t = (-220 \text{ rad/s})(14 \text{ s}) = -3100 \text{ rad}$$

In general, when the angular acceleration is constant, the angular displacement can be obtained from

$$\theta = \overline{\omega}t = \tfrac{1}{2}(\omega_0 + \omega)t \tag{8.6}$$

This equation and Equation 8.4 provide a complete description of rotational motion under the condition of constant angular acceleration. Equation 8.4 (with $t_0 = 0$ s) and Equation 8.6 are compared with the analogous results for linear motion in the first two rows of Table 8.1. The purpose of this comparison is to emphasize that the mathematical forms of Equations 8.4 and 2.4 are identical, as are the forms of Equations 8.6 and 2.7. Of course, the symbols used for the rotational variables are different from those used for the linear variables, as Table 8.2 indicates.

TABLE 8.1	The Equations of Kinematics for Rotational and Linear Motion		
Rotational Motion (α = constant)		**Linear Motion (a = constant)**	
$\omega = \omega_0 + \alpha t$	(8.4)	$v = v_0 + at$	(2.4)
$\theta = \tfrac{1}{2}(\omega_0 + \omega)t$	(8.6)	$x = \tfrac{1}{2}(v_0 + v)t$	(2.7)
$\theta = \omega_0 t + \tfrac{1}{2}\alpha t^2$	(8.7)	$x = v_0 t + \tfrac{1}{2}at^2$	(2.8)
$\omega^2 = \omega_0^2 + 2\alpha\theta$	(8.8)	$v^2 = v_0^2 + 2ax$	(2.9)

TABLE 8.2	Symbols Used in Rotational and Linear Kinematics	
Rotational Motion	**Quantity**	**Linear Motion**
θ	Displacement	x
ω_0	Initial velocity	v_0
ω	Final velocity	v
α	Acceleration	a
t	Time	t

In Chapter 2, Equations 2.4 and 2.7 are used to derive the remaining two equations of kinematics (Equations 2.8 and 2.9). These additional equations convey no new information but are convenient to have when solving problems. Similar derivations can be carried out here. The results are listed as Equations 8.7 and 8.8 below and in Table 8.1; they can be inferred directly from their counterparts in linear motion by making the substitution of symbols indicated in Table 8.2:

$$\theta = \omega_0 t + \tfrac{1}{2}\alpha t^2 \tag{8.7}$$
$$\omega^2 = \omega_0^2 + 2\alpha\theta \tag{8.8}$$

The four equations in the left column of Table 8.1 are called the **equations of rotational kinematics for constant angular acceleration**. The following example illustrates that they are used in the same fashion as the equations of linear kinematics.

EXAMPLE 5 | Blending with a Blender

The blades of an electric blender are whirling with an angular velocity of +375 rad/s while the "puree" button is pushed in, as **Figure 8.8** shows. When the "blend" button is pressed, the blades accelerate and reach a greater angular velocity after the blades have rotated through an angular displacement of +44.0 rad. The angular acceleration has a constant value of +1740 rad/s². Find the final angular velocity of the blades.

Reasoning The three known variables are listed in the following table, along with a question mark indicating that a value for the final angular velocity ω is being sought.

θ	α	ω	ω_0	t
+44.0 rad	+1740 rad/s²	?	+375 rad/s	

We can use Equation 8.8, because it relates the angular variables $\theta, \alpha, \omega,$ and ω_0.

> **Problem-Solving Insight** Each equation of rotational kinematics contains four of the five kinematic variables, $\theta, \alpha, \omega, \omega_0,$ and t. Therefore, it is necessary to have values for three of these variables if one of the equations is to be used to determine a value for an unknown variable.

Axis of rotation

FIGURE 8.8 The angular velocity of the blades in an electric blender changes each time a different push button is chosen.

Puree   Blend

Solution From Equation 8.8 ($\omega^2 = \omega_0^2 + 2\alpha\theta$) it follows that

$$\omega = +\sqrt{\omega_0^2 + 2\alpha\theta} = +\sqrt{(375 \text{ rad/s})^2 + 2(1740 \text{ rad/s}^2)(44.0 \text{ rad})}$$

$$= \boxed{+542 \text{ rad/s}}$$

The negative root is disregarded, since the blades do not reverse their direction of rotation.

> **Problem-Solving Insight** *The equations of rotational kinematics can be used with any self-consistent set of units for $\theta, \alpha, \omega, \omega_0,$ and t.* Radians are used in Example 5 only because data are given in terms of radians. Had the data for $\theta, \alpha,$ and ω_0 been provided in rev, rev/s², and rev/s, respectively, then Equation 8.8 could have been used to determine the answer for ω directly in rev/s. In any case, the reasoning strategy for applying the kinematics equations is summarized as follows.

REASONING STRATEGY Applying the Equations of Rotational Kinematics

1. Make a drawing to represent the situation being studied, showing the direction of rotation.

2. Decide which direction of rotation is to be called positive (+) and which direction is to be called negative (−). In this text we choose the counterclockwise direction to be positive and the clockwise direction to be negative, but this is arbitrary. However, do not change your decision during the course of a calculation.

3. In an organized way, write down the values (with appropriate + and − signs) that are given for any of the five rotational kinematic variables ($\theta, \alpha, \omega, \omega_0,$ and t). Be on the alert for implied data, such as the phrase "starts from rest," which means that the value of the initial angular velocity is $\omega_0 = 0$ rad/s. The data table in Example 5 is a good way to keep track of this information. In addition, identify the variable(s) that you are being asked to determine.

4. Before attempting to solve a problem, verify that the given information contains values for at least three of the five kinematic variables. Once the three variables are identified for which values are known, the appropriate relation from Table 8.1 can be selected.

5. When the rotational motion is divided into segments, the final angular velocity of one segment is the initial angular velocity for the next segment.

6. Keep in mind that there may be two possible answers to a kinematics problem. Try to visualize the different physical situations to which the answers correspond.

Check Your Understanding

(The answers are given at the end of the book.)

5. The blades of a ceiling fan start from rest and, after two revolutions, have an angular speed of 0.50 rev/s. The angular acceleration of the blades is constant. What is the angular speed after eight revolutions?

6. Equation 8.7 ($\theta = \omega_0 t + \frac{1}{2}\alpha t^2$) is being used to solve a problem in rotational kinematics. Which one of the following sets of values for the variables ω_0, α, and t *cannot* be substituted directly into this equation to calculate a value for θ? **(a)** $\omega_0 = 1.0$ rad/s, $\alpha = 1.8$ rad/s^2, and $t = 3.8$ s **(b)** $\omega_0 = 0.16$ rev/s, $\alpha = 1.8$ rad/s^2, and $t = 3.8$ s **(c)** $\omega_0 = 0.16$ rev/s, $\alpha = 0.29$ rev/s^2, and $t = 3.8$ s

8.4 Angular Variables and Tangential Variables

In the familiar ice-skating stunt known as crack-the-whip, a number of skaters attempt to maintain a straight line as they skate around the one person (the pivot) who remains in place. **Figure 8.9** shows each skater moving on a circular arc and includes the corresponding velocity vector at the instant portrayed in the picture. For every individual skater, the vector is drawn tangent to the appropriate circle and, therefore, is called the **tangential velocity** $\vec{v}_T$. The magnitude of the tangential velocity is referred to as the **tangential speed**.

THE PHYSICS OF . . . "crack-the-whip." Of all the skaters involved in the stunt, the one farthest from the pivot has the hardest job. Why? Because, in keeping the line straight, this skater covers more distance than anyone else. To accomplish this, he must skate faster than anyone else and, thus, must have the largest tangential speed. In fact, the line remains straight only if each person skates with the correct tangential speed. The skaters closer to the pivot must move with smaller tangential speeds than those farther out, as indicated by the magnitudes of the tangential velocity vectors in **Figure 8.9**.

With the aid of **Figure 8.10**, it is possible to show that the tangential speed of any skater is directly proportional to the skater's distance r from the pivot, assuming a given angular speed for the rotating line. When the line rotates as a rigid unit for a time t, it sweeps out the angle θ shown in the drawing. The distance s through which a skater moves along a circular arc can be calculated from $s = r\theta$ (Equation 8.1), provided that θ is measured in radians. Dividing both sides of this equation by t, we find that $s/t = r(\theta/t)$. The term s/t is the tangential speed v_T (e.g., in meters/second) of the skater, while θ/t is the angular speed ω (in radians/second) of the line:

$$v_T = r\omega \qquad (\omega \text{ in rad/s}) \tag{8.9}$$

In this expression, the terms v_T and ω refer to the magnitudes of the tangential and angular velocities, respectively, and are numbers without algebraic signs.

It is important to emphasize that the angular speed ω in Equation 8.9 must be expressed in radian measure (e.g., in rad/s); no other units, such as revolutions per second, are acceptable. This restriction arises because the equation was derived by using the definition of radian measure, $s = r\theta$.

The real challenge for the crack-the-whip skaters is to keep the line straight while making it pick up angular speed—that is, while giving it an angular acceleration. To make the angular speed of the line increase, each skater must increase his tangential speed, since the two speeds are related according to $v_T = r\omega$. Of course, the fact that a skater must skate faster and faster means that he must accelerate, and his tangential

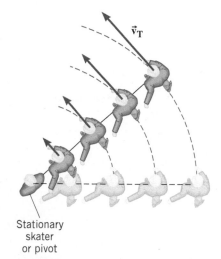

FIGURE 8.9 When doing the stunt known as crack-the-whip, each skater along the radial line moves on a circular arc. The tangential velocity $\vec{v}_T$ of each skater is represented by an arrow that is tangent to each arc.

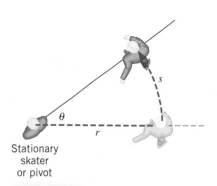

FIGURE 8.10 During a time t, the line of skaters sweeps through an angle θ. An individual skater, located at a distance r from the stationary skater, moves through a distance s on a circular arc.

acceleration a_T can be related to the angular acceleration α of the line. If the time is measured relative to $t_0 = 0$ s, the definition of linear acceleration is given by Equation 2.4 as $a_T = (v_T - v_{T0})/t$, where v_T and v_{T0} are the final and initial tangential speeds, respectively. Substituting $v_T = r\omega$ for the tangential speed shows that

$$a_T = \frac{v_T - v_{T0}}{t} = \frac{(r\omega) - (r\omega_0)}{t} = r\left(\frac{\omega - \omega_0}{t}\right)$$

Since $\alpha = (\omega - \omega_0)/t$ according to Equation 8.4, it follows that

$$a_T = r\alpha \qquad (\alpha \text{ in rad/s}^2) \qquad\qquad \textbf{(8.10)}$$

This result shows that, for a given value of α, the tangential acceleration a_T is proportional to the radius r, so the skater farthest from the pivot must have the largest tangential acceleration. In this expression, the terms a_T and α refer to the magnitudes of the numbers involved, without reference to any algebraic sign. Moreover, as is the case for ω in $v_T = r\omega$, only radian measure can be used for α in Equation 8.10.

There is an advantage to using the angular velocity ω and the angular acceleration α to describe the rotational motion of a rigid object. The advantage is that these angular quantities describe the motion of the *entire object*. In contrast, the tangential quantities v_T and a_T describe only the motion of a single point on the object, and Equations 8.9 and 8.10 indicate that different points located at different distances r have different tangential velocities and accelerations. Example 6 stresses this advantage.

EXAMPLE 6 | A Helicopter Blade

A helicopter blade has an angular speed of $\omega = 6.50$ rev/s and has an angular acceleration of $\alpha = 1.30$ rev/s^2. For points 1 and 2 on the blade in **Figure 8.11**, find the magnitudes of **(a)** the tangential speeds and **(b)** the tangential accelerations.

Reasoning Since the radius r for each point and the angular speed ω of the helicopter blade are known, we can find the tangential speed v_T for each point by using the relation $v_T = r\omega$. However, since this equation can be used only with radian measure, the angular speed ω must be converted to rad/s from rev/s. In a similar manner, the tangential acceleration a_T for points 1 and 2 can be found using $a_T = r\alpha$, provided the angular acceleration α is expressed in rad/s^2 rather than in rev/s^2.

> **Problem-Solving Insight** When using Equations 8.9 and 8.10 to relate tangential and angular quantities, remember that the angular quantity is always expressed in radian measure. These equations are not valid if angles are expressed in degrees or revolutions.

Solution **(a)** Converting the angular speed ω to rad/s from rev/s, we obtain

$$\omega = \left(6.50\,\frac{\text{rev}}{\text{s}}\right)\left(\frac{2\pi\text{ rad}}{1\text{ rev}}\right) = 40.8\,\frac{\text{rad}}{\text{s}}$$

The tangential speed for each point is

Point 1 $v_T = r\omega = (3.00\text{ m})(40.8\text{ rad/s})$
$$= \boxed{122\text{ m/s } (273\text{ mph})} \qquad\qquad \textbf{(8.9)}$$

Point 2 $v_T = r\omega = (6.70\text{ m})(40.8\text{ rad/s})$
$$= \boxed{273\text{ m/s } (611\text{ mph})} \qquad\qquad \textbf{(8.9)}$$

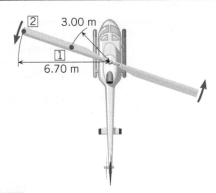

FIGURE 8.11 Points 1 and 2 on the rotating blade of the helicopter have the same angular speed and acceleration, but they have *different* tangential speeds and accelerations.

The rad unit, being dimensionless, does not appear in the final answers.

(b) Converting the angular acceleration α to rad/s^2 from rev/s^2, we find

$$\alpha = \left(1.30\,\frac{\text{rev}}{\text{s}^2}\right)\left(\frac{2\pi\text{ rad}}{1\text{ rev}}\right) = 8.17\,\frac{\text{rad}}{\text{s}^2}$$

The tangential accelerations can now be determined:

Point 1 $a_T = r\alpha = (3.00\text{ m})(8.17\text{ rad/s}^2) = \boxed{24.5\text{ m/s}^2}$ **(8.10)**

Point 2 $a_T = r\alpha = (6.70\text{ m})(8.17\text{ rad/s}^2) = \boxed{54.7\text{ m/s}^2}$ **(8.10)**

EXAMPLE 7 | BIO Wing Speed of the Rufous Hummingbird

Rufous hummingbirds have a top linear speed of about 50 mi/h, and fly over 2000 miles during their migratory transits. However, what stands out compared to other birds is their ability to dart and hover. This is due in part to their small body mass (3.4 g) combined with their high wingbeat frequency (41 beat cycles per second). If the length of a single hummingbird wing is 47 mm, and the angular displacement of the wing from the top to the bottom of the stroke (i.e., half of a full wingbeat) is 105°, determine **(a)** the average angular speed (in rad/s) of the wing and **(b)** the average linear speed (in m/s) of the wingtip.

Reasoning **(a)** Since we know the wingbeat frequency in cycle/s and the angular displacement from the top to the bottom of the stroke, we can calculate the average angular speed using Equation 8.2. **(b)** Once we have the average angular speed, then we can find the average linear speed (i.e., the tangential speed) of the wingtip using Equation 8.9.

Solution **(a)** The total angular displacement during a full cycle is equal to $2 \times 105° = 210°$. Since the wingbeat frequency is 41 cycles/s, the magnitude of the total angular displacement in one second is given by

$$\Delta\theta = (41 \text{ cycles})(210°) = 8610°$$

Next, we can write this in radians by multiplying by the proper conversion factor as follows:

$$\Delta\theta = (8610 \text{ deg})\left(\frac{2\pi \text{ rad}}{360 \text{ deg}}\right) = 150 \text{ rad}$$

The average angular speed is then given by Equation 8.2 as

$$\overline{\omega} = \frac{\Delta\theta}{\Delta t} = \frac{150 \text{ rad}}{1 \text{ s}} = \boxed{150 \text{ rad/s}}$$

This is the average angular speed, which means that much higher values are achieved during a complete wingbeat cycle. Peak angular velocities of hummingbird wings can exceed 200 rad/s! Compare this to the angular speed of the helicopter blades (40.8 rad/s) in Example 6!

(b) The average tangential speed of the wingtips can be calculated from the average angular speed and the wing length (r) using Equation 8.9:

$$\overline{v}_{\text{T}} = r\overline{\omega} = (0.047 \text{ m})(150 \text{ rad/s}) = \boxed{7.1 \text{ m/s}},$$

where we have modified the equation by placing bars over the variables to specify that they are average values. This is equal to 16 mi/h. Again, this is an average value, so the maximum speed is much larger. Compare this to the speed of the tips of the helicopter blades (611 mi/h) in Example 6. Although the angular speed of the hummingbird wings is much greater than that of the helicopter blades, the helicopter blades are much longer, and this is what accounts for the much greater tangential speed of the helicopter blade tips.

Check Your Understanding

(The answers are given at the end of the book.)

7. A thin rod rotates at a constant angular speed. In case A the axis of rotation is perpendicular to the rod at its center. In case B the axis is perpendicular to the rod at one end. In which case, if either, are there points on the rod that have the same tangential speeds?

8. It is possible to build a clock in which the tips of the hour hand and the second hand move with the same tangential speed. This is normally never done, however. Why? **(a)** The length of the hour hand would be 3600 times greater than the length of the second hand. **(b)** The hour hand and the second hand would have the same length. **(c)** The length of the hour hand would be 3600 times smaller than the length of the second hand.

9. The earth rotates once per day about its axis, which is perpendicular to the plane of the equator and passes through the north geographic pole. Where on the earth's surface should you stand in order to have the smallest possible tangential speed?

10. A building is located on the earth's equator. As the earth rotates about its axis, which floor of the building has the greatest tangential speed? **(a)** The first floor **(b)** The tenth floor **(c)** The twentieth floor

11. The blade of a lawn mower is rotating at an angular speed of 17 rev/s. The tangential speed of the outer edge of the blade is 32 m/s. What is the radius of the blade?

8.5 Centripetal Acceleration and Tangential Acceleration

When an object picks up speed as it moves around a circle, it has a tangential acceleration, as discussed in the previous section. In addition, the object also has a centripetal acceleration, as emphasized in Chapter 5. That chapter deals with **uniform circular motion**, in which a particle moves at a constant tangential speed on a circular path. The tangential speed v_T is the magnitude of the tangential velocity vector. Even when the magnitude of the tangential velocity is constant, an acceleration is present, since the direction of the velocity changes continually. Because the resulting acceleration points toward the center of the circle, it is called the centripetal acceleration. **Interactive Figure 8.12a** shows the centripetal acceleration $\vec{a}_c$ for a model airplane flying in uniform circular motion on a guide wire. The magnitude of $\vec{a}_c$ is

$$a_c = \frac{v_T^2}{r} \tag{5.2}$$

The subscript "T" has now been included in this equation as a reminder that it is the tangential speed that appears in the numerator.

The centripetal acceleration can be expressed in terms of the angular speed ω by using $v_T = r\omega$ (Equation 8.9):

$$a_c = \frac{v_T^2}{r} = \frac{(r\omega)^2}{r} = r\omega^2 \qquad (\omega \text{ in rad/s}) \tag{8.11}$$

Only radian measure (rad/s) can be used for ω in this result, since the relation $v_T = r\omega$ presumes radian measure.

While considering uniform circular motion in Chapter 5, we ignored the details of how the motion is established in the first place. In **Interactive Figure 8.12b**, for instance, the engine of the plane produces a thrust in the tangential direction, and this force leads to a tangential acceleration. In response, the tangential speed of the plane increases from moment to moment, until the situation shown in the drawing results. While the tangential speed is changing, the motion is called **nonuniform circular motion**.

Interactive Figure 8.12b illustrates an important feature of nonuniform circular motion. Since the direction and the magnitude of the tangential velocity are both changing, the airplane experiences two acceleration components simultaneously. The changing direction means that there is a centripetal acceleration $\vec{a}_c$. The magnitude of $\vec{a}_c$ at any moment can be calculated using the value of the instantaneous angular speed and the radius: $a_c = r\omega^2$. The fact that the magnitude of the tangential velocity is changing means that there is also a tangential acceleration $\vec{a}_T$. The magnitude of $\vec{a}_T$ can be determined from the angular acceleration α according to $a_T = r\alpha$, as the previous section explains. If the magnitude F_T of the net tangential force and the mass m are known, a_T also can be calculated using Newton's second law, $F_T = ma_T$. **Interactive Figure 8.12b** shows the two acceleration components. The total acceleration is given by the vector sum of $\vec{a}_c$ and $\vec{a}_T$. Since $\vec{a}_c$ and $\vec{a}_T$ are perpendicular, the magnitude of the total acceleration $\vec{a}$ can be obtained from the Pythagorean theorem as $a = \sqrt{a_c^2 + a_T^2}$, while the angle ϕ in the drawing can be determined from $\tan \phi = a_T/a_c$. The next example applies these concepts to a discus thrower.

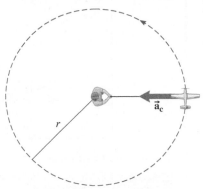

(a) Uniform circular motion

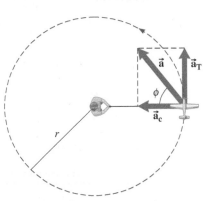

(b) Nonuniform circular motion

INTERACTIVE FIGURE 8.12   (a) If a model airplane flying on a guide wire has a constant tangential speed, the motion is uniform circular motion, and the plane experiences only a centripetal acceleration $\vec{a}_c$. (b) Nonuniform circular motion occurs when the tangential speed changes. Then there is a tangential acceleration $\vec{a}_T$ in addition to the centripetal acceleration.

Analyzing Multiple-Concept Problems

EXAMPLE 8 | A Discus Thrower

Discus throwers often warm up by throwing the discus with a twisting motion of their bodies. **Figure 8.13a** illustrates a top view of such a warm-up throw. Starting from rest, the thrower accelerates the discus to a final angular velocity of +15.0 rad/s in a time of 0.270 s before releasing it. During the acceleration, the discus moves on a circular arc of radius 0.810 m. Find the magnitude a of the total acceleration of the discus just before the discus is released.

Reasoning   Since the tangential speed of the discus increases as the thrower turns, the discus simultaneously experiences a tangential acceleration $\vec{a}_T$ and a centripetal acceleration $\vec{a}_c$ that are oriented at right angles to each other (see the drawing). The magnitude a of the total acceleration is $a = \sqrt{a_c^2 + a_T^2}$, where a_c and a_T are the magnitudes of the centripetal and tangential accelerations. The magnitude of the centripetal acceleration can be evaluated from Equation 8.11 ($a_c = r\omega^2$),

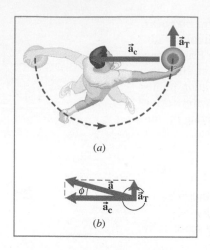

(a)

(b)

and the magnitude of the tangential acceleration follows from Equation 8.10 ($a_T = r\alpha$). The angular acceleration α can be found from its definition in Equation 8.4.

FIGURE 8.13 (a) A discus thrower and the centripetal acceleration $\vec{a}_c$ and tangential acceleration $\vec{a}_T$ of the discus. (b) The total acceleration $\vec{a}$ of the discus just before the discus is released is the vector sum of $\vec{a}_c$ and $\vec{a}_T$.

Knowns and Unknowns   The data for this problem are:

Description	Symbol	Value	Comment
Explicit Data			
Final angular velocity	ω	+15.0 rad/s	Positive, because discus moves counterclockwise (see drawing).
Time	t	0.270 s	
Radius of circular arc	r	0.810 m	
Implicit Data			
Initial angular velocity	ω_0	0 rad/s	Discus starts from rest.
Unknown Variable			
Magnitude of total acceleration	a	?	

Modeling the Problem

STEP 1   Total Acceleration   Figure 8.13b shows the two perpendicular components of the acceleration of the discus at the moment of release. The centripetal acceleration $\vec{a}_c$ arises because the discus is traveling on a circular path; this acceleration always points toward the center of the circle (see Section 5.2). The tangential acceleration $\vec{a}_T$ arises because the tangential velocity of the discus is increasing; this acceleration is tangent to the circle (see Section 8.4). Since $\vec{a}_c$ and $\vec{a}_T$ are perpendicular, we can use the Pythagorean theorem to find the magnitude a of the total acceleration, as given by Equation 1 at the right. Values for a_c and a_T will be obtained in the next two steps.

$$a = \sqrt{a_c^2 + a_T^2} \qquad (1)$$

$$\boxed{?} \quad \boxed{?}$$

STEP 2   Centripetal Acceleration   The magnitude a_c of the centripetal acceleration can be evaluated from Equation 8.11 as

$$\boxed{a_c = r\omega^2}$$

where r is the radius and ω is the angular speed of the discus. Both r and ω are known, so we can substitute this expression for a_c into Equation 1, as indicated at the right. In Step 3, we will evaluate the magnitude a_T of the tangential acceleration.

$$a = \sqrt{a_c^2 + a_T^2} \qquad (1)$$

$$\boxed{a_c = r\omega^2} \quad \boxed{?}$$

STEP 3   Tangential Acceleration   According to Equation 8.10, the magnitude a_T of the tangential acceleration is $a_T = r\alpha$, where r is the radius of the path and α is the magnitude of the angular acceleration. The angular acceleration is defined (see Equation 8.4) as the change $\omega - \omega_0$ in the angular velocity divided by the time t, or $\alpha = (\omega - \omega_0)/t$. Thus, the tangential acceleration can be written as

$$\boxed{a_T = r\alpha = r\left(\frac{\omega - \omega_0}{t}\right)}$$

All the variables on the right side of this equation are known, so we can substitute this expression for a_T into Equation 1 (see the right column).

$$a = \sqrt{a_c^2 + a_T^2} \qquad (1)$$

$$\boxed{a_c = r\omega^2} \quad \boxed{a_T = r\left(\frac{\omega - \omega_0}{t}\right)}$$

Solution   Algebraically combining the results of the three steps, we have:

STEP 1   STEP 2   STEP 3

$$a = \sqrt{a_c^2 + a_T^2} = \sqrt{r^2\omega^4 + a_T^2} = \sqrt{r^2\omega^4 + r^2\left(\frac{\omega - \omega_0}{t}\right)^2}$$

The magnitude of the total acceleration of the discus is

$$a = \sqrt{r^2\omega^4 + r^2\left(\frac{\omega - \omega_0}{t}\right)^2}$$

$$= \sqrt{(0.810 \text{ m})^2(15.0 \text{ rad/s})^4 + (0.810 \text{ m})^2 \left(\frac{15.0 \text{ rad/s} - 0 \text{ rad/s}}{0.270 \text{ s}}\right)^2} = \boxed{188 \text{ m/s}^2}$$

Note that we can also determine the angle ϕ between the total acceleration of the discus and its centripetal acceleration (see **Figure 8.13b**). From trigonometry, we have that $\tan \phi = a_T/a_c$, so

$$\phi = \tan^{-1}\left(\frac{a_T}{a_c}\right) = \tan^{-1}\left[\frac{r\left(\frac{\omega - \omega_0}{t}\right)}{r\omega^2}\right] = \tan^{-1}\left[\frac{\left(\frac{15.0 \text{ rad/s} - 0 \text{ rad/s}}{0.270 \text{ s}}\right)}{(15.0 \text{ rad/s})^2}\right] = 13.9° \longleftarrow$$

Related Homework: Problems 41, 43, 56

Math Skills   The tangent function ($\tan \phi$) is defined as $\tan \phi = \dfrac{h_o}{h_a}$ (Equation 1.3).

As shown in **Figure 8.14a**, h_o is the length of the side of a right triangle opposite the angle ϕ, and h_a is the length of the side adjacent to the angle ϕ. As always, the first step in applying such a trigonometric function is to identify this angle and its associated right triangle. **Figure 8.13b** establishes the angle ϕ and is reproduced in **Figure 8.14b**. A comparison of the shaded right triangles in the drawing reveals that $h_o = a_T$ and $h_a = a_c$. Thus, it follows that $\tan \phi = \dfrac{h_o}{h_a} = \dfrac{a_T}{a_c}$, and ϕ is the angle whose tangent is $\dfrac{a_T}{a_c}$. This result can be expressed by using the inverse tangent function ($\tan^{-1}$):

$$\phi = \tan^{-1}\left(\frac{a_T}{a_c}\right)$$

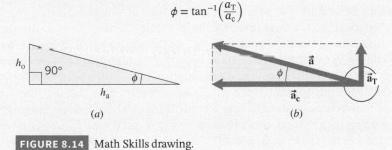

(a)   (b)

FIGURE 8.14   Math Skills drawing.

Check Your Understanding

(The answers are given at the end of the book.)

12. A car is up on a hydraulic lift at a garage. The wheels are free to rotate, and the drive wheels are rotating with a constant angular velocity. Which one of the following statements is true? **(a)** A point on the rim has no tangential and no centripetal acceleration. **(b)** A point on the rim has both a nonzero tangential acceleration and a nonzero centripetal acceleration. **(c)** A point on the rim has a nonzero tangential acceleration but no centripetal acceleration. **(d)** A point on the rim has no tangential acceleration but does have a nonzero centripetal acceleration.

13. Section 5.6 discusses how the uniform circular motion of a space station can be used to create artificial gravity. This can be done by adjusting the angular speed of the space station, so the centripetal acceleration at an astronaut's feet equals g, the magnitude of the acceleration due to the earth's gravity (see **Figure 5.19**). If such an adjustment is made, will the acceleration at the astronaut's head due to the artificial gravity be **(a)** greater than, **(b)** equal to, or **(c)** less than g?

14. A bicycle is turned upside down, the front wheel is spinning (see **CYU Figure 8.2**), and there is an angular acceleration. At the instant shown, there are six points on the wheel that have arrows associated with them. Which one of the following quantities could the arrows *not* represent? **(a)** Tangential velocity **(b)** Centripetal acceleration **(c)** Tangential acceleration

CYU FIGURE 8.2

15. A rotating object starts from rest and has a constant angular acceleration. Three seconds later the centripetal acceleration of a point on the object has a magnitude of 2.0 m/s². What is the magnitude of the centripetal acceleration of this point six seconds after the motion begins?

8.6 | Rolling Motion

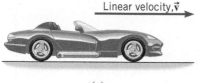

Linear velocity, $\vec{v}$

(a)

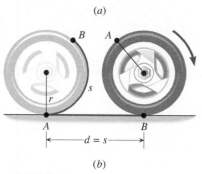

(b)

FIGURE 8.15 (*a*) An automobile moves with a linear speed v. (*b*) If the tires roll and do not slip, the distance d through which an axle moves equals the circular arc length s along the outer edge of a tire.

Rolling motion is a familiar situation that involves rotation, as **Figure 8.15** illustrates for the case of an automobile tire. The essence of rolling motion is that there is *no slipping* at the point of contact where the tire touches the ground. To a good approximation, the tires on a normally moving automobile roll and do not slip. In contrast, the squealing tires that accompany the start of a drag race are rotating, but they are not rolling while they rapidly spin and slip against the ground.

When the tires in **Figure 8.15** roll, there is a relationship between the angular speed at which the tires rotate and the linear speed (assumed constant) at which the car moves forward. In part *b* of the drawing, consider the points labeled *A* and *B* on the left tire. Between these points we apply a coat of red paint to the tread of the tire; the length of this circular arc of paint is *s*. The tire then rolls to the right until point *B* comes in contact with the ground. As the tire rolls, all the paint comes off the tire and sticks to the ground, leaving behind the horizontal red line shown in the drawing. The axle of the wheel moves through a linear distance *d*, which is equal to the length of the horizontal strip of paint. Since the tire does not slip, the distance *d* must be equal to the circular arc length *s*, measured along the outer edge of the tire: $d = s$. Dividing both sides of this equation by the elapsed time *t* shows that $d/t = s/t$. The term d/t is the speed at which the axle moves parallel to the ground—namely, the linear speed v of the car. The term s/t is the tangential speed v_T at which a point on the outer edge of the tire moves relative to the axle. In addition, v_T is related to the angular speed ω about the axle according to $v_\text{T} = r\omega$ (Equation 8.9). Therefore, it follows that

$$\underbrace{v}_{\substack{\text{Linear} \\ \text{speed}}} = \underbrace{v_\text{T} = r\omega}_{\substack{\text{Tangential} \\ \text{speed}}} \quad (\omega \text{ in rad/s}) \tag{8.12}$$

If the car in **Figure 8.15** has a linear acceleration $\vec{a}$ parallel to the ground, a point on the tire's outer edge experiences a tangential acceleration $\vec{a}_\text{T}$ relative to the axle. The same kind of reasoning as that used in the previous paragraph reveals that the magnitudes of these accelerations are the same and that they are related to the angular acceleration α of the wheel relative to the axle:

$$\underbrace{a}_{\substack{\text{Linear} \\ \text{acceleration}}} = \underbrace{a_\text{T} = r\alpha}_{\substack{\text{Tangential} \\ \text{acceleration}}} \quad (\alpha \text{ in rad/s}^2) \tag{8.13}$$

Equations 8.12 and 8.13 may be applied to any rolling motion, because the object does not slip against the surface on which it is rolling.

Check Your Understanding

(The answers are given at the end of the book.)

16. The speedometer of a truck is set to read the linear speed of the truck, but uses a device that actually measures the angular speed of the rolling tires that came with the truck. However, the owner replaces the tires with larger-diameter versions. Does the reading on the speedometer after the replacement give a speed that is **(a)** less than, **(b)** equal to, or **(c)** greater than the true linear speed of the truck?

17. Rolling motion is an example that involves rotation about an axis that is not fixed. Give three other examples of rotational motion about an axis that is not fixed.

8.7 | *The Vector Nature of Angular Variables

We have presented angular velocity and angular acceleration by taking advantage of the analogy between angular variables and linear variables. Like the linear velocity and the linear acceleration, the angular quantities are also vectors and have a direction as well as a magnitude. As yet, however, we have not discussed the directions of these vectors.

When a rigid object rotates about a fixed axis, it is the axis that identifies the motion, and the angular velocity vector points along this axis. **Interactive Figure 8.16** shows how to determine the direction using a *right-hand rule:*

> **Right-Hand Rule**   Grasp the axis of rotation with your right hand, so that your fingers circle the axis in the same sense as the rotation. Your extended thumb points along the axis in the direction of the angular velocity vector.

No part of the rotating object moves in the direction of the angular velocity vector.

Angular acceleration arises when the angular velocity changes, and the acceleration vector also points along the axis of rotation. The acceleration vector has the same direction as the *change* in the angular velocity. That is, when the magnitude of the angular velocity (which is the angular speed) is increasing, the angular acceleration vector points in the same direction as the angular velocity. Conversely, when the magnitude of the angular velocity is decreasing, the angular acceleration vector points in the direction opposite to the angular velocity.

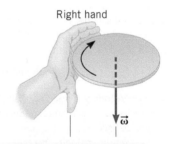

Right hand

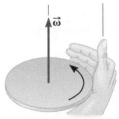

Right hand

INTERACTIVE FIGURE 8.16   The angular velocity vector $\vec{\omega}$ of a rotating object points along the axis of rotation. The direction along the axis depends on the sense of the rotation and can be determined with the aid of a right-hand rule (see the text).

EXAMPLE 9 | BIO The Physics of a Centrifuge

One popular piece of equipment in research laboratories is a centrifuge (**Figure 8.17a**). Typical applications of a centrifuge include separating immiscible liquids or suspended solids like blood, for example. Samples are placed in tubes (**Figure 8.17b**) and then rotated at high speeds around a fixed axis. The centripetal acceleration causes the denser substances and particles to separate and settle at the bottom of the tubes. Suppose a centrifuge starts from rest and reaches its maximum angular speed of 4510 rpm in 12.2 s. Through how many revolutions does the centrifuge spin during this time?

Reasoning   This is a standard rotational kinematics problem, with the following known variables:

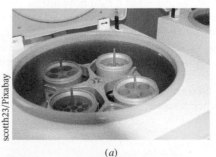

scotth23/Pixabay

(a)   (b)

FIGURE 8.17   (a) Common laboratory centrifuge. (b) Drawing of a typical sample tube used in the centrifuge. The tubes are rotated at high speed. The centripetal acceleration separates the materials in the tube by density, with the more dense particles settling at the bottom of the tube.

θ	α	ω	ω_0	t
?		+4510 rpm	0	12.2 s

Given the information in the table, we can simply use Equation 8.6 to calculate the angular displacement, θ.

Solution   In order to apply Equation 8.6, we need the value of the average angular speed, $\overline{\omega}$. We can calculate $\overline{\omega}$ using Equation 8.5:

$$\overline{\omega} = \tfrac{1}{2}(\omega_0 + \omega) = \tfrac{1}{2}(0 + 4510 \text{ rpm}) = 2260 \text{ rpm}.$$

Now we can solve for the angular displacement. However, before we can apply Equation 8.6, we need to have consistent units. We need to convert the time given from seconds to minutes:

$$t = (12.2 \text{ s})\left(\frac{1 \text{ min}}{60 \text{ s}}\right) = 0.203 \text{ min}.$$

Now we can calculate the angular displacement:

$$\theta = \overline{\omega}t = (2260 \text{ rpm})(0.203 \text{ min}) = \boxed{459 \text{ revolutions}}$$

Concept Summary

8.1 Rotational Motion and Angular Displacement   When a rigid body rotates about a fixed axis, the angular displacement is the angle swept out by a line passing through any point on the body and intersecting the axis of rotation perpendicularly. By convention, the angular displacement is positive if it is counterclockwise and negative if it is clockwise.

The radian (rad) is the SI unit of angular displacement. The angle θ in radians is defined in Equation 8.1 as the circular arc of length s traveled by a point on the rotating body divided by the radial distance r of the point from the axis.

$$(\theta \text{ in radians}) = \frac{s}{r} \qquad \text{(8.1)}$$

8.2 Angular Velocity and Angular Acceleration   The average angular velocity $\overline{\omega}$ is the angular displacement $\Delta\theta$ divided by the elapsed time Δt, according to Equation 8.2. As Δt approaches zero, the average angular velocity becomes equal to the instantaneous angular velocity ω. The magnitude of the instantaneous angular velocity is called the instantaneous angular speed.

The average angular acceleration $\overline{\alpha}$ is the change $\Delta\omega$ in the angular velocity divided by the elapsed time Δt, according to Equation 8.4. As Δt approaches zero, the average angular acceleration becomes equal to the instantaneous angular acceleration α.

$$\overline{\omega} = \frac{\Delta\theta}{\Delta t} \qquad \text{(8.2)}$$

$$\overline{\alpha} = \frac{\Delta\omega}{\Delta t} \qquad \text{(8.4)}$$

8.3 The Equations of Rotational Kinematics   The equations of rotational kinematics apply when a rigid body rotates with a constant angular acceleration about a fixed axis. These equations relate the angular displacement $\theta - \theta_0$, the angular acceleration α, the final angular velocity ω, the initial angular velocity ω_0, and the elapsed time $t - t_0$. Assuming that $\theta_0 = 0$ rad at $t_0 = 0$ s, the equations of rotational kinematics are written as in Equations 8.4 and 8.6–8.8. These equations may be used with any self-consistent set of units and are not restricted to radian measure.

$$\omega = \omega_0 + \alpha t \qquad \text{(8.4)}$$

$$\theta = \tfrac{1}{2}(\omega + \omega_0)t \qquad \text{(8.6)}$$

$$\theta = \omega_0 t + \tfrac{1}{2}\alpha t^2 \qquad \text{(8.7)}$$

$$\omega^2 = \omega_0^2 + 2\alpha\theta \qquad \text{(8.8)}$$

8.4 Angular Variables and Tangential Variables   When a rigid body rotates through an angle θ about a fixed axis, any point on the body moves on a circular arc of length s and radius r. Such a point has a tangential velocity (magnitude $= v_T$) and, possibly, a tangential acceleration (magnitude $= a_T$). The angular and tangential variables are related by Equations 8.1, 8.9, and 8.10. These equations refer to the magnitudes of the variables involved, without reference to positive or negative signs, and only radian measure can be used when applying them.

$$s = r\theta \qquad (\theta \text{ in rad}) \qquad \text{(8.1)}$$

$$v_T = r\omega \qquad (\omega \text{ in rad/s}) \qquad \text{(8.9)}$$

$$a_T = r\alpha \qquad (\alpha \text{ in rad/s}^2) \qquad \text{(8.10)}$$

8.5 Centripetal Acceleration and Tangential Acceleration   The magnitude a_c of the centripetal acceleration of a point on an object rotating with uniform or nonuniform circular motion can be expressed in terms of the radial distance r of the point from the axis and the angular speed ω, as shown in Equation 8.11. This point experiences a total acceleration $\vec{a}$ that is the vector sum of two perpendicular acceleration components, the centripetal acceleration $\vec{a}_c$ and the tangential acceleration $\vec{a}_T$; $\vec{a} = \vec{a}_c + \vec{a}_T$.

$$a_c = r\omega^2 \qquad (\omega \text{ in rad/s}) \qquad \text{(8.11)}$$

8.6 Rolling Motion   The essence of rolling motion is that there is no slipping at the point where the object touches the surface upon which it is rolling. As a result, the tangential speed v_T of a point on the outer edge of a rolling object, measured relative to the axis through the center of the object, is equal to the linear speed v with which the object moves parallel to the surface. In other words, we have Equation 8.12.

The magnitudes of the tangential acceleration a_T and the linear acceleration a of a rolling object are similarly related, as in Equation 8.13.

$$v = v_T = r\omega \qquad (\omega \text{ in rad/s}) \qquad \text{(8.12)}$$

$$a = a_T = r\alpha \qquad (\alpha \text{ in rad/s}^2) \qquad \text{(8.13)}$$

8.7 The Vector Nature of Angular Variables   The direction of the angular velocity vector is given by a right-hand rule. Grasp the axis of rotation with your right hand, so that your fingers circle the axis in the same sense as the rotation. Your extended thumb points along the axis in the direction of the angular velocity vector. The angular acceleration vector has the same direction as the change in the angular velocity.

Focus on Concepts

Additional questions are available for assignment in WileyPLUS.

Section 8.1   Rotational Motion and Angular Displacement

1. The moon is 3.85×10^8 m from the earth and has a diameter of 3.48×10^6 m. You have a pea (diameter = 0.50 cm) and a dime (diameter = 1.8 cm). You close one eye and hold each object at arm's length (71 cm) between your open eye and the moon. Which objects, if any, completely cover your view of the moon? Assume that the moon and both objects are sufficiently far from your eye that the given diameters are equal to arc lengths when calculating angles. (a) Both (b) Neither (c) Pea (d) Dime

Section 8.2   Angular Velocity and Angular Acceleration

2. The radius of the circle traced out by the second hand on a clock is 6.00 cm. In a time t the tip of the second hand moves through an arc length of 24.0 cm. Determine the value of t in seconds.

3. A rotating object has an angular acceleration of $\alpha = 0$ rad/s^2. Which one or more of the following three statements is consistent with a zero angular acceleration? A. The angular velocity is $\omega = 0$ rad/s at all times. B. The angular velocity is $\omega = 10$ rad/s at all times. C. The angular displacement θ has the same value at all times. (a) A, B, and C (b) A and B, but not C (c) A only (d) B only (e) C only

Section 8.3   The Equations of Rotational Kinematics

4. A rotating wheel has a constant angular acceleration. It has an angular velocity of 5.0 rad/s at time $t = 0$ s, and 3.0 s later has an angular velocity of 9.0 rad/s. What is the angular displacement of the wheel during the 3.0-s interval? (a) 15 rad (b) 21 rad (c) 27 rad (d) There is not enough information given to determine the angular displacement.

5. A rotating object starts from rest at $t = 0$ s and has a constant angular acceleration. At a time of $t = 7.0$ s the object has an angular velocity of $\omega = 16$ rad/s. What is its angular velocity at a time of $t = 14$ s?

Section 8.4   Angular Variables and Tangential Variables

6. A merry-go-round at a playground is a circular platform that is mounted parallel to the ground and can rotate about an axis that is perpendicular to the platform at its center. The angular speed of the merry-go-round is constant, and a child at a distance of 1.4 m from the axis has a tangential speed of 2.2 m/s. What is the tangential speed of another child, who is located at a distance of 2.1 m from the axis? (a) 1.5 m/s (b) 3.3 m/s (c) 2.2 m/s (d) 5.0 m/s (e) 0.98 m/s

7. A small fan has blades that have a radius of 0.0600 m. When the fan is turned on, the tips of the blades have a tangential acceleration of 22.0 m/s^2 as the fan comes up to speed. What is the angular acceleration α of the blades?

Section 8.5   Centripetal Acceleration and Tangential Acceleration

8. A wheel rotates with a constant angular speed ω. Which one of the following is true concerning the angular acceleration α of the wheel, the tangential acceleration a_T of a point on the rim of the wheel, and the centripetal acceleration a_c of a point on the rim?

(a) $\alpha = 0$ rad/s^2, $a_T = 0$ m/s^2, and $a_c = 0$ m/s^2

(b) $\alpha = 0$ rad/s^2, $a_T \neq 0$ m/s^2, and $a_c = 0$ m/s^2

(c) $\alpha \neq 0$ rad/s^2, $a_T = 0$ m/s^2, and $a_c = 0$ m/s^2

(d) $\alpha = 0$ rad/s^2, $a_T = 0$ m/s^2, and $a_c \neq 0$ m/s^2

(e) $\alpha \neq 0$ rad/s^2, $a_T \neq 0$ m/s^2, and $a_c \neq 0$ m/s^2

9. A platform is rotating with an angular speed of 3.00 rad/s and an angular acceleration of 11.0 rad/s^2. At a point on the platform that is 1.25 m from the axis of rotation, what is the magnitude of the *total* acceleration a?

Section 8.6   Rolling Motion

10. The radius of each wheel on a bicycle is 0.400 m. The bicycle travels a distance of 3.0 km. Assuming that the wheels do not slip, how many revolutions does each wheel make?

(a) 1.2×10^3 revolutions

(b) 2.4×10^2 revolutions

(c) 6.0×10^3 revolutions

(d) 8.4×10^{-4} revolutions

(e) Since the time of travel is not given, there is not enough information for a solution.

Problems

Additional questions are available for assignment in WileyPLUS.

SSM Student Solutions Manual
MMH Problem-solving help
GO Guided Online Tutorial
V-HINT Video Hints
CHALK Chalkboard Videos
BIO Biomedical application
E Easy
M Medium
H Hard
WS Worksheet
T Team Problem

Section 8.1   Rotational Motion and Angular Displacement,

Section 8.2   Angular Velocity and Angular Acceleration

1. **E** **SSM** **CHALK** A pitcher throws a curveball that reaches the catcher in 0.60 s. The ball curves because it is spinning at an average angular velocity of 330 rev/min (assumed constant) on its way to

the catcher's mitt. What is the angular displacement of the baseball (in radians) as it travels from the pitcher to the catcher?

2. **E GO** The table that follows lists four pairs of initial and final angles of a wheel on a moving car. The elapsed time for each pair of angles is 2.0 s. For each of the four pairs, determine the average angular velocity (magnitude and direction as given by the algebraic sign of your answer).

	Initial angle θ_0	Final angle θ
(a)	0.45 rad	0.75 rad
(b)	0.94 rad	0.54 rad
(c)	5.4 rad	4.2 rad
(d)	3.0 rad	3.8 rad

3. **E** The earth spins on its axis once a day and orbits the sun once a year ($365\frac{1}{4}$ days). Determine the average angular velocity (in rad/s) of the earth as it **(a)** spins on its axis and **(b)** orbits the sun. In each case, take the positive direction for the angular displacement to be the direction of the earth's motion.

4. **E** Our sun rotates in a circular orbit about the center of the Milky Way galaxy. The radius of the orbit is 2.2×10^{20} m, and the angular speed of the sun is 1.1×10^{-15} rad/s. How long (in years) does it take for the sun to make one revolution around the center?

5. **E GO** The initial angular velocity and the angular acceleration of four rotating objects at the same instant in time are listed in the table that follows. For each of the objects (a), (b), (c), and (d), determine the final angular *speed* after an elapsed time of 2.0 s.

	Initial angular velocity ω_0	Angular acceleration α
(a)	+12 rad/s	+3.0 rad/s²
(b)	+12 rad/s	−3.0 rad/s²
(c)	−12 rad/s	+3.0 rad/s²
(d)	−12 rad/s	−3.0 rad/s²

6. **E GO** The table that follows lists four pairs of initial and final angular velocities for a rotating fan blade. The elapsed time for each of the four pairs of angular velocities is 4.0 s. For each of the four pairs, find the average angular acceleration (magnitude and direction as given by the algebraic sign of your answer).

	Initial angular velocity ω_0	Final angular velocity ω
(a)	+2.0 rad/s	+5.0 rad/s
(b)	+5.0 rad/s	+2.0 rad/s
(c)	−7.0 rad/s	−3.0 rad/s
(d)	+4.0 rad/s	−4.0 rad/s

7. **E** Conceptual Example 2 provides some relevant background for this problem. A jet is circling an airport control tower at a distance of 18.0 km. An observer in the tower watches the jet cross in front of the moon. As seen from the tower, the moon subtends an angle of 9.04×10^{-3} radians. Find the distance traveled (in meters) by the jet as the observer watches the nose of the jet cross from one side of the moon to the other.

8. **E SSM** A Ferris wheel rotates at an angular velocity of 0.24 rad/s. Starting from rest, it reaches its operating speed with an average

angular acceleration of 0.030 rad/s². How long does it take the wheel to come up to operating speed?

9. **E GO** A floor polisher has a rotating disk that has a 15-cm radius. The disk rotates at a constant angular velocity of 1.4 rev/s and is covered with a soft material that does the polishing. An operator holds the polisher in one place for 45 s, in order to buff an especially scuffed area of the floor. How far (in meters) does a spot on the outer edge of the disk move during this time?

10. **M** The sun appears to move across the sky, because the earth spins on its axis. To a person standing on the earth, the sun subtends an angle of $\theta_{\text{sun}} = 9.28 \times 10^{-3}$ rad (see Conceptual Example 2). How much time (in seconds) does it take for the sun to move a distance equal to its own diameter?

11. **M GO** A propeller is rotating about an axis perpendicular to its center, as the drawing shows. The axis is parallel to the ground. An arrow is fired at the propeller, travels parallel to the axis, and passes through one of the open spaces between the propeller blades. The angular open spaces between the three propeller blades are each $\pi/3$ rad (60.0°). The vertical drop of the arrow may be ignored. There is a maximum value ω for the angular speed of the propeller, beyond which the arrow cannot pass through an open space without being struck by one of the blades. Find this maximum value when the arrow has the lengths L and speeds v shown in the following table.

	L	v
(a)	0.71 m	75.0 m/s
(b)	0.71 m	91.0 m/s
(c)	0.81 m	91.0 m/s

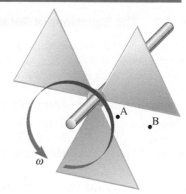

PROBLEM 11

12. **M CHALK** Two people start at the same place and walk around a circular lake in opposite directions. One walks with an angular speed of 1.7×10^{-3} rad/s, while the other has an angular speed of 3.4×10^{-3} rad/s. How long will it be before they meet?

13. **M** A space station consists of two donut-shaped living chambers, A and B, that have the radii shown in the drawing. As the station rotates, an astronaut in chamber A is moved 2.40×10^2 m along a circular arc. How far along a circular arc is an astronaut in chamber B moved during the same time?

$r_A = 3.20 \times 10^2$ m

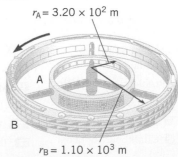

PROBLEM 13 $r_B = 1.10 \times 10^3$ m

14. [M] [V-HINT] The drawing shows a device that can be used to measure the speed of a bullet. The device consists of two rotating disks, separated by a distance of $d = 0.850$ m, and rotating with an angular speed of 95.0 rad/s. The bullet first passes through the left disk and then through the right disk. It is found that the angular displacement between the two bullet holes is $\theta = 0.240$ rad. From these data, determine the speed of the bullet.

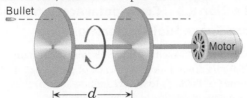

Bullet

Motor

PROBLEM 14

15. [M] [GO] An automatic dryer spins wet clothes at an angular speed of 5.2 rad/s. Starting from rest, the dryer reaches its operating speed with an average angular acceleration of 4.0 rad/s². How long does it take the dryer to come up to speed?

16. [M] [SSM] A stroboscope is a light that flashes on and off at a constant rate. It can be used to illuminate a rotating object, and if the flashing rate is adjusted properly, the object can be made to appear stationary. **(a)** What is the shortest time between flashes of light that will make a three-bladed propeller appear stationary when it is rotating with an angular speed of 16.7 rev/s? **(b)** What is the next shortest time?

17. [H] The drawing shows a golf ball passing through a windmill at a miniature golf course. The windmill has 8 blades and rotates at an angular speed of 1.25 rad/s. The opening between successive blades is equal to the width of a blade. A golf ball (diameter 4.50×10^{-2} m) has just reached the edge of one of the rotating blades (see the drawing). Ignoring the thickness of the blades, find the *minimum* linear speed with which the ball moves along the ground, such that the ball will not be hit by the next blade.

Golf ball

PROBLEM 17

Section 8.3   The Equations of Rotational Kinematics

18. [E] A figure skater is spinning with an angular velocity of $+15$ rad/s. She then comes to a stop over a brief period of time. During this time, her angular displacement is $+5.1$ rad. Determine **(a)** her average angular acceleration and **(b)** the time during which she comes to rest.

19. [E] [SSM] A gymnast is performing a floor routine. In a tumbling run she spins through the air, increasing her angular velocity from 3.00 to 5.00 rev/s while rotating through one-half of a revolution. How much time does this maneuver take?

20. [E] The angular speed of the rotor in a centrifuge increases from 420 to 1420 rad/s in a time of 5.00 s. **(a)** Obtain the angle through which the rotor turns. **(b)** What is the magnitude of the angular acceleration?

21. [E] A wind turbine is initially spinning at a constant angular speed. As the wind's strength gradually increases, the turbine experiences a constant angular acceleration of 0.140 rad/s². After making 2870 revolutions, its angular speed is 137 rad/s. **(a)** What is the initial angular velocity of the turbine? **(b)** How much time elapses while the turbine is speeding up?

22. [E] [GO] A car is traveling along a road, and its engine is turning over with an angular velocity of $+220$ rad/s. The driver steps on the accelerator, and in a time of 10.0 s the angular velocity increases to $+280$ rad/s. **(a)** What would have been the angular displacement of the engine if its angular velocity had remained constant at the initial value of $+220$ rad/s during the entire 10.0-s interval? **(b)** What would have been the angular displacement if the angular velocity had been equal to its final value of $+280$ rad/s during the entire 10.0-s interval? **(c)** Determine the actual value of the angular displacement during the 10.0-s interval.

23. [E] [SSM] The wheels of a bicycle have an angular velocity of $+20.0$ rad/s. Then, the brakes are applied. In coming to rest, each wheel makes an angular displacement of $+15.92$ revolutions. **(a)** How much time does it take for the bike to come to rest? **(b)** What is the angular acceleration (in rad/s²) of each wheel?

24. [M] [SSM] [MMH] A motorcyclist is traveling along a road and accelerates for 4.50 s to pass another cyclist. The angular acceleration of each wheel is $+6.70$ rad/s², and, just after passing, the angular velocity of each wheel is $+74.5$ rad/s, where the plus signs indicate counterclockwise directions. What is the angular displacement of each wheel during this time?

25. [M] [V-HINT] A top is a toy that is made to spin on its pointed end by pulling on a string wrapped around the body of the top. The string has a length of 64 cm and is wound around the top at a spot where its radius is 2.0 cm. The thickness of the string is negligible. The top is initially at rest. Someone pulls the free end of the string, thereby unwinding it and giving the top an angular acceleration of $+12$ rad/s². What is the final angular velocity of the top when the string is completely unwound?

26. [M] [MMH] The drive propeller of a ship starts from rest and accelerates at 2.90×10^{-3} rad/s² for 2.10×10^{3} s. For the next 1.40×10^{3} s the propeller rotates at a constant angular speed. Then it decelerates at 2.30×10^{-3} rad/s² until it slows (without reversing direction) to an angular speed of 4.00 rad/s. Find the total angular displacement of the propeller.

27. [M] [GO] The drawing shows a graph of the angular velocity of a rotating wheel as a function of time. Although not shown in the graph, the angular velocity continues to increase at the same rate until $t = 8.0$ s. What is the angular displacement of the wheel from 0 to 8.0 s?

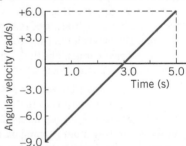

PROBLEM 27

28. [M] [V-HINT] At the local swimming hole, a favorite trick is to run horizontally off a cliff that is 8.3 m above the water. One diver runs off the edge of the cliff, tucks into a "ball," and rotates on the way down with an average angular speed of 1.6 rev/s. Ignore air resistance and determine the number of revolutions she makes while on the way down.

29. [M] [V-HINT] A spinning wheel on a fireworks display is initially rotating in a counterclockwise direction. The wheel has an angular acceleration of -4.00 rad/s². Because of this acceleration, the angular velocity of the wheel changes from its initial value to a final value of -25.0 rad/s. While this change occurs, the angular displacement of the wheel is zero. (Note the similarity to that of a ball being thrown vertically upward, coming to a momentary halt, and then falling downward to its initial position.) Find the time required for the change in the angular velocity to occur.

Section 8.4 Angular Variables and Tangential Variables

30. **E** **GO** A fan blade is rotating with a constant angular acceleration of $+12.0 \, \text{rad/s}^2$. At what point on the blade, as measured from the axis of rotation, does the magnitude of the tangential acceleration equal that of the acceleration due to gravity?

31. **E** An auto race takes place on a circular track. A car completes one lap in a time of 18.9 s, with an average tangential speed of 42.6 m/s. Find (**a**) the average angular speed and (**b**) the radius of the track.

32. **E** **SSM** A string trimmer is a tool for cutting grass and weeds; it utilizes a length of nylon "string" that rotates about an axis perpendicular to one end of the string. The string rotates at an angular speed of 47 rev/s, and its tip has a tangential speed of 54 m/s. What is the length of the rotating string?

33. **E** **V-HINT** In 9.5 s a fisherman winds 2.6 m of fishing line onto a reel whose radius is 3.0 cm (assumed to be constant as an approximation). The line is being reeled in at a constant speed. Determine the angular speed of the reel.

34. **E** The earth has a radius of 6.38×10^6 m and turns on its axis once every 23.9 h. (**a**) What is the tangential speed (in m/s) of a person living in Ecuador, a country that lies on the equator? (**b**) At what latitude (i.e., the angle θ in the drawing) is the tangential speed one-third that of a person living in Ecuador?

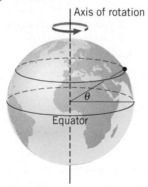

PROBLEM 34

35. **M** **SSM** A baseball pitcher throws a baseball horizontally at a linear speed of 42.5 m/s (about 95 mi/h). Before being caught, the baseball travels a horizontal distance of 16.5 m and rotates through an angle of 49.0 rad. The baseball has a radius of 3.67 cm and is rotating about an axis as it travels, much like the earth does. What is the tangential speed of a point on the "equator" of the baseball?

36. **M** **GO** A person lowers a bucket into a well by turning the hand crank, as the drawing illustrates. The crank handle moves with a constant tangential speed of 1.20 m/s on its circular path. The rope holding the bucket unwinds without slipping on the barrel of the crank. Find the linear speed with which the bucket moves down the well.

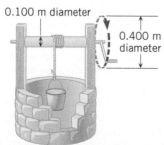

PROBLEM 36

37. **M** A thin rod (length = 1.50 m) is oriented vertically, with its bottom end attached to the floor by means of a frictionless hinge. The mass of the rod may be ignored, compared to the mass of an object

fixed to the top of the rod. The rod, starting from rest, tips over and rotates downward. (**a**) What is the angular speed of the rod just before it strikes the floor? (*Hint: Consider using the principle of conservation of mechanical energy.*) (**b**) What is the magnitude of the angular acceleration of the rod just before it strikes the floor?

Section 8.5 Centripetal Acceleration and Tangential Acceleration

38. **E** **SSM** A racing car travels with a constant tangential speed of 75.0 m/s around a circular track of radius 625 m. Find (**a**) the magnitude of the car's total acceleration and (**b**) the direction of its total acceleration relative to the radial direction.

39. **E** **GO** Two Formula One racing cars are negotiating a circular turn, and they have the same centripetal acceleration. However, the path of car A has a radius of 48 m, while that of car B is 36 m. Determine the ratio of the angular speed of car A to the angular speed of car B.

40. **E** **SSM** The earth orbits the sun once a year (3.16×10^7 s) in a nearly circular orbit of radius 1.50×10^{11} m. With respect to the sun, determine (**a**) the angular speed of the earth, (**b**) the tangential speed of the earth, and (**c**) the magnitude and direction of the earth's centripetal acceleration.

41. **E** **MMH** Review Multiple-Concept Example 8 in this chapter as an aid in solving this problem. In a fast-pitch softball game the pitcher is impressive to watch, as she delivers a pitch by rapidly whirling her arm around so that the ball in her hand moves on a circle. In one instance, the radius of the circle is 0.670 m. At one point on this circle, the ball has an angular acceleration of $64.0 \, \text{rad/s}^2$ and an angular speed of 16.0 rad/s. (**a**) Find the magnitude of the total acceleration (centripetal plus tangential) of the ball. (**b**) Determine the angle of the total acceleration relative to the radial direction.

42. **M** **GO** A rectangular plate is rotating with a constant angular speed about an axis that passes perpendicularly through one corner, as the drawing shows. The centripetal acceleration measured at corner A is n times as great as that measured at corner B. What is the ratio L_1/L_2 of the lengths of the sides of the rectangle when $n = 2.00$?

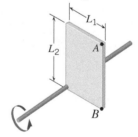

PROBLEM 42

43. **M** **V-HINT** Multiple-Concept Example 8 explores the approach taken in problems such as this one. The blades of a ceiling fan have a radius of 0.380 m and are rotating about a fixed axis with an angular velocity of $+1.50 \, \text{rad/s}$. When the switch on the fan is turned to a higher speed, the blades acquire an angular acceleration of $+2.00 \, \text{rad/s}^2$. After 0.500 s has elapsed since the switch was reset, what is (**a**) the total acceleration (in m/s^2) of a point on the tip of a blade and (**b**) the angle ϕ between the total acceleration $\vec{a}$ and the centripetal acceleration $\vec{a}_c$? (See **Interactive Figure 8.12b**.)

44. **M** **CHALK** **SSM** The sun has a mass of 1.99×10^{30} kg and is moving in a circular orbit about the center of our galaxy, the Milky Way. The radius of the orbit is 2.3×10^4 light-years (1 light-year $= 9.5 \times 10^{15}$ m), and the angular speed of the sun is 1.1×10^{-15} rad/s. (**a**) Determine the tangential speed of the sun. (**b**) What is the magnitude of the net force that acts on the sun to keep it moving around the center of the Milky Way?

Section 8.6   Rolling Motion

Note: All problems in this section assume that there is no slipping of the surfaces in contact during the rolling motion.

45. **E** **SSM** A motorcycle accelerates uniformly from rest and reaches a linear speed of 22.0 m/s in a time of 9.00 s. The radius of each tire is 0.280 m. What is the magnitude of the angular acceleration of each tire?

46. **E** An automobile tire has a radius of 0.330 m, and its center moves forward with a linear speed of $v = 15.0$ m/s. **(a)** Determine the angular speed of the wheel. **(b)** Relative to the axle, what is the tangential speed of a point located 0.175 m from the axle?

47. **E** **MMH** A car is traveling with a speed of 20.0 m/s along a straight horizontal road. The wheels have a radius of 0.300 m. If the car speeds up with a linear acceleration of 1.50 m/s² for 8.00 s, find the angular displacement of each wheel during this period.

48. **E** Suppose you are riding a stationary exercise bicycle, and the electronic meter indicates that the wheel is rotating at 9.1 rad/s. The wheel has a radius of 0.45 m. If you ride the bike for 35 min, how far would you have gone if the bike could move?

49. **M** **V-HINT** A motorcycle, which has an initial linear speed of 6.6 m/s, decelerates to a speed of 2.1 m/s in 5.0 s. Each wheel has a radius of 0.65 m and is rotating in a counterclockwise (positive) direction. What are **(a)** the constant angular acceleration (in rad/s²) and **(b)** the angular displacement (in rad) of each wheel?

50. **M** **GO** A dragster starts from rest and accelerates down a track. Each tire has a radius of 0.320 m and rolls without slipping. At a distance of 384 m, the angular speed of the wheels is 288 rad/s. Determine **(a)** the linear speed of the dragster and **(b)** the magnitude of the angular acceleration of its wheels.

51. **M** **GO** A bicycle is rolling down a circular portion of a path; this portion of the path has a radius of 9.00 m. As the drawing illustrates, the angular displacement of the bicycle is 0.960 rad. What is the angle (in radians) through which each bicycle wheel (radius = 0.400 m) rotates?

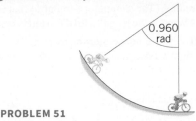

PROBLEM 51

52. **M** **SSM** The penny-farthing is a bicycle that was popular between 1870 and 1890. As the drawing shows, this type of bicycle has a large front wheel and a small rear wheel. During a ride, the front wheel (radius = 1.20 m) makes 276 revolutions. How many revolutions does the rear wheel (radius = 0.340 m) make?

PROBLEM 52

53. **M** **CHALK** **GO** A ball of radius 0.200 m rolls with a constant linear speed of 3.60 m/s along a horizontal table. The ball rolls off the edge and falls a vertical distance of 2.10 m before hitting the floor. What is the angular displacement of the ball while the ball is in the air?

Additional Problems   **Online**

54. **E** A flywheel has a constant angular deceleration of 2.0 rad/s². **(a)** Find the angle through which the flywheel turns as it comes to rest from an angular speed of 220 rad/s. **(b)** Find the time for the flywheel to come to rest.

55. **E** **SSM** An electric fan is running on HIGH. After the LOW button is pressed, the angular speed of the fan decreases to 83.8 rad/s in 1.75 s. The deceleration is 42.0 rad/s². Determine the initial angular speed of the fan.

56. **E** Refer to Multiple-Concept Example 8 for insight into this problem. During a tennis serve, a racket is given an angular acceleration of magnitude 160 rad/s². At the top of the serve, the racket has an angular speed of 14 rad/s. If the distance between the top of the racket and the shoulder is 1.5 m, find the magnitude of the total acceleration of the top of the racket.

57. **M** **SSM** The compact disc (or CD) is an outdated storage medium that was used prominently in the 1990s to store digital data, including music, on a spiral track. Music is put onto a CD with the assumption that, while the disc spins during playback, the music will be detected at a *constant tangential speed* at any point. Since $v_T = r\omega$, a CD rotates at a smaller angular speed for music near the outer edge and a larger angular speed for music near the inner part of the disc. For music at the outer edge ($r = 0.0568$ m), the angular speed is 3.50 rev/s. Find **(a)** the constant tangential speed at which music is detected and **(b)** the angular speed (in rev/s) for music at a distance of 0.0249 m from the center of a CD.

58. **M** **V-HINT** After 10.0 s, a spinning roulette wheel at a casino has slowed down to an angular velocity of +1.88 rad/s. During this time, the wheel has an angular acceleration of −5.04 rad/s². Determine the angular displacement of the wheel.

59. **M** **GO** At a county fair there is a betting game that involves a spinning wheel. As the drawing shows, the wheel is set into rotational motion with the beginning of the angular section labeled "1" at the marker at the top of the wheel. The wheel then decelerates and eventually comes to a halt on one of the numbered sections. The wheel in the drawing is divided into twelve sections, each of which is an angle of 30.0°. Determine the numbered section on which the wheel comes to a halt when the deceleration of the wheel has a magnitude of 0.200 rev/s² and the initial angular velocity is **(a)** +1.20 rev/s and **(b)** +1.47 rev/s.

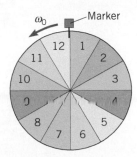

PROBLEM 59

60. M V-HINT MMH A racing car, starting from rest, travels around a circular turn of radius 23.5 m. At a certain instant, the car is still accelerating, and its angular speed is 0.571 rad/s. At this time, the total acceleration (centripetal plus tangential) makes an angle of 35.0° with respect to the radius. (The situation is similar to that in **Interactive Figure 8.12b**.) What is the magnitude of the total acceleration?

61. M GO SSM An automobile, starting from rest, has a linear acceleration to the right whose magnitude is 0.800 m/s² (see the figure). During the next 20.0 s, the tires roll and do not slip. The radius of each wheel is 0.330 m. At the end of this time, what is the angle through which each wheel has rotated?

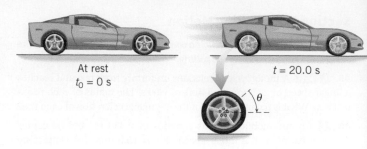

At rest
$t_0 = 0$ s

$t = 20.0$ s

θ

PROBLEM 61

Physics in Biology, Medicine, and Sports

62. M BIO **8.3**   A dentist causes the bit of a high-speed drill to accelerate from an angular speed of 1.05×10^4 rad/s to an angular speed of 3.14×10^4 rad/s. In the process, the bit turns through 1.88×10^4 rad. Assuming a constant angular acceleration, how long would it take the bit to reach its maximum speed of 7.85×10^4 rad/s, starting from rest?

63. E BIO **8.4**   Some bacteria are propelled by biological motors that spin hairlike flagella. A typical bacterial motor turning at a constant angular velocity has a radius of 1.5×10^{-8} m, and a tangential speed at the rim of 2.3×10^{-5} m/s. **(a)** What is the angular speed (the magnitude of the angular velocity) of this bacterial motor? **(b)** How long does it take the motor to make one revolution?

64. M BIO GO **8.5**   In a large centrifuge used for training pilots and astronauts, a small chamber is fixed at the end of a rigid arm that rotates in a horizontal circle. A trainee riding in the chamber of a centrifuge rotating with a constant angular speed of 2.5 rad/s experiences a centripetal acceleration of 3.2 times the acceleration due to gravity. In a second training exercise, the centrifuge speeds up from rest with a constant angular acceleration. When the centrifuge reaches an angular speed of 2.5 rad/s, the trainee experiences a total acceleration equal to 4.8 times the acceleration due to gravity. **(a)** How long is the arm of the centrifuge? **(b)** What is the angular acceleration of the centrifuge in the second training exercise?

65. M BIO **8.2**   On January 19, 2015, at the National Stadium in Warsaw, Poland, figure skater Olivia Rybicka-Oliver, at only 11 years old, shattered the Guinness World Record for the fastest spin on ice at 340 rpm (see the photo). **(a)** What was her angular speed in rad/s? **(b)** Assume she comes to rest from this maximum angular speed in 3.2 s and calculate the number of rotations she completes during this time.

66. M BIO **8.4**   Evan Nagao (see the photo) is an incredible talent who can make a yo-yo perform amazing tricks. He is a multiple former U.S. national champion who has even launched his own yo-yo brand. One basic yo-yo trick is to make it "sleep." One throws the yo-yo straight downward by quickly snapping the elbow and wrist. The yo-yo then spins at the bottom of the string very quickly. Assume the yo-yo begins from rest and is thrown downward in 5.00 ms, at which point it rotates at a constant 4650 rpm while hanging at the bottom of the string. **(a)** What was the average angular acceleration of the yo-yo in rad/s² during the throw? **(b)** If the diameter of the yo-yo is 7.00 cm, calculate the tangential speed of a point on its outer edge while it is sleeping.

PROBLEM 66

Ken McKay/ITV/Shutterstock.com

67. M BIO **8.3**   *Helicobacter pylori* (*H. pylori*) is a helically shaped bacterium that is usually found in the stomach. It burrows through the gastric mucous lining to establish an infection in the stomach's epithelial cells (see the photo). Approximately 90% of the people infected with *H. pylori* will never experience symptoms. Others may develop peptic ulcers and show symptoms of chronic gastritis. The method of motility of *H. pylori* is a prokaryotic flagellum attached to the back of the bacterium that rigidly rotates like a propeller on a ship. The flagellum is composed of proteins, is approximately 40.0 nm in diameter, and can reach rotation speeds as high as 1.50×10^3 rpm. If the speed of the bacterium is 10.0 μm/s, how far has it moved in the time it takes the flagellum to rotate through an angular displacement of 5.00×10^2 rad?

PROBLEM 65   Olivia Oliver

H. PYLORI CROSSING MUCUS LAYER OF STOMACH

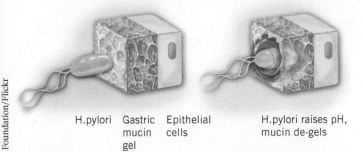

H.pylori   Gastric   Epithelial   H.pylori raises pH,
         mucin   cells   mucin de-gels
         gel

PROBLEM 67

68. **M** **BIO** **8.3** A stationary bicycle is a great option to improve one's fitness level. The rider's pedaling action rotates a weighted flywheel that simulates the resistance a person would feel while riding outside. The Keiser M3 stationary bike uses one of the industry's lightest flywheels at just 8 lb (see the photo). This allows the rider to reach maximum speed in less time. Assume a rider begins from rest and accelerates the flywheel at a constant rate to 110 rpm in 3.7 s. **(a)** What is the angular acceleration of the flywheel in rad/s^2, and **(b)** how many rotations does the wheel complete during this time?

Keiser Corporation

PROBLEM 68

69. **M** **BIO** **8.3** Wind dispersal, also called *anemochory*, is one of the most primitive ways that a plant spreads its seeds. Maple trees do this, in which they have winged seeds called *samaras* that spin like a helicopter as they drop toward the ground (see the photo). This causes the seeds to drop slowly, thereby increasing the likelihood that the wind carries them farther away from their parent tree. As the seed falls, it quickly reaches terminal velocity due to air resistance, and it rotates at a constant angular speed of 25 rad/s. If a seed falls from a height of 9.2 m and falls with a terminal speed of 1.1 m/s, how many rotations does the seed complete during its fall?

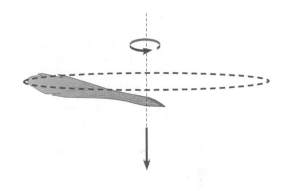

PROBLEM 69

Concepts and Calculations Problems

Online

In this chapter we have studied the concepts of angular displacement, angular velocity, and angular acceleration. We now review some important aspects of these quantities. Problem 70 illustrates that the angular acceleration and the angular velocity can have the same or opposite direction, depending on whether the angular speed is increasing or decreasing. Problem 71 reviews the two different types of acceleration, centripetal and tangential, that a car can have when it travels on a circular road.

70. **M** **CHALK** A rider on a mountain bike is traveling to the left in the figure. Each wheel has an angular velocity of +21.7 rad/s, where, as usual, the plus sign indicates that the wheel is rotating in the counterclockwise direction. **(a)** To pass another cyclist, the rider pumps harder, and the angular velocity of the wheels increases from +21.7 to +28.5 rad/s in a time of 3.50 s. **(b)** After passing the cyclist, the rider begins to coast, and the angular velocity of the wheels decreases from +28.5 to +15.3 rad/s in a time of 10.7 s. *Concepts:* (i) Is the angular acceleration positive or negative when the rider is passing the cyclist

and the angular speed of the wheels is increasing? (ii) Is the angular acceleration positive or negative when the rider is coasting and the angular speed of the wheels is decreasing? *Calculations:* In both instances, **(a)** and **(b)**, determine the magnitude and direction of the angular acceleration (assumed constant) of the wheels.

(a) Angular speed   (b) Angular speed
    increasing   decreasing

PROBLEM 70

71. M CHALK SSM Suppose you are driving a car in a counter-clockwise direction on a circular road whose radius is $r = 390$ m (see the figure). You look at the speedometer and it reads a steady 32 m/s (about 72 mi/h). *Concepts:* (i) Does an object traveling at a constant tangential speed (for example, $v_T = 32$ m/s) along a circular path have an acceleration? (ii) Is there a tangential acceleration $\vec{a}_T$ when the angular speed of an object changes (e.g., when the car's angular speed decreases to 4.9×10^{-2} rad/s)? *Calculations:* (a) What is the angular speed of the car? (b) Determine the acceleration (magnitude and direction) of the car. (c) To avoid a rear-end collision with the vehicle ahead, you apply the brakes and reduce your angular speed to 4.9×10^{-2} rad/s in a time of 4.0 s. What is the tangential acceleration (magnitude and direction) of the car?

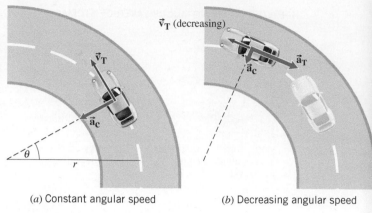

(a) Constant angular speed   (b) Decreasing angular speed

PROBLEM 71

Team Problems   Online

72. E T WS **A Rotating Scientific Satellite.** You and your team are designing a device that is to be placed into orbit to measure the earth's magnetic field. It is constructed from two identical spheres, each of diameter 0.550 m and mass 270.0 kg, connected to each other via a cable so that the distance between the sphere centers is 22.0 m. The composite object is then set into rotational motion, resembling a rotating dumbbell, where the cable both holds the spheres together and provides the centripetal force that keeps them on a circular path of diameter 22.0 m. Your task is to determine the maximum angular velocity of the device if the tension in the cable is not to exceed 15 500 N. Express your answer in both rad/s and rpm.

73. E T WS **Soft Braking.** A circular space station with a diameter of 3.20 km is spinning at a rate such that its crew on the outer rim experiences artificial gravity equal to that on the surface of the earth. But something has gone wrong, and you and your team have been assigned the task of reducing the station's rotation to zero so that crucial repairs can be made. The problem is that the crew cannot leave the outer rim while this happens. In order to bring the station to a halt in the least amount of time that will also be safe to the people inside, the linear acceleration at the rim where they are located cannot exceed $g = 9.80$ m/s². (a) What constant value of the angular acceleration (which will be negative, since the rate of rotation must decrease) will accomplish this? (b) How long will it take the space station to come to rest?

74. M T **Energy of a Bullet Dissipated by Plywood.** As part of a criminal investigation, you need to determine how much of a bullet's

energy is dissipated by a 0.500-inch piece of plywood. You construct a device that consists of three disks that are separated by a distance $d = 0.950$ m and rotate on a common axis. The bullet is fired through the first disk (a few inches above its center), which is composed of a light plastic that has a negligible effect on the speed of the bullet. The bullet then passes through the second disk, which is composed of 0.500-inch plywood. Finally, the bullet strikes the third disk, where it becomes embedded. The disks rotate with an angular velocity of $\omega = 92.0$ rad/s. The angular displacement between holes in the first and second disks is $\Delta\theta_{12} = 0.255$ rad, and the angular displacement between the holes in the second and third disks is $\Delta\theta_{23} = 0.273$ rad. If the mass of the bullet is 15.0 g, find (a) the initial speed of the bullet and (b) the energy dissipated by the 0.500-inch plywood.

75. M T **Three Wheels.** Three rubber wheels are mounted on axles so that their outer edges make tight contact with each other and their centers are on a line. The wheel on the far left axle is connected to a motor that rotates it at 25 rpm, and drives the wheel in contact with it on its right, which, in turn, drives the wheel on its right. The left wheel (Wheel 1) has a diameter of $d_1 = 0.20$ m, the middle wheel (Wheel 2) has $d_2 = 0.30$ m, and the far right wheel (Wheel 3) has $d_3 = 0.45$ m. (a) If Wheel 1 rotates clockwise, in which direction does Wheel 3 rotate? (b) What is the angular speed of Wheel 3, and what is the tangential speed on its outer edge? (c) What arrangement of the wheels gives the largest tangential speed on the outer edge of the wheel in the far right position (assuming the wheel in the far left position is driven at 25 rpm)?

A hurricane, or typhoon, is composed of large rotating masses of air with a very low pressure eye at its center. When the rotating air near the outer edge of the storm is forced inward due to the lower pressure, its moment of inertia decreases. As a consequence of this, its angular speed increases to conserve angular momentum. This effect produces the very high and dangerous winds in the eye wall of the storm (see photograph). Moment of inertia and conservation of angular momentum are two important topics in rotational dynamics, which we will study in this chapter.

skeeze/Pixabay

Rotational Dynamics

9.1 The Action of Forces and Torques on Rigid Objects

The mass of most rigid objects, such as a propeller or a wheel, is spread out and not concentrated at a single point. These objects can move in a number of ways. **Figure 9.1a** illustrates one possibility called translational

After reading this module, you should be able to...

9.1 Distinguish between torque and force.

9.2 Analyze rigid objects in equilibrium.

9.3 Determine the center of gravity of rigid objects.

9.4 Analyze rotational dynamics using moments of inertia.

9.5 Apply the relation between rotational work and energy.

9.6 Solve problems using the conservation of angular momentum.

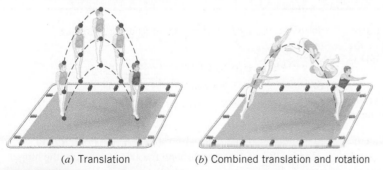

(a) Translation   (b) Combined translation and rotation

FIGURE 9.1 Examples of (a) translational motion and (b) combined translational and rotational motions.

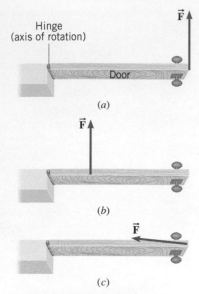

Hinge
(axis of rotation)

$\vec{F}$

Door

(a)

$\vec{F}$

(b)

$\vec{F}$

(c)

INTERACTIVE FIGURE 9.2 With a force of a given magnitude, a door is easier to open by (a) pushing at the outer edge than by (b) pushing closer to the axis of rotation (the hinge). (c) Pushing into the hinge makes it difficult to open the door.

motion, in which all points on the body travel on parallel paths (not necessarily straight lines). In pure translation there is no rotation of any line in the body. Because translational motion can occur along a curved line, it is often called curvilinear motion or linear motion. Another possibility is rotational motion, which may occur in combination with translational motion, as is the case for the somersaulting gymnast in **Figure 9.1***b*.

We have seen many examples of how a net force affects linear motion by causing an object to accelerate. We now need to take into account the possibility that a rigid object can also have an angular acceleration. A net external force causes linear motion to change, but what causes rotational motion to change? For example, something causes the rotational velocity of a speedboat's propeller to change when the boat accelerates. Is it simply the net force? As it turns out, it is not the net external force, but rather the net external torque that causes the rotational velocity to change. Just as greater net forces cause greater linear accelerations, greater net torques cause greater rotational or angular accelerations.

Interactive Figure 9.2 helps to explain the idea of torque. When you push on a door with a force $\vec{F}$, as in part *a*, the door opens more quickly when the force is larger. Other things being equal, a larger force generates a larger torque. However, the door does not open as quickly if you apply the same force at a point closer to the hinge, as in part *b*, because the force now produces less torque. Furthermore, if your push is directed nearly at the hinge, as in part *c*, you will have a hard time opening the door at all, because the torque is nearly zero. In summary, the torque depends on the magnitude of the force, on the point where the force is applied relative to the axis of rotation (the hinge in **Interactive Figure 9.2**), and on the direction of the force.

For simplicity, we deal with situations in which the force lies in a plane that is perpendicular to the axis of rotation. In **Figure 9.3**, for instance, the axis is perpendicular to the page and the force lies in the plane of the paper. The drawing shows the line of action and the lever arm of the force, two concepts that are important in the definition of torque. The **line of action** is an extended line drawn colinear with the force. The **lever arm** is the distance ℓ between the line of action and the axis of rotation, measured on a line that is perpendicular to both. The torque is represented by the symbol τ (Greek letter *tau*), and its magnitude is defined as the magnitude of the force times the lever arm:

DEFINITION OF TORQUE

Magnitude of torque = (Magnitude of the force) × (Lever arm) = $F\ell$ (9.1)

Direction: The torque τ is positive when the force tends to produce a counterclockwise rotation about the axis, and negative when the force tends to produce a clockwise rotation.

SI Unit of Torque: newton · meter (N · m)

Equation 9.1 indicates that forces of the same magnitude can produce *different* torques, depending on the value of the lever arm, and Example 1 illustrates this important feature.

EXAMPLE 1 | Different Lever Arms, Different Torques

In **Figure 9.3** a force (magnitude = 55 N) is applied to a door. However, the lever arms are different in the three parts of the drawing: **(a)** $\ell = 0.80$ m, **(b)** $\ell = 0.60$ m, and **(c)** $\ell = 0$ m. Find the torque in each case.

Reasoning In each case the lever arm is the perpendicular distance between the axis of rotation and the line of action of the force. In part *a* this perpendicular distance is equal to the width of the door. In parts *b* and *c*, however, the lever arm is less than the width. Because the lever arm is different in each case, the torque is different, even though the magnitude of the applied force is the same.

Solution Using Equation 9.1, we find the following values for the torques:

(a) $\tau = +F\ell = +(55\ \text{N})(0.80\ \text{m}) = \boxed{+44\ \text{N} \cdot \text{m}}$

(b) $\tau = +F\ell = +(55\ \text{N})(0.60\ \text{m}) = \boxed{+33\ \text{N} \cdot \text{m}}$

(c) $\tau = +F\ell = +(55\ \text{N})(0\ \text{m}) = \boxed{0\ \text{N} \cdot \text{m}}$

In parts *a* and *b* the torques are positive, since the forces tend to produce a counterclockwise rotation of the door. In part *c* the line of action of $\vec{F}$ passes through the axis of rotation (the hinge). Hence, the lever arm is zero, and the torque is zero.

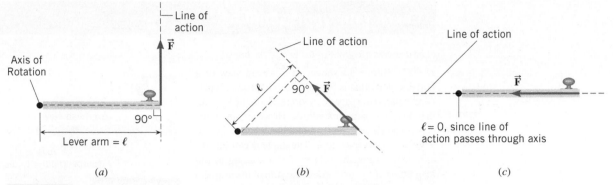

FIGURE 9.3 In this top view, the hinges of a door appear as a black dot (•) and define the axis of rotation. The line of action and lever arm ℓ are illustrated for a force applied to the door (*a*) perpendicularly and (*b*) at an angle. (*c*) The lever arm is zero because the line of action passes through the axis of rotation.

In our bodies, muscles and tendons produce torques about various joints. Example 2 illustrates how the Achilles tendon produces a torque about the ankle joint.

EXAMPLE 2 | BIO The Physics of the Achilles Tendon

Figure 9.4a shows the ankle joint and the Achilles tendon attached to the heel at point *P*. The tendon exerts a force $\vec{\mathbf{F}}$ (magnitude = 720 N), as **Figure 9.4b** indicates. Determine the torque (magnitude and direction) of this force about the ankle joint, which is located 3.6×10^{-2} m away from point *P*.

Reasoning To calculate the magnitude of the torque, it is necessary to have a value for the lever arm ℓ. However, the lever arm is not the given distance of 3.6×10^{-2} m. Instead, the lever arm is the perpendicular distance between the axis of rotation at the

ankle joint and the line of action of the force $\vec{\mathbf{F}}$. In **Figure 9.4b** this distance is indicated by the dashed red line.

Solution From the drawing, it can be seen that the lever arm is $\ell = (3.6 \times 10^{-2}$ m) cos 55°. The magnitude of the torque is

$$F\ell = (720 \text{ N})(3.6 \times 10^{-2} \text{ m}) \cos 55° = 15 \text{ N} \cdot \text{m} \qquad (9.1)$$

The force $\vec{\mathbf{F}}$ tends to produce a clockwise rotation about the ankle joint, so the torque is negative: $\boxed{\tau = -15 \text{ N} \cdot \text{m}}$.

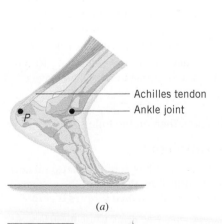

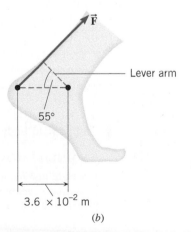

FIGURE 9.4 The force $\vec{\mathbf{F}}$ generated by the Achilles tendon produces a clockwise (negative) torque about the ankle joint.

Check Your Understanding

(The answers are given at the end of the book.)

1. CYU Figure 9.1 shows an overhead view of a horizontal bar that is free to rotate about an axis perpendicular to the page. Two forces act on the bar, and they have the same magnitude. However, one force is perpendicular to the bar, and the other makes an angle ϕ with respect to it. The angle ϕ can be 90°, 45°, or 0°. Rank the values of ϕ according to the magnitude of the net torque (the sum of the torques) that the two forces produce, largest net torque first.

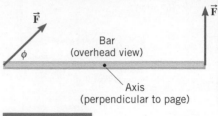

CYU FIGURE 9.1

2. Sometimes, even with a wrench, one cannot loosen a nut that is frozen tightly to a bolt. It is often possible to loosen the nut by slipping one end of a long pipe over the wrench handle and pushing at the other end of the pipe. With the aid of the pipe, does the applied force produce a smaller torque, a greater torque, or the same torque on the nut?

3. Is it possible **(a)** for a large force to produce a small, or even zero, torque and **(b)** for a small force to produce a large torque?

4. CYU Figure 9.2 shows a woman struggling to keep a stack of boxes balanced on a dolly. The woman's left foot is on the axle of the dolly. Assuming that the boxes are identical, which one creates the greatest torque with respect to the axle?

Howard Berman/Stone/Getty Images

CYU FIGURE 9.2

9.2 | Rigid Objects in Equilibrium

If a rigid body is in equilibrium, neither its linear motion nor its rotational motion changes. This lack of change leads to certain equations that apply for rigid-body equilibrium. For instance, an object whose linear motion is not changing has no acceleration $\vec{\mathbf{a}}$. Therefore, the net force $\Sigma\vec{\mathbf{F}}$ applied to the object must be zero, since $\Sigma\vec{\mathbf{F}} = m\vec{\mathbf{a}}$ and $\vec{\mathbf{a}} = 0$. For two-dimensional motion the x and y components of the net force are separately zero: $\Sigma F_x = 0$ and $\Sigma F_y = 0$. In calculating the net force, we include only forces from external agents, or *external forces.** In addition to linear motion, we must consider rotational motion, which also does not change under equilibrium conditions. This means that the net external torque acting on the object must be zero, because torque is what causes rotational motion to change. Using the symbol $\Sigma\tau$ to represent the net external torque (the sum of all positive and negative torques), we have

$$\Sigma\tau = 0 \qquad\qquad (9.2)$$

We define rigid-body equilibrium, then, in the following way.

EQUILIBRIUM OF A RIGID BODY

A rigid body is in equilibrium if it has zero translational acceleration and zero angular acceleration. In equilibrium, the sum of the externally applied forces is zero, and the sum of the externally applied torques is zero:

$$\Sigma F_x = 0 \quad\text{and}\quad \Sigma F_y = 0 \qquad (4.9\text{a and } 4.9\text{b})$$

$$\Sigma\tau = 0 \qquad\qquad (9.2)$$

---

*We ignore internal forces that one part of an object exerts on another part, because they occur in action–reaction pairs, each of which consists of oppositely directed forces of equal magnitude. The effect of one force cancels the effect of the other, as far as the acceleration of the entire object is concerned.

The reasoning strategy for analyzing the forces and torques acting on a body in equilibrium is given below. The first four steps of the strategy are essentially the same as those outlined in Section 4.11, where only forces are considered. Steps 5 and 6 have been added to account for any external torques that may be present. Example 3 illustrates how this reasoning strategy is applied to a diving board.

REASONING STRATEGY **Applying the Conditions of Equilibrium to a Rigid Body**

1. **Select the object to which the equations for equilibrium are to be applied.**

2. **Draw a free-body diagram that shows all the external forces acting on the object.**

3. **Choose a convenient set of x, y axes and resolve all forces into components that lie along these axes.**

4. **Apply the equations that specify the balance of forces at equilibrium: $\Sigma F_x = 0$ and $\Sigma F_y = 0$.**

5. **Select a convenient axis of rotation. The choice is arbitrary. Identify the point where each external force acts on the object, and calculate the torque produced by each force about the chosen axis. Set the sum of the torques equal to zero: $\Sigma\tau = 0$.**

6. **Solve the equations in Steps 4 and 5 for the desired unknown quantities.**

EXAMPLE 3 | A Diving Board

A woman whose weight is 530 N is poised at the right end of a diving board with a length of 3.90 m. The board has negligible weight and is bolted down at the left end, while being supported 1.40 m away by a fulcrum, as **Figure 9.5a** shows. Find the forces $\vec{F}_1$ and $\vec{F}_2$ that the bolt and the fulcrum, respectively, exert on the board.

Reasoning Part b of the figure shows the free-body diagram of the diving board. Three forces act on the board: $\vec{F}_1$, $\vec{F}_2$, and the force due to the diver's weight $\vec{W}$. In choosing the directions of $\vec{F}_1$ and $\vec{F}_2$, we have used our intuition: $\vec{F}_1$ points downward because the bolt must pull in that direction to counteract the tendency of the board to rotate clockwise about the fulcrum; $\vec{F}_2$ points upward, because the board pushes downward against the fulcrum, which, in reaction, pushes upward on the board. Since the board is stationary, it is in equilibrium.

Solution Since the board is in equilibrium, the sum of the vertical forces must be zero:

$$\Sigma F_y = -F_1 + F_2 - W = 0 \qquad \textbf{(4.9b)}$$

Similarly, the sum of the torques must be zero, $\Sigma\tau = 0$. For calculating torques, we select an axis that passes through the left end of the board and is perpendicular to the page. (We will see shortly that this choice is arbitrary.) The force $\vec{F}_1$ produces no torque since it passes through the axis and, therefore, has a zero lever arm, while $\vec{F}_2$ creates a counterclockwise (positive) torque, and $\vec{W}$ produces a clockwise (negative) torque. The free-body diagram shows the lever arms for the torques:

$$\Sigma\tau = +F_2\ell_2 - W\ell_w = 0 \qquad \textbf{(9.2)}$$

Solving this equation for F_2 yields

$$F_2 = \frac{W\ell_w}{\ell_2} = \frac{(530\ \text{N})(3.90\ \text{m})}{1.40\ \text{m}} = \boxed{1480\ \text{N}}$$

This value for F_2, along with $W = 530$ N, can be substituted into Equation 4.9b to show that $\boxed{F_1 = 950\ \text{N}}$.

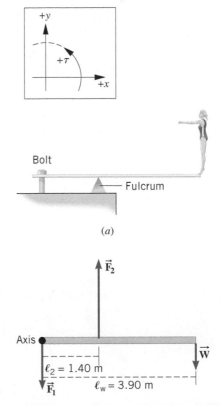

(a)

(b) Free-body diagram of the diving board

FIGURE 9.5 (a) A diver stands at the end of a diving board. (b) The free-body diagram for the diving board. The box at the top left shows the positive x and y directions for the forces, as well as the positive direction (counterclockwise) for the torques.

In Example 3 the sum of the external torques is calculated using an axis that passes through the left end of the diving board.

> **Problem-Solving Insight**   The choice of the axis is completely arbitrary, because if an object is in equilibrium, it is in equilibrium with respect to any axis whatsoever.

Thus, the sum of the external torques is zero, no matter where the axis is placed. One usually chooses the location so that the lines of action of one or more of the unknown forces pass through the axis. Such a choice simplifies the torque equation, because the torques produced by these forces are zero. For instance, in Example 3 the torque due to the force $\vec{F}_1$ does not appear in Equation 9.2, because the lever arm of this force is zero.

In a calculation of torque, the lever arm of the force must be determined relative to the axis of rotation. In Example 3 the lever arms are obvious, but sometimes a little care is needed in determining them, as in the next example.

EXAMPLE 4 | Fighting a Fire

In **Animated Figure 9.6a** an 8.00-m ladder of weight $W_L = 355$ N leans against a smooth vertical wall. The term "smooth" means that the wall can exert only a normal force directed perpendicular to the wall and cannot exert a frictional force parallel to it. A firefighter, whose weight is $W_F = 875$ N, stands 6.30 m up from the bottom of the ladder. Assume that the ladder's weight acts at the ladder's center, and neglect the hose's weight. Find the forces that the wall and the ground exert on the ladder.

Reasoning   **Animated Figure 9.6b** shows the free-body diagram of the ladder. The following forces act on the ladder:

1. Its weight $\vec{W}_L$
2. A force due to the weight $\vec{W}_F$ of the firefighter
3. The force $\vec{P}$ applied to the top of the ladder by the wall and directed perpendicular to the wall
4. The forces $\vec{G}_x$ and $\vec{G}_y$, which are the horizontal and vertical components of the force exerted by the ground on the bottom of the ladder

The ground, unlike the wall, is not smooth, so that the force $\vec{G}_x$ is produced by static friction and prevents the ladder from slipping. The force $\vec{G}_y$ is the normal force applied to the ladder by the ground. The ladder is in equilibrium, so the sum of these forces and the sum of the torques produced by them must be zero.

Solution   Since the net force acting on the ladder is zero, we have

$$\Sigma F_x = G_x - P = 0 \tag{4.9a}$$

$$\Sigma F_y = G_y - W_L - W_F = 0 \tag{4.9b}$$

Solving Equation 4.9b gives

$$G_y = W_L + W_F = 355 \text{ N} + 875 \text{ N} = \boxed{1230 \text{ N}}$$

Equation 4.9a cannot be solved as it stands, because it contains two unknown variables. However, we know that the net torque acting on an object in equilibrium is zero. In calculating torques, we use an axis at the left end of the ladder, directed perpendicular to the page, as **Animated Figure 9.6c** indicates. This axis is convenient, because $\vec{G}_x$ and $\vec{G}_y$ produce no torques about it, their

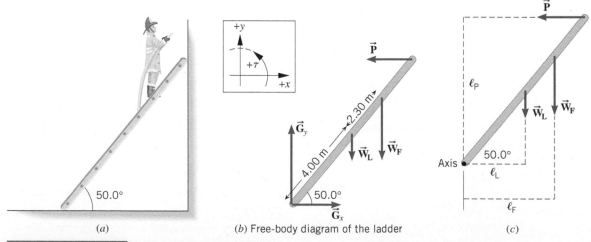

(a)   (b) Free-body diagram of the ladder   (c)

ANIMATED FIGURE 9.6   (a) A ladder leaning against a smooth wall. (b) The free-body diagram for the ladder. (c) Three of the forces that act on the ladder and their lever arms. The axis of rotation is at the lower end of the ladder and is perpendicular to the page.

lever arms being zero. Consequently, these forces will not appear in the equation representing the balance of torques. The lever arms for the remaining forces are shown in **Animated Figure 9.6c** as red dashed lines of length ℓ. The following list summarizes these forces, their lever arms, and the torques:

Force	Lever Arm	Torque
$W_L = 355$ N	$\ell_L = (4.00 \text{ m}) \cos 50.0°$	$-W_L\ell_L$
P	$\ell_P = (8.00 \text{ m}) \sin 50.0°$	$+P\ell_P$
$W_F = 875$ N	$\ell_F = (6.30 \text{ m}) \cos 50.0°$	$-W_F\ell_F$

Setting the sum of the torques equal to zero gives

$$\Sigma\tau = -W_L\ell_L - W_F\ell_F + P\ell_P = 0 \qquad (9.2)$$

Solving this equation for P gives

$$P = \frac{W_L\ell_L + W_F\ell_F}{\ell_P}$$

$$= \frac{(355 \text{ N})(4.00 \text{ m})\cos 50.0° + (875 \text{ N})(6.30 \text{ m})\cos 50.0°}{(8.00 \text{ m})\sin 50.0°} = \boxed{727 \text{ N}}$$

Substituting $P = 727$ N into Equation 4.9a reveals that

$$G_x = P = \boxed{727 \text{ N}}$$

Math Skills Sometimes it is necessary to use trigonometry to determine the lever arms from the distances given in a problem. As examples, consider the lever arms for the forces $\vec{W}_L$ and $\vec{P}$ (see **Animated Figure 9.6c**). To determine the lever arm ℓ_L for the force $\vec{W}_L$, we will use the cosine function. Equation 1.2 defines the cosine of the angle θ as $\cos\theta = \dfrac{h_a}{h}$, where h_a is the side of a right triangle adjacent

to the angle θ and h is the hypotenuse of the right triangle, as shown by the smaller triangle in **Figure 9.7a**. By comparing this triangle with the smaller triangle in **Figure 9.7b** it can be seen that $\theta = 50.0°$ and that $h_a = \ell_L$ and $h = 4.00$ m. As a result we have

$$\cos 50.0° = \frac{h_a}{h} = \frac{\ell_L}{4.00 \text{ m}} \quad \text{or} \quad \ell_L = (4.00 \text{ m})\cos 50.0°$$

To determine the lever arm ℓ_P for the force $\vec{P}$, we will use the sine function. Equation 1.1 defines the sine of the angle θ as $\sin\theta = \dfrac{h_o}{h}$, where h_o is the side of a right triangle opposite the angle θ and h is the hypotenuse of the right triangle, as shown by the larger triangle in **Figure 9.7a**. By comparing this triangle with the larger triangle in **Figure 9.7b** it can be seen that $\theta = 50.0°$ and that $h_o = \ell_P$ and $h = 8.00$ m. As a result we have

$$\sin 50.0° = \frac{h_o}{h} = \frac{\ell_P}{8.00 \text{ m}} \quad \text{or} \quad \ell_P = (8.00 \text{ m})\sin 50.0°$$

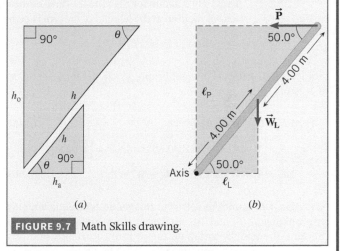

(a)   (b)

FIGURE 9.7 Math Skills drawing.

To a large extent the directions of the forces acting on an object in equilibrium can be deduced using intuition. Sometimes, however, the direction of an unknown force is not obvious, and it is inadvertently drawn reversed in the free-body diagram. This kind of mistake causes no difficulty.

Problem-Solving Insight Choosing the direction of an unknown force backward in the free-body diagram simply means that the value determined for the force will be a negative number.

EXAMPLE 5 | BIO The Physics of Bodybuilding

A bodybuilder holds a dumbbell of weight $\vec{W}_d$ as in **Figure 9.8a**. His arm is horizontal and weighs $W_a = 31.0$ N. The deltoid muscle is assumed to be the only muscle acting and is attached to the arm as shown. The maximum force $\vec{M}$ that the deltoid muscle can supply has a magnitude of 1840 N. **Figure 9.8b** shows the distances that locate where the various forces act on the arm. What is the weight of the heaviest dumbbell that can be held, and what are the horizontal and vertical force components, $\vec{S}_x$ and $\vec{S}_y$, that the shoulder joint applies to the left end of the arm?

Reasoning **Figure 9.8b** is the free-body diagram for the arm. Note that $\vec{S}_x$ is directed to the right, because the deltoid

muscle pulls the arm in toward the shoulder joint, and the joint pushes back in accordance with Newton's third law. The direction of the force $\vec{S}_y$, however, is less obvious, and we are alert for the possibility that the direction chosen in the free-body diagram is backward. If so, the value obtained for $\vec{S}_y$ will be negative.

Problem-Solving Insight When a force is negative, such as the vertical force $S_y = -297$ N in this example, it means that the direction of the force is opposite to that chosen originally.

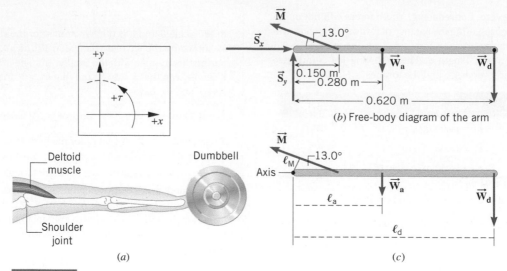

FIGURE 9.8 (*a*) The fully extended, horizontal arm of a bodybuilder supports a dumbbell. (*b*) The free-body diagram for the arm. (*c*) Three of the forces that act on the arm and their lever arms. The axis of rotation at the left end of the arm is perpendicular to the page. Force vectors are not to scale.

Solution The arm is in equilibrium, so the net force acting on it is zero:

$$\Sigma F_x = S_x - M\cos 13.0° = 0 \qquad (4.9a)$$

or $S_x = M\cos 13.0° = (1840 \text{ N})\cos 13.0° = \boxed{1790 \text{ N}}$

$$\Sigma F_y = S_y + M\sin 13.0° - W_a - W_d = 0 \qquad (4.9b)$$

Equation 4.9b cannot be solved at this point, because it contains two unknowns, S_y and W_d. However, since the arm is in equilibrium, the torques acting on the arm must balance, and this fact provides another equation. To calculate torques, we choose an axis through the left end of the arm and perpendicular to the page. With this axis, the torques due to $\vec{S}_x$ and $\vec{S}_y$ are zero, because the line of action of each force passes through the axis and the lever arm of each force is zero. The list below summarizes the remaining forces, their lever arms (see **Figure 9.8c**), and the torques.

Force	Lever Arm	Torque
$W_a = 31.0$ N	$\ell_a = 0.280$ m	$-W_a\ell_a$
W_d	$\ell_d = 0.620$ m	$-W_d\ell_d$
$M = 1840$ N	$\ell_M = (0.150 \text{ m})\sin 13.0°$	$+M\ell_M$

The condition specifying a zero net torque is

$$\Sigma\tau = -W_a\ell_a - W_d\ell_d + M\ell_M = 0 \qquad (9.2)$$

Solving this equation for W_d yields

$$W_d = \frac{-W_a\ell_a + M\ell_M}{\ell_d}$$

$$= \frac{-(31.0 \text{ N})(0.280 \text{ m}) + (1840 \text{ N})(0.150 \text{ m})\sin 13.0°}{0.620 \text{ m}} = \boxed{86.1 \text{ N}}$$

Substituting this value for W_d into Equation 4.9b and solving for S_y gives $\boxed{S_y = -297 \text{ N}}$. The minus sign indicates that the choice of direction for S_y in the free-body diagram is wrong. In reality, S_y has a magnitude of 297 N but is directed downward, not upward.

Check Your Understanding

(*The answers are given at the end of the book.*)

5. Three forces (magnitudes either *F* or 2*F*) in **CYU Figure 9.3** act on each of the thin, square sheets shown in the drawing. In parts A and B of the drawing, the force labeled $2\vec{F}$ acts at the center of the sheet. When considering angular acceleration, use an axis of rotation that is perpendicular to the plane of a sheet at its center. Determine in which drawing **(a)** the translational acceleration is equal to zero, but the angular acceleration is not equal to zero; **(b)** the translational acceleration is not equal to zero, but the angular acceleration is equal to zero; and **(c)** both the translational and angular accelerations are zero.

6. The free-body diagram in **CYU Figure 9.4** shows the forces that act on a thin rod. The three forces are drawn to scale and lie in the plane of the paper. Are these forces sufficient to keep the rod in equilibrium, or are additional forces necessary?

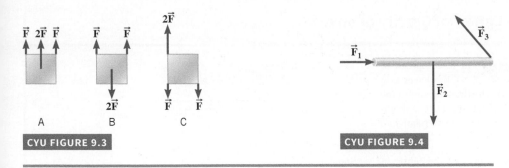

CYU FIGURE 9.3

CYU FIGURE 9.4

9.3 Center of Gravity

Often, it is important to know the torque produced by the weight of an *extended* body. In Examples 4 and 5, for instance, it is necessary to determine the torques caused by the weight of the ladder and the arm, respectively. In both cases the weight is considered to act at a definite point for the purpose of calculating the torque. This point is called the **center of gravity** (abbreviated "cg").

> **DEFINITION OF CENTER OF GRAVITY**
>
> **The center of gravity of a rigid body is the point at which its weight can be considered to act when the torque due to the weight is being calculated.**

When an object has a symmetrical shape and its weight is distributed uniformly, the center of gravity lies at its geometrical center. For instance, **Figure 9.9** shows a thin, uniform, horizontal rod of length L attached to a vertical wall by a hinge. The center of gravity of the rod is located at the geometrical center. The lever arm for the weight $\vec{W}$ is $L/2$, and the magnitude of the torque is $W(L/2)$. In a similar fashion, the center of gravity of any symmetrically shaped and uniform object, such as a sphere, disk, cube, or cylinder, is located at its geometrical center. However, this does not mean that the center of gravity must lie within the object itself. The center of gravity of a compact disc recording, for instance, lies at the center of the hole in the disc and is, therefore, "outside" the object.

Suppose we have a group of objects, with known weights and centers of gravity, and it is necessary to know the center of gravity for the group as a whole. As an example, **Figure 9.10a** shows a group composed of two parts: a horizontal uniform board (weight $\vec{W}_1$) and a uniform box (weight $\vec{W}_2$) near the left end of the board. The center of gravity can be determined by calculating the net torque created by the board and box about an axis that is picked arbitrarily to be at the right end of the board. Part *a* of the figure shows the weights $\vec{W}_1$ and $\vec{W}_2$ and their corresponding lever arms x_1 and x_2. The net torque is $\Sigma\tau = W_1x_1 + W_2x_2$. It is also possible to calculate the net torque by treating the total weight $\vec{W}_1 + \vec{W}_2$ as if it were located at the center of gravity and had the lever arm x_{cg}, as part *b* of the drawing indicates: $\Sigma\tau = (W_1 + W_2)x_{cg}$. The two values for the net torque must be the same, so that

$$W_1x_1 + W_2x_2 = (W_1 + W_2)x_{cg}$$

This expression can be solved for x_{cg}, which locates the center of gravity relative to the axis:

Center of gravity
$$x_{cg} = \frac{W_1x_1 + W_2x_2 + \cdots}{W_1 + W_2 + \cdots} \qquad (9.3)$$

The notation "$+ \cdots$" indicates that Equation 9.3 can be extended to account for any number of weights distributed along a horizontal line. **Figure 9.10c** illustrates that the group can be balanced by a single external force (due to the index finger), if the line of action of the force passes through the center of gravity, and if the force is equal in magnitude, but opposite in direction, to the weight of the group. Example 6 demonstrates how to calculate the center of gravity for the human arm.

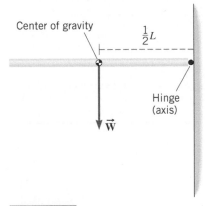

FIGURE 9.9 A thin, uniform, horizontal rod of length L is attached to a vertical wall by a hinge. The center of gravity of the rod is at its geometrical center.

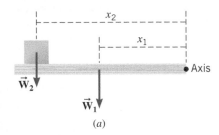

(a)

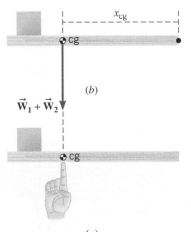

(b)

(c)

FIGURE 9.10 (*a*) A box rests near the left end of a horizontal board. (*b*) The total weight ($\vec{W}_1 + \vec{W}_2$) acts at the center of gravity of the group. (*c*) The group can be balanced by applying an external force (due to the index finger) at the center of gravity.

EXAMPLE 6 | BIO The Center of Gravity of an Arm

The horizontal arm illustrated in **Figure 9.11** is composed of three parts: the upper arm (weight $W_1 = 17$ N), the lower arm ($W_2 = 11$ N), and the hand ($W_3 = 4.2$ N). The drawing shows the center of gravity of each part, measured with respect to the shoulder joint. Find the center of gravity of the entire arm, relative to the shoulder joint.

Reasoning and Solution The coordinate x_{cg} of the center of gravity is given by

$$x_{cg} = \frac{W_1 x_1 + W_2 x_2 + W_3 x_3}{W_1 + W_2 + W_3} \quad (9.3)$$

$$= \frac{(17\text{ N})(0.13\text{ m}) + (11\text{ N})(0.38\text{ m}) + (4.2\text{ N})(0.61\text{ m})}{17\text{ N} + 11\text{ N} + 4.2\text{ N}} = \boxed{0.28\text{ m}}$$

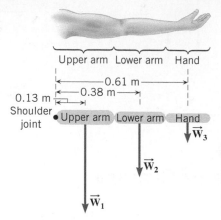

FIGURE 9.11 The three parts of a human arm, and the weight and center of gravity for each.

The center of gravity plays an important role in determining whether a group of objects remains in equilibrium as the weight distribution within the group changes. A change in the weight distribution causes a change in the position of the center of gravity, and if the change is too great, the group will not remain in equilibrium. Conceptual Example 7 discusses a shift in the center of gravity that led to an embarrassing result.

CONCEPTUAL EXAMPLE 7 | Overloading a Cargo Plane

Figure 9.12a shows a stationary cargo plane with its front landing gear 9 meters off the ground. This accident occurred because the plane was overloaded toward the rear. How did a shift in the center of gravity of the loaded plane cause the accident?

Reasoning and Solution **Figure 9.12b** shows a drawing of a correctly loaded plane, with the center of gravity located between the front and the rear landing gears. The weight $\vec{W}$ of the plane and cargo acts downward at the center of gravity, and the normal

(a)

Kevork Djansazian/AP Images

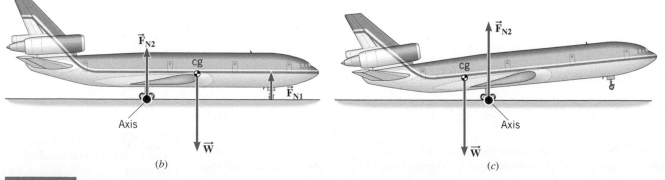

(b)   (c)

FIGURE 9.12 (a) This stationary cargo plane is sitting on its tail at Los Angeles International Airport, after being overloaded toward the rear. (b) In a correctly loaded plane, the center of gravity is between the front and the rear landing gears. (c) When the plane is overloaded toward the rear, the center of gravity shifts behind the rear landing gear, and the accident in part a occurs.

forces $\vec{F}_{N1}$ and $\vec{F}_{N2}$ act upward at the front and at the rear landing gear, respectively. With respect to an axis at the rear landing gear, the counterclockwise torque due to $\vec{F}_{N1}$ balances the clockwise torque due to $\vec{W}$, and the plane remains in equilibrium. **Figure 9.12c** shows the plane with too much cargo loaded toward the rear, just after the plane has begun to rotate counterclockwise. Because of the overloading, the center of gravity has shifted behind the rear landing gear. The torque due to $\vec{W}$ is now counterclockwise and is

not balanced by any clockwise torque. Due to the unbalanced counterclockwise torque, the plane rotates until its tail hits the ground, which applies an upward force to the tail. The clockwise torque due to this upward force balances the counterclockwise torque due to $\vec{W}$, and the plane comes again into an equilibrium state, this time with the front landing gear 9 meters off the ground.

Related Homework: Problems 12, 15, 55

As we have seen in Example 7, the center of gravity plays an important role in the equilibrium orientation of airplanes. It also plays a similar role in the design of vehicles that can be safely driven with minimal risk to the passengers. Example 8 discusses this role in the context of sport utility vehicles.

Analyzing Multiple-Concept Problems

EXAMPLE 8 | The Physics of the Static Stability Factor and Rollover

Figure 9.13a shows a sport utility vehicle (SUV) that is moving away from you and negotiating a horizontal turn. The radius of the turn is 16 m, and its center is on the right in the drawing. The center of gravity of the vehicle is 0.94 m above the ground and, as

an approximation, is assumed to be located midway between the wheels on the left and right sides. The separation between these wheels is the track width and is 1.7 m. What is the greatest speed at which the SUV can negotiate the turn without rolling over?

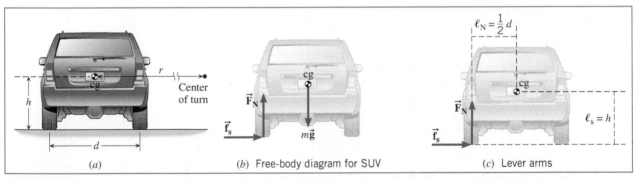

(a) (b) Free-body diagram for SUV (c) Lever arms

FIGURE 9.13 (a) A sport utility vehicle (SUV) is shown moving away from you, following a road curving to the right. The radius of the turn is r. The SUV's center of gravity ($\odot$) is located at a height h above the ground, and its wheels are separated by a distance d. (b) This free-body diagram shows the SUV at the instant just before rollover toward the outside of the turn begins. The forces acting on it are its weight $m\vec{g}$, the total normal force $\vec{F}_N$ (acting only on the left-side tires), and the total force of static friction $\vec{f}_s$ (also acting only on the left-side tires). (c) The lever arms ℓ_N (for $\vec{F}_N$) and ℓ_s (for $\vec{f}_s$) are for an axis passing through the center of gravity and perpendicular to the page.

Reasoning The free-body diagram in **Figure 9.13b** shows the SUV at the instant just before it begins to roll over toward the outside of the turn, which is on the left side of the drawing. At this moment the right-side wheels have just lost contact with the ground, so no forces are acting on them. The forces acting on the SUV are its weight $m\vec{g}$, the total normal force $\vec{F}_N$ (acting

only on the left-side tires), and the total force of static friction $\vec{f}_s$ (also acting only on the left-side tires). Since the SUV is moving around the turn, it has a centripetal acceleration and must be experiencing a centripetal force. The static frictional force $\vec{f}_s$ alone provides the centripetal force. The speed v at which the SUV (mass m) negotiates the turn of radius r is related to the magnitude F_c of the centripetal force by $F_c = mv^2/r$ (Equation 5.3). After applying this relation, we will consider rollover by analyzing the torques acting on the SUV with respect to an axis through the center of gravity and perpendicular to the drawing. The normal force $\vec{F}_N$ will play a role in this analysis, and it will be evaluated by using the fact that it must balance the weight of the vehicle.

Knowns and Unknowns We are given the following data:

Description	Symbol	Value	Comment
Radius of turn	r	16 m	
Location of center of gravity	h	0.94 m	Height above ground.
Track width	d	1.7 m	Distance between left- and right-side wheels.
Unknown Variable			
Speed of SUV	v	?	

Modeling the Problem

STEP 1 Speed and Centripetal Force According to Equation 5.3, the magnitude of the centripetal force (provided solely by the static frictional force of magnitude f_s) is related to the speed v of the car by

$$F_c = f_s = \frac{mv^2}{r}$$

Solving for the speed gives Equation 1 at the right. To use this result, we must have a value for f_s, which we obtain in Step 2.

$$v = \sqrt{\frac{rf_s}{m}} \quad \boxed{?} \tag{1}$$

STEP 2 Torques **Figure 9.13b** shows the free-body diagram for the SUV at the instant just before rollover begins, when the total normal force $\vec{F}_N$ and the total static frictional force $\vec{f}_s$ are acting only on the left-side wheels, the right-side wheels having just lost contact with the ground. At this moment, the sum of the torques due to these two forces is zero for an axis through the center of gravity and perpendicular to the page. Each torque has a magnitude given by Equation 9.1 as the magnitude of the force times the lever arm for the force, and part c of the drawing shows the lever arms. No lever arm is shown for the weight, because it passes through the axis and, therefore, contributes no torque. The force $\vec{f}_s$ produces a positive torque, since it causes a counterclockwise rotation about the chosen axis, while the force $\vec{F}_N$ produces a negative torque, since it causes a clockwise rotation. Thus, we have

$$f_s \ell_s - F_N \ell_N = 0 \quad \text{or} \quad \boxed{f_s = \frac{F_N \ell_N}{\ell_s} = \frac{F_N d}{2h}}$$

This result for f_s can be substituted into Equation 1, as illustrated at the right. We now proceed to Step 3, to obtain a value for F_N.

$$v = \sqrt{\frac{rf_s}{m}} \tag{1}$$

$$f_s = \frac{F_N d}{2h} \quad \boxed{?} \tag{2}$$

STEP 3 The Normal Force The SUV does not accelerate in the vertical direction, so the normal force must balance the car's weight mg, or

$$\boxed{F_N = mg}$$

The substitution of this result into Equation 2 is shown at the right.

$$v = \sqrt{\frac{rf_s}{m}} \tag{1}$$

$$f_s = \frac{F_N d}{2h} \tag{2}$$

$$F_N = mg$$

Solution The results of each step can be combined algebraically to show that

$$\overset{\text{STEP 1}}{\downarrow} \quad \overset{\text{STEP 2}}{\downarrow} \quad \overset{\text{STEP 3}}{\downarrow}$$
$$v = \sqrt{\frac{rf_s}{m}} = \sqrt{\frac{rF_N d}{2hm}} = \sqrt{\frac{rmgd}{2hm}}$$

The final expression is independent of the mass of the SUV, since m has been eliminated algebraically. The greatest speed at which the SUV can negotiate the turn without rolling over is

$$v = \sqrt{rg \underbrace{\left(\frac{d}{2h}\right)}_{\text{SSF}}} = \sqrt{(16\text{ m})(9.80\text{ m/s}^2)\left[\frac{1.7\text{ m}}{2(0.94\text{ m})}\right]} = \boxed{12\text{ m/s}}$$

At this speed, the SUV will negotiate the turn just on the verge of rolling over. In the final result and beneath the term $\frac{d}{2h}$, we have included the label SSF, which stands for *static stability factor*. The SSF provides one measure of how susceptible a vehicle is to rollover. Higher values of the SSF are better, because they lead to larger values for v, the greatest speed at which the vehicle can negotiate a turn without rolling over. Note that greater values for the track width d (more widely separated wheels) and smaller values for the height h (center of gravity closer to the ground) lead to higher values of the SSF. A Formula One racing car, for example, with its low center of gravity and large track width, is much less prone to rolling over than is an SUV, when both are driven at the same speed around the same curve.

The center of gravity of an object with an irregular shape and a nonuniform weight distribution can be found by suspending the object from two different points P_1 and P_2, one at a time. **Figure 9.14a** shows the object at the moment of release, when its weight $\vec{W}$, acting at the center of gravity, has a nonzero lever arm ℓ relative to the axis shown in the drawing. At this instant the weight produces a torque about the axis. The tension force $\vec{T}$ applied to the object by the suspension cord produces no torque because its line of action passes through the axis. Hence, in part a there is a net torque applied to the object, and the object begins to rotate. Friction eventually brings the object to rest as in part b, where the center of gravity lies directly below the point of suspension. In such an orientation, the line of action of the weight passes through the axis, so there is no longer any net torque. In the absence of a net torque the object remains at rest. By suspending the object from a second point P_2 (see **Figure 9.14c**), a second line through the object can be established, along which the center of gravity must also lie. The center of gravity, then, must be at the intersection of the two lines.

The center of gravity is closely related to the center-of-mass concept discussed in Section 7.5. To see why they are related, let's replace each occurrence of the weight in Equation 9.3 by $W = mg$, where m is the mass of a given object and g is the magnitude of the acceleration due to gravity at the location of the object. Suppose that g has the same value everywhere the objects are located. Then it can be algebraically eliminated from each term on the right side of Equation 9.3. The resulting equation, which contains only masses and distances, is the same as Equation 7.10, which defines the location of the center of mass. Thus, the two points are identical. For ordinary-sized objects, like cars and boats, the center of gravity coincides with the center of mass.

The center of gravity of the human body will vary depending on the body's shape. Therefore, it figures prominently in athletic events, such as that discussed within the context of center of mass in Example 10 of Chapter 7. The center of gravity of the body also impacts everyday activities, such as lifting a heavy object, as the following example illustrates.

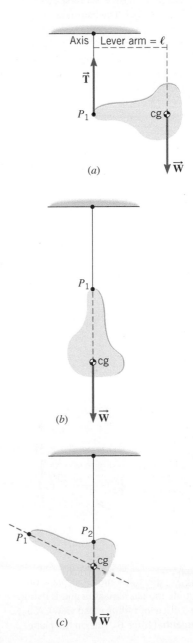

FIGURE 9.14 The center of gravity (cg) of an object can be located by suspending the object from two different points, P_1 and P_2, one at a time.

CONCEPTUAL EXAMPLE 9 | BIO The Importance of Proper Lifting Technique

Over one million people injure their back in the workplace annually, and an estimated 80% of the adult population will injure their back during their lifetime. Americans spend more than $50 billion a year on treating back pain. Many of these injuries could be avoided by using proper lifting techniques. What is the best technique to use to reduce the risk of a back injury?

Reasoning and Solution **Figure 9.15a** shows proper body position when picking up a heavy object. The lifter keeps their spine straight, bends at the knees, and keeps the object close to their body. The weight of the person's upper body ($\vec{\mathbf{W}}_U$) and the weight of the box ($\vec{\mathbf{W}}_B$) produce clockwise torques about the axis at the hip joint, with their forces acting at their respective centers of gravity. To maintain this position, the lower back muscles must apply a force to the spine ($\vec{\mathbf{F}}_B$). This force produces the counterclockwise torque necessary to balance the two torques from the weights. **Figure 9.15b** shows a very poor technique for lifting a heavy object. The back is no longer straight, but rounded. The person is leaning more forward and holding the box farther from their body. This increases the lever arm, and thus the torques, for both weight forces. In order to compensate and maintain equilibrium about the hip joint, the force from the lower back muscles must increase, placing them at a greater risk of injury.

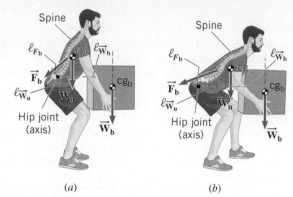

(a)   (b)

FIGURE 9.15 (a) Proper technique in lifting a heavy box. The person keeps their back straight, bends their knees, and holds the box close to their body. The lift is made with the legs, with very little rotation about the hip joint. (b) Poor lifting technique. The person is leaning too far forward, so that the back is more rounded and less vertical, as well as holding the box further from their body. This greatly increases the force required by the muscles in the lower back. Notice too how the person in (b) has to slide their left foot forward, in order to keep a point of support below their center of gravity.

Check Your Understanding

(*The answers are given at the end of the book.*)

7. Starting in the spring, fruit begins to grow on the outer end of a branch on a pear tree. As the fruit grows, does the center of gravity of the pear-growing branch **(a)** move toward the pears at the end of the branch, **(b)** move away from the pears, or **(c)** not move at all?

8. CYU **Figure 9.5** shows a wine rack for a single bottle of wine that seems to defy common sense as it balances on a tabletop. Where is the center of gravity of the combined wine rack and bottle of wine located?

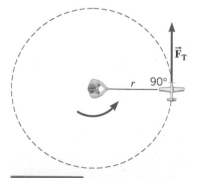

Wine rack

Rack is not attached to the tabletop.

CYU FIGURE 9.5

(a) At the neck of the bottle where it passes through the wine rack **(b)** Directly above the point where the wine rack touches the tabletop **(c)** At a location to the right of where the wine rack touches the tabletop

9. Bob and Bill have the same weight and wear identical shoes. When they both keep their feet flat on the floor and their bodies straight, Bob can lean forward farther than Bill can before falling. Other things being equal, whose center of gravity is closer to the ground when both are standing erect?

FIGURE 9.16 A model airplane on a guideline has a mass m and is flying on a circle of radius r (top view). A net tangential force $\vec{\mathbf{F}}_T$ acts on the plane.

9.4 Newton's Second Law for Rotational Motion About a Fixed Axis

The goal of this section is to put Newton's second law into a form suitable for describing the rotational motion of a rigid object about a fixed axis. We begin by considering a particle moving on a circular path. **Figure 9.16** presents a good approximation of this situation by using a small model plane on a guideline of negligible mass. The plane's engine produces a net external tangential force F_T that gives the plane a tangential acceleration a_T. In accord with Newton's second law, it follows that $F_T = ma_T$. The torque τ produced by this force is $\tau = F_T r$, where the radius r of the circular path is also the lever arm. As a result, the torque is $\tau = ma_T r$. However, the tangential acceleration is related to the angular acceleration α according to $a_T = r\alpha$ (Equation 8.10), where α must be expressed in rad/s^2. With this substitution for a_T, the torque becomes

$$\tau = \underbrace{(mr^2)}_{\substack{\text{Moment of} \\ \text{inertia } I}}\alpha \tag{9.4}$$

Equation 9.4 is the form of Newton's second law we have been seeking. It indicates that the net external torque τ is directly proportional to the angular acceleration α. The constant of proportionality is $I = mr^2$, which is called the **moment of inertia of the particle**. The SI unit for moment of inertia is kg · m^2.

If all objects were single particles, it would be just as convenient to use the second law in the form $F_T = ma_T$ as in the form $\tau = I\alpha$. The advantage in using $\tau = I\alpha$ is that it can be applied to any rigid body rotating about a fixed axis, and not just to a particle. To illustrate how this advantage arises, **Figure 9.17a** shows a flat sheet of material that rotates about an axis perpendicular to the sheet. The sheet is composed of a number of mass particles, $m_1, m_2, \ldots, m_N$, where N is very large. Only four particles are shown for the sake of clarity. Each particle behaves in the same way as the model airplane in **Figure 9.16** and obeys the relation $\tau = (mr^2)\alpha$:

$$\tau_1 = (m_1 r_1{}^2)\alpha$$
$$\tau_2 = (m_2 r_2{}^2)\alpha$$
$$.$$
$$.$$
$$.$$
$$\tau_N = (m_N r_N{}^2)\alpha$$

In these equations each particle has the same angular acceleration α, since the rotating object is assumed to be rigid. Adding together the N equations and factoring out the common value of α, we find that

$$\underbrace{\Sigma\tau}_{\substack{\text{Net} \\ \text{external torque}}} = \underbrace{(\Sigma mr^2)}_{\substack{\text{Moment} \\ \text{of inertia}}}\alpha \tag{9.5}$$

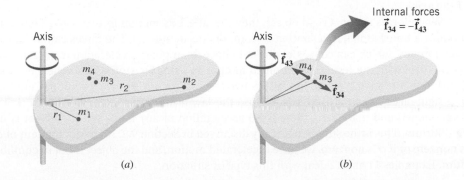

FIGURE 9.17 (a) A rigid body consists of a large number of particles, four of which are shown. (b) The internal forces that particles 3 and 4 exert on each other obey Newton's law of action and reaction.

where the expression $\Sigma\tau = \tau_1 + \tau_2 + \cdots + \tau_N$ is the sum of the external torques, and $\Sigma mr^2 = m_1r_1^2 + m_2r_2^2 + \cdots + m_Nr_N^2$ represents the sum of the individual moments of inertia. The latter quantity is the **moment of inertia I of the body**:

Moment of inertia of a body
$$I = m_1r_1^2 + m_2r_2^2 + \cdots + m_Nr_N^2 = \Sigma mr^2 \qquad (9.6)$$

In this equation, r is the perpendicular radial distance of each particle from the axis of rotation. Combining Equation 9.6 with Equation 9.5 gives the following result:

ROTATIONAL ANALOG OF NEWTON'S SECOND LAW FOR A RIGID BODY ROTATING ABOUT A FIXED AXIS

$$\text{Net external torque} = \begin{pmatrix} \text{Moment of} \\ \text{inertia} \end{pmatrix} \times \begin{pmatrix} \text{Angular} \\ \text{acceleration} \end{pmatrix}$$

$$\Sigma\tau = I\alpha \qquad (9.7)$$

Requirement: α must be expressed in rad/s².

The version of Newton's second law given in Equation 9.7 applies only for rigid bodies. The word "rigid" means that the distances r_1, r_2, r_3, etc. that locate each particle m_1, m_2, m_3, etc. (see **Figure 9.17a**) do not change during the rotational motion. In other words, a rigid body is one that does not change its shape while undergoing an angular acceleration in response to an applied net external torque.

The form of the second law for rotational motion, $\Sigma\tau = I\alpha$, is similar to the equation for translational (linear) motion, $\Sigma F = ma$, and is valid only in an inertial frame. The moment of inertia I plays the same role for rotational motion that the mass m does for translational motion. Thus, I is a measure of the rotational inertia of a body. When using Equation 9.7, α must be expressed in rad/s², because the relation $a_T = r\alpha$ (which requires radian measure) was used in the derivation.

When calculating the sum of torques in Equation 9.7, it is necessary to include only the *external torques*, those applied by agents outside the body. The torques produced by internal forces need not be considered, because they always combine to produce a net torque of zero. Internal forces are those that one particle within the body exerts on another particle. They always occur in pairs of oppositely directed forces of equal magnitude, in accord with Newton's third law (see m_3 and m_4 in **Figure 9.17b**). The forces in such a pair have the same line of action, so they have identical lever arms and produce torques of equal magnitudes. One torque is counterclockwise, while the other is clockwise, the net torque from the pair being zero.

It can be seen from Equation 9.6 that the moment of inertia depends on both the mass of each particle and its distance from the axis of rotation. The farther a particle is from the axis, the greater is its contribution to the moment of inertia. Therefore, although a rigid object possesses a unique total mass, it does not have a unique moment of inertia, as indicated by Example 10. This example shows how the moment of inertia can change when the axis of rotation changes. The procedure illustrated in Example 10 can be extended using integral calculus to evaluate the moment of inertia of a rigid object with a continuous mass distribution, and **Table 9.1** gives some typical results. These results depend on the total mass of the object, its shape, and the location and orientation of the axis.

When forces act on a rigid object, they can affect its motion in two ways. They can produce a translational acceleration a (components a_x and a_y). The forces can also produce torques, which can cause the object to have an angular acceleration α. In general, we can deal with the resulting combined motion by using Newton's second law. For the translational motion, we use the law in the form $\Sigma F = ma$. For the rotational motion of a rigid object about a fixed axis, we use the law in the form $\Sigma\tau = I\alpha$. When a (both components) and α are zero, there is no acceleration of any kind, and the object is in equilibrium. This is the situation already discussed in Section 9.2. If any component of a is nonzero or if α is nonzero, we have accelerated motion, and the object is not in equilibrium. Examples 11 and 12 deal with this type of situation.

TABLE 9.1 Moments of Inertia I for Various Rigid Objects of Mass M

Thin-walled hollow cylinder or hoop

$I = MR^2$

Solid cylinder or disk

$I = \frac{1}{2}MR^2$

Thin rod, axis perpendicular to rod and passing through center

$I = \frac{1}{12}ML^2$

Thin rod, axis perpendicular to rod and passing through one end

$I = \frac{1}{3}ML^2$

Solid sphere, axis through center

$I = \frac{2}{5}MR^2$

Solid sphere, axis tangent to surface

$I = \frac{7}{5}MR^2$

Thin-walled spherical shell, axis through center

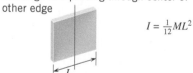

$I = \frac{2}{3}MR^2$

Thin rectangular sheet, axis parallel to one edge and passing through center of other edge

$I = \frac{1}{12}ML^2$

Thin rectangular sheet, axis along one edge

$I = \frac{1}{3}ML^2$

EXAMPLE 10 | The Moment of Inertia Depends on Where the Axis Is

Two particles each have a mass M and are fixed to the ends of a thin rigid rod, whose mass can be ignored. The length of the rod is L. Find the moment of inertia when this object rotates relative to an axis that is perpendicular to the rod at **(a)** one end and **(b)** the center. (See **Interactive Figure 9.18**.)

Reasoning   When the axis of rotation changes, the distance r between the axis and each particle changes. In determining the moment of inertia using $I = \Sigma mr^2$, we must be careful to use the distances that apply for each axis.

> **Problem-Solving Insight**   The moment of inertia depends on the location and orientation of the axis relative to the particles that make up the object.

Solution   **(a)** Particle 1 lies on the axis, as part a of the drawing shows, and has a zero radial distance: $r_1 = 0$. In contrast, particle 2 moves on a circle whose radius is $r_2 = L$. Noting that $m_1 = m_2 = M$, we find that the moment of inertia is

$$I = \Sigma mr^2 = m_1 r_1^2 + m_2 r_2^2 = M(0)^2 + M(L)^2 = \boxed{ML^2} \quad \text{(9.6)}$$

(b) Part b of the drawing shows that particle 1 no longer lies on the axis but now moves on a circle of radius $r_1 = L/2$. Particle 2 moves on a circle with the same radius, $r_2 = L/2$. Therefore,

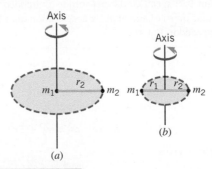

INTERACTIVE FIGURE   9.18   Two particles, masses m_1 and m_2, are attached to the ends of a massless rigid rod. The moment of inertia of this object is different, depending on whether the rod rotates about an axis through (a) the end or (b) the center of the rod.

$$I = \Sigma mr^2 = m_1 r_1^2 + m_2 r_2^2 = M(L/2)^2 + M(L/2)^2 = \boxed{\tfrac{1}{2}ML^2}$$

This value differs from the value in part (a) because the axis of rotation is different, and the distances of the particles from the axis are different.

Analyzing Multiple-Concept Problems

EXAMPLE 11 | The Torque of an Electric Saw Motor

The motor in an electric saw brings the circular blade from rest up to the rated angular velocity of 80.0 rev/s in 240.0 rev. One type of blade has a moment of inertia of 1.41×10^{-3} kg · m². What net torque (assumed constant) must the motor apply to the blade?

Reasoning   Newton's second law for rotational motion, $\Sigma\tau = I\alpha$ (Equation 9.7), can be used to find the net torque $\Sigma\tau$. However,

when using the second law, we will need a value for the angular acceleration α, which can be obtained by using one of the equations of rotational kinematics. In addition, we must remember that the value for α must be expressed in rad/s², not rev/s², because Equation 9.7 requires radian measure.

Knowns and Unknowns   The given data are summarized in the following table:

Description	Symbol	Value	Comment
Explicit Data			
Final angular velocity	ω	80.0 rev/s	Must be converted to rad/s.
Angular displacement	θ	240.0 rev	Must be converted to rad.
Moment of inertia	I	1.41×10^{-3} kg · m²	
Implicit Data			
Initial angular velocity	ω_0	0 rad/s	Blade starts from rest.
Unknown Variable			
Net torque applied to blade	$\Sigma\tau$	?	

Modeling the Problem

STEP 1 Newton's Second Law for Rotation Newton's second law for rotation (Equation 9.7) specifies the net torque $\Sigma\tau$ applied to the blade in terms of the blade's moment of inertia I and angular acceleration α. In Step 2, we will obtain the value for α that is needed in Equation 9.7.

$$\Sigma\tau = I\alpha \qquad (9.7)$$

$$\boxed{?}$$

STEP 2 Rotational Kinematics As the data table indicates, we have data for the angular displacement θ, the final angular velocity ω, and the initial angular velocity ω_0. With these data, Equation 8.8 from the equations of rotational kinematics can be used to determine the angular acceleration α:

$$\omega^2 = \omega_0^2 + 2\alpha\theta \qquad (8.8)$$

Solving for α gives

$$\alpha = \frac{\omega^2 - \omega_0^2}{2\theta}$$

which can be substituted into Equation 9.7, as shown at the right.

$$\Sigma\tau = I\alpha \qquad (9.7)$$

$$\boxed{\alpha = \frac{\omega^2 - \omega_0^2}{2\theta}}$$

Solution The results of each step can be combined algebraically to show that

$$\underset{\Sigma\tau}{} \overset{\text{STEP 1}}{=} \underset{I\alpha}{} \overset{\text{STEP 2}}{=} I\left(\frac{\omega^2 - \omega_0^2}{2\theta}\right)$$

In this result for $\Sigma\tau$, we must use radian measure for the variables ω, ω_0, and θ. To convert from revolutions (rev) to radians (rad), we will use the fact that 1 rev = 2π rad. Thus, the net torque applied by the motor to the blade is

$$\Sigma\tau = I\left(\frac{\omega^2 - \omega_0^2}{2\theta}\right)$$

$$= (1.41 \times 10^{-3}\,\text{kg} \cdot \text{m}^2)\left\{ \frac{\left[\left(80.0\,\frac{\text{rev}}{\text{s}}\right)\left(\frac{2\pi\,\text{rad}}{1\,\text{rev}}\right)\right]^2 - (0\,\text{rad/s})^2}{2(240.0\,\text{rev})\left(\frac{2\pi\,\text{rad}}{1\,\text{rev}}\right)} \right\} = \boxed{0.118\,\text{N} \cdot \text{m}}$$

Related Homework: Problems 23, 27, 29, 67

BIO **THE PHYSICS OF . . . wheelchairs.** To accelerate a wheelchair, the rider applies a force to a handrail on each wheel. The magnitude of the torque generated by the force is the product of the force-magnitude and the lever arm. As **Figure 9.19** illustrates, the lever arm is the radius of the circular rail, which is designed to be as large as possible. Thus, a relatively large torque can be generated for a given force, allowing for rapid acceleration.

JACKIE JOHNSTON/AP Images

FIGURE 9.19 A rider applies a force $\vec{F}$ to the circular handrail. The magnitude of the torque produced by this force is the product of the force-magnitude and the lever arm ℓ about the axis of rotation.

Example 11 shows how Newton's second law for rotational motion is used when design considerations demand an adequately large angular acceleration. There are also situations in which it is desirable to have as little angular acceleration as possible, and Conceptual Example 12 deals with one of them.

CONCEPTUAL EXAMPLE 12 | The Physics of Archery and Bow Stabilizers

Archers can shoot with amazing accuracy, especially using modern bows such as the one in **Figure 9.20**. Notice the bow stabilizer, a long, thin rod that extends from the front of the bow and has a relatively massive cylinder at the tip. Advertisements claim that the stabilizer helps to steady the archer's aim. Which of the following explains why this is true? The addition of the stabilizer **(a)** decreases the bow's moment of inertia, making it easier for the archer to hold the bow steady; **(b)** has nothing to do with the bow's moment of inertia; **(c)** increases the bow's moment of inertia, making it easier for the archer to hold the bow steady.

Reasoning An axis of rotation (the black dot) has been added to **Figure 9.20**. This axis passes through the archer's left shoulder and is perpendicular to the plane of the paper. Any angular acceleration of the archer's body about this axis will lead to a rotation of the bow and, thus, will degrade the archer's aim. The angular acceleration will depend on any unbalanced torques that occur while the archer's tensed muscles try to hold the drawn bow, as well as the bow's moment of inertia.

Answers (a) and (b) are incorrect. Adding a stabilizer to the bow increases its mass. According to the definition of the moment of inertia of a body (Equation 9.6), the increase in mass leads to an increase in the bow's moment of inertia.

Answer (c) is correct. Newton's second law for rotational motion states that the angular acceleration α of the bow is given by $\alpha = (\Sigma\tau)/I$ (Equation 9.7), where $\Sigma\tau$ is the net torque acting on the bow and I is its moment of inertia. The stabilizer increases I, especially the relatively massive cylinder at the tip, since it is so far from the axis. Note that the moment of inertia is in the denominator on the right side of this equation. Therefore, to the extent that I is larger, a given net torque $\Sigma\tau$ will create a smaller angular acceleration and, hence, less disturbance of the aim.

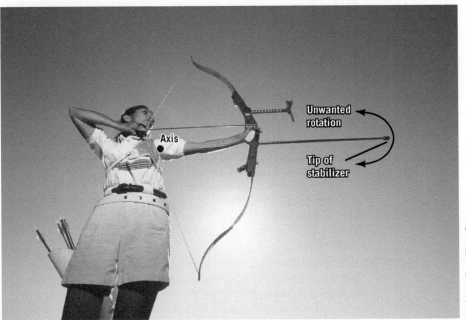

David Madison/Getty Images

FIGURE 9.20 The long, thin rod extending from the front of the bow is a stabilizer. The stabilizer helps to steady the archer's aim, as Conceptual Example 12 discusses.

We have seen that Newton's second law for rotational motion, $\Sigma\tau = I\alpha$, has the same form as the law for translational motion, $\Sigma F = ma$, so each rotational variable has a translational analog: torque τ and force F are analogous quantities, as are moment of inertia I and mass m, and angular acceleration α and linear acceleration a. The other physical concepts developed for studying translational motion, such as kinetic energy and momentum, also have rotational analogs. For future reference, **Table 9.2** itemizes these concepts and their rotational analogs.

TABLE 9.2 **Analogies Between Rotational and Translational Concepts**

Physical Concept	Rotational	Translational
Displacement	θ	s
Velocity	ω	v
Acceleration	α	a
The cause of acceleration	Torque τ	Force F
Inertia	Moment of inertia I	Mass m
Newton's second law	$\Sigma\tau = I\alpha$	$\Sigma F = ma$
Work	$\tau\theta$	Fs
Kinetic energy	$\frac{1}{2}I\omega^2$	$\frac{1}{2}mv^2$
Momentum	$L = I\omega$	$p = mv$

Check Your Understanding

(The answers are given at the end of the book.)

10. Three massless rods (A, B, and C) are free to rotate about an axis at their left end (see **CYU Figure 9.6**). The same force $\vec{\mathbf{F}}$ is applied to the right end of each rod. Objects with different masses are attached to the rods, but the total mass (3m) of the objects is the same for each rod. Rank the angular acceleration of the rods, largest to smallest.

11. A flat triangular sheet of uniform material is shown in **CYU Figure 9.7**. There are three possible axes of rotation, each perpendicular to the sheet and passing through one corner, A, B, or C. For which axis is the greatest net external torque required to bring the triangle up to an angular speed of 10.0 rad/s in 10.0 s, starting from rest? Assume that the net torque is kept constant while it is being applied.

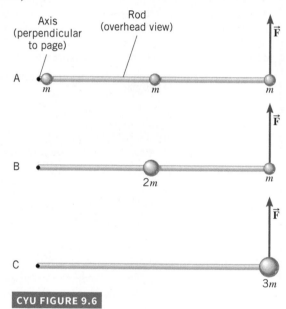

CYU FIGURE 9.6

12. At a given instant an object has an angular velocity. It also has an angular acceleration due to torques that are present. Therefore, the angular velocity is changing. Does the angular velocity at this instant increase, decrease, or remain the same **(a)** if additional torques are applied so as to make the net torque suddenly equal to zero and **(b)** if all the torques are suddenly removed?

13. The space probe in **CYU Figure 9.8** is initially moving with a constant translational velocity and zero angular velocity. **(a)** When the two engines are fired, each generating a thrust of magnitude T, does the translational velocity increase, decrease, or remain the same? **(b)** Does the angular velocity increase, decrease, or remain the same?

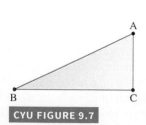

CYU FIGURE 9.7

CYU FIGURE 9.8

9.5 | Rotational Work and Energy

Work and energy are among the most fundamental and useful concepts in physics. Chapter 6 discusses their application to translational motion. These concepts are equally useful for rotational motion, provided they are expressed in terms of angular variables.

The work W done by a constant force that points in the same direction as the displacement is $W = Fs$ (Equation 6.1), where F and s are the magnitudes of the force and the displacement, respectively. To see how this expression can be rewritten using angular variables, consider **Figure 9.21**. Here a rope is wrapped around a wheel and is under a constant tension F. If the rope is pulled out a distance s, the wheel rotates through an angle $\theta = s/r$ (Equation 8.1), where r is the radius of the wheel and θ is in radians. Thus, $s = r\theta$, and the work done by the tension force in turning the wheel is $W = Fs = Fr\theta$. However, Fr is the torque τ applied to the wheel by the tension, so the rotational work can be written as follows:

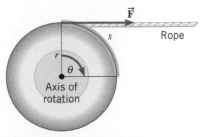

FIGURE 9.21 The force $\vec{F}$ does work in rotating the wheel through the angle θ.

DEFINITION OF ROTATIONAL WORK

The rotational work W_R done by a constant torque τ in turning an object through an angle θ is

$$W_R = \tau\theta \qquad (9.8)$$

Requirement: θ must be expressed in radians.

SI Unit of Rotational Work: joule (J)

Section 6.2 discusses the work–energy theorem and kinetic energy. There we saw that the work done on an object by a net external force causes the translational kinetic energy ($\frac{1}{2}mv^2$) of the object to change. In an analogous manner, the rotational work done by a net external torque causes the rotational kinetic energy to change. A rotating body possesses kinetic energy, because its constituent particles are moving. If the body is rotating with an angular speed ω, the tangential speed v_T of a particle at a distance r from the axis is $v_T = r\omega$ (Equation 8.9). **Figure 9.22** shows two such particles. If a particle's mass is m, its kinetic energy is $\frac{1}{2}mv_T^2 = \frac{1}{2}mr^2\omega^2$. The kinetic energy of the entire rotating body, then, is the sum of the kinetic energies of the particles:

FIGURE 9.22 The rotating wheel is composed of many particles, two of which are shown.

$$\text{Rotational KE} = \Sigma\left(\tfrac{1}{2}mr^2\omega^2\right) = \tfrac{1}{2}\underbrace{(\Sigma mr^2)}_{\substack{\text{Moment of}\\\text{inertia, }I}}\omega^2$$

In this result, the angular speed ω is the same for all particles in a rigid body and, therefore, has been factored outside the summation. According to Equation 9.6, the term in parentheses is the moment of inertia, $I = \Sigma mr^2$, so the rotational kinetic energy takes the following form:

DEFINITION OF ROTATIONAL KINETIC ENERGY

The rotational kinetic energy KE_R of a rigid object rotating with an angular speed ω about a fixed axis and having a moment of inertia I is

$$KE_R = \tfrac{1}{2}I\omega^2 \qquad (9.9)$$

Requirement: ω must be expressed in rad/s.

SI Unit of Rotational Kinetic Energy: joule (J)

Kinetic energy is one part of an object's total mechanical energy. The total mechanical energy is the sum of the kinetic and potential energies and obeys the principle of conservation of mechanical energy (see Section 6.5). Specifically, we need to remember that translational and rotational motion can occur simultaneously. When a bicycle coasts down a hill, for instance, its tires are both translating and rotating. An object such as a rolling bicycle tire has both translational and rotational kinetic energies, so that the total mechanical energy is

$$\underbrace{E}_{\substack{\text{Total mechanical}\\\text{energy}}} = \underbrace{\tfrac{1}{2}mv^2}_{\substack{\text{Translational}\\\text{kinetic energy}}} + \underbrace{\tfrac{1}{2}I\omega^2}_{\substack{\text{Rotational}\\\text{kinetic energy}}} + \underbrace{mgh}_{\substack{\text{Gravitational}\\\text{potential energy}}}$$

Here m is the mass of the object, v is the translational speed of its center of mass, I is its moment of inertia about an axis through the center of mass, ω is its angular speed, and h is the height of the object's center of mass relative to an arbitrary zero level. Mechanical energy is conserved if W_{nc}, the net work done by external nonconservative forces and external torques, is zero. If the total mechanical energy is conserved as an object moves, its final total mechanical energy E_f equals its initial total mechanical energy E_0: $E_f = E_0$.

Example 13 illustrates the effect of combined translational and rotational motion in the context of how the total mechanical energy of a cylinder is conserved as it rolls down an incline.

EXAMPLE 13 | Rolling Cylinders

A thin-walled hollow cylinder (mass = m_h, radius = r_h) and a solid cylinder (mass = m_s, radius = r_s) start from rest at the top of an incline (**Figure 9.23**). Both cylinders start at the same vertical height h_0 and roll down the incline without slipping. All heights are measured relative to an arbitrarily chosen zero level that passes through the center of mass of a cylinder when it is at the bottom of the incline (see the drawing). Ignoring energy losses due to retarding forces, determine which cylinder has the greatest translational speed on reaching the bottom.

Reasoning Only the conservative force of gravity does work on the cylinders, so the total mechanical energy is conserved as they roll down. The total mechanical energy E at any height h above the zero level is the sum of the translational kinetic energy ($\frac{1}{2}mv^2$), the rotational kinetic energy ($\frac{1}{2}I\omega^2$), and the gravitational potential energy (mgh):

$$E = \tfrac{1}{2}mv^2 + \tfrac{1}{2}I\omega^2 + mgh$$

As the cylinders roll down, potential energy is converted into kinetic energy, but the kinetic energy is shared between the translational form ($\frac{1}{2}mv^2$) and the rotational form ($\frac{1}{2}I\omega^2$). The object with more of its kinetic energy in the translational form will have the greater translational speed at the bottom of the incline. We expect the solid cylinder to have the greater translational speed, because more of its mass is located near the rotational axis and, thus, possesses less rotational kinetic energy.

Solution The total mechanical energy E_f at the bottom ($h_f = 0$ m) is the same as the total mechanical energy E_0 at the top ($h = h_0$, $v_0 = 0$ m/s, $\omega_0 = 0$ rad/s):

$$\tfrac{1}{2}mv_f^2 + \tfrac{1}{2}I\omega_f^2 + mgh_f = \tfrac{1}{2}mv_0^2 + \tfrac{1}{2}I\omega_0^2 + mgh_0$$

$$\tfrac{1}{2}mv_f^2 + \tfrac{1}{2}I\omega_f^2 = mgh_0$$

Since each cylinder rolls without slipping, the final rotational speed ω_f and the final translational speed v_f of its center of mass are related by Equation 8.12, $\omega_f = v_f/r$, where r is the radius of the cylinder. Substituting this expression for ω_f into the energy-conservation equation and solving for v_f yields

$$v_f = \sqrt{\frac{2mgh_0}{m + \dfrac{I}{r^2}}} \qquad \longleftarrow$$

Setting $m = m_h$, $r = r_h$, and $I = m_h r_h^2$ for the hollow cylinder and then setting $m = m_s$, $r = r_s$, and $I = \frac{1}{2}m_s r_s^2$ for the solid cylinder (see **Table 9.1**), we find that the two cylinders have the following translational speeds at the bottom of the incline:

Hollow cylinder $\quad v_f = \sqrt{gh_0}$

Solid cylinder $\quad v_f = \sqrt{\dfrac{4gh_0}{3}} = 1.15\sqrt{gh_0}$

The solid cylinder has the greater translational speed at the bottom and, thus, arrives there first.

Math Skills To obtain the desired expression for the translational speed v_f, we proceed as follows. The energy-conservation equation is $\frac{1}{2}mv_f^2 + \frac{1}{2}I\omega_f^2 = mgh_0$. Substituting $\omega_f = \frac{v_f}{r}$ (Equation 8.12) into this equation gives

$$\tfrac{1}{2}mv_f^2 + \tfrac{1}{2}I\left(\frac{v_f}{r}\right)^2 = mgh_0 \quad \text{or} \quad \tfrac{1}{2}\left(m + \frac{I}{r^2}\right)v_f^2 = mgh_0$$

Multiplying both sides of the right-hand equation by 2 and dividing both sides by $m + \dfrac{I}{r^2}$ shows that

$$\frac{2\tfrac{1}{2}\left(\cancel{m + \dfrac{I}{r^2}}\right)v_f^2}{\cancel{m + \dfrac{I}{r^2}}} = \frac{2mgh_0}{m + \dfrac{I}{r^2}} \quad \text{or} \quad v_f^2 = \frac{2mgh_0}{m + \dfrac{I}{r^2}}$$

Taking the square root of both sides of the right-hand equation shows that

$$v_f = \sqrt{\frac{2mgh_0}{m + \dfrac{I}{r^2}}}$$

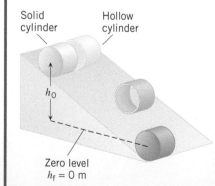

Solid cylinder   Hollow cylinder

h_0

Zero level
$h_f = 0$ m

FIGURE 9.23 A hollow cylinder and a solid cylinder start from rest and roll down the incline plane. The conservation of mechanical energy can be used to show that the solid cylinder, having the greater translational speed, reaches the bottom first.

Related Homework: Problems 42, 77

Check Your Understanding

(The answers are given at the end of the book.)

14. Two uniform solid balls are placed side by side at the top of an incline plane and, starting from rest, are allowed to roll down the incline. Which ball, if either, has the greater translational speed at the bottom if **(a)** they have the same radii, but one is more massive than the other and **(b)** they have the same mass, but one has a larger radius?

15. A thin sheet of plastic is uniform and has the shape of an equilateral triangle. Consider two axes for rotation. Both are perpendicular to the plane of the triangle, axis A passing through the center of the triangle and axis B passing through one corner. If the angular speed ω about each axis is the same, for which axis does the triangle have the greater rotational kinetic energy?

16. A hoop, a solid cylinder, a spherical shell, and a solid sphere are placed at rest at the top of an incline. All the objects have the same radius. They are then released at the same time. What is the order in which they reach the bottom (fastest first)?

9.6 | Angular Momentum

In Chapter 7 the linear momentum p of an object is defined as the product of its mass m and linear velocity v; that is, $p = mv$. For rotational motion the analogous concept is called the **angular momentum** L. The mathematical form of angular momentum is analogous to that of linear momentum, with the mass m and the linear velocity v being replaced with their rotational counterparts, the moment of inertia I and the angular velocity ω.

DEFINITION OF ANGULAR MOMENTUM

The angular momentum L of a body rotating about a fixed axis is the product of the body's moment of inertia I and its angular velocity ω with respect to that axis:

$$L = I\omega \tag{9.10}$$

Requirement: ω must be expressed in rad/s.

SI Unit of Angular Momentum: $\text{kg} \cdot \text{m}^2/\text{s}$

Linear momentum is an important concept in physics because the total linear momentum of a system is conserved when the sum of the average external forces acting on the system is zero. Then, the final total linear momentum P_f and the initial total linear momentum P_0 are the same: $P_f = P_0$. In the case of angular momentum, a similar line of reasoning indicates that when the sum of the average external torques is zero, the final and initial angular momenta are the same: $L_f = L_0$, which is the **principle of conservation of angular momentum**.

PRINCIPLE OF CONSERVATION OF ANGULAR MOMENTUM

The total angular momentum of a system remains constant (is conserved) if the net average external torque acting on the system is zero.

Example 14 illustrates an interesting consequence of the conservation of angular momentum.

CONCEPTUAL EXAMPLE 14 | BIO The Physics of a Spinning Skater

In **Figure 9.24a** an ice skater is spinning with both arms and a leg outstretched. In **Figure 9.24b** she pulls her arms and leg inward. As a result of this maneuver, her angular velocity ω increases dramatically. Why? Neglect any air resistance and assume that friction between her skates and the ice is negligible. **(a)** A net external torque acts on the skater, causing ω to increase. **(b)** No net external torque acts on her; she is simply obeying the conservation of angular momentum. **(c)** Due to the movements of her arms and legs, a net internal torque acts on the skater, causing her angular momentum and ω to increase.

Reasoning Considering the skater as the system, we will use the conservation of angular momentum as a guide; it indicates that only a net external torque can cause the total angular momentum of a system to change.

Answer (a) is incorrect. There is no net external torque acting on the skater, because air resistance and the friction between her skates and the ice are negligible.

Answer (c) is incorrect. The movements of her arms and legs do produce internal torques. However, only *external* torques, not internal torques, can change the angular momentum of a system.

Answer (b) is correct. Since any air resistance and friction are negligible, the net external torque acting on the skater is zero, and the skater's angular momentum is conserved as she pulls her arms and leg inward. However, angular momentum is the product of the moment of inertia I and the angular velocity ω (see Equation 9.10). By moving the mass of her arms and leg inward, the skater decreases the distance r of the mass from the axis of rotation and, consequently, decreases her moment of inertia

FIGURE 9.24 (*a*) A skater spins slowly on one skate, with both arms and one leg outstretched. (*b*) As she pulls her arms and leg in toward the rotational axis, her angular velocity ω increases.

I ($I = \Sigma mr^2$). Since the product of I and ω is constant, then ω must increase as I decreases. Thus, as she pulls her arms and leg inward, she spins with a larger angular velocity.

Related Homework: Problem 48

The next example involves a satellite and illustrates another application of the principle of conservation of angular momentum.

EXAMPLE 15 | The Physics of a Satellite in an Elliptical Orbit

An artificial satellite is placed into an elliptical orbit about the earth, as illustrated in **Figure 9.25**. Telemetry data indicate that its point of closest approach (called the *perigee*) is $r_P = 8.37 \times 10^6$ m from the center of the earth, and its point of greatest distance (called the *apogee*) is $r_A = 25.1 \times 10^6$ m from the center of the earth. The speed of the satellite at the perigee is $v_P = 8450$ m/s. Find its speed v_A at the apogee.

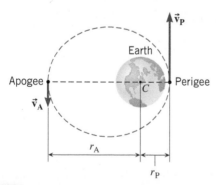

FIGURE 9.25 A satellite is moving in an elliptical orbit about the earth. The gravitational force exerts no torque on the satellite, so the angular momentum of the satellite is conserved.

Reasoning The only force of any significance that acts on the satellite is the gravitational force of the earth. However, at any instant, this force is directed toward the center of the earth and passes through the axis about which the satellite instantaneously rotates. Therefore, the gravitational force exerts *no torque* on the satellite (the lever arm is zero). Consequently, the net average external torque acting on the satellite is zero, and the angular momentum of the satellite remains constant at all times.

Solution Since the angular momentum is the same at the apogee (A) and the perigee (P), it follows that $I_A\omega_A = I_P\omega_P$. Furthermore, the orbiting satellite can be considered a point mass, so its moment of inertia is $I = mr^2$ (see Equation 9.4). In addition, the angular speed ω of the satellite is related to its tangential speed v_T by $\omega = v_T/r$ (Equation 8.9). If Equation 9.4 and Equation 8.9 are used at the apogee and perigee, the conservation of angular momentum gives the following result:

$$I_A\omega_A = I_P\omega_P \quad \text{or} \quad (mr_A^2)\left(\frac{v_A}{r_A}\right) = (mr_P^2)\left(\frac{v_P}{r_P}\right)$$

$$v_A = \frac{r_P v_P}{r_A} = \frac{(8.37 \times 10^6 \text{ m})(8450 \text{ m/s})}{25.1 \times 10^6 \text{ m}} = \boxed{2820 \text{ m/s}}$$

The answer is independent of the mass of the satellite. The satellite behaves just like the skater in **Figure 9.24**, because its speed is smaller at the apogee, where the moment of inertia is greater, and greater at the perigee, where the moment of inertia is smaller.

The result in Example 15 indicates that a satellite does not have a constant speed in an elliptical orbit. The speed changes from a maximum at the perigee to a minimum at the apogee; the closer the satellite comes to the earth, the faster it travels. Planets moving around the sun in elliptical orbits exhibit the same kind of behavior, and Johannes

Kepler (1571–1630) formulated his famous second law based on observations of such characteristics of planetary motion. Kepler's second law states that, in a given amount of time, a line joining any planet to the sun sweeps out the same amount of area no matter where the planet is on its elliptical orbit, as **Figure 9.26** illustrates. The conservation of angular momentum can be used to show why the law is valid, by means of a calculation similar to that in Example 15.

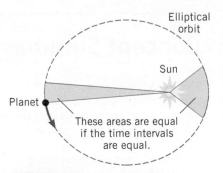

FIGURE 9.26 Kepler's second law of planetary motion states that a line joining a planet to the sun sweeps out equal areas in equal time intervals.

Check Your Understanding

(*The answers are given at the end of the book.*)

17. A woman is sitting on the spinning seat of a piano stool with her arms folded. Ignore any friction in the spinning stool. What happens to her **(a)** angular velocity and **(b)** angular momentum when she extends her arms outward?

18. Review Conceptual Example 14 as an aid in answering this question. Suppose the ice cap at the South Pole were to melt and the water were distributed uniformly over the earth's oceans. Would the earth's angular velocity increase, decrease, or remain the same?

19. Conceptual Example 14 provides background for this question. A cloud of interstellar gas is rotating. Because the gravitational force pulls the gas particles together, the cloud shrinks, and, under the right conditions, a star may ultimately be formed. Would the angular velocity of the star be less than, equal to, or greater than the angular velocity of the rotating gas?

20. A person is hanging motionless from a vertical rope over a swimming pool. She lets go of the rope and drops straight down. After letting go, is it possible for her to curl into a ball and start spinning?

EXAMPLE 16 | BIO The Physics of the Rotational Kinetic Energy of a Figure Skater

In Conceptual Example 14 we saw how a spinning figure skater can change her angular velocity by changing her moment of inertia. This occurs because, in the absence of external torques, the skater's angular momentum is conserved. But what about her rotational kinetic energy? Is that also conserved? Assume the skater in **Figure 9.24a** has a moment of inertia I_0 and is rotating with an angular velocity ω_0. When she brings her arms in (**Figure 9.24b**) assume her rotational velocity doubles. Calculate the ratio of her final rotational kinetic energy to her initial rotational kinetic energy (KE_{Rf}/KE_{R0}) and answer the question of whether or not her rotational kinetic energy is conserved.

Reasoning We can calculate the skater's rotational kinetic energy, both before and after she brings her arms in, by using Equation 9.9. To simplify the ratio of her rotational kinetic energies, we can apply conservation of angular momentum.

Solution Applying Equation 9.9, the ratio of her rotational kinetic energy will be given by the following:

$$\frac{KE_{R_f}}{KE_{R_0}} = \frac{\frac{1}{2}I_f\omega_f^2}{\frac{1}{2}I_0\omega_0^2}.$$

We can simplify this expression and rewrite it in the following way:

$$\frac{KE_{R_f}}{KE_{R_0}} = \frac{\frac{1}{2}I_f\omega_f^2}{\frac{1}{2}I_0\omega_0^2} = \frac{(I_f\omega_f)\omega_f}{(I_0\omega_0)\omega_0}.$$

The product of $I\omega$ in the numerator and denominator represents the angular momentum, L, after and before the skater brings her arms in, respectively. We know these two quantities are equal,

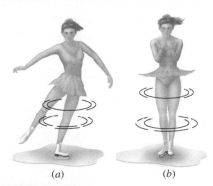

(a)   (b)

FIGURE 9.24 (REPEATED) (*a*) A skater spins slowly on one skate, with both arms and one leg outstretched. (*b*) As she pulls her arms and leg in toward the rotational axis, her angular velocity ω increases.

since the angular momentum in this problem is conserved. Therefore, the ratio of her rotational kinetic energies reduces to the ratio of her final and initial angular velocities:

$$\frac{KE_{R_f}}{KE_{R_0}} = \frac{\omega_f}{\omega_0}.$$

Here, we are told that her angular velocity doubles. Thus, $\omega_f = 2\omega_0$, and $\dfrac{KE_{R_f}}{KE_{R_0}} = \dfrac{2\omega_0}{\omega_0} = \boxed{2}$. While her angular momentum during this maneuver is conserved, her rotational kinetic energy is not. Where does the extra rotational kinetic energy come from? It comes from the work that she does internally to move her arms closer to the axis of rotation.

Concept Summary

9.1 The Action of Forces and Torques on Rigid Objects The line of action of a force is an extended line that is drawn colinear with the force. The lever arm ℓ is the distance between the line of action and the axis of rotation, measured on a line that is perpendicular to both.

The torque of a force has a magnitude that is given by the magnitude F of the force times the lever arm ℓ. The magnitude of the torque τ is given by Equation 9.1, and τ is positive when the force tends to produce a counterclockwise rotation about the axis, and negative when the force tends to produce a clockwise rotation.

$$\text{Magnitude of torque} = F\ell \tag{9.1}$$

9.2 Rigid Objects in Equilibrium A rigid body is in equilibrium if it has zero translational acceleration and zero angular acceleration. In equilibrium, the net external force and the net external torque acting on the body are zero, according to Equations 4.9a, 4.9b, and 9.2.

$$\Sigma F_x = 0 \quad \text{and} \quad \Sigma F_y = 0 \tag{4.9a and 4.9b}$$

$$\Sigma \tau = 0 \tag{9.2}$$

9.3 Center of Gravity The center of gravity of a rigid object is the point where its entire weight can be considered to act when calculating the torque due to the weight. For a symmetrical body with uniformly distributed weight, the center of gravity is at the geometrical center of the body. When a number of objects whose weights are $W_1, W_2, \ldots$ are distributed along the x axis at locations $x_1, x_2, \ldots$, the center of gravity x_{cg} is given by Equation 9.3. The center of gravity is identical to the center of mass, provided the acceleration due to gravity does not vary over the physical extent of the objects.

$$x_{cg} = \frac{W_1 x_1 + W_2 x_2 + \cdots}{W_1 + W_2 + \cdots} \tag{9.3}$$

9.4 Newton's Second Law for Rotational Motion About a Fixed Axis The moment of inertia I of a body composed of N particles is given by Equation 9.6, where m is the mass of a particle and r is the perpendicular distance of the particle from the axis of rotation.

$$I = m_1 r_1^2 + m_2 r_2^2 + \cdots + m_N r_N^2 = \Sigma m r^2 \tag{9.6}$$

For a rigid body rotating about a fixed axis, Newton's second law for rotational motion is stated as in Equation 9.7, where $\Sigma \tau$ is the net external torque applied to the body, I is the moment of inertia of the body, and α is its angular acceleration.

$$\Sigma \tau = I\alpha \quad (\alpha \text{ in rad/s}^2) \tag{9.7}$$

9.5 Rotational Work and Energy The rotational work W_R done by a constant torque τ in turning a rigid body through an angle θ is specified by Equation 9.8.

The rotational kinetic energy KE_R of a rigid object rotating with an angular speed ω about a fixed axis and having a moment of inertia I is specified by Equation 9.9.

The total mechanical energy E of a rigid body is the sum of its translational kinetic energy ($\frac{1}{2}mv^2$), its rotational kinetic energy ($\frac{1}{2}I\omega^2$), and its gravitational potential energy (mgh), according to Equation 1, where m is the mass of the object, v is the translational speed of its center of mass, I is its moment of inertia about an axis through the center of mass, ω is its angular speed, and h is the height of the object's center of mass relative to an arbitrary zero level.

The total mechanical energy is conserved if the net work done by external nonconservative forces and external torques is zero. When the total mechanical energy is conserved, the final total mechanical energy E_f equals the initial total mechanical energy E_0: $E_f = E_0$.

$$W_R = \tau\theta \quad (\theta \text{ in radians}) \tag{9.8}$$

$$KE_R = \frac{1}{2}I\omega^2 \quad (\omega \text{ in rad/s}) \tag{9.9}$$

$$E = \frac{1}{2}mv^2 + \frac{1}{2}I\omega^2 + mgh \tag{1}$$

9.6 Angular Momentum The angular momentum of a rigid body rotating with an angular velocity ω about a fixed axis and having a moment of inertia I with respect to that axis is given by Equation 9.10.

$$L = I\omega \quad (\omega \text{ in rad/s}) \tag{9.10}$$

The principle of conservation of angular momentum states that the total angular momentum of a system remains constant (is conserved) if the net average external torque acting on the system is zero. When the total angular momentum is conserved, the final angular momentum L_f equals the initial angular momentum L_0: $L_f = L_0$.

Focus on Concepts

Online

Additional questions are available for assignment in WileyPLUS.

Section 9.1 The Action of Forces and Torques on Rigid Objects

1. The wheels on a moving bicycle have both translational (or linear) and rotational motions. What is meant by the phrase "a rigid body, such as a bicycle wheel, is in equilibrium"? (a) The body cannot have translational or rotational motion of any kind. (b) The body can have translational motion, but it cannot have rotational motion. (c) The body cannot have translational motion, but it can have rotational motion. (d) The body can have translational and rotational motions, as long as its translational acceleration and angular acceleration are zero.

Section 9.2   Rigid Objects in Equilibrium

2. The drawing illustrates an overhead view of a door and its axis of rotation. The axis is perpendicular to the page. There are four forces acting on the door, and they have the same magnitude. Rank the torque τ that each force produces, largest to smallest.

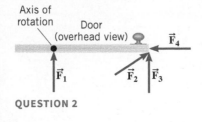

QUESTION 2

(a) τ_4, τ_3, τ_2, τ_1 (b) τ_3, τ_2, τ_1 and τ_4 (a two-way tie) (c) τ_2, τ_4, τ_3, τ_1 (d) τ_1, τ_4, τ_3, τ_2 (e) τ_2, τ_3 and τ_4 (a two-way tie), τ_1

3. Five hockey pucks are sliding across frictionless ice. The drawing shows a top view of the pucks and the three forces that act on each one. As shown, the forces have different magnitudes (F, $2F$, or $3F$), and are applied at different points on the pucks. Only one of the five pucks can be in equilibrium. Which puck is it? (a) 1 (b) 2 (c) 3 (d) 4 (e) 5

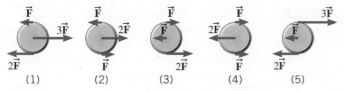

QUESTION 3

4. The drawing shows a top view of a square box lying on a frictionless floor. Three forces, which are drawn to scale, act on the box. Consider an angular acceleration with respect to an axis through the center of the box (perpendicular to the page). Which one of the following statements is correct? (a) The box will have a translational acceleration but not an angular acceleration. (b) The box will have both a translational and an angular acceleration. (c) The box will have an

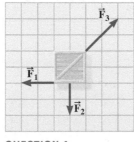

QUESTION 4

angular acceleration but not a translational acceleration. (d) The box will have neither a translational nor an angular acceleration. (e) It is not possible to determine whether the box will have a translational or an angular acceleration.

Section 9.4   Newton's Second Law for Rotational Motion About a Fixed Axis

5. The drawing shows three objects rotating about a vertical axis. The mass of each object is given in terms of m_0, and its perpendicular distance from the axis is specified in terms of r_0. Rank the three objects according to their moments of inertia, largest to smallest. (a) A, B, C (b) A, C, B (c) B, A, C (d) B, C, A (e) C, A, B

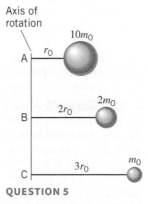

QUESTION 5

6. Two blocks are placed at the ends of a horizontal massless board, as in the drawing. The board is kept from rotating and rests on a support that serves as an axis of rotation. The moment of inertia

of this system relative to the axis is 12 kg · m². Determine the magnitude of the angular acceleration when the system is allowed to rotate.

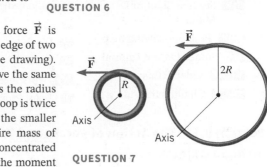

QUESTION 6

7. The same force $\vec{F}$ is applied to the edge of two hoops (see the drawing). The hoops have the same mass, whereas the radius of the larger hoop is twice the radius of the smaller one. The entire mass of each hoop is concentrated at its rim, so the moment of inertia is $I = Mr^2$, where M is the mass and r is the radius. Which hoop has the greater angular acceleration, and how many times as great is it compared to the angular acceleration of the other hoop? (a) The smaller hoop; two times as great (b) The smaller hoop; four times as great (c) The larger hoop; two times as great (d) The larger hoop; four times as great (e) Both have the same angular acceleration.

QUESTION 7

Section 9.5   Rotational Work and Energy

8. Two hoops, starting from rest, roll down identical inclined planes. The work done by nonconservative forces, such as air resistance, is zero ($W_{nc} = 0$ J). Both have the same

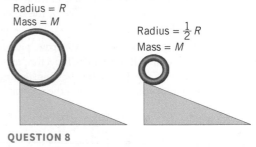

QUESTION 8

mass M, but, as the drawing shows, one hoop has twice the radius of the other. The moment of inertia for each hoop is $I = Mr^2$, where r is its radius. Which hoop, if either, has the greater total kinetic energy (translational plus rotational) at the bottom of the incline? (a) The larger hoop (b) The smaller hoop (c) Both have the same total kinetic energy.

Section 9.6   Angular Momentum

9. Under what condition(s) is the angular momentum of a rotating body, such as a spinning ice skater, conserved? (a) Each external force acting on the body must be zero. (b) Each external force and each external torque acting on the body must be zero. (c) Each external force may be nonzero, but the sum of the forces must be zero. (d) Each external torque may be nonzero, but the sum of the torques must be zero.

10. An ice skater is spinning on frictionless ice with her arms extended outward. She then pulls her arms in toward her body, reducing her moment of inertia. Her angular momentum is conserved, so as she reduces her moment of inertia, her angular velocity increases and she spins faster. Compared to her initial rotational kinetic energy, her final rotational kinetic energy is (a) the same (b) larger, because her angular speed is larger (c) smaller, because her moment of inertia is smaller.

Problems

Additional questions are available for assignment in WileyPLUS.

SSM Student Solutions Manual	**BIO** Biomedical application
MMH Problem-solving help	**E** Easy
GO Guided Online Tutorial	**M** Medium
V-HINT Video Hints	**H** Hard
CHALK Chalkboard Videos	**WS** Worksheet
	T Team Problem

Section 9.1   The Action of Forces and Torques on Rigid Objects

1. E SSM MMH The wheel of a car has a radius of 0.350 m. The engine of the car applies a torque of 295 N · m to this wheel, which does not slip against the road surface. Since the wheel does not slip, the road must be applying a force of static friction to the wheel that produces a countertorque. Moreover, the car has a constant velocity, so this countertorque balances the applied torque. What is the magnitude of the static frictional force?

2. E The steering wheel of a car has a radius of 0.19 m, and the steering wheel of a truck has a radius of 0.25 m. The same force is applied in the same direction to each steering wheel. What is the ratio of the torque produced by this force in the truck to the torque produced in the car?

3. E SSM You are installing a new spark plug in your car, and the manual specifies that it be tightened to a torque that has a magnitude of 45 N · m. Using the data in the drawing, determine the magnitude F of the force that you must exert on the wrench.

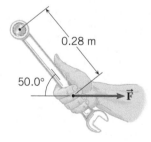

PROBLEM 3

4. E GO Two children hang by their hands from the same tree branch. The branch is straight, and grows out from the tree trunk at an angle of 27.0° above the horizontal. One child, with a mass of 44.0 kg, is hanging 1.30 m along the branch from the tree trunk. The other child, with a mass of 35.0 kg, is hanging 2.10 m from the tree trunk. What is the magnitude of the net torque exerted on the branch by the children? Assume that the axis is located where the branch joins the tree trunk and is perpendicular to the plane formed by the branch and the trunk.

5. E SSM The drawing shows a jet engine suspended beneath the wing of an airplane. The weight $\vec{W}$ of the engine is 10 200 N and acts as shown in the drawing. In flight the engine produces a thrust $\vec{T}$ of 62 300 N that is parallel to the ground. The rotational axis in the drawing is perpendicular to the plane of the paper. With respect to this axis, find the magnitude of the torque due to (a) the weight and (b) the thrust.

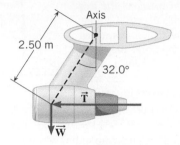

PROBLEM 5

6. E A square, 0.40 m on a side, is mounted so that it can rotate about an axis that passes through the center of the square. The axis is perpendicular to the plane of the square. A force of 15 N lies in this plane and is applied to the square. What is the magnitude of the maximum torque that such a force could produce?

7. M SSM A pair of forces with equal magnitudes, opposite directions, and different lines of action is called a "couple." When a couple acts on a rigid object, the couple produces a torque that does *not* depend on the location of the axis. The drawing shows a couple acting on a tire wrench, each force being perpendicular to the wrench. Determine an expression for the torque produced by the couple when the axis is perpendicular to the tire and passes through (a) point A, (b) point B, and (c) point C. Express your answers in terms of the magnitude F of the force and the length L of the wrench.

PROBLEM 7

8. M GO One end of a meter stick is pinned to a table, so the stick can rotate freely in a plane parallel to the tabletop. Two forces, both parallel to the tabletop, are applied to the stick in such a way that the net torque is zero. The first force has a magnitude of 2.00 N and is applied perpendicular to the length of the stick at the free end. The second force has a magnitude of 6.00 N and acts at a 30.0° angle with respect to the length of the stick. Where along the stick is the 6.00-N force applied? Express this distance with respect to the end of the stick that is pinned.

9. M V-HINT MMH A rod is lying on the top of a table. One end of the rod is hinged to the table so that the rod can rotate freely on the tabletop. Two forces, both parallel to the tabletop, act on the rod at the same place. One force is directed perpendicular to the rod and has a magnitude of 38.0 N. The second force has a magnitude of 55.0 N and is directed at an angle θ with respect to the rod. If the sum of the torques due to the two forces is zero, what must be the angle θ?

10. H A rotational axis is directed perpendicular to the plane of a square and is located as shown in the drawing. Two forces, $\vec{F}_1$ and $\vec{F}_2$, are applied to diagonally opposite corners, and act along the sides of the square, first as shown in part *a* and then as shown in part *b* of the drawing. In each case the net torque produced by the forces is zero. The square is one meter on a side, and the magnitude of $\vec{F}_2$ is three times that of $\vec{F}_1$. Find the distances *a* and *b* that locate the axis.

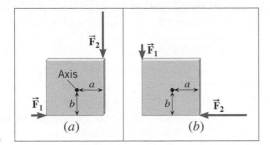

PROBLEM 10

Section 9.2 Rigid Objects in Equilibrium,

Section 9.3 Center of Gravity

11. **E SSM MMH** A hiker, who weighs 985 N, is strolling through the woods and crosses a small horizontal bridge. The bridge is uniform, weighs 3610 N, and rests on two concrete supports, one at each end. He stops one-fifth of the way along the bridge. What is the magnitude of the force that a concrete support exerts on the bridge **(a)** at the near end and **(b)** at the far end?

12. **E** Conceptual Example 7 provides useful background for this problem. Workers have loaded a delivery truck in such a way that its center of gravity is only slightly forward of the rear axle, as shown in the drawing. The mass of the truck and its contents is 7460 kg. Find the magnitudes of the forces exerted by the ground on **(a)** the front wheels and **(b)** the rear wheels of the truck.

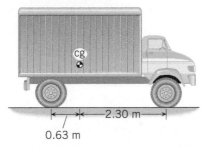

PROBLEM 12

13. **E GO** The drawing shows a rectangular piece of wood. The forces applied to corners B and D have the same magnitude of 12 N and are directed parallel to the long and short sides of the rectangle. The long side of the rectangle is twice as long as the short side. An axis of rotation is shown perpendicular to the plane of the rectangle at its center. A third force (not shown in the drawing) is applied to corner A, directed along the short side of the rectangle (either toward B or away from B), such that the piece of wood is at equilibrium. Find the magnitude and direction of the force applied to corner A.

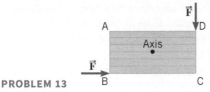

PROBLEM 13

14. **E GO** The wheels, axle, and handles of a wheelbarrow weigh 60.0 N. The load chamber and its contents weigh 525 N. The drawing shows these two forces in two different wheelbarrow designs. To support the wheelbarrow in equilibrium, the man's hands apply a force $\vec{F}$

to the handles that is directed vertically upward. Consider a rotational axis at the point where the tire contacts the ground, directed perpendicular to the plane of the paper. Find the magnitude of the man's force for both designs.

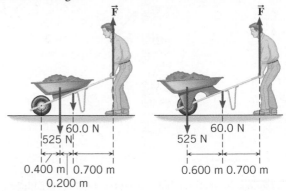

PROBLEM 14

15. **E MMH** Review Conceptual Example 7 as background material for this problem. A jet transport has a weight of 1.00×10^6 N and is at rest on the runway. The two rear wheels are 15.0 m behind the front wheel, and the plane's center of gravity is 12.6 m behind the front wheel. Determine the normal force exerted by the ground on **(a)** the front wheel and on **(b)** *each* of the two rear wheels.

16. **E GO** The drawing shows a uniform horizontal beam attached to a vertical wall by a frictionless hinge and supported from below at an angle $\theta = 39°$ by a brace that is attached to a pin. The beam has a weight of 340 N. Three

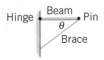

PROBLEM 16

additional forces keep the beam in equilibrium. The brace applies a force $\vec{P}$ to the right end of the beam that is directed upward at the angle θ with respect to the horizontal. The hinge applies a force to the left end of the beam that has a horizontal component $\vec{H}$ and a vertical component $\vec{V}$. Find the magnitudes of these three forces.

17. **M SSM CHALK** A uniform board is leaning against a smooth vertical wall. The board is at an angle θ above the horizontal ground. The coefficient of static friction between the ground and the lower end of the board is 0.650. Find the smallest value for the angle θ, such that the lower end of the board does not slide along the ground.

18. **M V-HINT** The drawing shows a bicycle wheel resting against a small step whose height is $h = 0.120$ m. The weight and radius of the wheel are $W = 25.0$ N and $r = 0.340$ m, respectively. A horizontal force $\vec{F}$ is applied to the axle of the wheel. As the magnitude of $\vec{F}$ increases, there comes a time when the wheel just begins to rise up and loses contact with the ground. What is the magnitude of the force when this happens?

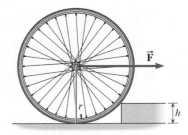

PROBLEM 18

19. **M SSM CHALK** A 1220-N uniform beam is attached to a vertical wall at one end and is supported by a cable at the other end. A 1960-N crate hangs from the far end of the beam. Using the data shown in the drawing, find **(a)** the magnitude of the tension in the wire and **(b)** the

magnitudes of the horizontal and vertical components of the force that the wall exerts on the left end of the beam.

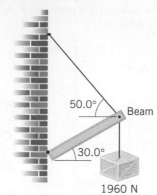

PROBLEM 19

20. **M** **V-HINT** A wrecking ball (weight = 4800 N) is supported by a boom, which may be assumed to be uniform and has a weight of 3600 N. As the drawing shows, a support cable runs from the top of the boom to the tractor. The angle between the support cable and the horizontal is 32°, and the angle between the boom and the horizontal is 48°. Find **(a)** the tension in the support cable and **(b)** the magnitude of the force exerted on the lower end of the boom by the hinge at point *P*.

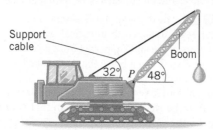

PROBLEM 20

21. **H** A man drags a 72-kg crate across the floor at a constant velocity by pulling on a strap attached to the bottom of the crate. The crate is tilted 25° above the horizontal, and the strap is inclined 61° above the horizontal. The center of gravity of the crate coincides with its geometrical center, as indicated in the drawing. Find the magnitude of the tension in the strap.

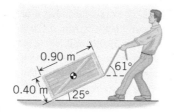

PROBLEM 21

22. **H** The drawing shows an A-shaped stepladder. Both sides of the ladder are equal in length. This ladder is standing on a frictionless horizontal surface, and only the crossbar (which has a negligible mass) of the "A" keeps the ladder from collapsing. The ladder is uniform and has a mass of 20.0 kg. Determine the tension in the crossbar of the ladder.

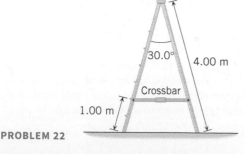

PROBLEM 22

Section 9.4 Newton's Second Law for Rotational Motion About a Fixed Axis

23. **E** Consult Multiple-Concept Example 11 to review an approach to problems such as this. A CD has a mass of 17 g and a radius of 6.0 cm. When inserted into a player, the CD starts from rest and accelerates to an angular velocity of 21 rad/s in 0.80 s. Assuming the CD is a uniform solid disk, determine the net torque acting on it.

24. **E** A clay vase on a potter's wheel experiences an angular acceleration of 8.00 rad/s² due to the application of a 10.0-N · m net torque. Find the total moment of inertia of the vase and potter's wheel.

25. **E** A solid circular disk has a mass of 1.2 kg and a radius of 0.16 m. Each of three identical thin rods has a mass of 0.15 kg. The rods are attached perpendicularly to the plane of the disk at its outer edge to form a three-legged stool (see the drawing). Find the moment of inertia of the stool with respect to an axis that is perpendicular to the plane of the disk at its center. (*Hint: When considering the moment of inertia of each rod, note that all of the mass of each rod is located at the same perpendicular distance from the axis.*)

PROBLEM 25

26. **E** A ceiling fan is turned on and a net torque of 1.8 N · m is applied to the blades. The blades have a total moment of inertia of 0.22 kg · m². What is the angular acceleration of the blades?

27. **E** **GO** Multiple-Concept Example 11 provides one model for solving this type of problem. Two wheels have the same mass and radius of 4.0 kg and 0.35 m, respectively. One has the shape of a hoop and the other the shape of a solid disk. The wheels start from rest and have a constant angular acceleration with respect to a rotational axis that is perpendicular to the plane of the wheel at its center. Each turns through an angle of 13 rad in 8.0 s. Find the net external torque that acts on each wheel.

28. **E** A 9.75-m ladder with a mass of 23.2 kg lies flat on the ground. A painter grabs the top end of the ladder and pulls straight upward with a force of 245 N. At the instant the top of the ladder leaves the ground, the ladder experiences an angular acceleration of 1.80 rad/s² about an axis passing through the bottom end of the ladder. The ladder's center of gravity lies halfway between the top and bottom ends. **(a)** What is the net torque acting on the ladder? **(b)** What is the ladder's moment of inertia?

29. **E** **SSM** **MMH** Multiple-Concept Example 11 offers useful background for problems like this. A cylinder is rotating about an axis that passes through the center of each circular end piece. The cylinder has a radius of 0.0830 m, an angular speed of 76.0 rad/s, and a moment of inertia of 0.615 kg · m². A brake shoe presses against the surface of the cylinder and applies a tangential frictional force to it. The frictional force reduces the angular speed of the cylinder by a factor of two during a time of 6.40 s. **(a)** Find the magnitude of the angular deceleration of the cylinder. **(b)** Find the magnitude of the force of friction applied by the brake shoe.

30. **M** **V-HINT** A 15.0-m length of hose is wound around a reel, which is initially at rest. The moment of inertia of the reel is 0.44 kg · m², and its radius is 0.160 m. When the reel is turning, friction at the axle exerts a torque of magnitude 3.40 N · m on the reel. If the hose is pulled so that the tension in it remains a constant 25.0 N, how long does it take to completely unwind the hose from the reel? Neglect the mass and thickness of the hose on the reel, and assume that the hose unwinds without slipping.

31. **E** **GO** The drawing shows two identical systems of objects; each consists of the same three small balls connected by massless rods. In both systems the axis is perpendicular to the page, but it is located at a different place, as shown. The same force of magnitude F is applied to the same ball in each system (see the drawing). The masses of the balls are $m_1 = 9.00$ kg, $m_2 = 6.00$ kg, and $m_3 = 7.00$ kg. The magnitude of the force is $F = 424$ N. **(a)** For each of the two systems, determine the moment of inertia about the given axis of rotation. **(b)** Calculate the torque (magnitude and direction) acting on each system. **(c)** Both systems start from rest, and the direction of the force moves with the system and always points along the 4.00-m rod. What is the angular velocity of each system after 5.00 s?

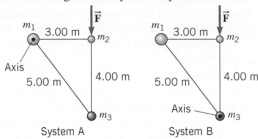

PROBLEM 31

32. **M** **GO** The drawing shows the top view of two doors. The doors are uniform and identical. Door A rotates about an axis through its left edge, and door B rotates about an axis through its center. The same force $\vec{F}$ is applied perpendicular to each door at its right edge, and the force remains perpendicular as the door turns. No other force affects the rotation of either door. Starting from rest, door A rotates through a certain angle in 3.00 s. How long does it take door B (also starting from rest) to rotate through the same angle?

PROBLEM 32

33. **M** **SSM** A stationary bicycle is raised off the ground, and its front wheel ($m = 1.3$ kg) is rotating at an angular velocity of 13.1 rad/s (see the drawing). The front brake is then applied for 3.0 s, and the wheel slows down to 3.7 rad/s. Assume that all the mass of the wheel is concentrated in the rim, the radius of which is 0.33 m. The coefficient of kinetic friction between each brake pad and the rim is $\mu_k = 0.85$. What is the magnitude of the normal force that *each* brake pad applies to the rim?

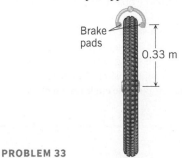

PROBLEM 33

34. **M** The **parallel axis theorem** provides a useful way to calculate the moment of inertia I about an arbitrary axis. The theorem states that $I = I_{cm} + Mh^2$, where I_{cm} is the moment of inertia of the object relative to an axis that passes through the center of mass and is parallel to the axis of interest. M is the total mass of the object, and h is the perpendicular distance between the two axes. Use this theorem and information to determine an expression for the moment of inertia of a solid cylinder of radius R relative to an axis that lies on the surface of the cylinder and is perpendicular to the circular ends.

35. **H** The crane shown in the drawing is lifting a 180-kg crate upward with an acceleration of 1.2 m/s². The cable from the crate passes over

a solid cylindrical pulley at the top of the boom. The pulley has a mass of 130 kg. The cable is then wound onto a hollow cylindrical drum that is mounted on the deck of the crane. The mass of the drum is 150 kg, and its radius is 0.76 m. The engine applies a counterclockwise torque to the drum in order to wind up the cable. What is the magnitude of this torque? Ignore the mass of the cable.

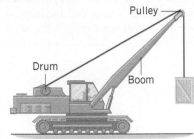

PROBLEM 35

Section 9.5 Rotational Work and Energy

36. **H** Calculate the kinetic energy that the earth has because of **(a)** its rotation about its own axis and **(b)** its motion around the sun. Assume that the earth is a uniform sphere and that its path around the sun is circular. For comparison, the total energy used in the United States in one year is about 1.1×10^{20} J.

37. **E** **SSM** Three objects lie in the x, y plane. Each rotates about the z axis with an angular speed of 6.00 rad/s. The mass m of each object and its perpendicular distance r from the z axis are as follows: (1) $m_1 = 6.00$ kg and $r_1 = 2.00$ m, (2) $m_2 = 4.00$ kg and $r_2 = 1.50$ m, (3) $m_3 = 3.00$ kg and $r_3 = 3.00$ m. **(a)** Find the tangential speed of each object. **(b)** Determine the total kinetic energy of this system using the expression $KE = \frac{1}{2}m_1v_1^2 + \frac{1}{2}m_2v_2^2 + \frac{1}{2}m_3v_3^2$. **(c)** Obtain the moment of inertia of the system. **(d)** Find the rotational kinetic energy of the system using the relation $KE_R = \frac{1}{2}I\omega^2$ to verify that the answer is the same as the answer to **(b)**.

38. **E** **GO** Two thin rods of length L are rotating with the same angular speed ω (in rad/s) about axes that pass perpendicularly through one end. Rod A is massless but has a particle of mass 0.66 kg attached to its free end. Rod B has a mass of 0.66 kg, which is distributed uniformly along its length. The length of each rod is 0.75 m, and the angular speed is 4.2 rad/s. Find the kinetic energies of rod A with its attached particle and of rod B.

39. **E** **SSM** A flywheel is a solid disk that rotates about an axis that is perpendicular to the disk at its center. Rotating flywheels provide a means for storing energy in the form of rotational kinetic energy and are being considered as a possible alternative to batteries in electric cars. The gasoline burned in a 300-mile trip in a typical midsize car produces about 1.2×10^9 J of energy. How fast would a 13-kg flywheel with a radius of 0.30 m have to rotate to store this much energy? Give your answer in rev/min.

40. **E** A helicopter has two blades (see Figure 8.11); each blade has a mass of 240 kg and can be approximated as a thin rod of length 6.7 m. The blades are rotating at an angular speed of 44 rad/s. **(a)** What is the total moment of inertia of the two blades about the axis of rotation? **(b)** Determine the rotational kinetic energy of the spinning blades.

41. **M** **MMH** A solid sphere is rolling on a surface. What fraction of its total kinetic energy is in the form of rotational kinetic energy about the center of mass?

42. **M** **GO** Review Example 13 before attempting this problem. A marble and a cube are placed at the top of a ramp. Starting from rest at the same height, the marble rolls without slipping and the cube slides (no kinetic friction) down the ramp. Determine the ratio of the center-of-mass speed of the cube to the center-of-mass speed of the marble at the bottom of the ramp.

43. **M** **V-HINT** Starting from rest, a basketball rolls from the top of a hill to the bottom, reaching a translational speed of 6.6 m/s. Ignore

frictional losses. **(a)** What is the height of the hill? **(b)** Released from rest at the same height, a can of frozen juice rolls to the bottom of the same hill. What is the translational speed of the frozen juice can when it reaches the bottom?

44. **M** **CHALK** **SSM** A bowling ball encounters a 0.760-m vertical rise on the way back to the ball rack, as the drawing illustrates. Ignore frictional losses and assume that the mass of the ball is distributed uniformly. The translational speed of the ball is 3.50 m/s at the bottom of the rise. Find the translational speed at the top.

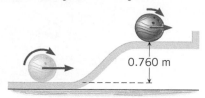

0.760 m

PROBLEM 44

45. **H** A tennis ball, starting from rest, rolls down the hill in the drawing. At the end of the hill the ball becomes airborne, leaving at an angle of 35° with respect to the ground. Treat the ball as a thin-walled spherical shell, and determine the range *x*.

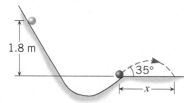

1.8 m

35°

x

PROBLEM 45

Section 9.6 Angular Momentum

46. **E** **SSM** Two disks are rotating about the same axis. Disk A has a moment of inertia of 3.4 kg · m² and an angular velocity of +7.2 rad/s. Disk B is rotating with an angular velocity of −9.8 rad/s. The two disks are then linked together without the aid of any external torques, so that they rotate as a single unit with an angular velocity of −2.4 rad/s. The axis of rotation for this unit is the same as that for the separate disks. What is the moment of inertia of disk B?

47. **E** When some stars use up their fuel, they undergo a catastrophic explosion called a *supernova*. This explosion blows much or all of the star's mass outward, in the form of a rapidly expanding spherical shell. As a simple model of the supernova process, assume that the star is a solid sphere of radius *R* that is initially rotating at 2.0 revolutions per day. After the star explodes, find the angular velocity, in revolutions per day, of the expanding supernova shell when its radius is 4.0*R*. Assume that all of the star's original mass is contained in the shell.

48. **E** **GO** Conceptual Example 14 provides useful background for this problem. A playground carousel is free to rotate about its center

on frictionless bearings, and air resistance is negligible. The carousel itself (without riders) has a moment of inertia of 125 kg · m². When one person is standing on the carousel at a distance of 1.50 m from the center, the carousel has an angular velocity of 0.600 rad/s. However, as this person moves inward to a point located 0.750 m from the center, the angular velocity increases to 0.800 rad/s. What is the person's mass?

49. **E** **MMH** A thin rod has a length of 0.25 m and rotates in a circle on a frictionless tabletop. The axis is perpendicular to the length of the rod at one of its ends. The rod has an angular velocity of 0.32 rad/s and a moment of inertia of 1.1×10^{-3} kg · m². A bug standing on the axis decides to crawl out to the other end of the rod. When the bug (mass = 4.2×10^{-3} kg) gets where it's going, what is the angular velocity of the rod?

50. **E** **GO** As seen from above, a playground carousel is rotating counterclockwise about its center on frictionless bearings. A person standing still on the ground grabs onto one of the bars on the carousel very close to its outer edge and climbs aboard. Thus, this person begins with an angular speed of zero and ends up with a nonzero angular speed, which means that he underwent a counterclockwise angular acceleration. The carousel has a radius of 1.50 m, an initial angular speed of 3.14 rad/s, and a moment of inertia of 125 kg · m². The mass of the person is 40.0 kg. Find the final angular speed of the carousel after the person climbs aboard.

51. **M** **GO** A thin, uniform rod is hinged at its midpoint. To begin with, one-half of the rod is bent upward and is perpendicular to the other half. This bent object is rotating at an angular velocity of 9.0 rad/s about an axis that is perpendicular to the left end of the rod and parallel to the rod's upward half (see the drawing). Without the aid of external torques, the rod suddenly assumes its straight shape. What is the angular velocity of the straight rod?

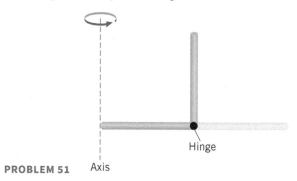

Hinge

PROBLEM 51 Axis

52. **H** A small 0.500-kg object moves on a frictionless horizontal table in a circular path of radius 1.00 m. The angular speed is 6.28 rad/s. The object is attached to a string of negligible mass that passes through a small hole in the table at the center of the circle. Someone under the table begins to pull the string downward to make the circle smaller. If the string will tolerate a tension of no more than 105 N, what is the radius of the smallest possible circle on which the object can move?

Additional Problems   Online

53. **E** A solid disk rotates in the horizontal plane at an angular velocity of 0.067 rad/s with respect to an axis perpendicular to the disk at its center. The moment of inertia of the disk is 0.10 kg · m². From above, sand is dropped straight down onto this rotating disk, so that a thin uniform ring of sand is formed at a distance of 0.40 m from the axis. The sand in the ring has a mass of 0.50 kg. After all the sand is in place, what is the angular velocity of the disk?

54. **E** A solid cylindrical disk has a radius of 0.15 m. It is mounted to an axle that is perpendicular to the circular end of the disk at its

center. When a 45-N force is applied tangentially to the disk, perpendicular to the radius, the disk acquires an angular acceleration of 120 rad/s². What is the mass of the disk?

55. **E** **GO** Review Conceptual Example 7 before starting this problem. A uniform plank of length 5.0 m and weight 225 N rests horizontally on two supports, with 1.1 m of the plank hanging over the right support (see the drawing). To what distance *x* can a person who weighs 450 N walk on the overhanging part of the plank before it just begins to tip?

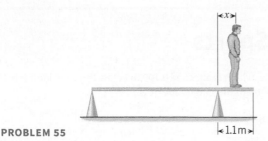

PROBLEM 55

56. E SSM A rotating door is made from four rectangular sections, as indicated in the drawing. The mass of each section is 85 kg. A person pushes on the outer edge of one section with a force of $F = 68$ N that is directed perpendicular to the section. Determine the magnitude of the door's angular acceleration.

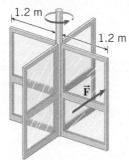

PROBLEM 56

57. M V-HINT A block (mass = 2.0 kg) is hanging from a massless cord that is wrapped around a pulley (moment of inertia = 1.1×10^{-3} kg · m²), as the drawing shows. Initially the pulley is prevented from rotating and the block is stationary. Then, the pulley is allowed to rotate as the block falls. The cord does not slip relative to the pulley as the block falls. Assume that the radius of the cord around the pulley remains constant at a value of 0.040 m during the block's descent. Find the angular acceleration of the pulley and the tension in the cord.

PROBLEM 57

58. M GO A thin, rigid, uniform rod has a mass of 2.00 kg and a length of 2.00 m. **(a)** Find the moment of inertia of the rod relative to an axis that is perpendicular to the rod at one end. **(b)** Suppose all the mass of the rod were located at a single point. Determine the perpendicular distance of this point from the axis in part (a), such that this point particle has the same moment of inertia as the rod does. This distance is called the **radius of gyration** of the rod.

59. M GO Two identical wheels are moving on horizontal surfaces. The center of mass of each has the same linear speed. However, one wheel is rolling, while the other is sliding on a frictionless surface without rolling. Each wheel then encounters an incline plane. One continues to roll up the incline, while the other continues to slide up. Eventually they come to a momentary halt, because the gravitational force slows them down. Each wheel is a disk of mass 2.0 kg. On the horizontal surfaces the center of mass of each wheel moves with a linear speed of 6.0 m/s. **(a)** What is the total kinetic energy of each wheel? **(b)** Determine the maximum height reached by each wheel as it moves up the incline.

60. H MMH By means of a rope whose mass is negligible, two blocks are suspended over a pulley, as the drawing shows. The pulley can be treated as a uniform solid cylindrical disk. The downward acceleration of the 44.0-kg block is observed to be exactly one-half the acceleration due to gravity. Noting that the tension in the rope is not the same on each side of the pulley, find the mass of the pulley.

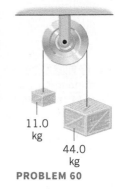

11.0 kg

44.0 kg

PROBLEM 60

61. M GO SSM A crate of mass 451 kg is being lifted by the mechanism shown in part a of the figure. The two cables are wrapped around their respective pulleys, which have radii of 0.600 and 0.200 m. The pulleys are fastened together to form a dual pulley and turn as a single unit about the center axle, relative to which the combined moment of inertia is 46.0 kg · m². The cables roll on the dual pulley without slipping. A tension of magnitude 2150 N is maintained in the cable attached to the motor. Find the angular acceleration of the dual pulley and the tension in the cable attached to the crate.

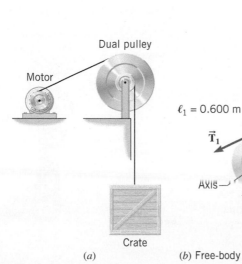

PROBLEM 61 (a)

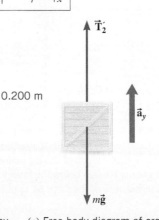

(b) Free-body diagram of pulley (c) Free-body diagram of crate

Physics in Biology, Medicine, and Sports

62. **E** **BIO** **9.3** A person is standing on a level floor. His head, upper torso, arms, and hands together weigh 438 N and have a center of gravity that is 1.28 m above the floor. His upper legs weigh 144 N and have a center of gravity that is 0.760 m above the floor. Finally, his lower legs and feet together weigh 87.0 N and have a center of gravity that is 0.250 m above the floor. Relative to the floor, find the location of the center of gravity for his entire body.

63. **E** **BIO** **9.3** The drawing shows a person (weight, $W = 584$ N) doing push-ups. Find the normal force exerted by the floor on *each* hand and *each* foot, assuming that the person holds this position.

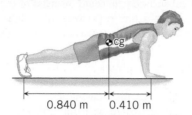

PROBLEM 63

0.840 m   0.410 m

64. **E** **BIO** **MMH** **9.3** A person exerts a horizontal force of 190 N in the test apparatus shown in the drawing. Find the horizontal force $\vec{M}$ (magnitude and direction) that his flexor muscle exerts on his forearm.

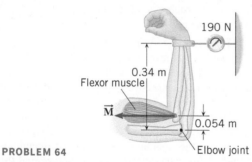

PROBLEM 64

65. **E** **BIO** **CHALK** **9.3** A man holds a 178-N ball in his hand, with the forearm horizontal (see the drawing). He can support the ball in this position because of the flexor muscle force $\vec{M}$, which is applied perpendicular to the forearm. The forearm weighs 22.0 N and has a center of gravity as indicated. Find **(a)** the magnitude of $\vec{M}$ and **(b)** the magnitude and direction of the force applied by the upper arm bone to the forearm at the elbow joint.

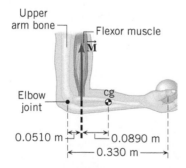

PROBLEM 65

66. **M** **BIO** **GO** **9.3** A person is sitting with one leg outstretched and stationary, so that it makes an angle of 30.0° with the horizontal, as the drawing indicates. The weight of the leg below the knee is 44.5 N, with the center of gravity located below the knee joint. The leg is being held in this position because of the force $\vec{M}$ applied by the quadriceps

muscle, which is attached 0.100 m below the knee joint (see the drawing). Obtain the magnitude of $\vec{M}$.

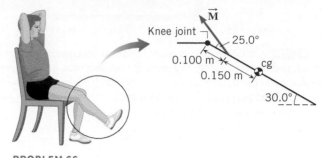

PROBLEM 66

67. **E** **BIO** **CHALK** **9.4** Multiple-Concept Example 11 reviews the approach and some of the concepts that are pertinent to this problem. The drawing shows a model for the motion of the human forearm in throwing a dart. Because of the force $\vec{M}$ applied by the triceps muscle, the forearm can rotate about an axis at the elbow joint. Assume that the forearm has the dimensions shown in the drawing and a moment of inertia of 0.065 kg · m² (including the effect of the dart) relative to the axis at the elbow. Assume also that the force $\vec{M}$ acts perpendicular to the forearm. Ignoring the effect of gravity and any frictional forces, determine the magnitude of the force $\vec{M}$ needed to give the dart a tangential speed of 5.0 m/s in 0.10 s, starting from rest.

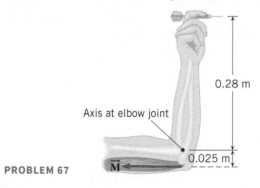

PROBLEM 67

68. **E** **BIO** **9.1** The drawing shows a lower leg being exercised. It has a 49-N weight attached to the foot and is extended at an angle θ with respect to the vertical. Consider a rotational axis at the knee. **(a)** When $\theta = 90.0°$, find the magnitude of the torque that the weight creates. **(b)** At what angle θ does the magnitude of the torque equal 15 N · m?

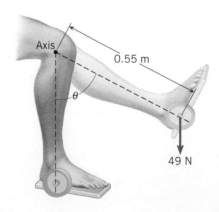

PROBLEM 68

69. M BIO V-HINT **9.2** The drawing shows an outstretched arm (0.61 m in length) that is parallel to the floor. The arm is pulling downward against the ring attached to the pulley system, in order to hold the 98-N weight stationary. To pull the arm downward, the latissimus dorsi muscle applies the force $\vec{M}$ in the drawing, at a point that is 0.069 m from the shoulder joint and oriented at an angle of 29°. The arm has a weight of 47 N and a center of gravity (cg) that is located 0.28 m from the shoulder joint. Find the magnitude of $\vec{M}$.

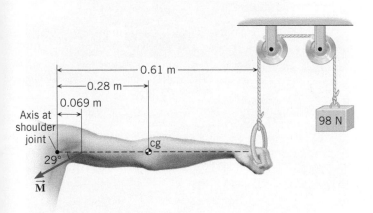

PROBLEM 69

70. M BIO **9.2** The figure shows the finishing position of the arm of a bodybuilder who is performing a standing lateral raise with a 5.00-kg dumbbell. This is an excellent exercise that isolates the lateral deltoid muscle. The weight of the arm itself is 42.5 N, and it acts at point *A* in the figure. Given the information shown, and the fact that $\theta = 35.0°$, **(a)** find the magnitude of the tension force in the deltoid ($\vec{F}_d$) and **(b)** the normal force from the shoulder joint ($\vec{F}_s$) acting on the humerus bone that holds the arm in this equilibrium position.

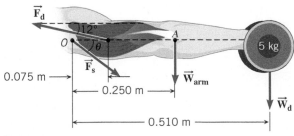

PROBLEM 70

71. E BIO **9.2** Contraction of the quadriceps muscle on the front of the upper leg results in lower-leg extension. This action creates tension in the quadriceps tendon, which is transmitted to the patellar tendon through the patella, or kneecap bone. Tension in the patellar tendon applies a force to the tibia bone in the lower leg, producing a rotation about the knee joint. Imagine an athlete is performing an exercise on a leg extension machine. At the moment shown in the drawing, she is holding the weight at rest. Use the information in the figure, and the fact that the magnitude of $\vec{F}$ is 100 N, to calculate the magnitude of the tension in the patellar tendon ($\vec{T}_p$) at this position. Neglect the weight of the lower leg.

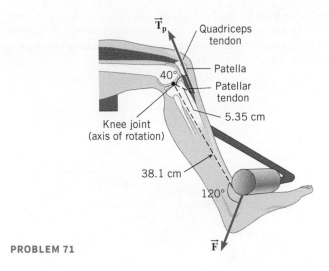

PROBLEM 71

72. M BIO **9.4** Raise your arm straight away from your side until it is in a horizontal position, as shown in the figure. Now relax your shoulder muscles and let your arm "free-fall" as it rotates back to your side. Calculate the tangential acceleration of your fingertips right as your arm begins to fall. Write your answer in terms of *g*, the acceleration of falling objects near the surface of the earth. Treat your arm as a uniform rod that rotates about an axis at one end (hinged at your shoulder). How does your answer compare with *g*?

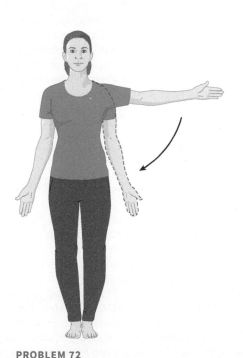

PROBLEM 72

73. E BIO **9.5** The title of the *fastest spinning animal on earth* may go to the *flattie spider* (see the figure). They can anchor one of their legs and pivot around it at incredible speeds to attack prey in any direction. Some of these spiders can spin with angular speeds of 3000°/s. **(a)** How many rotations does a flattie spider complete in the blink of an eye (300 ms)? **(b)** Assume the spider is rotating at a constant angular speed of 3000°/s. What is the spider's rotational kinetic energy? Assume the moment of inertia of the spider while pivoting about one leg is 9.4×10^{-5} kg · m².

magnitude of the tension in the elephant's trunk $\vec{T}$ ($m = 4540$ kg, $r_{FN} = r_{fs} = 133$ cm, $r_{cg} = 76.0$ cm, $r_T = 229$ cm).

Yu Zeng

PROBLEM 73

Michael Poliza/Caters News Agency

PROBLEM 74

74. **H** **BIO** **9.2** African elephants are the largest land animals. They consume approximately 10% of their body weight in food each day, which for an adult male, can be 1000 lb of vegetation! Their diet consists mostly of grasses, bamboo, tree bark, and fruit. They also like to dine on tree leaves. To reach these leaves, they often stand on their hind legs and extend their trunks (see the figure). The elephant in the figure is in equilibrium. The location of the elephant's center of gravity is shown, and the axis of rotation has been chosen to correspond to the hip joint. The forces in the elephant's free-body diagram are shown, and there is a static friction force between the elephant's back feet and the ground. Use the following information and calculate the

---

Concepts and Calculations Problems

Online

In this chapter we have seen that a rotational or angular acceleration results when a net external torque acts on an object. In contrast, when a net external force acts on an object, it leads to translational or linear acceleration, as discussed in Chapter 4. Torque and force are, then, fundamentally different concepts, and Problem 75 focuses on this fact. For rotational motion, the moment of inertia and the net external torque determine the angular acceleration of a rotating object, according to Newton's second law. Newton's second law together with the equations of rotational kinematics are useful in accounting for a wide variety of rotational motion. Problem 76 utilizes these concepts in a situation where a rotating object is slowing down.

75. **M** **CHALK** The figure shows a uniform crate resting on a horizontal surface. The crate has a square cross section and a weight of $W = 580$ N, which is uniformly distributed. At the bottom right edge of the surface is a small obstruction that prevents the crate from sliding when a horizontal pushing force $\vec{P}$ is applied to the left side. However, if this force is great enough, the crate will begin to tip and rotate over the obstruction. *Concepts:* (i) What causes the tipping—the force $\vec{P}$ or the torque that it creates? (ii) Where should $\vec{P}$ be applied so that a minimum force will be necessary to provide the necessary torque? In

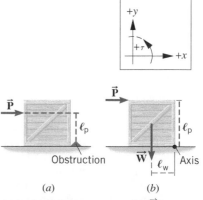

(a) A horizontal pushing force $\vec{P}$ is applied to a uniform crate, which has a square cross section and a weight $\vec{W}$. The crate rests on the ground, up against a small obstruction. (b) Some of the forces acting on the crate and their lever arms.

PROBLEM 75

other words, should the lever arm be a minimum or a maximum? *Calculations:* Determine the minimum pushing force that leads to tipping.

76. [M] [CHALK] [SSM] Two spheres are each rotating at an angular speed of 24 rad/s about axes that pass through their centers. Each has a radius of 0.20 m and a mass of 1.5 kg. However, as the figure shows, one is solid and the other is a thin-walled spherical shell. Suddenly, a net external torque due to friction (magnitude = 0.12 N · m) begins to act on each sphere and slows the motion down. *Concepts:* (i) Which sphere has the greater moment of inertia and why? (ii) Which sphere has the angular acceleration (a deceleration) with the smaller magnitude? (iii) Which sphere takes a longer time to come to a halt? *Calculations:* How long does it take each sphere to come to a halt?

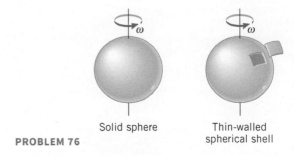

Solid sphere   Thin-walled spherical shell

PROBLEM 76

Team Problems

Online

77. [M] [T] [WS] **A Car Race.** You and your team are designing a small wooden car for a race where the cars are released from rest at the top of an inclined plane, and the first to get to the bottom wins. The rules for the design of the cars are as follows: (1) the body is made of a single piece of carved wood that weighs 0.75 kg, and (2) the combined mass of the wheels, regardless of shape, must be 0.12 kg, making the total mass of the car 0.87 kg. The four wheels, however, must be identical to each other. The car must also be exactly 18.0 cm in length, with no other specifications on the shape. You are free to use any type of wood that you like for the body, and any material, including metals, for the wheels, as long as the constituents conform to the mass requirements. You recognize that you should choose the most dense wood available (such as hickory) so as to minimize the cross-sectional area of your car body and reduce air resistance. Your choice for the wheels, however, needs some thought. You do not have time to construct your own wheels, but you have some available that all meet the mass requirements. Your choices are as follows: (1) solid wood disks of radius $r = 2.50$ cm, (2) solid aluminum (Al) disks ($r = 1.25$ cm), (3) solid lead (Pb) disks ($r = 1.25$ cm, but thinner than the Al disks, since Pb is denser than Al), (4) Al hollow cylinders ($r = 1.25$ cm), (5) Pb hollow cylinders ($r = 1.25$ cm), or (6) Pb solid disks ($r = 1.00$ cm). Which set of wheels do you choose, and why? Make a mathematical argument to justify your choice. Consult Example 13 for useful insight into this problem.

78. [E] [T] [WS] **A Spinning Space Station.** You and your team are on a space station in deep space that is shaped like a ring and is spinning about its central axis. The crew resides on the inner surface of the outer edge and experiences artificial gravity with an acceleration of 9.80 m/s². The station has a diameter of 750.0 m, and a mass of 3.70×10^7 kg. However, there is a plan to attach thousands of tungsten radiation shields by setting them moving toward the ring from above the plane of rotation. When they come into contact with the ring, they will adhere to it via precisely positioned magnets located on the shields and on the ring. With no external torques acting on the system, the shields will attach simultaneously, be uniformly distributed, and have a combined total mass of 9.98×10^6 kg. It can be assumed that the station is a thin ring, and has the same radius before and after the shields are added. You and your team must decide whether the addition of the shields will affect the station's

artificial gravity and, if so, what must be done to correct it. Find the new value of the centripetal acceleration (i.e., the new artificial gravity) felt by the crew in the ring before any corrections are made.

79. [M] [T] **Smugglers.** Rumor has it that a company has been smuggling gold out of the country using sealed, cylindrical barrels with hollow walls. They pour molten gold into the hollows, and then fill the remainder of the barrel's internal volume with packing peanuts. The total mass of the gold-walled barrel was designed so that it exactly matches those used to transport a volatile chemical that cannot be exposed to air (and therefore the barrel cannot be opened and checked). The X-ray machine usually used to screen containers is suspiciously damaged and not available. (a) There are 20 barrels total, and they are all identical: mass $m = 50.0$ kg, height $h = 1.20$ m, and diameter $D = 0.250$ m. How do you determine which ones have walls filled with gold (and are essentially hollow on the interior except for packing peanuts) and those completely filled with the volatile chemical (a tightly-packed powder) where the mass is uniformly distributed? Hint: apply the concepts of moment of inertia. Assume that, in the case of the gold-filled barrels, the entire mass is concentrated at the outer wall of the barrel and, in the case of the barrels filled with the chemical, the mass is distributed evenly throughout the volume of the cylinder. You can neglect the circular bottoms and the lids of the barrels, and assume there is no slipping. (b) What is the acceleration of the center of mass of each of the barrels as they roll down a 30.0° inclined plane? (c) How much time does it take each barrel to roll 10.0 m down the 30.0° plane?

80. [M] [T] **A Ride Inside a Tractor Tire.** You and your friends plan to roll down a hill on the inside of 600.0-pound tractor tire (diameter $D = 1.80$ m). The hill is inclined at an angle of 25.0° and you initially plan to start from a distance $L = 100.0$ m up the hill, but decide to first check whether it will be safe. (a) Assuming the masses of the tire and your 105-pound body are concentrated at the outer rim of a thin-walled cylinder/hoop, what is the effective acceleration your body experiences at the bottom of the hill where your angular speed is greatest, that is, how many "g's" will you experience? Assuming the human body can withstand a g-force of 8.00 g's (1 g = 9.80 m/s²), is it safe to make the ride from 100.0 m up the hill? (b) What is the maximum starting distance (L_{max}) up the hill that is safe?

CHAPTER **10**

Sarah Reinertsen is a professional athlete who holds numerous world records in her disability division. Her athletic performance is made possible by a high-tech prosthetic leg made of carbon fiber, which flexes and stores elastic potential energy, like a spring does. The elastic potential energy stored by a spring is one of the topics in this chapter.

Don Bartletti/Los Angeles Times/Getty Images

Simple Harmonic Motion and Elasticity

10.1 The Ideal Spring and Simple Harmonic Motion

Springs are familiar objects that have many applications, ranging from push-button switches on electronic components, to automobile suspension systems, to mattresses. In use, they can be stretched or compressed. For example, the top drawing in **Figure 10.1** shows a spring being stretched. Here a hand applies a pulling force F_x^{Applied} to the spring. The subscript x reminds us that F_x^{Applied} lies along the x axis (not shown in the drawing), which is parallel to the length of the spring. In response, the spring stretches and undergoes a displacement of x from its original, or "unstrained," length. The bottom drawing in **Figure 10.1** illustrates the spring being compressed. Now the hand applies a pushing force to the spring, and it again undergoes a displacement from its unstrained length.

Experiment reveals that for relatively small displacements, the force F_x^{Applied} required to stretch or compress a spring is directly proportional

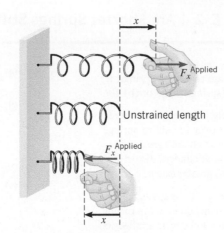

FIGURE 10.1 An ideal spring obeys the equation $F_x^{\text{Applied}} = kx$, where F_x^{Applied} is the force applied to the spring, x is the displacement of the spring from its unstrained length, and k is the spring constant.

to the displacement x, or $F_x^{\text{Applied}} \propto x$. As is customary, this proportionality may be converted into an equation by introducing a proportionality constant k:

$$F_x^{\text{Applied}} = kx \qquad\qquad \textbf{(10.1)}$$

The constant k is called the **spring constant**, and Equation 10.1 shows that it has the dimensions of force per unit length (N/m). A spring that behaves according to $F_x^{\text{Applied}} = kx$ is said to be an **ideal spring**. Example 1 illustrates one application of such a spring.

EXAMPLE 1 | The Physics of a Tire Pressure Gauge

When a tire pressure gauge is pressed against a tire valve, as in **Figure 10.2**, the air in the tire pushes against a plunger attached to a spring. Suppose the spring constant of the spring is $k = 320$ N/m and the bar indicator of the gauge extends 2.0 cm when the gauge is pressed against the tire valve. What force does the air in the tire apply to the spring?

Reasoning We assume that the spring is an ideal spring, so that the relation $F_x^{\text{Applied}} = kx$ is obeyed. The spring constant k is known, as is the displacement x. Therefore, we can determine the force applied to the spring.

Solution The force needed to compress the spring is given by Equation 10.1 as

$$F_x^{\text{Applied}} = kx = (320 \text{ N/m})(0.020 \text{ m}) = \boxed{6.4 \text{ N}}$$

Thus, the exposed length of the bar indicator indicates the force that the air pressure in the tire exerts on the spring. We will see later that pressure is force per unit area, so force is pressure times area. Since the area of the plunger surface is fixed, the bar indicator can be marked in units of pressure.

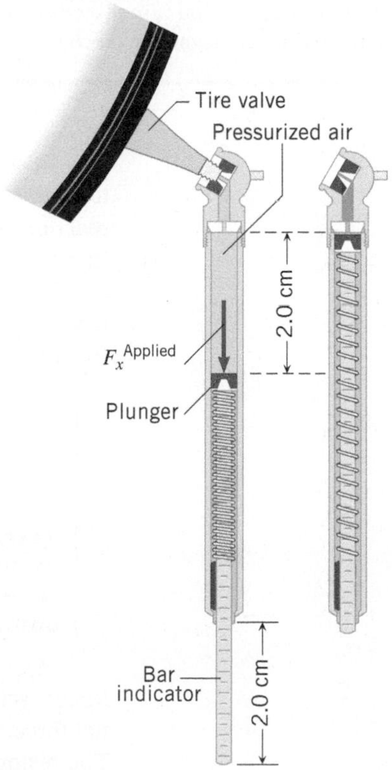

FIGURE 10.2 In a tire pressure gauge, the pressurized air from the tire exerts a force F_x^{Applied} that compresses a spring.

Sometimes the spring constant k is referred to as the **stiffness** of the spring, because a large value for k means the spring is "stiff," in the sense that a large force is required to stretch or compress it. Conceptual Example 2 examines what happens to the stiffness of a spring when the spring is cut into two shorter pieces.

CONCEPTUAL EXAMPLE 2 | Are Shorter Springs Stiffer Springs?

Figure 10.3a shows a 10-coil spring that has a spring constant k. When this spring is cut in half, so there are two 5-coil springs, is the spring constant of each of the shorter springs (a) $\frac{1}{2}k$ or (b) $2k$?

Reasoning When the length of a spring is increased or decreased, the change in length is distributed over the entire spring. Greater forces are required to cause changes that are a greater fraction of the spring's initial length, since such changes distort the atomic structure of the spring material to a greater extent.

Answer (a) is incorrect. When a force F_x^{Applied} is applied to the 10-coil spring, as in **Figure 10.3a**, the displacement of the spring from its unstrained length is x. If this same force were applied to a 5-coil spring that had a spring constant of $\frac{1}{2}k$, the displacement would be $2x$ because the spring is only half as stiff. Since this displacement would be a larger fraction of the length of the 5-coil spring, it would require a force greater than F_x^{Applied}, not equal to F_x^{Applied}. Therefore, the spring constant cannot be $\frac{1}{2}k$.

Answer (b) is correct. As indicated in **Figure 10.3a**, the displacement of the spring from its unstrained length is x when a force F_x^{Applied} is applied to the 10-coil spring. **Figure 10.3b** shows the spring divided in half between the fifth and sixth coils (counting from the right). The spring is in equilibrium, so the net force acting on the right half (coils 1–5) must be zero. Thus, as part b shows, a force of $-F_x^{\text{Applied}}$ must act on coil 5 in order to balance the force F_x^{Applied} that acts on coil 1. It is the adjacent coil 6 that exerts the force $-F_x^{\text{Applied}}$, and Newton's action–reaction law now comes into play. It tells us that coil 5, in response, exerts an oppositely directed force of equal magnitude on coil 6. In other words,

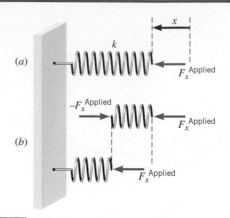

FIGURE 10.3 (a) The 10-coil spring has a spring constant k. The applied force is F_x^{Applied}, and the displacement of the spring from its unstrained length is x. (b) The spring in part a is divided in half, so that the forces acting on the two 5-coil springs can be analyzed.

the force F_x^{Applied} is also exerted on the left half of the spring, as part b also indicates. As a result, the left half compresses by an amount that is one-half the displacement x experienced by the 10-coil spring. We conclude, then, that the 5-coil spring must be twice as stiff as the 10-coil spring. In general, the spring constant is inversely proportional to the number of coils in the spring, so shorter springs are stiffer springs, other things being equal.

Related Homework: Problems 10, 18

To stretch or compress a spring, a force must be applied to it. In accord with Newton's third law, the spring exerts an oppositely directed force of equal magnitude. This reaction force is applied by the spring to the agent that does the pulling or pushing. In other words, the reaction force is applied to the object attached to the spring. The reaction force is also called a "restoring force," for a reason that will be clarified shortly. The restoring force of an ideal spring is obtained from the relation $F_x^{\text{Applied}} = kx$ by including the minus sign required by Newton's action–reaction law, as indicated in Equation 10.2.

> **HOOKE'S LAW* RESTORING FORCE OF AN IDEAL SPRING**
>
> The restoring force of an ideal spring is
>
> $$F_x = -kx \tag{10.2}$$
>
> where k is the spring constant and x is the displacement of the spring from its unstrained length. The minus sign indicates that the restoring force always points in a direction opposite to the displacement of the spring from its unstrained length.

In Chapter 4 we encountered four types of forces: the gravitational force, the normal force, frictional forces, and the tension force. These forces can contribute to the net external force, which Newton's second law relates to the mass and acceleration of an object. The restoring force of a spring can also contribute to the net external force. Once again, we see the unifying theme of Newton's second law, in that individual forces contribute to the net force, which, in turn, is responsible for the acceleration. Newton's second law plays a central role in describing the motion of objects attached to springs.

---

*As we will see in Section 10.8, Equation 10.2 is similar to a relationship first discovered by Robert Hooke (1635–1703).

Interactive Figure 10.4 helps to explain why the phrase "restoring force" is used. In the picture, an object of mass m is attached to a spring on a frictionless table. In part A, the spring has been stretched to the right, so it exerts the leftward-pointing force F_x. When the object is released, this force pulls it to the left, restoring it toward its equilibrium position. However, consistent with Newton's first law, the moving object has inertia and coasts beyond the equilibrium position, compressing the spring as in part B. The force exerted by the spring now points to the right and, after bringing the object to a momentary halt, acts to restore the object to its equilibrium position. But the object's inertia again carries it beyond the equilibrium position, this time stretching the spring and leading to the restoring force F_x shown in part C. The back-and-forth motion illustrated in the drawing then repeats itself, continuing forever, since no friction acts on the object or the spring.

When the restoring force has the mathematical form given by $F_x = -kx$, the type of friction-free motion illustrated in Interactive Figure 10.4 is designated as "simple harmonic motion."

By attaching a pen to the object and moving a strip of paper past it at a steady rate, we can record the position of the vibrating object as time passes. **Figure 10.5** illustrates the resulting graphical record of simple harmonic motion. The maximum excursion from equilibrium is the **amplitude A** of the motion. The shape of this graph is characteristic of simple harmonic motion and is called "sinusoidal," because it has the shape of a trigonometric sine or cosine function.

The restoring force also leads to simple harmonic motion when the object is attached to a vertical spring, just as it does when the spring is horizontal. When the spring is vertical, however, the weight of the object causes the spring to stretch, and the motion occurs with respect to the equilibrium position of the object on the stretched spring, as **Animated Figure 10.6** indicates. The amount of initial stretching d_0 due to the weight can be calculated by equating the weight to the magnitude of the restoring force that supports it; thus, $mg = kd_0$, which gives $d_0 = mg/k$.

INTERACTIVE FIGURE 10.4 The restoring force F_x (see blue arrows) produced by an ideal spring always points opposite to the displacement x (see black arrows) of the spring and leads to a back-and-forth motion of the object.

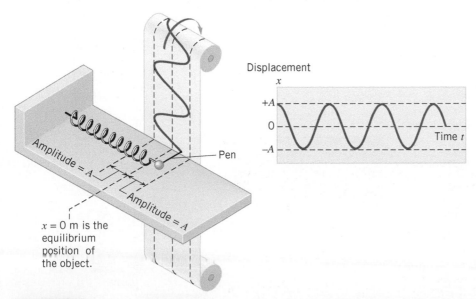

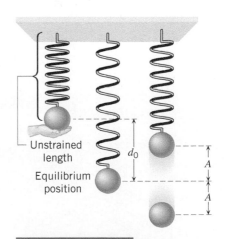

FIGURE 10.5 When an object moves in simple harmonic motion, a graph of its position as a function of time has a sinusoidal shape with an amplitude A. A pen attached to the object records the graph.

ANIMATED FIGURE 10.6 The weight of an object on a vertical spring stretches the spring by an amount d_0. Simple harmonic motion of amplitude A occurs with respect to the equilibrium position of the object on the stretched spring.

Check Your Understanding

(The answers are given at the end of the book.)

1. A steel ball is dropped onto a very hard floor. Over and over again, the ball rebounds to its original height (assuming that no energy is lost during the collision with the floor). Is the motion of the ball simple harmonic motion?

2. CYU Figure 10.1 shows identical springs that are attached to a box in two different ways. Initially, the springs are unstrained. The box is then pulled to the right and released. In each case the initial displacement of the box is the same. At the moment of release, which box, if either, experiences the greater net force due to the springs?

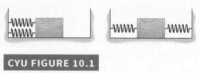

CYU FIGURE 10.1

3. A 0.42-kg block is attached to the end of a horizontal ideal spring and rests on a frictionless surface. The block is pulled so that the spring stretches for 2.1 cm relative to its unstrained length. When the block is released, it moves with an acceleration of 9.0 m/s². What is the spring constant of the spring?

4. Two people pull on a horizontal spring that is attached to an immovable wall. Then, they detach it from the wall and pull on opposite ends of the horizontal spring. They pull just as hard in each case. In which situation, if either, does the spring stretch more?

10.2 | Simple Harmonic Motion and the Reference Circle

Simple harmonic motion can be described in terms of displacement, velocity, and acceleration, and the model in **Figure 10.7** is helpful in explaining these characteristics. The model consists of a small ball attached to the top of a rotating turntable. The ball is moving in uniform circular motion (see Section 5.1) on a path known as the **reference circle**. As the ball moves, its shadow falls on a strip of film, which is moving upward at a steady rate and recording where the shadow is. A comparison of the film with the paper in **Figure 10.5** reveals the same kind of patterns, suggesting that the model is useful.

Displacement

Figure 10.8 shows the reference circle (radius = A) and indicates how to determine the displacement of the shadow on the film. The ball starts on the x axis at $x = +A$

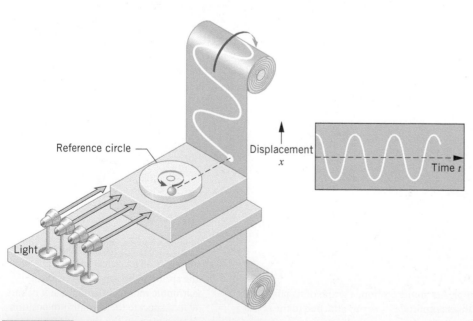

FIGURE 10.7 The ball mounted on the turntable moves in uniform circular motion, and its shadow, projected on a moving strip of film, executes simple harmonic motion.

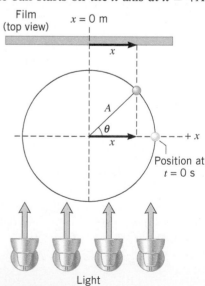

FIGURE 10.8 A top view of a ball on a turntable. The ball's shadow on the film has a displacement x that depends on the angle θ through which the ball has moved on the reference circle.

and moves through the angle θ in a time t. The circular motion is uniform, so the ball moves with a constant angular speed ω (in rad/s), and the angle has a value (in rad) of $\theta = \omega t$. The displacement x of the shadow is just the projection of the radius A onto the x axis:

$$x = A \cos \theta = A \cos \omega t \qquad \text{(10.3)}$$

Figure 10.9 shows a graph of this equation. As time passes, the shadow of the ball oscillates between the values of $x = +A$ and $x = -A$, corresponding to the limiting values of $+1$ and -1 for the cosine of an angle. The radius A of the reference circle, then, is the amplitude of the simple harmonic motion.

As the ball moves one revolution or cycle around the reference circle, its shadow executes one cycle of back-and-forth motion. For any object in simple harmonic motion, the time required to complete one cycle is the **period** T, as **Figure 10.9** indicates. The value of T depends on the angular speed ω of the ball because the greater the angular speed, the shorter the time it takes to complete one revolution. We can obtain the relationship between ω and T by recalling that $\omega = \Delta\theta/\Delta t$ (Equation 8.2), where $\Delta\theta$ is the angular displacement of the ball and Δt is the time interval. For one cycle, $\Delta\theta = 2\pi$ rad and $\Delta t = T$, so that

$$\omega = \frac{2\pi}{T} \qquad (\omega \text{ in rad/s}) \qquad \text{(10.4)}$$

Instead of the period, we often speak of the **frequency** f of the motion, the frequency being the number of cycles of the motion per second. For example, if an object on a spring completes 10 cycles in one second, the frequency is $f = 10$ cycles/s. The period T, or the time for one cycle, would be $\frac{1}{10}$ s. Thus, frequency and period are related according to

$$f = \frac{1}{T} \qquad \text{(10.5)}$$

Usually, one cycle per second is referred to as one hertz (Hz), the unit being named after Heinrich Hertz (1857–1894). One thousand cycles per second is called one kilohertz (kHz). Thus, five thousand cycles per second, for instance, can be written as 5 kHz.

Using the relationships $\omega = 2\pi/T$ and $f = 1/T$, we can relate the angular speed ω (in rad/s) to the frequency f (in cycles/s or Hz):

$$\omega = \frac{2\pi}{T} = 2\pi f \qquad (\omega \text{ in rad/s}) \qquad \text{(10.6)}$$

Because ω is directly proportional to the frequency f, ω is often called the **angular frequency**.

Velocity

The reference circle model can also be used to determine the velocity of an object in simple harmonic motion. **Figure 10.10** shows the tangential velocity $\vec{v}_T$ of the ball on the reference circle. The drawing indicates that the velocity $\vec{v}_x$ of the shadow is just the x component of the vector $\vec{v}_T$; that is, $v_x = -v_T \sin\theta$, where $\theta = \omega t$. The minus sign is necessary because $\vec{v}_x$ points to the left, in the direction of the negative x axis. Since the tangential speed v_T is related to the angular speed ω by $v_T = r\omega$ (Equation 8.9) and since $r = A$, it follows that $v_T = A\omega$. Therefore, the velocity in simple harmonic motion is given by

$$v_x = -v_T \sin\theta = -A\omega \sin\omega t \qquad (\omega \text{ in rad/s}) \qquad \text{(10.7)}$$

This velocity is *not* constant, but varies between maximum and minimum values as time passes. When the shadow changes direction at either end of the oscillatory motion, the velocity

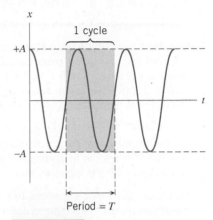

FIGURE 10.9 For simple harmonic motion, the graph of displacement x versus time t is a sinusoidal curve. The period T is the time required for one complete motional cycle.

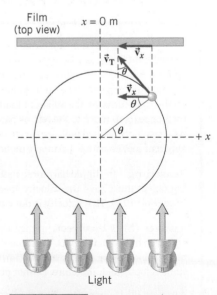

FIGURE 10.10 The velocity $\vec{v}_x$ of the ball's shadow is the x component of the tangential velocity $\vec{v}_T$ of the ball on the reference circle.

is momentarily zero. When the shadow passes through the $x = 0$ m position, the velocity has a maximum magnitude of $A\omega$, since the sine of an angle is between +1 and −1:

$$v_{max} = A\omega \quad (\omega \text{ in rad/s}) \tag{10.8}$$

Both the amplitude A and the angular frequency ω determine the maximum velocity, as Example 3 emphasizes.

EXAMPLE 3 | The Physics of a Loudspeaker Diaphragm

The diaphragm of a loudspeaker moves back and forth in simple harmonic motion to create sound, as in **Figure 10.11**. The frequency of the motion is $f = 1.0$ kHz and the amplitude is $A = 0.20$ mm. **(a)** What is the maximum speed of the diaphragm? **(b)** Where in the motion does this maximum speed occur?

Reasoning The maximum speed of an object vibrating in simple harmonic motion is $v_{max} = A\omega$ (ω in rad/s), according to Equation 10.8. The angular frequency ω is related to the frequency f by $\omega = 2\pi f$, according to Equation 10.6.

Solution **(a)** Using Equations 10.8 and 10.6, we find that the maximum speed of the vibrating diaphragm is

$$v_{max} = A\omega = A(2\pi f) = (0.20 \times 10^{-3}\text{m})(2\pi)(1.0 \times 10^3 \text{Hz})$$
$$= \boxed{1.3 \text{ m/s}}$$

(b) The speed of the diaphragm is zero when the diaphragm momentarily comes to rest at either end of its motion: $x = +A$ and

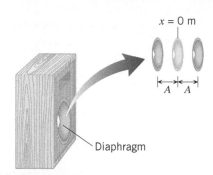

FIGURE 10.11 The diaphragm of a loudspeaker generates a sound by moving back and forth in simple harmonic motion.

$x = -A$. Its maximum speed occurs midway between these two positions, or at $x = 0$ m.

Simple harmonic motion is not just any kind of vibratory motion. It is a very specific kind and, among other things, must have the velocity given by Equation 10.7. For instance, advertising signs often use a "moving light" display to grab your attention. Conceptual Example 4 examines the back-and-forth motion in one such display, to see whether it is simple harmonic motion.

CONCEPTUAL EXAMPLE 4 | Moving Lights

Over the entrance to a restaurant is mounted a strip of equally spaced light bulbs, as **Figure 10.12a** illustrates. Starting at the left end, each bulb turns on in sequence for one-half second. Thus, a lighted bulb appears to move from left to right. After the last bulb on the right turns on, the apparent motion reverses. The lighted bulb then appears to move to the left, as part b of the drawing indicates. As a result, the lighted bulb appears to oscillate back and forth. Is the apparent motion simple harmonic motion? **(a)** No **(b)** Yes

Reasoning In simple harmonic motion the velocity of the moving object must have the velocity specified by $v = -A\omega \sin \omega t$ (Equation 10.7). This velocity is not constant as time passes.

Answer (b) is incorrect. Since the bulbs are equally spaced and each one remains lit for the same amount of time, the apparent motion in **Figure 10.12a** (or in **Figure 10.12b**) occurs at a constant velocity and, therefore, cannot be simple harmonic motion.

Answer (a) is correct. The apparent motion is not simple harmonic motion. If it were, the speed would be zero at each end of

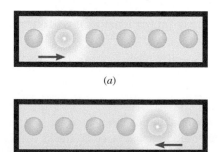

FIGURE 10.12 The motion of a lighted bulb is from (a) left to right and then from (b) right to left.

the sign and would increase to a maximum speed at the center, consistent with Equation 10.7. However, the speed is constant because the bulbs are equally spaced and each remains on for the same amount of time.

Acceleration

In simple harmonic motion, the velocity is not constant; consequently, there must be an acceleration. This acceleration can also be determined with the aid of the reference-circle model. As **Figure 10.13** shows, the ball on the reference circle moves in uniform circular motion, and, therefore, has a centripetal acceleration $\vec{a}_c$ that points toward the center of the circle. The acceleration $\vec{a}_x$ of the shadow is the x component of the centripetal acceleration; $a_x = -a_c \cos \theta$. The minus sign is needed because the acceleration of the shadow points to the left. Recalling that the centripetal acceleration is related to the angular speed ω by $a_c = r\omega^2$ (Equation 8.11) and using $r = A$, we find that $a_c = A\omega^2$. With this substitution and the fact that $\theta = \omega t$, the acceleration in simple harmonic motion becomes

$$a_x = -a_c \cos \theta = -A\omega^2 \cos \omega t \qquad (\omega \text{ in rad/s}) \qquad \textbf{(10.9)}$$

The acceleration, like the velocity, does *not* have a constant value as time passes. The maximum magnitude of the acceleration is

$$a_{\text{max}} = A\omega^2 \qquad (\omega \text{ in rad/s}) \qquad \textbf{(10.10)}$$

Although both the amplitude A and the angular frequency ω determine the maximum value, the frequency has a particularly strong effect, because it is squared. Example 5 shows that the acceleration can be remarkably large in a practical situation.

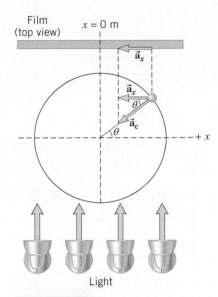

FIGURE 10.13 The acceleration $\vec{a}_x$ of the ball's shadow is the x component of the centripetal acceleration $\vec{a}_c$ of the ball on the reference circle.

EXAMPLE 5 | The Loudspeaker Revisited—The Maximum Acceleration

The loudspeaker diaphragm in **Figure 10.11** is vibrating at a frequency of $f = 1.0$ kHz, and the amplitude of the motion is $A = 0.20$ mm. **(a)** What is the maximum acceleration of the diaphragm, and **(b)** where does this maximum acceleration occur?

Reasoning The maximum acceleration of an object vibrating in simple harmonic motion is $a_{\text{max}} = A\omega^2$ (ω in rad/s), according to Equation 10.10. Equation 10.6 shows that the angular frequency ω is related to the frequency f by $\omega = 2\pi f$.

Solution **(a)** Using Equations 10.10 and 10.6, we find that the maximum acceleration of the vibrating diaphragm is

$$a_{\text{max}} = A\omega^2 = A(2\pi f)^2 = (0.20 \times 10^{-3} \text{m})[2\pi(1.0 \times 10^3 \text{Hz})]^2$$

$$= \boxed{7.9 \times 10^3 \text{m/s}^2}$$

This is an incredible acceleration, being more than 800 times the acceleration due to gravity, and the diaphragm must be built to withstand it.

(b) The maximum acceleration occurs when the force acting on the diaphragm is a maximum. The maximum force arises when the diaphragm is at the ends of its path, where the displacement is greatest. Thus, the maximum acceleration occurs at $x = +A$ and $x = -A$ in **Figure 10.11**.

Math Skills The angular frequency ω and the vibration frequency f are not the same thing, so you will need to be careful to distinguish between them when solving problems. The two quantities are proportional, but they differ by a factor of 2π. The relation between them is

$$\omega = 2\pi f \qquad \textbf{(10.6)}$$

You can always tell which quantity is being referred to in a problem by looking at the given data. A value for ω is given in radians per second, whereas a value for f is given in hertz (cycles per second).

Frequency of Vibration

With the aid of Newton's second law ($\Sigma F_x = ma_x$), it is possible to determine the frequency at which an object of mass m vibrates on a spring. We assume that the mass of the spring itself is negligible and that the only force acting on the object in the horizontal direction is due to the spring—that is, the Hooke's law restoring force. Thus, the net force is $\Sigma F_x = -kx$, and Newton's second law becomes $-kx = ma_x$, where a_x is the acceleration of the object. The displacement and acceleration of an oscillating spring are, respectively, $x = A \cos \omega t$ (Equation 10.3) and $a_x = -A\omega^2 \cos \omega t$ (Equation 10.9). Substituting these expressions for x and a_x into the relation $-kx = ma_x$, we find that

$$-k(A \cos \omega t) = m(-A\omega^2 \cos \omega t)$$

which yields

$$\omega = \sqrt{\frac{k}{m}} \qquad (\omega \text{ in rad/s}) \qquad \textbf{(10.11)}$$

In this expression, the angular frequency ω must be in radians per second. Larger spring constants k and smaller masses m result in larger frequencies. Example 6 illustrates an application of Equation 10.11.

Analyzing Multiple-Concept Problems

EXAMPLE 6 | BIO The Physics of a Body-Mass Measurement Device

Astronauts who spend a long time in orbit measure their body masses as part of their health-maintenance programs. On earth, it is simple to measure body weight W with a scale and convert it to mass m using the magnitude g of the acceleration due to gravity, since $W = mg$. However, this procedure does not work in orbit, because both the scale and the astronaut are in free fall and cannot press against each other (see Conceptual Example 13 in Chapter 5). Instead, astronauts use a body-mass measurement device, as **Figure 10.14** illustrates. The device consists of a spring-mounted chair in which the astronaut sits. The chair is then started oscillating in simple harmonic motion. The period of the motion is measured electronically and is automatically converted into a value of the astronaut's mass, after the mass of the chair is taken into account. The spring used in one such device has a spring constant of 606 N/m, and the mass of the chair is 12.0 kg. The measured oscillation period is 2.41 s. Find the mass of the astronaut.

Reasoning Since the astronaut and chair are oscillating in simple harmonic motion, the total mass m_{total} of

the two is related to the angular frequency ω of the motion by $\omega = \sqrt{k/m_{\text{total}}}$, or $m_{\text{total}} = k/\omega^2$. The angular frequency, in turn, is related to the period of the motion by $\omega = 2\pi/T$. These two relations will enable us to find the total mass of the astronaut and chair. From this result, we will subtract the mass of the chair (m_{chair}) to obtain the mass of the astronaut (m_{astro}).

Knowns and Unknowns The data for this problem are:

Description	Symbol	Value
Spring constant	k	606 N/m
Mass of chair	m_{chair}	12.0 kg
Period of oscillation	T	2.41 s
Unknown Variable		
Mass of astronaut	m_{astro}	?

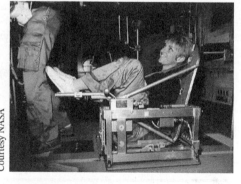

Courtesy NASA

FIGURE 10.14 Astronaut Millie Hughes-Fulford trains to use a body-mass measurement device developed for determining mass in orbit.

Modeling the Problem

STEP 1 Angular Frequency of Vibration The mass m_{astro} of the astronaut is equal to the total mass m_{total} of the astronaut and chair minus the mass m_{chair} of the chair:

$$m_{\text{astro}} = m_{\text{total}} - m_{\text{chair}} \qquad \textbf{(1)}$$

The mass of the chair is known. Since the astronaut and chair oscillate in simple harmonic motion, the angular frequency ω of the oscillation is related to the spring constant k of the spring and the total mass m_{total} by Equation 10.11:

$$\omega = \sqrt{\frac{k}{m_{\text{total}}}} \quad \text{or} \quad m_{\text{total}} = \frac{k}{\omega^2}$$

Substituting this result for m_{total} into Equation 1 gives Equation 2 at the right. The spring constant and mass of the chair are known, but the angular frequency ω is not; we will evaluate it in Step 2.

$$m_{\text{astro}} = \frac{k}{\omega^2} - m_{\text{chair}} \qquad \textbf{(2)}$$

?

STEP 2 Angular Frequency and Period   The angular frequency ω of the oscillating motion is inversely related to its period T by

$$\boxed{\omega = \frac{2\pi}{T}} \qquad\qquad (10.4)$$

All the variables on the right side of this equation are known, and we substitute it into Equation 2, as indicated in the right column.

$$m_{\text{astro}} = \frac{k}{\omega^2} - m_{\text{chair}} \qquad (2)$$

$$\uparrow$$

$$\boxed{\omega = \frac{2\pi}{T}} \qquad (10.4)$$

Solution   Algebraically combining the results of the steps above, we have:

STEP 1   STEP 2

$$m_{\text{astro}} = \frac{k}{\omega^2} - m_{\text{chair}} = \frac{k}{\left(\frac{2\pi}{T}\right)^2} - m_{\text{chair}} = \frac{kT^2}{4\pi^2} - m_{\text{chair}}$$

$$= \frac{(606\ \text{N/m})(2.41\ \text{s})^2}{4\pi^2} - 12.0\ \text{kg} = \boxed{77.2\ \text{kg}}$$

Related Homework: Problem 68

THE PHYSICS OF . . . detecting and measuring small amounts of chemicals.   Example 6 indicates that the mass of the vibrating object influences the frequency of simple harmonic motion. Electronic sensors are being developed that take advantage of this effect in detecting and measuring small amounts of chemicals. These sensors utilize tiny quartz crystals that vibrate when an electric current passes through them. If a crystal is coated with a substance that absorbs a particular chemical, then the mass of the coated crystal increases as the chemical is absorbed and, according to the relation $f = \frac{1}{2\pi}\sqrt{k/m}$ (Equations 10.6 and 10.11), the frequency of the simple harmonic motion decreases. The change in frequency is detected electronically, and the sensor is calibrated to give the mass of the absorbed chemical.

Check Your Understanding

(*The answers are given at the end of the book.*)

5. CYU Figure 10.2 shows plots of the displacement x versus the time t for three objects undergoing simple harmonic motion. Which object—I, II, or III—has the greatest maximum velocity?

6. In **Figure 10.13** the shadow moves in simple harmonic motion. Where on the path of the shadow is the acceleration equal to zero?

7. A particle is oscillating in simple harmonic motion. The time required for the particle to travel through one complete cycle is equal to the period of the motion, no matter what the amplitude is. But how can this be, since larger amplitudes mean that the particle travels farther?

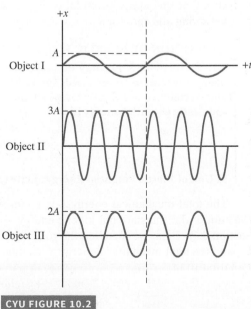

CYU FIGURE 10.2

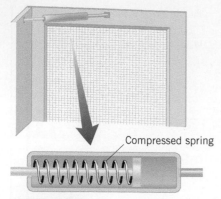

FIGURE 10.15 A door-closing unit. The elastic potential energy stored in the compressed spring is used to close the door.

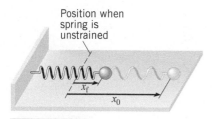

FIGURE 10.16 When the object is released, its displacement changes from an initial value of x_0 to a final value of x_f.

10.3 # Energy and Simple Harmonic Motion

We saw in Chapter 6 that an object above the surface of the earth has gravitational potential energy. Therefore, when the object is allowed to fall, like the hammer of the pile driver in Figure 6.13, it can do work. A spring also has potential energy when the spring is stretched or compressed, which we refer to as **elastic potential energy**. Because of elastic potential energy, a stretched or compressed spring can do work on an object that is attached to the spring.

THE PHYSICS OF . . . a door-closing unit. For instance, **Figure 10.15** shows a door-closing unit that is often found on screen doors. When the door is opened, a spring inside the unit is compressed and has elastic potential energy. When the door is released, the compressed spring expands and does the work of closing the door.

To find an expression for the elastic potential energy, we will determine the work done by the spring force on an object. **Figure 10.16** shows an object attached to one end of a stretched spring. When the object is released, the spring contracts and pulls the object from its initial position x_0 to its final position x_f. The work W done by a constant force is given by Equation 6.1 as $W = (F \cos \theta)s$, where F is the magnitude of the force, s is the magnitude of the displacement ($s = x_0 - x_f$), and θ is the angle between the force and the displacement. The magnitude of the spring force is not constant, however. Equation 10.2 gives the spring force as $F_x = -kx$, and as the spring contracts, the magnitude of this force changes from kx_0 to kx_f. In using Equation 6.1 to determine the work, we can account for the changing magnitude by using an average magnitude $\overline{F}_x$ in place of the constant magnitude F_x. Because the dependence of the spring force on x is linear, the magnitude of the average force is just one-half the sum of the initial and final values, or $\overline{F}_x = \frac{1}{2}(kx_0 + kx_f)$. The work $W_{elastic}$ done by the average spring force is, then,

$$W_{elastic} = (\overline{F}_x \cos \theta)s = \tfrac{1}{2}(kx_0 + kx_f)(\cos 0°)(x_0 - x_f)$$

$$W_{elastic} = \underbrace{\tfrac{1}{2}kx_0^2}_{\substack{\text{Initial elastic} \\ \text{potential energy}}} - \underbrace{\tfrac{1}{2}kx_f^2}_{\substack{\text{Final elastic} \\ \text{potential energy}}} \qquad \textbf{(10.12)}$$

In the calculation above, θ is 0°, since the spring force has the same direction as the displacement. Equation 10.12 indicates that the work done by the spring force is equal to the difference between the initial and final values of the quantity $\frac{1}{2}kx^2$. The quantity $\frac{1}{2}kx^2$ is analogous to the quantity mgh, which we identified in Section 6.3 as the gravitational potential energy. Here, the quantity $\frac{1}{2}kx^2$ is the elastic potential energy. Equation 10.13 indicates that the elastic potential energy is a maximum for a fully stretched or compressed spring and zero for a spring that is neither stretched nor compressed ($x = 0$ m).

DEFINITION OF ELASTIC POTENTIAL ENERGY

The elastic potential energy $PE_{elastic}$ is the energy that a spring has by virtue of being stretched or compressed. For an ideal spring that has a spring constant k and is stretched or compressed by an amount x relative to its unstrained length, the elastic potential energy is

$$PE_{elastic} = \tfrac{1}{2}kx^2 \qquad \textbf{(10.13)}$$

SI Unit of Elastic Potential Energy: joule (J)

The total mechanical energy E is a familiar idea that we originally defined to be the sum of the translational kinetic energy and the gravitational potential energy (see Section 6.5). Then, we included the rotational kinetic energy (see Section 9.5). We now expand the total mechanical energy to include the elastic potential energy, as shown in Equation 10.14.

$$\underbrace{E}_{\substack{\text{Total} \\ \text{mechanical} \\ \text{energy}}} = \underbrace{\tfrac{1}{2}mv^2}_{\substack{\text{Translational} \\ \text{kinetic} \\ \text{energy}}} + \underbrace{\tfrac{1}{2}I\omega^2}_{\substack{\text{Rotational} \\ \text{kinetic} \\ \text{energy}}} + \underbrace{mgh}_{\substack{\text{Gravitational} \\ \text{potential} \\ \text{energy}}} + \underbrace{\tfrac{1}{2}kx^2}_{\substack{\text{Elastic} \\ \text{potential} \\ \text{energy}}} \qquad \textbf{(10.14)}$$

As Section 6.5 discusses, the total mechanical energy is conserved when external nonconservative forces (such as friction) do no net work; that is, when $W_{\text{nc}} = 0$ J. Then, the final and initial values of E are the same: $E_f = E_0$. The principle of conservation of total mechanical energy is the subject of the next example.

EXAMPLE 7 | An Object on a Horizontal Spring

Interactive Figure 10.17 shows an object of mass $m = 0.200$ kg that is vibrating on a horizontal frictionless table. The spring has a spring constant of $k = 545$ N/m. The spring is stretched initially to $x_0 = 4.50$ cm and is then released from rest (see part A of the drawing). Determine the final translational speed v_f of the object when the final displacement of the spring is **(a)** $x_f = 2.25$ cm and **(b)** $x_f = 0$ cm.

Reasoning The conservation of mechanical energy indicates that, in the absence of friction (a nonconservative force), the final and initial total mechanical energies are the same:

$$E_f = E_0$$

$$\tfrac{1}{2}mv_f^2 + \tfrac{1}{2}I\omega_f^2 + mgh_f + \tfrac{1}{2}kx_f^2 = \tfrac{1}{2}mv_0^2 + \tfrac{1}{2}I\omega_0^2 + mgh_0 + \tfrac{1}{2}kx_0^2$$

Since the object is moving on a horizontal table, the final and initial heights are the same: $h_f = h_0$. The object is not rotating, so its angular speed is zero: $\omega_f = \omega_0 = 0$ rad/s. Also, as the problem states, the initial translational speed of the object is zero, $v_0 = 0$ m/s. With these substitutions, the conservation-of-energy equation becomes

$$\tfrac{1}{2}mv_f^2 + \tfrac{1}{2}kx_f^2 = \tfrac{1}{2}kx_0^2$$

from which we can obtain v_f:

$$v_f = \sqrt{\frac{k}{m}\left(x_0^2 - x_f^2\right)}$$

Solution **(a)** Since $x_0 = 0.0450$ m and $x_f = 0.0225$ m, the final translational speed is

$$v_f = \sqrt{\frac{545 \text{ N/m}}{0.200 \text{ kg}}[(0.0450 \text{ m})^2 - (0.0225 \text{ m})^2]} = \boxed{2.03 \text{ m/s}}$$

The total mechanical energy at this point is composed partly of translational kinetic energy ($\tfrac{1}{2}mv_f^2 = 0.414$ J) and partly of elastic potential energy ($\tfrac{1}{2}kx_f^2 = 0.138$ J), as indicated in part B of

Interactive Figure 10.17. The total mechanical energy E is the sum of these two energies: $E = 0.414$ J $+ 0.138$ J $= 0.552$ J. Because the total mechanical energy remains constant during the motion, this value equals the initial total mechanical energy when the object is stationary and the energy is entirely elastic potential energy ($E_0 = \tfrac{1}{2}kx_0^2 = 0.552$ J).

(b) When $x_0 = 0.0450$ m and $x_f = 0$ m, we have

$$v_f = \sqrt{\frac{k}{m}\left(x_0^2 - x_f^2\right)} = \sqrt{\frac{545 \text{ N/m}}{0.200 \text{ kg}}[(0.0450 \text{ m})^2 - (0 \text{ m})^2]}$$

$$= \boxed{2.35 \text{ m/s}}$$

Now the total mechanical energy is due entirely to the translational kinetic energy ($\tfrac{1}{2}mv_f^2 = 0.552$ J), since the elastic potential energy is zero (see part C of **Interactive Figure 10.17**). Note that the total mechanical energy is the same as it is in Solution part (a). In the absence of friction, the simple harmonic motion of a spring converts the different types of energy between one form and another, the total always remaining the same.

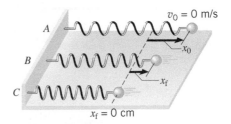

INTERACTIVE FIGURE 10.17 The total mechanical energy of this system is entirely elastic potential energy (A), partly elastic potential energy and partly kinetic energy (B), and entirely kinetic energy (C).

Conceptual Example 8 takes advantage of energy conservation to illustrate what happens to the maximum speed, amplitude, and angular frequency of a simple harmonic oscillator when its mass is changed suddenly at a certain point in the motion.

CONCEPTUAL EXAMPLE 8 | Changing the Mass of a Simple Harmonic Oscillator

Figure 10.18a shows a box of mass m attached to a spring that has a force constant k. The box rests on a horizontal, frictionless surface. The spring is initially stretched to $x = A$ and then released from rest. The box executes simple harmonic motion that is characterized by a maximum speed v_x^{max}, an amplitude A, and an angular frequency ω. When the box is passing through the point where the spring is unstrained ($x = 0$ m), a second box of

the same mass m and speed v_x^{max}, is attached to it, as in part b of the drawing. Discuss what happens to **(a)** the maximum speed, **(b)** the amplitude, and **(c)** the angular frequency of the subsequent simple harmonic motion.

Reasoning and Solution **(a)** The maximum speed of an object in simple harmonic motion occurs when the object is passing through the point where the spring is unstrained ($x = 0$ m), as in **Figure 10.18b**. Since the second box is attached at this point with the same speed, *the maximum speed of the two-box system remains the same as that of the one-box system.*

(b) At the same speed, the maximum kinetic energy of the two boxes is twice that of a single box, since the mass is twice as much. Subsequently, when the two boxes move to the left and compress the spring, their kinetic energy is converted into elastic potential energy. Since the two boxes have twice as much kinetic energy as one box alone, the two will have twice as much elastic potential energy when they come to a halt at the extreme left. Here, we are using the principle of conservation of mechanical energy, which applies since friction is absent. But the elastic potential energy is proportional to the amplitude squared (A^2) of the motion, *so the amplitude of the two-box system is $\sqrt{2}$ times as great as that of the one-box system.*

(c) The angular frequency ω of a simple harmonic oscillator is $\omega = \sqrt{k/m}$ (Equation 10.11). Since the mass of the two-box system is twice the mass of the one-box system, *the angular frequency of the two-box system is $\sqrt{2}$ times as small as that of the one-box system.*

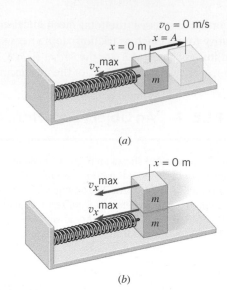

(a)

(b)

FIGURE 10.18 (a) A box of mass m, starting from rest at $x = A$, undergoes simple harmonic motion about $x = 0$ m. (b) When $x = 0$ m, a second box, with the same mass and speed, is attached.

In the previous two examples, gravitational potential energy plays no role because the spring is horizontal. The next example illustrates that gravitational potential energy must be taken into account when a spring is oriented vertically.

EXAMPLE 9 | A Falling Ball on a Vertical Spring

A 0.20-kg ball is attached to a vertical spring, as in **Figure 10.19**. The spring constant of the spring is 28 N/m. The ball, supported initially so that the spring is neither stretched nor compressed, is released from rest. In the absence of air resistance, how far does the ball fall before being brought to a momentary stop by the spring?

Reasoning Since air resistance is absent, only the conservative forces of gravity and the spring act on the ball. Therefore, the principle of conservation of mechanical energy applies:

$$E_f = E_0$$

$$\tfrac{1}{2}mv_f^2 + \tfrac{1}{2}I\omega_f^2 + mgh_f + \tfrac{1}{2}ky_f^2 = \tfrac{1}{2}mv_0^2 + \tfrac{1}{2}I\omega_0^2 + mgh_0 + \tfrac{1}{2}ky_0^2$$

Note that we have replaced x with y in the elastic-potential-energy terms ($\tfrac{1}{2}ky_f^2$ and $\tfrac{1}{2}ky_0^2$), in recognition of the fact that the spring is moving in the vertical direction. The problem states that the final and initial translational speeds of the ball are zero: $v_f = v_0 = 0$ m/s. The ball and spring do not rotate; therefore, the final and initial angular speeds are also zero: $\omega_f = \omega_0 = 0$ rad/s. As **Figure 10.19** indicates, the initial height of the ball is h_0, and the final height is $h_f = 0$ m. In addition, the spring is unstrained ($y_0 = 0$ m) to begin with, and so it has no elastic potential energy initially. With these substitutions, the conservation-of-mechanical-energy equation reduces to

$$\tfrac{1}{2}ky_f^2 = mgh_0$$

This result shows that the initial gravitational potential energy (mgh_0) is converted into elastic potential energy ($\tfrac{1}{2}ky_f^2$). When the ball falls to its lowest point, its displacement is $y_f = -h_0$, where the minus sign indicates that the displacement is downward.

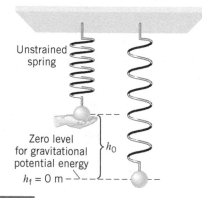

FIGURE 10.19 The ball is supported initially so that the spring is unstrained. After being released from rest, the ball falls through the distance h_0 before being momentarily stopped by the spring.

Substituting this result into the equation above and solving for h_0 yields $h_0 = 2mg/k$.

> **Problem-Solving Insight** When evaluating the total mechanical energy E, always include a potential energy term for every conservative force acting on the system. In Example 9 there are two such terms, gravitational and elastic.

Solution The distance that the ball falls before coming to a momentary halt is

$$h_0 = \frac{2mg}{k} = \frac{2(0.20 \text{ kg})(9.8 \text{ m/s}^2)}{28 \text{ N/m}} = \boxed{0.14 \text{ m}}$$

Check Your Understanding

(*The answers are given at the end of the book.*)

8. Is more elastic potential energy stored in a spring when the spring is compressed by one centimeter than when it is stretched by the same amount?

9. A block is attached to the end of a horizontal ideal spring and rests on a frictionless surface. The block is pulled so that the spring stretches relative to its unstrained length. In each of the following three cases, the spring is stretched initially by the same amount. Rank the amplitudes of the resulting simple harmonic motion in decreasing order (largest first). **(a)** The block is released from rest. **(b)** The block is given an initial speed v_0. **(c)** The block is given an initial speed $\frac{1}{2}v_0$.

10. A block is attached to a horizontal spring and slides back and forth on a frictionless horizontal surface. A second identical block is suddenly attached to the first block. The attachment is accomplished by joining the blocks at one extreme end of the oscillation cycle. The velocities of the blocks are exactly matched at the instant of joining. How do the **(a)** amplitude, **(b)** frequency, and **(c)** maximum speed of the simple harmonic motion change?

10.4 | The Pendulum

A **simple pendulum** consists of a particle of mass m, attached to a frictionless pivot P by a cable of length L and negligible mass. When the particle is pulled away from its equilibrium position by an angle θ and released, it swings back and forth as **Figure 10.20** shows. By attaching a pen to the bottom of the swinging particle and moving a strip of paper beneath it at a steady rate, we can record the position of the particle as time passes. The graphical record reveals a pattern that is similar (but not identical) to the sinusoidal pattern for simple harmonic motion.

Gravity causes the back-and-forth rotation about the axis at P. The rotation speeds up as the particle approaches the lowest point and slows down on the upward part of the swing. Eventually the angular speed is reduced to zero, and the particle swings back. As Section 9.4 discusses, a net torque is required to change the angular speed. The gravitational force $m\vec{g}$ produces this torque. (The tension $\vec{T}$ in the cable creates no torque, because it points directly at the pivot P and, therefore, has a zero lever arm.) According to Equation 9.1, the magnitude of the torque τ is the product of the magnitude mg of the gravitational force and the lever arm ℓ, so that $\tau = -(mg)\ell$. The minus sign is included since the torque is a restoring torque; it acts to reduce the angle θ [the angle θ is positive (counterclockwise), while the torque is negative (clockwise)]. The lever arm ℓ is the perpendicular distance between the line of action of $m\vec{g}$ and the pivot P. In **Figure 10.20**, ℓ is very nearly equal to the arc length s of the circular path when the angle θ is small (about 10° or less). Furthermore, if θ is expressed in radians, the arc length and the radius L of the circular path are related, according to $s = L\theta$ (Equation 8.1). It follows, then, that $\ell \approx s = L\theta$, and the gravitational torque is

$$\tau \approx \underbrace{-mgL}_{k'}\theta$$

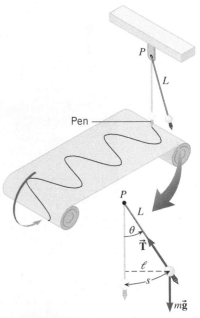

FIGURE 10.20 A simple pendulum swinging back and forth about the pivot P. If the angle θ is small (about 10° or less), the swinging is approximately simple harmonic motion.

In the equation above, the term mgL has a constant value k', independent of θ. *For small angles,* then, the torque that restores the pendulum to its vertical equilibrium position is proportional to the angular displacement θ. The expression $\tau = -k'\theta$ has the same form as the Hooke's law restoring force for an ideal spring, $F = -kx$. Therefore, we expect the frequency of the back-and-forth movement of the pendulum to be given by an equation analogous to Equation 10.11 $(\omega = 2\pi f = \sqrt{k/m})$. In place of the spring constant k, the constant $k' = mgL$ will appear, and, as usual in rotational motion, in place of the mass m, the moment of inertia I will appear:

$$\omega = 2\pi f = \sqrt{\frac{mgL}{I}} \qquad \text{(small angles only)} \qquad \textbf{(10.15)}$$

The moment of inertia of a particle of mass m, rotating at a radius $r = L$ about an axis, is given by $I = mL^2$ (Equation 9.6). Substituting this expression for I into Equation 10.15 reveals that for a simple pendulum

Simple pendulum $\omega = 2\pi f = \sqrt{\dfrac{g}{L}}$ (small angles only)   **(10.16)**

The mass of the particle has been eliminated algebraically from this expression, so only the length L and the magnitude g of the acceleration due to gravity determine the frequency of a simple pendulum. If the angle of oscillation is large, the pendulum does not exhibit simple harmonic motion, and Equation 10.16 does not apply. Equation 10.16 provides the basis for using a pendulum to keep time, as Example 10 demonstrates.

EXAMPLE 10 | Keeping Time

Figure 10.21 shows a clock that uses a pendulum to keep time. Determine the length of a simple pendulum that will swing back and forth in simple harmonic motion with a period of 1.00 s.

Reasoning   When a simple pendulum is swinging back and forth in simple harmonic motion, its frequency f is given by Equation 10.16 as $f = \dfrac{1}{2\pi}\sqrt{g/L}$, where g is the magnitude of the acceleration due to gravity and L is the length of the pendulum. We also

know from Equation 10.5 that the frequency is the reciprocal of the period T, so $f = 1/T$. Thus, the equation above becomes $1/T = \dfrac{1}{2\pi}\sqrt{g/L}$. We can solve this equation for the length L.

Solution   The length of the pendulum is

$$L = \frac{T^2 g}{4\pi^2} = \frac{(1.00 \text{ s})^2 (9.80 \text{ m/s}^2)}{4\pi^2} = \boxed{0.248 \text{ m}}$$

FIGURE 10.21   This pendulum clock keeps time as the pendulum swings back and forth.

Robert Mathena/Fundamental Photographs

It is not necessary that the object in **Figure 10.20** be a particle at the end of a cable. It may be a rigid extended object, in which case the pendulum is called a **physical pendulum**. For small oscillations, Equation 10.15 still applies, but the moment of inertia I is no longer mL^2. The proper value for the rigid object must be used. (See Section 9.4 for a discussion of moment of inertia.) In addition, the length L for a physical pendulum is the distance between the axis at P and the center of gravity of the object. The next example deals with an important type of physical pendulum.

Analyzing Multiple-Concept Problems

EXAMPLE 11 | BIO The Physics of Walking

When we walk, our legs alternately swing forward about the hip joint as a pivot. In this motion the leg is acting approximately as a physical pendulum. Treating the leg as a thin uniform rod of length 0.80 m, find the time it takes for the leg to swing forward.

Reasoning The time it takes for the leg to swing forward is one-half of the period T of the pendulum motion, which is related to the frequency f of the motion by $T = 1/f$ (Equation 10.5). The frequency of a physical pendulum is given by $f = \frac{1}{2\pi}\sqrt{mgL/I}$ (Equation 10.15), where m and I are, respectively, its mass and moment of inertia, and L is the distance

between the pivot at the hip joint and the center of gravity of the leg. By combining these two relations we will be able to find the time it takes for the leg to swing forward.

Knowns and Unknowns The data for this problem are:

Description	Symbol	Value
Length of leg	L_{leg}	0.80 m
Unknown Variable		
Time for leg to swing forward	t	?

Modeling the Problem

STEP 1 Period and Frequency The time t it takes for the leg to swing forward is one-half the period T, or

$$t = \tfrac{1}{2}T$$

The period, in turn, is related to the frequency f by $T = 1/f$ (see Equation 10.5). Substituting this expression for T into the equation above for t gives Equation 1 at the right. The frequency of oscillation will be evaluated in Step 2.

$$t = \tfrac{1}{2}\left(\tfrac{1}{f}\right) \quad (1)$$

$$\boxed{?}$$

STEP 2 The Frequency of a Physical Pendulum According to Equation 10.15, the frequency of a physical pendulum is given by

$$f = \frac{1}{2\pi}\sqrt{\frac{mgL}{I}} \quad (10.15)$$

Since the leg is being approximated as a thin uniform rod, the distance L between the pivot and the center of gravity is one-half the length L_{leg} of the leg, so $L = \tfrac{1}{2}L_{leg}$. Using this relation in Equation 10.15 gives Equation 2 at the right, which we then substitute into Equation 1, as shown. The moment of inertia I will be evaluated in Step 3.

$$t = \tfrac{1}{2}\left(\tfrac{1}{f}\right) \quad (1)$$

$$\boxed{f = \frac{1}{2\pi}\sqrt{\frac{mg(\tfrac{1}{2}L_{leg})}{I}}} \quad (2)$$

$$\boxed{?}$$

STEP 3 Moment of Inertia The leg is being approximated as a thin uniform rod of length L_{leg} that rotates about an axis that is perpendicular to one end. The moment of inertia I for such an object is given in Table 9.1 as

$$\boxed{I = \tfrac{1}{3}mL_{leg}^2}$$

We substitute this expression for I into Equation 2, as shown in the right column.

$$t = \tfrac{1}{2}\left(\tfrac{1}{f}\right) \quad (1)$$

$$\boxed{f = \frac{1}{2\pi}\sqrt{\frac{mg(\tfrac{1}{2}L_{leg})}{I}}} \quad (2)$$

$$\boxed{I = \tfrac{1}{3}mL_{leg}^2}$$

Solution Algebraically combining the results of the steps above, we have:

$$\overset{\text{STEP 1}}{\downarrow}\overset{\text{STEP 2}}{\downarrow}\qquad\qquad\overset{\text{STEP 3}}{\downarrow}$$

$$t = \tfrac{1}{2}\left(\tfrac{1}{f}\right) = \tfrac{1}{2}\left[\frac{1}{\frac{1}{2\pi}\sqrt{\frac{mg(\tfrac{1}{2}L_{leg})}{I}}}\right] = \tfrac{1}{2}\left[\frac{1}{\frac{1}{2\pi}\sqrt{\frac{mg(\tfrac{1}{2}L_{leg})}{\tfrac{1}{3}mL_{leg}^2}}}\right]$$

$$= \pi\sqrt{\frac{2L_{leg}}{3g}} = \pi\sqrt{\frac{2(0.80\text{ m})}{3(9.80\text{ m/s}^2)}} = \boxed{0.73\text{ s}}$$

Note that the mass m is algebraically eliminated from the final result.

Check Your Understanding

(The answers are given at the end of the book.)

11. Suppose that a grandfather clock (a simple pendulum) is running slowly; that is, the time it takes to complete each cycle is greater than it should be. Should you **(a)** shorten or **(b)** lengthen the pendulum to make the clock keep correct time?

12. Consult **Concept Simulation 10.2** to review the concept that is important here. In principle, the motions of a simple pendulum and an object on an ideal spring can both be used to provide the basic time interval or period used in a clock. Which of the two kinds of clocks becomes more inaccurate when carried to the top of a high mountain?

13. Concept Simulation 10.2 deals with the concept on which this question is based. Suppose you were kidnapped and held prisoner by space invaders in a completely isolated room, with nothing but a watch and a pair of shoes (including shoe laces of known length). How could you determine whether you were on earth or on the moon?

14. Two people are sitting on identical playground swings. One is pulled back 4° from the vertical and the other is pulled back 8°. They are both released at the same instant. Will they both come back to their starting points at the same time? Assume simple-pendulum motion.

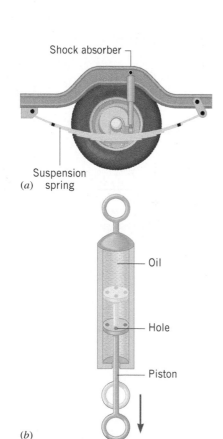

FIGURE 10.22 (*a*) A shock absorber mounted in the suspension system of an automobile and (*b*) a simplified, cutaway view of the shock absorber.

10.5 | Damped Harmonic Motion

In simple harmonic motion, an object oscillates with a constant amplitude, because there is no mechanism for dissipating energy. In reality, however, friction or some other energy-dissipating mechanism is always present. In the presence of energy dissipation, the oscillation amplitude decreases as time passes, and the motion is no longer simple harmonic motion. Instead, it is referred to as **damped harmonic motion**, the decrease in amplitude being called "damping."

THE PHYSICS OF . . . a shock absorber. One widely used application of damped harmonic motion is in the suspension system of an automobile. **Figure 10.22*a*** shows a shock absorber attached to a main suspension spring of a car. A shock absorber is designed to introduce damping forces, which reduce the vibrations associated with a bumpy ride. As part *b* of the drawing shows, a shock absorber consists of a piston in a reservoir of oil. When the piston moves in response to a bump in the road, holes in the piston head permit the piston to pass through the oil. Viscous forces that arise during this movement cause the damping.

Figure 10.23 illustrates the different degrees of damping that can exist. As applied to the example of a car's suspension system, these graphs show the vertical position of the chassis after it has been pulled upward by an amount A_0 at time $t_0 = 0$ s and then released. Part *a* of the figure compares undamped or simple harmonic motion in curve 1 (red) to slightly damped motion in curve 2 (green). In damped harmonic motion, the chassis oscillates with decreasing amplitude and eventually comes to rest. As the degree of damping is increased from curve 2 to curve 3 (gold), the car makes fewer oscillations before coming to a halt.

Figure 10.23*b* shows that as the degree of damping is increased further, there comes a point when the car does not oscillate at all after it is released but, rather, settles directly back to its equilibrium position, as in curve 4 (blue). The smallest degree of damping that completely eliminates the oscillations is termed "critical damping," and the motion is said to be *critically damped*. **Figure 10.23*b*** also shows that the car takes the longest time to return to its equilibrium position in curve 5 (purple), where the degree of damping is above the value for critical damping. When the damping exceeds the critical value, the motion is said to be *overdamped*. In contrast, when the damping

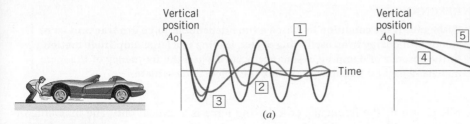

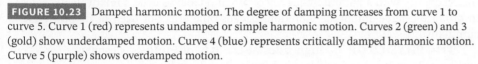

FIGURE 10.23 Damped harmonic motion. The degree of damping increases from curve 1 to curve 5. Curve 1 (red) represents undamped or simple harmonic motion. Curves 2 (green) and 3 (gold) show underdamped motion. Curve 4 (blue) represents critically damped harmonic motion. Curve 5 (purple) shows overdamped motion.

is less than the critical level, the motion is said to be *underdamped* (curves 2 and 3). Typical automobile shock absorbers are designed to produce underdamped motion somewhat like that in curve 3.

Check Your Understanding

(*The answer is given at the end of the book.*)

15. The shock absorbers on a car are badly in need of replacement and introduce very little damping. Does the number of occupants in the car affect the vibration frequency of the car's suspension system?

10.6 Driven Harmonic Motion and Resonance

In damped harmonic motion, a mechanism such as friction dissipates or reduces the energy of an oscillating system, with the result that the amplitude of the motion decreases in time. This section discusses the opposite effect—namely, the increase in amplitude that results when energy is continually added to an oscillating system.

To set an object on an ideal spring into simple harmonic motion, some agent must apply a force that stretches or compresses the spring initially. Suppose that this force is applied at all times, not just for a brief initial moment. The force could be provided, for example, by a person who simply pushes and pulls the object back and forth. The resulting motion is known as **driven harmonic motion**, because the additional force drives or controls the behavior of the object to a large extent. The additional force is identified as the **driving force**.

Figure 10.24 illustrates one particularly important example of driven harmonic motion. Here, the driving force has the same frequency as the spring system and always points in the direction of the object's velocity. The frequency of the spring system is $f = \frac{1}{2\pi} \sqrt{k/m}$ and is called a **natural frequency** because it is the frequency at which the spring system naturally oscillates. Since the driving force and the velocity always have the same direction, positive work is done on the object at all times, and the total mechanical energy of the system increases. As a result, the amplitude of the vibration becomes larger and will increase without limit if there is no damping force to dissipate the energy being added by the driving force. The situation depicted in **Figure 10.24** is known as **resonance**.

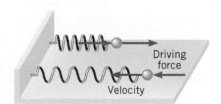

FIGURE 10.24 Resonance occurs when the frequency of the driving force (blue arrows) matches a frequency at which the object naturally vibrates. The red arrows represent the velocity of the object.

> **RESONANCE**
>
> **Resonance is the condition in which a time-dependent force can transmit large amounts of energy to an oscillating object, leading to a large-amplitude motion. In the absence of damping, resonance occurs when the frequency of the force matches a natural frequency at which the object will oscillate.**

The role played by the frequency of a driving force is a critical one. The matching of this frequency with a natural frequency of vibration allows even a relatively weak force to produce a large-amplitude vibration, because the effect of each push–pull cycle is cumulative.

THE PHYSICS OF . . . high tides at the Bay of Fundy. Resonance can occur with any object that can oscillate, and springs need not be involved. The greatest tides in the world occur in the Bay of Fundy, which lies between the Canadian provinces of New Brunswick and Nova Scotia. **Figure 10.25** shows the enormous difference between the water level at high and low tides, a difference that in some locations averages about 15 m. This phenomenon is partly due to resonance. The time, or period, that it takes for the tide to flow into and ebb out of a bay depends on the size of the bay, the topology of the bottom, and the configuration of the shoreline. The ebb and flow of the water in the Bay of Fundy has a period of 12.5 hours, which is very close to the lunar tidal period of 12.42 hours. The tide then "drives" water into and out of the Bay of Fundy at a frequency (once per 12.42 hours) that nearly matches the natural frequency of the bay (once per 12.5 hours). The result is the extraordinary high tide in the bay. (You can create a similar effect in a bathtub full of water by moving back and forth in synchronism with the waves you're causing.)

Andrew J. Martinez/Science Source

Andrew J. Martinez/Science Source

FIGURE 10.25 The Bay of Fundy, Canada, at (*left*) high tide and (*right*) low tide. In some places the water level changes by almost 15 m.

Check Your Understanding

(*The answer is given at the end of the book.*)

16. A car travels at a constant speed over a road that contains a series of equally spaced bumps. The spacing between bumps is d. The mass of the car is m, and the spring constant of the car's suspension springs is k. Because of resonance, a particularly jarring ride results. Ignoring the effect of the car's shock absorbers, derive an expression for the car's speed v in terms of d, m, and k, as well as some numerical constants.

Elastic Deformation

Stretching, Compression, and Young's Modulus

We have seen that a spring returns to its original shape when the force compressing or stretching it is removed. In fact, all materials become distorted in some way when they are squeezed or stretched, and many of them, such as rubber, return to their original shape when the squeezing or stretching is removed. Such materials are said to be "elastic." From an atomic viewpoint, elastic behavior has its origin in the forces that atoms exert on each other, and **Figure 10.26** symbolizes these forces as springs. It is because of these atomic-level "springs" that a material tends to return to its initial shape once the forces that cause the deformation are removed.

The interatomic forces that hold the atoms of a solid together are particularly strong, so considerable force must be applied to stretch or compress a solid object. Experiments have shown that the magnitude of the force can be expressed by the following relation, provided that the amount of stretch or compression is small compared to the original length of the object:

$$F = Y\left(\frac{\Delta L}{L_0}\right)A \qquad (10.17)$$

As **Figure 10.27** shows, F denotes the magnitude of the stretching force applied perpendicularly to the surface at the end, A is the cross-sectional area of the rod, ΔL is the increase in length, and L_0 is the original length. An analogous picture applies in the case of a force that causes compression. The term Y is a proportionality constant called *Young's modulus*, after Thomas Young (1773–1829). Solving Equation 10.17 for Y shows that Young's modulus has units of force per unit area (N/m^2).

> **Problem-Solving Insight** It should be noted that the magnitude of the force in Equation 10.17 is proportional to the fractional increase (or decrease) in length $\Delta L/L_0$, rather than the absolute change in length ΔL.

The magnitude of the force is also proportional to the cross-sectional area A, which need not be circular, but can have any shape (e.g., rectangular).

BIO **THE PHYSICS OF . . . surgical implants.** Table 10.1 reveals that the value of Young's modulus depends on the nature of the material. The values for metals are much larger than those for bone, for example. Equation 10.17 indicates that, for a given force, the material with the greater value of Y undergoes the smaller change in length. This difference between the changes in length is the reason why surgical implants (e.g., artificial hip joints), which are often made from stainless steel or titanium alloys, can lead to chronic deterioration of the bone that is in contact with the implanted prosthesis.

BIO **THE PHYSICS OF . . . bone structure.** Forces that are applied as in **Figure 10.27** and cause stretching are called "tensile" forces, because they create a tension in the material, much like the tension in a rope. Equation 10.17 also applies when the force compresses the material along its length. In this situation, the force is applied in a direction opposite to that shown in **Figure 10.27**, and ΔL stands for the amount by which the original length L_0 decreases. **Table 10.1** indicates, for example, that bone has different values of Young's modulus for compression and tension, the value for tension being greater. Such differences are related to the structure of the material. The solid part of bone consists of collagen fibers (a protein material) distributed throughout hydroxyapatite (a mineral). The collagen acts like the steel rods in reinforced concrete and increases the value of Y for tension relative to the value of Y for compression.

Most solids have Young's moduli that are rather large, reflecting the fact that a large force is needed to change the length of a solid object by even a small amount, as Example 12 illustrates.

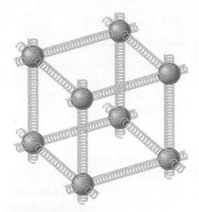

FIGURE 10.26 The forces between atoms act like springs. The atoms are represented by red spheres, and the springs between some atoms have been omitted for clarity.

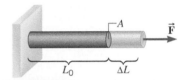

FIGURE 10.27 In this diagram, $\vec{F}$ denotes the stretching force, A the cross-sectional area, L_0 the original length of the rod, and ΔL the amount of stretch.

TABLE 10.1	Values for the Young's Modulus of Solid Materials
Material	**Young's Modulus Y (N/m^2)**
Aluminum	6.9×10^{10}
Bone	
Compression	9.4×10^{9}
Tension	1.6×10^{10}
Brass	9.0×10^{10}
Brick	1.4×10^{10}
Copper	1.1×10^{11}
Mohair	2.9×10^{9}
Nylon	3.7×10^{9}
Pyrex glass	6.2×10^{10}
Steel	2.0×10^{11}
Teflon	3.7×10^{9}
Titanium	1.2×10^{11}
Tungsten	3.6×10^{11}

EXAMPLE 12 | BIO The Physics of Bone Compression

A circus performer supports the combined weight (1080 N) of a number of colleagues (see **Figure 10.28**). Each thighbone (femur) of this performer has a length of 0.55 m and an effective cross-sectional area of 7.7×10^{-4} m^2. Determine the amount by which each thighbone compresses under the extra weight.

Reasoning The additional weight supported by each thighbone is $F = \frac{1}{2}(1080 \text{ N}) = 540$ N, and **Table 10.1** indicates that Young's modulus for bone compression is 9.4×10^9 N/m^2. Since the length and cross-sectional area of the thighbone are also known, we can use Equation 10.17 to find the amount by which the additional weight compresses the thighbone.

Solution The amount of compression ΔL of each thighbone is

$$\Delta L = \frac{FL_0}{YA} = \frac{(540 \text{ N})(0.55 \text{ m})}{(9.4 \times 10^9 \text{ N/m}^2)(7.7 \times 10^{-4} \text{ m}^2)}$$

$$= \boxed{4.1 \times 10^{-5} \text{ m}}$$

This is a very small change, the fractional decrease being $\Delta L/L_0 = 0.000\ 075$.

FIGURE 10.28 The entire weight of the balanced group is supported by the legs of the performer who is lying on his back.

UWE LEIN/AP Images

Just as Example 12 showed us that the fractional change in the length of the human femur changes very little in response to a large compressive force, Example 13 now illustrates that human ligaments both are strong and can endure relatively large elastic deformations.

EXAMPLE 13 | BIO The Iliofemoral Ligament

Also known as the "Y-ligament," the iliofemoral ligament in the hip joins the ilium to the femur (see **Figure 10.29**). It is the strongest ligament in the human body, and can withstand stretching forces exceeding 770.0 lb! It has a Young's modulus of 5.00×10^7 N/m^2 and an effective cross-sectional area of about 5.50 cm^2 in an average male adult. Assuming a uniform cross-sectional area and an unstrained length of 9.40 cm, determine **(a)** how far the ligament stretches beyond its unstrained length when subjected to a 770.0-lb stretching force, and **(b)** the percent change in its length under this load.

Reasoning (a) We are given the force F, the unstrained length L_0, the effective cross-sectional area A, and Young's modulus Y of the ligament material. Assuming that the effective cross-sectional area is uniform, and that the force is a tensile force (i.e., a "stretching" force), we can calculate the change in length (ΔL) using Equation 10.17. **(b)** To find the percent change in length, we need to divide the value of ΔL calculated in part (a) by the unstrained length L_0, and multiply the result by 100.

Solution (a) This is an elastic deformation problem where the deformation results from a stretching force. We can therefore apply Equation 10.17, but we first have to convert the force F into units of N, L_0 into m, and the cross-sectional area A into m^2:

$$F = (770.0 \text{ lb})\left(4.45 \frac{\text{N}}{\text{lb}}\right) = 3427 \text{ N}$$

$$L_0 = (9.40 \text{ cm})\left(\frac{\text{m}}{100 \text{ cm}}\right) = 9.40 \times 10^{-2} \text{ m}$$

and

$$A = (5.50 \text{ cm}^2)\left(\frac{1 \text{ m}}{100 \text{ cm}}\right)\left(\frac{1 \text{ m}}{100 \text{ cm}}\right) = 5.50 \times 10^{-4} \text{ m}^2$$

Solving Equation 10.17 for ΔL and plugging in the known values gives

$$\Delta L = \frac{FL_0}{YA} = \frac{(3427 \text{ N})(0.0940 \text{ m})}{(5.00 \times 10^7 \text{ N/m}^2)(5.50 \times 10^{-4} \text{ m}^2)} = \boxed{0.0117 \text{ m}}$$

or about 1.17 cm.

(b) The percent change is given by the change in length divided by the unstrained length, multiplied by 100:

$$\% \text{ change in length} = \left(\frac{\Delta L}{L_0}\right) \times 100 = \left(\frac{0.0117 \text{ m}}{0.0940 \text{ m}}\right) \times 100 = \boxed{12.4\%}$$

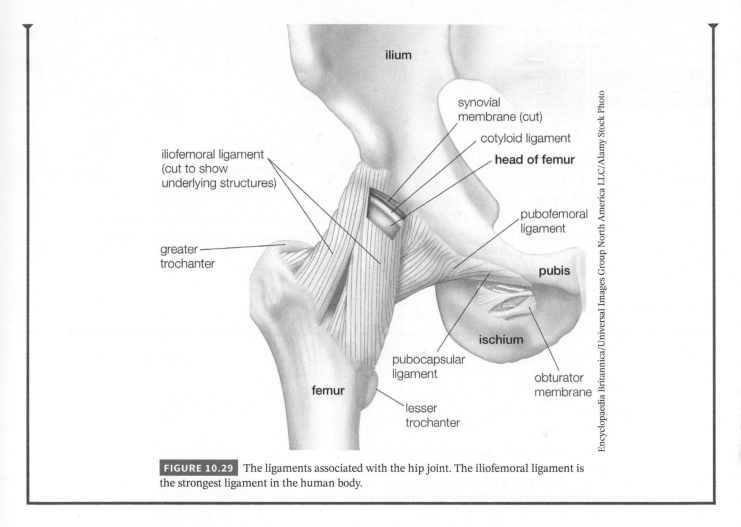

FIGURE 10.29 The ligaments associated with the hip joint. The iliofemoral ligament is the strongest ligament in the human body.

Shear Deformation and the Shear Modulus

It is possible to deform a solid object in a way other than by stretching or compressing it. For instance, place a book on a rough table and push on the top cover, as in **Figure 10.30a**. Notice that the top cover, and the pages below it, become shifted relative to the stationary bottom cover. The resulting deformation is called a **shear deformation** and occurs because of the combined effect of the force $\vec{F}$ applied (by the hand) to the top of the book and the force $-\vec{F}$ applied (by the table) to the bottom of the book. In general, shearing forces cause a solid object to change its shape. In **Figure 10.30** the directions of the forces are parallel to the covers of the book, each of which has an area A, as illustrated in part b of the drawing. These two forces have equal magnitudes, but opposite directions, so the book remains in equilibrium. Equation 10.18 gives the magnitude F of the force needed to produce an amount of shear ΔX for an object with thickness L_0:

$$F = S\left(\frac{\Delta X}{L_0}\right)A \qquad (10.18)$$

This equation is very similar to Equation 10.17. The constant of proportionality S is called the **shear modulus** and, like Young's modulus, has units of force per unit area (N/m^2). The value of S depends on the nature of the material, and **Table 10.2** gives some representative values. Example 14 illustrates how to determine the shear modulus of a familiar dessert.

TABLE 10.2	Values for the Shear Modulus of Solid Materials
Material	**Shear Modulus S (N/m²)**
Aluminum	2.4×10^{10}
Bone	1.2×10^{10}
Brass	3.5×10^{10}
Copper	4.2×10^{10}
Lead	5.4×10^{9}
Nickel	7.3×10^{10}
Steel	8.1×10^{10}
Tungsten	1.5×10^{11}

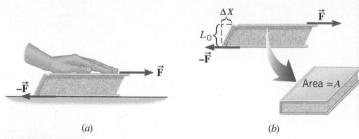

(a)   (b)

FIGURE 10.30 (a) An example of a shear deformation. The shearing forces $\vec{F}$ and $-\vec{F}$ are applied parallel to the top and bottom covers of the book. (b) The shear deformation is ΔX. The area of each cover is A, and the thickness of the book is L_0.

EXAMPLE 14 | J-E-L-L-O

A block of Jell-O is resting on a plate. **Figure 10.31a** gives the dimensions of the block. You are bored, impatiently waiting for dinner, and push tangentially across the top surface with a force of $F = 0.45$ N, as in part b of the drawing. The top surface moves a distance $\Delta X = 6.0 \times 10^{-3}$ m relative to the bottom surface. Use this idle gesture to measure the shear modulus of Jell-O.

Reasoning The finger applies a force that is parallel to the top surface of the Jell-O block. The shape of the block changes, because the top surface moves a distance ΔX relative to the bottom surface. The magnitude of the force required to produce this change in shape is given by Equation 10.18 as $F = S(\Delta X/L_0)A$. We know the values for all the variables in this relation except S, which, therefore, can be determined.

Solution Solving Equation 10.18 for the shear modulus S, we find that $S = FL_0/(A\ \Delta X)$, where $A = (0.070$ m$)(0.070$ m$)$ is the area of the top surface, and $L_0 = 0.030$ m is the thickness of the block:

$$S = \frac{FL_0}{A\ \Delta X} = \frac{(0.45\ \text{N})(0.030\ \text{m})}{(0.070\ \text{m})(0.070\ \text{m})(6.0 \times 10^{-3}\text{m})} = \boxed{460\ \text{N/m}^2}$$

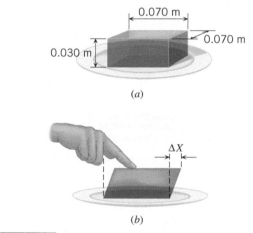

(a)

(b)

FIGURE 10.31 (a) A block of Jell-O and (b) a shearing force applied to it.

Jell-O can be deformed easily, so its shear modulus is significantly less than that of a more rigid material like steel (see **Table 10.2**).

Although Equations 10.17 and 10.18 are algebraically similar, they refer to different kinds of deformations. The tensile force in **Figure 10.27** is perpendicular to the surface whose area is A, whereas the shearing force in **Figure 10.30** is parallel to that surface. Furthermore, the ratio $\Delta L/L_0$ in Equation 10.17 is different from the ratio $\Delta X/L_0$ in Equation 10.18. The distances ΔL and L_0 are parallel, whereas ΔX and L_0 are perpendicular. Young's modulus refers to a *change in length* of one dimension of a solid object as a result of tensile or compressive forces. The shear modulus refers to a *change in shape* of a solid object as a result of shearing forces.

Volume Deformation and the Bulk Modulus

When a compressive force is applied along one dimension of a solid, the length of that dimension decreases. It is also possible to apply compressive forces so that the size of every dimension (length, width, and depth) decreases, leading to a decrease in volume, as **Figure 10.32** illustrates. This kind of overall compression occurs, for example, when an object is submerged in a liquid, and the liquid presses inward everywhere on the object. The forces acting in such situations are applied perpendicular to every surface, and it is more convenient to speak of the perpendicular force per unit area, rather than the amount of any one force in particular. The magnitude of the perpendicular force per unit area is called the **pressure P**.

DEFINITION OF PRESSURE

The pressure P is the magnitude F of the force acting perpendicular to a surface divided by the area A over which the force acts:

$$P = \frac{F}{A} \qquad (10.19)$$

Pressure is a scalar, not a vector, quantity.

SI Unit of Pressure: N/m^2 = pascal (Pa)

Equation 10.19 indicates that the SI unit for pressure is the unit of force divided by the unit of area, or newton/meter^2 (N/m^2). This unit of pressure is often referred to as a *pascal* (Pa), named after the French scientist Blaise Pascal (1623–1662).

Suppose we change the pressure on an object by an amount ΔP, where ΔP represents the final pressure P minus the initial pressure P_0: $\Delta P = P - P_0$. Because of this change in pressure, the volume of the object changes by an amount $\Delta V = V - V_0$, where V and V_0 are the final and initial volumes, respectively. Such a pressure change occurs, for example, when a swimmer dives deeper into the water. Experiment reveals that the change ΔP in pressure needed to change the volume by an amount ΔV is directly proportional to the fractional change $\Delta V/V_0$ in the volume:

$$\Delta P = -B\left(\frac{\Delta V}{V_0}\right) \qquad (10.20)$$

This relation is analogous to Equations 10.17 and 10.18, except that the area A in those equations does not appear here explicitly; the area is already taken into account by the concept of pressure (magnitude of the force per unit area). The proportionality constant B is known as the **bulk modulus**. The minus sign occurs because an increase in pressure (ΔP positive) always creates a decrease in volume (ΔV negative), and B is given as a positive quantity. Like Young's modulus and the shear modulus, the bulk modulus has units of force per unit area (N/m^2), and its value depends on the nature of the material. **Table 10.3** gives representative values of the bulk modulus.

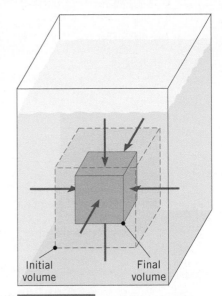

FIGURE 10.32 The arrows denote the forces that push perpendicularly on every surface of an object immersed in a liquid. The magnitude of the force per unit area is the pressure. When the pressure increases, the volume of the object decreases.

Initial volume   Final volume

TABLE 10.3	Values for the Bulk Modulus of Solid and Liquid Materials
Material	**Bulk Modulus B [N/m^2 (=Pa)]**
Solids	
Aluminum	7.1×10^{10}
Brass	6.7×10^{10}
Copper	1.3×10^{11}
Diamond	4.43×10^{11}
Lead	4.2×10^{10}
Nylon	6.1×10^{9}
Osmium	4.62×10^{11}
Pyrex glass	2.6×10^{10}
Steel	1.4×10^{11}
Liquids	
Ethanol	8.9×10^{8}
Oil	1.7×10^{9}
Water	2.2×10^{9}

Check Your Understanding

(The answers are given at the end of the book.)

17. Young's modulus for steel is greater than that for a particular unknown material. What does this mean about how these materials compress when used in construction? **(a)** Steel compresses much more easily than the unknown material does. **(b)** The unknown material compresses more easily than steel does. **(c)** Young's modulus has nothing to do with compression, so not enough information is given for an answer.

18. Two rods are made from the same material. One has a circular cross section, and the other has a square cross section. The circle just fits within the square. When the same force is applied to stretch these rods, they each stretch by the same amount. Which rod, if either, is longer?

19. A trash compactor crushes empty aluminum cans, thereby reducing the total volume, so that $\Delta V/V_0 = -0.75$ in Equation 10.20. Can the value given in **Table 10.3** for the bulk modulus of aluminum be used to calculate the change ΔP in pressure generated in the trash compactor?

20. Both sides of the relation $F = S(\Delta X/L_0)A$ (Equation 10.18) can be divided by the area A to give F/A on the left side. Can this F/A term be called a pressure, such as the pressure that appears in $\Delta P = -B(\Delta V/V_0)$ (Equation 10.20)?

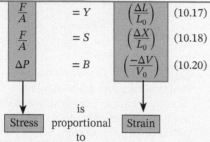

TABLE 10.4 Stress and Strain Relations for Elastic Behavior

$\dfrac{F}{A}$	$= Y$	$\left(\dfrac{\Delta L}{L_0}\right)$	(10.17)
$\dfrac{F}{A}$	$= S$	$\left(\dfrac{\Delta X}{L_0}\right)$	(10.18)
ΔP	$= B$	$\left(\dfrac{-\Delta V}{V_0}\right)$	(10.20)

Stress   is proportional to   Strain

10.8 Stress, Strain, and Hooke's Law

Equations 10.17, 10.18, and 10.20 specify the amount of force needed for a given amount of elastic deformation, and they are repeated in **Table 10.4** to emphasize their common features. The left side of each equation is the magnitude of the force per unit area required to cause an elastic deformation. In general, the ratio of the magnitude of the force to the area is called the **stress**. The right side of each equation involves the change in a quantity (ΔL, ΔX, or ΔV) divided by a quantity (L_0 or V_0) relative to which the change is compared. The terms $\Delta L/L_0$, $\Delta X/L_0$, and $\Delta V/V_0$ are unitless ratios, and each is referred to as the **strain** that results from the applied stress. In the case of stretch and compression, the strain is the fractional change in length, whereas in volume deformation it is the fractional change in volume. In shear deformation the strain refers to a change in shape of the object. Experiments show that these three equations, with constant values for Young's modulus, the shear modulus, and the bulk modulus, apply to a wide range of materials. Therefore, stress and strain are directly proportional to one another, a relationship first discovered by Robert Hooke (1635–1703) and now referred to as **Hooke's law**.

> **HOOKE'S LAW FOR STRESS AND STRAIN**
>
> **Stress is directly proportional to strain.**
>
> **SI Unit of Stress:** newton per square meter (N/m²) = pascal (Pa)
>
> **SI Unit of Strain:** Strain is a unitless quantity.

In reality, materials obey Hooke's law only up to a certain limit, as **Figure 10.33** shows. As long as stress remains proportional to strain, a plot of stress versus strain is a straight line. The point on the graph where the material begins to deviate from straight-line behavior is called the "proportionality limit." Beyond the proportionality limit stress and strain are no longer directly proportional. However, if the stress does not exceed the "elastic limit" of the material, the object will return to its original size and shape once the stress is removed. The "elastic limit" is the point beyond which the object no longer returns to its original size and shape when the stress is removed; the object remains permanently deformed.

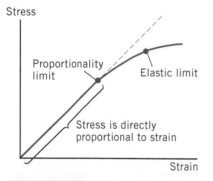

FIGURE 10.33 Hooke's law (stress is directly proportional to strain) is valid only up to the proportionality limit of a material. Beyond this limit, Hooke's law no longer applies. Beyond the elastic limit, the material remains deformed even when the stress is removed.

Check Your Understanding

(*The answer is given at the end of the book.*)

21. The block in **CYU Figure 10.3** rests on the ground. Which face—A, B, or C—experiences the largest stress and which face experiences the smallest stress when the block is resting on it?

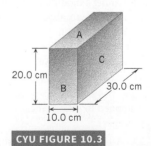

CYU FIGURE 10.3

EXAMPLE 15 | BIO The Physics of the Mechanical Properties of Bone

Human bones also follow Hooke's Law for Stress and Strain, as indicated in Figure 10.34. This is essentially Figure 10.33, where we plot the stress versus strain (under compression) specifically for a human femur—the largest bone in the body. In Section 6.9, we plotted the average force versus displacement for a *variable* force. The area under the curve was equal to the work done by the force. Now let's apply a similar analysis to Figure 10.34. Calculate the area under the curve where Hooke's law is valid (shaded region), and use unit analysis to describe what this area represents.

Reasoning As we did in Section 6.9, we need to calculate the area of the shaded region. Since we are to consider only the range where Hooke's law is valid, the shaded region will have the shape of a triangle whose area is equal to $\frac{1}{2}$(base × height).

Solution Using the data in Figure 10.34 we calculate the area of the shaded region:

$$\text{Area} = \tfrac{1}{2}(\text{base} \times \text{height}) = \tfrac{1}{2}(0.010)(200 \times 10^6 \text{ Pa}) = \boxed{1.00 \times 10^6 \text{ Pa}}$$

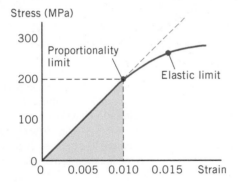

FIGURE 10.34 Stress versus strain for a human femur bone. Hooke's Law is obeyed up to the proportionality limit.

Since stress has the units of pressure, and strain is unitless, our answer also has units of pressure, or Pa. However, we can gain a greater understanding of the significance of the area under the curve if we write pressure as the stress $P = \frac{F}{A}$ and the strain as $\frac{\Delta L}{L_0}$. Now when we calculate the area under the curve in general, we have:

$$\text{Area} = \tfrac{1}{2}\left(\frac{F}{A}\right)\left(\frac{\Delta L}{L_0}\right)$$

Here, the force is given by Hooke's law:

$$F = kx = k\,\Delta L$$

If we substitute this into our equation for the area above, we get the following:

$$\text{Area} = \frac{\tfrac{1}{2}k(\Delta L)^2}{AL_0}$$

The numerator is equivalent to Equation 10.13 and represents the elastic potential energy. The denominator is equal to an area times a length, which has the units of volume. Therefore, in general, the area under the curve is equivalent to energy per unit volume, or energy density:

$$\text{Area} = \frac{\text{PE [J]}}{\text{Vol [m}^3]}$$

For our simple example above, the result indicates that one million J/m^3 of elastic potential energy is stored in the bone under compression. Of course, the stored energy in a human femur is much smaller than one million joules, as the change in volume under compression is much smaller than 1 m^3. However, if the energy delivered to the bone is greater than the energy it can absorb, the bone will fail, and a fracture will occur.

Concept Summary

10.1 The Ideal Spring and Simple Harmonic Motion The force that must be applied to stretch or compress an ideal spring is given by Equation 10.1, where k is the spring constant and x is the displacement of the spring from its unstrained length.

$$F_x^{\text{Applied}} = kx \qquad (10.1)$$

A spring exerts a restoring force on an object attached to the spring. The restoring force F_x produced by an ideal spring is given by Equation 10.2, where the minus sign indicates that the restoring force points opposite to the displacement of the spring.

$$F_x = -kx \qquad (10.2)$$

Simple harmonic motion is the oscillatory motion that occurs when a restoring force of the form $F_x = -kx$ acts on an object. A graphical record of position versus time for an object in simple harmonic motion is sinusoidal. The amplitude A of the motion is the maximum distance that the object moves away from its equilibrium position.

10.2 Simple Harmonic Motion and the Reference Circle The period T of simple harmonic motion is the time required to complete one cycle of the motion, and the frequency f is the number of cycles per second that occurs. Frequency and period are related, according to Equation 10.5. The frequency f (in Hz) is related to the angular frequency ω (in rad/s), according to Equation 10.6.

$$f = \frac{1}{T} \qquad (10.5)$$

$$\omega = 2\pi f \qquad (\omega \text{ in rad/s}) \qquad (10.6)$$

The maximum speed $v_{\max}$ of an object moving in simple harmonic motion is given by Equation 10.8, where A is the amplitude of the motion.

$$v_{\max} = A\omega \qquad (\omega \text{ in rad/s}) \qquad (10.8)$$

The maximum acceleration a_{max} of an object moving in simple harmonic motion is given by Equation 10.10.

$$a_{max} = A\omega^2 \qquad (\omega \text{ in rad/s}) \qquad \textbf{(10.10)}$$

The angular frequency of simple harmonic motion is given by Equation 10.11.

$$\omega = \sqrt{\frac{k}{m}} \qquad (\omega \text{ in rad/s}) \qquad \textbf{(10.11)}$$

10.3 Energy and Simple Harmonic Motion The elastic potential energy of an object attached to an ideal spring is given by Equation 10.13. The total mechanical energy E of such a system is the sum of its translational and rotational kinetic energies, gravitational potential energy, and elastic potential energy, according to Equation 10.14. If external nonconservative forces like friction do no net work, the total mechanical energy of the system is conserved, as indicated by Equation 1.

$$PE_{elastic} = \tfrac{1}{2}kx^2 \qquad \textbf{(10.13)}$$

$$E = \tfrac{1}{2}mv^2 + \tfrac{1}{2}I\omega^2 + mgh + \tfrac{1}{2}kx^2 \qquad \textbf{(10.14)}$$

$$E_f = E_0 \qquad \textbf{(1)}$$

10.4 The Pendulum A simple pendulum is a particle of mass m attached to a frictionless pivot by a cable whose length is L and whose mass is negligible. The small-angle ($\leq 10°$) back-and-forth swinging of a simple pendulum is simple harmonic motion, but large-angle movement is not. The frequency f of the small-angle motion is given by Equation 10.16.

$$2\pi f = \sqrt{\frac{g}{L}} \qquad (\text{small angles only}) \qquad \textbf{(10.16)}$$

A physical pendulum consists of a rigid object, with moment of inertia I and mass m, suspended from a frictionless pivot. For small-angle displacements, the frequency f of simple harmonic motion for a physical pendulum is given by Equation 10.15, where L is the distance between the axis of rotation and the center of gravity of the rigid object.

$$2\pi f = \sqrt{\frac{mgL}{I}} \qquad (\text{small angles only}) \qquad \textbf{(10.15)}$$

10.5 Damped Harmonic Motion Damped harmonic motion is motion in which the amplitude of oscillation decreases as time passes. Critical damping is the minimum degree of damping that eliminates any oscillations in the motion as the object returns to its equilibrium position.

10.6 Driven Harmonic Motion and Resonance Driven harmonic motion occurs when a driving force acts on an object along with the restoring force. Resonance is the condition under which the driving force can transmit large amounts of energy to an oscillating object, leading to large-amplitude motion. In the absence of damping, resonance occurs when the frequency of the driving force matches a natural frequency at which the object oscillates.

10.7 Elastic Deformation One type of elastic deformation is stretch and compression. The magnitude F of the force required to stretch or compress an object of length L_0 and cross-sectional area A by an amount ΔL (see **Figure 10.27**) is given by Equation 10.17, where Y is a constant called Young's modulus.

$$F = Y\left(\frac{\Delta L}{L_0}\right)A \qquad \textbf{(10.17)}$$

Another type of elastic deformation is shear. The magnitude F of the shearing force required to create an amount of shear ΔX for an object of thickness L_0 and cross-sectional area A is (see **Figure 10.30**) given by Equation 10.18, where S is a constant called the shear modulus.

$$F = S\left(\frac{\Delta X}{L_0}\right)A \qquad \textbf{(10.18)}$$

A third type of elastic deformation is volume deformation, which has to do with pressure. The pressure P is the magnitude F of the force acting perpendicular to a surface divided by the area A over which the force acts, according to Equation 10.19. The SI unit for pressure is N/m^2, a unit known as a pascal (Pa): 1 Pa = 1 N/m^2. The change ΔP in pressure needed to change the volume V_0 of an object by an amount ΔV (see **Figure 10.32**) is given by Equation 10.20, where B is a constant known as the bulk modulus.

$$P = \frac{F}{A} \qquad \textbf{(10.19)}$$

$$\Delta P = -B\left(\frac{\Delta V}{V_0}\right) \qquad \textbf{(10.20)}$$

10.8 Stress, Strain, and Hooke's Law Stress is the magnitude of the force per unit area applied to an object and causes strain. For stretch/compression, the strain is the fractional change $\Delta L/L_0$ in length. For shear, the strain reflects the change in shape of the object and is given by $\Delta X/L_0$ (see **Figure 10.30**). For volume deformation, the strain is the fractional change in volume $\Delta V/V_0$. Hooke's law states that stress is directly proportional to strain.

Focus on Concepts

Online

Additional questions are available for assignment in WileyPLUS.

Section 10.1 The Ideal Spring and Simple Harmonic Motion

1. Which one of the following graphs correctly represents the restoring force F of an ideal spring as a function of the displacement x of the spring from its unstrained length?

(a)   (b)   (c)   (d)   (e)

QUESTION 1

Section 10.2   Simple Harmonic Motion and the Reference Circle

2. You have two springs. One has a greater spring constant than the other. You also have two objects, one with a greater mass than the other. Which object should be attached to which spring, so that the resulting spring–object system has the greatest possible period of oscillation? (**a**) The object with the greater mass should be attached to the spring with the greater spring constant. (**b**) The object with the greater mass should be attached to the spring with the smaller spring constant. (**c**) The object with the smaller mass should be attached to the spring with the smaller spring constant. (**d**) The object with the smaller mass should be attached to the spring with the greater spring constant.

3. An object is oscillating in simple harmonic motion with an amplitude A and an angular frequency ω. What should you do to increase the maximum speed of the motion? (**a**) Reduce both A and ω by 10%. (**b**) Increase A by 10% and reduce ω by 10%. (**c**) Reduce A by 10% and increase ω by 10%. (**d**) Increase both A and ω by 10%.

Section 10.3   Energy and Simple Harmonic Motion

4. The kinetic energy of an object attached to a horizontal ideal spring is denoted by KE and the elastic potential energy by PE. For the simple harmonic motion of this object the maximum kinetic energy and the maximum elastic potential energy during an oscillation cycle are KE_{max} and PE_{max}, respectively. In the absence of friction, air resistance, and any other nonconservative forces, which of the following equations applies to the object–spring system?

A. KE + PE = constant
B. $KE_{max} = PE_{max}$
(**a**) A, but not B (**b**) B, but not A (**c**) A and B (**d**) Neither A nor B

5. A block is attached to a horizontal spring. On top of this block rests another block. The two-block system slides back and forth in simple harmonic motion on a frictionless horizontal surface. At one extreme end of the oscillation cycle, where the blocks come to a momentary halt before reversing the direction of their motion, the top block is suddenly lifted vertically upward, without changing the zero velocity of the bottom block. The simple harmonic motion then continues. What happens to the amplitude and the angular frequency of the ensuing motion? (**a**) The amplitude remains the same, and the angular frequency increases. (**b**) The amplitude increases, and the angular frequency remains the same. (**c**) Both the amplitude and the angular frequency remain the same. (**d**) Both the amplitude and the angular frequency decrease. (**e**) Both the amplitude and the angular frequency increase.

Section 10.4   The Pendulum

6. Five simple pendulums are shown in the drawings. The lengths of the pendulums are drawn to scale, and the masses are either m or $2m$, as shown. Which pendulum has the smallest angular frequency of oscillation?   (**a**) A   (**b**) B   (**c**) C   (**d**) D   (**e**) E

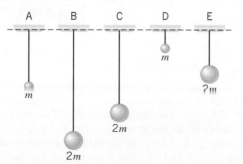

QUESTION 6

Section 10.5   Damped Harmonic Motion

7. An object on a spring is oscillating in simple harmonic motion. Suddenly, friction appears and causes the energy of the system to be dissipated. The system now exhibits _____. (**a**) driven harmonic motion (**b**) Hooke's-law type of motion (**c**) damped harmonic motion

Section 10.6   Driven Harmonic Motion and Resonance

8. An external force (in addition to the spring force) is continually applied to an object of mass m attached to a spring that has a spring constant k. The frequency of this external force is such that resonance occurs. Then the frequency of this external force is doubled, and the force is applied to one of the spring systems shown in the drawing. With which system would resonance occur?   (**a**) A   (**b**) B   (**c**) C   (**d**) D   (**e**) E

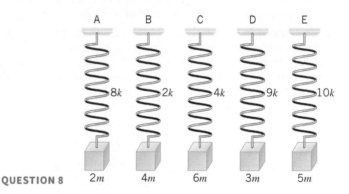

QUESTION 8

Section 10.7   Elastic Deformation

9. Drawings A and B show two cylinders that are identical in all respects, except that one is hollow. Identical forces are applied to each cylinder in order to stretch them. Which cylinder, if either, stretches more? (**a**) A and B both stretch by the same amount. (**b**) A stretches more than B. (**c**) B stretches more than A. (**d**) Insufficient information is given for an answer.

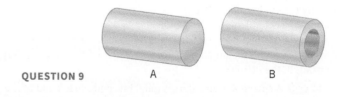

QUESTION 9

Section 10.8   Stress, Strain, and Hooke's Law

10. A material has a shear modulus of $5.0 \times 10^9 \, \text{N/m}^2$. A shear stress of $8.5 \times 10^6 \, \text{N/m}^2$ is applied to a piece of the material. What is the resulting shear strain?

Problems

Additional questions are available for assignment in WileyPLUS.

SSM Student Solutions Manual

MMH Problem-solving help

GO Guided Online Tutorial

V-HINT Video Hints

CHALK Chalkboard Videos

BIO Biomedical application

E Easy

M Medium

H Hard

WS Worksheet

T Team Problem

Section 10.1 The Ideal Spring and Simple Harmonic Motion

1. **E** **SSM** A hand exerciser utilizes a coiled spring. A force of 89.0 N is required to compress the spring by 0.0191 m. Determine the force needed to compress the spring by 0.0508 m.

2. **E** The drawing shows three identical springs hanging from the ceiling. Nothing is attached to the first spring, whereas a 4.50-N block hangs from the second spring. A block of unknown weight hangs from the third spring. From the drawing, determine (a) the spring constant (in N/m) and (b) the weight of the block hanging from the third spring.

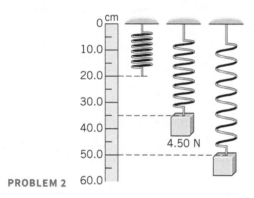

PROBLEM 2

3. **E** In a room that is 2.44 m high, a spring (unstrained length = 0.30 m) hangs from the ceiling. A board whose length is 1.98 m is attached to the free end of the spring. The board hangs straight down, so that its 1.98-m length is perpendicular to the floor. The weight of the board (104 N) stretches the spring so that the lower end of the board just extends to, but does not touch, the floor. What is the spring constant of the spring?

4. **E** **GO** A spring lies on a horizontal table, and the left end of the spring is attached to a wall. The other end is connected to a box. The box is pulled to the right, stretching the spring. Static friction exists between the box and the table, so when the spring is stretched only by a small amount and the box is released, the box does not move. The mass of the box is 0.80 kg, and the spring has a spring constant of 59 N/m. The coefficient of static friction between the box and the table on which it rests is $\mu_s = 0.74$. How far can the spring be stretched from its unstrained position without the box moving when it is released?

5. **E** **SSM** A person who weighs 670 N steps onto a spring scale in the bathroom, and the spring compresses by 0.79 cm. (a) What is the spring constant? (b) What is the weight of another person who compresses the spring by 0.34 cm?

6. **E** A spring ($k = 830$ N/m) is hanging from the ceiling of an elevator, and a 5.0-kg object is attached to the lower end. By how much does the spring stretch (relative to its unstrained length) when the elevator is accelerating upward at $a = 0.60$ m/s^2?

7. **E** **CHALK** A 0.70-kg block is hung from and stretches a spring that is attached to the ceiling. A second block is attached to the first one, and the amount that the spring stretches from its unstrained length triples. What is the mass of the second block?

8. **M** **V-HINT** A uniform 1.4-kg rod that is 0.75 m long is suspended at rest from the ceiling by two springs, one at each end of the rod. Both springs hang straight down from the ceiling. The springs have identical lengths when they are unstretched. Their spring constants are 59 N/m and 33 N/m. Find the angle that the rod makes with the horizontal.

9. **M** **SSM** In 0.750 s, a 7.00-kg block is pulled through a distance of 4.00 m on a frictionless horizontal surface, starting from rest. The block has a constant acceleration and is pulled by means of a horizontal spring that is attached to the block. The spring constant of the spring is 415 N/m. By how much does the spring stretch?

10. **M** **GO** Review Conceptual Example 2 as an aid in solving this problem. An object is attached to the lower end of a 100-coil spring that is hanging from the ceiling. The spring stretches by 0.160 m. The spring is cut into two identical springs of 50 coils each. As the drawing shows, each spring is attached between the ceiling and the object. By how much does each spring stretch?

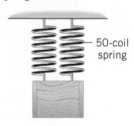

PROBLEM 10

11. **M** **CHALK** **SSM** A small ball is attached to one end of a spring that has an unstrained length of 0.200 m. The spring is held by the other end, and the ball is whirled around in a horizontal circle at a speed of 3.00 m/s. The spring remains nearly parallel to the ground during the motion and is observed to stretch by 0.010 m. By how much would the spring stretch if it were attached to the ceiling and the ball allowed to hang straight down, motionless?

12. **M** **MMH** To measure the static friction coefficient between a 1.6-kg block and a vertical wall, the setup shown in the drawing is used. A spring (spring constant = 510 N/m) is attached to the block. Someone pushes on the end of the spring in a direction perpendicular to the wall until the block does not slip downward. The spring is compressed by 0.039 m. What is the coefficient of static friction?

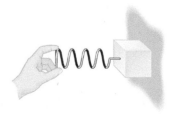

PROBLEM 12

13. **H** A 30.0-kg block is resting on a flat horizontal table. On top of this block is resting a 15.0-kg block, to which a horizontal spring is attached, as the drawing illustrates. The spring constant of the spring is 325 N/m. The coefficient of kinetic friction between the lower block and the table is 0.600, and the coefficient of static friction between the two blocks is 0.900. A horizontal force $\vec{F}$ is applied

to the lower block as shown. This force is increasing in such a way as to keep the blocks moving at a *constant speed*. At the point where the upper block begins to slip on the lower block, determine (a) the amount by which the spring is compressed and (b) the magnitude of the force $\vec{\mathbf{F}}$.

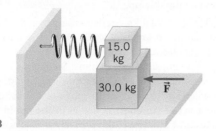

PROBLEM 13

14. 🅗 A 15.0-kg block rests on a horizontal table and is attached to one end of a massless, horizontal spring. By pulling horizontally on the other end of the spring, someone causes the block to accelerate uniformly and reach a speed of 5.00 m/s in 0.500 s. In the process, the spring is stretched by 0.200 m. The block is then pulled at a *constant speed* of 5.00 m/s, during which time the spring is stretched by only 0.0500 m. Find (a) the spring constant of the spring and (b) the coefficient of kinetic friction between the block and the table.

Section 10.2 Simple Harmonic Motion and the Reference Circle

15. 🅔 The fan blades on a jet engine make one thousand revolutions in a time of 50.0 ms. Determine (a) the period (in seconds) and (b) the frequency (in Hz) of the rotational motion. (c) What is the angular frequency of the blades?

16. 🅔 **MMH** A block of mass $m = 0.750$ kg is fastened to an unstrained horizontal spring whose spring constant is $k = 82.0$ N/m. The block is given a displacement of $+0.120$ m, where the $+$ sign indicates that the displacement is along the $+x$ axis, and then released from rest. (a) What is the force (magnitude and direction) that the spring exerts on the block just before the block is released? (b) Find the angular frequency ω of the resulting oscillatory motion. (c) What is the maximum speed of the block? (d) Determine the magnitude of the maximum acceleration of the block.

17. 🅔 **MMH** An 0.80-kg object is attached to one end of a spring, as in Figure 10.5, and the system is set into simple harmonic motion. The displacement x of the object as a function of time is shown in the drawing. With the aid of these data, determine (a) the amplitude A of the motion, (b) the angular frequency ω, (c) the spring constant k, (d) the speed of the object at $t = 1.0$ s, and (e) the magnitude of the object's acceleration at $t = 1.0$ s.

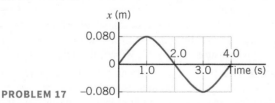

PROBLEM 17

18. 🅔 Refer to Conceptual Example 2 as an aid in solving this problem. A 100-coil spring has a spring constant of 420 N/m. It is cut into four shorter springs, each of which has 25 coils. One end of a 25-coil spring is attached to a wall. An object of mass 46 kg is attached to the other end of the spring, and the system is set into horizontal oscillation. What is the angular frequency of the motion?

19. 🅜 **GO** Objects of equal mass are oscillating up and down in simple harmonic motion on two different vertical springs. The spring

constant of spring 1 is 174 N/m. The motion of the object on spring 1 has twice the amplitude as the motion of the object on spring 2. The magnitude of the maximum velocity is the same in each case. Find the spring constant of spring 2.

20. 🅜 **SSM** **MMH** **CHALK** A spring stretches by 0.018 m when a 2.8-kg object is suspended from its end. How much mass should be attached to this spring so that its frequency of vibration is $f = 3.0$ Hz?

21. 🅜 **V-HINT** An object attached to a horizontal spring is oscillating back and forth along a frictionless surface. The maximum speed of the object is 1.25 m/s, and its maximum acceleration is 6.89 m/s². How much time elapses between an instant when the object's speed is at a maximum and the next instant when its acceleration is at a maximum?

22. 🅜 **GO** **MMH** A vertical spring (spring constant $= 112$ N/m) is mounted on the floor. A 0.400-kg block is placed on top of the spring and pushed down to start it oscillating in simple harmonic motion. The block is not attached to the spring. (a) Obtain the frequency (in Hz) of the motion. (b) Determine the amplitude at which the block will lose contact with the spring.

23. 🅗 A tray is moved horizontally back and forth in simple harmonic motion at a frequency of $f = 2.00$ Hz. On this tray is an empty cup. Obtain the coefficient of static friction between the tray and the cup, given that the cup begins slipping when the amplitude of the motion is 5.00×10^{-2} m.

Section 10.3 Energy and Simple Harmonic Motion

24. 🅔 A pen contains a spring with a spring constant of 250 N/m. When the tip of the pen is in its retracted position, the spring is compressed 5.0 mm from its unstrained length. In order to push the tip out and lock it into its writing position, the spring must be compressed an additional 6.0 mm. How much work is done by the spring force to ready the pen for writing? Be sure to include the proper algebraic sign with your answer.

25. 🅔 **GO** The drawing shows three situations in which a block is attached to a spring. The position labeled "0 m" represents the unstrained position of the spring. The block is moved from an initial position x_0 to a final position x_f, the magnitude of the displacement being denoted by the symbol s. Suppose the spring has a spring constant of $k = 46.0$ N/m. Using the data provided in the drawing, determine the total work done by the restoring force of the spring for each situation.

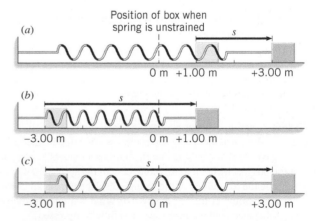

PROBLEM 25

26. 🅔 A spring is hung from the ceiling. A 0.450-kg block is then attached to the free end of the spring. When released from rest, the block drops 0.150 m before momentarily coming to rest, after which it moves back upward. (a) What is the spring constant of the spring? (b) Find the angular frequency of the block's vibrations.

27. 🅔 A 3.2-kg block is hanging stationary from the end of a vertical spring that is attached to the ceiling. The elastic potential energy of this spring-block system is 1.8 J. What is the elastic potential energy of the system when the 3.2-kg block is replaced by a 5.0-kg block?

28. **E** **SSM** A vertical spring with a spring constant of 450 N/m is mounted on the floor. From directly above the spring, which is unstrained, a 0.30-kg block is dropped from rest. It collides with and sticks to the spring, which is compressed by 2.5 cm in bringing the block to a momentary halt. Assuming air resistance is negligible, from what height (in cm) above the compressed spring was the block dropped?

29. **E** In preparation for shooting a ball in a pinball machine, a spring ($k = 675$ N/m) is compressed by 0.0650 m relative to its unstrained length. The ball ($m = 0.0585$ kg) is at rest against the spring at point A. When the spring is released, the ball slides (without rolling). It leaves the spring and arrives at point B, which is 0.300 m higher than point A. Ignore friction, and find the ball's speed at point B.

30. **E** **MMH** A heavy-duty stapling gun uses a 0.140-kg metal rod that rams against the staple to eject it. The rod is attached to and pushed by a stiff spring called a "ram spring" ($k = 32\,000$ N/m). The mass of this spring may be ignored. The ram spring is compressed by 3.0×10^{-2} m from its unstrained length and then released from rest. Assuming that the ram spring is oriented vertically and is still compressed by 0.8×10^{-2} m when the downward-moving ram hits the staple, find the speed of the ram at the instant of contact.

31. **E** A rifle fires a 2.10×10^{-2}-kg pellet straight upward, because the pellet rests on a compressed spring that is released when the trigger is pulled. The spring has a negligible mass and is compressed by 9.10×10^{-2} m from its unstrained length. The pellet rises to a maximum height of 6.10 m above its position on the compressed spring. Ignoring air resistance, determine the spring constant.

32. **E** **SSM** A 1.00×10^{-2}-kg block is resting on a horizontal frictionless surface and is attached to a horizontal spring whose spring constant is 124 N/m. The block is shoved parallel to the spring axis and is given an initial speed of 8.00 m/s, while the spring is initially unstrained. What is the amplitude of the resulting simple harmonic motion?

33. **M** **V-HINT** An 86.0-kg climber is scaling the vertical wall of a mountain. His safety rope is made of nylon that, when stretched, behaves like a spring with a spring constant of 1.20×10^3 N/m. He accidentally slips and falls freely for 0.750 m before the rope runs out of slack. How much is the rope stretched when it breaks his fall and momentarily brings him to rest?

34. **M** **V-HINT** A horizontal spring is lying on a frictionless surface. One end of the spring is attached to a wall, and the other end is connected to a movable object. The spring and object are compressed by 0.065 m, released from rest, and subsequently oscillate back and forth with an angular frequency of 11.3 rad/s. What is the speed of the object at the instant when the spring is *stretched* by 0.048 m relative to its unstrained length?

35. **M** **GO** A spring is resting vertically on a table. A small box is dropped onto the top of the spring and compresses it. Suppose the spring has a spring constant of 450 N/m and the box has a mass of 1.5 kg. The speed of the box just before it makes contact with the spring is 0.49 m/s. (a) Determine the magnitude of the spring's displacement at an instant when the acceleration of the box is zero. (b) What is the magnitude of the spring's displacement when the spring is fully compressed?

36. **M** **CHALK** **SSM** **MMH** A spring is compressed by 0.0620 m and is used to launch an object horizontally with a speed of 1.50 m/s. If the object were attached to the spring, at what angular frequency (in rad/s) would it oscillate?

37. **M** **GO** A 0.60-kg metal sphere oscillates at the end of a vertical spring. As the spring stretches from 0.12 to 0.23 m (relative to its unstrained length), the speed of the sphere decreases from 5.70 to 4.80 m/s. What is the spring constant of the spring?

38. **M** A 1.1-kg object is suspended from a vertical spring whose spring constant is 120 N/m. (a) Find the amount by which the spring is stretched from its unstrained length. (b) The object is pulled straight down by an additional distance of 0.20 m and released from rest. Find the speed with which the object passes through its original position on the way up.

39. **H** **SSM** A 70.0-kg circus performer is fired from a cannon that is elevated at an angle of 40.0° above the horizontal. The cannon uses strong elastic bands to propel the performer, much in the same way that a slingshot fires a stone. Setting up for this stunt involves stretching the bands by 3.00 m from their unstrained length. At the point where the performer flies free of the bands, his height above the floor is the same as the height of the net into which he is shot. He takes 2.14 s to travel the horizontal distance of 26.8 m between this point and the net. Ignore friction and air resistance and determine the effective spring constant of the firing mechanism.

40. **H** A 1.00×10^{-2}-kg bullet is fired horizontally into a 2.50-kg wooden block attached to one end of a massless horizontal spring ($k = 845$ N/m). The other end of the spring is fixed in place, and the spring is unstrained initially. The block rests on a horizontal, frictionless surface. The bullet strikes the block perpendicularly and quickly comes to a halt within it. As a result of this completely inelastic collision, the spring is compressed along its axis and causes the block/bullet to oscillate with an amplitude of 0.200 m. What is the speed of the bullet?

Section 10.4 The Pendulum

41. **E** A simple pendulum is made from a 0.65-m-long string and a small ball attached to its free end. The ball is pulled to one side through a small angle and then released from rest. After the ball is released, how much time elapses before it attains its greatest speed?

42. **E** **GO** **MMH** Astronauts on a distant planet set up a simple pendulum of length 1.2 m. The pendulum executes simple harmonic motion and makes 100 complete vibrations in 280 s. What is the magnitude of the acceleration due to gravity on this planet?

43. **E** **GO** The length of a simple pendulum is 0.79 m and the mass of the particle (the "bob") at the end of the cable is 0.24 kg. The pendulum is pulled away from its equilibrium position by an angle of 8.50° and released from rest. Assume that friction can be neglected and that the resulting oscillatory motion is simple harmonic motion. (a) What is the angular frequency of the motion? (b) Using the position of the bob at its lowest point as the reference level, determine the total mechanical energy of the pendulum as it swings back and forth. (c) What is the bob's speed as it passes through the lowest point of the swing?

44. **E** **CHALK** A spiral staircase winds up to the top of a tower in an old castle. To measure the height of the tower, a rope is attached to the top of the tower and hung down the center of the staircase. However, nothing is available with which to measure the length of the rope. Therefore, at the bottom of the rope a small object is attached so as to form a simple pendulum that just clears the floor. The period of the pendulum is measured to be 9.2 s. What is the height of the tower?

45. **E** **GO** Two physical pendulums (not simple pendulums) are made from meter sticks that are suspended from the ceiling at one end. The sticks are uniform and are identical in all respects, except that one is made of wood (mass = 0.17 kg) and the other of metal (mass = 0.85 kg). They are set into oscillation and execute simple harmonic motion. Determine the period of (a) the wood pendulum and (b) the metal pendulum.

46. **H** A small object oscillates back and forth at the bottom of a frictionless hemispherical bowl, as the drawing illustrates. The radius of the bowl is R, and the angle θ is small enough that the object oscillates in simple harmonic motion. Derive an expression for the angular frequency ω of the motion. Express your answer in terms of R and g, the magnitude of the acceleration due to gravity.

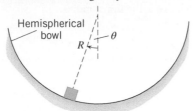

Hemispherical bowl

R θ

PROBLEM 46

Section 10.7   Elastic Deformation,

Section 10.8   Stress, Strain, and Hooke's Law

47. **E** **SSM** A tow truck is pulling a car out of a ditch by means of a steel cable that is 9.1 m long and has a radius of 0.50 cm. When the car just begins to move, the tension in the cable is 890 N. How much has the cable stretched?

48. **E** Two stretched cables both experience the same stress. The first cable has a radius of 3.5×10^{-3} m and is subject to a stretching force of 270 N. The radius of the second cable is 5.1×10^{-3} m. Determine the stretching force acting on the second cable.

49. **E** **SSM** **CHALK** The pressure increases by 1.0×10^4 N/m² for every meter of depth beneath the surface of the ocean. At what depth does the volume of a Pyrex glass cube, 1.0×10^{-2} m on an edge at the ocean's surface, decrease by 1.0×10^{-10} m³?

50. **E** A 59-kg water skier is being pulled by a nylon tow rope that is attached to a boat. The unstretched length of the rope is 12 m, and its cross-sectional area is 2.0×10^{-5} m². As the skier moves, a resistive force (due to the water) of magnitude 130 N acts on her; this force is directed opposite to her motion. What is the change in the length of the rope when the skier has an acceleration whose magnitude is 0.85 m/s²?

51. **E** A solid steel cylinder is standing (on one of its ends) vertically on the floor. The length of the cylinder is 3.6 m and its radius is 65 cm. When an object is placed on top of the cylinder, the cylinder compresses by an amount of 5.7×10^{-7} m. What is the weight of the object?

52. **E** The drawing shows a 160-kg crate hanging from the end of a steel bar. The length of the bar is 0.10 m, and its cross-sectional area is 3.2×10^{-4} m². Neglect the weight of the bar itself and determine **(a)** the shear stress on the bar and **(b)** the vertical deflection ΔY of the right end of the bar.

PROBLEM 52

53. **E** A copper cube, 0.30 m on a side, is subjected to two shearing forces, each of which has a magnitude $F = 6.0 \times 10^6$ N (see the drawing). Find the angle θ (in degrees), which is one measure of how the shape of the block has been altered by shear deformation.

PROBLEM 53

54. **E** **SSM** Two metal beams are joined together by four rivets, as the drawing indicates. Each rivet has a radius of 5.0×10^{-3} m and is to

be exposed to a shearing stress of no more than 5.0×10^8 Pa. What is the maximum tension $\vec{T}$ that can be applied to each beam, assuming that each rivet carries one-fourth of the total load?

PROBLEM 54

55. **E** A copper cylinder and a brass cylinder are stacked end to end, as in the drawing. Each cylinder has a radius of 0.25 cm. A compressive force of $F = 6500$ N is applied to the right end of the brass cylinder. Find the amount by which the length of the stack decreases.

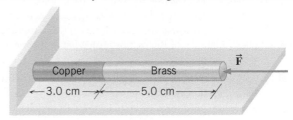

Copper   Brass $\vec{F}$

←3.0 cm→ ←5.0 cm→

PROBLEM 55

56. **E** A piece of aluminum is surrounded by air at a pressure of 1.01×10^5 Pa. The aluminum is placed in a vacuum chamber where the pressure is reduced to zero. Determine the fractional change $\Delta V/V_0$ in the volume of the aluminum.

57. **E** **GO** One end of a piano wire is wrapped around a cylindrical tuning peg and the other end is fixed in place. The tuning peg is turned so as to stretch the wire. The piano wire is made from steel ($Y = 2.0 \times 10^{11}$ N/m²). It has a radius of 0.80 mm and an unstrained length of 0.76 m. The radius of the tuning peg is 1.8 mm. Initially, there is no tension in the wire, but when the tuning peg is turned, tension develops. Find the tension in the wire when the tuning peg is turned through two revolutions. Ignore the radius of the wire compared to the radius of the tuning peg.

58. **M** **GO** A piece of mohair (see Table 10.1) taken from an Angora goat has a radius of 31×10^{-6} m. What is the least number of identical pieces of mohair needed to suspend a 75-kg person, so the strain experienced by each piece is less than 0.010? Assume that the tension is the same in all the pieces.

59. **M** **GO** **CHALK** A square plate is 1.0×10^{-2} m thick, measures 3.0×10^{-2} m on a side, and has a mass of 7.2×10^{-2} kg. The shear modulus of the material is 2.0×10^{10} N/m². One of the square faces rests on a flat horizontal surface, and the coefficient of static friction between the plate and the surface is 0.91. A force is applied to the top of the plate, as in **Figure 10.30a**. Determine **(a)** the maximum possible amount of shear stress, **(b)** the maximum possible amount of shear strain, and **(c)** the maximum possible amount of shear deformation ΔX (see **Figure 10.30b**) that can be created by the applied force just before the plate begins to move.

60. **H** A 6.8-kg bowling ball is attached to the end of a nylon cord with a cross-sectional area of 3.4×10^{-5} m². The other end of the cord is fixed to the ceiling. When the bowling ball is pulled to one side and released from rest, it swings downward in a circular arc. At the instant it reaches its lowest point, the bowling ball is 1.4 m lower than the point from which it was released, and the cord is stretched 2.7×10^{-3} m from its unstrained length. What is the unstrained length of the cord? (*Hint: When calculating any quantity other than the strain, ignore the increase in the length of the cord.*)

61. **H** **SSM** A solid brass sphere is subjected to a pressure of 1.0×10^5 Pa due to the earth's atmosphere. On Venus the pressure due to the atmosphere is 9.0×10^6 Pa. By what fraction $\Delta r/r_0$ (including the algebraic sign) does the radius of the sphere change when it is exposed to the Venusian atmosphere? Assume that the change in radius is very small relative to the initial radius.

Additional Problems

Online

62. **E** A loudspeaker diaphragm is producing a sound for 2.5 s by moving back and forth in simple harmonic motion. The angular frequency of the motion is 7.54×10^4 rad/s. How many times does the diaphragm move back and forth?

63. **E** A person bounces up and down on a trampoline, while always staying in contact with it. The motion is simple harmonic motion, and it takes 1.90 s to complete one cycle. The height of each bounce above the equilibrium position is 45.0 cm. Determine **(a)** the amplitude and **(b)** the angular frequency of the motion. **(c)** What is the maximum speed attained by the person?

64. **E** A simple pendulum is swinging back and forth through a small angle, its motion repeating every 1.25 s. How much longer should the pendulum be made in order to increase its period by 0.20 s?

65. **E** An archer, about to shoot an arrow, is applying a force of +240 N to a drawn bowstring. The bow behaves like an ideal spring whose spring constant is 480 N/m. What is the displacement of the bowstring?

66. **M** **V-HINT** A block rests on a frictionless horizontal surface and is attached to a spring. When set into simple harmonic motion, the block oscillates back and forth with an angular frequency of 7.0 rad/s. The drawing indicates the position of the block when the spring is unstrained. This position is labeled "$x = 0$ m." The drawing also shows a small bottle located 0.080 m to the right of this position. The block is pulled to the right, stretching the spring by 0.050 m, and is then thrown to the left. In order for the block to knock over the bottle, it must be thrown with a speed exceeding v_0. Ignoring the width of the block, find v_0.

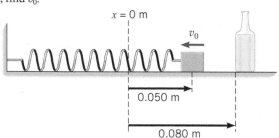

PROBLEM 66

67. **M** **GO** A vertical ideal spring is mounted on the floor and has a spring constant of 170 N/m. A 0.64-kg block is placed on the spring in two different ways. **(a)** In one case, the block is placed on the spring and not released until it rests stationary on the spring in its equilibrium position. Determine the amount (magnitude only) by which the spring is compressed. **(b)** In a second situation, the block is released from rest immediately after being placed on the spring and falls downward until it comes to a momentary halt. Determine the amount (magnitude only) by which the spring is now compressed.

68. **M** Multiple-Concept Example 6 reviews the principles that play roles in this problem. A bungee jumper, whose mass is 82 kg, jumps from a tall platform. After reaching his lowest point, he continues to oscillate up and down, reaching the low point two more times in 9.6 s. Ignoring air resistance and assuming that the bungee cord is an ideal spring, determine its spring constant.

69. **M** **GO** When an object of mass m_1 is hung on a vertical spring and set into vertical simple harmonic motion, it oscillates at a frequency of 12.0 Hz. When another object of mass m_2 is hung on the spring along with the first object, the frequency of the motion is 4.00 Hz. Find the ratio m_2/m_1 of the masses.

70. **H** **SSM** The drawing shows a top view of a frictionless horizontal surface, where there are two springs with particles of mass m_1 and m_2 attached to them. Each spring has a spring constant of 120 N/m. The particles are pulled to the right and then released from the positions shown in the drawing. How much time passes before the particles are side by side for the first time at $x = 0$ m if **(a)** $m_1 = m_2 = 3.0$ kg and **(b)** $m_1 = 3.0$ kg and $m_2 = 27$ kg?

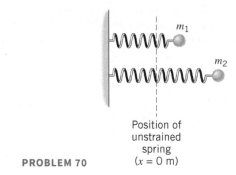

Position of
unstrained
spring
$(x = 0 \text{ m})$

PROBLEM 70

71. **M** **GO** A helicopter is using a steel cable to lift a 2100-kg jeep (see the figure). The unstretched length of the cable is 16 m, and its radius is 5.0×10^{-3} m. By what amount does the cable stretch when the jeep is hoisted straight upward with an acceleration of +1.5 m/s²?

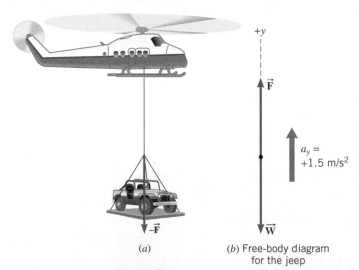

(a) The jeep applies a force $-\vec{\mathbf{F}}$ to the lower end of the cable, thereby stretching it. *(b)* The free-body diagram for the jeep, showing the two forces acting on it.

PROBLEM 71

Physics in Biology, Medicine, and Sports

72. **E** **BIO** **SSM** **10.2** When responding to sound, the human eardrum vibrates about its equilibrium position. Suppose an eardrum is vibrating with an amplitude of 6.3×10^{-7} m and a maximum speed of 2.9×10^{-3} m/s. **(a)** What is the frequency (in Hz) of the eardrum's vibration? **(b)** What is the maximum acceleration of the eardrum?

73. **E** **BIO** **V-HINT** **10.7** When subjected to a force of compression, the length of a bone decreases by 2.7×10^{-5} m. When this same bone is subjected to a tensile force of the same magnitude, by how much does it stretch?

74. **M** **BIO** **10.8** A gymnast does a one-arm handstand. The humerus, which is the upper arm bone (between the elbow and the shoulder joint), may be approximated as a 0.30-m-long cylinder with an outer radius of 1.00×10^{-2} m and a hollow inner core with a radius of 4.0×10^{-3} m. Excluding the arm, the mass of the gymnast is 63 kg. **(a)** What is the compressional strain of the humerus? **(b)** By how much is the humerus compressed?

75. **M** **BIO** **GO** **10.8** Depending on how you fall, you can break a bone easily. The severity of the break depends on how much energy the bone absorbs in the accident, and to evaluate this let us treat the bone as an ideal spring. The maximum applied force of compression that one man's thighbone can endure without breaking is 7.0×10^4 N. The minimum effective cross-sectional area of the bone is 4.0×10^{-4} m^2, its length is 0.55 m, and Young's modulus is $Y = 9.4 \times 10^9$ N/m^2. The mass of the man is 65 kg. He falls straight down without rotating, strikes the ground stiff-legged on one foot, and comes to a halt without rotating. To see that it is easy to break a thighbone when falling in this fashion, find the maximum distance through which his center of gravity can fall without his breaking a bone.

76. **E** **BIO** **10.8** The femur is a bone in the leg whose minimum cross-sectional area is about 4.0×10^{-4} m^2. A compressional force in excess of 6.8×10^4 N will fracture this bone. **(a)** Find the maximum stress that this bone can withstand. **(b)** What is the strain that exists under a maximum-stress condition?

77. **E** **BIO** **10.7** Between each pair of vertebrae in the spinal column is a cylindrical disc of cartilage. Typically, this disc has a radius of about 3.0×10^{-2} m and a thickness of about 7.0×10^{-3} m. The shear modulus of cartilage is 1.2×10^7 N/m^2. Suppose that a shearing force of magnitude 11 N is applied parallel to the top surface of the disc while the bottom surface remains fixed in place. How far does the top surface move relative to the bottom surface?

78. **H** **BIO** **10.7** A cylindrically shaped piece of collagen (a substance found in the body in connective tissue) is being stretched by a force that increases from 0 to 3.0×10^{-2} N. The length and radius of the collagen are, respectively, 2.5 and 0.091 cm, and Young's modulus is 3.1×10^6 N/m^2. **(a)** If the stretching obeys Hooke's law, what is the spring constant k for collagen? **(b)** How much work is done by the variable force that stretches the collagen? (See Section 6.9 for a discussion of the work done by a variable force.)

79. **E** **BIO** **10.7** Elastic therapeutic tape, also called kineso tape or KT-tape, consists of an elastic cotton weave with an adhesive backing. It is placed directly on the skin to provide support to injured or painful joints, pulled muscles, and other ailments (see the photo). While there is no definitive scientific evidence that KT-tape offers significant health benefits, its popularity surged after many athletes wore it in the 2008 Olympic Games. Standard KT-tape is 5.08 cm wide and 0.180 mm thick. One end of a 10.0-cm length of KT-tape is hung from the ceiling, and a 1.00-kg mass is attached to the other end, causing the tape to stretch by 2.14 mm. What is the Young's modulus of the tape?

PROBLEM 79

80. **H** **BIO** **10.7** The silk of Darwin's bark spider (*Caerostris darwini*), named after the naturalist Charles Darwin, is the strongest biological material ever discovered (see the photo). In fact, many of the mechanical properties of this silk supersede those of steel. While the Young's modulus for steel (2.00×10^{11} N/m^2) is larger than that for silk (1.15×10^{10} N/m^2), the density of the silk is much less, such that the strength-to-density ratio of the silk is about five times that of steel. Consider a 10.0-cm unstrained length of spider silk with a circular cross section and a radius of 2.50×10^{-6} m. When this segment supports the weight of a dragonfly ($m = 3.00 \times 10^{-3}$ kg), **(a)** what is the percent change in the length of the silk? **(b)** What force would have to be applied to a steel filament of the same dimensions in order to produce the same percent change in length?

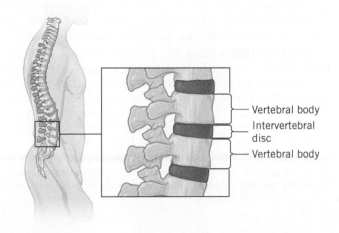

PROBLEM 80

81. **E** **BIO** **10.8** The intervertebral discs in the human spine (see the figure) are multilayered elastic structures composed of collagen that act like shock absorbers between the vertebrae. While lifting heavy loads improperly, such as lifting a heavy box with your back arched instead of using your legs, the compressive forces on the discs can be quite large. In one case, the compressive force on a disc is 3100 N. If the disc is compressed a distance of 1.5 mm, what is the effective spring constant of the disc?

— Vertebral body
— Intervertebral disc
— Vertebral body

PROBLEM 81

82. **M** **BIO** **10.4** Most injuries to the shoulder will require rehab and physical therapy. Pendulum exercises are a great way to increase strength as well as to improve flexibility and range of motion in an injured shoulder. Here, the patient bends forward until their back is horizontal. The injured arm is hung vertically downward and is allowed to swing freely back and forth like a pendulum (see the figure). Treat the arm (length of 69 cm) in the figure as a physical pendulum—a uniform rod rotating about an axis through one of its ends—and calculate the period of the arm's small oscillations.

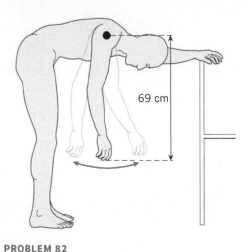

69 cm

PROBLEM 82

83. **M** **BIO** **10.3** Skeletal muscles are attached to bones by tendons. Tendons are composed of elastic tissues that, for small stretching displacements, behave like a spring and follow Hooke's law. The largest and strongest tendon in the human body is the Achilles tendon (see the figure), which attaches the muscles of the lower leg to the heel bone (calcaneus). While the Achilles tendon is incredibly strong, it is often under high tension and has a low blood supply. This makes the tendon prone to injuries, such as tears and ruptures. In fact, 1 out of every 10 000 people rupture their Achilles tendon each year. One unfortunate soul had his Achilles tendon rupture when the tension in the tendon was 28 500 N. If the effective spring constant of the tendon is 4.60×10^6 N/m, **(a)** how far did the tendon stretch before it ruptured? **(b)** How much elastic potential energy was stored in the tendon when it ruptured?

Anatomy of the Lower Leg

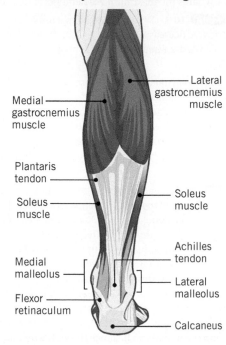

Medial gastrocnemius muscle

Lateral gastrocnemius muscle

Plantaris tendon

Soleus muscle

Soleus muscle

Achilles tendon

Medial malleolus

Flexor retinaculum

Lateral malleolus

Calcaneus

PROBLEM 83

Concepts and Calculations Problems

Online

This chapter has examined an important kind of vibratory motion known as simple harmonic motion. Specifically, it has discussed how the motion's displacement, velocity, and acceleration vary with time, and explained what determines the frequency of the motion. In addition, we saw that the elastic force is conservative, so that the total mechanical energy is conserved if nonconservative forces, such as friction and air resistance, are absent. We conclude now with some problems that review important features of simple harmonic motion.

84. **M** **CHALK** **SSM** A 75-kg diver is standing at the end of a diving board while it is vibrating up and down in simple harmonic motion, as indicated in the figure. The diving board has an effective spring constant of $k = 4100$ N/m, and the vertical distance between the highest and lowest points in the motion is 0.30 m. *Concepts:* (i) How is the amplitude A related to the vertical distance between the highest and lowest points of the diver's motion? (ii) Starting from the top, where is the diver located one-quarter of a period later, and what can be said about his speed at this point? (iii) If the amplitude were to double, would the period also double? Explain. *Calculations:* **(a)** What is the amplitude of the motion? **(b)** Starting when the diver is at the highest point, what is his speed one-quarter of a period later? **(c)** If the vertical distance between his highest and lowest points were changed to 0.10 m, what would be the time required for the diver to make one complete motional cycle?

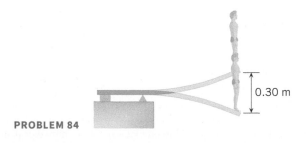

0.30 m

PROBLEM 84

85. **M** **CHALK** A 68.0-kg bungee jumper is standing on a tall platform ($h_0 = 46.0$ m), as indicated in the figure. The bungee cord has a natural

length of $L_0 = 9.00$ m and, when stretched, behaves like an ideal spring with a spring constant of $k = 66.0$ N/m. The jumper falls from rest, and it is assumed that the only forces acting on him are his weight and, for the latter part of the descent, the elastic force of the bungee cord. *Concepts:* (i) Can we use the conservation of mechanical energy to find his speed at any point along the descent? Explain your answer. (ii) What type of energy does he have when he is standing on the platform? (iii) What types of energy does he have at point A? (iv) What types of energy does he have at point B? *Calculations:* What is his speed when he is at the following heights above the water: (a) $h_A = 37.0$ m, and (b) $h_B = 15.0$ m?

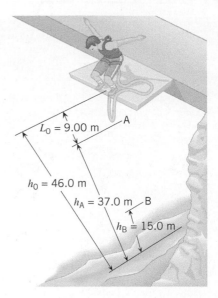

PROBLEM 85

Team Problems

Online

86. **E** **T** **WS** **Moving a Fragile Particle Detector.** You and your team are assigned to design a structure that protects a heavy, fragile detector as it is moved from one lab to another. The idea is to suspend it by springs inside a large cubical box. You decide to use four identical springs, one attached to each of the four upper corners of the box. The opposite end of each spring is connected to a single hook on the top of the instrument that will hang from the springs, suspending it inside the box. The box has sides of length $L = 1.00$ m, and the springs have an unstrained length of 70.0 cm. Assuming that you want the hook on the top of the instrument to which all of the springs are connected to be located at the geometric center of the box when the 78.0-kg instrument is suspended, what should you choose for the spring constant of the springs?

87. **E** **T** **WS** **The Spring and the Pendulum.** A steel ball of mass $m = 6.25$ kg hangs from the end of a vertical spring of spring constant $k = 112$ N/m. You and your team have been given the task of designing a simple pendulum that utilizes the same mass m and has the same frequency of oscillation f as the spring–mass system. (a) What is the frequency of oscillation of the spring–mass system? (b) Find the required length L and mass m of the pendulum so that it has the same frequency of oscillation as that found in part (a). (c) What is the relationship between the length of the pendulum and the equilibrium displacement of the spring from its unstrained length when the 6.25-kg mass hangs from it? *Hint: Assume that the frequency of oscillation of a vertical spring system is the same as that of a horizontal spring system.*

88. **M** **T** **Bungee Cord Escape.** You are running from pirates on a tropical island somewhere in the Caribbean. You have somehow become separated from the rest of your group and now find yourself on the edge of a cliff with your pursuers less than 10 minutes behind you. According to a sign posted on the guardrail at the cliff's edge, the drop to the beach below is $h = 140.0$ feet. Your team members (waiting for you on the beach, near your boat) have a rope, but there is no time for anyone to climb the cliff to save you. You break into a deserted cabin nearby, and rummage around for a rope. Instead, you find a brand

new, still-in-package, bungee cord that must have been intended for tourists jumping from a nearby bridge. You figure you might be able to attach it to the guardrail and jump to the beach, letting go at the bottom before it reverses your motion. You read the bungee cord specifications on the package: the total length of the cord is $L_0 = 100.0$ m, the maximum elastic deformation is 200% (i.e., it can safely triple its length), and the elastic constant is $k = 75.0$ N/m. (a) If you weigh 170 lb, how far is the bungee cord designed to let you fall before it stops you and reverses your direction? Will this afford you a safe landing? (b) You realize that you don't have to hang from the very end of the bungee, but rather from some point in the middle. How far from the attached end should you grasp the unstretched bungee cord so that you land softly on the beach? Will you be able to perform the jump and stay under the elastic deformation limit?

89. **M** **T** **A Light-Beam Metronome.** You are given the task of opening an antiquated "light lock," which is unlocked by shining red light pulses of a certain frequency into a light sensor on the lock. You are given a red laser pointer, a spring of unstretched length $L = 15.0$ cm and spring constant $k = 7.20$ N/m, a sheet of steel ($\rho = 7.60$ g/cm^3) that is 0.125 inches thick, and some tools. You come up with the idea to take a piece of the steel sheet (of mass m), cut a slot in it, and hang it from the spring. If you shine the laser through the slot and onto the sensor, and then stretch the spring and let it go, the steel plate will oscillate and cause the beam to pass through the slot periodically. (a) Assuming the beam is passing through the slot (and onto the lock's sensor) when the spring-mass system is in equilibrium, how is the frequency at which the light pulses hit the sensor related to the frequency of the spring/mass (i.e., steel plate) system. (b) Based on your answer for (a), what should be the frequency of the spring/mass system if the unlocking frequency is 3.50 Hz? (c) What should be the mass m of the steel plate? (d) Calculate some reasonable dimensions for the steel plate (i.e., they should be consistent with the mass that is required for the spring-mass system). You may assume the material removed to make the slot in the steel plate is of negligible mass.

CHAPTER 11

The air is a fluid, and this chapter examines the forces and pressures that fluids exert when they are at rest and when they are in motion. In a tornado the air is moving very rapidly and, as we will see, moving air has a lower pressure than that of stationary air. This difference in air pressure is one of the reasons that tornadoes, such as the one in this photograph, are so destructive.

Design Pics Inc/Alamy Stock Photo

Fluids

11.1 | Mass Density

Fluids are materials that can flow, and they include both gases and liquids. Air is the most common gas, and flows from place to place as wind. Water is the most familiar liquid, and flowing water has many uses, from generating hydroelectric power to white-water rafting. The **mass density** of a liquid or gas is one of the important factors that determine its behavior as a fluid. As indicated below, the mass density is the mass per unit volume and is denoted by the Greek letter rho (ρ).

> **DEFINITION OF MASS DENSITY**
>
> The mass density ρ is the mass m of a substance divided by its volume V:
>
> $$\rho = \frac{m}{V} \tag{11.1}$$
>
> SI Unit of Mass Density: kg/m^3

Equal volumes of different substances generally have different masses, so the density depends on the nature of the material, as **Table 11.1** indicates. Gases have the smallest densities because gas molecules are relatively far apart and a gas contains a large fraction of empty space. In contrast, the molecules are much more tightly packed in liquids and solids, and the tighter packing leads to larger densities.

TABLE 11.1 Mass Densities[a] of Common Substances

Substance	Mass Density ρ (kg/m³)	Substance	Mass Density ρ (kg/m³)
Solids		**Liquids**	
Aluminum	2700	Blood (whole, 37 °C)	1060
Brass	8470	Ethyl alcohol	806
Concrete	2200	Mercury	13 600
Copper	8890	Oil (hydraulic)	800
Diamond	3520	Water (4 °C)	1.000×10^3
Gold	19 300	**Gases**	
Ice	917	Air	1.29
Iron (steel)	7860	Carbon dioxide	1.98
Lead	11 300	Helium	0.179
Quartz	2660	Hydrogen	0.0899
Silver	10 500	Nitrogen	1.25
Wood (yellow pine)	550	Oxygen	1.43

[a]Unless otherwise noted, densities are given at 0 °C and 1 atm pressure.

The densities of gases are very sensitive to changes in temperature and pressure. However, for the range of temperatures and pressures encountered in this text, the densities of liquids and solids do not differ much from the values in **Table 11.1**.

It is the mass of a substance, not its weight, that enters into the definition of density. In situations where weight is needed, it can be calculated from the mass density, the volume, and the acceleration due to gravity, as Example 1 illustrates.

EXAMPLE 1 | Blood as a Fraction of Body Weight

The body of a man whose weight is 690 N contains about 5.2×10^{-3} m³ (5.5 qt) of blood. **(a)** Find the blood's weight and **(b)** express it as a percentage of the body weight.

Reasoning To find the weight W of the blood, we need the mass m, since $W = mg$ (Equation 4.5), where g is the magnitude of the acceleration due to gravity. According to **Table 11.1**, the density of blood is 1060 kg/m³, so the mass of the blood can be found by using the given volume of 5.2×10^{-3} m³ in Equation 11.1.

Solution **(a)** According to Equation 4.5, the blood's weight is $W = mg$. Equation 11.1 can be solved for m to show that the mass is $m = \rho V$. Substituting this result into Equation 4.5 gives

$$W = mg = (\rho V)g = (1060 \text{ kg/m}^3)(5.2 \times 10^{-3} \text{ m}^3)(9.80 \text{ m/s}^2)$$
$$= \boxed{54 \text{ N}}$$

(b) The percentage of body weight contributed by the blood is

$$\text{Percentage} = \frac{54 \text{ N}}{690 \text{ N}} \times 100 = \boxed{7.8\%}$$

A convenient way to compare densities is to use the concept of **specific gravity**. The specific gravity of a substance is its density divided by the density of a standard reference material, usually chosen to be water at 4 °C.

$$\text{Specific gravity} = \frac{\text{Density of substance}}{\text{Density of water at 4 °C}} = \frac{\text{Density of substance}}{1.000 \times 10^3 \text{ kg/m}^3} \quad (11.2)$$

Being the ratio of two densities, specific gravity has no units. For example, **Table 11.1** reveals that diamond has a specific gravity of 3.52, since the density of diamond is 3.52 times the density of water at 4 °C.

The next two sections deal with the important concept of pressure. We will see that the density of a fluid is one factor determining the pressure that a fluid exerts.

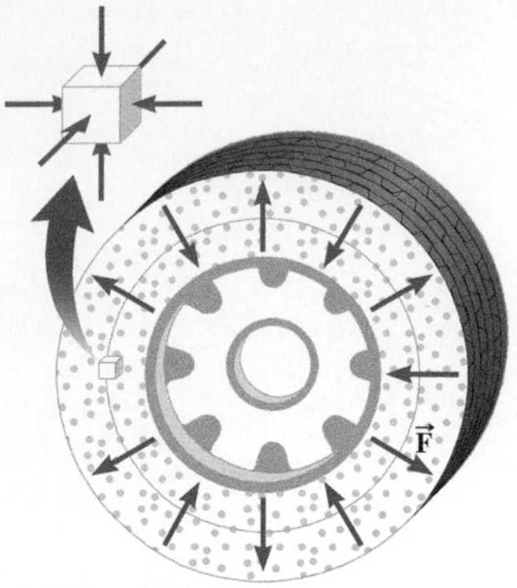

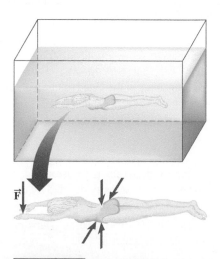

FIGURE 11.1 In colliding with the inner walls of the tire, the air molecules (blue dots) exert a force on every part of the wall surface. If a small cube were inserted inside the tire, the cube would experience forces (blue arrows) acting perpendicular to each of its six faces.

11.2 | Pressure

People who have fixed a flat tire know something about pressure. The final step in the job is to reinflate the tire to the proper pressure. The underinflated tire is soft because it contains an insufficient number of air molecules to push outward against the rubber and give the tire that solid feel. When air is added from a pump, the number of molecules and the collective force they exert are increased. The air molecules within a tire are free to wander throughout its entire volume, and in the course of their wandering they collide with one another and the inner walls of the tire. The collisions with the walls allow the air to exert a force against every part of the wall surface, as **Figure 11.1** shows. The pressure P exerted by a fluid is defined in Section 10.7 (Equation 10.19) as the magnitude F of the force acting perpendicular to a surface divided by the area A over which the force acts:

$$P = \frac{F}{A} \tag{11.3}$$

The SI unit for pressure is a newton/meter^2 (N/m^2), a combination that is referred to as a *pascal* (Pa). A pressure of 1 Pa is a very small amount. Many common situations involve pressures of approximately 10^5 Pa, an amount referred to as one *bar* of pressure. Alternatively, force can be measured in pounds and area in square inches, so another unit for pressure is pounds per square inch (lb/in.^{2}), often abbreviated as "psi."

Because of its pressure, the air in a tire applies a force to any surface with which the air is in contact. Suppose, for instance, that a small cube is inserted inside the tire. As **Figure 11.1** shows, the air pressure causes a force to act perpendicularly on each face of the cube. In a similar fashion, a liquid such as water also exerts pressure. A swimmer, for example, feels the water pushing perpendicularly inward everywhere on her body, as **Figure 11.2** illustrates. In general, a static fluid cannot produce a force parallel to a surface, for if it did, the surface would apply a reaction force to the fluid, consistent with Newton's action–reaction law. In response, the fluid would flow and would not then be static.

While fluid pressure can generate a force, which is a vector quantity, ***pressure itself is not a vector***. In the definition of pressure, $P = F/A$, the symbol F refers only to the magnitude of the force, so that pressure has no directional characteristic. The force generated by the pressure of a static fluid is always perpendicular to the surface that the fluid contacts, as Example 2 illustrates.

FIGURE 11.2 Water applies a force perpendicular to each surface within the water, including the walls and bottom of the swimming pool, and all parts of the swimmer's body.

EXAMPLE 2 | The Force on a Swimmer

Suppose that the pressure acting on the back of a swimmer's hand is 1.2×10^5 Pa, a realistic value near the bottom of the diving end of a pool. The surface area of the back of the hand is 8.4×10^{-3} m². **(a)** Determine the magnitude of the force that acts on it. **(b)** Discuss the direction of the force.

Reasoning From the definition of pressure in Equation 11.3, we can see that the magnitude of the force is the pressure times the area. The direction of the force is always perpendicular to the surface that the water contacts.

> **Problem-Solving Insight** Force is a vector, but pressure is not.

Solution **(a)** A pressure of 1.2×10^5 Pa is 1.2×10^5 N/m². From Equation 11.3, we find

$$F = PA = (1.2 \times 10^5\,\text{N/m}^2)(8.4 \times 10^{-3}\,\text{m}^2) = \boxed{1.0 \times 10^3\,\text{N}}$$

This is a rather large force, about 230 lb.

(b) In **Figure 11.2**, the hand (palm downward) is oriented parallel to the bottom of the pool. Since the water pushes perpendicularly against the back of the hand, the force $\vec{F}$ is directed downward in the drawing. This downward-acting force is balanced by an upward-acting force on the palm, so that the hand is in equilibrium. If the hand were rotated by 90°, the directions of these forces would also be rotated by 90°, always being perpendicular to the hand.

A person need not be under water to experience the effects of pressure. Walking about on land, we are at the bottom of the earth's atmosphere, which is a fluid and pushes inward on our bodies just like the water in a swimming pool. As **Figure 11.3** indicates, there is enough air above the surface of the earth to create the following pressure at sea level:

Atmospheric pressure at sea level 1.013×10^5 Pa = 1 atmosphere

This pressure corresponds to 14.70 lb/in.² and is referred to as one *atmosphere* (atm), a significant amount of pressure. Look, for instance, in **Figure 11.3** at the results of pumping out the air from within a gasoline can. With no internal air to push outward, the inward push of the external air is unbalanced and is strong enough to crumple the can.

THE PHYSICS OF . . . lynx paws. In contrast to the situation in **Figure 11.3**, reducing the pressure is sometimes beneficial. Lynx, for example, are well suited for hunting on snow because of their oversize paws (see **Figure 11.4**). The large paws act as snowshoes and distribute the weight over a large area. Thus, they reduce the weight per unit area, or the pressure that the cat applies to the surface, which helps to keep it from sinking into the snow.

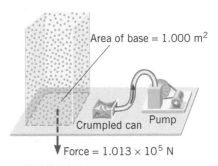

FIGURE 11.3 Atmospheric pressure at sea level is 1.013×10^5 Pa, which is sufficient to crumple a can if the inside air is pumped out.

Michael Lane/123RF

FIGURE 11.4 Lynx have large paws that act as natural snowshoes.

Check Your Understanding

(The answers are given at the end of the book.)

1. **BIO** As you climb a mountain, your ears "pop" because of the changes in atmospheric pressure. In which direction, outward or inward, does your eardrum move **(a)** as you climb up and **(b)** as you climb down?

2. A bottle of juice is sealed under partial vacuum, with a lid on which a red dot or "button" is painted. Around the button the following phrase is printed: "Button pops up when seal is broken." Why does the button remain pushed in when the seal is intact? **(a)** The pressure inside the bottle is greater than the pressure outside the bottle. **(b)** The pressure inside the bottle is less than the pressure outside the bottle. **(c)** There is a greater force acting on the interior surface of the seal than acts on the exterior surface.

3. A method for resealing a partially full bottle of wine under a vacuum uses a specially designed rubber stopper to close the bottle. A simple pump is attached to the stopper, and to remove air from the bottle, the plunger of the pump is pulled up and then released. After about 15 pull-and-release cycles the wine is under a partial vacuum. On the fifteenth pull-and-release cycle, does it require **(a)** more force, **(b)** less force, or **(c)** the same force to pull the plunger up than it did on the first cycle?

11.3 | Pressure and Depth in a Static Fluid

The deeper an underwater swimmer goes, the more strongly the water pushes on his body and the greater is the pressure that he experiences. To determine the relation between pressure and depth, we turn to Newton's second law ($\Sigma\vec{F} = m\vec{a}$). In using the second law, we will focus on two external forces that act on the fluid. One is the gravitational force—that is, the weight of the fluid. The other is the collisional force that is responsible for fluid pressure, as Section 11.2 discusses. Since the fluid is at rest, its acceleration is zero ($\vec{a} = 0$ m/s^2), and it is in equilibrium. By applying the second law in the form $\Sigma\vec{F} = 0$, we will derive a relation between pressure and depth. This relation is especially important because it leads to Pascal's principle (Section 11.5) and Archimedes' principle (Section 11.6), both of which are essential in describing the properties of static fluids.

Interactive Figure 11.5 shows a container of fluid and focuses attention on one column of the fluid. The free-body diagram in the figure shows all the vertical forces acting on the column. On the top face (area $= A$), the fluid pressure P_1 generates a downward force whose magnitude is P_1A. Similarly, on the bottom face, the pressure P_2 generates an upward force of magnitude P_2A. The pressure P_2 is greater than the pressure P_1 because the bottom face supports the weight of more fluid than the upper one does. In fact, the excess weight supported by the bottom face is exactly the weight of the fluid within the column. As the free-body diagram indicates, this weight is mg, where m is the mass of the fluid and g is the magnitude of the acceleration due to gravity. Since the column is in equilibrium, we can set the sum of the vertical forces equal to zero and find that

$$\Sigma F_y = P_2A - P_1A - mg = 0 \quad \text{or} \quad P_2A = P_1A + mg$$

The mass m is related to the density ρ and the volume V of the column by $m = \rho V$ (Equation 11.1). Since the volume is the cross-sectional area A times the vertical dimension h, we have $m = \rho Ah$. With this substitution, the condition for equilibrium becomes $P_2A = P_1A + \rho Ahg$. The area A can be eliminated algebraically from this expression, with the result that

$$P_2 = P_1 + \rho gh \tag{11.4}$$

Equation 11.4 indicates that if the pressure P_1 is known at a higher level, the larger pressure P_2 at a deeper level can be calculated by adding the increment ρgh. In determining the pressure increment ρgh, we assumed that the density ρ is the same at any vertical distance h or, in other words, the fluid is incompressible. The assumption is reasonable for liquids, since the bottom layers can support the upper layers with little compression. In a gas, however, the lower layers are compressed markedly by the weight of the upper layers, with the result that the density varies with vertical distance. For example, the density of our atmosphere is larger near the earth's surface than it is at higher altitudes.

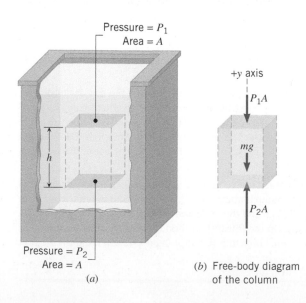

(a)

(b) Free-body diagram of the column

INTERACTIVE FIGURE 11.5
(a) A container of fluid in which one column of the fluid is outlined. The fluid is at rest. (b) The free-body diagram, showing the vertical forces acting on the column.

When applied to gases, the relation $P_2 = P_1 + \rho g h$ can be used only when h is small enough that any variation in ρ is negligible.

A significant feature of Equation 11.4 is that the pressure increment $\rho g h$ is affected by the vertical distance h, but not by any horizontal distance within the fluid. Conceptual Example 3 helps to clarify this feature.

CONCEPTUAL EXAMPLE 3 | The Hoover Dam

Lake Mead is the largest wholly artificial reservoir in the United States and was formed after the completion of the Hoover Dam in 1936. As **Figure 11.6a** suggests, the water in the reservoir backs up behind the dam for a considerable distance (about 200 km or 120 miles). Suppose that all the water were removed, except for a relatively narrow vertical column in contact with the dam. **Figure 11.6b** shows a side view of this hypothetical situation, in which the water against the dam has the same depth as in **Figure 11.6a**. How would the dam needed to contain the water in this hypothetical reservoir compare with the Hoover Dam? Would it need to be **(a)** less massive or **(b)** equally massive?

Reasoning Imagine a small square in the inner face of the dam, located beneath the water. The magnitude of the force on this square is the product of its area and the pressure of the water, according to Equation 11.3. However, the relation $P_2 = P_1 + \rho g h$ (Equation 11.4) indicates that the pressure at a given point depends on the vertical distance h that the small square is below the water. Thus, the force exerted by the water at any given location depends on the depth at that location.

Answer (a) is incorrect. Since our hypothetical reservoir contains much less water than Lake Mead, it is tempting to say that a less massive structure would be required. Note from the Reasoning section that the force exerted on a given section of the dam depends on how far below the surface the section is located. The horizontal distance of the water backed up behind the dam does not appear in Equation 11.4, and, therefore, has no effect on the pressure and, hence, on the force.

Answer (b) is correct. The force exerted on a given section of the dam depends only on how far that section is located vertically below the surface (see the Reasoning section). Certainly, as one goes deeper and deeper, the water pressure and force become greater. But no matter how deep one goes, the force that the water applies on a given section of the dam does not depend on the amount of water backed up behind the dam. Thus, the dam for our imaginary reservoir would sustain the same forces that the Hoover Dam sustains and, therefore, would need to be equally massive.

Adam G. Sylvester/Science Source

(a)

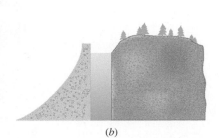

(b)

FIGURE 11.6 (a) The Hoover Dam in Nevada and Lake Mead behind it. (b) A hypothetical reservoir formed by removing most of the water from Lake Mead. Conceptual Example 3 compares the dam needed for this hypothetical reservoir with the Hoover Dam.

The next example deals further with the relationship between pressure and depth given by Equation 11.4.

EXAMPLE 4 | The Swimming Hole

Figure 11.7 shows the cross section of a swimming hole. Points *A* and *B* are both located at a distance of *h* = 5.50 m below the surface of the water. Find the pressure at each of these two points.

Reasoning The pressure at point *B* is the same as that at point *A*, since both are located at the *same vertical distance* beneath the surface and only the vertical distance *h* affects the pressure increment ρgh in Equation 11.4. To understand this important feature more clearly, consider the path $AA'B'B$ in **Figure 11.7**. The pressure decreases on the way up along the vertical segment AA' and increases by the same amount on the way back down along segment $B'B$. Since no change in pressure occurs along the horizontal segment $A'B'$, the pressure is the same at *A* and *B*.

> **Problem-Solving Insight** The pressure at any point in a fluid depends on the vertical distance *h* of the point beneath the surface. However, for a given vertical distance, the pressure is the same, no matter where the point is located horizontally in the fluid.

Solution The pressure acting on the surface of the water is the atmospheric pressure of 1.01×10^5 Pa. Using this value as P_1 in Equation 11.4, we can determine a value for the pressure P_2 at

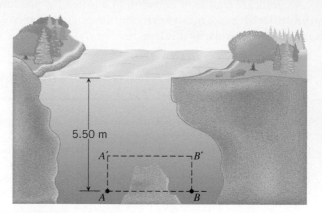

FIGURE 11.7 The pressures at points *A* and *B* are the same, since both points are located at the same vertical distance of 5.50 m beneath the surface of the water.

either point *A* or *B*, both of which are located 5.50 m under the water. **Table 11.1** gives the density of water as 1.000×10^3 kg/m^3.

$$P_2 = P_1 + \rho gh$$
$$P_2 = 1.01 \times 10^5 \text{ Pa} + (1.000 \times 10^3 \text{ kg/m}^3)(9.80 \text{ m/s}^2)(5.50 \text{ m})$$
$$= \boxed{1.55 \times 10^5 \text{ Pa}}$$

Figure 11.8 shows an irregularly shaped container of liquid. Reasoning similar to that used in Example 4 leads to the conclusion that the pressure is the same at points *A*, *B*, *C*, and *D*, since each is at the same vertical distance *h* beneath the surface. In effect, the arteries in our bodies are also an irregularly shaped "container" for the blood. The next example examines the blood pressure at different places in this "container."

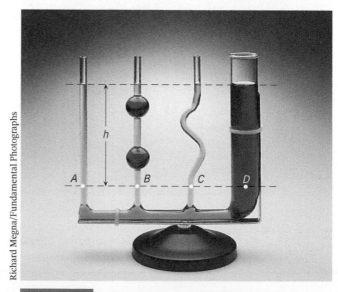

Richard Megna/Fundamental Photographs

FIGURE 11.8 Since points *A*, *B*, *C*, and *D* are at the same distance *h* beneath the liquid surface, the pressure at each of them is the same.

EXAMPLE 5 | BIO The Physics of Blood Pressure

Blood in the arteries is flowing, but as a first approximation, the effects of this flow can be ignored and the blood treated as a static fluid. Estimate the amount by which the blood pressure P_2 in the anterior tibial artery at the foot exceeds the blood pressure P_1 in the aorta at the heart when a person is **(a)** reclining horizontally as in **Figure 11.9a** and **(b)** standing as in **Figure 11.9b**.

Reasoning and Solution **(a)** When the body is horizontal, there is little or no vertical separation between the feet and the heart. Since $h = 0$ m,

$$P_2 - P_1 = \rho g h = \boxed{0 \text{ Pa}} \qquad \textbf{(11.4)}$$

(b) When an adult is standing up, the vertical separation between the feet and the heart is about 1.35 m, as **Figure 11.9b** indicates. **Table 11.1** gives the density of blood as 1060 kg/m³, so that

$$P_2 - P_1 = \rho g h = (1060 \text{ kg/m}^3)(9.80 \text{ m/s}^2)(1.35 \text{ m})$$

$$= \boxed{1.40 \times 10^4 \text{ Pa}}$$

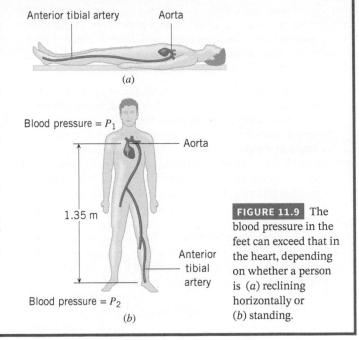

FIGURE 11.9 The blood pressure in the feet can exceed that in the heart, depending on whether a person is (a) reclining horizontally or (b) standing.

Sometimes fluid pressure places limits on how a job can be done. Conceptual Example 6 and Example 7 illustrate how fluid pressure restricts the height to which water can be pumped, as well as placing a limit on the length of a straw.

CONCEPTUAL EXAMPLE 6 | The Physics of Pumping Water

Figure 11.10 shows two methods for pumping water from a well. In one method, the pump is submerged in the water at the bottom of the well, while in the other, it is located at ground level. If the well is shallow, either technique can be used. However, if the well is very deep, only one of the methods works. Which pumping method works, **(a)** the submerged pump or **(b)** the pump located at ground level?

Reasoning To answer this question, we need to examine the nature of the job done by the pump in each place. The pump at the bottom of the well pushes water up the pipe, while the pump at ground level does not push water at all. Instead, the ground-level pump removes air from the pipe, creating a partial vacuum within it. (It's acting just like you do when drinking through a straw. You draw some of the air out of the straw, and the external air pressure pushes the liquid up into it, as discussed in Example 7.)

Answer (b) is incorrect. As the pump at ground level removes air from the pipe, the pressure above the water within the pipe is reduced (see point A in the drawing). The greater air pressure outside the pipe (see point B) pushes water up the pipe. However, even the strongest pump can only remove *all* of the air. Once the air is completely removed, an increase in pump strength does not increase the height to which the water is pushed by the external air pressure. Thus, the ground-level pump can only cause water to rise to a certain maximum height and cannot be used for very deep wells.

Answer (a) is correct. For a very deep well, the column of water becomes very tall, and the pressure at the bottom of the pipe becomes large, due to the pressure increment $\rho g h$ in the relation

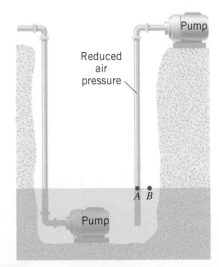

FIGURE 11.10 A water pump can be placed at the bottom of a well or at ground level. Conceptual Example 6 discusses the two placements.

$P_2 = P_1 + \rho g h$ (Equation 11.4). However, as long as the pump can push with sufficient strength to overcome the large pressure, it can shove the next increment of water into the pipe, so the method can be used for very deep wells.

Related Homework: Problem 20

EXAMPLE 7 | BIO The Physics of Drinking Straws—How Long Can They Be?

Consider for a moment how your favorite cold beverage ends up in your mouth, when you drink it with a straw. Many people would say that you "suck" the liquid up the straw with your mouth. Based on the fact that the liquid moves up the straw due to a difference in pressure, what is the maximum length of a vertical drinking straw?

Reasoning and Solution   **Figure 11.11a** shows that when an open-ended drinking straw is placed in a liquid, the liquid quickly fills the straw to the same height as the top of the liquid in the glass. One atmosphere of pressure is pushing down on the top of the liquid in the glass, as well as the liquid in the open straw. In **Figure 11.11b**, the person drinking from the straw forms a seal around the straw with their lips and then inhales through their mouth by contracting their diaphragm muscle. This increases the volume of the lungs and lowers the air pressure inside them. Since the air in the straw and lungs share the same space, the air pressure in the straw above the liquid is also reduced. As a result, the higher atmospheric pressure outside of the straw pushes the liquid up the straw. The height that the column of liquid in the straw rises above the liquid in the glass depends on the difference in pressure between the top of the column (P_2) and the liquid in the glass (P_1). If the pressure difference is great enough, the liquid is pushed up into the mouth, where it can be swallowed. The height of the column of liquid in the straw can be determined from Equation 11.4:

$$P_2 = P_1 + \rho g h \Rightarrow h = \frac{P_2 - P_1}{\rho g}$$

The maximum value of h will be determined by the maximum value of the pressure difference, $P_2 - P_1$. The value of P_2 is 1 atmosphere, or 1.103×10^5 Pa, and the smallest pressure for P_1 is, in principle, 0 Pa, which assumes that all of the air above the liquid in the straw is removed. Thus, under the best of circumstances, the greatest pressure difference can only be 1 atmosphere. Using this information, we can calculate the maximum value for h.

$$h = \frac{1.01 \times 10^5 \text{ Pa} - 0 \text{ Pa}}{(1.000 \times 10^3 \text{ kg/m}^3)(9.80 \text{ m/s}^2)} = \boxed{10.3 \text{ m}}$$

Here, we have assumed that the density of the liquid in the glass is equal to that of fresh water. This is a fairly long straw at over 33 feet! Since the act of breathing in only lowers the pressure in the straw, but does not create the condition of zero pressure, we would expect the maximum length of the straw to be somewhat less than 10.3 m. Simple experiments, which you could recreate for yourself, show that liquid can be drawn up to a height of at least 6 m. This result also holds for the situation described in Example 6 for the case in which the pump is located at ground level (see Figure 11.10). The maximum depth from which the ground-level can pump fresh water is 10.3 m.

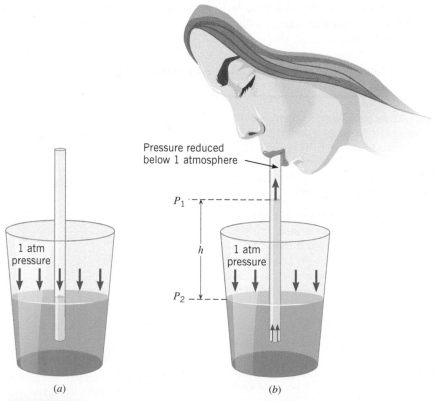

Pressure reduced
below 1 atmosphere

P_1

h

P_2

1 atm
pressure

1 atm
pressure

(a)

(b)

FIGURE 11.11 (a) An open-ended straw is placed in a glass of liquid. The liquid rises in the straw to match the height of the top of the liquid in the glass. The pressure at the top of the liquid is 1 atmosphere both inside and outside of the straw. (b) When the drinker inhales, the air pressure above the liquid inside the straw is reduced below atmospheric pressure. The higher atmospheric pressure outside pushes the liquid up the straw. The height of the liquid column in the straw depends on the pressure difference between the top of the liquid in the glass and the top of the liquid in the straw.

Related Homework: Problem 20

Check Your Understanding

(The answers are given at the end of the book.)

4. A scuba diver is swimming under water, and **CYU Figure 11.1** shows a plot of the water pressure acting on the diver as a function of time. In each of the three regions, **(a)** A → B, **(b)** B → C, and **(c)** C → D, does the depth of the diver increase, decrease, or remain constant?

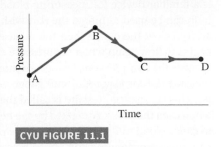

CYU FIGURE 11.1

5. A 15-meter-high tank is closed and completely filled with water. A valve is then opened at the bottom of the tank and water begins to flow out. When the water stops flowing, will the tank be completely empty, or will there still be a noticeable amount of water in it?

6. Could you use a straw to sip a drink on the moon? **(a)** Yes. It would be no different than drinking with a straw on earth. **(b)** No, because there is no air on the moon and, therefore, no air pressure to push the liquid up the straw. **(c)** Yes, and it is easier on the moon because the acceleration due to gravity on the moon is only $\frac{1}{6}$ of that on the earth.

7. A scuba diver is below the surface of the water when a storm approaches, dropping the air pressure above the water. Would a sufficiently sensitive pressure gauge attached to his wrist register this drop in air pressure? Assume that the diver's wrist does not move as the storm approaches.

11.4 | Pressure Gauges

One of the simplest pressure gauges is the mercury barometer used for measuring atmospheric pressure. This device is a tube sealed at one end, filled completely with mercury, and then inverted, so that the open end is under the surface of a pool of mercury (see **Figure 11.12**). Except for a negligible amount of mercury vapor, the space above the mercury in the tube is empty, and the pressure P_1 is nearly zero there. The pressure P_2 at point A at the bottom of the mercury column is the same as the pressure at point B— namely, atmospheric pressure—because these two points are at the same level. With $P_1 = 0$ Pa and $P_2 = P_{atm}$, it follows from Equation 11.4 that $P_{atm} = 0$ Pa $+ \rho gh$. Thus, the atmospheric pressure can be determined from the height h of the mercury in the tube, the density ρ of mercury, and the acceleration due to gravity. Usually, weather forecasters report the pressure in terms of the height h, expressing it in millimeters or inches of mercury. For instance, using $P_{atm} = 1.013 \times 10^5$ Pa and $\rho = 13.6 \times 10^3$ kg/m³ for the density of mercury, we find that $h = P_{atm}/(\rho g) = 760$ mm (29.9 inches).* Slight variations from this value occur, depending on weather conditions and altitude.

Figure 11.13 shows another kind of pressure gauge, the open-tube manometer. The phrase "open-tube" refers to the fact that one side of the U-tube is open to atmospheric pressure. The tube contains a liquid, often mercury, and its other side is connected to the container whose pressure P_2 is to be measured. When the pressure in the container is equal to the atmospheric pressure, the liquid levels in both sides of the U-tube are the same. When the pressure in the container is greater than atmospheric pressure, as in **Figure 11.13**, the liquid in the tube is pushed downward on the left side and upward on the right side. The relation $P_2 = P_1 + \rho gh$ can be used to determine the container pressure. Atmospheric pressure exists at the top of the right column, so that $P_1 = P_{atm}$. The pressure P_2 is the same at points A and B, so we find that $P_2 = P_{atm} + \rho gh$, or

$$P_2 - P_{atm} = \rho gh$$

The height h is proportional to $P_2 - P_{atm}$, which is called the **gauge pressure**. The gauge pressure is the amount by which the container pressure differs from atmospheric pressure. The actual value for P_2 is called the **absolute pressure**.

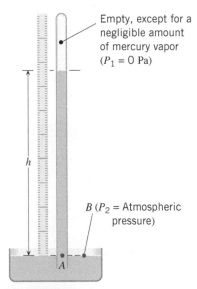

FIGURE 11.12 A mercury barometer.

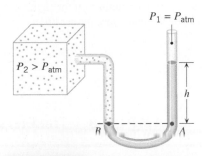

FIGURE 11.13 The U-shaped tube is called an open-tube manometer and can be used to measure the pressure P_2 in a container.

---

*A pressure of one millimeter of mercury is sometimes referred to as one *torr*, to honor the inventor of the barometer, Evangelista Torricelli (1608–1647). Thus, one atmosphere of pressure is 760 torr.

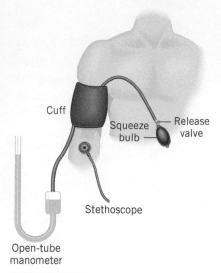

Cuff

Squeeze bulb — Release valve

Stethoscope

Open-tube manometer

FIGURE 11.14 A sphygmomanometer is used to measure blood pressure.

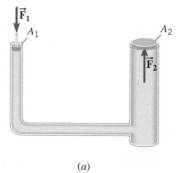

(a)

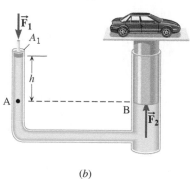

(b)

FIGURE 11.15 (a) An external force $\vec{F}_1$ is applied to the piston on the left. As a result, a force $\vec{F}_2$ is exerted on the cap on the chamber on the right. (b) The familiar hydraulic car lift.

Problem-Solving Insight When solving problems that deal with pressure, be sure to note the distinction between gauge pressure and absolute pressure.

BIO **THE PHYSICS OF . . . a sphygmomanometer.** The sphygmomanometer is a familiar device for measuring blood pressure. As **Figure 11.14** illustrates, a squeeze bulb can be used to inflate the cuff with air, which cuts off the flow of blood through the artery below the cuff. When the release valve is opened, the cuff pressure drops. Blood begins to flow again when the pressure created by the heart at the peak of its beating cycle exceeds the cuff pressure. Using a stethoscope to listen for the initial flow, the operator can measure the corresponding cuff gauge pressure with, for example, an open-tube manometer. This cuff gauge pressure is called the *systolic* pressure. Eventually, there comes a point when even the pressure created by the heart at the low point of its beating cycle is sufficient to cause blood to flow. Identifying this point with the stethoscope, the operator can measure the corresponding cuff gauge pressure, which is referred to as the *diastolic* pressure. The systolic and diastolic pressures are reported in millimeters of mercury, and values of less than 120 and 80, respectively, are typical of a young, healthy heart.

11.5 # Pascal's Principle

As we have seen, the pressure in a fluid increases with depth, due to the weight of the fluid above the point of interest. A completely enclosed fluid may be subjected to an additional pressure by the application of an external force. For example, **Figure 11.15a** shows two interconnected cylindrical chambers. The chambers have different diameters and, together with the connecting tube, are completely filled with a liquid. The larger chamber is sealed at the top with a cap, while the smaller one is fitted with a movable piston. Consider the pressure P_1 at a point immediately beneath the piston. According to the definition of pressure, it is the magnitude F_1 of the external force divided by piston area A_1, or $P_1 = F_1/A_1$. If it is necessary to know the pressure P_2 *at any deeper place in the liquid*, we just add to the value of P_1 the increment $\rho g h$, which takes into account the depth h below the piston: $P_2 = P_1 + \rho g h$. The important feature here is this: The pressure P_1 adds to the pressure $\rho g h$ due to the depth of the liquid at any point, whether that point is in the smaller chamber, the connecting tube, or the larger chamber. Therefore, if the applied pressure P_1 is increased or decreased, the pressure at any other point within the confined liquid changes correspondingly. This behavior is described by **Pascal's principle**.

PASCAL'S PRINCIPLE

Any change in the pressure applied to a completely enclosed fluid is transmitted undiminished to all parts of the fluid and the enclosing walls.

The usefulness of the arrangement in **Figure 11.15a** becomes apparent when we calculate the force F_2 applied by the liquid to the cap on the right side. The area of the cap is A_2 and the pressure there is P_2. As long as the tops of the left and right chambers are at the same level, the pressure increment $\rho g h$ is zero, so that the relation $P_2 = P_1 + \rho g h$ becomes $P_2 = P_1$. Consequently, $F_2/A_2 = F_1/A_1$, and

$$F_2 = F_1\left(\frac{A_2}{A_1}\right) \tag{11.5}$$

If area A_2 is larger than area A_1, a large force $\vec{F}_2$ can be applied to the cap on the right chamber, starting with a smaller force $\vec{F}_1$ on the left. Depending on the ratio of the areas A_2/A_1, the force $\vec{F}_2$ can be large indeed, as in the familiar hydraulic car lift shown in **Figure 11.15b**. In this device the force $\vec{F}_2$ is not applied to a cap that seals the larger chamber, but, rather, to a movable plunger that lifts a car. Examples 8 and 9 deal with a hydraulic car lift.

EXAMPLE 8 | The Physics of a Hydraulic Car Lift

In the hydraulic car lift shown in **Figure 11.15b**, the input piston on the left has a radius of $r_1 = 0.0120$ m and a negligible weight. The output plunger on the right has a radius of $r_2 = 0.150$ m. The combined weight of the car and the plunger is 20 500 N. Since the output force has a magnitude of $F_2 = 20\,500$ N, it supports the car. Suppose that the bottom surfaces of the piston and plunger are at the same level, so that $h = 0$ m in **Figure 11.15b**. What is the magnitude F_1 of the input force needed so that $F_2 = 20\,500$ N?

Reasoning When the bottom surfaces of the piston and plunger are at the same level, as in **Figure 11.15a**, Equation 11.5 applies, and we can use it to determine F_1.

> **Problem-Solving Insight** Note that the relation $F_1 = F_2(A_1/A_2)$, which results from Pascal's principle, applies only when the points 1 and 2 lie at the same depth ($h = 0$ m) in the fluid.

Solution According to Equation 11.5, we have

$$F_2 = F_1\left(\frac{A_2}{A_1}\right) \quad \text{or} \quad F_1 = F_2\left(\frac{A_1}{A_2}\right)$$

Using $A = \pi r^2$ for the circular areas of the piston and plunger, we find that

$$F_1 = F_2\left(\frac{A_1}{A_2}\right) = F_2\left(\frac{\pi r_1^2}{\pi r_2^2}\right) = (20\ 500\ \text{N})\frac{(0.0120\ \text{m})^2}{(0.150\ \text{m})^2} = \boxed{131\ \text{N}}$$

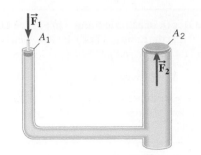

(a)

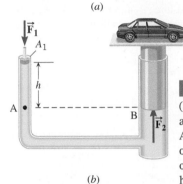

(b)

FIGURE 11.15 (REPEATED)
(a) An external force $\vec{F}_1$ is applied to the piston on the left. As a result, a force $\vec{F}_2$ is exerted on the cap on the chamber on the right. (b) The familiar hydraulic car lift.

Analyzing Multiple-Concept Problems

EXAMPLE 9 | A Car Lift Revisited

The data are the same as in Example 8. Suppose now, however, that the bottom surfaces of the piston and plunger are at *different* levels, such that $h = 1.10$ m in **Figure 11.15b**. The car lift uses hydraulic oil that has a density of $8.00 \times 10^2\ \text{kg/m}^3$. What is the magnitude F_1 of the input force that is now needed to produce an output force having a magnitude of $F_2 = 20\ 500$ N?

Reasoning Since the bottom surfaces of the piston and plunger are at different levels, Equation 11.5 no longer

applies. Our approach will be based on the definition of pressure (Equation 11.3) and the relation between pressure and depth given in Equation 11.4. It is via Equation 11.4 that we will take into account the fact that the bottom of the output plunger is $h = 1.10$ m below the bottom of the input piston.

Knowns and Unknowns We have the following data:

Description	Symbol	Value	Comment
Radius of input piston	r_1	0.0120 m	
Radius of output piston	r_2	0.150 m	
Magnitude of output force	F_2	20 500 N	
Density of hydraulic oil	ρ	$8.00 \times 10^2\ \text{kg/m}^3$	
Level difference between input piston and output plunger	h	1.10 m	See **Figure 11.15b**.
Unknown Variables			
Magnitude of input force	F_1	?	

Modeling the Problem

STEP 1 Force and Pressure The input force F_1 is determined by the pressure P_1 acting on the bottom surface of the input piston, which has an area A_1. According to Equation 11.3, the input force is $F_1 = P_1 A_1$, where $A_1 = \pi r_1^2$ because the piston has a circular cross section with a radius r_1. With this expression for the area A_1, the input force can be written as in Equation 1 at the right. We have a value for r_1, but P_1 is unknown, so we turn to Step 2 in order to obtain it.

$$F_1 = P_1 \pi r_1^2 \qquad (1)$$

?

STEP 2 Pressure and Depth in a Static Fluid In **Figure 11.15b**, the bottom surface of the plunger at point B is at the same level as point A, which is at a depth h beneath the input piston. Equation 11.4 applies, so that

$$P_2 = P_1 + \rho gh \quad \text{or} \quad \boxed{P_1 = P_2 - \rho gh}$$

This result for P_1 can be substituted into Equation 1, as shown at the right. Now it is necessary to have a value for P_2, the pressure at the bottom surface of the output plunger, which we obtain in Step 3.

$$F_1 = P_1 \pi r_1^2 \qquad (1)$$

$$\boxed{P_1 = P_2 - \rho gh} \qquad (2)$$

$$\boxed{?}$$

STEP 3 Pressure and Force By using Equation 11.3, we can relate the unknown pressure P_2 to the given output force F_2 and the area A_2 of the bottom surface of the output plunger: $P_2 = F_2/A_2$. The area A_2 is a circle with a radius r_2, so $A_2 = \pi r_2^2$. Thus, the pressure P_2 is

$$\boxed{P_2 = \frac{F_2}{\pi r_2^2}}$$

At the right, this result is substituted into Equation 2.

$$F_1 = P_1 \pi r_1^2 \qquad (1)$$

$$\boxed{P_1 = P_2 - \rho gh} \qquad (2)$$

$$\boxed{P_2 = \frac{F_2}{\pi r_2^2}}$$

Solution Combining the results of each step algebraically, we find that

STEP 1 STEP 2 STEP 3

$$F_1 = P_1 \pi r_1^2 = (P_2 - \rho gh)\pi r_1^2 = \left(\frac{F_2}{\pi r_2^2} - \rho gh \right)\pi r_1^2$$

Thus, we find that the necessary input force is

$$F_1 = \frac{F_2 r_1^2}{r_2^2} - \rho gh \pi r_1^2$$

$$= \frac{(20\,500\ \text{N})(0.0120\ \text{m})^2}{(0.150\ \text{m})^2} - (8.00 \times 10^2\ \text{kg/m}^3)(9.80\ \text{m/s}^2)(1.10\ \text{m})\pi(0.0120\ \text{m})^2$$

$$= \boxed{127\ \text{N}}$$

The answer here is less than the answer in Example 8 because the weight of the 1.10-m column of hydraulic oil provides some of the input force to support the car and plunger.

Related Homework: Problem 31

In a device such as a hydraulic car lift, the same amount of work is done by both the input and output forces in the absence of friction. The larger output force $\vec{\mathbf{F}}_2$ moves through a smaller distance, while the smaller input force $\vec{\mathbf{F}}_1$ moves through a larger distance. The work, being the product of the magnitude of the force and the distance, is the same in either case since mechanical energy is conserved (assuming that friction is negligible).

An enormous variety of clever devices use hydraulic fluids, just as the car lift does. In **Figure 11.16**, for instance, a hydraulic fluid multiplies a small input force into the large output force required to operate the digging scoop of the excavator.

FIGURE 11.16 Excavators such as this one are a familiar sight at construction jobs. They use hydraulic fluids to generate the large output forces necessary for digging. The shiny cylinder attached to the digging scoop is a sure sign that a hydraulic fluid is at work.

11.6 | Archimedes' Principle

Anyone who has tried to push a beach ball under the water has felt how the water pushes back with a strong upward force. This upward force is called the **buoyant force**, and all fluids apply such a force to objects that are immersed in them. The buoyant force exists because fluid pressure is larger at greater depths.

In **Interactive Figure 11.17** a cylinder of height h is being held under the surface of a liquid. The pressure P_1 on the top face generates the downward force P_1A, where A is the area of the face. Similarly, the pressure P_2 on the bottom face generates the upward force P_2A. Since the pressure is greater at greater depths, the upward force exceeds the downward force. Consequently, the liquid applies to the cylinder a net upward force, or buoyant force, whose magnitude F_B is

$$F_B = P_2A - P_1A = (P_2 - P_1)A = \rho g h A$$

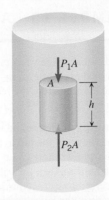

INTERACTIVE FIGURE 11.17 The fluid applies a downward force P_1A to the top face of the submerged cylinder and an upward force P_2A to the bottom face.

We have substituted $P_2 - P_1 = \rho g h$ from Equation 11.4 into this result. In so doing, we find that the buoyant force equals $\rho g h A$. The quantity hA is the volume of liquid that the cylinder moves aside or displaces in being submerged, and ρ denotes the density of the liquid, not the density of the material from which the cylinder is made. Therefore, $\rho h A$ gives the mass m of the displaced fluid, so that the buoyant force equals mg, the weight of the displaced fluid. The phrase "weight of the displaced fluid" refers to the weight of the fluid that would spill out if the container were filled to the brim before the cylinder is inserted into the liquid. The buoyant force is not a new type of force. It is just the name given to the net upward force exerted by the fluid on the object.

The shape of the object in **Interactive Figure 11.17** is not important. No matter what its shape, the buoyant force pushes it upward in accord with **Archimedes' principle**. It was an impressive accomplishment that the Greek scientist Archimedes (ca. 287–212 BC) discovered the essence of this principle so long ago.

> **ARCHIMEDES' PRINCIPLE**
>
> **Any fluid applies a buoyant force to an object that is partially or completely immersed in it; the magnitude of the buoyant force equals the weight of the fluid that the object displaces:**
>
> $$\underbrace{F_B}_{\substack{\text{Magnitude of} \\ \text{buoyant force}}} = \underbrace{W_{\text{fluid}}}_{\substack{\text{Weight of} \\ \text{displaced fluid}}} \qquad (11.6)$$

The effect that the buoyant force has depends on its strength compared with the strengths of the other forces that are acting. For example, if the buoyant force is strong enough to balance the force of gravity, an object will float in a fluid. **Figure 11.18** explores this possibility. In part a, a block that weighs 100 N displaces some liquid, and the liquid applies a buoyant force F_B to the block, according to Archimedes' principle. Nevertheless, if the block were released, it would fall further into the liquid because the buoyant force is not sufficiently strong to balance the weight of the block. In part b, however, enough of the block is submerged to provide a buoyant force that can balance the 100-N weight, so the block is in equilibrium and floats when released. If the buoyant force were not large enough to balance the weight, even with the block completely submerged, the block would sink. Even if an object sinks, there is still a buoyant force acting on it; it's just that the buoyant force is not large enough to balance the weight. Example 9 provides additional insight into what determines whether an object floats or sinks in a fluid.

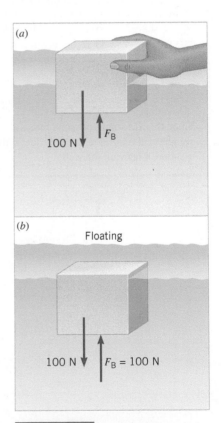

FIGURE 11.18 (a) An object of weight 100 N is being immersed in a liquid. The deeper the object is, the more liquid it displaces, and the stronger the buoyant force is. (b) The buoyant force matches the 100-N weight, so the object floats.

EXAMPLE 10 | A Swimming Raft

A solid, square pinewood raft measures 4.0 m on a side and is 0.30 m thick. **(a)** Determine whether the raft floats in water, and **(b)** if so, how much of the raft is beneath the surface (see the distance h in **Figure 11.19**).

Reasoning To determine whether the raft floats, we will compare the weight of the raft to the maximum possible buoyant force and see whether there could be enough buoyant force to balance the weight. If so, then the value of the distance h can be obtained by utilizing the fact that the floating raft is in equilibrium, with the magnitude of the buoyant force equaling the raft's weight.

Solution **(a)** The weight of the raft is $W_{raft} = m_{pine}g$ (Equation 4.5), where m_{pine} is the mass of the raft and can be calculated as $m_{pine} = \rho_{pine}V_{pine}$ (Equation 11.1). The pinewood's density is $\rho_{pine} = 550$ kg/m³ (**Table 11.1**), and its volume is $V_{pine} = 4.0$ m × 4.0 m × 0.30 m = 4.8 m³. Thus, we find the weight of the raft to be

$$W_{raft} = m_{pine}g = (\rho_{pine}V_{pine})g$$
$$= (550 \text{ kg/m}^3)(4.8 \text{ m}^3)(9.80 \text{ m/s}^2) = 26\ 000 \text{ N}$$

The maximum possible buoyant force occurs when the entire raft is under the surface, displacing a volume of water that is $V_{water} = V_{pine} = 4.8$ m³. According to Archimedes' principle, the weight of this volume of water is the maximum buoyant force F_B^{MAX}. It can be obtained using the density of water:

$$F_B^{MAX} = \rho_{water}V_{water}g$$
$$= (1.000 \times 10^3 \text{ kg/m}^3)(4.8 \text{ m}^3)(9.80 \text{ m/s}^2) = 47\ 000 \text{ N}$$

Since the maximum possible buoyant force exceeds the 26 000-N weight of the raft, the raft will float only partially submerged at a distance h beneath the water.

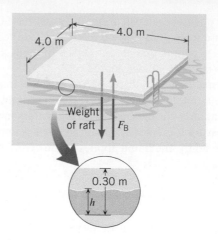

FIGURE 11.19 A raft floating with a distance h beneath the water.

(b) We now find the value of h. The buoyant force balances the raft's weight, so $F_B = 26\ 000$ N. However, according to Equation 11.6, the magnitude of the buoyant force is also the weight of the displaced water, so $F_B = 26\ 000$ N $= W_{water}$. Using the density of water, we can also express the weight of the displaced water as $W_{water} = \rho_{water}V_{water}g$, where the water volume is $V_{water} = 4.0$ m × 4.0 m × h. As a result,

$$26\ 000 \text{ N} = W_{water} = \rho_{water}(4.0 \text{ m} \times 4.0 \text{ m} \times h)g$$

$$h = \frac{26\ 000 \text{ N}}{\rho_{water}(4.0 \text{ m} \times 4.0 \text{ m})g}$$

$$= \frac{26\ 000 \text{ N}}{(1.000 \times 10^3 \text{ kg/m}^3)(4.0 \text{ m} \times 4.0 \text{ m})(9.80 \text{ m/s}^2)}$$

$$= \boxed{0.17 \text{ m}}$$

In order to decide whether the raft will float in part (a) of Example 10, we compared the weight of the raft $(\rho_{pine}V_{pine})g$ to the maximum possible buoyant force $(\rho_{water}V_{water})g = (\rho_{water}V_{pine})g$. The comparison depends only on the densities ρ_{pine} and ρ_{water}. The take-home message is this: **Any object that is *solid throughout* will float in the liquid if its density is less than the density of the liquid, and will sink in the liquid if its density is greater than the density of the liquid.** For instance, at 0 °C ice has a density of 917 kg/m³, whereas water has a density of 1000 kg/m³. Therefore, ice floats in water. In the case that the object and liquid have equal densities, the weight of the object and the buoyant force have equal magnitudes, and the object will suspend motionless in the medium, that is, it will neither sink nor float. In such a case, the object is said to have **neutral buoyancy** in the liquid.

Although a solid piece of a high-density material like steel will sink in water, such materials can, nonetheless, be used to make floating objects. A supertanker, for example, floats because it is *not* solid metal. It contains enormous amounts of empty space and, because of its shape, displaces enough water to balance its own large weight. Conceptual Example 11 focuses on an interesting aspect of a floating ship.

CONCEPTUAL EXAMPLE 11 | How Much Water Is Needed to Float a Ship?

A ship floating in the ocean is a familiar sight. But is all that water really necessary? Can an ocean vessel float in the amount of water that a swimming pool contains, for instance?

Reasoning and Solution In principle, a ship can float in much less than the amount of water in a swimming pool. To see why,

look at **Figure 11.20**. Part *a* shows the ship floating in the ocean because it contains empty space within its hull and displaces enough water to balance its own weight. Part *b* shows the water that the ship displaces, which, according to Archimedes' principle, has a weight that equals the ship's weight. Although this wedge-shaped portion of water represents the water *displaced* by the ship, it is not the amount that must be present to float the ship, as part *c* illustrates. This part of the drawing shows a canal, the cross section of which matches the shape in part *b*. **All that is needed, in principle, is a thin section of water that separates the hull of the floating ship from the sides of the canal.** This thin section of water could have a very small volume indeed.

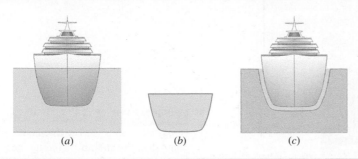

(*a*)   (*b*)   (*c*)

FIGURE 11.20 (*a*) A ship floating in the ocean. (*b*) This is the water that the ship displaces. (*c*) The ship floats here in a canal that has a cross section similar in shape to that in part *b*.

THE PHYSICS OF . . . a state-of-charge battery indicator. Archimedes' principle is used in some car batteries to alert the owner that recharging is necessary, via a state-of-charge indicator, such as the one illustrated in **Figure 11.21**. The battery includes a viewing port that looks down through a plastic rod, which extends into the battery acid. Attached to the end of this rod is a "cage" containing a green ball. The cage has holes in it that allow the acid to enter. When the battery is charged, the density of the acid is great enough that its buoyant force makes the ball rise to the top of the cage, to just beneath the plastic rod. The viewing port shows a green dot. As the battery discharges, the density of the acid decreases. Since the buoyant force is the weight of the acid displaced by the ball, the buoyant force also decreases. As a result, the ball sinks into one of the two chambers oriented at an angle beneath the plastic rod (see **Figure 11.21**). With the ball no longer visible, the viewing port shows a dark or black dot, warning that the battery charge is low.

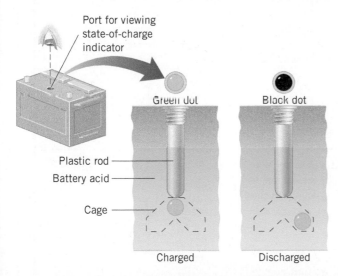

Port for viewing state-of-charge indicator

Green dot   Black dot

Plastic rod
Battery acid

Cage

Charged   Discharged

FIGURE 11.21 A state-of-charge indicator for a car battery.

Archimedes' principle has allowed us to determine how an object can float in a liquid. This principle also applies to gases, as the next example illustrates.

Analyzing Multiple-Concept Problems

EXAMPLE 12 | The Physics of a Goodyear Airship

Normally, a Goodyear airship, such as the one in **Figure 11.22**, contains about $5.40 \times 10^3 \text{ m}^3$ of helium (He), whose density is 0.179 kg/m^3. Find the weight W_L of the load that the airship can carry in equilibrium at an altitude where the density of air is 1.20 kg/m^3.

Reasoning The airship and its load are in equilibrium. Thus, the buoyant force F_B applied to the airship by the surrounding air balances the weight W_{He} of the helium and the weight W_L of the load, including the solid parts of the airship. The free-body diagram in **Figure 11.22b** shows these forces. The buoyant force is the weight of the air that the ship displaces, according to Archimedes' principle. The weight of air or helium is given as $W = mg$ (Equation 4.5). Using the density ρ and volume V, we can express the mass m as $m = \rho V$ (Equation 11.1).

F_B

W_{He}

W_L

(a)

(b) Free-body diagram of the airship

FIGURE 11.22 (a) A helium-filled Goodyear airship. (b) The free-body diagram of the airship, including the load weight W_L, the weight W_{He} of the helium, and the buoyant force F_B.

Knowns and Unknowns The following table summarizes the data:

Description	Symbol	Value
Volume of helium in airship	V_{He}	$5.40 \times 10^3 \text{ m}^3$
Density of helium	ρ_{He}	0.179 kg/m^3
Density of air	ρ_{air}	1.20 kg/m^3
Unknown Variables		
Weight of load	W_L	?

Modeling the Problem

STEP 1 Equilibrium Because the forces in the free-body diagram (see **Figure 11.22b**) balance at equilibrium, we have

$$W_{He} + W_L = F_B$$

Rearranging this result gives Equation 1 at the right. Neither the buoyant force F_B nor the weight W_{He} of the helium is known. We will determine them in Steps 2 and 3.

$$W_L = F_B - W_{He} \qquad \text{(1)}$$

STEP 2 Weight and Density According to Equation 4.5, the weight is given by $W = mg$. On the other hand, the density ρ is defined as the mass m divided by the volume V (see Equation 11.1), so we know that $m = \rho V$. Thus, the weight of the helium in the airship is

$$\boxed{W_{He} = m_{He}g = \rho_{He}V_{He}g}$$

The result can be substituted into Equation 1, as shown at the right. We turn now to Step 3 to evaluate the buoyant force.

$$W_L = F_B - W_{He} \qquad \text{(1)}$$
$$\boxed{W_{He} = \rho_{He}V_{He}g}$$

STEP 3 Archimedes' Principle The buoyant force is given by Archimedes' principle as the weight of the air displaced by the airship. Thus, following the approach in Step 2, we can write the buoyant force as follows:

$$F_B = W_{air} = \rho_{air}V_{air}g$$

In this result, V_{air} is very nearly the same as V_{He}, since the volume occupied by the materials of the ship's outer structure is negligible compared to V_{He}. Assuming that $V_{air} = V_{He}$, we see that the expression for the buoyant force becomes

$$\boxed{F_B = \rho_{air}V_{He}g}$$

Substitution of this value for the buoyant force into Equation 1 is shown at the right.

$$W_L = F_B - W_{He} \qquad \text{(1)}$$
$$\boxed{W_{He} = \rho_{He}V_{He}g}$$
$$\boxed{F_B = \rho_{air}V_{He}g}$$

Solution   Combining the results of each step algebraically, we find that

$$\underset{\downarrow}{\text{STEP 1}} \qquad \underset{\downarrow}{\text{STEP 2}} \qquad \underset{\downarrow}{\text{STEP 3}}$$

$$W_L = F_B - W_{He} = F_B - \rho_{He}V_{He}g = \rho_{air}V_{He}g - \rho_{He}V_{He}g$$

The weight of the load that the airship can carry at an altitude where $\rho_{air} = 1.20 \text{ kg/m}^3$ is, then,

$$W_L = (\rho_{air} - \rho_{He})V_{He}g = (1.20 \text{ kg/m}^3 - 0.179 \text{ kg/m}^3)(5.40 \times 10^3 \text{ m}^3)(9.80 \text{m/s}^2)$$

$$= \boxed{5.40 \times 10^4 \text{ N}}$$

Related Homework: Problem 43

Check Your Understanding

(The answers are given at the end of the book.)

8. A glass is filled to the brim with water and has an ice cube floating in it. When the ice cube melts, what happens? **(a)** Water spills out of the glass. **(b)** The water level in the glass drops. **(c)** The water level in the glass does not change.

9. A steel beam is suspended completely under water by a cable that is attached to one end of the beam, so it hangs vertically. Another identical beam is also suspended completely under water, but by a cable that is attached to the center of the beam, so it hangs horizontally. Which beam, if either, experiences the greater buoyant force? Neglect any change in water density with depth.

10. A glass beaker, filled to the brim with water, is resting on a scale. A solid block is then placed in the water, causing some of it to spill over. The water that spills is wiped away, and the beaker is still filled to the brim. How do the initial and final readings on the scale compare if the block is made from **(a)** wood (whose density is less than that of water) and **(b)** iron (whose density is greater than that of water)?

11. On a distant planet the acceleration due to gravity is less than it is on earth. Would you float more easily in water on this planet than on earth?

12. As a person dives toward the bottom of a swimming pool, the pressure increases noticeably. Does the buoyant force acting on her also increase? Neglect any change in water density with depth.

13. A pot is partially filled with water, in which a plastic cup is floating. Inside the floating cup is a small block of lead. When the lead block is removed from the cup and placed into the water, it sinks to the bottom. When this happens, does the water level in the pot **(a)** rise, **(b)** fall, or **(c)** remain the same?

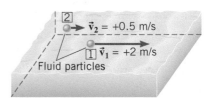

FIGURE 11.23 Two fluid particles in a stream. At different locations in the stream the particle velocities may be different, as indicated by $\vec{v}_1$ and $\vec{v}_2$.

11.7 | Fluids in Motion

Fluids can move or flow in many ways. Water may flow smoothly and slowly in a quiet stream or violently over a waterfall. The air may form a gentle breeze or a raging tornado. To deal with such diversity, it helps to identify some of the basic types of fluid flow.

Fluid flow can be steady or unsteady. In **steady flow** the velocity of the fluid particles at any point is constant as time passes. For instance, in **Figure 11.23** a fluid particle flows with a velocity of $\vec{v}_1 = +2$ m/s past point 1. In steady flow every particle passing through this point has this same velocity. At another location the velocity may be different, as in a river, which usually flows fastest near its center and slowest near its banks. Thus, at point 2 in the figure, the fluid velocity is $\vec{v}_2 = +0.5$ m/s, and if the flow is steady, all particles passing through this point have a velocity of $+0.5$ m/s. **Unsteady flow** exists whenever the velocity at a point in the fluid changes as time passes. **Turbulent flow** is an extreme kind of unsteady flow and occurs when there are sharp obstacles or bends in the path of a fast-moving fluid, as in the rapids in **Figure 11.24**. In turbulent flow, the velocity at a point changes erratically from moment to moment, both in magnitude and in direction.

Frans Lemmens/Corbis Unreleased/Getty Images

FIGURE 11.24 The flow of water in white-water rapids is an example of turbulent flow.

Fluid flow can be compressible or incompressible. Most liquids are nearly incompressible; that is, the density of a liquid remains almost constant as the pressure changes. To a good approximation, then, liquids flow in an incompressible manner. In contrast, gases are highly compressible. However, there are situations in which the density of a flowing gas remains constant enough that the flow can be considered incompressible.

Fluid flow can be viscous or nonviscous. A viscous fluid, such as honey, does not flow readily and is said to have a large viscosity.* In contrast, water is less viscous and flows more readily; water has a smaller viscosity than honey. The flow of a viscous fluid is an energy-dissipating process. The viscosity hinders neighboring layers of fluid from sliding freely past one another. A fluid with zero viscosity flows in an unhindered manner with no dissipation of energy. Although no real fluid has zero viscosity at normal temperatures, some fluids have negligibly small viscosities. An incompressible, nonviscous fluid is called an **ideal fluid**.

When the flow is steady, **streamlines** are often used to represent the trajectories of the fluid particles. A streamline is a line drawn in the fluid such that a tangent to the streamline at any point is parallel to the fluid velocity at that point. **Figure 11.25** shows the velocity vectors at three points along a streamline. The fluid velocity can vary (in both magnitude and direction) from point to point along a streamline, but at any given point, the velocity is constant in time, as required by the condition of steady flow. In fact, steady flow is often called **streamline flow**.

Figure 11.26a illustrates a method for making streamlines visible by using small tubes to release a colored dye into the moving liquid. The dye does not immediately mix with the liquid and is carried along a streamline. In the case of a flowing gas, such as that in a wind tunnel, streamlines are often revealed by smoke streamers, as part *b* of the figure shows.

In steady flow, the pattern of streamlines is steady in time, and, as **Figure 11.26a** indicates, no two streamlines cross one another. If they did cross, every particle arriving at the crossing point could go one way or the other. This would mean that the velocity at the crossing point would change from moment to moment, a condition that does not exist in steady flow.

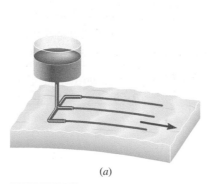

FIGURE 11.25 At any point along a streamline, the velocity vector of the fluid particle at that point is tangent to the streamline.

(a)   (b)

FIGURE 11.26 (a) In the steady flow of a liquid, a colored dye reveals the streamlines. (b) A smoke streamer reveals a streamline pattern for the air flowing around this skier, as he tests his position for air resistance in a wind tunnel.

Check Your Understanding

(*The answer is given at the end of the book.*)

14. In steady flow, the velocity $\vec{v}$ of a fluid particle at any point is constant in time. On the other hand, the fluid in a pipe accelerates when it moves from a larger-diameter section of the pipe into a smaller-diameter section, so the velocity is increasing during the transition. Does the condition of steady flow rule out such an acceleration?

*See Section 11.11 for a discussion of viscosity.

11.8 # The Equation of Continuity

Have you ever used your thumb to control the water flowing from the end of a hose, as in **Figure 11.27**? If so, you have seen that the water velocity increases when your thumb reduces the cross-sectional area of the hose opening. This kind of fluid behavior is described by the **equation of continuity**. This equation expresses the following simple idea: If a fluid enters one end of a pipe at a certain rate (e.g., 5 kilograms per second), then fluid must also leave at the same rate, assuming that there are no places between the entry and exit points to add or remove fluid. The mass of fluid per second (e.g., 5 kg/s) that flows through a tube is called the **mass flow rate**.

Animated **Figure 11.28** shows a small mass of fluid or fluid element (dark blue) moving along a tube. Upstream at position 2, where the tube has a cross-sectional area A_2, the fluid has a speed v_2 and a density ρ_2. Downstream at location 1, the corresponding quantities are A_1, v_1, and ρ_1. During a small time interval Δt, the fluid at point 2 moves a distance of $v_2 \Delta t$, as the drawing shows. The volume of fluid that has flowed past this point is the cross-sectional area times this distance, or $A_2 v_2 \Delta t$. The mass Δm_2 of this fluid element is the product of the density and volume: $\Delta m_2 = \rho_2 A_2 v_2 \Delta t$. Dividing Δm_2 by Δt gives the mass flow rate (the mass per second):

$$\text{Mass flow rate at position 2} = \frac{\Delta m_2}{\Delta t} = \rho_2 A_2 v_2 \qquad (11.7a)$$

Similar reasoning leads to the mass flow rate at position 1:

$$\text{Mass flow rate at position 1} = \frac{\Delta m_1}{\Delta t} = \rho_1 A_1 v_1 \qquad (11.7b)$$

Since no fluid can cross the sidewalls of the tube, the mass flow rates at positions 1 and 2 must be equal. However, these positions were selected arbitrarily, so the mass flow rate has the same value everywhere in the tube, an important result known as the **equation of continuity**. The equation of continuity is an expression of the fact that mass is conserved (i.e., neither created nor destroyed) as the fluid flows.

> **EQUATION OF CONTINUITY**
>
> The mass flow rate ($\rho A v$) has the same value at every position along a tube that has a single entry and a single exit point for fluid flow. For two positions along such a tube
>
> $$\rho_1 A_1 v_1 = \rho_2 A_2 v_2 \qquad (11.8)$$
>
> where ρ = fluid density (kg/m^3)
> A = cross-sectional area of tube (m^2)
> v = fluid speed (m/s)
>
> **SI Unit of Mass Flow Rate: kg/s**

The density of an incompressible fluid does not change during flow, so that $\rho_1 = \rho_2$, and the equation of continuity reduces to

Incompressible fluid $$A_1 v_1 = A_2 v_2 \qquad (11.9)$$

The quantity $A v$ represents the volume of fluid per second (measured in m^3/s, for instance) that passes through the tube and is referred to as the **volume flow rate Q**:

$$Q = \text{Volume flow rate} = A v \qquad (11.10)$$

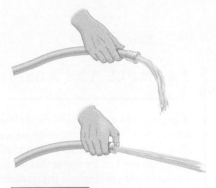

FIGURE 11.27 When the end of a hose is partially closed off, thus reducing its cross-sectional area, the fluid velocity increases.

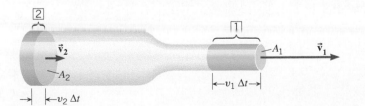

ANIMATED FIGURE 11.28 In general, a fluid flowing in a tube that has different cross-sectional areas A_1 and A_2 at positions 1 and 2 also has different velocities $\vec{v}_1$ and $\vec{v}_2$ at these positions.

Equation 11.9 shows that where the tube's cross-sectional area is large, the fluid speed is small, and, conversely, where the tube's cross-sectional area is small, the speed is large. Example 13 explores this behavior in more detail for the hose in **Figure 11.27**.

EXAMPLE 13 | A Garden Hose

A garden hose has an unobstructed opening with a cross-sectional area of 2.85×10^{-4} m², from which water fills a bucket in 30.0 s. The volume of the bucket is 8.00×10^{-3} m³ (about two gallons). Find the speed of the water that leaves the hose through **(a)** the unobstructed opening and **(b)** an obstructed opening with half as much area.

Reasoning   If we can determine the volume flow rate Q, we can obtain the speed of the water from Equation 11.10 as $v = Q/A$, since the area A is given. We can find the volume flow rate from the volume of the bucket and its fill time.

Solution   **(a)** The volume flow rate Q is equal to the volume of the bucket divided by the fill time. Therefore, the speed of the water is

$$v = \frac{Q}{A} = \frac{(8.00 \times 10^{-3}\,\text{m}^3)/(30.0\,\text{s})}{2.85 \times 10^{-4}\,\text{m}^2} = \boxed{0.936\ \text{m/s}} \qquad \textbf{(11.10)}$$

(b) Water can be considered incompressible, so the equation of continuity can be applied in the form $A_1 v_1 = A_2 v_2$. When the opening of the hose is unobstructed, its area is A_1, and the speed of the water is that determined in part (a)—namely, $v_1 = 0.936$ m/s. Since $A_2 = \frac{1}{2} A_1$, we find that

$$v_2 = \left(\frac{A_1}{A_2}\right) v_1 = \left(\frac{A_1}{\frac{1}{2} A_1}\right)(0.936\ \text{m/s}) = \boxed{1.87\ \text{m/s}} \qquad \textbf{(11.9)}$$

The next example applies the equation of continuity to the flow of blood.

EXAMPLE 14 | BIO The Physics of a Clogged Artery

In the condition known as atherosclerosis, a deposit, or atheroma, forms on the arterial wall and reduces the opening through which blood can flow. In the carotid artery in the neck, blood flows three times faster through a partially blocked region than it does through an unobstructed region. Determine the ratio of the effective radii of the artery at the two places.

Reasoning   Blood, like most liquids, is incompressible, and the equation of continuity in the form of $A_1 v_1 = A_2 v_2$ (Equation 11.9) can be applied. In applying this equation, we use the fact that the area of a circle is πr^2.

> **Problem-Solving Insight**   The equation of continuity in the form $A_1 v_1 = A_2 v_2$ applies only when the density of the fluid is constant. If the density is not constant, the equation of continuity is $\rho_1 A_1 v_1 = \rho_2 A_2 v_2$.

Solution   From Equation 11.9, it follows that

$$\underbrace{(\pi r_U^2) v_U}_{\substack{\text{Unobstructed} \\ \text{volume flow rate}}} = \underbrace{(\pi r_O^2) v_O}_{\substack{\text{obstructed} \\ \text{volume flow rate}}}$$

The ratio of the radii is

$$\frac{r_U}{r_O} = \sqrt{\frac{v_O}{v_U}} = \sqrt{3} = \boxed{1.7}$$

Check Your Understanding

(The answers are given at the end of the book.)

15. Water flows from left to right through the five sections (A, B, C, D, E) of the pipe shown in **CYU Figure 11.2**. In which section(s) does the water speed increase, decrease, and remain constant? Treat the water as an incompressible fluid.

	Speed Increases	**Speed Decreases**	**Speed Is Constant**
(a)	A, B	D, E	C
(b)	D	B	A, C, E
(c)	D, E	A, B	C
(d)	B	D	A, C, E
(e)	A, B	C, D	E

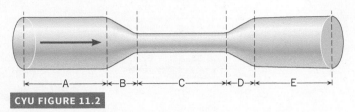

CYU FIGURE 11.2

16. In case A, water falls downward from a faucet. In case B, water shoots upward, as in a fountain. In each case, as the water moves downward or upward, it has a cross-sectional area that **(a)** does not change, **(b)** becomes larger in A and smaller in B, **(c)** becomes smaller in A and larger in B, **(d)** becomes larger in both cases, **(e)** becomes smaller in both cases.

11.9 Bernoulli's Equation

For *steady flow,* the speed, pressure, and elevation of an *incompressible and nonviscous* fluid are related by an equation discovered by Daniel Bernoulli (1700–1782). To derive **Bernoulli's equation**, we will use the work–energy theorem. This theorem, which is introduced in Chapter 6, states that the net work W_{nc} done on an object by external non-conservative forces is equal to the change in the total mechanical energy of the object (see Equation 6.8). As mentioned earlier, the pressure within a fluid is caused by colli-sional forces, which are nonconservative. Therefore, when a fluid is accelerated because of a difference in pressures, work is being done by nonconservative forces ($W_{nc} \neq 0$ J), and this work changes the total mechanical energy of the fluid from an initial value of E_0 to a final value of E_f. The total mechanical energy is not conserved. We will now see how the work–energy theorem leads directly to Bernoulli's equation.

To begin with, let us make two observations about a moving fluid. First, whenever a fluid is flowing in a horizontal pipe and encounters a region of reduced cross-sectional area, the pressure of the fluid drops, as the pressure gauges in **Figure 11.29a** indicate. The reason for this follows from Newton's second law. When moving from the wider region 2 to the narrower region 1, the fluid speeds up or accelerates, consistent with the conservation of mass (as expressed by the equation of continuity). According to the second law, the accelerating fluid must be subjected to an unbalanced force. However, there can be an unbalanced force only if the pressure in region 2 exceeds the pressure in region 1. We will see that the difference in pressures is given by Bernoulli's equation. The second observation is that if the fluid moves to a higher elevation, the pressure at the lower level is greater than the pressure at the higher level, as in **Figure 11.29b**. The basis for this observation is our previous study of static fluids, and Bernoulli's equation will confirm it, provided that the cross-sectional area of the pipe does not change.

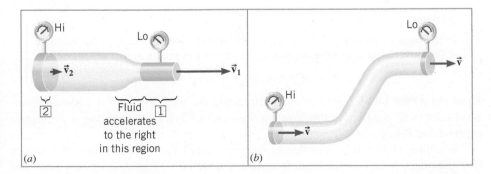

FIGURE 11.29 (*a*) In this horizontal pipe, the pressure in region 2 is greater than that in region 1. The difference in pressures leads to the net force that accelerates the fluid to the right. (*b*) When the fluid changes elevation, the pressure at the bottom is greater than the pressure at the top, assuming that the cross-sectional area of the pipe is constant.

To derive Bernoulli's equation, consider **Figure 11.30a**. This drawing shows a fluid element of mass m, upstream in region 2 of a pipe. Both the cross-sectional area and the elevation are different at different places along the pipe. The speed, pressure, and elevation in this region are v_2, P_2, and y_2, respectively. Downstream in region 1 these variables have the values v_1, P_1, and y_1. As Chapter 6 discusses, an object moving under the influ-ence of gravity has a total mechanical energy E that is the sum of the kinetic energy KE and the gravitational potential energy PE: $E = \text{KE} + \text{PE} = \frac{1}{2}mv^2 + mgy$. When work W_{nc}

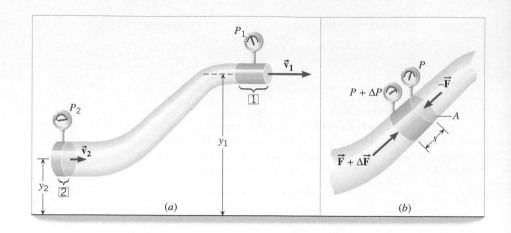

FIGURE 11.30 (*a*) A fluid element (dark blue) moving through a pipe whose cross-sectional area and elevation change. (*b*) The fluid element experiences a force $-\vec{\mathbf{F}}$ on its top surface due to the fluid above it, and a force $\vec{\mathbf{F}} + \Delta\vec{\mathbf{F}}$ on its bottom surface due to the fluid below it.

is done on the fluid element by external nonconservative forces, the total mechanical energy changes. According to the work–energy theorem, the work equals the change in the total mechanical energy:

$$W_{\text{nc}} = E_1 - E_2 = \underbrace{\left(\tfrac{1}{2}mv_1^2 + mgy_1\right)}_{\substack{\text{Total mechanical} \\ \text{energy in region 1}}} - \underbrace{\left(\tfrac{1}{2}mv_2^2 + mgy_2\right)}_{\substack{\text{Total mechanical} \\ \text{energy in region 2}}} \qquad \textbf{(6.8)}$$

Figure 11.30*b* helps us understand how the work W_{nc} arises. On the top surface of the fluid element, the surrounding fluid exerts a pressure P. This pressure gives rise to a force of magnitude $F = PA$, where A is the cross-sectional area. On the bottom surface, the surrounding fluid exerts a slightly greater pressure, $P + \Delta P$, where ΔP is the pressure difference between the ends of the element. As a result, the force on the bottom surface has a magnitude of $F + \Delta F = (P + \Delta P)A$. The magnitude of the *net* force pushing the fluid element up the pipe is $\Delta F = (\Delta P)A$. When the fluid element moves through its own length s, the work done is the product of the magnitude of the net force and the distance: Work $= (\Delta F)s = (\Delta P)As$. The quantity As is the volume V of the element, so the work is $(\Delta P)V$. The total work done on the fluid element in moving it from region 2 to region 1 is the sum of the small increments of work $(\Delta P)V$ done as the element moves along the pipe. This sum amounts to $W_{\text{nc}} = (P_2 - P_1)V$, where $P_2 - P_1$ is the pressure difference between the two regions. With this expression for W_{nc}, the work–energy theorem becomes

$$W_{\text{ne}} = (P_2 - P_1)V = \left(\tfrac{1}{2}mv_1^2 + mgy_1\right) - \left(\tfrac{1}{2}mv_2^2 + mgy_2\right)$$

By dividing both sides of this result by the volume V, recognizing that m/V is the density ρ of the fluid, and rearranging terms, we obtain Bernoulli's equation.

> **BERNOULLI'S EQUATION**
>
> In the steady flow of a nonviscous, incompressible fluid of density ρ, the pressure P, the fluid speed v, and the elevation y at any two points (1 and 2) are related by
>
> $$P_1 + \tfrac{1}{2}\rho v_1^2 + \rho g y_1 = P_2 + \tfrac{1}{2}\rho v_2^2 + \rho g y_2 \qquad \textbf{(11.11)}$$

Since the points 1 and 2 were selected arbitrarily, the term $P + \tfrac{1}{2}\rho v^2 + \rho g y$ has a constant value at all positions in the flow. For this reason, Bernoulli's equation is sometimes expressed as $P + \tfrac{1}{2}\rho v^2 + \rho g y = \text{constant}$.

Equation 11.11 can be regarded as an extension of the earlier result that specifies how the pressure varies with depth in a static fluid ($P_2 = P_1 + \rho g h$), the terms $\tfrac{1}{2}\rho v_1^2$ and $\tfrac{1}{2}\rho v_2^2$ accounting for the effects of fluid speed. Bernoulli's equation reduces to the result for static fluids when the speed of the fluid is the same everywhere ($v_1 = v_2$), as it is when the cross-sectional area remains constant. Under such conditions, Bernoulli's equation is $P_1 + \rho g y_1 = P_2 + \rho g y_2$. After rearrangement, this result becomes

$$P_2 = P_1 + \rho g(y_1 - y_2) = P_1 + \rho g h$$

which is the result (Equation 11.4) for static fluids.

11.10 | Applications of Bernoulli's Equation

When a moving fluid is contained in a horizontal pipe, all parts of it have the same elevation ($y_1 = y_2$), and Bernoulli's equation simplifies to

$$P_1 + \tfrac{1}{2}\rho v_1{}^2 = P_2 + \tfrac{1}{2}\rho v_2{}^2 \qquad\qquad (11.12)$$

Thus, the quantity $P + \tfrac{1}{2}\rho v^2$ remains constant throughout a horizontal pipe; if v increases, P decreases, and vice versa. This is the result that we deduced qualitatively from Newton's second law (Section 11.9). Conceptual Example 15 illustrates it.

CONCEPTUAL EXAMPLE 15 | Tarpaulins and Bernoulli's Equation

A tarpaulin is a piece of canvas that is used to cover a cargo, like that pulled by the truck in **Figure 11.31**. Whenever the truck stops, the tarpaulin lies flat. Why does it bulge outward whenever the truck is speeding down the highway? **(a)** The air rushing over the outside surface of the canvas creates a higher pressure than does the stationary air inside the cargo area. **(b)** The air rushing over the outside surface of the canvas creates a lower pressure than does the stationary air inside the cargo area. **(c)** The air inside the cargo area heats up, thus increasing the pressure on the tarp and pushing it outward.

Reasoning When the truck is stationary, the air outside and inside the cargo area is stationary, so the pressure is the same in both places. This pressure applies the same force to the outer and inner surfaces of the canvas, with the result that the tarpaulin lies flat. When the truck is moving, the outside air rushes over the top surface of the canvas, and the pressure generated by the moving air is different than the pressure of the stationary air.

Answer (a) is incorrect. A higher pressure outside and a lower pressure in the cargo area would cause the tarpaulin to sink inward, not bulge outward.

Answer (c) is incorrect. A heating effect would not disappear every time the truck stops and reappear only when the truck is moving.

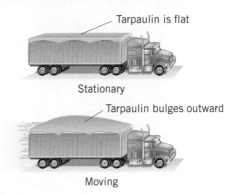

FIGURE 11.31 The tarpaulin that covers the cargo is flat when the truck is stationary but bulges outward when the truck is moving.

Answer (b) is correct. In accord with Bernoulli's equation (Equation 11.12), the moving air outside the canvas has a lower pressure than does the stationary air inside the cargo area. The greater inside pressure generates a greater force on the inner surface of the canvas, and the tarpaulin bulges outward.

Related Homework: Problems 54, 64

THE PHYSICS OF . . . household plumbing. The impact of fluid flow on pressure is widespread. **Figure 11.32**, for instance, illustrates how household plumbing takes into account the implications of Bernoulli's equation. The U-shaped section of pipe beneath the sink is called a "trap," because it traps water, which serves as a barrier to prevent sewer gas

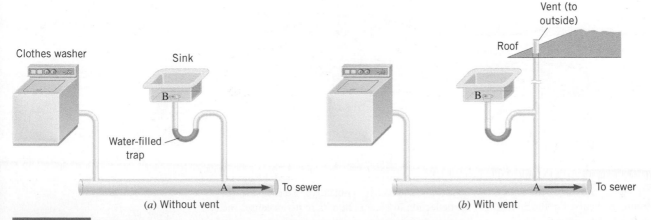

FIGURE 11.32 In a household plumbing system, a vent is necessary to equalize the pressures at points A and B, thus preventing the trap from being emptied. An empty trap allows sewer gas to enter the house.

from leaking into the house. Part *a* of the drawing shows poor plumbing. When water from the clothes washer rushes through the sewer pipe, the high-speed flow causes the pressure at point A to drop. The pressure at point B in the sink, however, remains at the higher atmospheric pressure. As a result of this pressure difference, the water is pushed out of the trap and into the sewer line, leaving no protection against sewer gas. A correctly designed system is vented to the outside of the house, as in **Figure 11.32b**. The vent ensures that the pressure at A remains the same as that at B (atmospheric pressure), even when water from the clothes washer is rushing through the pipe. Thus, the purpose of the vent is to prevent the trap from being emptied, not to provide an escape route for sewer gas.

THE PHYSICS OF . . . airplane wings. One of the most spectacular examples of how fluid flow affects pressure is the dynamic lift on airplane wings. **Figure 11.33a** shows a wing moving to the right, with the air flowing leftward past the wing. Hence, according to Bernoulli's equation, the pressures on the top of and beneath the wing are both lower than atmospheric pressure.* However, the pressure above the wing is reduced relative to the pressure under the wing. This is due to the wing's shape, which causes the air to travel faster (more reduction in pressure) over the curved top surface and more slowly (less reduction in pressure) over the flatter bottom. Thus, the wing is lifted upward. Part *b* of the figure shows the wing of an airplane.

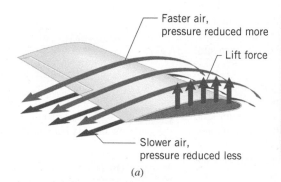

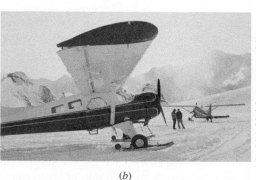

FIGURE 11.33 (*a*) Air flowing around an airplane wing. The wing is moving to the right. (*b*) The end of this wing has roughly the shape indicated in part *a*.

THE PHYSICS OF . . . a curveball. The curveball, one of a baseball pitcher's main weapons, is another illustration of the effects of fluid flow. **Figure 11.34a** shows a baseball moving to the right with no spin. The view is from above, looking down toward the ground. Here, air flows with the same speed around both sides of the ball, and the pressure is reduced on both sides by the same amount relative to atmospheric pressure. No net force exists to make the ball curve to either side. However, when the ball is given a spin, the air close to its surface is dragged around with it, and the situation changes. The speed of the air on one-half of the ball is increased, and the pressure there is even more reduced relative to atmospheric pressure. On the other half of the ball, the speed of the air is decreased,

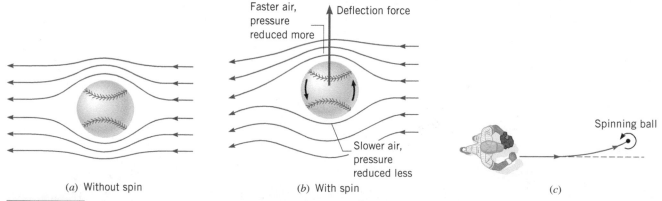

FIGURE 11.34 These views of a baseball are from above, looking down toward the ground, with the ball moving to the right. (*a*) Without spin, the ball does not curve to either side. (*b*) A spinning ball curves in the direction of the deflection force. (*c*) The spin in part *b* causes the ball to curve as shown here.

*The application of Bernoulli's equation to an airplane wing is only an approximation, as air is both viscous and compressible.

leading to a lesser reduction in pressure than occurs without spin. Part *b* of the picture shows the effects of a counterclockwise spin. The baseball experiences a net deflection force and curves on its way from the pitcher's mound to the plate, as part *c* shows.*

As a final application of Bernoulli's equation, **Figure 11.35a** shows a large tank from which water is emerging through a small pipe near the bottom. Bernoulli's equation can be used to determine the speed (called the efflux speed) at which the water leaves the pipe, as the next example shows.

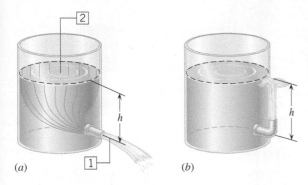

(a) (b)

FIGURE 11.35 (*a*) Bernoulli's equation can be used to determine the speed of the liquid leaving the small pipe. (*b*) An ideal fluid (no viscosity) will rise to the fluid level in the tank after leaving a vertical outlet nozzle.

EXAMPLE 16 | Efflux Speed

The tank in **Figure 11.35a** is open to the atmosphere at the top. Find an expression for the speed of the liquid leaving the pipe at the bottom.

Reasoning We assume that the liquid behaves as an ideal fluid. Therefore, we can apply Bernoulli's equation, and in preparation for doing so, we locate two points in the liquid in **Figure 11.35a**. Point 1 is just outside the efflux pipe, and point 2 is at the top surface of the liquid. The pressure at each of these points is equal to the atmospheric pressure, a fact that will be used to simplify Bernoulli's equation.

Solution Since the pressures at points 1 and 2 are the same, we have $P_1 = P_2$, and Bernoulli's equation becomes $\frac{1}{2}\rho v_1^2 + \rho g y_1 = \frac{1}{2}\rho v_2^2 + \rho g y_2$. The density ρ can be eliminated algebraically from this result, which can then be solved for the square of the efflux speed v_1:

$$v_1^2 = v_2^2 + 2g(y_2 - y_1) = v_2^2 + 2gh$$

We have substituted $h = y_2 - y_1$ for the height of the liquid above the efflux tube. If the tank is very large, the liquid level changes only slowly, and the speed at point 2 can be set equal to zero, so that $\boxed{v_1 = \sqrt{2gh}}$.

Math Skills To obtain the desired expression for v_1^2, we start with Bernoulli's equation in the following form:

$$\tfrac{1}{2}\rho v_1^2 + \rho g y_1 = \tfrac{1}{2}\rho v_2^2 + \rho g y_2$$

To solve this equation for v_1^2, we first isolate the term $\frac{1}{2}\rho v_1^2$ on the left side of the equals sign. To do this, we subtract $\rho g y_1$ from both sides of the equals sign:

$$\tfrac{1}{2}\rho v_1^2 + \rho g y_1 - \rho g y_1 = \tfrac{1}{2}\rho v_2^2 + \rho g y_2 - \rho g y_1 \quad \text{or}$$

$$\tfrac{1}{2}\rho v_1^2 = \tfrac{1}{2}\rho v_2^2 + \rho g y_2 - \rho g y_1$$

Note that the density ρ appears in each term of the result. Therefore, it can be eliminated algebraically:

$$\tfrac{1}{2}\rho v_1^2 = \tfrac{1}{2}\rho v_2^2 + \rho g y_2 - \rho g y_1 \quad \text{or} \quad \tfrac{1}{2}v_1^2 = \tfrac{1}{2}v_2^2 + g y_2 - g y_1$$

We can now simplify the resulting equation by factoring out the term g that appears on the right side in the terms $g y_2$ and $g y_1$:

$$\tfrac{1}{2}v_1^2 = \tfrac{1}{2}v_2^2 + g(y_2 - y_1)$$

Multiplying both sides of the equals sign by 2 yields

$$2\tfrac{1}{2}v_1^2 = 2\tfrac{1}{2}v_2^2 + 2g(y_2 - y_1) \quad \text{or} \quad v_1^2 = v_2^2 + 2g(y_2 - y_1)$$

In Example 16 the liquid is assumed to be an ideal fluid, and the speed with which it leaves the pipe is the same as if the liquid had freely fallen through a height *h* (see Equation 2.9 with *x* = *h* and *a* = *g*). This result is known as **Torricelli's theorem**. If the outlet pipe were pointed directly upward, as in **Figure 11.35b**, the liquid would rise to a height *h* equal to the fluid level above the pipe. However, if the liquid is not an ideal fluid, its viscosity cannot be neglected. Then, the efflux speed would be less than that given by Bernoulli's equation, and the liquid would rise to a height less than *h*. Example 17 illustrates how the Bernoulli equation can be used to analyze blood flow in the case of an abdominal aortic aneurysm.

*In the jargon used in baseball, the pitch shown in **Figure 11.34** parts *b* and *c* is called a "slider."

EXAMPLE 17 | BIO The Physics of an Abdominal Aortic Aneurysm

Example 14 demonstrated how the velocity of blood in an artery changes, as clogging reduces the effective cross-sectional area of the artery. Here we show that an *increase* in the cross-sectional area of an artery can be equally as dangerous. **Figure 11.36** represents a person who has a potentially fatal condition known as an abdominal aortic aneurysm (AAA). The aorta is the largest blood vessel in the body that carries blood away from the heart, and factors such as smoking, high blood pressure, and genetic history can lead to a weakening of the arterial wall that results in the formation of a large bulge (aneurysm) in the artery (**Figure 11.36a**). If the wall weakens too much, the blood pressure can cause it to rupture, leading to a significant loss of blood within the circulatory system. Because the cross-sectional area of the artery increases in the region of the aneurysm, the pressure will be different there. We can model the normal aorta and the aneurysm as two circular tubes with different diameters (**Figure 11.36b**). Assume the diameter of a normal aorta in an adult is $d_1 = 2.0$ cm, and the diameter of the aneurysm is $d_2 = 3.5$ cm. The velocity of the blood in the normal part of the aorta is $v_1 = 0.40$ m/s. Use the equation of continuity and Bernoulli's equation to calculate the difference in pressure ($\Delta P = P_2 - P_1$) between the aneurysm and the normal aorta. Assume the blood is an ideal fluid and the patient is lying down horizontally, so that the normal portion of the aorta and the aneurysm are at the same height.

Reasoning We can first apply the Equation of Continuity (Equation 11.8) in order to find v_2—the speed of the blood in the region of the aneurysm. Once we know v_2, then we can apply the Bernoulli equation (Equation 11.11) to find the pressure difference ΔP.

Solution We begin with Equation 11.8: $\rho_1 A_1 v_1 = \rho_2 A_2 v_2$. Here, the density of the fluid is constant, and each area is circular, or equal to $\pi(d/2)^2$. Therefore, Equation 11.8 reduces to the following: $d_1^2 v_1 = d_2^2 v_2$. Thus, $v_2 = (d_1/d_2)^2 v_1$. Now we apply the Bernoulli equation: $P_1 + \frac{1}{2}\rho v_1^2 + \rho g y_1 = P_2 + \frac{1}{2}\rho v_2^2 + \rho g y_2$. We now solve this for $\Delta P = P_2 - P_1$, where we note that $y_1 = y_2$: $\Delta P = \frac{1}{2}\rho(v_1^2 v_2^2)$. We can substitute in our expression above for v_2 and obtain our final result:

$$\Delta P = \frac{1}{2}\rho v_1^2(1 - (d_1/d_2)^4)$$

$$= \frac{1}{2}(1060 \text{ kg/m}^3)(0.40 \text{ m/s})^2(1 - (2.0 \text{ cm}/3.5 \text{ cm})^4) = \boxed{76\text{Pa}}$$

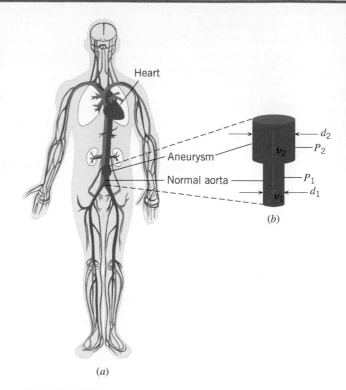

(b)

(a)

FIGURE 11.36 (a) Diagram of the human circulatory system showing the most common location of an AAA. (b) The normal aorta and aneurysm represented as circular tubes with different cross-sectional areas. The different areas lead to different blood velocities and pressures in each region.

Because the velocity of the blood slows down in the aneurysm due to its larger cross-sectional area, the blood pressure increases. This creates a dangerous cycle: As the aneurysm continues to grow, the pressure keeps increasing. If left untreated, the increasing pressure can rupture the arterial wall.

Related Homework: Problem 77

Check Your Understanding

(The answers are given at the end of the book.)

17. Fluid is flowing from left to right through a pipe (see **CYU Figure 11.3**). Points A and B are at the same elevation, but the cross-sectional areas of the pipe are different. Points B and C are at different elevations, but the cross-sectional areas are the same. Rank the pressures at the three points, highest to lowest: **(a)** A and B (a tie), C **(b)** C, A and B (a tie) **(c)** B, C, A **(d)** C, B, A **(e)** A, B, C

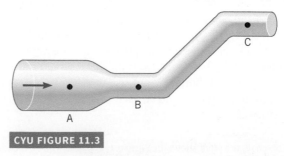

CYU FIGURE 11.3

18. Have you ever had a large truck (traveling in your direction) pass you from behind? You probably noticed that your car was pulled toward the truck as it passed. How does the speed of the air (and, hence, its pressure) between your car and the truck compare to the speed of the air on the opposite side of your car? The speed of the air between the two vehicles is **(a)** less and produces a smaller air pressure **(b)** less and produces a greater air pressure **(c)** greater and produces a smaller air pressure **(d)** greater and produces a greater air pressure.

19. Hold two sheets of paper by adjacent corners, so that they hang downward. The sheets should be parallel and slightly separated, so that you can see the floor through the gap between them. Blow air strongly down through the gap. What happens to the sheets? **(a)** Nothing **(b)** The sheets move further apart. **(c)** The sheets come closer together.

20. Suppose that you are a right-handed batter in a baseball game, so the ball is moving from your left to your right. You are caught unprepared, looking directly at the ball as it passes by for a strike. If the ball curves upward on its way to the plate, which way is it spinning? **(a)** Clockwise **(b)** Counterclockwise

21. You are sitting on a stationary train next to an *open* window, and the pressure of the air in your inner ear is equal to the pressure outside your ear. The train starts up, and as it accelerates to a high speed, your ears "pop." Your eardrums respond to a decrease or increase in the air pressure by "popping" outward or inward, respectively. Assume that the air pressure in your inner ear has not had time to change, so it remains the same as when the train was stationary. Do your eardrums "pop" **(a)** outward or **(b)** inward?

22. Sometimes the weather conditions at an airport give rise to air that has an unusually low density. What effect does such a low-density condition have on a plane's ability to generate the required lift force for takeoff? **(a)** It has no effect. **(b)** It makes it easier. **(c)** It makes it harder.

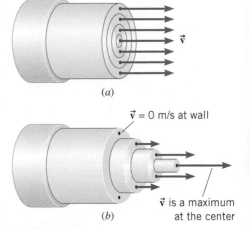

$\vec{v}$ = 0 m/s at wall

$\vec{v}$ is a maximum
(b)   at the center

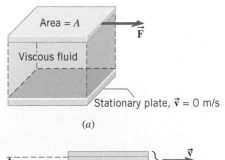

FIGURE 11.37 (*a*) In ideal (nonviscous) fluid flow, all fluid particles across the pipe have the same velocity. (*b*) In viscous flow, the speed of the fluid is zero at the surface of the pipe and increases to a maximum along the center axis.

11.11 | *Viscous Flow

In an ideal fluid there is no viscosity to hinder the fluid layers as they slide past one another. Within a pipe of uniform cross section, every layer of an ideal fluid moves with the same velocity, even the layer next to the wall, as **Figure 11.37a** shows. When viscosity is present, the fluid layers have different velocities, as part *b* of the drawing illustrates. The fluid at the center of the pipe has the greatest velocity. In contrast, the fluid layer next to the wall surface does not move at all, because it is held tightly by intermolecular forces. So strong are these forces that if a solid surface moves, the adjacent fluid layer moves along with it and remains at rest *relative* to the moving surface.

To help introduce viscosity in a quantitative fashion, **Figure 11.38a** shows a viscous fluid between two parallel plates. The top plate is free to move, while the bottom one is stationary. If the top plate is to move with a velocity $\vec{v}$ relative to the bottom plate, a force $\vec{F}$ is required. For a highly viscous fluid, like honey, a large force is needed; for a less viscous fluid, like water, a smaller one will do. As part *b* of the drawing suggests, we may imagine the fluid to be composed of many thin horizontal layers. When the top plate moves, the intermediate fluid layers slide over each other. The velocity of each layer is different, changing uniformly from $\vec{v}$ at the top plate to zero at the bottom plate. The resulting flow is called **laminar flow**, since a thin layer is often referred to as a *lamina*. As each layer moves, it is subjected to viscous forces from its neighbors. The purpose of the force $\vec{F}$ is to compensate for the effect of these forces, so that any layer can move with a constant velocity.

The amount of force required in **Figure 11.38a** depends on several factors. Larger areas A, being in contact with more fluid, require larger forces, so that the force is proportional to the contact area ($F \propto A$). For a given area, greater speeds require larger forces, with the result that the force is proportional to the speed ($F \propto v$). The force is also inversely proportional to the perpendicular distance y between the top and bottom plates ($F \propto 1/y$). The larger the distance y, the smaller is the force required to achieve a given speed with a given contact area. These three proportionalities can be expressed simultaneously in the following way: $F \propto Av/y$. Equation 11.13 expresses this relationship with the aid of a proportionality constant η (Greek letter *eta*), which is called the **coefficient of viscosity** or simply the **viscosity**.

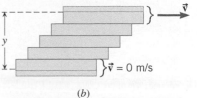

FIGURE 11.38 (*a*) A force $\vec{F}$ is applied to the top plate, which is in contact with a viscous fluid. (*b*) Because of the force $\vec{F}$, the top plate and the adjacent layer of fluid move with a constant velocity $\vec{v}$.

FORCE NEEDED TO MOVE A LAYER OF VISCOUS FLUID WITH A CONSTANT VELOCITY

The magnitude of the tangential force $\vec{F}$ required to move a fluid layer at a constant speed v, when the layer has an area A and is located a perpendicular distance y from an immobile surface, is given by

$$F = \frac{\eta A v}{y} \qquad (11.13)$$

where η is the coefficient of viscosity.

SI Unit of Viscosity: Pa · s

Common or CGS Unit of Viscosity: poise (P)

By solving Equation 11.13 for the viscosity, $\eta = Fy/(vA)$, it can be seen that the SI unit for viscosity is N · m/[(m/s) · m²] = Pa · s. Another common unit for viscosity is the *poise* (P), which is used in the CGS system of units and is named after the French physician Jean Poiseuille (1799–1869; pronounced, approximately, as Pwah-zoy′). The following relation exists between the two units:

$$1 \text{ poise (P)} = 0.1 \text{ Pa} \cdot \text{s}$$

Values of viscosity depend on the nature of the fluid. Under ordinary conditions, the viscosities of liquids are significantly *larger* than those of gases. Moreover, the viscosities of either liquids or gases depend markedly on temperature. Usually, the viscosities of liquids decrease as the temperature is increased. Anyone who has heated honey or oil, for example, knows that these fluids flow much more freely at an elevated temperature. In contrast, the viscosities of gases increase as the temperature is raised. An ideal fluid has $\eta = 0$ P.

Viscous flow occurs in a wide variety of situations, such as oil moving through a pipeline or a liquid being forced through the needle of a hypodermic syringe. **Figure 11.39** identifies the factors that determine the volume flow rate Q (in m³/s) of the fluid. First, a difference in pressures $P_2 - P_1$ must be maintained between any two locations along the pipe for the fluid to flow. In fact, Q is proportional to $P_2 - P_1$, a greater pressure difference leading to a larger flow rate. Second, a long pipe offers greater resistance to the flow than a short pipe does, and Q is inversely proportional to the length L. Third, high-viscosity fluids flow less readily than low-viscosity fluids, and Q is inversely proportional to the viscosity η. Finally, the volume flow rate is larger in a pipe of larger radius, other things being equal. The dependence on the radius R is a surprising one, Q being proportional to the fourth power of the radius, or R^4. If, for instance, the pipe radius is reduced to one-half of its original value, the volume flow rate is reduced to one-sixteenth of its original value, assuming the other variables remain constant. The mathematical relation for Q in terms of these parameters was discovered by Poiseuille and is known as **Poiseuille's law.**

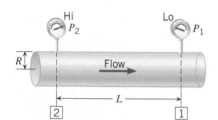

FIGURE 11.39 For viscous flow, the difference in pressures $P_2 - P_1$, the radius R and length L of the tube, and the viscosity η of the fluid influence the volume flow rate.

POISEUILLE'S LAW

A fluid whose viscosity is η, flowing through a pipe of radius R and length L, has a volume flow rate Q given by

$$Q = \frac{\pi R^4 (P_2 - P_1)}{8\eta L} \qquad (11.14)$$

where P_1 and P_2 are the pressures at the ends of the pipe.

THE PHYSICS OF . . . pipeline pumping stations. The fact that Q is inversely proportional to L in Equation 11.14 has important consequences for long pipelines, such as the Alaskan pipeline (see **Figure 11.40**). Solving Equation 11.14 for P_1 shows that

$$P_1 = P_2 - \frac{Q8\eta L}{\pi R^4}$$

Thus, the pressure P_1 at the downstream end of a length of pipe is less than the pressure P_2 at the upstream end. Long pipelines must have pumping stations at various places along the line to compensate for the drop in pressure.

Example 18 illustrates the use of Poiseuille's law.

FIGURE 11.40 As oil flows along the Alaskan pipeline, the pressure drops because oil is a viscous fluid. Pumping stations are located along the pipeline to compensate for the drop in pressure.

EXAMPLE 18 | BIO The Physics of a Hypodermic Syringe

A hypodermic syringe is filled with a solution whose viscosity is 1.5×10^{-3} Pa · s. As **Figure 11.41** shows, the plunger area of the syringe is 8.0×10^{-5} m^2, and the length of the needle is 0.025 m. The internal radius of the needle is 4.0×10^{-4} m. The gauge pressure in the vein that is injected is 1900 Pa (14 mm of mercury). What force must be applied to the plunger, so that 1.0×10^{-6} m^3 of solution can be injected in 3.0 s?

Reasoning The necessary force is the pressure applied to the plunger times the area of the plunger. Since viscous flow is occurring, the pressure is different at different points along the syringe. However, the barrel of the syringe is so wide that little pressure difference is required to sustain the flow up to point 2, where the fluid encounters the narrow needle. Consequently, the pressure applied to the plunger is nearly equal to the pressure P_2 at point 2. To find this pressure, we apply Poiseuille's law to the needle. Poiseuille's law indicates that $P_2 - P_1 = 8\eta LQ/(\pi R^4)$. We note that the pressure P_1 is given as a gauge pressure, which, in this case, is the amount of pressure in excess of atmospheric pressure. This causes no difficulty, because we need to find the amount of force in excess of the force applied to the plunger by the atmosphere. The volume flow rate Q can be obtained from the time needed to inject the known volume of solution.

Solution The volume flow rate is $Q = (1.0 \times 10^{-6}$ m$^3)/(3.0$ s$) = 3.3 \times 10^{-7}$ m^3/s. According to Poiseuille's law (Equation 11.14), the required pressure difference is

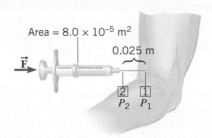

FIGURE 11.41 The difference in pressure $P_2 - P_1$ required to sustain the fluid flow through a hypodermic needle can be found with the aid of Poiseuille's law.

$$P_2 - P_1 = \frac{8\eta LQ}{\pi R^4} = \frac{8(1.5 \times 10^{-3}\,\text{Pa} \cdot \text{s})(0.025\,\text{m})(3.3 \times 10^{-7}\,\text{m}^3/\text{s})}{\pi(4.0 \times 10^{-4}\,\text{m})^4}$$

$$= 1200\,\text{Pa}$$

Since $P_1 = 1900$ Pa, the pressure P_2 must be $P_2 = 1200$ Pa $+ 1900$ Pa $= 3100$ Pa. The force that must be applied to the plunger is this pressure times the plunger area:

$$F = (3100\,\text{Pa})(8.0 \times 10^{-5}\,\text{m}^2) = \boxed{0.25\,\text{N}}$$

Concept Summary

11.1 Mass Density Fluids are materials that can flow, and they include gases and liquids. The mass density ρ of a substance is its mass m divided by its volume V, according to Equation 11.1.

$$\rho = \frac{m}{V} \tag{11.1}$$

The specific gravity of a substance is its mass density divided by the density of water at 4 °C (1.000×10^3 kg/m^3), according to Equation 11.2.

$$\text{Specific gravity} = \frac{\text{Density of substance}}{1.000 \times 10^3\,\text{kg/m}^3} \tag{11.2}$$

11.2 Pressure The pressure P exerted by a fluid is the magnitude F of the force acting perpendicular to a surface embedded in the fluid divided by the area A over which the force acts, as shown by Equation 11.3. The SI unit for measuring pressure is the pascal (Pa); 1 Pa = 1 N/m^2. One atmosphere of pressure is 1.013×10^5 Pa or 14.7 lb/in.^{2}

$$P = \frac{F}{A} \tag{11.3}$$

11.3 Pressure and Depth in a Static Fluid In the presence of gravity, the upper layers of a fluid push downward on the layers beneath, with the result that fluid pressure is related to depth. In an incompressible static fluid whose density is ρ, the relation is given by Equation 11.4,

where P_1 is the pressure at one level, P_2 is the pressure at a level that is h meters deeper, and g is the magnitude of the acceleration due to gravity.

$$P_2 = P_1 + \rho g h \tag{11.4}$$

11.4 Pressure Gauges Two basic types of pressure gauges are the mercury barometer and the open-tube manometer.

The gauge pressure is the amount by which a pressure P differs from atmospheric pressure. The absolute pressure is the actual value for P.

11.5 Pascal's Principle Pascal's principle states that any change in the pressure applied to a completely enclosed fluid is transmitted undiminished to all parts of the fluid and the enclosing walls.

11.6 Archimedes' Principle The buoyant force is the upward force that a fluid applies to an object that is partially or completely immersed in it. Archimedes' principle states that the magnitude of the buoyant force equals the weight of the fluid that the partially or completely immersed object displaces, as indicated by Equation 11.6.

$$\underbrace{F_B}_{\substack{\text{Magnitude of} \\ \text{buoyant force}}} = \underbrace{W_{\text{fluid}}}_{\substack{\text{Weight of} \\ \text{displaced fluid}}} \tag{11.6}$$

11.7 Fluids in Motion/11.8 The Equation of Continuity In steady flow, the velocity of the fluid particles at any point is constant as time passes.

An incompressible, nonviscous fluid is known as an ideal fluid.

The mass flow rate of a fluid with a density ρ, flowing with a speed v in a pipe of cross-sectional area A, is the mass per second (kg/s) flowing past a point and is given by Equation 11.7.

$$\text{Mass flow rate} = \rho A v \qquad (11.7)$$

The equation of continuity expresses the fact that mass is conserved: what flows into one end of a pipe flows out the other end, assuming there are no additional entry or exit points in between. Expressed in terms of the mass flow rate, the equation of continuity is given by Equation 11.8, where the subscripts 1 and 2 denote any two points along the pipe.

$$\rho_1 A_1 v_1 = \rho_2 A_2 v_2 \qquad (11.8)$$

If a fluid is incompressible, the density at any two points is the same, $\rho_1 = \rho_2$. For an incompressible fluid, the equation of continuity is expressed as in Equation 11.9. The product Av is known as the volume flow rate Q (in m³/s), according to Equation 11.10.

$$A_1 v_1 = A_2 v_2 \qquad (11.9)$$

$$Q = \text{Volume flow rate} = Av \qquad (11.10)$$

11.9 Bernoulli's Equation/11.10 Applications of Bernoulli's Equation In the steady flow of an ideal fluid whose density is ρ, the pressure P, the fluid speed v, and the elevation y at any two points (1 and 2) in the fluid are related by Bernoulli's equation (see Equation 11.11). When the flow is horizontal ($y_1 = y_2$), Bernoulli's equation indicates that higher fluid speeds are associated with lower fluid pressures.

$$P_1 + \tfrac{1}{2}\rho v_1^2 + \rho g y_1 = P_2 + \tfrac{1}{2}\rho v_2^2 + \rho g y_2 \qquad (11.11)$$

11.11 Viscous Flow The magnitude F of the tangential force required to move a fluid layer at a constant speed v, when the layer has an area A and is located a perpendicular distance y from an immobile surface, is given by Equation 11.13, where η is the coefficient of viscosity.

$$F = \frac{\eta A v}{y} \qquad (11.13)$$

A fluid whose viscosity is η, flowing through a pipe of radius R and length L, has a volume flow rate Q given by Poiseuille's law (see Equation 11.14), where P_1 and P_2 are the pressures at the downstream and upstream ends of the pipe, respectively.

$$Q = \frac{\pi R^4 (P_2 - P_1)}{8 \eta L} \qquad (11.14)$$

Focus on Concepts

Online

Additional questions are available for assignment in WileyPLUS.

Section 11.3 Pressure and Depth in a Static Fluid

1. The drawing shows three containers filled to the same height with the same fluid. In which container, if any, is the pressure at the bottom greatest? (a) Container A, because its bottom has the greatest surface area. (b) All three containers have the same pressure at the bottom. (c) Container A, because it has the greatest volume of fluid. (d) Container B, because it has the least volume of fluid. (e) Container C, because its bottom has the least surface area.

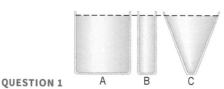

QUESTION 1   A   B   C

2. Two liquids, 1 and 2, are in equilibrium in a U-tube that is open at both ends, as in the drawing. The liquids do not mix, and liquid 1 rests on top of liquid 2. How is the density ρ_1 of liquid 1 related to the density ρ_2 of liquid 2? (a) ρ_1 is equal to ρ_2 because the liquids are in equilibrium. (b) ρ_1 is greater than ρ_2. (c) ρ_1 is less than ρ_2. (d) There is not enough information to tell which liquid has the greater density.

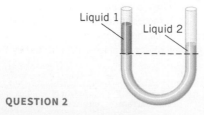

Liquid 1
Liquid 2

QUESTION 2

Section 11.6 Archimedes' Principle

3. A beaker is filled to the brim with water. A solid object of mass 3.00 kg is lowered into the beaker so that the object is fully submerged in the water (see the drawing). During this process, 2.00 kg of water flows over the rim and out of the beaker. What is the buoyant force that acts on the submerged object, and, when released, does the object rise, sink, or remain in place? (a) 29.4 N; the object rises. (b) 29.4 N; the object sinks. (c) 19.6 N; the object rises. (d) 19.6 N; the object sinks. (e) 19.6 N; the object remains in place.

Object

QUESTION 3

4. Three solid objects are floating in a liquid, as in the drawing. They have different weights and volumes, but have the same thickness (the dimension perpendicular to the page). Rank the objects according to their density, largest first. (a) A, B, C (b) A, C, B (c) B, A, C (d) B, C, A (e) C, A, B

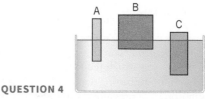

A   B   C

QUESTION 4

Section 11.8 The Equation of Continuity

5. A hollow pipe is submerged in a stream of water so that the length of the pipe is parallel to the velocity of the water. If the water speed doubles and the cross-sectional area of the pipe triples, what happens to the volume flow rate of the water passing through the pipe? (a) The

volume flow rate does not change. (**b**) The volume flow rate increases by a factor of 2. (**c**) The volume flow rate increases by a factor of 3. (**d**) The volume flow rate increases by a factor of 4. (**e**) The volume flow rate increases by a factor of 6.

6. In the drawing, water flows from a wide section of a pipe to a narrow section. In which part of the pipe is the volume flow rate the greatest? (**a**) The wide section (**b**) The narrow section (**c**) The volume flow rate is the same in both sections of the pipe.

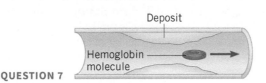

QUESTION 6

Section 11.9 Bernoulli's Equation

7. Blood flows through a section of a horizontal artery that is partially blocked by a deposit along the artery wall. A hemoglobin molecule moves from the narrow region into the wider region. What happens to the pressure acting on the molecule? (**a**) The pressure increases. (**b**) The pressure decreases. (**c**) There is no change in the pressure.

QUESTION 7

8. Water is flowing down through the pipe shown in the drawing. Point A is at a higher elevation than B and C are. The cross-sectional areas are the same at A and B but are wider at C. Rank the pressures at the three points, largest first. (**a**) P_A, P_B, P_C (**b**) P_C, P_B, P_A (**c**) P_B, P_C, P_A

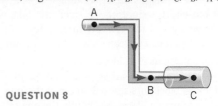

QUESTION 8

Section 11.11 Viscous Flow

9. A *viscous* fluid is flowing through two horizontal pipes. The pressure difference $P_2 - P_1$ between the ends of each pipe is the same. The pipes have the same radius, although one is twice as long as the other. How does the volume flow rate Q_B in the longer pipe compare with the rate Q_A in the shorter pipe? (**a**) Q_B is the same as Q_A. (**b**) Q_B is twice as large as Q_A. (**c**) Q_B is four times as large as Q_A. (**d**) Q_B is one-half as large as Q_A. (**e**) Q_B is one-fourth as large as Q_A.

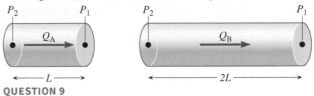

QUESTION 9

Problems

Online

Additional questions are available for assignment in WileyPLUS.

Section 11.1 Mass Density

1. **E** **SSM** One of the concrete pillars that support a house is 2.2 m tall and has a radius of 0.50 m. The density of concrete is about 2.2×10^3 kg/m³. Find the weight of this pillar in pounds (1 N = 0.2248 lb).

2. **E** A cylindrical storage tank has a radius of 1.22 m. When filled to a height of 3.71 m, it holds 14 300 kg of a liquid industrial solvent. What is the density of the solvent?

3. **E** **SSM** Accomplished silver workers in India can pound silver into incredibly thin sheets, as thin as 3.00×10^{-7} m (about one-hundredth of the thickness a sheet of paper). Find the area of such a sheet that can be formed from 1.00 kg of silver.

4. **E** Neutron stars consist only of neutrons and have unbelievably high densities. A typical mass and radius for a neutron star might be 2.7×10^{28} kg and 1.2×10^3 m. (**a**) Find the density of such a star. (**b**) If a dime ($V = 2.0 \times 10^{-7}$ m³) were made from this material, how much would it weigh (in pounds)?

5. **E** One end of a wire is attached to a ceiling, and a solid brass ball is tied to the lower end. The tension in the wire is 120 N. What is the radius of the brass ball?

6. **M** **GO** Planners of an experiment are evaluating the design of a sphere of radius R that is to be filled with helium (0 °C, 1 atm pressure). Ultrathin silver foil of thickness T will be used to make the sphere, and the designers claim that the mass of helium in the sphere will equal the mass of silver used. Assuming that T is much less than R, calculate the ratio T/R for such a sphere.

7. **M** **SSM** A bar of gold measures 0.15 m × 0.050 m × 0.050 m. How many gallons of water have the same mass as this bar?

8. **M** A full can of black cherry soda has a mass of 0.416 kg. It contains 3.54×10^{-4} m³ of liquid. Assuming that the soda has the same density as water, find the volume of aluminum used to make the can.

9. **M** **V-HINT** **CHALK** A hypothetical spherical planet consists entirely of iron. What is the period of a satellite that orbits this planet just above its surface? Consult **Table 11.1** as necessary.

10. **H** An antifreeze solution is made by mixing ethylene glycol ($\rho = 1116$ kg/m³) with water. Suppose that the specific gravity of such a solution is 1.0730. Assuming that the total volume of the solution is the sum of its parts, determine the volume percentage of ethylene glycol in the solution.

Section 11.2 Pressure

11. **E** **SSM** An airtight box has a removable lid of area 1.3×10^{-2} m² and negligible weight. The box is taken up a mountain where the air

pressure outside the box is 0.85×10^5 Pa. The inside of the box is completely evacuated. What is the magnitude of the force required to pull the lid off the box?

12. **E** A person who weighs 625 N is riding a 98-N mountain bike. Suppose that the entire weight of the rider and bike is supported equally by the two tires. If the pressure in each tire is 7.60×10^5 Pa, what is the area of contact between each tire and the ground?

13. **E** **MMH** A solid concrete block weighs 169 N and is resting on the ground. Its dimensions are 0.400 m × 0.200 m × 0.100 m. A number of identical blocks are stacked on top of this one. What is the smallest number of whole blocks (including the one on the ground) that can be stacked so that their weight creates a pressure of at least two atmospheres on the ground beneath the first block?

14. **E** United States currency is printed using intaglio presses that generate a printing pressure of 8.0×10^4 lb/in.² A $20 bill is 6.1 in. by 2.6 in. Calculate the magnitude of the force that the printing press applies to one side of the bill.

15. **E** **SSM** A glass bottle of soda is sealed with a screw cap. The absolute pressure of the carbon dioxide inside the bottle is 1.80×10^5 Pa. Assuming that the top and bottom surfaces of the cap each have an area of 4.10×10^{-4} m², obtain the magnitude of the force that the screw thread exerts on the cap in order to keep it on the bottle. The air pressure outside the bottle is one atmosphere.

16. **M** **GO** A 58-kg skier is going down a slope oriented 35° above the horizontal. The area of each ski in contact with the snow is 0.13 m². Determine the pressure that each ski exerts on the snow.

17. **M** **GO** A suitcase (mass $m = 16$ kg) is resting on the floor of an elevator. The part of the suitcase in contact with the floor measures 0.50 m × 0.15 m. The elevator is moving upward with an acceleration of magnitude 1.5 m/s². What pressure (in excess of atmospheric pressure) is applied to the floor beneath the suitcase?

18. **M** **CHALK** A cylinder is fitted with a piston, beneath which is a spring, as in the drawing. The cylinder is open to the air at the top. Friction is absent. The spring constant of the spring is 3600 N/m. The piston has a negligible mass and a radius of 0.024 m. **(a)** When the air beneath the piston is completely pumped out, how much does the atmospheric pressure cause the spring to compress? **(b)** How much work does the atmospheric pressure do in compressing the spring?

PROBLEM 18

Section 11.3 Pressure and Depth in a Static Fluid,

Section 11.4 Pressure Gauges

19. **E** The Mariana trench is located in the floor of the Pacific Ocean at a depth of about 11 000 m below the surface of the water. The density of seawater is 1025 kg/m³. **(a)** If an underwater vehicle were to explore such a depth, what force would the water exert on the vehicle's observation window (radius = 0.10 m)? **(b)** For comparison, determine the weight of a jetliner whose mass is 1.2×10^5 kg.

20. **E** Review Conceptual Example 6 as an aid in understanding this problem. Consider the pump on the right side of **Figure 11.10**, which acts to reduce the air pressure in the pipe. The air pressure outside the pipe is one atmosphere. Find the maximum depth from which this pump can extract water from the well.

21. **E** **GO** A meat baster consists of a squeeze bulb attached to a plastic tube. When the bulb is squeezed and released, with the open end of the tube under the surface of the basting sauce, the sauce rises in the tube to a distance h, as the drawing shows. Using 1.013×10^5 Pa

for the atmospheric pressure and 1200 kg/m³ for the density of the sauce, find the absolute pressure in the bulb when the distance h is **(a)** 0.15 m and **(b)** 0.10 m.

PROBLEM 21

22. **E** **SSM** The main water line enters a house on the first floor. The line has a gauge pressure of 1.90×10^5 Pa. **(a)** A faucet on the second floor, 6.50 m above the first floor, is turned off. What is the gauge pressure at this faucet? **(b)** How high could a faucet be before no water would flow from it, even if the faucet were open?

23. **E** **SSM** A water tower is a familiar sight in many towns. The purpose of such a tower is to provide storage capacity and to provide sufficient pressure in the pipes that deliver the water to customers. The drawing shows a spherical reservoir that contains 5.25×10^5 kg of water when full. The reservoir is vented to the atmosphere at the top. For a full reservoir, find the gauge pressure that the water has at the faucet in **(a)** house A and **(b)** house B. Ignore the diameter of the delivery pipes.

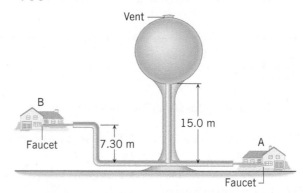

PROBLEM 23

24. **M** **GO** **CHALK** A mercury barometer reads 747.0 mm on the roof of a building and 760.0 mm on the ground. Assuming a constant value of 1.29 kg/m³ for the density of air, determine the height of the building.

25. **M** **SSM** A 1.00-m-tall container is filled to the brim, partway with mercury and the rest of the way with water. The container is open to the atmosphere. What must be the depth of the mercury so that the absolute pressure on the bottom of the container is twice the atmospheric pressure?

26. **M** **CHALK** Two identical containers are open at the top and are connected at the bottom via a tube of negligible volume and a valve that is closed. Both containers are filled initially to the same height of 1.00 m, one with water, the other with mercury, as the drawing indicates. The valve is then opened. Water and mercury are immiscible. Determine the fluid level in the left container when equilibrium is reestablished.

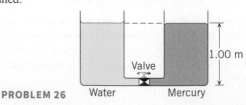

PROBLEM 26

27. **M** **V-HINT** The vertical surface of a reservoir dam that is in contact with the water is 120 m wide and 12 m high. The air pressure is one atmosphere. Find the magnitude of the total force acting on this surface in a completely filled reservoir. (*Hint: The pressure varies linearly with depth, so you must use an average pressure.*)

28. **H** A house has a roof with the dimensions shown in the drawing. Determine the magnitude and direction of the net force that the atmosphere applies to the roof when the outside pressure drops suddenly by 75.0 mm of mercury before the air pressure in the attic can adjust. Express your answer in **(a)** newtons and **(b)** pounds.

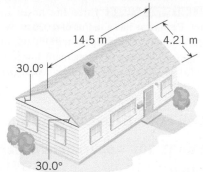

PROBLEM 28

Section 11.5 Pascal's Principle

29. **E** **SSM** The atmospheric pressure above a swimming pool changes from 755 to 765 mm of mercury. The bottom of the pool is a rectangle (12 m × 24 m). By how much does the force on the bottom of the pool increase?

30. **E** A barber's chair with a person in it weighs 2100 N. The output plunger of a hydraulic system begins to lift the chair when the barber's foot applies a force of 55 N to the input piston. Neglect any height difference between the plunger and the piston. What is the ratio of the radius of the plunger to the radius of the piston?

31. **E** **MMH** Multiple-Concept Example 9 presents an approach to problems of this kind. The hydraulic oil in a car lift has a density of 8.30×10^2 kg/m^3. The weight of the input piston is negligible. The radii of the input piston and output plunger are 7.70×10^{-3} m and 0.125 m, respectively. What input force F is needed to support the 24 500-N combined weight of a car and the output plunger, when **(a)** the bottom surfaces of the piston and plunger are at the same level, and **(b)** the bottom surface of the output plunger is 1.30 m *above* that of the input piston?

32. **E** **GO** In the process of changing a flat tire, a motorist uses a hydraulic jack. She begins by applying a force of 45 N to the input piston, which has a radius r_1. As a result, the output plunger, which has a radius r_2, applies a force to the car. The ratio r_2/r_1 has a value of 8.3. Ignore the height difference between the input piston and output plunger and determine the force that the output plunger applies to the car.

33. **M** **V-HINT** A dump truck uses a hydraulic cylinder, as the drawing illustrates. When activated by the operator, a pump injects hydraulic oil into the cylinder at an absolute pressure of 3.54×10^6 Pa and drives the output plunger, which has a radius of 0.150 m. Assuming that the plunger remains perpendicular to the floor of the load bed, find the torque that the plunger creates about the axis identified in the drawing.

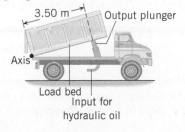

PROBLEM 33

34. **M** **GO** **CHALK** The drawing shows a hydraulic chamber with a spring (spring constant = 1600 N/m) attached to the input piston and a rock of mass 40.0 kg resting on the output plunger. The piston and plunger are nearly at the same height, and each has a negligible mass. By how much is the spring compressed from its unstrained position?

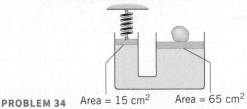

PROBLEM 34   Area = 15 cm^2 Area = 65 cm^2

35. **M** **SSM** The drawing shows a hydraulic system used with disc brakes. The force $\vec{F}$ is applied perpendicularly to the brake pedal. The pedal rotates about the axis shown in the drawing and causes a force to be applied perpendicularly to the input piston (radius = 9.50×10^{-3} m) in the master cylinder. The resulting pressure is transmitted by the brake fluid to the output plungers (radii = 1.90×10^{-2} m), which are covered with the brake linings. The linings are pressed against both sides of a disc attached to the rotating wheel. Suppose that the magnitude of $\vec{F}$ is 9.00 N. Assume that the input piston and the output plungers are at the same vertical level, and find the force applied to each side of the rotating disc.

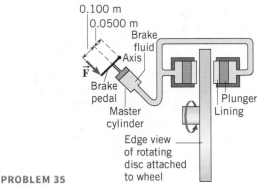

PROBLEM 35

Section 11.6 Archimedes' Principle

36. **E** The density of ice is 917 kg/m^3, and the density of seawater is 1025 kg/m^3. A swimming polar bear climbs onto a piece of floating ice that has a volume of 5.2 m^3. What is the weight of the heaviest bear that the ice can support without sinking completely beneath the water?

37. **E** **SSM** A 0.10-m × 0.20-m × 0.30-m block is suspended from a wire and is completely under water. What buoyant force acts on the block?

38. **E** **GO** A hydrometer is a device used to measure the density of a liquid. It is a cylindrical tube weighted at one end, so that it floats with the heavier end downward. The tube is contained inside a large "medicine dropper," into which the liquid is drawn using the squeeze bulb (see the drawing). For use with your car, marks are put on the tube so that the level at which it floats indicates whether the liquid is battery acid (more dense) or antifreeze (less dense). The hydrometer has a weight of $W = 5.88 \times 10^{-2}$ N and a cross-sectional area of $A = 7.85 \times 10^{-5}$ m^2. How far from the bottom of the tube should the mark be put that denotes **(a)** battery acid ($\rho = 1280$ kg/m^3) and **(b)** antifreeze ($\rho = 1073$ kg/m^3)?

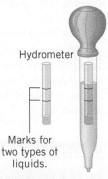

Hydrometer

Marks for two types of liquids.

PROBLEM 38

39. **E** A duck is floating on a lake with 25% of its volume beneath the water. What is the average density of the duck?

40. **E** A paperweight, when weighed in air, has a weight of $W = 6.9$ N. When completely immersed in water, however, it has a weight of $W_{in\ water} = 4.3$ N. Find the volume of the paperweight.

41. **E** **SSM** An 81-kg person puts on a life jacket, jumps into the water, and floats. The jacket has a volume of 3.1×10^{-2} m³ and is completely submerged under the water. The volume of the person's body that is under water is 6.2×10^{-2} m³. What is the density of the life jacket?

42. **E** **V-HINT** A lost shipping container is found resting on the ocean floor and completely submerged. The container is 6.1 m long, 2.4 m wide, and 2.6 m high. Salvage experts attach a spherical balloon to the top of the container and inflate it with air pumped down from the surface. When the balloon's radius is 1.5 m, the shipping container just begins to rise toward the surface. What is the mass of the container? Ignore the mass of the balloon and the air within it. Do *not* neglect the buoyant force exerted on the shipping container by the water. The density of seawater is 1025 kg/m³.

43. **M** **CHALK** **SSM** Refer to Multiple-Concept Example 12 to see a problem similar to this one. What is the smallest number of whole logs ($\rho = 725$ kg/m³, radius = 0.0800 m, length = 3.00 m) that can be used to build a raft that will carry four people, each of whom has a mass of 80.0 kg?

44. **M** **GO** A hot-air balloon is accelerating upward under the influence of two forces, its weight and the buoyant force. For simplicity, consider the weight to be only that of the hot air within the balloon, thus ignoring the balloon fabric and the basket. The hot air inside the balloon has a density of $\rho_{hot\ air} = 0.93$ kg/m³, and the density of the cool air outside is $\rho_{cool\ air} = 1.29$ kg/m³. What is the acceleration of the rising balloon?

45. **M** A hollow cubical box is 0.30 m on an edge. This box is floating in a lake with one-third of its height beneath the surface. The walls of the box have a negligible thickness. Water from a hose is poured into the open top of the box. What is the depth of the water in the box just at the instant that water from the lake begins to pour into the box from the lake?

46. **M** **GO** To verify her suspicion that a rock specimen is hollow, a geologist weighs the specimen in air and in water. She finds that the specimen weighs twice as much in air as it does in water. The density of the solid part of the specimen is 5.0×10^3 kg/m³. What fraction of the specimen's apparent volume is solid?

47. **H** **SSM** A solid cylinder (radius = 0.150 m, height = 0.120 m) has a mass of 7.00 kg. This cylinder is floating in water. Then oil ($\rho = 725$ kg/m³) is poured on top of the water until the situation shown in the drawing results. How much of the height of the cylinder is in the oil?

PROBLEM 47

48. **H** One kilogram of glass ($\rho = 2.60 \times 10^3$ kg/m³) is shaped into a hollow spherical shell that just barely floats in water. What are the inner and outer radii of the shell? Do not assume that the shell is thin.

Section 11.8 The Equation of Continuity

49. **E** A fuel pump sends gasoline from a car's fuel tank to the engine at a rate of 5.88×10^{-2} kg/s. The density of the gasoline is 735 kg/m³, and the radius of the fuel line is 3.18×10^{-3} m. What is the speed at which the gasoline moves through the fuel line?

50. **E** A room has a volume of 120 m³. An air-conditioning system is to replace the air in this room every twenty minutes, using ducts that have a square cross section. Assuming that air can be treated as an incompressible fluid, find the length of a side of the square if the air speed within the ducts is (a) 3.0 m/s and (b) 5.0 m/s.

51. **E** **GO** Water flows straight down from an open faucet. The cross-sectional area of the faucet is 1.8×10^{-4} m², and the speed of the water is 0.85 m/s as it leaves the faucet. Ignoring air resistance, find the cross-sectional area of the water stream at a point 0.10 m below the faucet.

52. **M** **GO** **MMH** **CHALK** Three fire hoses are connected to a fire hydrant. Each hose has a radius of 0.020 m. Water enters the hydrant through an underground pipe of radius 0.080 m. In this pipe the water has a speed of 3.0 m/s. (a) How many kilograms of water are poured onto a fire in one hour by all three hoses? (b) Find the water speed in each hose.

Section 11.9 Bernoulli's Equation,

Section 11.10 Applications of Bernoulli's Equation

53. **E** **SSM** Prairie dogs are burrowing rodents. They do not suffocate in their burrows, because the effect of air speed on pressure creates sufficient air circulation. The animals maintain a difference in the shapes of two entrances to the burrow, and because of this difference, the air ($\rho = 1.29$ kg/m³) blows past the openings at different speeds, as the drawing indicates. Assuming that the openings are at the same vertical level, find the difference in air pressure between the openings and indicate which way the air circulates.

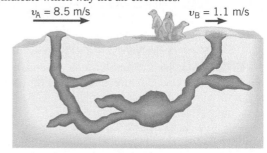

$v_A = 8.5$ m/s $v_B = 1.1$ m/s

PROBLEM 53

54. **E** Review Conceptual Example 15 before attempting this problem. The truck in that example is traveling at 27 m/s. The density of air is 1.29 kg/m³. By how much does the pressure inside the cargo area beneath the tarpaulin exceed the outside pressure?

55. **E** **SSM** An airplane wing is designed so that the speed of the air across the top of the wing is 251 m/s when the speed of the air below the wing is 225 m/s. The density of the air is 1.29 kg/m³. What is the lifting force on a wing of area 24.0 m²?

56. **E** **GO** Water flowing out of a horizontal pipe emerges through a nozzle. The radius of the pipe is 1.9 cm, and the radius of the nozzle is 0.48 cm. The speed of the water in the pipe is 0.62 m/s. Treat the water as an ideal fluid, and determine the absolute pressure of the water in the pipe.

57. **E** **MMH** A small crack occurs at the base of a 15.0-m-high dam. The effective crack area through which water leaves is 1.30×10^{-3} m². (a) Ignoring viscous losses, what is the speed of water flowing through the crack? (b) How many cubic meters of water per second leave the dam?

58. **E** Water is circulating through a closed system of pipes in a two-floor apartment. On the first floor, the water has a gauge pressure of 3.4×10^5 Pa and a speed of 2.1 m/s. However, on the second floor, which is 4.0 m higher, the speed of the water is 3.7 m/s. The speeds are different because the pipe diameters are different. What is the gauge pressure of the water on the second floor?

59. 🄴 🄶🄾 A ship is floating on a lake. Its hold is the interior space beneath its deck; the hold is empty and is open to the atmosphere. The hull has a hole in it, which is below the water line, so water leaks into the hold. The effective area of the hole is 8.0×10^{-3} m^2 and is located 2.0 m beneath the surface of the lake. What volume of water per second leaks into the ship?

60. 🄼 🅂🅂🄼 A Venturi meter is a device that is used for measuring the speed of a fluid within a pipe. The drawing shows a gas flowing at speed v_2 through a horizontal section of pipe whose cross-sectional area is $A_2 = 0.0700$ m^2. The gas has a density of $\rho = 1.30$ kg/m^3. The Venturi meter has a cross-sectional area of $A_1 = 0.0500$ m^2 and has been substituted for a section of the larger pipe. The pressure difference between the two sections is $P_2 - P_1 = 120$ Pa. Find **(a)** the speed v_2 of the gas in the larger, original pipe and **(b)** the volume flow rate Q of the gas.

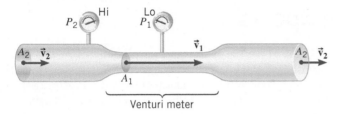

PROBLEM 60

61. 🄼 🅅-🄷🄸🄽🅃 🄲🄷🄰🄻🄺 A hand-pumped water gun is held level at a height of 0.75 m above the ground and fired. The water stream from the gun hits the ground a horizontal distance of 7.3 m from the muzzle. Find the gauge pressure of the water gun's reservoir at the instant when the gun is fired. Assume that the speed of the water in the reservoir is zero and that the water flow is steady. Ignore both air resistance and the height difference between the reservoir and the muzzle.

62. 🄼 🄲🄷🄰🄻🄺 A liquid is flowing through a horizontal pipe whose radius is 0.0200 m. The pipe bends straight upward through a height of 10.0 m and joins another horizontal pipe whose radius is 0.0400 m. What volume flow rate will keep the pressures in the two horizontal pipes the same?

63. 🄼 🅅-🄷🄸🄽🅃 An airplane has an effective wing surface area of 16 m^2 that is generating the lift force. In level flight the air speed over the top of the wings is 62.0 m/s, while the air speed beneath the wings is 54.0 m/s. What is the weight of the plane?

64. 🄼 🄼🄼🄷 The construction of a flat rectangular roof (5.0 m × 6.3 m) allows it to withstand a maximum net outward force that is 22 000 N. The density of the air is 1.29 kg/m^3. At what wind speed will this roof blow outward?

65. 🄼 🄶🄾 A pump and its horizontal intake pipe are located 12 m beneath the surface of a large reservoir. The speed of the water in the intake pipe causes the pressure there to decrease, in accord with

Bernoulli's principle. Assuming nonviscous flow, what is the maximum speed with which water can flow through the intake pipe?

66. 🄷 🅂🅂🄼 A uniform rectangular plate is hanging vertically downward from a hinge that passes along its left edge. By blowing air at 11.0 m/s *over the top of the plate only*, it is possible to keep the plate in a horizontal position, as illustrated in part *a* of the drawing. To what value should the air speed be reduced so that the plate is kept at a 30.0° angle with respect to the vertical, as in part *b* of the drawing? (*Hint: Apply Bernoulli's equation in the form of Equation 11.12.*)

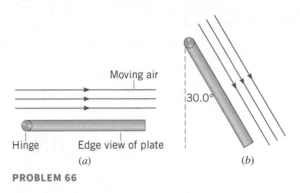

Moving air

Hinge   Edge view of plate
(a)

30.0°

(b)

PROBLEM 66

Section 11.11   Viscous Flow

67. 🄴 A pipe is horizontal and carries oil that has a viscosity of 0.14 Pa · s. The volume flow rate of the oil is 5.3×10^{-5} m^3/s. The length of the pipe is 37 m, and its radius is 0.60 cm. At the output end of the pipe the pressure is atmospheric pressure. What is the absolute pressure at the input end?

68. 🄼 🄶🄾 🄲🄷🄰🄻🄺 Two hoses are connected to the same outlet using a Y-connector, as the drawing shows. The hoses A and B have the same length, but hose B has the larger radius. Each is open to the atmosphere at the end where the water exits. Water flows through both hoses as a viscous fluid, and Poiseuille's law [$Q = \pi R^4 (P_2 - P_1)/(8\eta L)$] applies to each. In this law, P_2 is the pressure upstream, P_1 is the pressure downstream, and Q is the volume flow rate. The ratio of the radius of hose B to the radius of hose A is $R_B/R_A = 1.50$. Find the ratio of the speed of the water in hose B to the speed in hose A.

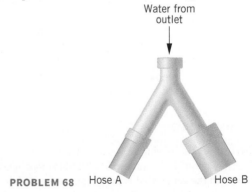

Water from outlet

PROBLEM 68   Hose A   Hose B

Additional Problems

Online

69. 🄴 🅂🅂🄼 Measured along the surface of the water, a rectangular swimming pool has a length of 15 m. Along this length, the flat bottom of the pool slopes downward at an angle of 11° below the horizontal, from one end to the other. By how much does the pressure at the bottom of the deep end exceed the pressure at the bottom of the shallow end?

70. 🄴 A water bed for sale has dimensions of 1.83 m × 2.13 m × 0.229 m. The floor of the bedroom will tolerate an additional weight of no more than 6660 N. Find the weight of the water in the bed and determine whether the bed should be purchased.

71. 🄴 🄶🄾 An underground pump initially forces water through a horizontal pipe at a flow rate of 740 gallons per minute. After several years of operation, corrosion and mineral deposits have reduced the inner radius of the pipe to 0.19 m from 0.24 m, but the pressure difference between the ends of the pipe is the same as it was initially. Find the final flow rate in the pipe in gallons per minute. Treat water as a viscous fluid.

72. **E** (a) The mass and the radius of the sun are, respectively, 1.99×10^{30} kg and 6.96×10^8 m. What is its density? (b) If a solid object is made from a material that has the same density as the sun, would it sink or float in water? Why? (c) Would a solid object sink or float in water if it were made from a material whose density was the same as that of the planet Saturn (mass $= 5.7 \times 10^{26}$ kg, radius $= 6.0 \times 10^7$ m)? Provide a reason for your answer.

73. **M** **SSM** A water line with an internal radius of 6.5×10^{-3} m is connected to a shower head that has 12 holes. The speed of the water in the line is 1.2 m/s. (a) What is the volume flow rate in the line?

(b) At what speed does the water leave one of the holes (effective hole radius $= 4.6 \times 10^{-4}$ m) in the head?

74. **M** **V-HINT** A log splitter uses a pump with hydraulic oil to push a piston, which is attached to a chisel. The pump can generate a pressure of 2.0×10^7 Pa in the hydraulic oil, and the piston has a radius of 0.050 m. In a stroke lasting 25 s, the piston moves 0.60 m. What is the power needed to operate the log splitter's pump?

75. **M** An object is solid throughout. When the object is completely submerged in ethyl alcohol, its apparent weight is 15.2 N. When completely submerged in water, its apparent weight is 13.7 N. What is the volume of the object?

76. **M** **GO** A gold prospector finds a solid rock that is composed solely of quartz and gold. The mass and volume of the rock are, respectively, 12.0 kg and 4.00×10^{-3} m³. Find the mass of the gold in the rock.

Physics in Biology, Medicine, and Sports

77. **M** **GO** **SSM** **BIO** **11.10**   Review Example 17 as an aid in understanding this problem. An aneurysm is an abnormal enlargement of a blood vessel such as the aorta. Because of the aneurysm, the normal cross-sectional area A_1 of the aorta increases to a value of $A_2 = 1.7A_1$. The speed of the blood ($\rho = 1060$ kg/m³) through a normal portion of the aorta is $v_1 = 0.40$ m/s. Assuming that the aorta is horizontal (the person is lying down), determine the amount by which the pressure P_2 in the enlarged region exceeds the pressure P_1 in the normal region.

78. **M** **BIO** **V-HINT** **11.2**   As the initially empty urinary bladder fills with urine and expands, its internal pressure increases by 3300 Pa, which triggers the micturition reflex (the feeling of the need to urinate). The drawing shows a horizontal, square section of the bladder wall with an edge length of 0.010 m. Because the bladder is stretched, four tension forces of equal magnitude T act on the square section, one at each edge, and each force is directed at an angle θ below the horizontal. What is the magnitude T of the tension force acting on one edge of the section when the internal bladder pressure is 3300 Pa and each of the four tension forces is directed 5.0° below the horizontal?

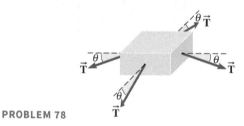

PROBLEM 78

79. **E** **BIO** **11.4**   The drawing shows an intravenous feeding. With the distance shown, nutrient solution ($\rho = 1030$ kg/m³) can just barely enter the blood in the vein. What is the gauge pressure of the venous blood? Express your answer in millimeters of mercury.

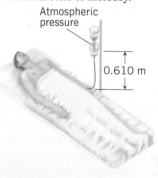

Atmospheric pressure

0.610 m

PROBLEM 79

80. **E** **BIO** **11.3**   The human lungs can function satisfactorily up to a limit where the pressure difference between the outside and inside of the lungs is one-twentieth of an atmosphere. If a diver uses a snorkel for breathing, how far below the water can she swim? Assume the diver is in salt water whose density is 1025 kg/m³.

81. **E** **BIO** **11.3**   At a given instant, the blood pressure in the heart is 1.6×10^4 Pa. If an artery in the brain is 0.45 m above the heart, what is the pressure in the artery? Ignore any pressure changes due to blood flow.

82. **E** **BIO** **SSM** **11.8**   A patient recovering from surgery is being given fluid intravenously. The fluid has a density of 1030 kg/m³, and 9.5×10^{-4} m³ of it flows into the patient every six hours. Find the mass flow rate in kg/s.

83. **E** **BIO** **MMH** **11.8**   (a) The volume flow rate in an artery supplying the brain is 3.6×10^{-6} m³/s. If the radius of the artery is 5.2 mm, determine the average blood speed. (b) Find the average blood speed at a constriction in the artery if the constriction reduces the radius by a factor of 3. Assume that the volume flow rate is the same as that in part (a).

84. **M** **BIO** **V-HINT** **11.8**   The aorta carries blood away from the heart at a speed of about 40 cm/s and has a radius of approximately 1.1 cm. The aorta branches eventually into a large number of tiny capillaries that distribute the blood to the various body organs. In a capillary, the blood speed is approximately 0.07 cm/s, and the radius is about 6×10^{-4} cm. Treat the blood as an incompressible fluid, and use these data to determine the approximate number of capillaries in the human body.

85. **E** **BIO** **11.10**   The blood speed in a normal segment of a horizontal artery is 0.11 m/s. An abnormal segment of the artery is narrowed down by an arteriosclerotic plaque to one-fourth the normal cross-sectional area. What is the difference in blood pressures between the normal and constricted segments of the artery?

86. **E** **BIO** **11.11**   In the human body, blood vessels can dilate, or increase their radii, in response to various stimuli, so that the volume flow rate of the blood increases. Assume that the pressure at either end of a blood vessel, the length of the vessel, and the viscosity of the blood remain the same, and determine the factor $R_{dilated}/R_{normal}$ by which the radius of a vessel must change in order to double the volume flow rate of the blood through the vessel.

87. **E** **BIO** **GO** **11.11**   A blood transfusion is being set up in an emergency room for an accident victim. Blood has a density of 1060 kg/m³ and a viscosity of 4.0×10^{-3} Pa · s. The needle being used has a

length of 3.0 cm and an inner radius of 0.25 mm. The doctor wishes to use a volume flow rate through the needle of 4.5×10^{-8} m³/s. What is the distance h above the victim's arm where the level of the blood in the transfusion bottle should be located? As an approximation, assume that the level of the blood in the transfusion bottle and the point where the needle enters the vein in the arm have the same pressure of one atmosphere. (In reality, the pressure in the vein is slightly above atmospheric pressure.)

88. **E** **BIO** **11.9** One way to administer an inoculation is with a "gun" that shoots the vaccine through a narrow opening. No needle is necessary, for the vaccine emerges with sufficient speed to pass directly into the tissue beneath the skin. The speed is high, because the vaccine ($\rho = 1100$ kg/m³) is held in a reservoir where a high pressure pushes it out. The pressure on the surface of the vaccine in one gun is 4.1×10^6 Pa above the atmospheric pressure outside the narrow opening. The dosage is small enough that the vaccine's surface in the reservoir is nearly stationary during an inoculation. The vertical height between the vaccine's surface in the reservoir and the opening can be ignored. Find the speed at which the vaccine emerges.

89. **E** **BIO** **11.3** If a scuba diver descends too quickly into the sea, the internal pressure on each eardrum remains at atmospheric pressure, while the external pressure increases due to the increased water depth. At sufficient depths, the difference between the external and internal pressures can rupture an eardrum. Eardrums can rupture when the pressure difference is as little as 35 kPa. What is the depth at which this pressure difference could occur? The density of seawater is 1025 kg/m³.

90. **E** **BIO** **11.3** The Mariana snailfish (see the photo) holds the record for the world's deepest living fish. The snailfish has been found in the Mariana Trench at a depth of 26 500 feet below the water's surface. (a) Relative to atmospheric pressure at the surface of the ocean, what is the water pressure at the depth where the snailfish lives? Use the density of sea water (1030 kg/m³) in your calculation. (b) How many total pounds of force would be applied to a human head at this depth? Treat the head as a sphere with a radius of 9.80 cm.

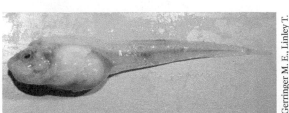

PROBLEM 90

91. **M** **BIO** **11.1** Astronauts living and working on the International Space Station in orbit above the earth will experience many negative physiological effects from microgravity. One of the more serious ones is a decrease in bone density. Our bones are composed of living tissue that is constantly being broken down and replaced with new bone. On the earth's surface, these two processes occur at roughly the same rate, so our bone density remains fairly constant for long periods of time. In space, the muscles and bones are not subjected to the constant weights, forces, and stresses that come with living in gravity. This lack of constant stress leads to a slower rate of new bone growth and a decrease in overall bone density (see the figure). The longest and strongest bone in the human body—the femur—has a volume of 118 cm³ and a density of 1.94 g/cm³. If an astronaut on the space station experiences a net loss of 20.0% of the bone mass in their femur after 5 months, what is the bone density at that time?

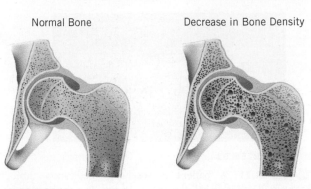

Normal Bone   Decrease in Bone Density

PROBLEM 91

92. **M** **BIO** **11.3** The tympanic membrane, or eardrum, is a structure that separates the external and middle parts of the ear (see the figure). It is sensitive to and vibrates in response to changes in air pressure and transmits these vibrations to other structures in the inner ear that lead to the sensation of hearing. Under normal conditions, the pressures on the inside and outside of the tympanic membrane are kept approximately equal. The auditory tube, also called the Eustachian tube, is responsible for this equilibration. However, rapid changes in external pressure can cause large pressure differentials on the tympanic membrane, causing it to rupture. A differential force across the eardrum membrane as little as 5.0 N can cause a rupture. (a) If the cross-sectional area of the membrane is 1.0 cm², what is the maximum tolerable pressure difference between the external and inner ear? (b) Based on your answer in part (a), to what maximum depth could a person dive in fresh water before rupturing an eardrum? Experienced divers can actually go much deeper than the answer to part (b) by using techniques to equilibrate the pressure on the inside and outside of the membrane as they dive.

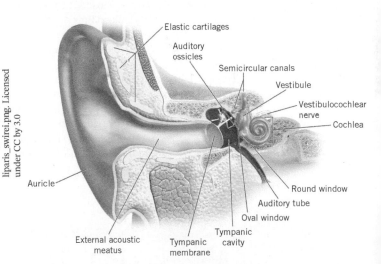

The Anatomy of the Ear

PROBLEM 92

93. **M** **BIO** **11.6** The blue whale is the largest animal known to have ever existed (see the photo). It can be almost 30 m long with a mass of 1.73×10^5 kg. Imagine a blue whale is completely immersed and at rest under the surface of salt water. What volume of salt water is displaced by the whale? Calculate your answer in gallons. Take the density of salt water to be 1030 kg/m³.

Give your answer in cm³. Assume the height of the column of medication in the IV bag and connecting tube remains constant during the 10.0 minutes.

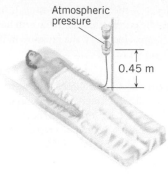

Atmospheric pressure

0.45 m

PROBLEM 94

94. [M] [BIO] **11.11** A patient is receiving medication through an intravenous infusion, as the figure shows. The medication has a viscosity of 2.2×10^{-3} N · s/m² and a density of 1050 kg/m³. The needle being used has a length of 0.035 m, has an inner radius of 5.0×10^{-3} m, and has been inserted into a vein that has a gauge pressure of 2.0×10^3 Pa. The IV bag hangs 0.45 m above the location of the needle. What volume of medication enters the vein in 10.0 minutes?

Concepts and Calculations Problems

Online

Pressure plays an important role in the behavior of fluids. As we have seen in this chapter, pressure is the magnitude of the force acting perpendicular to a surface divided by the area of the surface. Pressure should not be confused, however, with the force itself. Problem 95 serves to emphasize that pressure and force are different concepts. Problem 96 focuses on the essence of Archimedes' principle and its application to the situation of a buoyant force acting on a submerged object.

95. [M] [CHALK] The figure shows a rear view of a loaded two-wheeled wheelbarrow on a horizontal surface. It has balloon tires and a weight $W = 684$ N, which is uniformly distributed. The left tire has a contact area with the ground of $A_L = 6.6 \times 10^{-4}$ m², whereas the right tire is underinflated and has a contact area of $A_R = 9.9 \times 10^{-4}$ m². *Concepts:* (i) Force is a vector. Therefore both a direction and a magnitude are needed to specify it. Are both a direction and magnitude needed to specify a pressure? (ii) How is the force each tire applies to the ground related to the force the ground applies to each tire? (iii) Do the left and right tires apply the same force to the ground? Explain. (iv) Do the left and right tires apply the same pressure to the ground? *Calculations:* Find the force and pressure that each tire applies to the ground.

96. [M] [CHALK] [SSM] A father (weight $W = 830$ N) and his daughter (weight $W = 340$ N) are spending the day at the lake. They are each sitting on a beach ball that is just submerged beneath the water (see the figure). *Concepts:* (i) Each ball is in equilibrium, being stationary and having no acceleration. Thus, the net force acting on each ball is zero. What balances the downward-acting weight in each case? (ii) In which case is the buoyant force greater? (iii) In this situation, what determines the magnitude of the buoyant force? (iv) Which beach ball has the larger radius? *Calculations:* Ignoring the weight of the air in each ball, and the volumes of their legs that are under the water, find the radius of each ball.

PROBLEM 96

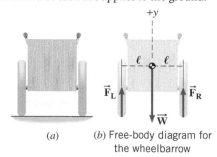

(a)   *(b)* Free-body diagram for the wheelbarrow

$\vec{F}_L$ $+y$ $\vec{F}_R$ $\vec{W}$

ℓ ℓ

(a) Rear view of a wheelbarrow with balloon tires. *(b)* Free-body diagram of the wheelbarrow, showing its weight $\vec{W}$ and the forces $\vec{F}_L$ and $\vec{F}_R$ that the ground applies, respectively, to the left and right tires.

PROBLEM 95

Team Problems

97. M T WS **A Fire Hose.** You and your team need to put out a fire on a rooftop that is currently too high for the water stream emerging from a fire hose. With no nozzle attached, and the 4.00-in.-diameter hose aimed at a 60.0° angle above the horizontal, the peak of the stream's trajectory barely makes it to the bottom of the third floor, which is 20.0 feet above the ground. Assuming the flow rate does not change, what diameter nozzle (in cm) would be sufficient to get water to the roof of the five-story building (i.e., 60.0 feet above the ground) when the hose is directed 60.0° above the horizontal? Treat the water trajectory like projectile motion, and neglect air resistance.

98. E T WS **A Malfunctioning Submarine.** You and your team are inside a small research submarine in the Weddell Sea off the coast of western Antarctica, searching for undersea tunnels leading inland, when the sub suffers serious malfunctions. First, power fails, and you are left with only emergency batteries that will provide dim light and run the air purifying system for no more than 2 hours. Next, the valves to the ballast tanks (i.e., the mechanism used to control the buoyancy of the craft) inadvertently opened and the tanks completely filled with seawater, making the sub sink to the seabed, 70 feet below the surface. All safety devices, such as inflatable buoys that could be sent to warn those on the surface of the emergency, are nonfunctional. You find the submarine's operation manual and learn that there is a manual pump that can be used to expel water from the ballast tanks at a rate of about 1.0 gallon per minute (gpm). In the manual, you find that the mass of the submarine, with crew and supplies, and with empty ballast tanks, is 9250 kg. The shape of the sub is roughly cylindrical, with a length of 5.0 m and a diameter of 1.6 m. The ballast tanks are contained inside the submarine's body, and account for 15% of its total volume. Using the manual pump, how long will it take to remove enough water to float the submarine? The density of seawater is 1040 kg/m³.

99. M T **Crossing a River.** You and your team come to a slow flowing river that you need to cross. The nearest bridge is 20 miles to the north, too dangerous and too far to trek with your group. You explore the area down river and discover an abandoned shed with a stash of 55-gallon drums (all empty and weighing 35.0 lb each) and a stack of 10-foot planks: You will build a raft. The six members of your team have a combined weight of 925 pounds (assuming everyone was truthful). You also have a four-wheeler (all-terrain vehicle), which weighs 450.0 lb, and other gear, which adds another 315 lb. Your simple raft design is as follows: a platform of planks with barrels strapped to the bottom. You estimate that the planks weigh about 45.0 lb each, and you will need 20 of them to make a platform that can accommodate everything for the one-way trip. You measure the dimensions of the cylindrical drums and find they have diameter $D = 22.5$ inches and height $h = 33.5$ inches (note: not exactly "55 gallons"). **(a)** What is the minimum number of barrels that you will need so that the raft will float when fully loaded? **(b)** What is the minimum number of barrels you will need if you want the platform to be at least 6 inches above the water when fully loaded?

100. M T **A Spring Gun.** A hydraulic press is used to compress a spring that will then be used to project a 25.0-kg steel ball. The system is similar, but smaller in scale, to the car jack illustrated in **Figure 11.15**. In this case, the smaller cylinder has a diameter of $d_1 = 0.750$ cm and has a manually operated plunger. The larger cylinder has a diameter of $d_2 = 10.0$ cm, and its piston compresses the spring. The idea is that the gun operator pulls a lever that pushes the plunger on the small cylinder, which transmits a pressure to the larger piston that, in turn, exerts a force on the spring and compresses it. Once the spring is compressed, the steel ball is loaded and the spring is released, ejecting the ball. **(a)** If a force of 950.0 N is exerted on the primary (smaller) piston to compress the spring 1.25 m from its equilibrium (uncompressed) position, what is the spring constant k of the spring? **(b)** What is the velocity of the steel ball just after it is ejected? **(c)** Neglecting air resistance, what is the maximum range of this "spring gun"?

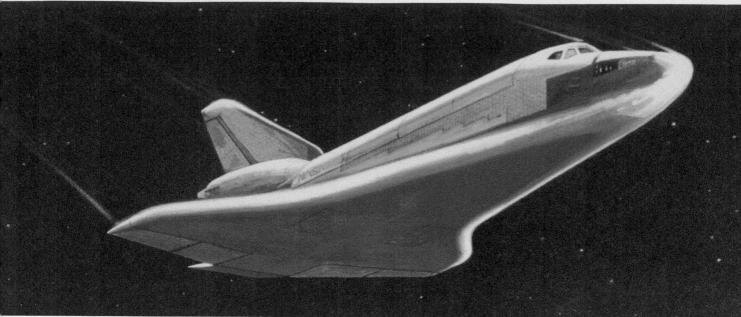

As this artist's rendition shows, the underside of the Space Shuttle consists of special ceramic tiles that shield the Shuttle's interior and flight crew from the high temperatures and heat produced upon reentering the Earth's atmosphere after a mission in space. In 2003 the Space Shuttle *Columbia's* seven-member crew was killed when the shuttle disintegrated on reentry after suffering damage to the heat shield during launch.

NASA/Science Source

LEARNING OBJECTIVES

After reading this module, you should be able to...

12.1 Convert temperature scales.

12.2 Define the Kelvin scale and absolute zero.

12.3 Identify types of thermometers.

12.4 Analyze linear thermal expansion.

12.5 Analyze volume thermal expansion.

12.6 Define heat and internal energy.

12.7 Solve specific heat problems.

12.8 Solve heat problems with phase changes.

12.9 Analyze phase equilibrium.

12.10 Solve humidity problems.

Temperature and Heat

12.1 Common Temperature Scales

To measure temperature we use a thermometer. Many thermometers make use of the fact that materials usually expand with increasing temperature. For example, **Figure 12.1** shows the common mercury-in-glass thermometer, which consists of a mercury-filled glass bulb connected to a capillary tube. When the mercury is heated, it expands into the capillary tube, the amount of expansion being proportional to the change in temperature. The outside of the glass is marked with an appropriate scale for reading the temperature.

A number of different temperature scales have been devised, two popular choices being the **Celsius** (formerly, centigrade) and **Fahrenheit scales**. **Figure 12.1** illustrates these scales. Historically,* both scales were defined by assigning two temperature points on the scale and then dividing the distance between them into a number of equally spaced intervals. One

*Today, the Celsius and Fahrenheit scales are defined in terms of the Kelvin temperature scale; Section 12.2 discusses the Kelvin scale.

point was chosen to be the temperature at which ice melts under one atmosphere of pressure (the "ice point"), and the other was the temperature at which water boils under one atmosphere of pressure (the "steam point"). On the Celsius scale, an ice point of 0 °C (0 degrees Celsius) and a steam point of 100 °C were selected. On the Fahrenheit scale, an ice point of 32 °F (32 degrees Fahrenheit) and a steam point of 212 °F were chosen. The Celsius scale is used worldwide, while the Fahrenheit scale is used mostly in the United States, often in home medical thermometers.

There is a subtle difference in the way the temperature of an object is reported, as compared to a *change* in its temperature. For example, the temperature of the human body is about 37 °C, where the symbol °C stands for "degrees Celsius." However, the *change* between two temperatures is specified in "Celsius degrees" (C°)—not in "degrees Celsius." Thus, if the body temperature rises to 39 °C, the change in temperature is 2 Celsius degrees or 2 C°, not 2 °C.

As **Figure 12.1** indicates, the separation between the ice and steam points on the Celsius scale is divided into 100 Celsius degrees, while on the Fahrenheit scale the separation is divided into 180 Fahrenheit degrees. Therefore, the size of the Celsius degree is larger than that of the Fahrenheit degree by a factor of $\frac{180}{100}$, or $\frac{9}{5}$. Examples 1 and 2 illustrate how to convert between the Celsius and Fahrenheit scales using this factor.

The reasoning strategy used in Examples 1 and 2 for converting between different temperature scales is summarized below.

REASONING STRATEGY   **Converting Between Different Temperature Scales**

1. **Determine the magnitude of the difference between the stated temperature and the ice point on the initial scale.**

2. **Convert this number of degrees from one scale to the other scale by using the appropriate conversion factor. For conversion between the Celsius and Fahrenheit scales, the factor is based on the fact that 1 C° = $\frac{9}{5}$ F°**

3. **Add or subtract the number of degrees on the new scale to or from the ice point on the new scale.**

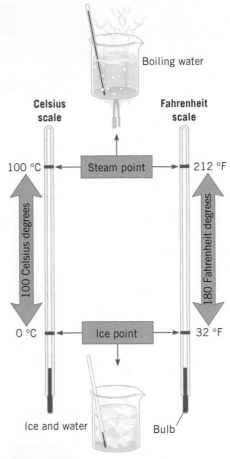

FIGURE 12.1 The Celsius and Fahrenheit temperature scales.

EXAMPLE 1 | Converting from a Fahrenheit to a Celsius Temperature

A healthy person has an oral temperature of 98.6 °F. What would this reading be on the Celsius scale?

Reasoning and Solution A temperature of 98.6 °F is 66.6 Fahrenheit degrees above the ice point of 32.0 °F. Since 1 C° = $\frac{9}{5}$ F°, the difference of 66.6 F° is equivalent to

$$(66.6 \text{ F°})\left(\frac{1 \text{ C}}{\frac{9}{5} \text{ F°}}\right) = 37.0 \text{ C°}$$

Thus, the person's temperature is 37.0 Celsius degrees above the ice point. Adding 37.0 Celsius degrees to the ice point of 0 °C on the Celsius scale gives a Celsius temperature of $\boxed{37.0 \text{ °C}}$.

EXAMPLE 2 | Converting from a Celsius to a Fahrenheit Temperature

A time and temperature sign on a bank indicates that the outdoor temperature is −20.0 °C. Find the corresponding temperature on the Fahrenheit scale.

Reasoning and Solution The temperature of −20.0 °C is 20.0 Celsius degrees *below* the ice point of 0 °C. This number of Celsius degrees corresponds to

$$(20.0 \text{ C°})\left(\frac{\frac{9}{5} \text{ F°}}{1 \text{ C°}}\right) = 36.0 \text{ F°}$$

The temperature, then, is 36.0 Fahrenheit degrees below the ice point. Subtracting 36.0 Fahrenheit degrees from the ice point of 32.0 °F on the Fahrenheit scale gives a Fahrenheit temperature of $\boxed{−4.0 \text{ °F}}$.

Check Your Understanding

(The answer is given at the end of the book.)

1. On a new temperature scale the steam point is 348 °X, and the ice point is 112 °X. What is the temperature on this scale that corresponds to 28.0 °C?

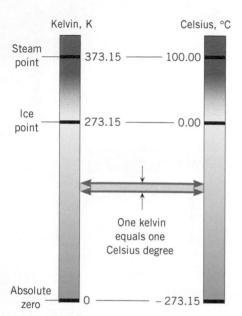

FIGURE 12.2 on the left shows:

Kelvin, K   Celsius, °C

Steam point   373.15 —— 100.00

Ice point   273.15 —— 0.00

One kelvin equals one Celsius degree

Absolute zero   0 —— −273.15

FIGURE 12.2 A comparison of the Kelvin and Celsius temperature scales.

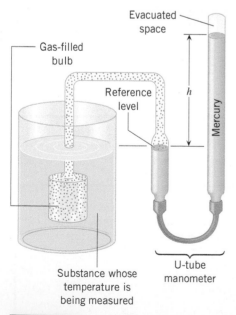

Evacuated space

Gas-filled bulb

Reference level

Mercury

h

Substance whose temperature is being measured

U-tube manometer

FIGURE 12.3 A constant-volume gas thermometer.

12.2 The Kelvin Temperature Scale

Although the Celsius and Fahrenheit scales are widely used, the **Kelvin temperature scale** has greater scientific significance. It was introduced by the Scottish physicist William Thompson (Lord Kelvin, 1824–1907), and in his honor each degree on the scale is called a kelvin (K). By international agreement, the symbol K is not written with a degree sign (°), nor is the word "degrees" used when quoting temperatures. For example, a temperature of 300 K (not 300 °K) is read as "three hundred kelvins," not "three hundred degrees kelvin." The kelvin is the SI base unit for temperature.

Figure 12.2 compares the Kelvin and Celsius scales. The size of one kelvin is identical to the size of one Celsius degree because there are one hundred divisions between the ice and steam points on both scales. As we will discuss shortly, experiments have shown that there exists a lowest possible temperature, below which no substance can be cooled. This lowest temperature is defined to be the zero point on the Kelvin scale and is referred to as *absolute zero*.

The ice point (0 °C) occurs at 273.15 K on the Kelvin scale. Thus, the Kelvin temperature T and the Celsius temperature T_c are related by

$$T = T_c + 273.15 \qquad (12.1)$$

The number 273.15 in Equation 12.1 is an experimental result, obtained in studies that utilize a gas-based thermometer.

When a gas confined to a fixed volume is heated, its pressure increases. Conversely, when the gas is cooled, its pressure decreases. For example, the air pressure in automobile tires can rise by as much as 20% after the car has been driven and the tires have become warm. The change in gas pressure with temperature is the basis for the **constant-volume gas thermometer**.

A constant-volume gas thermometer consists of a gas-filled bulb to which a pressure gauge is attached, as in **Figure 12.3**. The gas is often hydrogen or helium at a low density, and the pressure gauge can be a U-tube manometer filled with mercury. The bulb is placed in thermal contact with the substance whose temperature is being measured. The volume of the gas is held constant by raising or lowering the *right* column of the U-tube manometer in order to keep the mercury level in the *left* column at the same reference level. The absolute pressure of the gas is proportional to the height h of the mercury on the right. As the temperature changes, the pressure changes and can be used to indicate the temperature, once the constant-volume gas thermometer has been calibrated.

Suppose that the absolute pressure of the gas in **Figure 12.3** is measured at different temperatures. If the results are plotted on a pressure-versus-temperature graph, a straight line is obtained, as in **Figure 12.4**. If the straight line is extended or extrapolated to lower and lower temperatures, the line crosses the temperature axis at −273.15 °C. In reality, no gas can be cooled to this temperature, because all gases liquify before reaching it. However, helium and hydrogen liquify at such low temperatures that they are often used in the thermometer. This kind of graph can be obtained for different amounts and types of low-density gases. In all cases, a straight line is found that extrapolates to

−273.15 °C on the temperature axis, which suggests that the value of −273.15 °C has fundamental significance. The significance of this number is that it is the **absolute zero point** for temperature measurement. The phrase "absolute zero" means that temperatures lower than −273.15 °C cannot be reached by continually cooling a gas or any other substance. If lower temperatures could be reached, then further extrapolation of the straight line in **Figure 12.4** would suggest that negative absolute gas pressures could exist. Such a situation would be impossible, because a negative absolute gas pressure has no meaning. Thus, the Kelvin scale is chosen so that its zero temperature point is the lowest temperature attainable.

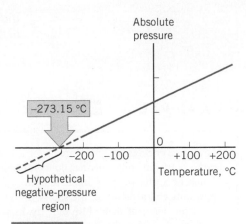

FIGURE 12.4 A plot of absolute pressure versus temperature for a low-density gas at constant volume. The graph is a straight line and, when extrapolated (dashed line), crosses the temperature axis at −273.15 °C.

12.3 | Thermometers

All thermometers make use of the change in some physical property with temperature. A property that changes with temperature is called a **thermometric property**. For example, the thermometric property of the mercury thermometer is the length of the mercury column, while in the constant-volume gas thermometer it is the pressure of the gas. Several other important thermometers and their thermometric properties will now be discussed.

The *thermocouple* is a thermometer used extensively in scientific laboratories. It consists of thin wires of different metals, welded together at the ends to form two junctions, as **Figure 12.5** illustrates. Often the metals are copper and constantan (a copper–nickel alloy). One of the junctions, called the "hot" junction, is placed in thermal contact with the object whose temperature is being measured. The other junction, termed the "reference" junction, is kept at a known constant temperature (usually an ice–water mixture at 0 °C). The thermocouple generates a voltage that depends on the *difference in temperature* between the two junctions. This voltage is the thermometric property and is measured by a voltmeter, as the drawing indicates. With the aid of calibration tables, the temperature of the hot junction can be obtained from the voltage. Thermocouples are used to measure temperatures as high as 2300 °C or as low as −270 °C.

Most substances offer resistance to the flow of electricity, and this resistance changes with temperature. As a result, electrical resistance provides another thermometric property. *Electrical resistance thermometers* are often made from platinum wire, because platinum has excellent mechanical and electrical properties in the temperature range from −270 °C to +700 °C. The resistance of platinum wire is known as a function of temperature. Thus,

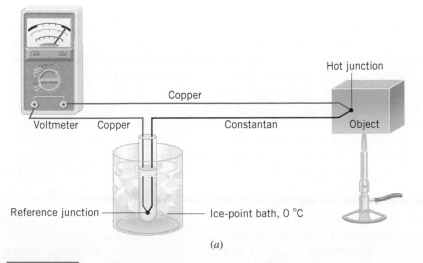

(a)

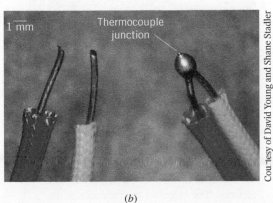

(b)

FIGURE 12.5 (a) A thermocouple is made from two different types of wires, copper and constantan in this case. (b) A thermocouple junction between two different wires.

the temperature of a substance can be determined by placing the resistance thermometer in thermal contact with the substance and measuring the resistance of the wire.

Almost anything that has a predictable response to temperature can be used as a thermometer. As Example 3 shows, this even applies to the chirp of the snowy tree cricket.

EXAMPLE 3 | BIO A Cricket Thermometer

Snowy tree crickets are green, are about 1.5 cm long, and have an orange mark and long antennae on their heads. They can be found in most of North America, and have been extensively studied. One interesting characteristic of the snowy tree cricket is the temperature dependence of their chirp rate: the higher the ambient temperature, the higher the chirp rate. This temperature dependence can serve as a crude audible thermometer. On a particularly cold night ($T = 13$ °C), suppose the total number of chirps recorded in your back yard in one minute was 119. The next day, a warm front comes through, the nighttime temperature increases to 26 °C, and the number of chirps recorded in one minute is 258. Assuming a linear relationship between the chirp rate and the temperature, and that the number of crickets does not change, what is the temperature on the next night when 204 chirps are counted in one minute?

Reasoning   Since we assume a linear relationship between the number of chirps per minute N and the temperature T, we can apply the point-slope formula to obtain the slope, and then a linear equation that relates T and N.

Solution   The point-slope formula for a line passing through points (x_1, y_1) and (x_2, y_2) is given by $y_2 - y_1 = m(x_2 - x_1)$, where m is the slope, y is the dependent variable N, and x is the independent variable T. Writing this as $N_2 - N_1 = m(T_2 - T_1)$ and solving for the slope, we have

$$m = \frac{N_2 - N_1}{T_2 - T_1} = \frac{258 \text{ chirps/min} - 119 \text{ chirps/min}}{26\,°C - 13\,°C} = 10.7 \frac{\text{chirps}}{\text{min} \cdot °C}$$

The equation of a line is given by $y = mx + b$, which we write as $N = mT + b$, where b is the N-axis intercept. Since we know the slope (m), we can solve this equation for b, and plug in one of the two sets of known values (say $N = 119$ and $T = 13$ °C) to obtain the following:

$$b = N - mT = 119 \text{ chirps/min} - [10.7 \text{ chirps/(min} \cdot °C)](13\,°C)$$
$$= -20.1 \text{ chirps/min}$$

Note that this negative number of chirps has no physical meaning, but is just the y-intercept of the line. Solving $N = mT + b$ for T, and substituting the values for m and b, we have

$$T = \frac{N - b}{m} = \frac{N}{m} - \frac{b}{m} = \frac{N}{[10.7 \text{ chirps/(min} \cdot °C)]}$$
$$- \frac{-20.1 \text{ chirps/min}}{[10.7 \text{ chirps/(min} \cdot °C)]}$$

or

$$T = \left(0.0935 \frac{\text{min} \cdot °C}{\text{chirps}}\right)N + 1.88\,°C$$

Substituting $N = 204$ chirps/min into this equation, and taking 2 significant figures, we have

$$T = \left(0.0935 \frac{\text{min} \cdot °C}{\text{chirps}}\right)\left(204 \frac{\text{chirps}}{\text{min}}\right) + 1.88\,°C = \boxed{21\,°C}$$

This result is only an approximation, and does not work, for instance, at temperatures below about 10 °C or above 38 °C, where snowy tree crickets stop chirping altogether.

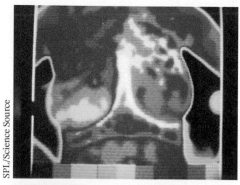

SPL/Science Source

FIGURE 12.6 The presence of breast cancer can be detected in a thermogram via the elevated temperatures associated with the malignant tissue. The healthy breast on the left side of the thermogram registers predominantly blue (see the temperature coding bar to the right of the image), indicating temperatures that are not unusually elevated. The breast on the right side of the thermogram, however, registers the colors yellow, orange, and red, which indicate the elevated temperatures due to breast cancer.

BIO   **THE PHYSICS OF . . . thermography.**   Radiation emitted by an object can also be used to indicate temperature. At low to moderate temperatures, the predominant radiation emitted is infrared. As the temperature is raised, the intensity of the radiation increases substantially. In one interesting application, an infrared camera registers the intensity of the infrared radiation produced at different locations on the human body. The camera is connected to a color monitor that displays the different infrared intensities as different colors. This "thermal painting" is called a *thermograph* or *thermogram*. Thermography is an important diagnostic tool in medicine. For example, breast cancer is indicated in the thermogram in **Figure 12.6** by the elevated temperatures associated with malignant tissue. In another application, **Figure 12.7** shows thermographic images of a smoker's forearms before (left) and 5 minutes after (right) he has smoked a cigarette. After smoking, the forearms are cooler due to the effect of nicotine, which causes vasoconstriction (narrowing of the blood vessels) and reduces blood flow, a result that can lead to a higher risk from blood clotting. Temperatures in these images range from over 34 °C to about 28 °C and are indicated in decreasing order by the colors white, red, yellow, green, and blue.

Oceanographers and meteorologists also use thermograms extensively, to map the temperature distribution on the surface of the earth. For example, **Figure 12.8** shows a satellite image of the sea-surface temperature of the Pacific Ocean. The region depicted in red is the 1997/98 El Niño, a large area of

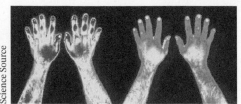

FIGURE 12.7 Thermogram showing a smoker's forearms before (*left*) and 5 minutes after (*right*) he has smoked a cigarette. Temperatures decrease from white (over 34 °C) to yellow, to red, to green, to blue (about 28 °C).

FIGURE 12.8 A thermogram of the 1997/98 El Niño (red), a large region of abnormally high temperatures in the Pacific Ocean.

the ocean, approximately twice the width of the United States, where temperatures reached abnormally high values. This El Niño caused major weather changes in certain regions of the earth.

12.4 | Linear Thermal Expansion

Normal Solids

Have you ever found the metal lid on a glass jar too tight to open? One solution is to run hot water over the lid, which loosens it because the metal expands more than the glass does. To varying extents, most materials expand when heated and contract when cooled. The increase in any one dimension of a solid is called **linear expansion**, linear in the sense that the expansion occurs along a line. **Figure 12.9** illustrates the linear expansion of a rod whose length is L_0 when the temperature is T_0. When the temperature of the rod increases to $T_0 + \Delta T$, the length becomes $L_0 + \Delta L$, where ΔT and ΔL are the changes in temperature and length, respectively. Conversely, when the temperature decreases to $T_0 - \Delta T$, the length decreases to $L_0 - \Delta L$.

For modest temperature changes, experiments show that the change in length is directly proportional to the change in temperature ($\Delta L \propto \Delta T$). In addition, the change in length is proportional to the initial length of the rod, a fact that can be understood with the aid of **Figure 12.10**. Part *a* of the drawing shows two identical rods. Each rod has a length L_0 and expands by ΔL when the temperature increases by ΔT. Part *b* shows the two heated rods combined into a single rod, for which the total expansion is the sum of the expansions of each part—namely, $\Delta L + \Delta L = 2\Delta L$. Clearly, the amount of expansion doubles if the rod is twice as long to begin with. In other words, the change in length is directly proportional to the original length ($\Delta L \propto L_0$). Equation 12.2 expresses the fact that ΔL is proportional to both L_0 and ΔT ($\Delta L \propto L_0 \Delta T$) by using a proportionality constant α, which is called the **coefficient of linear expansion**.

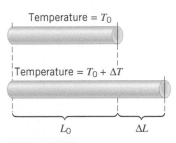

FIGURE 12.9 When the temperature of a rod is raised by ΔT, the length of the rod increases by ΔL.

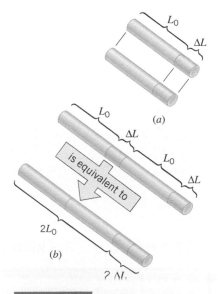

FIGURE 12.10 (*a*) Each of two identical rods expands by ΔL when heated. (*b*) When the rods are combined into a single rod of length $2L_0$, the "combined" rod expands by $2\,\Delta L$.

LINEAR THERMAL EXPANSION OF A SOLID

The length L_0 of an object changes by an amount ΔL when its temperature changes by an amount ΔT:

$$\Delta L = \alpha L_0\, \Delta T \qquad (12.2)$$

where α is the coefficient of linear expansion.

Common Unit for the Coefficient of Linear Expansion: $\frac{1}{C°} = (C°)^{-1}$

TABLE 12.1   **Coefficients of Thermal Expansion for Solids and Liquids**[a]

| Substance | Coefficient of Thermal Expansion $(C°)^{-1}$ | |
	Linear (α)	Volume (β)
Solids		**Liquids**
Aluminum	23×10^{-6}	69×10^{-6}
Brass	19×10^{-6}	57×10^{-6}
Concrete	12×10^{-6}	36×10^{-6}
Copper	17×10^{-6}	51×10^{-6}
Glass (common)	8.5×10^{-6}	26×10^{-6}
Glass (Pyrex)	3.3×10^{-6}	9.9×10^{-6}
Gold	14×10^{-6}	42×10^{-6}
Iron or steel	12×10^{-6}	36×10^{-6}
Lead	29×10^{-6}	87×10^{-6}
Nickel	13×10^{-6}	39×10^{-6}
Quartz (fused)	0.50×10^{-6}	1.5×10^{-6}
Silver	19×10^{-6}	57×10^{-6}
Liquids[b]		
Benzene	—	1240×10^{-6}
Carbon tetrachloride	—	1240×10^{-6}
Ethyl alcohol	—	1120×10^{-6}
Gasoline	—	950×10^{-6}
Mercury	—	182×10^{-6}
Methyl alcohol	—	1200×10^{-6}
Water	—	207×10^{-6}

[a]The values for α and β pertain to a temperature near 20°C.

[b]Since liquids do not have fixed shapes, the coefficient of linear expansion is not defined for them.

Solving Equation 12.2 for α shows that $\alpha = \Delta L/(L_0 \Delta T)$. Since the length units of ΔL and L_0 algebraically cancel, the coefficient of linear expansion α has the unit of $(C°)^{-1}$ when the temperature difference ΔT is expressed in Celsius degrees (C°). Different materials with the same initial length expand and contract by different amounts as the temperature changes, so the value of α depends on the nature of the material. **Table 12.1** shows some typical values. Coefficients of linear expansion also vary somewhat depending on the range of temperatures involved, but the values in **Table 12.1** are adequate approximations. Example 4 deals with a situation in which a dramatic effect due to thermal expansion can be observed, even though the change in temperature is small.

EXAMPLE 4 | Buckling of a Sidewalk

A concrete sidewalk is constructed between two buildings on a day when the temperature is 25 °C. The sidewalk consists of two slabs, each three meters in length and of negligible thickness (**Figure 12.11a**). As the temperature rises to 38 °C, the slabs expand, but no space is provided for thermal expansion. The buildings do not move, so the slabs buckle upward. Determine the vertical distance y in part b of the drawing.

Reasoning The expanded length of each slab is equal to its original length plus the change in length ΔL due to the rise in temperature. We know the original length, and Equation 12.2 can be used to find the change in length. Once the expanded length has been determined, the Pythagorean theorem can be employed to find the vertical distance y in **Figure 12.11b**.

Solution The change in temperature is $\Delta T = 38\ ^\circ C - 25\ ^\circ C = 13\ C^\circ$, and the coefficient of linear expansion for concrete is given in **Table 12.1**. The change in length of each slab associated with this temperature change is

$$\Delta L = \alpha L_0\, \Delta T = [12 \times 10^{-6}(C^\circ)^{-1}](3.0\ \text{m})(13\ C^\circ)$$
$$= 0.000\ 47\ \text{m} \qquad\qquad \textbf{(12.2)}$$

The expanded length of each slab is, thus, 3.000 47 m. We can now determine the vertical distance y by applying the Pythagorean theorem to the right triangle in **Figure 12.11b**:

$$y = \sqrt{(3.000\ 47\ \text{m})^2 - (3.000\ 00\ \text{m})^2} = \boxed{0.053\ \text{m}}$$

Math Skills The calculation of the answer for y involves two numbers, each of which has six significant figures. Yet the answer has only two significant figures. This reduction in significant figures occurs because the calculation includes a subtraction step. ***Be on the lookout for subtractions, because they may reduce the number of significant figures in your answers.*** For instance, in the present case we have $y = \sqrt{(3.000\ 47\ \text{m})^2 - (3.000\ 00\ \text{m})^2}$, which has the form of $y = \sqrt{a^2 - b^2} = \sqrt{(a+b)(a-b)}$. Therefore, we can write the expression for y in the following way:

$$y = \sqrt{(3.000\ 47\ \text{m})^2 - (3.000\ 00\ \text{m})^2}$$
$$= \sqrt{(3.000\ 47\ \text{m} + 3.000\ 00\ \text{m})(3.000\ 47\ \text{m} - 3.000\ 00\ \text{m})}$$

In this result, the subtraction reduces the number of significant figures from six to two:

$$3.000\ 47\ \text{m} - 3.000\ 00\ \text{m} = 0.000\ 47\ \text{m}$$

As a result, our expression for y becomes

$$y = \sqrt{(6.000\ 47\ \text{m})(0.000\ 47\ \text{m})} = \boxed{0.053\ \text{m}}$$

which limits the answer to two significant figures.

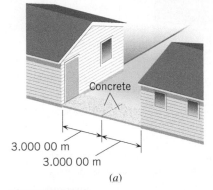

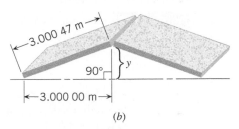

3.000 47 m

90°

y

3.000 00 m

(a)

(b)

FIGURE 12.11 (a) Two concrete slabs completely fill the space between the buildings. (b) When the temperature increases, each slab expands, causing the sidewalk to buckle.

The buckling of a sidewalk is one consequence of not providing sufficient room for thermal expansion. To eliminate such problems, engineers incorporate expansion joints or spaces at intervals along bridge roadbeds, as **Figure 12.12** shows.

BIO **THE PHYSICS OF . . . an antiscalding device.** Although Example 4 shows how thermal expansion can cause problems, there are also times when it can be useful. For instance, each year thousands of children are taken to emergency rooms suffering from burns caused by scalding tap water. Such accidents can be reduced with the aid of the antiscalding device shown in **Figure 12.13**. This device screws onto the end of a faucet and quickly shuts off the flow of water when it becomes too hot. As the water temperature rises, the actuator spring expands and pushes the plunger forward, shutting off the flow. When the water cools, the spring contracts and the water flow resumes.

Thermal Stress

If the concrete slabs in **Figure 12.11** had not buckled upward, they would have been subjected to immense forces from the buildings. The forces needed to prevent a solid object from expanding must be strong enough to counteract any change in length that

David R. Frazier Photolibrary, Inc./Alamy Stock Photo

FIGURE 12.12 An expansion joint in a bridge.

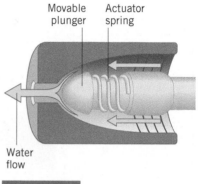

Movable plunger   Actuator spring

Water flow

FIGURE 12.13 An antiscalding device.

would occur due to a change in temperature. Although the change in temperature may be small, the forces—and hence the stresses—can be enormous. They can, in fact, lead to serious structural damage. Example 5 illustrates just how large the stresses can be.

Analyzing Multiple-Concept Problems

EXAMPLE 5 | The Physics of Thermal Stress

A steel beam is used in the roadbed of a bridge. The beam is mounted between two concrete supports when the temperature is 23 °C, with no room provided for thermal expansion (see **Figure 12.14**). What compressional stress must the concrete supports apply to each end of the beam, if they are to keep the beam from expanding when the temperature rises to 42 °C? Assume that the distance between the concrete supports does not change as the temperature rises.

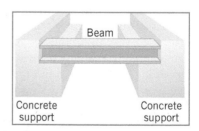

Beam

Concrete support   Concrete support

FIGURE 12.14 A steel beam is mounted between concrete supports with no room provided for thermal expansion.

Reasoning When the temperature rises by an amount ΔT, the natural tendency of the beam is to expand. If the beam were free to expand, it would lengthen by an amount $\Delta L = \alpha L_0 \, \Delta T$ (Equation 12.2). However, the concrete supports prevent this expansion from occurring by exerting a compressional force on each end of the beam. The magnitude F of this force depends on ΔL through the relation $F = YA(\Delta L/L_0)$ (Equation 10.17), where Y is Young's modulus for steel and A

is the cross-sectional area of the beam. According to the discussion in Section 10.8, the compressional stress is equal to the magnitude of the compressional force divided by the cross-sectional area, or Stress $= F/A$.

Knowns and Unknowns The data for this problem are listed in the table:

Description	Symbol	Value
Initial temperature	T_0	23 °C
Final temperature	T	42 °C
Unknown Variables		
Stress	—	?

Modeling the Problem

STEP 1 Stress and Force The compressional stress is defined as the magnitude F of the compressional force divided by the cross-sectional area A of the beam (see Section 10.8), or

$$\text{Stress} = \frac{F}{A}$$

According to Equation 10.17, the magnitude of the compressional force that the concrete supports exert on each end of the steel beam is given by

$$F = YA\left(\frac{\Delta L}{L_0}\right)$$

where Y is Young's modulus, ΔL is the change in length, and L_0 is the original length of the beam. Substituting this expression for F into the definition of stress gives Equation 1 in the right column. Young's modulus Y for steel is known (see Table 10.1), but we do not know either ΔL or L_0. However, ΔL is proportional to L_0, so we will focus on ΔL in Step 2.

$$\text{Stress} = Y\frac{\Delta L}{L_0} \qquad (1)$$

$$\boxed{?}$$

STEP 2 Linear Thermal Expansion If it were free to do so, the beam would expand by an amount $\Delta L = \alpha L_0\, \Delta T$ (Equation 12.2), where α is the coefficient of linear expansion. The change in temperature is the final temperature T minus the initial temperature T_0, or $\Delta T = T - T_0$. Thus, the beam would have expanded by an amount

$$\boxed{\Delta L = \alpha L_0(T - T_0)}$$

In this expression the variables T and T_0 are known, and α is available in **Table 12.1**. We substitute this relation for ΔL into Equation 1, as indicated in the right column.

$$\text{Stress} = Y\frac{\Delta L}{L_0} \qquad (1)$$

$$\boxed{\Delta L = \alpha L_0\,(T - T_0)}$$

Solution Algebraically combining the results of each step, we have

$$\text{Stress} \overset{\text{STEP 1}}{=} Y\frac{\Delta L}{L_0} \overset{\text{STEP 2}}{=} Y\frac{\alpha \cancel{L_0}(T - T_0)}{\cancel{L_0}} = Y\alpha(T - T_0)$$

Note that the original length L_0 of the beam is eliminated algebraically from this result. Taking the value of $Y = 2.0 \times 10^{11}$ N/m^2 from Table 10.1 and $\alpha = 12 \times 10^{-6}$ (C°)$^{-1}$ from **Table 12.1**, we find that

$$\text{Stress} = Y\alpha(T - T_0)$$

$$= (2.0 \times 10^{11}\ \text{N/m}^2)[12 \times 10^{-6}\ (\text{C}°)^{-1}](42\ °\text{C} - 23\ °\text{C}) = \boxed{4.6 \times 10^7\ \text{N/m}^2}$$

This is a large stress, equivalent to nearly one million pounds per square foot.

Related Homework: Problems 17, 20

The Bimetallic Strip

A **bimetallic strip** is made from two thin strips of metal that have *different* coefficients of linear expansion, as **Animated Figure 12.15a** shows. Often brass [$\alpha = 19 \times 10^{-6}$ (C°)$^{-1}$] and steel [$\alpha = 12 \times 10^{-6}$ (C°)$^{-1}$] are selected. The two pieces are welded or riveted together. When the bimetallic strip is heated, the brass, having the larger value of α, expands more than the steel. Since the two metals are bonded together,

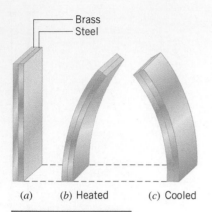

(a)   (b) Heated   (c) Cooled

(a) A bimetallic strip and how it behaves when (b) heated and (c) cooled.

the bimetallic strip bends into an arc as in part b, with the longer brass piece having a larger radius than the steel piece. When the strip is cooled, the bimetallic strip bends in the opposite direction, as in part c.

THE PHYSICS OF . . . an automatic coffee maker. Bimetallic strips are frequently used as adjustable automatic switches in electrical appliances. **Figure 12.16** shows an automatic coffee maker that turns off when the coffee is brewed to the selected strength. In part a, while the brewing cycle is on, electricity passes through the heating coil that heats the water. The electricity can flow because the contact mounted on the bimetallic strip touches the contact mounted on the "strength" adjustment knob, thus providing a continuous path for the electricity. When the bimetallic strip gets hot enough to bend away, as in part b of the drawing, the contacts separate. The electricity stops because it no longer has a continuous path along which to flow, and the brewing cycle is shut off. Turning the "strength" knob adjusts the brewing time by adjusting the distance through which the bimetallic strip must bend for the contact points to separate.

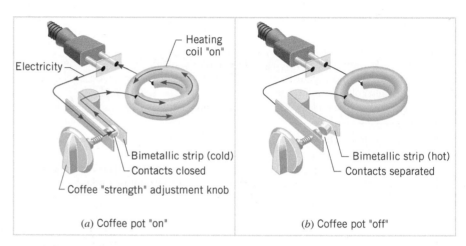

(a) Coffee pot "on"   (b) Coffee pot "off"

FIGURE 12.16 A bimetallic strip controls whether this coffee pot is (a) "on" (strip cold, straight) or (b) "off" (strip hot, bent).

The Expansion of Holes

An interesting example of linear expansion occurs when there is a hole in a piece of solid material. We know that the material itself expands when heated, but what about the hole? Does it expand, contract, or remain the same? Conceptual Example 6 provides some insight into the answer to this question.

CONCEPTUAL EXAMPLE 6 | Do Holes Expand or Contract When the Temperature Increases?

Interactive Figure 12.17a shows eight square tiles that are attached together and arranged to form a square pattern with a hole in the center. If the tiles are heated, does the size of the hole **(a)** decrease or **(b)** increase?

Reasoning We can analyze this problem by disassembling the pattern into separate tiles, heating them, and then reassembling the pattern. What happens to each of the individual tiles can be explained using what we know about linear expansion.

Answer (a) is incorrect. When a tile is heated both its length and width expand. It is tempting to think, therefore, that the

hole in the pattern decreases as the surrounding tiles expand into it. However, this is not correct, because any one tile is prevented from expanding into the hole by the expansion of the tiles next to it.

Answer (b) is correct. Since each tile expands upon heating, the pattern also expands, and the hole along with it, as shown in Interactive Figure 12.17b. In fact, if we had a ninth tile that was identical to the others and heated it to the same extent, it would fit exactly into the hole, as Interactive Figure 12.17c indicates. Thus, not only does the hole expand, it does so exactly

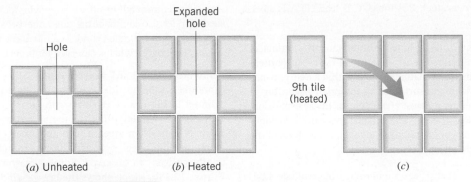

Hole

Expanded hole

9th tile (heated)

(a) Unheated   (b) Heated   (c)

INTERACTIVE FIGURE 12.17 (a) The tiles are arranged to form a square pattern with a hole in the center. (b) When the tiles are heated, the pattern, including the hole in the center, expands. (c) The expanded hole is the same size as a heated tile.

as each of the tiles does. Since the ninth tile is made of the same material as the others, we see that the hole expands just as if it were made of the material of the surrounding tiles. The thermal expansion of the hole and the surrounding material is analogous to a photographic enlargement: everything is enlarged, including holes.

Related Homework: Problem 10

Instead of the separate tiles in Example 6, we could have used a square plate with a square hole in the center. The hole in the plate would have expanded just like the hole in the pattern of tiles. Furthermore, the same conclusion applies to a hole of any shape. Thus, it follows that

Problem-Solving Insight A hole in a piece of solid material expands when heated and contracts when cooled, just as if it were filled with the material that surrounds it.

The equation $\Delta L = \alpha L_0 \, \Delta T$ can be used to find the change in any linear dimension of the hole, such as its radius or diameter, if the hole is circular. Example 7 illustrates this type of linear expansion.

EXAMPLE 7 | A Heated Engagement Ring

A gold engagement ring has an inner diameter of 1.5×10^{-2} m and a temperature of 27 °C. The ring falls into a sink of hot water whose temperature is 49 °C. What is the change in the diameter of the hole in the ring?

Reasoning The hole expands as if it were filled with gold, so the change in the diameter is given by $\Delta L = \alpha L_0 \Delta T$, where $\alpha = 14 \times 10^{-6}$ (C°)$^{-1}$ is the coefficient of linear expansion for gold

(Table 12.1), L_0 is the original diameter, and ΔT is the change in temperature.

Solution The change in the ring's diameter is

$$\Delta L = \alpha L_0 \, \Delta T = [14 \times 10^{-6} \, (\text{C}°)^{-1}](1.5 \times 10^{-2} \, \text{m})(49 \, °\text{C} - 27 \, °\text{C})$$

$$= \boxed{4.6 \times 10^{-6} \, \text{m}}$$

The previous two examples illustrate that holes expand like the surrounding material when heated. Therefore, holes in materials with larger coefficients of linear expansion expand more than those in materials with smaller coefficients of linear expansion. Conceptual Example 8 explores this aspect of thermal expansion.

CONCEPTUAL EXAMPLE 8 | Expanding Cylinders

Figure 12.18 shows a cross-sectional view of three cylinders, A, B, and C. One is made from lead, one from brass, and one from steel. All three have the same temperature, and they barely fit inside each other. As the cylinders are heated to the same higher temperature, C falls off, while A becomes tightly wedged against B. Which cylinder is made from which material? **(a)** A is brass, B is lead, C is steel **(b)** A is lead, B is brass, C is steel **(c)** A is lead, B is steel, C is brass **(d)** A is brass, B is steel, C is lead **(e)** A is steel, B is brass, C is lead **(f)** A is steel, B is lead, C is brass

Reasoning We will consider how the outer and inner diameters of each cylinder change as the temperature is raised. In particular, with regard to the inner diameter we note that a hole expands as if it were filled with the surrounding material. According to **Table 12.1**, lead has the greatest coefficient of linear expansion, followed by brass, and then by steel. Thus, the outer and inner diameters of the lead cylinder change the most, while those of the steel cylinder change the least.

Answers (a), (b), (e), and (f) are incorrect. Since the steel cylinder expands the least, it cannot be the outer one, for if it were, the greater expansion of the middle cylinder would prevent the

steel cylinder from falling off, as outer cylinder C actually does. The steel cylinder also cannot be the inner one, because then the greater expansion of the middle cylinder would allow the steel cylinder to fall out, contrary to what is observed for inner cylinder A.

Answers (c) and (d) are correct. Since the steel cylinder cannot be on the outside or on the inside, it must be the middle cylinder B. **Figure 12.18a** shows the lead cylinder as the outer cylinder C. It will fall off as the temperature is raised, since lead expands more than steel. The brass inner cylinder A expands more than the steel cylinder that surrounds it and becomes tightly wedged, as observed. Similar reasoning applies also to **Figure 12.18b**, which shows the brass cylinder as the outer cylinder and the lead cylinder as the inner one, since both brass and lead expand more than steel.

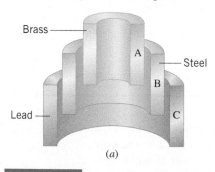 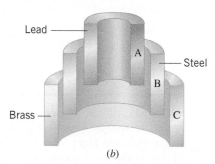

(a)   (b)

FIGURE 12.18 Conceptual Example 8 discusses the arrangements of the three cylinders shown in cutaway views in parts *a* and *b*.

Check Your Understanding

(The answers are given at the end of the book.)

2. A rod is hung from an aluminum frame, as **CYU Figure 12.1** shows. The rod and the frame have the same temperature, and there is a small gap between the rod and the floor. The frame and rod are then heated uniformly. Will the rod ever touch the floor if the rod is made from **(a)** aluminum, **(b)** lead, **(c)** brass?

3. A simple pendulum is made using a long, thin metal wire. When the temperature drops, does the period of the pendulum increase, decrease, or remain the same?

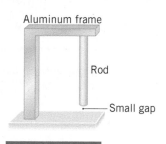

CYU FIGURE 12.1

4. For added strength, many highways and buildings are constructed with reinforced concrete (concrete that is reinforced with embedded steel rods). **Table 12.1** shows that the coefficient of linear expansion α for concrete is the same as that for steel. Why is it important that these two coefficients be the same?

5. One type of cooking pot is made from stainless steel and has a copper coating over the outside of the bottom. At room temperature the bottom of this pot is flat, but when heated the bottom is not flat. When the bottom of this pot is heated, is it bowed outward or inward?

6. A metal ball has a diameter that is slightly *greater* than the diameter of a hole that has been cut into a metal plate. The coefficient of linear expansion for the metal from which the ball is made is *greater* than that for the metal of the plate. Which one or more of the following procedures can be used to make the ball pass through the hole? **(a)** Raise the temperatures of the ball and the plate by the same amount. **(b)** Lower the temperatures of the ball and the plate by the same amount. **(c)** Heat the ball and cool the plate. **(d)** Cool the ball and heat the plate.

7. A hole is cut through an aluminum plate. A brass ball has a diameter that is slightly *smaller* than the diameter of the hole. The plate and the ball have the same temperature at all times. Should the plate and ball both be heated or both be cooled to *prevent* the ball from falling through the hole?

Volume Thermal Expansion

The volume of a normal material increases as the temperature increases. Most solids and liquids behave in this fashion. By analogy with linear thermal expansion, the change in volume ΔV is proportional to the change in temperature ΔT and to the initial volume V_0, provided the change in temperature is not too large. These two proportionalities can be converted into Equation 12.3 with the aid of a proportionality constant β, known as the **coefficient of volume expansion**. The algebraic form of this equation is similar to that for linear expansion, $\Delta L = \alpha L_0 \, \Delta T$.

VOLUME THERMAL EXPANSION

The volume V_0 of an object changes by an amount ΔV when its temperature changes by an amount ΔT:

$$\Delta V = \beta V_0 \, \Delta T \qquad (12.3)$$

where β is the coefficient of volume expansion.

Common Unit for the Coefficient of Volume Expansion: $(\text{C}°)^{-1}$

The unit for β, like that for α, is $(\text{C}°)^{-1}$. Values for β depend on the nature of the material, and **Table 12.1** lists some examples measured near 20 °C. The values of β for liquids are substantially larger than those for solids, because liquids typically expand more than solids, given the same initial volumes and temperature changes. **Table 12.1** also shows that, for most solids, the coefficient of volume expansion is three times as much as the coefficient of linear expansion: $\beta = 3\alpha$.

If a cavity exists within a solid object, the volume of the cavity increases when the object expands, just as if the cavity were filled with the surrounding material. The expansion of the cavity is analogous to the expansion of a hole in a sheet of material. Accordingly, the change in volume of a cavity can be found using the relation $\Delta V = \beta V_0 \, \Delta T$, where β is the coefficient of volume expansion of the material that surrounds the cavity. Example 9 illustrates this point.

EXAMPLE 9 | The Physics of the Overflow of an Automobile Radiator

A small plastic container, called the coolant reservoir, catches the radiator fluid that overflows when an automobile engine becomes hot (see **Figure 12.19**). The radiator is made of copper, and the coolant has a coefficient of volume expansion of $\beta = 4.10 \times 10^{-4} \, (\text{C}°)^{-1}$. If the radiator is filled to its 15-quart capacity when the engine is cold (6.0 °C), how much overflow will spill into the reservoir when the coolant reaches its operating temperature of 92 °C?

Reasoning When the temperature increases, both the coolant and the radiator expand. If they were to expand by the same amount, there would be no overflow. However, the liquid coolant expands more than the radiator, and the overflow volume is the amount of coolant expansion *minus* the amount of the radiator cavity expansion.

Solution When the temperature increases by 86 C°, the coolant expands by an amount

$$\Delta V = \beta V_0 \, \Delta T = [4.10 \times 10^{-4} \, (\text{C}°)^{-1}](15 \text{ quarts})(86 \text{ C}°) \quad \textbf{(12.3)}$$

$$= 0.53 \text{ quarts}$$

The radiator cavity expands as if it were filled with copper [$\beta = 51 \times 10^{-6} \, (\text{C}°)^{-1}$; see **Table 12.1**]. The expansion of the radiator cavity is

$$\Delta V = \beta V_0 \, \Delta T = [51 \times 10^{-6} \, (\text{C}°)^{-1}](15 \text{ quarts})(86 \text{ C}°)$$

$$= 0.066 \text{ quarts}$$

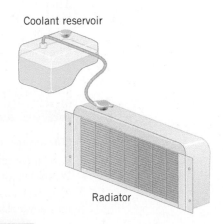

Coolant reservoir

Radiator

FIGURE 12.19 An automobile radiator and a coolant reservoir for catching the overflow from the radiator.

The overflow volume is 0.53 quarts − 0.066 quarts = $\boxed{0.46 \text{ quarts}}$.

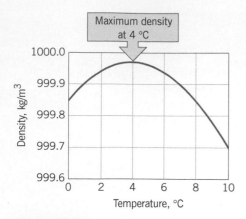

FIGURE 12.20 The density of water in the temperature range from 0 to 10 °C. At 4 °C water has a maximum density of 999.973 kg/m³. (This value is equivalent to the often-quoted density of 1.000 00 grams per milliliter.)

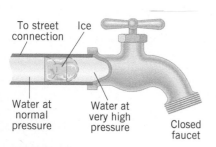

FIGURE 12.21 As water freezes and expands, enormous pressure is applied to the liquid water between the ice and the closed faucet.

Although most substances expand when heated, a few do not. For instance, if water at 0 °C is heated, its volume *decreases* until the temperature reaches 4 °C. Above 4 °C water behaves normally, and its volume increases as the temperature increases. Because a given mass of water has a minimum volume at 4 °C, the density (mass per unit volume) of water is greatest at 4 °C, as **Figure 12.20** shows.

BIO **THE PHYSICS OF . . . ice formation and the survival of aquatic life.** The fact that water has its greatest density at 4 °C, rather than at 0 °C, has important consequences for the way in which a lake freezes. When the air temperature drops, the surface layer of water is chilled. As the temperature of the surface layer drops toward 4 °C, this layer becomes more dense than the warmer water below. The denser water sinks and pushes up the deeper and warmer water, which in turn is chilled at the surface. This process continues until the temperature of the entire lake reaches 4 °C. Further cooling of the surface water below 4 °C makes it *less dense* than the deeper layers; consequently, the surface layer does not sink but stays on top. Continued cooling of the top layer to 0 °C leads to the formation of ice that floats on the water, because ice has a smaller density than water at any temperature. Below the ice, however, the water temperature remains above 0 °C. The sheet of ice acts as an insulator that reduces the loss of heat from the lake, especially if the ice is covered with a blanket of snow, which is also an insulator. As a result, lakes usually do not freeze solid, even during prolonged cold spells, so fish and other aquatic life can survive.

THE PHYSICS OF . . . bursting water pipes. The fact that the density of ice is smaller than the density of water has an important consequence for homeowners, who have to contend with the possibility of bursting water pipes during severe winters. Water often freezes in a section of pipe exposed to unusually cold temperatures. The ice can form an immovable plug that prevents the subsequent flow of water, as **Figure 12.21** illustrates. When water (larger density) turns to ice (smaller density), its volume expands by 8.3%. Therefore, when more water freezes at the left side of the plug, the expanding ice pushes liquid back into the pipe leading to the street connection, and no damage is done. However, when ice forms on the right side of the plug, the expanding ice pushes liquid to the right—but it has nowhere to go if the faucet is closed. As ice continues to form and expand, the water pressure between the plug and faucet rises. Even a small increase in the amount of ice produces a large increase in the pressure. This situation is analogous to the thermal stress discussed in Multiple-Concept Example 5, where a small change in the length of the steel beam produces a large stress on the concrete supports. The entire section of pipe to the right of the blockage experiences the same elevated pressure, according to Pascal's principle (Section 11.5). Therefore, the pipe can burst at any point where it is structurally weak, even within the heated space of the building. If you should lose heat during the winter, there is a simple way to prevent pipes from bursting. Simply open the faucet so it drips a little. The excessive pressure will be relieved.

Check Your Understanding

(The answers are given at the end of the book.)

8. Suppose that liquid mercury and glass both had the same coefficient of volume expansion. Would a mercury-in-glass thermometer still work?

9. Is the buoyant force provided by warmer water (above 4 °C) greater than, less than, or equal to the buoyant force provided by cooler water (also above 4 °C)?

12.6 Heat and Internal Energy

An object with a high temperature is said to be hot, and the word "hot" brings to mind the word "heat." **Heat** flows from a hotter object to a cooler object when the two are placed in contact. It is for this reason that a cup of hot coffee feels hot to the touch, while a glass of ice water feels cold. When the person in **Figure 12.22a** touches the coffee cup, heat flows from the hotter cup into the cooler hand. When the person touches the glass in part *b* of the drawing, heat again flows from hot to cold, in this case from the warmer hand into the colder glass. The response of the nerves in the hand to the arrival or departure of heat prompts the brain to identify the coffee cup as being hot and the glass as being cold.

What exactly is heat? As the following definition indicates, heat is a form of energy, energy in transit from hot to cold.

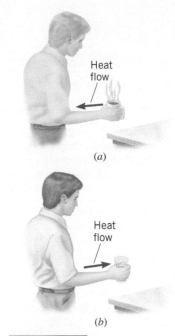

> **DEFINITION OF HEAT**
>
> **Heat is energy that flows from a higher-temperature object to a lower-temperature object because of the difference in temperatures.**
>
> **SI Unit of Heat: joule (J)**

Being a kind of energy, heat is measured in the same units used for work, kinetic energy, and potential energy. Thus, the SI unit for heat is the joule.

The heat that flows from hot to cold in **Figure 12.22** originates in the **internal energy** of the hot substance. The internal energy of a substance is the sum of the molecular kinetic energy (due to random motion of the molecules), the molecular potential energy (due to forces that act between the atoms of a molecule and between molecules), and other kinds of molecular energy. When heat flows in circumstances where no work is done, the internal energy of the hot substance decreases and the internal energy of the cold substance increases. Although heat may originate in the internal energy supply of a substance, *it is not correct to say that a substance contains heat.* The substance has internal energy, not heat. The word "heat" only refers to the energy actually in transit from hot to cold.

The next two sections consider some effects of heat. For instance, when preparing spaghetti, the first thing that a cook does is to heat the water. Heat from the stove causes the internal energy of the water to increase. Associated with this increase is a rise in temperature. After a while, the temperature reaches 100 °C, and the water begins to boil. During boiling, the added heat causes the water to change from a liquid to a vapor phase (steam). The next section investigates how the addition (or removal) of heat causes the temperature of a substance to change. Then, Section 12.8 discusses the relationship between heat and phase change, such as that which occurs when water boils.

FIGURE 12.22 Heat is energy in transit from hot to cold. (*a*) Heat flows from the hotter coffee cup to the colder hand. (*b*) Heat flows from the warmer hand to the colder glass of ice water.

12.7 | Heat and Temperature Change: Specific Heat Capacity

Solids and Liquids

Greater amounts of heat are needed to raise the temperature of solids or liquids to higher values. A greater amount of heat is also required to raise the temperature of a greater mass of material. Similar comments apply when the temperature is lowered, except that heat must be removed. For limited temperature ranges, experiment shows that the heat Q is directly proportional to the change in temperature ΔT and to the mass m. These two proportionalities are expressed below in Equation 12.4, with the help of a proportionality constant c that is referred to as the **specific heat capacity** of the material.

> **HEAT SUPPLIED OR REMOVED IN CHANGING THE TEMPERATURE OF A SUBSTANCE**
>
> **The heat Q that must be supplied or removed to change the temperature of a substance of mass m by ΔT degrees is**
>
> $$Q = cm\,\Delta T \tag{12.4}$$
>
> **where c is the specific heat capacity of the substance.**
>
> **Common Unit for Specific Heat Capacity: J/(kg · C°)**

Solving Equation 12.4 for the specific heat capacity shows that $c = Q/(m\,\Delta T)$, so the unit for specific heat capacity is J/(kg · C°). **Table 12.2** reveals that the value of the

TABLE 12.2	Specific Heat Capacities[a] of Some Solids and Liquids
Substance	**Specific Heat Capacity, c J/(kg · C°)**
Solids	
Aluminum	9.00×10^2
Copper	387
Glass	840
Human body (37 °C, average)	3500
Ice (−15 °C)	2.00×10^3
Iron or steel	452
Lead	128
Silver	235
Liquids	
Benzene	1740
Ethyl alcohol	2450
Glycerin	2410
Mercury	139
Water (15 °C)	4186

[a]Except as noted, the values are for 25 °C and 1 atm of pressure.

specific heat capacity depends on the nature of the material. Examples 10 and 11 illustrate the use of Equation 12.4.

EXAMPLE 10 | BIO A Hot Jogger

In a half hour, a 65-kg jogger can generate 8.0×10^5 J of heat. This heat is removed from the jogger's body by a variety of means, including the body's own temperature-regulating mechanisms. If the heat were not removed, how much would the jogger's body temperature increase?

Reasoning   The increase in body temperature depends on the amount of heat Q generated by the jogger, her mass m, and the specific heat capacity c of the human body. Since numerical values are known for these three variables, we can determine the potential rise in temperature by using Equation 12.4.

Solution   Table 12.2 gives the average specific heat capacity of the human body as 3500 J/(kg · C°). With this value, Equation 12.4 shows that

$$\Delta T = \frac{Q}{cm} = \frac{8.0 \times 10^5 \text{ J}}{[3500 \text{ J/(kg} \cdot \text{C}°)](65 \text{ kg})} = \boxed{3.5 \text{ C}°}$$

An increase in body temperature of 3.5 °C could be life-threatening. One way in which the jogger's body prevents it from occurring is to remove excess heat by perspiring. In contrast, cats, such as the

FIGURE 12.23   Cats, such as this lion, often pant to get rid of excess heat.

one in **Figure 12.23**, do not perspire but often pant to remove excess heat.

EXAMPLE 11 | Taking a Hot Shower

Cold water at a temperature of 15 °C enters a heater, and the resulting hot water has a temperature of 61 °C. A person uses 120 kg of hot water in taking a shower. **(a)** Find the energy needed to heat the water. **(b)** Assuming that the utility company charges $0.10 per kilowatt · hour for electrical energy, determine the cost of heating the water.

Reasoning   The amount Q of heat needed to raise the water temperature can be found from the relation $Q = cm \Delta T$, since the specific heat capacity, mass, and temperature change of the water are known. To determine the cost of this energy, we multiply the cost per unit of energy ($0.10 per kilowatt · hour) by the amount of energy used, expressed in energy units of kilowatt · hours.

Solution   (a) From equation 12.4, the amount of heat needed to heat the water is

$$Q = cm \Delta T = [4186 \text{ J/(kg} \cdot \text{C}°)](120 \text{ kg})(61 \text{ °C} - 15 \text{ °C})$$
$$= \boxed{2.3 \times 10^7 \text{ J}}$$

(b) The kilowatt · hour (kWh) is the unit of energy that utility companies use in your electric bill. To calculate the cost, we need to determine the number of joules in one kilowatt · hour. Recall that 1 kilowatt is 1000 watts (1 kW = 1000 W), 1 watt is 1 joule per second (1 W = 1 J/s; see Table 6.3), and 1 hour is equal to 3600 seconds (1 h = 3600 s). Thus,

$$1 \text{ kWh} = (1 \text{ kWh})\left(\frac{1000 \text{ W}}{1 \text{ kW}}\right)\left(\frac{1 \text{ J/s}}{1 \text{ W}}\right)\left(\frac{3600 \text{ s}}{1 \text{ h}}\right) = 3.60 \times 10^6 \text{ J}$$

The number of kilowatt · hours of energy used to heat the water is

$$(2.3 \times 10^7 \text{ J})\left(\frac{1 \text{ kWh}}{3.60 \times 10^6 \text{ J}}\right) = 6.4 \text{ kWh}$$

At a cost of $0.10 per kWh, the bill for the heat is $\boxed{\$0.64}$ or 64 cents.

Gases

As we will see in Section 15.6, the value of the specific heat capacity depends on whether the pressure or volume is held constant while energy in the form of heat is added to or removed from a substance. The distinction between constant pressure and constant volume is usually not important for solids and liquids but is significant for gases. In Section 15.6, we will see that a greater value for the specific heat capacity is obtained for a gas at constant pressure than for a gas at constant volume.

Heat Units Other than the Joule

There are three heat units other than the joule in common use. One kilocalorie (1 kcal) was defined historically as the amount of heat needed to raise the temperature of one kilogram of water by one Celsius degree.* With $Q = 1.00$ kcal, $m = 1.00$ kg, and $\Delta T = 1.00$ C°, the equation $Q = cm\,\Delta T$ shows that such a definition is equivalent to a specific heat capacity for water of $c = 1.00$ kcal/(kg · C°). Similarly, one calorie (1 cal) was defined as the amount of heat needed to raise the temperature of one gram of water by one Celsius degree, which yields a value of $c = 1.00$ cal/(g · C°). (Nutritionists use the word "Calorie," with a capital C, to specify the energy content of foods; this use is unfortunate, since 1 Calorie = 1000 calories = 1 kcal.) The British thermal unit (Btu) is the other commonly used heat unit and was defined historically as the amount of heat needed to raise the temperature of one pound of water by one Fahrenheit degree.

It was not until the time of James Joule (1818–1889) that the relationship between energy in the form of work (in units of joules) and energy in the form of heat (in units of kilocalories) was firmly established. Joule's experiments revealed that the performance of mechanical work, like rubbing your hands together, can make the temperature of a substance rise, just as the absorption of heat can. His experiments and those of later workers have shown that

$$1\text{ kcal} = 4186\text{ joules}\quad\text{or}\quad 1\text{ cal} = 4.186\text{ joules}$$

Because of its historical significance, this conversion factor is known as the **mechanical equivalent of heat**.

Calorimetry

In Section 6.8 we encountered the principle of conservation of energy, which states that energy can be neither created nor destroyed, but can only be converted from one form to another. There we dealt with kinetic and potential energies. In this chapter we have expanded our concept of energy to include heat, which is energy that flows from a higher-temperature object to a lower-temperature object because of the difference in temperature. No matter what its form, whether kinetic energy, potential energy, or heat, energy can be neither created nor destroyed. This fact governs the way objects at different temperatures come to an equilibrium temperature when they are placed in contact. If there is no heat loss to the external surroundings, the amount of heat gained by the cooler objects equals the amount of heat lost by the hotter ones, a process that is consistent with the conservation of energy. Just this kind of process occurs within a thermos. A perfect thermos would prevent any heat from leaking out or in. However, energy in the form of heat can flow *between* materials inside the thermos to the extent that they have different temperatures; for example, between ice cubes and warm tea. The transfer of energy continues until a common temperature is reached at thermal equilibrium.

The kind of heat transfer that occurs within a thermos of iced tea also occurs within a calorimeter, which is the experimental apparatus used in a technique known as **calorimetry**. **Figure 12.24** shows that, like a thermos, a calorimeter is essentially an insulated container. It can be used to determine the specific heat capacity of a substance, as the next example illustrates.

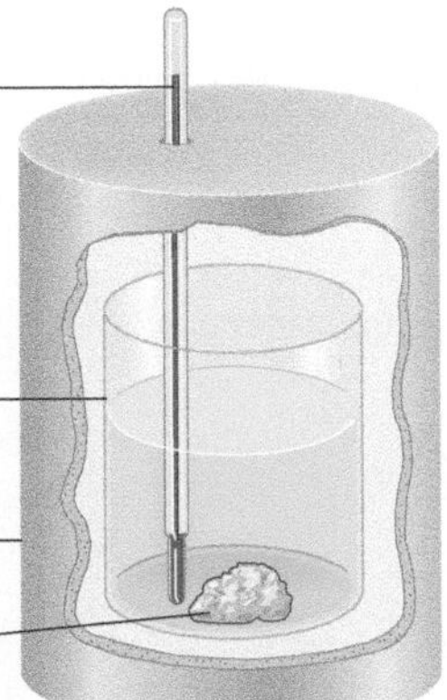

FIGURE 12.24 A calorimeter can be used to measure the specific heat capacity of an unknown material.

*From 14.5 to 15.5 °C.

EXAMPLE 12 | Measuring the Specific Heat Capacity

The calorimeter cup in **Figure 12.24** is made from 0.15 kg of aluminum and contains 0.20 kg of water. Initially, the water and the cup have a common temperature of 18.0 °C. A 0.040-kg mass of unknown material is heated to a temperature of 97.0 °C and then added to the water. The temperature of the water, the cup, and the unknown material is 22.0 °C after thermal equilibrium is reestablished. Ignoring the small amount of heat gained by the thermometer, find the specific heat capacity of the unknown material.

Reasoning   Since energy is conserved and there is negligible heat flow between the calorimeter and the outside surroundings, the amount of heat gained by the cold water and the aluminum cup as they warm up is equal to the amount of heat lost by the unknown material as it cools down. Each amount of heat can be calculated using the relation $Q = cm\,\Delta T$, if we are careful to write the change in temperature ΔT as the higher temperature minus the lower temperature. The equation "Heat gained = Heat lost" contains a single unknown quantity, the desired specific heat capacity.

> **Problem-Solving Insight**   In the equation "Heat gained = Heat lost," both sides must have the same algebraic sign. Therefore, when calculating

heat contributions for use in this equation, write any temperature changes as the higher minus the lower temperature.

Solution

$$\underbrace{(cm\,\Delta T)_{\text{Al}} + (cm\,\Delta T)_{\text{water}}}_{\substack{\text{Heat gained by}\\ \text{aluminum and water}}} = \underbrace{(cm\,\Delta T)_{\text{unknown}}}_{\substack{\text{Heat lost by}\\ \text{unknown material}}}$$

$$c_{\text{unknown}} = \frac{c_{\text{Al}}\, m_{\text{Al}}\, \Delta T_{\text{Al}} + c_{\text{water}}\, m_{\text{water}}\, \Delta T_{\text{water}}}{m_{\text{unknown}}\, \Delta T_{\text{unknown}}}$$

The changes in temperature of the substances are $\Delta T_{\text{Al}} = \Delta T_{\text{water}} = 22.0\ °C - 18.0\ °C = 4.0\ C°$, and $\Delta T_{\text{unknown}} = 97.0\ °C - 22.0\ °C = 75.0\ C°$. **Table 12.2** contains values for the specific heat capacities of aluminum and water. Substituting these data into the equation above, we find that

$$c_{\text{unknown}} = \frac{[9.00 \times 10^2\ \text{J/(kg} \cdot \text{C°)}](0.15\ \text{kg})(4.0\ \text{C°})}{(0.040\ \text{kg})(75.0\ \text{C°})}$$

$$+ \frac{[4186\ \text{J/(kg} \cdot \text{C)}](0.20\ \text{kg})(4.0\ \text{C°})}{(0.040\ \text{kg})(75.0\ \text{C°})}$$

$$= \boxed{1300\ \text{J/(kg} \cdot \text{C°)}}$$

Check Your Understanding

(The answers are given at the end of the book.)

10. Two different objects are supplied with equal amounts of heat. Which one or more of the following statements explain why their temperature changes would not necessarily be the same? **(a)** The objects have the same mass but are made from materials that have different specific heat capacities. **(b)** The objects are made from the same material but have different masses. **(c)** The objects have the same mass and are made from the same material.

11. Two objects are made from the same material but have different masses. The two are placed in contact, and neither one loses any heat to the environment. Which object experiences the temperature change with the greater magnitude, or does each object experience a temperature change of the same magnitude?

12. Consider an object of mass m that experiences a change ΔT in its temperature. Various possibilities for these variables are listed in **CYU Table 12.1**. Rank these possibilities in descending order (largest first) according to how much heat is needed to bring about the change in temperature.

CYU TABLE 12.1

	m (kg)	ΔT (C°)
(a)	2.0	15
(b)	1.5	40
(c)	3.0	25
(d)	2.5	20

12.8 | Heat and Phase Change: Latent Heat

Surprisingly, there are situations in which the addition or removal of heat does not cause a temperature change. Consider a well-stirred glass of iced tea that has come to thermal equilibrium. Even though heat enters the glass from the warmer room, the temperature of the tea does not rise above 0 °C as long as ice cubes are present.

Apparently the heat is being used for some purpose other than raising the temperature. In fact, the heat is being used to melt the ice, and only when all of it is melted will the temperature of the liquid begin to rise.

An important point illustrated by the iced tea example is that there is more than one type or phase of matter. For instance, some of the water in the glass is in the solid phase (ice) and some is in the liquid phase. The gas or vapor phase is the third familiar phase of matter. In the gas phase, water is referred to as water vapor or steam. All three phases of water are present in the scene depicted in **Figure 12.25**, although the water vapor in the air is not visible in the photograph.

Matter can change from one phase to another, and heat plays a role in the change. **Figure 12.26** summarizes the various possibilities. A solid can **melt** or **fuse** into a liquid if heat is added, while the liquid can **freeze** into a solid if heat is removed. Similarly, a liquid can **evaporate** into a gas if heat is supplied, while the gas can **condense** into a liquid if heat is taken away. Rapid evaporation, with the formation of vapor bubbles within the liquid, is called **boiling**. Finally, a solid can sometimes change directly into a gas if heat is provided. We say that the solid **sublimes** into a gas. Examples of sublimation are (1) solid carbon dioxide, CO_2 (dry ice), turning into gaseous CO_2 and (2) solid naphthalene (moth balls) turning into naphthalene fumes. Conversely, if heat is removed under the right conditions, the gas will condense directly into a solid.

Figure 12.27 displays a graph that indicates what typically happens when heat is added to a material that changes phase. The graph records temperature versus heat added and refers to water at the normal atmospheric pressure of 1.01×10^5 Pa. The water starts off as ice at the subfreezing temperature of -30 °C. As heat is added, the temperature of the ice increases, in accord with the specific heat capacity of ice [2000 J/(kg · C°)]. Not until the temperature reaches the normal melting/freezing point of 0 °C does the water begin to change phase. Then, when heat is added, the solid changes into the liquid, the temperature staying at 0 °C until *all the ice has melted*. Once all the material is in the liquid phase, additional heat causes the temperature to increase again, now in accord with the specific heat capacity of liquid water [4186 J/(kg · C°)]. When the temperature reaches the normal boiling/condensing point of 100 °C, the water begins to change from the liquid to the gas phase and continues to do so as long as heat is added. The temperature remains at 100 °C *until all liquid is gone*. When all of the material is in the gas phase, additional heat once again causes the temperature to rise, this time according to the specific heat capacity of water vapor at constant atmospheric pressure [2020 J/(kg · C°)]. Conceptual Example 13 applies the information in **Figure 12.27** to a familiar situation.

FIGURE 12.25 This antarctic scene shows resting crabeater seals. The three phases of water are present: solid ice is floating in liquid water, and water vapor is present in the air (invisible) and in the clouds in the sky.

Martin Schneiter/123RF

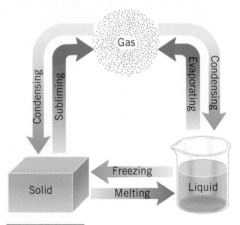

FIGURE 12.26 Three familiar phases of matter—solid, liquid, and gas—and the phase changes that can occur between any two of them.

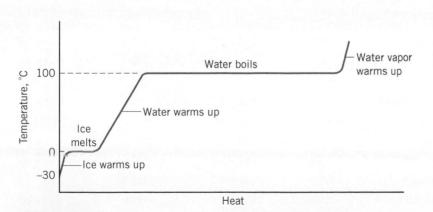

FIGURE 12.27 The graph shows the way the temperature of water changes as heat is added, starting with ice at -30 °C. The pressure is atmospheric pressure.

CONCEPTUAL EXAMPLE 13 | Saving Energy

Suppose you are cooking spaghetti, and the instructions say "boil the pasta in water for ten minutes." To cook spaghetti in an open pot using the least amount of energy, should you **(a)** turn up the burner to its fullest so the water vigorously boils or **(b)** turn down the burner so the water barely boils?

Reasoning The spaghetti needs to cook at the temperature of boiling water for ten minutes. In an open pot the pressure is atmospheric pressure, and water boils at 100 °C, regardless of whether it is vigorously boiling or just barely boiling. To convert water into steam requires energy in the form of heat from the burner, and the

greater the amount of water converted, the greater the amount of energy needed.

Answer (a) is incorrect. Causing the water to boil vigorously just wastes energy unnecessarily. All it accomplishes is to convert more water into steam.

Answer (b) is correct. Keeping the water just barely boiling uses the least amount of energy to keep the spaghetti at 100 °C, because it minimizes the amount of water converted into steam.

When a substance changes from one phase to another, the amount of heat that must be added or removed depends on the type of material and the nature of the phase change. The heat per kilogram associated with a phase change is referred to as **latent heat:**

> **HEAT SUPPLIED OR REMOVED IN CHANGING THE PHASE OF A SUBSTANCE**
>
> The heat Q that must be supplied or removed to change the phase of a mass m of a substance is
>
> $$Q = mL \qquad (12.5)$$
>
> where L is the latent heat of the substance.
>
> **SI Unit of Latent Heat:** J/kg

The **latent heat of fusion** L_f refers to the change between solid and liquid phases, the **latent heat of vaporization** L_v applies to the change between liquid and gas phases, and the **latent heat of sublimation** L_s refers to the change between solid and gas phases.

Table 12.3 gives some typical values of latent heats of fusion and vaporization. For instance, the latent heat of fusion for water is $L_f = 3.35 \times 10^5$ J/kg. Thus, 3.35×10^5 J of heat must be supplied to melt one kilogram of ice at 0 °C into liquid water at 0 °C; conversely, this amount of heat must be removed from one kilogram of liquid water at 0 °C to freeze the liquid into ice at 0 °C.

TABLE 12.3 Latent Heats[a] of Fusion and Vaporization

Substance	Melting Point (°C)	Latent Heat of Fusion, L_f (J/kg)	Boiling Point (°C)	Latent Heat of Vaporization, L_v (J/kg)
Ammonia	−77.8	33.2×10^4	−33.4	13.7×10^5
Benzene	5.5	12.6×10^4	80.1	3.94×10^5
Copper	1083	20.7×10^4	2566	47.3×10^5
Ethyl alcohol	−114.4	10.8×10^4	78.3	8.55×10^5
Gold	1063	6.28×10^4	2808	17.2×10^5
Lead	327.3	2.32×10^4	1750	8.59×10^5
Mercury	−38.9	1.14×10^4	356.6	2.96×10^5
Nitrogen	−210.0	2.57×10^4	−195.8	2.00×10^5
Oxygen	−218.8	1.39×10^4	−183.0	2.13×10^5
Water	0.0	33.5×10^4	100.0	22.6×10^5

[a]The values pertain to 1 atm pressure.

BIO **THE PHYSICS OF . . . steam burns.** In comparison to the latent heat of fusion, **Table 12.3** indicates that the latent heat of vaporization for water has the much larger value of $L_v = 22.6 \times 10^5$ J/kg. When water boils at 100 °C, 22.6×10^5 J of heat must be supplied for each kilogram of liquid turned into steam. And when steam condenses at 100 °C, this amount of heat is released from each kilogram of steam that changes back into liquid. Liquid water at 100 °C is hot enough by itself to cause a bad burn, and the additional effect of the large latent heat can cause severe tissue damage if condensation occurs on the skin.

THE PHYSICS OF . . . high-tech clothing. By taking advantage of the latent heat of fusion, designers can now engineer clothing that can absorb or release heat to help maintain a comfortable and approximately constant temperature close to your body. As the photograph in **Figure 12.28** shows, the fabric in this type of clothing is coated with microscopic balls of heat-resistant plastic that contain a substance known as a "phase-change material" (PCM). When you are enjoying your favorite winter sport, for example, it is easy to become overheated. The PCM prevents this by melting, absorbing excess body heat in the process. When you are taking a break and cooling down, however, the PCM freezes and releases heat to keep you warm. The temperature range over which the PCM can maintain a comfort zone is related to its melting/freezing temperature, which is determined by its chemical composition.

Examples 14 and 15 illustrate how to take into account the effect of latent heat when using the conservation-of-energy principle.

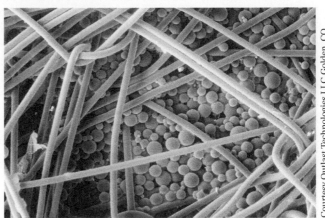

FIGURE 12.28 This highly magnified image shows a high-tech fabric that can automatically adjust in reaction to your body heat and help maintain a constant temperature next to your skin. See the text discussion.

Courtesy Outlast Technologies LLC,Golden, CO

EXAMPLE 14 | Ice-Cold Lemonade

Ice at 0 °C is placed in a Styrofoam cup containing 0.32 kg of lemonade at 27 °C. The specific heat capacity of lemonade is virtually the same as that of water; that is, $c = 4186$ J/(kg · C°). After the ice and lemonade reach an equilibrium temperature, some ice still remains. The latent heat of fusion for water is $L_f = 3.35 \times 10^5$ J/kg. Assume that the mass of the cup is so small that it absorbs a negligible amount of heat, and ignore any heat lost to the surroundings. Determine the mass of ice that has melted.

Reasoning According to the principle of energy conservation, the amount of heat gained by the melting ice equals the amount of heat lost by the cooling lemonade. According to Equation 12.5, the amount of heat gained by the melting ice is $Q = mL_f$, where m is the mass of the melted ice, and L_f is the latent heat of fusion for water. The amount of heat lost by the lemonade is given by

$Q = cm \, \Delta T$, where ΔT is the higher temperature of 27 °C minus the lower equilibrium temperature. The equilibrium temperature is 0 °C, because there is some ice remaining, and ice is in equilibrium with liquid water when the temperature is 0 °C.

Solution

$$\underbrace{(mL_f)_{\text{ice}}}_{\substack{\text{Heat gained} \\ \text{by ice}}} = \underbrace{(cm \, \Delta T)_{\text{lemonade}}}_{\substack{\text{Heat lost by} \\ \text{lemonade}}}$$

The mass m_{ice} of ice that has melted is

$$m_{\text{ice}} = \frac{(cm \, \Delta T)_{\text{lemonade}}}{L_f}$$

$$= \frac{[4186 \text{ J/(kg} \cdot \text{C°)}](0.32 \text{ kg})(27 \text{ °C} - 0 \text{ °C})}{3.35 \times 10^5 \text{ J/kg}} = \boxed{0.11 \text{ kg}}$$

EXAMPLE 15 | Getting Ready for a Party

A 7.00-kg glass bowl [$c = 840$ J/(kg · C°)] contains 16.0 kg of punch at 25.0 °C. Two-and-a-half kilograms of ice [$c = 2.00 \times 10^3$ J/(kg · C°)] are added to the punch. The ice has an initial temperature of −20.0 °C, having been kept in a very cold freezer. The punch may be treated as if it were water [$c = 4186$ J/(kg · C°)], and it may be

assumed that there is no heat flow between the punch bowl and the external environment. The latent heat of fusion for water is 3.35×10^5 J/kg. When thermal equilibrium is reached, all the ice has melted, and the final temperature of the mixture is above 0 °C. Determine this temperature.

Reasoning The final temperature can be found by using the conservation of energy principle: the amount of heat gained is equal to the amount of heat lost. Heat is gained (a) by the ice in warming up to the melting point, (b) by the ice in changing phase from a solid to a liquid, and (c) by the liquid that results from the ice warming up to the final temperature; heat is lost (d) by the punch and (e) by the bowl in cooling down. The amount of heat gained or lost by each component in changing temperature can be determined from the relation $Q = cm \, \Delta T$, where ΔT is the higher temperature minus the lower temperature. The amount of heat gained when water changes phase from a solid to a liquid at 0 °C is $Q = mL_f$, where m is the mass of water and L_f is the latent heat of fusion.

Solution The amount of heat gained or lost by each component is listed as follows:

(a) Heat gained when ice warms to 0.0 °C $= [2.00 \times 10^3 \text{ J/(kg} \cdot \text{C°)}](2.50 \text{ kg})$ $[0.0 \text{ °C} - (-20.0 \text{ °C})]$

(b) Heat gained when ice melts at 0.0 °C $= (2.50 \text{ kg})(3.35 \times 10^5 \text{ J/kg})$

(c) Heat gained when melted ice (liquid) warms to temperature T $= [4186 \text{ J/(kg} \cdot \text{C°)}](2.50 \text{ kg})$ $(T - 0.0 \text{ °C})$

(d) Heat lost when punch cools to temperature T $= [4186 \text{ J/(kg} \cdot \text{C°)}](16.0 \text{ kg})$ $(25.0 \text{ °C} - T)$

(e) Heat lost when bowl cools to temperature T $= [840 \text{ J/(kg} \cdot \text{C°)}](7.00 \text{ kg})$ $(25.0 \text{ °C} - T)$

Setting the amount of heat gained equal to the amount of heat lost gives:

$$\underbrace{(a) + (b) + (c)}_{\text{Heat gained}} = \underbrace{(d) + (e)}_{\text{Heat lost}}$$

This equation can be solved to show that $\boxed{T = 11 \text{ °C}}$.

Math Skills Carrying out the multiplications listed for the terms (a), (b), (c), (d), and (e), we find that the heat-gained-equals-heat-lost equation can be written as follows:

$$\underbrace{(1.00 \times 10^5)}_{(a)} + \underbrace{(8.38 \times 10^5)}_{(b)} + \underbrace{(1.05 \times 10^4)T}_{(c)}$$

$$= \underbrace{\boxed{(1.67 \times 10^6)} - (6.70 \times 10^4) \, T}_{(d)}$$

$$+ \underbrace{\boxed{(1.47 \times 10^5)} - (5.88 \times 10^3) \, T}_{(e)}$$

Units have been omitted for the sake of clarity. Combining terms shows that

$$(9.38 \times 10^5) + (1.05 \times 10^4)T = (1.82 \times 10^6)$$
$$- (7.29 \times 10^4)T \qquad \textbf{(1)}$$

In combining terms, we have paid particular attention to the fact that not all of the exponential powers of ten are the same. For example, in the boxes on the right-hand side of our starting equation, we have

$$(1.67 \times 10^6) + (1.47 \times 10^5) = 1.82 \times 10^6$$

We do not have

$$(1.67 \times 10^6) + (1.47 \times 10^6) = 3.14 \times 10^6 \qquad \textbf{\textit{or}}$$
$$(1.67 \times 10^5) + (1.47 \times 10^5) = 3.14 \times 10^5$$

Rearranging and further combining terms in Equation 1 yields

$$(8.34 \times 10^4)T = 8.8 \times 10^5 \qquad \text{or} \qquad T = \boxed{11 \text{ °C}}$$

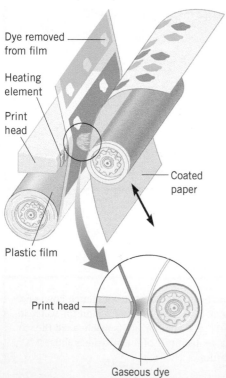

Dye removed from film

Heating element

Print head

Coated paper

Plastic film

Print head

Gaseous dye

THE PHYSICS OF . . . a dye-sublimation color printer. An interesting application of the phase change between a solid and a gas is found in one kind of color printer used with computers. A dye-sublimation printer uses a thin plastic film coated with separate panels of cyan (blue), yellow, and magenta pigment or dye. A full spectrum of colors is produced by using combinations of tiny spots of these dyes. As **Figure 12.29** shows, the coated film passes in front of a print head that extends across the width of the paper and contains 2400 heating elements. When a heating element is turned on, the dye in front of it absorbs heat and goes from a solid to a gas—it sublimes—with no liquid phase in between. A coating on the paper absorbs the gaseous dye on contact, producing a small spot of color. The intensity of the spot is controlled by the heating element, since each element can produce 256 different temperatures; the hotter the element, the greater the amount of dye transferred to the paper. The paper makes three separate passes across the print head, once for each of the dyes. The final result is an image of near-photographic quality. Some printers also employ a fourth pass, in which a clear plastic coating is deposited over the photograph. This coating makes the print waterproof and also helps to prevent premature fading.

FIGURE 12.29 A dye-sublimation printer. As the plastic film passes in front of the print head, the heat from a given heating element causes one of three pigments or dyes on the film to sublime from a solid to a gas. The gaseous dye is absorbed onto the coated paper as a dot of color. The size of the dots on the paper has been exaggerated for clarity.

Check Your Understanding

(The answers are given at the end of the book.)

13. Fruit blossoms are permanently damaged at temperatures of about −4 °C (a hard freeze). Orchard owners sometimes spray a film of water over the blossoms to protect them when a hard freeze is expected. Why does this technique offer protection?

14. When ice cubes are used to cool a drink, both their mass and temperature are important in how effective they are. **CYU Table 12.2** lists several possibilities for the mass and temperature of the ice cubes used to cool one particular drink. Rank the possibilities in descending order (best first) according to their cooling effectiveness. Note that the latent heat of phase change and the specific heat capacity must be considered.

CYU TABLE 12.2

	Mass of ice cubes	Temperature of ice cubes
(a)	m	−6.0 °C
(b)	$\frac{1}{2}m$	−12 °C
(c)	$2m$	−3.0 °C

12.9 | *Equilibrium Between Phases of Matter

Under specific conditions of temperature and pressure, a substance can exist at equilibrium in more than one phase at the same time. Consider **Interactive Figure 12.30**, which shows a container kept at a constant temperature by a large reservoir of heated sand. Initially the container is evacuated. Part *a* shows it just after it has been partially filled with a liquid and a few fast-moving molecules are escaping the liquid and forming a vapor phase. These molecules pick up the required energy (the latent heat of vaporization) during collisions with neighboring molecules in the liquid. However, the reservoir of heated sand replenishes the energy carried away, thus maintaining the constant temperature. At first, the movement of molecules is predominantly from liquid to vapor, although some molecules in the vapor phase do reenter the liquid. As the molecules accumulate in the vapor, the number reentering the liquid eventually equals the number entering the vapor, at which point equilibrium is established, as in part *b*. From this point on, the concentration of molecules in the vapor phase does not change, and the vapor pressure remains constant. The pressure of the vapor that coexists in equilibrium with the liquid is called the **equilibrium vapor pressure** of the liquid.

The equilibrium vapor pressure does not depend on the volume of space above the liquid. If more space were provided, more liquid would vaporize, until equilibrium was reestablished at the same vapor pressure, assuming the same temperature is maintained. In fact, the equilibrium vapor pressure depends only on the temperature of the liquid; a higher temperature causes a higher pressure, as the graph in **Figure 12.31** indicates for the specific case of water. Only when the temperature and vapor pressure correspond to a

Constant-temperature heated sand

(a)

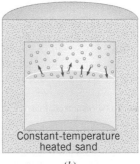

Constant-temperature heated sand

(b)

INTERACTIVE FIGURE 12.30
(*a*) Some of the molecules begin entering the vapor phase in the evacuated space above the liquid. (*b*) Equilibrium is reached when the number of molecules entering the vapor phase equals the number returning to the liquid.

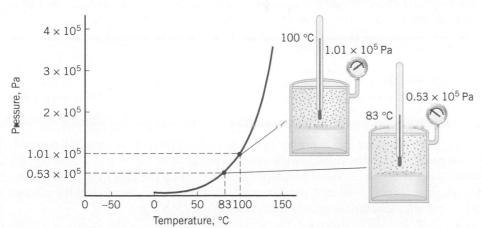

FIGURE 12.31 A plot of the equilibrium vapor pressure versus temperature is called the vapor pressure curve or the vaporization curve, the example shown being that for the liquid/vapor equilibrium of water.

point on the curved line, which is called the **vapor pressure curve** or the **vaporization curve**, do liquid and vapor phases coexist at equilibrium.

To illustrate the use of a vaporization curve, let's consider what happens when water boils in a pot that is *open to the air*. Assume that the air pressure acting on the water is 1.01×10^5 Pa (one atmosphere). When boiling occurs, bubbles of water vapor form throughout the liquid, rise to the surface, and break. For these bubbles to form and rise, the pressure of the vapor inside them must at least equal the air pressure acting on the surface of the water. According to **Figure 12.31**, a value of 1.01×10^5 Pa corresponds to a temperature of 100 °C. Consequently, water boils at 100 °C at one atmosphere of pressure. In general, a *liquid boils at the temperature for which its vapor pressure equals the external pressure*. Water will not boil, then, at sea level if the temperature is only 83 °C, because at this temperature the vapor pressure of water is only 0.53×10^5 Pa (see **Figure 12.31**), a value less than the external pressure of 1.01×10^5 Pa. However, water does boil at 83 °C on a mountain at an altitude of just under five kilometers, because the atmospheric pressure there is 0.53×10^5 Pa.

The fact that water can boil at a temperature less than 100 °C leads to an interesting phenomenon that Conceptual Example 16 discusses.

CONCEPTUAL EXAMPLE 16 | How to Boil Water That Is Cooling Down

Figure 12.32a shows water boiling in an open flask. Shortly after the flask is removed from the burner, the boiling stops. A cork is then placed in the neck of the flask to seal it, and water is poured over the neck of the flask, as in part *b* of the drawing. To restart the boiling, should the water poured over the neck be **(a)** cold or **(b)** hot—but not boiling?

Reasoning When the open flask is removed from the burner, the water begins to cool and the pressure above its surface is one atmosphere (1.01×10^5 Pa). Boiling quickly stops, because water cannot boil when its temperature is less than 100 °C and the pressure above its surface is one atmosphere. To restart the boiling, it is necessary either to reheat the water to 100 °C or reduce the pressure above the water in the corked flask to something less than one atmosphere so that boiling can occur at a temperature less than 100 °C.

Answer (b) is incorrect. Certainly, pouring hot water over the corked flask will reheat the water. However, since the water being poured is not boiling, its temperature must be less than 100 °C. Therefore, it cannot reheat the water within the flask to 100 °C and restart the boiling.

Answer (a) is correct. When cold water is poured over the corked flask, it causes some of the water vapor inside to condense. Consequently, the pressure above the liquid in the flask drops. When it drops to the value of the vapor pressure of the

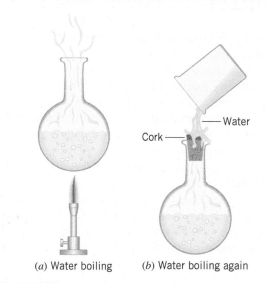

(a) Water boiling   (b) Water boiling again

FIGURE 12.32 (*a*) Water is boiling at a temperature of 100 °C and a pressure of one atmosphere. (*b*) The water boils at a temperature that is less than 100 °C, because the cool water reduces the pressure above the water in the flask.

water in the flask at its current temperature (which is now less than 100 °C), the boiling restarts.

THE PHYSICS OF . . . spray cans. The operation of spray cans is based on the equilibrium between a liquid and its vapor. **Figure 12.33a** shows that a spray can contains a liquid propellant that is mixed with the product (such as hair spray). Inside the can, propellant vapor forms over the liquid. A propellant is chosen that has an equilibrium vapor pressure that is greater than atmospheric pressure at room temperature. Consequently, when the nozzle of the can is pressed, as in part *b* of the drawing, the vapor pressure forces the liquid propellant and product up the tube in the can and out the nozzle as a spray. When the nozzle is released, the coiled spring reseals the can and the propellant vapor builds up once again to its equilibrium value.

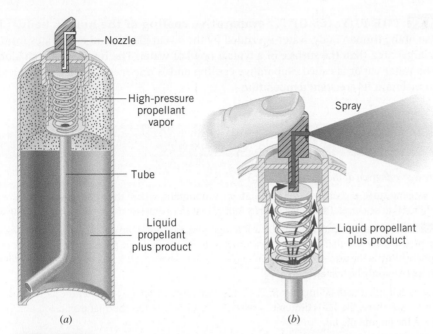

FIGURE 12.33 (*a*) A closed spray can containing liquid and vapor in equilibrium. (*b*) An open spray can.

As is the case for liquid/vapor equilibrium, a solid can be in equilibrium with its liquid phase only at specific conditions of temperature and pressure. For each temperature, there is a single pressure at which the two phases can coexist in equilibrium. A plot of the equilibrium pressure versus equilibrium temperature is referred to as the **fusion curve**, and **Figure 12.34***a* shows a typical curve for a normal substance. A normal substance expands on melting (e.g., carbon dioxide and sulfur). Since higher pressures make it more difficult for such materials to expand, a higher melting temperature is needed for a higher pressure, and the fusion curve slopes upward to the right. Part *b* of the picture illustrates the fusion curve for water, one of the few substances that contract when they melt. Higher pressures make it easier for such substances to melt. Consequently, a lower melting temperature is associated with a higher pressure, and the fusion curve slopes downward to the right.

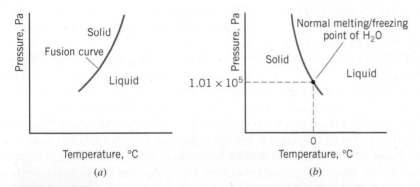

FIGURE 12.34 (*a*) The fusion curve for a normal substance that expands on melting. (*b*) The fusion curve for water, one of the few substances that contract on melting.

It should be noted that just because two phases can coexist in equilibrium does not necessarily mean that they will. Other factors may prevent it. For example, water in an *open* bowl may never come into equilibrium with water vapor if air currents are present. Under such conditions the liquid, perhaps at a temperature of 25 °C, attempts to establish the corresponding equilibrium vapor pressure of 3.2×10^3 Pa. If air currents continually blow the water vapor away, however, equilibrium will never be established, and eventually the water will evaporate completely. Each kilogram of water that goes into the vapor phase takes along the latent heat of vaporization. Because of this heat loss, the remaining liquid would become cooler, except for the fact that the surroundings replenish the loss.

BIO **THE PHYSICS OF . . . evaporative cooling of the human body.**   In the case of the human body, water is exuded by the sweat glands and evaporates from a much larger area than the surface of a typical bowl of water. The removal of heat along with the water vapor is called evaporative cooling and is one mechanism that the body uses to maintain its constant temperature.

Check Your Understanding

(*The answers are given at the end of the book.*)

15.   A camping stove is used to boil water high on a mountain, where the atmospheric pressure is lower than it is at sea level. Does it necessarily follow that the same stove can boil water at sea level?

16.   **BIO** Medical instruments are sterilized at a high temperature in an autoclave, which is essentially a pressure cooker that heats the instruments in water under a pressure greater than one atmosphere. Why is the water in an autoclave able to reach a very high temperature, while water in an open pot can only be heated to 100 °C?

17.   A jar is half filled with boiling water. The lid is then screwed on the jar. After the jar has cooled to room temperature, the lid is difficult to remove. Why? Ignore the thermal expansion and contraction of the jar and the lid.

18.   A bottle of carbonated soda (sealed and under a pressure greater than one atmosphere) is left outside in subfreezing temperatures, although the soda remains liquid. When the soda is brought inside and opened immediately, it suddenly freezes. Why?

19.   When a bowl of water is placed in a closed container and the water vapor is pumped away rapidly enough, why does the remaining liquid turn into ice?

12.10   *Humidity

Air is a mixture of gases, including nitrogen, oxygen, and water vapor. The total pressure of the mixture is the sum of the partial pressures of the component gases. The **partial pressure** of a gas is the pressure it would exert if it alone occupied the entire volume at the same temperature as the mixture. The partial pressure of water vapor in air depends on weather conditions. It can be as low as zero or as high as the equilibrium vapor pressure of water at the given temperature.

THE PHYSICS OF . . . relative humidity.   To provide an indication of how much water vapor is in the air, weather forecasters usually give the **relative humidity**. If the relative humidity is too low, the air contains such a small amount of water vapor that skin and mucous membranes tend to dry out. If the relative humidity is too high, especially on a hot day, we become very uncomfortable and our skin feels "sticky." Under such conditions, the air holds so much water vapor that the water exuded by sweat glands cannot evaporate efficiently. The relative humidity is defined as the ratio (expressed as a percentage) of the actual partial pressure of water vapor in the air to the equilibrium vapor pressure at a given temperature.

$$\text{Percent relative humidity} = \frac{\text{Partial pressure of water vapor}}{\text{Equilibrium vapor pressure of water at the existing temperature}} \times 100 \qquad (12.6)$$

The term in the denominator on the right of Equation 12.6 is given by the vaporization curve of water and is the pressure of the water vapor in equilibrium with the liquid. At a given temperature, the partial pressure of the water vapor in the air cannot exceed this value. If it did, the vapor would not be in equilibrium with the liquid and would condense as dew or rain to reestablish equilibrium.

When the partial pressure of the water vapor equals the equilibrium vapor pressure of water at a given temperature, the relative humidity is 100%. In such a situation, the vapor is said to be *saturated* because it is present in the maximum amount, as it would be above a pool of liquid at equilibrium in a closed container. If the relative humidity is less than 100%, the water vapor is said to be *unsaturated*. Example 17 demonstrates how to find the relative humidity.

EXAMPLE 17 | Relative Humidities

One day, the partial pressure of water vapor in the air is 2.0×10^3 Pa. Using the vaporization curve for water in **Figure 12.35**, determine the relative humidity if the temperature is **(a)** 32 °C and **(b)** 21 °C.

Reasoning and Solution **(a)** According to **Figure 12.35**, the equilibrium vapor pressure of water at 32 °C is 4.8×10^3 Pa. Equation 12.6 reveals that the relative humidity is

$$\text{Relative humidity at 32 °C} = \frac{2.0 \times 10^3 \text{ Pa}}{4.8 \times 10^3 \text{ Pa}} \times 100 = \boxed{42\%}$$

(b) A similar calculation shows that

$$\text{Relative humidity at 21 °C} = \frac{2.0 \times 10^3 \text{ Pa}}{2.5 \times 10^3 \text{ Pa}} \times 100 = \boxed{80\%}$$

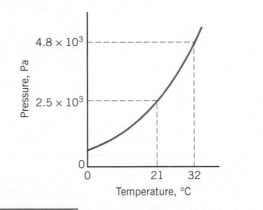

FIGURE 12.35 The vaporization curve of water.

THE PHYSICS OF . . . fog formation. When air containing a given amount of water vapor is cooled, a temperature is reached in which the partial pressure of the vapor equals the equilibrium vapor pressure. This temperature is known as the **dew point**. For instance, **Figure 12.36** shows that if the partial pressure of water vapor is 3.2×10^3 Pa, the dew point is 25 °C. This partial pressure would correspond to a relative humidity of 100%, if the ambient temperature were equal to the dew-point temperature. Hence, the dew point is the temperature below which water vapor in the air condenses in the form of liquid drops (dew or fog). The closer the actual temperature is to the dew point, the closer the relative humidity is to 100%. Thus, for fog to form, the air temperature must drop below the dew point. Similarly, water condenses on the outside of a cold glass when the temperature of the air next to the glass falls below the dew point.

THE PHYSICS OF . . . a home dehumidifier. The cold coils in a home dehumidifier function very much in the same way that the cold glass does. The coils are kept cold by a circulating refrigerant (see **Figure 12.37**). When the air blown across them by the fan cools below the dew point, water vapor condenses in the form of droplets, which collect in a receptacle.

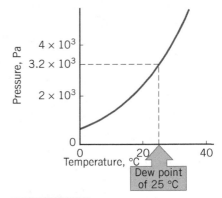

FIGURE 12.36 On the vaporization curve of water, the dew point is the temperature that corresponds to the actual partial pressure of water vapor in the air.

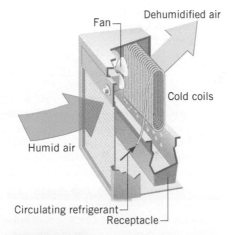

FIGURE 12.37 The cold coils of a dehumidifier cool the air blowing across them to below the dew point, and water vapor condenses out of the air.

Check Your Understanding

(The answers are given at the end of the book.)

20. A bowl of water is covered tightly and allowed to sit at a constant temperature of 23 °C for a long time. What is the relative humidity in the space between the surface of the water and the cover?

21. Is it possible for dew to form on Tuesday night and not on Monday night, even though Monday night is the cooler night?

22. Two rooms in a mansion have the same temperature. One of these rooms contains an indoor swimming pool. On a cold day the windows of one of the two rooms are "steamed up." Which room is it likely to be? Explain.

EXAMPLE 18 | BIO Heating Up Your Brain

The human brain is a remarkable organ. It only requires approximately 20 W of power to function normally. For comparison, a man-made computer processor with the same computational power as your brain would consume 10–20 MW of power! With that being said, the brain is the most energy-hungry organ in the human body. It only accounts for 1/50 of the body's total weight, but it requires 1/5 (20%) of the body's total energy. This power output from the brain produces heat, which raises the temperature of the brain and the surrounding bone and tissues. The brain contains approximately 1.2 kg of water, which accounts for 77% of its total mass. If a human brain was operating at a constant power output of 20 W for one hour, what would be the increase in its temperature? Assume it is composed of only the 1.2 kg of water, and no heat transfer occurs between the brain and its surroundings.

Reasoning We can calculate the change in temperature (ΔT) of the water inside the brain by knowing the total heat (Q) produced

in 1 hour. Equation 12.4 then provides the relationship between Q and ΔT.

Solution We can relate the power output of the brain to the total heat produced, since $P = Q/t$. Therefore, $Q = Pt$. Now we apply Equation 12.4: $Q = cm\,\Delta T$. Substituting in for Q and solving for ΔT, we get the following:

$$\Delta T = Pt/mc = (20\ \text{W})(60\ \text{s})/(1.2\ \text{kg})(4186\ \text{J}/(\text{kg}\cdot\text{C}^\circ)) = \boxed{0.24\ ^\circ\text{C}}$$

Although this is a small temperature change, we have done the calculation assuming the entire brain has a high specific heat capacity equal to that of water. The actual rise in temperature will be higher if we consider the rest of the brain that is not water. Studies have shown that after behavioral stimuli, it is not unusual to observe a 2–3 °C temperature increase in the brain. This is a significant change and must be controlled by the body's temperature regulation system that consists of the circulatory system and the surface of the skin.

Concept Summary

12.1 Common Temperature Scales   On the Celsius temperature scale, there are 100 equal divisions between the ice point (0 °C) and the steam point (100 °C). On the Fahrenheit temperature scale, there are 180 equal divisions between the ice point (32 °F) and the steam point (212 °F).

12.2 The Kelvin Temperature Scale   For scientific work, the Kelvin temperature scale is the scale of choice. One kelvin (K) is equal in size to one Celsius degree. However, the temperature T on the Kelvin scale differs from the temperature T_c on the Celsius scale by an additive constant of 273.15, as indicated by Equation 12.1. The lower limit of temperature is called absolute zero and is designated as 0 K on the Kelvin scale.

$$T = T_c + 273.15 \qquad (12.1)$$

12.3 Thermometers   The operation of any thermometer is based on the change in some physical property with temperature; this physical property is called a thermometric property. Examples of thermometric properties are the length of a column of mercury, electrical voltage, and electrical resistance.

12.4 Linear Thermal Expansion   Most substances expand when heated. For linear expansion, an object of length L_0 experiences a change ΔL in length when the temperature changes by ΔT, as shown in Equation 12.2, where α is the coefficient of linear expansion.

$$\Delta L = \alpha L_0\,\Delta T \qquad (12.2)$$

For an object held rigidly in place, a thermal stress can occur when the object attempts to expand or contract. The stress can be large, even for small temperature changes.

When the temperature changes, a hole in a plate of solid material expands or contracts as if the hole were filled with the surrounding material.

12.5 Volume Thermal Expansion   For volume expansion, the change ΔV in the volume of an object of volume V_0 is given by Equation 12.3, where β is the coefficient of volume expansion. When the temperature changes, a cavity in a piece of solid material expands or contracts as if the cavity were filled with the surrounding material.

$$\Delta V = \beta V_0\,\Delta T \qquad (12.3)$$

12.6 Heat and Internal Energy The internal energy of a substance is the sum of the kinetic, potential, and other kinds of energy that the molecules of the substance have. Heat is energy that flows from a higher-temperature object to a lower-temperature object because of the difference in temperatures. The SI unit for heat is the joule (J).

12.7 Heat and Temperature Change: Specific Heat Capacity The heat Q that must be supplied or removed to change the temperature of a substance of mass m by an amount ΔT is given by Equation 12.4, where c is a constant known as the specific heat capacity.

$$Q = cm\,\Delta T \qquad\qquad \textbf{(12.4)}$$

When materials are placed in thermal contact within a perfectly insulated container, the principle of energy conservation requires that the amount of heat lost by warmer materials equals the amount of heat gained by cooler materials.

Heat is sometimes measured with a unit called the kilocalorie (kcal). The conversion factor between kilocalories and joules is known as the mechanical equivalent of heat: 1 kcal = 4186 joules.

12.8 Heat and Phase Change: Latent Heat Heat must be supplied or removed to make a material change from one phase to another. The heat Q that must be supplied or removed to change the phase of a mass m of a substance is given by Equation 12.5, where L is the latent heat of the substance and has SI units of J/kg. The latent heats of fusion, vaporization, and sublimation refer, respectively, to the solid/liquid, the liquid/vapor, and the solid/vapor phase changes.

$$Q = mL \qquad\qquad \textbf{(12.5)}$$

12.9 Equilibrium Between Phases of Matter The equilibrium vapor pressure of a substance is the pressure of the vapor phase that is in equilibrium with the liquid phase. For a given substance, vapor pressure depends only on temperature. For a liquid, a plot of the equilibrium vapor pressure versus temperature is called the vapor pressure curve or vaporization curve.

The fusion curve gives the combinations of temperature and pressure for equilibrium between solid and liquid phases.

12.10 Humidity The relative humidity is defined as in Equation 12.6.

$$\begin{array}{c}\text{Percent}\\\text{relative}\\\text{humidity}\end{array} = \dfrac{\begin{array}{c}\text{Partial pressure}\\\text{of water vapor}\end{array}}{\begin{array}{c}\text{Equilibrium vapor pressure of}\\\text{water at the existing temperature}\end{array}} \times 100 \qquad \textbf{(12.6)}$$

The dew point is the temperature below which the water vapor in the air condenses. On the vaporization curve of water, the dew point is the temperature that corresponds to the actual pressure of water vapor in the air.

Focus on Concepts

Online

Additional questions are available for assignment in WileyPLUS.

Section 12.2   The Kelvin Temperature Scale

1. Which one of the following statements correctly describes the Celsius and the Kelvin temperature scales? (a) The size of the degree on the Celsius scale is larger than that on the Kelvin scale by a factor of 9/5. (b) Both scales assign the same temperature to the ice point, but they assign different temperatures to the steam point. (c) Both scales assign the same temperature to the steam point, but they assign different temperatures to the ice point. (d) The Celsius scale assigns the same values to the ice and the steam points that the Kelvin scale assigns. (e) The size of the degree on each scale is the same.

Section 12.4   Linear Thermal Expansion

2. The drawing shows two thin rods, one made from aluminum [$\alpha = 23 \times 10^{-6}$ (C°)$^{-1}$] and the other from steel [$\alpha = 12 \times 10^{-6}$ (C°)$^{-1}$]. Each rod has the same length and the same initial temperature and is attached

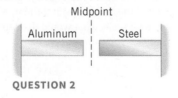

QUESTION 2

at one end to an immovable wall, as shown. The temperatures of the rods are increased, both by the same amount, until the gap between the rods is closed. Where do the rods meet when the gap is closed? (a) The rods meet exactly at the midpoint. (b) The rods meet to the right of the midpoint. (c) The rods meet to the left of the midpoint.

3. A ball is slightly too large to fit through a hole in a flat plate. The drawing shows two arrangements of this situation. In Arrangement I the ball is made from metal A and the plate from metal B. When both the ball and the plate are cooled by the same number of Celsius degrees, the ball passes through the hole. In Arrangement II the ball is also made from metal A,

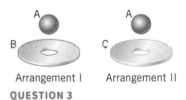

Arrangement I   Arrangement II

QUESTION 3

but the plate is made from metal C. Here, the ball passes through the hole when both the ball and the plate are heated by the same number of Celsius degrees. Rank the coefficients of linear thermal expansion of metals A, B, and C in descending order (largest first): (a) α_B, α_A, α_C (b) α_B, α_C, α_A (c) α_C, α_B, α_A (d) α_C, α_A, α_B (e) α_A, α_B, α_C

Section 12.5   Volume Thermal Expansion

4. A solid sphere and a solid cube are made from the same material. The sphere would just fit within the cube, if it could. Both begin at the same temperature, and both are heated to the same temperature. Which object, if either, has the greater change in volume? (a) The sphere. (b) The cube. (c) Both have the same change in volume. (d) Insufficient information is given for an answer.

5. A container can be made from steel [$\beta = 36 \times 10^{-6}$ (C°)$^{-1}$] or lead [$\beta = 87 \times 10^{-6}$ (C°)$^{-1}$]. A liquid is poured into the container, filling it to the brim. The liquid is either water [$\beta = 207 \times 10^{-6}$ (C°)$^{-1}$] or ethyl alcohol [$\beta = 1120 \times 10^{-6}$ (C°)$^{-1}$]. When the full container is heated, some liquid spills out. To keep the overflow to a minimum, from what material should the container be made and what should the liquid be? (a) Lead, water (b) Steel, water (c) Lead, ethyl alcohol (d) Steel, ethyl alcohol

Section 12.7   Heat and Temperature Change: Specific Heat Capacity

6. Which of the following cases (if any) requires the greatest amount of heat? In each case the material is the same. (**a**) 1.5 kg of the material is to be heated by 7.0 C°. (**b**) 3.0 kg of the material is to be heated by 3.5 C°. (**c**) 0.50 kg of the material is to be heated by 21 C°. (**d**) 0.75 kg of the material is to be heated by 14 C°. (**e**) The amount of heat required is the same in each of the four previous cases.

7. The following three hot samples have the same temperature. The same amount of heat is removed from each sample. Which one experiences the smallest drop in temperature, and which one experiences the largest drop?

> Sample A. 4.0 kg of water $[c = 4186 \text{ J/(kg} \cdot \text{C}°)]$
>
> Sample B. 2.0 kg of oil $[c = 2700 \text{ J/(kg} \cdot \text{C}°)]$
>
> Sample C. 9.0 kg of dirt $[c = 1050 \text{ J/(kg} \cdot \text{C}°)]$

(**a**) C smallest and A largest (**b**) B smallest and C largest (**c**) A smallest and B largest (**d**) C smallest and B largest (**e**) B smallest and A largest

Section 12.8   Heat and Phase Change: Latent Heat

8. The latent heat of fusion for water is 33.5×10^4 J/kg, while the latent heat of vaporization is 22.6×10^5 J/kg. What mass m of water must be frozen in order to release the amount of heat that 1.00 kg of steam releases when it condenses?

Section 12.9   Equilibrium Between Phases of Matter

9. Which one or more of the following techniques can be used to freeze water?

> **A.** Cooling the water below its normal freezing point of 0 °C at the normal atmospheric pressure of 1.01×10^5 Pa
>
> **B.** Cooling the water below its freezing point of −1 °C at a pressure greater than 1.01×10^5 Pa
>
> **C.** Rapidly pumping away the water vapor above the liquid in an insulated container (The insulation prevents heat flowing from the surroundings into the remaining liquid.)

(**a**) Only A   (**b**) Only B   (**c**) Only A and B   (**d**) A, B, and C   (**e**) Only C

Section 12.10   Humidity

10. Which of the following three statements concerning relative humidity values of 30% and 40% are true? Note that when the relative humidity is 30%, the air temperature may be different than it is when the relative humidity is 40%.

> **A.** It is possible that at a relative humidity of 30% there is a smaller partial pressure of water vapor in the air than there is at a relative humidity of 40%.
>
> **B.** It is possible that there is the same partial pressure of water vapor in the air at 30% and at 40% relative humidity.
>
> **C.** It is possible that at a relative humidity of 30% there is a greater partial pressure of water vapor in the air than there is at a relative humidity of 40%.

(**a**) A, B, and C   (**b**) Only A and B   (**c**) Only A and C   (**d**) Only B and C (**e**) Only A

Problems   Online

Additional questions are available for assignment in WileyPLUS.

Note: *For problems in this set, use the values of α and β in* **Table 12.1**, *and the values of c, L_f and L_v in* **Tables 12.2 and 12.3,** *unless stated otherwise.*

SSM Student Solutions Manual	**BIO** Biomedical application
MMH Problem-solving help	**E** Easy
GO Guided Online Tutorial	**M** Medium
V-HINT Video Hints	**H** Hard
CHALK Chalkboard Videos	**WS** Worksheet
	T Team Problem

Section 12.1   Common Temperature Scales,

Section 12.2   The Kelvin Temperature Scale,

Section 12.3   Thermometers

1. **E** **SSM** Suppose you are hiking down the Grand Canyon. At the top, the temperature early in the morning is a cool 3 °C. By late afternoon, the temperature at the bottom of the canyon has warmed to a sweltering 34 °C. What is the *difference* between the higher and lower temperatures in (**a**) Fahrenheit degrees and (**b**) kelvins?

2. **E** You are sick, and your temperature is 312.0 kelvins. Convert this temperature to the Fahrenheit scale.

3. **E** **CHALK** On the moon the surface temperature ranges from 375 K during the day to 1.00×10^2 K at night. What are these temperatures on the (**a**) Celsius and (**b**) Fahrenheit scales?

4. **E** **GO** **CHALK** The drawing shows two thermometers, A and B, whose temperatures are measured in °A and °B. The ice and boiling points of water are also indicated. (**a**) Using the data in the drawing, determine the number of B degrees on the B scale that correspond to 1 A° on the A scale. (**b**) If the temperature of a substance reads +40.0 °A on the A scale, what would that temperature read on the B scale?

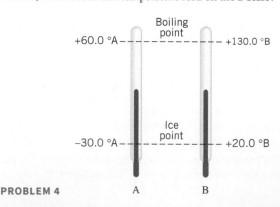

PROBLEM 4   A   B

5. **M** **GO** A copper–constantan thermocouple generates a voltage of 4.75×10^{-3} volts when the temperature of the hot junction is 110.0 °C and the reference junction is kept at a temperature of 0.0 °C. If the voltage is proportional to the difference in temperature between the junctions, what is the temperature of the hot junction when the voltage is 1.90×10^{-3} volts?

6. **M** **V-HINT** If a nonhuman civilization were to develop on Saturn's largest moon, Titan, its scientists might well devise a temperature scale based on the properties of methane, which is much more abundant on the surface than water is. Methane freezes at −182.6 °C on Titan, and boils at −155.2 °C. Taking the boiling point of methane as 100.0 °M (degrees Methane) and its freezing point as 0 °M, what temperature on the Methane scale corresponds to the absolute zero point of the Kelvin scale?

Section 12.4 Linear Thermal Expansion

7. **E** A steel section of the Alaskan pipeline had a length of 65 m and a temperature of 18 °C when it was installed. What is its change in length when the temperature drops to a frigid −45 °C?

8. **E** **SSM** A steel aircraft carrier is 370 m long when moving through the icy North Atlantic at a temperature of 2.0 °C. By how much does the carrier lengthen when it is traveling in the warm Mediterranean Sea at a temperature of 21 °C?

9. **E** The Eiffel Tower is a steel structure whose height increases by 19.4 cm when the temperature changes from −9 to +41 °C. What is the approximate height (in meters) at the lower temperature?

10. **E** Conceptual Example 6 provides background for this problem. A hole is drilled through a copper plate whose temperature is 11 °C. **(a)** When the temperature of the plate is increased, will the radius of the hole be larger or smaller than the radius at 11 °C? Why? **(b)** When the plate is heated to 110 °C, by what fraction $\Delta r/r_0$ will the radius of the hole change?

11. **E** A commonly used method of fastening one part to another part is called "shrink fitting." A steel rod has a diameter of 2.0026 cm, and a flat plate contains a hole whose diameter is 2.0000 cm. The rod is cooled so that it just fits into the hole. When the rod warms up, the enormous thermal stress exerted by the plate holds the rod securely to the plate. By how many Celsius degrees should the rod be cooled?

12. **E** **SSM** When the temperature of a coin is raised by 75 C°, the coin's diameter increases by 2.3×10^{-5} m. If the original diameter of the coin is 1.8×10^{-2} m, find the coefficient of linear expansion.

13. **E** One January morning in 1943, a warm chinook wind rapidly raised the temperature in Spearfish, South Dakota, from below freezing to +12.0 °C. As the chinook died away, the temperature fell to −20.0 °C in 27.0 minutes. Suppose that a 19-m aluminum flagpole were subjected to this temperature change. Find the average speed at which its height would decrease, assuming the flagpole responded instantaneously to the changing temperature.

14. **E** **GO** One rod is made from lead and another from quartz. The rods are heated and experience the same change in temperature. The change in length of each rod is the same. If the initial length of the lead rod is 0.10 m, what is the initial length of the quartz rod?

15. **M** **GO** A thin rod consists of two parts joined together. One-third of it is silver and two-thirds is gold. The temperature decreases by 26 C°. Determine the fractional decrease $\dfrac{\Delta L}{L_{0,\,Silver} + L_{0,\,Gold}}$ in the rod's length, where $L_{0,\,Silver}$ and $L_{0,\,Gold}$ are the initial lengths of the silver and gold rods.

16. **M** **CHALK** The brass bar and the aluminum bar in the drawing are each attached to an immovable wall. At 28 °C the air gap between the rods is 1.3×10^{-3} m. At what temperature will the gap be closed?

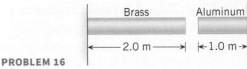

PROBLEM 16

17. **M** **V-HINT** Multiple-Concept Example 5 reviews the concepts that are involved in this problem. A ruler is accurate when the temperature is 25 °C. When the temperature drops to −14 °C, the ruler shrinks and no longer measures distances accurately. However, the ruler can be made to read correctly if a force of magnitude 1.2×10^3 N is applied to each end so as to stretch it back to its original length. The ruler has a cross-sectional area of 1.6×10^{-5} m², and it is made from a material whose coefficient of linear expansion is 2.5×10^{-5} (C°)$^{-1}$. What is Young's modulus for the material from which the ruler is made?

18. **M** **SSM** A simple pendulum consists of a ball connected to one end of a thin brass wire. The period of the pendulum is 2.0000 s. The temperature rises by 140 C°, and the length of the wire increases. Determine the period of the heated pendulum.

19. **M** **V-HINT** As the drawing shows, two thin strips of metal are bolted together at one end; both have the same temperature. One is steel, and the other is aluminum. The steel strip is 0.10% longer than the aluminum strip. By how much should the temperature of the strips be increased, so that the strips have the same length?

PROBLEM 19

20. **M** **GO** Consult Multiple-Concept Example 5 for insight into solving this problem. A copper rod is fastened securely at both ends to immovable supports. When this rod is stretched, it will rupture when a tensile stress of 2.3×10^7 N/m² is applied at each end. The rod just fits between the supports, so initially there is no stress applied to the rod. The rod is then cooled. What is the magnitude $|\Delta T|$ of the change in temperature of the rod when it ruptures?

21. **M** **GO** A ball and a thin plate are made from different materials and have the same initial temperature. The ball does not fit through a hole in the plate, because the diameter of the ball is slightly larger than the diameter of the hole. However, the ball will pass through the hole when the ball and the plate are both heated to a common higher temperature. In each of the arrangements in the drawing the diameter of the ball is 1.0×10^{-5} m larger than the diameter of the hole in the thin plate, which has a diameter of 0.10 m. The initial temperature of each arrangement is 25.0 °C. At what temperature will the ball fall through the hole in each arrangement?

PROBLEM 21

22. **H** An 85.0-N backpack is hung from the middle of an aluminum wire, as the drawing shows. The temperature of the wire then drops by 20.0 C°. Find the tension in the wire at the lower temperature. Assume that the distance between the supports does not change, and ignore any thermal stress.

PROBLEM 22

Section 12.5 Volume Thermal Expansion

23. **E** A flask is filled with 1.500 L (L = liter) of a liquid at 97.1 °C. When the liquid is cooled to 15.0 °C, its volume is only 1.383 L, however. Neglect the contraction of the flask and use **Table 12.1** to identify the liquid.

24. **E** **MMH** A thin spherical shell of silver has an inner radius of 2.0×10^{-2} m when the temperature is 18 °C. The shell is heated to 147 °C. Find the change in the interior volume of the shell.

25. **E** A test tube contains 2.54×10^{-4} m^3 of liquid carbon tetrachloride at a temperature of 75.0 °C. The test tube and the carbon tetrachloride are cooled to a temperature of −13.0 °C, which is above the freezing point of carbon tetrachloride. Find the volume of carbon tetrachloride in the test tube at −13.0 °C.

26. **E** **SSM** A copper kettle contains water at 24 °C. When the water is heated to its boiling point of 100.0 °C, the volume of the kettle expands by 1.2×10^{-5} m^3. Determine the volume of the kettle at 24 °C.

27. **E** Suppose you are selling apple cider for two dollars a gallon when the temperature is 4.0 °C. The coefficient of volume expansion of the cider is 280×10^{-6} (C°)$^{-1}$. How much more money (in pennies) would you make per gallon by refilling the container on a day when the temperature is 26 °C? Ignore the expansion of the container.

28. **E** **MMH** During an all-night cram session, a student heats up a one-half liter (0.50×10^{-3} m^3) glass (Pyrex) beaker of cold coffee. Initially, the temperature is 18 °C, and the beaker is filled to the brim. A short time later when the student returns, the temperature has risen to 92 °C. The coefficient of volume expansion of coffee is the same as that of water. How much coffee (in cubic meters) has spilled out of the beaker?

29. **E** **MMH** Many hot-water heating systems have a reservoir tank connected directly to the pipeline, to allow for expansion when the water becomes hot. The heating system of a house has 76 m of copper pipe whose inside radius is 9.5×10^{-3} m. When the water and pipe are heated from 24 to 78 °C, what must be the minimum volume of the reservoir tank to hold the overflow of water?

30. **E** **SSM** Suppose that the steel gas tank in your car is completely filled when the temperature is 17 °C. How many gallons will spill out of the twenty-gallon tank when the temperature rises to 35 °C?

31. **E** **GO** An aluminum can is filled to the brim with a liquid. The can and the liquid are heated so their temperatures change by the same amount. The can's initial volume at 5 °C is 3.5×10^{-4} m^3. The coefficient of volume expansion for aluminum is 69×10^{-6} (C°)$^{-1}$. When the can and the liquid are heated to 78 °C, 3.6×10^{-6} m^3 of liquid spills over. What is the coefficient of volume expansion of the liquid?

32. **M** **V-HINT** A spherical brass shell has an interior volume of 1.60×10^{-3} m^3. Within this interior volume is a solid steel ball that has a volume of 0.70×10^{-3} m^3. The space between the steel ball and the inner surface of the brass shell is filled completely with mercury. A small hole is drilled through the brass, and the temperature of the arrangement is increased by 12 C°. What is the volume of the mercury that spills out of the hole?

33. **M** **GO** At the bottom of an old mercury-in-glass thermometer is a 45-mm^3 reservoir filled with mercury. When the thermometer was placed under your tongue, the warmed mercury would expand into a very narrow cylindrical channel, called a capillary, whose radius was 1.7×10^{-2} mm. Marks were placed along the capillary that indicated the temperature. Ignore the thermal expansion of the glass and determine how far (in mm) the mercury would expand into the capillary when the temperature changed by 1.0 C°.

34. **M** **SSM** The bulk modulus of water is $B = 2.2 \times 10^9$ N/m^2. What change in pressure ΔP (in atmospheres) is required to keep water from expanding when it is heated from 15 to 25 °C?

35. **M** **GO** **CHALK** The density of mercury is 13 600 kg/m^3 at 0 °C. What would its density be at 166 °C?

36. **H** **SSM** Two identical thermometers made of Pyrex glass contain, respectively, identical volumes of mercury and methyl alcohol. If the expansion of the glass is taken into account, how many times greater is the distance between the degree marks on the methyl alcohol thermometer than the distance on the mercury thermometer?

37. **H** The column of mercury in a barometer (see Figure 11.12) has a height of 0.760 m when the pressure is one atmosphere and the temperature is 0.0 °C. Ignoring any change in the glass containing the mercury, what will be the height of the mercury column for the same one atmosphere of pressure when the temperature rises to 38.0 °C on a hot day? (*Hint: The pressure in the barometer is given by Pressure = ρgh, and the density ρ of the mercury changes when the temperature changes.*)

Section 12.6 Heat and Internal Energy,

Section 12.7 Heat and Temperature Change: Specific Heat Capacity

38. **E** **SSM** Ideally, when a thermometer is used to measure the temperature of an object, the temperature of the object itself should not change. However, if a significant amount of heat flows from the object to the thermometer, the temperature will change. A thermometer has a mass of 31.0 g, a specific heat capacity of $c = 815$ J/(kg · C°), and a temperature of 12.0 °C. It is immersed in 119 g of water, and the final temperature of the water and thermometer is 41.5 °C. What was the temperature of the water before the insertion of the thermometer?

39. **E** **SSM** Blood can carry excess energy from the interior to the surface of the body, where the energy is dispersed in a number of ways. While a person is exercising, 0.6 kg of blood flows to the body's surface and releases 2000 J of energy. The blood arriving at the surface has the temperature of the body's interior, 37.0 °C. Assuming that blood has the same specific heat capacity as water, determine the temperature of the blood that leaves the surface and returns to the interior.

40. **E** An ice chest at a beach party contains 12 cans of soda at 5.0 °C. Each can of soda has a mass of 0.35 kg and a specific heat capacity of 3800 J/(kg · C°). Someone adds a 6.5-kg watermelon at 27 °C to the chest. The specific heat capacity of watermelon is nearly the same as that of water. Ignore the specific heat capacity of the chest and determine the final temperature T of the soda and watermelon.

41. **E** A piece of glass has a temperature of 83.0 °C. Liquid that has a temperature of 43.0 °C is poured over the glass, completely covering it, and the temperature at equilibrium is 53.0 °C. The mass of the glass and the liquid is the same. Ignoring the container that holds the glass and liquid and assuming that the heat lost to or gained from the surroundings is negligible, determine the specific heat capacity of the liquid.

42. **E** **GO** Two bars of identical mass are at 25 °C. One is made from glass and the other from another substance. The specific heat capacity of glass is 840 J/(kg · C°). When identical amounts of heat are supplied to each, the glass bar reaches a temperature of 88 °C, while the other bar reaches 250.0 °C. What is the specific heat capacity of the other substance?

43. **E** **SSM** At a fabrication plant, a hot metal forging has a mass of 75 kg and a specific heat capacity of 430 J/(kg · C°). To harden it, the forging is immersed in 710 kg of oil that has a temperature of 32 °C and a specific heat capacity of 2700 J/(kg · C°). The final temperature of the oil and forging at thermal equilibrium is 47 °C. Assuming that heat flows only between the forging and the oil, determine the initial temperature of the forging.

44. **M** **MMH** **CHALK** A 0.35-kg coffee mug is made from a material that has a specific heat capacity of 920 J/(kg · C°) and contains 0.25 kg of water. The cup and water are at 15 °C. To make a cup of coffee, a small electric heater is immersed in the water and brings it to a boil in three minutes. Assume that the cup and water always have the same temperature and determine the minimum power rating of this heater.

45. **M** **GO** Three portions of the same liquid are mixed in a container that prevents the exchange of heat with the environment. Portion A has a mass m and a temperature of 94.0 °C, portion B also has a mass m but a temperature of 78.0 °C, and portion C has a mass m_C and a temperature of 34.0 °C. What must be the mass of portion C so that the final temperature T_f of the three-portion mixture is $T_f = 50.0$ °C? Express your answer in terms of m; for example, $m_C = 2.20\ m$.

46. **M** **V-HINT** The heating element of a water heater in an apartment building has a maximum power output of 28 kW. Four residents of the building take showers at the same time, and each receives heated water at a volume flow rate of 14×10^{-5} m³/s. If the water going into the heater has a temperature of 11 °C, what is the maximum possible temperature of the hot water that each showering resident receives?

47. **M** **SSM** A rock of mass 0.20 kg falls from rest from a height of 15 m into a pail containing 0.35 kg of water. The rock and water have the same initial temperature. The specific heat capacity of the rock is 1840 J/(kg · C°). Ignore the heat absorbed by the pail itself, and determine the rise in the temperature of the rock and water.

48. **H** A steel rod ($\rho = 7860$ kg/m³) has a length of 2.0 m. It is bolted at both ends between immobile supports. Initially there is no tension in the rod, because the rod just fits between the supports. Find the tension that develops when the rod loses 3300 J of heat.

Section 12.8 Heat and Phase Change: Latent Heat

49. **E** **SSM** How much heat must be added to 0.45 kg of aluminum to change it from a solid at 130 °C to a liquid at 660 °C (its melting point)? The latent heat of fusion for aluminum is 4.0×10^5 J/kg.

50. **E** Suppose that the amount of heat removed when 3.0 kg of water freezes at 0.0 °C were removed from ethyl alcohol at its freezing/melting point of −114.4 °C. How many kilograms of ethyl alcohol would freeze?

51. **E** To help prevent frost damage, fruit growers sometimes protect their crop by spraying it with water when overnight temperatures are expected to go below freezing. When the water turns to ice during the night, heat is released into the plants, thereby giving a measure of protection against the cold. Suppose a grower sprays 7.2 kg of water at 0 °C onto a fruit tree. (a) How much heat is released by the water when it freezes? (b) How much would the temperature of a 180-kg tree rise if it absorbed the heat released in part (a)? Assume that the specific heat capacity of the tree is 2.5×10^3 J/(kg · C°) and that no phase change occurs within the tree itself.

52. **E** **GO** (a) Objects A and B have the same mass of 3.0 kg. They melt when 3.0×10^4 J of heat is added to object A and when 9.0×10^4 J is added to object B. Determine the latent heat of fusion for the substance from which each object is made. (b) Find the heat required to melt object A when its mass is 6.0 kg.

53. **E** **SSM** Find the mass of water that vaporizes when 2.10 kg of mercury at 205 °C is added to 0.110 kg of water at 80.0 °C.

54. **E** A mass $m = 0.054$ kg of benzene vapor at its boiling point of 80.1 °C is to be condensed by mixing the vapor with water at 41 °C. What is the minimum mass of water required to condense all of the benzene vapor? Assume that the mixing and condensation take place in a perfectly insulating container.

55. **E** A certain quantity of steam has a temperature of 100.0 °C. To convert this steam into ice at 0.0 °C, energy in the form of heat must be removed from the steam. If this amount of energy were used to accelerate the ice from rest, what would be the linear speed of the ice? For comparison, bullet speeds of about 700 m/s are common.

56. **E** A thermos contains 150 cm³ of coffee at 85 °C. To cool the coffee, you drop two 11-g ice cubes into the thermos. The ice cubes are initially at 0 °C and melt completely. What is the final temperature of the coffee? Treat the coffee as if it were water.

57. **M** **V-HINT** A snow maker at a resort pumps 130 kg of lake water per minute and sprays it into the air above a ski run. The water droplets freeze in the air and fall to the ground, forming a layer of snow. If all the water pumped into the air turns to snow, and the snow cools to the ambient air temperature of −7.0 °C, how much heat does the snow-making process release each minute? Assume that the temperature of the lake water is 12.0 °C, and use 2.00×10^3 J/(kg · C°) for the specific heat capacity of snow.

58. **M** **SSM** **CHALK** **MMH** A 42-kg block of ice at 0 °C is sliding on a horizontal surface. The initial speed of the ice is 7.3 m/s and the final speed is 3.5 m/s. Assume that the part of the block that melts has a very small mass and that all the heat generated by kinetic friction goes into the block of ice. Determine the mass of ice that melts into water at 0 °C.

59. **M** **GO** Water at 23.0 °C is sprayed onto 0.180 kg of molten gold at 1063 °C (its melting point). The water boils away, forming steam at 100.0 °C and leaving solid gold at 1063 °C. What is the minimum mass of water that must be used?

60. **M** **SSM** An unknown material has a normal melting/freezing point of −25.0 °C, and the liquid phase has a specific heat capacity of 160 J/(kg · C°). One-tenth of a kilogram of the solid at −25.0 °C is put into a 0.150-kg aluminum calorimeter cup that contains 0.100 kg of glycerin. The temperature of the cup and the glycerin is initially 27.0 °C. All the unknown material melts, and the final temperature at equilibrium is 20.0 °C. The calorimeter neither loses energy to nor gains energy from the external environment. What is the latent heat of fusion of the unknown material?

61. **M** **GO** When it rains, water vapor in the air condenses into liquid water, and energy is released. (a) How much energy is released when 0.0254 m (one inch) of rain falls over an area of 2.59×10^6 m² (one square mile)? (b) If the average energy needed to heat one home for a year is 1.50×10^{11} J, how many homes could be heated for a year with the energy determined in part (a)?

62. **M** **SSM** It is claimed that if a lead bullet goes fast enough, it can melt completely when it comes to a halt suddenly and all its kinetic energy is converted into heat via friction. Find the minimum speed of a lead bullet (initial temperature = 30.0 °C) for such an event to happen.

63. **M** **V-HINT** Equal masses of two different liquids have the same temperature of 25.0 °C. Liquid A has a freezing point of −68.0 °C and a specific heat capacity of 1850 J/(kg · C°). Liquid B has a freezing point of −96.0 °C and a specific heat capacity of 2670 J/(kg · C°). The same amount of heat must be removed from each liquid in order to freeze it into a solid at its respective freezing point. Determine the difference $L_{f,A} - L_{f,B}$ between the latent heats of fusion for these liquids.

64. **M** **CHALK** Occasionally, huge icebergs are found floating on the ocean's currents. Suppose one such iceberg is 120 km long, 35 km wide, and 230 m thick. **(a)** How much heat would be required to melt this iceberg (assumed to be at 0 °C) into liquid water at 0 °C? The density of ice is 917 kg/m³. **(b)** The annual energy consumption by the United States is about 1.1×10^{20} J. If this energy were delivered to the iceberg every year, how many years would it take before the ice melted?

Section 12.9 Equilibrium Between Phases of Matter,

Section 12.10 Humidity

65. **E** **CHALK** What atmospheric pressure would be required for carbon dioxide to boil at a temperature of 20 °C? See the vapor pressure curve for carbon dioxide in the drawing.

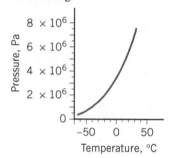

PROBLEM 65

66. **E** **SSM** **CHALK** At a temperature of 10 °C the percent relative humidity is R_{10}, and at 40 °C it is R_{40}. At each of these temperatures the partial pressure of water vapor in the air is the same. Using the

vapor pressure curve for water that accompanies this problem, determine the ratio R_{10}/R_{40} of the two humidity values.

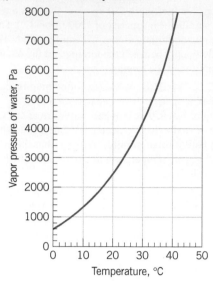

PROBLEM 66

67. **E** What is the relative humidity on a day when the temperature is 30 °C and the dew point is 10 °C? Use the vapor pressure curve that accompanies Problem 66.

68. **E** **GO** The vapor pressure of water at 10 °C is 1300 Pa. **(a)** What percentage of atmospheric pressure is this? Take atmospheric pressure to be 1.013×10^5 Pa. **(b)** What percentage of the total air pressure at 10 °C is due to water vapor when the relative humidity is 100%? **(c)** The vapor pressure of water at 35 °C is 5500 Pa. What is the relative humidity at this temperature if the partial pressure of water in the air has not changed from what it was at 10 °C when the relative humidity was 100%?

69. **E** **GO** The temperature of 2.0 kg of water is 100.0 °C, but the water is not boiling, because the external pressure acting on the water surface is 3.0×10^5 Pa. Using the vapor pressure curve for water given in Figure 12.31, determine the amount of heat that must be added to the water to bring it to the point where it just begins to boil.

Additional Problems   Online

70. **E** An aluminum baseball bat has a length of 0.86 m at a temperature of 17 °C. When the temperature of the bat is raised, the bat lengthens by 0.000 16 m. Determine the final temperature of the bat.

71. **E** A 0.200-kg piece of aluminum that has a temperature of −155 °C is added to 1.5 kg of water that has a temperature of 3.0 °C. At equilibrium the temperature is 0.0 °C. Ignoring the container and assuming that the heat exchanged with the surroundings is negligible, determine the mass of water that has been frozen into ice.

72. **E** A thick, vertical iron pipe has an inner diameter of 0.065 m. A thin aluminum disk, heated to a temperature of 85 °C, has a diameter that is 3.9×10^{-5} m greater than the pipe's inner diameter. The disk is laid on top of the open upper end of the pipe, perfectly centered on it, and allowed to cool. What is the temperature of the aluminum disk when the disk falls into the pipe? Ignore the temperature change of the pipe.

73. **E** **SSM** A lead object and a quartz object each have the same initial volume. The volume of each increases by the same amount, because the temperature increases. If the temperature of the lead object increases by 4.0 C°, by how much does the temperature of the quartz object increase?

74. **E** **GO** If the price of electrical energy is $0.10 per kilowatt · hour, what is the cost of using electrical energy to heat the water in a swimming pool (12.0 m × 9.00 m × 1.5 m) from 15 to 27 °C?

75. **M** Concrete sidewalks are always laid in sections, with gaps between each section. For example, the drawing shows three identical 2.4-m sections, the outer two of which are against immovable walls. The two identical gaps between the sections are provided so that thermal expansion will not create the thermal stress that could lead to cracks. What is the minimum gap width necessary to account for an increase in temperature of 32 C°?

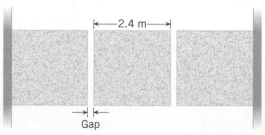

PROBLEM 75

76. **M** **GO** A constant-volume gas thermometer (see Figures 12.3 and 12.4) has a pressure of 5.00×10^3 Pa when the gas temperature is 0.00 °C. What is the temperature (in °C) when the pressure is 2.00×10^3 Pa?

77. **M** When 4200 J of heat are added to a 0.15-m-long silver bar, its length increases by 4.3×10^{-3} m. What is the mass of the bar?

78. **H** A wire is made by attaching two segments together, end to end. One segment is made of aluminum and the other is steel. The effective coefficient of linear expansion of the two-segment wire is 19×10^{-6} (C°)$^{-1}$. What fraction of the length is aluminum?

79. **H** An insulated container is partly filled with oil. The lid of the container is removed, 0.125 kg of water heated to 90.0 °C is poured in, and the lid is replaced. As the water and the oil reach equilibrium, the volume of the oil increases by 1.20×10^{-5} m^3. The density of the oil is 924 kg/m^3, its specific heat capacity is 1970 J/(kg · C°), and its coefficient of volume expansion is 721×10^{-6} (C°)$^{-1}$. What is the temperature when the oil and the water reach equilibrium?

80. **H** A steel bicycle wheel (without the rubber tire) is rotating freely with an angular speed of 18.00 rad/s. The temperature of the wheel changes from −100.0 to +300.0 °C. No net external torque acts on the wheel, and the mass of the spokes is negligible. **(a)** Does the angular speed increase or decrease as the wheel heats up? Why? **(b)** What is the angular speed at the higher temperature?

81. **M** **GO** **SSM** The figure shows a swimming pool on a sunny day. If the water absorbs 2.00×10^9 J of heat from the sun, what is the change in volume of the water?

PROBLEM 81

Physics in Biology, Medicine, and Sports

82. **E** **BIO** **SSM** **12.2** Dermatologists often remove small precancerous skin lesions by freezing them quickly with liquid nitrogen, which has a temperature of 77 K. What is this temperature on the **(a)** Celsius and **(b)** Fahrenheit scales?

83. **E** **BIO** **MMH** **12.7** When resting, a person has a metabolic rate of about 3.0×10^5 joules per hour. The person is submerged neck-deep into a tub containing 1.2×10^3 kg of water at 21.00 °C. If the heat from the person goes only into the water, find the water temperature after half an hour.

84. **E** **BIO** **GO** **12.6** When you drink cold water, your body must expend metabolic energy in order to maintain normal body temperature (37 °C) by warming up the water in your stomach. Could drinking ice water, then, substitute for exercise as a way to "burn calories?" Suppose you expend 430 kilocalories during a brisk hour-long walk. How many liters of ice water (0 °C) would you have to drink in order to use up 430 kilocalories of metabolic energy? For comparison, the stomach can hold about 1 liter.

85. **M** **BIO** **MMH** **12.6** One ounce of a well-known breakfast cereal contains 110 Calories (1 food Calorie = 4186 J). If 2.0% of this energy could be converted by a weight lifter's body into work done in lifting a barbell, what is the heaviest barbell that could be lifted a distance of 2.1 m?

86. **E** **BIO** **12.8** The latent heat of vaporization of H$_2$O at body temperature (37.0 °C) is 2.42×10^6 J/kg. To cool the body of a 75-kg jogger [average specific heat capacity = 3500 J/(kg · C°)] by 1.5 C°, how many kilograms of water in the form of sweat have to be evaporated?

87. **E** **BIO** **12.10** Suppose that air in the human lungs has a temperature of 37 °C, and the partial pressure of water vapor has a value of 5.5×10^3 Pa. What is the relative humidity in the lungs? Consult the vapor pressure curve for water that accompanies Problem 75.

88. **E** **BIO** **12.7** A person eats a container of strawberry yogurt. The Nutritional Facts label states that it contains 240 Calories (1 Calorie = 4186 J). What mass of perspiration would one have to lose to get rid of this energy? At body temperature, the latent heat of vaporization of water is 2.42×10^6 J/kg.

89. **M** **BIO** **12.8** The human brain accounts for 1/5 of the body's total power output, with an average power of 20.0 W. How long, in hours, would it take the heat produced by the brain to vaporize a cup of water with an initial temperature of 25.0 °C into steam at 1.00×10^2 °C? Assume the mass of the water in the cup is 237 g and all of the heat produced by the brain is transferred to the water.

90. **E** **BIO** **12.2** Since 1871, the normal internal body temperature for human beings has been 98.6 °F. This average was based on over a million measurements analyzed by a German physician, Carl Reinhold August Wunderlich. However, in 2020 researchers at Stanford University published a report claiming normal body temperature has dropped since the Industrial Revolution. Analyzing almost 700 000 temperature readings from 1862 to 2017, the researchers found the average body temperature has decreased by 0.03 °C per decade. The new average body temperature was found to be 97.9 °F. What is this temperature in **(a)** Celsius, and **(b)** Kelvin?

91. **M** **BIO** **12.6** The Peloton spin cycle is an upscale stationary bike that you can purchase to use in your home. It features a 22-inch sweat-proof screen, where the rider interacts with a live instructor, effectively recreating a studio spin class at home. The Peloton is great for improving cardiovascular fitness. If the Peloton instructor maintains an average power output of 435 W for 30.0 minutes, how many Calories does the instructor burn during this time?

92. **M** **BIO** **12.7** Imagine a physics lecture hall with 100 students who are settling in for a 1-hour lecture. At the start of the lecture, the temperature of the air in the room is a comfortable 21.0 °C. Unfortunately, the room's air conditioner breaks right as the lecture begins. Each student has an average power output of about 60.0 W at room temperature. Imagine the energy released by each student goes into heating just the air in the room, which has a volume of 9.50×10^2 m^3 and a density of 1.20 kg/m^3. Assume the volume of the air remains constant and the specific heat capacity of the air is 718 J/(kg · °C). Calculate the room's temperature at the end of the lecture. In reality, a significant portion of the heat produced would be absorbed by the walls, ceiling, floors, chairs, desks, and so on, which we are neglecting.

93. **M** **BIO** **12.7** Fever, or pyrexia, is the body's natural response to infection, which leads to an increase in the body's setpoint temperature. This was one of the most common symptoms experienced by people infected with the COVID-19 virus during the pandemic in 2020. Normal body temperature was recently established to be 97.9 °F (36.6 °C), and during a high fever, the body temperature may rise to 103 °F (39.4 °C) or more. This heat is produced by the body ramping up its metabolic rate and burning more Calories. How many Calories would a 75.0-kg person have to consume in order to raise their temperature from 36.6 °C to 39.4 °C ? Take the average specific heat capacity of the human body to be 3470 J/(kg · C°).

Concepts and Calculations Problems

Online

Problem 94 provides insight on the variables involved when the length and volume of an object change due to a temperature change. Problem 95 discusses how different factors affect the temperature change of an object to which heat is being added.

94. **M** **CHALK** The figure shows three rectangular blocks made from the same material. The initial dimensions of each are expressed as multiples of D, where $D = 2.00$ cm. The blocks are heated and their temperatures increase by 35.0 C°. The coefficients of linear and volume expansion are $\alpha = 1.50 \times 10^{-5}$ (C°)$^{-1}$ and $\beta = 4.50 \times 10^{-5}$ (C°)$^{-1}$, respectively. *Concepts:* (i) Does the change in the vertical height of a block depend only on its height, or does it also depend on its width and depth? Without doing any calculations, rank the blocks according to their change in height, largest first. (ii) Does the change in the volume of a block depend only on its height, or does it also depend on its width and depth? Without doing any calculations, rank the blocks according to their change in volume, largest first. *Calculations:* Determine the change in the (a) vertical heights and (b) volumes of the blocks.

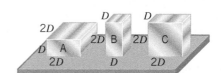

PROBLEM 94

95. **M** **CHALK** **SSM** Objects A and B in the figure are made from copper, but the mass of object B is three times that of object A. Object C is made from glass and has the same mass as object B. The *same amount* of heat Q is supplied to each one: $Q = 14$ J. *Concepts:* (i) Which object, A or B, experiences a greater rise in temperature, and why? (ii) Which object, B or C, experiences a greater rise in temperature, and why? *Calculations:* Determine the rise in temperature of each block.

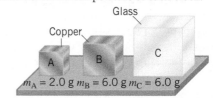

PROBLEM 95

Team Problems

Online

96. **M** **T** **WS** **Rule of Mixtures.** Two barrels each contain a liquid that is a mixture of water and acetone, a clear reagent that mixes well with water. One barrel contains 10.0% water and 90.0% acetone by volume, and the other 10.0% acetone and 90.0% water by volume. The problem is that you do not know which is which and, although acetone has a distinct odor, you cannot tell them apart by smelling them. Although the density of acetone ($\rho_A = 784$ kg/m^3) is less than that of water ($\rho_W = 1.00 \times 10^3$ kg/m^3), you do not have a scale with which to measure their weights. However, you have a 1000-mL beaker marked with lines every 100 mL, a 1400-mL calorimeter vessel, a small freezer, a set of brass masses from an old balance scale, and a thermometer. You look up the specific heat capacities and find that they are very different for water ($c_{p,w} = 4128$ J/kg · C°) and acetone ($c_{p,A} = 2160$ J/kg · C°), and devise an idea to take advantage of that difference in order to identify the mixtures. You and your team plan to use the "rule of mixtures," which states that the effective specific heat capacity of a mixture is the weighted average of the specific heats of the components, that is,

$$c_{p,T} = \left(\frac{m_1}{m_T}\right)c_{p1} + \left(\frac{m_2}{m_T}\right)c_{p2}$$

where $c_{p,T}$ is the effective specific heat of the total mixture, c_{p1} and c_{p2} are the specific heats of the two liquid components, m_1 and m_2 are the respective masses of the components, and m_T is the mass of the entire mixture (i.e., $m_T = m_1 + m_2$). You put a 1.00×10^3-g brass mass ($c_{p,B} = 920.0$ J/kg · C°) in the freezer ($T = -11.0$ °C) and wait for it to equilibrate. Next, using the beaker, you measure out 1.00×10^3 mL of one of the mixtures and pour it into the calorimeter. The mixtures both start at room temperature (28.2 °C). Finally, you take the brass weight out of the freezer, submerge it in the liquid in the calorimeter, close the lid, and insert the thermometer into the liquid through a hole in the lid. You wait for the temperature to equilibrate, record the temperature, and then repeat the process for the other mixture. (a) Which mixture will equilibrate at a lower temperature? Explain your answer. (b) What are the respective masses of acetone and water in 1.00×10^3 mL of each mixture, and what are the total masses of each mixture? (c) What are the effective specific heats of each mixture? (d) Calculate the expected final temperature of each mixture using this procedure. The calorimeter vessel is made of Styrofoam, and you can neglect any heat transfer to or from the vessel.

97. **E** **T** **WS** **A Giant Frisbee.** A solid copper disk of radius 10.50 m is launched toward the sun like a gigantic Frisbee with a constant linear velocity of 350 m/s and a rotation rate of exactly 438.7 rpm about its central axis. At the time of launch, the disk was very cold ($-191.1\ °C$). As it approaches the sun, however, it will get very hot. You and your team have been brought in to determine whether the increased temperature will affect the motion of the disk. **(a)** Will the rate of rotation be affected when the disk becomes hot? If so, how and why? **(b)** What will be the rate of rotation (in rad/s) when the copper disk reaches a temperature of 785.9 °C? There are no external torques acting on the disk.

98. **M** **T** **A Crude Thermometer.** You and your team are given the task of constructing a crude thermometer that covers a temperature range from 0 °C to 60.0 °C. You have at your disposal an aluminum rod of length $L_0 = 2.00$ m (at $T = 0\ °C$) and diameter $D = 0.500$ cm, and a broken clock that is missing its hour hand, and its longer minute hand is hanging loose and pointing in the six o'clock direction. The long hand of the clock is 8.20 inches long and pivots loosely at the clock's center. You get the idea to mount the rod horizontally so that one end butts against a wall and the other end pushes against the dangling minute hand of the clock. A temperature-induced change in the length of the rod will then be reflected in a change in the angle that the minute hand makes with the vertical (i.e., relative to the six o'clock position). **(a)** If you want the full temperature range to span the angle between the 6 and 7 markings (uniformly spaced) on the clock, how far from the central pivot should the end of the rod make contact with the minute hand? **(b)** The current temperature in the room is 65.0 °F. At what angle relative to the vertical (the six o'clock position) should the minute hand point at this temperature if it is to point directly at 6 when $T = 0\ °C$? **(c)** What would be the angular range of your "clock-thermometer" if the rod were made of steel, rather than aluminum? Assume that it is placed at the same position on the minute hand as determined in (a).

99. **M** **T** **A Submerged Machine.** You and your team are testing a device that is to be submerged in the cold waters of Antarctica. It is designed to rotate a small wheel at a very precise rate. One component of the device consists of a steel wheel (diameter $D_{steel} = 3.0000$ cm at $T = 78.400\ °F$) and a large aluminum wheel that drives it (diameter $D_{Al} = 150.00$ cm at $T = 78.400\ °F$). **(a)** If the aluminum wheel rotates at 35.000 RPM, at what rate does the smaller, steel wheel turn while both wheels are at $T = 78.400\ °F$? **(b)** You submerge the device in a large vat of water held at 32.800 °C, simulating its working environment in the Antarctic seas. Assuming the larger, aluminum driving wheel still rotates at 35.000 RPM, at what rate does the smaller, steel wheel rotate when the device is submerged in the cold water? **(c)** How should you adjust the rate of the aluminum driving wheel so that the steel wheel rotates at the same rate as it had at $T = 78.400\ °F$? Note: $\alpha_{Al} = 23 \times 10^{-6}$ and $\alpha_{steel} = 12 \times 10^{-6}$.

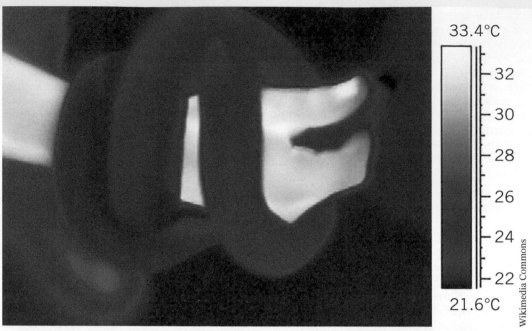

CHAPTER 13

33.4°C

─ 32

─ 30

─ 28

─ 26

─ 24

─ 22

21.6°C

Wikimedia Commons

All objects with a temperature above absolute zero (−273.15 °C) emit infrared or thermal radiation. The image above, known as a thermal image, or thermogram, is made with a special camera that is sensitive to this type of radiation. A colored scale represents the relative temperature of the objects in the image. Warmer objects, such as people, most animals, and other sources of heat, stand out in contrast to the cooler background and colder objects. The image above shows a snake wrapped around a human hand and wrist. The snake, being a coldblooded animal, matches its temperature to its environment and appears much cooler than the warm hand that is holding it. In this chapter, we will study radiation, as well as the other processes of heat transfer.

LEARNING OBJECTIVES

After reading this module, you should be able to...

13.1 Define convection.

13.2 Solve conduction problems.

13.3 Solve radiation problems.

13.4 Analyze heat transfer applications.

The Transfer of Heat

13.1 Convection

When heat is transferred to or from a substance, the internal energy of the substance can change, as we saw in Chapter 12. This change in internal energy is accompanied by a change in temperature or a change in phase. The transfer of heat affects us in many ways. For instance, within our homes, furnaces distribute heat on cold days, and air conditioners remove it on hot days. Our bodies constantly transfer heat in one direction or another, to prevent the adverse effects of hypo- and hyperthermia. And virtually all our energy originates in the sun and is transferred to us over a distance of 150 million kilometers through the void of space. Today's sunlight provides the energy to drive photosynthesis in the plants that provide our food and, hence, metabolic energy. Ancient sunlight nurtured the organic matter that became the fossil fuels of oil, natural gas, and coal. This chapter examines the three processes by which heat is transferred: convection, conduction, and radiation.

When part of a fluid is warmed, such as the air above a fire, the volume of that part of the fluid expands, and the density decreases.

FIGURE 13.1 A plume of smoke originates from an oil burn near the site of the Deepwater Horizon oil spill disaster in the Gulf of Mexico in 2010. The plume rises hundreds of meters into the air because of convection.

FIGURE 13.2 Convection currents are set up when a pot of water is heated.

According to Archimedes' principle (see Section 11.6), the surrounding cooler and denser fluid exerts a buoyant force on the warmer fluid and pushes it upward. As warmer fluid rises, the surrounding cooler fluid replaces it. This cooler fluid, in turn, is warmed and pushed upward. Thus, a continuous flow is established, which carries along heat. Whenever heat is transferred by the bulk movement of a gas or a liquid, the heat is said to be transferred by **convection**. The fluid flow itself is called a **convection current**.

> **CONVECTION**
> **Convection is the process in which heat is carried from place to place by the bulk movement of a fluid.**

The smoke rising from a fire, like the one in **Figure 13.1**, is one visible result of convection. **Figure 13.2** shows the less visible example of convection currents in a pot of water being heated on a gas burner. The currents distribute the heat from the burning gas to all parts of the water. Conceptual Example 1 deals with some of the important roles that convection plays in the home.

CONCEPTUAL EXAMPLE 1 | The Physics of Heating and Cooling by Convection

Hot water baseboard heating units are frequently used in homes, and a cooling coil is a major component of a refrigerator. The locations of these heating and cooling devices are different because each is designed to maximize the production of convection currents. Where should the heating unit and the cooling coil be located? **(a)** Heating unit near the floor of the room and cooling coil near the top of the refrigerator **(b)** Heating unit near the ceiling of the room and cooling coil near the bottom of the refrigerator

Reasoning An important goal for the heating system is to distribute heat throughout a room. The analogous goal for the cooling coil is to remove heat from all of the space within a refrigerator. In each case, the heating or cooling device must be positioned so that convection makes the goal achievable.

Answer (b) is incorrect. If the heating unit were placed near the ceiling of the room, warm air from the unit would remain there, because warm air does not fall (it rises). Thus, there would be very little natural movement (or convection) of air to distribute

the heat throughout the room. If the cooling coil were located near the bottom of the refrigerator, the cool air would remain there, because cool air does not rise (it sinks). There would be very little convection to carry the heat from other parts of the refrigerator to the coil for removal.

Answer (a) is correct. The air above the baseboard unit is heated, like the air above a fire. Buoyant forces from the surrounding cooler air push the warm air upward. Cooler air near the ceiling is displaced downward and then warmed by the baseboard heating unit, causing the convection current illustrated in **Animated Figure 13.3a**. Within the refrigerator, air in contact with the top-mounted coil is cooled, its volume decreases, and its density increases. The surrounding warmer and less dense air cannot provide sufficient buoyant force to support the cooler air, which sinks downward. In the process, warmer air near the bottom of the refrigerator is displaced upward and is then cooled by the coil, establishing the convection current shown in **Animated Figure 13.3b**.

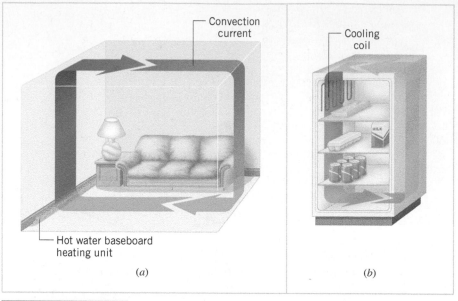

ANIMATED FIGURE 13.3 (*a*) Air warmed by the baseboard heating unit is pushed to the top of the room by the cooler and denser air. (*b*) Air cooled by the cooling coil sinks to the bottom of the refrigerator. In both (*a*) and (*b*) a convection current is established.

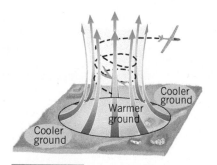

FIGURE 13.4 Updrafts, or thermals, are caused by the convective movement of air that the ground has warmed.

THE PHYSICS OF . . . thermals. Another example of convection occurs when the ground, heated by the sun's rays, warms the neighboring air. Surrounding cooler and denser air pushes the heated air upward. The resulting updraft or "thermal" can be quite strong, depending on the amount of heat that the ground can supply. As **Figure 13.4** illustrates, these thermals can be used by glider pilots to gain considerable altitude. Birds such as eagles utilize thermals in a similar fashion.

THE PHYSICS OF . . . an inversion layer. Sometimes meteorological conditions cause a layer to form in the atmosphere where the temperature increases with increasing altitude. Such a layer is called an **inversion layer** because its temperature profile is inverted compared to the usual situation, wherein the air temperature decreases with increasing altitude. In the usual situation, upward convection currents occur and are important for dispersing pollutants from industrial sources and automobile exhaust systems. An inversion layer, in contrast, arrests the normal upward convection currents, causing a stagnant-air condition in which the concentration of pollutants increases substantially. This condition leads to a smog layer that can often be seen hovering over large cities.

We have been discussing **natural convection**, in which a temperature difference causes the density at one place in a fluid to be different from the density at another. Sometimes, natural convection is inadequate to transfer sufficient amounts of heat. In such cases **forced convection** is often used, and an external device such as a pump or a fan mixes the warmer and cooler portions of the fluid.

BIO **THE PHYSICS OF . . . rapid thermal exchange.** **Figure 13.5** shows an application of forced convection that is revolutionizing the way in which the effects of overheating are being treated. Athletes, for example, are especially prone to overheating, and the device illustrated in **Figure 13.5** is appearing more and more frequently at athletic events. The technique is known as rapid thermal exchange and takes advantage of specialized blood vessels called arteriovenous anastomoses (AVAs) that are found in the palms of the hands (and soles of the feet). These blood vessels are used to help dissipate unwanted heat from the body. The device in the drawing consists of a small chamber containing a curved metal plate, through which cool water is circulated from a refrigerated supply. The overheated athlete inserts his hand into the chamber and places his palm on the plate. The chamber seals around the wrist and is evacuated slightly to reduce the air pressure and thereby promote circulation of blood through the hand. Forced convection plays two roles in this treatment. It causes the water to circulate through the metal plate and remove heat from the blood in the AVAs. Also, the cooled blood returns through veins to the heart, which pumps it throughout the body, thus lowering the body temperature and relieving the effects of overheating.

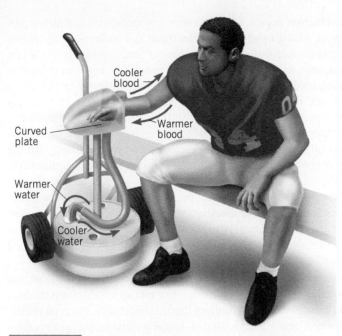

Cooler
blood

Curved
plate

Warmer
blood

Warmer
water

Cooler
water

FIGURE 13.5 An overheated athlete uses a rapid-thermal-exchange device to cool down. He places the palm of his hand on a curved metal plate within a slightly evacuated chamber. Forced convection circulates cool water through the plate, which cools the blood flowing through the hand. The cooled blood returns through veins to the heart, which circulates it throughout the body.

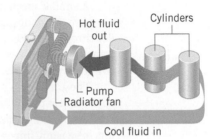

Hot fluid
out

Cylinders

Pump

Radiator fan

Cool fluid in

FIGURE 13.6 The forced convection generated by a pump circulates radiator fluid through an automobile engine to remove excess heat.

Figure 13.6 shows the application of forced convection in an automobile engine. As in the previous application, forced convection occurs in two ways. First, a pump circulates radiator fluid (water and antifreeze) through the engine to remove excess heat from the combustion process. Second, a radiator fan draws air through the radiator. Heat is transferred from the hotter radiator fluid to the cooler air, thereby cooling the fluid.

BIO **THE PHYSICS OF . . . the windchill factor.** Forced convection also plays the principal role in the windchill factor that is often mentioned in weather reports. The wind mixes the cold ambient air with the warm layer of air that immediately surrounds the exposed portions of your body. The forced convection removes heat from exposed body surfaces, thereby making you feel colder than you would if there were no wind.

Check Your Understanding

(*The answer is given at the end of the book.*)

1. The transfer of heat by convection is *smallest* in **(a)** solids, **(b)** liquids, **(c)** gases.

13.2 Conduction

Anyone who has fried a hamburger in an all-metal skillet knows that the metal handle becomes hot. Somehow, heat is transferred from the burner to the handle. Clearly, heat is not being transferred by the bulk movement of the metal or the surrounding air, so convection can be ruled out. Instead, heat is transferred directly through the metal by a process called **conduction**.

> **CONDUCTION**
> Conduction is the process whereby heat is transferred directly through a material, with any bulk motion of the material playing no role in the transfer.

One mechanism for conduction occurs when the atoms or molecules in a hotter part of the material vibrate or move with greater energy than those in a cooler part. By means of collisions, the more energetic molecules pass on some of their energy to their less energetic neighbors. For example, imagine a gas filling the space between two walls that face each other and are maintained at different temperatures. Molecules strike the hotter wall, absorb energy from it, and rebound with a greater kinetic energy than when they arrived. As these more energetic molecules collide with their less energetic neighbors, they transfer some of their energy to them. Eventually, this energy is passed on until it reaches the molecules next to the cooler wall. These molecules, in turn, collide with the wall, giving up some of their energy to it in the process. Through such molecular collisions, heat is conducted from the hotter to the cooler wall.

A similar mechanism for the conduction of heat occurs in metals. Metals are different from most substances in having a pool of electrons that are more or less free to wander throughout the metal. These free electrons can transport energy and allow metals to transfer heat very well. The free electrons are also responsible for the excellent electrical conductivity that metals have.

Those materials that conduct heat well are called **thermal conductors**, and those that conduct heat poorly are known as **thermal insulators**. Most metals are excellent thermal conductors; wood, glass, and most plastics are common thermal insulators. Thermal insulators have many important applications. Virtually all new housing construction incorporates thermal insulation in attics and walls to reduce heating and cooling costs. And the wooden or plastic handles on many pots and pans reduce the flow of heat to the cook's hand.

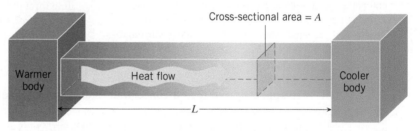

INTERACTIVE FIGURE 13.7 Heat is conducted through the bar when the ends of the bar are maintained at different temperatures. The heat flows from the warmer to the cooler end.

To illustrate the factors that influence the conduction of heat, **Interactive Figure 13.7** displays a rectangular bar. The ends of the bar are in thermal contact with two bodies, one of which is kept at a constant higher temperature, while the other is kept at a constant lower temperature. Although not shown for the sake of clarity, the sides of the bar are insulated, so the heat lost through them is negligible. The amount of heat Q conducted through the bar from the warmer end to the cooler end depends on a number of factors:

1. Q is proportional to the time t during which conduction takes place ($Q \propto t$). More heat flows in longer time periods.

2. Q is proportional to the temperature difference ΔT between the ends of the bar ($Q \propto \Delta T$). A larger difference causes more heat to flow. No heat flows when both ends have the same temperature and $\Delta T = 0$ C°.

3. Q is proportional to the cross-sectional area A of the bar ($Q \propto A$). **Interactive Figure 13.8** helps to explain this fact by showing two identical bars (insulated sides not shown) placed between the warmer and cooler bodies. Clearly, twice as much heat flows through two bars as through one, because the cross-sectional area has been doubled.

4. Q is inversely proportional to the length L of the bar ($Q \propto 1/L$). Greater lengths of material conduct less heat. To experience this effect, put two insulated mittens (the

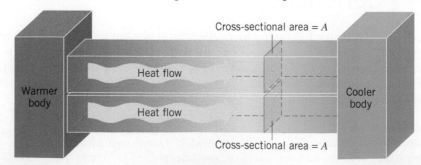

INTERACTIVE FIGURE 13.8 Twice as much heat flows through two identical bars as through one.

pot holders that cooks keep near the stove) on the *same hand*. Then, touch a hot pot and notice that it feels cooler than when you wear only one mitten, signifying that less heat passes through the greater thickness ("length") of material.

These proportionalities can be stated together as $Q \propto (A\,\Delta T)t/L$. Equation 13.1 expresses this result with the aid of a proportionality constant k, which is called the **thermal conductivity**.

CONDUCTION OF HEAT THROUGH A MATERIAL
The heat Q conducted during a time t through a bar of length L and cross-sectional area A is

$$Q = \frac{(kA\,\Delta T)t}{L} \qquad (13.1)$$

where ΔT is the temperature difference between the ends of the bar (the higher temperature minus the lower temperature) and k is the thermal conductivity of the material.

SI Unit of Thermal Conductivity: $J/(s \cdot m \cdot C°)$

Since $k = QL/(tA\,\Delta T)$, the SI unit for thermal conductivity is $J \cdot m/(s \cdot m^2 \cdot C°)$ or $J/(s \cdot m \cdot C°)$. The SI unit of power is the joule per second (J/s), or watt (W), so the thermal conductivity is also given in units of $W/(m \cdot C°)$.

Different materials have different thermal conductivities, and Table 13.1 gives some representative values. Because metals are such good thermal conductors, they have large thermal conductivities. In comparison, liquids and gases generally have small thermal conductivities. In fact, in most fluids the heat transferred by conduction is negligible compared to that transferred by convection when there are strong convection currents. Air, for instance, with its small thermal conductivity, is an excellent thermal insulator when confined to small spaces where no appreciable convection currents can be established. Goose down, Styrofoam, and wool derive their fine insulating properties in part from the small dead-air spaces within them, as Figure 13.9 illustrates.

THE PHYSICS OF . . . dressing warmly. We also take advantage of dead-air spaces when we dress "in layers" during very cold weather and put on several layers of relatively thin clothing rather than one thick layer. The air trapped between the layers acts as an excellent insulator.

Example 2 deals with the role that conduction through body fat plays in regulating body temperature.

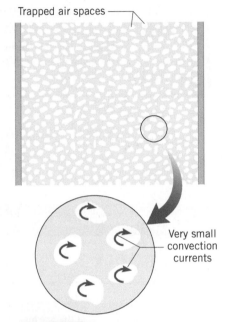

Trapped air spaces

Very small convection currents

FIGURE 13.9 Styrofoam is an excellent thermal insulator because it contains many small, dead-air spaces. These small spaces inhibit heat transfer by convection currents, and air itself has a very low thermal conductivity.

TABLE 13.1 **Thermal Conductivities**[a] **of Selected Materials**

Substance	Thermal Conductivity, k $[J/(s \cdot m \cdot C°)]$	Substance	Thermal Conductivity, k $[J/(s \cdot m \cdot C°)]$
Metals		**Other Materials**	
Aluminum	240	Asbestos	0.090
Brass	110	Body fat	0.20
Copper	390	Concrete	1.1
Iron	79	Diamond	2450
Lead	35	Glass	0.80
Silver	420	Goose down	0.025
Steel (stainless)	14	Ice (0 °C)	2.2
Gases		Styrofoam	0.010
Air	0.0256	Water	0.60
Hydrogen (H_2)	0.180	Wood (oak)	0.15
Nitrogen (N_2)	0.0258	Wool	0.040
Oxygen (O_2)	0.0265		

[a]Except as noted, the values pertain to temperatures near 20 °C.

EXAMPLE 2 | BIO The Physics of Heat Transfer in the Human Body

When excessive heat is produced within the body, it must be transferred to the skin and dispersed if the temperature at the body interior is to be maintained at the normal value of 37.0 °C. One possible mechanism for transfer is conduction through body fat. Suppose that heat travels through 0.030 m of fat in reaching the skin, which has a total surface area of 1.7 m² and a temperature of 34.0 °C. Find the amount of heat that reaches the skin in half an hour (1800 s).

Reasoning and Solution   In Table 13.1 the thermal conductivity of body fat is given as $k = 0.20$ J/(s · m · C°). According to Equation 13.1,

$$Q = \frac{(kA\,\Delta T)t}{L}$$

$$Q = \frac{[0.20\text{ J/(s}\cdot\text{m}\cdot\text{C}°)](1.7\text{ m}^2)(37.0\text{ °C} - 34.0\text{ °C})(1800\text{ s})}{0.030\text{ m}}$$

$$= \boxed{6.1 \times 10^4\text{ J}}.$$

For comparison, a jogger can generate over ten times this amount of heat in a half hour. Thus, conduction through body fat is not a particularly effective way of removing excess heat. Heat transfer via blood flow to the skin is more effective and has the added advantage that the body can vary the blood flow as needed (see Problem 36).

Example 3 uses Equation 13.1 to determine what the temperature is at a point between the warmer and cooler ends of the bar in Interactive Figure 13.7.

Analyzing Multiple-Concept Problems

EXAMPLE 3 | The Temperature at a Point Between the Ends of a Bar

In Interactive Figure 13.7 the temperatures at the ends of the bar are 85.0 °C at the warmer end and 27.0 °C at the cooler end. The bar has a length of 0.680 m. What is the temperature at a point that is 0.220 m from the cooler end of the bar?

Reasoning   The point in question is closer to the cooler end than to the warmer end of the bar. It might be expected, therefore, that the temperature at this point is less than halfway between 27.0 °C and 85.0 °C. We will demonstrate that this is, in fact, the case, by applying Equation 13.1. This expression applies because no heat escapes through the insulated sides of the bar, and we will use it twice to determine the desired temperature.

Knowns and Unknowns   The available data are as follows:

Description	Symbol	Value
Temperature at warmer end	T_W	85.0 °C
Temperature at cooler end	T_C	27.0 °C
Length of bar	L	0.680 m
Distance from cooler end	D	0.220 m
Unknown Variable		
Temperature at distance D from cooler end	T	?

Modeling the Problem

STEP 1 The Conduction of Heat   The heat Q conducted in a time t past the point in question (which is a distance D from the cooler end of the bar) is given by Equation 13.1 as

$$Q = \frac{kA(T - T_C)t}{D}$$

where k is the thermal conductivity of the material from which the bar is made, A is the bar's cross-sectional area, and T and T_C are, respectively, the temperature at the point in question and at the cooler end of the bar. Solving for T gives Equation 1 at the right. The variables Q, k, A, and t are unknown, so we proceed to Step 2 to deal with them.

$$T = T_C + \frac{QD}{kAt} \qquad (1)$$

STEP 2 The Conduction of Heat Revisited The heat Q that is conducted from the point in question to the cooler end of the bar originates at the warmer end of the bar. Thus, since no heat is lost through the sides, we may apply Equation 13.1 a second time to obtain an expression for Q:

$$Q = \frac{kA(T_W - T_C)t}{L}$$

where T_W and T_C are, respectively, the temperatures at the warmer and cooler ends of the bar, which has a length L. This expression for Q can be substituted into Equation 1, as indicated at the right. The terms k, A, and t remain to be dealt with. Fortunately, however, values for them are unnecessary, because they can be eliminated algebraically from the final calculation.

$$T = T_C + \frac{QD}{kAt} \qquad (1)$$

$$Q = \frac{kA(T_W - T_C)t}{L}$$

Solution Combining the results of each step algebraically, we find that

$$\overset{\text{STEP 1}}{\underset{\downarrow}{}} \quad \overset{\text{STEP 2}}{\underset{\downarrow}{}}$$

$$T = T_C + \frac{QD}{kAt} = T_C + \frac{\left[\frac{kA(T_W - T_C)t}{L}\right]D}{kAt}$$

Simplifying this result gives

$$T = T_C + \frac{\frac{\cancel{kA}(T_W - T_C)\cancel{t}}{L}D}{\cancel{kAt}} = T_C + \frac{(T_W - T_C)D}{L}$$

$$= 27.0\ °C + \frac{(85.0\ °C - 27.0\ °C)(0.220\ m)}{0.680\ m} = \boxed{45.8\ °C}$$

As expected, this temperature is less than halfway between 27.0 °C and 85.0 °C.

Related Homework: Problem 32

Virtually all homes contain insulation in the walls to reduce heat loss. Example 4 illustrates how to determine this loss with and without insulation.

EXAMPLE 4 | The Physics of Layered Insulation

One wall of a house consists of 0.019-m-thick plywood backed by 0.076-m-thick insulation, as **Figure 13.10** shows. The temperature at the inside surface is 25.0 °C, while the temperature at the outside surface is 4.0 °C, both being constant. The thermal conductivities of the insulation and the plywood are, respectively, 0.030 and 0.080 J/(s · m · C°), and the area of the wall is 35 m². Find the heat conducted through the wall in one hour **(a)** with the insulation and **(b)** without the insulation.

Reasoning The temperature T at the insulation–plywood interface (see **Figure 13.10**) must be determined before the heat conducted through the wall can be obtained. In calculating this temperature, we use the fact that no heat is accumulating in the wall because the inner and outer temperatures are constant. Therefore, the heat conducted through the insulation must equal the heat conducted through the plywood during the same time; that is, $Q_{insulation} = Q_{plywood}$. Each of the Q values can be expressed as $Q = (kA\ \Delta T)t/L$, according to Equation 13.1, leading to an expression that can be solved for the interface temperature. Once a value for T is available, Equation 13.1 can be used to obtain the heat conducted through the wall.

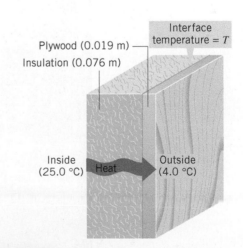

FIGURE 13.10 Heat flows through the insulation and plywood from the warmer inside to the cooler outside.

Problem-Solving Insight When heat is conducted through a multi-layered material (such as the plywood/insulation in this example) and the high and low temperatures are constant, the heat conducted through each layer is the same.

Solution **(a)** Using Equation 13.1 and the fact that $Q_{insulation} = Q_{plywood}$, we find that

$$\left[\frac{(kA\,\Delta T)t}{L}\right]_{insulation} = \left[\frac{(kA\,\Delta T)t}{L}\right]_{plywood}$$

$$\frac{[0.030\,\text{J/(s}\cdot\text{m}\cdot\text{C}°)]A(25.0\,°\text{C} - T)t}{0.076\,\text{m}} = \frac{[0.080\,\text{J/(s}\cdot\text{m}\cdot\text{C}°)]A(T - 4.0\,°\text{C})t}{0.019\,\text{m}}$$

Note that on each side of the equals sign we have written ΔT as the higher temperature minus the lower temperature. Eliminating the area A and time t algebraically and solving this equation for T reveals that the temperature at the insulation–plywood interface is $T = 5.8\,°\text{C}$.

The heat conducted through the wall is either $Q_{insulation}$ or $Q_{plywood}$, since the two quantities are equal. Choosing $Q_{insulation}$ and using $T = 5.8\,°\text{C}$ in Equation 13.1, we find that

$$Q_{insulation} = \frac{[0.030\,\text{J/(s}\cdot\text{m}\cdot\text{C}°)](35\,\text{m}^2)(25.0\,°\text{C} - 5.8\,°\text{C})(3600\,\text{s})}{0.076\,\text{m}}$$

$$= \boxed{9.5 \times 10^5\,\text{J}}$$

(b) It is straightforward to use Equation 13.1 to calculate the amount of heat that would flow through the plywood in one hour if the insulation were absent:

$$Q_{plywood} = \frac{[0.080\,\text{J/(s}\cdot\text{m}\cdot\text{C}°)](35\,\text{m}^2)(25.0\,°\text{C} - 4.0\,°\text{C})(3600\,\text{s})}{0.019\,\text{m}}$$

$$= \boxed{110 \times 10^5\,\text{J}}$$

Without insulation, the heat loss is increased by a factor of about 12.

Math Skills Without units (omitted for the sake of clarity), the equation that needs to be solved for the temperature T is

$$\frac{0.030\cancel{A}(25.0 - T)\cancel{t}}{0.076} = \frac{0.080\cancel{A}(T - 4.0)\cancel{t}}{0.019} \quad \text{or}$$

$$\frac{0.030(25.0 - T)}{0.076} = \frac{0.080(T - 4.0)}{0.019}$$

The terms A and t appear on both sides of the equation as factors in the numerator and, therefore, have been eliminated algebraically. Rearranging the result slightly and carrying out the indicated divisions, we obtain

$$\frac{0.030}{0.076}(25.0 - T) = \frac{0.080}{0.019}(T - 4.0) \quad \text{or}$$

$$0.394(25.0 - T) = 4.21(T - 4.0)$$

Expanding the terms on each side of the equals sign and rearranging the result shows that

$$9.85 - 0.394T = 4.21T - 16.84 \quad \text{or} \quad 26.69 = 4.604T$$

In these results, in order to avoid round-off errors, we have carried along more significant figures than only the two justified by the original data. However, in solving the final equation for T we round off to two significant figures and obtain

$$T = \frac{26.69}{4.604} = 5.8\,°\text{C}.$$

THE PHYSICS OF . . . protecting fruit plants from freezing. Fruit growers sometimes protect their crops by spraying them with water when overnight temperatures are expected to drop below freezing. Some fruit crops, like the strawberries in **Figure 13.11**, can withstand temperatures down to freezing (0 °C), but not below freezing. When water is sprayed on the plants, it can freeze and release heat (see Section 12.8),

Pierre Ducharme/Reuters/Newscom

FIGURE 13.11 When the temperature dips below freezing, strawberry growers spray their plants with water to put a coat of ice on them. Because of the heat released when the water freezes and because of the relatively small thermal conductivity of ice, this procedure protects the plants against subfreezing temperatures. The photograph shows berries that survived a temperature of −3 °C with little or no damage.

some of which goes into warming the plant. In addition, water and ice have relatively small thermal conductivities, as **Table 13.1** indicates. Thus, they also protect the crop by acting as thermal insulators that reduce heat loss from the plants.

Although a layer of ice may be beneficial to strawberries, it is not so desirable inside a refrigerator, as Conceptual Example 5 discusses.

CONCEPTUAL EXAMPLE 5 | An Iced-up Refrigerator

In a refrigerator, heat is removed by a cold refrigerant fluid that circulates within a tubular space embedded inside a metal plate, as **Figure 13.12** illustrates. A good refrigerator cools food as quickly as possible. Which arrangement works best: **(a)** an aluminum plate coated with ice, **(b)** an aluminum plate without ice, **(c)** a stainless steel plate coated with ice, or **(d)** a stainless steel plate without ice?

Reasoning   Figure 13.12 (see the blow-ups) shows the metal cooling plate with and without a layer of ice. Without ice, heat passes by conduction through the metal plate to the refrigerant fluid within. For a given temperature difference across the thickness of the metal, the rate of heat transfer depends on the thermal conductivity of the metal. When the plate becomes coated with ice, any heat that is removed by the refrigerant fluid must first be transferred by conduction through the ice before it encounters the metal plate.

Answers (a), (c), and (d) are incorrect.   For answers (a) and (c), the relation $Q = \dfrac{(kA\,\Delta T)t}{L}$ (Equation 13.1) indicates that the heat conducted per unit time (Q/t) is inversely proportional to the thickness L of the ice. As ice builds up, the heat removed per unit time by the cooling plate decreases. Thus, when covered with ice, the cooling plate—regardless of whether it's made from aluminum or stainless steel—does not work as well as a plate that is ice-free. Answer (d)—the stainless steel plate without ice—is incorrect, because heat is transferred more readily through a plate that has a greater thermal conductivity, and stainless steel has a smaller thermal conductivity than does aluminum (see **Table 13.1**).

Answer (b) is correct.   The relation $Q = \dfrac{(kA\,\Delta T)t}{L}$ (Equation 13.1) shows that the heat conducted per unit time (Q/t) is directly proportional to the thermal conductivity k of the metal plate. Since the

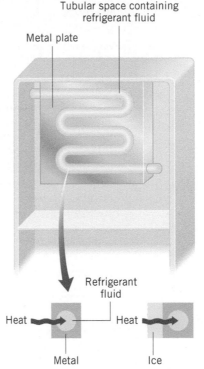

FIGURE 13.12   In a refrigerator, cooling is accomplished by a cold refrigerant fluid that circulates through a tubular space embedded within a metal plate. Sometimes the plate becomes coated with a layer of ice.

thermal conductivity of aluminum is more than 17 times greater than the thermal conductivity of stainless steel (see **Table 13.1**), aluminum is the preferred plate. The aluminum plate arrangement works best without an ice buildup. When ice builds up, the heat removed per unit time decreases because of the increased thickness of material through which the heat must pass.

Related Homework: Problem 10

Check Your Understanding

(The answers are given at the end of the book.)

2. A poker used in a fireplace is held at one end, while the other end is in the fire. In terms of being cooler to the touch, should a poker be made from **(a)** a high-thermal-conductivity material, **(b)** a low-thermal-conductivity material, or **(c)** can either type be used?

3. Several days after a snowstorm, the outdoor temperature remains below freezing. The roof on one house is uniformly covered with snow. On a neighboring house, however, the snow on the roof has completely melted. Which house is better insulated?

4. Concrete walls often contain steel reinforcement bars. Does the steel **(a)** enhance, **(b)** degrade, or **(c)** have no effect on the insulating value of the concrete wall? (Consult **Table 13.1**.)

5. To keep your hands as warm as possible during skiing, should you wear mittens or gloves? (Mittens, except for the thumb, do not have individual finger compartments.) Assume that the mittens and gloves are the same size and are made of the same material. You should wear: **(a)** gloves,

(Continued)

because the individual finger compartments mean that the gloves have a smaller thermal conductivity; **(b)** gloves, because the individual finger compartments mean that the gloves have a larger thermal conductivity; **(c)** mittens, because they have less surface area exposed to the cold air.

6. A water pipe is buried slightly beneath the ground. The ground is covered with a thick layer of snow, which contains a lot of small dead-air spaces within it. The air temperature suddenly drops to well below freezing. The accumulation of snow **(a)** has no effect on whether the water in the pipe freezes, **(b)** causes the water in the pipe to freeze more quickly than if the snow were not there, **(c)** helps prevent the water in the pipe from freezing.

7. Some animals have hair strands that are hollow, air-filled tubes. Others have hair strands that are solid. Which kind, if either, would be more likely to give an animal an advantage for surviving in very cold climates?

8. Two bars are placed between plates whose temperatures are T_{hot} and T_{cold} (see **CYU Figure 13.1**). The thermal conductivity of bar 1 is six times that of bar 2 ($k_1 = 6k_2$), but bar 1 has only one-third the cross-sectional area ($A_1 = \frac{1}{3}A_2$). Ignore any heat loss through the sides of the bars. What can you conclude about the amounts of heat Q_1 and Q_2, respectively, that bar 1 and bar 2 conduct in a given amount of time? **(a)** $Q_1 = \frac{1}{4}Q_2$ **(b)** $Q_1 = \frac{1}{8}Q_2$ **(c)** $Q_1 = 2Q_2$ **(d)** $Q_1 = 4Q_2$ **(e)** $Q_1 = Q_2$

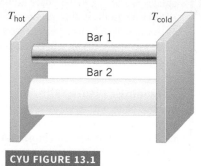

CYU FIGURE 13.1

9. A piece of Styrofoam and a piece of wood are joined together to form a layered slab. The two pieces have the same thickness and cross-sectional area, but the Styrofoam has the smaller thermal conductivity. The temperature of the exposed Styrofoam surface is greater than the temperature of the exposed wood surface, both temperatures being constant. Is the temperature of the Styrofoam–wood interface **(a)** closer to the higher temperature of the exposed Styrofoam surface, **(b)** closer to the lower temperature of the exposed wood surface, or **(c)** halfway between the two temperatures?

13.3 Radiation

Energy from the sun is brought to earth by large amounts of visible light waves, as well as by substantial amounts of infrared and ultraviolet waves. These waves are known as electromagnetic waves, a class that also includes the microwaves used for cooking and the radio waves used for AM and FM broadcasts. The sunbather in **Figure 13.13** feels hot because her body absorbs energy from the sun's electromagnetic waves. Anyone who has stood by a roaring fire or put a hand near an incandescent light bulb has experienced a similar effect. Thus, fires and light bulbs also emit electromagnetic waves, and when the energy of such waves is absorbed, it can have the same effect as heat.

The process of transferring energy via electromagnetic waves is called **radiation**, and, unlike convection or conduction, it does not require a material medium. Electromagnetic waves from the sun, for example, travel through the void of space during their journey to earth.

FIGURE 13.13 Suntans are produced by ultraviolet rays.

StockLite/Shutterstock.com

> **RADIATION**
> Radiation is the process in which energy is transferred by means of electromagnetic waves.

All bodies continuously radiate energy in the form of electromagnetic waves. Even an ice cube radiates energy, although so little of it is in the form of visible light that an ice cube cannot be seen in the dark. Likewise, the human body emits insufficient visible light to be seen in the dark. However, as **Figures 12.6** and **12.7** illustrate, the infrared waves radiating from the body can be detected in the dark by electronic cameras. Generally, an object does not emit much visible light until the temperature of the object exceeds about 1000 K. Then a characteristic red glow appears, like that of a heating coil on an electric stove. When its temperature reaches about 1700 K, an object begins to glow white-hot, like the tungsten filament in an incandescent light bulb.

In the transfer of energy by radiation, the absorption of electromagnetic waves is just as important as their emission. The surface of an object plays a significant role in determining how much radiant energy the object will absorb or emit. The two blocks in sunlight in **Interactive Figure 13.14**, for example, are identical, except that one has a rough surface coated with lampblack (a fine black soot), while the other has a highly polished silver surface. As the thermometers indicate, the temperature of the black block rises at a much faster rate than that of the silvery block. This is because lampblack absorbs about 97% of the incident radiant energy, while the silvery surface absorbs only about 10%. The remaining part of the incident energy is reflected in each case. We see the lampblack as black in color because it reflects so little of the light falling on it, while the silvery surface looks like a mirror because it reflects so much light. Since the color black is associated with nearly complete absorption of visible light, the term **perfect blackbody** or, simply, **blackbody** is used when referring to an object that absorbs *all* the electromagnetic waves falling on it.

All objects emit and absorb electromagnetic waves simultaneously. When a body has the same constant temperature as its surroundings, the amount of radiant energy being absorbed must balance the amount being emitted in a given interval of time. The block coated with lampblack absorbs and emits the same amount of radiant energy, and the silvery block does too. In either case, if absorption were greater than emission, the block would experience a net gain in energy. As a result, the temperature of the block would rise and not be constant. Similarly, if emission were greater than absorption, the temperature would fall. Since absorption and emission are balanced, *a material that is a good absorber, like lampblack, is also a good emitter, and a material that is a poor absorber, like polished silver, is also a poor emitter.* A perfect blackbody, being a perfect absorber, is also a perfect emitter.

THE PHYSICS OF . . . summer clothing. The fact that a black surface is both a good absorber and a good emitter is the reason people are uncomfortable wearing dark clothes during the summer. Dark clothes absorb a large fraction of the sun's radiation and then reemit it in all directions. About one-half of the emitted radiation is directed inward toward the body and creates the sensation of warmth. Light-colored clothes, in contrast, are cooler to wear, because they absorb and reemit relatively little of the incident radiation.

THE PHYSICS OF . . . a white sifaka lemur warming up. The use of light colors for comfort also occurs in nature. Most lemurs, for instance, are nocturnal and have dark fur like the lemur shown in **Figure 13.15a**. Since they are active at night, the dark fur poses no disadvantage in absorbing excessive sunlight. **Figure 13.15b** shows a species of lemur called the white sifaka, which lives in semiarid regions where there is little shade. The white color of the fur may help in thermoregulation, by reflecting sunlight, but during the cool mornings, reflection of sunlight would hinder warming up. However, these lemurs have black skin and only sparse fur on their bellies, and to warm up in the morning, they turn their dark bellies toward the sun. The dark color enhances the absorption of sunlight.

The amount of radiant energy Q emitted by a perfect blackbody is proportional to the radiation time interval t ($Q \propto t$). The longer the time, the greater is the amount of energy radiated. Experiment shows that Q is also proportional to the surface area A ($Q \propto A$). An object with a large surface area radiates more energy than one with a small

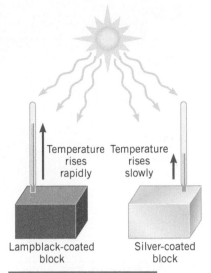

Temperature rises rapidly   Temperature rises slowly

Lampblack-coated block   Silver-coated block

INTERACTIVE FIGURE 13.14 The temperature of the block coated with lampblack rises faster than the temperature of the block coated with silver because the black surface absorbs radiant energy from the sun at the greater rate.

(a)   (b)

FIGURE 13.15 (a) Most lemurs, like this one, are nocturnal and have dark fur. (b) The species of lemur called the white sifaka, however, is active during the day and has white fur.

surface area, other things being equal. Finally, experiment reveals that Q is proportional to the *fourth power of the Kelvin temperature* T ($Q \propto T^4$), so the emitted energy increases markedly with increasing temperature. If, for example, the Kelvin temperature of an object doubles, the object emits 2^4 or 16 times more energy. Combining these factors into a single proportionality, we see that $Q \propto T^4At$. This proportionality is converted into an equation by inserting a proportionality constant σ, known as the *Stefan–Boltzmann constant*. It has been found experimentally that $\sigma = 5.67 \times 10^{-8}$ J/(s $\cdot$ m^2 $\cdot$ K^4):

$$Q = \sigma T^4 At$$

The relationship above holds only for a perfect emitter. Most objects are not perfect emitters, however. Suppose that an object radiates only about 80% of the visible light energy that a perfect emitter would radiate, so Q (for the object) $= (0.80)\sigma T^4 At$. The factor such as the 0.80 in this equation is called the **emissivity** e and is a dimensionless number between zero and one. The emissivity is the ratio of the energy an object actually radiates to the energy the object would radiate if it were a perfect emitter. For visible light, the value of e for the human body, for instance, varies between about 0.65 and 0.80, the smaller values pertaining to lighter skin colors. For infrared radiation, e is nearly one for all skin colors. For a perfect blackbody emitter, $e = 1$. Including the factor e on the right side of the expression $Q = \sigma T^4 At$ leads to the **Stefan–Boltzmann law of radiation**.

> **THE STEFAN–BOLTZMANN LAW OF RADIATION**
>
> The radiant energy Q, emitted in a time t by an object that has a Kelvin temperature T, a surface area A, and an emissivity e, is given by
>
> $$Q = e\sigma T^4 At \qquad (13.2)$$
>
> where $\sigma = 5.67 \times 10^{-8}$ J/(s $\cdot$ m^2 $\cdot$ K^4) is the Stefan–Boltzmann constant.

In Equation 13.2, the Stefan–Boltzmann constant σ is a universal constant in the sense that its value is the same for all bodies, regardless of the nature of their surfaces. The emissivity e, however, depends on the condition of the surface. Example 6 shows how the Stefan–Boltzmann law can be used to determine the size of a star.

EXAMPLE 6 | A Supergiant Star

The supergiant star Betelgeuse has a surface temperature of about 2900 K (about one-half that of our sun) and emits a radiant power (in joules per second, or watts) of approximately 4×10^{30} W (about 10 000 times as great as that of our sun). Assuming that Betelgeuse is a perfect emitter (emissivity $e = 1$) and spherical, find its radius.

Reasoning According to the Stefan–Boltzmann law, the power emitted is $Q/t = e\sigma T^4 A$. A star with a relatively small temperature T can have a relatively large radiant power Q/t only if its surface area A is large. As we will see, Betelgeuse has a very large surface area, so its radius is enormous.

> **Problem-Solving Insight** First solve an equation for the unknown in terms of the known variables. Then substitute numbers for the known variables, as this example shows.

Solution Solving the Stefan–Boltzmann law for the surface area, we find

$$A = \frac{Q/t}{e\sigma T^4}$$

But the surface area of a sphere is $A = 4\pi r^2$, so $r = \sqrt{A/(4\pi)}$. Therefore, we have

$$r = \sqrt{\frac{Q/t}{4\pi e \sigma T^4}} = \sqrt{\frac{4 \times 10^{30} \text{ W}}{4\pi(1)[5.67 \times 10^{-8} \text{ J/(s} \cdot \text{m}^2 \cdot \text{ K}^4)](2900 \text{ K})^4}}$$

$$= \boxed{3 \times 10^{11} \text{m}}$$

For comparison, Mars orbits the sun at a distance of 2.28×10^{11} m. Betelgeuse is certainly a "supergiant."

The next example explains how to apply the Stefan–Boltzmann law when an object, such as a wood stove, simultaneously emits and absorbs radiant energy.

EXAMPLE 7 | An Unused Wood-Burning Stove

An unused wood-burning stove has a constant temperature of 18 °C (291 K), which is also the temperature of the room in which the stove stands. The stove has an emissivity of 0.900 and a surface area of 3.50 m^2. What is the *net* radiant power generated by the stove?

Reasoning Power is the change in energy per unit time (Equation 6.10b), or Q/t, which, according to the Stefan–Boltzmann law, is $Q/t = e\sigma T^4 A$ (Equation 13.2). In this problem, however, we need to find the *net* power produced by the stove. The net power is the power the stove emits minus the power the stove absorbs. The power the stove absorbs comes from the walls, ceiling, and floor of the room, all of which emit radiation.

Solution Remembering that temperature must be expressed in kelvins when using the Stefan–Boltzmann law, we find that

$$\text{Power emitted by unheated stove at 18 °C} = \frac{Q}{t} = e\sigma T^4 A \qquad (13.2)$$

$$= (0.900)[5.67 \times 10^{-8}\,\text{J/(s} \cdot \text{m}^2 \cdot \text{K}^4)](291\,\text{K})^4(3.50\,\text{m}^2) = 1280\,\text{W}$$

The fact that the unheated stove emits 1280 W of power and yet maintains a constant temperature means that the stove also absorbs 1280 W of radiant power from its surroundings. Thus, the *net* power generated by the unheated stove is zero:

$$\underset{\substack{\text{Power emitted}\\\text{by stove at}\\18\,°C}}{\underbrace{1280\,\text{W}}} - \underset{\substack{\text{Power emitted by}\\\text{room at 18 °C and}\\\text{absorbed by stove}}}{\underbrace{1280\,\text{W}}} = \boxed{0\,\text{W}}$$

Problem-Solving Insight In applying the Stefan–Boltzmann law (Equation 13.2), the temperature must be expressed in kelvins.

When an object has a higher temperature than its surroundings, the object emits a net radiant power $P_{net} = (Q/t)_{net}$. The net power is the power the object emits minus the power it absorbs. Applying the Stefan–Boltzmann law leads to the following expression for P_{net} when the temperature of the object is T and the temperature of the environment is T_0:

$$P_{net} = e\sigma A(T^4 - T_0^4) \qquad (13.3)$$

Check Your Understanding

(*The answers are given at the end of the book.*)

10. **BIO** One way that heat is transferred from place to place inside the human body is by the flow of blood. Which one of the following heat transfer processes—forced convection, conduction, or radiation—best describes this action of the blood?

11. Two strips of material, A and B, are identical, except that they have emissivities of 0.4 and 0.7, respectively. The strips are heated to the same temperature and have a red glow. A brighter glow signifies that more energy per second is being radiated. Which strip has the brighter glow?

12. One day during the winter the sun has been shining all day. Toward sunset a light snow begins to fall. It collects without melting on a cement playground, but melts instantly on contact with a black asphalt road adjacent to the playground. Why the difference? **(a)** Being black, asphalt has a higher emissivity than cement, so the asphalt absorbs more radiant energy from the sun during the day and, consequently, warms above the freezing point. **(b)** Being black, asphalt has a lower emissivity than cement, so it absorbs more radiant energy from the sun during the day and, consequently, warms above the freezing point.

13. **BIO** If you were stranded in the mountains in cold weather, it would help to minimize energy losses from your body if you curled up into the tightest ball possible. Which factor in the relation $Q = e\sigma T^4 At$ (Equation 13.2) are you using to the best advantage by curling into a ball? **(a)** e **(b)** σ **(c)** T **(d)** A **(e)** t

14. Two identical cubes have the same temperature. One of them, however, is cut in two and the pieces are separated (see **CYU Figure 13.2**). The radiant energy emitted by the cube cut into two pieces is $Q_{two\ pieces}$ and that emitted by the uncut cube is Q_{cube}. What is true about the radiant energy emitted in a given time?

(a) $Q_{two\ pieces} = 2Q_{cube}$ **(b)** $Q_{two\ pieces} = \frac{4}{3}Q_{cube}$

(c) $Q_{two\ pieces} = Q_{cube}$ **(d)** $Q_{two\ pieces} = \frac{1}{2}Q_{cube}$

(e) $Q_{two\ pieces} = \frac{1}{3}Q_{cube}$

Cube cut into   Uncut cube
two pieces

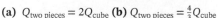
CYU FIGURE 13.2

15. Two objects have the same size and shape. Object A has an emissivity of 0.3, and object B has an emissivity of 0.6. Each radiates the same power. How is the Kelvin temperature T_A of A related to the Kelvin temperature T_B of B? **(a)** $T_A = T_B$ **(b)** $T_A = 2T_B$ **(c)** $T_A = \frac{1}{2}T_B$ **(d)** $T_A = \sqrt{2}T_B$ **(e)** $T_A = \sqrt[4]{2}T_B$

13.4 | Applications

THE PHYSICS OF . . . rating thermal insulation by R values. To keep heating and air conditioning bills to a minimum, it pays to use good thermal insulation in your home. Insulation inhibits convection between inner and outer walls and minimizes heat transfer by conduction. With respect to conduction, the logic behind home insulation ratings comes directly from Equation 13.1. According to this equation, the heat per unit time Q/t flowing through a thickness of material is $Q/t = kA\,\Delta T/L$. Keeping the value for Q/t to a minimum means using materials that have small thermal conductivities k and large thicknesses L. Construction engineers, however, prefer to use Equation 13.1 in the slightly different form shown below:

$$\frac{Q}{t} = \frac{A\,\Delta T}{L/k}$$

The term L/k in the denominator is called the *R value*. An *R* value expresses in a single number the combined effects of thermal conductivity and thickness. Larger *R* values reduce the heat per unit time flowing through the material and, therefore, mean better insulation. It is also convenient to use *R* values to describe layered slabs formed by sandwiching together a number of materials with different thermal conductivities and different thicknesses. The *R* values for the individual layers can be added to give a single *R* value for the entire slab. It should be noted, however, that *R* values are expressed using units of feet, hours, F°, and BTU for thickness, time, temperature, and heat, respectively.

THE PHYSICS OF . . . regulating the temperature of an orbiting satellite. When it is in the earth's shadow, an orbiting satellite is shielded from the intense electromagnetic waves emitted by the sun. But when it moves out of the earth's shadow, the satellite experiences the full effect of these waves. As a result, the temperature within a satellite would decrease and increase sharply during an orbital period and sensitive electronic circuitry would suffer, unless precautions are taken. To minimize temperature fluctuations, satellites are often covered with a highly reflecting and, hence, poorly absorbing metal foil, as **Figure 13.16** shows. By reflecting much of the sunlight, the foil minimizes temperature rises. Being a poor absorber, the foil is also a poor emitter and reduces radiant energy losses. Reducing these losses keeps the temperature from falling excessively.

THE PHYSICS OF . . . a thermos bottle. A thermos bottle, sometimes referred to as a Dewar flask, reduces the rate at which hot liquids cool down or cold liquids warm up. A thermos usually consists of a double-walled glass vessel with silvered inner walls (see **Interactive Figure 13.17**) and minimizes heat transfer via convection, conduction, and radiation. The space between the walls is evacuated to reduce energy losses due to conduction and convection. The silvered surfaces reflect most of the radiant energy that would otherwise enter or leave the liquid in the thermos. Finally, little heat is lost through the glass or the rubberlike gaskets and stopper, because these materials have relatively small thermal conductivities.

THE PHYSICS OF . . . a halogen cooktop stove. Halogen cooktops use radiant energy to heat pots and pans. A halogen cooktop uses several quartz–iodine lamps, like the ones in ultra-bright automobile headlights. These lamps are electrically powered and mounted below a ceramic top (see **Figure 13.18**). The electromagnetic energy they radiate

StockTrek Images/Getty Images

FIGURE 13.16 The highly reflective metal foil covering this satellite (the Hubble Space Telescope) minimizes temperature changes.

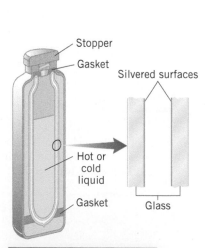

INTERACTIVE FIGURE 13.17 A thermos bottle minimizes energy transfer due to convection, conduction, and radiation.

Stopper
Gasket
Silvered surfaces
Hot or cold liquid
Gasket
Glass

FIGURE 13.18 In a halogen cooktop, quartz–iodine lamps emit a large amount of electromagnetic energy that is absorbed directly by a pot or pan.

Ceramic top
Quartz–iodine lamp

passes through the ceramic top and is absorbed directly by the bottom of the pot. Consequently, the pot heats up very quickly, rivaling the time of a pot on an open gas burner.

EXAMPLE 8 | BIO Staying Cool While Exercising

As warm-blooded animals, humans regulate their internal body temperature through the metabolism of food. These baseline metabolic processes occur continuously in the body, even when the body is at rest or sleeping. However, while exercising, the body's muscles convert even more of the chemical energy stored in food into mechanical work. Furthermore, the body's metabolic energy conversion is rather poor, where 60% of the available energy is converted to heat. As an example, let's say a jogger is doing work with a power output of 250 W. She is running on a day when the air temperature is 18 °C (291 K), and her skin temperature is 34 °C (307 K). How much water must she evaporate per hour by sweating if she is to maintain a skin temperature of 34 °C? Assume the surface area of her skin is 1.4 m^2 with an emissivity of 0.65, and she only loses heat by radiation.

Reasoning Heat is carried away from her skin by radiation, and we can calculate the rate of heat loss by using Equation 13.3. Any excess heat produced by physical activity must be removed by water evaporation. By knowing the quantity of excess heat, we can calculate the amount of water we need to evaporate by using Equation 12.5 and the latent heat of fusion for water.

Solution We begin by calculating the radiation heat loss from the skin due to the difference in the skin and air temperature. Applying Equation 13.3, we have:

$$P_{net} = e\sigma A(T^4 - T_0^4) = (0.65)[(5.67 \times 10^{-8} \text{ J})/(\text{s} \cdot \text{m}^2 \cdot \text{K}^4)](1.4 \text{ m}^2)$$
$$[(307 \text{ K})^4 - (291 \text{ K})^4] = 88 \text{ W}.$$

Any excess power output above 88 W will produce heat that has to be removed to maintain her current skin temperature. Her total power output that produces heat will be 60% of 250 W, or (0.60)(250 W) = 150 W. Therefore, the additional power output is 150 W – 88 W = 62 W, or 62 J/s. Thus, in one hour, or 3600 s, this is an extra $Q = (62 \text{ J/s})(3600 \text{ s}) = 2.2 \times 10^5$ J of heat that must be removed by evaporation. Finally, we apply Equation 12.5 to calculate the mass of water evaporated:

$$m = Q/L = 2.2 \times 10^5 \text{ J}/22.6 \times 10^5 \text{ J} = \boxed{0.097 \text{ kg}}$$

The process of radiation is important in regulating the temperature of our bodies, as the previous example discussed. The radiation emitted by human beings can also indicate potential health issues, as the following example shows.

EXAMPLE 9 | BIO The Physics of Thermography—Revisited

In Section 12.3, the medical diagnostic technique of thermography was introduced. The intensity of the radiant heat emitted by different parts of the body depends on temperature, in accordance with the Stefan–Boltzmann law. By measuring the radiated power from the body, color-contrast images like Figure 12.6 can be created. The sensitivity of thermography has increased significantly, to where it now can often diagnose medical conditions, such as breast cancer, before more traditional techniques, like mammography. In fact, the minimum resolvable temperature difference of modern thermal cameras is tens of millikelvins. Consider a tumor just below the surface of the skin that creates an increased skin temperature of 41.0 °C over an area of 6.25 cm^2. What is the difference in the net power radiated by this region and a region of the same area with a normal skin temperature of 36.5 °C? Assume that the emissivity of the skin from both areas is 0.650.

Reasoning The net power radiated from the skin can be calculated with Equation 13.3. This relationship can be used twice—once for each area of the skin with its respective temperature. The difference in these two values gives us the desired result.

Solution We apply Equation 13.3 to calculate the net power radiated from the hotter area of the skin P_{net-H} and the cooler area P_{net-C}. We calculate the difference in these two quantities, remembering to express the temperatures in kelvins and to convert the area to m^2:

$$P_{net-H} - P_{net-C} = e\sigma A(T_H^4 - T_0^4) - e\sigma A(T_C^4 - T_0^4) = e\sigma A(T_H^4 - T_C^4)$$
$$= (0.650)[5.67 \times 10^{-8} \text{ J}/(\text{s} \cdot \text{m}^2 \cdot \text{K}^4)](6.25 \times 10^{-4} \text{ m}^2)[(314 \text{ K})^4$$
$$- (309.5 \text{ K})^4] = \boxed{1.26 \times 10^{-2} \text{ W}}$$

Notice in the final expression that the temperature of the ambient air in the room (T_0) is eliminated, since we are calculating the *difference* in the radiated net power.

Concept Summary

13.1 Convection Convection is the process in which heat is carried from place to place by the bulk movement of a fluid. During natural convection, the warmer, less dense part of a fluid is pushed upward by the buoyant force provided by the surrounding cooler and denser part.

Forced convection occurs when an external device, such as a fan or a pump, causes the fluid to move.

13.2 Conduction Conduction is the process whereby heat is transferred directly through a material, with any bulk motion of the material playing no role in the transfer. Materials that conduct heat well, such as most metals, are known as thermal conductors. Materials that conduct heat poorly, such as wood, glass, and most plastics, are referred to as thermal insulators. The heat Q conducted during a time t through a bar of length L and cross-sectional area A is given by Equation 13.1, where ΔT is the temperature difference between the ends of the bar and k is the thermal conductivity of the material.

$$Q = \frac{(kA\,\Delta T)t}{L} \qquad (13.1)$$

13.3 Radiation Radiation is the process in which energy is transferred by means of electromagnetic waves. All objects, regardless of their temperature, simultaneously absorb and emit electromagnetic waves. Objects that are good absorbers of radiant energy are also good emitters, and objects that are poor absorbers are also poor emitters. An object that absorbs all the radiation incident upon it is called a perfect blackbody. A perfect blackbody, being a perfect absorber, is also a perfect emitter.

The radiant energy Q emitted during a time t by an object whose surface area is A and whose Kelvin temperature is T is given by the Stefan–Boltzmann law of radiation (see Equation 13.2), where $\sigma = 5.67 \times 10^{-8}$ J/(s · m² · K⁴) is the Stefan–Boltzmann constant and e is the emissivity, a dimensionless number characterizing the surface of the object. The emissivity lies between 0 and 1, being zero for a nonemitting surface and one for a perfect blackbody.

$$Q = e\sigma T^4 A t \qquad (13.2)$$

The net radiant power is the power an object emits minus the power it absorbs. The net radiant power P_{net} emitted by an object with a temperature T located in an environment with a temperature T_0 is given by Equation 13.3.

$$P_{net} = e\sigma A(T^4 - T_0^4) \qquad (13.3)$$

Focus on Concepts

Online

Additional questions are available for assignment in WileyPLUS.

Section 13.2 Conduction

1. The heat conducted through a bar depends on which of the following?

A. The coefficient of linear expansion

B. The thermal conductivity

C. The specific heat capacity

D. The length of the bar

E. The cross-sectional area of the bar

(a) A, B, and D (b) A, C, and D (c) B, C, D, and E (d) B, D, and E (e) C, D, and E

2. Two bars are conducting heat from a region of higher temperature to a region of lower temperature. The bars have identical lengths and cross-sectional areas, but are made from different materials. In the drawing they are placed "in parallel" between the two temperature regions in arrangement A, whereas they are placed end to end in arrangement B. In which arrangement is the heat that is conducted the greatest? (a) The heat conducted is the same in both arrangements. (b) Arrangement A (c) Arrangement B (d) It is not possible to determine which arrangement conducts more heat.

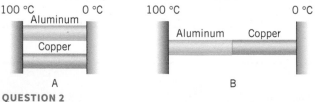

QUESTION 2

3. The drawing shows a composite slab consisting of three materials through which heat is conducted from left to right. The materials have identical thicknesses and cross-sectional areas. Rank the materials according to their thermal conductivities, largest first. (a) k_1, k_2, k_3 (b) k_1, k_3, k_2 (c) k_2, k_1, k_3 (d) k_2, k_3, k_1 (e) k_3, k_2, k_1

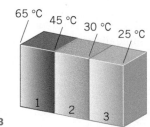

QUESTION 3

4. The long single bar on the left in the drawing has a thermal conductivity of 240 J/(s · m · C°). The ends of the bar are at temperatures of 400 and 200 °C, and the temperature of its midpoint is halfway between these two temperatures, or 300 °C. The two bars on the right are half as long as the bar on the left, and the thermal conductivities of these bars are different (see the drawing). All of the bars have the same cross-sectional area. What can be said about the temperature at the point where the two bars on the right are joined together? (a) The temperature at the point where the two bars are joined together is 300 °C. (b) The temperature at the point where the two bars are joined together is greater than 300 °C. (c) The temperature at the point where the two bars are joined together is less than 300 °C.

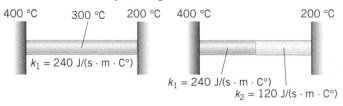

QUESTION 4

Section 13.3 Radiation

5. Three cubes are made from the same material. As the drawing indicates, they have different sizes and temperatures. Rank the cubes according to the radiant energy they emit per second, largest first. (a) A, B, C (b) A, C, B (c) B, A, C (d) B, C, A (e) C, B, A

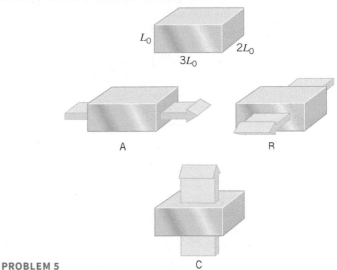

QUESTION 5

6. An astronaut on a space station has two objects that are identical in all respects, except that one is painted black and the other is painted silver. Initially, they are at the same temperature. When taken from inside the space station and placed in outer space, which object, if either, cools down at a faster rate? (a) The object painted black (b) The object painted silver (c) Both objects cool down at the same rate. (d) It is not possible to determine which object cools down at the faster rate.

7. The emissivity e of object B is $\frac{1}{16}$ that of object A, although both objects are identical in size and shape. If the objects radiate the same energy per second, what is the ratio T_B/T_A of their Kelvin temperatures? (a) $\frac{1}{16}$ (b) $\frac{1}{4}$ (c) $\frac{1}{2}$ (d) 2 (e) 4

Problems

Online

Additional questions are available for assignment in WileyPLUS.

Note: *For problems in this set, use the values for thermal conductivities given in* **Table 13.1** *unless stated otherwise.*

- **SSM** Student Solutions Manual
- **MMH** Problem-solving help
- **GO** Guided Online Tutorial
- **V-HINT** Video Hints
- **CHALK** Chalkboard Videos

- **BIO** Biomedical application
- **E** Easy
- **M** Medium
- **H** Hard
- **WS** Worksheet
- **T** Team Problem

Section 13.2 Conduction

1. **E** In an electrically heated home, the temperature of the ground in contact with a concrete basement wall is 12.8 °C. The temperature at the inside surface of the wall is 20.0 °C. The wall is 0.10 m thick and has an area of 9.0 m². Assume that one kilowatt · hour of electrical energy costs $0.10. How many hours are required for one dollar's worth of energy to be conducted through the wall?

2. **E** **SSM** **MMH** A person's body is covered with 1.6 m² of wool clothing. The thickness of the wool is 2.0×10^{-3} m. The temperature at the outside surface of the wool is 11 °C, and the skin temperature is 36 °C. How much heat per second does the person lose due to conduction?

3. **E** **GO** Two objects are maintained at constant temperatures, one hot and one cold. Two identical bars can be attached end to end, as in part *a* of the drawing, or one on top of the other, as in part *b*. When either of these arrangements is placed between the hot and the cold objects for the same amount of time, heat Q flows from left to right. Find the ratio Q_a/Q_b.

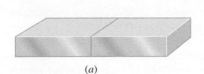

(a)

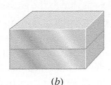

(b)

PROBLEM 3

4. **E** **SSM** One end of an iron poker is placed in a fire where the temperature is 502 °C, and the other end is kept at a temperature of 26 °C. The poker is 1.2 m long and has a radius of 5.0×10^{-3} m. Ignoring the heat lost along the length of the poker, find the amount of heat conducted from one end of the poker to the other in 5.0 s.

5. **E** **GO** The block in the drawing has dimensions $L_0 \times 2L_0 \times 3L_0$, where $L_0 = 0.30$ m. The block has a thermal conductivity of 250 J/(s · m · C°). In drawings A, B, and C, heat is conducted through the block in three different directions; in each case the temperature of the warmer surface is 35 °C and that of the cooler surface is 19 °C. Determine the heat that flows in 5.0 s for each case.

PROBLEM 5

6. **E** **GO** A closed box is filled with dry ice at a temperature of −78.5 °C, while the outside temperature is 21.0 °C. The box is cubical, measuring 0.350 m on a side, and the thickness of the walls is 3.00×10^{-2} m. In one day, 3.10×10^6 J of heat is conducted through the six walls. Find the thermal conductivity of the material from which the box is made.

7. **E** **MMH** One end of a brass bar is maintained at 306 °C, while the other end is kept at a constant, but lower, temperature. The cross-sectional area of the bar is 2.6×10^{-4} m². Because of insulation, there is negligible heat loss through the sides of the bar. Heat flows through the bar, however, at the rate of 3.6 J/s. What is the temperature of the bar at a point 0.15 m from the hot end?

8. **E** **GO** A wall in a house contains a single window. The window consists of a single pane of glass whose area is 0.16 m² and whose thickness is 2.0 mm. Treat the wall as a slab of the insulating material Styrofoam whose area and thickness are 18 m² and 0.10 m, respectively. Heat is lost via conduction through the wall and the window. The temperature difference between the inside and outside is the same for the wall and the window. Of the total heat lost by the wall and the window, what is the percentage lost by the window?

9. **E** **V-HINT** **MMH** A composite rod is made from stainless steel and iron and has a length of 0.50 m. The cross section of this composite rod is shown in the drawing and consists of a square within a circle. The square cross section of the steel is 1.0 cm on a side. The temperature at one end of the rod is 78 °C, while it is 18 °C at the other end. Assuming that no heat exits through the cylindrical outer surface, find the total amount of heat conducted through the rod in two minutes.

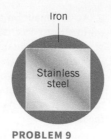
Iron

Stainless steel

PROBLEM 9

10. **M** **CHALK** Review Conceptual Example 5 before attempting this problem. To illustrate the effect of ice on the aluminum cooling plate, consider the drawing shown here and the data that it contains. Ignore any limitations due to significant figures. **(a)** Calculate the heat per second per square meter that is conducted through the ice–aluminum combination.

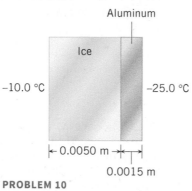
Aluminum

Ice

−10.0 °C   −25.0 °C

$\leftarrow$ 0.0050 m $\rightarrow$

0.0015 m

PROBLEM 10

(b) Calculate the heat per second per square meter that would be conducted through the aluminum if the ice were not present. Notice how much larger the answer is in (b) as compared to (a).

11. **M** A cubical piece of heat-shield tile from the space shuttle measures 0.10 m on a side and has a thermal conductivity of 0.065 J/(s · m · C°). The outer surface of the tile is heated to a temperature of 1150 °C, while the inner surface is maintained at a temperature of 20.0 °C. **(a)** How much heat flows from the outer to the inner surface of the tile in five minutes? **(b)** If this amount of heat were transferred to two liters (2.0 kg) of liquid water, by how many Celsius degrees would the temperature of the water rise?

12. **M** **GO** Two pots are identical except that the flat bottom of one is aluminum, whereas that of the other is copper. Water in these pots is boiling away at 100.0 °C at the same rate. The temperature of the heating element on which the aluminum bottom is sitting is 155.0 °C. Assume that heat enters the water only through the bottoms of the pots and find the temperature of the heating element on which the copper bottom rests.

13. **M** **GO** A pot of water is boiling under one atmosphere of pressure. Assume that heat enters the pot only through its bottom, which is copper and rests on a heating element. In two minutes, the mass of water boiled away is $m = 0.45$ kg. The radius of the pot bottom is $R = 6.5$ cm, and the thickness is $L = 2.0$ mm. What is the temperature T_E of the heating element in contact with the pot?

14. **M** **V-HINT** In a house the temperature at the surface of a window is 25 °C. The temperature outside at the window surface is 5.0 °C. Heat is lost through the window via conduction, and the heat lost per second has a certain value. The temperature outside begins to fall, while the conditions inside the house remain the same. As a result, the heat lost per second increases. What is the temperature at the outside window surface when the heat lost per second doubles?

15. **H** The drawing shows a solid cylindrical rod made from a center cylinder of lead and an outer concentric jacket of copper. Except for its ends, the rod is insulated (not shown), so

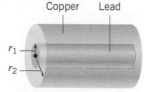
Copper   Lead

r_1
r_2

PROBLEM 15

that the loss of heat from the curved surface is negligible. When a temperature difference is maintained between its ends, this rod conducts one-half the amount of heat that it would conduct if it were solid copper. Determine the ratio of the radii r_1/r_2.

Section 13.3 Radiation

16. **E** **GO** Light bulb 1 operates with a filament temperature of 2700 K, whereas light bulb 2 has a filament temperature of 2100 K. Both filaments have the same emissivity, and both bulbs radiate the same power. Find the ratio A_1/A_2 of the filament areas of the bulbs.

17. **E** **SSM** The amount of radiant power produced by the sun is approximately 3.9×10^{26} W. Assuming the sun to be a perfect blackbody sphere with a radius of 6.96×10^8 m, find its surface temperature (in kelvins).

18. **E** In an old house, the heating system uses radiators, which are hollow metal devices through which hot water or steam circulates. In one room the radiator has a dark color (emissivity = 0.75). It has a temperature of 62 °C. The new owner of the house paints the radiator a lighter color (emissivity = 0.50). Assuming that it emits the same radiant power as it did before being painted, what is the temperature (in degrees Celsius) of the newly painted radiator?

19. **E** **SSM** A person is standing outdoors in the shade where the temperature is 28 °C. **(a)** What is the radiant energy absorbed per second by his head when it is covered with hair? The surface area of the hair (assumed to be flat) is 160 cm² and its emissivity is 0.85. **(b)** What would be the radiant energy absorbed per second by the same person if he were bald and the emissivity of his head were 0.65?

20. **E** The Kelvin temperature of an object is T_1, and the object radiates a certain amount of energy per second. The Kelvin temperature of the object is then increased to T_2, and the object radiates twice the energy per second that it radiated at the lower temperature. What is the ratio T_2/T_1?

21. **E** A baking dish is removed from a hot oven and placed on a cooling rack. As the dish cools down to 35 °C from 175 °C, its net radiant power decreases to 12.0 W. What was the net radiant power of the baking dish when it was first removed from the oven? Assume that the temperature in the kitchen remains at 22 °C as the dish cools down.

22. **E** **GO** Sirius B is a white star that has a surface temperature (in kelvins) that is four times that of our sun. Sirius B radiates only 0.040 times the power radiated by the sun. Our sun has a radius of 6.96×10^8 m. Assuming that Sirius B has the same emissivity as the sun, find the radius of Sirius B.

23. **E** **GO** A solar collector is placed in direct sunlight where it absorbs energy at the rate of 880 J/s for each square meter of its surface. The emissivity of the solar collector is $e = 0.75$. What equilibrium temperature does the collector reach? Assume that the only energy loss is due to the emission of radiation.

24. **M** **GO** Liquid helium is stored at its boiling-point temperature of 4.2 K in a spherical container ($r = 0.30$ m). The container is a perfect blackbody radiator. The container is surrounded by a spherical shield whose temperature is 77 K. A vacuum exists in the space between the container and the shield. The latent heat of vaporization for helium is 2.1×10^4 J/kg. What mass of liquid helium boils away through a venting valve in one hour?

25. **M** **CHALK** A solid sphere has a temperature of 773 K. The sphere is melted down and recast into a cube that has the same emissivity and emits the same radiant power as the sphere. What is the cube's temperature?

Additional Problems

26. **E** **SSM** Due to a temperature difference ΔT, heat is conducted through an aluminum plate that is 0.035 m thick. The plate is then replaced by a stainless steel plate that has the same temperature difference and cross-sectional area. How thick should the steel plate be so that the same amount of heat per second is conducted through it?

27. **E** **GO** A copper pipe with an outer radius of 0.013 m runs from an outdoor wall faucet into the interior of a house. The temperature of the faucet is 4.0 °C, and the temperature of the pipe, at 3.0 m from the faucet, is 25 °C. In fifteen minutes, the pipe conducts a total of 270 J of heat to the outdoor faucet from the house interior. Find the inner radius of the pipe. Ignore any water inside the pipe.

28. **E** How many days does it take for a perfect blackbody cube (0.0100 m on a side, 30.0 °C) to radiate the same amount of energy that a one-hundred-watt light bulb uses in one hour?

29. **E** **GO** An object is inside a room that has a constant temperature of 293 K. Via radiation, the object emits three times as much power as it absorbs from the room. What is the temperature (in kelvins) of the object? Assume that the temperature of the object remains constant.

30. **E** The concrete wall of a building is 0.10 m thick. The temperature inside the building is 20.0 °C, while the temperature outside is 0.0 °C. Heat is conducted through the wall. When the building is unheated, the inside temperature falls to 0.0 °C, and heat conduction ceases. However, the wall does emit radiant energy when its temperature is 0.0 °C. The radiant energy emitted per second per square meter at 0.0 °C is the same as the heat lost per second per square meter due to conduction when the temperature inside the building is 20.0 °C. What is the emissivity of the wall?

31. **M** **GO** Part *a* of the drawing shows a rectangular bar whose dimensions are $L_0 \times 2L_0 \times 3L_0$. The bar is at the same constant

temperature as the room (not shown) in which it is located. The bar is then cut, lengthwise, into two identical pieces, as shown in part *b* of the drawing. The temperature of each piece is the same as that of the original bar. (a) What is the ratio of the power absorbed by the two bars in part *b* of the drawing to the single bar in part *a*? (b) Suppose that the temperature of the single bar in part *a* is 450.0 K. What would the temperature (in kelvins) of the room and the two bars in part *b* have to be so that the two bars absorb the same power as the single bar in part *a*?

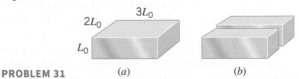

PROBLEM 31   (*a*)   (*b*)

32. **M** **V-HINT** Multiple-Concept Example 3 discusses an approach to problems such as this. The ends of a thin bar are maintained at different temperatures. The temperature of the cooler end is 11 °C, while the temperature at a point 0.13 m from the cooler end is 23 °C and the temperature of the warmer end is 48 °C. Assuming that heat flows only along the length of the bar (the sides are insulated), find the length of the bar.

33. **M** **CHALK** **GO** A copper rod has a length of 1.5 m and a cross-sectional area of 4.0×10^{-4} m². One end of the rod is in contact with boiling water and the other with a mixture of ice and water. What is the mass of ice per second that melts? Assume that no heat is lost through the side surface of the rod.

34. **M** **GO** **SSM** A wood-burning stove (emissivity = 0.900 and surface area = 3.50 m²) is being used to heat a room. The fire keeps the stove surface at a constant 198 °C (471 K) and the room at a constant 29 °C (302 K). Determine the *net* radiant power generated by the stove.

Physics in Biology, Medicine, and Sports

35. **E** **BIO** **13.2** The amount of heat per second conducted from the blood capillaries beneath the skin to the surface is 240 J/s. The energy is transferred a distance of 2.0×10^{-3} m through a body whose surface area is 1.6 m². Assuming that the thermal conductivity is that of body fat, determine the temperature difference between the capillaries and the surface of the skin.

36. **E** **BIO** **SSM** **13.2** In the conduction equation $Q = (kA\,\Delta T)t/L$, the combination of terms kA/L is called the *conductance*. The human body has the ability to vary the conductance of the tissue beneath the skin by means of vasoconstriction and vasodilation, in which the flow of blood to the veins and capillaries is decreased and increased, respectively. The conductance can be adjusted over a range such that the tissue beneath the skin is equivalent to a thickness of 0.080 mm of Styrofoam or 3.5 mm of air. By what factor (high/low) can the body adjust the conductance?

37. **E** **BIO** **SSM** **13.3** A person eats a dessert that contains 260 Calories. (This "Calorie" unit, with a capital C, is the one used by nutritionists; 1 Calorie = 4186 J. See Section 12.7.) The skin temperature of this individual is 36 °C and that of her environment is 21 °C. The emissivity of her skin is 0.75 and its surface area is 1.3 m². How much

time would it take for her to emit a *net* radiant energy from her body that is equal to the energy contained in this dessert?

38. **E** **BIO** **13.3** A person's body is producing energy internally due to metabolic processes. If the body loses more energy than metabolic processes are generating, its temperature will drop. If the drop is severe, it can be life-threatening. Suppose that a person is unclothed and energy is being lost via radiation from a body surface area of 1.40 m², which has a temperature of 34 °C and an emissivity of 0.700. Also suppose that metabolic processes are producing energy at a rate of 115 J/s. What is the temperature of the coldest room in which this person could stand and not experience a drop in body temperature?

39. **E** **BIO** **MMH** **13.3** Suppose the skin temperature of a naked person is 34 °C when the person is standing inside a room whose temperature is 25 °C. The skin area of the individual is 1.5 m². (a) Assuming the emissivity is 0.80, find the net loss of radiant power from the body. (b) Determine the number of food Calories of energy (1 food Calorie = 4186 J) that are lost in one hour due to the net loss rate obtained in part (a). Metabolic conversion of food into energy replaces this loss.

40. **BIO** **M** **13.3** A person infected with the SARS-CoV-2 virus is running a fever with an internal body temperature of 103 °F (39.4 °C).

This raises the person's skin temperature to 36.5 °C. The person is standing unclothed in a room with an ambient temperature of 21.0 °C. If the temperature of the person's skin is maintained at 36.5 °C solely by the energy obtained from the metabolic conversion of food, how many Calories would this person have to consume per hour? Assume heat is lost only by radiation, and the surface area and emissivity of the person's skin are 1.50 m^2 and 0.800, respectively.

41. BIO M 13.3 Hypothermia is a dangerous condition in which the core temperature of the body falls to an unsafe level. In humans, this occurs around 35 °C. Typically, the cause of hypothermia is exposure to cold. Extremely cold conditions, like being outside without protection in below-freezing temperatures or falling into cold water, are certainly dangerous. However, even less extreme conditions can be serious. In fact, any situation that leads to the body's heat loss can result in hypothermia. Furthermore, the outside temperature may not be the most significant factor that contributes to hypothermia. Consider a very cold day, where the ambient air temperature is 4.20 °C, and an unclothed person is outside with a skin temperature of 39.5 °C. Find the ratio of the rate of heat loss of the person completely submerged in water to the rate of heat loss of the same person standing in the open air. Assume the water has the same temperature as the ambient air, there are no wind or water currents, and the heat is lost through the same thickness of water and air. The thermal conductivities of air and water at 4.20 °C are 2.42×10^{-2} W/(m · C°) and 5.74×10^{-1} W/(m · C°), respectively.

42. BIO M 13.3 Shape memory polymers are a class of materials that can remember their original shape and transition between shapes when stretched, twisted, or compressed. They have recently found uses in orthopedics and as other surgical and dental implants. One of the most promising new materials is PEEK (polyether ether ketone). It is a very good electrical and thermal insulator. As a dental implant material, it provides the patient a comfortable implant that does not conduct heat or alter taste, which can be a constant bother with traditional materials (see the photos). Imagine a PEEK implant that is 1.50 cm in length with a circular diameter of 0.25 cm. An ice cube is in contact with the end of the implant. How much heat is transferred along the length of the implant in 35 seconds, if the change in

temperature across the implant is 27 °C? The thermal conductivity of PEEK is 0.25 W/(m · C°).

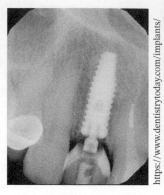

Extraction of a tooth followed by the insertion of a PEEK implant. The image of the X-ray shows the implant inserted into a tapered implant body that is secured in the jaw bone.

PROBLEM 42

43. BIO M 13.3 What is the rate of heat transfer in J/s through the body's skin and the fat layer just beneath its surface? Treat the skin and fat as a single layer that is 1.20 cm thick. Assume the temperature of the inner surface of the layer corresponds to internal body temperature, or 36.7 °C, and the temperature of the outer surface of the layer corresponds to skin temperature, or 34.0 °C. Take the total surface area of the layer to be 1.50 m^2 and use a value of 0.243 W/(m · C°) for the average thermal conductivity of the layer.

44. BIO M 13.3 Many surgeries require the use of tools that cut through or remove parts of skeletal bones. These procedures can locally generate a lot of heat. For example, thermal bone necrosis will begin if the bone is kept at 47.0 °C for at least one minute. One of the most important factors that determines how hot the bone gets is its thermal conductivity. Imagine a bone saw cutting through a 2.50-cm length of cylindrical bone with a radius of 1.20 cm. How much heat is conducted along the bone in 1.00 minute of cutting, if the temperature in the bone at the location of the cut is 47.0 °C, and the temperature at the other end of the length of bone is 37 °C? The thermal conductivity of the bone is 0.625 W/(m · C°). Assume all of the heat created by the saw is transferred to the bone by conduction.

Concepts and Calculations Problems   Online

Heat conduction is governed by Equation 13.1, as we have seen. Problem 45 illustrates a familiar application of this relation in the kitchen. It also gives us the opportunity to review the idea of latent heat of vaporization. Problem 46 deals with a case in which heat loss by radiation leads to freezing of water. It stresses the importance of the area from which the radiation occurs, and reviews the concept of the latent heat of fusion.

45. M CHALK Two pots are identical, except that in one case the flat bottom is aluminum and in the other it is copper. Each pot contains the same amount of boiling water and sits on a heating element that has a temperature of 155 °C. In the aluminum pot, the water boils away completely in 360 s. *Concepts:* (i) Is the heat needed to boil away the water in the aluminum-bottom pot less than, greater than, or the same as the heat needed in the copper-bottom pot? (ii) One of the factors in Equation 13.1 that influences the amount of heat conducted through the bottom of each pot is the temperature difference ΔT between the upper and lower surfaces of the pot's bottom. Is this temperature difference for the aluminum-bottom pot less than, greater than, or the same as that for the copper-bottom pot? (iii) Is

the time required to boil away the water completely in the copper-bottom pot less than, greater than, or the same as that required for the aluminum-bottom pot? *Calculations:* Find the time it takes the water in the copper-bottom pot to boil away completely.

46. M CHALK SSM One half of a kilogram of liquid water at 273 K (0 °C) is placed outside on a day when the temperature is 261 K (−12 °C). Assume that the heat is lost from the water only by means of radiation and that the emissivity of the radiating surface is 0.60. Consider two cases: when the surface area of the water is **(a)** 0.035 m^2 (as it might be in a cup) and **(b)** 1.5 m^2 (as it could be if the water were spilled out to form a thin sheet). *Concepts:* (i) In case (a) is the heat that must be removed to freeze the water less than, greater than, or the same as in case (b)? (ii) The loss of heat by radiation depends on the temperature of the radiating object. Does the temperature of the water change as it freezes? (iii) The water both loses and gains heat by radiation. How, then, can heat transfer by radiation lead to freezing of the water? (iv) Will it take longer for the water to freeze in case (a) when the area is smaller or in case (b) when the area is larger? *Calculations:* For each case, (a) and (b), determine the time it takes the water to freeze into ice at 0 °C.

Team Problems

47. **E** **T** **WS** **Flying Close to the Sun.** A space probe you designed is approaching the sun, and you and your team need to determine whether it needs to deviate from its current course. The probe has a large square detector 10.0 meters on a side (area = $1.00 \times 10^2 \text{ m}^2$) that consists of a thin sheet composed of strong carbon fibers. The carbon sheet can withstand high temperatures with a melting point of 3600 °C, but aluminum brackets that support the carbon sheet have a relatively lower melting temperature of 660.3 °C. Based on the size and temperature of the star, at the closest distance of approach the sheet will absorb energy at a rate of 48 250 J/s for each square meter of its surface. Assuming that the aluminum brackets are in thermal contact with the sheet and will reach the same temperature, and that the emissivity of the carbon-fiber sheet is $e = 0.880$, what equilibrium temperature will the sheet and aluminum brackets reach? Should you deviate from the current course? You can assume that the only energy loss is due to the emission of radiation, and you can neglect any effects of the small aluminum brackets.

48. **M** **T** **WS** **The Cold Night Sky.** You and your team are on an expedition on a high-altitude mesa and need ice to cool a small vial that contains a perishable biological sample that you have collected. Even though you only need a few grams of ice in order to keep the small, highly insulated sample container cold enough for the trip home the next day, you do not have a freezer with which to make ice. In addition, even though you are at high altitude, the nighttime temperature will reach a low of 4.00 °C, well above the freezing point of water (0 °C). The weather report not only calls for that low temperature (4.00 °C) to be sustained for 6.0 hours during the night, but also that there will be a perfectly clear sky with no moon visible. You then recall something you once read about how ancient Persians made ice in the desert by pouring shallow puddles of water on flat stones in the evening and waking up to find them frozen, even though the ambient temperature never dropped below freezing. You know that at your elevation, and with perfectly clear conditions, the night sky acts like a blackbody with a temperature of −31.0 °C. You take a thick slab of Styrofoam and carve a square cavity into its top surface that is 10.0 cm on a side and 2.0 cm deep. You set the Styrofoam, which acts as an excellent thermal insulator, on flat ground. Next, you pour in 25.0 mL (25.0 g) of water, which covers the bottom of the cavity. **(a)** Starting at a temperature of 4.00 °C, how much heat must be removed from the water to make it freeze? **(b)** If the emissivity of water is $e = 0.950$, what is the rate at which heat is being removed from the water through radiation? **(c)** How long (in hours) will it take the water to freeze with the starting temperature equal to 4.00 °C?

Will the water freeze before the ambient temperature starts to increase, that is, in less than 6.0 hours? (*Hint: You will need to apply Equation 13.3. As an approximation, you can take the starting temperature of the water to be T = 0 °C, since this is the temperature of the water at which most of the heat (i.e., the latent heat of the phase change) is removed.*)

49. **M** **T** **The Diamond Ring Solution.** The processing chip on the computer that controls the navigation equipment on your spacecraft is overheating. Unless you fix the problem, the chip will be damaged and the navigation system will shut down. You open the panel and find that the small copper disk that was supposed to bridge the gap between the smooth top of the chip and the cooling plate is missing, leaving a 2.0 mm gap between them. In this configuration, the heat cannot escape the chip at the required rate. You notice by the thin smudge of thermal grease (a highly thermally conductive material used to promote good thermal contact between surfaces) that the missing copper disk was 2.0 mm thick and had a diameter of 1.0 cm. You know that the chip is designed to run below 75 °C, and the copper cooling plate is held at a constant 5.0 °C. **(a)** What was the rate of heat flow from the chip to the copper plate when the original copper disk was in place and the chip was at its maximum operating temperature? **(b)** The only material that you have available on board to bridge the gap between the chip and copper plate is lead. If the cross-sectional area of the lead piece you plan to wedge into the gap is 1.0 cm², what is the rate of heat flow from the chip to the copper plate? Does it match the value calculated in part (a)? **(c)** While brainstorming for other possible solutions to your problem, you happen to glance down at the engagement ring on your finger: a large, glittering diamond. The top and bottom surfaces are flat and nearly rectangular ($L = 0.75$ cm and $W = 0.50$ cm), and the thickness looks to be about 2.0 mm, just right to bridge the gap. You pry the diamond out of its holder and press it into the gap. What is the rate of heat flow now? Good enough?

50. **M** **T** **Indirect Cooling With Liquid Nitrogen.** You are designing a system to cool an insulated silver plate of dimensions 2.00 cm × 2.00 cm × 0.500 cm. One end of a thermally insulated copper wire (diameter $D = 2.50$ mm and length $L = 15.0$ cm) is dipped into a vat of liquid nitrogen ($T = 77.2$ K), and the other end is attached to the bottom of the silver plate. **(a)** If the silver plate starts at room temperature (72.0 °F), what is the initial rate of heat flow between the plate and the liquid nitrogen reservoir? **(b)** Assuming the rate of heat flow calculated in part (a), estimate the temperature of the silver plate after 30.0 seconds.

CHAPTER 14

A scuba diver carries his air supply in the tank on his back. To the extent that the air in a scuba tank behaves like an ideal gas, its pressure, volume, and temperature are related by the ideal gas law. We will see that it is possible to use this law to estimate how long a diver, using a tank of a given size, can stay under the water at a given depth.

Greg Amptman/Shutterstock.com

LEARNING OBJECTIVES

After reading this module, you should be able to...

14.1 Express the amount of a substance in moles.

14.2 Apply the ideal gas law.

14.3 Apply the kinetic theory of gases to ideal gases.

14.4 Solve diffusion problems.

The Ideal Gas Law and Kinetic Theory

14.1 Molecular Mass, the Mole, and Avogadro's Number

Often, we wish to compare the mass of one atom with another. To facilitate the comparison, a mass scale known as the **atomic mass scale** has been established. To set up this scale, a reference value (along with a unit) is chosen for one of the elements. The unit is called the **atomic mass unit** (symbol: u). By international agreement, the reference element is chosen to be the most abundant type or isotope* of carbon, which is called carbon-12. Its atomic mass† is defined to be exactly twelve atomic mass units, or 12 u. The relationship between the atomic mass unit and the kilogram is

$$1 \text{ u} = 1.6605 \times 10^{-27} \text{ kg}$$

*Isotopes are discussed in Section 31.1.
†In chemistry the expression "atomic weight" is frequently used in place of "atomic mass."

The atomic masses of all the elements are listed in the periodic table, part of which is shown in **Interactive Figure 14.1**. The complete periodic table is given on the inside of the back cover of the textbook. In general, the masses listed are average values and take into account the various isotopes of an element that exist naturally. For brevity, the unit "u" is often omitted from the table. For example, a magnesium atom (Mg) has an average atomic mass of 24.305 u, whereas a lithium atom (Li) has an average atomic mass of 6.941 u. Thus, atomic magnesium is more massive than atomic lithium by a factor of (24.305 u)/(6.941 u) = 3.502. In the periodic table, the atomic mass of carbon (C) is given as 12.011 u, rather than exactly 12 u. This is because a small amount (about 1%) of the naturally occurring material is an isotope called carbon-13. The value of 12.011 u is an average that reflects the small contribution of carbon-13.

The molecular mass of a molecule is the sum of the atomic masses of its atoms. For instance, the elements hydrogen and oxygen have atomic masses of 1.007 94 u and 15.9994 u, respectively, so the molecular mass of a water molecule (H_2O) is, therefore, 2(1.007 94 u) + 15.9994 u = 18.0153 u.

Macroscopic amounts of materials contain large numbers of atoms or molecules. Even in a small volume of gas, 1 cm³, for example, the number is enormous. It is convenient to express such large numbers in terms of a single unit, the **gram-mole**, or simply the **mole** (symbol: mol). *One gram-mole of a substance contains as many particles (atoms or molecules) as there are atoms in 12 grams of the isotope carbon-12.* Experiment shows that 12 grams of carbon-12 contain 6.022×10^{23} atoms. The number of atoms per mole is known as **Avogadro's number N_A**, after the Italian scientist Amedeo Avogadro (1776–1856):

$$N_A = 6.022 \times 10^{23}\,\text{mol}^{-1}$$

Thus, the number of moles n contained in any sample is the number of particles N in the sample divided by the number of particles per mole N_A (Avogadro's number):

$$n = \frac{N}{N_A}$$

Although defined in terms of carbon atoms, the concept of a mole can be applied to any collection of objects by noting that one mole contains Avogadro's number of objects. Thus, one mole of atomic sulfur contains 6.022×10^{23} sulfur atoms, one mole of water contains 6.022×10^{23} H_2O molecules, and one mole of golf balls contains 6.022×10^{23} golf balls. The mole is the SI base unit for expressing "the amount of a substance."

The number n of moles contained in a sample can also be found from its mass. To see how, multiply and divide the right-hand side of the previous equation by the mass $m_{particle}$ of a single particle, expressed in grams:

$$n = \frac{m_{particle}N}{m_{particle}N_A} = \frac{m}{\text{Mass per mole}}$$

The numerator $m_{particle}N$ is the mass of a particle times the number of particles in the sample, which is the mass m of the sample expressed in grams. The denominator $m_{particle}N_A$ is the mass of a particle times the number of particles per mole, which is the mass per mole, expressed in grams per mole.

Problem-Solving Insight The mass per mole (in g/mol) of any substance has the same numerical value as the atomic or molecular mass of the substance (in atomic mass units).

To understand this fact, consider the carbon-12 and sodium atoms as examples. The mass per mole of carbon-12 is 12 g/mol, since, by definition, 12 grams of carbon-12 contain one mole of atoms. On the other hand, the mass per mole of sodium (Na) is 22.9898 g/mol for the following reason: as indicated in **Interactive Figure 14.1**, a sodium atom is more massive than a carbon-12 atom by the ratio of their atomic masses, (22.9898 u)/(12 u) = 1.915 82. Therefore, the mass per mole of sodium is 1.915 82 times as great as that of carbon-12, which means equivalently that (1.915 82)(12 g/mol) = 22.9898 g/mol. Thus, the numerical value of the mass per mole of sodium (22.9898) is the same as the numerical value of its atomic mass.

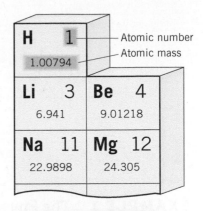

INTERACTIVE FIGURE 14.1   A portion of the periodic table showing the atomic number and atomic mass of each element. In the periodic table it is customary to omit the symbol "u" denoting the atomic mass unit.

Since one gram-mole of a substance contains Avogadro's number of particles (atoms or molecules), the mass $m_{particle}$ of a particle (in grams) can be obtained by dividing the mass per mole (in g/mol) by Avogadro's number:

$$m_{particle} = \frac{\text{Mass per mole}}{N_A}$$

Example 1 illustrates how to use the concepts of the mole, atomic mass, and Avogadro's number to determine the number of atoms and molecules present in two famous gemstones.

EXAMPLE 1 | The Physics of Gemstones

Figure 14.2a shows the Hope diamond (44.5 carats), which is almost pure carbon. **Figure 14.2b** shows the Rosser Reeves ruby (138 carats), which is primarily aluminum oxide (Al_2O_3). One carat is equivalent to a mass of 0.200 g. Determine **(a)** the number of carbon atoms in the diamond and **(b)** the number of Al_2O_3 molecules in the ruby.

Reasoning The number N of atoms (or molecules) in a sample is the number of moles n times the number of atoms per mole N_A (Avogadro's number); $N = nN_A$. We can determine the number of moles by dividing the mass of the sample m by the mass per mole of the substance.

Solution **(a)** The Hope diamond's mass is m = (44.5 carats) $[(0.200 \text{ g})/(1 \text{ carat})]$ = 8.90 g. Since the average atomic mass of naturally occurring carbon is 12.011 u (see the periodic table on the inside of the back cover), the mass per mole of this substance is 12.011 g/mol. The number of moles of carbon in the Hope diamond is

$$n = \frac{m}{\text{Mass per mole}} = \frac{8.90 \text{ g}}{12.011 \text{ g/mol}} = 0.741 \text{ mol}$$

The number of carbon atoms in the Hope diamond is

$$N = nN_A = (0.741 \text{ mol})(6.022 \times 10^{23} \text{ atoms/mol})$$

$$= \boxed{4.46 \times 10^{23} \text{ atoms}}$$

(b) The mass of the Rosser Reeves ruby is m = (138 carats)$[(0.200 \text{ g})/(1 \text{ carat})]$ = 27.6 g. The molecular mass of an aluminum oxide molecule (Al_2O_3) is the sum of the atomic masses of its atoms, which are 26.9815 u for aluminum and 15.9994 u for oxygen (see the periodic table on the inside of the back cover):

$$\text{Molecular mass} = \underbrace{2(26.9815 \text{ u})}_{\substack{\text{Mass of 2} \\ \text{aluminum atoms}}} + \underbrace{3(15.9994 \text{ u})}_{\substack{\text{Mass of 3} \\ \text{oxygen atoms}}} = 101.9612 \text{ u}$$

Thus, the mass per mole of Al_2O_3 is 101.9612 g/mol. Calculations like those in part (a) reveal that the Rosser Reeves ruby contains 0.271 mol or $\boxed{1.63 \times 10^{23} \text{ molecules of } Al_2O_3}$.

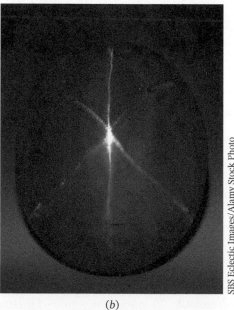

(a)   *(b)*

Chip Clark/Smithsonian Institute

SBS Eclectic Images/Alamy Stock Photo

FIGURE 14.2 (*a*) The Hope diamond surrounded by 16 smaller diamonds. (*b*) The Rosser Reeves ruby. Both gems are on display at the Smithsonian Institution in Washington, D.C.

Check Your Understanding

(*The answers are given at the end of the book.*)

1. Consider one mole of hydrogen (H_2) and one mole of oxygen (O_2). Which, if either, has the greater number of molecules and which, if either, has the greater mass?

2. The molecules of substances A and B are composed of different atoms. However, the two substances have the same mass densities. Consider the possibilities for the molecular masses of the two types of molecules and decide whether 1 m^3 of substance A contains the same number of molecules as 1 m^3 of substance B.

3. A gas mixture contains equal masses of the monatomic gases argon (atomic mass = 39.948 u) and neon (atomic mass = 20.179 u). These two are the only gases present. Of the total number of atoms in the mixture, what percentage is neon?

14.2 | The Ideal Gas Law

An **ideal gas** is an idealized model for real gases that have sufficiently low densities. The condition of low density means that the molecules of the gas are so far apart that they do not interact (except during collisions that are effectively elastic). The ideal gas law expresses the relationship between the absolute pressure, the Kelvin temperature, the volume, and the number of moles of the gas.

In discussing the constant-volume gas thermometer, Section 12.2 has already explained the relationship between the absolute pressure and Kelvin temperature of a low-density gas. This thermometer utilizes a small amount of gas (e.g., hydrogen or helium) placed inside a bulb and kept at a constant volume. Since the density is low, the gas behaves as an ideal gas. Experiment reveals that a plot of gas pressure versus temperature is a straight line, as in Figure 12.4. This plot is redrawn in **Animated Figure 14.3**, with the change that the temperature axis is now labeled in kelvins rather than in degrees Celsius. The graph indicates that the absolute pressure P is directly proportional to the Kelvin temperature T ($P \propto T$), for a fixed volume and a fixed number of molecules.

The relation between absolute pressure and the number of molecules of an ideal gas is simple. Experience indicates that it is possible to increase the pressure of a gas by adding more molecules; this is exactly what happens when a tire is pumped up. When the volume and temperature of a low-density gas are kept constant, doubling the number of molecules doubles the pressure. Thus, the absolute pressure of an ideal gas at constant temperature and constant volume is proportional to the number of molecules or, equivalently, to the number of moles n of the gas ($P \propto n$).

To see how the absolute pressure of a gas depends on the volume of the gas, look at the partially filled balloon in **Figure 14.4a**. This balloon is "soft," because the pressure of the air is low. However, if all the air in the balloon is squeezed into a smaller "bubble," as in part *b* of the figure, the "bubble" has a tighter feel. This tightness indicates that the pressure in the smaller volume is high enough to stretch the rubber substantially. Thus, it is possible to increase the pressure of a gas by reducing its volume, and if the number of molecules and the temperature are kept constant, the absolute pressure of an ideal gas is inversely proportional to its volume V ($P \propto 1/V$).

The three relations just discussed for the absolute pressure of an ideal gas can be expressed as a single proportionality, $P \propto nT/V$. This proportionality can be written as an equation by inserting a proportionality constant R, called the **universal gas constant**. Experiments have shown that $R = 8.31$ J/(mol · K) for any real gas with a density sufficiently low to ensure ideal gas behavior. The resulting equation is called the **ideal gas law**.

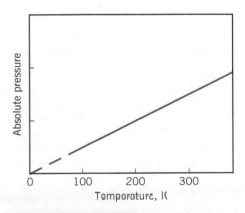

ANIMATED FIGURE 14.3 The pressure inside a constant-volume gas thermometer is directly proportional to the Kelvin temperature, a proportionality that is characteristic of an ideal gas.

(a) (b)

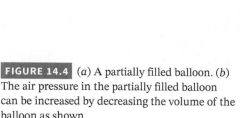

FIGURE 14.4 (a) A partially filled balloon. (b) The air pressure in the partially filled balloon can be increased by decreasing the volume of the balloon as shown.

Andy Washnik

Andy Washnik

IDEAL GAS LAW

The absolute pressure P of an ideal gas is directly proportional to the Kelvin temperature T and the number of moles n of the gas and is inversely proportional to the volume V of the gas: $P = R(nT/V)$. In other words,

$$PV = nRT \qquad (14.1)$$

where R is the universal gas constant and has the value of 8.31 J/(mol · K).

Sometimes, it is convenient to express the ideal gas law in terms of the total number of particles N, instead of the number of moles n. To obtain such an expression, we multiply and divide the right side of Equation 14.1 by Avogadro's number $N_A = 6.022 \times 10^{23}$ particles/mol* and recognize that the product nN_A is equal to the total number N of particles:

$$PV = nRT = nN_A\left(\frac{R}{N_A}\right)T = N\left(\frac{R}{N_A}\right)T$$

The constant term R/N_A is referred to as **Boltzmann's constant**, in honor of the Austrian physicist Ludwig Boltzmann (1844–1906), and is represented by the symbol k:

$$k = \frac{R}{N_A} = \frac{8.31 \text{ J/(mol · K)}}{6.022 \times 10^{23} \text{ mol}^{-1}} = 1.38 \times 10^{-23} \text{ J/K}$$

With this substitution, the ideal gas law becomes

$$PV = NkT \qquad (14.2)$$

Example 2 presents an application of the ideal gas law.

EXAMPLE 2 | BIO The Physics of Oxygen in the Lungs

In the lungs, a thin respiratory membrane separates tiny sacs of air (absolute pressure = 1.00×10^5 Pa) from the blood in the capillaries. These sacs are called alveoli, and it is from them that oxygen enters the blood. The average radius of the alveoli is 0.125 mm, and the air inside contains 14% oxygen. Assuming that the air behaves as an ideal gas at body temperature (310 K), find the number of oxygen molecules in one of the sacs.

Reasoning The pressure and temperature of the air inside an alveolus are known, and its volume can be determined since we know the radius. Thus, the ideal gas law in the form $PV = NkT$ can be used directly to find the number N of air particles inside one of the sacs. The number of oxygen molecules is 14% of the number of air particles.

Problem-Solving Insight In the ideal gas law, the temperature T must be expressed on the Kelvin scale. The Celsius and Fahrenheit scales cannot be used.

Solution The volume of a spherical sac is $V = \frac{4}{3}\pi r^3$, where r is the radius. Solving Equation 14.2 for the number of air particles, we have

$$N = \frac{PV}{kT} = \frac{(1.00 \times 10^5 \text{ Pa})\left[\frac{4}{3}\pi(0.125 \times 10^{-3} \text{ m})^3\right]}{(1.38 \times 10^{-23} \text{ J/K})(310 \text{ K})} = 1.9 \times 10^{14}$$

The number of oxygen molecules is 14% of this value, or $0.14N = \boxed{2.7 \times 10^{13}}$.

*"Particles" is not an SI unit and is often omitted. Then, particles/mol = 1/mol = mol⁻¹.

With the aid of the ideal gas law, it can be shown that one mole of an ideal gas occupies a volume of 22.4 liters at a temperature of 273 K (0 °C) and a pressure of one atmosphere (1.013×10^5 Pa). These conditions of temperature and pressure are known as **standard temperature and pressure (STP)**. Conceptual Example 3 discusses the physics of beer bubbles, and Example 4 applies the ideal gas law to hyperbaric medicine.

CONCEPTUAL EXAMPLE 3 | The Physics of Rising Beer Bubbles

If you look carefully at the bubbles rising in a glass of beer (see Figure 14.5), you'll see them grow in size as they move upward, often doubling in volume by the time they reach the surface. Beer bubbles contain mostly carbon dioxide (CO_2), a gas that is dissolved in the beer because of the fermentation process. Which variable describing the gas is responsible for the growth of the rising bubbles? **(a)** The Kelvin temperature T **(b)** The absolute pressure P **(c)** The number of moles n

Reasoning   The variables T, P, and n are related to the volume V of a bubble by the ideal gas law ($V = nRT/P$). We assume that this law applies and use it to guide our thinking. According to this law, an increase in temperature, a decrease in pressure, or an increase in the number of moles could account for the growth in size of the upward-moving bubbles.

Answers (a) and (b) are incorrect.   Temperature can be eliminated immediately, since it is constant throughout the beer. Pressure cannot be dismissed so easily. As a bubble rises, its depth decreases, and so does the fluid pressure that a bubble experiences. Since volume is inversely proportional to pressure according to the ideal gas law, at least part of the bubble growth is due to the decreasing pressure of the surrounding beer. However, some bubbles *double in volume* on the way up. To account for the doubling, there would need to be two atmospheres of pressure at the bottom of the glass, compared to the one atmosphere at the top. The pressure increment due to depth is $\rho g h$ (see Equation 11.4), so an extra pressure of one atmosphere at the bottom would mean 1.01×10^5 Pa $= \rho g h$. Solving for h with ρ equal to the density of water reveals that $h = 10.3$ m. Since most beer glasses are only about 0.2 m tall, we can rule out a change in pressure as the major cause of the change in volume.

Answer (c) is correct.   The process of elimination brings us to the conclusion that the number of moles of CO_2 in a bubble must somehow be increasing on the way up. This is, in fact, the case. Each bubble acts as a nucleation site for CO_2 molecules dissolved in the surrounding beer, so as a bubble moves upward, it accumulates carbon dioxide and grows larger.

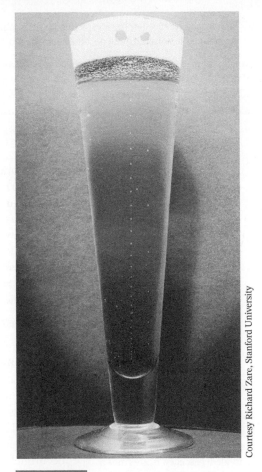

Courtesy Richard Zare, Stanford University

FIGURE 14.5 The bubbles in a glass of beer grow larger as they move upward.

Related Homework: Problem 22

EXAMPLE 4 | BIO Hyperbaric Medicine

Hyperbaric Medicine   A hyperbaric chamber like that shown in Figure 14.6 provides a high-pressure, high-oxygen environment to treat medical conditions such as carbon monoxide poisoning, cancer, burns, and wounds that are difficult to heal. One of the more prominent uses, however, is to treat decompression sickness that sometimes occurs in diving accidents when a person is brought to the surface too quickly. The process involves placing a patient in a chamber and increasing the pressure, while supplying them with pure oxygen, often through a tightly fitted mask. The pressures can reach as high as six times the pressure at sea level, that is, 6 atmospheres! Suppose a cylindrical hyperbaric chamber has a radius $r = 1.10$ m and a length $L = 2.70$ m and is to be filled with pure nitrogen (N_2) gas to a pressure $P = 6.00$ atm. The temperature of the interior of the chamber is set to 22.0 °C. Assuming that nitrogen gas (N_2) behaves like an ideal gas with a molar mass of 28.0 g/mol, determine the mass of nitrogen inside the chamber in kg.

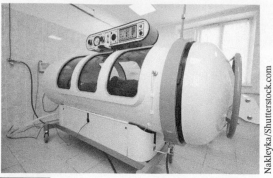

FIGURE 14.6 A hyperbaric chamber for the treatment of ailments such as carbon monoxide poisoning and decompression sickness.

Reasoning The pressure and temperature are known, and the volume of the chamber can be calculated since its dimensions are given. To find the mass m, we can apply the ideal gas law in the form $PV = nRT$, and use the definition of the number of moles $n = m/m_{mol}$, where m_{mol} is the molar mass of nitrogen (N_2).

Solution Solving Equation 14.1 (the ideal gas law) for n, and substituting for the volume of the chamber ($V = \pi r^2 L$), we have

$$n = \frac{PV}{RT} = \frac{\pi r^2 LP}{RT}$$

The radius of the chamber is $r = D/2 = 0.550$ m. Note that we must also convert the pressure to Pa (6.00 atm = 6.08×10^5 Pa) and the temperature to Kelvin (22.0 °C = 295 K). Since $n = m/m_{mol}$, and the molar mass of nitrogen is $m_{mol} = 28.0$ g/mol, we have

$$m = \left(\frac{\pi r^2 LP}{RT}\right) m_{mol}$$

$$= \left(\frac{\pi(0.550 \text{ m})^2(2.70 \text{ m})(6.08 \times 10^5 \text{ Pa})}{(8.31 \text{ J} \cdot \text{mol}^{-1} \cdot \text{K}^{-1})(295 \text{ K})}\right)(28.0 \text{ g/mol})$$

$$= 17.8 \times 10^3 \text{ g} = \boxed{17.8 \text{ kg}}$$

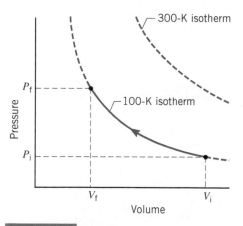

FIGURE 14.7 A pressure-versus-volume plot for a gas at a constant temperature is called an isotherm. For an ideal gas, each isotherm is a plot of the equation $P = nRT/V = \text{constant}/V$.

Historically, the work of several investigators led to the formulation of the ideal gas law. The Irish scientist Robert Boyle (1627–1691) discovered that at a constant temperature, the absolute pressure of a fixed mass (fixed number of moles) of a low-density gas is inversely proportional to its volume ($P \propto 1/V$). This fact is often called Boyle's law and can be derived from the ideal gas law by noting that $P = nRT/V = \text{constant}/V$ when n and T are constants. Alternatively, if an ideal gas changes from an initial pressure and volume (P_i, V_i) to a final pressure and volume (P_f, V_f), it is possible to write $P_i V_i = nRT$ and $P_f V_f = nRT$. Since the right sides of these equations are equal, we may equate the left sides to give the following concise way of expressing **Boyle's law**:

Constant T, constant n $\qquad P_i V_i = P_f V_f \qquad$ **(14.3)**

Figure 14.7 illustrates how pressure and volume change according to Boyle's law for a fixed number of moles of an ideal gas at a constant temperature of 100 K. The gas begins with an initial pressure and volume of P_i and V_i and is compressed. The pressure increases as the volume decreases, according to $P = nRT/V$, until the final pressure and volume of P_f and V_f are reached. The curve that passes through the initial and final points is called an **isotherm**, meaning "same temperature." If the temperature had been 300 K, rather than 100 K, the compression would have occurred along the 300-K isotherm. Different isotherms do not intersect. Example 5 deals with an application of Boyle's law to scuba diving.

Analyzing Multiple-Concept Problems

EXAMPLE 5 | The Physics of Scuba* Diving

When a scuba diver descends to greater depths, the water pressure increases. The air pressure inside the body cavities (e.g., lungs, sinuses) must be maintained at the same pressure as that of the surrounding water; otherwise the cavities would collapse. A special valve automatically adjusts the pressure of the air coming from the scuba tank to ensure that the air pressure equals the water pressure at all times. The scuba gear in **Figure 14.8a** consists of a 0.0150-m³ tank filled with compressed air at an absolute pressure of 2.02×10^7 Pa. Assume that the diver consumes air at the rate of 0.0300 m³ per minute and that the temperature of the air does not change as the diver goes deeper into the water. How long (in minutes) can a diver stay under water at a depth of 10.0 m? Take the density of seawater to be 1025 kg/m³.

Reasoning The time (in minutes) that a scuba diver can remain under water is equal to the volume of air that is

---

*The word is an acronym for *self*-contained *u*nderwater *b*reathing *a*pparatus.

FIGURE 14.8 (a) The air pressure inside the body cavities of a scuba diver must be maintained at the same value as the pressure of the surrounding water. (b) The pressure P_2 at a depth h is greater than the pressure P_1 at the surface.

available divided by the volume per minute consumed by the diver. The volume of air available to the diver depends on the volume and pressure of the air in the scuba tank, as well as the pressure of the air inhaled by the diver, according to Boyle's law. The pressure of the air inhaled equals the water pressure that acts on the diver. This pressure can be found from a knowledge of the diver's depth beneath the surface of the water.

Knowns and Unknowns   The data for this problem are listed below:

Description	Symbol	Value	Comment
Explicit Data			
Volume of air in tank	V_i	$0.0150\ \text{m}^3$	
Pressure of air in tank	P_i	$2.02 \times 10^7\ \text{Pa}$	
Rate of air consumption	C	$0.0300\ \text{m}^3/\text{min}$	
Mass density of seawater	ρ	$1025\ \text{kg/m}^3$	
Depth of diver	h	$10.0\ \text{m}$	
Implicit Data			
Air pressure at surface of water	P_1	$1.01 \times 10^5\ \text{Pa}$	Atmospheric pressure at sea level (see Section 11.2).
Unknown Variable			
Time that diver can remain at 10.0-m depth	t	?	

Modeling the Problem

STEP 1 Duration of the Dive   The air inside the scuba tank has an initial pressure of P_i and a volume of V_i (the volume of the tank). A scuba diver does *not* breathe the air *directly* from the tank, because the tank pressure of 2.02×10^7 Pa is nearly 200 times atmospheric pressure and would cause his lungs to explode. Instead, a valve on the tank adjusts the pressure of the air being sent to the diver so it equals the surrounding water pressure P_f. The time t (in minutes) that the diver can remain under water is equal to the total volume of air consumed by the diver divided by the rate C (in cubic meters per minute) at which the air is consumed:

$$t = \frac{\text{Total volume of air consumed}}{C}$$

The total volume of air consumed is the volume V_f available at the breathing pressure P_f minus the volume V_i of the scuba tank, because this amount of air always remains behind in the tank. Thus, we have Equation 1 at the right. The volume V_i and the rate C are known, but the final volume V_f is not, so we turn to Step 2 to evaluate it.

$$t = \frac{V_f - V_i}{C} \qquad (1)$$

STEP 2 Boyle's Law Since the temperature of the air remains constant, the air volume V_f available to the diver at the pressure P_f is related to the initial pressure P_i and volume V_i of air in the tank by Boyle's law $P_iV_i = P_fV_f$ (Equation 14.3). Solving for V_f yields

$$V_f = \frac{P_iV_i}{P_f}$$

This expression for V_f can be substituted into Equation 1, as indicated at the right. The initial pressure P_i and volume V_i are given. However, we still need to determine the pressure P_f of the air inhaled by the diver, and we will evaluate it in the next step.

$$t = \frac{V_f - V_i}{C} \quad (1)$$

$$V_f = \frac{P_iV_i}{P_f} \quad (2)$$

$$?$$

STEP 3 Pressure and Depth in a Static Fluid Figure 14.8b shows the diver at a depth h below the surface of the water. The absolute pressure P_2 at this depth is related to the pressure P_1 at the surface of the water by Equation 11.4: $P_2 = P_1 + \rho gh$, where ρ is the mass density of seawater and g is the magnitude of the acceleration due to gravity. Since P_1 is the air pressure at the surface of the water, it is atmospheric pressure. Recall that the valve on the scuba tank adjusts the pressure P_f of the air inhaled by the diver to be equal to the pressure P_2 of the surrounding water. Thus, $P_2 = P_f$, and Equation 11.4 becomes

$$P_f = P_1 + \rho gh$$

We now substitute this expression for P_f into Equation 2, as indicated in the right column.

$$t = \frac{V_f - V_i}{C} \quad (1)$$

$$V_f = \frac{P_iV_i}{P_f} \quad (2)$$

$$P_f = P_1 + \rho gh$$

Problem-Solving Insight When using the ideal gas law, either directly or in the form of Boyle's law, remember that the pressure P must be the absolute pressure, not the gauge pressure.

Solution Algebraically combining the results of the three steps, we have

STEP 1 **STEP 2** **STEP 3**

$$t = \frac{V_f - V_i}{C} = \frac{\frac{P_iV_i}{P_f} - V_i}{C} = \frac{\frac{P_iV_i}{P_1 + \rho gh} - V_i}{C}$$

The time that the diver can remain at a depth of 10.0 m is

$$t = \frac{\frac{P_iV_i}{P_1 + \rho gh} - V_i}{C}$$

$$= \frac{\frac{(2.02 \times 10^7\,\text{Pa})(0.0150\,\text{m}^3)}{1.01 \times 10^5\,\text{Pa} + (1025\,\text{kg/m}^3)(9.80\,\text{m/s}^2)(10.0\,\text{m})} - 0.0150\,\text{m}^3}{0.0300\,\text{m}^3/\text{min}} = \boxed{49.6\,\text{min}}$$

Note that at a fixed consumption rate C, greater values for h lead to smaller values for t. In other words, a deeper dive must have a shorter duration.

Related Homework: Problems 18, 22

Another investigator whose work contributed to the formulation of the ideal gas law was the Frenchman Jacques Charles (1746–1823). He discovered that at a constant pressure, the volume of a fixed mass (fixed number of moles) of a low-density gas is directly proportional to the Kelvin temperature ($V \propto T$). This relationship is known as Charles' law and can be obtained from the ideal gas law by noting that $V = nRT/P = (\text{constant})T$, if n and P are constant. Equivalently, when an ideal gas changes from an initial volume and temperature (V_i, T_i) to a final volume and temperature (V_f, T_f), it is possible to write $V_i/T_i = nR/P$ and $V_f/T_f = nR/P$. Thus, one way of stating **Charles' law** is

Constant P, constant n $\qquad \dfrac{V_i}{T_i} = \dfrac{V_f}{T_f}$ (14.4)

Check Your Understanding

(The answers are given at the end of the book.)

4. A tightly sealed house has a large ceiling fan that blows air out of the house and into the attic. The owners turn the fan on and forget to open any windows or doors. What happens to the air pressure in the house after the fan has been on for a while, and does it become easier or harder for the fan to do its job?

5. Above the liquid in a can of hair spray is a gas at a relatively high pressure. The label on the can includes the warning "DO NOT STORE AT HIGH TEMPERATURES." Why is the warning given?

6. What happens to the pressure in a tightly sealed house when the electric furnace turns on and runs for a while?

7. **BIO** When you climb a mountain, your eardrums "pop" outward as the air pressure decreases. When you come down, they pop inward as the pressure increases. At the sea coast, you swim through a completely submerged passage and emerge into a pocket of air trapped within a cave. As the tide comes in, the water level in the cave rises, and your eardrums pop. Is this popping analogous to what happens as you climb up or climb down a mountain?

8. Atmospheric pressure decreases with increasing altitude. Given this fact, explain why helium-filled weather balloons are underinflated when they are launched from the ground. Assume that the temperature does not change much as the balloon rises.

9. A slippery cork is being pressed into an almost full (but not 100% full) bottle of wine. When released, the cork slowly slides back out. However, if half the wine is removed from the bottle before the cork is inserted, the cork does not slide out. Explain.

10. Consider equal masses of three monatomic gases: argon (atomic mass = 39.948 u), krypton (atomic mass = 83.80 u), and xenon (atomic mass = 131.29 u). The pressure and volume of each gas is the same. Which gas has the greatest and which the smallest temperature?

14.3 | Kinetic Theory of Gases

As useful as it is, the ideal gas law provides no insight as to how pressure and temperature are related to properties of the molecules themselves, such as their masses and speeds. To show how such microscopic properties are related to the pressure and temperature of an ideal gas, this section examines the dynamics of molecular motion. The pressure that a gas exerts on the walls of a container is due to the force exerted by the gas molecules when they collide with the walls. Therefore, we will begin by combining the notion of collisional forces exerted by a fluid (Section 11.2) with Newton's second and third laws of motion (Sections 4.3 and 4.5). These concepts will allow us to obtain an expression for the pressure in terms of microscopic properties. We will then combine this with the ideal gas law to show that the average translational kinetic energy $\overline{\text{KE}}$ of a particle in an ideal gas is $\overline{\text{KE}} = \frac{3}{2}kT$ where k is Boltzmann's constant and T is the Kelvin temperature. In the process, we will also see that the internal energy U of a monatomic ideal gas is $U = \frac{3}{2}nRT$ where n is the number of moles and R is the universal gas constant.

The Distribution of Molecular Speeds

A macroscopic container filled with a gas at standard temperature and pressure contains a large number of particles (atoms or molecules). These particles are in constant, random motion, colliding with each other and with the walls of the container. In the course of one second, a particle undergoes many collisions, and each one changes the particle's speed and direction of motion. As a result, the atoms or molecules have different speeds. It is possible, however, to speak about an average particle speed. At any given instant, some particles have speeds less than, some near, and some greater than the average. For conditions of low gas density, the distribution of speeds within a large collection of molecules at a constant temperature was calculated by the Scottish physicist James Clerk

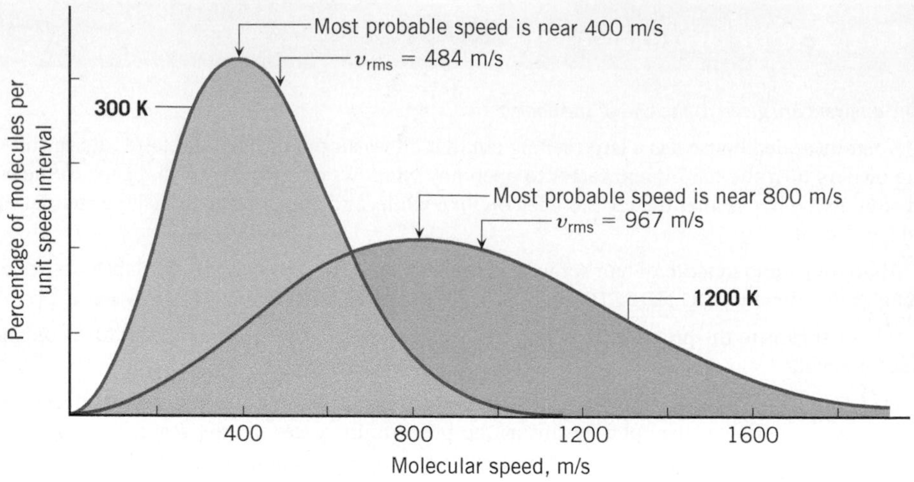

Most probable speed is near 400 m/s

v_{rms} = 484 m/s

300 K

Most probable speed is near 800 m/s

v_{rms} = 967 m/s

1200 K

Percentage of molecules per unit speed interval

Molecular speed, m/s

400   800   1200   1600

FIGURE 14.9 The Maxwell distribution curves for molecular speeds in oxygen gas at temperatures of 300 and 1200 K.

Maxwell (1831–1879). **Figure 14.9** displays the Maxwell speed distribution curves for O_2 gas at two different temperatures. When the temperature is 300 K, the maximum in the curve indicates that the most probable speed is about 400 m/s. At a temperature of 1200 K, the distribution curve shifts to the right, and the most probable speed increases to about 800 m/s. One particularly useful type of average speed, known as the rms speed and written as v_{rms}, is also shown in the drawing. When the temperature of the oxygen gas is 300 K the rms speed is 484 m/s, and it increases to 967 m/s when the temperature rises to 1200 K. The meaning of the rms speed and the reason why it is so important will be discussed shortly.

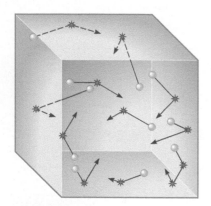

FIGURE 14.10 The pressure that a gas exerts is caused by the collisions of its molecules with the walls of the container.

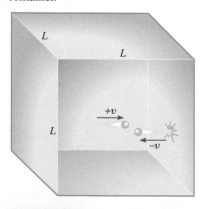

FIGURE 14.11 A gas particle is shown colliding elastically with the right wall of the container and rebounding from it.

Kinetic Theory

If a ball is thrown against a wall, it exerts a force on the wall. As **Figure 14.10** suggests, gas particles do the same thing, except that their masses are smaller and their speeds are greater. The number of particles is so great and they strike the wall so often that the effect of their individual impacts appears as a continuous force. Dividing the magnitude of this force by the area of the wall gives the pressure exerted by the gas.

To calculate the force, consider an ideal gas composed of N identical particles in a cubical container whose sides have length L. Except for elastic* collisions, these particles do not interact. **Figure 14.11** focuses attention on one particle of mass m as it strikes the right wall perpendicularly and rebounds elastically. While approaching the wall, the particle has a velocity $+v$ and linear momentum $+mv$ (see Section 7.1 for a review of linear momentum). The particle rebounds with a velocity $-v$ and momentum $-mv$, travels to the left wall, rebounds again, and heads back toward the right. The time t between collisions with the right wall is the round-trip distance $2L$ divided by the speed of the particle; that is, $t = 2L/v$. According to Newton's second law of motion, in the form of the impulse-momentum, the average force exerted on the particle by the wall is given by the change in the particle's momentum per unit time:

$$\text{Average force} = \frac{\text{Final momentum} - \text{Initial momentum}}{\text{Time between successive collisions}} \qquad (7.4)$$

$$= \frac{(-mv) - (+mv)}{2L/v} = \frac{-mv^2}{L}$$

According to Newton's law of action–reaction, the force applied to the wall by the particle is equal in magnitude to this value, but oppositely directed (i.e., $+mv^2/L$).

*The term "elastic" is used here to mean that *on the average*, in a large number of particles, there is no gain or loss of translational kinetic energy because of collisions.

Math Skills It is important to note that the average of the squared speed $\overline{v^2}$ is *not the same thing* as the square of the average speed, which is $\overline{v}^2$. To see why these two quantities are not equal, consider two particles with the following speeds: v_A is the speed of particle A and v_B is the speed of particle B. To compute the average of the squared speed and the square of the average speed, we proceed in the usual fashion when computing averages. We calculate each average by adding the two values (one for particle A and one for particle B) and then dividing the result by 2.

$$\overline{v^2} = \frac{v_A^2 + v_B^2}{2}$$

$$\overline{v}^2 = \left(\frac{v_A + v_B}{2}\right)^2 = \frac{v_A^2 + 2v_A v_B + v_B^2}{4}$$

Clearly, $\overline{v^2}$ is not equal to $\overline{v}^2$.

The magnitude F of the *total* force exerted on the right wall is equal to the number of particles that collide with the wall during the time t multiplied by the average force exerted by each particle. Since the N particles move randomly in three dimensions, one-third of them on the average strike the right wall during the time t. Therefore, the total force is

$$F = \left(\frac{N}{3}\right)\left(\frac{m\overline{v^2}}{L}\right)$$

In this result v^2 has been replaced by $\overline{v^2}$, *the average value* of the squared speed. The collection of particles possesses a Maxwell distribution of speeds, so an average value for v^2 must be used, rather than a value for any individual particle. The square root of the quantity $\overline{v^2}$ is called the **root-mean-square speed**, or, for short, the **rms speed**; $v_{rms} = \sqrt{\overline{v^2}}$. With this substitution, the total force becomes

$$F = \left(\frac{N}{3}\right)\left(\frac{m v_{rms}^2}{L}\right)$$

Pressure is force per unit area, so the pressure P acting on a wall of area L^2 is

$$P = \frac{F}{L^2} = \left(\frac{N}{3}\right)\left(\frac{m v_{rms}^2}{L^3}\right) = \left(\frac{N}{3}\right)\left(\frac{m v_{rms}^2}{V}\right)$$

where $V = L^3$ is the volume of the box. This result can be written as

$$PV = \tfrac{2}{3}N\left(\tfrac{1}{2}m v_{rms}^2\right) \tag{14.5}$$

Equation 14.5 relates the macroscopic properties of the gas—its pressure and volume—to the microscopic properties of the constituent particles—their mass and speed. Since the term $\tfrac{1}{2}m v_{rms}^2$ is the average translational kinetic energy $\overline{KE}$ of an individual particle, it follows that

$$PV = \tfrac{2}{3}N(\overline{KE})$$

This result is similar to the ideal gas law, $PV = NkT$ (Equation 14.2). Both equations have identical terms on the left, so the terms on the right must be equal: $\tfrac{2}{3}N(\overline{KE}) = NkT$. Therefore,

$$\overline{KE} = \tfrac{1}{2}m v_{rms}^2 = \tfrac{3}{2}kT \tag{14.6}$$

Equation 14.6 is significant, because it allows us to interpret temperature in terms of the motion of gas particles. This equation indicates that the Kelvin temperature is directly proportional to the average translational kinetic energy per particle in an ideal gas, no matter what the pressure and volume are. On the average, the particles have greater kinetic energies when the gas is hotter than when it is cooler. Conceptual Example 6 discusses a common misconception about the relation between kinetic energy and temperature.

CONCEPTUAL EXAMPLE 6 | Does a Single Particle Have a Temperature?

Each particle in a gas has kinetic energy. Furthermore, the equation $\frac{1}{2}mv_{rms}^2 = \frac{3}{2}kT$ establishes the relationship between the average kinetic energy per particle and the temperature of an ideal gas. Is it valid, then, to conclude that a single particle has a temperature?

Reasoning and Solution We know that a gas contains an enormous number of particles that are traveling with a distribution of speeds, such as those indicated by the graphs in Figure 14.9. Therefore, the particles do not all have the same kinetic energy, but possess a distribution of kinetic energies ranging from very nearly zero to extremely large values. If each particle had a temperature that was associated with its kinetic energy, there would

be a whole range of different temperatures within the gas. This is not so, for a gas at thermal equilibrium has only one temperature (see Section 15.2), a temperature that would be registered by a thermometer placed in the gas. Thus, temperature is a property that characterizes the gas as a whole, a fact that is inherent in the relation $\frac{1}{2}mv_{rms}^2 = \frac{3}{2}kT$. The term v_{rms} is a kind of *average particle speed*. Therefore, $\frac{1}{2}mv_{rms}^2$ is the *average kinetic energy* per particle and is characteristic of the gas as a whole. Since the Kelvin temperature is proportional to $\frac{1}{2}mv_{rms}^2$ it is also a characteristic of the gas as a whole and cannot be ascribed to each gas particle individually. Thus, *a single gas particle does not have a temperature*.

If two ideal gases have the same temperature, the relation $\frac{1}{2}mv_{rms}^2 = \frac{3}{2}kT$ indicates that the average kinetic energy of each kind of gas particle is the same. In general, however, the rms speeds of the different particles are not the same, because the masses may be different. Air, for example, is primarily a mixture of nitrogen N_2 (molecular mass = 28.0 u) and oxygen O_2 (molecular mass = 32.0 u). If we assume that each behaves like an ideal gas and apply Equation 14.6, we find the rms speeds of nitrogen and oxygen at room temperature (293 K) are 511 m/s and 478 m/s, respectively. As a comparison, the speed of sound in air at the same temperature is 343 m/s.

The equation $\overline{KE} = \frac{3}{2}kT$ has also been applied to particles much larger than atoms or molecules. The English botanist Robert Brown (1773–1858) observed through a microscope that pollen grains suspended in water move on very irregular, zigzag paths. This Brownian motion can also be observed with other particle suspensions, such as fine smoke particles in air. In 1905, Albert Einstein (1879–1955) showed that Brownian motion could be explained as a response of the large suspended particles to impacts from the moving molecules of the fluid medium (e.g., water or air). As a result of the impacts, the suspended particles have the same average translational kinetic energy as the fluid molecules—namely, $\overline{KE} = \frac{3}{2}kT$. Unlike the molecules, however, the particles are large enough to be seen through a microscope and, because of their relatively large mass, have a comparatively small average speed.

The Internal Energy of a Monatomic Ideal Gas

Chapter 15 deals with the science of thermodynamics, in which the concept of internal energy plays an important role. Using the results just developed for the average translational kinetic energy, we conclude this section by expressing the internal energy of a monatomic ideal gas in a form that is suitable for use later on.

The internal energy of a substance is the sum of the various kinds of energy that the atoms or molecules of the substance possess. A monatomic ideal gas is composed of single atoms. These atoms are assumed to be so small that the mass is concentrated at a point, with the result that the moment of inertia I about the center of mass is negligible. Thus, the rotational kinetic energy $\frac{1}{2}I\omega^2$ is also negligible. Vibrational kinetic and potential energies are absent, because the atoms are not connected by chemical bonds and, except for elastic collisions, do not interact. As a result, the internal energy U is the total translational kinetic energy of the N atoms that constitute the gas: $U = N(\frac{1}{2}mv_{rms}^2)$.

Since $\frac{1}{2}mv_{rms}^2 = \frac{3}{2}kT$ according to Equation 14.6, the internal energy can be written in terms of the Kelvin temperature as

$$U = N\left(\tfrac{3}{2}kT\right)$$

Usually, U is expressed in terms of the number of moles n, rather than the number of atoms N. Using the fact that Boltzmann's constant is $k = R/N_A$, where R is the universal gas constant and N_A is Avogadro's number, and realizing that $N/N_A = n$, we find that

Monatomic ideal gas $U = \tfrac{3}{2}nRT$ **(14.7)**

Thus, the internal energy depends on the number of moles and the Kelvin temperature of the gas. In fact, it can be shown that the internal energy is proportional to the Kelvin temperature for *any type* of ideal gas (e.g., monatomic, diatomic, etc.). For example, when hot-air balloonists turn on the burner, they increase the temperature, and hence the internal energy per mole, of the air inside the balloon (see **Figure 14.12**).

Example 7 illustrates how Equation 14.7 can be used to calculate how much internal energy is contained in the air inside human lungs.

FIGURE 14.12 When the burner is turned on to heat the air within a hot-air balloon, the temperature of the air rises. Air behaves approximately as an ideal gas, for which the internal energy per mole is proportional to the Kelvin temperature.

MUSTAFA QURAISHI/AP Images

EXAMPLE 7 | BIO The Energy in Your Lungs

How much internal energy is contained in the air within your lungs? The result might surprise you. The volume of human adult lungs is approximately 6 liters (6.0×10^{-3} m^3), and the pressure in your lungs is approximately 1 atm (1.0×10^5 Pa). Assume the temperature of the air is close to internal body temperature or 37 °C (310 K). To simplify the calculation, we will assume your lungs are filled with helium gas, which is something many of you might have done to make your voice sound funny!

Reasoning Since we are using helium gas in our calculation, which is monatomic, we can apply Equation 14.7 to calculate its internal energy. The internal energy depends on n, the number of moles of the gas. This quantity can be determined by applying the ideal gas law (Equation 14.1) with the information given in the problem.

Solution We begin by calculating the number of moles of helium gas in the lungs using Equation 14.1:

$$n = \frac{PV}{RT} = \frac{(1.0 \times 10^5\,\text{Pa})(6.0 \times 10^{-3}\,\text{m}^3)}{[8.31\,\text{J/(mol} \cdot \text{K)}](310\,\text{K})} = 0.23\ \text{mol}.$$

Using this value for n, we can apply Equation 14.7 to calculate the internal energy of the gas:

$$U = \tfrac{3}{2}nRT = \tfrac{3}{2}(0.23\ \text{mol})[8.31\,\text{J/(mol} \cdot \text{K)}](310\,\text{K}) = \boxed{890\ \text{J}}$$

Is this a large energy? It is! This value is equivalent to the kinetic energy of a baseball moving at 110 m/s, or almost 250 mph!

Check Your Understanding

(The answers are given at the end of the book.)

11. The kinetic theory of gases assumes that, for a given collision time, a gas molecule rebounds with the same speed after colliding with the wall of a container. If the speed after the collision were less than the speed before the collision, the duration of the collision remaining the same, would the pressure of the gas be greater than, equal to, or less than the pressure predicted by kinetic theory?

12. If the temperature of an ideal gas were doubled from 50 to 100 °C, would the average translational kinetic energy of the gas particles also double?

13. The pressure of a monatomic ideal gas doubles, while the volume decreases to one-half its initial value. Does the internal energy of the gas increase, decrease, or remain unchanged?

14. The atoms in a container of helium (He) have the same translational rms speed as the atoms in a container of argon (Ar). Treating each gas as an ideal gas, decide which, if either, has the greater temperature.

15. The pressure of a monatomic ideal gas is doubled, while its volume is reduced by a factor of four. What is the ratio of the new rms speed of the atoms to the initial rms speed?

14.4 | *Diffusion

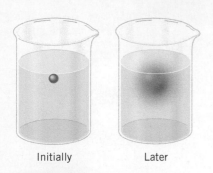

FIGURE 14.13 A drop of ink placed in water eventually becomes completely dispersed because of diffusion.

Initially   Later

You can smell the fragrance of a perfume at some distance from an open bottle because perfume molecules leave the space above the liquid in the bottle, where they are relatively concentrated, and spread out into the air, where they are less concentrated. During their journey, they collide with other molecules, so their paths resemble the zigzag paths characteristic of Brownian motion. The process in which molecules move from a region of higher concentration to one of lower concentration is called **diffusion**. Diffusion also occurs in liquids and solids, and **Figure 14.13** illustrates ink diffusing through water. However, compared to the rate of diffusion in gases, the rate is generally smaller in liquids and even smaller in solids. The host medium, such as the air or water in the examples above, is referred to as the **solvent**, while the diffusing substance, like the perfume molecules or the ink in **Figure 14.13**, is known as the **solute**. Relatively speaking, diffusion is a slow process, even in a gas. Conceptual Example 8 illustrates why.

CONCEPTUAL EXAMPLE 8 | Why Diffusion Is Relatively Slow

The fragrance from an open bottle of perfume takes several seconds or sometimes even minutes to reach the other side of a room by the process of diffusion. Which of the following accounts for the fact that diffusion is relatively slow? **(a)** The nature of Brownian motion **(b)** The relatively slow translational rms speeds that characterize gas molecules at room temperature

Reasoning The important characteristic of the paths followed by objects in Brownian motion is their zigzag shapes.

Answer (b) is incorrect. A gas molecule near room temperature has a translational rms speed of hundreds of meters per second. Such speeds are not slow. It would take a molecule traveling at such a speed just a fraction of a second to cross an ordinary room.

Answer (a) is correct. When a perfume molecule diffuses through air, it makes millions of collisions each second with air molecules. As **Figure 14.14** illustrates, the velocity of the molecule changes abruptly because of each collision, but between collisions, it moves in a straight line. Although it does move very fast between collisions, a perfume molecule wanders only slowly away from the bottle because of the zigzag path. It would take a

long time indeed to diffuse in this manner across a room. Usually, however, convection currents are present and carry the fragrance to the other side of the room in a matter of seconds or minutes.

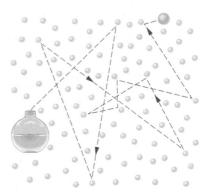

FIGURE 14.14 A perfume molecule collides with millions of air molecules during its journey, so the path has a zigzag shape. Although the air molecules are shown as stationary, they are also moving.

BIO **THE PHYSICS OF . . . drug delivery systems.** Diffusion is the basis for drug delivery systems that bypass the need to administer medication orally or via injections. **Interactive Figure 14.15** shows one such system, the transdermal patch. The word "transdermal" means "across the skin." Such patches, for example, are used to deliver nicotine in programs designed to help you stop smoking. The patch is attached to the skin using an adhesive, and the backing of the patch contains the drug within a reservoir. The concentration of the drug in the reservoir is relatively high, just like the concentration of perfume molecules above the liquid in a bottle. The drug diffuses slowly through a control membrane and directly into the skin, where its concentration is relatively low. Diffusion carries it into the blood vessels present in the skin. The purpose of the control membrane is to limit the rate of diffusion, which can also be adjusted in the reservoir by dissolving the drug in a neutral material to lower its initial concentration. Another diffusion-controlled drug delivery system utilizes capsules that are inserted surgically beneath the skin. Some contraceptives are administered in this fashion. The drug in the capsule diffuses slowly into the bloodstream over extended periods that can be as long as a year.

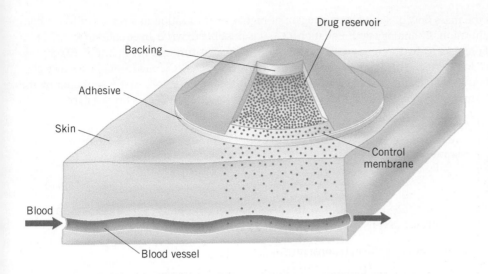

INTERACTIVE FIGURE 14.15 Using diffusion, a transdermal patch delivers a drug directly into the skin, where it enters blood vessels. The backing contains the drug within the reservoir, and the control membrane limits the rate of diffusion into the skin.

The diffusion process can be described in terms of the arrangement in **Interactive Figure 14.16a**. A hollow channel of length L and cross-sectional area A is filled with a fluid. The left end of the channel is connected to a container in which the solute concentration C_2 is relatively high, while the right end is connected to a container in which the solute concentration C_1 is lower. These concentrations are defined as the total mass of the solute molecules divided by the volume of the solution (e.g., 0.1 kg/m³). Because of the difference in concentration between the ends of the channel, $\Delta C = C_2 - C_1$, there is a net diffusion of the solute from the left end to the right end.

Interactive Figure 14.16a is similar to **Interactive Figure 13.7** for the conduction of heat along a bar, which, for convenience, is reproduced in **Interactive Figure 14.16b**. When the ends of the bar are maintained at different temperatures, T_2 and T_1, the heat Q conducted along the bar in a time t is

$$Q = \frac{(kA\,\Delta T)t}{L} \qquad (13.1)$$

where $\Delta T = T_2 - T_1$, and k is the thermal conductivity. Whereas conduction is the flow of heat from a region of higher temperature to a region of lower temperature, diffusion

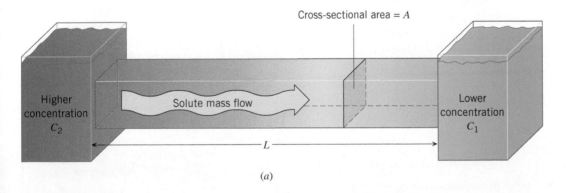

(a)

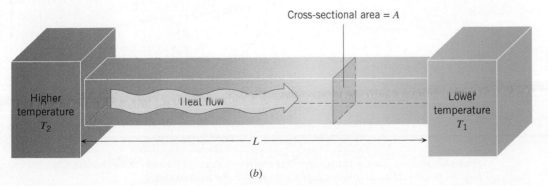

(b)

INTERACTIVE FIGURE 14.16 (a) Solute diffuses through the channel from the region of higher concentration to the region of lower concentration. (b) Heat is conducted along a bar whose ends are maintained at different temperatures.

is the mass flow of solute from a region of higher concentration to a region of lower concentration. By analogy with Equation 13.1, it is possible to write an equation for diffusion: (1) replace Q by the mass m of solute that is diffusing through the channel, (2) replace $\Delta T = T_2 - T_1$ by the difference in concentrations $\Delta C = C_2 - C_1$, and (3) replace k by a constant known as the diffusion constant D. The resulting equation, first formulated by the German physiologist Adolf Fick (1829–1901), is referred to as **Fick's law of diffusion**.

> **FICK'S LAW OF DIFFUSION**
>
> **The mass m of solute that diffuses in a time t through a solvent contained in a channel of length L and cross-sectional area A is***
>
> $$m = \frac{(DA\,\Delta C)t}{L} \tag{14.8}$$
>
> **where ΔC is the concentration difference between the ends of the channel and D is the diffusion constant.**
>
> **SI Unit for the Diffusion Constant: m^2/s**

It can be verified from Equation 14.8 that the diffusion constant has units of m^2/s, the exact value depending on the nature of the solute and the solvent. For example, the diffusion constant for ink in water is different from that for ink in benzene. Example 9 illustrates an important application of Fick's law.

EXAMPLE 9 | BIO The Physics of Water Loss from Plant Leaves

Large amounts of water can be given off by plants. It has been estimated, for instance, that a single sunflower plant can lose up to a pint of water a day during the growing season. **Figure 14.17** shows a cross-sectional view of a leaf. Inside the leaf, water passes from the liquid phase to the vapor phase at the walls of the mesophyll cells. The water vapor then diffuses through the intercellular air spaces and eventually exits the leaf through small openings, called stomatal pores. The diffusion constant for water vapor in air is $D = 2.4 \times 10^{-5}$ m^2/s. A stomatal pore has a cross-sectional area of about $A = 8.0 \times 10^{-11}$ m^2 and a length of about $L = 2.5 \times 10^{-5}$ m. The concentration of water vapor on the interior side of a pore is roughly $C_2 = 0.022$ kg/m^3, whereas the concentration on the outside is approximately $C_1 = 0.011$ kg/m^3. Determine the mass of water vapor that passes through a stomatal pore in one hour.

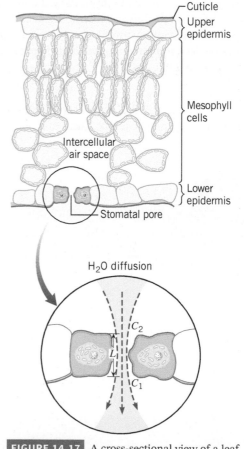

Reasoning and Solution   Fick's law of diffusion shows that

$$m = \frac{(DA\,\Delta C)t}{L} \tag{14.8}$$

$$= \frac{(2.4 \times 10^{-5}\,m^2/s)(8.0 \times 10^{-11}\,m^2)(0.022\,kg/m^3 - 0.011\,kg/m^3)(3600\,s)}{2.5 \times 10^{-5}\,m}$$

$$= \boxed{3.0 \times 10^{-9}\,kg}$$

This amount of water may not seem significant. However, a single leaf may have as many as a million stomatal pores, so the water lost by an entire plant can be substantial.

FIGURE 14.17   A cross-sectional view of a leaf. Water vapor diffuses out of the leaf through a stomatal pore.

---

*Fick's law assumes that the temperature of the solvent is constant throughout the channel. Experiments indicate that the diffusion constant depends strongly on the temperature.

Check Your Understanding

(The answers are given at the end of the book.)

16. **BIO** In the lungs, oxygen in very small sacs called alveoli diffuses into the blood. The diffusion occurs directly through the walls of the sacs, which have a thickness L. The total effective area A across which diffusion occurs is the sum of the individual areas (each quite small) of the various sac walls. Considering the fact that the mass m of oxygen that enters the blood per second needs to be large and referring to Fick's law of diffusion, what can you deduce about L and about the total number of sacs present in the lungs?

17. The same solute is diffusing through the same solvent in each of three cases. For each case, CYU Table 14.1 gives the length and cross-sectional area of the diffusion channel. The concentration difference between the ends of the diffusion channel is the same in each case. Rank the diffusion rates (in kg/s) in descending order (largest first).

CYU TABLE 14.1

Case	Length	Cross-Sectional Area
(a)	$\frac{1}{2}L$	A
(b)	L	$\frac{1}{2}A$
(c)	$\frac{1}{3}L$	$2A$

Concept Summary

14.1 Molecular Mass, the Mole, and Avogadro's Number Each element in the periodic table is assigned an atomic mass. One atomic mass unit (u) is exactly one-twelfth the mass of an atom of carbon-12. The molecular mass of a molecule is the sum of the atomic masses of its atoms.

The number of moles n contained in a sample is equal to the number of particles N (atoms or molecules) in the sample divided by the number of particles per mole N_A, as shown in Equation 1, where N_A is called Avogadro's number and has a value of $N_A = 6.022 \times 10^{23}$ particles per mole.

$$n = \frac{N}{N_A} \tag{1}$$

The number of moles n is also equal to the mass m of the sample (expressed in grams) divided by the mass per mole (expressed in grams per mole), as shown in Equation 2. The mass per mole (in g/mol) of a substance has the same numerical value as the atomic or molecular mass of one of its particles (in atomic mass units).

$$n = \frac{m}{\text{Mass per mole}} \tag{2}$$

The mass $m_{particle}$ of a particle (in grams) can be obtained by dividing the mass per mole (in g/mol) by Avogadro's number, according to Equation 3.

$$m_{particle} = \frac{\text{Mass per mole}}{N_A} \tag{3}$$

14.2 The Ideal Gas Law The ideal gas law relates the absolute pressure P, the volume V, the number of moles n, and the Kelvin temperature T of an ideal gas, according to Equation 14.1, where $R = 8.31$ J/(mol · K) is the universal gas constant. An alternative form of the ideal gas law is given by Equation 14.2, where N is the number of particles and $k = \frac{R}{N_A}$ is Boltzmann's constant. A real gas behaves as an ideal gas when its density is low enough that its particles do not interact, except via elastic collisions.

$$PV = nRT \tag{14.1}$$

$$PV = NkT \tag{14.2}$$

A form of the ideal gas law that applies when the number of moles and the temperature are constant is known as Boyle's law. Using the subscripts "i" and "f" to denote, respectively, initial and final conditions, we can write Boyle's law as in Equation 14.3. A form of the ideal gas law that applies when the number of moles and the pressure are constant is called Charles' law and is given by Equation 14.4.

$$P_i V_i = P_f V_f \tag{14.3}$$

$$\frac{V_i}{T_i} = \frac{V_f}{T_f} \tag{14.4}$$

14.3 Kinetic Theory of Gases The distribution of particle speeds in an ideal gas at constant temperature is the Maxwell speed distribution (see Figure 14.9). The kinetic theory of gases indicates that the Kelvin temperature T of an ideal gas is related to the average translational kinetic energy $\overline{KE}$ of a particle, according to Equation 14.6, where v_{rms} is the root-mean-square speed of the particles.

$$\overline{KE} = \tfrac{1}{2}mv_{rms}^2 = \tfrac{3}{2}kT \tag{14.6}$$

The internal energy U of n moles of a monatomic ideal gas is given by Equation 14.7. The internal energy of any type of ideal gas (e.g., monatomic, diatomic) is proportional to its Kelvin temperature.

$$U = \tfrac{3}{2}nRT \tag{14.7}$$

14.4 Diffusion Diffusion is the process whereby solute molecules move through a solvent from a region of higher solute concentration to a region of lower solute concentration. Fick's law of diffusion states that the mass m of solute that diffuses in a time t through the solvent in a channel of length L and cross-sectional area A is given by Equation 14.8, where ΔC is the solute concentration difference between the ends of the channel and D is the diffusion constant.

$$m = \frac{(DA\,\Delta C)t}{L} \tag{14.8}$$

Focus on Concepts

Online

Additional questions are available for assignment in WileyPLUS.

Section 14.1 Molecular Mass, the Mole, and Avogadro's Number

1. All but one of the following statements are true. Which one is not true? **(a)** A mass (in grams) equal to the molecular mass (in atomic mass units) of a pure substance contains the same number of molecules, no matter what the substance is. **(b)** One mole of any pure substance contains the same number of molecules. **(c)** Ten grams of a pure substance contains twice as many molecules as five grams of the substance. **(d)** Ten grams of a pure substance contains the same number of molecules, no matter what the substance is. **(e)** Avogadro's number of molecules of a pure substance and one mole of the substance have the same mass.

2. A mixture of ethyl alcohol (molecular mass = 46.1 u) and water (molecular mass = 18.0 u) contains one mole of molecules. The mixture contains 20.0 g of ethyl alcohol. What mass m of water does it contain?

Section 14.2 The Ideal Gas Law

3. For an ideal gas, each of the following unquestionably leads to an increase in the pressure of the gas, except one. Which one is it? **(a)** Increasing the temperature and decreasing the volume, while keeping the number of moles of the gas constant **(b)** Increasing the temperature, the volume, and the number of moles of the gas **(c)** Increasing the temperature, while keeping the volume and the number of moles of the gas constant **(d)** Increasing the number of moles of the gas, while keeping the temperature and the volume constant **(e)** Decreasing the volume, while keeping the temperature and the number of moles of the gas constant.

4. The cylinder in the drawing contains 3.00 mol of an ideal gas. By moving the piston, the volume of the gas is reduced to one-fourth its initial value, while the temperature is held constant. How many moles Δn of the gas must be allowed to escape through the valve, so that the pressure of the gas does not change?

Valve

QUESTION 4

5. Carbon monoxide is a gas at 0 °C and a pressure of 1.01×10^5 Pa. It is a diatomic gas, each of its molecules consisting of one carbon atom (atomic mass = 12.0 u) and one oxygen atom (atomic mass = 16.0 u). Assuming that carbon monoxide is an ideal gas, calculate its density ρ.

Section 14.3 Kinetic Theory of Gases

6. If the speed of every atom in a monatomic ideal gas were doubled, by what factor would the Kelvin temperature of the gas be multiplied? **(a)** 4 **(b)** 2 **(c)** 1 **(d)** $\frac{1}{2}$ **(e)** $\frac{1}{4}$

7. The atomic mass of a nitrogen atom (N) is 14.0 u, while that of an oxygen atom (O) is 16.0 u. Three diatomic gases have the same temperature: nitrogen (N_2), oxygen (O_2), and nitric oxide (NO). Rank these gases in ascending order (smallest first), according to the values of their translational rms speeds: **(a)** O_2, N_2, NO **(b)** NO, N_2, O_2 **(c)** N_2, NO, O_2 **(d)** O_2, NO, N_2 **(e)** N_2, O_2, NO

8. The pressure of a monatomic ideal gas is doubled, while the volume is cut in half. By what factor is the internal energy of the gas multiplied? **(a)** $\frac{1}{4}$ **(b)** $\frac{1}{2}$ **(c)** 1 **(d)** 2 **(e)** 4

Section 14.4 Diffusion

9. The following statements concern how to increase the rate of diffusion (in kg/s). All but one statement are always true. Which one is not necessarily true? **(a)** Increase the cross-sectional area of the diffusion channel, keeping constant its length and the difference in solute concentrations between its ends. **(b)** Increase the difference in solute concentrations between the ends of the diffusion channel, keeping constant its cross-sectional area and its length. **(c)** Decrease the length of the diffusion channel, keeping constant its cross-sectional area and the difference in solute concentrations between its ends. **(d)** Increase the cross-sectional area of the diffusion channel, and decrease its length, keeping constant the difference in solute concentrations between its ends. **(e)** Increase the cross-sectional area of the diffusion channel, increase the difference in solute concentrations between its ends, and increase its length.

10. The diffusion rate for a solute is 4.0×10^{-11} kg/s in a solvent-filled channel that has a cross-sectional area of 0.50 cm^2 and a length of 0.25 cm. What would be the diffusion rate m/t in a channel with a cross-sectional area of 0.30 cm^2 and a length of 0.10 cm?

Problems

Online

Additional questions are available for assignment in WileyPLUS.

Note: *The pressures referred to in these problems are absolute pressures, unless indicated otherwise.*

SSM Student Solutions Manual	**MMH** Problem-solving help

GO Guided Online Tutorial	**M** Medium
V-HINT Video Hints	**H** Hard
CHALK Chalkboard Videos	**WS** Worksheet
BIO Biomedical application	**T** Team Problem
E Easy	

Section 14.1 Molecular Mass, the Mole, and Avogadro's Number

1. **E** **SSM** Hemoglobin has a molecular mass of 64 500 u. Find the mass (in kg) of one molecule of hemoglobin.

2. **E** A mass of 135 g of a certain element is known to contain 30.1×10^{23} atoms. What is the element?

3. **E** **GO** A certain element has a mass per mole of 196.967 g/mol. What is the mass of a single atom in (a) atomic mass units and (b) kilograms? (c) How many moles of atoms are in a 285-g sample?

4. **E** **GO** The chlorophyll-*a* molecule ($C_{55}H_{72}MgN_4O_5$) is important in photosynthesis. (a) Determine its molecular mass (in atomic mass units). (b) What is the mass (in grams) of 3.00 moles of chlorophyll-*a* molecules?

5. **M** **V-HINT** A runner weighs 580 N (about 130 lb), and 71% of this weight is water. (a) How many moles of water are in the runner's body? (b) How many water molecules (H_2O) are there?

6. **M** **GO** Consider a mixture of three different gases: 1.20 g of argon (molecular mass = 39.948 g/mol), 2.60 g of neon (molecular mass = 20.180 g/mol), and 3.20 g of helium (molecular mass = 4.0026 g/mol). For this mixture, determine the percentage of the total number of atoms that corresponds to each of the components.

7. **M** **V-HINT** **CHALK** A cylindrical glass of water (H_2O) has a radius of 4.50 cm and a height of 12.0 cm. The density of water is 1.00 g/cm³. How many moles of water molecules are contained in the glass?

Section 14.2 The Ideal Gas Law

8. **E** **SSM** It takes 0.16 g of helium (He) to fill a balloon. How many grams of nitrogen (N_2) would be required to fill the balloon to the same pressure, volume, and temperature?

9. **E** A 0.030-m³ container is initially evacuated. Then, 4.0 g of water is placed in the container, and, after some time, all the water evaporates. If the temperature of the water vapor is 388 K, what is its pressure?

10. **E** An ideal gas at 15.5 °C and a pressure of 1.72×10^5 Pa occupies a volume of 2.81 m³. (a) How many moles of gas are present? (b) If the volume is raised to 4.16 m³ and the temperature raised to 28.2 °C, what will be the pressure of the gas?

11. **E** **GO** Four closed tanks, A, B, C, and D, each contain an ideal gas. The table gives the absolute pressure and volume of the gas in each tank. In each case, there is 0.10 mol of gas. Using this number and the data in the table, compute the temperature of the gas in each tank.

	A	B	C	D
Absolute pressure (Pa)	25.0	30.0	20.0	2.0
Volume (m³)	4.0	5.0	5.0	75

12. **E** A Goodyear blimp typically contains 5400 m³ of helium (He) at an absolute pressure of 1.1×10^5 Pa. The temperature of the helium is 280 K. What is the mass (in kg) of the helium in the blimp?

13. **E** A clown at a birthday party has brought along a helium cylinder, with which he intends to fill balloons. When full, each balloon contains 0.034 m³ of helium at an absolute pressure of 1.2×10^5 Pa. The cylinder contains helium at an absolute pressure of 1.6×10^7 Pa and has a volume of 0.0031 m³. The temperature of the helium in the tank and in the balloons is the same and remains constant. What is the maximum number of balloons that can be filled?

14. **E** **GO** The volume of an ideal gas is held constant. Determine the ratio P_2/P_1 of the final pressure to the initial pressure when the temperature of the gas rises (a) from 35.0 to 70.0 K and (b) from 35.0 to 70.0 °C.

15. **E** Two ideal gases have the same mass density and the same absolute pressure. One of the gases is helium (He), and its temperature is 175 K. The other gas is neon (Ne). What is the temperature of the neon?

16. **E** On the sunlit surface of Venus, the atmospheric pressure is 9.0×10^6 Pa, and the temperature is 740 K. On the earth's surface the atmospheric pressure is 1.0×10^5 Pa, while the surface temperature can reach 320 K. These data imply that Venus has a "thicker" atmosphere at its surface than does the earth, which means that the number of molecules per unit volume (N/V) is greater on the surface of Venus than on the earth. Find the ratio $(N/V)_{Venus}/(N/V)_{Earth}$.

17. **E** **V-HINT** A tank contains 0.85 mol of molecular nitrogen (N_2). Determine the mass (in grams) of nitrogen that must be *removed* from the tank in order to lower the pressure from 38 to 25 atm. Assume that the volume and temperature of the nitrogen in the tank do not change.

18. **M** **V-HINT** Multiple-Concept Example 5 reviews the principles that play roles in this problem. A primitive diving bell consists of a cylindrical tank with one end open and one end closed. The tank is lowered into a freshwater lake, open end downward. Water rises into the tank, compressing the trapped air, whose temperature remains constant during the descent. The tank is brought to a halt when the distance between the surface of the water in the tank and the surface of the lake is 40.0 m. Atmospheric pressure at the surface of the lake is 1.01×10^5 Pa. Find the fraction of the tank's volume that is filled with air.

19. **M** **GO** A tank contains 11.0 g of chlorine gas (Cl_2) at a temperature of 82 °C and an absolute pressure of 5.60×10^5 Pa. The mass per mole of Cl_2 is 70.9 g/mol. (a) Determine the volume of the tank. (b) Later, the temperature of the tank has dropped to 31 °C and, due to a leak, the pressure has dropped to 3.80×10^5 Pa. How many grams of chlorine gas have leaked out of the tank?

20. **M** **CHALK** **MMH** The dimensions of a room are 2.5 m × 4.0 m × 5.0 m. Assume that the air in the room is composed of 79% nitrogen (N_2) and 21% oxygen (O_2). At a temperature of 22 °C and a pressure of 1.01×10^5 Pa, what is the mass (in grams) of the air?

21. **M** **GO** The drawing shows an ideal gas confined to a cylinder by a massless piston that is attached to an ideal spring. Outside the cylinder is a vacuum. The cross-sectional area of the piston is $A = 2.50 \times 10^{-3}$ m². The initial pressure, volume, and temperature of the gas are, respectively, P_0, $V_0 = 6.00 \times 10^{-4}$ m³, and $T_0 = 273$ K, and the spring is initially stretched by an amount $x_0 = 0.0800$ m with respect to its unstrained length. The gas is heated, so that its final pressure, volume, and temperature are P_f, V_f, and T_f, and the spring is stretched by an amount $x_f = 0.1000$ m with respect to its unstrained length. What is the final temperature of the gas?

— Piston

PROBLEM 21

22. **M** **SSM** Multiple-Concept Example 5 and Conceptual Example 3 are pertinent to this problem. A bubble, located 0.200 m beneath the surface in a glass of beer, rises to the top. The air pressure at the top is 1.01×10^5 Pa. Assume that the density of beer is the same as that of fresh water. If the temperature and number of moles of CO_2 in the bubble remain constant as the bubble rises, find the ratio of the bubble's volume at the top to its volume at the bottom.

23. **M** **V-HINT** The relative humidity is 55% on a day when the temperature is 30.0 °C. Using the graph that accompanies Problem 66 in Chapter 12, determine the number of moles of water vapor per cubic meter of air.

24. **M** **MMH** One assumption of the ideal gas law is that the atoms or molecules themselves occupy a negligible volume. Verify that this assumption is reasonable by considering gaseous xenon (Xe). Xenon has an atomic radius of 2.0×10^{-10} m. For STP conditions, calculate the percentage of the total volume occupied by the atoms.

25. **H** A spherical balloon is made from an amount of material whose mass is 3.00 kg. The thickness of the material is negligible compared to the 1.50-m radius of the balloon. The balloon is filled with helium (He) at a temperature of 305 K and just floats in air, neither rising nor falling. The density of the surrounding air is 1.19 kg/m³. Find the absolute pressure of the helium gas.

26. **H** The mass of a hot-air balloon and its occupants is 320 kg (excluding the hot air inside the balloon). The air outside the balloon has a pressure of 1.01×10^5 Pa and a density of 1.29 kg/m³. To lift off, the air inside the balloon is heated. The volume of the heated balloon is 650 m³. The pressure of the heated air remains the same as the pressure of the outside air. To what temperature (in kelvins) must the air be heated so that the balloon just lifts off? The molecular mass of air is 29 u.

Section 14.3 Kinetic Theory of Gases

27. **E** **SSM** **MMH** Very fine smoke particles are suspended in air. The translational rms speed of a smoke particle is 2.8×10^{-3} m/s, and the temperature is 301 K. Find the mass of a particle.

28. **E** **GO** Four tanks A, B, C, and D are filled with monatomic ideal gases. For each tank, the mass of an individual atom and the rms speed of the atoms are expressed in terms of m and v_{rms}, respectively (see the table). Suppose that $m = 3.32 \times 10^{-26}$ kg, and $v_{rms} = 1223$ m/s. Find the temperature of the gas in each tank.

	A	B	C	D
Mass	m	m	$2m$	$2m$
Rms speed	v_{rms}	$2v_{rms}$	v_{rms}	$2v_{rms}$

29. **E** **GO** Suppose a tank contains 680 m³ of neon (Ne) at an absolute pressure of 1.01×10^5 Pa. The temperature is changed from 293.2 to 294.3 K. What is the increase in the internal energy of the neon?

30. **E** Two moles of an ideal gas are placed in a container whose volume is 8.5×10^{-3} m³. The absolute pressure of the gas is 4.5×10^5 Pa. What is the average translational kinetic energy of a molecule of the gas?

31. **E** Two gas cylinders are identical. One contains the monatomic gas argon (Ar), and the other contains an equal mass of the monatomic gas krypton (Kr). The pressures in the cylinders are the same, but the temperatures are different. Determine the ratio $\dfrac{\overline{KE}_{Krypton}}{\overline{KE}_{Argon}}$ of the average kinetic energy of a krypton atom to the average kinetic energy of an argon atom.

32. **E** A container holds 2.0 mol of gas. The total average kinetic energy of the gas molecules in the container is equal to the kinetic

energy of an 8.0×10^{-3}-kg bullet with a speed of 770 m/s. What is the Kelvin temperature of the gas?

33. **M** **CHALK** **MMH** The temperature near the surface of the earth is 291 K. A xenon atom (atomic mass = 131.29 u) has a kinetic energy equal to the average translational kinetic energy and is moving straight up. If the atom does not collide with any other atoms or molecules, how high up will it go before coming to rest? Assume that the acceleration due to gravity is constant throughout the ascent.

34. **M** **GO** Compressed air can be pumped underground into huge caverns as a form of energy storage. The volume of a cavern is 5.6×10^5 m³, and the pressure of the air in it is 7.7×10^6 Pa. Assume that air is a diatomic ideal gas whose internal energy U is given by $U = \frac{5}{2}nRT$. If one home uses 30.0 kW · h of energy per day, how many homes could this internal energy serve for one day?

35. **H** A cubical box with each side of length 0.300 m contains 1.000 moles of neon gas at room temperature (293 K). What is the average rate (in atoms/s) at which neon atoms collide with one side of the container? The mass of a single neon atom is 3.35×10^{-26} kg.

Section 14.4 Diffusion

36. **E** A tube has a length of 0.015 m and a cross-sectional area of 7.0×10^{-4} m². The tube is filled with a solution of sucrose in water. The diffusion constant of sucrose in water is 5.0×10^{-10} m²/s. A difference in concentration of 3.0×10^{-3} kg/m³ is maintained between the ends of the tube. How much time is required for 8.0×10^{-13} kg of sucrose to be transported through the tube?

37. **E** **GO** The diffusion constant for the amino acid glycine in water has a value of 1.06×10^{-9} m²/s. In a 2.0-cm-long tube with a cross-sectional area of 1.5×10^{-4} m², the mass rate of diffusion is $m/t = 4.2 \times 10^{-14}$ kg/s, because the glycine concentration is maintained at a value of 8.3×10^{-3} kg/m³ at one end of the tube and at a lower value at the other end. What is the lower concentration?

38. **E** **SSM** A large tank is filled with methane gas at a concentration of 0.650 kg/m³. The valve of a 1.50-m pipe connecting the tank to the atmosphere is inadvertently left open for twelve hours. During this time, 9.00×10^{-4} kg of methane diffuses out of the tank, leaving the concentration of methane in the tank essentially unchanged. The diffusion constant for methane in air is 2.10×10^{-5} m²/s. What is the cross-sectional area of the pipe? Assume that the concentration of methane in the atmosphere is zero.

39. **M** **V-HINT** **CHALK** Carbon tetrachloride (CCl_4) is diffusing through benzene (C_6H_6), as the drawing illustrates. The concentration of CCl_4 at the left end of the tube is maintained at 1.00×10^{-2} kg/m³, and the diffusion constant is 20.0×10^{-10} m²/s. The CCl_4 enters the tube at a mass rate of 5.00×10^{-13} kg/s. Using these data and those shown in the drawing, find (a) the mass of CCl_4 per second that passes point A and (b) the concentration of CCl_4 at point A.

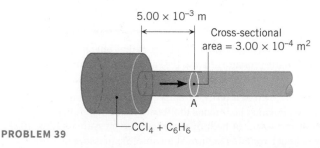

5.00×10^{-3} m

Cross-sectional area = 3.00×10^{-4} m²

A

$CCl_4 + C_6H_6$

PROBLEM 39

Additional Problems

40. **E** At the start of a trip, a driver adjusts the absolute pressure in her tires to be 2.81×10^5 Pa when the outdoor temperature is 284 K. At the end of the trip she measures the pressure to be 3.01×10^5 Pa. Ignoring the expansion of the tires, find the air temperature inside the tires at the end of the trip.

41. **E** **GO** When you push down on the handle of a bicycle pump, a piston in the pump cylinder compresses the air inside the cylinder. When the pressure in the cylinder is greater than the pressure inside the inner tube to which the pump is attached, air begins to flow from the pump to the inner tube. As a biker slowly begins to push down the handle of a bicycle pump, the pressure inside the cylinder is 1.0×10^5 Pa, and the piston in the pump is 0.55 m above the bottom of the cylinder. The pressure inside the inner tube is 2.4×10^5 Pa. How far down must the biker push the handle before air begins to flow from the pump to the inner tube? Ignore the air in the hose connecting the pump to the inner tube, and assume that the temperature of the air in the pump cylinder does not change.

42. **E** **SSM** In a diesel engine, the piston compresses air at 305 K to a volume that is one-sixteenth of the original volume and a pressure that is 48.5 times the original pressure. What is the temperature of the air after the compression?

43. **E** **GO** When a gas is diffusing through air in a diffusion channel, the diffusion rate is the number of gas atoms per second diffusing from one end of the channel to the other end. The faster the atoms move, the greater is the diffusion rate, so the diffusion rate is proportional to the rms speed of the atoms. The atomic mass of ideal gas A is 1.0 u, and that of ideal gas B is 2.0 u. For diffusion through the same channel under the same conditions, find the ratio of the diffusion rate of gas A to the diffusion rate of gas B.

44. **E** Near the surface of Venus, the rms speed of carbon dioxide molecules (CO_2) is 650 m/s. What is the temperature (in kelvins) of the atmosphere at that point?

45. **M** **GO** At the normal boiling point of a material, the liquid phase has a density of 958 kg/m^3, and the vapor phase has a density of 0.598 kg/m^3. The average distance between neighboring molecules in the vapor phase is d_{vapor}. The average distance between neighboring molecules in the liquid phase is d_{liquid}. Determine the ratio d_{vapor}/d_{liquid}. (*Hint: Assume that the volume of each phase is filled with many cubes, with one molecule at the center of each cube.*)

46. **M** **GO** Helium (He), a monatomic gas, fills a 0.010-m^3 container. The pressure of the gas is 6.2×10^5 Pa. How long would a 0.25-hp engine have to run (1 hp = 746 W) to produce an amount of energy equal to the internal energy of this gas?

47. **H** A gas fills the right portion of a horizontal cylinder whose radius is 5.00 cm. The initial pressure of the gas is 1.01×10^5 Pa. A frictionless movable piston separates the gas from the left portion of the cylinder, which is evacu-

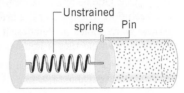

PROBLEM 47

ated and contains an ideal spring, as the drawing shows. The piston is initially held in place by a pin. The spring is initially unstrained, and the length of the gas-filled portion is 20.0 cm. When the pin is removed and the gas is allowed to expand, the length of the gas-filled chamber doubles. The initial and final temperatures are equal. Determine the spring constant of the spring.

48. **M** **GO** **SSM** Air is primarily a mixture of nitrogen N_2 (molecular mass = 28.0 u) and oxygen O_2 (molecular mass = 32.0 u). Assume that each behaves like an ideal gas and determine the rms speed of the nitrogen and oxygen molecules when the air temperature is 293 K.

Physics in Biology, Medicine, and Sports

49. **E** **BIO** **14.1** Manufacturers of headache remedies routinely claim that their own brands are more potent pain relievers than the competing brands. Their way of making the comparison is to compare the number of molecules in the standard dosage. Tylenol uses 325 mg of acetaminophen ($C_8H_9NO_2$) as the standard dose, whereas Advil uses 2.00×10^2 mg of ibuprofen ($C_{13}H_{18}O_2$). Find the number of molecules of pain reliever in the standard doses of **(a)** Tylenol and **(b)** Advil.

50. **E** **BIO** **SSM** **14.1** The active ingredient in the allergy medication Claritin contains carbon (C), hydrogen (H), chlorine (Cl), nitrogen (N), and oxygen (O). Its molecular formula is $C_{22}H_{23}ClN_2O_2$. The standard adult dosage utilizes 1.572×10^{19} molecules of this species. Determine the mass (in grams) of the active ingredient in the standard dosage.

51. **M** **BIO** **SSM** **14.1** The preparation of homeopathic "remedies" involves the repeated dilution of solutions containing an active ingredient such as arsenic trioxide (As_2O_3). Suppose one begins with 18.0 g of arsenic trioxide dissolved in water, and repeatedly dilutes the

solution with pure water, each dilution reducing the amount of arsenic trioxide remaining in the solution by a factor of 100. Assuming perfect mixing at each dilution, what is the maximum number of dilutions one may perform so that at least one molecule of arsenic trioxide remains in the diluted solution? For comparison, homeopathic "remedies" are commonly diluted 15 or even 30 times.

52. **E** **BIO** **CHALK** **SSM** **14.2** A young male adult takes in about 5.0×10^{-4} m^3 of fresh air during a normal breath. Fresh air contains approximately 21% oxygen. Assuming that the pressure in the lungs is 1.0×10^5 Pa and that air is an ideal gas at a temperature of 310 K, find the number of oxygen molecules in a normal breath.

53. **E** **BIO** **14.4** Insects do not have lungs as we do, nor do they breathe through their mouths. Instead, they have a system of tiny tubes, called tracheae, through which oxygen diffuses into their bodies. The tracheae begin at the surface of an insect's body and penetrate into the interior. Suppose that a trachea is 1.9 mm long with a cross-sectional area of 2.1×10^{-9} m^2. The concentration of oxygen in the air outside the insect is 0.28 kg/m^3, and the diffusion constant is 1.1×10^{-5} m^2/s. If the mass per second of oxygen diffusing through a trachea is 1.7×10^{-12} kg/s, find the oxygen concentration at the interior end of the tube.

54. **E** **BIO** **SSM** **14.2** Oxygen for hospital patients is kept in special tanks, where the oxygen has a pressure of 65.0 atmospheres and a temperature of 288 K. The tanks are stored in a separate room, and the oxygen is pumped to the patient's room, where it is administered at a pressure of 1.00 atmosphere and a temperature of 297 K. What volume does 1.00 m^3 of oxygen in the tanks occupy at the conditions in the patient's room?

55. **M** **BIO** **14.1** *Remdesivir* is a broad-spectrum antiviral medication. It has been used to treat certain RNA viruses, such as SARS and MERS. As of early 2020, the drug was being tested as a specific treatment for the SARS-CoV-2 virus. The chemical formula of the drug is $C_{27}H_{35}N_6O_8P$. One common delivery method of the drug is by intravenous solution. The mass of the drug in the dose is 1.00×10^2 mg. (a) How many moles of Remdesivir are there in a standard dose? (b) How many molecules of Remdesivir are contained in this dose?

56. **E** **BIO** **14.2** The lungs of a blue whale are enormous. Their total volume is capable of holding 13 000 gallons (49.2 m^3) of air. Nevertheless, relative to body size, human lungs are actually larger. Human lungs occupy 7% of the body's internal cavity, while the blue whale's lungs occupy only 3%. However, a whale's lungs are highly efficient. With every breath, they are able to pass 90% of the inhaled oxygen into their blood, as compared to just 10% to 15% for humans. Furthermore, a blue whale's blood is 60% hemoglobin, compared to only 30% in humans. These and other adaptations allow the whales to store twice as much oxygen per volume of blood than humans. These abilities allow the whale to remain under water for extended periods of time—some as long as 90 minutes! Assume with each breath, the blue whale takes in 10.3 m^3 of oxygen. Treat the oxygen in the air as an ideal gas at a temperature of 20.0 °C. The partial pressure of oxygen in the atmosphere at sea level is 21.3 kPa. Calculate what mass of oxygen is transferred to the whale's blood with every breath.

57. **M** **BIO** **14.3** The air we breathe into our lungs is a mixture of gases. While the atmosphere is mostly composed of nitrogen and oxygen, there are other gases too. One of these is a monatomic gas that has an rms speed of 428 m/s in your lungs at 20.0 °C. What gas is this?

58. **M** **BIO** **14.3** The earth's atmosphere is composed of a mixture of gases. The oxygen (O_2) we need to sustain our lives makes up only 21.0% of the air we breathe. On a hot summer day, when the temperature outside is 39.0 °C, with what rms speed do the oxygen molecules strike the surface of your skin?

59. **H** **BIO** **14.4** While the majority of the symptoms experienced by people infected with the SARS-CoV-2 (COVID-19) virus are mild, some develop the very serious condition of ARDS (acute respiratory distress syndrome), which results in a rapid and widespread

inflammation of the lungs. The major function of the lungs—the process of breathing—facilitates the vital exchange of gases between the environment and the body. The rate of exchange of these gases is largely governed by diffusion, which takes place in the extensive network of tiny sacs in the lungs called alveoli—hundreds of millions of them! In fact, the surface area of the alveoli in the human lungs is so large that if every alveoli were laid flat, they would cover a surface area as large as 70 to 80 m^2. The diffusion of oxygen and carbon dioxide takes place over this large surface. Conditions like ARDS, and other diseases in the lungs, can reduce this surface area, resulting in an insufficient amount of oxygen being transferred to the blood during respiration, a condition known as hypoxemia (see the photo), which is why patients are given pure oxygen to breathe. ARDS results in considerable fluid accumulation in the lungs, which floods the alveoli and inhibits respiration. ARDS patients often have rapid breathing and are gasping for breath. This requires considerable physical exertion, and many patients quickly become exhausted. At this point, patients may be moved into an intensive care unit and placed on a mechanical ventilator to help them breathe. This is a very serious condition, and the overall prognosis for someone with ARDS is poor, with a mortality rate near 40%. For those who survive, they often experience a lower quality of life, due to irreversible damage.

Consider a small section (5.2 μm^2) of the respiratory membrane that covers each alveolus. It is composed of a layer of squamous epithelial cells with a thickness of 1.0 μm. If the concentration of oxygen on one side of the membrane is 4.5×10^{-17} mol/μm^3 and 7.8×10^{-18} mol/μm^3 on the other side, what mass of oxygen (in micrograms) is transported across the membrane each second? The diffusion constant for oxygen across the membrane is 1.7×10^{-9} m^2/s.

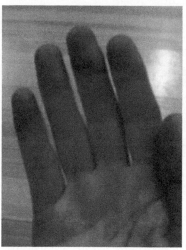

https://medicineonlinepk.blogspot.com/2013/10/cyanosis-causes-mechanisms.html

A common indicator of hypoxemia (or hypoxia) is a condition called cyanosis, which results in blue/purple coloring of the skin, often at peripheral areas, such as fingers, toes, and lips. The coloring is a result of an increase in the concentration of deoxygenated hemoglobin in the blood.
PROBLEM 59

Concepts and Calculations Problems

Online

This chapter introduces the ideal gas law, which is a relation between the pressure, volume, temperature, and number of moles of an ideal gas. In Section 10.1, we examined how the compression of a spring depends on the forced applied to it. Problem 60

reviews how a gas produces a force and why an ideal gas at different temperatures causes a spring to compress by different amounts. The kinetic theory of gases is important because it allows us to understand the relation between the macroscopic properties of a gas, such as pressure and temperature, and the microscopic properties of molecules, such as speed and mass. Problem 61 reveals the essential features of this theory.

60. M CHALK The figure shows three identical chambers containing a piston and a spring whose spring constant is $k = 5.8 \times 10^4$ N/m. The chamber in part a is completely evacuated, and the piston just touches its left end. In this position, the spring is unstrained. In part b of the drawing, 0.75 mol of ideal gas 1 is introduced into the space between the left end of the chamber and the piston, and the spring compresses by $x_1 = 15$ cm. In part c, 0.75 mol of ideal gas 2 is introduced into the space between the left end of the chamber and the piston, and the spring compresses by $x_2 = 24$ cm. *Concepts:* (i) Which gas exerts a greater force on the piston? (ii) How is the force required to compress a spring related to the displacement of the spring from the unstrained position? (iii) Which gas exerts the greater pressure on the piston? (iv) Which gas has the greater temperature? *Calculations:* Find the temperature of each gas.

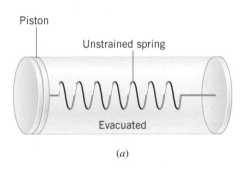

Piston

Unstrained spring

Evacuated

(*a*)

PROBLEM 60

61. M CHALK SSM In outer space the density of matter is extremely low, about one atom per cm³. The matter is mainly hydrogen atoms ($m = 1.67 \times 10^{-27}$ kg) whose rms speed is 260 m/s. A cubical box, 2.0 m on a side, is placed in outer space, and the hydrogen atoms are allowed to enter. *Concepts:* (i) Why do hydrogen atoms exert a force on the wall of the box? (ii) Do the atoms generate a pressure on the walls of the box? (iii) Do hydrogen atoms in outer space have a temperature? If so, how is the temperature related to the microscopic properties of the atoms? *Calculations:* (a) What is the magnitude of the force that the atoms exert on one wall of the box? (b) Determine the pressure that the atoms exert. (c) Does outer space have a temperature and, if so, what is it?

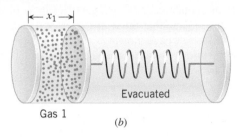

$\leftarrow x_1 \rightarrow$

Evacuated

Gas 1

(*b*)

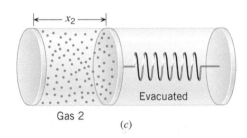

$\leftarrow x_2 \rightarrow$

Evacuated

Gas 2

(*c*)

Team Problems

Online

62. E T WS **Danger on a Spacecraft.** You and your team are on a spacecraft and discover that an external argon gas storage tank is about to fail, and you need to reduce the pressure by releasing some of the gas into space. The pressure gauge for this tank is no longer operable, but you know that the last pressure reading for the tank was 27.2 atm, and it should still be at that pressure since it has not leaked, and nothing else has changed. You and your team must reduce the pressure to 18.0 atm using only a mass flow meter that tells you how much mass of the gas flows through an exit valve. You also know that the volume of the tank is 4.35 m³ and that the temperature is 140.5 K, which has not changed since the last pressure measurement. (a) What is the mass of argon (in kg) in the tank before you open the valve? (b) What mass of argon gas (in kg) should you release from the tank in order to reduce the pressure to 18.0 atm? Assume that the gas behaves like an ideal gas and that the volume and temperature of the tank do not change.

63. M T WS **Dalton's Law of Partial Pressures.** You and your team are on a space station that has been damaged. The air in one entire section of the station has leaked through a crack, and the

volume is now at vacuum (i.e., $P = 0$ atm). The leak has since been repaired, and you and your team have been given the task of making this part of the station livable again by refilling it with one atmosphere of breathable air: a mixture of nitrogen (75.0%) and oxygen (25.0%), where the percentages are by weight (wt%). The problem is that you have separate sources for nitrogen and oxygen, and have to determine how much of each must be introduced into the void in order to have the proper mix of gases at a pressure of one atmosphere (1.013×10^5 Pa). One of your team members suggests that you use *Dalton's Rule of Partial Pressures*, which states that, for ideal gases, the total pressure of a gas mixture is the sum of the partial pressures that each amount of the constituent gases would have in the same volume and at the same temperature. For instance, in this case, the total pressure $P_T = P_N(75.0$ wt%$) + P_O(25.0$ wt%$) = 1.00$ atm, where P_N would be the pressure with only the given amount of nitrogen in the mixture, and P_O would be the pressure with only the given amount of oxygen in the mixture. If the volume of the section that needs to be filled is 98.0 m³, and is to be held at a comfortable temperature of 23.0 °C, determine the masses (in kg) of nitrogen and oxygen that must be introduced into the evacuated chamber. Assume that N_2 and O_2 behave like ideal gases with molar masses of 28.0 g/mol (N_2) and 32.0 g/mol (O_2).

64. Ⓜ Ⓣ **A Gas Dosing Device.** You and your team are given the task of designing a crude system to portion out a specific amount (moles) of krypton (Kr) gas. You brainstorm and come up with the idea of a device that consists of a cylinder ($D = 11.0$ cm) with a spring-loaded piston. The side of the piston with the spring is at atmospheric pressure. Krypton gas is introduced into the cylinder on the other side of the piston (without the spring), and the piston moves and compresses from its equilibrium position. When the spring is in its equilibrium position, the piston is at the very end of the cylinder, and the gas volume is zero. **(a)** If the spring ($k = 8.20 \times 10^4$ N/m) compresses 12.5 cm when the Kr gas is introduced into the cylinder, how many moles of Kr are contained in the cylinder? Treat it as an ideal, monatomic gas, and assume that the temperature of the gas has equilibrated with the outside environment ($T = 72.0$ °F) when you measure the compressed distance. **(b)** How far should the spring compress for 1.50 moles of Kr gas (at $T = 72.0$ °F)?

65. Ⓜ Ⓣ **An Underwater Gas Piston.** You and your team have been given the task of preparing a gas piston device that is to be deployed to the bottom of the Southern Sea, beneath the Brunt Ice Shelf off the coast of Antarctica. The inner volume of the cylinder has a diameter of 9.00 cm and a length of 25.0 cm, and is fitted with a movable piston that compresses the argon (Ar) gas that it contains. Before deployment, the piston is located and held at the very top of the cylinder, providing the maximum volume for the gas. When the device is submerged, however, the piston will move inward due to the water pressure created by the depth, and the change in temperature of the gas. The device is to be taken to a depth of 350.0 m in salt water (density $\rho = 1.03$ g/cm³) at a temperature of 33.0 °F. If the maximum allowable compression distance of the piston is 13.0 cm, what is the minimum gas pressure in the cylinder that must be attained when it is filled at the surface (at atmospheric pressure and $T = 59.0$ °F) to prevent the maximum displacement of the piston from exceeding its maximum (13.0 cm) when submerged to a depth of 350.0 m?

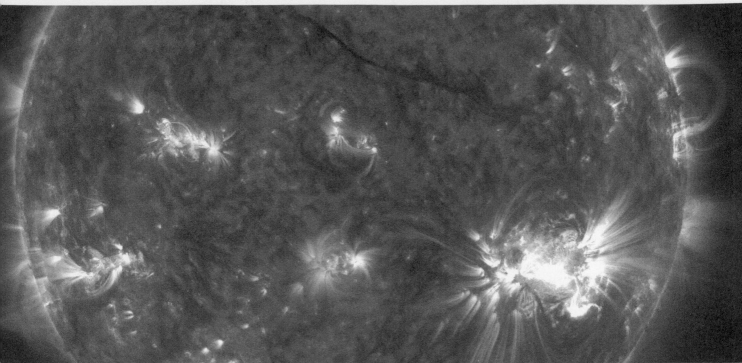

The sun fuses 600 million tons of hydrogen into helium every second, and about 4 million tons of this matter is converted into energy. This is an irreversible process that transfers heat from a hot source, the sun, to a cold source, space. Some of this energy will reach the earth and other planets in the solar system, allowing the formation of highly ordered and complex biological systems, like plants and animals. However, during this process, the total disorder, or entropy, of the universe increases. In this chapter, we will study the laws that govern heat transfer and its relationship to work and energy in physical systems.

NASA/SDO

Thermodynamics

15.1 Thermodynamic Systems and Their Surroundings

We have studied heat (Chapter 12) and work (Chapter 6) as separate topics. Often, however, they occur simultaneously. In an automobile engine, for instance, fuel is burned at a relatively high temperature, some of its internal energy is used for doing the work of driving the pistons up and down, and the excess heat is removed by the cooling system to prevent overheating. **Thermodynamics** is the branch of physics that is built upon the fundamental laws that heat and work obey.

In thermodynamics the collection of objects on which attention is being focused is called the **system**, while everything else in the environment is called the **surroundings**. For example, the system in an automobile engine could be the burning gasoline, whereas the surroundings would then include the pistons, the exhaust system, the radiator, and the outside air. The system and its surroundings are separated by walls of some kind. Walls that permit heat to flow through

LEARNING OBJECTIVES

After reading this module, you should be able to...

15.1 Identify thermodynamic systems and their surroundings.

15.2 Define the zeroth law of thermodynamics.

15.3 Define the first law of thermodynamics.

15.4 Analyze thermal processes.

15.5 Analyze thermal processes in an ideal gas.

15.6 Distinguish between different forms of heat capacity.

15.7 Define the second law of thermodynamics.

15.8 Analyze heat engines.

15.9 Analyze Carnot engines.

15.10 Analyze the operation of refrigerators, air conditioners, and heat pumps.

15.11 Calculate the change in entropy in a thermodynamic process.

15.12 Define the third law of thermodynamics.

Steve Vidler/SuperStock

FIGURE 15.1 The air in one of these hot-air balloons is one example of a thermodynamic system.

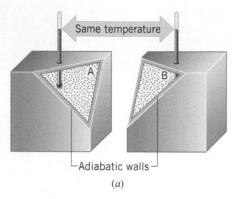

(a)

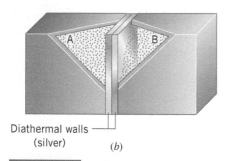

Diathermal walls (silver)   (b)

FIGURE 15.2 (a) Systems A and B are surrounded by adiabatic walls and register the same temperature on a thermometer. (b) When system A is put into thermal contact with system B through diathermal walls, no net flow of heat occurs between the systems.

them, such as those of the engine block, are called **diathermal walls**. Perfectly insulating walls that do not permit heat to flow between the system and its surroundings are known as **adiabatic walls**.

To understand what the laws of thermodynamics have to say about the relationship between heat and work, it is necessary to describe the physical condition or **state of a system**. We might be interested, for instance, in the hot air within one of the balloons in **Figure 15.1**. The hot air itself would be the system, and the skin of the balloon provides the walls that separate this system from the surrounding cooler air. The state of the system would be specified by giving values for the pressure, volume, temperature, and mass of the hot air.

As this chapter discusses, there are four laws of thermodynamics. We begin with the one known as the zeroth law and then consider the remaining three.

15.2 | The Zeroth Law of Thermodynamics

The zeroth law of thermodynamics deals with the concept of **thermal equilibrium**. Two systems are said to be in thermal equilibrium if there is no net flow of heat between them when they are brought into thermal contact. For instance, you are definitely *not* in thermal equilibrium with the water in Lake Michigan in January. Just dive into it, and you will find out how quickly your body loses heat to the frigid water. To help explain the central idea of the zeroth law of thermodynamics, **Figure 15.2a** shows two systems labeled A and B. Each is within a container whose adiabatic walls are made from insulation that prevents the flow of heat, and each has the same temperature, as indicated by a thermometer. In part *b*, one wall of each container is replaced by a thin silver sheet, and the two sheets are touched together. Silver has a large thermal conductivity, so heat flows through it readily and the silver sheets behave as diathermal walls. Even though the diathermal walls would permit it, no net flow of heat occurs in part *b*, indicating that the two systems are in thermal equilibrium. There is no net flow of heat because the two systems have the same temperature. We see, then, that ***temperature is the indicator of thermal equilibrium in the sense that there is no net flow of heat between two systems in thermal contact that have the same temperature.***

In **Figure 15.2** the thermometer plays an important role. System A is in equilibrium with the thermometer, and so is system B. In each case, the thermometer registers the same temperature, thereby indicating that the two systems are equally hot. Consequently, systems A and B are found to be in thermal equilibrium with each other. In effect, the thermometer is a third system. The fact that system A and system B are each in thermal equilibrium with this third system at the same temperature means that they are in thermal equilibrium with each other. This finding is an example of the **zeroth law of thermodynamics.**

THE ZEROTH LAW OF THERMODYNAMICS

Two systems individually in thermal equilibrium with a third system* are in thermal equilibrium with each other.

The zeroth law establishes temperature as the indicator of thermal equilibrium and implies that all parts of a system must be in thermal equilibrium if the system is to have a definable single temperature. In other words, there can be no flow of heat within a system that is in thermal equilibrium.

15.3 | The First Law of Thermodynamics

The atoms and molecules of a substance have kinetic and potential energy. These and other kinds of molecular energy constitute the internal energy of a substance. When a substance participates in a process involving energy in the form of work and heat, the internal

---

*The state of the third system is the same when it is in thermal equilibrium with either of the two systems. In **Figure 15.2**, for example, the mercury level is the same in the thermometer in either system.

energy of the substance can change. The relationship between work, heat, and changes in the internal energy is known as the first law of thermodynamics. We will now see that the first law of thermodynamics is an expression of the conservation of energy.

Suppose that a system gains heat Q and that this is the only effect occurring. Consistent with the law of conservation of energy, the internal energy of the system increases from an initial value of U_i to a final value of U_f, the change being $\Delta U = U_f - U_i = Q$. In writing this equation, we use the following convention.

> **Problem-Solving Insight**   Heat Q is positive when the system gains heat and negative when the system loses heat.

The internal energy of a system can also change because of work. If a system does work W on its surroundings and there is no heat flow, energy conservation indicates that the internal energy of the system decreases from U_i to U_f, the change now being $\Delta U = U_f - U_i = -W$. The minus sign is included with the work because we employ the following convention.

> **Problem-Solving Insight**   Work is positive when it is done by the system and negative when it is done on the system.

A system can gain or lose energy simultaneously in the form of heat Q and work W. The change in internal energy due to both factors is given by Equation 15.1. Thus, the **first law of thermodynamics** is just the conservation-of-energy principle applied to heat, work, and the change in the internal energy.

> **THE FIRST LAW OF THERMODYNAMICS**
>
> The internal energy of a system changes from an initial value U_i to a final value of U_f due to heat Q and work W:
>
> $$\Delta U = U_f - U_i = Q - W \qquad (15.1)$$
>
> Q is positive when the system gains heat and negative when it loses heat. W is positive when work is done by the system and negative when work is done on the system.

Examples 1 and 2 illustrate the use of Equation 15.1 and the sign conventions for Q and W.

EXAMPLE 1 | Positive and Negative Work

Interactive Figure 15.3 illustrates a system and its surroundings. In part a, the system gains 1500 J of heat from its surroundings, and 2200 J of work is done *by* the system on the surroundings. In part b, the system also gains 1500 J of heat, but 2200 J of work is done *on* the system by the surroundings. In each case, determine the change in the internal energy of the system.

Reasoning   In Interactive Figure 15.3a the system loses more energy in doing work than it gains in the form of heat, so the internal energy of the system decreases. Thus, we expect the change in the internal energy, $\Delta U = U_f - U_i$, to be negative. In part b of the drawing, the system gains energy in the form of both heat and work. The internal energy of the system increases, and we expect ΔU to be positive.

> **Problem-Solving Insight**   When using the first law of thermodynamics, as expressed by Equation 15.1, be careful to follow the proper sign conventions for the heat Q and the work W.

Solution   (a) The heat is positive, $Q = +1500$ J, since it is gained by the system. The work is positive, $W = +2200$ J, since it is done by the system. According to the first law of thermodynamics

$$\Delta U = Q - W = (+1500 \text{ J}) - (+2200 \text{ J}) = \boxed{-700 \text{ J}} \qquad (15.1)$$

The minus sign for ΔU indicates that the internal energy has decreased, as expected.

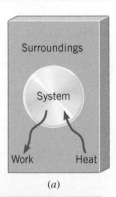 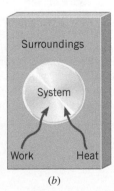

(a)   *(b)*

INTERACTIVE FIGURE 15.3   (a) The system gains energy in the form of heat but loses energy because work is done by the system. (b) The system gains energy in the form of heat and also gains energy because work is done on the system.

(b) The heat is positive, $Q = +1500$ J, since it is gained by the system. But the work is negative, $W = -2200$ J, since it is done *on* the system. Thus,

$$\Delta U = Q - W = (+1500 \text{ J}) - (-2200 \text{ J}) = \boxed{+3700 \text{ J}} \qquad (15.1)$$

The plus sign for ΔU indicates that the internal energy has increased, as expected.

EXAMPLE 2 | BIO Refueling Your Lost Internal Energy

Suppose you spend 1 hour jogging and do work with a power output of 170 W. During this time, you also give off 4.2×10^5 J of heat to your surroundings. How many bananas would you have to eat to replace the internal energy you lost by jogging? Assume one banana contains 100 Calories.

Reasoning We can use the first law of thermodynamics (Equation 15.1) to calculate the change in the jogger's internal energy. This depends on the heat (Q) lost or gained by the system and the work (W) done on or by the system. Once we know the jogger's change in internal energy, then we can use the mechanical equivalent of heat to calculate the Calories (bananas) she must eat to replace this lost energy.

Solution We begin by using the first law of thermodynamics (Equation 15.1): $\Delta U = Q - W$. The heat from the system (the jogger) is given in the problem, and it will be negative, since it is lost by the system. The work (W) will be positive, since it is done by the system. We can calculate the work using the power output and the time she runs:

$$W = Pt = (170 \text{ W})(3600 \text{ s}) = 6.1 \times 10^5 \text{ J}$$

Calculating the change in the internal energy, we get:

$$\Delta U = -4.2 \times 10^5 \text{ J} - 6.1 \times 10^5 \text{ J} = -10.3 \times 10^5 \text{ J}$$

As expected, her change in internal energy is negative, and it's this quantity that must be replaced by the food she eats. Each banana contains 100 Calories, which is equal to 100 kcal. This is equivalent to $(100 \text{ kcal}) \times (4186 \text{ J/kcal}) = 4.186 \times 10^5$ J per banana. The number of bananas required to replace her lost internal energy will be:

$$(10.3 \times 10^5 \text{ J})/(4.186 \times 10^5 \text{ J/banana}) = \boxed{2.5 \text{ bananas}}$$

There is a lot of energy content in food!

In the first law of thermodynamics, the internal energy U, heat Q, and work W are energy quantities, and each is expressed in energy units such as joules. However, there is a fundamental difference between U, on the one hand, and Q and W on the other. The next example sets the stage for explaining this difference.

EXAMPLE 3 | An Ideal Gas

The temperature of three moles of a monatomic ideal gas is reduced from $T_i = 540$ K to $T_f = 350$ K by two different methods. In the first method 5500 J of heat flows into the gas, whereas in the second, 1500 J of heat flows into it. In each case find **(a)** the change in the internal energy and **(b)** the work done by the gas.

Reasoning Since the internal energy of a monatomic ideal gas is $U = \frac{3}{2}nRT$ (Equation 14.7) and since the number of moles n is fixed, only a change in temperature T can alter the internal energy. Because the change in T is the same in both methods, the change in U is also the same. From the given temperatures, the change ΔU in internal energy can be determined. Then, the first law of thermodynamics can be used with ΔU and the given heat values to calculate the work for each of the methods.

Solution **(a)** Using Equation 14.7 for the internal energy of a monatomic ideal gas, we find for each method of adding heat that

$$\Delta U = \frac{3}{2}nR(T_f - T_i) = \frac{3}{2}(3.0 \text{ mol})[8.31 \text{ J/(mol} \cdot \text{K)}](350 \text{ K} - 540 \text{ K})$$

$$= \boxed{-7100 \text{ J}}$$

(b) Since ΔU is now known and the heat is given in each method, we can use Equation 15.1 ($\Delta U = Q - W$) to determine the work:

1st method $\quad W = Q - \Delta U = 5500 \text{ J} - (-7100 \text{ J}) = \boxed{12\,600 \text{ J}}$

2nd method $\quad W = Q - \Delta U = 1500 \text{ J} - (-7100 \text{ J}) = \boxed{8600 \text{ J}}$

In each method the gas does work, but it does more in the first method.

To understand the difference between U and either Q or W, consider the value for ΔU in Example 3. In both methods ΔU is the same. Its value is determined once the initial and final temperatures are specified because the internal energy of an ideal gas depends only on the temperature. Temperature is one of the variables (along with pressure and volume) that define the state of a system.

> **Problem-Solving Insight** The internal energy depends only on the state of a system, not on the method by which the system arrives at a given state.

In recognition of this characteristic, internal energy is referred to as a ***function of state***.* In contrast, heat and work are not functions of state because they have different values for each different method used to make the system change from one state to another, as in Example 3.

....

*The fact that an ideal gas is used in Example 3 does not restrict our conclusion. Had a real (nonideal) gas or other material been used, the only difference would have been that the expression for the internal energy would have been more complicated. It might have involved the volume V, as well as the temperature T, for instance.

Check Your Understanding

(The answer is given at the end of the book.)

1. A gas is enclosed within a chamber that is fitted with a frictionless piston. The piston is then pushed in, thereby compressing the gas. Which statement below regarding this process is consistent with the first law of thermodynamics? **(a)** The internal energy of the gas will increase. **(b)** The internal energy of the gas will decrease. **(c)** The internal energy of the gas will not change. **(d)** The internal energy of the gas may increase, decrease, or remain the same, depending on the amount of heat that the gas gains or loses.

15.4 | Thermal Processes

A system can interact with its surroundings in many ways, and the heat and work that come into play always obey the first law of thermodynamics. This section introduces four common thermal processes. In each case, the process is assumed to be **quasi-static**, which means that it occurs slowly enough that a uniform pressure and temperature exist throughout all regions of the system at all times.

An isobaric process is one that occurs at constant pressure. For instance, **Figure 15.4** shows a substance (solid, liquid, or gas) contained in a chamber fitted with a frictionless piston. The pressure P experienced by the substance is always the same, because it is determined by the external atmosphere and the weight of the piston and the block resting on it. Heating the substance makes it expand and do work W in lifting the piston and block through the displacement $\vec{s}$. The work can be calculated from $W = Fs$ (Equation 6.1), where F is the magnitude of the force and s is the magnitude of the displacement. The force is generated by the pressure P acting on the bottom surface of the piston (area $= A$), according to $F = PA$ (Equation 10.19). With this substitution for F, the work becomes $W = (PA)s$. But the product $A \cdot s$ is the change in volume of the material, $\Delta V = V_f - V_i$, where V_f and V_i are the final and initial volumes, respectively. Thus, the relation is

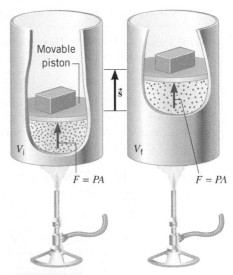

Isobaric process $W = P\,\Delta V = P(V_f - V_i)$ **(15.2)**

Consistent with our sign convention, this result predicts a positive value for the work done *by a system* when it expands isobarically (V_f exceeds V_i). Equation 15.2 also applies to an isobaric compression (V_f less than V_i). Then, the work is negative, since work must be done *on the system* to compress it. Example 4 emphasizes that $W = P\,\Delta V$ applies to any system, solid, liquid, or gas, as long as the pressure remains constant while the volume changes.

FIGURE 15.4 The substance in the chamber is expanding isobarically because the pressure is held constant by the external atmosphere and the weight of the piston and the block.

Analyzing Multiple-Concept Problems

EXAMPLE 4 | Isobaric Expansion of Water

One gram of water is placed in the cylinder in **Figure 15.4**, and the pressure is maintained at 2.0×10^5 Pa. The temperature of the water is raised by 31 C°. In one case, the water is in the liquid phase, expands by the small amount of 1.0×10^{-8} m³, and has a specific heat capacity of 4186 J/(kg · C°). In another case, the water is in the gas phase, expands by the much greater amount of 7.1×10^{-5} m³, and has a specific heat capacity of 2020 J/(kg · C°). Determine the change in the internal energy of the water in each case.

Reasoning The change ΔU in the internal energy is given by the first law of thermodynamics as $\Delta U = Q - W$ (Equation 15.1). The heat Q may be evaluated as $Q = cm\,\Delta T$ (Equation 12.4). Finally, since the process occurs at a constant pressure (isobaric), the work W may be found using $W = P\,\Delta V$ (Equation 15.2).

Knowns and Unknowns The following table summarizes the given data:

Description	Symbol	Value	Comment
Mass of water	m	1.0 g	0.0010 kg
Pressure on water	P	2.0×10^5 Pa	Pressure is constant.
Increase in temperature	ΔT	31 C°	
Increase in volume of liquid	ΔV_{liquid}	1.0×10^{-8} m³	Expansion occurs.
Specific heat capacity of liquid	c_{liquid}	4186 J/(kg · C°)	
Increase in volume of gas	ΔV_{gas}	7.1×10^{-5} m³	Expansion occurs.
Specific heat capacity of gas	c_{gas}	2020 J/(kg · C°)	

Unknown Variables

Change in internal energy of liquid	ΔU_{liquid}	?	
Change in internal energy of gas	ΔU_{gas}	?	

Modeling the Problem

STEP 1 The First Law of Thermodynamics   The change ΔU in the internal energy is given by the first law of thermodynamics, as shown at the right. In Equation 15.1, neither the heat Q nor the work W is known, so we turn to Steps 2 and 3 to evaluate them.

$$\Delta U = Q - W \qquad \text{(15.1)}$$

STEP 2 Heat and Specific Heat Capacity   According to Equation 12.4, the heat Q needed to raise the temperature of a mass m of material by an amount ΔT is

$$Q = cm\,\Delta T$$

where c is the material's specific heat capacity. Data are available for all of the terms on the right side of this expression, which can be substituted into Equation 15.1, as shown at the right. The remaining unknown variable in Equation 15.1 is the work W, and we evaluate it in Step 3.

$$\Delta U = Q - W \qquad \text{(15.1)}$$
$$Q = cm\,\Delta T \qquad \text{(12.4)}$$

STEP 3 Work Done at Constant Pressure   Under constant-pressure, or isobaric, conditions, the work W done is given by Equation 15.2 as

$$W = P\,\Delta V$$

where P is the pressure acting on the material and ΔV is the change in the volume of the material. Substitution of this expression into Equation 15.1 is shown at the right.

$$\Delta U = Q - W \qquad \text{(15.1)}$$
$$Q = cm\,\Delta T \qquad \text{(12.4)}$$
$$W = P\,\Delta V \qquad \text{(15.1)}$$

Solution   Combining the results of each step algebraically, we find that

$$\underset{\downarrow}{\text{STEP 1}} \quad \underset{\downarrow}{\text{STEP 2}} \quad \underset{\downarrow}{\text{STEP 3}}$$
$$\Delta U = Q - W = cm\,\Delta T - W = cm\,\Delta T - P\,\Delta V$$

Applying this result to the liquid and to the gaseous water gives

$$\Delta U_{\text{liquid}} = c_{\text{liquid}}\, m\,\Delta T - P\,\Delta V_{\text{liquid}}$$

$$= [4186 \text{ J/(kg · C°)}](0.0010 \text{ kg})(31 \text{ C°}) - (2.0 \times 10^5 \text{ Pa})(1.0 \times 10^{-8} \text{ m}^3)$$

$$= 130 \text{ J} - 0.0020 \text{ J} = \boxed{130 \text{ J}}$$

$$\Delta U_{\text{gas}} = c_{\text{gas}}\, m\,\Delta T - P\,\Delta V_{\text{gas}}$$

$$= [2020 \text{ J/(kg · C°)}](0.0010 \text{ kg})(31 \text{ C°}) - (2.0 \times 10^5 \text{ Pa})(7.1 \times 10^{-5} \text{ m}^3)$$

$$= 63 \text{ J} - 14 \text{ J} = \boxed{49 \text{ J}}$$

For the liquid, virtually all the 130 J of heat serves to change the internal energy, since the volume change and the corresponding work of expansion are so small. In contrast, a significant fraction of the 63 J of heat added to the gas causes work of expansion to be done, so that only 49 J is left to change the internal energy.

Related Homework: Problem 15

It is often convenient to display thermal processes graphically. For instance, **Interactive Figure 15.5** shows a plot of pressure versus volume for an isobaric expansion. Since the pressure is constant, the graph is a horizontal straight line, beginning at the initial volume V_i and ending at the final volume V_f. In terms of such a plot, the work $W = P(V_f - V_i)$ is the area under the graph, which is the shaded rectangle of height P and width $V_f - V_i$.

Another common thermal process is an *isochoric process, one that occurs at constant volume*. **Figure 15.6a** illustrates an isochoric process in which a substance (solid, liquid, or gas) is heated. The substance would expand if it could, but the rigid container keeps the volume constant, so the pressure–volume plot shown in **Figure 15.6b** is a vertical straight line. Because the volume is constant, the pressure inside rises, and the substance exerts more and more force on the walls. Although enormous forces can be generated in the closed container, no work is done ($W = 0$ J), since the walls do not move. Consistent with zero work being done, the area under the vertical straight line in **Figure 15.6b** is zero. Since no work is done, the first law of thermodynamics indicates that the heat in an isochoric process serves only to change the internal energy: $\Delta U = Q - W = Q$.

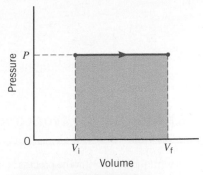

INTERACTIVE FIGURE 15.5 For an isobaric process, a pressure–volume plot is a horizontal straight line. The work done [$W = P(V_f - V_i)$] is the colored rectangular area under the graph.

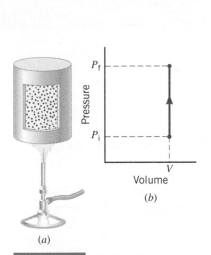

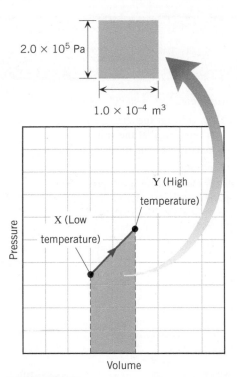

FIGURE 15.6 (*a*) The substance in the chamber is being heated isochorically because the rigid chamber keeps the volume constant. (*b*) The pressure–volume plot for an isochoric process is a vertical straight line. The area under the graph is zero, indicating that no work is done.

FIGURE 15.7 The colored area gives the work done by the gas for the process from X to Y.

A third important thermal process is an *isothermal process, one that takes place at constant temperature*. The next section illustrates the important features of an isothermal process when the system is an ideal gas.

Last, there is the *adiabatic process, one that occurs without the transfer of heat*. Since there is no heat transfer, Q equals zero, and the first law indicates that $\Delta U = Q - W = -W$. Thus, when work is done by a system adiabatically, W is positive and the internal energy decreases by exactly the amount of the work done. When work is done on a system adiabatically, W is negative and the internal energy increases correspondingly. The next section discusses an adiabatic process for an ideal gas.

A process may be complex enough that it is not recognizable as one of the four just discussed. For instance, **Figure 15.7** shows a process for a gas in which the pressure, volume, and temperature are changed along the straight line from X to Y. With the aid of integral calculus, the following can be proved.

Problem-Solving Insight   The area under a pressure–volume graph is the work for any kind of process.

Thus, the area representing the work has been colored in **Figure 15.7**. The volume increases, so that work is done by the gas. This work is positive by convention, as is the area. In contrast, if a process reduces the volume, work is done on the gas, and this work is negative by convention. Correspondingly, the area under the pressure–volume graph would be assigned a negative value. In Example 5, we determine the work for the case shown in **Figure 15.7**.

EXAMPLE 5 | Work and the Area Under a Pressure–Volume Graph

Determine the work for the process in which the pressure, volume, and temperature of a gas are changed along the straight line from X to Y in **Figure 15.7**.

Reasoning The work is given by the area (in color) under the straight line between X and Y. Since the volume increases, work is done by the gas on the surroundings, so the work is positive. The

area can be found by counting squares in **Figure 15.7** and multiplying by the area per square.

Solution We estimate that there are 8.9 colored squares in the drawing. The area of one square is $(2.0 \times 10^5 \, \text{Pa})(1.0 \times 10^{-4} \, \text{m}^3) = 2.0 \times 10^1$ J, so the work is

$$W = +(8.9 \text{ squares})(2.0 \times 10^1 \, \text{J/square}) = \boxed{+180 \, \text{J}}$$

Check Your Understanding

(*The answers are given at the end of the book.*)

2. Is it possible for the temperature of a substance to rise without heat flowing into the substance? **(a)** Yes, provided that the volume of the substance does not change. **(b)** Yes, provided that the substance expands and does positive work. **(c)** Yes, provided that work is done on the substance and it contracts.

3. **CYU Figure 15.1** shows a pressure-versus-volume plot for a three-step process: A → B, B → C, and C → A. For each step, the work can be positive, negative, or zero. Which answer in **CYU Table 15.1** correctly describes the work for the three steps?

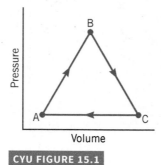

Pressure vs. Volume

CYU FIGURE 15.1

CYU TABLE 15.1 **Work Done by the System**

	A → B	B → C	C → A
(a)	Positive	Negative	Negative
(b)	Positive	Positive	Negative
(c)	Negative	Negative	Positive
(d)	Positive	Negative	Zero
(e)	Negative	Positive	Zero

4. **CYU Figure 15.2** shows a pressure–volume graph in which a gas expands at constant pressure from *A* to *B*, and then goes from *B* to *C* at constant volume. Complete **CYU Table 15.2** by deciding whether each of the four unspecified quantities is positive (+), negative (−), or zero (0).

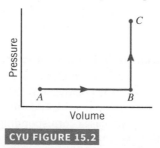

Pressure vs. Volume

CYU FIGURE 15.2

CYU TABLE 15.2

	ΔU	Q	W
$A \to B$	+	?	?
$B \to C$	?	+	?

5. When a solid melts at constant pressure, the volume of the resulting liquid does not differ much from the volume of the solid. According to the first law of thermodynamics, how does the internal energy of the liquid compare to the internal energy of the solid? The internal energy of the liquid is **(a)** greater than, **(b)** the same as, **(c)** less than the internal energy of the solid.

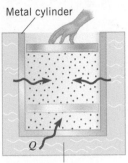

Metal cylinder

Hot water at temperature T

(a)

15.5 | Thermal Processes Using an Ideal Gas

Isothermal Expansion or Compression

When a system performs work isothermally, the temperature remains constant. In **Animated Figure 15.8a**, for instance, a metal cylinder contains n moles of an ideal gas, and the large mass of hot water maintains the cylinder and gas at a constant Kelvin temperature T. The piston is held in place initially so the volume of the gas is V_i. As the external force applied to the piston is reduced quasi-statically, the pressure decreases as the gas expands to the final volume V_f. **Animated Figure 15.8b** gives a plot of pressure ($P = nRT/V$) versus volume for the process. The solid red line in the graph is called an isotherm (meaning "constant temperature") and represents the relation between pressure and volume when the temperature is held constant. The work W done by the gas is *not* given by $W = P\,\Delta V = P(V_f - V_i)$ because the pressure is not constant. Nevertheless, the work is equal to the area under the graph. The techniques of integral calculus lead to the following result* for W:

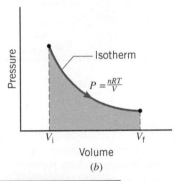

(b)

ANIMATED FIGURE 15.8 *(a)* The ideal gas in the cylinder is expanding isothermally at temperature T. The force holding the piston in place is reduced slowly, so the expansion occurs quasi-statically. *(b)* The work done by the gas is given by the colored area.

Isothermal expansion or compression of an ideal gas

$$W = nRT \ln\left(\frac{V_f}{V_i}\right) \qquad (15.3)$$

Where does the energy for this work originate? Since the internal energy of any ideal gas is proportional to the Kelvin temperature ($U = \frac{3}{2}nRT$ for a monatomic ideal gas, for example), the internal energy remains constant throughout an isothermal process, and the change in internal energy is zero. As a result, the first law of thermodynamics becomes $\Delta U = 0 = Q - W$. In other words, $Q = W$, and the energy for the work originates in the hot water. Heat flows into the gas from the water, as **Animated Figure 15.8a** illustrates. If the gas is compressed isothermally, Equation 15.3 still applies, and heat flows out of the gas into the water. Example 6 deals with the isothermal expansion of an ideal gas.

EXAMPLE 6 | Isothermal Expansion of an Ideal Gas

Two moles of the monatomic gas argon expand isothermally at 298 K, from an initial volume of $V_i = 0.025$ m^3 to a final volume of $V_f = 0.050$ m^3. Assuming that argon behaves as an ideal gas, find **(a)** the work done by the gas, **(b)** the change in the internal energy of the gas, and **(c)** the heat supplied to the gas.

Reasoning and Solution **(a)** The work done by the gas can be found from Equation 15.3:

$$W = nRT \ln\left(\frac{V_f}{V_i}\right) = (2.0 \text{ mol})[8.31 \text{ J/(mol} \cdot \text{K)}](298 \text{ K}) \ln\left(\frac{0.050 \text{ m}^3}{0.025 \text{ m}^3}\right)$$

$$= \boxed{+3400 \text{ J}}$$

(b) The internal energy of a monatomic ideal gas is $U = \frac{3}{2}nRT$ (Equation 14.7) and does not change when the temperature is constant. Therefore, $\boxed{\Delta U = 0 \text{ J}}$.

(c) The heat Q supplied can be determined from the first law of thermodynamics:

$$Q = \Delta U + W = 0 \text{ J} + 3400 \text{ J} = \boxed{+3400 \text{ J}} \qquad (15.1)$$

Adiabatic Expansion or Compression

When a system performs work adiabatically, no heat flows into or out of the system. Figure 15.9a shows an arrangement in which n moles of an ideal gas do work under adiabatic conditions, expanding quasi-statically from an initial volume V_i to a final volume

---

*In this result, "ln" denotes the natural logarithm to the base $e = 2.71828$. The natural logarithm is related to the common logarithm to the base ten by $\ln(V_f/V_i) = 2.303 \log(V_f/V_i)$.

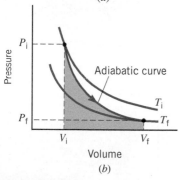

Metal cylinder

Insulating material

(a)

FIGURE 15.9 (a) The ideal gas in the cylinder is expanding adiabatically. The force holding the piston in place is reduced slowly, so the expansion occurs quasi-statically. (b) A plot of pressure versus volume yields the adiabatic curve shown in red, which intersects the isotherms (blue) at the initial temperature T_i and the final temperature T_f. The work done by the gas is given by the colored area.

V_f. The arrangement is similar to that in **Animated Figure 15.8** for isothermal expansion. However, a different amount of work is done here, because the cylinder is now surrounded by insulating material that prevents the flow of heat from occurring, so $Q = 0$ J. According to the first law of thermodynamics, the change in internal energy is $\Delta U = Q - W = -W$. Since the internal energy of an ideal monatomic gas is $U = \frac{3}{2}nRT$ (Equation 14.7), it follows directly that $\Delta U = U_f - U_i = \frac{3}{2}nR(T_f - T_i)$, where T_i and T_f are the initial and final Kelvin temperatures. With this substitution, the relation $\Delta U = -W$ becomes

Adiabatic expansion or compression of a monatomic ideal gas

$$W = \frac{3}{2}nR(T_i - T_f) \tag{15.4}$$

When an ideal gas expands adiabatically, it does positive work, so W is positive in Equation 15.4. Therefore, the term $T_i - T_f$ is also positive, so the final temperature of the gas must be less than the initial temperature. The internal energy of the gas is reduced to provide the necessary energy to do the work, and because the internal energy is proportional to the Kelvin temperature, the temperature decreases. **Figure 15.9b** shows a plot of pressure versus volume for an adiabatic process. The adiabatic curve (red) intersects the isotherms (blue) at the higher initial temperature $[T_i = P_iV_i/(nR)]$ and also at the lower final temperature $[T_f = P_fV_f/(nR)]$. The colored area under the adiabatic curve represents the work done.

The reverse of an adiabatic expansion is an adiabatic compression (W is negative), and Equation 15.4 indicates that the final temperature of the ideal gas exceeds the initial temperature. The energy provided by the agent doing the work increases the internal energy of the gas. As a result, the gas becomes hotter.

The equation that gives the adiabatic curve (red) between the initial pressure and volume (P_i, V_i) and the final pressure and volume (P_f, V_f) in **Figure 15.9b** can be derived using integral calculus. The result is

Adiabatic expansion or compression of an ideal gas

$$P_iV_i^\gamma = P_fV_f^\gamma \tag{15.5}$$

where the exponent γ (Greek gamma) is the ratio of the specific heat capacities at constant pressure and constant volume, $\gamma = c_P/c_V$. Equation 15.5 applies in conjunction with the ideal gas law, because *each point* on the adiabatic curve satisfies the relation $PV = nRT$.

Math Skills Suppose it is necessary to solve Equation 15.5 for the variable V_i. If the value for γ were $\gamma = 2$ (which it never is for an ideal gas), we would begin by taking the square root of both sides of the equals sign in Equation 15.5 and would deal with the square root of V_i^2 (or V_f^2) as follows:

$$\sqrt{V_i^2} = (V_i^2)^{1/2} = V_i^{2(1/2)} = V_i^1 = V_i$$

We will see in the next section that $\gamma = \frac{5}{3}$ for a monatomic ideal gas, and we deal with such a case in a similar way. We begin by raising both sides of Equation 15.5 to the power $1/\gamma$:

$$P_iV_i^\gamma = P_fV_f^\gamma \quad \text{or} \quad (P_iV_i^\gamma)^{1/\gamma} = (P_fV_f^\gamma)^{1/\gamma}$$

Simplifying the terms in the right-hand equation, we have

$$(P_iV_i^\gamma)^{1/\gamma} = (P_fV_f^\gamma)^{1/\gamma} \quad \text{or} \quad P_i^{1/\gamma}V_i^{\gamma(1/\gamma)} = P_f^{1/\gamma}V_f^{\gamma(1/\gamma)} \quad \text{or} \quad P_i^{1/\gamma}V_i = P_f^{1/\gamma}V_f$$

Dividing both sides of the far-right equation by $P_i^{1/\gamma}$ gives

$$\frac{P_i^{1/\gamma}V_i}{P_i^{1/\gamma}} = \frac{P_f^{1/\gamma}V_f}{P_i^{1/\gamma}} \quad \text{or} \quad V_i = \frac{P_f^{1/\gamma}V_f}{P_i^{1/\gamma}} \quad \text{or} \quad V_i = \left(\frac{P_f}{P_i}\right)^{1/\gamma}V_f$$

For $\gamma = \frac{5}{3}$, the exponent of $1/\gamma$ in the final expression for V_i is $\frac{1}{\gamma} = \frac{3}{5}$.

Table 15.1 summarizes the work done in the four types of thermal processes that we have been considering. For each process it also shows how the first law of thermodynamics depends on the work and other variables.

TABLE 15.1 Summary of Thermal Processes

Type of Thermal Process	Work Done	First Law of Thermodynamics ($\Delta U = Q - W$)
Isobaric (constant pressure)	$W = P(V_f - V_i)$	$\Delta U = Q - \underbrace{P(V_f - V_i)}_{W}$
Isochoric (constant volume)	$W = 0 \text{ J}$	$\Delta U = Q - \underbrace{0 \text{ J}}_{Q}$
Isothermal (constant temperature)	$W = nRT \ln\left(\frac{V_f}{V_i}\right)$ (for an ideal gas)	$\underbrace{0 \text{ J}}_{\Delta U \text{ for an ideal gas}} = \underbrace{Q - nRT \ln\left(\frac{V_f}{V_i}\right)}_{W}$
Adiabatic (no heat flow)	$W = \frac{3}{2}nR(T_i - T_f)$ (for a monatomic ideal gas)	$\Delta U = \underbrace{0 \text{ J}}_{Q} - \underbrace{\frac{3}{2}nR(T_i - T_f)}_{W}$

Check Your Understanding

(The answers are given at the end of the book.)

6. One hundred joules of heat is added to a gas, and the gas expands at constant pressure. Is it possible that the internal energy increases by 100 J? **(a)** Yes **(b)** No; the increase in the internal energy is less than 100 J, since work is done by the gas. **(c)** No; the increase in the internal energy is greater than 100 J, since work is done by the gas.

7. A gas is compressed isothermally, and its internal energy increases. Is the gas an ideal gas? **(a)** No, because if the temperature of an ideal gas remains constant, its internal energy must also remain constant. **(b)** No, because if the temperature of an ideal gas remains constant, its internal energy must decrease. **(c)** Yes, because if the temperature of an ideal gas remains constant, its internal energy must increase.

8. A material undergoes an isochoric process that is also adiabatic. Is the internal energy of the material at the end of the process **(a)** greater than, **(b)** less than, or **(c)** the same as it was at the start?

9. **CYU Figure 15.3** shows an arrangement for an adiabatic free expansion or "throttling" process. The process is adiabatic because the entire arrangement is contained within perfectly insulating walls. The gas in chamber A rushes suddenly into chamber B through a hole in the partition. Chamber B is initially evacuated, so the gas expands there under zero external pressure and the work ($W = P \,\Delta V$) it does is zero. Assume that the gas is an ideal gas. How does the final temperature of the gas after expansion compare to its initial temperature? The final temperature is **(a)** greater than, **(b)** less than, **(c)** the same as the initial temperature.

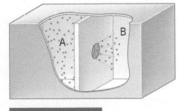

CYU FIGURE 15.3

15.6 | Specific Heat Capacities

In this section the first law of thermodynamics is used to gain an understanding of the factors that determine the specific heat capacity of a material. Remember, when the temperature of a substance changes as a result of heat flow, the change in temperature ΔT and the amount of heat Q are related according to $Q = cm \,\Delta T$ (Equation 12.4). In this expression c denotes the specific heat capacity in units of J/(kg · C°), and m is the mass in kilograms. Now, however, it is more convenient to express the amount of material as the number of moles n, rather than the number of kilograms. Therefore, we replace the expression $Q = cm \,\Delta T$ with the following analogous expression:

$$Q = Cn \,\Delta T \qquad\qquad (15.6)$$

where the capital letter C (as opposed to the lowercase c) refers to the **molar specific heat capacity** in units of J/(mol · K). In addition, the unit for measuring the temperature change ΔT is the kelvin (K) rather than the Celsius degree (C°), and $\Delta T = T_f - T_i$, where T_f and T_i are the final and initial temperatures. For gases it is necessary to distinguish between the molar specific heat capacities C_P and C_V, which apply, respectively, to conditions of constant pressure and constant volume. With the help of the first law of thermodynamics and an ideal gas as an example, it is possible to see why C_P and C_V differ.

To determine the molar specific heat capacities, we must first calculate the heat Q needed to raise the temperature of an ideal gas from T_i to T_f. According to the first law, $Q = \Delta U + W$. We also know that the internal energy of a monatomic ideal gas is $U = \frac{3}{2} nRT$ (Equation 14.7). As a result, $\Delta U = U_f - U_i = \frac{3}{2} nR(T_f - T_i)$. When the heating process occurs at constant pressure, the work done is given by Equation 15.2: $W = P \Delta V = P(V_f - V_i)$. For an ideal gas, $PV = nRT$, so the work becomes $W = nR(T_f - T_i)$. On the other hand, when the volume is constant, $\Delta V = 0$ m³, and the work done is zero. The calculation of the heat is summarized below:

$$Q = \Delta U + W$$
$$Q_{\text{constant pressure}} = \tfrac{3}{2} nR(T_f - T_i) + nR(T_f - T_i) = \tfrac{5}{2} nR(T_f - T_i)$$
$$Q_{\text{constant volume}} = \tfrac{3}{2} nR(T_f - T_i) + 0$$

The molar specific heat capacities can now be determined, since Equation 15.6 indicates that $C = Q/[n(T_f - T_i)]$:

Constant pressure for a monatomic ideal gas
$$C_P = \frac{Q_{\text{constant pressure}}}{n(T_f - T_i)} = \tfrac{5}{2} R \qquad (15.7)$$

Constant volume for a monatomic ideal gas
$$C_V = \frac{Q_{\text{constant volume}}}{n(T_f - T_i)} = \tfrac{3}{2} R \qquad (15.8)$$

The ratio γ of the specific heats is

Monatomic ideal gas
$$\gamma = \frac{C_P}{C_V} = \frac{\tfrac{5}{2} R}{\tfrac{3}{2} R} = \frac{5}{3} \qquad (15.9)$$

For real monatomic gases near room temperature, experimental values of C_P and C_V give ratios very close to the theoretical value of $\gamma = \tfrac{5}{3}$.

Many gases are not monatomic. Instead, they consist of molecules formed from more than one atom. The oxygen in our atmosphere, for example, is a diatomic gas, because it consists of molecules formed from two oxygen atoms. Similarly, atmospheric nitrogen is a diatomic gas consisting of molecules formed from two nitrogen atoms. Whereas the individual atoms in a monatomic ideal gas can exhibit only translational motion, the molecules in a diatomic ideal gas can exhibit translational and rotational motion, as well as vibrational motion at sufficiently high temperatures. The result of such additional motions is that Equations 15.7–15.9 do not apply to a diatomic ideal gas. Instead, if the temperature is sufficiently moderate that the diatomic molecules do not vibrate, the molar specific heat capacities of a diatomic ideal gas are $C_P = \tfrac{7}{2} R$ and $C_V = \tfrac{5}{2} R$, with the result that $\gamma = \frac{C_P}{C_V} = \tfrac{7}{5}$.

The difference between C_P and C_V arises because work is done when the gas expands in response to the addition of heat under conditions of constant pressure, whereas no work is done under conditions of constant volume. For a monatomic ideal gas, C_P exceeds C_V by an amount equal to R, the ideal gas constant:

$$C_p - C_V = R \qquad (15.10)$$

In fact, it can be shown that Equation 15.10 applies to any kind of ideal gas—monatomic, diatomic, etc.

Check Your Understanding

(The answers are given at the end of the book.)

10. Suppose that a material contracts when it is heated. Following the same line of reasoning used in the text to reach Equations 15.7 and 15.8, deduce the relationship between the specific heat capacity at constant pressure (C_P) and the specific heat capacity at constant volume (C_V). Which of the following describes the relationship? **(a)** $C_P = C_V$ **(b)** C_P is greater than C_V **(c)** C_P is less than C_V.

11. You want to heat a gas so that its temperature will be as high as possible. Should you heat the gas under conditions of **(a)** constant pressure or **(b)** constant volume? **(c)** It does not matter what the conditions are.

15.7 | The Second Law of Thermodynamics

Ice cream melts when left out on a warm day. A cold can of soda warms up on a hot day at a picnic. Ice cream and soda never become colder when left in a hot environment, for heat always flows spontaneously from hot to cold, and never from cold to hot. The spontaneous flow of heat is the focus of one of the most profound laws in all of science, the **second law of thermodynamics**.

> **THE SECOND LAW OF THERMODYNAMICS: THE HEAT FLOW STATEMENT**
>
> Heat flows spontaneously from a substance at a higher temperature to a substance at a lower temperature and does not flow spontaneously in the reverse direction.

It is important to realize that the second law of thermodynamics deals with a different aspect of nature than does the first law of thermodynamics. The second law is a statement about the natural tendency of heat to flow from hot to cold, whereas the first law deals with energy conservation and focuses on both heat and work. A number of important devices depend on heat and work in their operation, and to understand such devices both laws are needed. For instance, an automobile engine is a type of heat engine because it uses heat to produce work. In discussing heat engines, Sections 15.8 and 15.9 will bring together the first and second laws to analyze engine efficiency. Then, in Section 15.10 we will see that refrigerators, air conditioners, and heat pumps also utilize heat and work and are closely related to heat engines. The way in which these three appliances operate also depends on both the first and second laws of thermodynamics. Before moving on to our study of heat engines, we discuss an important thermodynamic system, where heat from a hotter source, the Sun, is transferred to a colder object, planet earth.

THE PHYSICS OF . . . global climate change. Earth, the Sun, and the surrounding space constitute a thermodynamic system, and our planet is warming. The surface temperature of Earth is determined primarily by the radiation energy absorbed from the Sun. Energy produced by the decay of radioactive elements (Chapter 31) in Earth's interior also heat the surface, but this contributes only a small fraction, about 1/1000, of the heating provided by the Sun. Since Earth's surface is in thermal equilibrium with its surroundings, it must be emitting radiant energy into space at the same rate at which it is absorbing it, as described in Section 13.3. The rate of the emitted radiation depends on the surface temperature (Equation 13.2), and if the absorption and emission of radiation were the only two factors in determining the surface temperature of Earth, its average temperature would be −18 °C, which is far too cold to support life as we know it. However, since the average temperature of Earth's surface is closer to 15 °C, there must be another mechanism that contributes to warming. This is where Earth's atmosphere comes into play. Gases in Earth's atmosphere, such as carbon dioxide (CO_2), water vapor

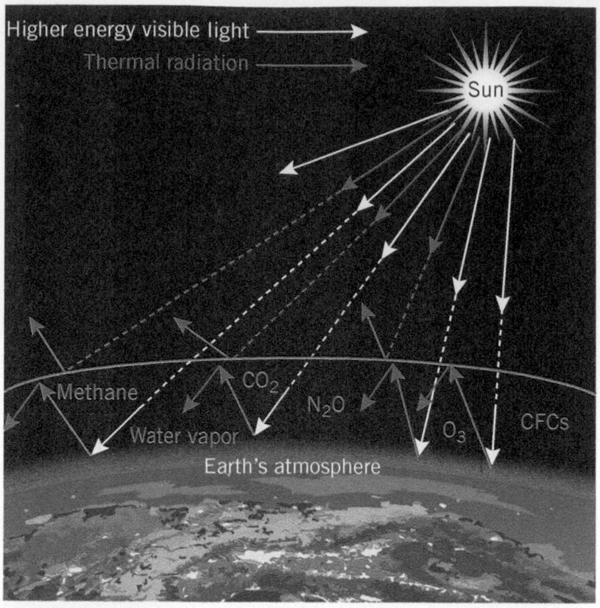

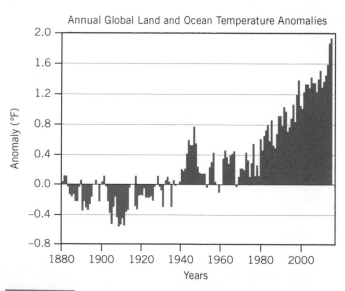

FIGURE 15.10 An atmosphere populated with greenhouse gases absorbs more of the thermal radiation emitted from the oceans and land masses and redirects a portion of it back to the earth, thereby warming its surface.

(H_2O), chloroflourocarbons (CF_2Cl_2), methane (CH_4), nitrous oxide (N_2O), and ozone (O_3)-the so-called **greenhouse gases**-allow the Sun's shorter wavelength visible light to pass through and reach Earth, while the longer wavelength thermal radiation is absorbed and reemitted, much of it back into space (see **Figure 15.10**). The sunlight that does reach Earth warms its surface, and then the surface reemits a portion of this energy as thermal radiation. Just as the greenhouse gases in the atmosphere prevented the thermal radiation from the Sun from reaching Earth's surface, they now prevent the thermal radiation emitted by Earth's surface from leaving the atmosphere. This effectively acts like a blanket over Earth, trapping additional heat and raising the surface temperature. This process, called the **greenhouse effect**, is what produces the moderate temperature on Earth's surface.

Venus, a planet similar in size and density to Earth, is nearly 30% closer to the Sun than Earth is, and yet its average surface temperature, based solely on the absorption and emission of the Sun's radiant energy (neglecting atmospheric effects), should be −41 °C. This is even colder than the value predicted for Earth, and it results from the fact that Venus reflects a greater fraction of the Sun's energy than Earth does. Nevertheless, the average surface temperature on Venus is 460 °C, which makes it the hottest planet in the solar system! The atmosphere of Venus is composed of greater than 96% CO_2 (compared to just 0.04% on Earth), and this high concentration of greenhouse gases effectively traps the heat from the Sun's radiant energy, resulting in the planet's high surface temperature. The conditions on Venus correspond to a **runaway greenhouse effect**, which occurs when the concentration of greenhouse gases in a planet's atmosphere are so high that it blocks the thermal radiation from the planet and prevents the planet from cooling.

Since the late 1880's, the average surface temperature of Earth has increased by over 1.8 °F (**Figure 15.11**). While pollution, overpopulation, and deforestation have all contributed to the effects of this global warming, the most significant cause of the increasing temperature is the greenhouse effect. And while it is true that the average surface temperature of Earth and the level of carbon dioxide in the atmosphere have naturally fluctuated over time, since 1750 the level of CO_2 has increased by over 45%. Ice-core data over the last 800 000 years show the average level of CO_2 in the atmosphere was about 225 ppm (parts per million), while never spiking above 300 ppm. Just in the last 270 years, the level has increased from 280 ppm to over 400 ppm (**Figure 15.12**). In other words, human beings as a species have never lived on Earth during a time when the CO_2 levels in the atmosphere were this high. Human activities after the Industrial Revolution, primarily in the form of the extraction and burning of fossil fuels, such as oil, coal, and natural gas, are responsible for the vast majority of the increase in CO_2. To date, there are no alternative explanations, natural or otherwise, that can reasonably account for the rapid change in climate. The Intergovernmental Panel on Climate Change (IPCC) concluded that "human influence on climate has been the dominant cause of observed global warming since the mid-20th century".[1]

An increase of 1.8 °F does not seem like much, so why should we be concerned? The last decade (2000–2019) was the hottest decade on record. In fact, the past five years have been the hottest five years on record for the second year running. With this rise in surface temperature, vast amounts of the polar ice caps, sea ice, glaciers, and permafrost are melting, which is leading to a rise in sea levels. One conservative estimate suggests a 4.2-foot rise in sea level for every 1 °F increase in average surface

FIGURE 15.11 The annual average change in the surface temperature of the earth relative to 1880. Reprinted from USGRCP and Wuebbles *et al.* (eds.), *Climate Science Special Report: Fourth National Climate Assessment,* Vol. 1 (Washington, DC: USGRCP), 2017.)

[1]Allen *et al.*, "Chapter 1: Framing and Context," IPCC SR15, 2018, pp. 49–91.

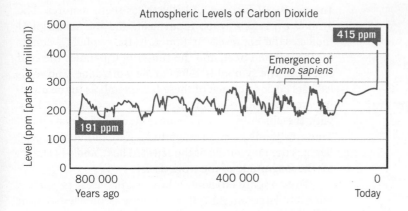

FIGURE 15.12 The level of carbon dioxide gas in the atmosphere over time. The average concentration was about 225 ppm dating back 800 000 years. However, after the year 1750, the level increased, and it is presently over 400 ppm- a level never before reached in the history of human beings. Adapted from NOAA Climate.gov based on data from Lüthi, *et al.*, 2008, via NOAA NCEI Paleoclimatology Program.

temperature.[2] Widespread coastal flooding is expected in the future, with 150 million people being displaced during high tide by the year 2050.[3] A warmer climate also results in more extreme weather, with heavy rain and flooding, more intense hurricanes, tornados, and snow storms, as well as longer and more widespread periods of drought. These changes in weather will ultimately displace our coastal populations and affect our food supply, both in agriculture and livestock production, as well as have negative impacts on our health.

The effects of global warming cannot be stopped completely, but the impacts of climate change can be mitigated by reducing our emission of greenhouse gases. For example, in order to keep the mean surface temperature of Earth from rising more than 2.7 °F (1.5 °C), we need to bring our CO_2 emissions to zero by 2050. Climate experts agree that an increase in temperature above 2.7 °F would result in catastrophic changes in climate. This is a huge challenge for humankind, and the fate of countless other animal species and entire eco-systems will depend on our ability to act. We will need to reduce our dependence on fossil fuels and develop alternative sources of renewable energy, such as wind and solar. Nuclear energy offers certain advantages over fossil fuels, but it also has significant disadvantages, such as the creation and storage of nuclear waste. However, more than just using different sources of energy, combating global climate change will require us to change our behavior, in terms of how we live, work, travel, and play. Fortunately, we still have time, but the window for action is quickly closing. This will not be solved by one country alone, but will require a concerted effort of all nations. The Paris Agreement, which was created under the United Nations Framework Convention on Climate Change, is a multinational effort aimed at limiting the negative impacts of global warming. It proposed both energy and climate policy with the goals of 1) reducing CO_2 emissions by 20%, 2) increasing the renewable energy market by 20%, and 3) improving the efficiency of existing energy technologies by 20%. The agreement was signed in April 2016 by 196 nations. The energy carried by sunlight and its impact on the greenhouse effect is discussed more in Section 24.4.

15.8 | Heat Engines

THE PHYSICS OF . . . a heat engine.   A **heat engine** is any device that uses heat to perform work. It has three essential features:

1. Heat is supplied to the engine at a relatively high input temperature from a place called the *hot reservoir.*

2. Part of the input heat is used to perform work by the *working substance* of the engine, which is the material within the engine that actually does the work (e.g., the gasoline–air mixture in an automobile engine).

3. The remainder of the input heat is rejected to a place called the *cold reservoir,* which has a temperature lower than the input temperature.

---

[2]U.S. Global Change Research Program (USGCRP), "Climate Science Special Report: Chapter 12. Sea Level Rise," retrieved December 27, 2018, from science2017.globalchange.gov.
[3]Kulp, S. A., and Strauss, B. H., "New Elevation Data Triple Estimates of Global Vulnerability to Sea-Level Rise and Coastal Flooding," *Nature Communications*, Vol. 10, no. 1, p. 4844 (2019).

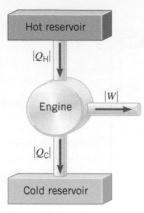

Hot reservoir

$|Q_\text{H}|$

Engine

$|W|$

$|Q_\text{C}|$

Cold reservoir

INTERACTIVE FIGURE 15.13 This schematic representation of a heat engine shows the input heat (magnitude = $|Q_\text{H}|$) that originates from the hot reservoir, the work (magnitude = $|W|$) that the engine does, and the heat (magnitude = $|Q_\text{C}|$) that the engine rejects to the cold reservoir.

Interactive Figure 15.13 illustrates these features. The symbol Q_H refers to the input heat, and the subscript H indicates the hot reservoir. Similarly, the symbol Q_C stands for the rejected heat, and the subscript C denotes the cold reservoir. The symbol W refers to the work done. The vertical bars enclosing each of these three symbols in the drawing are included to emphasize that we are concerned here with the absolute values, or magnitudes, of the symbols. Thus, $|Q_\text{H}|$ indicates the magnitude of the input heat, $|Q_\text{C}|$ denotes the magnitude of the rejected heat, and $|W|$ stands for the magnitude of the work done.

To be highly efficient, a heat engine must produce a relatively large amount of work from as little input heat as possible. Therefore, the **efficiency** e of a heat engine is defined as the ratio of the magnitude of the work $|W|$ done by the engine to the magnitude of the input heat $|Q_\text{H}|$:

$$e = \frac{|W|}{|Q_\text{H}|} \qquad (15.11)$$

If the input heat were converted entirely into work, the engine would have an efficiency of 1.00, since $|W| = |Q_\text{H}|$; such an engine would be 100% efficient. ***Efficiencies are often quoted as percentages obtained by multiplying the ratio $|W|/|Q_\text{H}|$ by a factor of 100***. Thus, an efficiency of 68% would mean that a value of 0.68 is used for the efficiency in Equation 15.11.

An engine, like any device, must obey the principle of conservation of energy. Some of the engine's input heat $|Q_\text{H}|$ is converted into work $|W|$, and the remainder $|Q_\text{C}|$ is rejected to the cold reservoir. If there are no other losses in the engine, the principle of energy conservation requires that

$$|Q_\text{H}| = |W| + |Q_\text{C}| \qquad (15.12)$$

Solving this equation for $|W|$ and substituting the result into Equation 15.11 leads to the following alternative expression for the efficiency e of a heat engine:

$$e = \frac{|Q_\text{H}| - |Q_\text{C}|}{|Q_\text{H}|} = 1 - \frac{|Q_\text{C}|}{|Q_\text{H}|} \qquad (15.13)$$

Example 7 illustrates how the concepts of efficiency and energy conservation are applied to a heat engine.

Math Skills The symbols $|Q_\text{H}|$, $|Q_\text{C}|$, and $|W|$ refer to absolute values or magnitudes only. It is essential to remember that these symbols never have negative values assigned to them when they appear in equations. For example, the value assigned to $|W|$ is the same if $W = -1830$ J or $W = +1830$ J.

This follows because

$$|W| = |-1830 \text{ J}| = 1830 \text{ J}$$

and

$$|W| = |+1830 \text{ J}| = 1830 \text{ J}$$

Analyzing Multiple-Concept Problems

EXAMPLE 7 | An Automobile Engine

An automobile engine has an efficiency of 22.0% and produces 2510 J of work. How much heat is rejected by the engine?

Reasoning Energy conservation indicates that the amount of heat rejected to the cold reservoir is the part of the input heat that is *not* converted into work. The work is given, and the input heat can be obtained since the efficiency of the engine is also given.

Knowns and Unknowns The following data are available:

Description	Symbol	Value
Efficiency of engine	e	22.0% (0.220)
Magnitude of work	$|W|$	2510 J
Unknown Variable		
Magnitude of rejected heat	$|Q_\text{C}|$	?

Modeling the Problem

STEP 1 The Conservation of Energy According to the energy-conservation principle, the magnitudes of the input heat $|Q_H|$, the work done $|W|$, and the rejected heat $|Q_C|$ are related according to $|Q_H| = |W| + |Q_C|$ (Equation 15.12). Solving for $|Q_C|$ gives Equation 1 at the right. In this result, $|W|$ is known, but $|Q_H|$ is not, although it will be evaluated in Step 2.

$$|Q_C| = |Q_H| - |W| \qquad (1)$$

$$\uparrow$$

$$\boxed{?}$$

STEP 2 Engine Efficiency Equation 15.11 gives the engine efficiency as $e = |W|/|Q_H|$. Solving for $|Q_H|$, we find that

$$\boxed{|Q_H| = \frac{|W|}{e}}$$

which can be substituted into Equation 1 as shown in the right column.

$$|Q_C| = |Q_H| - |W| \qquad (1)$$

$$\uparrow$$

$$\boxed{|Q_H| = \frac{|W|}{e}}$$

> **Problem-Solving Insight** When efficiency is stated as a percentage (e.g., 22.0%), it must be converted to a decimal fraction (e.g., 0.220) before being used in an equation.

Solution Combining the results of each step algebraically, we find that

$$\overset{\text{STEP 1}}{\downarrow} \qquad \overset{\text{STEP 2}}{\downarrow}$$

$$|Q_C| = |Q_H| - |W| = \left|\frac{|W|}{e}\right| - |W|$$

The magnitude of the rejected heat, then, is

$$|Q_C| = |W|\left(\frac{1}{e} - 1\right) = (2510 \text{ J})\left(\frac{1}{0.220} - 1\right) = \boxed{8900 \text{ J}}$$

Related Homework: Problem 41

In Example 7, less than one-quarter of the input heat is converted into work because the efficiency of the automobile engine is only 22.0%. If the engine were 100% efficient, all the input heat would be converted into work. Unfortunately, nature does not permit 100%-efficient heat engines to exist, as the next section discusses.

15.9 | Carnot's Principle and the Carnot Engine

What is it that allows a heat engine to operate with maximum efficiency? The French engineer Sadi Carnot (1796–1832) proposed that a heat engine has maximum efficiency when the processes within the engine are reversible. *A reversible process is one in which both the system and its environment can be returned to exactly the states they were in before the process occurred.*

In a reversible process, *both* the system and its environment can be returned to their initial states. Therefore, a process that involves an energy-dissipating mechanism, such as friction, cannot be reversible because the energy wasted due to friction would alter the system or the environment or both. There are also reasons other than friction why a process may not be reversible. For instance, the spontaneous flow of heat from a hot substance to a cold substance is irreversible, even though friction is not present. For heat to flow in the reverse direction, work must be done, as we will see in Section 15.10. The agent doing such work must be located in the environment of the hot and cold substances, and, therefore, the environment must change while the heat is moved back from cold to hot. Since the system and the environment cannot *both* be returned to their initial states, the process of spontaneous heat flow is irreversible. In fact, all spontaneous processes are irreversible, such as the explosion of an unstable chemical or the bursting of a bubble. When the word "reversible" is used in connection with engines, it does not just mean a gear that allows the engine to operate a device in reverse. All cars have a reverse gear, for instance, but no automobile engine is thermodynamically reversible, since friction exists no matter which way the car moves.

Today, the idea that the efficiency of a heat engine is a maximum when the engine operates reversibly is referred to as **Carnot's principle**.

> **CARNOT'S PRINCIPLE: AN ALTERNATIVE STATEMENT OF THE SECOND LAW OF THERMODYNAMICS**
>
> No irreversible engine operating between two reservoirs at constant temperatures can have a greater efficiency than a reversible engine operating between the same temperatures. Furthermore, all reversible engines operating between the same temperatures have the same efficiency.

Carnot's principle is quite remarkable, for no mention is made of the working substance of the engine. It does not matter whether the working substance is a gas, a liquid, or a solid. As long as the process is reversible, the efficiency of the engine is a maximum. However, Carnot's principle does *not* state, or even imply, that a reversible engine has an efficiency of 100%.

It can be shown that if Carnot's principle were not valid, it would be possible for heat to flow spontaneously from a cold substance to a hot substance, in violation of the second law of thermodynamics. In effect, then, Carnot's principle is another way of expressing the second law.

No real engine operates reversibly. Nonetheless, the idea of a reversible engine provides a useful standard for judging the performance of real engines. **Figure 15.14** shows a reversible engine, called a **Carnot engine**, that is particularly useful as an idealized model. An important feature of a Carnot engine is that all input heat (magnitude = $|Q_H|$) originates from a hot reservoir *at a single temperature* T_H and all rejected heat (magnitude = $|Q_C|$) goes into a cold reservoir *at a single temperature* T_C.

Carnot's principle implies that the efficiency of a reversible engine is independent of the working substance of the engine, and therefore can depend only on the temperatures of the hot and cold reservoirs. Since efficiency is $e = 1 - |Q_C|/|Q_H|$ according to Equation 15.13, the ratio $|Q_C|/|Q_H|$ can depend only on the reservoir temperatures. This observation led Lord Kelvin to propose a **thermodynamic temperature scale**. He proposed that the thermodynamic temperatures of the cold and hot reservoirs be defined such that their ratio is equal to $|Q_C|/|Q_H|$. Thus, the thermodynamic temperature scale is related to the heats absorbed and rejected by a Carnot engine, and is independent of the working substance. If a reference temperature is properly chosen, it can be shown that the thermodynamic temperature scale is identical to the Kelvin scale introduced in Section 12.2 and used in the ideal gas law. As a result, the ratio of the magnitude of the rejected heat $|Q_C|$ to the magnitude of the input heat $|Q_H|$ is

$$\frac{|Q_C|}{|Q_H|} = \frac{T_C}{T_H} \qquad (15.14)$$

where the temperatures T_C and T_H *must be expressed in kelvins.*

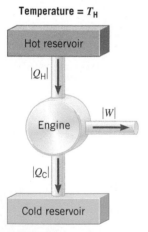

Temperature = T_H

Hot reservoir

$|Q_H|$

Engine → $|W|$

$|Q_C|$

Cold reservoir

Temperature = T_C

FIGURE 15.14 A Carnot engine is a reversible engine in which all input heat (magnitude = $|Q_H|$) originates from a hot reservoir at a single temperature T_H, and all rejected heat (magnitude = $|Q_C|$) goes into a cold reservoir at a single temperature T_C.

Math Skills When determining the efficiency of a Carnot engine with Equation 15.15, be sure the temperatures T_C and T_H of the cold and hot reservoirs are expressed in kelvins; degrees Celsius or Fahrenheit will lead to errors. Suppose, for example, that a Carnot engine has reservoir temperatures of $T_C = 254$ K (-19 °C) and $T_H = 294$ K (21 °C). According to Equation 15.15, the efficiency of this engine is

$$e_{Carnot} = 1 - \frac{T_C}{T_H} = 1 - \frac{254\ K}{294\ K} = 0.136 \quad \text{or} \quad 13.6\%$$

For comparison, the incorrect calculation using Celsius degrees is shown below:

Incorrect Calculation $\quad e_{Carnot} = 1 - \dfrac{T_C}{T_H} = 1 - \dfrac{-19\ °C}{21\ °C} = 1.90 \quad \text{or} \quad 190\%$

This result makes no sense, since it implies an impossible efficiency of greater than 100%. The trouble with the Celsius scale in the calculation is that it allows values that are below zero, or negative. The Kelvin scale, in contrast, is chosen so that its zero temperature point is the lowest temperature attainable. Therefore, it cannot give efficiencies greater than 100% when used in Equation 15.15.

The efficiency e_{Carnot} of a Carnot engine can be written in a particularly useful way by substituting Equation 15.14 into Equation 15.13 for the efficiency, $e = 1 - |Q_C|/|Q_H|$:

Efficiency of a Carnot engine $\qquad\qquad e_{Carnot} = 1 - \dfrac{T_C}{T_H} \qquad\qquad$ **(15.15)**

This relation gives the *maximum possible efficiency* for a heat engine operating between two Kelvin temperatures T_C and T_H, and the next example illustrates its application.

EXAMPLE 8 | The Physics of Extracting Work from a Warm Ocean

Water near the surface of a tropical ocean has a temperature of 298.2 K (25.0 °C), whereas water 700 m beneath the surface has a temperature of 280.2 K (7.0 °C). It has been proposed that the warm water be used as the hot reservoir and the cool water as the cold reservoir of a heat engine. Find the maximum possible efficiency for such an engine.

Reasoning The maximum possible efficiency is the efficiency that a Carnot engine would have (Equation 15.15) operating between temperatures of $T_H = 298.2$ K and $T_C = 280.2$ K.

Solution Using $T_H = 298.2$ K and $T_C = 280.2$ K in Equation 15.15, we find that

$$e_{Carnot} = 1 - \frac{T_C}{T_H} = 1 - \frac{280.2\ K}{298.2\ K} = \boxed{0.060\ (6.0\%)}$$

In Example 8 the maximum possible efficiency is only 6.0%. The small efficiency arises because the Kelvin temperatures of the hot and cold reservoirs are so close. A greater efficiency is possible only when there is a greater difference between the reservoir temperatures. However, there are limits on how large the efficiency of a heat engine can be, as Conceptual Example 9 discusses.

CONCEPTUAL EXAMPLE 9 | Natural Limits on the Efficiency of a Heat Engine

Consider a hypothetical engine that receives 1000 J of heat as input from a hot reservoir and delivers 1000 J of work, rejecting no heat to a cold reservoir whose temperature is above 0 K. Which law of thermodynamics does this engine violate? **(a)** The first law **(b)** The second law **(c)** Both the first and second laws

Reasoning The first law of thermodynamics is an expression of energy conservation. The second law states that no irreversible engine operating between two reservoirs at constant temperatures can have a greater efficiency than a reversible engine operating between the same two reservoirs. The efficiency of such a reversible engine is e_{Carnot}, the efficiency of a Carnot engine.

Answers (a) and (c) are incorrect. From the point of view of energy conservation, nothing is wrong with an engine that converts 1000 J of heat into 1000 J of work. Energy has been neither created nor destroyed; it has only been transformed from one form (heat) into another form (work). Therefore, this engine does not violate the first law of thermodynamics.

Answer (b) is correct. Since all of the input heat is converted into work, the efficiency of the engine is 1, or 100%. But Equation 15.15, which is based on the second law of thermodynamics, indicates that the maximum possible efficiency is $e_{Carnot} = 1 - T_C/T_H$, where T_C and T_H are the temperatures of the cold and hot reservoirs, respectively. Since we are told that T_C is above 0 K, it is clear that the ratio T_C/T_H is greater than zero, so the maximum possible efficiency is less than 1 (or less than 100%). The hypothetical engine, therefore, violates the second law of thermodynamics, which limits the efficiencies of heat engines to values less than 100%.

THE PHYSICS OF . . . thermal pollution. Example 9 has emphasized that *even a perfect heat engine has an efficiency that is less than 1.0 or 100%.* In this regard, we note that the maximum possible efficiency, as given by Equation 15.15, approaches 1.0 when T_C approaches absolute zero (0 K). However, experiments have shown that it is not possible to cool a substance to absolute zero (see Section 15.12), so nature does not allow a 100%-efficient heat engine to exist. As a result, there will always be heat rejected to a cold reservoir whenever a heat engine is used to do work, even if friction and other irreversible processes are eliminated completely. This rejected heat is a form of thermal pollution. The second law of thermodynamics requires that at least some thermal pollution be generated whenever heat engines are used to perform work. This kind of thermal pollution can be reduced only if society reduces its dependence on heat engines to do work.

Check Your Understanding

(The answers are given at the end of the book.)

12. The second law of thermodynamics, in the form of Carnot's principle, indicates that the most efficient heat engine operating between two temperatures is a reversible one. Does this mean that a reversible engine operating between the temperatures of 600 and 400 K must be more efficient than an *irreversible* engine operating between 700 and 300 K?

13. Three reversible engines, A, B, and C, use the same cold reservoir for their exhaust heats. However, they use different hot reservoirs that have the following temperatures: (A) 1000 K, (B) 1100 K, and (C) 900 K. Rank these engines in order of increasing efficiency (smallest efficiency first). **(a)** A, C, B **(b)** C, B, A **(c)** B, A, C **(d)** C, A, B

14. Suppose that you wish to improve the efficiency of a Carnot engine. Which answer describes the best way? **(a)** Lower the Kelvin temperature of the cold reservoir by a factor of four. **(b)** Raise the Kelvin temperature of the hot reservoir by a factor of four. **(c)** Cut the Kelvin temperature of the cold reservoir in half and double the Kelvin temperature of the hot reservoir. **(d)** All three choices give the same improvement in efficiency.

15. Consider a hypothetical device that takes 10 000 J of heat from a hot reservoir and 5000 J of heat from a cold reservoir (whose temperature is greater than 0 K) and produces 15 000 J of work. What can be said about this device? **(a)** It violates the first law of thermodynamics but not the second law. **(b)** It violates the second law of thermodynamics but not the first law. **(c)** It violates both the first and second laws of thermodynamics. **(d)** It does not violate either the first or the second law of thermodynamics.

15.10 Refrigerators, Air Conditioners, and Heat Pumps

The natural tendency of heat is to flow from hot to cold, as indicated by the second law of thermodynamics. However, if work is used, heat can be *made* to flow from cold to hot, against its natural tendency. Refrigerators, air conditioners, and heat pumps are, in fact, devices that do just that. As **Figure 15.15** illustrates, these devices use work (magnitude = |W|) to extract heat (magnitude = |Q_C|) from the cold reservoir and deposit heat (magnitude = |Q_H|) into the hot reservoir. Generally speaking, such a process is called a **refrigeration process**. A comparison of the left and right sides of this drawing shows that the directions of the arrows symbolizing heat and work in a refrigeration process are opposite to those in an engine process. Nonetheless, energy is conserved during a refrigeration process, just as it is in an engine process, so |Q_H| = |W| + |Q_C|. Moreover, if the process occurs reversibly, we have ideal devices that are called Carnot refrigerators, Carnot air conditioners, and Carnot heat pumps. For these ideal devices, the relation |Q_C|/|Q_H| = T_C/T_H (Equation 15.14) applies, just as it does for a Carnot engine.

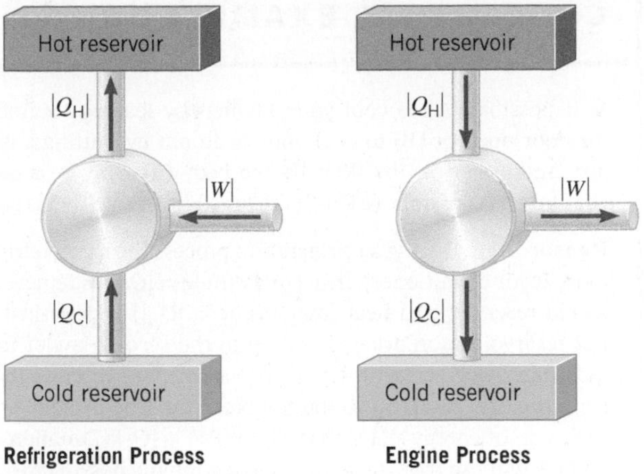

THE PHYSICS OF . . . refrigerators. In a **refrigerator**, the interior of the unit is the cold reservoir, while the warmer exterior is the hot reservoir. As **Figure 15.16** illustrates, the refrigerator takes heat from the food inside and deposits it into the kitchen, along with the energy needed to do the work of making the heat flow from cold to hot. For this reason, the outside surfaces (usually the sides and back) of most refrigerators are warm to the touch while the units operate.

THE PHYSICS OF . . . air conditioners. An **air conditioner** is like a refrigerator, except that the room itself is the cold reservoir and the outdoors is the hot reservoir. **Figure 15.17** shows a window unit, which cools a room by removing heat and depositing it outside, along with the work used to make the heat flow from cold to hot. Conceptual Example 10 considers a common misconception about refrigerators and air conditioners.

Refrigeration Process **Engine Process**

FIGURE 15.15 In the refrigeration process on the left, work (magnitude = |W|) is used to remove heat (magnitude = |Q_C|) from the cold reservoir and deposit heat (magnitude = |Q_H|) into the hot reservoir. Compare this with the engine process on the right.

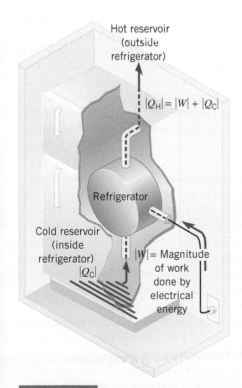

FIGURE 15.16 A refrigerator.

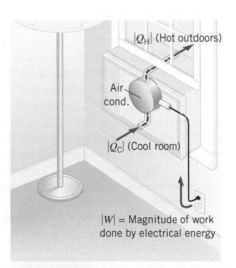

FIGURE 15.17 A window air conditioner removes heat from a room, which is the cold reservoir, and deposits heat outdoors, which is the hot reservoir.

CONCEPTUAL EXAMPLE 10 | You Can't Beat the Second Law of Thermodynamics

Is it possible (A) to cool your kitchen by leaving the refrigerator door open or (B) to cool your bedroom by putting a window air conditioner on the floor by the bed? **(a)** Only A is possible. **(b)** Only B is possible. **(c)** Both are possible. **(d)** Neither is possible.

Reasoning   During a refrigeration process (be it in a refrigerator or in an air conditioner), heat (magnitude = $|Q_C|$) is removed from a cold reservoir and heat (magnitude = $|Q_H|$) is deposited into a hot reservoir. Moreover, according to the second law of thermodynamics, work (magnitude = $|W|$) is required to move this heat from the cold reservoir to the hot reservoir. The principle of conservation of energy states that $|Q_H| = |W| + |Q_C|$ (Equation 15.12), and we will use this as a guide in assessing the possibilities.

Answers (a), (b), and (c) are incorrect.   If you wanted to cool your kitchen by leaving the refrigerator door open, the refrigerator would have to take heat from directly in front of the open door and pump *less* heat out the back of the unit and into the kitchen (since the refrigerator is supposed to be cooling the entire kitchen). Likewise, if you tried to cool your entire bedroom by placing the air conditioner on the floor by the bed, the air conditioner would have

to take heat (magnitude = $|Q_C|$) from directly in front of the unit and deposit *less* heat (magnitude = $|Q_H|$) out the back. According to the second law of thermodynamics this cannot happen, since $|Q_H| = |W| + |Q_C|$; that is, $|Q_H|$ is greater than (not less than) $|Q_C|$ because $|W|$ is greater than zero.

Answer (d) is correct.   The heat (magnitude = $|Q_C|$) removed from the air directly in front of the open refrigerator is deposited back into the kitchen at the rear of the unit. Moreover, according to the second law of thermodynamics, work (magnitude = $|W|$) is needed to move that heat from cold to hot, and the energy from this work is also deposited into the kitchen as additional heat. Thus, the open refrigerator puts into the kitchen an amount of heat $|Q_H| = |W| + |Q_C|$, which is *more* than it removes from in front of the open refrigerator. Thus, rather than cooling the kitchen, the open refrigerator warms it up. Putting an air conditioner on the floor to cool your bedroom is similarly a no-win game. The heat pumped out the back of the air conditioner and into the bedroom is greater than the heat pulled into the front of the unit. Consequently, the air conditioner actually warms the bedroom.

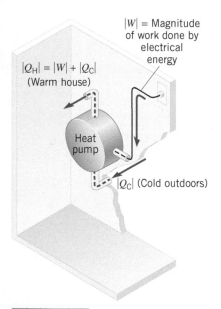

$|Q_H| = |W| + |Q_C|$
(Warm house)

$|W|$ = Magnitude of work done by electrical energy

Heat pump

$|Q_C|$ (Cold outdoors)

FIGURE 15.18   In a heat pump the cold reservoir is the wintry outdoors, and the hot reservoir is the inside of the house.

The quality of a refrigerator or air conditioner is rated according to its coefficient of performance. Such appliances perform well when they remove a relatively large amount of heat (magnitude = $|Q_C|$) from a cold reservoir by using as small an amount of work (magnitude = $|W|$) as possible. Therefore, the coefficient of performance is defined as the ratio of $|Q_C|$ to $|W|$, and the greater this ratio is, the better the performance is:

Refrigerator or air conditioner   Coefficient of performance $= \dfrac{|Q_C|}{|W|}$ **(15.16)**

Commercially available refrigerators and air conditioners have coefficients of performance in the range 2 to 6, depending on the temperatures involved. The coefficients of performance for these real devices are less than those for ideal, or Carnot, refrigerators and air conditioners.

In a sense, refrigerators and air conditioners operate like pumps. They pump heat "uphill" from a lower temperature to a higher temperature, just as a water pump forces water uphill from a lower elevation to a higher elevation. It would be appropriate to call them heat pumps. However, the name "heat pump" is reserved for the device illustrated in **Figure 15.18**, which is a home heating appliance.

THE PHYSICS OF . . . heat pumps.   The **heat pump** uses work (magnitude = $|W|$) to make heat (magnitude = $|Q_C|$) from the wintry outdoors (the cold reservoir) flow up the temperature "hill" into a warm house (the hot reservoir). According to the conservation of energy, the heat pump deposits inside the house an amount of heat $|Q_H| = |W| + |Q_C|$. The air conditioner and the heat pump do closely related jobs. The air conditioner refrigerates the inside of the house and heats up the outdoors, whereas the heat pump refrigerates the outdoors and heats up the inside. These jobs are so closely related that most heat pump systems serve in a dual capacity, being equipped with a switch that converts them from heaters in the winter into air conditioners in the summer.

Heat pumps are popular for home heating in today's energy-conscious world, and it is easy to understand why. Suppose that 1000 J of energy is available for home heating. **Figure 15.19** shows that a conventional electric heating system uses this 1000 J to heat a coil of wire, just as in a toaster. A fan blows air across the hot coil, and forced convection carries the 1000 J of heat into the house. In contrast, the heat pump in **Figure 15.18** does not use the 1000 J directly as heat. Instead, it uses the 1000 J to do the work (magnitude = $|W|$) of pumping heat (magnitude = $|Q_C|$) from the cooler outdoors into the warmer house and, in so doing, delivers an amount of energy $|Q_H| = |W| + |Q_C|$. With $|W| = 1000$ J, this becomes $|Q_H| = 1000$ J $+ |Q_C|$, so that the heat pump delivers more than 1000 J of heat, whereas the conventional electric heating system delivers only 1000 J.

It is also possible to specify a coefficient of performance for heat pumps. However, unlike refrigerators and air conditioners, the job of a heat pump is to heat, not to cool. As a result, the coefficient of performance of a heat pump is the ratio of the magnitude of the heat $|Q_H|$ delivered into the house to the magnitude of the work $|W|$ required to deliver it:

Heat pump $$\text{Coefficient of performance} = \frac{|Q_H|}{|W|}$$ (15.17)

The coefficient of performance depends on the indoor and outdoor temperatures. Commercial units have coefficients of about 3 to 4 under favorable conditions.

Check Your Understanding

(The answers are given at the end of the book.)

16. Each drawing in **CYU Figure 15.4** represents a hypothetical heat engine or a hypothetical heat pump and shows the corresponding heats and work. Only one of these hypothetical situations is allowed in nature. Which is it?

FIGURE 15.19 This conventional electric heating system is delivering 1000 J of heat to the living room.

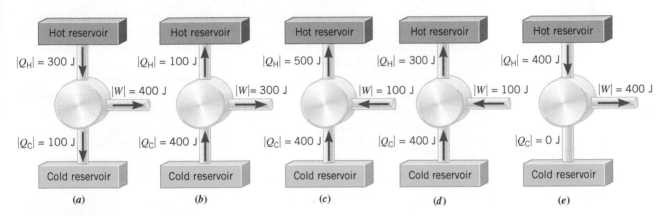

(a)   (b)   (c)   (d)   (e)

CYU FIGURE 15.4

17. A refrigerator is kept in a garage that is not heated in the cold winter or air-conditioned in the hot summer. Does it cost more for this refrigerator to make a kilogram of ice cubes in the winter or in the summer? **(a)** In the summer **(b)** In the winter **(c)** It costs the same in both seasons.

18. The coefficient of performance of a heat pump that is removing heat from the cold outdoors **(a)** must always be less than one, **(b)** can be either less than or greater than one, **(c)** must always be greater than one.

19. A kitchen air conditioner and a refrigerator both remove heat from a cold reservoir and deposit it in a hot reservoir. The air conditioner _____ the kitchen, whereas the refrigerator _____ the kitchen. **(a)** cools, cools **(b)** cools, warms **(c)** warms, warms **(d)** warms, cools

20. On a summer day a window air conditioner cycles on and off, according to how the temperature within the room changes. When are you more likely to be able to fry an egg on the outside part of the unit? **(a)** When the unit is on **(b)** When the unit is off **(c)** It does not matter whether the unit is on or off.

15.11 Entropy

A Carnot engine has the maximum possible efficiency for its operating conditions because the processes occurring within it are reversible. Irreversible processes, such as friction, cause real engines to operate at less than maximum efficiency, for they reduce our ability to use heat to perform work. As an extreme example, imagine that a hot object is placed in thermal contact with a cold object, so heat flows spontaneously, and hence irreversibly, from hot to cold. Eventually both objects reach the same temperature, and $T_C = T_H$. A Carnot engine using these two objects as heat reservoirs is unable to do work, because the efficiency of the engine is zero $[e_{Carnot} = 1 - (T_C/T_H) = 0]$. In general, irreversible processes cause us to lose some, but not necessarily all, of the ability to perform work. This partial loss can be expressed in terms of a concept called **entropy**.

To introduce the idea of entropy we recall the relation $|Q_C|/|Q_H| = T_C/T_H$ (Equation 15.14) that applies to a Carnot engine. It is possible to rearrange this equation as $|Q_C|/T_C = |Q_H|/T_H$, which focuses attention on the heat Q divided by the Kelvin temperature T. The quantity Q/T is called the change in the entropy ΔS:

$$\Delta S = \left(\frac{Q}{T}\right)_R \tag{15.18}$$

In this expression the temperature T must be in kelvins, and the subscript R refers to the word "reversible." It can be shown that Equation 15.18 applies to any process in which heat enters (Q is positive) or leaves (Q is negative) a system reversibly at a constant temperature. Such is the case for the heat that flows into and out of the reservoirs of a Carnot engine. Equation 15.18 indicates that the SI unit for entropy is a joule per kelvin (J/K).

Entropy, like internal energy, is a function of the state or condition of the system. Only the state of a system determines the entropy S that a system has. Therefore, the change in entropy ΔS is equal to the entropy of the final state of the system minus the entropy of the initial state.

We can now describe what happens to the entropy of a Carnot engine. As the engine operates, the entropy of the hot reservoir decreases, since heat of magnitude $|Q_H|$ departs reversibly at a Kelvin temperature T_H. The corresponding change in the entropy is $\Delta S_H = -|Q_H|/T_H$, where the minus sign is needed to indicate a decrease, since the symbol $|Q_H|$ denotes only the magnitude of the heat. In contrast, the entropy of the cold reservoir increases by an amount $\Delta S_C = +|Q_C|/T_C$, for the rejected heat reversibly enters the cold reservoir at a Kelvin temperature T_C. The total change in entropy is

$$\Delta S_C + \Delta S_H = +\frac{|Q_C|}{T_C} - \frac{|Q_H|}{T_H} = 0$$

because $|Q_C|/T_C = |Q_H|/T_H$ according to Equation 15.14.

The fact that the total change in entropy is zero for a Carnot engine is a specific illustration of a general result. It can be proved that when *any* reversible process occurs, the change in the entropy of the universe is zero; $\Delta S_{universe} = 0$ J/K for a reversible process. The word "universe" means that $\Delta S_{universe}$ takes into account the entropy changes of all parts of the system and all parts of the environment. ***Reversible processes do not alter the total entropy of the universe***. To be sure, the entropy of one part of the universe may change because of a reversible process, but if so, the entropy of another part changes in the opposite way by the same amount.

What happens to the entropy of the universe when an *irreversible* process occurs is more complex, because the expression $\Delta S = (Q/T)_R$ does not apply directly. However, if a system changes irreversibly from an initial state to a final state, this expression can be used to calculate ΔS indirectly, as **Figure 15.20** indicates. We imagine a hypothetical reversible process that causes the system to change between *the same initial and final states* and then find ΔS for this reversible process. The value obtained for ΔS also applies to the irreversible process that actually occurs, since only the nature of the initial and final states, and not the path between them, determines ΔS. Example 11 illustrates this indirect method and shows that spontaneous (irreversible) processes increase the entropy of the universe.

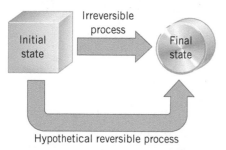

ΔS for irreversible process $=$ ΔS for hypothetical reversible process

FIGURE 15.20 Although the relation $\Delta S = (Q/T)_R$ applies to reversible processes, it can be used as part of an indirect procedure to find the entropy change for an irreversible process. This drawing illustrates the procedure discussed in the text.

EXAMPLE 11 | The Entropy of the Universe Increases

Figure 15.21 shows 1200 J of heat flowing spontaneously through a copper rod from a hot reservoir at 650 K to a cold reservoir at 350 K. Determine the amount by which this irreversible process changes the entropy of the universe, assuming that no other changes occur.

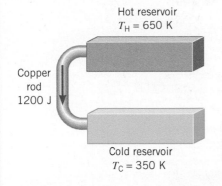

Hot reservoir
$T_H = 650$ K

Copper rod
1200 J

Cold reservoir
$T_C = 350$ K

FIGURE 15.21 Heat flows spontaneously from a hot reservoir to a cold reservoir.

Reasoning The hot-to-cold heat flow is irreversible, so the relation $\Delta S = (Q/T)_R$ is applied to a hypothetical process whereby the 1200 J of heat is taken reversibly from the hot reservoir and added reversibly to the cold reservoir.

Solution The total entropy change of the universe is the algebraic sum of the entropy changes for each reservoir:

$$\Delta S_{universe} = \underbrace{-\frac{1200\ J}{650\ K}}_{\substack{\text{Entropy lost by} \\ \text{hot reservoir}}} + \underbrace{\frac{1200\ J}{350\ K}}_{\substack{\text{Entropy gained} \\ \text{by cold reservoir}}} = \boxed{+1.6\ J/K}$$

The irreversible process causes the entropy of the universe to increase by 1.6 J/K.

Example 11 is a specific illustration of a general result: ***Any irreversible process increases the entropy of the universe.*** In other words, $\Delta S_{universe} > 0$ J/K for an irreversible process. Reversible processes do not alter the entropy of the universe, whereas irreversible processes cause the entropy to increase. Therefore, the entropy of the universe continually increases, like time itself, and entropy is sometimes called "time's arrow." It can be shown that this behavior of the entropy of the universe provides a completely general statement of the second law of thermodynamics, which applies not only to heat flow but also to all kinds of other processes.

THE SECOND LAW OF THERMODYNAMICS STATED IN TERMS OF ENTROPY

The total entropy of the universe does not change when a reversible process occurs ($\Delta S_{universe} = 0$ J/K) and increases when an irreversible process occurs ($\Delta S_{universe} > 0$ J/K).

When an irreversible process occurs and the entropy of the universe increases, the energy available for doing work decreases, as the next example illustrates.

EXAMPLE 12 | Energy Unavailable for Doing Work

Suppose that 1200 J of heat is used as input for an engine under two different conditions. In **Figure 15.22a** the heat is supplied by a hot reservoir whose temperature is 650 K. In part *b* of the drawing, the heat flows irreversibly through a copper rod into a second reservoir whose temperature is 350 K and then enters the engine. In either case, a 150-K reservoir is used as the cold reservoir. For each case, determine the maximum amount of work that can be obtained from the 1200 J of heat.

Reasoning According to Equation 15.11, the work (magnitude = $|W|$) obtained from the engine is the product of its efficiency e and the input heat (magnitude = $|Q_H|$), or $|W| = e|Q_H|$. For a given input heat, the maximum amount of work is obtained when the efficiency is a maximum; that is, when the engine is a Carnot engine. The efficiency of a Carnot engine is given by Equation 15.15 as $e_{Carnot} = 1 - T_C/T_H$. Therefore, the efficiency

may be determined from the Kelvin temperatures of the hot and cold reservoirs.

Solution

Before irreversible heat flow

$$e_{Carnot} = 1 - \frac{T_C}{T_H} = 1 - \frac{150\ K}{650\ K} = 0.77$$

$$|W| = (e_{Carnot})(1200\ J) = (0.77)(1200\ J) = \boxed{920\ J}$$

After irreversible heat flow

$$e_{Carnot} = 1 - \frac{T_C}{T_H} = 1 - \frac{150\ K}{350\ K} = 0.57$$

$$|W| = (e_{Carnot})(1200\ J) = (0.57)(1200\ J) = \boxed{680\ J}$$

When the 1200 J of input heat is taken from the 350-K reservoir instead of the 650-K reservoir, the efficiency of the Carnot engine is smaller. As a result, less work (680 J versus 920 J) can be extracted from the input heat.

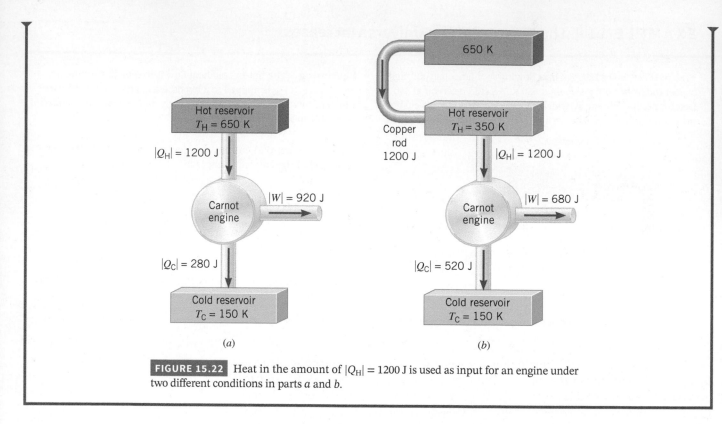

FIGURE 15.22 Heat in the amount of $|Q_H| = 1200$ J is used as input for an engine under two different conditions in parts *a* and *b*.

Example 12 shows that 240 J less work (920 J − 680 J) can be performed when the input heat is obtained from the hot reservoir with the lower temperature. In other words, the irreversible process of heat flow through the copper rod causes energy to become unavailable for doing work in the amount of $W_{unavailable} = 240$ J. Example 11 shows that this irreversible process simultaneously causes the entropy of the universe to increase by an amount $\Delta S_{universe} = +1.6$ J/K. These values for $W_{unavailable}$ and $\Delta S_{universe}$ are in fact related. If you multiply $\Delta S_{universe}$ by 150 K, which is the lowest Kelvin temperature in Example 12, you obtain $W_{unavailable} = (150\text{ K}) \times (1.6\text{ J/K}) = 240$ J. This illustrates the following general result:

$$W_{unavailable} = T_0 \Delta S_{universe} \tag{15.19}$$

where T_0 is the Kelvin temperature of the coldest heat reservoir. Since irreversible processes cause the entropy of the universe to increase, they cause energy to be degraded, in the sense that part of the energy becomes unavailable for the performance of work. In contrast, there is no penalty when reversible processes occur, because for them $\Delta S_{universe} = 0$ J/K, and there is no loss of work.

Entropy can also be interpreted in terms of order and disorder. As an example, consider a block of ice (**Figure 15.23**) with each of its H_2O molecules fixed rigidly in place in a highly structured and ordered arrangement. In comparison, the puddle of water into which the ice melts is disordered and unorganized, because the molecules in a liquid are free to move from place to place. Heat is required to melt the ice and produce the disorder. Moreover, heat flow into a system increases the entropy of the system, according to $\Delta S = (Q/T)_R$. We associate an increase in entropy, then, with an increase in disorder. Conversely, we associate a decrease

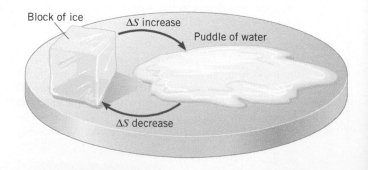

FIGURE 15.23 A block of ice is an example of an ordered system relative to a puddle of water.

in entropy with a decrease in disorder or a greater degree of order. Example 13 illustrates an order-to-disorder change and the increase of entropy that accompanies it.

EXAMPLE 13 | Order to Disorder

Find the change in entropy that results when a 2.3-kg block of ice melts slowly (reversibly) at 273 K (0 °C).

Reasoning Since the phase change occurs reversibly at a constant temperature, the change in entropy can be found by using Equation 15.18, $\Delta S = (Q/T)_R$, where Q is the heat absorbed by the melting ice. This heat can be determined by using the relation $Q = mL_f$ (Equation 12.5), where m is the

mass and $L_f = 3.35 \times 10^5$ J/kg is the latent heat of fusion of water (see Table 12.3).

Solution Using Equation 15.18 and Equation 12.5, we find that the change in entropy is

$$\Delta S = \left(\frac{Q}{T}\right)_R = \frac{mL_f}{T} = \frac{(2.3 \text{ kg})(3.35 \times 10^5 \text{ J/kg})}{273 \text{ K}} = \boxed{+2.8 \times 10^3 \text{ J/K}}$$

a result that is positive, since the ice absorbs heat as it melts.

Figure 15.24 shows another order-to-disorder change that can be described in terms of entropy.

Martin Hunter/Getty Images

FIGURE 15.24 With the aid of explosives, demolition experts caused this building to go from the ordered state (lower entropy), top photograph, to the disordered state (higher entropy), bottom photograph.

Check Your Understanding

(The answers are given at the end of the book.)

21. Two equal amounts of water are mixed together in an insulated container, and no work is done in the process. The initial temperatures of the water are different, but the mixture reaches a uniform temperature. Do the internal energy and entropy of the water increase, decrease, or remain constant as a result of the mixing process (refer to **CYU Table 15.3**)?

CYU TABLE 15.3

	Internal Energy of the Water	Entropy of the Water
(a)	Increases	Increases
(b)	Decreases	Decreases
(c)	Remains constant	Decreases
(d)	Remains constant	Increases
(e)	Remains constant	Remains constant

22. An event happens somewhere in the universe and, as a result, the entropy of an object changes by −5 J/K. Consistent with the second law of thermodynamics, which one (or more) of the following is a possible value for the entropy change for the rest of the universe? **(a)** −5 J/K **(b)** 0 J/K **(c)** +5 J/K **(d)** +10 J/K

23. In each of the following cases, which has the greater entropy, a handful of popcorn kernels or the popcorn that results from them; a salad before or after it has been tossed; and a messy apartment or a neat apartment?

24. A glass of water contains a teaspoon of dissolved sugar. After a while, the water evaporates, leaving behind sugar crystals. The entropy of the sugar crystals is less than the entropy of the dissolved sugar because the sugar crystals are in a more ordered state. Why doesn't this process violate the second law of thermodynamics? **(a)** Because, considering what happens to the water, the total entropy of the universe also decreases. **(b)** Because, considering what happens to the water, the total entropy of the universe increases. **(c)** Because the second law does not apply to this situation.

25. A builder uses lumber to construct a building, which is unfortunately destroyed in a fire. Thus, the lumber existed at one time or another in three different states: (A) as unused building material, (B) as a building, and (C) as a burned-out shell of a building. Rank these three states in order of decreasing entropy (largest first). **(a)** C, B, A **(b)** A, B, C **(c)** C, A, B **(d)** A, C, B **(e)** B, A, C

15.12 The Third Law of Thermodynamics

To the zeroth, first, and second laws of thermodynamics we add the third (and last) law. The **third law of thermodynamics** indicates that it is impossible to reach a temperature of absolute zero.

> **THE THIRD LAW OF THERMODYNAMICS**
>
> **It is not possible to lower the temperature of any system to absolute zero ($T = 0$ K) in a finite number of steps.**

This law, like the second law, can be expressed in a number of ways, but a discussion of them is beyond the scope of this text. The third law is needed to explain a number of experimental observations that cannot be explained by the other laws of thermodynamics.

Concept Summary

15.1 Thermodynamic Systems and Their Surroundings A thermodynamic system is the collection of objects on which attention is being focused, and the surroundings are everything else in the environment. The state of a system is the physical condition of the system, as described by values for physical parameters, often pressure, volume, and temperature.

15.2 The Zeroth Law of Thermodynamics Two systems are in thermal equilibrium if there is no net flow of heat between them when they are brought into thermal contact. Temperature is the indicator of thermal equilibrium in the sense that there is no net flow of heat between two systems in thermal contact that have the same temperature. The zeroth law of thermodynamics states that two systems individually in thermal equilibrium with a third system are in thermal equilibrium with each other.

15.3 The First Law of Thermodynamics The first law of thermodynamics states that due to heat Q and work W, the internal energy of a system changes from its initial value of U_i to a final value of U_f according to Equation 15.1. In this equation Q is positive when the system gains heat and negative when it loses heat. W is positive when work is done by the system and negative when work is done on the system. The first law of thermodynamics is the conservation-of-energy principle applied to heat, work, and the change in the internal energy.

$$\Delta U = U_f - U_i = Q - W \qquad (15.1)$$

The internal energy is called a function of state because it depends only on the state of the system and not on the method by which the system came to be in a given state.

15.4 Thermal Processes A thermal process is quasi-static when it occurs slowly enough that a uniform pressure and temperature exist throughout the system at all times. An isobaric process is one that occurs at constant pressure. The work W done when a system changes at a constant pressure P from an initial volume V_i to a final volume V_f is given by Equation 15.2. An isochoric process is one that takes place at constant volume, and no work is done in such a process. An isothermal process is one that takes place at constant temperature. An adiabatic process is one that takes place without the transfer of heat. The work done in any kind of quasi-static process is given by the area under the corresponding pressure–volume graph.

$$W = P\,\Delta V = P(V_f - V_i) \qquad (15.2)$$

15.5 Thermal Processes Using an Ideal Gas When n moles of an ideal gas change quasi-statically from an initial volume V_i to a final volume V_f at a constant Kelvin temperature T, the work done is given by Equation 15.3, and the process is said to be isothermal.

$$W = nRT \ln\left(\frac{V_f}{V_i}\right) \qquad (15.3)$$

When n moles of a monatomic ideal gas change quasi-statically and adiabatically from an initial temperature T_i to a final temperature T_f, the work done is given by Equation 15.4. During an adiabatic process, and in addition to the ideal gas law, an ideal gas obeys Equation 15.5, where $\gamma = c_P/c_V$ is the ratio of the specific heat capacities at constant pressure and constant volume.

$$W = \tfrac{3}{2}nR(T_i - T_f) \qquad (15.4)$$

$$P_i V_i^{\gamma} = P_f V_f^{\gamma} \qquad (15.5)$$

15.6 Specific Heat Capacities The molar specific heat capacity C of a substance determines how much heat Q is added or removed when the temperature of n moles of the substance changes by an amount ΔT, according to Equation 15.6. For a monatomic ideal gas,

the molar specific heat capacities at constant pressure and constant volume are given by Equations 15.7 and 15.8, respectively, where R is the ideal gas constant. For a diatomic ideal gas at moderate temperatures that do not allow vibration to occur, these values are $C_P = \tfrac{7}{2}R$ and $C_V = \tfrac{5}{2}R$. For any type of ideal gas, the difference between C_P and C_V is given by Equation 15.10.

$$Q = Cn\,\Delta T \qquad (15.6)$$

$$C_P = \tfrac{5}{2}R \qquad (15.7)$$

$$C_V = \tfrac{3}{2}R \qquad (15.8)$$

$$C_P - C_V = R \qquad (15.10)$$

15.7 The Second Law of Thermodynamics The second law of thermodynamics can be stated in a number of equivalent forms. In terms of heat flow, the second law declares that heat flows spontaneously from a substance at a higher temperature to a substance at a lower temperature and does not flow spontaneously in the reverse direction.

15.8 Heat Engines A heat engine produces work (magnitude $= |W|$) from input heat (magnitude $= |Q_H|$) that is extracted from a heat reservoir at a relatively high temperature. The engine rejects heat (magnitude $= |Q_C|$) into a reservoir at a relatively low temperature. The efficiency e of a heat engine is given by Equation 15.11.

$$e = \frac{\text{Work done}}{\text{Input heat}} = \frac{|W|}{|Q_H|} \qquad (15.11)$$

The conservation of energy requires that $|Q_H|$ must be equal to $|W|$ plus $|Q_C|$, as in Equation 15.12. By combining Equation 15.12 with Equation 15.11, the efficiency of a heat engine can also be written as shown in Equation 15.13.

$$|Q_H| = |W| + |Q_C| \qquad (15.12)$$

$$e = 1 - \frac{|Q_C|}{|Q_H|} \qquad (15.13)$$

15.9 Carnot's Principle and the Carnot Engine A reversible process is one in which *both* the system and its environment can be returned to exactly the states they were in before the process occurred.

Carnot's principle is an alternative statement of the second law of thermodynamics. It states that no irreversible engine operating between two reservoirs at constant temperatures can have a greater efficiency than a reversible engine operating between the same temperatures. Furthermore, all reversible engines operating between the same temperatures have the same efficiency.

A Carnot engine is a reversible engine in which all input heat (magnitude $= |Q_H|$) originates from a hot reservoir at a single Kelvin temperature T_H and all rejected heat (magnitude $= |Q_C|$) goes into a cold reservoir at a single Kelvin temperature T_C. For a Carnot engine, Equation 15.14 applies. The efficiency e_{Carnot} of a Carnot engine is the maximum efficiency that an engine operating between two fixed temperatures can have and is given by Equation 15.15.

$$\frac{|Q_C|}{|Q_H|} = \frac{T_C}{T_H} \qquad (15.14)$$

$$e_{\text{Carnot}} = 1 - \frac{T_C}{T_H} \qquad (15.15)$$

15.10 Refrigerators, Air Conditioners, and Heat Pumps Refrigerators, air conditioners, and heat pumps are devices that utilize work (magnitude $= |W|$) to make heat (magnitude $= |Q_C|$) flow from a lower Kelvin temperature T_C to a higher Kelvin temperature T_H. In the process (the refrigeration process) they deposit heat (magnitude $= |Q_H|$)

at the higher temperature. The principle of the conservation of energy requires that $|Q_H| = |W| + |Q_C|$.

$$\text{Coefficient of performance of a refrigerator} = \frac{|Q_C|}{|W|} \qquad (15.16)$$

If the refrigeration process is ideal, in the sense that it occurs reversibly, the devices are called Carnot devices and the relation $|Q_C|/|Q_H| = T_C/T_H$ (Equation 15.14) holds.

The coefficient of performance of a refrigerator or an air conditioner is given by Equation 15.16. The coefficient of performance of a heat pump, however, is given by Equation 15.17.

$$\text{Coefficient of performance of a heat pump} = \frac{|Q_H|}{|W|} \qquad (15.17)$$

15.11 Entropy The change in entropy ΔS for a process in which heat Q enters or leaves a system reversibly at a constant Kelvin temperature T is given by Equation 15.18, where the subscript R stands for "reversible."

$$\Delta S = \left(\frac{Q}{T}\right)_R \qquad (15.18)$$

The second law of thermodynamics can be stated in a number of equivalent forms. In terms of entropy, the second law states that the total entropy of the universe does not change when a reversible process occurs ($\Delta S_{\text{universe}} = 0$ J/K) and increases when an irreversible process occurs ($\Delta S_{\text{universe}} > 0$ J/K).

Irreversible processes cause energy to be degraded in the sense that part of the energy becomes unavailable for the performance of work. The energy $W_{\text{unavailable}}$ that is unavailable for doing work because of an irreversible process is shown in Equation 15.19, where $\Delta S_{\text{universe}}$ is the total entropy change of the universe and T_0 is the Kelvin temperature of the coldest reservoir into which heat can be rejected.

$$W_{\text{unavailable}} = T_0 \, \Delta S_{\text{universe}} \qquad (15.19)$$

Increased entropy is associated with a greater degree of disorder and decreased entropy with a lesser degree of disorder (more order).

15.12 The Third Law of Thermodynamics The third law of thermodynamics states that it is not possible to lower the temperature of any system to absolute zero ($T = 0$ K) in a finite number of steps.

Focus on Concepts

Additional questions are available for assignment in WileyPLUS.

Section 15.3 The First Law of Thermodynamics

1. The first law of thermodynamics states that the change ΔU in the internal energy of a system is given by $\Delta U = Q - W$, where Q is the heat and W is the work. Both Q and W can be positive or negative numbers. Q is a positive number if _____, and W is a positive number if _____. (a) the system *loses* heat; work is done *by* the system (b) the system *loses* heat; work is done *on* the system (c) the system *gains* heat; work is done *by* the system (d) the system *gains* heat; work is done *on* the system

Section 15.4 Thermal Processes

2. The drawing shows the expansion of three ideal gases. Rank the gases according to the work they do, largest to smallest. (a) A, B, C (b) A and B (a tie), C (c) B and C (a tie), A (d) B, C, A (e) C, A, B

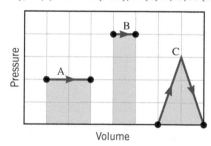

QUESTION 2

3. The pressure–volume graph shows three paths in which a gas expands from an initial state A to a final state B. The change $\Delta U_{A \to B}$ in internal energy is the same for each of the paths. Rank the paths according to the heat Q added to the gas, largest to smallest. (a) 1, 2, 3 (b) 1, 3, 2 (c) 2, 1, 3 (d) 3, 1, 2 (e) 3, 2, 1

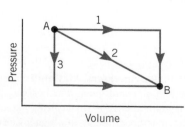

QUESTION 3

Section 15.5 Thermal Processes Using an Ideal Gas

4. An ideal monatomic gas expands isothermally from A to B, as the graph shows. What can be said about this process? (a) The gas does no work. (b) No heat enters or leaves the gas. (c) The first law of thermodynamics does not apply to an isothermal process. (d) The ideal gas law is not valid during an isothermal process. (e) There is no change in the internal energy of the gas.

QUESTION 4

5. A monatomic ideal gas is thermally insulated, so no heat can flow between it and its surroundings. Is it possible for the temperature of the gas to rise? (a) Yes. The temperature can rise if work is done *by* the gas. (b) No. The only way that the temperature can rise is if heat is added to the gas. (c) Yes. The temperature can rise if work is done *on* the gas.

Section 15.8 Heat Engines

6. A heat engine takes heat Q_H from a hot reservoir and uses part of this energy to perform work W. Assuming that Q_H cannot be changed, how can the efficiency of the engine be improved? (a) Increase the work W; the heat Q_C rejected to the cold reservoir increases as a result. (b) Increase the work W; the heat Q_C rejected to the cold reservoir remains unchanged. (c) Increase the work W; the heat Q_C rejected to the cold reservoir decreases as a result. (d) Decrease the work W; the heat Q_C rejected to the cold reservoir remains unchanged. (e) Decrease the work W; the heat Q_C rejected to the cold reservoir decreases as a result.

Section 15.9   Carnot's Principle and the Carnot Engine

7. The three Carnot engines shown in the drawing operate with hot and cold reservoirs whose temperature differences are 100 K. Rank the efficiencies of the engines, largest to smallest. (a) All engines have the same efficiency. (b) A, B, C (c) B, A, C (d) C, B, A (e) C, A, B

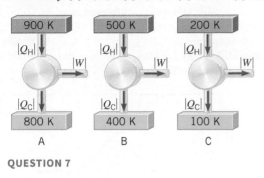

QUESTION 7

Section 15.10   Refrigerators, Air Conditioners, and Heat Pumps

8. A refrigerator operates for a certain time, and the work done by the electrical energy during this time is $W = 1000$ J. What can be said about the heat delivered to the room containing the refrigerator? (a) The heat delivered to the room is less than 1000 J. (b) The heat delivered to the room is equal to 1000 J. (c) The heat delivered to the room is greater than 1000 J.

Section 15.11   Entropy

9. Heat is transferred from the sun to the earth via electromagnetic waves (see Chapter 24). Because of this transfer, the entropy of the sun _____, the entropy of the earth _____, and the entropy of the sun–earth system _____. (a) increases, decreases, decreases (b) decreases, increases, increases (c) increases, increases, increases (d) increases, decreases, increases (e) decreases, increases, decreases.

Problems

Online

SSM Student Solutions Manual

MMH Problem-solving help

GO Guided Online Tutorial

V-HINT Video Hints

CHALK Chalkboard Videos

BIO Biomedical application

E Easy

M Medium

H Hard

WS Worksheet

T Team Problem

Section 15.3   The First Law of Thermodynamics

1. **E** In moving out of a dormitory at the end of the semester, a student does 1.6×10^4 J of work. In the process, his internal energy decreases by 4.2×10^4 J. Determine each of the following quantities (including the algebraic sign): (a) W (b) ΔU (c) Q

2. **E** The internal energy of a system changes because the system gains 165 J of heat and performs 312 J of work. In returning to its initial state, the system loses 114 J of heat. During this return process, (a) what work is involved, and (b) is the work done by the system or on the system?

3. **E SSM** A system does 164 J of work on its environment and gains 77 J of heat in the process. Find the change in the internal energy of (a) the system and (b) the environment.

4. **E GO** A system does 4.8×10^4 J of work, and 7.6×10^4 J of heat flows into the system during the process. Find the change in the internal energy of the system.

5. **E V-HINT** Three moles of an ideal monatomic gas are at a temperature of 345 K. Then, 2438 J of heat is added to the gas, and 962 J of work is done on it. What is the final temperature of the gas?

Section 15.4   Thermal Processes

6. **E** A system undergoes a two-step process. In the first step, the internal energy of the system increases by 228 J when 166 J of work is done on the system. In the second step, the internal energy of the

system increases by 115 J when 177 J of work is done on the system. For the overall process, find the heat. What type of process is the overall process? Explain.

7. **E SSM** When a .22-caliber rifle is fired, the expanding gas from the burning gunpowder creates a pressure behind the bullet. This pressure causes the force that pushes the bullet through the barrel. The barrel has a length of 0.61 m and an opening whose radius is 2.8×10^{-3} m. A bullet (mass = 2.6×10^{-3} kg) has a speed of 370 m/s after passing through this barrel. Ignore friction and determine the average pressure of the expanding gas.

8. **E GO** A system gains 2780 J of heat at a constant pressure of 1.26×10^5 Pa, and its internal energy increases by 3990 J. What is the change in the volume of the system, and is it an increase or a decrease?

9. **E V-HINT** A system gains 1500 J of heat, while the internal energy of the system increases by 4500 J and the volume decreases by 0.010 m³. Assume that the pressure is constant and find its value.

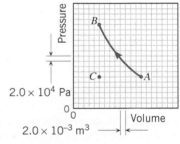

PROBLEM 10

10. **E** The volume of a gas is changed along the curved line between A and B in the drawing. Do not assume that the curved line is an isotherm or that the gas is ideal. (a) Find the magnitude of the work for the process, and (b) determine whether the work is positive or negative.

11. **E SSM** (a) Using the data presented in the accompanying pressure–volume graph, estimate the magnitude of the

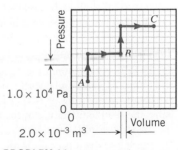

PROBLEM 11

work done when the system changes from A to B to C along the path shown. **(b)** Determine whether the work is done by the system or on the system and, hence, whether the work is positive or negative.

12. **E** Sections 14.2 and 14.3 provide useful information for this problem. When a monatomic ideal gas expands at a constant pressure of 2.6×10^5 Pa, the volume of the gas increases by 6.2×10^{-3} m^3. **(a)** Determine the heat that flows into or out of the gas. **(b)** Specify the direction of the flow.

13. **E** **CHALK** A gas is contained in a chamber such as that in Figure 15.4. Suppose that the region outside the chamber is evacuated and the total mass of the block and the movable piston is 135 kg. When 2050 J of heat flows into the gas, the internal energy of the gas increases by 1730 J. What is the distance s through which the piston rises?

14. **M** A piece of aluminum has a volume of 1.4×10^{-3} m^3. The coefficient of volume expansion for aluminum is $\beta = 69 \times 10^{-6}$ (C°)$^{-1}$. The temperature of this object is raised from 20 to 320 °C. How much work is done by the expanding aluminum if the air pressure is 1.01×10^5 Pa?

15. **M** **SSM** Refer to Multiple-Concept Example 4 to see how the concepts pertinent to this problem are used. The pressure of a gas remains constant while the temperature, volume, and internal energy of the gas increase by 53.0 C°, 1.40×10^{-3} m^3, and 939 J, respectively. The mass of the gas is 24.0 g, and its specific heat capacity is 1080 J/(kg · C°). Determine the pressure.

16. **M** **GO** Refer to the drawing that accompanies Problem 11. When a system changes from A to B along the path shown on the pressure-versus-volume graph, it gains 2700 J of heat. What is the change in the internal energy of the system?

17. **H** Water is heated in an open pan where the air pressure is one atmosphere. The water remains a liquid, which expands by a small amount as it is heated. Determine the ratio of the work done by the water to the heat absorbed by the water.

Section 15.5 Thermal Processes Using an Ideal Gas

18. **E** Six grams of helium (molecular mass = 4.0 u) expand isothermally at 370 K and do 9600 J of work. Assuming that helium is an ideal gas, determine the ratio of the final volume of the gas to the initial volume.

19. **E** **SSM** Five moles of a monatomic ideal gas expand adiabatically, and its temperature decreases from 370 to 290 K. Determine **(a)** the work done (including the algebraic sign) by the gas, and **(b)** the change in its internal energy.

20. **E** **GO** Three moles of neon expand isothermally to 0.250 from 0.100 m^3. Into the gas flows 4.75×10^3 J of heat. Assuming that neon is an ideal gas, find its temperature.

21. **E** The temperature of a monatomic ideal gas remains constant during a process in which 4700 J of heat flows out of the gas. How much work (including the proper + or − sign) is done?

22. **E** **GO** One-half mole of a monatomic ideal gas expands adiabatically and does 610 J of work. By how many kelvins does its temperature change? Specify whether the change is an increase or a decrease.

23. **E** **SSM** A monatomic ideal gas has an initial temperature of 405 K. This gas expands and does the same amount of work whether the expansion is adiabatic or isothermal. When the expansion is adiabatic, the final temperature of the gas is 245 K. What is the ratio of the final to the initial volume when the expansion is isothermal?

24. **E** **V-HINT** Heat is added isothermally to 2.5 mol of a monatomic ideal gas. The temperature of the gas is 430 K. How much heat must be added to make the volume of the gas double?

25. **M** **MMH** **CHALK** A diesel engine does not use spark plugs to ignite the fuel and air in the cylinders. Instead, the temperature required to ignite the fuel occurs because the pistons compress the air in the cylinders. Suppose that air at an initial temperature of 21 °C is compressed adiabatically to a temperature of 688 °C. Assume the air to be an ideal gas for which $\gamma = \frac{7}{5}$. Find the compression ratio, which is the ratio of the initial volume to the final volume.

26. **M** A monatomic ideal gas expands from point A to point B along the path shown in the drawing. **(a)** Determine the work done by the gas. **(b)** The temperature of the gas at point A is 185 K. What is its temperature at point B? **(c)** How much heat has been added to or removed from the gas during the process?

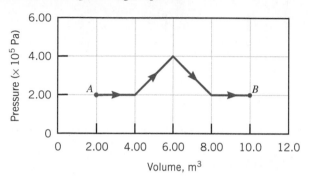

PROBLEM 26

27. **M** **SSM** The drawing refers to one mole of a monatomic ideal gas and shows a process that has four steps, two isobaric (A to B, C to D) and two isochoric (B to C, D to A). Complete the following table by calculating ΔU, W, and Q (including the algebraic signs) for each of the four steps.

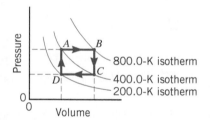

PROBLEM 27

	ΔU	W	Q
A to B			
B to C			
C to D			
D to A			

28. **M** **GO** A monatomic ideal gas ($\gamma = \frac{5}{3}$) is contained within a perfectly insulated cylinder that is fitted with a movable piston. The initial pressure of the gas is 1.50×10^5 Pa. The piston is pushed so as to compress the gas, with the result that the Kelvin temperature doubles. What is the final pressure of the gas?

29. **M** The pressure and volume of an ideal monatomic gas change from A to B to C, as the drawing shows. The curved line between A and C is an isotherm. **(a)** Determine the total heat for the process and **(b)** state whether the flow of heat is into or out of the gas.

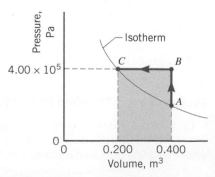

PROBLEM 29

30. **H** The work done by one mole of a monatomic ideal gas ($\gamma = \frac{5}{3}$) in expanding adiabatically is 825 J. The initial temperature and volume of the gas are 393 K and 0.100 m^3. Obtain **(a)** the final temperature and **(b)** the final volume of the gas.

31. **H** **SSM** The drawing shows an adiabatically isolated cylinder that is divided initially into two identical parts by an adiabatic partition. Both sides contain one mole of a monatomic ideal gas ($\gamma = \frac{5}{3}$), with the initial temperature being 525 K on the left and 275 K on the right. The partition is then allowed to move slowly (i.e., quasi-statically) to the right, until the pressures on each side of the partition are the same. Find the final temperatures on the **(a)** left and **(b)** right.

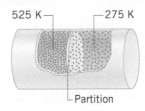

525 K ─┐   ┌─ 275 K

└─ Partition

PROBLEM 31

Section 15.6   Specific Heat Capacities

32. **E** Argon is a monatomic gas whose atomic mass is 39.9 u. The temperature of eight grams of argon is raised by 75 K under conditions of constant pressure. Assuming that argon behaves as an ideal gas, how much heat is required?

33. **E** **SSM** The temperature of 2.5 mol of a monatomic ideal gas is 350 K. The internal energy of this gas is doubled by the addition of heat. How much heat is needed when it is added at **(a)** constant volume and **(b)** constant pressure?

34. **E** **GO** Under constant-volume conditions, 3500 J of heat is added to 1.6 moles of an ideal gas. As a result, the temperature of the gas increases by 75 K. How much heat would be required to cause the same temperature change under constant-pressure conditions? Do not assume anything about whether the gas is monatomic, diatomic, etc.

35. **E** **SSM** Heat is added to two identical samples of a monatomic ideal gas. In the first sample the heat is added while the volume of the gas is kept constant, and the heat causes the temperature to rise by 75 K. In the second sample, an identical amount of heat is added while the pressure (but not the volume) of the gas is kept constant. By how much does the temperature of this sample increase?

36. **E** **GO** A monatomic ideal gas in a rigid container is heated from 217 K to 279 K by adding 8500 J of heat. How many moles of gas are there in the container?

37. **E** Three moles of a monatomic ideal gas are heated at a constant volume of 1.50 m^3. The amount of heat added is 5.24×10^3 J. **(a)** What is the change in the temperature of the gas? **(b)** Find the change in its internal energy. **(c)** Determine the change in pressure.

38. **M** **V-HINT** A monatomic ideal gas expands at constant pressure. **(a)** What percentage of the heat being supplied to the gas is used to increase the internal energy of the gas? **(b)** What percentage is used for doing the work of expansion?

39. **M** **CHALK** Suppose a monatomic ideal gas is contained within a vertical cylinder that is fitted with a movable piston. The piston is frictionless and has a negligible mass. The area of the piston is 3.14×10^{-2} m^2, and the pressure outside the cylinder is 1.01×10^5 Pa. Heat (2093 J) is removed from the gas. Through what distance does the piston drop?

40. **M** **V-HINT** A monatomic ideal gas is heated while at a constant volume of 1.00×10^{-3} m^3, using a ten-watt heater. The pressure of the gas increases by 5.0×10^4 Pa. How long was the heater on?

Section 15.8   Heat Engines

41. **E** Multiple-Concept Example 7 provides a review of the concepts that play roles here. An engine has an efficiency of 64% and produces 5500 J of work. Determine **(a)** the input heat and **(b)** the rejected heat.

42. **E** **GO** Engine 1 has an efficiency of 0.18 and requires 5500 J of input heat to perform a certain amount of work. Engine 2 has an efficiency of 0.26 and performs the same amount of work. How much input heat does the second engine require?

43. **E** Due to a tune-up, the efficiency of an automobile engine increases by 5.0%. For an input heat of 1300 J, how much more work does the engine produce after the tune-up than before?

44. **M** **SSM** **CHALK** Due to design changes, the efficiency of an engine increases from 0.23 to 0.42. For the same input heat $|Q_\text{H}|$, these changes increase the work done by the more efficient engine and reduce the amount of heat rejected to the cold reservoir. Find the ratio of the heat rejected to the cold reservoir for the improved engine to that for the original engine.

45. **H** Engine A receives three times more input heat, produces five times more work, and rejects two times more heat than engine B. Find the efficiency of **(a)** engine A and **(b)** engine B.

Section 15.9   Carnot's Principle and the Carnot Engine

46. **E** A Carnot engine operates with an efficiency of 27.0% when the temperature of its cold reservoir is 275 K. Assuming that the temperature of the hot reservoir remains the same, what must be the temperature of the cold reservoir in order to increase the efficiency to 32.0%?

47. **E** An engine has a hot-reservoir temperature of 950 K and a cold-reservoir temperature of 620 K. The engine operates at three-fifths maximum efficiency. What is the efficiency of the engine?

48. **E** **SSM** A Carnot engine has an efficiency of 0.700, and the temperature of its cold reservoir is 378 K. **(a)** Determine the temperature of its hot reservoir. **(b)** If 5230 J of heat is rejected to the cold reservoir, what amount of heat is put into the engine?

49. **E** An engine does 18 500 J of work and rejects 6550 J of heat into a cold reservoir whose temperature is 285 K. What would be the smallest possible temperature of the hot reservoir?

50. **E** **GO** A Carnot engine has an efficiency of 0.40. The Kelvin temperature of its hot reservoir is quadrupled, and the Kelvin temperature of its cold reservoir is doubled. What is the efficiency that results from these changes?

51. **E** **MMH** A Carnot engine operates between temperatures of 650 and 350 K. To improve the efficiency of the engine, it is decided either to raise the temperature of the hot reservoir by 40 K or to lower the temperature of the cold reservoir by 40 K. Which change gives the greater improvement? Justify your answer by calculating the efficiency in each case.

52. **M** The hot reservoir for a Carnot engine has a temperature of 890 K, while the cold reservoir has a temperature of 670 K. The heat input for this engine is 4800 J. The 670-K reservoir also serves as the hot reservoir for a second Carnot engine. This second engine uses the rejected heat of the first engine as input and extracts additional work from it. The rejected heat from the second engine goes into a reservoir that has a temperature of 420 K. Find the total work delivered by the two engines.

53. M SSM Suppose that the gasoline in a car engine burns at 631 °C, while the exhaust temperature (the temperature of the cold reservoir) is 139 °C and the outdoor temperature is 27 °C. Assume that the engine can be treated as a Carnot engine (a gross oversimplification). In an attempt to increase mileage performance, an inventor builds a second engine that functions between the exhaust and outdoor temperatures and uses the exhaust heat to produce additional work. Assume that the inventor's engine can also be treated as a Carnot engine. Determine the ratio of the total work produced by both engines to that produced by the first engine alone.

54. M V-HINT A power plant taps steam superheated by geothermal energy to 505 K (the temperature of the hot reservoir) and uses the steam to do work in turning the turbine of an electric generator. The steam is then converted back into water in a condenser at 323 K (the temperature of the cold reservoir), after which the water is pumped back down into the earth where it is heated again. The output power (work per unit time) of the plant is 84 000 kilowatts. Determine (a) the maximum efficiency at which this plant can operate and (b) the minimum amount of rejected heat that must be removed from the condenser every twenty-four hours.

Section 15.10 Refrigerators, Air Conditioners, and Heat Pumps

55. E SSM A Carnot air conditioner maintains the temperature in a house at 297 K on a day when the temperature outside is 311 K. What is the coefficient of performance of the air conditioner?

56. E GO The inside of a Carnot refrigerator is maintained at a temperature of 277 K, while the temperature in the kitchen is 299 K. Using 2500 J of work, how much heat can this refrigerator remove from its inside compartment?

57. E A refrigerator operates between temperatures of 296 and 275 K. What would be its maximum coefficient of performance?

58. E GO Two Carnot air conditioners, A and B, are removing heat from different rooms. The outside temperature is the same for both rooms, 309.0 K. The room serviced by unit A is kept at a temperature of 294.0 K, while the room serviced by unit B is kept at 301.0 K. The heat removed from either room is 4330 J. For both units, find the magnitude of the work required and the magnitude of the heat deposited outside.

59. E The water in a deep underground well is used as the cold reservoir of a Carnot heat pump that maintains the temperature of a house at 301 K. To deposit 14 200 J of heat in the house, the heat pump requires 800 J of work. Determine the temperature of the well water.

60. E GO A Carnot engine has an efficiency of 0.55. If this engine were run backward as a heat pump, what would be the coefficient of performance?

61. E A Carnot refrigerator is used in a kitchen in which the temperature is kept at 301 K. This refrigerator uses 241 J of work to remove 2561 J of heat from the food inside. What is the temperature inside the refrigerator?

62. M V-HINT The wattage of a commercial ice maker is 225 W and is the rate at which it does work. The ice maker operates just like a refrigerator or an air conditioner and has a coefficient of performance of 3.60. The water going into the unit has a temperature of 15.0 °C, and the ice maker produces ice cubes at 0.0 °C. Ignoring the work needed to keep stored ice from melting, find the maximum amount (in kg) of ice that the unit can produce in one day of continuous operation.

63. M SSM CHALK A Carnot refrigerator transfers heat from its inside (6.0 °C) to the room air outside (20.0 °C). (a) Find the coefficient of performance of the refrigerator. (b) Determine the magnitude of the minimum work needed to cool 5.00 kg of water from 20.0 to 6.0 °C when it is placed in the refrigerator.

Section 15.11 Entropy

64. E Consider three engines that each use 1650 J of heat from a hot reservoir (temperature = 550 K). These three engines reject heat to a cold reservoir (temperature = 330 K). Engine I rejects 1120 J of heat. Engine II rejects 990 J of heat. Engine III rejects 660 J of heat. One of the engines operates reversibly, and two operate irreversibly. However, of the two irreversible engines, one violates the second law of thermodynamics and could not exist. For each of the engines determine the total entropy change of the universe, which is the sum of the entropy changes of the hot and cold reservoirs. On the basis of your calculations, identify which engine operates reversibly, which operates irreversibly and could exist, and which operates irreversibly and could not exist.

65. E Heat Q flows spontaneously from a reservoir at 394 K into a reservoir at 298 K. Because of the spontaneous flow, 2800 J of energy is rendered unavailable for work when a Carnot engine operates between the reservoir at 298 K and a reservoir at 248 K. Find Q.

66. E SSM Find the change in entropy of the H_2O molecules when (a) three kilograms of ice melts into water at 273 K and (b) three kilograms of water changes into steam at 373 K. (c) On the basis of the answers to parts (a) and (b), discuss which change creates more disorder in the collection of H_2O molecules.

67. E GO On a cold day, 24 500 J of heat leaks out of a house. The inside temperature is 21 °C, and the outside temperature is −15 °C. What is the increase in the entropy of the universe that this heat loss produces?

68. M MMH (a) After 6.00 kg of water at 85.0 °C is mixed in a perfect thermos with 3.00 kg of ice at 0.0 °C, the mixture is allowed to reach equilibrium. When heat is added to or removed from a solid or liquid of mass m and specific heat capacity c, the change in entropy can be shown to be $\Delta S = mc \ln(T_f/T_i)$, where T_i and T_f are the initial and final Kelvin temperatures. Using this expression and the change in entropy for melting, find the change in entropy that occurs. (b) Should the entropy of the universe increase or decrease as a result of the mixing process? Give your reasoning and state whether your answer in part (a) is consistent with your answer here.

69. M V-HINT CHALK The sun is a sphere with a radius of 6.96×10^8 m and an average surface temperature of 5800 K. Determine the amount by which the sun's thermal radiation increases the entropy of the entire universe each second. Assume that the sun is a perfect blackbody, and that the average temperature of the rest of the universe is 2.73 K. Do not consider the thermal radiation absorbed by the sun from the rest of the universe.

70. M GO An irreversible engine operates between temperatures of 852 and 314 K. It absorbs 1285 J of heat from the hot reservoir and does 264 J of work. (a) What is the change $\Delta S_{universe}$ in the entropy of the universe associated with the operation of this engine? (b) If the engine were reversible, what would be the magnitude $|W|$ of the work it would have done, assuming that it operated between the same temperatures and absorbed the same heat as the irreversible engine? (c) Using the results of parts (a) and (b), find the difference between the work produced by the reversible and irreversible engines.

Additional Problems

71. **E** The pressure of a monatomic ideal gas ($\gamma = \frac{5}{3}$) doubles during an adiabatic compression. What is the ratio of the final volume to the initial volume?

72. **E** **SSM** One-half mole of a monatomic ideal gas absorbs 1200 J of heat while 2500 J of work is done by the gas. **(a)** What is the temperature change of the gas? **(b)** Is the change an increase or a decrease?

73. **E** A gas, while expanding under isobaric conditions, does 480 J of work. The pressure of the gas is 1.6×10^5 Pa, and its initial volume is 1.5×10^{-3} m³. What is the final volume of the gas?

74. **E** **V-HINT** A lawnmower engine with an efficiency of 0.22 rejects 9900 J of heat every second. What is the magnitude of the work that the engine does in one second?

75. **E** **SSM** A process occurs in which the entropy of a system increases by 125 J/K. During the process, the energy that becomes unavailable for doing work is zero. **(a)** Is this process reversible or irreversible? Give your reasoning. **(b)** Determine the change in the entropy of the surroundings.

76. **E** A Carnot heat pump operates between an outdoor temperature of 265 K and an indoor temperature of 298 K. Find its coefficient of performance.

77. **E** The temperatures indoors and outdoors are 299 and 312 K, respectively. A Carnot air conditioner deposits 6.12×10^5 J of heat outdoors. How much heat is removed from the house?

78. **E** **GO** Carnot engine A has an efficiency of 0.60, and Carnot engine B has an efficiency of 0.80. Both engines utilize the same hot reservoir, which has a temperature of 650 K and delivers 1200 J of heat to each engine. Find the magnitude of the work produced by each engine and the temperatures of the cold reservoirs that they use.

79. **E** **SSM** The pressure and volume of a gas are changed along the path *ABCA*. Using the data shown in the graph, determine the work done (including the algebraic sign) in each segment of the path: **(a)** *A* to *B*, **(b)** *B* to *C*, and **(c)** *C* to *A*.

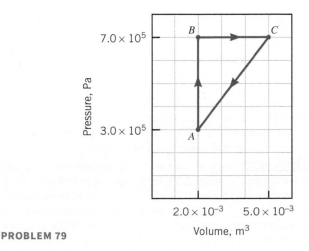

PROBLEM 79

80. **M** **GO** An ideal gas is taken through the three processes (A → B, B → C, and C → A) shown in the drawing. In general, for each process the internal energy *U* of the gas can change because heat *Q* can be added to or removed from the gas and work *W* can be done by the gas or on the gas. For the three processes shown in the drawing, fill in the five missing entries in the following table.

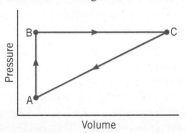

PROBLEM 80

Process	ΔU	Q	W
A → B	**(b)**	+561 J	**(a)**
B → C	+4303 J	**(c)**	+3740 J
C → A	**(d)**	**(e)**	−2867 J

81. **H** **SSM** An engine has an efficiency e_1. The engine takes input heat of magnitude $|Q_H|$ from a hot reservoir and delivers work of magnitude $|W_1|$. The heat rejected by this engine is used as input heat for a second engine, which has an efficiency e_2 and delivers work of magnitude $|W_2|$. The overall efficiency of this two-engine device is the magnitude of the total work delivered ($|W_1| + |W_2|$) divided by the magnitude $|Q_H|$ of the input heat. Find an expression for the overall efficiency e in terms of e_1 and e_2.

82. **M** **GO** **SSM** An ideal, or Carnot, heat pump is used to heat a house to a temperature of 294 K (21 °C). How much work must the pump do to deliver 3350 J of the heat into the house on a day when the outdoor temperature is 273 K (0 °C), and on another day when the temperature is 252 K (−21 °C)?

Physics in Biology, Medicine, and Sports

83. **E** **BIO** **GO** 15.3 In a game of football outdoors on a cold day, a player will begin to feel exhausted after using approximately 8.0×10^5 J of internal energy. (a) One player, dressed too lightly for the weather, has to leave the game after losing 6.8×10^5 J of heat. How much work has he done? (b) Another player, wearing clothes that offer better protection against heat loss, is able to remain in the game long enough to do 2.1×10^5 J of work. What is the magnitude of the heat that he has lost?

84. **M** **BIO** **CHALK** **SSM** 15.3 In exercising, a weight lifter loses 0.150 kg of water through evaporation, the heat required to evaporate the water coming from the weight lifter's body. The work done in lifting weights is 1.40×10^5 J. (a) Assuming that the latent heat of vaporization of perspiration is 2.42×10^6 J/kg, find the change in the internal energy of the weight lifter. (b) Determine the minimum number of nutritional Calories of food (1 nutritional Calorie = 4186 J) that must be consumed to replace the loss of internal energy.

85. **E** **BIO** **SSM** 15.8 Heat engines take input energy in the form of heat, use some of that energy to do work, and exhaust the remainder. Similarly, a person can be viewed as a heat engine that takes an input of internal energy, uses some of it to do work, and gives off the rest as heat. Suppose that a trained athlete can function as a heat engine with an efficiency of 0.11. (a) What is the magnitude of the internal energy that the athlete uses in order to do 5.1×10^4 J of work? (b) Determine the magnitude of the heat the athlete gives off.

86. **M** **BIO** **GO** 15.8 A 52-kg mountain climber, starting from rest, climbs a vertical distance of 730 m. At the top, she is again at rest. In the process, her body generates 4.1×10^6 J of energy via metabolic processes. In fact, her body acts like a heat engine, the efficiency of which is given by Equation 15.11 as $e = |W|/|Q_H|$, where $|W|$ is the magnitude of the work she does and $|Q_H|$ is the magnitude of the input heat. Find her efficiency as a heat engine.

87. **M** **BIO** 15.3 A weightlifter drinks a protein shake that contains 2.00×10^2 Calories. She then performs multiple repetitions on the bench press and does 2.75×10^5 J of work. After her workout, her net change in internal energy is $+1.50 \times 10^5$ J. During her workout, she loses heat to the environment, which results in the vaporization of perspiration from the surface of her skin. What mass of water did she lose due to perspiration? Assume the latent heat of vaporization of the perspiration is 2.42×10^6 J/kg.

88. **M** **BIO** 15.3 A long-distance runner performs work in the form of exercise with an average power output of 650 W. In 35 minutes of running, his body loses 5.2×10^5 J of heat. (a) What was the decrease in the internal energy of the runner? (b) How many calories would the runner have to ingest to replace the energy lost in part (a)?

89. **H** **BIO** 15.8 Imagine a person lifts 50 bags of landscaping rock, each with a mass of 35.0 kg, onto a table that is 1.00 m high. During this activity, 6.20×10^4 J of heat is exhausted by the body through perspiration. (a) Treating the body as a heat engine, what is its efficiency? (b) How much energy was extracted from the body's internal heat reservoir, which is created by the metabolism of food?

90. **M** **BIO** 15.3 During an intense 30-minute fat-burning cardio workout class, the instructor maintains an average power output of 2.0×10^2 W, and she loses heat to the environment at an average rate of 2.5×10^2 W. Ignoring heat loss due to perspiration, how much fat will the instructor lose during the workout? The energy content of fat is 9.3 kcal/g.

Concepts and Calculations Problems

Online

The first law of thermodynamics is basically a restatement of the conservation-of-energy principle in terms of heat and work. Problem 91 emphasizes this fact by showing that the conservation principle and the first law provide the same approach to the problem. It also provides a review of the concept of the latent heat of sublimation and the ideal gas law. Problem 92 reviews some of the central features of heat engines, as well as kinetic energy and the work–energy theorem.

91. **M** **CHALK** The sublimation of zinc (mass per mole = 0.0654 kg/mol) takes place at a temperature of 6.00×10^2 K, and the latent heat of sublimation is 1.99×10^6 J/kg. The pressure remains constant during the sublimation. Assume that the zinc vapor can be treated as a monatomic ideal gas and that the volume of solid zinc is negligible compared to the corresponding vapor. *Concepts:* (i) What is sublimation, and what is the latent heat of sublimation? (ii) When a solid phase changes to a gas phase, does the volume of the material increase or decrease, and by how much? (iii) As the material changes from a solid to a gas, does it do work on the environment, or does the environment do work on it? How much work is involved? (iv) In this problem we begin with heat Q and realize that it is used for two purposes: First, it makes the solid change into a gas, which entails a change ΔU in the internal energy of the material, $\Delta U = U_{gas} - U_{solid}$. Second, it allows the expanding material to do work W on the environment. According to the conservation-of-energy principle, how is Q related to ΔU and W? (v) According to the first law of thermodynamics, how is Q related to ΔU and W? *Calculations:* What is the change in the internal energy of zinc when 1.50 kg of zinc sublimates?

92. **M** **CHALK** **SSM** Each of two Carnot engines uses the same cold reservoir at a temperature of 275 K for its exhaust heat. Each engine receives 1450 J of input heat. The work from either of these engines is used to drive a pulley arrangement that uses a rope to accelerate a 125-kg crate from rest along a horizontal frictionless surface, as shown in the figure. With engine 1 the crate attains a speed of 2.00 m/s, while with engine 2 it attains a speed of 3.00 m/s. *Concepts:* (i) With which engine is the change in the crate's energy greater? (ii) Which engine does more work? Explain your answer. (iii) For which engine is the temperature of the hot reservoir greater? *Calculations:* Find the temperature of the hot reservoir for each engine.

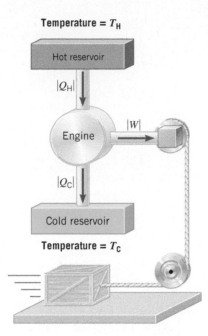

Temperature = T_H

Hot reservoir

$|Q_H|$

Engine $|W|$

$|Q_C|$

Cold reservoir

Temperature = T_C

PROBLEM 92

Team Problems

Online

93. E T WS Relieve the Pressure. A pressure relief valve is designed to prevent a gas containment vessel (a metal cylinder) from exploding due to overpressurization. The idea is that when the internal pressure exceeds a certain level, the valve automatically opens to let out some of the gas, and then closes again when the pressure is reduced to a safe level. You and your team have been assigned to test a pressure relief valve on an insulated neon gas storage tank to make sure that it is operable. The tank has a volume of 2.75 m³ and is currently at a pressure of 5.20×10^5 Pa, and the pressure relief valve is set to open at a pressure of 8.50×10^5 Pa. To increase the pressure, you turn on a 250.0-watt heater located inside the tank. How long after you turn on the heater should the relief valve open? Neon is a monatomic gas.

94. E T WS A Water Chiller. You and your team have been assigned to set up a special medical X-ray device that requires chilled water to be circulated through it in order that it not overheat. The problem is that a water chiller was not included with the device and you need to determine the type of chiller that is needed, and then obtain one. You read in the operation manual of the device that the temperature of the water going into the X-ray system must be no greater than 18.0 °C, and must circulate at a rate of no less than 8.00 liters per minute. The temperature of the water entering the chiller is 29.0 °C. If the coefficient of performance (COP) of the water chiller is 3.65, what is the minimum wattage required for a chiller that will do the job, where the wattage is the rate at which the chiller can do work? You can treat the water chiller like a typical refrigeration system.

95. T M A High-Performance Engine. You and your team are tasked with evaluating the parameters of a high-performance engine. The engine compresses the fuel-air mixture in the cylinder to make it ignite rather than igniting it with a spark plug (this is how diesel engines operate). You are given the *compression ratio* for the cylinders in the engine, which is the ratio of the initial volume of the cylinder (before the piston compresses the gas) to the final volume of the cylinder (after the gas is compressed): $V_i{:}V_f$. Assume that the fuel-gas mixture enters the cylinder at a temperature of 22.0 °C, and that the gas behaves like an ideal gas with $\gamma = 7/5$. **(a)** If the compression ratio is 15.4:1, what is the final temperature of the gas if the compression is adiabatic? **(b)** With everything else the same as in (a), what is the final temperature of the gas if the compression ratio is increased to 17.0:1? **(c)** Everything else being the same, for which compression ratio do you think the engine runs more efficiently? Give a qualitative argument for your answer.

96. T M A Gas Lift. A monatomic gas is contained in a long, vertical cylinder of inner radius $r = 14.0$ cm. A movable piston of negligible mass is inserted in the cylinder at a height $h = 1.20$ m. Initially, the gas inside the piston is at ambient temperature (23.5 °C) and standard pressure (1 atm). You place a 250.0-kg mass on the top of the piston and allow it to isothermally compress the gas below it. **(a)** What is the final pressure of the gas? **(b)** How far is the piston displaced from the equilibrium position? **(c)** Assuming an isobaric expansion, to what temperature should the gas inside the cylinder be heated in order to lift the mass 15.0 cm above the original position of the cylinder (i.e., to a height of 1.35 m)?

CHAPTER **16**

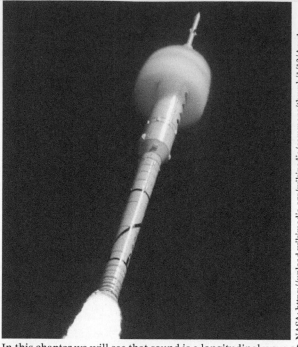

NASA; https://upload.wikimedia.org/wikipedia/commons/thumb/1/13/Ares1-X%2828OCT2009%29.jpg/180px-Ares1-X%2828OCT2009%29.jpg

In this chapter we will see that sound is a longitudinal wave of pressure fluctuations and travels through air at a speed of 343 m/s when the temperature is 20 °C. Sound is produced by a vibrating object, such as the surfaces of a rocket in flight. When a rocket approaches and then exceeds the speed of sound, it is said to break through the sound barrier. The so-called barrier is formed due to sound waves previously emitted by the rocket at speeds less than the speed of sound. Here, an Aries 1-X rocket emerges from a cloud called a vapor cone or vapor egg during launch. These often occur several seconds after the rocket breaks the sound barrier and passes through what flight control refers to as "Max Q," which is the time during the launch that the vehicle experiences maximum aerodynamic pressure. After this time, the pressure around the rocket decreases, which is accompanied by a drop in temperature. Under the right atmospheric conditions, water vapor will condense on the surface of the shock wave.

LEARNING OBJECTIVES

After reading this module, you should be able to...

16.1 Distinguish between transverse and longitudinal waves.

16.2 Calculate speed, frequency, and wavelength for a wave.

16.3 Calculate the speed of a wave on a string.

16.4 Use wave functions to describe a wave.

16.5 Examine sound waves.

16.6 Calculate the speed of a sound wave.

16.7 Analyze sound intensity.

16.8 Calculate sound intensity level.

16.9 Solve Doppler effect problems.

16.10 Describe applications of sound in medicine.

16.11 Examine the sensitivity of the human ear using Fletcher–Munson curves.

Waves and Sound

16.1 The Nature of Waves

There are two features common to all waves:

1. A wave is a traveling disturbance.
2. A wave carries energy from place to place.

Consider a water wave, for instance. In **Figure 16.1** the wave created by the motorboat travels across the lake and disturbs the fisherman. However, *there is no bulk flow of water* outward from the motorboat. The wave is not a bulk movement of water such as a river, but, rather, a disturbance traveling on the surface of the lake. Part of the wave's energy in **Figure 16.1** is transferred to the fisherman and his boat.

We will consider two basic types of waves, transverse and longitudinal. **Interactive Figure 16.2** illustrates how a transverse wave can

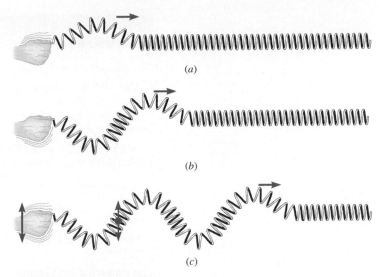

(*a*) An upward pulse moves to the right, followed by (*b*) a downward pulse. (*c*) When the end of the Slinky is moved up and down continuously, a transverse wave is produced.

FIGURE 16.1 The wave created by the motorboat travels across the lake and disturbs the fisherman.

be generated using a Slinky, a remarkable toy in the form of a long, loosely coiled spring. If one end of the Slinky is jerked up and down, as in part *a*, an upward pulse is sent traveling toward the right. If the end is then jerked down and up, as in part *b*, a downward pulse is generated and also moves to the right. If the end is continually moved up and down in simple harmonic motion, an entire wave is produced. As part *c* illustrates, the wave consists of a series of alternating upward and downward sections that propagate to the right, disturbing the vertical position of the Slinky in the process. To focus attention on the disturbance, a colored dot is attached to the Slinky in part *c* of the drawing. As the wave advances, the dot is displaced up and down in simple harmonic motion. The motion of the dot occurs perpendicular, or transverse, to the direction in which the wave travels. Thus, *a transverse wave is one in which the disturbance occurs perpendicular to the direction of travel of the wave*. Radio waves, light waves, and microwaves are transverse waves. Transverse waves also travel on the strings of instruments such as guitars and banjos.

A longitudinal wave can also be generated with a Slinky, and **Interactive Figure 16.3** demonstrates how. When one end is pushed forward along its length (i.e., longitudinally) and then pulled back to its starting point, as in part *a*, a region where the coils are compressed together is sent traveling to the right. If the end is pulled backward and then pushed forward to its starting point, as in part *b*, a region where the coils are stretched apart is formed and also moves to the right. If the end is continually moved back and forth in simple harmonic motion, an entire wave is created. As part *c* shows, the wave consists of a series of alternating compressed and stretched regions that travel to the right and disturb the coils. A colored dot is once again attached to the Slinky to emphasize the vibratory nature of the disturbance. In response to the wave, the dot moves back and forth in simple harmonic motion along the line of travel of the wave. Thus, *a longitudinal wave is one in which the disturbance occurs parallel to the line of travel of the wave*. A sound wave is a longitudinal wave.

A water wave is neither transverse nor longitudinal, since the motion of the water particles is not strictly perpendicular or strictly parallel to the line along which the wave travels. Instead, the motion includes both transverse and longitudinal components, since the water particles at the surface move on nearly circular paths, as **Figure 16.4** indicates.

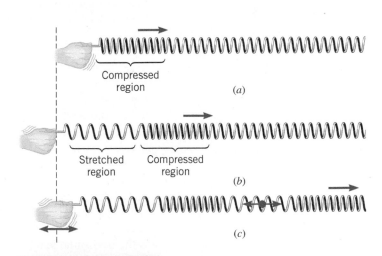

(*a*) A compressed region moves to the right, followed by (*b*) a stretched region. (*c*) When the end of the Slinky is moved back and forth continuously, a longitudinal wave is produced.

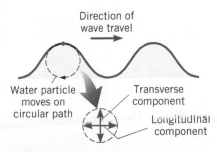

FIGURE 16.4 A water wave is neither transverse nor longitudinal, since water particles at the surface move clockwise on nearly circular paths as the wave moves from left to right.

Check Your Understanding

(The answers are given at the end of the book.)

1. Considering the nature of a water wave (see **Figure 16.4**), which of the following statements correctly describes how a fishing float moves on the surface of a lake when a wave passes beneath it? **(a)** It bobs up and down vertically. **(b)** It moves back and forth horizontally. **(c)** It moves in a vertical plane, exhibiting both motions described in (a) and (b) simultaneously.

2. Suppose that the longitudinal wave in **Interactive Figure 16.3c** moves to the right at a speed of 1 m/s. Does one coil of the Slinky move a distance of 1 mm to the right in a time of 1 ms?

16.2 | Periodic Waves

The transverse and longitudinal waves that we have been discussing are called **periodic waves** because they consist of cycles or patterns that are produced over and over again by the source. In **Figures 16.2** and **16.3** the repetitive patterns occur as a result of the simple harmonic motion of the left end of the Slinky, so that every segment of the Slinky vibrates in simple harmonic motion. Sections 10.1 and 10.2 discuss the simple harmonic motion of an object on a spring and introduce the concepts of cycle, amplitude, period, and frequency. This same terminology is used to describe periodic waves, such as the sound waves we hear (discussed later in this chapter) and the light waves we see (discussed in Chapter 24).

Figure 16.5 uses a graphical representation of a transverse wave on a Slinky to review the terminology. One **cycle** of a wave is shaded in color in both parts of the drawing. A wave is a series of many cycles. In part *a* the vertical position of the Slinky is plotted on the vertical axis, and the corresponding distance along the length of the Slinky is plotted on the horizontal axis. Such a graph is equivalent to a photograph of the wave taken at one instant in time and shows the disturbance that exists at each point along the Slinky's length. As marked on this graph, the **amplitude** *A* is the maximum excursion of a particle of the medium (i.e., the Slinky) in which the wave exists from the particle's undisturbed position. The amplitude is the distance between a crest, or highest point on the wave pattern, and the undisturbed position; it is also the distance between a trough, or lowest point on the wave pattern, and the undisturbed position. The **wavelength** *λ* is the horizontal length of one cycle of the wave, as shown in **Figure 16.5a**. The wavelength is also the horizontal distance between two successive crests, two successive troughs, or any two successive equivalent points on the wave.

Part *b* of **Figure 16.5** shows a graph in which time, rather than distance, is plotted on the horizontal axis. This graph is obtained by observing a single point on the Slinky. As the wave passes, the point under observation oscillates up and down in simple harmonic motion. As indicated on the graph, the **period** *T* is the time required for one complete up/down cycle, just as it is for an object vibrating on a spring. The period *T* is related to the **frequency** *f*, just as it is for any example of simple harmonic motion:

$$f = \frac{1}{T} \tag{10.5}$$

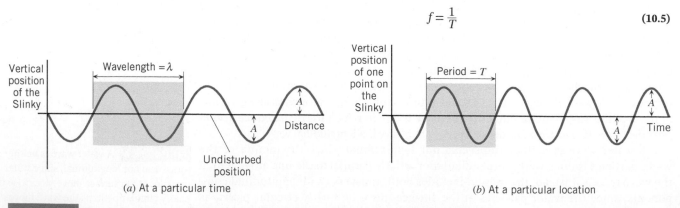

(a) At a particular time

(b) At a particular location

FIGURE 16.5 One cycle of the wave is shaded in color, and the amplitude of the wave is denoted as *A*.

The period is commonly measured in seconds, and frequency is measured in cycles per second, or hertz (Hz). If, for instance, one cycle of a wave takes one-tenth of a second to pass an observer, then ten cycles pass the observer per second, as Equation 10.5 indicates [$f = 1/(0.1 \text{ s}) = 10 \text{ cycles/s} = 10 \text{ Hz}$].

A simple relation exists between the period, the wavelength, and the speed of any periodic wave, a relation that **Figure 16.6** helps to introduce. Imagine waiting at a railroad crossing, while a freight train moves by at a constant speed v. The train consists of a long line of identical boxcars, each of which has a length λ and requires a time T to pass, so the speed is $v = \lambda/T$. This same equation applies for a wave and relates the speed of the wave to the wavelength λ and the period T. Since the frequency of a wave is $f = 1/T$, the expression for the speed is

$$v = \frac{\lambda}{T} = f\lambda \qquad (16.1)$$

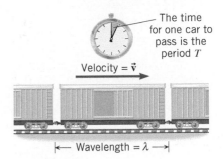

The time for one car to pass is the period T

Velocity = $\vec{v}$

$\longmapsto$ Wavelength = λ $\longrightarrow$|

FIGURE 16.6 A train moving at a constant speed serves as an analogy for a traveling wave.

The terminology just discussed and the fundamental relations $f = 1/T$ and $v = f\lambda$ apply to longitudinal as well as to transverse waves. Example 1 shows how the wavelength of a wave is determined by the wave speed and the frequency established by the source.

EXAMPLE 1 | The Wavelengths of Radio Waves

AM and FM radio waves are transverse waves consisting of electric and magnetic disturbances traveling at a speed of 3.00×10^8 m/s. A station broadcasts an AM radio wave whose frequency is 1230×10^3 Hz (1230 kHz on the dial) and an FM radio wave whose frequency is 91.9×10^6 Hz (91.9 MHz on the dial). Find the distance between adjacent crests in each wave.

Reasoning The distance between adjacent crests is the wavelength λ. Since the speed of each wave is $v = 3.00 \times 10^8$ m/s and the frequencies are known, the relation $v = f\lambda$ can be used to determine the wavelengths.

Problem-Solving Insight The equation $v = f\lambda$ applies to any kind of periodic wave.

Solution

AM
$$\lambda = \frac{v}{f} = \frac{3.00 \times 10^8 \text{ m/s}}{1230 \times 10^3 \text{ Hz}} = \boxed{244 \text{ m}}$$

FM
$$\lambda = \frac{v}{f} = \frac{3.00 \times 10^8 \text{ m/s}}{91.9 \times 10^6 \text{ Hz}} = \boxed{3.26 \text{ m}}$$

Notice that the wavelength of an AM radio wave is longer than two and one-half football fields!

Check Your Understanding

(The answer is given at the end of the book.)

3. A sound wave (a periodic longitudinal wave) from a loudspeaker travels from air into water. The frequency of the wave does not change, because the loudspeaker producing the sound determines the frequency. The speed of sound in air is 343 m/s, whereas the speed in fresh water is 1482 m/s. When the sound wave enters the water, does its wavelength increase, decrease, or remain the same?

16.3 | The Speed of a Wave on a String

The properties of the material* or medium through which a wave travels determine the speed of the wave. For example, **Figure 16.7** shows a transverse wave on a string and draws attention to four string particles that have been drawn as colored dots. As the wave moves to the right, each particle is displaced, one after the other, from its undisturbed

---

*Electromagnetic waves (discussed in Chapter 24) can move through a vacuum, as well as through materials such as glass and water.

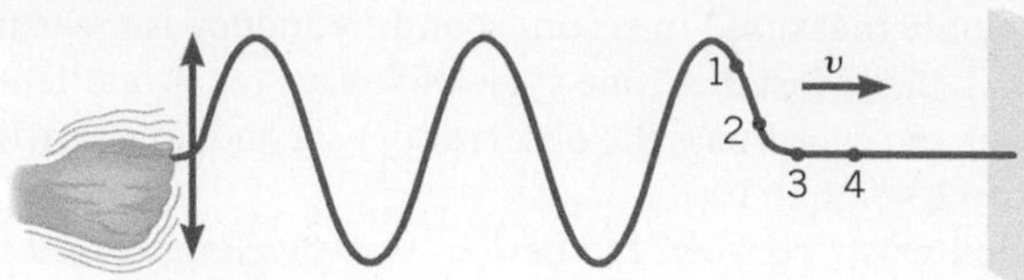

FIGURE 16.7 As a transverse wave moves to the right with speed v, each string particle is displaced, one after the other, from its undisturbed position.

position. In the drawing, particles 1 and 2 have already been displaced upward, while particles 3 and 4 are not yet affected by the wave. Particle 3 will be next to move because the section of string immediately to its left (i.e., particle 2) will pull it upward.

Figure 16.7 leads us to conclude that the speed with which the wave moves to the right depends on how quickly one particle of the string is accelerated upward in response to the net pulling force exerted by its adjacent neighbors. In accord with Newton's second law, a stronger net force results in a greater acceleration, and, thus, a faster-moving wave. The ability of one particle to pull on its neighbors depends on how tightly the string is stretched—that is, on the tension (see Section 4.10 for a review of tension). The greater the tension, the greater the pulling force the particles exert on each other and the faster the wave travels, other things being equal. Along with the tension, a second factor influences the wave speed. According to Newton's second law, the inertia or mass of particle 3 in **Figure 16.7** also affects how quickly it responds to the upward pull of particle 2. For a given net pulling force, a smaller mass has a greater acceleration than a larger mass. Therefore, other things being equal, a wave travels faster on a string whose particles have a small mass, or, as it turns out, on a string that has a small mass per unit length. The mass per unit length is called the *linear density* of the string. It is the mass m of the string divided by its length L, or m/L. Effects of the tension F and the mass per unit length are evident in the following expression for the speed v of a small-amplitude wave on a string:

$$v = \sqrt{\frac{F}{m/L}} \qquad (16.2)$$

The motion of transverse waves along a string is important in the operation of musical instruments, such as the guitar, the violin, and the piano. In these instruments, the strings are either plucked, bowed, or struck to produce transverse waves. Example 2 discusses the speed of the waves on the strings of a guitar.

EXAMPLE 2 | The Physics of Waves on Guitar Strings

Transverse waves travel on each string of an electric guitar after the string is plucked (see **Figure 16.8**). The length of each string between its two fixed ends is 0.628 m, and the mass is 0.208 g for the highest pitched E string and 3.32 g for the lowest pitched E string. Each string is under a tension of 226 N. Find the speeds of the waves on the two strings.

Reasoning The speed of a wave on a guitar string, as expressed by Equation 16.2, depends on the tension F in the string and its linear density m/L. Since the tension is the same for both strings, and smaller linear densities give rise to greater speeds, we expect the wave speed to be greatest on the string with the smallest linear density.

Solution The speeds of the waves are given by Equation 16.2 as

High-pitched E $\quad v = \sqrt{\dfrac{F}{m/L}} = \sqrt{\dfrac{226 \text{ N}}{(0.208 \times 10^{-3} \text{ kg})/(0.628 \text{ m})}} = \boxed{826 \text{ m/s}}$

Low-pitched E $\quad v = \sqrt{\dfrac{F}{m/L}} = \sqrt{\dfrac{226 \text{ N}}{(3.32 \times 10^{-3} \text{ kg})/(0.628 \text{ m})}} = \boxed{207 \text{ m/s}}$

Notice how fast the waves move: the speeds correspond to 1850 and 463 mi/h.

Transverse vibration of the string

FIGURE 16.8 Plucking a guitar string generates transverse waves.

Conceptual Example 3 offers additional insight into the nature of a wave as a traveling disturbance.

CONCEPTUAL EXAMPLE 3 | Wave Speed Versus Particle Speed

As indicated in **Figure 16.9**, the speed of a transverse wave on a string is v_{wave}, and the speed at which a string particle moves is $v_{particle}$. Which of the following statements is correct? **(a)** The speeds v_{wave} and $v_{particle}$ are identical. **(b)** The speeds v_{wave} and $v_{particle}$ are different.

Reasoning   A wave moves on a string at a speed v_{wave} that is determined by the properties of the string and has a constant value everywhere on the string at all times, assuming that these properties are the same everywhere on the string. Each particle on the string, however, moves in simple harmonic motion, assuming that the source generating the wave (e.g., the hand in **Interactive Figure 16.2c**) moves in simple harmonic motion. Each particle has a speed $v_{particle}$ that is characteristic of simple harmonic motion.

Answer (a) is incorrect.   The speed v_{wave} has a constant value at all times. In contrast, $v_{particle}$ is not constant at all times, because it is the speed that characterizes simple harmonic motion and that speed varies as time passes. Thus, the two speeds are not identical.

Answer (b) is correct.   The speed v_{wave} is determined by the tension F and the mass per unit length m/L of the string, according to $v_{wave} = \sqrt{\dfrac{F}{m/L}}$ (see Equation 16.2). The speed $v_{particle}$ is characteristic of simple harmonic motion, according to $v_{particle} = A\omega \sin \omega t$ (Equation 10.7 without the minus sign, since we deal

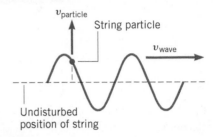

FIGURE 16.9   A transverse wave on a string is moving to the right with a constant speed v_{wave}. A string particle moves up and down in simple harmonic motion about the undisturbed position of the string. A string particle moves with a speed $v_{particle}$.

here only with speed, which is the magnitude of the velocity). The particle speed depends on the amplitude A and the angular frequency ω of the simple harmonic motion, as well as the time t; the speed is greatest when the particle is passing through the undisturbed position of the string, and it is zero when the particle has its maximum displacement. Thus, the two speeds are different, because v_{wave} depends on the properties of the string and $v_{particle}$ depends on the properties of the source creating the wave.

Related Homework: Problem 19

Check Your Understanding

(The answers are given at the end of the book.)

4. One end of each of two identical strings is attached to a wall. Each string is being pulled equally tight by someone at the other end. A transverse pulse is sent traveling along string A. A bit later an identical pulse is sent traveling along string B. What, if anything, can be done to make the pulse on string B catch up with and pass the pulse on string A?

5. In Section 4.10 the concept of a *massless* rope is discussed. Considering Equation 16.2, would it take any time for a transverse wave to travel the length of a truly massless rope?

6. A wire is strung tightly between two immovable posts. Review Section 12.4 and decide whether the speed of a transverse wave on this wire would increase, decrease, or remain the same when the temperature increases. Ignore any change in the mass per unit length of the wire.

7. Examine Conceptual Example 3 before addressing this question. A wave moves on a string with a constant velocity. Does this mean that the particles of the string always have zero acceleration?

8. A rope of mass m is hanging down from the ceiling. Nothing is attached to the loose end of the rope. As a transverse wave travels upward on the rope, does the speed of the wave increase, decrease, or remain the same?

9. String I and string II have the same length. However, the mass of string I is twice the mass of string II, and the tension in string I is eight times the tension in string II. A wave of the same amplitude and frequency travels on each of these strings. Which of the drawings in **CYU Figure 16.1** correctly shows the waves: **(a)** A **(b)** B **(c)** C?

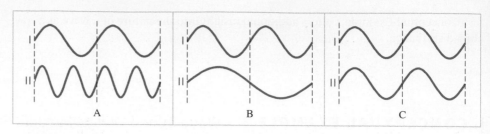

Math Skills Since the phase angles in Equation 16.3 and Equation 16.4 are measured in radians, *a calculator must be set to its radian mode when it is used to evaluate the functions* $\sin(2\pi f t - 2\pi x/\lambda)$ *or* $\sin(2\pi f t + 2\pi x/\lambda)$. Suppose, for instance, that $2\pi f t - 2\pi x/\lambda = 0.500$ radians, which corresponds to 28.6°. If your calculator is set to its radian mode when you evaluate sin 0.500, the correct value of 0.479 is displayed. However, with the calculator set to its degree mode, an incorrect value of 0.00873 is shown.

16.4 | *The Mathematical Description of a Wave

When a wave travels through a medium, it displaces the particles of the medium from their undisturbed positions. Suppose that a particle is located at a distance x from a coordinate origin. We would like to know the displacement y of this particle from its undisturbed position at any time t as the wave passes. For periodic waves that result from simple harmonic motion of the source, the expression for the displacement involves a sine or cosine, a fact that is not surprising. After all, in Chapter 10 simple harmonic motion is described using sinusoidal equations, and the graphs for a wave in **Figure 16.5** look like a plot of displacement versus time for an object oscillating on a spring (see **Figure 10.5**).

Our approach will be to present the expression for the displacement and then show graphically that it gives a correct description. Equation 16.3 represents the displacement of a particle caused by a wave traveling in the $+x$ direction (to the right), with an amplitude A, frequency f, and wavelength λ. Equation 16.4 applies to a wave moving in the $-x$ direction (to the left).

Wave motion toward $+x$ $$y = A \sin\left(2\pi f t - \frac{2\pi x}{\lambda}\right)$$ (16.3)

Wave motion toward $-x$ $$y = A \sin\left(2\pi f t + \frac{2\pi x}{\lambda}\right)$$ (16.4)

These equations apply to transverse or longitudinal waves and assume that $y = 0$ m when $x = 0$ m and $t = 0$ s. The term $(2\pi f t - 2\pi x/\lambda)$ in Equation 16.3 or the term $(2\pi f t + 2\pi x/\lambda)$ in Equation 16.4 is called the **phase angle** of the wave. In either case the phase angle is measured in *radians*, not in degrees.

Consider a transverse wave moving in the $+x$ direction along a string. A string particle located at the origin ($x = 0$ m) exhibits simple harmonic motion with a phase angle of $2\pi f t$; that is, its displacement as a function of time is $y = A \sin (2\pi f t)$. A particle located at a distance x also exhibits simple harmonic motion, but its phase angle in Equation 16.3 is

$$2\pi f t - \frac{2\pi x}{\lambda} = 2\pi f\left(t - \frac{x}{f\lambda}\right) = 2\pi f\left(t - \frac{x}{v}\right)$$

The quantity x/v is the time needed for the wave to travel the distance x. In other words, the simple harmonic motion that occurs at x is delayed by the time interval x/v compared to the motion at the origin.

Figure 16.10 shows the displacement y plotted as a function of position x along the string at a series of time intervals separated by one-fourth of the period T ($t = 0$ s, $\frac{1}{4}T$, $\frac{2}{4}T$, $\frac{3}{4}T$, T). These graphs are constructed by substituting the corresponding value for t into Equation 16.3, remembering that $f = 1/T$, and then calculating y at a series of values for x. The graphs are like photographs taken at various times as the wave moves to the right. For reference, the colored square on each graph marks the place on the wave that

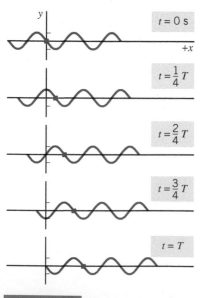

FIGURE 16.10 Equation 16.3 is plotted here at a series of times separated by one-fourth of the period T. The colored square in each graph marks the place on the wave that is located at $x = 0$ m when $t = 0$ s. As time passes, the wave moves to the right.

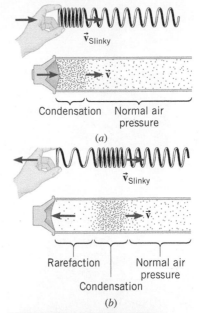

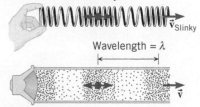

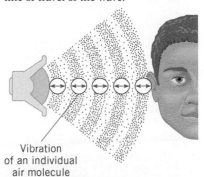

is located at $x = 0$ m when $t = 0$ s. As time passes, the colored square moves to the right, along with the wave. In a similar manner, it can be shown that Equation 16.4 represents a wave moving in the $-x$ direction.

16.5 The Nature of Sound

Longitudinal Sound Waves

Sound is a longitudinal wave that is created by a vibrating object, such as a guitar string, the human vocal cords, or the diaphragm of a loudspeaker. Moreover, sound can be created or transmitted only in a medium, such as a gas, liquid, or solid. As we will see, the particles of the medium must be present for the disturbance of the wave to move from place to place. Sound cannot exist in a vacuum.

THE PHYSICS OF . . . a loudspeaker diaphragm. To see how sound waves are produced and why they are longitudinal, consider the vibrating diaphragm of a loudspeaker. When the diaphragm moves outward, it compresses the air directly in front of it, as in **Interactive Figure 16.11a**. This compression causes the air pressure to rise slightly. The region of increased pressure is called a **condensation**, and it travels away from the speaker at the speed of sound. The condensation is analogous to the compressed region of coils in a longitudinal wave on a Slinky, which is included in **Interactive Figure 16.11a** for comparison. After producing a condensation, the diaphragm reverses its motion and moves inward, as in part *b* of the drawing. The inward motion produces a region known as a **rarefaction**, where the air pressure is slightly less than normal. The rarefaction is similar to the stretched region of coils in a longitudinal Slinky wave. Following immediately behind the condensation, the rarefaction also travels away from the speaker at the speed of sound. **Figure 16.12** further emphasizes the similarity between a sound wave and a longitudinal Slinky wave. As the wave passes, the colored dots attached both to the Slinky and to an air molecule execute simple harmonic motion about their undisturbed positions. The colored arrows on either side of the dots indicate that the simple harmonic motion occurs parallel to the line of travel. The drawing also shows that the wavelength λ is the distance between the centers of two successive condensations; λ is also the distance between the centers of two successive rarefactions.

Figure 16.13 illustrates a sound wave spreading out in space after being produced by a loudspeaker. When the condensations and rarefactions arrive at the ear, they force the eardrum to vibrate at the same frequency as the speaker diaphragm. The vibratory motion of the eardrum is interpreted by the brain as sound. It should be emphasized that sound is not a mass movement of air, like the wind. As the condensations and rarefactions of the sound wave travel outward from the vibrating diaphragm in **Figure 16.13**, the individual air molecules are not carried along with the wave. Rather, each molecule executes simple harmonic motion about a fixed location. In doing so, one molecule collides with its neighbor and passes the condensations and rarefactions forward. The neighbor, in turn, repeats the process.

The Frequency of a Sound Wave

Each cycle of a sound wave includes one condensation and one rarefaction, and the **frequency** is the number of cycles per second that passes by a given location. For example, if the diaphragm of a speaker vibrates back and forth in simple harmonic motion at a frequency of 1000 Hz, then 1000 condensations, each followed by a rarefaction, are generated every second, thus forming a sound wave whose frequency is also 1000 Hz. A sound with a single frequency is called a **pure tone**. Experiments have shown that a healthy young person hears all sound frequencies from approximately 20 to 20 000 Hz (20 kHz). The ability to hear the high frequencies decreases with age, however, and a normal middle-aged adult hears frequencies only up to 12–14 kHz.

INTERACTIVE FIGURE 16.11 (*a*)When the speaker diaphragm moves outward, it creates a condensation. (*b*) When the diaphragm moves inward, it creates a rarefaction. The condensation and rarefaction on the Slinky are included for comparison. In reality, the velocity of the wave on the Slinky $\vec{v}_{Slinky}$ is much smaller than the velocity of sound in air $\vec{v}$. For simplicity, the two waves are shown here to have the same velocity.

FIGURE 16.12 Both the wave on the Slinky and the sound wave are longitudinal. The colored dots attached to the Slinky and to an air molecule vibrate back and forth parallel to the line of travel of the wave.

FIGURE 16.13 Condensations and rarefactions travel from the speaker to the listener, but the individual air molecules do not move with the wave. A given molecule vibrates back and forth about a fixed location.

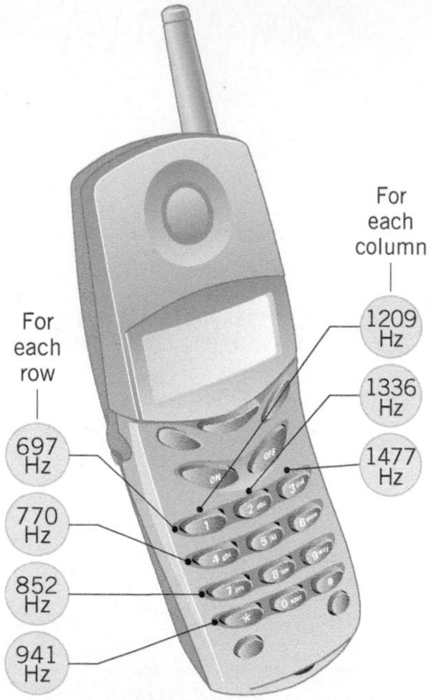

THE PHYSICS OF . . . a touch-tone telephone. Pure tones are used in touch-tone telephones, such as the one shown in Figure 16.14. These phones simultaneously produce two pure tones when each button is pressed, a different pair of tones for each different button. The tones are transmitted electronically to the central telephone office, where they activate switching circuits that complete the call. For example, the drawing indicates that pressing the "5" button produces pure tones of 770 and 1336 Hz simultaneously. These frequencies are characteristic of the second row and second column of buttons, respectively. Similarly, the "9" button generates tones of 852 and 1477 Hz.

Sound can be generated whose frequency lies below 20 Hz or above 20 kHz, although humans normally do not hear it. Sound waves with frequencies below 20 Hz are said to be **infrasonic**, while those with frequencies above 20 kHz are referred to as **ultrasonic**. Some species of bats known as microbats use ultrasonic frequencies up to 120 kHz for locating prey and for navigating (Figure 16.15), while rhinoceroses use infrasonic frequencies as low as 5 Hz to call one another (Figure 16.16).

Frequency is an objective property of a sound wave because frequency can be measured with an electronic frequency counter. A listener's perception of frequency, however, is subjective. The brain interprets the frequency detected by the ear primarily in terms of the subjective quality called **pitch**. A pure tone with a large (high) frequency is interpreted as a high-pitched sound, while a pure tone with a small (low) frequency is interpreted as a low-pitched sound. For instance, a piccolo produces high-pitched sounds, and a tuba produces low-pitched sounds.

FIGURE 16.14 A touch-tone telephone and a schematic showing the two pure tones produced when each button is pressed.

FIGURE 16.15 Some bats use ultrasonic sound waves for locating prey and for navigating. This bat has captured a katydid.

FIGURE 16.16 Rhinoceroses call to one another using infrasonic sound waves.

The Pressure Amplitude of a Sound Wave

Figure 16.17 illustrates a pure-tone sound wave traveling in a tube. Attached to the tube is a series of gauges that indicate the pressure variations along the wave. The graph shows that the air pressure varies sinusoidally along the length of the tube. Although this graph has the appearance of a transverse wave, remember that the sound itself is a longitudinal wave. The graph also shows the **pressure amplitude** of the wave, which is the magnitude of the maximum change in pressure, measured relative to the undisturbed or atmospheric pressure. The pressure fluctuations in a sound wave are normally very small. For instance, in a typical conversation between two people the pressure amplitude is about 3×10^{-2} Pa, certainly a small amount compared with the atmospheric pressure of $1.01 \times 10^{+5}$ Pa. The ear is remarkable in being able to detect such small changes.

Loudness is an attribute of sound that depends primarily on the amplitude of the wave: the larger the amplitude, the louder the sound. The pressure amplitude is an objective property of a sound wave, since it can be measured. Loudness, on the other hand, is subjective. Each individual determines what is loud, depending on the acuteness of his or her hearing.

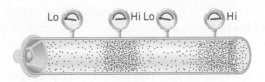

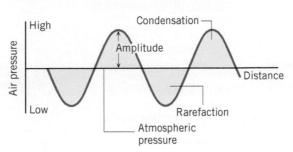

FIGURE 16.17 A sound wave is a series of alternating condensations and rarefactions. The graph shows that the condensations are regions of higher than normal air pressure, and the rarefactions are regions of lower than normal air pressure.

Check Your Understanding

(The answer is given at the end of the book.)

10. In a traveling sound wave, are there any particles that are *always* at rest as the wave passes by?

16.6 | The Speed of Sound

Gases

Sound travels through gases, liquids, and solids at considerably different speeds, as Table 16.1 reveals. Near room temperature, the speed of sound in air is 343 m/s (767 mi/h) and is markedly greater in liquids and solids. For example, sound travels more than four times faster in water and more than seventeen times faster in steel than it does in air. In general, sound travels slowest in gases, faster in liquids, and fastest in solids.

Like the speed of a wave on a guitar string, the speed of sound depends on the properties of the medium. In a gas, it is only when molecules collide that the condensations and rarefactions of a sound wave can move from place to place. It is reasonable, then, to expect the speed of sound in a gas to have the same order of magnitude as the average molecular speed between collisions. For an ideal gas this average speed is the translational rms speed given by Equation 14.6: $v_{rms} = \sqrt{3kT/m}$, where T is the Kelvin temperature, m is the mass of a molecule, and k is Boltzmann's constant. Although the expression for v_{rms} overestimates the speed of sound, it does give the correct dependence on Kelvin temperature and particle mass. Careful analysis shows that the speed of sound in an ideal gas is given by

Ideal gas
$$v = \sqrt{\frac{\gamma kT}{m}}$$
(16.5)

where $\gamma = c_P/c_V$ is the ratio of the specific heat capacity at constant pressure c_P to the specific heat capacity at constant volume c_V.

The factor γ is introduced in Section 15.5, where the adiabatic compression and expansion of an ideal gas are discussed. In Section 15.6 it is shown that γ has the value of $\gamma = \frac{5}{3}$ for ideal monatomic gases and a value of $\gamma = \frac{7}{5}$ for ideal diatomic gases. The value of γ appears in Equation 16.5 because the condensations and rarefactions of a sound wave are formed by adiabatic compressions and expansions of the gas. The regions that

TABLE 16.1	Speed of Sound in Gases, Liquids, and Solids
Substance	**Speed (m/s)**
Gases	
Air (0 °C)	331
Air (20 °C)	343
Carbon dioxide (0 °C)	259
Oxygen (0 °C)	316
Helium (0 °C)	965
Liquids	
Chloroform (20 °C)	1004
Ethyl alcohol (20 °C)	1162
Mercury (20 °C)	1450
Fresh water (20 °C)	1482
Seawater (20 °C)	1522
Solids	
Copper	5010
Glass (Pyrex)	5640
Lead	1960
Steel	5960

are compressed (the condensations) become slightly warmed, and the regions that are expanded (the rarefactions) become slightly cooled. However, no appreciable heat flows from a condensation to an adjacent rarefaction because the distance between the two (half a wavelength) is relatively large for most audible sound waves and a gas is a poor thermal conductor. Thus, the compression and expansion process is adiabatic. Example 4 illustrates the use of Equation 16.5.

Analyzing Multiple-Concept Problems

EXAMPLE 4 | The Physics of an Ultrasonic Ruler

Figure 16.18 shows an ultrasonic ruler that is used to measure the distance to a target, such as a wall. To initiate the measurement, the ruler generates a pulse of ultrasonic sound that travels to the wall and, much like an echo, reflects from it. The reflected pulse returns to the ruler, which measures the time it takes for the round-trip. Using a preset value for the speed of sound, the unit determines the distance to the wall and displays it on a digital readout. Suppose that the round-trip travel time is 20.0 ms on a day when the air temperature is 32 °C. Assuming that air is an ideal diatomic gas

$(\gamma = \frac{7}{5})$ and that the average molecular mass of air is 28.9 u, find the distance between the ultrasonic ruler and the wall.

Reasoning   The distance between the ruler and the wall is equal to the speed of sound multiplied by the time it takes for the sound pulse to reach the wall. The speed v of sound can be determined from a knowledge of the air temperature T and the average mass m of an air molecule by using the relation $v = \sqrt{\gamma k T / m}$ (Equation 16.5). The time can be deduced from the given data.

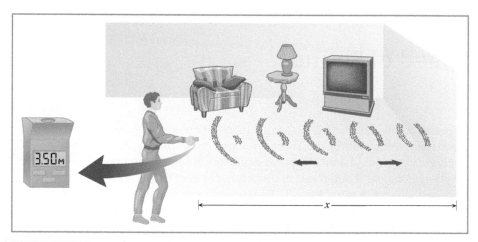

FIGURE 16.18   An ultrasonic ruler uses sound with a frequency greater than 20 kHz to measure the distance x to the wall. The blue arcs and blue arrow denote the outgoing sound wave, and the red arcs and red arrow denote the wave reflected from the wall.

Knowns and Unknowns   The data for this problem are listed below:

Description	Symbol	Value	Comment
Round-trip time of sound	t_{RT}	20.0 ms	20.0 ms = 20.0×10^{-3} s
Air temperature	T_c	32 °C	
Ratio of specific heats for air	γ	$\frac{7}{5}$	
Average molecular mass of air	m	28.9 u	Must convert "u" to kilograms.
Unknown Variable			
Distance between ruler and wall	x	?	

Modeling the Problem

STEP 1 Kinematics   Since sound moves at a constant speed, the distance x it travels is the product of its speed v and the time t, or $x = vt$. The time for the sound to reach the wall is one-half the round-trip time t_{RT}, so $t = \frac{1}{2}t_{RT}$. Thus, the distance to the wall is

$$x = v\left(\tfrac{1}{2}t_{RT}\right)$$

The round-trip time t_{RT} is known, but the speed of sound in air at 32 °C is not. We will find an expression for this speed in Step 2.

$$x = v\left(\tfrac{1}{2}t_{RT}\right) \qquad (1)$$

$$\boxed{?}$$

STEP 2 Speed of Sound   Since the air is assumed to be an ideal gas, the speed v of sound is related to the Kelvin temperature T and the average mass m of an air molecule by

$$v = \sqrt{\frac{\gamma k T}{m}} \qquad (16.5)$$

where γ is the ratio of the specific heat capacity of air at constant pressure to that at constant volume (see Section 15.5), and k is Boltzmann's constant. The temperature in this expression must be the Kelvin temperature of the air, which is related to its Celsius temperature T_c by $T = T_c + 273.15$ (Equation 12.1). Thus, the speed of sound in air is

$$\boxed{v = \sqrt{\frac{\gamma k (T_c + 273.15)}{m}}}$$

This expression for v can be substituted into Equation 1, as shown on the right.

$$x = v\left(\tfrac{1}{2}t_{RT}\right) \qquad (1)$$

$$\boxed{v = \sqrt{\frac{\gamma k (T_c + 273.15)}{m}}}$$

> **Problem-Solving Insight**   When using the equation $v = \sqrt{\gamma k T/m}$ to calculate the speed of sound in an ideal gas, be sure to express the temperature T in kelvins and not in degrees Celsius or Fahrenheit.

Solution   Combining the results of the modeling steps, we have

STEP 1   STEP 2

$$x = v\left(\tfrac{1}{2}t_{RT}\right) = \sqrt{\frac{\gamma k (T_c + 273.15)}{m}}\left(\tfrac{1}{2}t_{RT}\right)$$

Since the average mass of an air molecule is given in atomic mass units (28.9 u), we must convert it to kilograms by using the conversion factor $1\,u = 1.6605 \times 10^{-27}$ kg (see Section 14.1). Thus,

$$m = (28.9\,u)\left(\frac{1.6605 \times 10^{-27}\,\text{kg}}{1\,u}\right) = 4.80 \times 10^{-26}\,\text{kg}$$

The distance from the ultrasonic ruler to the wall is

$$x = \sqrt{\frac{\gamma k (T_c + 273.15)}{m}}\left(\tfrac{1}{2}t_{RT}\right)$$

$$= \sqrt{\frac{\tfrac{7}{5}(1.38 \times 10^{-23}\,\text{J/K})(32\,°\text{C} + 273.15)}{4.80 \times 10^{-2}\,\text{kg}}}\left[\tfrac{1}{2}(20.0 \times 10^{-3}\,\text{s})\right] = \boxed{3.50\,\text{m}}$$

Related Homework: Problem 46

THE PHYSICS OF . . . sonar.   Sonar (**so**und **na**vigation **r**anging) is a technique for determining water depth and locating underwater objects, such as reefs, submarines, and schools of fish. The core of a sonar unit consists of an ultrasonic transmitter and receiver mounted on the bottom of a ship. The transmitter emits a short pulse of ultrasonic sound, and at a later time the reflected pulse returns and is detected by the receiver. The distance to the object is determined from the electronically measured round-trip time of the pulse and a knowledge of the speed of sound in water; the distance registers automatically on an appropriate meter. Such a measurement is similar to the distance measurement discussed for the ultrasonic ruler in Example 4.

Conceptual Example 5 illustrates how the speed of sound in air can be used to estimate the distance to a thunderstorm, using a handy rule of thumb.

CONCEPTUAL EXAMPLE 5 | Lightning, Thunder, and a Rule of Thumb

In a thunderstorm, lightning and thunder occur nearly simultaneously. The light waves from the lightning travel at a speed of $v_{light} = 3.0 \times 10^8$ m/s, whereas the sound waves from the thunder travel at $v_{sound} = 343$ m/s. There is a rule of thumb for estimating how far away a storm is. After you see a lightning flash, count the seconds until you hear the thunder; divide the number of seconds by five to get the approximate distance (in miles) to the storm. In this rule, which of the two speeds plays a role? **(a)** Both v_{sound} and v_{light} **(b)** Only v_{sound} **(c)** Only v_{light}

Reasoning At a distance of one mile from a storm, the observer in **Figure 16.19** detects either type of wave only after a time that is equal to the distance divided by the speed at which the wave travels. This fact will guide our analysis.

Answers (b) and (c) are incorrect. The rule involves the time that passes *between* seeing the lightning flash and hearing the thunder, not just the time at which either type of wave is detected. Therefore, both the speeds v_{light} and v_{sound} must play a role in the rule.

Answer (a) is correct. Light from the flash travels so rapidly that it reaches the observer almost instantaneously; its travel time for one mile (1.6×10^3 m) is only

$$t_{light} = \frac{1.6 \times 10^3 \, m}{v_{light}} = \frac{1.6 \times 10^3 \, m}{3.0 \times 10^8 \, m/s} = 5 \times 10^{-6} \, s$$

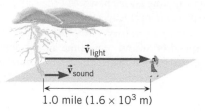

FIGURE 16.19 A lightning bolt from a thunderstorm generates a flash of light and sound (thunder). The speed of light is much greater than the speed of sound. Therefore, the light reaches the person first, followed later by the sound.

In comparison, the sound of the thunder travels very slowly; its travel time for one mile is

$$t_{sound} = \frac{1.6 \times 10^3 \, m}{v_{sound}} = \frac{1.6 \times 10^3 \, m}{343 \, m/s} = 5 \, s$$

Since t_{light} is negligible compared to t_{sound}, the time between seeing the lightning flash and hearing the thunder is about 5 s for every mile of distance from the storm.

Liquids

In a liquid, the speed of sound depends on the density ρ and the *adiabatic* bulk modulus B_{ad} of the liquid:

Liquid
$$v = \sqrt{\frac{B_{ad}}{\rho}}$$
(16.6)

The bulk modulus is introduced in Section 10.7 in a discussion of the volume deformation of liquids and solids. There it is tacitly assumed that the temperature remains constant while the volume of the material changes; that is, the compression or expansion is isothermal. However, the condensations and rarefactions in a sound wave occur under *adiabatic* rather than isothermal conditions. Thus, the adiabatic bulk modulus B_{ad} must be used when calculating the speed of sound in liquids. Values of B_{ad} will be provided as needed in this text.

Table 16.1 gives some data for the speed of sound in liquids. In seawater, for instance, the speed is 1522 m/s, which is more than four times as great as the speed in air. The speed of sound is an important parameter in the measurement of distance, as discussed for the ultrasonic ruler in Example 4.

BIO **THE PHYSICS OF . . . cataract surgery.** Accurate distance measurements using ultrasonic sound also play an important role in medicine, where the sound often travels through liquid-like materials in the body. A routine preoperative procedure in cataract surgery, for example, uses an ultrasonic probe called an A-scan to measure the length of the eyeball in front of the lens, the thickness of the lens, and the length of the eyeball between the lens and the retina (see **Figure 16.20**). The measurement is similar to that discussed in Example 4 and relies on the fact that the speed of sound in the material in front of and behind the lens of the eye is 1532 m/s, whereas that within the lens is 1641 m/s. In cataract surgery, the cataractous lens is removed and often replaced with an implanted artificial lens. Data provided by the A-scan facilitate the design of the lens implant (its size and the optical correction that it introduces).

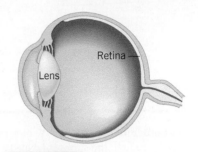

FIGURE 16.20 A cross-sectional view of the human eye.

Solid Bars

When sound travels through a long, slender, solid bar, the speed of the sound depends on the properties of the medium according to

Long, slender, solid bar
$$v = \sqrt{\frac{Y}{\rho}}$$
(16.7)

where Y is Young's modulus (defined in Section 10.7) and ρ is the density.

Check Your Understanding

(*The answers are given at the end of the book.*)

11. Do you expect an echo to return to you more quickly on a hot day or a cold day, other things being equal?

12. Carbon monoxide (CO), hydrogen (H_2), and nitrogen (N_2) may be treated as ideal gases. Each has the same temperature and nearly the same value for the ratio of the specific heat capacities at constant pressure and constant volume. In which two of the three gases is the speed of sound approximately the same?

13. Jell-O starts out as a liquid and then sets to a gel. As the Jell-O sets and becomes more solid, does the speed of sound in this material increase, decrease, or remain the same?

16.7 | Sound Intensity

Sound waves carry energy that can be used to do work, like forcing the eardrum to vibrate. In an extreme case such as a sonic boom, the energy can be sufficient to cause damage to windows and buildings. The amount of energy transported per second by a sound wave is called the **power** of the wave and is measured in SI units of joules per second (J/s) or watts (W).

When a sound wave leaves a source, such as the loudspeaker in **Figure 16.21**, the power spreads out and passes through imaginary surfaces that have increasingly larger areas. For instance, the same sound power passes through the surfaces labeled 1 and 2 in the drawing. However, the power is spread out over a greater area in surface 2. We will bring together the ideas of sound power and the area through which the power passes and, in the process, will formulate the concept of sound intensity. The idea of wave intensity is not confined to sound waves. It will recur, for example, in Chapter 24 when we discuss another important type of waves, electromagnetic waves.

The **sound intensity** I is defined as the sound power P that passes perpendicularly through a surface divided by the area A of that surface:

$$I = \frac{P}{A}$$
(16.8)

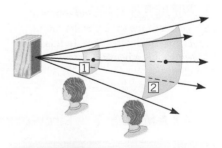

FIGURE 16.21 The power carried by a sound wave spreads out after leaving a source, such as a loudspeaker. Thus, the power passes perpendicularly through surface 1 and then through surface 2, which has the larger area.

The unit of sound intensity is power per unit area, or W/m^2. The next example illustrates how the sound intensity changes as the distance from a loudspeaker changes.

EXAMPLE 6 | Sound Intensities

In **Figure 16.21**, 12×10^{-5} W of sound power passes perpendicularly through the surfaces labeled 1 and 2. These surfaces have areas of $A_1 = 4.0$ m^2 and $A_2 = 12$ m^2. Determine the sound intensity at each surface and discuss why listener 2 hears a quieter sound than listener 1.

Reasoning The sound intensity I is the sound power P passing perpendicularly through a surface divided by the area A of that surface. Since the same sound power passes through both surfaces and surface 2 has the greater area, the sound intensity is less at surface 2.

> **Problem-Solving Insight** Sound intensity I and sound power P are different concepts. They are related, however, since intensity equals power per unit area.

Solution The sound intensity at each surface follows from Equation 16.8:

Surface 1
$$I_1 = \frac{P}{A_1} = \frac{12 \times 10^{-5}\,\text{W}}{4.0\,\text{m}^2} = \boxed{3.0 \times 10^{-5}\,\text{W/m}^2}$$

Surface 2
$$I_2 = \frac{P}{A_2} = \frac{12 \times 10^{-5}\,\text{W}}{12\,\text{m}^2} = \boxed{1.0 \times 10^{-5}\,\text{W/m}^2}$$

The sound intensity is less at the more distant surface, where the same power passes through a threefold greater area. The ear of a listener, with its fixed area, intercepts less power where the intensity, or power per unit area, is smaller. Thus, listener 2 intercepts less of the sound power than listener 1. With less power striking the ear, the sound is quieter.

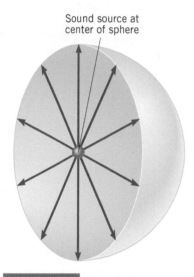

Sound source at center of sphere

FIGURE 16.22 The sound source at the center of the sphere emits sound uniformly in all directions. In this drawing, only a hemisphere is shown for clarity.

For a 1000-Hz tone, the smallest sound intensity that the human ear can detect is about 1×10^{-12} W/m^2; this intensity is called the **threshold of hearing**. On the other extreme, continuous exposure to intensities greater than 1 W/m^2 can be painful and can result in permanent hearing damage. The human ear is remarkable for the wide range of intensities to which it is sensitive.

If a source emits sound *uniformly in all directions,* the intensity depends on distance in a simple way. **Figure 16.22** shows such a source at the center of an imaginary sphere (for clarity only a hemisphere is shown). The radius of the sphere is r. Since all the radiated power P passes through the spherical surface of area $A = 4\pi r^2$, the intensity at a distance r is

Spherically uniform radiation
$$I = \frac{P}{4\pi r^2} \tag{16.9}$$

From this we see that the intensity of a source that radiates sound uniformly in all directions varies as $1/r^2$. For example, if the distance increases by a factor of two, the sound intensity decreases by a factor of $2^2 = 4$. Example 7 illustrates the effect of the $1/r^2$ dependence of intensity on distance.

EXAMPLE 7 | Fireworks

During a fireworks display, a rocket explodes high in the air above the observers. Assume that the sound spreads out uniformly in all directions and that reflections from the ground can be ignored. When the sound reaches listener 2 in **Figure 16.23**, who is $r_2 = 640$ m away from the explosion, the sound has an intensity of $I_2 = 0.10$ W/m^2. What is the sound intensity detected by listener 1, who is $r_1 = 160$ m away from the explosion?

Reasoning Listener 1 is four times closer to the explosion than listener 2. Therefore, the sound intensity detected by listener 1 is $4^2 = 16$ times greater than the sound intensity detected by listener 2.

> **Problem-Solving Insight** Equation 16.9 can be used only when the sound spreads out uniformly in all directions and there are no reflections of the sound waves.

FIGURE 16.23 If an explosion in a fireworks display radiates sound uniformly in all directions, the intensity at any distance r is $I = P/(4\pi r^2)$, where P is the sound power of the explosion.

Solution The ratio of the sound intensities can be found using Equation 16.9:

$$\frac{I_1}{I_2} = \frac{\dfrac{P}{4\pi r_1^{\,2}}}{\dfrac{P}{4\pi r_2^{\,2}}} = \frac{r_2^{\,2}}{r_1^{\,2}} = \frac{(640 \text{ m})^2}{(160 \text{ m})^2} = 16$$

As a result, $I_1 = (16)I_2 = (16)(0.10 \text{ W/m}^2) = \boxed{1.6 \text{ W/m}^2}$.

Equation 16.9 is valid only when no walls, ceilings, floors, etc. are present to reflect the sound and cause it to pass through the same surface more than once. Conceptual Example 8 demonstrates why this is so. Example 9 illustrates the application of Equation 16.9 to analyze the limitations of echolocation used by bats to locate prey.

CONCEPTUAL EXAMPLE 8 | Reflected Sound and Sound Intensity

Suppose that the person singing in the shower in **Figure 16.24** produces a sound power P. Sound reflects from the surrounding shower stall. At a distance r in front of the person, does the expression $I = P/(4\pi r^2)$ (Equation 16.9) **(a)** overestimate, **(b)** underestimate, or **(c)** give the correct total sound intensity?

Reasoning In arriving at Equation 16.9, it was assumed that the sound spreads out uniformly from the source and passes only once through the imaginary surface that surrounds it (see **Figure 16.22**). In **Figure 16.24**, only part of this imaginary surface (colored blue) is shown, but nonetheless, if Equation 16.9 is to apply, the same assumption must hold.

Answers (a) and (c) are incorrect. Equation 16.9 cannot overestimate the sound intensity, because it assumes that the sound passes through the imaginary surface only once and, hence, does not take into account the reflected sound within the shower stall. For the same reason, neither can Equation 16.9 give the correct sound intensity.

Answer (b) is correct. **Figure 16.24** illustrates three paths by which the sound passes through the imaginary surface. The "direct" sound travels along a path from its source directly to the surface. It is the intensity of this sound that is given by $I = P/(4\pi r^2)$. The remaining paths are two of the many that characterize the sound reflected from the shower stall. The *total* sound

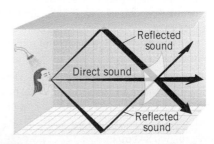

FIGURE 16.24 When someone sings in the shower, the sound power passing through part of an imaginary spherical surface (shown in blue) is the sum of the direct sound power and the reflected sound power.

power that passes through the surface is the sum of the direct and reflected powers. Thus the total sound intensity at a distance r from the source is greater than the intensity of the direct sound alone, so Equation 16.9 underestimates the sound intensity from the singing. People like to sing in the shower because their voices sound so much louder due to the enhanced intensity caused by the reflected sound.

Related Homework: Problem 61

EXAMPLE 9 | BIO Ultrasonic Waves and the Echolocation of Bats

A bat like the one shown in **Figure 16.15** emits ultrasonic pulses through its larynx at frequencies up to 120 kHz. The emitted sound waves scatter from objects and then "echo" back to the bat, whose brain then analyzes the time of travel, frequency, and intensity of the echo in order to echolocate obstacles and prey. The properties of the emitted sound, and the sensitivity of the bat's ears, set the size and distance limits of what the bat can detect. For instance, only objects of linear physical dimensions greater than or equal to the wavelength of the sound can be detected. Also, if the cross-sectional area of the target object is too small, the intensity of the echo will be undetectable by the bat's ears. Consider a stationary bat that emits a 65.0-kHz sound wave of power $P_0 = 1.90 \times 10^{-5}$ W, measured at a point just in front of the bat's mouth. At this frequency, the lowest intensity the bat can detect is 1.20×10^{-12} W/m². At a distance $r = 1.50$ m, what is the smallest diameter D of a spherical object that the bat can detect based on **(a)** the frequency f and **(b)** the intensity I_0 of the sound waves the bat emits? **(c)** Is it the wavelength or the intensity of the sound the bat emits that limits the size of an object than can be detected at a distance of 1.50 m? Assume that the speed of sound in air is 343 m/s, and that all of the power striking the object is reflected from it, that is, no energy is absorbed.

Reasoning **(a)** Knowing the speed of sound in air, and the frequency of the waves, the wavelength can be determined using Equation 16.1, $v = f\lambda$. The wavelength sets the lower size limit of what can be detected, and the wavelength will be the minimum diameter of a spherical object that can be detected based on the frequency of the waves. **(b)** Assuming that the pulse emitted by the bat is a spherical wave, we can calculate the intensity of the emitted pulse at the location of the object using Equation 16.9. We can then calculate the reflected power in terms of the unknown cross-sectional area of the object, and then the intensity of this echo at the bat's location, again using Equation 16.9. Finally, we can determine the cross-sectional area of the target that delivers the minimum detectable power to the bat and, finally, solve for the diameter of that circular area. **(c)** Comparing the results of (a) and (b) will reveal whether it is the emitted frequency or emitted intensity that limits the minimum size of an object that the bat can detect.

Solution **(a)** The wavelength λ of the emitted sound waves will be equal to the minimum diameter (D) of the spherical object that the bat can detect. Solving Equation 16.1 for the wavelength, we have

$$\lambda = \frac{v}{f} = \frac{343 \text{ m/s}}{65.0 \times 10^3 \text{ 1/s}} = \boxed{5.28 \times 10^{-3} \text{ m}}$$

Therefore, for the given frequency, D can be no smaller than 5.28×10^{-3} m.

FIGURE 16.15 (REPEATED) Some bats use ultrasonic sound waves for locating prey and for navigating. This bat has captured a katydid.

MerlinTuttle.org/Science Source

(b) Assuming that the sound waves emitted by the bat behave like spherical waves, following Equation 16.9, the intensity I_r at the location of the target object a distance r from the bat is given by

$$I_r = \frac{P_0}{4\pi r^2}$$

where P_0 is the power emitted by the bat. From this equation, it can be seen that the intensity of the wave decreases as the distance r between the source (the bat) and the object increases, as expected. The power reflected by the object will be the incident intensity I_r impinging upon the object multiplied by the cross-sectional area A of the object, or

$$P_r = I_r A = \left(\frac{P_0}{4\pi r^2}\right) A \qquad (1)$$

Assuming that the sound waves reflected from the object propagate as spherical waves, the intensity of the reflected waves will also diminish as they travel back to the bat. Therefore, Equation 16.9 must be applied again, this time treating the reflection as the source. The final intensity I_F detected by the bat is then given by

$$I_F = \frac{P_r}{4\pi r^2}$$

Substituting Equation 1 for P_r, we have

$$I_F = \frac{P_r}{4\pi r^2} = \frac{P_0 A}{(4\pi r^2)^2} = \frac{P_0 \pi \left(\frac{D}{2}\right)^2}{(4\pi r^2)^2}$$

where D is the diameter of the spherical object. Solving for D, we have

$$D = \left(\frac{64\pi r^4 I_F}{P_0}\right)^{\frac{1}{2}} = \left[\frac{64\pi (1.50 \text{ m})^4 (1.20 \times 10^{-12} \text{ W/m}^2)}{1.90 \times 10^{-5} \text{ W}}\right]^{\frac{1}{2}}$$

$$= \boxed{8.02 \times 10^{-3} \text{ m}}$$

(c) Since the minimum size of an object that can be detected by the bat at a distance of 1.50 m is limited to $D = 5.28 \times 10^{-3}$ m due to the wavelength of the sound, and to $D = 8.02 \times 10^{-3}$ m due to the emitted intensity, it is the **emitted intensity** that limits the minimum size of an object that the bat can detect.

Check Your Understanding

(The answers are given at the end of the book.)

14. BIO Some animals rely on an acute sense of hearing for survival, and the visible parts of the ears on such animals are often relatively large. How does this anatomical feature help to increase the sensitivity of the animal's hearing for low-intensity sounds?

15. A source is emitting sound uniformly in all directions. There are no reflections anywhere. A *flat* surface faces the source. Is the sound intensity the same at all points on the surface?

16.8 | Decibels

The **decibel** (dB) is a measurement unit used when comparing two sound intensities. The simplest method of comparison would be to compute the ratio of the intensities. For instance, we could compare $I = 8 \times 10^{-12}$ W/m^2 to $I_0 = 1 \times 10^{-12}$ W/m^2 by computing $I/I_0 = 8$ and stating that I is eight times as great as I_0. However, because of the way in which the human hearing mechanism responds to intensity, it is more appropriate to use a logarithmic scale for the comparison. For this purpose, the **intensity level** β (expressed in decibels) is defined as follows:

$$\beta = (10 \text{ dB}) \log\left(\frac{I}{I_0}\right) \qquad (16.10)$$

where "log" denotes the common logarithm to the base ten. I_0 is the intensity of the reference level to which I is being compared and is sometimes the threshold of hearing; that is, $I_0 = 1.00 \times 10^{-12}$ W/m^2. With the aid of a calculator, the intensity level can be evaluated for the values of I and I_0 given above:

$$\beta = (10 \text{ dB}) \log\left(\frac{8 \times 10^{-12} \text{ W/m}^2}{1 \times 10^{-12} \text{ W/m}^2}\right) = (10 \text{ dB}) \log 8 = (10 \text{ dB})(0.9) = 9 \text{ dB}$$

This result indicates that I is 9 decibels greater than I_0. Although β is called the "intensity level," it is *not* an intensity and does *not* have intensity units of W/m^2. In fact, the decibel, like the radian, is dimensionless.

Notice that if both I and I_0 are at the threshold of hearing, then $I = I_0$, and the intensity level is 0 dB according to Equation 16.10:

$$\beta = (10 \text{ dB}) \log\left(\frac{I_0}{I_0}\right) = (10 \text{ dB}) \log 1 = 0 \text{ dB}$$

since log 1 = 0. Thus,

Problem-Solving Insight An intensity level of zero decibels does not mean that the sound intensity I is zero; it means that $I = I_0$.

Intensity levels can be measured with a sound-level meter, such as the one in **Figure 16.25**. The intensity level β is displayed on its scale, assuming that the threshold of hearing is 0 dB. **Table 16.2** lists the intensities I and the associated intensity levels β for some common sounds, using the threshold of hearing as the reference level.

When a sound wave reaches a listener's ear, the sound is interpreted by the brain as loud or soft, depending on the intensity of the wave. Greater intensities give rise to

Math Skills In Equation 16.10, $\log\left(\frac{I}{I_0}\right)$ refers to the common logarithm and *not* the natural logarithm, which is $\ln\left(\frac{I}{I_0}\right)$. Since most calculators provide single-button access to both types of logarithm, *be careful to hit the "log" button and not the "ln" button when you evaluate the logarithm.* For example, if $\frac{I}{I_0} = 10.00$, it follows that log (10.00) = 1.000. But if you mistakenly hit the "ln" button, your calculator will display 2.303, and the value that you determine for the intensity level β in Equation 16.10 will be incorrect. When using Equation 16.10, also note that the ratio $\frac{I}{I_0}$ cannot be a negative number, since the logarithm of a negative number is not defined.

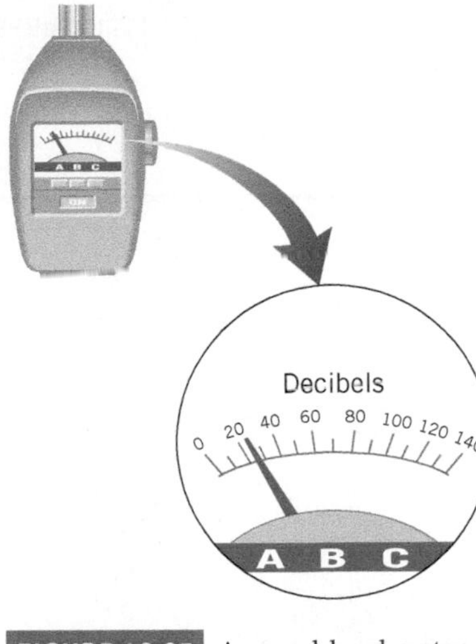

FIGURE 16.25 A sound-level meter and a close-up view of its decibel scale.

TABLE 16.2	Typical Sound Intensities and Intensity Levels Relative to the Threshold of Hearing	
	Intensity I (W/m^2)	**Intensity Level β (dB)**
Threshold of hearing	1.0×10^{-12}	0
Rustling leaves	1.0×10^{-11}	10
Whisper	1.0×10^{-10}	20
Normal conversation (1 meter)	3.2×10^{-6}	65
Inside car in city traffic	1.0×10^{-4}	80
Car without muffler	1.0×10^{-2}	100
Live rock concert	1.0	120
Threshold of pain	10	130

louder sounds. However, the relation between intensity and loudness is not a simple proportionality, because doubling the intensity does *not* double the loudness, as we will now see.

Suppose you are sitting in front of a stereo system that is producing an intensity level of 90 dB. If the volume control on the amplifier is turned up slightly to produce a 91-dB level, you would just barely notice the change in loudness. *Hearing tests have revealed that a one-decibel (1-dB) change in the intensity level corresponds to approximately the smallest change in loudness that an average listener with normal hearing can detect.* Since 1 dB is the smallest perceivable increment in loudness, a change of 3 dB—say, from 90 to 93 dB—is still a rather small change in loudness. Example 10 determines the factor by which the sound intensity must be increased to achieve such a change.

EXAMPLE 10 | Comparing Sound Intensities

Audio system 1 produces an intensity level of $\beta_1 = 90.0$ dB, and system 2 produces an intensity level of $\beta_2 = 93.0$ dB. The corresponding intensities (in W/m²) are I_1 and I_2. Determine the ratio I_2/I_1.

Reasoning Intensity levels are related to intensities by logarithms (see Equation 16.10), and it is a property of logarithms (see Appendix D) that $\log A - \log B = \log (A/B)$. Subtracting the two intensity levels and using this property, we find that

$$\beta_2 - \beta_1 = (10\text{ dB}) \log\left(\frac{I_2}{I_0}\right) - (10\text{ dB}) \log\left(\frac{I_1}{I_0}\right) = (10\text{ dB}) \log\left(\frac{I_2/I_0}{I_1/I_0}\right)$$

$$= (10\text{ dB}) \log\left(\frac{I_2}{I_1}\right)$$

Solution Using the result just obtained, we find

$$93.0\text{ dB} - 90.0\text{ dB} = (10\text{ dB}) \log\left(\frac{I_2}{I_1}\right)$$

$$0.30 = \log\left(\frac{I_2}{I_1}\right) \quad \text{or} \quad \frac{I_2}{I_1} = 10^{0.30} = \boxed{2.0}$$

Doubling the intensity changes the loudness by only a small amount (3 dB) and does not double it, so there is no simple proportionality between intensity and loudness.

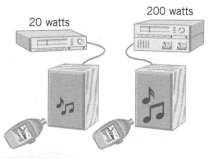

FIGURE 16.26 In spite of its tenfold greater power, the 200-watt audio system has only about double the loudness of the 20-watt system, when both are set for maximum volume.

To double the loudness of a sound, the intensity must be increased by more than a factor of two. *Experiment shows that if the intensity level increases by 10 dB, the new sound seems approximately twice as loud as the original sound.* For instance, a 70-dB intensity level sounds about twice as loud as a 60-dB level, and an 80-dB intensity level sounds about twice as loud as a 70-dB level. The factor by which the sound intensity must be increased to double the loudness can be determined as in Example 10:

$$\beta_2 - \beta_1 = 10.0\text{ dB} = (10\text{ dB})\left[\log\left(\frac{I_2}{I_0}\right) - \log\left(\frac{I_1}{I_0}\right)\right]$$

Solving this equation as in Example 10 reveals that $I_2/I_1 = 10.0$. Thus, increasing the sound intensity by a factor of ten will double the perceived loudness. Consequently, with both audio systems in **Figure 16.26** set at maximum volume, the 200-watt system will sound only twice as loud as the much cheaper 20-watt system.

Check Your Understanding

(The answers are given at the end of the book.)

16. If two people talk simultaneously and each creates an intensity level of 65 dB at a certain point, does the total intensity level at this point equal 130 dB?

17. Two observation points are located at distances r_1 and r_2 from a source of sound. The sound spreads out uniformly from the source, and there are no reflecting surfaces in the environment. The sound intensity level at distance r_2 is 6 dB less than the level at distance r_1. **(a)** What is the ratio I_2/I_1 of the sound intensities at the two distances? **(b)** What is the ratio r_2/r_1 of the distances?

16.9 The Doppler Effect

Have you ever heard an approaching fire truck and noticed the distinct change in the sound of the siren as the truck passes? The effect is similar to what you get when you put together the two syllables "eee" and "yow" to produce "eee-yow." While the truck approaches, the pitch of the siren is relatively high ("eee"), but as the truck passes and moves away, the pitch suddenly drops ("yow"). Something similar, but less familiar, occurs when an observer moves toward or away from a stationary source of sound. Such phenomena were first identified in 1842 by the Austrian physicist Christian Doppler (1803–1853) and are collectively referred to as the Doppler effect.

To explain why the Doppler effect occurs, we will bring together concepts that we have discussed previously—namely, the velocity of an object and the wavelength and frequency of a sound wave (Section 16.5). We will combine the effects of the velocities of the source and observer of the sound with the definitions of wavelength and frequency. In so doing, we will see that the **Doppler effect** is the change in frequency or pitch of the sound detected by an observer because the sound source and the observer have different velocities with respect to the medium of sound propagation.

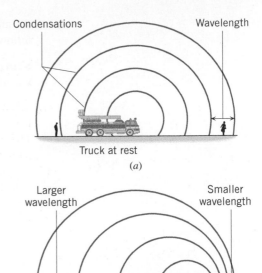

Condensations **Wavelength**

Truck at rest
(a)

Larger wavelength **Smaller wavelength**

Truck moving
(b)

Moving Source

To see how the Doppler effect arises, consider the sound emitted by a siren on the stationary fire truck in **Animated Figure 16.27a**. Like the truck, the air is assumed to be stationary with respect to the earth. Each solid blue arc in the drawing represents a condensation of the sound wave. Since the sound pattern is symmetrical, listeners standing in front of or behind the truck detect the same number of condensations per second and, consequently, hear the same frequency. Once the truck begins to move, the situation changes, as part *b* of the picture illustrates. Ahead of the truck, the condensations are now closer together, resulting in a decrease in the wavelength of the sound. This "bunching-up" occurs because the moving truck "gains ground" on a previously emitted condensation before emitting the next one. Since the condensations are closer together, the observer standing in front of the truck senses more of them arriving per second than she does when the truck is stationary. The increased rate of arrival corresponds to a greater sound frequency, which the observer hears as a higher pitch. Behind the moving truck, the condensations are farther apart than they are when the truck is stationary. This increase in the wavelength occurs because the truck pulls away from condensations emitted toward the rear. Consequently, fewer condensations per second arrive at the ear of an observer behind the truck, corresponding to a smaller sound frequency or lower pitch.

If the stationary siren in **Animated Figure 16.27a** emits a condensation at the time $t = 0$ s, it will emit the next one at time T, where T is the period of the wave. The distance between these two condensations is the wavelength λ of the sound produced by the stationary source, as **Figure 16.28a** indicates. When the truck is moving with a speed v_s (the subscript "s" stands for the "source" of sound) toward a stationary observer, the siren also emits condensations at $t = 0$ s and at time T. However, prior to emitting the second condensation, the truck moves closer to the observer by a distance $v_s T$, as **Figure 16.28b** shows. As a result, the distance between successive condensations is no longer the wavelength λ created by the stationary siren, but, rather, a wavelength λ' that is shortened by the amount $v_s T$:

$$\lambda' = \lambda - v_s T$$

Let's denote the frequency perceived by the stationary observer as f_o, where the subscript "o" stands for "observer." According to Equation 16.1, f_o is equal to the speed of sound v divided by the shortened wavelength λ':

$$f_o = \frac{v}{\lambda'} = \frac{v}{\lambda - v_s T}$$

But for the stationary siren, we have $\lambda = v/f_s$ and $T = 1/f_s$, where f_s is the frequency of the sound emitted by the source (not the frequency f_o perceived by the observer). With

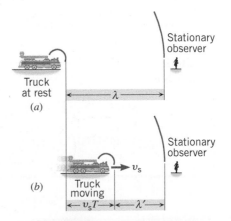

Stationary observer

Truck at rest
(a)
λ

Stationary observer

(b) Truck moving v_s
$\vdash v_s T \dashv \vdash \lambda' \dashv$

FIGURE 16.28 (a) When the fire truck is stationary, the distance between successive condensations is one wavelength λ. (b) When the truck moves with a speed v_s, the wavelength of the sound in front of the truck is shortened to λ'.

the aid of these substitutions for λ and T, the expression for f_o can be arranged to give the following result:

Source moving toward
stationary observer
$$f_o = f_s\left(\frac{1}{1 - \frac{v_s}{v}}\right)$$
(16.11)

Since the term $1 - v_s/v$ is in the denominator in Equation 16.11 and is less than one, the frequency f_o heard by the observer is *greater* than the frequency f_s emitted by the source. The difference between these two frequencies, $f_o - f_s$, is called the **Doppler shift**, and its magnitude depends on the ratio of the speed of the source v_s to the speed of sound v.

When the siren moves away from, rather than toward, the observer, the wavelength λ' becomes *greater* than λ according to

$$\lambda' = \lambda + v_s T$$

Notice the presence of the "+" sign in this equation, in contrast to the "−" sign that appeared earlier. The same reasoning that led to Equation 16.11 can be used to obtain an expression for the observed frequency f_o:

Source moving away from
stationary observer
$$f_o = f_s\left(\frac{1}{1 + \frac{v_s}{v}}\right)$$
(16.12)

The denominator $1 + v_s/v$ in Equation 16.12 is greater than one, so the frequency f_o heard by the observer is *less* than the frequency f_s emitted by the source. The next example illustrates how large the Doppler shift is in a familiar situation.

EXAMPLE 11 | The Sound of a Passing Train

A high-speed train is traveling at a speed of 44.7 m/s (100 mi/h) when the engineer sounds the 415-Hz warning horn. The speed of sound is 343 m/s. What are the frequency and wavelength of the sound, as perceived by a person standing at a crossing, when the train is **(a)** approaching and **(b)** leaving the crossing?

Reasoning When the train approaches, the person at the crossing hears a sound whose frequency is greater than 415 Hz because of the Doppler effect. As the train moves away, the person hears a frequency that is less than 415 Hz. We may use Equations 16.11 and 16.12, respectively, to determine these frequencies. In either case, the observed wavelength can be obtained according to Equation 16.1 as the speed of sound divided by the observed frequency.

Solution **(a)** When the train approaches, the observed frequency is

$$f_o = f_s\left(\frac{1}{1 - \frac{v_s}{v}}\right)$$

$$= (415\ \text{Hz})\left(\frac{1}{1 - \frac{44.7\ \text{m/s}}{343\ \text{m/s}}}\right) = \boxed{477\ \text{Hz}}$$
(16.11)

The observed wavelength is

$$\lambda' = \frac{v}{f_o} = \frac{343\ \text{m/s}}{477\ \text{Hz}} = \boxed{0.719\ \text{m}}$$
(16.1)

(b) When the train leaves the crossing, the observed frequency is

$$f_o = f_s\left(\frac{1}{1 + \frac{v_s}{v}}\right)$$

$$= (415\ \text{Hz})\left(\frac{1}{1 + \frac{44.7\ \text{m/s}}{343\ \text{m/s}}}\right) = \boxed{367\ \text{Hz}}$$
(16.12)

In this case, the observed wavelength is

$$\lambda' = \frac{v}{f_o} = \frac{343\ \text{m/s}}{367\ \text{Hz}} = \boxed{0.935\ \text{m}}$$
(16.1)

Moving Observer

Figure 16.29 shows how the Doppler effect arises when the sound source is stationary and the observer moves. As in the case of the moving source, we assume that the air is stationary. The observer moves with a speed v_o ("o" stands for "observer") toward the stationary source and covers a distance $v_o t$ in a time t. During this time, the moving observer encounters all the condensations that he would if he were stationary, *plus an additional number*. The additional number of condensations encountered is the distance $v_o t$ divided by the distance λ between successive condensations, or $v_o t/\lambda$. Thus, the additional number of condensations encountered per second by the moving observer is v_0/λ. Since a stationary observer would hear a frequency f_s emitted by the source, the moving observer hears a greater frequency f_0 given by

$$f_0 = f_s + \frac{v_0}{\lambda} = f_s\left(1 + \frac{v_0}{f_s\lambda}\right)$$

Using the fact that $v = f_s\lambda$, where v is the speed of sound, we find that

Observer moving toward stationary source
$$f_0 = f_s\left(1 + \frac{v_0}{v}\right) \tag{16.13}$$

An observer moving *away from* a stationary source moves in the same direction as the sound wave and, as a result, intercepts *fewer* condensations per second than a stationary observer does. In this case, the moving observer hears a smaller frequency f_0 that is given by

Observer moving away from stationary source
$$f_0 = f_s\left(1 - \frac{v_0}{v}\right) \tag{16.14}$$

It should be noted that the physical mechanism producing the Doppler effect is different when the source moves and the observer is stationary from when the observer moves and the source is stationary. When the source moves, as in **Figure 16.28b**, the wavelength of the sound perceived by the observer changes from λ to λ'. When the wavelength changes, the stationary observer hears a different frequency f_0 than the frequency produced by the source. On the other hand, when the observer moves and the source is stationary, the *wavelength λ does not change* (see, for example, **Figure 16.29**). Instead, the moving observer intercepts a different number of wave condensations per second than does a stationary observer and therefore detects a different frequency f_0.

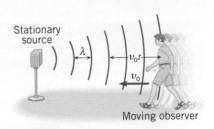

FIGURE 16.29 An observer moving with a speed v_0 toward the stationary source intercepts more wave condensations per unit of time than does a stationary observer.

General Case

It is possible for *both* the sound source and the observer to move with respect to the medium of sound propagation. If the medium is stationary, Equations 16.11–16.14 may be combined to give the observed frequency f_0 as

Source and observer both moving
$$f_0 = f_s\left(\frac{1 \pm \frac{v_0}{v}}{1 \mp \frac{v_s}{v}}\right) \tag{16.15}$$

In the numerator, the plus sign applies when the observer moves toward the source, and the minus sign applies when the observer moves away from the source. In the denominator, the minus sign is used when the source moves toward the observer, and the plus sign is used when the source moves away from the observer. The symbols v_0, v_s, and v denote numbers without an algebraic sign because the direction of travel has been taken into account by the plus and minus signs that appear directly in this equation.

NEXRAD

THE PHYSICS OF . . . NEXRAD. NEXRAD stands for **Nex**t Generation Weather **Rad**ar and is a nationwide system used by the National Weather Service to provide dramatically improved early warning of severe storms, such as the tornado in **Figure 16.30**. The system is based on radar waves, which are a type of electromagnetic wave (see Chapter 24) and, like sound waves, can exhibit the Doppler effect. The Doppler effect is at the heart of NEXRAD. As the drawing illustrates, a tornado is a swirling mass of air and water droplets. Radar pulses are sent out by a NEXRAD unit, whose protective covering is shaped like a soccer ball. The waves reflect from the water droplets and return to the unit, where the frequency is observed and compared to the outgoing frequency. For instance, droplets at point A in the drawing are moving toward the unit, and the radar waves reflected from them have their frequency Doppler-shifted to higher values. Droplets at point B, however, are moving away from the unit, and the frequency of the waves reflected from these droplets is Doppler-shifted to lower values. Computer processing of the Doppler frequency shifts leads to color-enhanced views on display screens (see

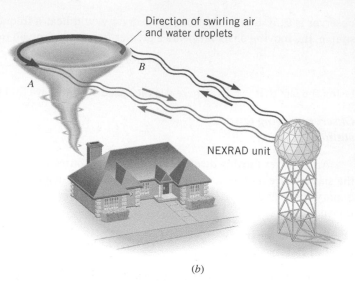

(a)   (b)

FIGURE 16.30 (a) A tornado is one of nature's most dangerous storms. (b) The National Weather Service uses the NEXRAD system, which is based on Doppler-shifted radar, to identify the storms that are likely to spawn tornadoes.

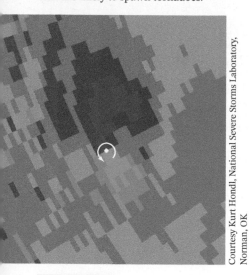

Courtesy Kurt Hondl, National Severe Storms Laboratory, Norman, OK

FIGURE 16.31 This color-enhanced view of a tornado shows circulating winds detected by a NEXRAD station, which is located below and to the right of the figure. The white dot and arrow indicate the storm center and the direction of wind circulation.

Figure 16.31). These views reveal the direction and magnitude of the wind velocity and can identify, from distances up to 140 mi, the swirling air masses that are likely to spawn tornadoes. The equations that specify the Doppler frequency shifts of radar waves are different from those given for sound waves by Equations 16.11–16.15. The reason for the difference is that radar waves propagate from one place to another by a different mechanism than that of sound waves (see Section 24.5).

Check Your Understanding

(The answers are given at the end of the book.)

18. At a swimming pool, a music fan up on a diving platform is listening to a radio. As the radio is playing a tone that has a constant frequency f_s, it is accidentally knocked off the platform. Describe the Doppler effect heard by **(a)** the person on the platform and **(b)** a person down below in the water. In each case, state whether the observed frequency f_o is greater or smaller than f_s and describe how f_o changes (if it changes) as the radio falls.

19. When a car is at rest, its horn emits a frequency of 600 Hz. A person standing in the middle of the street with this car behind him hears the horn with a frequency of 580 Hz. Does he need to jump out of the way?

20. A source of sound produces the same frequency under water as it does in air. This source has the same velocity in air as it does under water. Consider the ratio f_o/f_s of the observed frequency f_o to the source frequency f_s. Is this ratio greater in air or under water when the source **(a)** approaches and **(b)** moves away from the observer?

21. Two cars, one behind the other, are traveling in the same direction at the same speed. Does either driver hear the other's horn at a frequency that is different from the frequency heard when both cars are at rest?

22. When a truck is stationary, its horn produces a frequency of 500 Hz. You are driving your car, and this truck is following behind. You hear its horn at a frequency of 520 Hz. **(a)** Refer to Equation 16.15 and decide which algebraic sign should be used in the numerator and which in the denominator. **(b)** Which driver, if either, is driving faster?

16.10 Applications of Sound in Medicine

BIO **THE PHYSICS OF . . . ultrasonic imaging.** When ultrasonic waves are used in medicine for diagnostic purposes, high-frequency sound pulses are produced by a transmitter and directed into the body. As in sonar, reflections occur. They occur each time a pulse encounters a boundary between two tissues that have different densities or a boundary between a tissue and the adjacent fluid. By scanning ultrasonic waves across the body and detecting the echoes generated from various internal locations, it is possible to obtain an image, or sonogram, of the inner anatomy. Ultrasonic imaging is employed extensively in obstetrics to examine the developing fetus (**Figure 16.32**). The fetus, surrounded by the amniotic sac, can be distinguished from other anatomical features so that fetal size, position, and possible abnormalities can be detected. Ultrasound is also used in other medically related areas. For instance, tumors in the liver, kidney, brain, and pancreas can be detected with ultrasound. Yet another application involves monitoring the real-time movement of pulsating structures, such as heart valves ("echocardiography") and large blood vessels.

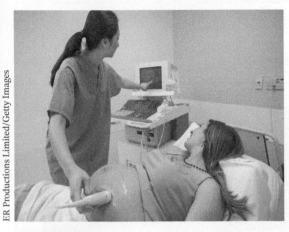
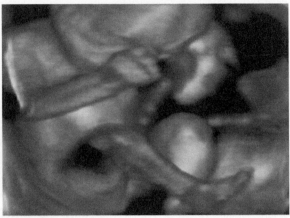

FIGURE 16.32 An ultrasonic scanner can be used to produce an image of the fetus as it develops in the uterus. Conventional scanning produces two-dimensional images. Three-dimensional scanning uses computer technology to produce images that are more detailed than the 2-D variety. The image on the right, for example, was obtained using 3-D scanning in the second trimester of pregnancy and illustrates the kind of detail that is achievable.

When ultrasound is used to form images of internal anatomical features or foreign objects in the body, the wavelength of the sound wave must be about the same size as, or smaller than, the object to be located. Therefore, high frequencies in the range from 1 to 15 MHz (1 MHz = 1 megahertz = 1×10^6 Hz) are the norm. For instance, the wavelength of 5-MHz ultrasound is $\lambda = v/f = 0.3$ mm, if a value of 1540 m/s is used for the speed of sound through tissue. A sound wave with a frequency higher than 5 MHz and a correspondingly shorter wavelength is required for locating objects smaller than 0.3 mm.

BIO THE PHYSICS OF . . . the cavitron ultrasonic surgical aspirator. Neuro-surgeons use a device called a **c**avitron **u**ltrasonic **s**urgical **a**spirator (CUSA) to remove brain tumors once thought to be inoperable. Ultrasonic sound waves cause the slender tip of the CUSA probe (see **Figure 16.33**) to vibrate at approximately 23 kHz. The probe shatters any section of the tumor that it touches, and the fragments are flushed out of the brain with a saline solution. Because the tip of the probe is small, the surgeon can selectively remove small bits of malignant tissue without damaging the surrounding healthy tissue.

BIO THE PHYSICS OF . . . bloodless surgery with HIFU. Another application of ultrasound is in a new type of bloodless surgery, which can eliminate abnormal cells, such as those in benign hyperplasia of the prostate gland. This technique is known as HIFU (**h**igh-**i**ntensity **f**ocused **u**ltrasound). It is analogous to focusing the sun's electromagnetic waves by using a magnifying glass and producing a small region where the energy carried by the waves can cause localized heating. Ultrasonic waves can be used in a similar fashion. The waves enter directly through the skin and come into focus inside the body over a region that is sufficiently well defined to be surgically useful. Within this region the energy of the waves causes localized heating, leading to a temperature of about 56 °C (normal body temperature is 37 °C), which is sufficient to kill abnormal cells. The killed cells are eventually removed by the body's natural processes.

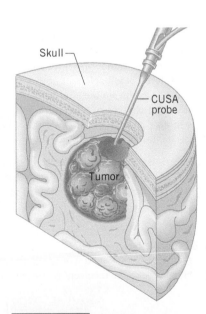

FIGURE 16.33 Neurosurgeons use a cavitron ultrasonic surgical aspirator (CUSA) to "cut out" brain tumors without adversely affecting the surrounding healthy tissue.

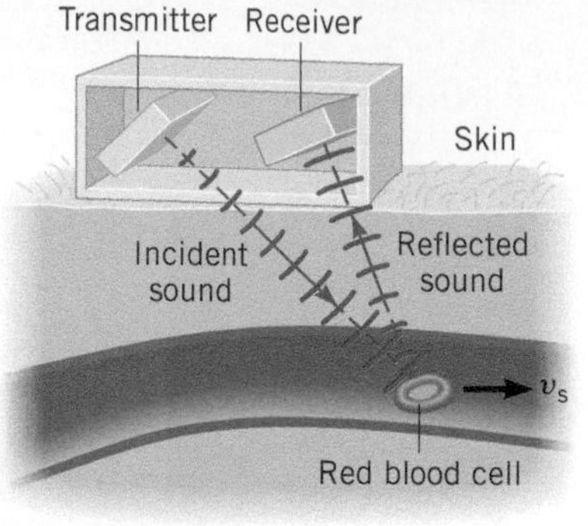

Transmitter Receiver

Skin

Incident
sound

Reflected
sound

v_s

Red blood cell

FIGURE 16.34 A Doppler flow meter measures the speed of red blood cells.

BIO **THE PHYSICS OF . . . the Doppler flow meter.** The Doppler flow meter is a particularly interesting medical application of the Doppler effect. This device measures the speed of blood flow, using transmitting and receiving elements that are placed directly on the skin, as in **Figure 16.34**. The transmitter emits a continuous sound whose frequency is typically about 5 MHz. When the sound is reflected from the red blood cells, its frequency is changed in a kind of Doppler effect because the cells are moving. The receiving element detects the reflected sound, and an electronic counter measures its frequency. From the change in frequency the speed of the blood flow can be determined. Typically, the change in frequency is around 600 Hz for flow speeds of about 0.1 m/s. The Doppler flow meter can be used to locate regions where blood vessels have narrowed, since greater flow speeds occur in the narrowed regions, according to the equation of continuity (see Section 11.8). In addition, the Doppler flow meter can be used to detect the motion of a fetal heart as early as 8–10 weeks after conception.

EXAMPLE 12 | **BIO** The Physics of Color Doppler Ultrasound

As discussed in Section 16.10, the technique of medical ultrasound can be used to image many different internal structures of the body. When combined with the Doppler effect, the technique of Color Doppler Ultrasound can provide information on the velocity of fluids, such as blood, moving in these structures. The Doppler information is converted by a computer into a false-color image that is overlaid on the ultrasound image of the structure, such as a blood vessel. The colors indicate the speed and direction of the blood flow (**Figure 16.35**). As an example, consider an ultrasound image created with sound waves at a frequency of 5.00 MHz. The sound waves travel at a speed of 1540 m/s through the body and reflect off of red blood cells flowing toward the transmitter. A sound receiver, also located in the transmitter, detects the reflected sound waves with a frequency that is 980 Hz higher than the sound sourced by the transmitter. Calculate the speed of the blood moving toward the transmitter.

Reasoning This is a straightforward application of the Doppler effect equations for sound, where we have to take into account two cases. First, as the sound wave is sent from the transmitter toward the red blood cells, we have the case of the observer (the blood) moving toward a stationary source (the transmitter/receiver), or Equation 16.13. Next, the sound wave is reflected off the red blood cell back toward the receiver. Thus, the blood cell acts as the source, and we have the case of a source moving toward a stationary observer, or Equation 16.11. Therefore, we will need to use both Equation 16.13 and Equation 16.11 to calculate the speed of the blood (v_B).

Solution The sound will be Doppler shifted on the way to the blood cells and also Doppler shifted upon reflection. For the sound traveling from the transmitter to the red blood cells, we use Equation 16.13:

$$f_o = f_s\left(1 + \frac{v_o}{v}\right) = f_s\left(1 + \frac{v_B}{v}\right)$$

where we have represented the speed of the observer (v_o) as v_B, the speed of the blood. For the sound traveling back from the red blood cells to the receiver, we use Equation 16.11:

$$f_o = f_s\left(\frac{1}{1 - \frac{v_s}{v}}\right) = f_s\left(\frac{1}{1 - \frac{v_B}{v}}\right)$$

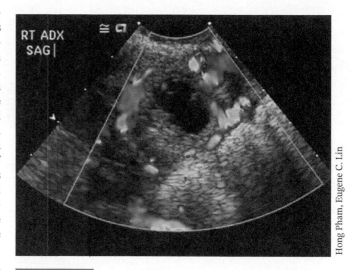

Hong Pham, Eugene C. Lin

FIGURE 16.35 Color Doppler ultrasound image showing the red and blue color overlay that provides information on the speed and direction of blood flow within internal body structures. Blue indicates blood flow toward the ultrasonic transducer, and red indicates flow away. This particular image shows increased vascularity in the right fallopian tube of an expecting mother, which, unfortunately, is evidence of an ectopic pregnancy.

where now, since the source of the sound is the reflection off of the red blood cells, we replace v_s with v_B. We can combine these two expressions by recognizing that the observed frequency in Equation 16.13 becomes the source frequency in Equation 16.11. Making this substitution, we have:

$$f_o = f_s\left(\frac{1 + \frac{v_B}{v}}{1 - \frac{v_B}{v}}\right)$$

We can now rearrange this equation and solve for v_B:

$$\frac{f_o}{f_s}\left(1 - \frac{v_B}{v}\right) = \left(1 + \frac{v_B}{v}\right) \Rightarrow v_B = \frac{v(f_o - f_s)}{(f_o + f_s)} = \frac{(1540\ \text{m/s})(980\ \text{Hz})}{(10\ 000\ 980\ \text{Hz})}$$

$$= \boxed{0.15\ \text{m/s}}$$

16.11 | *The Sensitivity of the Human Ear

BIO THE PHYSICS OF . . . hearing.   Although the ear is capable of detecting sound intensities as small as 1×10^{-12} W/m², it is *not* equally sensitive to all frequencies, as **Figure 16.36** shows. This figure displays a series of graphs known as the **Fletcher–Munson curves**, after H. Fletcher and M. Munson, who first determined them in 1933. In these graphs the audible sound frequencies are plotted on the horizontal axis, and the sound intensity levels (in decibels) are plotted on the vertical axis. Each curve is a *constant loudness* curve because it shows the sound intensity level needed at each frequency to make the sound appear to have the same loudness. For example, the lowest (red) curve represents the threshold of hearing. It shows the intensity levels at which sounds of different frequencies just become audible. The graph indicates that the intensity level of a 100-Hz sound must be about 37 dB greater than the intensity level of a 1000-Hz sound to be at the threshold of hearing. Therefore, the ear is *less sensitive* to a 100-Hz sound than it is to a 1000-Hz sound. In general, **Figure 16.36** reveals that the ear is most sensitive in the range of about 1–5 kHz, and becomes progressively less sensitive at higher and lower frequencies.

Each curve in **Figure 16.36** represents a different loudness, and each is labeled according to its intensity level at 1000 Hz. For instance, the curve labeled "60" represents all sounds that have the same loudness as a 1000-Hz sound whose intensity level is 60 dB. These constant-loudness curves become flatter as the loudness increases, the relative flatness indicating that the ear is nearly equally sensitive to all frequencies when the sound is loud. Thus, when you listen to loud sounds, you hear the low frequencies, the middle frequencies, and the high frequencies about equally well. However, when you listen to quiet sounds, the high and low frequencies seem to be absent, because the ear is relatively insensitive to these frequencies under such conditions.

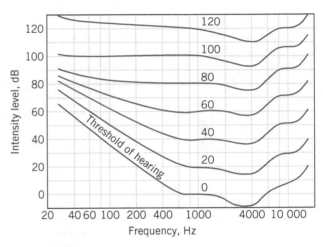

FIGURE 16.36 Each curve represents the intensity levels at which sounds of various frequencies have the same loudness. The curves are labeled by their intensity levels at 1000 Hz and are known as the Fletcher–Munson curves.

Concept Summary

16.1 The Nature of Waves   A wave is a traveling disturbance and carries energy from place to place. In a transverse wave, the disturbance occurs perpendicular to the direction of travel of the wave. In a longitudinal wave, the disturbance occurs parallel to the line along which the wave travels.

16.2 Periodic Waves   A periodic wave consists of cycles or patterns that are produced over and over again by the source of the wave. The amplitude of the wave is the maximum excursion of a particle of the medium from the particle's undisturbed position. The wavelength λ is the distance along the length of the wave between two successive equivalent points, such as two crests or two troughs. The period T is the time required for the wave to travel a distance of one wavelength. The frequency f (in hertz) is the number of wave cycles per second that passes an observer and is the reciprocal of the period (in seconds), as shown in Equation 10.5. The speed v of a wave is related to its wavelength and frequency according to Equation 16.1.

$$f = \frac{1}{T} \tag{10.5}$$

$$v = f\lambda \tag{16.1}$$

16.3 The Speed of a Wave on a String   The speed of a wave depends on the properties of the medium in which the wave travels. For a transverse wave on a string that has a tension F and a mass per unit length m/L, the wave speed is given by Equation 16.2. The mass per unit length is also called the linear density.

$$v = \sqrt{\frac{F}{m/L}} \tag{16.2}$$

16.4 The Mathematical Description of a Wave   When a wave of amplitude A, frequency f, and wavelength λ moves in the $+x$ direction through a medium, the wave causes a displacement y of a particle at position x according to Equation 16.3. For a wave moving in the $-x$ direction, the displacement y is given by Equation 16.4.

$$y = A \sin\left(2\pi ft - \frac{2\pi x}{\lambda}\right) \qquad (16.3)$$

$$y = A \sin\left(2\pi ft + \frac{2\pi x}{\lambda}\right) \qquad (16.4)$$

16.5 The Nature of Sound Sound is a longitudinal wave that can be created only in a medium; it cannot exist in a vacuum. Each cycle of a sound wave includes one condensation (a region of greater than normal pressure) and one rarefaction (a region of less than normal pressure).

A sound wave with a single frequency is called a pure tone. Frequencies less than 20 Hz are called infrasonic. Frequencies greater than 20 kHz are called ultrasonic. The brain interprets the frequency detected by the ear primarily in terms of the subjective quality known as pitch. A high-pitched sound is one with a large frequency (e.g., piccolo). A low-pitched sound is one with a small frequency (e.g., tuba).

The pressure amplitude of a sound wave is the magnitude of the maximum change in pressure, measured relative to the undisturbed pressure. The pressure amplitude is associated with the subjective quality of loudness. The larger the pressure amplitude, the louder the sound.

16.6 The Speed of Sound The speed of sound v depends on the properties of the medium. In an ideal gas, the speed of sound is given by Equation 16.5, where $\gamma = c_P/c_V$ is the ratio of the specific heat capacities at constant pressure and constant volume, k is Boltzmann's constant, T is the Kelvin temperature, and m is the mass of a molecule of the gas. In a liquid, the speed of sound is given by Equation 16.6, where B_{ad} is the adiabatic bulk modulus and ρ is the mass density. In a solid that has a Young's modulus of Y and the shape of a long slender bar, the speed of sound is given by Equation 16.7.

$$v = \sqrt{\frac{\gamma kT}{m}} \qquad (16.5)$$

$$v = \sqrt{\frac{B_{ad}}{\rho}} \qquad (16.6)$$

$$v = \sqrt{\frac{Y}{\rho}} \qquad (16.7)$$

16.7 Sound Intensity The intensity I of a sound wave is the power P that passes perpendicularly through a surface divided by the area A of the surface, as shown in Equation 16.8. The SI unit for intensity

is watts per square meter (W/m²). The smallest sound intensity that the human ear can detect is known as the threshold of hearing and is about 1×10^{-12} W/m² for a 1-kHz sound. When a source radiates sound uniformly in all directions and no reflections are present, the intensity of the sound is inversely proportional to the square of the distance from the source, according to Equation 16.9.

$$I = \frac{P}{A} \qquad (16.8)$$

$$I = \frac{P}{4\pi r^2} \qquad (16.9)$$

16.8 Decibels The intensity level β (in decibels) is used to compare a sound intensity I to the sound intensity I_0 of a reference level, as indicated by Equation 16.10. The decibel, like the radian, is dimensionless. An intensity level of zero decibels means that $I = I_0$. One decibel is approximately the smallest change in loudness that an average listener with healthy hearing can detect. An increase of ten decibels in the intensity level corresponds approximately to a doubling of the loudness of the sound.

$$\beta = (10 \text{ dB}) \log\left(\frac{I}{I_0}\right) \qquad (16.10)$$

16.9 The Doppler Effect The Doppler effect is the change in frequency detected by an observer because the sound source and the observer have different velocities with respect to the medium of sound propagation. If the observer and source move with speeds v_o and v_s, respectively, and if the medium is stationary, the frequency f_o detected by the observer is given by Equation 16.15, where f_s is the frequency of the sound emitted by the source and v is the speed of sound. In the numerator, the plus sign applies when the observer moves toward the source, and the minus sign applies when the observer moves away from the source. In the denominator, the minus sign is used when the source moves toward the observer, and the plus sign is used when the source moves away from the observer.

$$f_o = f_s \left(\frac{1 \pm \dfrac{v_o}{v}}{1 \mp \dfrac{v_s}{v}}\right) \qquad (16.15)$$

Focus on Concepts

Online

Additional questions are available for assignment in WileyPLUS.

Section 16.1 The Nature of Waves

1. Domino toppling is an event in which a large number of dominoes are lined up close together and then allowed to topple, one after the other. The disturbance that propagates along the line of dominoes is _____. (a) partly transverse and partly longitudinal (b) transverse (c) longitudinal

Section 16.2 Periodic Waves

2. A transverse wave on a string has an amplitude A. A tiny spot on the string is colored red. As one cycle of the wave passes by, what is the total distance traveled by the red spot? (a) A (b) $2A$ (c) $\frac{1}{2}A$ (d) $4A$ (e) $\frac{1}{4}A$

Section 16.3 The Speed of a Wave on a String

3. As a wave moves through a medium at a speed v, the particles of the medium move in simple harmonic motion about their undisturbed positions. The maximum speed of the simple harmonic motion is v_{max}. When the amplitude of the wave doubles, _____. (a) v doubles, but v_{max} remains the same (b) v remains unchanged, but v_{max} doubles (c) both v and v_{max} remain unchanged (d) both v and v_{max} double

4. A rope is attached to a hook in the ceiling and is hanging straight down. The rope has a mass m, and nothing is attached to the free end of the rope. As a transverse wave travels down the rope from the top, _____. (a) the speed of the wave does not change (b) the speed of the wave increases (c) the speed of the wave decreases

Section 16.4 The Mathematical Description of a Wave

5. The equation that describes a transverse wave on a string is

$$y = (0.0120 \text{ m}) \sin[(483 \text{ rad/s})t - (3.00 \text{ rad/m})x]$$

where y is the displacement of a string particle and x is the position of the particle on the string. The wave is traveling in the $+x$ direction. What is the speed v of the wave?

Section 16.5 The Nature of Sound

6. As the amplitude of a sound wave in air decreases to zero, _____. (a) nothing happens to the condensations and rarefactions of the wave (b) the condensations and rarefactions of the wave occupy more and more distance along the direction in which the wave is traveling (c) the condensations of the wave disappear, but nothing happens to the rarefactions (d) nothing happens to the condensations of the wave, but the rarefactions disappear (e) both the condensations and the rarefactions of the wave disappear

Section 16.6 The Speed of Sound

7. An echo is sound that returns to you after being reflected from a distant surface (e.g., the side of a cliff). Assuming that the distances involved are the same, an echo under water and an echo in air return to you _____. (a) at different times, the echo under water returning more slowly (b) at different times, the echo under water returning more quickly (c) at the same time

8. A horn on a boat sounds a warning, and the sound penetrates the water. How does the frequency of the sound in the air compare to its frequency in the water? How does the wavelength in the air compare to the wavelength in the water? (a) The frequency in the air is smaller than the frequency in the water, and the wavelength in the air is greater than the wavelength in the water. (b) The frequency in the air is greater than the frequency in the water, and the wavelength in the air is smaller than the wavelength in the water. (c) The frequency in the air is the same as the frequency in the water, and the wavelength in the air is the same as the wavelength in the water. (d) The frequency in the air is the same as the frequency in the water, and the wavelength in the air is smaller than the wavelength in the water. (e) The frequency in the air is the same as the frequency in the water, and the wavelength in the air is greater than the wavelength in the water.

Section 16.7 Sound Intensity

9. A source emits sound uniformly in all directions. There are no reflections of the sound. At a distance of 12 m from the source, the intensity of the sound is 5.0×10^{-3} W/m^2. What is the total sound power P emitted by the source?

Section 16.8 Decibels

10. A source emits sound uniformly in all directions. There are no reflections of the sound. At a distance r_1 from the source, the sound is 7.0 dB louder than it is at a distance r_2 from the source. What is the ratio r_1/r_2?

Section 16.9 The Doppler Effect

11. A red car and a blue car can move along the same straight one-lane road. Both cars can move only at one speed when they move (e.g., 60 mph). The driver of the red car sounds his horn. In which one of the following situations does the driver of the blue car hear the highest horn frequency? (a) Both cars are moving at the same speed, and they are moving apart. (b) Both cars are moving in the same direction at the same speed. (c) Both cars are moving at the same speed, and they are moving toward each other. (d) The red car is moving toward the blue car, which is stationary. (e) The blue car is moving toward the red car, which is stationary.

12. What happens to the Doppler effect in air (i.e., the shift in frequency of a sound wave) as the temperature increases? (a) It is greater at higher temperatures, but only in the case of a moving source and a stationary observer. (b) It is greater at higher temperatures, but only in the case of a moving observer and a stationary source. (c) It is greater at higher temperatures than at lower temperatures. (d) It is less at higher temperatures than at lower temperatures. (e) The Doppler effect does not change as the temperature increases.

Problems

Online

Additional questions are available for assignment in WileyPLUS.

SSM Student Solutions Manual

MMH Problem-solving help

GO Guided Online Tutorial

V-HINT Video Hints

CHALK Chalkboard Videos

BIO Biomedical application

E Easy

M Medium

H Hard

WS Worksheet

T Team Problem

Section 16.1 The Nature of Waves,

Section 16.2 Periodic Waves

1. **E SSM** Light is an electromagnetic wave and travels at a speed of 3.00×10^8 m/s. The human eye is most sensitive to yellow-green light, which has a wavelength of 5.45×10^{-7} m. What is the frequency of this light?

2. **E** Consider the freight train in **Figure 16.6**. Suppose that 15 boxcars pass by in a time of 12.0 s and each has a length of 14.0 m.

(a) What is the frequency at which each boxcar passes? **(b)** What is the speed of the train?

3. **E** A woman is standing in the ocean, and she notices that after a wave crest passes, five more crests pass in a time of 50.0 s. The distance between two successive crests is 32 m. Determine, if possible, the wave's **(a)** period, **(b)** frequency, **(c)** wavelength, **(d)** speed, and **(e)** amplitude. If it is not possible to determine any of these quantities, then so state.

4. **E** Tsunamis are fast-moving waves often generated by underwater earthquakes. In the deep ocean their amplitude is barely noticeable, but upon reaching shore, they can rise up to the astonishing height of a six-story building. One tsunami, generated off the Aleutian islands in Alaska, had a wavelength of 750 km and traveled a distance of 3700 km in 5.3 h. **(a)** What was the speed (in m/s) of the wave? For reference, the speed of a 747 jetliner is about 250 m/s. Find the wave's **(b)** frequency and **(c)** period.

5. **E** **SSM** In **Interactive Figure 16.2c** the hand moves the end of the Slinky up and down through two complete cycles in one second. The wave moves along the Slinky at a speed of 0.50 m/s. Find the distance between two adjacent crests on the wave.

6. **E** **GO** **CHALK** A person fishing from a pier observes that four wave crests pass by in 7.0 s and estimates the distance between two successive crests to be 4.0 m. The timing starts with the first crest and ends with the fourth. What is the speed of the wave?

7. **E** Using the data in the graphs that accompany this problem, determine the speed of the wave.

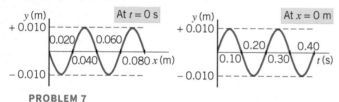

PROBLEM 7

8. **M** A 3.49-rad/s ($33\frac{1}{3}$ rpm) record has a 5.00-kHz tone cut in the groove. If the groove is located 0.100 m from the center of the record (see the drawing), what is the wavelength in the groove?

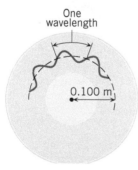

One wavelength

0.100 m

PROBLEM 8

9. **M** **V-HINT** **CHALK** The speed of a transverse wave on a string is 450 m/s, and the wavelength is 0.18 m. The amplitude of the wave is 2.0 mm. How much time is required for a particle of the string to move through a total distance of 1.0 km?

10. **M** **GO** A jetskier is moving at 8.4 m/s in the direction in which the waves on a lake are moving. Each time he passes over a crest, he feels a bump. The bumping frequency is 1.2 Hz, and the crests are separated by 5.8 m. What is the wave speed?

Section 16.3 The Speed of a Wave on a String

11. **E** The mass of a string is 5.0×10^{-3} kg, and it is stretched so that the tension in it is 180 N. A transverse wave traveling on this string

has a frequency of 260 Hz and a wavelength of 0.60 m. What is the length of the string?

12. **E** **SSM** The middle C string on a piano is under a tension of 944 N. The period and wavelength of a wave on this string are 3.82 ms and 1.26 m, respectively. Find the linear density of the string.

13. **E** A wire is stretched between two posts. Another wire is stretched between two posts that are twice as far apart. The tension in the wires is the same, and they have the same mass. A transverse wave travels on the shorter wire with a speed of 240 m/s. What would be the speed of the wave on the longer wire?

14. **E** **MMH** To measure the acceleration due to gravity on a distant planet, an astronaut hangs a 0.055-kg ball from the end of a wire. The wire has a length of 0.95 m and a linear density of 1.2×10^{-4} kg/m. Using electronic equipment, the astronaut measures the time for a transverse pulse to travel the length of the wire and obtains a value of 0.016 s. The mass of the wire is negligible compared to the mass of the ball. Determine the acceleration due to gravity.

15. **E** Two wires are parallel, and one is directly above the other. Each has a length of 50.0 m and a mass per unit length of 0.020 kg/m. However, the tension in wire A is 6.00×10^2 N, and the tension in wire B is 3.00×10^2 N. Transverse wave pulses are generated simultaneously, one at the left end of wire A and one at the right end of wire B. The pulses travel toward each other. How much time does it take until the pulses pass each other?

16. **E** **SSM** The drawing shows two transverse waves traveling on strings. The linear density of each string is 0.065 kg/m. The tension is provided by a 26-N block that is hanging from the string. Find the speed of the wave in part **(a)** and in part **(b)** of the drawing.

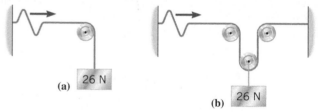

(a) 26 N

(b) 26 N

PROBLEM 16

17. **M** **GO** A steel cable has a cross-sectional area 2.83×10^{-3} m² and is kept under a tension of 1.00×10^4 N. The density of steel is 7860 kg/m³. *Note that this value is not the linear density of the cable.* At what speed does a transverse wave move along the cable?

18. **M** **GO** The drawing shows a graph of two waves traveling to the right at the same speed. **(a)** Using the data in the drawing, determine the wavelength of each wave. **(b)** The speed of the waves is 12 m/s; calculate the frequency of each one. **(c)** What is the maximum speed for a particle attached to each wave?

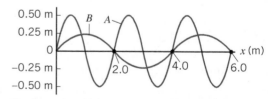

PROBLEM 18

19. **M** **V-HINT** Review Conceptual Example 3 before starting this problem. The amplitude of a transverse wave on a string is 4.5 cm. The ratio of the maximum particle speed to the speed of the wave is 3.1. What is the wavelength (in cm) of the wave?

20. **M** **CHALK** The drawing shows a frictionless incline and pulley. The two blocks are connected by a wire (mass per unit length = 0.0250 kg/m) and remain stationary. A transverse wave on the wire has a speed of 75.0 m/s. Neglecting the weight of the wire relative to the tension in the wire, find the masses m_1 and m_2 of the blocks.

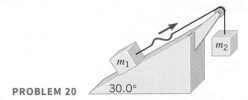

PROBLEM 20

21. **H** The drawing shows a 15.0-kg ball being whirled in a circular path on the end of a string. The motion occurs on a frictionless, horizontal table. The angular speed of the ball is $\omega = 12.0$ rad/s. The string has a mass of 0.0230 kg. How much time does it take for a wave on the string to travel from the center of the circle to the ball?

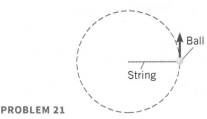

PROBLEM 21

Section 16.4   The Mathematical Description of a Wave

(Note: The phase angles $(2\pi ft - 2\pi x/\lambda)$ and $(2\pi ft + 2\pi x/\lambda)$ are measured in radians, not degrees.)

22. **E GO** A wave traveling along the x axis is described mathematically by the equation $y = 0.17 \sin (8.2\pi t + 0.54\pi x)$, where y is the displacement (in meters), t is in seconds, and x is in meters. What is the speed of the wave?

23. **E V-HINT** A wave has the following properties: amplitude = 0.37 m, period = 0.77 s, wave speed = 12 m/s. The wave is traveling in the $-x$ direction. What is the mathematical expression (similar to Equation 16.3 or 16.4) for the wave?

24. **E** The drawing shows a graph that represents a transverse wave on a string. The wave is moving in the $+x$ direction with a speed of 0.15 m/s. Using the information contained in the graph, write the mathematical expression (similar to Equation 16.3 or 16.4) for the wave.

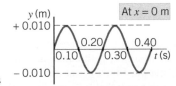

PROBLEM 24

25. **E SSM** A wave traveling in the $+x$ direction has an amplitude of 0.35 m, a speed of 5.2 m/s, and a frequency of 14 Hz. Write the equation of the wave in the form given by either Equation 16.3 or 16.4.

26. **M GO CHALK** A transverse wave is traveling on a string. The displacement y of a particle from its equilibrium position is given by $y = (0.021$ m$) \sin (25t - 2.0 x)$. Note that the phase angle $25t - 2.0x$ is in radians, t is in seconds, and x is in meters. The linear density of the string is 1.6×10^{-2} kg/m. What is the tension in the string?

27. **M SSM** The tension in a string is 15 N, and its linear density is 0.85 kg/m. A wave on the string travels toward the $-x$ direction; it has an amplitude of 3.6 cm and a frequency of 12 Hz. What are the (a) speed and (b) wavelength of the wave? (c) Write down a mathematical expression (like Equation 16.3 or 16.4) for the wave, substituting numbers for the variables A, f, and λ.

28. **H** A transverse wave on a string has an amplitude of 0.20 m and a frequency of 175 Hz. Consider the particle of the string at $x = 0$ m. It begins with a displacement of $y = 0$ m when $t = 0$ s, according to Equation 16.3 or 16.4. How much time passes between the first two instants when this particle has a displacement of $y = 0.10$ m?

Section 16.5   The Nature of Sound,

Section 16.6   The Speed of Sound

29. **E SSM** For research purposes a sonic buoy is tethered to the ocean floor and emits an infrasonic pulse of sound (speed = 1522 m/s). The period of this sound is 71 ms. Determine the wavelength of the sound.

30. **E CHALK** To navigate, a porpoise emits a sound wave that has a wavelength of 1.5 cm. The speed at which the wave travels in seawater is 1522 m/s. Find the period of the wave.

31. **E MMH** At what temperature is the speed of sound in helium (ideal gas, $\gamma = 1.67$, atomic mass = 4.003 u) the same as its speed in oxygen at 0 °C?

32. **E** Have you ever listened for an approaching train by kneeling next to a railroad track and putting your ear to the rail? Young's modulus for steel is $Y = 2.0 \times 10^{11}$ N/m², and the density of steel is $\rho = 7860$ kg/m³. On a day when the temperature is 20 °C, how many times greater is the speed of sound in the rail than in the air?

33. **E SSM** The speed of a sound in a container of hydrogen at 201 K is 1220 m/s. What would be the speed of sound if the temperature were raised to 405 K? Assume that hydrogen behaves like an ideal gas.

34. **E GO** Suppose you are part of a team that is trying to break the sound barrier with a jet-powered car, which means that it must travel faster than the speed of sound in air. In the morning, the air temperature is 0 °C, and the speed of sound is 331 m/s. What speed must your car exceed if it is to break the sound barrier when the temperature has risen to 43 °C in the afternoon? Assume that air behaves like an ideal gas.

35. **E** As the drawing illustrates, a siren can be made by blowing a jet of air through 20 equally spaced holes in a rotating disk. The time it takes for successive holes to move past the air jet is the period of the sound. The siren is to produce a 2200-Hz tone. What must be the angular speed ω (in rad/s) of the disk?

PROBLEM 35

36. **E** At a height of ten meters above the surface of a freshwater lake, a sound pulse is generated. The echo from the bottom of the lake returns to the point of origin 0.110 s later. The air and water temperature are 20 °C. How deep is the lake?

37. **E** An observer stands 25 m behind a marksman practicing at a rifle range. The marksman fires the rifle horizontally, the speed of the bullets is 840 m/s, and the air temperature is 20 °C. How far does each bullet travel before the observer hears the report of the rifle? Assume that the bullets encounter no obstacles during this interval, and ignore both air resistance and the vertical component of the bullets' motion.

38. **E GO** An ultrasonic ruler, such as the one discussed in Example 4 in Section 16.6, displays the distance between the ruler and an object, such as a wall. The ruler sends out a pulse of ultrasonic sound

and measures the time it takes for the pulse to reflect from the object and return. The ruler uses this time, along with a preset value for the speed of sound in air, to determine the distance. Suppose that you use this ruler under water, rather than in air. The actual distance from the ultrasonic ruler to an object is 25.0 m. The adiabatic bulk modulus and density of seawater are $B_{ad} = 2.37 \times 10^9$ Pa and $\rho = 1025$ kg/m^3, respectively. Assume that the ruler uses a preset value of 343 m/s for the speed of sound in air. Determine the distance reading that the ruler displays.

39. **E** An explosion occurs at the end of a pier. The sound reaches the other end of the pier by traveling through three media: air, fresh water, and a slender metal handrail. The speeds of sound in air, water, and the handrail are 343, 1482, and 5040 m/s, respectively. The sound travels a distance of 125 m in each medium. (a) Through which medium does the sound arrive first, second, and third? (b) After the first sound arrives, how much later do the second and third sounds arrive?

40. **M** **V-HINT** A sound wave travels twice as far in neon (Ne) as it does in krypton (Kr) in the same time interval. Both neon and krypton can be treated as monatomic ideal gases. The atomic mass of neon is 20.2 u, and the atomic mass of krypton is 83.8 u. The temperature of the krypton is 293 K. What is the temperature of the neon?

41. **M** **SSM** A hunter is standing on flat ground between two vertical cliffs that are directly opposite one another. He is closer to one cliff than to the other. He fires a gun and, after a while, hears three echoes. The second echo arrives 1.6 s after the first, and the third echo arrives 1.1 s after the second. Assuming that the speed of sound is 343 m/s and that there are no reflections of sound from the ground, find the distance between the cliffs.

42. **M** **GO** A monatomic ideal gas ($\gamma = 1.67$) is contained within a box whose volume is 2.5 m^3. The pressure of the gas is 3.5×10^5 Pa. The total mass of the gas is 2.3 kg. Find the speed of sound in the gas.

43. **M** **V-HINT** A long slender bar is made from an unknown material. The length of the bar is 0.83 m, its cross-sectional area is 1.3×10^{-4} m^2, and its mass is 2.1 kg. A sound wave travels from one end of the bar to the other end in 1.9×10^{-4} s. From which one of the materials listed in Table 10.1 is the bar most likely to be made?

44. **M** **CHALK** As the drawing shows, one microphone is located at the origin, and a second microphone is located on the +y axis. The microphones are separated by a distance of $D = 1.50$ m. A source of sound is located on the +x axis, its distances from microphones 1 and 2 being L_1 and L_2, respectively. The speed of sound is 343 m/s. The sound reaches microphone 1 first, and then, 1.46 ms later, it reaches microphone 2. Find the distances L_1 and L_2.

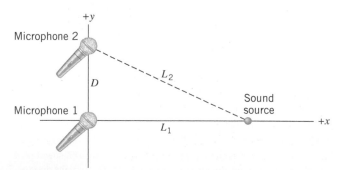

45. **M** **SSM** When an earthquake occurs, two types of sound waves are generated and travel through the earth. The primary, or P, wave has a speed of about 8.0 km/s and the secondary, or S, wave has a speed of about 4.5 km/s. A seismograph, located some distance away, records the arrival of the P wave and then, 78 s later, records the

arrival of the S wave. Assuming that the waves travel in a straight line, how far is the seismograph from the earthquake?

46. **M** **V-HINT** Consult Multiple-Concept Example 4 in order to review a model for solving this type of problem. Suppose that you are standing by the side of a road in the Sahara desert where the temperature has reached a hot 56 °C (130 °F). A truck, traveling at a constant speed, passes by. After 4.00 s have elapsed, you use the ultrasonic ruler discussed in Example 4 to measure the distance to the truck. A sound pulse leaves the ultrasonic ruler and returns 0.120 s later. Assume that the average molecular mass of air is 28.9 u, air is an ideal diatomic gas ($\gamma = \frac{7}{5}$), and the truck moves a negligible distance in the time it takes for the sound pulse to reach it. Determine how fast the truck is moving.

Section 16.7 Sound Intensity

47. **E** **SSM** At a distance of 3.8 m from a siren, the sound intensity is 3.6×10^{-2} W/m^2. Assuming that the siren radiates sound uniformly in all directions, find the total power radiated.

48. **E** **GO** A source of sound is located at the center of two concentric spheres, parts of which are shown in the drawing. The source emits sound uniformly in all directions. On the spheres are drawn three small patches that may or may not have equal areas. However, the same sound power passes through each patch. The source produces 2.3 W of sound power, and the radii of the concentric spheres are $r_A = 0.60$ m and $r_B = 0.80$ m. (a) Determine the sound intensity at each of the three patches. (b) The sound power that passes through each of the patches is 1.8×10^{-3} W. Find the area of each patch.

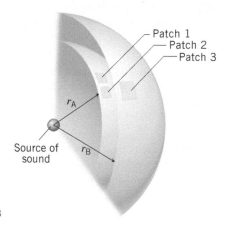

49. **E** Suppose that a public address system emits sound uniformly in all directions and that there are no reflections. The intensity at a location 22 m away from the sound source is 3.0×10^{-4} W/m^2. What is the intensity at a spot that is 78 m away?

50. **E** **SSM** A loudspeaker has a circular opening with a radius of 0.0950 m. The electrical power needed to operate the speaker is 25.0 W. The average sound intensity at the opening is 17.5 W/m^2. What percentage of the electrical power is converted by the speaker into sound power?

51. **E** **GO** A man stands at the midpoint between two speakers that are broadcasting an amplified static hiss uniformly in all directions. The speakers are 30.0 m apart and the total power of the sound coming from each speaker is 0.500 W. Find the total sound intensity that the man hears (a) when he is at his initial position halfway between the speakers, and (b) after he has walked 4.0 m directly toward one of the speakers.

52. **M** **SSM** **MMH** A dish of lasagna is being heated in a microwave oven. The effective area of the lasagna that is exposed to the microwaves is 2.2×10^{-2} m^2. The mass of the lasagna is 0.35 kg, and its specific heat capacity is 3200 J/(kg · C°). The temperature rises by 72 °C in 8.0 minutes. What is the intensity of the microwaves in the oven?

53. **M** **GO** **CHALK** Two sources of sound are located on the *x* axis, and each emits power uniformly in all directions. There are no reflections. One source is positioned at the origin and the other at $x = +123$ m. The source at the origin emits four times as much power as the other source. Where on the *x* axis are the two sounds equal in intensity? Note that there are two answers.

Section 16.8   Decibels

54. **E** A woman stands a distance *d* from a loud motor that emits sound uniformly in all directions. The sound intensity at her position is an uncomfortable 3.2×10^{-3} W/m^2. There are no reflections. At a position twice as far from the motor, what are (a) the sound intensity and (b) the sound intensity level relative to the threshold of hearing?

55. **E** The volume control on a surround-sound amplifier is adjusted so the sound intensity level at the listening position increases from 23 to 61 dB. What is the ratio of the final sound intensity to the original sound intensity?

56. **E** **GO** Using an intensity of 1×10^{-12} W/m^2 as a reference, the threshold of hearing for an average young person is 0 dB. Person 1 and person 2, who are not average, have thresholds of hearing that are $\beta_1 = -8.00$ dB and $\beta_2 = +12.0$ dB. What is the ratio I_1/I_2 of the sound intensity I_1 when person 1 hears the sound at his own threshold of hearing compared to the sound intensity I_2 when person 2 hears the sound at his own threshold of hearing?

57. **E** A listener doubles his distance from a source that emits sound uniformly in all directions. There are no reflections. By how many decibels does the sound intensity level change?

58. **E** Sound is passing perpendicularly through an open window whose dimensions are 1.1 m × 0.75 m. The sound intensity level is 95 dB above the threshold of hearing. How much sound *energy* comes through the window in one hour?

59. **E** The bellow of a territorial bull hippopotamus has been measured at 115 dB above the threshold of hearing. What is the sound intensity?

60. **M** **CHALK** **SSM** When one person shouts at a football game, the sound intensity level at the center of the field is 60.0 dB. When all the people shout together, the intensity level increases to 109 dB. Assuming that each person generates the same sound intensity at the center of the field, how many people are at the game?

61. **M** **GO** Review Conceptual Example 8 as background for this problem. A loudspeaker is generating sound in a room. At a certain point, the sound waves coming directly from the speaker (without reflecting from the walls) create an intensity level of 75.0 dB. The waves reflected from the walls create, by themselves, an intensity level of 72.0 dB at the same point. What is the total intensity level? (*Hint: The answer is not 147.0 dB.*)

62. **M** **V-HINT** A portable radio is sitting at the edge of a balcony 5.1 m above the ground. The unit is emitting sound uniformly in all directions. By accident, it falls from rest off the balcony and continues to play on the way down. A gardener is working in a flower bed directly below the falling unit. From the instant the unit begins to fall, how much time is required for the sound intensity level heard by the gardener to increase by 10.0 dB?

Section 16.9   The Doppler Effect

63. **E** A bird is flying directly toward a stationary bird-watcher and emits a frequency of 1250 Hz. The bird-watcher, however, hears a frequency of 1290 Hz. What is the speed of the bird, expressed as a percentage of the speed of sound?

64. **E** **SSM** From a vantage point very close to the track at a stock car race, you hear the sound emitted by a moving car. You detect a frequency that is 0.86 times as small as the frequency emitted by the car when it is stationary. The speed of sound is 343 m/s. What is the speed of the car?

65. **E** Dolphins emit clicks of sound for communication and echolocation. A marine biologist is monitoring a dolphin swimming in seawater where the speed of sound is 1522 m/s. When the dolphin is swimming directly away at 8.0 m/s, the marine biologist measures the number of clicks occurring per second to be at a frequency of 2500 Hz. What is the difference (in Hz) between this frequency and the number of clicks per second actually emitted by the dolphin?

66. **E** **MMH** A convertible moves toward you and then passes you; all the while, its loudspeakers are producing a sound. The speed of the car is a constant 9.00 m/s, and the speed of sound is 343 m/s. What is the ratio of the frequency you hear while the car is approaching to the frequency you hear while the car is moving away?

67. **E** The security alarm on a parked car goes off and produces a frequency of 960 Hz. The speed of sound is 343 m/s. As you drive toward this parked car, pass it, and drive away, you observe the frequency to change by 95 Hz. At what speed are you driving?

68. **M** **V-HINT** A car is parked 20.0 m directly south of a railroad crossing. A train is approaching the crossing from the west, headed directly east at a speed of 55.0 m/s. The train sounds a short blast of its 289-Hz horn when it reaches a point 20.0 m west of the crossing. What frequency does the car's driver hear when the horn blast reaches the car? The speed of sound in air is 343 m/s. (*Hint: Assume that only the component of the train's velocity that is directed toward the car affects the frequency heard by the driver.*)

69. **M** **GO** A loudspeaker in a parked car is producing sound whose frequency is 20 510 Hz. A healthy young person with normal hearing is standing nearby on the sidewalk but cannot hear the sound because the frequency is too high. When the car is moving, however, this person can hear the sound. (a) Is the car moving toward or away from the person? Why? (b) If the speed of sound is 343 m/s, what is the minimum speed of the moving car?

70. **M** **SSM** **MMH** Two trucks travel at the same speed. They are far apart on adjacent lanes and approach each other essentially head-on. One driver hears the horn of the other truck at a frequency that is 1.14 times the frequency he hears when the trucks are stationary. The speed of sound is 343 m/s. At what speed is each truck moving?

71. **M** **GO** A wireless transmitting microphone is mounted on a small platform that can roll down an incline, directly away from a loudspeaker that is mounted at the top of the incline. The loudspeaker broadcasts a tone that has a fixed frequency of 1.000×10^4 Hz, and the speed of sound is 343 m/s. At a time of 1.5 s following the release of the platform, the microphone detects a frequency of 9939 Hz. At a time of 3.5 s following the release of the platform, the microphone detects a frequency of 9857 Hz. What is the acceleration (assumed constant) of the platform?

72. **M** **V-HINT** A car is accelerating while its horn is sounding. Just after the car passes a stationary person, the person hears a frequency of 966.0 Hz. Fourteen seconds later, the frequency heard by the person has decreased to 912.0 Hz. When the car is stationary, its horn emits a sound whose frequency is 1.00×10^3 Hz. The speed of sound is 343 m/s. What is the acceleration of the car?

73. **M** **GO** The siren on an ambulance is emitting a sound whose frequency is 2450 Hz. The speed of sound is 343 m/s. **(a)** If the ambulance is stationary and you (the "observer") are sitting in a parked car, what are the wavelength and the frequency of the sound you hear? **(b)** Suppose that the ambulance is moving toward you at a speed of 26.8 m/s. Determine the wavelength and the frequency of the sound

you hear. **(c)** If the ambulance is moving toward you at a speed of 26.8 m/s and you are moving toward it at a speed of 14.0 m/s, find the wavelength and frequency of the sound you hear.

74. **M** **SSM** Two submarines are under water and approaching each other head-on. Sub A has a speed of 12 m/s and sub B has a speed of 8 m/s. Sub A sends out a 1550-Hz sonar wave that travels at a speed of 1522 m/s. **(a)** What is the frequency detected by sub B? **(b)** Part of the sonar wave is reflected from sub B and returns to sub A. What frequency does sub A detect for this reflected wave?

Additional Problems

Online

75. **E** A recording engineer works in a soundproofed room that is 44.0 dB quieter than the outside. If the sound intensity that leaks into the room is 1.20×10^{-10} W/m², what is the intensity outside?

76. **E** You are flying in an ultralight aircraft at a speed of 39 m/s. An eagle, whose speed is 18 m/s, is flying directly toward you. Each of the given speeds is relative to the ground. The eagle emits a shrill cry whose frequency is 3400 Hz. The speed of sound is 330 m/s. What frequency do you hear?

77. **E** **SSM** Suppose that the linear density of the A string on a violin is 7.8×10^{-4} kg/m. A wave on the string has a frequency of 440 Hz and a wavelength of 65 cm. What is the tension in the string?

78. **E** A car driving along a highway at a speed of 23 m/s strays onto the shoulder. Evenly spaced parallel grooves called "rumble strips" are carved into the pavement of the shoulder. Rolling over the rumble strips causes the car's wheels to oscillate up and down at a frequency of 82 Hz. How far apart are the centers of adjacent rumble-strip grooves?

79. **E** **SSM** When Gloria wears her hearing aid, the sound intensity level increases by 30.0 dB. By what factor does the sound intensity increase?

80. **E** The average sound intensity inside a busy neighborhood restaurant is 3.2×10^{-5} W/m². How much energy goes into each ear (area = 2.1×10^{-3} m²) during a one-hour meal?

81. **E** **SSM** Suppose that the amplitude and frequency of the transverse wave in **Interactive Figure 16.2c** are, respectively, 1.3 cm and 5.0 Hz. Find the *total vertical distance* (in cm) through which the colored dot moves in 3.0 s.

82. **E** A bat emits a sound whose frequency is 91 kHz. The speed of sound in air at 20.0 °C is 343 m/s. However, the air temperature is 35 °C, so the speed of sound is not 343 m/s. Assume that air behaves like an ideal gas, and find the wavelength of the sound.

83. **E** **SSM** You are riding your bicycle directly away from a stationary source of sound and hear a frequency that is 1.0% lower than the emitted frequency. The speed of sound is 343 m/s. What is your speed?

84. **E** **GO** Argon (molecular mass = 39.9 u) is a monatomic gas. Assuming that it behaves like an ideal gas at 298 K ($\gamma = 1.67$), find **(a)** the rms speed of argon atoms and **(b)** the speed of sound in argon.

85. **E** **SSM** The sound intensity level at a rock concert is 115 dB, while that at a jazz fest is 95 dB. Determine the ratio of the sound intensity at the rock concert to the sound intensity at the jazz fest.

86. **E** An amplified guitar has a sound intensity level that is 14 dB greater than the same unamplified sound. What is the ratio of the amplified intensity to the unamplified intensity?

87. **E** In a discussion person A is talking 1.5 dB louder than person B, and person C is talking 2.7 dB louder than person A. What is the ratio of the sound intensity of person C to the sound intensity of person B?

88. **E** In **Interactive Figure 16.3c** the colored dot exhibits simple harmonic motion as the longitudinal wave passes. The wave has an amplitude of 5.4×10^{-3} m and a frequency of 4.0 Hz. Find the maximum acceleration of the dot.

89. **M** **GO** **(a)** A uniform rope of mass m and length L is hanging straight down from the ceiling. A small-amplitude transverse wave is sent up the rope from the bottom end. Derive an expression that gives the speed v of the wave on the rope in terms of the distance y above the bottom end of the rope and the magnitude g of the acceleration due to gravity. **(b)** Use the expression that you have derived to calculate the speeds at distances of 0.50 m and 2.0 m above the bottom end of the rope.

90. **M** **V-HINT** A spider hangs from a strand of silk whose radius is 4.0×10^{-6} m. The density of the silk is 1300 kg/m³. When the spider moves, waves travel along the strand of silk at a speed of 280 m/s. Ignore the mass of the silk strand, and determine the mass of the spider.

91. **M** **V-HINT** A member of an aircraft maintenance crew wears protective earplugs that reduce the sound intensity by a factor of 350. When a jet aircraft is taking off, the sound intensity level experienced by the crew member is 88 dB. What sound intensity level would the crew member experience if he removed the protective earplugs?

92. **M** **GO** **SSM** A speedboat, starting from rest, moves along a straight line away from a dock. The boat has a constant acceleration of +3.00 m/s² (see the figure). Attached to the dock is a siren that is producing a 755-Hz tone. If the air temperature is 20 °C, what is the frequency of the sound heard by a person on the boat when the boat's displacement from the dock is +45.0 m?

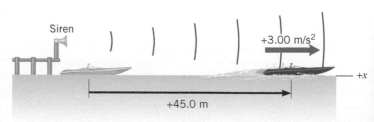

Siren   +3.00 m/s²

+45.0 m

+x

PROBLEM 92

Physics in Biology, Medicine, and Sports

93. **E** **BIO** **16.7** A typical adult ear has a surface area of 2.1×10^{-3} m². The sound intensity during a normal conversation is about 3.2×10^{-6} W/m² at the listener's ear. Assume that the sound strikes the surface of the ear perpendicularly. How much power is intercepted by the ear?

94. **M** **BIO** **V-HINT** **16.7** Deep ultrasonic heating is used to promote the healing of torn tendons. It is produced by applying ultrasonic sound over the affected area of the body. The sound transducer (generator) is circular with a radius of 1.8 cm, and it produces a sound intensity of 5.9×10^3 W/m². How much time is required for the transducer to emit 4800 J of sound energy?

95. **E** **BIO** **SSM** **16.8** A middle-aged man typically has poorer hearing than a middle-aged woman. In one case a woman can just begin to hear a musical tone, while a man can just begin to hear the tone only when its intensity level is increased by 7.8 dB relative to the just-audible intensity level for the woman. What is the ratio of the sound intensity just detected by the man to the sound intensity just detected by the woman?

96. **E** **BIO** **V-HINT** **16.8** Hearing damage may occur when a person is exposed to a sound intensity level of 90.0 dB (relative to the threshold of hearing) for a period of 9.0 hours. One particular eardrum has an area of 2.0×10^{-4} m². How much sound energy is incident on this eardrum during this time?

97. **M** **BIO** **16.8** Tortoises do have ears, but they are very different than human ears. They have no outer ear. Their eardrums are located along the side of their head behind their eyes and are covered with a flap of scales (see the photo). The physiology of their ears makes them sensitive to lower frequency sounds that travel well through water and also along the ground as vibrations. Consider a tortoise that hears a 72-dB sound from a boat propeller under water with a frequency of 1.0×10^2 Hz.* If the eardrum of the tortoise is approximately circular with a radius of 3.2 mm, how much power is delivered to the eardrum?

Shelly/totallytortoise.com

PROBLEM 97

*Sound intensity levels underwater are calculated slightly differently than in air, which we won't take into account for this calculation. In fact, the threshold of hearing for animals underwater can be many times higher than in air.

98. **E** **BIO** **16.6** Frequency ranges for hearing vary dramatically across the animal kingdom. For example, turtles are sensitive to lower frequencies (20 to 1000 Hz), humans have a greater range (20 to 20 000 Hz), but the champion of frequency range is the porpoise (75 to 150 000 Hz). What is the wavelength of a 150 000-Hz sound wave in seawater at a temperature of 20 °C?

99. **M** **BIO** **16.10** According to Section 16.10, the smallest detail that can be resolved by medical ultrasound is essentially equal to one wavelength of the sound. **(a)** If the speed of a 2.20-MHz ultrasonic wave in human tissue is 1540 m/s, what is the size of the smallest structure that can be observed? **(b)** How far below the surface of the skin can the sound waves penetrate? Assume their maximum penetration depth is 2.00×10^2 wavelengths. **(c)** How long would it take the sound waves to leave the transmitter, reach their maximum depth, and return to the receiver at the skin's surface?

100. **E** **BIO** **16.8** Review Example 9 as background material for this problem. Dolphins, like bats, use echolocation to detect objects in their respective environments, including prey. This allows them to scan the physical world using sound. While bats emit short pulses of high frequency sound waves with a frequency of about 60 kHz, dolphins use high frequency ultrasonic waves with a frequency of about 10^6 Hz. They produce a series of whistles or clicks at rates of up to 2000 per second. The clicks echo off of objects and prey and travel back to the dolphin, where they are absorbed in the lower jawbone, which transmits the vibrations to the inner ear. While the nature of sound waves in both air and water is similar, there are significant differences in how the intensity level of the sound is determined. The sound intensity of a wave in W/m² can be directly compared between a sound wave in air and one in water. However, the sound intensity level in dB is not the same. The dB is a relative unit of measure, unlike W/m², which is absolute. The reference used for air (typically 1.00×10^{-12} W/m²) is different for water. This fact, along with the speed of the wave in water (about 5 times faster than in air), and the greater density of the water results in an intensity level in water being 61.5 dB greater than in air. In other words, $\beta_{air} = \beta_{water} - 61.5$ dB.

If a dolphin detects a reflected sound wave in water with an intensity level of 80.0 dB, what would be the **(a)** intensity level and **(b)** intensity of this wave in air?

101. **M** **BIO** **16.8** With the advent of tiny, powerful rare earth magnets, and improvements in rechargeable battery and wireless technologies, wireless earbuds, like Apple's AirPods (see the photo), enhance the enjoyment of listening to music, movies, and podcasts, and allow handsfree conversation. However, since the earbuds focus sound waves directly into the auditory canal at close distances, care must be taken to protect the eardrum from long-term exposure to high volumes (intensity levels). Hearing experts recommend listening at a maximum intensity level of 85 dB for no more than 8.0 hours a day. Assuming a constant intensity level of 85 dB, how much energy does the eardrum absorb during the 8.0 hours? Take the area of the eardrum to be 0.65 cm².

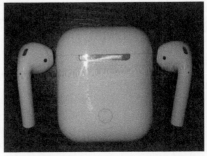

Courtesy of David Young

PROBLEM 101

Concepts and Calculations Problems

Online

One of the most important concepts we encountered in this chapter is the transverse wave. For instance, transverse waves travel along a guitar string when it is plucked or along a violin string when it is bowed. Problem 102 reviews how the travel speed depends on the properties of the string and on the tension in it. Problem 103 illustrates how the Doppler effect arises when an observer is moving away from or toward a stationary source of sound. In fact, we will see that it's possible for both situations to occur at the same time.

102. **M CHALK** The figure shows waves traveling on two strings. Each string is attached to a wall at one end and to a box that has a weight of 28.0 N at the other end. String 1 has a mass of 8.00 g and a length of 4.00 cm, and string 2 has a mass of 12.0 g and a length of 8.00 cm. *Concepts:* (i) Is the tension the same in each string? (ii) Is the speed of each wave the same? (iii) String 1 has a smaller mass and, hence, less inertia than string 2. Does this mean that the speed of the wave on string 1 is greater than the speed on string 2? *Calculations:* Determine the speed of the wave on each string.

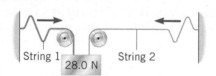

PROBLEM 102

103. **M CHALK SSM** A siren, mounted on a tower, emits a sound whose frequency is 2140 Hz. A person is driving a car away from the tower at a speed of 27.0 m/s. As the figure illustrates, the sound reaches the person by two paths: the sound reflected from the building in front of the car, and the sound coming directly from the siren. The speed of sound is 343 m/s. *Concepts:* (i) One way that the Doppler effect can arise is that the wavelength of the sound changes. For either the direct or reflected sound, does the wavelength change? (ii) Why does the driver hear a frequency for the reflected sound that is different from 2140 Hz, and is it greater or smaller than 2140 Hz? (iii) Why does the driver hear a frequency for the direct sound that is different from 2140 Hz, and is it greater or smaller than 2140 Hz? *Calculations:* What frequency does the person hear for the (a) reflected and (b) direct sound?

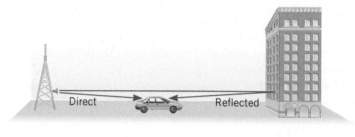

PROBLEM 103

Team Problems

Online

104. **E T WS** **A Pneumatic Warning Siren.** A pneumatic siren can be made by blowing a narrow jet of air through equally spaced holes on a rotating disk. The time it takes for successive holes to pass thought the jet stream is equal to the period of the resulting sound wave. You and your team are designing such a device to be used as a community warning alarm to be activated in the case of severe weather, or other dangerous situations. The siren should produce a tone as close as possible to 710 Hz, and people with average hearing should be able to detect it above normal conversation at a distance of at least 2.0 miles. The circular pattern of holes will be located 24 cm from the center of a thin aluminum disk, and the motor that rotates the disk runs at 1500 rpm. (a) What *even* integer number of equally spaced holes will result in a sound frequency closest to the desired 710 Hz? (b) What should be the minimum sound output power of the siren in watts (W) so that people with average hearing can detect it above normal conversation at a distance of 2.0 miles from the source? You can assume that the output is isotropic, that is, it emanates like a spherical wave, and that the sound intensity of normal conversion is 3.20×10^{-6} W/m² (from Table 16.2).

105. **E T WS** **Tuning the Waves.** You and your team are designing a system that allows you to control the speed of a transverse wave on a thin tungsten wire by adjusting its tension with a spring. One end of the wire is connected to an inner wall of a large wooden crate and the other end to one side of a spring. The opposite side of the spring is then connected to a screw that is mounted on the opposite wall of the crate. The screw can then be adjusted to stretch the spring, and therefore adjust the tension in the wire. The uniform tungsten wire is 75.6 cm long and has a mass of 3.15×10^{-3} kg. The spring has an unstretched length of 11.0 cm and a spring constant $k = 280.0$ N/m. In the lowest possible tension setting of the screw, that is, that which stretches the spring the least, the spring is stretched by 0.500 cm from

its equilibrium position. The screw can be turned to stretch it an additional 2.75 cm. What are the highest and lowest attainable speeds of transverse waves that can travel on the wire with this setup? You can neglect any stretching of wire.

106. **M WS** **Setting Safety Parameters.** You and your team are asked to determine the hearing safety parameters for a loud experiment. Since the sound will persist for only a short time, the largest allowed intensity level that will not result in permanent hearing damage is 135 dB. The experiment consists of a single source with a power output of 5.50×10^2 W that projects sound isotropically (i.e., uniformly in all directions). (a) At what distance should you set the safety perimeter (i.e., the radius of a sphere in meters) centered at the source? (b) How far from the source should you set the perimeter if the power of the source doubles?

107. **M WS** **A Mysterious Underwater Object.** You and your team are on a reconnaissance mission in a submarine exploring a mysterious object in the cold waters of the Weddell Sea, off the coast of Antarctica. The sonar indicates that the object, which had otherwise been moving erratically, has changed course and is now on a direct collision course with your sub. The captain issues an "all stop" order, bringing the sub to a halt relative to the water. The sonar operator "pings" the object, which amounts to sending a short blast of sound in the direction of the object. The emitted sound wave has a frequency of 1.550 kHz and a speed of 1552 m/s (the speed of sound in seawater). The sound reflects from the object and returns 2.582 s after it was emitted from your sub, and its frequency has shifted to 1.598 kHz. (a) How far from the sub was the object when the sound reflected from it? (b) What is the object's speed? (c) How long after you receive the return signal will it take the object to reach your submarine?

Each raindrop that strikes the water's surface creates waves that propagate outward in a circular pattern. When two or more of these sets of waves meet on the surface, they overlap, or interfere with each other, in a process called linear superposition. Where the waves meet crest-to-crest or trough-to-trough, the amplitude of the resultant wave increases, and where the waves meet trough-to-crest, the amplitude of the resultant wave decreases. These phenomena are known as constructive and destructive interference, respectively, and they are two important topics in this chapter.

Sandid/1642 images/Pixabay

The Principle of Linear Superposition and Interference Phenomena

17.1 The Principle of Linear Superposition

Often, two or more sound waves are present at the same place at the same time, as is the case with sound waves when everyone is talking at a party or when music plays from the speakers of a stereo system. To illustrate what happens when several waves pass simultaneously through the same region,

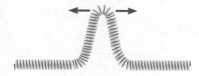

(a) Overlap begins

(b) Total overlap; the Slinky has twice the height of either pulse

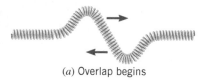

(c) The receding pulses

ANIMATED FIGURE 17.1 Two transverse "up" pulses passing through each other.

(a) Overlap begins

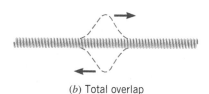

(b) Total overlap

(c) The receding pulses

ANIMATED FIGURE 17.2 Two transverse pulses, one "up" and one "down," passing through each other.

let's consider **Animated Figures 17.1** and **17.2**, which show two transverse pulses of equal heights moving toward each other along a Slinky. In **Animated Figure 17.1** both pulses are "up," whereas in **Animated Figure 17.2** one is "up" and the other is "down." Part *a* of each figure shows the two pulses beginning to overlap. The pulses merge, and the Slinky assumes a shape that is *the sum of the shapes of the individual pulses*. Thus, when the two "up" pulses overlap completely, as in **Animated Figure 17.1***b*, the Slinky has a pulse height that is twice the height of an individual pulse. Likewise, when the "up" pulse and the "down" pulse overlap exactly, as in **Animated Figure 17.2***b*, they momentarily cancel, and the Slinky becomes straight. In either case, the two pulses move apart after overlapping, and the Slinky once again conforms to the shapes of the individual pulses, as in part *c* of both figures.

The adding together of individual pulses to form a resultant pulse is an example of a more general concept called the **principle of linear superposition**.

> **THE PRINCIPLE OF LINEAR SUPERPOSITION**
>
> When two or more waves are present simultaneously at the same place, the resultant disturbance is the sum of the disturbances from the individual waves.

This principle can be applied to all types of waves, including sound waves, water waves, and electromagnetic waves such as light, radio waves, and microwaves. It embodies one of the most important concepts in physics, and the remainder of this chapter deals with examples related to it.

Check Your Understanding

(The answer is given at the end of the book.)

1. **CYU Figure 17.1** shows a graph of two pulses traveling toward each other at *t* = 0 s. Each pulse has a constant speed of 1 cm/s. When *t* = 2 s, what is the height of the resultant pulse at **(a)** *x* = 3 cm and **(b)** *x* = 4 cm?

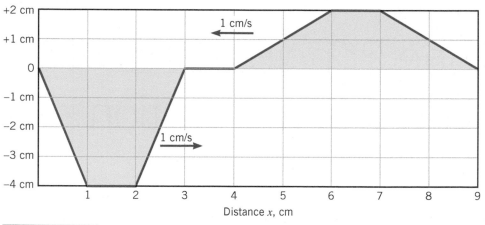

CYU FIGURE 17.1

17.2 | Constructive and Destructive Interference of Sound Waves

Suppose that the sounds from two speakers overlap in the middle of a listening area, as in **Interactive Figure 17.3**, and that each speaker produces a sound wave of the same amplitude and frequency. For convenience, the wavelength of the sound is chosen to be $\lambda = 1$ m. In addition, assume that the diaphragms of the speakers vibrate in phase; that is, they move outward together and inward together. If the distance of each speaker from the overlap point is the same (3 m in the drawing), the condensations (C) of one wave always meet the condensations of the other when the waves come together; similarly,

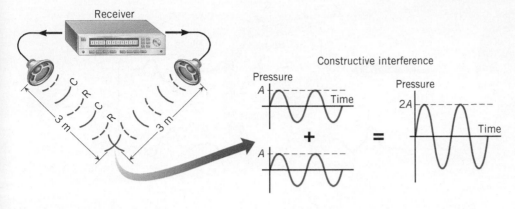

INTERACTIVE FIGURE 17.3 The speakers in this drawing vibrate in phase. As a result of constructive interference between the two sound waves (amplitude = A), a loud sound (amplitude = $2A$) is heard at an overlap point located equally distant from two in-phase speakers (C, condensation; R, rarefaction).

rarefactions (R) always meet rarefactions. According to the principle of linear super-position, the combined pattern is the sum of the individual patterns. As a result, the pressure fluctuations at the overlap point have twice the amplitude A that the individ-ual waves have, and a listener at this spot hears a louder sound than the sound coming from either speaker alone. When two waves always meet condensation-to-condensation and rarefaction-to-rarefaction (or crest-to-crest and trough-to-trough), they are said to be **exactly in phase** and to exhibit **constructive interference**.

Now consider what happens if one of the speakers is moved. The result is surprising. In **Interactive Figure 17.4**, the left speaker is moved away* from the overlap point by a distance equal to one-half of the wavelength, or 0.5 m. Therefore, at the overlap point, a condensation arriving from the left meets a rarefaction arriving from the right. Like-wise, a rarefaction arriving from the left meets a condensation arriving from the right. According to the principle of linear superposition, the net effect is a mutual cancella-tion of the two waves. The condensations from one wave offset the rarefactions from the other, leaving only a *constant air pressure*. A constant air pressure, devoid of condensa-tions and rarefactions, means that a listener detects no sound. When two waves always meet condensation-to-rarefaction (or crest-to-trough), they are said to be **exactly out of phase** and to exhibit **destructive interference**.

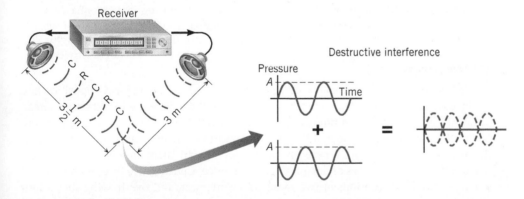

INTERACTIVE FIGURE 17.4 The speakers in this drawing vibrate in phase. However, the left speaker is one-half of a wavelength $\left(\frac{1}{2}\text{ m}\right)$ farther from the overlap point than the right speaker. Because of destructive interference, no sound is heard at the overlap point (C, condensation; R, rarefaction).

When two waves meet, they interfere constructively if they always meet exactly in phase and destructively if they always meet exactly out of phase. In either case, this means that the wave patterns do not shift relative to one another as time passes. Sources that produce waves in this fashion are called **coherent sources**.

THE PHYSICS OF . . . noise-canceling headphones. Destructive interfer-ence is the basis of a useful technique for reducing the loudness of undesirable sounds. For instance, **Figure 17.5** shows a pair of noise-canceling headphones. Small micro-phones are mounted inside the headphones and detect noise such as the engine noise

---

*When the left speaker is moved back, its sound intensity and, hence, its pressure amplitude decrease at the overlap point. In this chapter assume that the power delivered to the left speaker by the receiver is increased slightly to keep the amplitudes equal at the overlap point.

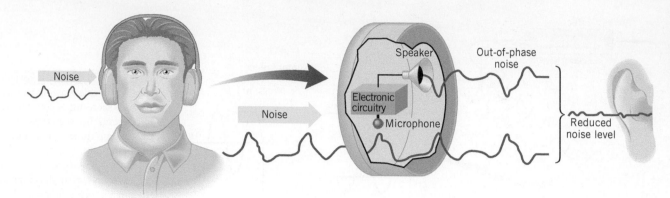

FIGURE 17.5 Noise-canceling headphones utilize destructive interference.

that an airplane pilot would hear. The headphones also contain circuitry to process the electronic signals from the microphones and reproduce the noise in a form that is exactly out of phase compared to the original. This out-of-phase version is played back through the headphone speakers and, because of destructive interference, combines with the original noise to produce a quieter background.

If the left speaker in **Interactive Figure 17.4** were moved away from the overlap point by *another* one-half wavelength $\left(3\frac{1}{2}\text{ m} + \frac{1}{2}\text{ m} = 4\text{ m}\right)$, the two waves would again be in phase, and constructive interference would occur. The listener would hear a loud sound because the left wave travels one whole wavelength $(\lambda = 1\text{ m})$ farther than the right wave and, at the overlap point, condensation meets condensation and rarefaction meets rarefaction. In general, the important issue is the *difference* in the path lengths traveled by each wave in reaching the overlap point:

> **Problem-Solving Insight** For two wave sources vibrating in phase, a difference in path lengths that is zero or an integer number (1, 2, 3, . . .) of wavelengths leads to constructive interference; a difference in path lengths that is a half-integer number $\left(\frac{1}{2}, 1\frac{1}{2}, 2\frac{1}{2}, \ldots\right)$ of wavelengths leads to destructive interference.

> **Problem-Solving Insight** For two wave sources vibrating out of phase, a difference in path lengths that is a half-integer number $\left(\frac{1}{2}, 1\frac{1}{2}, 2\frac{1}{2}, \ldots\right)$ of wavelengths leads to constructive interference; a difference in path lengths that is zero or an integer number (1, 2, 3, . . .) of wavelengths leads to destructive interference.

Interference effects can also be detected if the two speakers are fixed in position and the listener moves about the room. Consider **Figure 17.6**, where the sound waves spread outward from each of two in-phase speakers, as indicated by the concentric circular arcs. Each solid arc represents the middle of a condensation, and each dashed arc represents the middle of a rarefaction. Where the two waves overlap, there are places of constructive interference and places of destructive interference. Constructive interference occurs wherever two condensations or two rarefactions intersect, and the drawing shows four such places as solid dots. A listener stationed at any one of these locations hears a loud sound. On the other hand, destructive interference occurs wherever a condensation and a rarefaction intersect, such as the two open dots in the picture. A listener situated at a point of destructive interference hears no sound. At locations where neither constructive nor destructive interference occurs, the two waves partially reinforce or partially cancel, depending on the position relative to the speakers. Thus, it is possible for a listener to walk about the overlap region and hear marked variations in loudness.

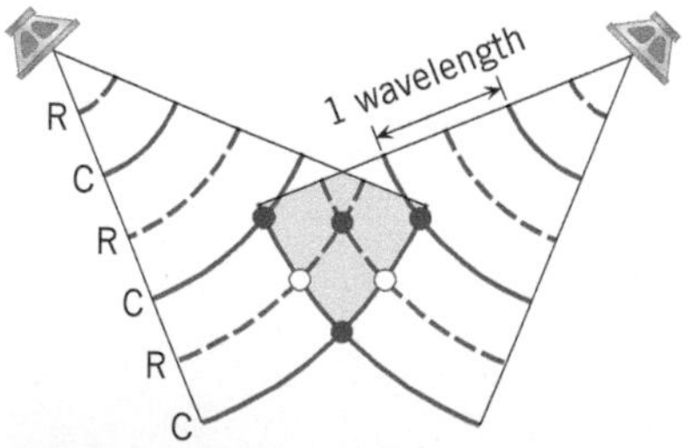

The individual sound waves from the speakers in **Figure 17.6** carry energy, and the energy delivered to the overlap region is the sum of the energies of the individual waves. This fact is consistent with the principle of conservation of energy, which we first encountered in Section 6.8. This principle states that energy can neither be created nor destroyed, but can only be converted from one form to another. One of the interesting consequences of interference is that the energy is redistributed, so there are places within the overlap region where the sound is loud and other places where there is no sound at all. Interference, so to speak, "robs Peter to pay Paul," but energy is always conserved in the process. Example 1 illustrates how to decide what a listener hears.

FIGURE 17.6 Two sound waves overlap in the shaded region. The solid lines denote the middle of the condensations (C), and the dashed lines denote the middle of the rarefactions (R). Constructive interference occurs at each solid dot (●) and destructive interference at each open dot (○).

EXAMPLE 1 | What Does a Listener Hear?

In **Figure 17.7** two in-phase loudspeakers, A and B, are separated by 3.20 m. A listener is stationed at point C, which is 2.40 m in front of speaker B. The triangle ABC is a right triangle. Both speakers are playing identical 214-Hz tones, and the speed of sound is 343 m/s. Does the listener hear a loud sound or no sound?

Reasoning The listener will hear either a loud sound or no sound, depending on whether the interference occurring at point C is constructive or destructive. To determine which it is, we need to find the difference in the distances traveled by the two sound waves that reach point C and see whether the difference is an integer or half-integer number of wavelengths. In either event, the wavelength can be found from the relation $\lambda = v/f$ (Equation 16.1).

> **Problem-Solving Insight** To decide whether two sources of sound produce constructive or destructive interference at a point, determine the difference in path lengths between each source and that point and compare it to the wavelength of the sound.

Solution Since the triangle ABC is a right triangle, the distance AC is given by the Pythagorean theorem as $\sqrt{(3.20\ \text{m})^2 + (2.40\ \text{m})^2} = 4.00$ m. The distance BC is given as 2.40 m. Thus, the difference in the travel distances for the waves is 4.00 m − 2.40 m = 1.60 m. The wavelength of the sound is

$$\lambda = \frac{v}{f} = \frac{343\ \text{m/s}}{214\ \text{Hz}} = 1.60\ \text{m} \qquad \textbf{(16.1)}$$

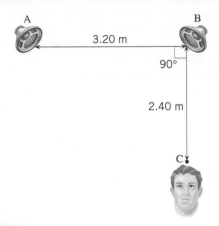

FIGURE 17.7 Example 1 discusses whether this setup leads to constructive or destructive interference at point C for 214-Hz sound waves.

Since the difference in the distances is an integer number ($n = 0, 1, 2 \ldots$) of wavelengths (here $n = 1$), constructive interference occurs at point C, and the ***listener hears a loud sound.***

It should be noted that, if the difference in the distances had not been equal to an integer (constructive interference) or an odd half-integer (destructive interference) number of wavelengths, the listener would hear a sound with a loudness somewhere between zero and the maximum which occurs as a result of constructive interference.

Up to this point, we have been assuming that the speaker diaphragms vibrate synchronously, or in phase; that is, they move outward together and inward together. This may not be the case, however, and Conceptual Example 2 considers what happens then.

CONCEPTUAL EXAMPLE 2 | Out-of-Phase Speakers

To make a speaker operate, two wires (one red and one black, for instance) must be connected between the speaker and the receiver (amplifier), as in **Figure 17.8**. Consider one of the speakers in **Interactive Figure 17.4**, where the red wire connects the red terminal of the speaker to the red terminal of the receiver. Similarly, the black wire connects the black terminal of the speaker to the black terminal of the receiver. For the other speaker, however, the wires connect a terminal of one color on the speaker to a terminal of a different color on the receiver. Since the two speakers are not connected to the receiver in exactly the same way, the two diaphragms will vibrate out of phase, one moving outward every time the other moves inward, and vice versa. A listener at the overlap point in **Interactive Figure 17.4** would now hear **(a)** no sound because destructive interference occurs **(b)** a loud sound because constructive interference occurs.

Reasoning Since the speaker diaphragms in **Interactive Figure 17.4** are now vibrating out of phase, one of them is moving exactly opposite to the way it was moving originally; let us assume that it is the left speaker. The effect of this change is that every condensation originating from the left speaker becomes a rarefaction, and every rarefaction becomes a condensation.

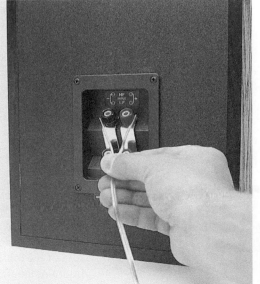

FIGURE 17.8 A loudspeaker is connected to a receiver (amplifier) by two wires.

Andy Washnik

Answer (a) is incorrect. When the two speakers in **Interactive Figure 17.4** are wired *in phase*, a condensation from one speaker always meets a rarefaction from the other at the overlap point, and destructive interference occurs. However, if one of the speakers were wired out of phase relative to the other, a condensation from one speaker would meet a condensation from the other, and destructive interference would *not* occur.

Answer (b) is correct. If the left speaker in **Interactive Figure 17.4** were connected out of phase with respect to the right speaker, a condensation from the right speaker would meet a condensation (not a rarefaction) from the left speaker at the overlap point. Similarly, a rarefaction from the right speaker would meet a rarefaction from the left speaker. The result is *constructive* interference, and a loud sound would be heard.

THE PHYSICS OF . . . wiring the speakers in an audio system. Instructions for connecting stereo or surround-sound systems specifically warn owners to avoid out-of-phase vibration of speaker diaphragms. If the wires and the terminals of the speakers and the receiver are not color-coded, you can check for problems in the following way. Play music with a lot of low-frequency bass tones. Set the receiver to its monaural mode, so the same sound comes out of both speakers being tested. If the diaphragms are in phase, the bass sound will either remain the same or become slightly louder as you slide the speakers together. If the diaphragms are out of phase, the bass sound will fade noticeably (due to destructive interference) when the speakers are right next to each other. In this event, simply interchange the wires to the terminals on one (not both) of the speakers.

The phenomena of constructive and destructive interference are exhibited by all types of waves, not just sound waves. We will encounter interference effects again in Chapter 27, in connection with light waves.

Check Your Understanding

(*The answers are given at the end of the book.*)

2. Does the principle of linear superposition imply that two sound waves, passing through the same place at the same time, always create a louder sound than is created by either wave alone?

3. Suppose that you are sitting at the overlap point between the two speakers in **Interactive Figure 17.4**. Because of destructive interference, you hear no sound, even though both speakers are emitting identical sound waves. One of the speakers is suddenly shut off. Will you now hear a sound? **(a)** No. **(b)** Yes. **(c)** Yes, but only if you move a distance of one wavelength closer to the speaker that is still producing sound.

4. Starting at the overlap point in **Interactive Figure 17.3**, you walk along a straight path that is *perpendicular* to the line between the speakers and passes through the midpoint of that line. As you walk, the loudness of the sound **(a)** changes from loud to faint to loud **(b)** changes from faint to loud to faint **(c)** does not change.

5. Starting at the overlap point in **Interactive Figure 17.3**, you walk along a path that is *parallel* to the line between the speakers. As you walk, the loudness of the sound **(a)** changes from loud to faint to loud **(b)** changes from faint to loud to faint **(c)** does not change.

17.3 Diffraction

Section 16.5 discusses the fact that sound is a pressure wave created by a vibrating object, such as a loudspeaker. The previous two sections of this chapter have examined what happens when two sound waves are present simultaneously at the same place; according to the principle of linear superposition, a resultant disturbance is formed from the sum of the individual waves. This principle reveals that overlapping sound waves exhibit interference effects, whereby the sound energy is redistributed within the overlap region. We will now use the principle of linear superposition to explore another interference effect, that of diffraction.

When a wave encounters an obstacle or the edges of an opening, it bends around them. For instance, a sound wave produced by a stereo system bends around the edges of an open doorway, as **Figure 17.9a** illustrates. If such bending did not occur, sound could be heard outside the room only at locations directly in front of the doorway, as part *b* of the drawing suggests. (It is assumed that no sound is transmitted directly through the walls.) The bending of a wave around an obstacle or the edges of an opening is called **diffraction**. All kinds of waves exhibit diffraction.

To demonstrate how the bending of waves arises, **Figure 17.10** shows an expanded view of **Figure 17.9a**. When the sound wave reaches the doorway, the air in the doorway is set into longitudinal vibration. In effect, each molecule of the air in the doorway becomes a source of a sound wave in its own right, and, for purposes of illustration, the drawing shows two of the molecules. Each produces a sound wave that expands outward in three dimensions, much like a water wave does in two dimensions when a stone is dropped into a pond. The sound waves generated by all the molecules in the doorway must be added together to obtain the total sound wave at any location outside the room, in accord with the principle of linear superposition. However, even considering only the waves from the two molecules in the picture, it is clear that the expanding wave patterns reach locations off to either side of the doorway. The net effect is a "bending," or diffraction, of the sound around the edges of the opening. Further insight into the origin of diffraction can be obtained with the aid of Huygens' principle (see Section 27.5).

When the sound waves generated by every molecule in the doorway are added together, it is found that there are places where the intensity is a maximum and places where it is zero, in a fashion similar to that discussed in the previous section. Analysis shows that at a great distance from the doorway the intensity is a maximum directly opposite the center of the opening. As the distance to either side of the center increases, the intensity decreases and reaches zero, then rises again to a maximum, falls again to zero, rises back to a maximum, and so on. Only the maximum at the center is a strong one. The other maxima are weak and become progressively weaker at greater distances from the center. In **Figure 17.10** the angle θ defines the location of the first minimum intensity point on either side of the center. Equation 17.1 gives θ in terms of the wavelength λ and the width D of the doorway and assumes that the doorway can be treated like a slit whose height is very large compared to its width:

Single slit—first minimum $$\sin \theta = \frac{\lambda}{D}$$ **(17.1)**

Waves also bend around the edges of openings other than single slits. Particularly important is the diffraction of sound by a circular opening, such as that in a loudspeaker. In this case, the angle θ is related to the wavelength λ and the diameter D of the opening by

Circular opening—first minimum $$\sin \theta = 1.22 \frac{\lambda}{D}$$ **(17.2)**

An important point to remember about Equations 17.1 and 17.2 is that the extent of the diffraction depends on the ratio of the wavelength to the size of the opening. If the ratio λ/D is small, then θ is small and little diffraction occurs. The waves are beamed in the forward direction as they leave an opening, much like the light from a flashlight. Such sound waves are said to have "narrow dispersion." Since high-frequency sound has a relatively small wavelength, it tends to have a narrow dispersion. On the other hand, for larger values of the ratio λ/D, the angle θ is larger. The waves spread out over a larger region and are said to have a "wide dispersion." Low-frequency sound, with its relatively large wavelength, typically has a wide dispersion.

In a stereo loudspeaker, a wide dispersion of the sound is desirable. Example 3 illustrates, however, that there are limitations to the dispersion that can be achieved, depending on the loudspeaker design.

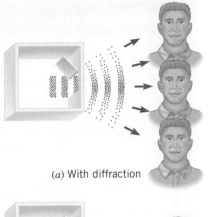

(a) With diffraction

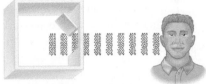

(b) Without diffraction

FIGURE 17.9 (a) The bending of a sound wave around the edges of the doorway is an example of diffraction. The source of the sound within the room is not shown. (b) If diffraction did not occur, the sound wave would not bend as it passed through the doorway.

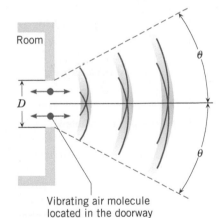

FIGURE 17.10 Each vibrating molecule of the air in the doorway generates a sound wave that expands outward and bends, or diffracts, around the edges of the doorway. Because of interference effects among the sound waves produced by all the molecules, the sound intensity is mostly confined to the region defined by the angle θ on either side of the doorway.

Analyzing Multiple-Concept Problems

EXAMPLE 3 | The Physics of Loudspeakers

A 1500-Hz sound and a 8500-Hz sound emerge from a loud-speaker through a circular opening that has a diameter of 0.30 m (see **Figure 17.11**). Assuming that the speed of sound in air is 343 m/s, find the diffraction angle θ for each sound.

Reasoning The diffraction angle θ depends on the ratio of the wavelength λ of the sound to the diameter D of the opening, according to $\sin \theta = 1.22(\lambda/D)$ (Equation 17.2). The wavelength for each sound can be obtained from the given frequencies and the speed of sound.

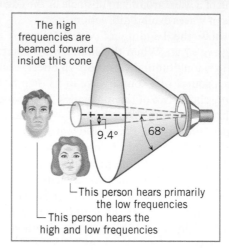

The high frequencies are beamed forward inside this cone

9.4°   68°

This person hears primarily the low frequencies

This person hears the high and low frequencies

FIGURE 17.11 Because the dispersion of high frequencies is less than the dispersion of low frequencies, you should be directly in front of the speaker to hear both the high and low frequencies equally well.

Knowns and Unknowns The following data are available:

Description	Symbol	Value
Sound frequency	f	1500 Hz or 8500 Hz
Diameter of speaker opening	D	0.30 m
Speed of sound	v	343 m/s
Unknown Variables		
Diffraction angle	θ	?

Modeling the Problem

STEP 1 The Diffraction Angle Equation 17.2 indicates that the diffraction angle is related to the ratio of the wavelength λ of the sound to the diameter D of the opening by $\sin \theta = 1.22(\lambda/D)$. Solving this expression for θ gives Equation 1 at the right. A value for D is given. To determine the value for λ, we turn to Step 2.

$$\theta = \sin^{-1}\left(1.22\frac{\lambda}{D}\right) \qquad (1)$$

?

STEP 2 Wavelength The wavelength λ is related to the frequency f and the speed v of the sound according to $v = f\lambda$ (Equation 16.1). Solving for λ gives

$$\lambda = \frac{v}{f}$$

which can be substituted into Equation 1 as shown at the right.

$$\theta = \sin^{-1}\left(1.22\frac{\lambda}{D}\right) \qquad (1)$$

$$\lambda = \frac{v}{f}$$

> **Problem-Solving Insight** When a wave passes through an opening, the extent of diffraction is greater when the ratio λ/D is greater, where λ is the wavelength of the wave and D is the width or diameter of the opening.

Solution Combining the results of each step algebraically, we find that

STEP 1   STEP 2

$$\theta = \sin^{-1}\left(1.22\frac{\lambda}{D}\right) = \sin^{-1}\left(1.22\frac{v/f}{D}\right)$$

Applying the above result to each of the sound frequencies shows that

1500-Hz sound $\theta = \sin^{-1}\left(1.22\dfrac{v}{fD}\right) = \sin^{-1}\left[1.22\dfrac{(343 \text{ m/s})}{(1500 \text{ Hz})(0.30 \text{ m})}\right] = \boxed{68°}$

8500-Hz sound $\theta = \sin^{-1}\left(1.22\dfrac{v}{fD}\right) = \sin^{-1}\left[1.22\dfrac{(343 \text{ m/s})}{(8500 \text{ Hz})(0.30 \text{ m})}\right] = \boxed{9.4°}$

Figure 17.11 illustrates these results. With a 0.30-m opening, the dispersion of the higher-frequency sound is limited to only 9.4°. To increase the dispersion, a smaller opening is needed. It is for this reason that loudspeaker designers use a small-diameter speaker called a *tweeter* to generate the high-frequency sound (see **Figure 17.12**).

Related Homework: Problems 12, 14

Tweeter

FIGURE 17.12 Small-diameter speakers (called tweeters) are used to produce high-frequency sound. The small diameter helps to promote a wider dispersion of the sound.

JPagetRFphotos/Alamy Stock Photo

As we have seen, diffraction is an interference effect, one in which some of the wave's energy is directed into regions that would otherwise not be accessible. Energy, of course, is conserved during this process, because energy is only redistributed during diffraction; no energy is created or destroyed.

Check Your Understanding

(The answers are given at the end of the book.)

6. At an open-air rock concert you are standing directly in front of a speaker. You hear the high-frequency sounds of a female vocalist as well as the low-frequency sounds of the rhythmic bass. As you walk to one side of the speaker, the sounds of the vocalist _____, and those of the rhythmic bass _____. **(a)** drop off noticeably; also drop off noticeably **(b)** drop off only slightly; drop off noticeably **(c)** drop off only slightly; also drop off only slightly **(d)** drop off noticeably; drop off only slightly

7. Refer to Example 1 in Section 16.2. Which type of radio wave, AM or FM, diffracts more readily around a given obstacle? **(a)** AM, because it has a greater wavelength **(b)** FM, because it has a greater wavelength **(c)** AM, because it has a greater frequency **(d)** FM, because it has a greater frequency

17.4 | Beats

In situations where waves with the *same frequency* overlap, we have seen how the principle of linear superposition leads to constructive and destructive interference and how it explains diffraction. We will see in this section that two overlapping waves with *slightly different frequencies* give rise to the phenomenon of beats. However, the principle of linear superposition again provides an explanation of what happens when the waves overlap.

A tuning fork has the property of producing a single-frequency sound wave when struck with a sharp blow. **Figure 17.13** shows sound waves coming from two tuning forks placed side by side. The tuning forks in the drawing are identical, and each is designed to produce a 440-Hz tone. However, a small piece of putty has been attached

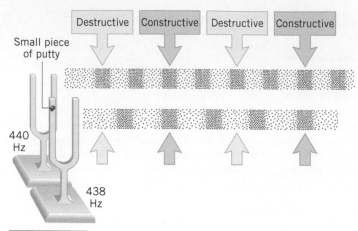

FIGURE 17.13 Two tuning forks have slightly different frequencies of 440 and 438 Hz. The phenomenon of beats occurs when the forks are sounded simultaneously. The sound waves are not drawn to scale.

to one fork, whose frequency is lowered to 438 Hz because of the added mass. When the forks are sounded simultaneously, the loudness of the resulting sound rises and falls periodically—faint, then loud, then faint, then loud, and so on. The periodic variations in loudness are called *beats* and result from the interference between two sound waves with slightly different frequencies.

For clarity, **Figure 17.13** shows the condensations and rarefactions of the sound waves separately. In reality, however, the waves spread out and overlap. In accord with the principle of linear superposition, the ear detects the combined total of the two. Notice that there are places where the waves interfere constructively and places where they interfere destructively. When a region of constructive interference reaches the ear, a loud sound is heard. When a region of destructive interference arrives, the sound intensity drops to zero (assuming each of the waves has the same amplitude). The number of times per second that the loudness rises and falls is the **beat frequency** and is the **difference** between the two sound frequencies. Thus, in the situation illustrated in Figure 17.13, an observer hears the sound loudness rise and fall at the rate of 2 times per second (440 Hz – 438 Hz).

Figure 17.14 helps to explain why the beat frequency is the difference between the two frequencies. The drawing displays graphical representations of the pressure patterns of a 10-Hz wave and a 12-Hz wave, along with the pressure pattern that results when the two overlap. These frequencies have been chosen for convenience, even though they lie below the audio range and are inaudible. Audible sound waves behave in exactly the same way. The top two drawings, in blue, show the pressure variations in a one-second interval of each wave. The third drawing, in red, shows the result of adding together the blue patterns according to the principle of linear superposition. Notice that the amplitude in the red drawing is not constant, as it is in the individual waves. Instead, the amplitude changes from a minimum to a maximum, back to a minimum, and so on. When such pressure variations reach the ear and occur in the audible frequency range, they produce a loud sound when the amplitude is a maximum and a faint sound when the amplitude is a minimum. Two loud–faint cycles, or beats, occur in the one-second interval shown in the drawing, corresponding to a beat frequency of 2 Hz. Thus, the beat frequency is the difference between the frequencies of the individual waves, or 12 Hz – 10 Hz = 2 Hz. Beats will occur whenever two frequencies are slightly different, as demonstrated in the following example.

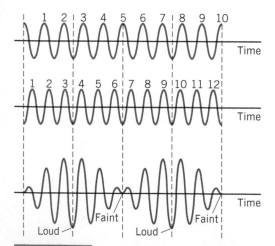

FIGURE 17.14 A 10-Hz and a 12-Hz sound wave, when added together, produce a wave with a beat frequency of 2 Hz. The drawings show the pressure patterns (in blue) of the individual waves and the pressure pattern (in red) that results when the two overlap. The time interval shown is one second.

EXAMPLE 4 | BIO The Physics of Detecting Small Differences in Frequency

The ability of our ears to distinguish between two pure tones of slightly different frequencies is known as **frequency discrimination**. Research has shown that for frequencies below 1000-Hz, we are able to distinguish a second tone with a change in frequency of only 1 to 3 Hz, assuming the second tone is emitted right after the first one and at nearly the same intensity. In general, our ability to discriminate two frequencies steadily decreases at higher frequencies, and studies involving people with peripheral hearing loss have shown that auditory pathology can actually improve frequency discrimination, as the brain undergoes neural reorganization after cochlear damage. With this being said, we are able to distinguish two slightly different frequencies *simultaneously*. The rise and fall in intensity of the beat frequency is very noticeable, especially when the difference in frequencies is small. For example, imagine the situation where you are at rest in your car, and you are honking the horn, which has a frequency of 525 Hz. Your friend is driving toward you in the same-model car and honking his horn with the same 525-Hz frequency. You hear a beat frequency of 2.00 Hz. How fast is your friend driving toward you? Assume the speed of sound in air is 343 m/s.

Reasoning   Since the two horns produce sounds of the same frequency, the beats indicate a shift in one of the observed frequencies.

Since your friend's car is moving and yours is stationary, the shift is due to the motion of his car and can be attributed to the Doppler effect. Equation 16.11 can be used to find the speed of the moving car from the Doppler shift. Here, we have the case of a source moving toward a stationary observer.

Solution   The Doppler effect will result in a higher observed frequency, since the sound source is moving toward the stationary observer. Since the beat frequency is equal to the Doppler shift we have the following:

$$f_o - f_s = 2 \text{ Hz} \Rightarrow f_o = f_s + 2 \text{ Hz} = 525 \text{ Hz} + 2 \text{ Hz} = 527 \text{ Hz.}$$

We can now apply Equation 16.11 and solve for the speed of the source (moving car):

$$f_o = f_s\left(\frac{1}{1-\frac{v_s}{v}}\right) \Rightarrow 1 - \frac{v_s}{v} = \frac{f_s}{f_o} \Rightarrow v_s = v\left(1 - \frac{f_s}{f_o}\right)$$

$$= (343 \text{ m/s})\left(1 - \frac{525 \text{ Hz}}{527 \text{ Hz}}\right) = \boxed{1.30 \text{ m/s}}$$

As a final note, car horns are often arranged in pairs—one with a frequency of 500 Hz and the second with a frequency between 405 – 420 Hz. The beat frequency produced by the two horns is more perceptible than a single tone, making it easier to detect, especially in an environment with a high level of ambient noise.

THE PHYSICS OF . . . tuning a musical instrument.   Musicians often tune their instruments by listening to a beat frequency. For instance, a guitar player plucks an out-of-tune string along with a tone that has the correct frequency. He then adjusts the string tension until the beats vanish, ensuring that the string is vibrating at the correct frequency.

Check Your Understanding

(The answers are given at the end of the book.)

8. Tuning fork A (frequency unknown) and tuning fork B (frequency = 384 Hz) together produce 6 beats in 2 seconds. When a small piece of putty is attached to tuning fork A, as in **Figure 17.13**, the beat frequency decreases. What is the frequency of tuning fork A before the putty is attached? **(a)** 378 Hz **(b)** 381 Hz **(c)** 387 Hz **(d)** 390 Hz

9. A tuning fork has a frequency of 440 Hz. The string of a violin and this tuning fork, when sounded together, produce a beat frequency of 1 Hz. From these two pieces of information alone, is it possible to determine the exact frequency of the violin string? **(a)** Yes; the frequency of the violin string is 441 Hz. **(b)** No, because the frequency of the violin string could be either 439 or 441 Hz. **(c)** Yes; the frequency of the violin string is 439 Hz.

10. When the regions of constructive and destructive interference in **Figure 17.13** move past a listener's ear, a beat frequency of 2 Hz is heard. Suppose that the tuning forks in the drawing are sounded under water and that the listener is also under water. The forks vibrate at 438 and 440 Hz, just as they do in air. However, sound travels four times faster in water than in air. The beat frequency heard by the underwater listener is **(a)** 16 Hz **(b)** 8 Hz **(c)** 4 Hz **(d)** 2 Hz.

17.5 | Transverse Standing Waves

A standing wave is another interference effect that can occur when two waves overlap. Standing waves can arise with transverse waves, such as those on a guitar string, and also with longitudinal sound waves, such as those in a flute. In any case, the principle

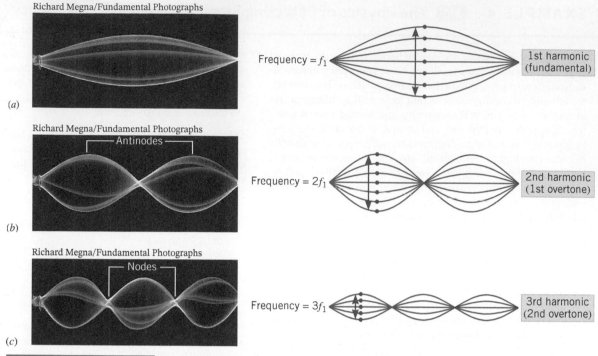

Richard Megna/Fundamental Photographs

(a)   Frequency = f_1 1st harmonic (fundamental)

Richard Megna/Fundamental Photographs
Antinodes

(b)   Frequency = $2f_1$ 2nd harmonic (1st overtone)

Richard Megna/Fundamental Photographs
Nodes

(c)   Frequency = $3f_1$ 3rd harmonic (2nd overtone)

INTERACTIVE FIGURE 17.15   Vibrating a string at certain unique frequencies sets up transverse standing wave patterns, such as the three shown in the photographs on the left. Each drawing on the right shows the various shapes that the string assumes at various times as it vibrates. The red dots attached to the strings focus attention on the maximum vibration that occurs at an antinode. In each of the drawings, one-half of a wave cycle is outlined in red. (Richard Megna/Fundamental Photographs)

of linear superposition provides an explanation of the effect, just as it does for diffraction and beats.

Interactive Figure 17.15 shows some of the essential features of transverse standing waves. In this figure the left end of each string is vibrated back and forth, while the right end is attached to a wall. Regions of the string move so fast that they appear only as a blur in the photographs. Each of the patterns shown is called a **transverse standing wave pattern**. Notice that the patterns include special places called nodes and antinodes. The **nodes** are places that do not vibrate at all, and the **antinodes** are places where maximum vibration occurs. To the right of each photograph is a drawing that helps us to visualize the motion of the string as it vibrates in a standing wave pattern. These drawings freeze the shape of the string at various times and emphasize the maximum vibration that occurs at an antinode with the aid of a red dot attached to the string.

Each standing wave pattern is produced at a unique frequency of vibration. These frequencies form a series, the smallest frequency f_1 corresponding to the one-loop pattern and the larger frequencies being integer multiples of f_1, as **Interactive Figure 17.15** indicates. Thus, if f_1 is 10 Hz, the frequency needed to establish the 2-loop pattern is $2f_1$ or 20 Hz, whereas the frequency needed to create the 3-loop pattern is $3f_1$ or 30 Hz, and so on. The frequencies in this series (f_1, $2f_1$, $3f_1$, etc.) are called **harmonics**. The lowest frequency f_1 is called the first harmonic, and the higher frequencies are designated as the second harmonic ($2f_1$), the third harmonic ($3f_1$), and so forth. The harmonic number (1st, 2nd, 3rd, etc.) corresponds to the number of loops in the standing wave pattern. The frequencies in this series are also referred to as the fundamental frequency, the first overtone, the second overtone, and so on. Thus, frequencies above the fundamental are **overtones** (see **Interactive Figure 17.15**).

Standing waves arise because identical waves travel on the string in *opposite directions* and combine in accord with the principle of linear superposition. A standing wave is said to be *standing* because it does not travel in one direction or the other, as do the individual waves that produce it. **Figure 17.16** shows why there are waves traveling in both directions on the string. At the top of the picture, one-half of a wave cycle (the remainder of the wave is omitted for clarity) is moving toward the wall on the right. When the half-cycle reaches the wall, it causes the string to pull upward on the wall. Consistent with Newton's action-reaction law, the wall pulls downward on the string, and a downward-pointing half-cycle is sent back toward the left. Thus, the wave reflects from the wall. Upon arriving back at the point of origin, the wave reflects again, this time from the hand vibrating the string. For small vibration amplitudes, the hand is essentially fixed and behaves as the wall does in causing reflections. Repeated reflections at both ends of the string create a multitude of wave cycles traveling in both directions.

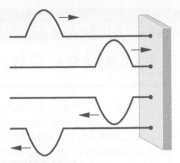

FIGURE 17.16 In reflecting from the wall, a forward-traveling half-cycle becomes a backward-traveling half-cycle that is inverted.

As each new cycle is formed by the vibrating hand, previous cycles that have reflected from the wall arrive and reflect again from the hand. Unless the timing is right, however, the new and the reflected cycles tend to offset one another, and a standing wave is not formed. Think about pushing someone on a swing and timing your pushes so that they reinforce one another. Such reinforcement in the case of the wave cycles leads to a large-amplitude standing wave. Suppose that the string has a length L and its left end is being vibrated at a frequency f_1. The time required to create a new wave cycle is the period T of the wave, where $T = 1/f_1$ (Equation 10.5). On the other hand, the time needed for a cycle to travel from the hand to the wall and back, a distance of $2L$, is $2L/v$, where v is the wave speed. Reinforcement between new and reflected cycles occurs if these two times are equal; that is, if $1/f_1 = 2L/v$. Thus, a standing wave is formed when the string is vibrated with a frequency of $f_1 = v/(2L)$.

Repeated reinforcement between newly created and reflected cycles causes a large-amplitude standing wave to develop on the string, *even when the hand itself vibrates with only a small amplitude.* Thus, the motion of the string is a resonance effect, analogous to that discussed in Section 10.6 for an object attached to a spring. The frequency f_1 at which resonance occurs is sometimes called a **natural frequency** of the string, similar to the frequency at which an object oscillates on a spring.

There is a difference between the resonance of the string and the resonance of a spring system, however. An object on a spring has only a single natural frequency, whereas the string has a *series* of natural frequencies. The series arises because a reflected wave cycle need not return to its point of origin in time to reinforce *every* newly created cycle. Reinforcement can occur, for instance, on *every other* new cycle, as it does if the string is vibrated at twice the frequency f_1, or $f_2 = 2f_1$. Likewise, if the vibration frequency is $f_3 = 3f_1$, reinforcement occurs on *every third* new cycle. Similar arguments apply for any frequency $f_n = nf_1$, where n is an integer. As a result, the series of natural frequencies that lead to standing waves on a string fixed at both ends is

String fixed at both ends $$f_n = n\left(\frac{v}{2L}\right) \qquad n = 1, 2, 3, 4, \ldots \qquad (17.3)$$

It is also possible to obtain Equation 17.3 in another way. In **Interactive Figure 17.15**, one-half of a wave cycle is outlined in red for each of the harmonics, to show that each loop in a standing wave pattern corresponds to one-half a wavelength. Since the two fixed ends of the string are nodes, the length L of the string must contain an integer number n of half-wavelengths: $L = n\left(\frac{1}{2}\lambda_n\right)$ or $\lambda_n = 2L/n$. Using this result for the wavelength in the relation $f_n\lambda_n = v$ shows that $f_n(2L/n) = v$, which can be rearranged to give Equation 17.3.

> **Problem-Solving Insight** The distance between two successive nodes (or between two successive antinodes) of a standing wave is equal to one-half of a wavelength.

Standing waves on a string are important in the way many musical instruments produce sound. For instance, a guitar string is stretched between two supports and, when plucked, vibrates according to the series of natural frequencies given by Equation 17.3. The next two examples show how this series of frequencies governs the design of a guitar.

Analyzing Multiple-Concept Problems

EXAMPLE 5 | Playing a Guitar

The heaviest string on an electric guitar has a linear density of 5.28×10^{-3} kg/m and is stretched with a tension of 226 N. This string produces the musical note E when vibrating along its entire length in a standing wave at the fundamental frequency of 164.8 Hz. **(a)** Find the length L of the string between its two fixed ends (see **Figure 17.17a**). **(b)** A guitar player wants the string to vibrate at a fundamental frequency of 2×164.8 Hz = 329.6 Hz, as it must if the musical note E is to be sounded one octave higher in pitch. To accomplish this, he presses the string against the proper fret before plucking the string. Find the distance L between the fret and the bridge of the guitar (see **Figure 17.17b**).

Reasoning The series of natural frequencies (including the fundamental) for a string fixed at both ends is given by $f_n = nv/(2L)$ (Equation 17.3), where $n = 1, 2, 3$, etc. This equation can be solved for the length L. The speed v at which waves travel can be obtained from the tension and the linear density of the string. The fundamental frequencies that are given correspond to $n = 1$.

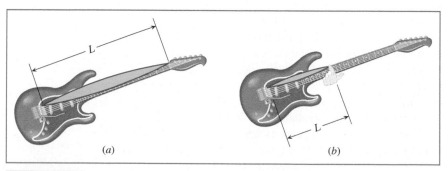

FIGURE 17.17 These drawings show the standing waves (in blue) that exist on a guitar string under different playing conditions.

Knowns and Unknowns The given data are summarized as follows:

Description	Symbol	Value	Comment
Explicit Data			
Linear density of string	m/L	5.28×10^{-3} kg/m	
Tension in string	F	226 N	
Natural frequency at which string vibrates	f_n	164.8 Hz or 329.6 Hz	These are fundamental frequencies.
Implicit Data			
Integer variable in series of natural frequencies	n	1	Fundamental frequencies are given.
Unknown Variable			
Length	L	?	

Modeling the Problem

STEP 1 Natural Frequencies According to Equation 17.3, the natural frequencies for a string fixed at both ends are given by $f_n = nv/(2L)$, where n takes on the integer values 1, 2, 3, etc., v is the speed of the waves on the string, and L is the length between the two fixed ends. Solving this expression for L gives Equation 1 at the right. In this equation, only the speed v is unknown. We will obtain a value for it in Step 2.

$$L = \frac{nv}{2f_n} \qquad (1)$$

$$\boxed{?}$$

STEP 2 Speed of the Waves on the String The speed v of the waves traveling on the string is given by Equation 16.2 as

$$v = \sqrt{\frac{F}{m/L}}$$

where F is the tension and m/L is the linear density, both of which are given. The substitution of this expression for the speed into Equation 1 is shown at the right.

$$L = \frac{nv}{2f_n} \qquad (1)$$

$$\boxed{v = \sqrt{\frac{F}{m/L}}}$$

Solution Combining the results of each step algebraically, we find that

$$\overset{\text{STEP 1}}{\underset{\downarrow}{}} \quad \overset{\text{STEP 2}}{\underset{\downarrow}{}}$$

$$L = \frac{nv}{2f_n} = \frac{n}{2f_n}\sqrt{\frac{F}{m/L}}$$

It is given that the two natural frequencies at which the string vibrates are fundamental frequencies. Thus, $n = 1$ and $f_n = f_1$. We now determine the desired lengths in parts (a) and (b).

(a) $L = \frac{n}{2f_n}\sqrt{\frac{F}{m/L}} = \frac{1}{2(164.8\ \text{Hz})}\sqrt{\frac{226\ \text{N}}{5.28 \times 10^{-3}\ \text{kg/m}}} = \boxed{0.628\ \text{m}}$

(b) $L = \frac{n}{2f_n}\sqrt{\frac{F}{m/L}} = \frac{1}{2(329.6\ \text{Hz})}\sqrt{\frac{226\ \text{N}}{5.28 \times 10^{-3}\ \text{kg/m}}} = \boxed{0.314\ \text{m}}$

The length in part (b) is one-half the length in part (a) because the fundamental frequency in part (b) is twice the fundamental frequency in part (a).

Related Homework: Problems 28, 31, 33

CONCEPTUAL EXAMPLE 6 | The Physics of the Frets on a Guitar

Figure 17.18 shows the frets on the neck of a guitar. They allow the player to produce a complete sequence of musical notes using a single string. Starting with the fret at the top of the neck, each successive fret shows where the player should press to get the next note in the sequence. Musicians call the sequence the chromatic scale, and every thirteenth note in it corresponds to one octave, or a doubling of the sound frequency. Which describes the spacing between the frets? It is **(a)** the same everywhere along the neck **(b)** greatest at the top of the neck and decreases with each additional fret further on down toward the bridge **(c)** smallest at the top of the neck and increases with each additional fret further on down toward the bridge.

Reasoning Our reasoning is based on the relation $f_1 = v/(2L)$ (Equation 17.3, with $n = 1$). The value of n is 1 because a string vibrates mainly at its fundamental frequency when plucked, as mentioned in Example 5. This equation shows that L, which is the length between a given fret and the bridge of the guitar, is inversely proportional to the fundamental frequency f_1, or $L = v/(2f_1)$. In Example 5 we found that the E string has a length of $L = 0.628$ m, corresponding to a fundamental frequency of $f_1 = 164.8$ Hz. We also found that the length between the bridge and the fret that must be pressed to double this frequency

to 2×164.8 Hz $= 329.6$ Hz is one-half of 0.628 m, or $L = 0.314$ m. To understand the spacing between frets as one moves down the neck, consider the fret that must be pressed to double the frequency again, from 329.6 Hz to 659.2 Hz. The length between the bridge and this fret would be one-half of 0.314 m, or $L = 0.157$ m. Thus, the distances of the three frets that we have been discussing are 0.628 m, 0.314 m, and 0.157 m, as indicated in **Figure 17.18**. The distance D_1 between the first two of these frets is $D_1 = 0.628$ m $-$ 0.314 m $=$ 0.314 m. Similarly, the distance between the second and third of these frets is $D_2 = 0.314$ m $-$ 0.157 m $=$ 0.157 m.

Answers (a) and (c) are incorrect. The distances between the frets are shown in **Figure 17.18**. Clearly, the distances D_1 and D_2 are not equal, nor are they smaller at the top of the neck and greater further on down.

Answer (b) is correct. Figure 17.18 shows that D_1 is greater than D_2. Thus, the spacing between the frets is greatest at the top of the neck and decreases with each additional fret further on down.

Related Homework: Problem 38

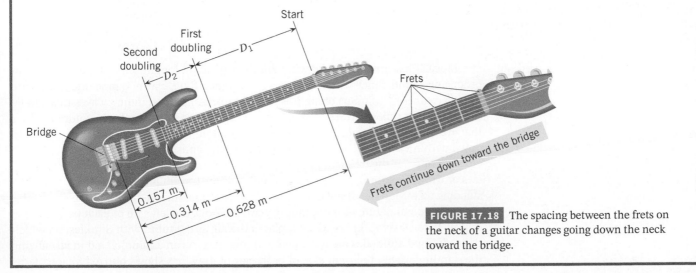

FIGURE 17.18 The spacing between the frets on the neck of a guitar changes going down the neck toward the bridge.

Check Your Understanding

(The answers are given at the end of the book.)

11. A standing wave that corresponds to the fourth harmonic is set up on a string that is fixed at both ends. **(a)** How many loops are in this standing wave? **(b)** How many nodes (excluding the nodes at the ends of the string) does this standing wave have? **(c)** Is there a node or an antinode at the midpoint of the string? **(d)** If the frequency of this standing wave is 440 Hz, what is the frequency of the lowest-frequency standing wave that could be set up on this string?

12. The tension in a guitar string is doubled. By what factor does the frequency of the vibrating string change? **(a)** It increases by a factor of 2. **(b)** It increases by a factor of $\sqrt{2}$. **(c)** It decreases by a factor of 2. **(d)** It decreases by a factor of $\sqrt{2}$.

13. A string is vibrating back and forth as in **Interactive Figure 17.15a**. The tension in the string is decreased by a factor of four, with the frequency and length of the string remaining the same. A new standing wave pattern develops on the string. How many loops are in this new pattern? **(a)** 5 **(b)** 4 **(c)** 3 **(d)** 2

14. A rope is hanging vertically straight down. The top end is being vibrated back and forth, and a standing wave with many loops develops on the rope, analogous (but not identical) to a standing wave on a horizontal rope. The rope has mass. The separation between successive nodes is **(a)** everywhere the same along the rope **(b)** greater near the top of the rope than near the bottom **(c)** greater near the bottom of the rope than near the top.

17.6 | Longitudinal Standing Waves

Standing wave patterns can also be formed from longitudinal waves. For example, when sound reflects from a wall, the forward- and backward-going waves can produce a standing wave. **Figure 17.19** illustrates the vibrational motion in a longitudinal standing wave on a Slinky. As in a transverse standing wave, there are nodes and antinodes. At the nodes the Slinky coils do not vibrate at all; that is, they have no displacement. At the antinodes the coils vibrate with maximum amplitude and, thus, have a maximum displacement. The red dots in **Figure 17.19** indicate the lack of vibration at a node and the maximum vibration at an antinode. The vibration occurs along the line of travel of the individual waves, as is to be expected for longitudinal waves. In a standing wave of sound, at the nodes and antinodes, the molecules or atoms of the medium behave as the red dots do.

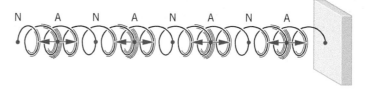

FIGURE 17.19 A longitudinal standing wave on a Slinky showing the displacement nodes (N) and antinodes (A).

Musical instruments in the wind family depend on longitudinal standing waves in producing sound. Since wind instruments (trumpet, flute, clarinet, pipe organ, etc.) are modified tubes or columns of air, it is useful to examine the standing waves that can be set up in such tubes. **Interactive Figure 17.20** shows two cylindrical columns of air that are open at both ends. Sound waves, originating from a tuning fork, travel up and down within each tube, since they reflect from the ends of the tubes, even though the ends are open. If the frequency f of the tuning fork matches one of the natural frequencies of the air column, the downward- and upward-traveling waves combine to form a standing wave, and the sound of the tuning fork becomes markedly louder. To emphasize the longitudinal nature of the standing wave patterns, the left side of each pair of drawings in **Interactive Figure 17.20** replaces the air in the tubes with Slinkies, on which the nodes and antinodes are indicated with red dots. As an additional aid in visualizing the standing waves, the right side of each pair of drawings shows blurred blue patterns

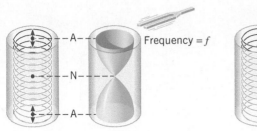
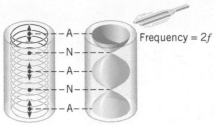

Frequency = f

Frequency = $2f$

INTERACTIVE FIGURE 17.20 A pictorial representation of longitudinal standing waves on a Slinky (left side of each pair) and in a tube of air (right side of each pair) that is open at both ends (A, antinode; N, node).

within each tube. These patterns symbolize the amplitude of the vibrating air molecules at various locations. Wherever the pattern is widest, the amplitude of vibration is greatest (a displacement antinode), and wherever the pattern is narrowest there is no vibration (a displacement node).

To determine the natural frequencies of the air columns in **Interactive Figure 17.20**, notice that there is a displacement antinode at each end of the open tube because the air molecules there are free to move.* As in a transverse standing wave, the distance between two successive antinodes is one-half of a wavelength, so the length L of the tube must be an integer number n of half-wavelengths: $L = n\left(\frac{1}{2}\lambda_n\right)$ or $\lambda_n = 2L/n$. Using this wavelength in the relation $f_n = v/\lambda_n$ shows that the natural frequencies f_n of the tube are

Tube open at both ends $f_n = n\left(\dfrac{v}{2L}\right)$ $n = 1, 2, 3, 4, \ldots$ **(17.4)**

At these frequencies, large-amplitude standing waves develop within the tube due to resonance. Example 7 illustrates how Equation 17.4 is involved when a flute is played.

EXAMPLE 7 | The Physics of a Flute

When all the holes are closed on one type of flute, the lowest note it can sound is a middle C (fundamental frequency = 261.6 Hz). The air temperature is 293 K, and the speed of sound is 343 m/s. Assuming the flute is a cylindrical tube open at both ends, determine the distance L in **Figure 17.21**—that is, the distance from the mouthpiece to the end of the tube. (This distance is approximate, since the antinode does not occur exactly at the mouthpiece.)

Reasoning   For a tube open at both ends, the series of natural frequencies (including the fundamental) is given by $f_n = nv/(2L)$ (Equation 17.4), where n takes on the integer values 1, 2, 3, etc. To obtain a value for L, we can solve this equation, since the given fundamental frequency corresponds to $n = 1$ and the speed v of sound is known.

Solution   Solving Equation 17.4 for the length L, we obtain

$$L = \frac{nv}{2f_n} = \frac{(1)(343 \text{ m/s})}{2(261.6 \text{ Hz})} = 0.656 \text{ m}$$

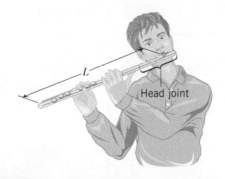

FIGURE 17.21   The length L of a flute between the mouthpiece and the end of the instrument determines the fundamental frequency of the lowest playable note.

Standing waves can also exist in a tube with only one end open, as the patterns in **Figure 17.22** indicate. Note the difference between these patterns and those in **Interactive Figure 17.20**. Here the standing waves have a displacement antinode at the open end and a displacement node at the closed end, where the air molecules are not free to move. Since the distance between a node and an adjacent antinode is one-fourth of a wavelength, the length L of the tube must be an odd number of

*In reality, the antinode does not occur exactly at the open end. However, if the tube's diameter is small compared to its length, little error is made in assuming that the antinode is located right at the end.

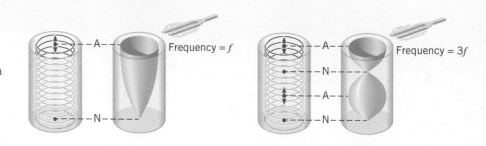

FIGURE 17.22 A pictorial representation of the longitudinal standing waves on a Slinky (left side of each pair) and in a tube of air (right side of each pair) that is open only at one end (A, antinode; N, node).

quarter-wavelengths: $L = 1\left(\frac{1}{4}\lambda\right)$ and $L = 3\left(\frac{1}{4}\lambda\right)$ for the two standing wave patterns in **Figure 17.22**. In general, then, $L = n\left(\frac{1}{4}\lambda_n\right)$, where n is any odd integer ($n = 1, 3, 5, \ldots$). From this result it follows that $\lambda_n = 4L/n$, and the natural frequencies f_n can be obtained from the relation $f_n = v/\lambda_n$:

Tube open at only one end $\qquad f_n = n\left(\dfrac{v}{4L}\right) \qquad n = 1, 3, 5, \ldots \qquad$ **(17.5)**

A tube open at only one end can develop standing waves only at the odd harmonic frequencies f_1, f_3, f_5, etc. In contrast, a tube open at both ends can develop standing waves at all harmonic frequencies f_1, f_2, f_3, etc. Moreover, the fundamental frequency f_1 of a tube open at only one end (Equation 17.5) is one-half that of a tube open at both ends (Equation 17.4). In other words, a tube open only at one end needs to be only one-half as long as a tube open at both ends in order to produce the *same* fundamental frequency.

Energy is also conserved when a standing wave is produced, either on a string or in a tube of air. The energy of the standing wave is the sum of the energies of the individual waves that comprise the standing wave. Once again, interference redistributes the energy of the individual waves to create locations of greatest energy (displacement antinodes) and locations of no energy (displacement nodes). Example 8 shows how the energy contained in standing waves set up in the human ear canal can lead to hearing loss.

EXAMPLE 8 | BIO The Physics of Hearing Loss—Standing Waves in the Ear

The human ear canal (**Figure 17.23**) essentially acts like a tube closed at one end. If the ear canal has a length of 2.3 cm, what are the fundamental wavelength and frequency for standing waves in the ear? Take the speed of sound to be 343 m/s. What is the significance of this result? Consult the graph in **Figure 16.36**.

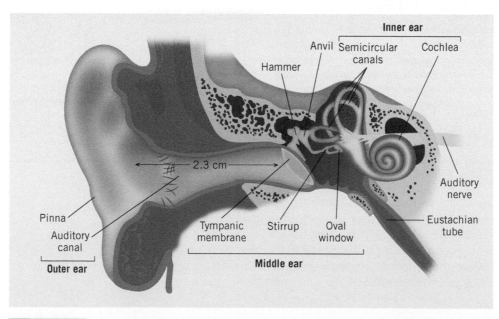

FIGURE 17.23 Diagram of the human ear showing the auditory canal as a tube closed at one end.

Reasoning We treat the auditory canal as a tube closed at one end. We can use Equation 17.5 to calculate the standing wave frequencies.

Solution The fundamental frequency is calculated for $n = 1$. Using this value in Equation 17.5, we get the following:

$$f_1 = (1)\left(\frac{343 \text{ m/s}}{(4)(0.023 \text{ m})}\right) = \boxed{3700 \text{ Hz}}$$

The fundamental wavelength is related to the fundamental frequency by

$$\lambda = \frac{v}{f} = \frac{343 \text{ m/s}}{3700 \text{ Hz}} = 9.3 \times 10^{-2} \text{ m} = \boxed{9.3 \text{ cm}}$$

Looking at the graph in **Figure 16.36**, we see that the intensity curves dip in the range of 2000 – 5000 Hz, with the ear having the greatest sensitivity at the fundamental frequency just below 4000 Hz. In fact, the ear can detect sound intensities below the threshold of hearing at 3700 Hz. Due to the resonance, sounds at this frequency are amplified. Damage to the ear can occur if exposed to high-intensity sounds for prolonged periods of time. People working in high-noise environments without ear protection often show signs of 3700-Hz hearing loss.

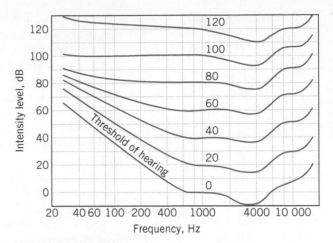

FIGURE 16.36 (REPEATED) Each curve represents the intensity levels at which sounds of various frequencies have the same loudness. The curves are labeled by their intensity levels at 1000 Hz and are known as Fletcher-Munson curves.

Check Your Understanding

(*The answers are given at the end of the book.*)

15. A cylindrical bottle, partially filled with water, is open at the top. When you blow across the top of the bottle a standing wave is set up inside it. Is there a node or an antinode **(a)** at the top of the bottle and **(b)** at the surface of the water? **(c)** If the standing wave is vibrating at its fundamental frequency, what is the distance between the top of the bottle and the surface of the water? Express your answer in terms of the wavelength λ of the standing wave. **(d)** If you take a sip from the bottle, is the fundamental frequency of the standing wave raised, lowered, or does it remain the same?

16. In **Interactive Figure 17.20** both tubes are filled with air, in which the speed of sound is v_{air}. Suppose, instead, that the tube near the tuning fork labeled "Frequency = $2f$" is filled not with air, but with another gas in which the speed of sound is v_{gas}. The frequency of each tuning fork remains unchanged. How should v_{gas} compare with v_{air} in order that the standing wave pattern in each tube has the same appearance? **(a)** $v_{gas} = \frac{1}{2}v_{air}$ **(b)** $v_{gas} = 2 v_{air}$ **(c)** $v_{gas} = \frac{1}{4}v_{air}$ **(d)** $v_{gas} = 4 v_{air}$

17. Standing waves can ruin the acoustics of a concert hall if there is excessive reflection of the sound waves that the performers generate. For example, suppose that a performer generates a 2093-Hz tone. If a large-amplitude standing wave is present, it is possible for a listener to move a distance of only 4.1 cm and hear the loudness of the tone change from loud to faint. What does the distance of 4.1 cm represent? **(a)** One wavelength of the sound **(b)** One-half the wavelength of the sound **(c)** One-fourth the wavelength of the sound

18. A wind instrument is brought into a warm house from the cold outdoors. What happens to the natural frequencies of the instrument? Neglect any change in the length of the instrument. **(a)** They increase. **(b)** They decrease. **(c)** They remain the same.

17.7 | *Complex Sound Waves

Musical instruments produce sound in a way that depends on standing waves. Examples 5 and 6 illustrate the role of transverse standing waves on the string of an electric guitar, while Example 7 stresses the role of longitudinal standing waves in the air column within a flute. In each example, sound is produced at the fundamental frequency of the instrument.

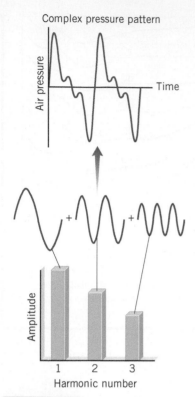

FIGURE 17.24 The topmost graph shows the pattern of pressure fluctuations such as a singer might produce. The pattern is the sum of the first three harmonics. The relative amplitudes of the harmonics correspond to the heights of the vertical bars in the bar graph.

In general, however, a musical instrument does not produce just the fundamental frequency when it plays a note, but simultaneously generates a number of harmonics as well. Different instruments, such as a violin and a trumpet, generate harmonics to different extents, and the harmonics give the instruments their characteristic sound qualities or timbres. Suppose, for instance, that a violinist and a trumpet player both sound concert A, a note whose fundamental frequency is 440 Hz. Even though both instruments are playing the same note, most people can distinguish the sound of the violin from that of the trumpet. The instruments sound different because the relative amplitudes of the harmonics (880 Hz, 1320 Hz, etc.) that the instruments create are different.

The sound wave corresponding to a note produced by a musical instrument or a singer is called a **complex sound wave** because it consists of a mixture of the fundamental and harmonic frequencies. The pattern of pressure fluctuations in a complex wave can be obtained by using the principle of linear superposition, as **Figure 17.24** indicates. This drawing shows a bar graph in which the heights of the bars give the relative amplitudes of the harmonics contained in a note such as a singer might produce. When the individual pressure patterns for each of the three harmonics are added together, they yield the complex pressure pattern shown at the top of the picture.*

THE PHYSICS OF . . . a spectrum analyzer. In practice, a bar graph such as that in **Figure 17.24** is determined with the aid of an electronic instrument known as a spectrum analyzer. When the note is produced, the complex sound wave is detected by a microphone that converts the wave into an electrical signal. The electrical signal, in turn, is fed into the spectrum analyzer, as **Figure 17.25** illustrates. The spectrum analyzer then determines the amplitude and frequency of each harmonic present in the complex wave and displays the results on its screen.

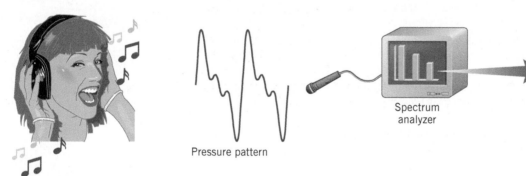

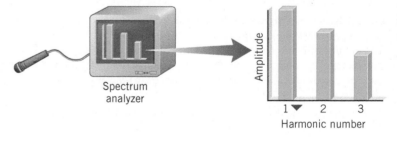

FIGURE 17.25 A microphone detects a complex sound wave produced by a singer's voice, and a spectrum analyzer determines the amplitude and frequency of each harmonic present in the wave.

Concept Summary

17.1 The Principle of Linear Superposition The principle of linear superposition states that when two or more waves are present simultaneously at the same place, the resultant disturbance is the sum of the disturbances from the individual waves.

17.2 Constructive and Destructive Interference of Sound Waves Constructive interference occurs at a point when two waves meet there crest-to-crest and trough-to-trough, thus reinforcing each

*In carrying out the addition, we assume that each individual pattern begins at zero at the origin when the time equals zero.

other. Destructive interference occurs when the waves meet crest-to-trough and cancel each other.

When waves meet crest-to-crest and trough-to-trough, they are exactly in phase. When they meet crest-to-trough, they are exactly out of phase.

For two wave sources vibrating in phase, a difference in path lengths that is zero or an integer number (1, 2, 3, . . .) of wavelengths leads to constructive interference; a difference in path lengths that is a half-integer number $\left(\frac{1}{2}, 1\frac{1}{2}, 2\frac{1}{2}, . . .\right)$ of wavelengths leads to destructive interference.

For two wave sources vibrating out of phase, a difference in path lengths that is a half-integer number $\left(\frac{1}{2}, 1\frac{1}{2}, 2\frac{1}{2}, . . .\right)$ of wavelengths leads to constructive interference; a difference in path lengths that is zero or an integer number (1, 2, 3, . . .) of wavelengths leads to destructive interference.

17.3 Diffraction Diffraction is the bending of a wave around an obstacle or the edges of an opening. The angle through which the wave bends depends on the ratio of the wavelength λ of the wave to the width D of the opening; the greater the ratio λ/D, the greater the angle.

When a sound wave of wavelength λ passes through an opening, the first place where the intensity of the sound is a minimum relative to that at the center of the opening is specified by the angle θ. If the opening is a rectangular slit of width D, such as a doorway, the angle is given by Equation 17.1. If the opening is a circular opening of diameter D, such as that in a loudspeaker, the angle is given by Equation 17.2.

$$\sin \theta = \frac{\lambda}{D} \qquad (17.1)$$

$$\sin \theta = 1.22 \frac{\lambda}{D} \qquad (17.2)$$

17.4 Beats Beats are the periodic variations in amplitude that arise from the linear superposition of two waves that have slightly different frequencies. When the waves are sound waves, the variations in amplitude cause the loudness to vary at the beat frequency, which is the difference between the frequencies of the waves.

17.5 Transverse Standing Waves A standing wave is the pattern of disturbance that results when oppositely traveling waves of the same frequency and amplitude pass through each other. A standing wave has places of minimum and maximum vibration called, respectively, nodes and antinodes. Under resonance conditions, standing waves can be established only at certain natural frequencies. The frequencies in this series (f_1, $2f_1$, $3f_1$, etc.) are called harmonics. The lowest frequency f_1 is called the first harmonic, the next frequency $2f_1$ is the second harmonic, and so on. For a string that is fixed at both ends and has a length L, the natural frequencies are specified by Equation 17.3, where v is the speed of the wave on the string and n is a positive integer.

$$f_n = n\left(\frac{v}{2L}\right) \quad n = 1, 2, 3, 4, . . . \qquad (17.3)$$

17.6 Longitudinal Standing Waves For a gas in a cylindrical tube open at both ends, the natural frequencies of vibration are specified by Equation 17.4, where v is the speed of sound in the gas and L is the length of the tube. For a gas in a cylindrical tube open at only one end, the natural frequencies of vibration are given by Equation 17.5.

$$f_n = n\left(\frac{v}{2L}\right) \quad n = 1, 2, 3, 4, . . . \qquad (17.4)$$

$$f_n = n\left(\frac{v}{4L}\right) \quad n = 1, 3, 5, 7, . . . \qquad (17.5)$$

17.7 Complex Sound Waves A complex sound wave consists of a mixture of a fundamental frequency and overtone frequencies.

Focus on Concepts

Online

Additional questions are available for assignment in WileyPLUS.

Section 17.1 The Principle of Linear Superposition

1. The drawing shows four moving pulses. Although shown as separated, the four pulses exactly overlap each other at the instant shown. Which combination of these pulses would produce a resultant pulse with the highest peak *and* the deepest valley at this instant? (a) 1 and 2 (b) 2, 3, and 4 (c) 2 and 3 (d) 1, 2, and 3 (e) 1 and 3

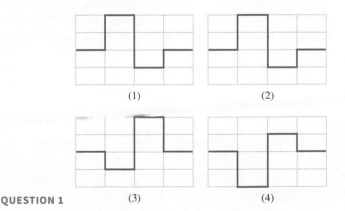

QUESTION 1

(1) (2) (3) (4)

Section 17.2 Constructive and Destructive Interference of Sound Waves

2. Two cellists, one seated directly behind the other in an orchestra, play the same note for the conductor, who is directly in front of them. Because of the separation between the cellists, destructive interference occurs at the conductor. This separation is the smallest that produces destructive interference. Would this separation increase, decrease, or remain the same if the cellists produced a note with a higher frequency? (a) The separation between the cellists would remain the same. (b) The separation would decrease because the wavelength of the sound is greater. (c) The separation would decrease because the wavelength of the sound is smaller. (d) The separation would increase because the wavelength of the sound is greater. (e) The separation would increase because the wavelength of the sound is smaller.

Section 17.3 Diffraction

3. A loudspeaker is producing sound of a certain wavelength. Which combination of the wavelength λ (expressed as a multiple of λ_0) and the speaker's diameter D (expressed as a multiple of D_0) would exhibit the greatest amount of diffraction when the sound leaves the speaker and enters the room? (a) $\lambda = \lambda_0$, $D = D_0$ (b) $\lambda = 2\lambda_0$, $D = D_0$ (c) $\lambda = \lambda_0$, $D = 2D_0$ (d) $\lambda = 2\lambda_0$, $D = 2D_0$ (e) $\lambda = 3\lambda_0$, $D = 2D_0$

4. Sound of a given frequency leaves a loudspeaker and spreads out due to diffraction. The speaker is placed in a room that contains either air or helium. The speed of sound in helium is about three times as great as the speed of sound in air. In which room, if either, does the sound exhibit the greater diffraction when leaving the speaker? **(a)** The greater diffraction occurs in the air-filled room, because the wavelength of the sound is smaller in that room. **(b)** The greater diffraction occurs in the air-filled room, because the wavelength of the sound is greater in that room. **(c)** The diffraction is the same in both rooms. **(d)** The greater diffraction occurs in the helium-filled room, because the wavelength of the sound is smaller in that room. **(e)** The greater diffraction occurs in the helium-filled room, because the wavelength of the sound is greater in that room.

Section 17.4   Beats

5. Two musicians are comparing their trombones. The first produces a tone that is known to be 438 Hz. When the two trombones play together they produce 6 beats every 2 seconds. Which statement is true about the second trombone? **(a)** It is producing either a 432-Hz sound or a 444-Hz sound. **(b)** It is producing either a 436-Hz sound or a 440-Hz sound. **(c)** It is producing a 444-Hz sound, and could be producing no other sound frequency. **(d)** It is producing either a 435-Hz sound or a 441-Hz sound. **(e)** It is producing a 441-Hz sound and could be producing no other sound frequency.

Section 17.5   Transverse Standing Waves

6. Two transverse standing waves are shown in the drawing. The strings have the same tension and length, but the bottom string is more massive. Which standing wave, if either, is vibrating at the higher frequency? **(a)** The top standing wave has the higher frequency, because the traveling waves have a smaller speed due to the smaller mass of the string. **(b)** The top standing wave has the higher frequency, because the traveling waves have a larger speed due to the

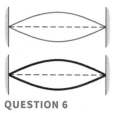

QUESTION 6

smaller mass of the string. **(c)** Both standing waves have the same frequency, because the frequency of vibration does not depend on the mass of the string. **(d)** The bottom standing wave has the higher frequency, because the traveling waves have a smaller speed due to the larger mass of the string. **(e)** The bottom standing wave has the higher frequency, because the traveling waves have a larger speed due to the larger mass of the string.

7. A standing wave on a string fixed at both ends is vibrating at its fourth harmonic. If the length, tension, and linear density are kept constant, what can be said about the wavelength and frequency of the fifth harmonic relative to the fourth harmonic? **(a)** The wavelength of the fifth harmonic is longer, and its frequency is higher. **(b)** The wavelength of the fifth harmonic is longer, and its frequency is lower. **(c)** The wavelength of the fifth harmonic is shorter, and its frequency is higher. **(d)** The wavelength of the fifth harmonic is shorter, and its frequency is lower.

Section 17.6   Longitudinal Standing Waves

8. A longitudinal standing wave is established in a tube that is open at both ends (see the drawing). The length of the tube is 0.80 m. What is the wavelength of the waves that make up the standing wave? **(a)** 0.20 m **(b)** 0.40 m **(c)** 0.80 m **(d)** 1.20 m **(e)** 1.60 m

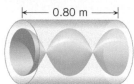

QUESTION 8

9. A longitudinal standing wave is established in a tube open at only one end (see the drawing). The frequency of the standing wave is 660 Hz, and the speed of sound in air is 343 m/s. What is the length of the tube? **(a)** 0.13 m **(b)** 0.26 m **(c)** 0.39 m **(d)** 0.52 m **(e)** 0.65 m

QUESTION 9

Problems

Online

Additional questions are available for assignment in WileyPLUS.

SSM Student Solutions Manual

MMH Problem-solving help

GO Guided Online Tutorial

V-HINT Video Hints

CHALK Chalkboard Videos

BIO Biomedical application

E Easy

M Medium

H Hard

WS Worksheet

T Team Problem

Section 17.1   The Principle of Linear Superposition,

Section 17.2   Constructive and Destructive Interference of Sound Waves

1. **E** In **Figure 17.7**, suppose that the separation between speakers A and B is 5.00 m and the speakers are vibrating in phase. They are playing identical 125-Hz tones, and the speed of sound is 343 m/s.

What is the largest possible distance between speaker B and the observer at C, such that he observes destructive interference?

2. **E** Two speakers, one directly behind the other, are each generating a 245-Hz sound wave. What is the smallest separation distance between the speakers that will produce destructive interference at a listener standing in front of them? The speed of sound is 343 m/s.

3. **E** **SSM** The drawing graphs a string on which two rectangular pulses are traveling at a constant speed of 1 cm/s at time $t = 0$ s. Using the principle of linear superposition, draw the shape of the string at $t = 1$ s, 2 s, 3 s, and 4 s.

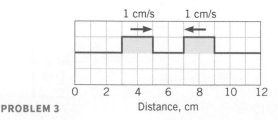

PROBLEM 3

4. ⓔ Loudspeakers A and B are vibrating in phase and are playing the same tone, which has a frequency of 250 Hz. They are set up as in **Figure 17.7**, and point C is located as shown there. However, the distance between the speakers and the distance between speaker B and point C have the same value d. The speed of sound is 343 m/s. What is the smallest value of d, such that constructive interference occurs at point C?

5. ⓔ 𝖲𝖲𝖬 Two waves are traveling in opposite directions on the same string. The displacements caused by the individual waves are given by $y_1 = (24.0 \text{ mm})\sin(9.00\pi t - 1.25\pi x)$ and $y_2 = (35.0 \text{ mm})\sin(2.88\pi t + 0.400\pi x)$. Note that the phase angles $(9.00\pi t - 1.25\pi x)$ and $(2.88\pi t + 0.400\pi x)$ are in radians, t is in seconds, and x is in meters. At $t = 4.00$ s, what is the net displacement (in mm) of the string at **(a)** $x = 2.16$ m and **(b)** $x = 2.56$ m? Be sure to include the algebraic sign (+ or –) with your answers.

6. ⓔ 𝖢𝖧𝖠𝖫𝖪 𝖦𝖮 Both drawings show the same square, each of which has a side of length $L = 0.75$ m. An observer O is stationed at one corner of each square. Two loudspeakers are located at corners of the square, as in either drawing 1 or drawing 2. The speakers produce the same single-frequency tone in either drawing and are in phase. The speed of sound is 343 m/s. Find the single smallest frequency that will produce both constructive interference in drawing 1 and destructive interference in drawing 2.

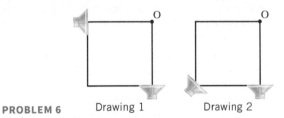

PROBLEM 6   Drawing 1   Drawing 2

7. ⓔ 𝖲𝖲𝖬 The drawing shows a loudspeaker A and point C, where a listener is positioned. A second loudspeaker B is located somewhere to the right of loudspeaker A. Both speakers vibrate in phase and are playing a 68.6-Hz tone. The speed of sound is 343 m/s. What is the closest to speaker A that speaker B can be located, so that the listener hears no sound?

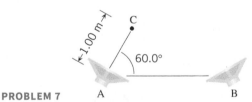

PROBLEM 7   A   B

8. ⓔ 𝖦𝖮 Suppose that the two speakers in **Figure 17.7** are separated by 2.50 m and are vibrating exactly *out of phase* at a frequency of 429 Hz. The speed of sound is 343 m/s. Does the observer at C observe constructive or destructive interference when his distance from speaker B is **(a)** 1.15 m and **(b)** 2.00 m?

9. Ⓜ 𝖬𝖬𝖧 Two loudspeakers on a concert stage are vibrating in phase. A listener is 50.5 m from the left speaker and 26.0 m from the right one. The listener can respond to all frequencies from 20 to 20 000 Hz, and the speed of sound is 343 m/s. What are the two lowest frequencies that can be heard loudly due to constructive interference?

10. Ⓜ 𝖦𝖮 𝖢𝖧𝖠𝖫𝖪 A listener is standing in front of two speakers that are producing sound of the same frequency and amplitude, except that they are vibrating out of phase. Initially, the distance between the listener and each speaker is the same (see the drawing). As the listener moves sideways, the sound intensity gradually changes. When the distance x in the drawing is 0.92 m, the change reaches the maximum amount (either loud to soft, or soft to loud). Using the data shown in the drawing and 343 m/s for the speed of sound, determine the frequency of the sound coming from the speakers.

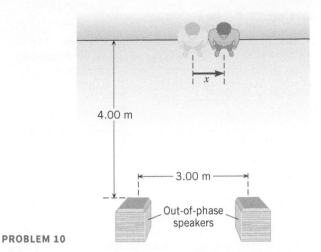

4.00 m

3.00 m

Out-of-phase speakers

PROBLEM 10

11. Ⓗ Speakers A and B are vibrating in phase. They are directly facing each other, are 7.80 m apart, and are each playing a 73.0-Hz tone. The speed of sound is 343 m/s. On the line between the speakers there are three points where constructive interference occurs. What are the distances of these three points from speaker A?

Section 17.3   Diffraction

12. ⓔ Consult Multiple-Concept Example 3 for background pertinent to this problem. A speaker has a diameter of 0.30 m. **(a)** Assuming that the speed of sound is 343 m/s, find the diffraction angle θ for a 2.0-kHz tone. **(b)** What speaker diameter D should be used to generate a 6.0-kHz tone whose diffraction angle is as wide as that for the 2.0-kHz tone in part (a)?

13. ⓔ 𝖲𝖲𝖬 Sound exits a diffraction horn loudspeaker through a rectangular opening like a small doorway. Such a loudspeaker is mounted outside on a pole. In winter, when the temperature is 273 K, the diffraction angle θ has a value of 15.0°. What is the diffraction angle for the same sound on a summer day when the temperature is 311 K?

14. ⓔ 𝖦𝖮 For one approach to problems such as this, see Multiple-Concept Example 3. Sound emerges through a doorway, as in **Figure 17.10**. The width of the doorway is 77 cm, and the speed of sound is 343 m/s. Find the diffraction angle θ when the frequency of the sound is **(a)** 5.0 kHz and **(b)** 5.0×10^2 Hz.

15. ⓔ 𝖦𝖮 The following two lists give the diameters and sound frequencies for three loudspeakers. Pair each diameter with a frequency, so that the diffraction angle is the same for each of the speakers, and then find the common diffraction angle. Take the speed of sound to be 343 m/s.

Diameter, D	Frequency, f
0.050 m	6.0 kHz
0.10 m	4.0 kHz
0.15 m	12.0 kHz

16. Ⓜ A 3.00-kHz tone is being produced by a speaker with a diameter of 0.175 m. The air temperature changes from 0 to 29 °C. Assuming air to be an ideal gas, find the *change* in the diffraction angle θ.

17. Ⓜ 𝖦𝖮 Sound (speed = 343 m/s) exits a diffraction horn loudspeaker through a rectangular opening like a small doorway. A person is sitting at an angle α off to the side of a diffraction horn that has a width D of 0.060 m. This individual does not hear a sound wave that has a frequency of 8100 Hz. When she is sitting at an angle $\alpha/2$, the frequency that she does not hear is different. What is this frequency?

Section 17.4 Beats

18. **E** **SSM** Two pure tones are sounded together. The drawing shows the pressure variations of the two sound waves, measured with respect to atmospheric pressure. What is the beat frequency?

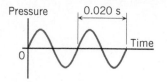
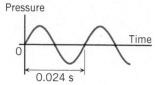

PROBLEM 18

19. **E** **GO** Two pianos each sound the same note simultaneously, but they are both out of tune. On a day when the speed of sound is 343 m/s, piano A produces a wavelength of 0.769 m, while piano B produces a wavelength of 0.776 m. How much time separates successive beats?

20. **E** **V-HINT** **MMH** A 440.0-Hz tuning fork is sounded together with an out-of-tune guitar string, and a beat frequency of 3 Hz is heard. When the string is tightened, the frequency at which it vibrates increases, and the beat frequency is heard to decrease. What was the original frequency of the guitar string?

21. **E** Two ultrasonic sound waves combine and form a beat frequency that is in the range of human hearing for a healthy young person. The frequency of one of the ultrasonic waves is 70 kHz. What are (a) the smallest possible and (b) the largest possible value for the frequency of the other ultrasonic wave?

22. **E** **SSM** Two out-of-tune flutes play the same note. One produces a tone that has a frequency of 262 Hz, while the other produces 266 Hz. When a tuning fork is sounded together with the 262-Hz tone, a beat frequency of 1 Hz is produced. When the same tuning fork is sounded together with the 266-Hz tone, a beat frequency of 3 Hz is produced. What is the frequency of the tuning fork?

23. **M** **GO** **CHALK** Two cars have identical horns, each emitting a frequency of $f_s = 395$ Hz. One of the cars is moving with a speed of 12.0 m/s toward a bystander waiting at a corner, and the other car is parked. The speed of sound is 343 m/s. What is the beat frequency heard by the bystander?

24. **M** **GO** A sound wave is traveling in seawater, where the adiabatic bulk modulus and density are 2.31×10^9 Pa and 1025 kg/m^3, respectively. The wavelength of the sound is 3.35 m. A tuning fork is struck under water and vibrates at 440.0 Hz. What would be the beat frequency heard by an underwater swimmer?

25. **H** Two loudspeakers are mounted on a merry-go-round whose radius is 9.01 m. When stationary, the speakers both play a tone whose frequency is 100.0 Hz. As the drawing illustrates, they are situated at opposite ends of a diameter. The speed of sound is 343.00 m/s, and the merry-go-round revolves once every 20.0 s. What is the beat frequency that is detected by the listener when the merry-go-round is near the position shown?

Merry-go-round
(top view)

PROBLEM 25

Listener

Section 17.5 Transverse Standing Waves

26. **E** The fundamental frequency of a string fixed at both ends is 256 Hz. How long does it take for a wave to travel the length of this string?

27. **E** A string that is fixed at both ends has a length of 2.50 m. When the string vibrates at a frequency of 85.0 Hz, a standing wave with five loops is formed. (a) What is the wavelength of the waves that travel on the string? (b) What is the speed of the waves? (c) What is the fundamental frequency of the string?

28. **E** **SSM** The approach to solving this problem is similar to that taken in Multiple-Concept Example 5. On a cello, the string with the largest linear density (1.56×10^{-2} kg/m) is the C string. This string produces a fundamental frequency of 65.4 Hz and has a length of 0.800 m between the two fixed ends. Find the tension in the string.

29. **E** **GO** Two wires, each of length 1.2 m, are stretched between two fixed supports. On wire A there is a second-harmonic standing wave whose frequency is 660 Hz. However, the same frequency of 660 Hz is the third harmonic on wire B. Find the speed at which the individual waves travel on each wire.

30. **E** **SSM** Suppose that the strings on a violin are stretched with the same tension and each has the same length between its two fixed ends. The musical notes and corresponding fundamental frequencies of two of these strings are G (196.0 Hz) and E (659.3 Hz). The linear density of the E string is 3.47×10^{-4} kg/m. What is the linear density of the G string?

31. **E** To review the concepts that play roles in this problem, consult Multiple-Concept Example 5. Sometimes, when the wind blows across a long wire, a low-frequency "moaning" sound is produced. This sound arises because a standing wave is set up on the wire, like a standing wave on a guitar string. Assume that a wire (linear density = 0.0140 kg/m) sustains a tension of 323 N because the wire is stretched between two poles that are 7.60 m apart. The lowest frequency that an average, healthy human ear can detect is 20.0 Hz. What is the lowest harmonic number n that could be responsible for the "moaning" sound?

32. **E** A string has a linear density of 8.5×10^{-3} kg/m and is under a tension of 280 N. The string is 1.8 m long, is fixed at both ends, and is vibrating in the standing wave pattern shown in the drawing. Determine the (a) speed, (b) wavelength, and (c) frequency of the traveling waves that make up the standing wave.

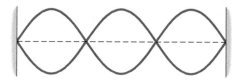

PROBLEM 32

33. **E** **GO** Multiple-Concept Example 5 deals with the same concepts as this problem. A 41-cm length of wire has a mass of 6.0 g. It is stretched between two fixed supports and is under a tension of 160 N. What is the fundamental frequency of this wire?

34. **M** **CHALK** **GO** A copper block is suspended from a wire, as in part 1 of the drawing. A container of mercury is then raised up around the block, as in part 2, so that 50.0% of the block's volume is submerged in the mercury. The density of copper is 8890 kg/m^3, and that of mercury is 13 600 kg/m^3. Find the ratio of the fundamental frequency of the wire in part 2 to the fundamental frequency of the wire in part 1.

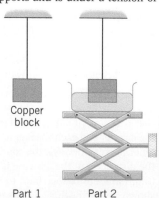

Copper
block

Part 1   Part 2
PROBLEM 34

35. **M** **V-HINT** The drawing shows two strings that have the same length and linear density. The left end of each string is attached to a wall, while the right end passes over a pulley and is connected to objects of different weights (W_A and W_B). Different standing waves are set up on each string, but their frequencies are the same. If W_A = 44 N, what is W_B?

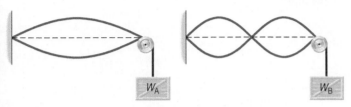

PROBLEM 35

36. **M** **SSM** The E string on an electric bass guitar has a length of 0.628 m and, when producing the note E, vibrates at a fundamental frequency of 41.2 Hz. Players sometimes add to their instruments a device called a "D-tuner." This device allows the E string to be used to produce the note D, which has a fundamental frequency of 36.7 Hz. The D-tuner works by extending the length of the string, keeping all other factors the same. By how much does a D-tuner extend the length of the E string?

37. **M** **GO** Standing waves are set up on two strings fixed at each end, as shown in the drawing. The two strings have the same tension and mass per unit length, but they differ in length by 0.57 cm. The waves on the shorter string propagate with a speed of 41.8 m/s, and the fundamental frequency of the shorter string is 225 Hz. Determine the beat frequency produced by the two standing waves.

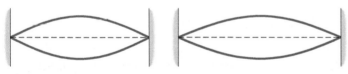

PROBLEM 37

38. **H** Review Conceptual Example 6 before attempting this problem. As the drawing shows, the length of a guitar string is 0.628 m. The frets are numbered for convenience. A performer can play a musical scale on a single string because the spacing *between the frets* is designed according to the following rule: When the string is pushed against any fret j, the fundamental frequency of the shortened string is larger by a factor of the twelfth root of two ($\sqrt[12]{2}$) than it is when the string is pushed against the fret $j - 1$. Assuming that the tension in the string is the same for any note, find the spacing **(a)** between fret 1 and fret 0 and **(b)** between fret 7 and fret 6.

PROBLEM 38

Section 17.6   Longitudinal Standing Waves,

Section 17.7   Complex Sound Waves

39. **E** **GO** A tube with a cap on one end, but open at the other end, has a fundamental frequency of 130.8 Hz. The speed of sound is 343 m/s **(a)** If the cap is removed, what is the new fundamental frequency of the tube? **(b)** How long is the tube?

40. **E** An organ pipe is open at both ends. It is producing sound at its third harmonic, the frequency of which is 262 Hz. The speed of sound is 343 m/s. What is the length of the pipe?

41. **E** The range of human hearing is roughly from twenty hertz to twenty kilohertz. Based on these limits and a value of 343 m/s for the speed of sound, what are the lengths of the longest and shortest pipes (open at both ends and producing sound at their fundamental frequencies) that you expect to find in a pipe organ?

42. **E** **MMH** The fundamental frequencies of two air columns are the same. Column A is open at both ends, while column B is open at only one end. The length of column A is 0.70 m. What is the length of column B?

43. **E** **SSM** A tube is open only at one end. A certain harmonic produced by the tube has a frequency of 450 Hz. The next higher harmonic has a frequency of 750 Hz. The speed of sound in air is 343 m/s. **(a)** What is the integer n that describes the harmonic whose frequency is 450 Hz? **(b)** What is the length of the tube?

44. **M** **V-HINT** A thin 1.2-m aluminum rod sustains a longitudinal standing wave with vibration antinodes at each end of the rod. There are no other antinodes. The density and Young's modulus of aluminum are, respectively, 2700 kg/m³ and 6.9×10^{10} N/m². What is the frequency of the rod's vibration?

45. **M** **CHALK** **SSM** A person hums into the top of a well and finds that standing waves are established at frequencies of 42, 70.0, and 98 Hz. The frequency of 42 Hz is not necessarily the fundamental frequency. The speed of sound is 343 m/s. How deep is the well?

46. **M** **GO** A vertical tube is closed at one end and open to air at the other end. The air pressure is 1.01×10^5 Pa. The tube has a length of 0.75 m. Mercury (mass density = 13 600 kg/m³) is poured into it to shorten the effective length for standing waves. What is the absolute pressure at the bottom of the mercury column, when the fundamental frequency of the shortened, air-filled tube is equal to the third harmonic of the original tube?

Additional Problems

Online

47. **E** A string is fixed at both ends and is vibrating at 130 Hz, which is its third harmonic frequency. The linear density of the string is 5.6×10^{-3} kg/m, and it is under a tension of 3.3 N. Determine the length of the string.

48. **E** **GO** One method for measuring the speed of sound uses standing waves. A cylindrical tube is open at both ends, and one end admits sound

from a tuning fork. A movable plunger is inserted into the other end at a distance L from the end of the tube where the tuning fork is. For a fixed frequency, the plunger is moved until the smallest value of L is measured that allows a standing wave to be formed. Suppose that the tuning fork produces a 485-Hz tone, and that the smallest value observed for L is 0.264 m. What is the speed of sound in the gas in the tube?

49. **E** **MMH** The drawing graphs a string on which two pulses (half up and half down) are traveling at a constant speed of 1 cm/s at *t* = 0 s. Using the principle of linear superposition, draw the shape of the string at *t* = 1 s, 2 s, 3 s, and 4 s.

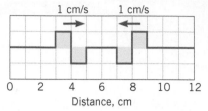

PROBLEM 49

50. **E** **SSM** The A string on a string bass vibrates at a fundamental frequency of 55.0 Hz. If the string's tension were increased by a factor of four, what would be the new fundamental frequency?

51. **M** **V-HINT** The two speakers in the drawing are vibrating in phase, and a listener is standing at point P. Does constructive or destructive interference occur at P when the speakers produce sound waves whose frequency is **(a)** 1466 Hz and **(b)** 977 Hz? Justify your answers with appropriate calculations. Take the speed of sound to be 343 m/s.

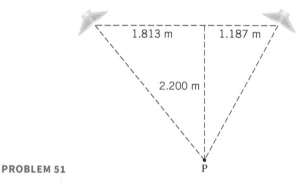

PROBLEM 51

52. **H** **SSM** The arrangement in the drawing shows a block (mass = 15.0 kg) that is held in position on a frictionless incline by a cord (length = 0.600 m). The mass per unit length of the cord is 1.20×10^{-2} kg/m, so the mass of the cord is negligible compared to the mass of the block. The cord is being vibrated at a frequency of 165 Hz (vibration source not shown in the drawing). What are the values of the angle θ between 15.0° and 90.0° at which a standing wave exists on the cord?

PROBLEM 52

53. **M** **GO** A flautist is playing a flute as discussed in Example 7, but now the temperature is 305 K instead of 293 K. As a result, the speed of sound is no longer 343 m/s. Therefore, with the length calculated in Example 7, the note middle C does not have the proper fundamental frequency of 261.6 Hz. In other words, the flute is out of tune. To adjust the tuning, the flautist can alter the flute's length by changing the extent to which the head joint (see **Figure 17.21**) is inserted into the main stem of the instrument. To what length must the flute be adjusted to play middle C at its proper frequency?

Physics in Biology, Medicine, and Sports

54. **E** **BIO** **SSM** **17.6** Sound enters the ear, travels through the auditory canal, and reaches the eardrum. The auditory canal is approximately a tube open at only one end. The other end is closed by the eardrum. A typical length for the auditory canal in an adult is about 2.9 cm. The speed of sound is 343 m/s. What is the fundamental frequency of the canal? (Interestingly, the fundamental frequency is in the frequency range where human hearing is most sensitive.)

55. **E** **BIO** **17.6** Divers working in underwater chambers at great depths must deal with the danger of nitrogen narcosis (the "bends"), in which nitrogen dissolves into the blood at toxic levels. One way to avoid this danger is for divers to breathe a mixture containing only helium and oxygen. Helium, however, has the effect of giving the voice a high-pitched quality, like that of Donald Duck's voice. To see why this occurs, assume for simplicity that the voice is generated by the vocal cords vibrating above a gas-filled cylindrical tube that is open only at one end. The quality of the voice depends on the harmonic frequencies generated by the tube; larger frequencies lead to higher-pitched voices. Consider two such tubes at 20 °C. One is filled with air, in which the speed of sound is 343 m/s. The other is filled with helium, in which the speed of sound is 1.00×10^3 m/s. To see the effect of helium on voice quality, calculate the ratio of the *n*th natural frequency of the helium-filled tube to the *n*th natural frequency of the air-filled tube.

56. **E** **BIO** **17.6** Human adults are most sensitive to sound waves that satisfy a standing wave condition for certain frequencies. Example 8 showed the fundamental frequency for humans is 3700 Hz.

How does this frequency compare to the fundamental frequency of an elephant's ear, whose auditory canal is 18 cm in length? Again, treat the auditory canal as a tube open only at one end. Use 340 m/s for the speed of sound. Compared to humans, you should find that an elephant's ears are sensitive to much lower frequency sounds.

57. **E** **BIO** **17.4** Binaural beat therapy is a form of sound therapy used to treat anxiety, stress, and sleep disorders. A patient wears headphones that apply slightly different frequencies of sound to the left and right ears. The frequencies are chosen such that the auditory part of the brain only hears a single tone, while the frequencies of the impulses traveling along the left and right auditory nerves are slightly different. There are multiple types of brain waves: alpha, beta, gamma, delta, and theta. Each of these operate over a slightly different frequency range. Synchronizing the beat frequency to those of the different types of brain waves may offer therapeutic benefits, although clinical trials have proved inconclusive. A person undergoing binaural beat therapy is wearing headphones that deliver a single tone with a frequency of 1.200×10^3 Hz into the left ear and a single tone with a frequency of 1.208×10^3 Hz into the right ear. Her therapy is targeting her brain's alpha waves (7 to 13 Hz) in an effort to promote relaxation. How many hours should she listen in order to complete 1.000×10^3 beat cycles?

58. **E** **BIO** **17.4** Sound passing through the vocal cords is modified by a set of resonating cavities (see the figure). These include the pharnyx, oral cavity (mouth), nasal cavity, and paranasal sinuses. The so-called nasal sounds of which examples in English would be "m," "n," and "ng," are created by having the air expelled through the vocal cords leave almost entirely through the nose and not the mouth. The pharynx and nasal cavity combine to form a resonating chamber that is open at one end (see the blue outline in the figure). **(a)** If the overall length of this chamber is 22.0 cm, what is the chamber's fundamental frequency? **(b)** What is the frequency of the chamber's third harmonic? The air being expelled through the vocal cords has been warmed by the body's temperature, so assume the speed of sound in air for these conditions is 353 m/s.

Science X™/Dr. Stephan A. Reber

PROBLEM 59

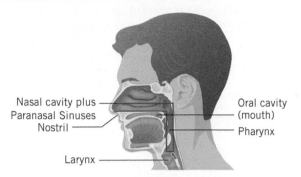

Nasal cavity plus
Paranasal Sinuses
Nostril

Oral cavity
(mouth)

Pharynx

Larynx

PROBLEM 58

59. **E** **BIO** **17.6** When a male American alligator is attempting to attract a female, he engages in a unique courtship ritual known as *water dancing*. With his head and tail raised up above the water, he lowers his body until his upper back is just below the water's surface. With his mouth closed, he then exhales while emitting a low frequency tone. When the frequency of the sound he emits matches the standing wave condition for his chest cavity, resonance occurs, and the water just above his upper back vibrates violently, producing quite the auditory and visual display (see the photo). Assume the alligator emits the fundamental frequency for resonance (58.3 Hz) and that his chest cavity can be approximated as a tube open only at one end. Use this information to estimate the length of his chest cavity, and take the speed of sound to be 343 m/s.

60. **M** **BIO** **17.3** Scientists have recently learned that elephants emit infrasonic sound waves with frequencies in the range of 15 to 35 Hz, as a means of communicating with each other. This technique was discovered after researchers observed large herds of African elephants of the savannah communicating over great distances, sometimes as far as 4 km. Infrasonic sounds were detected right before the occurrence of synchronized movements of the entire herd. Elephants living in the Central African rainforests also displayed similar behavior, even though the dense forest vegetation made it impossible for the elephants at the rear of the herd to see the leaders at the front. Diffraction allows the low frequency sound waves to effectively bend around the trees and vegetation, spanning the size of the herd (see the photo). While the range in the forest is reduced compared to the savannah, communications up to 3 km were observed. Imagine an elephant emitting a low frequency sound wave at 27.0 Hz. The wave passes through an opening in a grove of trees and foliage that is 25.0 m in diameter and approximately circular. How far has the wave spread out after traveling 1.00 km from the trees? Assume the speed of sound in air is 343 m/s.

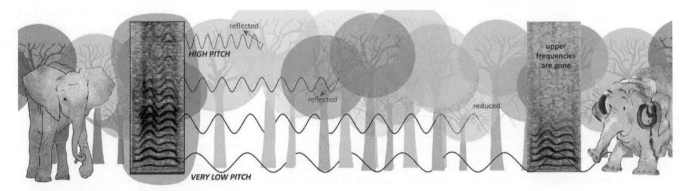

This graphic shows how lower frequency sounds are able to travel farther than higher frequency sounds, largely due to greater diffractive bending. Source: Elephant Listening Project, Cornell Labs. © Cornell University.

PROBLEM 60

Concepts and Calculations Problems

Online

Diffraction is the bending of a traveling wave around an obstacle or around the edges of an opening, and is one of the consequences of superposition. Problem 61 compares diffraction in two different media and reviews some of the fundamental properties of sound

waves. Problem 62 deals with standing waves of sound in a gas. One of the factors that affects the formation of standing waves is the speed at which the individual waves travel. Problem 62 reviews how the speed of sound depends on the properties of the gas.

61. **M** **CHALK** **SSM** A sound wave with a frequency of 15 kHz emerges through a circular opening that has a diameter of 0.20 m. *Concepts:* (i) The diffraction angle for a wave emerging through a circular opening is given by $\sin \theta = 1.22 \, \lambda/D$, where λ is the wavelength of the sound and D is the diameter of the opening. What is meant by the diffraction angle? (ii) How is the wavelength related to the frequency of the sound? (iii) Is the wavelength of the sound in air greater than, smaller than, or equal to the wavelength in water? Why? (Note: The speed of sound in air is 343 m/s and the speed of sound in water is 1482 m/s.) (iv) Is the diffraction angle of the sound in air greater than, smaller than, or equal to the diffraction angle in water? Explain. *Calculations:* Find the diffraction angle θ when the sound travels **(a)** in air and **(b)** in water.

62. **M** **CHALK** Two tubes of gas are identical and are open only at one end. One tube contains neon (Ne) and the other krypton (Kr). Both are monatomic gases, have the same temperature, and may be assumed to be ideal gases. The fundamental frequency of the tube containing neon is 481 Hz. *Concepts:* (i) For a gas-filled tube open only at one end, the fundamental frequency ($n = 1$) is $f_1 = v/(4L)$, where v is the speed of sound and L is the length of the tube. How is the speed related to the properties of the gas? (ii) All of the factors that affect the speed of sound in this problem are the same except for the atomic masses, which are given by 20.180 u for neon, and 83.80 u for krypton. Is the speed of sound in krypton greater than, smaller than, or equal to the speed of sound in neon? Why? (iii) Is the fundamental frequency of the tube containing krypton greater than, less than, or equal to the fundamental frequency of the tube containing neon? Explain. *Calculations:* What is the fundamental frequency of the tube containing krypton?

Team Problems

Online

63. **M** **T** **WS** **Finding the Dead Frequencies.** A football field is a rectangle of width 53.3 yards (48.8 m) and length 100 yards (91.4 m). The goal lines are on the extremes (100 yards apart) with the 50-yard line at the center. Yard lines are drawn every 5 yards and labeled every 10 yards, counting up from each goal line to the 50-yard line. To the left and right of the midfield line (that is, the 50-yard line) are the 40-yard lines, then the 30-yard lines, and so on. A new announcer box was constructed near the football field, centered on the 50-yard line, and 20.0 yards away from the sideline. Two loudspeakers are mounted on the upper left and right corners of the box, which is 12.0 yards wide. You and your team have been assigned to identify all frequencies between 0 Hz and 1.25 kHz for which dead spots occur due to destructive interference at the 35.0-yard markers on the opposite sideline. Assume that the speakers are in phase, and that the speed of sound in air is 343 m/s. You can neglect the height of the speakers above the ground.

64. **E** **T** **WS** **Finding a Faulty Valve.** You and your team are on a cargo ship and have encountered a problem. A bundle of 15 vertical pipes leads from the deck to the engine room. Each pipe is 15.0 cm in diameter, is 10.5 m long, and has a valve on its lower end, in a difficult to reach place. Unfortunately, one of the valves will not open. You try to look down the pipes using a flashlight, but cannot tell which pipe has the faulty valve. After some thinking, you get the idea to use a function generator with a speaker, a microphone, and some audio software to identify the pipe with the closed valve. The function generator produces sound at a single frequency (tone) that can be swept from 50.0 Hz to 20.0 kHz, and the speaker can be placed near the exposed open ends to send sound waves down the tubes. The microphone can then be placed near the open end to sense the intensity of the sound, which is the sum of the source sound plus the response of the tube, that is, it will be able to detect the resonances, or natural frequencies, of the tubes. The audio software will then reveal the precise frequencies at which these resonances occur. Assuming a pipe with an open valve behaves like a cylindrical tube with two open ends, and a pipe with a closed valve behaves like a cylindrical tube with only one open end, what are the first three resonant frequencies above 50.0 Hz that you should detect for a pipe with **(a)** an open valve, and **(b)** a closed valve? Assume that the speed of sound in air is 343 m/s.

65. **M** **T** **An Acoustic Remote Control.** You and your team are exploring an antiquated research facility in the mountains of southern Argentina that had been abandoned in the 1960s. You come to a giant locked door that has no visible handles or actuators, but you find a handheld device nearby that has two buttons, labeled "Open" and "Close." It looks like some kind of crude remote control, but when you push the buttons they just click and nothing happens. You open the device's top cover and inspect it. The internal mechanism resembles an old acoustic remote control called the "Space Command 600" that your parents had for their ancient TV. Pushing a button on the remote actuated a small hammer on the inside that struck the end of an aluminum rod, about 2 or 3 cm in length. The rod vibrated and emitted an ultrasonic sound wave that actuated an electrical circuit in the TV that was sensitive to that frequency. The TV remote had three buttons, and therefore three rods that vibrated at different frequencies. The first frequency turned the TV on and off, the second made the tuning dial click to the next station, and the third made the dial turn in the opposite direction. You pull off the cover of the device and find that it has places for two ¼-inch diameter rods, but both are missing. However, written on the inside of the cover of the device is the following: "Open = 95.50 kHz" and "Close = 102.5 kHz." Your team members search and eventually find a long piece of ¼-inch diameter aluminum rod. **(a)** To what lengths must you cut the rod in order to get the remote to work properly (i.e., so that the fundamental frequencies of the rods match those utilized by the remote)? **(b)** Suppose you had instead found a titanium rod. What lengths would be required in that case? (Young's moduli are $Y_{Al} = 6.9 \times 10^{10}$ N/m^2 and $Y_{Ti} = 1.2 \times 10^{11}$ N/m^2; the mass densities are $\rho_{Al} = 2700$ kg/m^3 and $\rho_{Ti} = 4500$ kg/m^3.)

66. **M** **T** **A Tunable Inclined Plane.** You and your team are designing a tunable, single-tone, acoustic emitter. One end of a string is connected to a post on the top of a frictionless inclined plane. The string supports a mass at rest below, on the surface of the plane. The angle θ of the incline relative to the horizontal is adjustable. The length of the string is $L = 35.0$ cm, and the mass per unit length of the string is $\lambda = 5.28 \times 10^{-3}$ kg/m. **(a)** When $\theta = 17.0°$, the plucked string is supposed to emit a fundamental frequency of $f = 212$ Hz. What value of the mass (M) is needed under these conditions? **(b)** What is the full frequency range of the device (i.e., by changing the angle)? **(c)** With the mass calculated in part (a), to what value should you set θ so that the fundamental frequency is 330 Hz?

Powers of Ten and Scientific Notation

In science, very large and very small decimal numbers are conveniently expressed in terms of powers of ten, some of which are listed below:

$$10^3 = 10 \times 10 \times 10 = 1000 \qquad 10^{-3} = \frac{1}{10 \times 10 \times 10}$$
$$= 0.001$$

$$10^2 = 10 \times 10 = 100 \qquad 10^{-2} = \frac{1}{10 \times 10} = 0.01$$

$$10^1 = 10 \qquad 10^{-1} = \frac{1}{10} = 0.1$$

$$10^0 = 1$$

Using powers of ten, we can write the radius of the earth in the following way, for example:

$$\text{Earth radius} = 6\ 380\ 000\ \text{m} = 6.38 \times 10^6\ \text{m}$$

The factor of ten raised to the sixth power is ten multiplied by itself six times, or one million, so the earth's radius is 6.38 million meters. Alternatively, the factor of ten raised to the sixth power indicates that the decimal point in the term 6.38 is to be moved six places *to the right* to obtain the radius as a number without powers of ten.

For numbers less than one, negative powers of ten are used. For instance, the Bohr radius of the hydrogen atom is

$$\text{Bohr radius} = 0.000\ 000\ 000\ 0529\ \text{m} = 5.29 \times 10^{-11}\ \text{m}$$

The factor of ten raised to the minus eleventh power indicates that the decimal point in the term 5.29 is to be moved eleven places *to the left* to obtain the radius as a number without powers of ten. Numbers expressed with the aid of powers of ten are said to be in ***scientific notation.***

Calculations that involve the multiplication and division of powers of ten are carried out as in the following examples:

$$(2.0 \times 10^6)(3.5 \times 10^3) = (2.0 \times 3.5) \times 10^{6+3} = 7.0 \times 10^9$$

$$\frac{9.0 \times 10^7}{2.0 \times 10^4} = \left(\frac{9.0}{2.0}\right) \times 10^7 \times 10^{-4}$$

$$= \left(\frac{9.0}{2.0}\right) \times 10^{7-4} = 4.5 \times 10^3$$

The general rules for such calculations are

$$\frac{1}{10^n} = 10^{-n} \tag{A-1}$$

$$10^n \times 10^m = 10^{n+m} \qquad \text{(Exponents added)} \tag{A-2}$$

$$\frac{10^n}{10^m} = 10^{n-m} \qquad \text{(Exponents subtracted)} \tag{A-3}$$

where n and m are any positive or negative number.

Scientific notation is convenient because of the ease with which it can be used in calculations. Moreover, scientific notation provides a convenient way to express the significant figures in a number, as Appendix B discusses.

APPENDIX B

Significant Figures

The number of *significant figures* in a number is the number of digits whose values are known with certainty. For instance, a person's height is measured to be 1.78 m, with the measurement error being in the third decimal place. All three digits are known with certainty, so that the number contains three significant figures. If a zero is given as the last digit to the right of the decimal point, the zero is presumed to be significant. Thus, the number 1.780 m contains four significant figures. As another example, consider a distance of 1500 m. This number contains only two significant figures, the one and the five. The zeros immediately to the left of the unexpressed decimal point are not counted as significant figures. However, zeros located between significant figures are significant, so a distance of 1502 m contains four significant figures.

Scientific notation is particularly convenient from the point of view of significant figures. Suppose it is known that a certain distance is fifteen hundred meters, to four significant figures. Writing the number as 1500 m presents a problem because it implies that only two significant figures are known. In contrast, the scientific notation of 1.500×10^3 m has the advantage of indicating that the distance is known to four significant figures.

When two or more numbers are used in a calculation, the number of significant figures in the answer is limited by the number of significant figures in the original data. For instance, a rectangular garden with sides of 9.8 m and 17.1 m has an area of (9.8 m)(17.1 m). A calculator gives 167.58 m² for this product. However, one of the original lengths is known only to two significant figures, so the final answer is limited to only two significant figures and should be rounded off to 170 m². In general, *when numbers are multiplied or divided, the number of significant figures in the final answer equals the smallest number of significant figures in any of the original factors.*

The number of significant figures in the answer to an addition or a subtraction is also limited by the original data. Consider the total distance along a biker's trail that consists of three segments with the distances shown as follows:

	2.5 km
	11 km
	5.26 km
Total	18.76 km

The distance of 11 km contains no significant figures to the right of the decimal point. Therefore, neither does the sum of the three distances, and the total distance should not be reported as 18.76 km. Instead, the answer is rounded off to 19 km. In general, *when numbers are added or subtracted, the last significant figure in the answer occurs in the last column (counting from left to right) containing a number that results from a combination of digits that are all significant.* In the answer of 18.76 km, the eight is the sum of 2 + 1 + 5, each digit being significant. However, the seven is the sum of 5 + 0 + 2, and the zero is not significant, since it comes from the 11-km distance, which contains no significant figures to the right of the decimal point.

Algebra

C.1 Proportions and Equations

Physics deals with physical variables and the relations between them. Typically, variables are represented by the letters of the English and Greek alphabets. Sometimes, the relation between variables is expressed as a proportion or inverse proportion. Other times, however, it is more convenient or necessary to express the relation by means of an equation, which is governed by the rules of algebra.

If two variables are ***directly proportional*** and one of them doubles, then the other variable also doubles. Similarly, if one variable is reduced to one-half its original value, then the other is also reduced to one-half its original value. In general, if x is directly proportional to y, then increasing or decreasing one variable by a given factor causes the other variable to change in the same way by the same factor. This kind of relation is expressed as $x \propto y$, where the symbol $\propto$ means "is proportional to."

Since the proportional variables x and y always increase and decrease by the same factor, the ratio of x to y must have a constant value, or $x/y = k$, where k is a constant, independent of the values for x and y. Consequently, a proportionality such as $x \propto y$ can also be expressed in the form of an equation: $x = ky$. The constant k is referred to as a ***proportionality constant.***

If two variables are ***inversely proportional*** and one of them increases by a given factor, then the other decreases by the same factor. An inverse proportion is written as $x \propto 1/y$. This kind of proportionality is equivalent to the following equation: $xy = k$, where k is a proportionality constant, independent of x and y.

C.2 Solving Equations

Some of the variables in an equation typically have known values, and some do not. It is often necessary to solve the equation so that a variable whose value is unknown is expressed in terms of the known quantities. ***In the process of solving an equation, it is permissible to manipulate the equation in any way, as long as a change made on one side of the equals sign is also made on the other side.*** For example,

consider the equation $v = v_0 + at$. Suppose values for v, v_0, and a are available, and the value of t is required. To solve the equation for t, we begin by subtracting v_0 from *both* sides:

$$
\begin{array}{rcl}
v & = & v_0 + at \\
-v_0 & = & -v_0 \\
\hline
v - v_0 & = & at
\end{array}
$$

Next, we divide both sides of $v - v_0 = at$ by the quantity a:

$$\frac{v - v_0}{a} = \frac{at}{a} = (1)t$$

On the right side, the a in the numerator divided by the a in the denominator equals one, so that

$$t = \frac{v - v_0}{a}$$

It is always possible to check the correctness of the algebraic manipulations performed in solving an equation by substituting the answer back into the original equation. In the previous example, we substitute the answer for t into $v = v_0 + at$:

$$v = v_0 + a\left(\frac{v - v_0}{a}\right) = v_0 + (v - v_0) = v$$

The result $v = v$ implies that our algebraic manipulations were done correctly.

Algebraic manipulations other than addition, subtraction, multiplication, and division may play a role in solving an equation. The same basic rule applies, however: Whatever is done to the left side of an equation must also be done to the right side. As another example, suppose it is necessary to express v_0 in terms of v, a, and x, where $v^2 = v_0^2 + 2ax$. By subtracting $2ax$ from both sides, we isolate v_0^2 on the right:

$$
\begin{array}{rcl}
v^2 & = & v_0^2 + 2ax \\
-2ax & = & -2ax \\
\hline
v^2 - 2ax & = & v_0^2
\end{array}
$$

To solve for v_0, we take the positive and negative square root of *both* sides of $v^2 - 2ax = v_0^2$:

$$v_0 = \pm\sqrt{v^2 - 2ax}$$

C.3 Simultaneous Equations

When more than one variable in a single equation is unknown, additional equations are needed if solutions are to be found for all of the unknown quantities. Thus, the equation $3x + 2y = 7$ cannot be solved by itself to give unique values for both x and y. However, if x and y also (i.e., simultaneously) obey the equation $x - 3y = 6$, then both unknowns can be found.

There are a number of methods by which such simultaneous equations can be solved. One method is to solve one equation for x in terms of y and substitute the result into the other equation to obtain an expression containing only the single unknown variable y. The equation $x - 3y = 6$, for instance, can be solved for x by adding $3y$ to each side, with the result that $x = 6 + 3y$. The substitution of this expression for x into the equation $3x + 2y = 7$ is shown below:

$$3x + 2y = 7$$
$$3(6 + 3y) + 2y = 7$$
$$18 + 9y + 2y = 7$$

We find, then, that $18 + 11y = 7$, a result that can be solved for y:

$$\begin{aligned} 18 + 11y &= \quad 7 \\ -18 \qquad &\quad -18 \\ \hline 11y &= -11 \end{aligned}$$

Dividing both sides of this result by 11 shows that $y = -1$. The value of $y = -1$ can be substituted in either of the original equations to obtain a value for x:

$$\begin{aligned} x - 3y &= 6 \\ x - 3(-1) &= 6 \\ x + 3 &= 6 \\ -3 \qquad\quad &\quad -3 \\ \hline x &= 3 \end{aligned}$$

Equations occur in physics that include the square of a variable. Such equations are said to be *quadratic* in that variable, and often can be put into the following form:

$$ax^2 + bx + c = 0 \qquad \text{(C-1)}$$

where a, b, and c are constants independent of x. This equation can be solved to give the **quadratic formula,** which is

$$x = \frac{-b \pm \sqrt{b^2 - 4ac}}{2a} \qquad \text{(C-2)}$$

The $\pm$ in the quadratic formula indicates that there are two solutions. For instance, if $2x^2 - 5x + 3 = 0$, then $a = 2$, $b = -5$, and $c = 3$. The quadratic formula gives the two solutions as follows:

Solution 1:
Plus sign

$$x = \frac{-b + \sqrt{b^2 - 4ac}}{2a}$$

$$= \frac{-(-5) + \sqrt{(-5)^2 - 4(2)(3)}}{2(2)}$$

$$= \frac{+5 + \sqrt{1}}{4} = \frac{3}{2}$$

Solution 2:
Minus sign

$$x = \frac{-b - \sqrt{b^2 - 4ac}}{2a}$$

$$= \frac{-(-5) - \sqrt{(-5)^2 - 4(2)(3)}}{2(2)}$$

$$= \frac{+5 - \sqrt{1}}{4} = 1$$

Exponents and Logarithms

Appendix A discusses powers of ten, such as 10^3, which means ten multiplied by itself three times, or $10 \times 10 \times 10$. The three is referred to as an **exponent.** The use of exponents extends beyond powers of ten. In general, the term y^n means the factor y is multiplied by itself n times. For example, y^2, or y squared, is familiar and means $y \times y$. Similarly, y^5 means $y \times y \times y \times y \times y$.

The rules that govern algebraic manipulations of exponents are the same as those given in Appendix A (see Equations A-1, A-2, and A-3) for powers of ten:

$$\frac{1}{y^n} = y^{-n} \tag{D-1}$$

$$y^n y^m = y^{n+m} \quad \text{(Exponents added)} \tag{D-2}$$

$$\frac{y^n}{y^m} = y^{n-m} \quad \text{(Exponents subtracted)} \tag{D-3}$$

To the three rules above we add two more that are useful. One of these is

$$y^n z^n = (yz)^n \tag{D-4}$$

The following example helps to clarify the reasoning behind this rule:

$$3^2 5^2 = (3 \times 3)(5 \times 5) = (3 \times 5)(3 \times 5) = (3 \times 5)^2$$

The other additional rule is

$$(y^n)^m = y^{nm} \quad \text{(Exponents multiplied)} \tag{D-5}$$

To see why this rule applies, consider the following example:

$$(5^2)^3 = (5^2)(5^2)(5^2) = 5^{2+2+2} = 5^{2 \times 3}$$

Roots, such as a square root or a cube root, can be represented with fractional exponents. For instance,

$$\sqrt{y} = y^{1/2} \quad \text{and} \quad \sqrt[3]{y} = y^{1/3}$$

In general, the nth root of y is given by

$$\sqrt[n]{y} = y^{1/n} \tag{D-6}$$

The rationale for Equation D-6 can be explained using the fact that $(y^n)^m = y^{nm}$. For instance, the fifth root of y is the number that, when multiplied by itself five times, gives back y. As shown below, the term $y^{1/5}$ satisfies this definition:

$$(y^{1/5})(y^{1/5})(y^{1/5})(y^{1/5})(y^{1/5}) = (y^{1/5})^5 = (y^{1/5})^{\times 5} = y$$

Logarithms are closely related to exponents. To see the connection between the two, note that it is possible to express any number y as another number B raised to the exponent x. In other words,

$$y = B^x \tag{D-7}$$

The exponent x is called the **logarithm** of the number y. The number B is called the **base number.** One of two choices for the base number is usually used. If $B = 10$, the logarithm is known as the *common logarithm,* for which the notation "log" applies:

Common logarithm $\quad y = 10^x \quad \text{or} \quad x = \log y \tag{D-8}$

If $B = e = 2.718 \ldots$, the logarithm is referred to as the *natural logarithm,* and the notation "ln" is used:

Natural logarithm $\quad y = e^z \quad \text{or} \quad z = \ln y \tag{D-9}$

The two kinds of logarithms are related by

$$\ln y = 2.3026 \log y \tag{D-10}$$

Both kinds of logarithms are often given on calculators.

The logarithm of the product or quotient of two numbers A and C can be obtained from the logarithms of the individual numbers according to the rules below. These rules are illustrated here for natural logarithms, but they are the same for any kind of logarithm.

$$\ln(AC) = \ln A + \ln C \tag{D-11}$$

$$\ln\left(\frac{A}{C}\right) = \ln A - \ln C \tag{D-12}$$

Thus, the logarithm of the product of two numbers is the sum of the individual logarithms, and the logarithm of the quotient of two numbers is the difference between the individual logarithms. Another useful rule concerns the logarithm of a number A raised to an exponent n:

$$\ln A^n = n \ln A \tag{D-13}$$

Rules D-11, D-12, and D-13 can be derived from the definition of the logarithm and the rules governing exponents.

Geometry and Trigonometry

E.1 Geometry

Angles

Two angles are equal if

1. They are vertical angles (see **Figure E1**).
2. Their sides are parallel (see **Figure E2**).

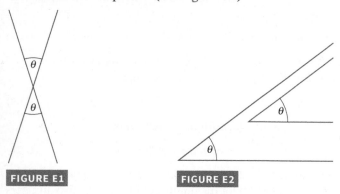

FIGURE E1

FIGURE E2

3. Their sides are mutually perpendicular (see **Figure E3**).

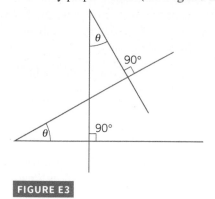

FIGURE E3

Triangles

1. The **sum of the angles** of any triangle is 180° (see **Figure E4**).

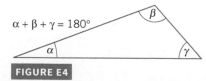

$$\alpha + \beta + \gamma = 180°$$

FIGURE E4

2. A **right triangle** has one angle that is 90°.
3. An **isosceles triangle** has two sides that are equal.
4. An **equilateral triangle** has three sides that are equal. Each angle of an equilateral triangle is 60°.

5. Two triangles are **similar** if two of their angles are equal (see **Figure E5**). The corresponding sides of similar triangles are proportional to each other:

$$\frac{a_1}{a_2} = \frac{b_1}{b_2} = \frac{c_1}{c_2}$$

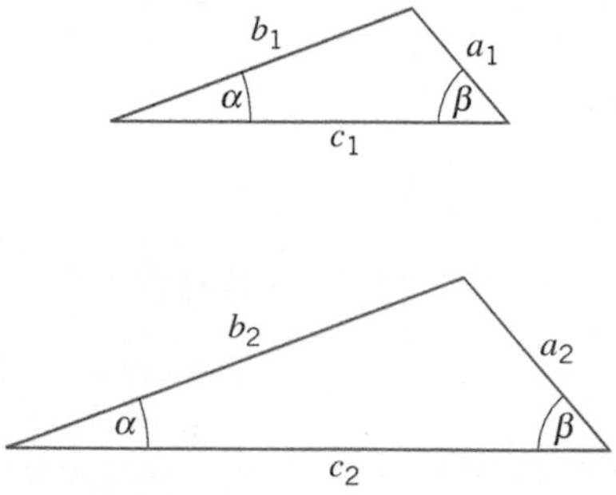

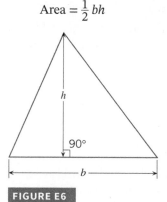

FIGURE E5

6. Two similar triangles are **congruent** if they can be placed on top of one another to make an exact fit.

Circumferences, Areas, and Volumes of Some Common Shapes

1. Triangle of base b and altitude h (see **Figure E6**):

$$\text{Area} = \frac{1}{2} bh$$

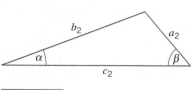

FIGURE E6

2. Circle of radius r:

$$\text{Circumference} = 2\pi r$$

$$\text{Area} = \pi r^2$$

3. Sphere of radius r:

$$\text{Surface area} = 4\pi r^2$$

$$\text{Volume} = \frac{4}{3}\pi r^3$$

4. Right circular cylinder of radius r and height h (see **Figure E7**):

$$\text{Surface area} = 2\pi r^2 + 2\pi rh$$
$$\text{Volume} = \pi r^2 h$$

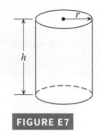

FIGURE E7

E.2 Trigonometry

Basic Trigonometric Functions

1. For a right triangle, the sine, cosine, and tangent of an angle θ are defined as follows (see **Figure E8**):

$$\sin\theta = \frac{\text{Side opposite }\theta}{\text{Hypotenuse}} = \frac{h_o}{h}$$

$$\cos\theta = \frac{\text{Side adjacent }\theta}{\text{Hypotenuse}} = \frac{h_a}{h}$$

$$\tan\theta = \frac{\text{Side opposite }\theta}{\text{Side adjacent to }\theta} = \frac{h_o}{h_a}$$

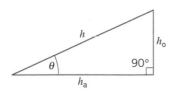

FIGURE E8

2. The secant ($\sec\theta$), cosecant ($\csc\theta$), and cotangent ($\cot\theta$) of an angle θ are defined as follows:

$$\sec\theta = \frac{1}{\cos\theta} \qquad \csc\theta = \frac{1}{\sin\theta} \qquad \cot\theta = \frac{1}{\tan\theta}$$

Triangles and Trigonometry

1. The ***Pythagorean theorem*** states that the square of the hypotenuse of a right triangle is equal to the sum of the squares of the other two sides (see **Figure E8**):

$$h^2 = h_o{}^2 + h_a{}^2$$

2. The ***law of cosines*** and the ***law of sines*** apply to any triangle, not just a right triangle, and they relate the angles and the lengths of the sides (see **Figure E9**):

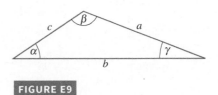

FIGURE E9

Law of cosines $c^2 = a^2 + b^2 - 2ab\cos\gamma$

Law of sines $\dfrac{a}{\sin\alpha} = \dfrac{b}{\sin\beta} = \dfrac{c}{\sin\gamma}$

Other Trigonometric Identities

1. $\sin(-\theta) = -\sin\theta$
2. $\cos(-\theta) = \cos\theta$
3. $\tan(-\theta) = -\tan\theta$
4. $(\sin\theta)/(\cos\theta) = \tan\theta$
5. $\sin^2\theta + \cos^2\theta = 1$
6. $\sin(\alpha \pm \beta) = \sin\alpha\cos\beta \pm \cos\alpha\sin\beta$

   If $\alpha = 90°$, $\sin(90° \pm \beta) = \cos\beta$

   If $\alpha = \beta$, $\sin 2\beta = 2\sin\beta\cos\beta$

7. $\cos(\alpha \pm \beta) = \cos\alpha\cos\beta \mp \sin\alpha\sin\beta$

   If $\alpha = 90°$, $\cos(90° \pm \beta) = \mp\sin\beta$

   If $\alpha = \beta$, $\cos 2\beta = \cos^2\beta - \sin^2\beta = 1 - 2\sin^2\beta$

Selected Isotopes[a]

Atomic No. Z	Element	Symbol	Atomic Mass No. A	Atomic Mass (u)	% Abundance, or Decay Mode if Radioactive	Half-Life (if Radioactive)
0	(Neutron)	n	1	1.008 665	β^-	10.37 min
1	Hydrogen	H	1	1.007 825	99.985	
	Deuterium	D	2	2.014 102	0.015	
	Tritium	T	3	3.016 050	β^-	12.33 yr
2	Helium	He	3	3.016 030	0.000 138	
			4	4.002 603	≈ 100	
3	Lithium	Li	6	6.015 121	7.5	
			7	7.016 003	92.5	
4	Beryllium	Be	7	7.016 928	EC, γ	53.29 days
			9	9.012 182	100	
5	Boron	B	10	10.012 937	19.9	
			11	11.009 305	80.1	
6	Carbon	C	11	11.011 432	β^+, EC	20.39 min
			12	12.000 000	98.90	
			13	13.003 355	1.10	
			14	14.003 241	β^-	5730 yr
7	Nitrogen	N	13	13.005 738	β^+, EC	9.965 min
			14	14.003 074	99.634	
			15	15.000 108	0.366	
8	Oxygen	O	15	15.003 065	β^+, EC	122.2 s
			16	15.994 915	99.762	
			18	17.999 160	0.200	
9	Fluorine	F	18	18.000 937	EC, β^+	1.8295 h
			19	18.998 403	100	
10	Neon	Ne	20	19.992 435	90.51	
			22	21.991 383	9.22	
11	Sodium	Na	22	21.994 434	β^+, EC, γ	2.602 yr
			23	22.989 767	100	
			24	23.990 961	β^-, γ	14.659 h
12	Magnesium	Mg	24	23.985 042	78.99	
13	Aluminum	Al	27	26.981 539	100	
14	Silicon	Si	28	27.976 927	92.23	
			31	30.975 362	β^-, γ	2.622 h
15	Phosphorus	P	31	30.973 762	100	
			32	31.973 907	β^-	14.282 days
16	Sulfur	S	32	31.972 070	95.02	
			35	34.969 031	β^-	87.51 days
17	Chlorine	Cl	35	34.968 852	75.77	
			37	36.965 903	24.23	
18	Argon	Ar	40	39.962 384	99.600	
19	Potassium	K	39	38.963 707	93.2581	
			40	39.963 999	β^-, EC, γ	1.277×10^9 yr

[a]Data for atomic masses are taken from *Handbook of Chemistry and Physics,* 66th ed., CRC Press, Boca Raton, FL. The masses are those for the neutral atom, including the Z electrons. Data for percent abundance, decay mode, and half-life are taken from E. Browne and R. Firestone, *Table of Radioactive Isotopes,* V. Shirley, Ed., Wiley, New York, 1986. α = alpha particle emission, β^- = negative beta emission, β^+ = positron emission, γ = γ-ray emission, EC = electron capture.

Appendix F Selected Isotopes (*continued*)

Atomic No. Z	Element	Symbol	Atomic Mass No. *A*	Atomic Mass (u)	% Abundance, or Decay Mode if Radioactive	Half-Life (if Radioactive)
20	Calcium	Ca	40	39.962 591	96.941	
21	Scandium	Sc	45	44.955 910	100	
22	Titanium	Ti	48	47.947 947	73.8	
23	Vanadium	V	51	50.943 962	99.750	
24	Chromium	Cr	52	51.940 509	83.789	
25	Manganese	Mn	55	54.938 047	100	
26	Iron	Fe	56	55.934 939	91.72	
27	Cobalt	Co	59	58.933 198	100	
			60	59.933 819	β^-, γ	5.271 yr
28	Nickel	Ni	58	57.935 346	68.27	
			60	59.930 788	26.10	
29	Copper	Cu	63	62.939 598	69.17	
			65	64.927 793	30.83	
30	Zinc	Zn	64	63.929 145	48.6	
			66	65.926 034	27.9	
31	Gallium	Ga	69	68.925 580	60.1	
32	Germanium	Ge	72	71.922 079	27.4	
			74	73.921 177	36.5	
33	Arsenic	As	75	74.921 594	100	
34	Selenium	Se	80	79.916 520	49.7	
35	Bromine	Br	79	78.918 336	50.69	
36	Krypton	Kr	84	83.911 507	57.0	
			89	88.917 640	β^-, γ	3.16 min
			92	91.926 270	β^-, γ	1.840 s
37	Rubidium	Rh	85	84.911 794	72.165	
38	Strontium	Sr	86	85.909 267	9.86	
			88	87.905 619	82.58	
			90	89.907 738	β^-	29.1 yr
			94	93.915 367	β^-, γ	1.235 s
39	Yttrium	Y	89	88.905 849	100	
40	Zirconium	Zr	90	89.904 703	51.45	
41	Niobium	Nb	93	92.906 377	100	
42	Molybdenum	Mo	98	97.905 406	24.13	
43	Technetium	Tc	98	97.907 215	β^-, γ	4.2×10^6 yr
44	Ruthenium	Ru	102	101.904 348	31.6	
45	Rhodium	Rh	103	102.905 500	100	
46	Palladium	Pd	106	105.903 478	27.33	
47	Silver	Ag	107	106.905 092	51.839	
			109	108.904 757	48.161	
48	Cadmium	Cd	114	113.903 357	28.73	
49	Indium	In	115	114.903 880	95.7; β^-	4.41×10^{14} yr
50	Tin	Sn	120	119.902 200	32.59	
51	Antimony	Sb	121	120.903 821	57.3	
52	Tellurium	Te	130	129.906 229	38.8; β^-	2.5×10^{21} yr
53	Iodine	I	127	126.904 473	100	
			131	130.906 114	β^-, γ	8.040 days

Appendix F Selected Isotopes (*continued*)

Atomic No. Z	Element	Symbol	Atomic Mass No. A	Atomic Mass (u)	% Abundance, or Decay Mode if Radioactive	Half-Life (if Radioactive)
54	Xenon	Xe	132	131.904 144	26.9	
			136	135.907 214	8.9	
			140	139.921 620	β^-, γ	13.6 s
55	Cesium	Cs	133	132.905 429	100	
			134	133.906 696	β^-, EC, γ	2.062 yr
56	Barium	Ba	137	136.905 812	11.23	
			138	137.905 232	71.70	
			141	140.914 363	β^-, γ	18.27 min
57	Lanthanum	La	139	138.906 346	99.91	
58	Cerium	Ce	140	139.905 433	88.48	
59	Praseodymium	Pr	141	140.907 647	100	
60	Neodymium	Nd	142	141.907 719	27.13	
61	Promethium	Pm	145	144.912 743	EC, α, γ	17.7 yr
62	Samarium	Sm	152	151.919 729	26.7	
63	Europium	Eu	153	152.921 225	52.2	
64	Gadolinium	Gd	158	157.924 099	24.84	
65	Terbium	Tb	159	158.925 342	100	
66	Dysprosium	Dy	164	163.929 171	28.2	
67	Holmium	Ho	165	164.930 319	100	
68	Erbium	Er	166	165.930 290	33.6	
69	Thulium	Tm	169	168.934 212	100	
70	Ytterbium	Yb	174	173.938 859	31.8	
71	Lutetium	Lu	175	174.940 770	97.41	
72	Hafnium	Hf	180	179.946 545	35.100	
73	Tantalum	Ta	181	180.947 992	99.988	
74	Tungsten (wolfram)	W	184	183.950 928	30.67	
75	Rhenium	Re	187	186.955 744	62.60; β^-	4.6×10^{10} yr
76	Osmium	Os	191	190.960 920	β^-, γ	15.4 days
			192	191.961 467	41.0	
77	Iridium	Ir	191	190.960 584	37.3	
			193	192.962 917	62.7	
78	Platinum	Pt	195	194.964 766	33.8	
79	Gold	Au	197	196.966 543	100	
			198	197.968 217	β^-, γ	2.6935 days
80	Mercury	Hg	202	201.970 617	29.80	
81	Thallium	Tl	205	204.974 401	70.476	
			208	207.981 988	β^-, γ	3.053 min
82	Lead	Pb	206	205.974 440	24.1	
			207	206.975 872	22.1	
			208	207.976 627	52.4	
			210	209.984 163	α, β^-, γ	22.3 yr
			211	210.988 735	β^-, γ	36.1 min
			212	211.991 871	β^-, γ	10.64 h
			214	213.999 798	β^-, γ	26.8 min
83	Bismuth	Bi	209	208.980 374	100	
			211	210.987 255	α, β^-, γ	2.14 min
			212	211.991 255	β^-, α, γ	1.0092 h

Appendix F Selected Isotopes (*continued*)

Atomic No. Z	Element	Symbol	Atomic Mass No. A	Atomic Mass (u)	% Abundance, or Decay Mode if Radioactive	Half-Life (if Radioactive)
84	Polonium	Po	210	209.982 848	α, γ	138.376 days
			212	211.988 842	α, γ	45.1 s
			214	213.995 176	α, γ	163.69 μs
			216	216.001 889	α, γ	150 ms
85	Astatine	At	218	218.008 684	α, β^-	1.6 s
86	Radon	Rn	220	220.011 368	α, γ	55.6 s
			222	222.017 570	α, γ	3.825 days
87	Francium	Fr	223	223.019 733	α, β^-, γ	21.8 min
88	Radium	Ra	224	224.020 186	α, γ	3.66 days
			226	226.025 402	α, γ	1.6×10^3 yr
			228	228.031 064	β^-, γ	5.75 yr
89	Actinium	Ac	227	227.027 750	α, β^-, γ	21.77 yr
			228	228.031 015	β^-, γ	6.13 h
90	Thorium	Th	228	228.028 715	α, γ	1.913 yr
			231	231.036 298	β^-, γ	1.0633 days
			232	232.038 054	100; α, γ	1.405×10^{10} yr
			234	234.043 593	β^-, γ	24.10 days
91	Protactinium	Pa	231	231.035 880	α, γ	3.276×10^4 yr
			234	234.043 303	β^-, γ	6.70 h
			237	237.051 140	β^-, γ	8.7 min
92	Uranium	U	232	232.037 130	α, γ	68.9 yr
			233	233.039 628	α, γ	1.592×10^5 yr
			235	235.043 924	0.7200; α, γ	7.037×10^8 yr
			236	236.045 562	α, γ	2.342×10^7 yr
			238	238.050 784	99.2745; α, γ	4.468×10^9 yr
			239	239.054 289	β^-, γ	23.47 min
93	Neptunium	Np	239	239.052 933	β^-, γ	2.355 days
94	Plutonium	Pu	239	239.052 157	α, γ	2.411×10^4 yr
			242	242.058 737	α, γ	3.763×10^5 yr
95	Americium	Am	243	243.061 375	α, γ	7.380×10^3 yr
96	Curium	Cm	245	245.065 483	α, γ	8.5×10^3 yr
97	Berkelium	Bk	247	247.070 300	α, γ	1.38×10^3 yr
98	Californium	Cf	249	249.074 844	α, γ	350.6 yr
99	Einsteinium	Es	254	254.088 019	α, γ, β^-	275.7 days
100	Fermium	Fm	253	253.085 173	EC, α, γ	3.00 days
101	Mendelevium	Md	255	255.091 081	EC, α	27 min
102	Nobelium	No	255	255.093 260	EC, α	3.1 min
103	Lawrencium	Lr	257	257.099 480	α, EC	646 ms
104	Rutherfordium	Rf	261	261.108 690	α	1.08 min
105	Dubnium	Db	262	262.113 760	α	34 s

Answers to Check Your Understanding

Chapter 1

CYU 1: (a) Yes. (b) No.

CYU 2: No.

CYU 3: a, b, c, f

CYU 4: No.

CYU 5: b, d

CYU 6: (a) 11 m (b) 5 m

CYU 7: No.

CYU 8: Yes.

CYU 9: (a) The magnitude of $\vec{B}$ is equal to the magnitude of $\vec{A}$.
(b) The direction of $\vec{B}$ is opposite to the direction of $\vec{A}$.

CYU 10: Vector $\vec{A}$ is perpendicular to vector $\vec{B}$.

CYU 11: Vector $\vec{A}$ points in the same direction as vector $\vec{B}$.

CYU 12: $\vec{A}$ and $\vec{D}$

CYU 13: (a) A_x is − and A_y is +
(b) B_x is + and B_y is −
(c) R_x is + and R_y is +

CYU 14: No.

CYU 15: Yes.

CYU 16: (a) $A_x = 0$ units and $A_y = +12$ units
(b) $A_x = −12$ units and $A_y = 0$ units
(c) $A_x = 0$ units and $A_y = −12$ units
(d) $A_x = +12$ units and $A_y = 0$ units

CYU 17: No.

CYU 18: a

Chapter 2

CYU 1: 0 m

CYU 2: scalar quantity

CYU 3: No.

CYU 4: a

CYU 5: average velocity = 2.7 m/s due east, average speed = 8.0 m/s

CYU 6: Yes.

CYU 7: No.

CYU 8: c

CYU 9: No.

CYU 10: No.

CYU 11: the rifle with the short barrel

CYU 12: 1.73 v

CYU 13: a

CYU 14: b

CYU 15: b

CYU 16: b

Chapter 3

CYU 1: b

CYU 2: c

CYU 3: a and c

CYU 4: (a) Yes; when the object is at its highest point.
(b) No.

CYU 5: No.

CYU 6: b

CYU 7: (a) when the ball is at its highest point in the trajectory
(b) at the initial and final positions of the motion

CYU 8: Yes.

CYU 9: Both bullets reach the ground at the same time.

CYU 10: (a) The displacement is greater for the stone thrown horizontally.
(b) The impact speed is greater for the stone thrown horizontally.
(c) The time of flight is the same for both stones.

CYU 11: No.

CYU 12: Ball A has the greater launch speed.

CYU 13: (a) +70 m/s (b) +30 m/s
(c) +40 m/s (d) −60 m/s

CYU 14: No.

CYU 15: The two times are the same.

CYU 16: (a) The range toward the front is the same as the range toward the rear.
(b) The range toward the front is greater than the range toward the rear.

CYU 17: swimmer A

Chapter 4

CYU 1: b

CYU 2: c

CYU 3: d

CYU 4: No, because two or more forces can cancel each other, leading to a net force of zero.

CYU 5: c

CYU 6: a and d

CYU 7: b

CYU 8: Yes, because the ratio of the two weights depends only on the masses of the objects, which are the same on the earth and on Mars.

CYU 9: a

CYU 10: d

CYU 11: No.

CYU 12: a

CYU 13: b

CYU 14: c

CYU 15: To pull, because the upward component of the pulling force reduces the normal force and, therefore, also reduces the force of kinetic friction acting on the sled.

CYU 16: 43°

CYU 17: c, a, b

CYU 18: a

CYU 19: a

CYU 20: b

CYU 21: d

CYU 22: No, because there must always be a vertical (upward) component of the tension force in the rope to balance the weight of the crate.

CYU 23: c

CYU 24: No, because the transfer described does not change the total mass being pulled by the engine.

CYU 25: a

Chapter 5

CYU 1: (a) The velocity is due south and the acceleration is due west.
(b) The velocity is due west and the acceleration is due north.

CYU 2: Yes, if you are going around a curve.

CYU 3: the person at the equator

CYU 4: a and b

CYU 5: *AB* or *DE*, *CD*, *BC*

CYU 6: (a) 4*r*
(b) 4*r*

CYU 7: No.

CYU 8: the same

CYU 9: edge of the turntable

CYU 10: car B

CYU 11: less than

CYU 12: (a) less than
(b) equal to

CYU 13: (a) Yes.
(b) Yes.
(c) Yes.
(d) Yes.

CYU 14: vertical

Chapter 6

CYU 1: b

CYU 2: d

CYU 3: d

CYU 4: a

CYU 5: No.

CYU 6: c

CYU 7: false

CYU 8: c

CYU 9: a

CYU 10: a, b, and c

CYU 11: b

CYU 12: d

CYU 13: b and d

CYU 14: e

CYU 15: c

CYU 16: a

CYU 17: No.

Chapter 7

CYU 1: No.

CYU 2: The total linear momentum is approximately zero because of the random directions and random speeds of the moving people.

CYU 3: (a) Yes.
(b) No.

CYU 4: (a) No.
(b) Yes.

CYU 5: b

CYU 6: (a) No.
(b) The impulse of the thrust is equal in magnitude and opposite in direction to the impulse of the force due to air resistance.

CYU 7: (a) No.
(b) No.

CYU 8: equal to

CYU 9: Yes.

CYU 10: a

CYU 11: decrease

CYU 12: (a) No.
(b) decrease

CYU 13: b, c, d

CYU 14: the cannonball

CYU 15: d

CYU 16: No. It is the total kinetic energy of the *system* that is the same before and after the collision.

CYU 17: c

CYU 18: nearer the heavier end

CYU 19: (a) zero
(b) Yes, opposite to the motion of the sunbather.

CYU 20: a

Chapter 8

CYU 1: Both axes lie in the plane of the paper. One passes through point *A* and is parallel to the line *BC*. The other passes through point *A* and the midpoint of the line *BC*.

CYU 2: B, C, A

CYU 3: No. The instantaneous angular speed of each blade is the same, but the blades are rotating in opposite directions.

CYU 4: c

CYU 5: 1.0 rev/s

CYU 6: b

CYU 7: Case A

CYU 8: a

CYU 9: at the north pole or at the south pole

CYU 10: c

CYU 11: 0.30 m

CYU 12: d

CYU 13: c

CYU 14: b

CYU 15: 8.0 m/s²

CYU 16: a

CYU 17: Among the many possible answers are the motions of a Frisbee through the air, the earth in its orbit, a twirling baton that has been thrown into the air, the blades on a moving lawn mower cutting the grass, and an ice skater performing a quadruple jump.

Chapter 9

CYU 1: 0°, 45°, 90°

CYU 2: greater torque

CYU 3: (a) Yes, if the lever arm is very small.
(b) Yes, if the lever arm is very large.

CYU 4: the box at the far right

CYU 5: (a) C
(b) A
(c) B

CYU 6: Additional forces are necessary.

CYU 7: a

CYU 8: b

CYU 9: Bob

CYU 10: A, B, C

CYU 11: axis B

CYU 12: (a) remains the same
(b) remains the same

CYU 13: (a) remains the same
(b) increases

CYU 14: (a) Both have the same translational speed.
(b) Both have the same translational speed.

CYU 15: axis B

CYU 16: solid sphere, solid cylinder, spherical shell, hoop

CYU 17: (a) decreases
(b) remains the same

CYU 18: decrease

CYU 19: greater than

CYU 20: No.

Chapter 10

CYU 1: No, because the force of gravity acting on the ball is constant, unlike the restoring force of simple harmonic motion.

CYU 2: Both boxes experience the same net force due to the springs.

CYU 3: 180 N/m

CYU 4: The spring stretches more when attached to the wall.

CYU 5: object II

CYU 6: at the position $x = 0$ m

CYU 7: The particle can cover the greater distance in the same time because at larger amplitudes the maximum speed is greater.

CYU 8: The same amount of energy is stored in both cases, since the elastic potential energy is proportional to the square of the displacement x.

CYU 9: b, c, a

CYU 10: The amplitude is unchanged. The frequency and maximum speed each decrease by a factor of $\sqrt{2}$.

CYU 11: a

CYU 12: the simple-pendulum clock, because its period depends on the acceleration due to gravity

CYU 13: Use a shoe and the shoe laces to make a simple pendulum whose period is related to the magnitude g of the acceleration due to gravity (see Equations 10.5 and 10.16). Measure the period of your pendulum and calculate g.

CYU 14: Yes, because for small angles the period of each person's motion is the same.

CYU 15: Yes, because the frequency depends on the mass of the car and its occupants (see Equations 10.6 and 10.11).

CYU 16: $v = \frac{d}{2\pi}\sqrt{\frac{k}{m}}$

CYU 17: b

CYU 18: The rod with the square cross section is longer.

CYU 19: No, because the value of B given in Table 10.3 applies to solid aluminum, not to a can that is mostly empty space.

CYU 20: No, because pressure involves a force that acts *perpendicular* to an area. In Equation 10.18 for shear deformation, the force acts parallel, not perpendicular, to the area A (see Figure 10.29).

CYU 21: Face B experiences the largest stress, and face C experiences the smallest stress.

Chapter 11

CYU 1: (a) outward
       (b) inward

CYU 2: b

CYU 3: a

CYU 4: (a) increase
       (b) decrease
       (c) remain constant

CYU 5: A noticeable amount of water will remain in the tank.

CYU 6: b

CYU 7: Yes; see Equation 11.4, in which P_2 is the pressure at his wrist and P_1 is the pressure above the water.

CYU 8: c

CYU 9: Both beams experience the same buoyant force.

CYU 10: (a) The readings are the same.
        (b) The final reading is greater than the initial reading.

CYU 11: No. You float because the weight of the water you displace equals your weight. Each weight is proportional to g, so its value makes no difference.

CYU 12: No. F_B depends only on the weight of the water she displaces, which doesn't change.

CYU 13: b

CYU 14: No.

CYU 15: d

CYU 16: c

CYU 17: e

CYU 18: c

CYU 19: c

CYU 20: b

CYU 21: a

CYU 22: c

Chapter 12

CYU 1: 178 °X

CYU 2: (a) No.   (b) Yes.   (c) No.

CYU 3: It decreases (see Equations 10.5 and 10.16).

CYU 4: With equal values for α, concrete and steel expand (contract) by the same amount as the temperature increases (decreases), thus minimizing problems with thermal stress.

CYU 5: The bottom is bowed outward, because it acts like a bimetallic strip.

CYU 6: b and d

CYU 7: cooled

CYU 8: No. When the temperature changes, the change in volume of the cavity within the glass would exactly compensate for the change in volume of the mercury, which would never rise or fall in the capillary tube of the thermometer.

CYU 9: Less than. The buoyant force is equal to the weight of the displaced water (see Section 11.6, Archimedes' principle), which is proportional to the water's density. Here, warmer water has a smaller density than cooler water does (see Figure 12.20).

CYU 10: a and b

CYU 11: the object with the smaller mass

CYU 12: c, b, d, a

CYU 13: Because heat is released when the water freezes at 0 °C (consistent with the latent heat of fusion of water), and this heat warms the blossoms.

CYU 14: c, a, b

CYU 15: No, because at sea level water boils at a higher temperature and the stove may not generate enough heat.

CYU 16: Because water in an open pot boils at 100 °C, thus preventing the temperature from rising further, whereas under the elevated pressure in the autoclave water has a boiling point above 100 °C.

CYU 17: Boiling water has a vapor pressure of one atmosphere, and the cool water in the sealed jar has a lower vapor pressure. The excess external pressure creates a net force pushing on the lid, making it hard to unscrew.

CYU 18: Under pressure in the sealed bottle, the soda has a freezing point lower than normal (see Figure 12.35b). The outside temperature is not cold enough to freeze it. When the bottle is opened, the pressure on the liquid decreases to one atmosphere, and the freezing point rises to its normal value. The liquid is now cold enough to freeze.

CYU 19: As the water vapor is removed, more forms in an attempt to reestablish equilibrium between liquid and vapor. When the pumping is rapid, the required latent heat is supplied mostly by the remaining liquid, which cools and eventually freezes.

CYU 20: 100%

CYU 21: Yes. The dew points on the two nights could be different, Tuesday's being higher than Monday's due to a greater partial pressure of water vapor in the air on Tuesday than on Monday.

CYU 22: The air above the swimming pool probably has a greater partial pressure of water vapor (due to inefficient humidity control) and, therefore, a higher dew point than that in the other room. Evidently, the temperature at the inner window-surfaces is below the dew point of the room with the swimming pool but above the dew point in the other room.

Chapter 13

CYU 1: a

CYU 2: b

CYU 3: the house with the snow on the roof

CYU 4: b

CYU 5: c

CYU 6: c

CYU 7: hollow, air-filled strands

CYU 8: c

CYU 9: b

CYU 10: forced convection

CYU 11: strip B

CYU 12: a

CYU 13: d

CYU 14: b

CYU 15: e

Chapter 14

CYU 1: Both have the same number of molecules, but oxygen has the greater mass.

CYU 2: In general, the number of molecules would be different. But they could be the same, if the molecular masses of the two types of molecules happen to be the same.

CYU 3: 66.4%

CYU 4: The ideal gas law gives the pressure as $P = nRT/V$, where T and V are constant. The fan reduces n in the house and increases it in the attic, so pressure decreases in the house and increases in the attic. The fan has a harder job pushing air out against the higher attic pressure.

CYU 5: The ideal gas law gives the gas pressure as $P = nRT/V$, where V and n are constant. As T increases, the pressure increases and could cause the can to burst.

CYU 6: The ideal gas law gives the gas pressure as $P = nRT/V$, where V and n are constant. As T increases, the pressure increases.

CYU 7: The ideal gas law gives the gas pressure as $P = nRT/V$, where T and n are constant. As V decreases due to the incoming tide, the pressure increases, and your ears pop inward, as if you were climbing down a mountain.

CYU 8: The ideal gas law gives the gas volume as $V = nRT/P$, where T and n are constant. As the pressure P decreases during the balloon's ascent, the volume increases. The balloon would overinflate if not underinflated to start with.

CYU 9: Boyle's law gives the final pressure in the bottle after the cork is pressed in: $P_f = P_i(V_i/V_f)$, where V_i/V_f is the volume of air above the wine before the cork is pressed in divided by the volume after the cork is pressed in. This ratio is much larger for the full bottle than for the half-full bottle, creating a pressure large enough to push the cork out.

CYU 10: Xenon has the greatest and argon the smallest temperature.

CYU 11: less than, which follows directly from the impulse–momentum theorem

CYU 12: No. The average kinetic energy is proportional to the Kelvin, not the Celsius, temperature.

CYU 13: It remains unchanged.

CYU 14: argon

CYU 15: $v_{\text{rms, new}}/v_{\text{rms, initial}} = 0.707$

CYU 16: L must be small and there must be many alveoli so that the total effective area A is large.

CYU 17: c, a, b

Chapter 15

CYU 1: d

CYU 2: c

CYU 3: b

CYU 4: $A \rightarrow B$: $Q = +$ and $W = +$
$B \rightarrow C$: $\Delta U = +$ and $W = 0$

CYU 5: a

CYU 6: b

CYU 7: a

CYU 8: c

CYU 9: c

CYU 10: c

CYU 11: b

CYU 12: No, because Carnot's principle only states that a reversible engine operating between two temperatures is more efficient than an irreversible engine operating between the *same temperatures*.

CYU 13: d

CYU 14: d

CYU 15: b

CYU 16: c

CYU 17: a

CYU 18: c

CYU 19: b

CYU 20: a

CYU 21: d

CYU 22: c and d

CYU 23: the popcorn that results from the kernels; a salad after it has been tossed; a messy apartment

CYU 24: b

CYU 25: c

Chapter 16

CYU 1: c

CYU 2: No. The coil moves back and forth in simple harmonic motion.

CYU 3: The wavelength increases.

CYU 4: The person pulling on string B should pull harder to increase the tension in the string.

CYU 5: In Equation 16.2, the speed would be infinitely large if m were zero, so it would take no time at all.

CYU 6: decrease

CYU 7: No, because the particles exhibit simple harmonic motion, in which the acceleration is not always zero.

CYU 8: increase

CYU 9: a

CYU 10: No, because each particle executes simple harmonic motion as the wave passes by.

CYU 11: hot day

CYU 12: CO and N_2

CYU 13: increase

CYU 14: Large outer ears intercept and direct more sound power into the auditory system than smaller ones do.

CYU 15: No, because not all points on the surface are at the same distance from the source.

CYU 16: No, because it is the intensities I_1 and I_2 that add to give a total intensity I_{total}. The intensity levels β_1 and β_2 do not add to give a total intensity level β_{total}.

CYU 17: (a) 1/4
(b) 2

CYU 18: (a) f_o is smaller than f_s, and f_o decreases during the fall.
(b) f_o is greater than f_s, and f_o increases during the fall.

CYU 19: No, because the observed frequency is less than the source frequency, so the car is moving away from him.

CYU 20: (a) greater in air
(b) greater under water

CYU 21: No, because there is no relative motion of the cars.

CYU 22: (a) minus sign in both places
(b) the truck driver

Chapter 17

CYU 1: (a) −3 cm
(b) −2 cm

CYU 2: No, because if the two sound waves have the same amplitude and frequency, they might cancel in a way analogous to that illustrated in Figure 17.2b and no sound will be heard.

CYU 3: b

CYU 4: c

CYU 5: a

CYU 6: d

CYU 7: a

CYU 8: c

CYU 9: b

CYU 10: d

CYU 11: (a) 4
(b) 3
(c) node
(d) 110 Hz

CYU 12: b

CYU 13: d

CYU 14: b

CYU 15: (a) antinode
(b) node
(c) $\frac{1}{4}\lambda$
(d) lowered

CYU 16: b

CYU 17: c ($\frac{1}{4}\lambda$ is the distance between an antinode and an adjacent node.)

CYU 18: a

Answers to Odd-Numbered Problems

Chapter 1

1. 124 m²
3. 10 159 m
5. 0.75 m²/s
7. 2.0 magnums
9. 6.65×10^5 m
11. [M]/[T]²
13. 25.9° south of west, 80.1 km
15. 0.25 m
17. 54.1 m
19. (a) 4500 m
    (b) 2.8 mi
21. 32°
23. due east, 200 N
    due west, 600 N
25. (a) 89 units
    (b) 89 units
27. 25 cm
29. (a) 8.6 units
    (b) 34.9° north of west
    (c) 8.6 units
    (d) 34.9° south of west
31. (a) 48° east of south
    (b) 48° west of south
33. (a) 465 N
    (b) –338 N
35. (a) 4.3 m, due east
    (b) 7.4 m, due south
37. 150 m/s
39. (a) 2.0×10^2 N
    (b) 41°
41. (a) 25° above the $+ x$ axis
    (b) 34.8 N
43. 3.00 m, 42.8° above the negative x axis
45. 0.532 m
47. (a) 2.7 km
    (b) 6.0×10^1 degrees, north of east
49. 6.88 km, 26.9°
51. (a) 5550 m
    (b) 8100 m
53. (a) 226 km
    (b) 51.3° north of east
55. 127 km, 74°
57. 18.7 N, 76°

59. 29.6 mL
61. 55 people
63. 9.72 cm
65. 71°
67. 20.4 m, 59.9°
69. $\theta_2 = 55.3°$, $\theta_1 = 95.5°$
71. (a) 2780 m
    (b) 37.0° south of west
    (c) 6.60°

Chapter 2

1. 9.1
3. 5×10^4 yr
5. (a) 7.07 km
    (b) 2.12 km 45.0° (north of east)
7. 64 s
9. 0.81 km
11. +16 m/s
13. (a) −0.40 m/s² (accelerating)
    (b) +0.80 m/s² (decelerating)
15. (a) $−3.94 \times 10^4$ m/s
    (b) $−5.79 \times 10^4$ m/s
17. (a) 2.8 m/s (b) 2.8 m/s
19. (a) due south
    (b) 9.4 m/s, due north
21. +30.0 m/s
23. 4.5 m
25. (a) 11.4 s (b) 44.7 m/s²
27. 3.1 m/s², directed southward
29. −1.0 m/s²
31. 39.2 m
33. 52.8 m
35. 0.87 m/s² (in the same direction as the car's velocity)
37. 1.7 s
39. 44.1 m/s
41. +39 m/s
43. −1.5 m/s²
45. 12 m
47. 3.33 m
49. 22 m
51. 3.43 s
53. 2.46 m
55. 0.40 s

57. −11 m/s
59. +2.7 m/s

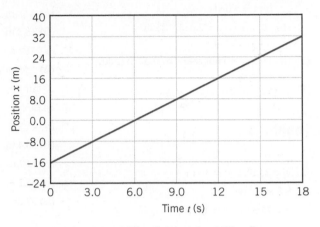

61. A: $−2.0 \times 10^1$ km/h B: $+1.0 \times 10^1$ km/h
    C: $+4.0 \times 10^1$ km/h
63. (a)

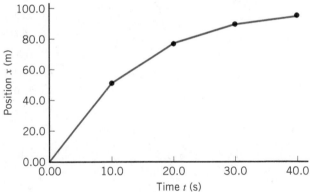

(b)

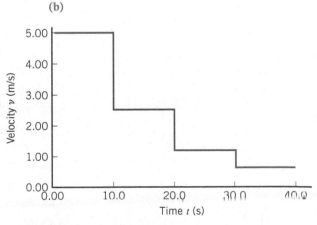

65. 109 m
67. 2.3 min
69. 5.63 s
71. (a) 2.2 m/s (b) 4.4 m/s (c) 0.021 m/s²

73. +62 m

75. 1.6×10^{-2} s

77. 283 m/s

79. 1.20 m/s^2

81. 7250 m

83. (a) 80 m/s [at $t = 2.0$ s]; 160 m/s [at $t = 4.0$ s]

   (b) 80 m/s [at $t = 2.0$ s]; 320 m/s [at $t = 4.0$ s]

85. (a) 11.5 m/s^2 (not safe) (b) 0.32 in (additional distance)

87. 3840 m, 50.6° west of south

Chapter 3

1. 2.8 m

3. $v_x = 11$ m/s, $v_y = 13$ m/s

5. 8.8×10^2 m

7. $x = 75.3$ km, $y = 143$ km

9. (a) 2.47 m/s^2

   (b) 2.24 m/s^2

11. (a) 2.99×10^4 m/s

    (b) 2.69×10^4 m/s

13. 8.6 m/s

15. (a) +1.6 m

    (b) +2.0 m

17. 9.4 mm

19. (a) 4.37 s

    (b) 93.5 m

21. 18 m/s

23. (a) 239 m/s, 57.1°

    (b) 239 m/s at an angle of 57.1° with respect to the horizontal

25. $a_x = 4.79$ m/s^2, $a_y = 7.59$ m/s^2

27. 30.0 m

29. 45.5 m/s

31. 62.8 m/s

33. (a) 85 m

    (b) 610 m

35. 4

37. 53 m

39. 4.52 m/s, 59.4°

41. 0.96 s

43. 3.4 m/s

45. 4

47. 27.2°

49. 190 s

51. 5.1 m/s, 51° north of west

53. 6.3 m/s, 18° north of east

55. 8.92°, south of west

57. 5.2 m/s, 52° west of south

59. 1130 m/s, 37.7°

61. 1.7 s

63. (a) 9.02 m

    (b) 1.05 s

65. 12 m

67. 14.9 m

69. 5.79 m/s

71. (a) 4.58 m

    (b) 18.3 m

73. (a) 51.1 s

    (b) 1.68 m/s

75. 23 cm

77. 34 m/s

79. (a) 41.8°

    (b) 36.4 m

    (c) 4.07 m

81. (a) 86.2 m

    (b) 37.0°

Chapter 4

1. 93 N

3. (a) +6.0 N (b) −24 N (c) −9.0 N

5. 32 N

7. 3560 N

9. (a) 3.6 N (b) 0.40 N

11. 30.9 m/s^2, 27.2° above the +x axis

13. 1.20 m/s^2, directed straight upward

15. 33 s

17. 18.4 N, 68° north of east

19. (a) 3.75 m/s^2 (b) 2.4×10^2 N

21. (a) 5.1×10^{-6} N (b) 5.1×10^{-6} N

23. 9.6×10^{-9} N

25 0.223 m/s^2

27. (a) 5.67×10^{-5} N (to the right)

    (b) 3.49×10^{-5} N (to the right)

    (c) 9.16×10^{-5} N (to the left)

29. 3.47×10^8 m

31. 178

33. (a) 980 N (b) 640 N

35. 7.3×10^2 N

37. 645 N

39. (a) 267 N, +x direction (b) 360 N

41. 2.9 m/s^2

43. 16.3 N

45. 68°

47. 1.00×10^2 N, 53.1° south of east

49. 929 N

51. 436 N, 5.7° above the horizontal

53. 406 N

55. +0.62 m/s^2

57. (a) 45 N (b) 37 N

59. 1320 N

61. (a) 914 N (b) 822 N

63. 5.03 m/s^2

65. 5.9×10^{-3} m/s^2

67. 0.14 m/s^2

69. (a) 1.1×10^4 N (b) 5.4×10^2 N

71. 4.0×10^2 N

73. (a) Case A: 2.9 N; Case B: 3.9 N

    (b) Case A: 0.97 m/s^2; Case B: 0.64 m/s^2

75. 8.17 s

77. 1.2 s

79. (a) 1.10×10^3 N (b) 931 N (c) 808 N

81. 39 N

83. 1730 N, directed due west

85. (a) 1.04×10^3 N (b) 1.04×10^3 N

    (c) 2.45 m/s^2 (d) 1.74×10^{-22} m/s^2

87. 0.265 m

89. (a) 0.60 m/s^2 (b) 104 N (left string); 230 N (right string)

91. 72 m/s

93. (a) 29.4 N (b) 44.1 N

95. 3090 N

97. −470 N

99. (a) 1100 N (b) 650 N

101. (a) 0.980 m/s^2, opposite to the direction of motion (b) 29.5 m

103. 42°

105. +2.3 m/s^2

107. 925 N (208 lb)

109. (a) $\mu_s = 0.22$ and $\theta = 10°$ (b) 12.4°

    (c) Yes. It is safe.

Chapter 5

1. 0.79 m/s^2

3. 0.71

5. 6.9 m/s^2

7. 332 m

9. (a) 464 m/s, 3.37×10^{-2} m/s^2

   (b) 402 m/s, 2.92×10^{-2} m/s^2

11. 1.5 m

13. 73 kg

15. 12 m/s

17. 0.187

19. (a) 0.91

    (b) 38 m/s

21. (a) 19 m/s

    (b) 23 m/s

23. 170 m

25. 184 m

27. 45 s

29. 3070 m/s

31. 5.92×10^3 m/s

33. 2.45×10^4 N

35. (a) 1.70×10^3 N

   (b) 1.66×10^3 N

37. 17 m/s

39. 8.48 m/s

41. 606 N

43. (a) 0.189 N

   (b) 4

45. 14.0 m/s

47. 28°

49. 816 m

51. 10 600 rev/min

53. 33 m/s

55. 0.38 m

57. 0.224 kg

59. 25 m/s, 18 m/s

61. 110 m

63. (a)

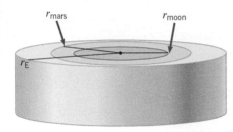

   (b) $r_{Mars} = 115$ m, $r_{moon} = 49.7$ m,
   $r_E = 300$ m

   (c) 34.8 s

Chapter 6

1. 2.2×10^3 J

3. -2.6×10^6 J

5. 42.8°

7. (a) More net work is done during the dive.

   (b) 6.8×10^7 J

9. 25°

11. 256 N

13. 2.5×10^7 J

15. 1450 kg

17. 3.2×10^3 J

19. 0.13

21. (a) 8.3×10^4 N

   (b) 9.1×10^3 J

   (c) -8.5×10^3 J (d) 4 m/s

23. (a) 10.0 m (b) 3.64 s

25. 5.24×10^5 J

27. (a) 27 J (b) 36 J (c) 8.8 J

   (d) The change in gravitational potential
   energy is -27 J $= -W$, where W is the
   work done by the weight.

29. (a) -3.0×10^4 J (b) The resistive force is
   not a conservative force.

31. (a) -1086 J (b) The skater is 2.01 m
   below the starting point.

33. 6.6 m/s

35. (a) 28.3 m/s (b) 28.3 m/s (c) 28.3 m/s

37. (a) 52.2 J (b) 48.8 m/s

39. 0.60 m

41. 43°

43. (a) 2.50 m (b) 1.98 m

45. 48°

47. -3.8 J

49. 8.6 m/s

51. 2.4 m/s

53. 4.17 m/s

55. (a) 3.3×10^4 W

   (b) 5.1×10^4 W

57. (a) 1.0×10^4 W (b) 13 hp

59. 49.6 J

61. 7.07 m/s

63. at $h = 20.0$ m: KE $= 0$ J;
   PE $= 392$ J; E $= 392$ J
   at $h = 10.0$ m: KE $= 196$ J;
   PE $= 196$ J; E $= 392$ J
   at $h = 0.0$ m: KE $= 392$ J;
   PE $= 0$ J; E $= 392$ J

65. 45 N

67. (a) -270 J (h) 140 N

69. 5.82×10^4 W (78 hp)

71. (a) 49.7 m/s (b) -4.89×10^4 J

73. 1.0×10^2 N

75. (a) 7110 J (b) 467 J

77. 61 J

79. $+1470$ J

81. (a) 14.4 m/s (b) 1.58 hp

83. (a) 479 m; It therefore has enough energy to lift the elevator and passenger
   (without friction).

   (b) $E_{grav} = 5.51 \times 10^5$ J and $E_{net} = 5.32 \times 10^5$ J. Since $E_{net} < E_{grav}$, the elevator
   will NOT make it to the top when the
   frictional force is included.

Chapter 7

1. $+69$ N, opposite to the velocity

3. -1.8 N, downward

5. -8.7 kg · m/s

7. (a) $+2.2 \times 10^3$ N

   (b) $+4.4 \times 10^3$ N

9. 960 N

11. 3.7 N · s

13. 1.4 m

15. (a) Bonzo has the larger mass, since he
   has the smaller recoil velocity.

   (b) 1.7

17. $+7.5$ m/s

19. 9.5%

21. -4.6 m/s

23. 0.34 s

25. 1.5 m

27. (a) 4.89 m/s

   (b) 1.22 m

29. $+9.3$ m/s

31. (a) -0.400 m/s, $+1.60$ m/s

   (b) $+0.800$ m/s

33. 7.4%

35. $+9.09$ m/s

37. 0.062 kg

39. (a) 73.0°

   (b) 4.28 m/s

41. 2.175×10^{-3} N

43. (a) 5.56 m/s

   (b) -2.83 m/s, $+2.73$ m/s

   (c) 0.409 m, 0.380 m

45. 1.6 m toward the origin

47. (a) 0 nm

   (b) $+0.0358$ nm

49. $-11\ 600$ N

51. 0.707

53. (a) $+2.4 \times 10^5$ N

   (b) $+4.8 \times 10^3$ N

   (c) Stiff-legged: $F_{Ground} = +2.4 \times 10^5$ N,
   Knees-bent: $F_{Ground} = +5.5 \times 10^3$ N

55. $x_{cm} = 0.11$ m, $y_{cm} = 0.20$ m

57. (a) 0.84 m/s

   (b) 29 N

59. 2500 N

61. (a) 1160 J, 580 kg · m/s

   (b) 1160 J, 140 kg · m/s

63. $x_{cm} = 11.2$ cm, $y_{cm} = 2.30$ cm

65. (a) 2.49 m/s

   (b) 3.3×10^{-3} m/s

   (ɵ) 3.00 m/s

Chapter 8

1. 21 rad

3. (a) $+7.3 \times 10^{-5}$ rad/s (b) $+2.0 \times 10^{-7}$ rad/s

5. (a) 18 rad/s (b) 6.0 rad/s (c) 6.0 rad/s
   (d) 18 rad/s

7. 163 m

9. 59 m

11. (a) 111 rad/s (b) 134 rad/s (c) 118 rad/s

13. 825 m

15. 1.3 s

17. 1.43×10^{-1} m/s

19. 0.125 s

21. (a) 117 rad/s (b) 140 s

23. (a) 10.0 s (b) -2.00 rad/s

25. 28 rad/s

27. 24 rad

29. 12.5 s

31. (a) 0.332 rad/s (b) 128 m

33. 9.1 rad/s

35. 4.63 m/s

37. 3.61 rad/s (b) 6.53 rad/s²

39. 0.87

41. (a) 177 m/s² (b) 14.0°

43. (a) 2.49 m/s² (b) 17.7°

45. 8.71 rad/s²

47. 693 rad

49. (a) -1.4 rad/s² (b) $+33$ rad

51. 21.6 rad

53. 11.8 rad

55. 157.3 rad/s

57. (a) 125 m/s (b) 7.98 rev/s

59. (a) 8 (b) 5

61. -485 rad

63. (a) 1500 rad/s (b) 4.2×10^{-3} s

65. (a) 36 rad/s (b) 9.0 rev

67. 31.8 μm

69. 33.3 rotations

71. (a) 8.2×10^{-2} rad/s (b) 2.6 m/s²
    (c) 3.2 m/s²

73. (a) -6.13×10^{-3} rad/s² (b) 12.8 s

75. (a) Clockwise (b) 1.12 rad/s; 0.26 m/s
    (c) The largest tangential speed occurs
    when the wheel with the largest di-
    ameter is in the drive-wheel position:
    $v = 0.59$ m/s

Chapter 9

1. 843 N

3. 2.1×10^2 N

5. (a) 13 500 N · m
   (b) 132 000 N · m

7. (a) FL
   (b) FL
   (c) FL

9. 43.7°

11. (a) 2590 N
    (b) 2010 N

13. 6.0 N pointing toward corner B

15. (a) 1.60×10^5 N
    (b) 4.20×10^5 N

17. 37.6°

19. (a) 2260 N
    (b) $S_x = 1450$ N, $S_y = 1450$ N

21. 430 N

23. 8.0×10^{-4} N · m

25. 0.027 kg · m²

27. Hoop: 0.20 N · m, Disk: 0.10 N · m

29. (a) 5.94 rad/s²
    (b) 44.0 N

31. (a) System A: 229 kg · m²,
    System B: 321 kg · m²
    (b) System A: -1270 N · m,
    System B: 0 N · m
    (c) System A: -27.7 rad/s, System B: 0
    rad/s

33. 0.78 N

35. 1700 N · m

37. (a) 12.0 m/s, 9.00 m/s, 18.0 m/s
    (b) 1.08×10^3 J
    (c) 60.0 kg · m³
    (d) 1.08×10^3 J

39. 6.1×10^5 rev/min

41. 2/7

43. (a) 3.7 m
    (b) 7.0 m/s

45. 2.0 m

47. 0.075 rev/d

49. 0.26 rad/s

51. 4.5 rad/s

53. 0.037 rad/s

55. 0.70 m

57. 180 rad/s², 5.0 N

59. (a) Rolling wheel: 54 J, Sliding wheel: 36 J
    (b) Rolling wheel: 2.8 m, Sliding wheel:
    1.8 m

61. 6.3 rad/s², 5.00×10^3 N

63. $F_{\text{Each Hand}} = 196$ N, $F_{\text{Each Foot}} = 96$ N

65. $M = 1.21 \times 10^3$ N, $F = 1.01 \times 10^3$ N,
    downward

67. 460 N

69. 1400 N

71. 4700 N

73. (a) 2.5 rotations
    (b) 0.129 J

75. 290 N

77. Choose solid cylinders over hallow ones.

79. (a) Roll the barrels down an inclined
    plane and time how long it takes,
    starting from rest, to roll a given
    distance. Due to their differing
    moments of inertia, they should take
    different times, and we can therefore
    identify which are solid and which
    are hollow cylinders.
    (b) $a_{\text{hallow}} = 2.45$ m/s², $a_{\text{solid}} = 3.27$ m/s²
    (c) $t_{\text{hallow}} = 2.86$ s, $t_{\text{solid}} = 2.47$ s

Chapter 10

1. 237 N

3. 650 N/m

5. (a) 8.5×10^4 N/m (b) 290 N

7. 1.4 kg

9. 0.240 m

11. 2.29×10^{-3} m

13. (a) 0.407 m (b) 397 N

15. (a) 50.0×10^{-6} s (b) 2.00×10^4 Hz
    (c) 1.26×10^5 rad/s

17. (a) 0.080 m (b) 1.6 rad (c) 2.0 N/m
    (d) $v = 0$ m/s (e) 0.20 m/s²

19. 696 N/m

21. 0.285 s

23. 0.806 s

25. (a) -1.84×10^2 J (b) $+1.84 \times 10^2$ J (c) 0 J

27. 4.4 J

29. 6.55 m/s

31. 303 N/m

33. 1.95 m

35. (a) 3.3×10^{-2} m (b) 7.6×10^{-2} m

37. 180 N/m

39. 2.37×10^3 N/m

41. 0.40 s

43. (a) 3.5 rad/s (b) 2.0×10^{-2} J (c) 0.41 m/s

45. (a) 1.64 s (b) 1.64 s

47. 5.2×10^{-4} m

49. 260 m

51. 4.2×10^4 N

53. 0.091°

55. 2.7×10^{-4} m

57. 1.2×10^4 N

59. (a) 710 N/m² (b) 3.6×10^{-8}
    (c) 3.6×10^{-10} m

61. -4.4×10^{-5}

63. (a) 0.450 m (b) 3.31 rad/s (c) 1.49 m/s

65. $+0.50$ m

67. (a) 0.037 m (b) 0.074 m

69. 8.00

71. 2.4×10^{-2} m

73. 1.6×10^{-5} m

75. 0.56 m

77. 2.3×10^{-6} m

79. 5.00×10^{7} m

81. 2.1×10^{6} N/m

83. (a) 6.20×10^{-3} m (b) 88.4 J

85. (a) 13 m/s (b) 11.7 m/s

87. (a) 0.67 Hz (b) 0.55 m

(c) $x = L$; i.e., the equilibrium displacement x of the spring is equal to the length L of the pendulum.

89. (a) For each full oscillation of the spring/mass, the beam will pass through the slot twice.

(b) 1.75 Hz (c) 0.60 kg

(d) $L = 8.27$ cm; $w = 3.00$ cm

Chapter 11

1. 8.3×10^{3} lb

3. 317 m^2

5. 7.0×10^{-2} m

7. 1.9 gal

9. 4240 s

11. 1.1×10^{3} N

13. 24

15. 32 N

17. 2400 Pa

19. (a) 3.5×10^{6} N

(b) 1.2×10^{6} N

21. (a) 9.95×10^{4} Pa

(b) 1.001×10^{5} Pa

23. (a) 2.45×10^{5} Pa

(b) 1.73×10^{5} Pa

25. 0.74 m

27. 2.3×10^{8} N

29. 3.8×10^{5} N

31. (a) 93.0 N

(b) 94.9 N

33. 8.50×10^{5} N · m

35. 108 N

37. 59 N

39. 250 kg/m^3

41. 390 kg/m^3

43. At least 20 logs are needed.

45. 0.20 m

47. 7.6×10^{-2} m

49. 2.52 m/s

51. 9.3×10^{-5} m^2

53. 46 Pa, (Enters at B and exits at A.)

55. 1.92×10^{5} N

57. (a) 17.1 m/s

(b) 2.22×10^{-2} m^3/s

59. 5.0×10^{-2} m^3/s

61. 1.7×10^{5} Pa

63. 9600 N

65. 21 m/s

67. 6.4×10^{5} Pa

69. 2.9×10^{4} Pa

71. 290 gal/min

73. (a) 1.6×10^{-4} m^3/s

(b) 2.0×10^{1} m/s

75. 7.9×10^{-4} m^3

77. 55 Pa

79. 46.2 mm Hg

81. 1.4×10^{4} Pa

83. (a) 4.2×10^{-2} m/s

(b) 0.38 m/s

85. 96 Pa

87. 0.34 m

89. 3.5 m

91. 1.55 g/cm^3

93. 44 400 gal

95. 2.0×10^{1} m/s

97. 7.72 cm

99. (a) $N_{barrels} = 6$

(b) $N_{barrels} = 8$

Chapter 12

1. (a) 56 F° (b) 31 K

3. (a) 102 °C (day); −173 °C (night)
(b) 215 °F (day); −2.80 × 10^2 °F (night)

5. 44.0 °C

7. 4.9×10^{-2} m

9. 3.2×10^{2} m

11. 110 C°

13. 8.6×10^{-6} m/s

15. 4.1×10^{-4}

17. 7.7×10^{10} N/m^2

19. 91 C°

21. Arrangement A: 31.7 °C; Arrangement B: 34.1 °C; Arrangement C: 30.4 °C

23. Gasoline

25. 2.26×10^{-4} m^3

27. 1 penny

29. 1.8×10^{-4} m^3

31. 2.1×10^{-4} (C°)$^{-1}$

33. 9.0 mm

35. 13 200 kg/m^3

37. 0.765 m

39. 36.2 °C

41. 2500 J/(kg · C°)

43. 940 °C

45. 4.50 m

47. 0.016 C°

49. 3.9×10^{5} J

51. (a) 2.4×10^{6} J (b) 5.3 C°

53. 9.49×10^{-3} kg

55. 2.46×10^{3} m/s

57. 5.2×10^{7} J

59. 4.38×10^{-3} kg

61. (a) 1.49×10^{14} J (b) 993 homes

63. 1.51×10^{5} J/kg

65. 5.5×10^{6} Pa

67. 33%

69. 2.8×10^{5} J

71. 0.027 kg

73. 230 C°

75. 1.4×10^{-3} m

77. 1.2×10^{-2} kg

79. 32.1 °C

81. 9.89×10^{-2} m^3

83. 21.03 °C

85. 4.4×10^{3} N

87. 87%

89. 8.47 hours

91. 187 Calories

93. 174 Calories

95. (Block A) 18 C°; (Block B) 6.0 C°; (Block C) 2.8 C°

97. (a) The rate of rotation decreases
(b) 44.5 rad.s

99. (a) 1750.00 rpm (b) 1749.50 rpm
(c) Set rotation rate of aluminum drive wheel to 35.010 rpm

Chapter 13

1. 14 h

3. 1/4

5. Case A: 4.0×10^{3} J, Case B: 9.0×10^{3} J, Case C: 3.6×10^{4} J

7. 287 °C

9. 85 J

11. (a) 2200 J
(b) 0.26 C°

13. 103.3 °C

15. 0.74

17. 5800 K

19. (a) 6.3 J/s
(b) 4.8 J/s

21. 275 W

23. 380 K

25. 732 K

27. 0.012 m

29. 386 K

31. (a) 1.27

(b) 424 K

33. 3.1×10^{-5} kg/s

35. 1.5 C°

37. 1.2×10^4 s

39. (a) 67 W

(b) 58 Calories

41. Ratio = 23.7

43. 139 J/s

45. 220 s

47. 718 °C

49. (a) 1072 J/s

(b) 123 J/s

(c) 3216 J/s

Chapter 14

1. 1.07×10^{-22} kg

3. (a) 196.967 u (b) 3.2706×10^{-25} kg

(c) 1.45 mol

5. (a) 2.3×10^3 mol (b) 1.4×10^{27}

7. 42.4 mol

9. 2.4×10^4 Pa

11. $T_A = 120$ K; $T_B = 180$ K; $T_C = 120$ K; $T_D = 180$ K

13. 12

15. 882 K

17. 8.1 g

19. (a) 8.17×10^{-4} m³ (b) 2.3 g

21. 3.70×10^2 K

23. 0.93 mol/m³

25. 6.19×10^5 Pa

27. 1.6×10^{-15} kg

29. 3.9×10^5 J

31. 2.098

33. 2820 m

35. 2.10×10^{26} atoms/s

37. 3.0×10^{-3} kg/m³

39. (a) 5.00×10^{-13} kg/s (b) 5.8×10^{-3} kg/m³

41. 0.32 m

43. 1.4

45. 11.7

47. 1.98×10^3 N/m

49. (a) 1.29×10^{21} (b) 5.84×10^{20}

51. 11

53. 0.14 kg/m³

55. (a) 1.66×10^{-4} moles

(b) 9.99×10^{19} molecules

57. Argon

59. 1.1×10^{-5} μg

61. (a) 1.5×10^{-16} N (b) 3.8×10^{-17} Pa

(c) 2.7 K

63. 87.4 kg (nitrogen); 29.2 kg (oxygen)

65. 1.84×10^6 Pa

Chapter 15

1. (a) $+1.6 \times 10^4$ J

(b) -4.2×10^4 J

(c) -2.6×10^4 J

3. (a) -87 J

(b) $+87$ J

5. 436 K

7. 1.2×10^7 Pa

9. 3.0×10^5 Pa

11. (a) $+3.0 \times 10^3$ J

(b) work is done by the system

13. 0.24 m

15. 3.1×10^5 Pa

17. 4.99×10^{-6}

19. (a) $+5.0 \times 10^3$ J

(b) -5.0×10^3 J

21. -4700 J

23. 1.81

25. 19.3

27. Step A to B: 4990 J, 3320 J, 8310 J

Step B to C: -4990 J, 0 J, -4990 J

Step C to D: -2490 J, -1660 J, -4150 J

Step D to A: 2490 J, 0 J, 2490 J

29. (a) -8.00×10^4 J

(b) Heat flows out of the gas.

31. (a) $T_{1f} = 477$ K

(b) $T_{2f} = 323$ K

33. (a) 1.1×10^4 J

(b) 1.8×10^4 J

35. 45 K

37. (a) 1.40×10^2 K

(b) 5.24×10^3 J

(c) 2.33×10^3 Pa

39. 0.264 m

41. (a) 8600 J

(b) 3100 J

43. 65 J

45. (a) 5/9

(b) 1/3

47. 0.21

49. 1090 K

51. lowering

53. 1.23

55. 21

57. 13

59. 284 K

61. 275 K

63. (a) 2.0×10^1

(b) 1.5×10^4 J

65. 1.4×10^4 J

67. 11.6 J/K

69. 1.4×10^{26} J/K

71. 0.66

73. 4.5×10^{-3} m³

75. (a) reversible

(b) -125 J/K

77. 5.86×10^5 J

79. (a) 0 J

(b) $+2.1 \times 10^3$ J

81. $e = e_1 + e_2 - e_1 e_2$

83. (a) $+1.2 \times 10^5$ J

(b) 5.9×10^5 J

85. (a) 4.6×10^5 J

(b) 4.1×10^5 J

87. 0.170 kg

89. (a) 21.7%

(b) 7.93×10^4 J

91. 2.87×10^6 J

93. 5450 s

95. (a) 881 K

(b) 916 K

(c) It is likely that the engine with the higher compression ratio is more efficient.

Chapter 16

1. 5.50×10^{14} Hz

3. (a) 10.0 s (b) 0.100 Hz (c) 32 m

(d) 3.2 m/s

(e) Not possible to determine amplitude

5. 0.25 m

7. 0.20 m/s

9. 5.0×10^1 m/s

11. 0.68 m

13. 340 m/s

15. 0.17 s

17. 21.2 m/s

19. 9.1 cm

21. 3.26×10^{-3} s

23. $y = (0.37 \text{ m}) \sin [(8.2 \text{ rad/s})t + (0.68 \text{ m}^{-1})x]$

25. $y = (0.35 \text{ m}) \sin [(88 \text{ rad/s})t - (17 \text{ m}^{-1})x]$

27. (a) 4.2 m/s (b) 0.35 m

(c) $y = (3.6 \times 10^{-2} \text{ m}) \sin [(75 \text{ rad/s})t + (18 \text{ m}^{-1})x]$

29. 110 m

31. 28.8 K

33. 1730 m/s

35. 690 rad/s

37. 61 m

39. (a) First in metal, second in water, third in air

(b) Second sound arrives 0.059 s later. Third sound arrives 0.339 s later.

41. 650 m

43. Tungsten

45. 8.0×10^5 m

47. 65 W

49. 2.4×10^{-5} W/m²

51. (a) 3.54×10^{-4} W/m²

(b) 4.39×10^{-4} W/m²

53. 82.0 m and 246 m

55. 6300

57. −6 dB

59. 0.316 W/m²

61. 76.8 dB

63. 3.1%

65. 13 Hz

67. 17 m/s

69. (a) Away (b) 8.7 m/s

71. 1.4 m/s²

73. (a) 0.140 m; 2450 Hz

(b) 0.129 m; 2660 Hz

(c) 0.129 m; 2770 Hz

75. 3.02×10^{-6} W/m²

77. 64 N

79. 1000

81. 78 cm

83. 3.4 m/s

85. 1.0×10^2

87. 2.6

89. (a) $v = (yg)^{1/2}$

(b) 2.2 m/s at $y = 0.50$ m; 4.4 m/s at $y = 2.0$ m

91. 113 dB

93. 6.7×10^{-9} W

95. 6.0

97. 5.1×10^{-10} W

99. (a) 0.700 mm (b) 14.0 cm (c) 1.8×10^{-4} s

101. 5.9×10^{-4} J

103. (a) 2310 Hz (b) 1970 Hz

105. Highest: 46.7 m/s; Lowest: 18.3 m/s

107. (a) 2004 m (b) 23.7 m/s (c) 83.2 s

Chapter 17

1. 8.42 m

3.

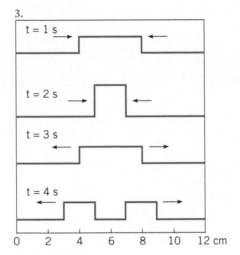

5. (a) +13.3 mm

(b) +48.8 mm

7. 3.89 m

9. 28 Hz and 42 Hz

11. 3.90 m, 1.55 m from speaker A, 6.25 m from speaker A

13. 16.0°

15. 44°

17. 1.5×10^4 Hz

19. 0.25 s

21. (a) 50 kHz

(b) 90 kHz

23. 14 Hz

25. 1.7 Hz

27. (a) 1.00 m

(b) 85.0 m/s

(c) 17.0 Hz

29. Wire A: 790 m/s, Wire B: 530 m/s

31. 2

33. 130 Hz

35. 11 N

37. 13 Hz

39. (a) 261.6 Hz

(b) 0.656 m

41. 20 Hz: 8.6 m, 20 kHz: 8.6×10^{-3} m

43. 0.57 m

45. 6.1 m

47. 0.28 m

49.

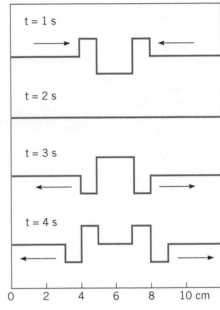

51. (a) destructive interference

(b) constructive interference

53. 0.669 m

55. 2.92

57. 20.83 minutes

59. 1.47 m

61. Air: 8.0°, Water: 37°

63. $(156.6)(m + \frac{1}{2})$ Hz

65. (a) $f = 95.50$ kHz: $L = 2.65$ cm, $f = 102.550$ kHz: $L = 2.47$ cm

(b) $f = 95.50$ kHz: $L = 2.70$ cm, $f = 102.550$ kHz: $L = 2.52$ cm

Fundamental Constants

Quantity	Symbol	Value*
Avogadro's number	N_A	$6.022\,140\,76 \times 10^{23}\,\text{mol}^{-1}$
Boltzmann's constant	k	$1.380\,649 \times 10^{-23}\,\text{J/K}$
Electron charge magnitude	e	$1.602\,176\,634 \times 10^{-19}\,\text{C}$
Permeability of free space	μ_0	$4\pi \times 10^{-7}\,\text{T} \cdot \text{m/A}$
Permittivity of free space	ε_0	$8.854\,187\,8128 \times 10^{-12}\,\text{C}^2/(\text{N} \cdot \text{m}^2)$
Planck's constant	h	$6.626\,070\,15 \times 10^{-34}\,\text{J} \cdot \text{s}$
Mass of electron	m_e	$9.109\,383\,7015 \times 10^{-31}\,\text{kg}$
Mass of neutron	m_n	$1.674\,927\,471 \times 10^{-27}\,\text{kg}$
Mass of proton	m_p	$1.672\,621\,923\,69 \times 10^{-27}\,\text{kg}$
Speed of light in vacuum	c	$2.997\,924\,58 \times 10^{8}\,\text{m/s}$
Universal gravitational constant	G	$6.674\,30 \times 10^{-11}\,\text{N} \cdot \text{m}^2/\text{kg}^2$
Universal gas constant	R	$8.314\,462\,618\,\text{J/(mol} \cdot \text{K)}$

*2018 CODATA recommended values.

Useful Physical Data

Acceleration due to earth's gravity	$9.80\,\text{m/s}^2 = 32.2\,\text{ft/s}^2$
Atmospheric pressure at sea level	$1.013 \times 10^5\,\text{Pa} = 14.70\,\text{lb/in.}^2$
Density of air (0 °C, 1 atm pressure)	$1.29\,\text{kg/m}^3$
Speed of sound in air (20 °C)	$343\,\text{m/s}$
Water	
Density (4 °C)	$1.000 \times 10^3\,\text{kg/m}^3$
Latent heat of fusion	$3.35 \times 10^5\,\text{J/kg}$
Latent heat of vaporization	$2.26 \times 10^6\,\text{J/kg}$
Specific heat capacity	$4186\,\text{J/(kg} \cdot \text{C}°)$
Earth	
Mass	$5.98 \times 10^{24}\,\text{kg}$
Radius (equatorial)	$6.38 \times 10^6\,\text{m}$
Mean distance from sun	$1.50 \times 10^{11}\,\text{m}$
Moon	
Mass	$7.35 \times 10^{22}\,\text{kg}$
Radius (mean)	$1.74 \times 10^6\,\text{m}$
Mean distance from earth	$3.85 \times 10^8\,\text{m}$
Sun	
Mass	$1.99 \times 10^{30}\,\text{kg}$
Radius (mean)	$6.96 \times 10^8\,\text{m}$

Frequently Used Mathematical Symbols

Symbol	Meaning
$=$	is equal to
$\neq$	is not equal to
$\propto$	is proportional to
$>$	is greater than
$<$	is less than
$\approx$	is approximately equal to
$\lvert x \rvert$	absolute value of x (always treated as a positive quantity)
Δ	the difference between two variables (e.g., ΔT is the final temperature minus the initial temperature)
Σ	the sum of two or more variables (e.g., $\sum\limits_{i=1}^{3} x_i = x_1 + x_2 + x_3$)

Conversion Factors

Length

1 in. = 2.54 cm

1 ft = 0.3048 m

1 mi = 5280 ft = 1.609 km

1 m = 3.281 ft

1 km = 0.6214 mi

1 angstrom (Å) = 10^{-10} m

Mass

1 slug = 14.59 kg

1 kg = 1000 grams = 6.852×10^{-2} slug

1 atomic mass unit (u) = 1.6605×10^{-27} kg

(1 kg has a weight of 2.205 lb where the acceleration due to gravity is 32.174 ft/s^2)

Time

1 d = 24 h = 1.44×10^3 min = 8.64×10^4 s

1 yr = 365.24 days = 3.156×10^7 s

Speed

1 mi/h = 1.609 km/h = 1.467 ft/s = 0.4470 m/s

1 km/h = 0.6214 mi/h = 0.2778 m/s = 0.9113 ft/s

Force

1 lb = 4.448 N

1 N = 10^5 dynes = 0.2248 lb

Work and Energy

1 J = 0.7376 ft · lb = 10^7 ergs

1 kcal = 4186 J

1 Btu = 1055 J

1 kWh = 3.600×10^6 J

1 eV = 1.602×10^{-19} J

Power

1 hp = 550 ft · lb/s = 745.7 W

1 W = 0.7376 ft · lb/s

Pressure

1 Pa = 1 N/m^2 = 1.450×10^{-4} lb/in.^{2}

1 lb/in.^{2} = 6.895×10^3 Pa

1 atm = 1.013×10^5 Pa = 1.013 bar = 14.70 lb/in.^{2} = 760 torr

Volume

1 liter = 10^{-3} m^3 = 1000 cm^3 = 0.03531 ft^3

1 ft^3 = 0.02832 m^3 = 7.481 U.S. gallons

1 U.S. gallon = 3.785×10^{-3} m^3 = 0.1337 ft^3

Angle

1 radian = 57.30°

1° = 0.01745 radian

Standard Prefixes Used to Denote Multiples of Ten

Prefix	Symbol	Factor
Tera	T	10^{12}
Giga	G	10^{9}
Mega	M	10^{6}
Kilo	k	10^{3}
Hecto	h	10^{2}
Deka	da	10^{1}
Deci	d	10^{-1}
Centi	c	10^{-2}
Milli	m	10^{-3}
Micro	μ	10^{-6}
Nano	n	10^{-9}
Pico	p	10^{-12}
Femto	f	10^{-15}

Basic Mathematical Formulas

Area of a circle = πr^2

Circumference of a circle = $2\pi r$

Surface area of a sphere = $4\pi r^2$

Volume of a sphere = $\frac{4}{3}\pi r^3$

Pythagorean theorem: $h^2 = h_o{}^2 + h_a{}^2$

Sine of an angle: $\sin \theta = h_o/h$

Cosine of an angle: $\cos \theta = h_a/h$

Tangent of an angle: $\tan \theta = h_o/h_a$

Law of cosines: $c^2 = a^2 + b^2 - 2ab \cos \gamma$

Law of sines: $a/\sin \alpha = b/\sin \beta = c/\sin \gamma$

Quadratic formula:

If $ax^2 + bx + c = 0$, then, $x = (-b \pm \sqrt{b^2 - 4ac})/(2a)$

SI Units

Quantity	Name of Unit	Symbol	Expression in Terms of Other SI Units
Length	meter	m	Base unit
Mass	kilogram	kg	Base unit
Time	second	s	Base unit
Electric current	ampere	A	Base unit
Temperature	kelvin	K	Base unit
Amount of substance	mole	mol	Base unit
Velocity	—	—	m/s
Acceleration	—	—	m/s^2
Force	newton	N	$kg \cdot m/s^2$
Work, energy	joule	J	$N \cdot m$
Power	watt	W	J/s
Impulse, momentum	—	—	$kg \cdot m/s$
Plane angle	radian	rad	m/m
Angular velocity	—	—	rad/s
Angular acceleration	—	—	rad/s^2
Torque	—	—	$N \cdot m$
Frequency	hertz	Hz	s^{-1}
Density	—	—	kg/m^3

Quantity	Name of Unit	Symbol	Expression in Terms of Other SI Units
Pressure, stress	pascal	Pa	N/m^2
Viscosity	—	—	$Pa \cdot s$
Electric charge	coulomb	C	$A \cdot s$
Electric field	—	—	N/C
Electric potential	volt	V	J/C
Resistance	ohm	Ω	V/A
Capacitance	farad	F	C/V
Inductance	henry	H	$V \cdot s/A$
Magnetic field	tesla	T	$N \cdot s/(C \cdot m)$
Magnetic flux	weber	Wb	$T \cdot m^2$
Specific heat capacity	—	—	$J/(kg \cdot K)$ or $J/(kg \cdot C°)$
Thermal conductivity	—	—	$J/(s \cdot m \cdot K)$ or $J/(s \cdot m \cdot C°)$
Entropy	—	—	J/K
Radioactive activity	becquerel	Bq	s^{-1}
Absorbed dose	gray	Gy	J/kg
Exposure	—	—	C/kg

The Greek Alphabet

Alpha	A	α	Iota	I	ι	Rho	P	ρ
Beta	B	β	Kappa	K	κ	Sigma	Σ	σ
Gamma	Γ	γ	Lambda	Λ	λ	Tau	T	τ
Delta	Δ	δ	Mu	M	μ	Upsilon	Υ	υ
Epsilon	E	ε	Nu	N	ν	Phi	Φ	ϕ
Zeta	Z	ζ	Xi	Ξ	ξ	Chi	X	χ
Eta	H	η	Omicron	O	o	Psi	ψ	ψ
Theta	Θ	θ	Pi	Π	π	Omega	Ω	ω

Periodic Table of the Elements

Legend (Key):
Symbol — Cl, Atomic number — 17, Atomic mass* — 35.453, Electron configuration — $3p^5$

Transition elements

Group I

Symbol	At. no.	Atomic mass	Config.
H	1	1.00794	$1s^1$
Li	3	6.941	$2s^1$
Na	11	22.9898	$3s^1$
K	19	39.0983	$4s^1$
Rb	37	85.4678	$5s^1$
Cs	55	132.905	$6s^1$
Fr	87	(223)	$7s^1$

Group II

Symbol	At. no.	Atomic mass	Config.
Be	4	9.01218	$2s^2$
Mg	12	24.305	$3s^2$
Ca	20	40.08	$4s^2$
Sr	38	87.62	$5s^2$
Ba	56	137.33	$6s^2$
Ra	88	(226)	$7s^2$

Transition elements — Period 4 (Sc–Zn)

Symbol	At. no.	Atomic mass	Config.
Sc	21	44.9559	$3d^14s^2$
Ti	22	47.87	$3d^24s^2$
V	23	50.9415	$3d^34s^2$
Cr	24	51.996	$3d^54s^1$
Mn	25	54.9380	$3d^54s^2$
Fe	26	55.845	$3d^64s^2$
Co	27	58.9332	$3d^74s^2$
Ni	28	58.69	$3d^84s^2$
Cu	29	63.546	$3d^{10}4s^1$
Zn	30	65.41	$3d^{10}4s^2$

Transition elements — Period 5 (Y–Cd)

Symbol	At. no.	Atomic mass	Config.
Y	39	88.9059	$4d^15s^2$
Zr	40	91.224	$4d^25s^2$
Nb	41	92.9064	$4d^45s^1$
Mo	42	95.94	$4d^55s^1$
Tc	43	(98)	$4d^55s^2$
Ru	44	101.07	$4d^75s^1$
Rh	45	102.906	$4d^85s^1$
Pd	46	106.42	$4d^{10}5s^0$
Ag	47	107.868	$4d^{10}5s^1$
Cd	48	112.41	$4d^{10}5s^2$

Transition elements — Period 6 (La, Hf–Hg)

Symbol	At. no.	Atomic mass	Config.
La	57	138.906	$5d^16s^2$
Hf	72	178.49	$5d^26s^2$
Ta	73	180.948	$5d^36s^2$
W	74	183.84	$5d^46s^2$
Re	75	186.207	$5d^56s^2$
Os	76	190.2	$5d^66s^2$
Ir	77	192.22	$5d^76s^2$
Pt	78	195.08	$5d^86s^2$
Au	79	196.967	$5d^{10}6s^1$
Hg	80	200.59	$5d^{10}6s^2$

Transition elements — Period 7 (Ac, Rf–Cn)

Symbol	At. no.	Atomic mass	Config.
Ac	89	(227)	$6d^17s^2$
Rf	104	(261)	$6d^27s^2$
Db	105	(262)	$6d^37s^2$
Sg	106	(266)	$6d^47s^2$
Bh	107	(264)	$6d^57s^2$
Hs	108	(277)	$6d^67s^2$
Mt	109	(268)	$6d^77s^2$
Ds	110	(281)	
Rg	111	(282)	
Cn	112	(285)	$5d^{10}6s^2$

Group III

Symbol	At. no.	Atomic mass	Config.
B	5	10.81	$2p^1$
Al	13	26.9815	$3p^1$
Ga	31	69.72	$4p^1$
In	49	114.82	$5p^1$
Tl	81	204.383	$6p^1$
Nh	113	(286)	

Group IV

Symbol	At. no.	Atomic mass	Config.
C	6	12.011	$2p^2$
Si	14	28.0855	$3p^2$
Ge	32	72.64	$4p^2$
Sn	50	118.71	$5p^2$
Pb	82	207.2	$6p^2$
Fl	114	(289)	

Group V

Symbol	At. no.	Atomic mass	Config.
N	7	14.0067	$2p^3$
P	15	30.9738	$3p^3$
As	33	74.9216	$4p^3$
Sb	51	121.76	$5p^3$
Bi	83	208.980	$6p^3$
Mc	115	(289)	

Group VI

Symbol	At. no.	Atomic mass	Config.
O	8	15.9994	$2p^4$
S	16	32.07	$3p^4$
Se	34	78.96	$4p^4$
Te	52	127.60	$5p^4$
Po	84	(209)	$6p^4$
Lv	116	(293)	

Group VII

Symbol	At. no.	Atomic mass	Config.
F	9	18.9984	$2p^5$
Cl	17	35.453	$3p^5$
Br	35	79.904	$4p^5$
I	53	126.904	$5p^5$
At	85	(210)	$6p^5$
Ts	117	(294)	

Group 0

Symbol	At. no.	Atomic mass	Config.
He	2	4.00260	$1s^2$
Ne	10	20.180	$2p^6$
Ar	18	39.948	$3p^6$
Kr	36	83.80	$4p^6$
Xe	54	131.29	$5p^6$
Rn	86	(222)	$6p^6$
Og	118	(294)	

Lanthanide series

Symbol	At. no.	Atomic mass	Config.
Ce	58	140.12	$4f^16s^2$
Pr	59	140.908	$4f^36s^2$
Nd	60	144.24	$4f^46s^2$
Pm	61	(145)	$4f^56s^2$
Sm	62	150.36	$4f^66s^2$
Eu	63	151.96	$4f^76s^2$
Gd	64	157.25	$5d^14f^76s^2$
Tb	65	158.925	$4f^96s^2$
Dy	66	162.50	$4f^{10}6s^2$
Ho	67	164.930	$4f^{11}6s^2$
Er	68	167.26	$4f^{12}6s^2$
Tm	69	168.934	$4f^{13}6s^2$
Yb	70	173.04	$4f^{14}6s^2$
Lu	71	174.967	$5d^14f^{14}6s^2$

Actinide series

Symbol	At. no.	Atomic mass	Config.
Th	90	232.038	$6d^27s^2$
Pa	91	231.036	$5f^26d^17s^2$
U	92	238.029	$5f^36d^17s^2$
Np	93	(237)	$5f^46d^17s^2$
Pu	94	(244)	$5f^66d^07s^2$
Am	95	(243)	$5f^76d^07s^2$
Cm	96	(247)	$5f^76d^17s^2$
Bk	97	(247)	$5f^96d^07s^2$
Cf	98	(251)	$5f^{10}6d^07s^2$
Es	99	(252)	$5f^{11}6d^07s^2$
Fm	100	(257)	$5f^{12}6d^07s^2$
Md	101	(258)	$5f^{13}6d^07s^2$
No	102	(259)	$6d^07s^2$
Lr	103	(262)	$6d^17s^2$

* Atomic mass values are averaged over isotopes according to the percentages that occur on the earth's surface. For unstable elements, the mass number of the most stable known isotope is given in parentheses. *Source: IUPAC Commission on Atomic Weights and Isotopic Abundances, 2001.*